Lecture Notes in Control and Information Sciences 410

Editors: M. Thoma, F. Allgöwer, M. Morari

Henri Bourlès and Bogdan Marinescu

Linear Time-Varying Systems

Algebraic-Analytic Approach

Authors

Henri Bourlès
SATIE, CNAM/ENS de Cachan
61 Avenue Du President Wilson
94230 Cachan
France
E-mail: henri.bourles@satie.ens-cachan.fr
henri.bourles@cnam.fr

Bogdan Marinescu
RTE EDF Transport
9, rue de la Porte de Buc
78000 Versailles, France
E-mail:
bogdan.marinescu@rte-france.com
and
SATIE-Ecole Normale Supérieure
de Cachan
61, avenue du Président Wilson
94230 Cachan, France
E-mail:
bogdan.marinescu@satie.ens-cachan.fr

ISBN 978-3-642-19726-0 e-ISBN 978-3-642-19727-7

DOI 10.1007/978-3-642-19727-7

Lecture Notes in Control and Information Sciences ISSN 0170-8643

Library of Congress Control Number: 2011925386

Typeset & Cover Design: Scientific Publishing Services Pvt. Ltd., Chennai, India.

Printed on acid-free paper

9 8 7 6 5 4 3 2 1

springer.com

à nos épouses Corinne et Isabelle, pour leur aide et pour leur patience qui à été mise à l'épreuve maintes fois

Preface

The aim of this book is to propose a new approach to analysis and control of linear time-varying (LTV) systems. More generally, systems over a ring of operators are considered. In the approach developed here, a system is defined in an intrinsic way, i.e., not by a particular representation (e.g., a transfer matrix or a state-space form) but as it is actually. The system equations, derived, e.g., from the laws of physics, are gathered to form an intrinsic mathematical object, namely a finitely presented module over a ring of operators. This module is nothing but the system equations when one takes into account the obvious fact that there are various equivalent ways to write these equations (in different orders, and using various variables). This is strongly connected with the engineering point of view, according to which a system is not a specific set of equations but an object of the material world which can be described by equivalent sets of equations. This viewpoint allowed us to formulate and solve efficiently several key problems of the theory of control in the case of linear time-varying systems. For example, the better-known control approach for LTV systems is probably gain scheduling [311], [211]. This method consists in approximating the LTV system by a finite sequence of linear time-invariant (LTI) systems, with suitable interpolations. It works quite well when the system coefficients are slowly varying. In a number of applications, however, this condition does not hold. The solution proposed in this book, based on algebraic analysis, does not require this assumption and is therefore more efficient.

"Algebraic analysis" (a term coined by the Japanese mathematician Mikio Sato) encompasses a variety of algebraic methods to study analytic objects. So is an LTV system; its algebraic structure, which is hidden when, e.g., gain scheduling is used, becomes very clear in the light of the methods presented in this book. Our approach is also useful for mathematicians since it shows how algebraic analysis can be applied to solve engineering problems.

Prerequisites

For most parts of the book, the prerequisites include (1) that background in algebra and analysis which is standard for people having, e.g., an engineering degree; (2) a certain knowledge of those problems and methods which are classical in control theory. For these two points, a good understanding of the textbook [43], by the first author, is sufficient and appropriate. Regarding purely algebraic aspects, the famous textbook [224], by S. MacLane and G. Birkhoff, can be a good complement.

Nevertheless, for a few parts of the book–such as Section 5.7–much more prerequisites (detailed in Chapter 1) are necessary. These parts are intended to researchers.

How to Read This Book

Chapters 1 to 4 present the mathematical background and form the first part of the book. Mathematicians interested in applications to control theory can directly jump to Chapter 5. This will be the exception. Most readers will have either to do the same but come back to the suitable parts of Chapters 1-4 when necessary, or to read the book from the beginning. Both methods can be efficient; their choice is a matter of taste and/or time. We recommend to complete at once the reading of this Preface by that of the Introductions of the various chapters.

In Part II, Chapter 5 presents the notion of system along with its behavior and structural properties. In Chapter 6, stability is investigated using the system poles. All other finite poles and zeros of an LTV system are also defined and studied. Chapter 7 deals with the structure at infinity and the impulsive behaviors of LTV systems.

Part III is devoted to applications in both fields of analysis and control. Modeling items are treated in Chapters 8 and 9, while open-loop and closed-loop control methods are presented in Chapters 10 and 11, respectively.

Some points of the classic (analytic) analysis of LTV systems are reviewed in Chapter 12 as complements to this new approach.

Each chapter contains a number of exercises, of various levels of difficulty: some of them are straightforward applications of the notions expounded in the body of the text; other ones – in general more difficult – propose complements. Hints are given when necessary.

Last, each chapter ends with historical notes.

Notation

Standard Notation

f.g.: finitely generated
f.p.: finitely presented
$\triangleq$: equal by definition
$\subseteq$, $\subsetneq$: inclusion, proper inclusion
Id_E : identity map of the set E
$A \backslash B$: complement of B in A ($B \subseteq A$)
$\mathbb{N} \triangleq \{0, 1, 2, ...\}$, $\mathbb{N}^\times \triangleq \{1, 2, ...\}$: the non-negative, the natural integers
$\mathbb{Z}$, $\mathbb{Q}$, $\mathbb{R}$, $\mathbb{C}$: the integer, rational, real, complex numbers
$\mathbf{k}$: $\mathbb{R}$ or $\mathbb{C}$
$\mathbb{T}$: torus $\mathbb{R}/\mathbb{Z}$
$\varphi : A \to B$: map φ with domain A and codomain B
$\varphi \bullet : x \mapsto \varphi(x)$: the right map which to x associates $\varphi(x)$
$\bullet \varphi : x \mapsto (x) \varphi$: the left map which to x associates $(x) \varphi$
$f|_A$: the restriction of the map f to a subset A of its domain
Y^X : set of all maps $X \to Y$
I_n : identity matrix of order n
${}^m S^p$: the $m \times p$ matrices over S (m rows, p columns)
${}^I S^J$, $S^{I \times J}$: the $I \times J$ matrices over S (I, J : possibly infinite sets of indices)
$S^p \triangleq {}^1 S^p$ (row), ${}^m S \triangleq {}^m S^1$ (column)
${}^I S^{(J)}$: the $I \times J$ matrices over S with finite number of nonzero columns
$\mathrm{Mat}_n(S) \triangleq {}^n S^n$: the $n \times n$ matrices over S
$\mathrm{diag}(\Delta)$: matrix with zero entries except the main diagonal Δ
$\mathrm{diag}(A_1, ..., A_n) = A_1 \oplus ... \oplus A_n$: diagonal sum of matrices
det, Tr : determinant, trace of a matrix
I_n : Identity matrix of order n
card : cardinal
$\cup$ union
$n! : 1 \times 2 \times ... \times n$
$\binom{n}{p}$: binomial coefficient $n!/(p!\,(n-p)!)$

sgn : signature of a permutation
$\bar{\lambda}$: conjugate of the complex number λ
$\Delta_{\mathrm{ij}} = \Delta_{\mathrm{i}}^{\mathrm{j}}$: Kronecker symbol

Chapter 1

$\prod_{i\in I} X_i$, X^I : product, power p. 4, 9
$\coprod_{i\in I} X_i$, $X^{(I)}$: coproduct, copower p. 5, 9
$\dot{\cup}$: disjoint union p. 5
$\mathrm{Ob}(\mathcal{C})$, $\mathrm{Mor}(\mathcal{C})$: class of all objects, of all morphisms, of $\mathcal{C}$ p. 6
$\mathcal{C}^{\mathrm{op}}$: opposite category p. 7
$Y \subseteq X$, $X \twoheadrightarrow Z$: subobject, quotient object of X p. 7
Set, **Mon** : category of sets, of monoids p. 8
Rng, **Fld** : category of rings, of fields p. 8
$_{\mathbf{R}}\mathbf{Mod}$, $\mathbf{Mod}_{\mathbf{R}}$: categories of left, right **R**-modules p. 8
$_{\mathbf{K}}\mathbf{Vsp}$, $\mathbf{Vsp}_{\mathbf{K}}$: categories of left, right **K**-vector spaces p. 8
Top : category of topological spaces p. 8
Ab : category of abelian groups p. 9
$\tilde{\rightarrow}$, $\simeq$: bijection or isomorphism p. , 12, 31
$\rightarrow$, $\hookrightarrow$, $\twoheadrightarrow$: morphism, monomorphism, epimorphism p. 7, 31
$\varinjlim F_i$: direct limit p. 14
$\varprojlim F_i$: inverse limit p. 16
$\bigoplus$: biproduct p. 18
$\mathrm{im}\, f$, $\ker f$: image, kernel of the (homo)morphism f p. 18, 31
$\mathrm{coker}\, f$, $\mathrm{coim}\, f$: cokernel, coimage of the morphism f p. 18
Ban, Frt : category of Banach spaces, of Fréchet spaces p. 20
HTVS : category of Hausdorff topological vector spaces p. 20
TVS, : category of topological vector spaces p. 20
LCS : category of locally convex topological vector spaces p. 20
$\mathrm{Hom}_{\mathcal{C}}(\bullet,\bullet)$: bifunctor Hom in the category $\mathcal{C}$ p. 21
$A \cap B$, $A + B$: intersection, union of two subobjects p. 25
$\mathcal{C}^{co}$: category of coherent objects in $\mathcal{C}$ p. 27
$\mathbf{M}^o$: opposite of the monoid $\mathbf{M}$ p. 28
$\mathbf{M}_0$, $\mathbf{M}^{\times}$ p. 28
$\mathbf{U}(\mathbf{M})$: the units of $\mathbf{M}$ p. 29
$(X)_l$, $(X)_r$, (X) : left, right, two-sided ideal generated by X p. 29
$\lhd_l, \lhd_r, \lhd$: left, right, two-sided ideal p. 29
$\mathbf{I}_l(\mathbf{M})$, $\mathbf{I}_r(\mathbf{M})$, $\mathbf{I}(\mathbf{M})$, $\mathbf{I}_l^{\alpha}(\mathbf{M})$, $\mathbf{I}_r^{\alpha}(\mathbf{M})$, $\mathbf{I}^{\alpha}(\mathbf{M})$ p. 29
$\mathbf{I}_l^{fg}(\mathbf{M})$, $\mathbf{I}_r^{fg}(\mathbf{M})$, $\mathbf{I}^{fg}(\mathbf{M})$ p. 29
$\mathbf{Z}(\mathbf{M})$: centre of $\mathbf{M}$ p. 30, 49
$\mathbf{Z}_{\mathbf{M}}(A)$, $\mathbf{N}_{\mathbf{M}}(A)$: centralizer, normalizer of A in $\mathbf{M}$ p. 30, 49
$[X]$: monoid generated by X p. 30
$\mathrm{GL}_n(\mathbf{R})$: general linear group over a ring $\mathbf{R}$ p. 30
$\mathbf{M}/\mathcal{R}$, $\mathcal{T}/\mathcal{R}$: quotient p. 6
$\mathrm{supp}\left((x_i)_{i\in I}\right)$: support of the family $(x_i)_{i\in I}$ p. 52

$\mathbf{M}^I$: power p. 32, 94
$\left[(a_i)_{i\in I}\right]_r$, $\left[(a_i)_{i\in I}\right]_l$, $\left[(a_i)_{i\in I}\right]$: lcrm, lclm, lcm p. 34
$\left((a_i)_{i\in I}\right)_l$, $\left((a_i)_{i\in I}\right)_r$, $\left((a_i)_{i\in I}\right)$: gcld, gcrd, gcd p. 34
$|c|$: length of c p. 36, 46, 44
$a \mid b$: a is a divisor of b p. 32
$a \parallel b$: a is a total divisor of b p. 35
X^* : free monoid on X p. 36
inf, sup : greatest lower bound, least upper bound p. 37
$\wedge$, $\vee$: meet, join p. 37
$|S|$: length of the poset S p. 39
ACC, ACC_α, DCC, DCC_α p. 39
$\dot{0}, \dot{1}$: least, greatest element p. 44
$(G:1)$: order of the group G p. 48
$\mathfrak{S}(E), \mathfrak{S}_n$: symmetric group of E, of degree n p. 48
$\langle X\rangle$, $\langle x_1,..,x_n\rangle$: group generated by X, by $x_1,..,x_n$ p. 49
$H \leq G$: subgroup of G p. 49
Alt_n, A_n : alternating group of degree n p. 49
G/H, $H\backslash G$: set of left, right cosets p. 49
$(G:H)$: index of H in G p. 50
$N \triangleleft G$: normal subgroup of G p. 50
G/N : quotient p. 50
$x \equiv x' \pmod N$ p. 50, 99
$\mathrm{SL}_n(\mathbf{R})$: special linear group over a commutative ring $\mathbf{R}$ p. 50
$\mathrm{Lat}(G)$: lattice of all normal subgroups of G p. 54
$|G|$: length of the group G p. 55
$h(G)$: Hirsch number of the group G p. 55
(x,y) : commutator of x and y in a group p. 56
(H,K) : subgroup generated by the commutators of H and K p. 56
$\mathbf{D}(G) = G^d$: derived group of G p. 56
G^{ab} : abelianization of the group G p. 57
$S[\mathfrak{T}]$: topological space p. 59
$\mathfrak{T}_d[S], \mathfrak{T}_f[S]$: discrete topology, finite topology p. 60
$\mathfrak{N}_X$: filter of neighborhoods of zero in X p. 59
$\sigma(X',X), \sigma(X,X')$: weak topologies p. 62
$\beta(X',X)$: strong topology p. 62
X'_β : strong dual p. 62
X_σ, X'_σ : weak spaces p. 62
${}^t f$: transpose, or adjoint, of the continuous linear map f p. 62
$\hat{\mathbf{G}}$: universal covering group of the Lie group $\mathbf{G}$ p. 69
$\pi_1(\mathbf{G})$: fundamental group of the Lie group $\mathbf{G}$ p. 69
$\mathcal{E}(X)$: the space of C^∞ functions on the manifold X p. 69
$\mathcal{E}'(X)$: the space of distributions with compact support p. 70
$\mathcal{D}(X)$: the space of C^∞ functions with compact support p. 69
$\mathcal{D}'(X)$: the space of distributions p. 70
$T \star S$: convolution product p. 70

$\mathcal{T}^{\infty}(\mathbf{G})$: space of punctual distributions over $\mathbf{G}$ p. 72
$\mathcal{T}_x^{\infty}(\mathbf{G})$: space of distributions with support $\subseteq \{x\}$ p. 72
$[X, Y]$: Lie bracket p. 72
$\mathfrak{g}$: Lie algebra of the Lie group $\mathbf{G}$ p. 72
$\exp_{\mathbf{G}}$: exponential p. 72
$\mathrm{U}(\mathfrak{g})$: universal enveloping algebra of $\mathfrak{g}$ p. 74
$\mathrm{Z}_{\mathbf{k}}(\mathfrak{g})$: centre of $\mathrm{U}(\mathfrak{g})$ p. 74
$\mathcal{O}(X)$: space of analytic functions on the real manifold X p. 81
$\mathcal{B}(X)$: space of hyperfunctions on X p. 81
$[\varphi] = [\varphi(z)]_{z=x}$: hyperfunction p. 81
$\mathrm{supp}\,[\varphi]$: support of the hyperfunction $[\varphi]$ p. 76
$\mathfrak{f.p.}(f)$: Hadamard's finite part of f p. 79
$A(n, \mathbb{R})$, $O(n, \mathbb{R})$, $\mathrm{SO}(n, \mathbb{R})$ p. 85
$\mathrm{Sp}(2n, \mathbf{k})$, $\mathfrak{S}_p(2n)$ p. 87
$\mathfrak{D}^n(\mathfrak{g})$, ad, $\mathrm{Der}(\mathfrak{g})$: p. 87

Chapter 2

$\mathbf{R}^o$: opposite of the ring $\mathbf{R}$ p. 93
$\mathrm{char}\,\mathbf{R}$: characteristic of the ring $\mathbf{R}$ p. 93
${}_{\mathbf{R}}\mathbf{R}$, $\mathbf{R}_{\mathbf{R}}$, ${}_{\mathbf{R}}\mathbf{R}_{\mathbf{R}}$: p. 96, 96
$N \lhd M$: submodule p. 96
$M_{[\mathbf{S}]}$ p. 96
${}_{\mathbf{R}}M_{\mathbf{S}}$: $(\mathbf{R}, \mathbf{S})$-bimodule p. 96
$\mathrm{End}_{\mathbf{R}}(M)$: endomorphism ring of M p. 96
$[a, b]$: commutator of a and b in an algebra p. 97
$\mathrm{Hom}_{\mathbf{R}}(M, N)$: module-homomorphisms $M \to N$ p. 98
$N \lhd M$: submodule of M p. 100
$\mathrm{Lat}_{\mathfrak{l}}(M)$: lattice of all submodules of the left module M p. 100
$\bigoplus_{i \in I} M_i$: direct sum p. 101
$[E]_{\mathbf{R}}$: module generated by E over $\mathbf{R}$ p. 102
$\dim_{\mathbf{K}}(V)$, $\dim(V)$: dimension of the $\mathbf{K}$-vector space V p. 104
$\langle \bullet, \bullet \rangle$: duality bracket p. 106
${}^t f$: transpose of the homomorphism f p. 106
$\bullet A$, $A\bullet$: p. 107
A^T : classical transpose of a matrix p. 108
I_d : identity functor p. 108
$A \equiv A'$: equivalent matrices p. 111
$A \equiv_l A'$, $A \equiv_r A'$: left-equivalent, right-equivalent matrices p. 111
$A \sim A'$: conjugate matrices p. 111
$A \bigotimes_{\mathbf{R}} B$: tensor product p. 112
$a \otimes b$, $f \otimes g$, $1 \otimes g$: p. 112
$M_{(\mathbf{S})}$: p. 113
$\mathbf{k}\{t\}$: ring of convergent power series p. 114, 147
$\mathbf{k}(\{t\})$: field of fractions of $\mathbf{k}\{t\}$ p. 114

$\mathfrak{E}(\mathbf{k})$: ring of entire functions p. 114, 172
$\mathcal{M}(\mathbf{k})$: field of meromorphic functions p. 114, 172
$\mathcal{PE}_N(\mathbf{k})$: ring of N-periodic entire functions p. 115
$\mathcal{PM}_N(\mathbf{k})$: ring of N-periodic meromorphic functions p. 115
$\mathbf{R}_\Delta$: subring of constants of $\mathbf{R}$ p. 116
$\operatorname{rad}\mathbf{R}$: Jacobson radical of $\mathbf{R}$ p. 117
$\mathfrak{a}\,\mathfrak{b}$: product of the ideals $\mathfrak{a}$ and $\mathfrak{b}$ p. 118
$\operatorname{Spec}\mathbf{R}$: prime spectrum of $\mathbf{R}$ p. 119
$\mathbf{R}[X]$: polynomial ring p. 131
$d^\circ(a)$: degree of the polynomial a p. 132
$\mathbf{K}(X)$: field of rational functions p. 133
$\mathbf{R}[X;\boldsymbol{\alpha},\Delta], \mathbf{R}[X;\Delta], \mathbf{R}[X;\boldsymbol{\alpha}]$: skew polynomial rings p. 135
$\mathbf{R}[\partial;\Delta]$, $\mathbf{R}[\mathbf{q};\boldsymbol{\alpha}]$: differential and difference operator ring p. 136
$\mathbf{R}[\mathbf{q}-1;\boldsymbol{\alpha},\Delta]$: differencing operator ring p. 136
$A_1(\mathbf{k})$: first Weyl algebra p. 137
$B_1(\mathbf{k})$: p. 137
$A_{1c}(\mathbf{k})$: p. 140
$S^{-1}\mathbf{R}\ (\mathbf{R}\,S^{-1})$: left (right) fractions with denominator in S p. 125
$\mathbf{Q}(\mathbf{R})$: field of fractions of the domain $\mathbf{R}$ p. 128
$\mathcal{T}_S(M)$, $\mathcal{T}(M)$: torsion submodule p. 129
$\hat{\mathbf{R}}$: completion of $\mathbf{R}$ in the $\mathfrak{a}$-adic topology p. 131
$A_1'(\mathbf{k})$: p. 141
$\mathbf{S} * G$: crossed product p. 141
$\mathbf{R}[[X^{-1};\boldsymbol{\alpha},\Delta]]$: ring of formal skew power series p. 148
$\omega(a)$: order of the formal power series a p. 148
$\sqrt{\mathfrak{a}}$: radical of the ideal $\mathfrak{a}$ p. 151
$\mathfrak{N}(\mathbf{R})$: nilradical of $\mathbf{R}$ p. 151
$\mathbf{A}^I$: affine space p. 153
$\mathcal{Z}_\mathfrak{a}$: algebraic set p. 153
$\mathrm{D}_n(\mathbf{R})$: p. 154
$\mathrm{E}_n(\mathbf{R})$: group of elementary matrices p. 154
$\operatorname{rk}(A)$: rank of the matrix A p. 390, 163
$A_{H,K}$: p. 157
$\operatorname{rk}_\mathbf{R}(M), \operatorname{rk}(M)$: rank of the free $\mathbf{R}$-module M p. 162
$\operatorname{rk}(f)$: rank of the homomorphism f p. 162
$\rho(A)$: inner rank of the matrix A p. 165
$\mathcal{H}_\infty$: Hardy space p. 169
$\mathbf{L}(c\mathbf{R},\mathbf{R})$: lattice of all right principal ideals containing $c\mathbf{R}$ p. 169
$\mathbf{R}_\mathfrak{p}$: local ring of $\mathbf{R}$ at $\mathfrak{p}$ p. 183
$A_o(\mathbb{C})$: p. 185
$A_n(\mathbf{k})$: n-th Weyl algebra p. 186
$B_n(\mathbf{k})$: p. 186
$\mathbf{R}\langle X\rangle$: free $\mathbf{R}$-ring p. 185
$\mathbf{R}\left[(X_i;\boldsymbol{\alpha}_i,\Delta_i)_{1\le i\le n}\right]$: multivariate skew polynomial ring p. 185

$\mathbf{R}\left[\left(X_i, X_i^{-1}; \boldsymbol{\alpha}_i, \Delta_i\right)_{1 \leq i \leq n}\right]$: multivariate Laurent polynomials p. 185
$\mathbf{K}\left(\left(X_i; \boldsymbol{\alpha}_i, \Delta_i\right)_{1 \leq i \leq n}\right)$: multivariate rational functions p. 186
$\mathbf{R}\left[\left[\left(X_i^{-1}; \boldsymbol{\alpha}_i, \Delta_i\right)_{1 \leq i \leq j+1}\right]\right]$: formal multivariate power series p. 186
$\mathbf{R}\left(\left(\left(X_i^{-1}; \boldsymbol{\alpha}_i, \Delta_i\right)_{1 \leq i \leq n}\right)\right)$: formal multivariate Laurent series p. 186

Chapter 3

${}_{\mathbf{A}}\mathbf{Mod}^{fg}$: category of finitely generated left **A**-modules p. 191
${}_{\mathbf{A}}\mathbf{Mod}^{fp}$: category of finitely presented left **A**-modules p. 191
$\mathrm{Ann}_l^{\mathbf{R}}(S)$, $\mathrm{Ann}_r^{\mathbf{R}}(S)$, $\mathrm{Ann}^{\mathbf{A}}(S)$: annihilator p. 194
$i(A)$: index of the matrix A p. 200
$\mathrm{Ass}(M)$: prime ideals associated with M p. 208
$\mathfrak{q}(\mathfrak{p})$: primary ideal with radical $\mathfrak{p}$ p. 208
$\bigotimes_{\mathbf{R}}$: bifunctor tensor product p. 213
ρ_* : change of ring p. 215
ρ_* : extension of ring induced along ρ p. 216
$\chi(M)$: characteristic of the module M p. 240
P^* : adjoint of the differential operator P p. 138
$\mathrm{pd}({}_{\mathbf{R}}M)$, $\mathrm{fd}({}_{\mathbf{R}}M)$: projective, flat dimension p. 242
$\mathrm{lgld}\,\mathbf{R}$, $\mathrm{rgld}\,\mathbf{R}$: left, right global dimension p. 243
$\mathrm{gld}\,\mathbf{R}$, $\mathrm{wgld}\,\mathbf{R}$: global, weak global dimension p. 243
$\mathcal{K}(\mathbf{R})$, $\dim(\mathbf{R})$: Krull dimension p. 243
${}^*\rho$: extension of ring coinduced along ρ p. 263
$M \subseteq_e E$: essential submodule of E p. 264
$N \subseteq_s M$: superfluous (or small) submodule of M p. 265

Chapter 4

$\mathbf{L}/\mathbf{K}$: extension $\mathbf{L} \supseteq \mathbf{K}$ p. 269
$[\mathbf{L} : \mathbf{K}]$: degree of the field extension $\mathbf{L}/\mathbf{K}$ p. 270
$\deg.\mathrm{tr}_{\mathbf{K}}\,\mathbf{L}$: transcendence degree of $\mathbf{L}/\mathbf{K}$ p. 271
$[a : \mathbf{K}]$: degree of the algebraic element a over $\mathbf{K}$ p. 273
$\mathrm{Gal}(\mathbf{L}/\mathbf{K})$: Galois group of the Galois extension $\mathbf{L}/\mathbf{K}$ p. 276
$\mathrm{Gal}_{\mathbf{K}}(f)$: Galois group of the polynomial f over $\mathbf{K}$ p. 276
${}^c a$: $(\boldsymbol{\alpha}, \Delta)$-conjugate of a by c p. 287
$\Delta^{\boldsymbol{\alpha},\Delta}(\mathrm{a})$: conjugacy class of a p. 287
$f_l(a)$, $f_r(a)$: left, right evaluation of f at a p. 288
$V_r(f)$, $V_l(f)$: set of left, right roots of f p. 289
f_Δ : minimal polynomial of the algebraic set Δ p. 292
$\mathrm{rk}(\Delta)$: rank of the algebraic set Δ p. 292
UnivR, UnivF : universal differential ring, field p. 302
$G_\Delta(\mathbf{L}/\mathbf{K})$: group of all $(\mathbf{K}, \Delta)$-automorphims of $\mathbf{L}$ p. 302
$\mathrm{Gal}_\Delta(\mathbf{L}/\mathbf{K})$: differential Galois group p. 302

$G_{\boldsymbol{\alpha}}(\mathbf{L}/\mathbf{K})$: group of all $(\mathbf{K},\boldsymbol{\alpha})$-automorphims of $\mathbf{L}$ p. 305
$\mathrm{Gal}_{\boldsymbol{\alpha}}(\mathbf{L}/\mathbf{K})$: difference Galois group p. 305

Chapter 5

$\mathcal{PE}(I)$: polynomial-exponential functions p. 325
$\mathcal{AG}$: arithmetic-geometric sequences p. 326
$\mathfrak{O}$: monoid of Ore functions p. 327
$(\mathfrak{O})$: set of Ore fields p. 327
$\mathcal{O}_{\infty}$: germs at infinity of analytic functions p. 330
$\mathcal{S}$: germs at infinity of complex-valued sequences p. 331
${}_{\mathbf{A}}\mathbf{TMod}$: category of topological modules p. 372
$\mathcal{H}(\mathbf{B})$: p. 383

Chapter 6

$\mathbf{O}_m$ the ring of formal power series over $\mathbf{K}_m$ p. 420
$\mathbf{O}$ the ring of formal power series over $\mathbf{K}$ p. 422
$v(a)$ the valuation of $a \in \mathbf{K}$ p. 419
M_{sp} the module of the poles of the system p. 427
M_{iz} the module of the invariant zeros p. 428
M_{idz} the module of the input-decoupling zeros p. 434
M_{odz} the module of output-decoupling zeros p. 434
M_{iodz} the module of input-output decoupling zeros p. 435
M_{op} the module of observable poles p. 436
M_{hm} the module of hidden modes p. 439
M_{tz} the module of transmission zeros p. 430
M_{tp} the module of transmission poles p. 430
M_{bz} the module of blocking zeros p. 433
$deg.tr.diff_{\mathbf{K}}\mathbf{L}$ the differential transcendence degree of the differential field extension $\mathbf{L}/\mathbf{K}$ p. 564
$\|\cdot\|_i, \|\cdot\|_{ip}$ the matrix induced norm p. 592
$a(T)$ full set of elementary Smith zeros of the torsion module T p. 428
$A(T) = \cup_i\{\Delta_{\widehat{\mathbf{K}}^{\infty}}(\mathrm{a}_1), ..., \Delta_{\widehat{\mathbf{K}}^{\infty}}(\mathrm{a}_\mathrm{n})\}$ where $\Delta_{\widehat{\mathbf{K}}^{\infty}}(\mathrm{a}_\mathrm{i})$ is the conjugacy class of a_i is the conjugacy class of a_i. 437
$\Delta_{\check{\mathbf{K}}}(\gamma_\mathrm{i})$ the conjugacy class of the pole γ_i. 424

Chapter 7

$\mathbf{K}, \mathbf{A}, \mathbf{S}$: rings p. 457
$\mathbf{B}, \mathbf{L}, \mathbf{Q}, \hat{\mathbf{K}}, \hat{\mathbf{S}}, \hat{\mathbf{L}}$: rings p. 459
TP_{∞}, TZ_{∞} : transmission poles, zeros, at infinity p. 460
$c_{\infty}(R(\partial))$: content at infinity of $R(\partial)$ p. 460
$\Delta_{\mathbf{M}}(G)$: MacMillan degree at infinity p. 463
${}_{\mathbf{S}}\mathbf{Mod}^{struc}$: full subcategory of ${}_{\mathbf{S}}\mathbf{Mod}$ p. 466
$\tilde{C}_{\mu}, \tilde{\Delta}^{(\mu-1)}, \tilde{\Delta}$: p. 466

$\tilde{\mathbf{E}}$: endomorphism ring of $\tilde{\Delta}$ p. 468
$\bar{\Delta}^{(\mu-1)}, \bar{\Delta}$: p. 472
$\bar{C}_\mu, C_\mu$: p. 473
$\mathfrak{B}, \mathfrak{B}_a$: p. 475
$\mathfrak{B}_\infty$: impulsive behavior p. 475
$\mathfrak{B}_{\infty,a}$: autonomous impulsive behavior p. 475
M^+ : impulsive system p. 475
$\mathcal{L}_+$: causal Laplace transform p. 478
Z_- : anticausal Z-transform p. 480

Contents

Part IV: Complements

Part I
Mathematical Tools

1
Categories, Divisibility and Groups

1.1 Introduction

Here we present the general mathematical notions necessary for the sequel. We begin with categories. For the applications to control theory, abelian categories of modules (**1.2.4.3**) are the most important. Therefore, in a first reading, beginners may limit themselves to that case. Functors (Subsect. 1.2.3) will be used throughout the book.

The section on categories is followed by three sections centered on divisibility and factorization. These notions are classical in, e.g., commutative rings, but we will constantly have to use them in the noncommutative case. Furthermore, since they only involve multiplication –and not addition–, we have preferred to present these notions in the framework of monoids. Our account culminates with the presentation of unique factorization noncommutative monoids.

Section 1.6 on groups is classical (apart from a few points, such as polycyclic by finite groups and Dedekind groups) and is included as an introduction to Lie groups (Sect. 1.8) and to Galois theory (Chapter 4).

Theoretically, the prerequisites to Sections 1.2-1.6 are only a basic knowledge of the theory of sets and of commutative rings. The case of Sections 1.7 (on topological vector spaces) and 1.8 (on Lie groups and Lie algebras) is quite different. Section 1.7 is a summary of the basic properties of topological spaces and of topological vector spaces, and Section 1.8 is an survey of Lie groups and Lie algebras. Both sections will probably be hard for the non-prepared reader, but they will only be needed for Section 5.7 on topological modules and behaviors. So, we advise the beginner to only read through them in a first reading.

Section 1.9 is devoted to Sato's hyperfunctions, an essential tool in algebraic analysis. If topological considerations are skipped, the prerequisites for this section are a basic knowledge of the functions of the complex variable: see ([43], sect. 12.4), and [208] for more details.

H. Bourlès and B. Marinescu: Linear Time-Varying Systems, LNCIS 410, pp. 3–90.
springerlink.com

1.2 Categories

We assume that the reader knows the bases of the Theory of Sets. Some points of this theory are recalled below.

1.2.1 Sets

1.2.1.1 Maps

Let X, Y be two sets and let $f : X \to Y$ be a map. Two notations can be used for f:

(i) $f : x \mapsto fx$ (or $f : x \mapsto f(x)$). In what follows, such a map is called a *right map* since it acts on the right. If Z is another set and $g : Y \longrightarrow Z$ is also a right map, then the composition

$$X \xrightarrow{f} Y \xrightarrow{g} Z \tag{1.1}$$

is gf, also written $g \circ f$. A right map f can be denoted by $f\bullet$.

(ii) $f : x \mapsto xf$ (or $f : x \mapsto (x) f$) [370], [79], [80]. Such a map is called a *left map*. If both f and g are left maps, the composition (1.1) is fg, also written $f \circ g$. To avoid confusions, a left map f is denoted by $\bullet f$.

Whether a map is a right or a left one is only a matter of notation. Notation (ii) is more natural although less usual. Nevertheless, in what follows all maps are right ones except when otherwise specified.

1.2.1.2 Products

Let $(X_i)_{i\in I}$ be a family of sets. The *product* of this family, denoted by $\prod_{i\in I} X_i$, is the set of all families $(x_i)_{i\in I}$ such that $x_i \in X_i$ for all $i \in I$. The map $\pi_j : \prod_{i\in I} X_i \to X_j$ defined by $\pi_j\left((x_i)_{i\in I}\right) = x_j$ is called the *canonical projection* with index j. The axiom of choice can be stated as follows:

Axiom 1. (Zermelo's axiom of choice). *Let $(X_i)_{i\in I}$ be a family of nonempty sets. There exists a family $(x_i)_{i\in I}$ such that $x_i \in X_i$ for all $i \in I$ (in other words, $\prod_{i\in I} X_i \neq \varnothing$).*

If for each $i \in I$, $X_i = X$, then $\prod_{i\in I} X_i = \prod_{i\in I} X$ is denoted by X^I and coincides with the set of all maps $I \to X$. The set X^I is called a *power* of X. If $I = \{1, ..., n\}$, X^I is also denoted by X^n and is called "X to the power n".

The product of two sets X_1 and X_2 is denoted by $X_1 \times X_2$ and is called the *Cartesian product* of X_1 and X_2, as is well-known.

1.2.1.3 Disjoint Unions

Let $(X_i)_{i\in I}$ a family of sets.

Lemma 2. *There exists a set Y with the following property: (i) Y is the union of a family $(X_i')_{i\in I}$ of* pairwise disjoint sets*; (ii) for each $i \in I$, there exists a bijection $\beta_i : X_i \longrightarrow X_i'$.*

Proof. Let $A = \bigcup_{i\in I} X_i$ and $i \in I$; then $x \mapsto (x, i)$ $(x \in X_i)$ is a bijection β_i on X_i onto a part X_i' of $A \times I$. Obviously, $X_i' \cap X_j'$ if $i \neq j$, thus the set $X = \bigcup_{i\in I} X_i'$ has the desired property. ∎

Let $j_i : X_i' \to \bigcup_{i\in I} X_i'$ be the inclusion and $\iota_i = j_i \circ \beta_i$.

Definition 3. *The set Y is called the* disjoint union *(or the* sum*) of the family $(X_i)_{i\in I}$; this set is denoted by $\dot{\bigcup} X_i$, and $\iota_i : X_i \to \dot{\bigcup} X_i$ is called the* canonical injection *with index i.*

Let us explain the difference between the *union* and the *disjoint union* through a simple example: let $X_1 = \{0, 3\}$ and $X_2 = \{3, 7\}$. Then $X_1 \cup X_2 = \{0, 3, 7\}$ whereas $X_1 \coprod X_2 = \{(1, 0), (1, 3), (2, 3), (2, 7)\}$; by an abuse of language, the latter set is often written $\{0, 3, 3, 7\}$.

Proposition 4. *If the family $(X_i)_{i\in I}$ is finite, then there exists a bijection $\alpha : \prod X_i \xrightarrow{\sim} \dot{\bigcup} X_i$.*

Proof. It is sufficient to consider the case when $\mathrm{card}(I) = 2$. Let $\alpha : X_1 \times X_2 \to X_1 \dot{\cup} X_2$ be given by $(x_1, x_2) \mapsto \{(1, x_1), (2, x_2)\}$; α is a bijection since its inverse is given by $\{(1, x_1), (2, x_2)\} \mapsto (x_1, x_2)$. ∎

1.2.1.4 Preordered Sets

Let I be a set; $\preceq$ is an *preorder relation* defined over I if the following properties hold for any elements $a, b, c \in I$: (i) if $a \preceq b$ and $b \preceq c$, then $a \preceq c$; (ii) $a \preceq a$. Then I, endowed with the preorder relation $\preceq$, is called a *preordered set*.

1.2.1.5 Equivalence Relations

The notion of binary relation in a set S is classical. We write $\mathcal{R}(x, x')$ to mean that the binary relation $\mathcal{R}$ is satisfied by the pair $(x, x') \in S \times S$. The graph of $\mathcal{R}$ is the subset of $S \times S$ consisting of all pairs (x, x') such that $\mathcal{R}(x, x')$.

When $\mathcal{R}$ is an equivalence relation, we write $x \equiv x' \pmod{\mathcal{R}}$ to mean that $\mathcal{R}(x, x')$ ([141], sect. 4.1). The quotient set $S/\mathcal{R}$ is the set of all equivalence classes and $\pi : S \ni x \mapsto \bar{x} \in S/\mathcal{R}$ be the *canonical surjection* (which to x associates its class $\bar{x}$, called the *canonical image* of x).

Let $\mathcal{R}, \mathcal{T}$ be equivalence relations in S. The equivalence relation $\mathcal{R}$ is said to be *finer* than $\mathcal{T}$, and $\mathcal{T}$ is said to be *coarser* than $\mathcal{R}$, if $x \equiv x' \,(\mathrm{mod}\, \mathcal{R})$ implies $x \equiv x' \,(\mathrm{mod}\, \mathcal{T})$.

Let $\mathcal{R}, \mathcal{T}$ be as above. Consider the binary relation $\mathcal{S}$ in $S/\mathcal{T}$ defined as follows: $\mathcal{S}(\bar{x}, \bar{y})$ if $x \equiv y \,(\mathrm{mod}\, \mathcal{R})$ for any $x, y \in S$ such that $x \in \bar{x}$ and $y \in \bar{y}$. This relation is *well-defined* (i.e., whether $\mathcal{S}(\bar{x}, \bar{y})$ holds only depends on the classes $\bar{x}$ and $\bar{y}$, not on the specific representatives x and y) and is an equivalence relation in $S/\mathcal{T}$.

Definition 5. *The above equivalence relation $\mathcal{S}$ is called the quotient of $\mathcal{T}$ by $\mathcal{R}$ and written $\mathcal{T}/\mathcal{R}$.*

1.2.2 General Notions on Categories

1.2.2.1 Categories

A *category* $\mathcal{C}$ consists of

(i) a class $\mathrm{Ob}(\mathcal{C})$, the objects of $\mathcal{C}$,

(ii) for each $X, Y \in \mathrm{Ob}(\mathcal{C})$, a set $\mathrm{Hom}_{\mathcal{C}}(X, Y)$, the *morphisms* (also called *arrows*) f from X to Y (X is called the *source* of f and Y its *target*),

(iii) for any $X, Y, Z \in \mathrm{Ob}(\mathcal{C})$, the composition $\mathrm{Hom}_{\mathcal{C}}(X, Y) \times \mathrm{Hom}_{\mathcal{C}}(Y, Z) \to \mathrm{Hom}_{\mathcal{C}}(X, Z)$.

One often writes $X \in \mathcal{C}$ instead of $X \in \mathrm{Ob}(\mathcal{C})$ and (when there is no risk of confusion) $f : X \to Y$ instead of $f \in \mathrm{Hom}_{\mathcal{C}}(X, X)$. The class of all morphisms of $\mathcal{C}$ is denoted by $\mathrm{Mor}(\mathcal{C})$.

The *cartesian product* $\mathcal{C}_1 \times \mathcal{C}_2$ of two categories is a category constructed as follows: its objects are the pairs $(X_1, X_2) \in \mathrm{Ob}(\mathcal{C}_1) \times \mathrm{Ob}(\mathcal{C}_2)$; its morphisms are the pairs (f_1, f_2) of morphisms $f_1 : X_1 \to Y_1$, $f_2 : X_2 \to Y_2$ with the composition

$$(f_1, f_2) \circ (g_1, g_2) = (f_1 \circ g_1, f_2 \circ g_2).$$

Two different notations can be used for the composition

$$X \xrightarrow{f} Y \xrightarrow{g} Z : \tag{1.2}$$

either $gf = g \circ f$, and then f and g are called *right morphisms*, or $fg = f \circ g$, and then f and g are called *left morphisms*; this is consistent with the terminology introduced for maps in (**1.2.1.1**). Nevertheless, a morphism $f : X \to Y$ is not a map when X and Y are not sets. A right morphism f *can be* written $f\bullet$, and to avoid confusions a left morphism f *is* written $\bullet f$. *Except when otherwise specified, all morphisms are right ones in what follows.*

By definition of a category, the following properties hold:

The composition of morphisms is associative; for any object X, there exists a morphism Id_X, the identity of X, such that for any $f : X \to Y$, $f \circ Id_X = Id_Y \circ f = f$, and Id_X belongs to $\mathrm{Hom}_{\mathcal{C}}(X, X)$;

$$\mathrm{Hom}_{\mathcal{C}}(X, Y) \cap \mathrm{Hom}_{\mathcal{C}}(X', Y') = \varnothing \text{ if } (X, Y) \neq (X', Y'). \tag{1.3}$$

A category $\mathcal{C}$ is said to be *small* if both $\mathrm{Ob}(\mathcal{C})$ and $\mathrm{Mor}(\mathcal{C})$ are sets, and *large* otherwise.

Consider a morphism $f : X \to Y$ of a category $\mathcal{C}$. For any $Z \in \mathcal{C}$, there exist well-defined maps $g \mapsto fg : \mathrm{Hom}_{\mathcal{C}}(Z, X) \longrightarrow \mathrm{Hom}_{\mathcal{C}}(Z, Y)$ and $g \mapsto gf : \mathrm{Hom}_{\mathcal{C}}(Y, Z) \longrightarrow \mathrm{Hom}_{\mathcal{C}}(X, Z)$; f is called a *monomorphism*, written $f : X \hookrightarrow Y$ (resp., an *epimorphism*, written $f : X \twoheadrightarrow Y$) if the first (resp., the second) map is always injective. In addition, f is called an *isomorphism* (written $f : X \tilde{\longrightarrow} Y$) if it has an inverse f^{-1} (such that $f^{-1} \circ f = Id_X$, $f \circ f^{-1} = Id_Y$). An isomorphism is a *bimorphism*, i.e., both a monomorphism and an epimorphism, but the converse does not hold true in general (**1.2.4.2**).

A category $\mathcal{D}$ is called a *subcategory* of $\mathcal{C}$ if $\mathrm{Ob}(\mathcal{D}) \subseteq \mathrm{Ob}(\mathcal{C})$ and for any $X, Y \in \mathrm{Ob}(\mathcal{D})$, $\mathrm{Hom}_{\mathcal{D}}(X, Y) \subseteq \mathrm{Hom}_{\mathcal{C}}(X, Y)$; the subcategory $\mathcal{D}$ is *full* if the latter inclusion is an equality, and then $\mathcal{D}$ is completely determined by $\mathrm{Ob}(\mathcal{D})$. The *opposite* of the category $\mathcal{C}$, written $\mathcal{C}^{\mathrm{op}}$, is defined by $\mathrm{Ob}(\mathcal{C}^{\mathrm{op}}) = \mathrm{Ob}(\mathcal{C})$ and $\mathrm{Hom}_{\mathcal{C}^{\mathrm{op}}}(X, Y) = \mathrm{Hom}_{\mathcal{C}}(Y, X)$; it is obtained from $\mathcal{C}$ by *dualization*, i.e., by "reversing the arrows".

Let X be an object of $\mathcal{C}$. The set P_X of all morphisms with target X has preorder $\leqq$ with $u \leqq v$ to mean that u factors through v (i.e., $u = vu'$ for some morphism u'). Two morphisms $u, v \in P_X$ are called *equivalent* if both $u \leqq v$ and $v \leqq u$. Dually, the set Q^X of all morphisms with source X has preorder $\geqq$ with $f \geqq g$ to mean that g factors through f (i.e., $g = g'f$ for some g'); f, g are called *equivalent* if both $f \geqq g$ and $g \geqq f$. A *subobject* of X is an equivalence class of monomorphisms belonging to P_X. Thus a representative of a subobject of X is a pair (Y, ι) such that $\iota : Y \to X$ is a monomorphism, called the *canonical injection* (or the *inclusion*) $Y \hookrightarrow X$. Dually, a *factor object* (or a *quotient object*) of X is an equivalence class of epimorphisms belonging to Q^X. A representative of a factor of X is a pair (Z, φ) such that $\varphi : X \to Z$ is an epimorphism, called the *canonical epimorphism* $X \twoheadrightarrow Z$. Abusing the language, the subobject (Y, ι) is denoted by Y ($Y \subseteq X$), and the factor object (Z, φ) is denoted by Z ($X \twoheadrightarrow Z$).

Remark 6. *Let (Y, ι_Y) be a representative of a subobject of X and let (Z, ι_Z) be a subobject of X (i.e., $Y \subseteq X$ and $Z \subseteq X$) such that ι_Z factors through ι_Y, i.e., $\iota_Z = \iota_Y \circ \iota_{Z,Y}$ where $\iota_{Z,Y} : Z \hookrightarrow Y$. Then $(Z, \iota_{Z,Y})$ is a representative of a subobject of Y and one can write $Z \subseteq Y$. Strictly speaking, the converse does not hold ([148], Sect. 1.1): if $(Z, \iota_{Z,Y})$ is a representative of a subobject of Y (so that $Z \subseteq Y$ and $Y \subseteq X$), then $(Z, \iota_{Z,Y})$ is not a representative of a subobject of X if $X \neq Y$. Nevertheless, $\iota_Z \triangleq \iota_Y \circ \iota_{Z,Y}$ is a monomorphism, so that abusing the language, and following Mitchell ([249], Sect. I.5), one* *can write $Z \subseteq X$. With this loose language, the relation $\subseteq$ becomes transitive and in both cases $Z \subseteq Y \subseteq X$. Although this is not fully correct, the risk of confusion is weak in practice.*

An object I in a category $\mathcal{C}$ is called *initial* if for any object X of $\mathcal{C}$ there exists a unique morphism $I \to X$. Dually, if I is such that for any object X of $\mathcal{C}$ there exists a unique morphism $X \to I$, I is said to be *terminal*. As easily seen, an initial (or terminal) object in a category $\mathcal{C}$, when it exists, is unique up to isomorphism. A *zero object* in $\mathcal{C}$ is an object which is both initial and terminal.

1.2.2.2 Examples

(1) The category denoted by **Set** is defined as follows: $\mathrm{Ob}(\mathbf{Set})$ is the class of all sets and for any $X, Y \in \mathrm{Ob}(\mathbf{Set})$, $\mathrm{Hom}_{\mathbf{Set}}(X, Y)$ is the set of all maps $X \to Y$. By a slight abuse of language, **Set** is called the *category of sets*, although a category is characterized by both its objects and its morphisms; here the morphisms are implied without ambiguity.

Lemma 7. *The category* **Set** *is a large category.*

Proof. If $\mathrm{Ob}(\mathbf{Set})$ is a set S, then the set $\mathcal{P}(S)$ of all parts of S belongs to S, thus $\mathrm{card}(\mathcal{P}(S)) \leq \mathrm{card}(S)$. But setting $\mathfrak{m} = \mathrm{card}(S)$, we have $\mathrm{card}(\mathcal{P}(S)) = 2^{\mathfrak{m}}$ ([26], §III.3, Prop. 12) and $\mathfrak{m} < 2^{\mathfrak{m}}$ by Cantor's theorem ([26], §III.3, Th. 2), thus we are led to a contradiction. ■

(2) Consider the category **Mon** of monoids (Subsect. 1.3.1 below): $\mathrm{Ob}(\mathbf{Mon})$ is the class of all monoids and for any $X, Y \in \mathrm{Ob}(\mathbf{Mon})$, $\mathrm{Hom}_{\mathbf{Mon}}(X, Y)$ is the set of all monoid-homomorphisms $X \to Y$.

(3) The category **Grp** of groups (Subsect. 1.6.1 below), the category **Rng** of rings, and that **Fld** of fields (Sect. 2.2 below) are likewise defined. So are also the categories ${}_{\mathbf{R}}\mathbf{Mod}$ (resp., $\mathbf{Mod}_{\mathbf{R}}$) of left (resp., right) **R**-modules and **R**-linear homomorphisms over a ring **R**, and, given a skew field **K**, the categories ${}_{\mathbf{K}}\mathbf{Vsp}$ (resp., $\mathbf{Vsp}_{\mathbf{K}}$) of left (resp., right) **K**-vector spaces and **K**-linear homomorphism (Sect. 2.2).

(4) The objects of the category **Top** of topological spaces (**1.7.1.2**) are all topological spaces, its morphisms are the continuous maps.

A *concrete category* is a subcategory of **Set**. All the above examples are concrete categories.

1.2.2.3 Products and Coproducts

The definition of a product in a category $\mathcal{C}$ (when such a product exists) is completely similar to that in the category **Set**: let $(X_i)_{i \in I}$ be a family of objects in a category $\mathcal{C}$. A *product* for this family is a family of morphisms $(\pi_i)_{i\in I}$, $\pi_i : X \to X_i$ with the property that for any family $(\alpha_i)_{i\in I}$, $\alpha_i : X'_i \to X_i$ there exists a unique morphism $\alpha : X' \to X$ such that $\pi_i\alpha = \alpha_i$ for all $i \in I$. The object X is denoted by $\prod_{i\in I} X_i$. The morphism $\pi_i : \prod_{i\in I} X_i \to X_i$ is called the *canonical projection* with index i (π_i is not necessarily an epimorphism). If for each $i \in I$, $X_i = X$, then $\prod_{i\in I} X_i = \prod_{i\in I} X$ is denoted by X^I which is called a *power* of X. If $I = \{1, ..., n\}$, X^I is also denoted by X^n and is called "X to the power n". The product of two objects X_1 and X_2 is denoted by $X_1 \times X_2$.

Arbitrary products exist in the category **Set** (**1.2.1.3**) and, as will be seen below, in the categories **Mon**, **Grp**, **Rng**, $_{\mathbf{R}}$**Mod** and **Mod**$_{\mathbf{R}}$. They also exist in **Top** ([34], §I.4) and in many other categories.

The *coproduct* of a family $(X_i)_{i\in I}$ of objects in a category $\mathcal{C}$ is defined dually to the product. Thus a coproduct for this family is a family of morphisms $(\iota_i)_{i\in I}$, $\iota_i : X_i \to X$ with the property that for any family $(\alpha_i)_{i\in I}$, $\alpha_i : X_i \to X'_i$ there exists a unique morphism $\alpha : X \to X'$ such that $\alpha\iota_i = \alpha_i$ for all $i \in I$. The object X is denoted by $\coprod_{i\in I} X_i$. The morphism $\iota_i : X_i \to \coprod_{i\in I} X_i$ is called the *canonical injection* with index i (ι_i is not necessarily a monomorphism). If for each $i \in I$, $X_i = X$, then $\coprod_{i\in I} X_i = \coprod_{i\in I} X$ is denoted by $X^{(I)}$ which is called a *copower* of X.

Arbitrary coproducts exist in the category **Set**, where they are called *disjoint unions* (**1.2.1.3**). They also exist in the category **Ab** of abelian groups (**1.6.4.3**), in $_{\mathbf{R}}$**Mod** and **Mod**$_{\mathbf{R}}$ (2.2.2.1), in **Top** (where a coproduct is a disjoint union of spaces) and in **Grp** (Sect. 1.6, Footnote 4).

Note that the existence of products in a category does not imply in general the existence of coproducts (for example, products, but not coproducts, exist in **Mon**) and the existence of coproducts does not imply in general the existence of products.

1.2.2.4 Generators and Cogenerators

Definition 8. *(i) An object U of a category $\mathcal{C}$ is a* generator *for $\mathcal{C}$ if the following condition holds: for any morphisms $f_1, f_2 : X \to Y$, there exists a morphism $\alpha : U \to X$ such that $f_1\alpha \neq f_2\alpha$ whenever $f_1 \neq f_2$.*
(ii) An object W of a category $\mathcal{C}$ is called a cogenerator *for $\mathcal{C}$ if it is a generator for $\mathcal{C}^{op}$, i.e., for any morphisms $f_1, f_2 : X \to Y$, there exists a morphism $\beta : Y \to W$ such that $\beta f_1 \neq \beta f_2$ whenever $f_1 \neq f_2$.*

Lemma 9. *If $\mathcal{C}$ has arbitrary* coproducts *(resp.,* products*), U is a* generator *(resp., W is a* cogenerator*) if, and only if every object is isomorphic to a factor of a copower $U^{(I)}$ (resp., to a subobject of a power W^I).*

Proof. We consider generators (the proof for cogenerators is obtained by dualization).

(a) Let $\mathcal{C}$ be a category with arbitrary coproducts, let X be an object of X, and let

$$I = \mathrm{Hom}_{\mathcal{C}}(U, X) = \{\alpha : \alpha \in \mathrm{Hom}_{\mathcal{C}}(U, X)\}.$$

By definition of a coproduct, we know that there exists $g : U^{(I)} \to X$ such that $g_{\alpha} = \alpha \ (\alpha \in I)$.

If g is not an epimorphism, the map

$$\mathrm{Hom}_{\mathcal{C}}(X, Y) \ni f \mapsto fg \in \mathrm{Hom}_{\mathcal{C}}\left(U^{(I)}, Y\right)$$

is not always injective, i.e., there exist an object Y and morphisms $f_1, f_2 : X \to Y$ such that

$$f_1 g = f_2 g \nRightarrow f_1 = f_2.$$

Now, $f_1 g = f_2 g$ means that $f_1 \alpha = f_2 \alpha$ for all $\alpha : U \to X$. Therefore, U is not a cogenerator.

(b) Conversely, suppose that there exists an epimorphism $g : U^{(I)} \twoheadrightarrow X$. Let $f_1, f_2 : X \to Y$ be distinct morphisms. Since $f_1 g \neq f_2 g$, there must exist an index $i \in I$ such that $f_1 \alpha \neq f_2 \alpha$ where $\alpha = g_i : U \to X$. Therefore, U is a cogenerator. ■

As will be seen (**3.2.1.1**), in the categories ${}_{\mathbf{R}}\mathbf{Mod}$ and $\mathbf{Mod}_{\mathbf{R}}$, the ring $\mathbf{R}$ (viewed as a module over itself) is a generator. The notion of cogenerator plays a crucial role in this book (**5.2.3.3**).

1.2.3 Functors

1.2.3.1 Functors and Subfunctors

Let $\mathcal{C}, \mathcal{D}$ be two categories. A *functor* (or, more specifically, a *covariant functor*) $\mathfrak{F} : \mathcal{C} \to \mathcal{D}$ consists of an "object-map" $\mathfrak{F} : \mathrm{Ob}(\mathcal{C}) \to \mathrm{Ob}(\mathcal{D})$ and for all $X, Y \in \mathcal{C}$, of an "arrow-map"[1] $\mathfrak{F} : \mathrm{Hom}_{\mathcal{C}}(X, Y) \to \mathrm{Hom}_{\mathcal{D}}(\mathfrak{F}(X), \mathfrak{F}(Y))$ such that:

$$\mathfrak{F}(Id_X) = Id_{\mathfrak{F}(X)},$$
$$\mathfrak{F}(f \circ g) = \mathfrak{F}(f) \circ \mathfrak{F}(g)$$

assuming that all morphisms in both $\mathcal{C}$ and $\mathcal{D}$ are right ones or that they are left ones.

A *contravariant functor* $\mathfrak{G} : \mathcal{C} \to \mathcal{D}$ is a covariant functor $\mathcal{C}^{\mathrm{op}} \to \mathcal{D}$. Thus, as opposed to a covariant functor, a contravariant functor "reverses the arrows" assuming that the morphisms in both $\mathcal{C}$ and $\mathcal{D}$ are right ones or that they are left ones:

$$\mathfrak{G}(Id_X) = Id_{\mathfrak{G}(X)},$$
$$\mathfrak{G}(f \circ g) = \mathfrak{G}(g) \circ \mathfrak{G}(f). \tag{1.4}$$

[1] The "object map" and the "arrow map" are often denoted by the same symbol.

Remark 10. *(i) If the morphisms in $\mathcal{C}$ are left (resp., right) ones and those in $\mathcal{D}$ are right (resp., left) ones, then (1.4) is changed to*

$$\mathfrak{G}(f \circ g) = \mathfrak{G}(f) \circ \mathfrak{G}(g). \tag{1.5}$$

(ii) Except when otherwise stated, all functors are covariant in the sequel. Let **P** *be a property satisfied by certain functors. A contravariant functor* $\mathfrak{G} : \mathcal{C} \to \mathcal{D}$ *is said to have the property* **P** *is the associated covariant functor* $\mathfrak{G} : \mathcal{C} \to \mathcal{D}$ *has this property.*

The *identity functor* of the category $\mathcal{C}$ is defined as: $Id_{\mathcal{C}}(f) = f$ $(f \in \mathrm{Mor}(\mathcal{C}))$, $Id_{\mathcal{C}}(X) = X$ $(X \in \mathrm{Ob}(\mathcal{C}))$. The *forgetful functor* $\mathfrak{U}$ is another simple example of functor: let $\mathcal{C}$ be a concrete category; $\mathfrak{U} : \mathcal{C} \to \mathbf{Set}$ is the functor described as follows: given any object X and any morphism f of $\mathcal{C}$, $\mathfrak{U}(X)$ is just the set of all elements belonging to X and $\mathfrak{U}(f)$ is the same map f, regarded just as a function between sets. Other kinds of forgetful functors can be considered. For example, there exists a forgetful functor $\mathfrak{U}' : \mathbf{Rgn} \to \mathbf{Ab}$ which assigns to each ring the underlying abelian group and to each ring-homomorphism the underlying $\mathbb{Z}$-linear homomorphism.

Let $\mathfrak{F} : \mathcal{C} \to \mathcal{D}$ be a functor and let $\mathcal{A}$ be a subcategory of $\mathcal{D}$ such that for any $X \in \mathrm{Ob}(\mathcal{C})$, $\mathfrak{F}(X) \in \mathrm{Ob}(\mathcal{A})$ and for any $f \in \mathrm{Mor}(\mathcal{C})$, $\mathfrak{F}(f) \in \mathrm{Mor}(\mathcal{A})$. Consider the functor $\mathfrak{G} : \mathcal{C} \to \mathcal{A}$ defined by $\mathfrak{G}(X) = \mathfrak{F}(X)$ for any $X \in \mathrm{Ob}(\mathcal{C})$, and $\mathfrak{G}(f) = \mathfrak{F}(f)$ for any $f \in \mathrm{Mor}(\mathcal{C})$. This functor is said to be *induced* by $\mathfrak{F}$; it is denoted by $\mathfrak{F} : \mathcal{C} \to \mathcal{D}$.

Let $\mathcal{C}$ be a concrete category and let $\mathfrak{F}$ be a functor $\mathcal{C} \to \mathbf{Set}$. A *subfunctor* of $\mathfrak{F}$ is a functor $\mathfrak{G} : \mathcal{C} \to \mathbf{Set}$ such that for any object $X \in \mathrm{Ob}(\mathcal{C})$, $\mathfrak{G}(X) \subseteq \mathfrak{F}(X)$ and for any morphism $u : X \to Y$ of $\mathcal{C}$, $\mathfrak{G}(u) : \mathfrak{G}(X) \to \mathfrak{G}(Y)$ coincides with $\mathfrak{F}(u)$ in $\mathfrak{G}(X)$.

Definition 11. *Let* $\mathfrak{F} : \mathcal{C} \to \mathcal{D}$ *be a functor.*

(i) $\mathfrak{F}$ *is* full *when for every pair of objects* X, X' *of* $\mathcal{C}$ *and for every morphism* $g : \mathfrak{F}(X) \to \mathfrak{F}(X')$ *in* $\mathcal{D}$*, there exists a morphism* $f : X \to X'$ *such that* $g = \mathfrak{F}(f)$*;* $\mathfrak{F}$ *is* faithful *when for every pair of objects* X, X' *of* $\mathcal{C}$ *and for every pair of morphisms* $f_1, f_2 : X \to X'$*, the equality* $\mathfrak{F}(f_1) = \mathfrak{F}(f_2)$ *implies* $f_1 = f_2$*;* $\mathfrak{F}$ *is* fully faithful *when* $\mathfrak{F}$ *is both full and faithful. For short,* $\mathfrak{F}$ *is faithful (resp., full, fully faithful) when its* arrow-map *is injective (resp., surjective, bijective).*
(ii) $\mathfrak{F}$ *is* injective *(resp.,* surjective, bijective*) if its object-map is injective (resp., surjective, bijective).*
(iii) $\mathfrak{F}$ *is an* isomorphism *if it is both bijective and fully faithful. An* antiisomorphism $\mathcal{C} \to \mathcal{D}$ *is an isomorphism* $\mathcal{C}^{op} \to \mathcal{D}$*.*
(iv) $\mathfrak{F}$ *is called an* equivalence *if there exist* $\mathfrak{G} : \mathcal{D} \to \mathcal{C}$ *such that* $\mathfrak{G} \circ \mathfrak{F}$ *(resp.,* $\mathfrak{F} \circ \mathfrak{G}$*) is* naturally isomorphic *to* $Id_{\mathcal{C}}$ *(resp.,* $Id_{\mathcal{D}}$*)* (**1.2.3.2**)*. A* duality $\mathcal{C} \to \mathcal{D}$ *is an equivalence* $\mathcal{C}^{op} \to \mathcal{D}$*.*

From the above, a subcategory $\mathcal{D}$ of $\mathcal{C}$ is full if, and only if the *inclusion functor* $\mathcal{D} \to \mathcal{C}$ (defined in an obvious way) is full. A full subcategory $\mathcal{D}$ of a given category C is determined by specifying just the class of its objects since its morphisms are all morphisms $X \to Y$ in $\mathcal{C}$ where $X, Y \in \mathrm{Ob}(\mathcal{D})$.

Let $\mathcal{C}_1, \mathcal{C}_2, \mathcal{D}$ be three categories. A functor $\mathfrak{F} : \mathcal{C}_1 \times \mathcal{C}_2 \to \mathcal{D}$ is called a *bifunctor*.

1.2.3.2 Natural Transformations

Let $\mathfrak{F} : \mathcal{C} \to \mathcal{D}$ and $\mathfrak{G} : \mathcal{C} \to \mathcal{D}$ be two functors. A *natural transformation* $t : \mathfrak{F} \to \mathfrak{G}$ is a class of morphisms $t_A : \mathfrak{F}(A) \to \mathfrak{G}(A)$, one for each $A \in \mathrm{Ob}(\mathcal{C})$, such that the diagram

$$\begin{array}{ccc} \mathfrak{F}(A) & \xrightarrow{\mathfrak{F}(f)} & \mathfrak{F}(B) \\ \downarrow t_A & & \downarrow t_B \\ \mathfrak{G}(A) & \xrightarrow{\mathfrak{G}(f)} & \mathfrak{G}(B) \end{array}$$

is commutative for every $f : A \to B$ in $\mathcal{C}$. A natural transformation t is also called a *morphism of functors*, and a *natural isomorphism* or an *isomorphism of functors* if each t_A is an isomorphism.

A classical and important example of natural isomorphism is given in Corollary 274 below (**2.2.5.5**).

1.2.3.3 Universal Element

Let C be a category and let $\mathfrak{F} : C \to \mathbf{Set}$ be a functor.

Definition 12. *A* universal element (u, R) *for* $\mathfrak{F}$ *is a pair* (u, R) *consisting of an object* $R \in \mathrm{Ob}(\mathcal{C})$ *and an element* $u \in \mathfrak{F}(R)$ *with the following property: for each object* $S \in \mathrm{Ob}(\mathcal{C})$ *and each element* $s \in \mathfrak{F}(S)$*, there exists a unique morphism* $h : R \to S$ *such that* $\mathfrak{F}(h)(u) = s$*. (For short,* u*, in place of the pair* (u, R)*, is called a universal element for* $\mathfrak{F}$*.) A* universal problem *consists in determining a universal element.*

This is depicted by the diagram below:

$$\begin{array}{ccc} R & u \in & \mathfrak{F}(R) \\ h\downarrow & \downarrow & \downarrow \mathfrak{F}(h) \\ S & s \in & \mathfrak{F}(S) \end{array}$$

Theorem 13. *(i) If* $u \in \mathfrak{F}(R)$ *is a universal element for the functor* $\mathfrak{F} : \mathcal{C} \to \mathcal{D}$*, then for any set* $S \in \mathrm{Ob}(\mathcal{C})$ *there exists a bijection*

$$\theta : S^R \cong \mathfrak{F}(R), \quad h \mapsto \mathfrak{F}(h)(u).$$

(ii) If $u \in \mathfrak{F}(R')$ *and* $u' \in \mathfrak{F}(R)$ *are two universal elements for the same functor* $\mathfrak{F}$*, then there exists an* isomorphism $\beta : R \xrightarrow{\sim} R'$ *such that* $\mathfrak{F}(\beta)(u) = u'$*.*

Proof. (i): Each $\mathfrak{F}(h)(u)$ is an element $s \in \mathfrak{F}(S)$ and conversely each $s \in \mathfrak{F}(S)$ is of the form $\mathfrak{F}(h)(u)$ for a unique $h \in S^R$ by Definition 12.

(ii): Since u is universal, there exists $h : R \to R'$ such that $u' = \mathfrak{F}(h)(u)$; similarly, since u' is universal, there exists $h' : R' \to R$ such that $u = \mathfrak{F}(h')(u')$. Therefore,

$$u = \mathfrak{F}(h')(\mathfrak{F}(h)(u)) = \mathfrak{F}(h') \circ (\mathfrak{F}(h)(u)) = \mathfrak{F}(h' \circ h)(u).$$

We also have $u = \mathfrak{F}(Id_R)(u)$, thus both $h' \circ h$ and Id_R transform u into u, thus $h' \circ h = Id_R$ by universality of u. By a similar rationale, $h \circ h' = Id_{R'}$ by universality of u', thus $h' = h^{-1}$ and $h = \beta$ is an isomorphism. ∎

An equalizer is an example of universal element ([224], sect. III.4). Let us define this notion without using functors, for the sake of simplicity.

Consider two morphism $f, g : Y \to Z$ and the following universal problem: find $h : X \to Y$ such that $fh = gh$ and for any $h' : Z' \to Y$ with $fh' = gh'$ there exists a unique $h'' : Z' \to Z$ with $hh'' = h'$.

Definition 14. *The pair (h, X) is an equalizer of (f, g). This equalizer is represented by the diagram below:*

$$Z \xrightarrow{h} Y \overset{f,g}{\rightrightarrows} X, \quad fh = gh.$$

An equalizer is necessarily a monomorphism. The notion of *coequalizer* is defined dually, and is an epimorphism. We say that a category $\mathcal{C}$ has equalizers (resp., has coequalizers) if each pair of morphisms of $\mathcal{C}$ has an equalizer (resp., a coequalizer).

In the category **Set**, consider $f, g : Y \to Z$. The equalizer of (f, g) is (X, h) where $X = \{y \in Y : f(y) = g(y)\}$ and $h : X \to Y$ is the inclusion. The coequalizer of (f, g) is $(\pi, \mathcal{R})$ where $\pi : Z \to Z/\mathcal{R}$ is the projection and $\mathcal{R}$ is the coarsest equivalence relation whose graph contains all pairs $(f(y), g(y))$ for $y \in Y$.

1.2.3.4 Direct (or inductive) Limits

A *filtering set* I is a preordered set (**1.2.1.4**) such that every pair of elements has an upper bound, i.e., for any two elements $i, j \in I$, there exists an element $k \in I$ (not necessarily distinct from i, j) with $i \preceq k$ and $j \preceq k$.

Let $\mathcal{C}$ be a category, let $(F_i)_{i \in I}$ be a family of objects of $\mathcal{C}$, and for each pair $(i, j) \in I \times I$ such that $i \preceq j$, let $\varphi^i_j : F_i \to F_j$ be a morphism for which

(1) $\varphi^i_i = Id_{F_i}$ for all $i \in I$;

(2) if $i \preceq j \preceq k$, $\varphi^i_k = \varphi^j_k \circ \varphi^i_j$.

Definition 15. $\mathfrak{D} = \{F_i, \varphi^i_j\}$ *is a* direct system *in $\mathcal{C}$ with index set I.*

(This can be represented by a commutative diagram that the reader is requested to draw.) A direct system is easily shown to be a covariant functor

$\mathcal{C} \to \mathcal{C}$ ([306], sect. 5.2). In the sequel, the expressions "direct limit" and "inductive limit" are synonymous; the corresponding notion is called "filtrant inductive limit" in ([183], Chap. 3).

Definition 16. *Let* $\mathfrak{D} = \{F_i, \varphi_j^i\}$ *be a direct system in* $\mathcal{C}$*. The* direct *(or* inductive*)* limit *of this system, written* $\varinjlim F_i$*, is (if it exists) a functor* $\mathcal{C} \to \mathcal{C}$ *consisting of an object and a family of morphisms* $\alpha_i : F_i \to \varinjlim F_i$ *with* $\alpha_i = \alpha_j \circ \varphi_j^i$ *such that for any object* X *and morphisms* $f_i : F_i \to X$*, there exists a morphism* $\beta : \varinjlim F_i \to X$ *such that* f_i *factors through* β *following* $f_i = \beta \circ \alpha_i$.

The direct limit is a universal element for the functor $\mathfrak{D}$ (Definition 12), thus, if it exists, it is unique up to isomorphism (Theorem 13(ii)). This is why we have spoken of *the* inductive limit instead of *an* inductive limit.

Proposition 17. *If* $\mathcal{C}$ *is a category with coequalizers and coproducts indexed by* I*, every direct system indexed by* I *in* $\mathcal{C}$ *has a direct limit.*

Proof. 1) Assume that $\mathcal{C} =$ **Set** or **Top** and proceed as follows: define

$$\varinjlim F_i = \left(\coprod_{i \in I} F_i \right) / \mathcal{R}$$

where $\mathcal{R}$ is the equivalence relation in $\coprod_{i \in I} F_i$ given by $x_i \equiv x_j \pmod{\mathcal{R}}$ (with $x_i \in F_i$, $x_j \in F_j$) if for any $(i,j) \in I \times I$ there exists an upper bound k of i and j such that $\varphi_k^i(x_i) = \varphi_k^j(x_j)$. One obtains from this definition canonical morphisms $\alpha_i : F_i \to \varinjlim F_i$ sending x_i to its equivalence class.

2) Assume that $\mathcal{C} =$ **Ab** or, more generally, that $\mathcal{C}$ is an abelian category (**1.2.4.3**) with arbitrary coproducts. Consider the morphisms α_i defined by the compositions below:

$$\alpha_i : F_i \longrightarrow \left(\coprod_{i \in I} F_i \right) \longrightarrow \left(\coprod_{i \in I} F_i \right) / \left(\sum_{i \preceq j} \operatorname{im} \left(\iota_j \varphi_j^i - \iota_i \right) \right)$$

where ι_i is the canonical injection $F_i \to \coprod_{i \in I} F_i$. The object on the right-hand side is $\varinjlim F_i$; this object, along with the morphisms $\alpha_i : F_i \to \varinjlim F_i$, is the direct limit of the family of objects $(F_i)_{i \in I}$ ([148], Prop.1.8).

3) For the general case, see ([282], Chap. I, Theorem 4.1) and, for the dual statement, ([224], sect. V.2, Theorem 1). ∎

When a category satisfies the condition in the above proposition for every inverse system $\mathfrak{I}$, it is called *complete*. Let us give some examples.

Example 18. *Let* $\preceq$ *be the trivial preorder relation in* I*, i.e.,* $i \preceq j$ *if, and only if* $i = j$*. Then there is no* φ_j^i *with* $i \neq j$ *and* $x_i \mathcal{R} x_j$ *if, and only if* $i = j$*, thus* $\mathcal{R}$ *is the equality. Therefore* $\varinjlim F_i = \coprod F_i$.

Example 19. *Let $I = \mathbb{N}$ and (F_i) be a sequence of subsets (resp., submonoids, subgroups, submodules) of a set (resp., a monoid, a group, a left* **R***-module) X such that $F_i \subseteq F_{i+1}$; let $\varphi_j^i : F_i \to F_j$ be the inclusion for $i \leq j$. Obviously, $\{F_i, \varphi_j^i\}$ is a direct system in* **Set** *(resp.,* **Mon**, **Grp**, ${}_{\mathbf{R}}$**Mod***) and the object $\varinjlim F_i$ is the set of classes $\bar{x}_i$ where $x_i \equiv x_j \pmod{\mathcal{R}}$ if, and only if there exists $k \geq \max(i,j)$ such that x_i and x_j belong to the same set F_k. Therefore, $\varinjlim F_i = \bigcup_{i\geq 0} F_i$. This is called the* directed union *of the sequence (F_i).*

Example 20. *Let X be a topological space, let Y be a set, let P be a nonempty part of X, and let $(V_i)_{i\in I}$ be a filtering family of neighborhoods of P (such that $i \preceq j$ if, and only if $V_j \subseteq V_i$). Let F_i be the set of all maps $V_i \to Y$ and, for any pair (i,j) such that $i \preceq j$, let $\varphi_j^i : F_i \to F_j$ be such that for any $f : V_i \to Y$, $\varphi_j^i(f) = f|_{V_j}$ (the restriction of f to V_j). Then $\mathfrak{D} = \{F_i, \varphi_j^i\}$ is a direct system in the category $\mathcal{C}$, the objects of which are the functions defined in a neighborhood of P and the morphisms of which are the restrictions to a neighborhood of P; $\varinjlim F_i$ is the set of* germs *of maps defined in a neighborhood of P.*

A germ can also defined as an equivalent class of functions defined in a neighborhood of P; the equivalence relation $\mathcal{R}$ is defined as follows: let U and W be two neighborhoods of P, and let $f : U \to Y$, $g : W \to Y$ be two functions; then $f \equiv g \pmod{\mathcal{R}}$ if, and only if f and g coincide in a neighborhood of P, and the class of f is its germ. Such a germ can be viewed– roughly speaking – as a function $f : U \to Y$ where U is a neighborhood of P which is "small enough".

If Y is a topological space, one can define germs of continuous functions; if both X and Y are differentiable manifolds, one can define germs of C^∞ functions (taking for the V_i open *neighborhoods), etc.*

1.2.3.5 Inverse (or projective) Limits

The definition and the construction of inverse limits is "dual" of that of direct limits, i.e., an inverse limit in a category $\mathcal{C}$ is a direct limit in the category $\mathcal{C}^{\mathrm{op}}$.

The definition of an inverse system is obtained by dualizing Definition 15. More specifically, let $\mathcal{C}$ be a category, let $(F_i)_{i\in I}$ be a family of objects of $\mathcal{C}$ and for each pair $(i,j) \in I \times I$ such that $i \preceq j$, let $\psi_i^j : F_j \to F_i$ be a morphism for which

(1') $\psi_i^i = Id_{F_i}$ for all $i \in I$;
(2') if $i \preceq j \preceq k$, $\psi_i^k = \psi_i^j \circ \psi_j^k$.

Definition 21. *$\mathfrak{I} = \left\{F_i, \psi_i^j\right\}$ is an* inverse system *in $\mathcal{C}$ with index set I.*

An inverse system can be proved to be a contravariant functor.

Definition 22. *Let* $\left\{F_i, \psi_i^j\right\}$ *be an inverse system in* $\mathcal{C}$. *The* inverse limit *(also called the* projective limit*) of this system, written* $\varprojlim F_i$, *is (if it exists) a functor* $\mathcal{C} \to \mathcal{C}$ *consisting of an object and a family of morphisms* $\alpha_i : \varprojlim F_i \to F_i$ *with* $\alpha_i = \psi_i^j \circ \alpha_j$ *such that for any object* X *and morphisms* $f_i : X \to F_i$, *there exists a morphism* $\beta : X \to \varprojlim F_i$ *such that* f_i *factors through* β *following* $f_i = \alpha_i \circ \beta$.

The inverse limit is a universal element for the functor $\mathfrak{I}$, thus, if it exists, it is unique up to isomorphism.

Proposition 23. *If* $\mathcal{C}$ *is a category with equalizers and products indexed by* I, *every inverse system indexed by* I *in* $\mathcal{C}$ *has an inverse limit.*

Proof. Dualize the proof of Proposition 17. Let us detail the case where $\mathcal{C} = \mathbf{Set}$ (the same construction is valid when $\mathcal{C}$ is **Mon**, **Grp**, **Rgn**, ${}_{\mathbf{R}}\mathbf{Mod}$ or **Top**). For each $i \in I$, let $\pi_j : \prod_{i \in I} F_i \to F_j$ be the canonical projection with index j and define

$$\varprojlim F_i = \left\{(x_i) \in \prod\nolimits_{i \in I} F_i : x_i = \psi_i^j(x_j) \quad \text{whenever } i \preceq j\right\}$$

and $\alpha_j : \varprojlim F_i \to F_j$ as the restriction $\pi_j|_\Gamma$ where $\Gamma \triangleq \varprojlim F_i$. One can easily check that this construction solves the universal problem. ■

When a category satisfies the condition in the above proposition for every direct system $\mathfrak{D}$, it is called *cocomplete*. Let us give some examples:

Example 24. *Let* $\preceq$ *be the trivial preorder relation in* I *(Example 18). Then* $\varprojlim F_i = \prod_{i \in I} F_i$.

Example 25. *If* $\left\{F_i, \psi_i^j\right\}$ *is an inverse system with index* I *and* I *has a top element* ∞, *then* $\varprojlim F_i = F_\infty$.

1.2.3.6 Image and Exactness of a Functor

Let $\mathfrak{F} : \mathcal{C} \to \mathcal{D}$ be a functor.

Definition 26. *The* image *of* $\mathfrak{F}$, *written* $\mathfrak{F}(\mathcal{C})$, *is a class consisting of two subclasses: that of all objects* $\mathfrak{F}(X)$, $X \in \mathrm{Ob}(\mathcal{C})$, *and that of all morphisms* $\mathfrak{F}(f)$, $f \in \mathrm{Mor}(\mathcal{C})$.

Lemma 27. *The image* $\mathfrak{F}(\mathcal{C})$ *is a subcategory of* $\mathcal{D}$ *if* $\mathfrak{F}$ *is injective.*

Proof. Let $A, A' \in \mathcal{C}$ and $f_1 : A_1 \to A$, $f_2 : A' \to A_2$ be such that $\mathfrak{F}(f_1)$ and $\mathfrak{F}(f_2)$ are composable in $\mathcal{D}$, so that $\mathfrak{F}(f_2) \circ \mathfrak{F}(f_1) \in \mathrm{Mor}(\mathcal{D})$. This means that $\mathfrak{F}(A) = \mathfrak{F}(A')$. If $\mathfrak{F}$ is injective, this implies $A = A'$, thus (since $\mathcal{C}$ is a category) f_1 and f_2 are composable, i.e., $f_2 \circ f_1 \in \mathrm{Mor}(\mathcal{C})$, and

$\mathfrak{F}(f_2) \circ \mathfrak{F}(f_1) = \mathfrak{F}(f_2 \circ f_1)$ is a morphism of $\mathfrak{F}(\mathcal{C})$. Therefore, $\mathfrak{F}(\mathcal{C})$ is a category. ∎

Assume that $\mathcal{C}$ and $\mathcal{D}$ have finite projective (resp., inductive) limits.

Definition 28. *The functor $\mathfrak{F}$ is* left *(resp.,* right*)* exact *if it commutes with finite projective (resp., inductive) limits; $\mathfrak{F}$ is* exact *if it is both left and right exact. A functor which is both faithful and exact is called* faithfully exact.

1.2.3.7 Adjoint Functors

Let $\mathcal{A}$, $\mathcal{C}$ be two categories and let $\mathfrak{F} : \mathcal{A} \to \mathcal{C}$, $\mathfrak{G} : \mathcal{C} \to \mathcal{A}$ be two functors.

Definition 29. *The ordered pair $(\mathfrak{F}, \mathfrak{G})$ is an* adjoint pair *(i.e., $\mathfrak{G}$ is a right adjoint to $\mathfrak{F}$ and $\mathfrak{F}$ is a left adjoint to $\mathfrak{G}$) if for each $A \in \mathrm{Ob}(\mathcal{A})$ and $C \in \mathrm{Ob}(\mathcal{C})$ there is a bijection*

$$\tau = \tau_{A,C} : \mathrm{Hom}_{\mathcal{C}}(\mathfrak{F}(A), C) \to \mathrm{Hom}_{\mathcal{D}}(A, \mathfrak{G}(C))$$

which is natural in each variable (**1.2.3.2**).

The importance of this notion is due to the following theorem ([183], Prop. 2.1.10):

Theorem 30. *Let $\mathcal{A}$ and $\mathcal{C}$ be categories, and let $\mathfrak{F} : \mathcal{A} \to \mathcal{C}$ and $\mathfrak{G} : \mathcal{C} \to \mathcal{A}$ be two functors. Assume that $(\mathfrak{F}, \mathfrak{G})$ is an adjoint pair.*
(i) If $\mathcal{A}$ has direct (resp., inverse) limits indexed by a filtering set of indices I (**1.2.3.4**), (**1.2.3.5**), *then $\mathfrak{F}$ (resp., $\mathfrak{G}$) commutes with direct (resp., inverse) limits indexed by I.*
(ii) In particular, if $\mathcal{A}$ has finite direct (resp., inverse) limits, $\mathfrak{F}$ is right exact (resp., $\mathfrak{G}$ is left exact).

Remark 31. *The converse of Theorem 30 holds true under certain additional conditions, according to* Freyd's Adjoint Functor Theorem*; see ([224], Sect. 5.6) for more details.*

1.2.4 Abelian Categories

1.2.4.1 Additive Categories

A *preadditive* category is a category $\mathcal{C}$ in which each set $\mathrm{Hom}_{\mathcal{C}}(X, Y)$ (X, $Y \in \mathrm{Ob}(\mathcal{C})$) is an abelian group and for which composition of morphisms is bilinear. Let $\mathcal{C}$ and $\mathcal{D}$ be preadditive categories; then a functor $\mathfrak{F} : \mathcal{C} \to \mathcal{D}$ is said to be *additive* if its arrow-map is $\mathbb{Z}$-linear.

Lemma 32. *Let $\mathcal{C}$, $\mathcal{D}$ be two categories, the former preadditive, the latter arbitrary, and let $\mathfrak{F} : \mathcal{C} \to \mathcal{D}$ be a functor. If $\mathfrak{F}$ is faithful, then $\mathfrak{F}$ is injective and its image $\mathfrak{F}(\mathcal{C})$ is a subcategory of $\mathcal{D}$. In addition, $\mathfrak{F}$ induces an isomorphism $\mathfrak{F} : \mathcal{C} \xrightarrow{\sim} \mathfrak{F}(\mathcal{C})$.*

Proof. For any object $X \in \mathcal{C}$ consider the zero map $O_X : X \to X$; if $\mathfrak{F}(X) = \mathfrak{F}(X')$ $(X, X' \in \mathcal{C})$, then $\mathfrak{F}(O_X) = \mathfrak{F}(O_{X'})$, thus $O_X = O_{X'}$ if $\mathfrak{F}$ is faithful, and this implies $X = X'$ by (1.3). Therefore, $\mathfrak{F}$ is injective, and $\mathfrak{F}(\mathcal{C})$ is a subcategory of $\mathcal{D}$ by Lemma 27. The induced functor $\mathfrak{F} : \mathcal{C} \to \mathfrak{F}(\mathcal{C})$ is still injective and faithful, and by construction (1.2.3.1) it is full and surjective, thus it is an isomorphism (Definition 11(iii)). ■

Item (1) below is a generalization of Proposition 4; its proof is detailed in ([80], Theorem 2.1.1):

Lemma and Definition 33. *(1) Let $\mathcal{C}$ be a preadditive category with a zero object 0 (1.2.2.1), let $(X_i)_{1\leq i\leq n}$ be a finite family of objects of $\mathcal{C}$, let X be an object of $\mathcal{C}$, and define $2n$ morphisms $\pi_i : X \to X_i$, $\iota_i : X_i \to X$, such that*

$$\pi_i \iota_j = \delta_{ij} Id_{X_j}.$$

The following conditions are equivalent:
(i) $\sum_{1\leq i\leq n} \iota_i \pi_i = Id_X$.
(ii)

$$X = \coprod_{1\leq i\leq n} X_i$$

with canonical injections $\iota_i : X_i \to X$.
(iii)

$$X = \prod_{1\leq i\leq n} X_i$$

with canonical projections $\pi_i : X \to X_i$.
(2) When the equivalent conditions of (1) hold, (X, π_i, ι_i) is called a biproduct *of the X_i and is denoted by*

$$X = \bigoplus_{1\leq i\leq n} X_i.$$

An *additive* category is a preadditive category which has a zero object 0 and a biproduct $\bigoplus$ of each pair of its objects.

Some additive categories have *arbitrary* products and coproducts, for example $_{\mathbf{R}}\mathbf{Mod}$. But not all additive categories have arbitrary products and coproducts, e.g., the category of *finite* abelian groups and $\mathbb{Z}$-linear homomorphisms.

Let $f : X \to Y$; the *kernel* of f, denoted by $\ker f$, is the equalizer of $(f, 0)$. Theforefore, $\ker f$ is a morphism $\kappa : K \to X$ such that $f \circ \kappa = 0$ and every morphism α such that $f \circ \alpha = 0$ factors uniquely through κ (i.e., $\alpha = \kappa \circ \alpha'$ for a unique α'); if it exists, $\ker f$ is unique up to isomorphism since it is a universal element–and, more specifically, a *universal morphism.* The *cokernel* of f, denoted by $\operatorname{coker} f$, is dually defined as the coequalizer of $(f, 0)$. Therefore, $\operatorname{coker} f$ is a morphism $\gamma : Y \to C$ such that $\gamma \circ f = 0$ and every morphism β such that $\beta \circ f = 0$ factors uniquely through γ (i.e., $\beta = \beta' \circ \gamma$ for a unique β'); if it exists, $\operatorname{coker} f$ is unique up to isomorphism since it is, like $\ker f$, a universal morphism. Since $\ker f$ is a monomorphism and $\operatorname{coker} f$ is an epimorphism, it can be convenient to think of $\ker f : K \hookrightarrow X$ as a subobject of X and of $\operatorname{coker} f : Y \twoheadrightarrow C$ as a quotient of Y. As easily seen, $\ker f$ is the largest subobject of X annihilated by f and $\operatorname{coker} f$ is the largest quotient of Y annihilating f. Assuming that $\ker f$ and $\operatorname{coker} f$ exist, we can also define the *image* and the *coimage* of f, respectively written $\operatorname{im} f$ and $\operatorname{coim} f$, as follows:

$$\operatorname{im} f = \ker \operatorname{coker} f, \quad \operatorname{coim} f = \operatorname{coker} \ker f. \tag{1.6}$$

If they exist, $\operatorname{im} f$ and $\operatorname{coim} f$ are unique up to isomorphism and are, respectively, a subobject of Y and a quotient of X.

1.2.4.2 Preabelian Categories

A *preabelian category* is an additive category in which each arrow has a kernel and a cokernel (thus an image and a coimage). As easily seen, a category $\mathcal{C}$ is preabelian if, and only if the opposite category $\mathcal{C}^{\mathrm{op}}$ is preabelian too.

Consider a morphism $f : X \to Y$. The following properties are equivalent: (i) f is a monomorphism (resp., an epimorphism); (ii) $\ker f = 0$ (resp., $\operatorname{coker} f = 0$); (iii) $\operatorname{coim} f \cong X$ (resp., $\operatorname{im} f = Y$). In addition, for any morphism $f : X \to Y$ we have

$$\ker \operatorname{coker} \ker f = \ker f, \quad \operatorname{coker} \ker \operatorname{coker} f = \operatorname{coker} f \tag{1.7}$$

as shown below (Example 100). In particular,

$$\operatorname{im} \ker g = \ker g, \quad \operatorname{coim} \operatorname{coker} u = \operatorname{coker} u. \tag{1.8}$$

Therefore by (1.6) we obtain:

Lemma 34. *(i) ι is a kernel if, and only if $\iota = \operatorname{im} \iota$.*
(ii) β is a cokernel if, and only if $\beta = \operatorname{coim} \beta$.
(iii) If $\iota = \operatorname{im} f$, then $\operatorname{coker} \iota = \operatorname{coker} f$ and $\operatorname{im} \iota = \operatorname{im} f = \iota$.

Consider the following diagram in a preabelian category, where $\pi : X \twoheadrightarrow \operatorname{coim} f$ is the canonical epimorphism and $\iota : \operatorname{im} f \hookrightarrow Y$ is the inclusion:

$$\begin{array}{ccccccc} \ker f & \xrightarrow{\kappa} & X & \xrightarrow{f} & Y & \xrightarrow{\gamma} & \operatorname{coker} f \\ & & \pi \downarrow & & \uparrow \iota & & \\ & & \operatorname{coim} f & \xrightarrow{\bar{f}} & \operatorname{im} f & & \end{array}$$

Lemma 35. *There exists a unique morphism $\bar{f} : \operatorname{coim} f \to \operatorname{im} f$, called the* induced morphism, *for which the above diagram is commutative.*

Proof. Since $f \circ \kappa = 0$ and $\pi = \operatorname{coim} f = \operatorname{coker} \kappa$, there exists $\tilde{f} : \operatorname{coim} f \to Y$ such that f factors uniquely through π, say $f = \tilde{f} \circ \pi$ where $\tilde{f} : \operatorname{coim} f \to Y$;

in addition, $\gamma \circ f = \gamma \circ \tilde{f} \circ \pi = 0$ and π is an epimorphism, thus $\gamma \circ \tilde{f} = 0$. Hence $\tilde{f}$ factors uniquely through $\ker \gamma = \operatorname{im} f$, i.e., there exists a unique morphism $\bar{f} : \operatorname{coim} f \to \operatorname{im} f$ such that $\tilde{f} = \iota \circ \bar{f}$ where $\iota : \operatorname{im} f \to Y$ is the inclusion. ■

Example 36. *(1) In the category* **TVS** *of all topological vector spaces over* $\mathbf{k}$ *(*$\mathbf{k} = \mathbb{R}$ *or* $\mathbb{C}$*) and* $\mathbf{k}$*-linear continuous maps (see Subsect. 1.7.1 below), let* $f : X \to Y$ *be a morphism (i.e., by definition, a* $\mathbf{k}$*-linear continuous map). Then* $\ker f = f^{-1}(\{0\})$, $\operatorname{im} f = f(X)$, $\operatorname{coker} f = Y/\operatorname{im} f$ *and* $\operatorname{coim} f = X/\ker f$. *Consider the induced morphism* $\bar{f} : \operatorname{coim} f \to \operatorname{coker} f$. *When* $\bar{f}$ *is an isomorphism,* f *is called a* strict morphism; *this happens if, and only if the following equivalent conditions are satisfied ([34], n°III.2.8):*
(a) f *is relatively open, i.e., the image* $f(U)$ *of any open subset of* X *is open in* $f(X)$;
(b) the image $f(U)$ *of any open neighborhood of* 0 *in* X *is a neighborhood of* 0 *in* $f(X)$.
(2) The same holds in the category **LCS** *of all locally convex topological* $\mathbf{k}$*-spaces (Subsect. 1.7.2 below), which is a full subcategory of* **TVS**.

Example 37. *(i) The category* **HTVS** *of all Hausdorff topological* $\mathbf{k}$*-vector spaces (*$\mathbf{k} = \mathbb{R}$ *or* $\mathbb{C}$*) and continuous* $\mathbf{k}$*-linear maps is a full subcategory of* **TVS**. *In* **HTVS**, *let* $f : X \to Y$ *be a morphism. Then* $\ker f = f^{-1}(\{0\})$, $\operatorname{im} f = \overline{f(X)}$ *(where* $\overline{f(X)}$ *is the closure of the space* $f(X)$ *in* Y*),* $\operatorname{coker} f = Y/\operatorname{im} f$ *and* $\operatorname{coim} f = X/\ker f$ *(see Exercise 195). When the induced morphism* $\bar{f}$ *is an isomorphism in* **HTVS**, f *is called a* strict morphism.
(ii) Let **HLCS** *be the category of Hausdorff locally convex* $\mathbf{k}$*-spaces;* **HLCS** *is a full subcategory of both* **LCS** *and* **HTVS**. *In* **HLCS**, *a morphism* $f : X \to Y$ *is strict if, and only if the following two conditions hold ([36], §IV.4, Prop. 2):*
(c_1*)* $\operatorname{im} {}^t f$ *is closed in the weak* dual* X'_σ *of* X;
(c_2*) for every equicontinuous part* A *of* X' *included in* $\operatorname{im} {}^t f$, *there exists an equicontinuous part* B *of* Y' *such that* $A = {}^t f(B)$.
A continuous $\mathbf{k}$*-linear map* $f : X_\sigma \to Y_\sigma$ *(where* X_σ *and* Y_σ *denote the* $\mathbf{k}$*-spaces* X *and* Y *endowed with the weakened topologies) is a strict morphism (and is called a* weak strict morphism*) if, and only if (*c_1*) holds ([36], §II.6, Corol. 3 of Prop. 7).*
(iii) The category **Frt** *of all Fréchet* $\mathbf{k}$*-spaces is a full subcategory of* **HTVS** *and the category* **Ban** *of all Banach* $\mathbf{k}$*-spaces is a full subcategory of* **Frt** *(see Subsect. 1.7.3 below). In* **Frt**, *the following conditions are equivalent, according to the* Banach-Schauder theorem *([36], n°IV.4.2):*
(α*)* $f : X \to Y$ *is a strict morphism;*
(β*)* $f(X)$ *is closed in* Y;
(γ*)* ${}^t f : Y'_\sigma \to X'_\sigma$ *is a strict morphism;*
(δ*)* f *is a weak strict morphism.*

Definition 38. *A preabelian category is called* semiabelian *if for any morphism f, the induced morphism $\bar{f}$ is a bimorphism* (**1.2.2.1**).

See Exercise 196. The terminology introduced in Examples 36 and 37 is now generalized.

Definition 39. *In a preabelian category $\mathcal{C}$, a morphism f is said to be* strict *if the induced morphism $\bar{f}$ is an isomorphism.*

Remark 40. *Let $f : X \to Y$ be a strict monomorphism (resp., epimorphism) in a preabelian category $\mathcal{C}$. Then f is a kernel (resp., a cokernel) of $\gamma : Y \twoheadrightarrow$* $\operatorname{coker} f$ *(resp., $\iota : \ker f \hookrightarrow X$). (See Exercise 198.)*

Definition 41. *In a preabelian category $\mathcal{C}$, a sequence*

$$X_1 \xrightarrow{f_1} X_2 \xrightarrow{f_2} X_3 \longrightarrow 0 \tag{1.9}$$

is exact if f_2 (or, more specifically, (f_2, X_3)) is a cokernel of f_1. Dually, a sequence

$$X_1 \xleftarrow{f_1} X_2 \xleftarrow{f_2} X_3 \longleftarrow 0 \tag{1.10}$$

is exact if f_2 (or, more specifically, (f_2, X_3)) is a kernel of f_1.

Left exactness and right exactness of a functor have been defined in an arbitrary category with finite direct and inverse limits (Definition 28). In a preabelian category, these notions can be expressed in terms of exact sequences, as shown by the theorem below ([183], proof of Prop. 8.3.18).

Theorem 42. *Let $\mathcal{C}$ be a preabelian category.*
(i) $\mathcal{C}$ has finite direct and inverse limits.
(ii) An additive covariant functor $\mathfrak{F}$ is left exact *if, and only if exactness of (1.10) implies exactness of*

$$\mathfrak{F}(X_1) \xleftarrow{\mathfrak{F}(f_1)} \mathfrak{F}(X_2) \xleftarrow{\mathfrak{F}(f_2)} \mathfrak{F}(X_3) \longleftarrow 0. \tag{1.11}$$

Such a functor is right exact *if, and only if exactness of (1.9) implies exactness of*

$$\mathfrak{F}(X_1) \xrightarrow{\mathfrak{F}(f_1)} \mathfrak{F}(X_2) \xrightarrow{\mathfrak{F}(f_2)} \mathfrak{F}(X_3) \longrightarrow 0. \tag{1.12}$$

(iii) Dually, a contravariant additive functor $\mathfrak{F}$ is left (resp., right) exact if, and only if exactness of (1.9) (resp., (1.10)) implies exactness of (1.11) (resp., (1.12)).
(iv) A functor (covariant or contravariant) is exact if is both left and right exact.

Theorem and Definition 43. *Let $\mathcal{C}$ be an additive category.*
(1) The bifunctor $\operatorname{Hom}_{\mathcal{C}}(\bullet, \bullet) : \mathcal{C} \times \mathcal{C} \to$ **Ab** *(where* **Ab** *denotes the category of abelian groups and $\mathbb{Z}$-linear maps) is defined as follows:*

(i) For any object X *of* $\mathcal{C}$, *the functor* $\mathfrak{A}_X = \mathrm{Hom}_{\mathcal{C}}(X, \bullet) : \mathcal{C} \to \mathbf{Ab}$ *is such that for any object* $Y \in \mathcal{C}$, $\mathfrak{A}_X(Y)$ *is the abelian group* $\mathrm{Hom}_{\mathcal{C}}(X, Y)$ *and for any morphism* $\gamma : Y \to Y'$ *of* $\mathcal{C}$, $\mathfrak{A}_X(\gamma)$ *is the left-composition by* γ, *i.e., for any morphism* $f : X \longrightarrow Y$, $\mathfrak{A}_X(\gamma)(f) = \gamma f$;
(ii) For any object Y *of* $\mathcal{C}$, *the functor* $\mathfrak{B}_Y = \mathrm{Hom}_{\mathcal{C}}(\bullet, Y)$ *is such that for any object* $X \in \mathcal{C}$, $\mathfrak{B}_Y(X)$ *is again the abelian group* $\mathrm{Hom}_{\mathcal{C}}(X, Y)$ *and for any morphism* $\alpha : X' \to X$, $\mathfrak{B}_Y(\alpha)$ *is the right-composition by* α, *i.e., for any morphism* $g : X \longrightarrow Y$, $\mathfrak{B}_Y(\alpha)(g) = g\alpha$.
See the two schemes below.

$$\begin{array}{ccc} Y & \xrightarrow{\gamma} & Y' \\ & \nwarrow^{f} & \uparrow \mathfrak{A}_X(\gamma)(f) \\ & & X \end{array} \qquad \begin{array}{ccc} X & \xleftarrow{\alpha} & X' \\ & \searrow^{g} & \downarrow \mathfrak{B}_Y(\alpha)(g) \\ & & Y \end{array}$$

(2) The functor $\mathfrak{A}_X = \mathrm{Hom}_{\mathcal{C}}(X, \bullet)$ *is covariant and the functor* $\mathfrak{B}_Y = \mathrm{Hom}_{\mathcal{C}}(\bullet, Y)$ *is contravariant. If* $\mathcal{C}$ *is preabelian, these two functors are left exact.*
(3) Assuming that $\mathcal{C}$ *has direct and inverse limits, in particular if it has arbitrary products (resp., coproducts), an object* X *is a generator (resp., a cogenerator) in* $\mathcal{C}$ (**1.2.2.4**) *if, and only if the functor* $\mathfrak{A}_X = \mathrm{Hom}_{\mathcal{C}}(X, \bullet)$ *(resp.,* $\mathfrak{B}_X = \mathrm{Hom}_{\mathcal{C}}(\bullet, X)$*) is faithful.*

Proof. (2): See ([183], Prop. 3.3.7); see also Exercise 200. (3) is clear. ∎

Definition 44. *In a preabelian category* $\mathcal{C}$, *an object* X *is called* projective *if* $\mathrm{Hom}_{\mathcal{C}}(X, \bullet)$ *transforms epimorphisms into surjections, and* injective *if* $\mathrm{Hom}_{\mathcal{C}}(\bullet, X)$ *transforms monomorphisms into surjections.*

Lemma 45. *(1) An object* I *is injective in a preabelian category* $\mathcal{C}$ *if, and only if it is projective in the category* $\mathcal{C}^{op}$.
(2) Specifically,
(i) An object P *is projective if, and only if, given a diagram with exact row as shown, there exists a map* $P \to A$ *to make the triangle commutative.*

$$\begin{array}{ccccc} & & P & & \\ & & \downarrow & & \\ A & \longrightarrow & B & \longrightarrow & 0 \end{array} \quad \Longrightarrow \quad \begin{array}{ccccc} & & P & & \\ & \swarrow & \downarrow & & \\ A & \longrightarrow & B & \longrightarrow & 0 \end{array}$$

(ii) Dually, an object I *is injective if, and only if, given a diagram with exact row as shown, there exists a map* $B \to I$ *to make the triangle commutative.*

$$\begin{array}{ccccc} 0 & \longrightarrow & A & \longrightarrow & B \\ & & \downarrow & & \\ & & I & & \end{array} \quad \Longrightarrow \quad \begin{array}{ccccc} 0 & \longrightarrow & A & \longrightarrow & B \\ & & \downarrow & \swarrow & \\ & & I & & \end{array}$$

(3) Assuming that $\mathcal{C}$ *has coproducts (resp., products) with index set* I *(possibly filtering),*

$$P = \coprod_{i \in I} P_i \quad (\text{resp., } I = \prod_{i \in I} I_i)$$

is projective (resp., injective) if, and only if each P_i *(resp.,* I_i*) is projective (resp., injective).*

Proof. (1) and (2) are clear. For (3), see ([249], Chap. II, Prop. 14.3). ∎

1.2.4.3 Abelian Categories

The proof of Item (1) below is easy:

Lemma and Definition 46. *Let* $\mathcal{C}$ *be a preabelian category.*
(1) The following conditions are equivalent:
(i) Every morphism is strict.
(ii) Every monomorphism is a kernel and every epimorphism is a cokernel.
(2) A preabelian category $\mathcal{C}$ *is* abelian *if it satisfies the above equivalent conditions.*

Therefore, every abelian category is a semiabelian category (Definition 38), but the converse does not hold.

Let $f : X \to Y$ and $g : Y \to Z$ be two morphisms in an abelian category $\mathcal{C}$ and consider the following diagram:

$$X \xrightarrow{f} Y \xrightarrow{g} Z.$$

Lemma and Definition 47. *(1) The following conditions are equivalent:*
(i) $\operatorname{im} f = \ker g$;
(ii) $\operatorname{coker} f = \operatorname{coim} g$.
(2) The pair of composable morphisms (f, g) *is said to be* exact *if the above equivalent conditions are satisfied.*

Proof. (1). (i)⇒(ii): If (i) holds, then $\ker \operatorname{coker} f = \ker g$ and $\operatorname{coker} \ker \operatorname{coker} f = \operatorname{coker} \ker g = \operatorname{coim} g$, thus (ii) holds by (1.7).

(ii)⇒(i): If (ii) holds, then $\operatorname{coker} f = \operatorname{coker} \ker g$ and $\ker \operatorname{coker} f = \ker \operatorname{coker} \ker g$, thus (i) holds by (1.7). ∎

Consider the following sequence of composable morphisms in an abelian category:

$$\ldots \longrightarrow X_{n-1} \xrightarrow{f_n} X_n \xrightarrow{f_{n+1}} X_{n+1} \longrightarrow \ldots \quad (n \geq 1)$$

Definition 48. *(i) The above sequence is said to be* exact at X_n *if* (f_n, f_{n+1}) *is exact, and it is said to be* exact *if each adjacent pair of morphisms is exact.*
(ii) Then, if $n = 4$ *and* $X_0 = X_4 = 0$, *this sequence is called a* short exact sequence.

(iii) A sequence

$$X \xrightarrow{f} Y \xrightarrow{g} Z$$

is called inexact *if* $\operatorname{im} f \neq \ker g$.

Proposition 49. *(1) A functor $\mathfrak{F}$ between abelian categories is faithful if, and only if it preserves inexact sequences.*
(2) An object X is projective (resp., injective) in an abelian category $\mathcal{C}$ if, and only if the functor $\operatorname{Hom}_{\mathcal{C}}(X, \bullet)$ *(resp.,* $\operatorname{Hom}_{\mathcal{C}}(\bullet, X)$*) is exact.*

Proof. (1): See ([80], Proposition 2.2.6). (2): See ([249], Sect. II.14). ∎

As easily shown, the categories $_{\mathbf{R}}\mathbf{Mod}$ and $\mathbf{Mod}_{\mathbf{R}}$ (detailed in Sect. 3.2 below) are abelian and can be considered as the "paradigms" of all abelian categories due to the following theorem, sometimes called the *Full Embedding Theorem* ([249], Chap. VI, Theorem 7.2), and stated below without proof:

Theorem 50. *(*Freyd-Mitchell theorem*). If $\mathcal{C}$ is a small abelian category, then there exist a ring* $\mathbf{R}$ *and a fully faithful exact functor $\mathfrak{F} : \mathcal{C} \to {_{\mathbf{R}}\mathbf{Mod}}$ (Definitions 11 & 28). Therefore, $\mathcal{C}$ is equivalent to a fully abelian subcategory of* $_{\mathbf{R}}\mathbf{Mod}$.

The above theorem can be used to generalize some results, classical in the category $_{\mathbf{R}}\mathbf{Mod}$, to any small abelian category. So are the *Five lemma*, the *Snake lemma* and the *Nine lemma* (also called the 3×3 lemma): see Exercise 201, as well as Exercise 674 and Theorem 644(ii) below.

We can now establish Noether's isomorphism theorems which are very important for the sequel. For this we need the notion of *lattice*, detailed a little further (Definition 108(i)).

Lemma 51. *Let X be an object of an abelian category $\mathcal{C}$. The family of all subobjects of X is a lattice.*

Proof. (i) Let $((A, \iota_A), (B, \iota_B))$ be a pair of subobjects of X of a category $\mathcal{C}$ "having pullbacks", e.g., an abelian category. As shown in ([224], Sect. V.7), there exists a subobject (C, ι_C) of X such that

$$\iota_C \leqq \iota_A, \quad \iota_C \leqq \iota_B$$

and whenever (D, ι_D) is a subobject of X such that

$$\iota_D \leqq \iota_A, \quad \iota_D \leqq \iota_B,$$

then $\iota_D \leqq \iota_C$. The subobject (C, ι_C) is called the *greatest lower bound*, or *meet*, of $((A, \iota_A), (B, \iota_B))$, and is denoted by

$$(C, \iota_C) = (A, \iota_A) \wedge (B, \iota_B);$$

C is also called (abusing the language) the *intersection* of A and B, and is denoted as

$$C = A \cap B. \tag{1.13}$$

(ii) If $\mathcal{C}$ is an "exact category"–a stronger condition than having pullbacks ([249], Sect. I.15)–, then, as shown in ([249], Chap. I, Corol. 15.3), every pair of subobjects $((A, \iota_A), (B, \iota_B))$ of the same object X has a *least upper bound,* also called their *joint*, and denoted by

$$(D, \iota_D) = (A, \iota_A) \vee (B, \iota_B);$$

D is also called (abusing the language) the *union* of A and B, (written $D = A \cup B$). In particular, dualizing (i) in an abelian category $\mathcal{C}$, every pair of factor objects has a greatest lower bound, and since ker and coker are order-reversing and inverses of each other (Example 100 below), every pair of subobjects has a least upper bound. ∎

The above can be generalized in an obvious way, defining the union and the intersection of an arbitrary family of subobjects of $X \in \mathcal{C}$ (although they do not necessarily exist). In an abelian category, it is often convenient to denote the union of A and B as a *sum*, i.e.,

$$D = A + B. \tag{1.14}$$

Theorem 52. *(*Noether's isomorphism theorems in an abelian category*). In an abelian category $\mathcal{C}$, the following hold:*
(1) Let $f : X \to Y$. Then

$$\boxed{X/\ker f \cong \operatorname{im} f}$$

(first isomorphism).
(2) Let $X \in \mathcal{C}$ and $Y, Z \subseteq X$. Then

$$\boxed{Y/(Y \cap Z) \cong (Y + Z)/Z}$$

(second isomorphism).
(3) Let $X \in \mathcal{C}$ and $Z \subseteq Y \subseteq X$. Then

$$\boxed{X/Y \cong (X/Z)/(Y/Z)}$$

(third isomorphism).

Proof. (1) is obvious, whereas (2) and (3) are consequences of the Nine lemma: see ([127], Sect. 2.66) or ([249], Sect. I.16). ∎

The proof of Item (2) below is left to the reader (it is detailed in [80], Corol. 2.1.5):

Lemma and Definition 53. *In an abelian category* $\mathcal{C}$*, consider a short exact sequence*

$$0 \longrightarrow X' \xrightarrow{\iota} X \xrightarrow{\varphi} X'' \longrightarrow 0. \tag{1.15}$$

(1) X *is called an extension of* X' *by* X''*, and* $(X', \iota) = \ker \varphi$*,* $(X'', \varphi) = \operatorname{coker} \iota$*.*
(2) The following conditions are equivalent:
(i) ι *is a* section*, i.e., there exists a morphism* $\rho : X \to X'$ *such that* $\rho\iota = Id_{X'}$*. (Then,* ρ *is a* retraction *associated with* ι*.)*
(ii) φ *is a* retraction*, i.e., there exists* $\sigma : X'' \to X$ *such that* $\varphi\sigma = Id_{X''}$*. (Then,* σ *is a* section *associated with* φ*.)*
(iii) $X \cong X' \bigoplus X''$ *with canonical injections* ι, σ *and canonical projections* ρ, φ*.*
(3) When the equivalent conditions of (2) hold, the short exact sequence (1.15) is said to be split*.*

Theorem 54. *Let* $\mathcal{C}$ *be an abelian category.*
(1) Let P *be an object of* $\mathcal{C}$*. The following conditions are equivalent:*
(i) P *is projective.*
(ii) Every short exact sequence

$$0 \longrightarrow A \xrightarrow{\iota} B \xrightarrow{\varphi} P \longrightarrow 0$$

with P *in the third place splits.*
(2) Dually, let I *be an object of* $\mathcal{C}$*. The following conditions are equivalent:*
(i) I *is injective.*
(ii) Every short exact sequence

$$0 \longrightarrow I \xrightarrow{\iota} B \xrightarrow{\varphi} C \longrightarrow 0$$

with I *in the first place splits.*

Proof. This theorem is proved in ([80], Theorems 2.2.8 & 2.2.9) using a direct approach. One can also proceed as follows: deduce (1) from Theorem 50 and Theorem 573 below; then deduce (2) from (1) by duality. ∎

1.2.4.4 Grothendieck Categories

A *Grothendieck category* is an abelian category with a generator and arbitrary coproducts, for which the following condition holds (Grothendieck's AB5 axiom [148]): if $A \in \mathcal{C}$, $(A_i)_{i \in I}$ is a directed increasing family of subobjects of A, and $B \subseteq A$, then

$$\textstyle\sum_{i \in I} (A_i \cap B) = \left(\sum_{i \in I} A_i\right) \cap B.$$

Definition 55. *Let $\mathcal{C}$ be a Grothendieck category.*
(i) An object A of $\mathcal{C}$ is said to be of finite type *if for each directed increasing family $(A_i)_{i \in I}$ of subobjects of A with $\sum_{i \in I} A_i = A$, there exists an index $j \in I$ such that $A_j = A$.*
(ii) An object A of $\mathcal{C}$ is called coherent *if (a) it is of finite type and (b) for each morphism $f : B \to A$ with B of finite type, $\ker f$ is also of finite type. The full subcategory of $\mathcal{C}$, the objects of which are coherent, is denoted by* Coh $(\mathcal{C})$.

1.3 Monoids

The reader already knows (at least) two examples of monoids. One of them is the set $\mathbb{N} \triangleq \{0, 1, 2, ...\}$ of all nonnegative integers, endowed with the usual multiplication. . The other one is the set $\mathbb{N}^{\times} \triangleq \{1, 2, ...\}$ of all natural integers, again endowed with the usual multiplication. In these monoids, the bases of arithmetic can be developed: one can define the notions of multiple and of divisor of a number, of greatest common multiple and least common divisor of several numbers, etc. If $xy = xz$ in $\mathbb{N}^{\times}$, then $y = z$. This does not hold in $\mathbb{N} \triangleq \{0, 1, 2, ...\}$ since x can be zero.

These simple examples are generalized in the sequel.

1.3.1 Monoids and Cancellation Monoids

1.3.1.1 Basic Notions

Definition 56. *A* monoid *is a set* $\mathbf{M}$ *equipped with an associative* internal law $\star$ *for which* $\mathbf{M}$ *has a* unit element *(i.e., an element e such that $x \star e = x = e \star x$ for any $x \in \mathbf{M}$); e is also called the* neutral element *of the internal law $\star$.*

The unit element of a monoid is unique. Indeed, if e, e' are unit elements, we have $e = e \star e' = e'$.

Definition 57. *(i) A* cancellation monoid $\mathbf{M}$ *is a monoid with the following property: for any $x, y, z \in \mathbf{M}$, $x \star y = x \star z$ or $y \star x = z \star x$ implies $y = z$.*
(ii) A monoid $\mathbf{M}$ *is said to be* commutative *if $x \star y = y \star x$ for any $x, y \in \mathbf{M}$.*

Remark 58. *In everything that follows, a monoid (as well as a group, a ring, etc.) is possibly noncommutative except when otherwise stated. Nevertheless, to avoid confusions, a* skew field *is called a* division ring, *as usual in English, and a field is a commutative division ring. See Definition 226 below.*

The internal law of a monoid $\mathbf{M}$ is often denoted as a multiplication (i.e., $x \star y$ is denoted by $x\,y$ or $x.y$) and the unit element by 1. Then, given two subsets A and B of $\mathbf{M}$, $A\,B$ is the set of all elements $a\,b$, $a \in A$, $b \in B$.

When there is no risk of confusion, the internal law of a *commutative monoid* $\mathbf{M}$ may be written additively (i.e., $x \star y$ is written $x+y$) and the unit element is then denoted by 0 (called the *zero element*); in this case $A+B$ ($A, B \subseteq \mathbf{M}$) is the set of all elements $a+b$, $a \in A$, $b \in B$.

Definition 59. *The* opposite *of a monoid* $\mathbf{M}$*, written* $\mathbf{M}^{\mathrm{op}}$*, is the set* $\mathbf{M}$ *endowed with the internal law* $(x, y) \mapsto x \,.\, y \triangleq y\,x$*.*

When a monoid $\mathbf{M}$ is commutative, the distinction between $\mathbf{M}$ and its opposite is irrelevant.

1.3.1.2 Rigid Monoids

Definition 60. *A cancellation monoid* $\mathbf{M}$ *is said to be* rigid *if whenever* $a\,b' = b\,a'$ $(a, b, a', b' \in \mathbf{M})$*, there exists* $z \in \mathbf{M}$ *such that either* $a = b\,z$ *or* $b = a\,z$*.*

See Exercise 203.

1.3.1.3 Zero in a Multiplicative Monoid

Definition 61. *A zero of a monoid* $\mathbf{M}$ *(multiplicatively noted) is an element denoted by* 0 *such that for any elements* $x, y \in \mathbf{M}$*,* $x\,0 = 0\,y = 0$*. If a monoid* $\mathbf{M}$ *has a zero, it is called a* monoid with zero*.*

A zero of a monoid with zero is unique. If $\mathbf{M}$ is a monoid without zero, one can prove that one zero can be added to $\mathbf{M}$, yielding a monoid with zero denoted by $\mathbf{M}_0$.

Notation 62. *For any monoid* $\mathbf{M}$*, one sets* $\mathbf{M}^{\times} = (\mathbf{M} \cup \{0\}) \setminus \{0\}$*.*

A monoid with zero is not a cancellation monoid; if $\mathbf{M}$ is a monoid with zero, $\mathbf{M}^{\times}$ is a submonoid without zero and *can be* a cancellation monoid.

1.3.2 Ideals a Monoid, and Other Notions

1.3.2.1 Submonoids, Ideals and Units

Definition 63. *(i) A* submonoid $\mathbf{H}$ *of a monoid* $\mathbf{M}$ *is a set* $\mathbf{H} \subseteq \mathbf{M}$*, containing the unit element* 1*, and such that* $x\,y \in \mathbf{H}$ *whenever* $x, y \in \mathbf{H}$*. A submonoid of* $\mathbf{M}$ *is* proper *if it is different from* $\mathbf{M}$*.*
(ii) A left *(resp.,* right*)* ideal *in* $\mathbf{M}$ *is a nonempty set* $\mathfrak{a} \subseteq \mathbf{M}$ *such that* $\mathbf{M}\,\mathfrak{a} = \mathfrak{a}$ *(resp.,* $\mathfrak{a}\,\mathbf{M} = \mathfrak{a}$*); a* two-sided ideal *is a left ideal which is a right ideal.* An ideal *means a two-sided ideal, except when otherwise stated. A left (resp., right, two-sided) ideal in* $\mathbf{M}$ *is* proper *if it is different from* $\mathbf{M}$*.*
(iii) Let $a \in \mathbf{M}$*; the left (resp., right) principal ideal generated by* a *is the set* $\mathbf{M}\,a$ *(resp.,* $a\,\mathbf{M}$*). The* principal ideal *generated by* a *(i.e.,* $\mathbf{M}\,a\,\mathbf{M}$*) is denoted by* (a)*.*

(iv) An element $x \in \mathbf{M}$ is left- (resp., right-) invertible if there exists $y \in \mathbf{M}$ such that $y\,x = 1$ (resp., $x\,y = 1$).
(v) A unit *of a monoid $\mathbf{M}$ is an element u which is invertible, i.e., both left- and right-invertible; the set of units of $\mathbf{M}$ is denoted by $\mathbf{U}(\mathbf{M})$ (an element which is not a unit is a* nonunit*).*
(vi) Two elements a and b are said to be associated *if $a = u\,b\,v$ for some $u, v \in \mathbf{U}(\mathbf{M})$; if $u = 1$ (resp., $v = 1$), a and b are right- (resp., left-) associated.*

An element $u \in \mathbf{M}$ is a unit if, and only if there exist $v, w \in \mathbf{M}$ such that $v\,u = u\,w = 1$; thus $v\,u\,w = v = w$, i.e., the left- and the right-inverse of a unit u coincide and are denoted by u^{-1}; $\mathbf{U}(\mathbf{M})$ is a group.

Definition 63(iii) can be generalized as follows: let X be a subset of a monoid $\mathbf{M}$; the intersection of all left ideals containing X is the smallest left ideal containing X; it is called the left ideal generated by X and written $(X)_l$. This definition holds with left and right interchanged and the right (resp., two-sided) ideal generated by X is denoted by $(X)_r$ (resp., (X)). If X is a finite set, the left, right and two-sided ideals generated by X are said to be *finitely generated.*

Notation 64. *(i) $\mathfrak{a} \lhd_l \mathbf{M}$ (resp., $\mathfrak{a} \lhd_r \mathbf{M}$, $\mathfrak{a} \lhd \mathbf{M}$) means that $\mathfrak{a}$ is a left (resp., right, two-sided) ideal in the monoid $\mathbf{M}$.*
(ii) The set of all left (resp., right, two-sided) ideals in the monoid $\mathbf{M}$ is denoted by $\boldsymbol{I}_l(\mathbf{M})$ (resp., $\boldsymbol{I}_r(\mathbf{M})$, $\boldsymbol{I}(\mathbf{M})$).
(iii) Let α be a cardinal; the set of all left (resp., right, two-sided) ideals in $\mathbf{M}$ generated by a set with cardinal α is denoted by $\boldsymbol{I}_l^{\alpha}(\mathbf{M})$ (resp., $\boldsymbol{I}_r^{\alpha}(\mathbf{M})$, $\boldsymbol{I}^{\alpha}(\mathbf{M})$).
(iv) The set of all left (resp., right, two-sided) finitely generated ideals in $\mathbf{M}$ is denoted by $\boldsymbol{I}_l^{fg}(\mathbf{M})$ (resp., $\boldsymbol{I}_r^{fg}(\mathbf{M})$, $\boldsymbol{I}^{fg}(\mathbf{M})$).

From the above, the set of all principal left (resp., right, two-sided) ideals in $\mathbf{M}$ is $\mathbf{I}_l^1(\mathbf{M})$ (resp., $\mathbf{I}_r^1(\mathbf{M})$, $\mathbf{I}^1(\mathbf{M})$).

Lemma 65. *Let c, c' be elements of a cancellation monoid $\mathbf{M}$. Then $c\,\mathbf{M} = c'\,\mathbf{M}$ if, and only if c and c' are right-associated.*

Proof. If $c\,\mathbf{M} = c'\,\mathbf{M}$, there exists $u, v \in \mathbf{M}$ such that $c = c'\,u$ and $c' = c\,v$. Therefore, $c = c\,v\,u$ and $c' = c'\,u\,v$; as a result, $v\,u = u\,v = 1$ which implies that u is a unit with inverse v. The converse is obvious. ∎

Definition 66. *Let $\mathbf{M}$ be a monoid.*
(i) A left (resp., right, two-sided) ideal $\mathfrak{m}$ in $\mathbf{M}$ is maximal *if it is a maximal element of the set of all* proper *left (resp., right, two-sided) ideals in $\mathbf{M}$ ordered by inclusion.*
(ii) A left (resp., right, two-sided) principal ideal $\mathfrak{m}$ in $\mathbf{M}$ is maximal *if it is a maximal element of the set of all* proper *left (resp., right, two-sided) principal ideals in $\mathbf{M}$ ordered by inclusion.*

1.3.2.2 Centralizer and Normalizer

Definition 67. *Two elements x, y of a monoid $\mathbf{M}$ are said to* commute *if $x\,y = y\,x$. The set $\mathbf{Z}(\mathbf{M})$ consisting those elements c which commute with all elements of $\mathbf{M}$ is called the* centre *of $\mathbf{M}$ and an element of $\mathbf{Z}(\mathbf{M})$ is said to be* central.

More general, let $A \subseteq \mathbf{M}$.

Definition 68. *(i) An element $x \in \mathbf{M}$ is said to* centralize *A if x commutes with all elements of A. The set of all $x \in \mathbf{M}$ which centralize A is called the* centralizer *of $\mathbf{A}$ in $\mathbf{M}$ and is denoted by $\mathbf{Z_M}(A)$.*
(ii) An element $x \in \mathbf{M}$ is said to normalize *A if $x\,A = A\,x$. The set of all $x \in \mathbf{M}$ which normalize A is called the* normalizer *of A in $\mathbf{M}$ and is denoted by $\mathbf{N_M}(A)$.*

1.3.2.3 Generator of a Monoid

Definition 69. *Let $X \subseteq \mathbf{M}$ where $\mathbf{M}$ is a monoid; the submonoid of $\mathbf{M}$* generated by *X, written $[X]$, is the smallest monoid containing X, i.e., the intersection of all monoids containing X; X is called a* generator *(or a* set of generators*) of $[X]$. The monoid M is said to be* finitely generated *if it is generated by finitely many elements.*

An element x belongs to $[X]$ if, and only if x can be written as a product of a finite number of elements of X (with the empty product equal to 1).

Example 70. *Consider the set $\mathbb{N} \triangleq \{0, 1, 2, ...\}$ of all non-negative integers (endowed with the multiplication) ; the set $\mathbb{N}^{\times} \triangleq \{1, 2, ...\}$ of all natural integers is a cancellation monoid and $\mathbf{U}(\mathbb{N}) = \mathbf{U}(\mathbb{N}^{\times}) = \{1\}$. In addition, $\mathbb{N}^{\times}$ is generated by the set of primes.*

Example 71. *The set $\mathrm{Mat}_n(\mathbf{R}) = {}^n\mathbf{R}^n$ of all square matrices of order $n \geq 1$ with entries in a ring $\mathbf{R}$, endowed with the multiplication of matrices, is a monoid with zero. The units of $\mathrm{Mat}_n(\mathbf{R})$ are the* invertible matrices *(with inverse in $\mathrm{Mat}_n(\mathbf{R})$); they form the* general linear group *of the square matrices of order n over $\mathbf{R}$, denoted by $\mathrm{GL}_n(\mathbf{R})$. If $n > 1$, $\mathrm{Mat}_n(\mathbf{R})$ is noncommutative (even if $\mathbf{R}$ is commutative) and $\mathrm{Mat}_n(\mathbf{R})^{\times}$ is not a cancellation monoid.*

1.3.3 Homomorphisms; Quotient Monoids

1.3.3.1 Homomorphisms

Let $\mathcal{C}$ be a concrete category, e.g., the category of monoids (resp., of groups, of rings). A morphism of $\mathcal{C}$ is called a monoid-homomorphism (resp., a group-homomorphism, a ring-homomorphism). (The expressions monoid-morphism, group-morphism, ring-morphism, etc., are also widely used in the mathematical community, and sound more modern. But, as pointed out by Godement [141], they do not make sense from the etymological point of view. Indeed, the word *homomorphism* is based on the greek roots *όμο –the same–* and *μορφή –shape*; thus a homomorphism is a structure-preserving map, and "homo" should not be removed. Likewise, the expression "morphism" is open to criticism in the theory of categories, but too widely accepted to be changed.) Similar expressions are used for monomorphisms, epimorphisms, and isomorphisms of $\mathcal{C}$. More specifically, in the category **Mon** (**1.2.2.2**), we have the following:

Definition 72. *Let* $\mathbf{M}$ *and* $\mathbf{N}$ *be two monoids.*
(i) A map $f : \mathbf{M} \rightarrow \mathbf{N}$ *is a* monoid-homomorphism *if* $f(a\,b) = f(a)\,f(b)$ *for any* $a, b \in \mathbf{M}$*, and* $f(1) = 1$*. This homomorphism is called a* monomorphism *(resp., an* epimorphism*, an* isomorphism*) if it is injective (resp., surjective, bijective), and it is then denoted as* $f : \mathbf{M} \hookrightarrow \mathbf{N}$ *(resp.,* $f : \mathbf{M} \twoheadrightarrow \mathbf{N}$*,* $f : \mathbf{M} \xrightarrow{\sim} \mathbf{N}$*). An* endomorphism *of* $\mathbf{M}$ *is a monoid-homomorphism* $\mathbf{M} \rightarrow \mathbf{M}$ *and an* automorphism *of* $\mathbf{M}$ *is an endomorphism which is an isomorphism.*
(ii) If there exists an isomorphism $f : \mathbf{M} \xrightarrow{\sim} \mathbf{N}$*, the monoids* $\mathbf{M}$ *and* $\mathbf{N}$ *are said to be* isomorphic *(written* $\mathbf{M} \cong \mathbf{N}$*).*
(iii) The image *of a homomorphism* $f : \mathbf{M} \rightarrow \mathbf{N}$ *is* $\operatorname{im} f \triangleq f(\mathbf{M})$ *and its* kernel *is* $\ker f \triangleq f^{-1}(\{1\}) = \{x \in \mathbf{M} : f(x) = 1\}$*.*

The proof of the following result is easy:

Proposition 73. *Let* $f : \mathbf{M} \rightarrow \mathbf{N}$ *be monoid-homomorphism.*
(i) $\operatorname{im} f$ *and* $\ker f$ *are submonoids of* $\mathbf{N}$ *and* $\mathbf{M}$*, respectively;*
(ii) $\mathbf{M}$ *is isomorphic to* $\operatorname{im} f$ *if, and only if* f *is a monomorphism.*

Remark 74. *By Proposition 73(ii), a monoid-monomorphism* f *is also called an* embedding *since, through* f*,* $\mathbf{M}$ *is embedded in* $\mathbf{N}$ *(by identifying* $\mathbf{M}$ *and* $f(\mathbf{M})$*).*

1.3.3.2 Quotient Monoids

Let $\mathcal{R}$ be an equivalence relation in the monoid $\mathbf{M}$ (1.2.1.5).

Definition 75. $\mathcal{R}$ *is said to be compatible with the monoid structure of* $\mathbf{M}$ *if the relations* $x \equiv x' \pmod{\mathcal{R}}$ *and* $y \equiv y' \pmod{\mathcal{R}}$ *imply* $x\,y \equiv x'\,y' \pmod{\mathcal{R}}$*.*

The proof of the following lemma is easy and left to the reader:

Lemma 76. *Let* $\mathcal{R}$ *be an equivalence relation in a monoid* $\mathbf{M}$*. The following conditions are equivalent:*
(i) $\mathcal{R}$ *is compatible with the monoid structure of* $\mathbf{M}$*;*
(ii) $\mathbf{M}/\mathcal{R}$ *is a monoid (called the* quotient monoid *of* $\mathbf{M}$ *by* $\mathcal{R}$*) endowed with the internal law given by:* $\pi(x)\,\pi(y) = \pi(x\,y)$ *for any* $x, y \in \mathbf{M}$*, where* π *is the canonical projection.*

Since π is obviously a monoid-epimorphism, we are led to the following:

Definition 77. *The above map* π *is called the* canonical epimorphism.

The following result can be easily proved ([26], n°II.6.7; [27], n°I.1.6). Item (ii) is connected with Noether's third isomorphism theorem (Theorem 52(3)):

Lemma 78. *Let* $\mathcal{R}$ *be an equivalence relation in* $\mathbf{M}$, *compatible with the monoid structure of* $\mathbf{M}$.
(i) An equivalence relation $\mathcal{S}$ *in* $\mathbf{M}/\mathcal{R}$ *is compatible with the monoid structure of* $\mathbf{M}/\mathcal{R}$ *if, and only if* $\mathcal{S}$ *is of the form* $\mathcal{T}/\mathcal{R}$ *where* $\mathcal{T}$ *is an equivalence relation in* $\mathbf{M}$, *compatible with the monoid structure of* $\mathbf{M}$, *and such that* $\mathcal{R}$ *is finer than* $\mathcal{T}$.
(ii) Then, the canonical epimorphism $\mathbf{M}/\mathcal{T} \twoheadrightarrow (\mathbf{M}/\mathcal{R}) / (\mathcal{T}/\mathcal{R})$ *is a monoid-isomorphism.*

Let $\mathbf{M}$ and $\mathbf{N}$ be two monoids and let $f : \mathbf{M} \to \mathbf{N}$ be a homomorphism. It is easily seen that the binary relation $\mathcal{R}$ defined in $\mathbf{M}$ as: $\mathcal{R}(x, y)$ if $f(x) = f(y)$, is an equivalence relation which is compatible with the monoid structure of $\mathbf{M}$. In addition, one can readily establish Item (1) below:

Lemma and Definition 79. *(1) There exists a unique homomorphism* $\bar{f}$: $\mathbf{M}/\mathcal{R} \to \operatorname{im} f$ *such that* $\bar{f}(\bar{x}) = f(x)$ *for any* $x \in \bar{x}$, *and* $\bar{f}$ *is an isomorphism.*
(2) The isomorphism $\bar{f}$ *is said to be* induced *by* f *(with respect to* $\mathcal{R}$*).*

1.3.4 Products of Monoids

Let $(\mathbf{M}_i)_{i \in I}$ be a family of monoids. Its product is the set of all families $(x_i)_{i \in I}$, $x_i \in \mathbf{M}_i$, and is denoted by $\prod_{i \in I} \mathbf{M}_i$ as in Subsect. 1.2.1.
If $\mathbf{M}_i = \mathbf{M}$ for all $i \in I$, then $\prod_{i \in I} \mathbf{M}_i$ is denoted by $\mathbf{M}^I$.

1.4 Divisibility

1.4.1 Divisors and Multiples

1.4.1.1 Divisors and Multiples: Definitions

Let $\mathbf{M}$ be a monoid.

Definition 80. *(1) Let* $a \in \mathbf{M}$ *and* $b \in \mathbf{M}$; a *is said to* left-divide b *(or to be a* left-divisor *of* b*), and* b *is said to be a* right-multiple *of* a, *if there exists* $c \in \mathbf{M}^{\times}$ *such that* $b = ac$. *Corresponding definitions hold with left and right interchanged.*
(2) If a *both left- and right-divides* b, *it is said to* divide b *(or to be a* divisor *of* b*), written* $a \mid b$, *and* b *is said to be a* multiple *of* a.

Clearly, $a \in \mathbf{M}$ is a left-divisor of $b \in \mathbf{M}$ if, and only if $b\mathbf{M} \subseteq a\mathbf{M}$; any element right-associated with a is also a left-divisor of b. Assuming that $\mathbf{M}$ is a monoid with zero, $a \in \mathbf{M}$ is a left-divisor of 0 if, and only if $0 \in a\mathbf{M}^{\times}$. 0 is a divisor of itself.

1.4.1.2 Least Common Multiples and Greatest Common Divisors

Let $(a_i)_{i\in I}$ be a nonempty family of elements of $\mathbf{M}$.

Definition 81. *(1) An element $b \in \mathbf{M}$ is said to be a* common right multiple *of $(a_i)_{i\in I}$ if it is a right-multiple of a_i for all $i \in I$, i.e.,*

$$b\,\mathbf{M} \subseteq \bigcap_{i\in I} a_i\,\mathbf{M};$$

b is said to be a least common right multiple *(*lcrm*) of $(a_i)_{i\in I}$ if (i) it is a common right multiple of that family and (ii) every common right multiple of $(a_i)_{i\in I}$ is a right multiple of b, i.e., for any $b' \in \mathbf{M}$ such that $b'\,\mathbf{M} \subseteq \bigcap_{i\in I} a_i\,\mathbf{M}$,*

$$b'\,\mathbf{M} \subseteq b\,\mathbf{M} \subseteq \bigcap_{i\in I} a_i\,\mathbf{M}.$$

(2) An element $b \in \mathbf{M}$ is said to be a common left divisor *of $(a_i)_{i\in I}$ if it is a left-divisor of a_i for all $i \in I$, i.e.,*

$$\bigcup_{i\in I} a_i\,\mathbf{M} \subseteq b\,\mathbf{M};$$

b is said to be a greatest common left divisor *(*gcld*) of $(a_i)_{i\in I}$ if (i) it is a common left divisor of that family and (ii) every common left divisor of $(a_i)_{i\in I}$ is a left divisor of b, i.e., for any $b' \in \mathbf{M}$ such that $\bigcup_{i\in I} a_i\,\mathbf{M} \subseteq b'\,\mathbf{M}$,*

$$\bigcup_{i\in I} a_i\,\mathbf{M} \subseteq b\,\mathbf{M} \subseteq b'\,\mathbf{M}.$$

Corresponding definitions hold with left and right interchanged; least common left multiples *(*lclm'*s) and* greatest common right divisors *(*gcrd'*s) are defined in this way.*
(3) An element $b \in \mathbf{M}$ is said to be a greatest common divisor *(*gcd*) of $(a_i)_{i\in I}$ if it is both a* gcld *and a* gcrd *of that family; similarly, b is said to be a* least common multiple *(*lcm*) of $(a_i)_{i\in I}$ if if it is both an* lclm *and an* lcrm *of that family.*

lcrm′s, gcld′s, lcm′s or gcd′s of a family or of a pair of elements do not exist in an arbitrary monoid: see Proposition 123, Theorem 131 and Remark 132 below.

Proposition 82. *In a cancellation monoid, the* lcrm *(resp., the* lclm*) and the* gcld *(resp., the* gcrd*) of a nonempty family of elements, if they exist, are unique up to right- (resp., left-) associates.*

Proof. Let $\mathbf{M}$ be a cancellation monoid and b, b' be lcrm′s or gcld′s of a family of elements. Then $b\,\mathbf{M} = b'\,\mathbf{M}$, thus b and b' are right-associated by Lemma 65. ■

In a cancellation monoid, let $(a_i)_{i\in I}$ be a family of elements; an lcrm (resp., lclm, lcm) of this family is denoted (if it exists) by $\left[(a_i)_{i\in I}\right]_r$ (resp., $\left[(a_i)_{i\in I}\right]_l$,

$\left[(a_i)_{i\in I}\right]$) and a gcld (resp., gcrd, gcd) by $\left((a_i)_{i\in I}\right)_l$ (resp., $\left((a_i)_{i\in I}\right)_r$, $\left((a_i)_{i\in I}\right)$) when this notation does not cause confusions.

1.4.2 Regular Elements and Total Divisibility

1.4.2.1 Regular and Invariant Elements; Atoms

Definition 83. *(i) If $a \in \mathbf{M}$ is not a left-divisor of 0 (Definition 80), it is called* right-regular *(a left-regular element is likewise defined), and* regular *if it is both left- and right-regular.*
(ii) An atom *(also called an* irreducible element*) is a regular element which is a nonunit and cannot be written as the product of two nonunits.*

Let $a \in \mathbf{M}$. If a can be expressed as a finite product of atoms, this factorization of a is said to be *complete* and a is said to be *atomic.*

Lemma 84. *Let $\mathbf{M}$ be a monoid and let $c \in \mathbf{M}$ be a regular element. The following conditions are equivalent:*
(i) c is an atom.
(ii) $\mathbf{M}c$ is a maximal left principal ideal in $\mathbf{M}$ (Definition 66).
(iii) $c\mathbf{M}$ is a maximal right principal ideal in $\mathbf{M}$.

Proof. (i)$\Rightarrow$(ii): Assume that (ii) does not hold. Then there exists a proper left ideal $\mathbf{M}\,b$ such that $\mathbf{M}\,c \subsetneq \mathbf{M}\,b$. Therefore, $c \in \mathbf{M}\,b$, and there exists $a \in \mathbf{M}$ such that $c = ab$. Since $\mathbf{M}\,c \neq \mathbf{M}\,b$, a is a nonunit, and since $\mathbf{M}\,b \neq \mathbf{M}$, b is a nonunit too. Therefore, (i) does not hold.

(ii)$\Rightarrow$(i): Assume that (i) does not hold. Then $c = ab$ where a and b are nonunits. Then $\mathbf{M}\,c \subsetneq \mathbf{M}\,b \subsetneq \mathbf{M}$, and (ii) does not hold.

Likewise, (i)$\Leftrightarrow$(iii). ■

Definition 85. *(i) A monoid $\mathbf{M}$ is said to be* atomic *if all elements of $\mathbf{M}^{\times}$ are regular and atomic.*
(ii) An element c of a monoid $\mathbf{M}$ is invariant *if (a) it is regular and (b) $c\,\mathbf{M} = \mathbf{M}\,c$.*

If c is invariant, it is clear that: (i) $c\,\mathbf{M} = \mathbf{M}\,c = \mathbf{M}\,c\,\mathbf{M}$; (ii) any associate of c is invariant.

Lemma 86. *Let c be an invariant element of a monoid $\mathbf{M}$.*
(i) The set of left-multiples of c coincides with the set of its right-multiples.
(ii) If $\mathbf{M}$ is a cancellation monoid, the set of left-divisors of c coincides with the set of its right-divisors.

Proof. (i) : If $a = b\,c$, there exists b' such that $b\,c = c\,b'$, thus $a = c\,b'$.

(ii) : Let $c = a\,b$. There exists b' such that $c\,b = b'\,c$, thus $a\,b\,b = b'\,a\,b$ whence $a\,b = b'\,a$ since $\mathbf{M}$ is a cancellation monoid. Therefore, $c = b'\,a$. ■

Definition 87. *If all elements of a monoid* $\mathbf{M}$ *are invariant,* $\mathbf{M}$ *is called an* invariant monoid.

According to the definitions, a commutative monoid $\mathbf{M}$ is invariant if, and only if all its elements are regular. The following is obvious:

Lemma 88. *Let* $\mathbf{M}$ *be a monoid such that any element* c *of* $\mathbf{M}$ *satisfies Property (b) in Definition 85 (this includes the case of invariant monoids and of commutative monoids). The following conditions are equivalent for an ideal* $\mathfrak{a}$*: (i)* $\mathfrak{a} \triangleleft_l \mathbf{M}$*; (ii)* $\mathfrak{a} \triangleleft_r \mathbf{M}$*; (iii)* $\mathfrak{a} \triangleleft \mathbf{M}$.

According to the above lemma, if $\mathbf{M}$ is an invariant monoid or a commutative monoid, it is useless to specify whether an ideal in $\mathbf{M}$ is a left, a right or a two-sided ideal.

1.4.2.2 Total Divisibility

Let $\mathbf{M}$ be a cancellation monoid.

Definition 89. *Let* $a, b \in \mathbf{M}$*;* a *is said to be a* total divisor *of* b *(written* $a \parallel b$*) if there exists an invariant element* c *such that* $a \mid c$ *and* $c \mid b$.

The following is clear:

Proposition 90. *(i) If* $a \parallel b$ *and* $b \parallel d$*, then* $a \parallel d$*. (ii)* $c \parallel c$ *if, and only if* c *is an invariant element.*

By the above proposition, $a \parallel b$ is a preorder relation but not an order relation, except if the monoid $\mathbf{M}$ is invariant.

1.4.3 Conical Monoids

Definition 91. *A monoid* $\mathbf{M}$ *is said to be* conical *if* $\mathbf{U}(\mathbf{M}) = \{1\}$.

Let $\mathbf{M}$ be an *invariant monoid.*

Consider the following equivalence relation $\mathcal{A}$ in $\mathbf{M}$: $x \equiv y \pmod{\mathcal{A}}$ if x and y are associated (Definition 63(vi)).

Lemma 92. *The equivalence relation* $\mathcal{A}$ *is compatible with the monoid structure of* $\mathbf{M}$.

Proof. According to Definition 75, we have to show that if $a \equiv a'$ and $b \equiv b'$, then $a\,b \equiv a'\,b'$. Let $a' = u\,a\,v \neq 0$ and $b' = u'\,b\,v' \neq 0$ where u, v, u', v' are units; then $a'\,b' = u\,a\,v\,u'\,b\,v'$. Since a is invariant, there exists $v'' \in \mathbf{M}^{\times}$ such that $a\,v = v''\,a$, thus $a = v''\,a\,v_1$ where $v_1 = v^{-1}$ is a unit. We have $a\,v_1 = v_1'\,a$, therefore $a = v''\,v_1'\,a$, and since a is regular, $1 = v''\,v_1'$, thus v'' is a unit. Similarly, $u'\,b = b\,u''$ where u'' is a unit. Therefore, $a'\,b' = u\,v''\,a\,b\,u''\,v'$ and $a'\,b' \equiv a\,b$. ∎

The quotient $\mathbf{M}/\mathcal{R}$ is denoted by $\mathbf{M}/\mathbf{U}(\mathbf{M})$. The only unit of $\mathbf{M}/\mathbf{U}(\mathbf{M})$ is 1. Therefore, $\mathbf{M}/\mathbf{U}(\mathbf{M})$ is a conical monoid.

Definition 93. *Let* $\mathbf{M}$ *be an invariant monoid;* $\mathbf{M}/\mathbf{U}(\mathbf{M})$ *is the conical monoid associated with* $\mathbf{M}$.

Example 94. *The set* $\mathbb{Z}^{\times}$ *of all integers except zero, with the usual multiplication, is a commutative monoid whose all elements are regular. Two elements* a, b *of* $\mathbb{Z}^{\times}$ *are associated if, and only if* $a = \pm b$. *The conical monoid associated with* $\mathbb{Z}$ *is isomorphic to (and is identified with)* $\mathbb{N}^{\times}$ *(Example 70).*

1.4.4 Free Monoids

Let X be a set and let X^* be the set of all finite sequences of elements of X (called *words* on X); any element of X^* can be written $w = (x_i)_{1 \leq i \leq n}$, $x_i \in X$; the empty word is written 1. The *length* of w (written $|w|$) is n (thus, $|1| = 0$). The multiplication in X^* is defined by juxtaposition: let $v = (x_i)_{n+1 \leq i \leq n+m}$; then $w\,v = (x_i)_{1 \leq i \leq n+m}$. It is easily verified that the multiplication is associative and that the unit element of X^* is 1, thus X^* is a monoid. An atom of X^* is a word of length 1.

Definition 95. X^* *is the* free monoid *on* X.

Assuming that the free monoid X^* is commutative, every element w of X^* can be written

$$w = \prod_{1 \leq i \leq |w|} x_i, \quad x_i \in X$$

and this factorization is unique in the following sense: if

$$w = \prod_{1 \leq i \leq |w|} y_i, \quad y_i \in X$$

then there exists a permutation $i \mapsto i'$ of $E = \{1, ..., |w|\}$ (i.e., a bijection $E \to E$) for which $y_i = x_{i'}$. This generalizes the unique decomposition of a natural integer into prime factors (see Subsect. 1.5.6).

1.5 Ordered Sets and Lattices

1.5.1 Posets

1.5.1.1 Basic Notions

Let S be a set; $\preceq$ is an *order relation* defined over S if (i) it is a preorder relation (**1.2.1.4**) and (ii) if $a \preceq b$ and $b \preceq a$, then $a = b$. If in addition $\preceq$ is such that (iii) for any elements $a, b \in S$, either $a \preceq b$ or $b \preceq a$, then $\preceq$ is called a *total order relation*.

A *poset* $(S, \preceq)$ is a set S equipped with an order relation $\preceq$; $(S, \preceq)$ is denoted by S when there is no confusion; $a \prec b$ means $a \preceq b$ and $a \neq b$.

1.5.1.2 Divisibility Monoids

Definition 96. *A* left-divisibility monoid *is a monoid* $\mathbf{M}$ *which is a poset and in which* $a \preceq b$ *means that* $\mathbf{M}\,b \subseteq \mathbf{M}\,a$. *The corresponding definition holds with left and right interchanged. A left- and right-divisibility monoid is a divisibility monoid.*

Let $\mathbf{M}$ be an *invariant monoid*. Then the following properties are equivalent:

(i) $\mathbf{M}\,b \subseteq \mathbf{M}\,a$,
(ii) $b\,\mathbf{M} \subseteq a\,\mathbf{M}$,
(iii) $\mathbf{M}\,b\,\mathbf{M} \subseteq \mathbf{M}\,a\,\mathbf{M}$,
(iv) $a \mid b$,
(v) $a \parallel b$.

Lemma 97. *If* $\mathbf{M}$ *is an invariant cancellation monoid, the conical monoid* $\mathbf{M}/\mathbf{U}(\mathbf{M})$ *associated with* $\mathbf{M}$ *is a divisibility monoid.*

Proof. Let $a, b \in \mathbf{M}$ be such that $a \mid b$ and $b \mid a$, e.g., $b = a\,q$ and $a = b\,q'$. Therefore $a = a\,q\,q'$, thus $q\,q' = 1$ and q, q' are units. As a result, a and b are associated and $\bar{a} = \bar{b}$ (where $\bar{x}$ is the canonical image of x in $\mathbf{M}/\mathbf{U}(\mathbf{M})$), i.e., divisibility is an order relation in $\mathbf{M}/\mathbf{U}(\mathbf{M})$. ∎

1.5.2 Bounds and Chains

1.5.2.1 Bounds and Isotone Maps

The facts below are detailed in [225] and [18].

An element m of a poset S is said to be *maximal* if $a \in S$ and $m \preceq a$ implies $m = a$. Let $P \subseteq S$; by the greatest (resp., smallest) element of P one means an element $m \in P$ (necessarily unique if it exists) such that $a \preceq m$ (resp., $m \preceq a$) for all $a \in P$; an element $m \in S$ is called an *upper bound* (resp., a *lower bound*) of P if $a \preceq m$ (resp., $m \preceq a$) for all $a \in P$. The *least upper bound* (resp., the *greatest lower bound*) of P, if it exists, is the greatest among all lower bounds (resp., the smallest among all upper bounds) of P and is denoted by $\sup P$ (resp., $\inf P$). If $P = \{a, b\}$, $\inf(a, b)$ and $\sup(a, b)$ are also denoted by $a \wedge b$ and $a \vee b$, and called *meet* and *join*, respectively. The element a is said to *cover* b if $b \prec a$ and there exists no element x such that $b \prec x \prec b$.

Let S, T be two posets; a map $\varphi : S \to T$ is said to be *order-preserving* (or *isotone*) if $\varphi(a) \preceq \varphi(b)$ for any $a, b \in S$ such that $a \preceq b$. (An isotone map is a morphism in the category of posets.) The relation $a \preceq b$ $(a, b \in S)$ is also written $b \succeq a$; $\succeq$ is an order relation over the set S and is called the *converse* of $\preceq$. The poset $(S, \succeq)$ is called the *dual* of $(S, \preceq)$ and is denoted by $\breve{S}$. An isotone map $\varphi : S \to \breve{T}$, viewed as a map $S \to T$, is *order-reversing* (or *antitone*). The map $\varphi : S \to \breve{S}$ such that $\varphi(x) = x$ for any $x \in S$ is called the *dualization* of S; by dualization, $a \wedge b$ is changed to $a \vee b$ and conversely.

The definition of a *Galois connection* is very useful:

Definition 98. *An isotone (resp., antitone) Galois connection between two posets* $(S, \preceq)$ *and* $(P, \preceq)$ *consists of two isotone (resp., antitone) functions* $f : S \to P$ *and* $g : P \to S$ *such that*

$$y \preceq f(x) \Longleftrightarrow x \preceq g(y) \quad (x \in S,\ y \in P). \tag{1.16}$$

In what follows, all Galois connections are antitone.

The term "Galois connection" comes from Galois' theory: see Remark 726 below (Subsect. 4.2.2).

Denote f by $x \mapsto x^{\perp}$ $(x \in S)$ and g by $y \mapsto y^{\perp}$ $(y \in P)$ (the risk of confusion is weak). Then (1.16) takes the form

$$\boxed{y \preceq x^{\perp} \Longleftrightarrow x \preceq y^{\perp} \quad (x \in S,\ y \in P).} \tag{1.17}$$

Proposition 99. *Consider two maps* $S \ni x \mapsto x^{\perp} \in P$ *and* $P \ni y \mapsto y^{\perp} \in S$.
(1) If these maps form a Galois connection, the following properties hold for all $x \in S,\ y \in P$:
(i) $x \preceq x^{\perp\perp}, \quad y \preceq y^{\perp\perp}$;
(ii) $x^{\perp} = x^{\perp\perp\perp}, \quad y^{\perp} = y^{\perp\perp\perp}$.
(2) Conversely, if (i) holds for all $x \in S,\ y \in P$, *then these two maps form a Galois connection.*

Proof. (1). By symmetry, it is sufficient to prove the properties for the variable x.

(i): Let $y = x^{\perp}$; then $y \preceq x^{\perp}$ and $x \preceq y^{\perp} = x^{\perp\perp}$.

(ii): From the above inequality we deduce that $x^{\perp\perp\perp} \preceq x^{\perp}$ since $f = (\bullet)^{\perp}$ is antitone; in addition, replacing y by $x^{\perp}$ in the second inequality of (i) we get $x^{\perp} \preceq x^{\perp\perp\perp}$, thus $x^{\perp} = x^{\perp\perp\perp}$.

(2). Assume that (i) holds for all $x \in S,\ y \in P$. Then whenever $y \preceq x^{\perp}$, one obtains $y^{\perp} \succeq x^{\perp\perp} \succeq x$, and by symmetry the maps considered form a Galois connection. ∎

Example 100. *Let* $\mathcal{C}$ *be a preabelian category, let* $X \in \mathcal{C}$, *and consider the preordered sets* P_X *and* Q^X *in* (**1.2.2.1**). *Choose a kernel for each* $u \in P_X$ *and a cokernel for each* $g \in Q^X$. *Then*

$$u \leqq \ker g \Leftrightarrow ug = 0 \Leftrightarrow \operatorname{coker} u \geqq g$$

which means that the functions

$$\ker : Q^X \to P_X, \quad \operatorname{coker} : P_X \to Q^X$$

constitute a Galois connection. This proves (1.7).

1.5.2.2 Chains and Intervals

A *chain* C of a poset S is a totally ordered subset of S, and its *length* is $|C| \triangleq \text{card}(C) - 1$ (assuming that C is nonempty). The *length* of S (written $|S|$) is the least upper bound of the lengths of its chains. Let $C = \{x_0, ..., x_m\}$ be a chain of S, of finite length $|C| = m$; a *refinement* C' of C is a chain of finite length such that $C \subseteq C'$ and C, C' have the same end-points (i.e., x_0 and x_m); this refinement is *proper* if $C \neq C'$. The chain $C = \{x_0, ..., x_m\}$ is said to be *connected* if x_i covers x_{i-1}, $1 \leq i \leq m$, i.e., if it has no *proper refinements*, i.e., if it is *maximal* between its two end-points (see Example 111 below).

Let $a, b \in S$; the *interval* $[a, b]$ is the set of all elements x such that $a \preceq x \preceq b$ (this interval if empty if the relation $a \preceq b$ is not satisfied). An interval is not a chain in general (see Example 111 below).

A poset S is said to be *well-ordered* if every nonempty subset of S has a least element.

1.5.3 Ascending and Descending Chain Conditions

1.5.3.1 Noetherian Posets, and Artinian Posets

Definition 101. *A poset S satisfies the* ascending chain condition *(ACC), or is* Noetherian, *when every nonempty subset of S has a maximal element. It satisfies the* descending chain condition *(DCC), or is* Artinian, *when its dual $\breve{S}$ satisfies ACC.*

Let $\mathbf{M}$ be a monoid and consider the sets $\mathbf{I}_l(\mathbf{M})$ and $\mathbf{I}_l^\alpha(\mathbf{M})$ for some cardinal α (Notation 64). These sets are posets ordered by inclusion.

Definition 102. *(i) The monoid $\mathbf{M}$ is said to satisfy left ACC (resp., ACC_α) if $\mathbf{I}_l(\mathbf{M})$ (resp., $\mathbf{I}_l^\alpha(\mathbf{M})$) satisfies ACC. This definition holds with ACC replaced by DCC, and with* left *replaced by* right *or* two-sided *(*mutatis mutandis*).*
(ii) The monoid $\mathbf{M}$ is said to be left Noetherian (resp., Artinian) if it satisfies left ACC (resp., DCC); this definition holds with left *replaced by* right *or* two-sided.

Proposition 103. *Let $\mathbf{M}$ be a cancellable monoid. If $\mathbf{M}$ satisfies left and right ACC_1, it is atomic.*

Proof. Let $c \in \mathbf{M}$. Then $c\mathbf{M} \neq \mathbf{M}$ if, and only if c is a nonunit by Lemma 65, and in this case by right ACC_1 there exists a maximal principal right ideal $p_1 \mathbf{M}$ such that $c\mathbf{M} \subseteq p_1 \mathbf{M} \subsetneq \mathbf{M}$. This means that $c = p_1 c_1$ and p_1 is an atom. Therefore, $\mathbf{M} c = \mathbf{M} p_1 c_1$ and since c and c_1 are not left-associated, $\mathbf{M} c \subsetneq \mathbf{M} c_1$ by Lemma 65. Continuing in this way, we get a strictly ascending chain of principal left ideals

$$\mathbf{M}c \subsetneq \mathbf{M}c_1 \subsetneq \mathbf{M}c_2 \subsetneq ...$$

which must terminate by left ACC_1. Therefore, every element of $\mathbf{M}$ is either a unit or a finite product of atoms, i.e., $\mathbf{M}$ is atomic. ■

Lemma 104. *A subset C of a poset S is well-ordered if, and only if it is an Artinian chain.*

Proof. Let $C \subseteq S$ be well-ordered and $a, b \in C$ (assuming that C is nonempty). The set $\{a, b\}$ has a least element, therefore $a \prec b$ or $a = b$ or $b \prec a$, i.e., C is a chain. Let X be a subset of C; X has a least element, thus C satisfies DCC. Conversely, an Artinian chain is obviously well-ordered. ■

The following result is less obvious:

Proposition 105. *Let S be a poset. The following properties are equivalent: (i) S is Noetherian; (ii) every chain of $\breve{S}$ is well-ordered; (iii) every well-ordered subset of S is finite.*

Proof. We will prove that (i)⇒(ii)⇒(iii)⇒(i).

(iii)⇒(i): If (i) fails, then S contains a nonempty subset X which does not have a maximal element. Let $x_1 \in X$; since x_1 is not maximal, there exists $x_2 \in X$ such that $x_2 \succ x_1$, etc., so that there exists an infinite ascending chain $x_1 \prec x_2 \prec ...$ and (iii) fails.

(ii)⇒(iii): If (iii) fails, S contains a well-ordered subset C which is infinite; C is an Artinian chain by Lemma 104. Let $C = \{x_0, x_1, ..\}$ with

$$x_0 \prec x_1 \prec ...$$

This set does not have a maximal element, thus it is not well-ordered in $\breve{S}$ and (ii) fails.

(i)⇒(ii): If (ii) fails, some chain $\breve{C}$ of $\breve{S}$ is not well-ordered, therefore some subchain does not have a least element and $\breve{S}$ does not satisfy DCC, i.e., S is not Noetherian and (i) fails. ■

Corollary 106. *A chain is finite if, and only if it is both Artinian and Noetherian.*

Proof. Let C be an Artinian chain; it is well-ordered by Lemma 104 and by Proposition 105 it is finite. The converse is obvious. ■

1.5.3.2 Zorn's Lemma

A poset S is said to be *inductively ordered* if every chain of S has an upper bound. As shown in, e.g., ([307], Theorem A.4), the following is equivalent to Zermelo's axiom of choice (Axiom 1):

Lemma 107. *(*Zorn's lemma*). Every nonempty inductively ordered set has a maximal element.*

1.5.4 Lattices

1.5.4.1 Definition and General Properties

Definition 108. *(i) A* lattice *L is a poset in which each pair (a,b) has a least upper bound $a \vee b$ and a greatest lower bound $a \wedge b$.*
(ii) A sublattice *N of L is a set $N \subseteq L$ which is a lattice for the same ordering.*
(iii) A complete lattice *is a poset in which each family $(a_i)_{i \in I}$ has a least upper bound $\bigvee_i a_i$ and a greatest lower bound $\bigwedge a_i$.*

As easily shown (see, e.g., [225], Chap. XIV, or [79], Sect. 3.1), in any lattice L, the following *modular inequality* is satisfied:

$$a \vee (b \wedge c) \preceq (a \vee b) \wedge c \text{ for all } a,b,c \in L \text{ such that } a \preceq c, \tag{1.18}$$

as well as the following *distributivity inequalities*:

$$a \vee (b \wedge c) \preceq (a \vee b) \wedge (a \vee c)\,, \tag{1.19}$$
$$a \wedge (b \vee c) \succeq (a \wedge b) \vee (a \wedge c) \tag{1.20}$$

for all $a,b,c \in L$. In addition, if (1.19) is an equality, then so is (1.20) and conversely, and then the first equality in (1.18) is an equality.

An interval and a chain of a lattice are sublattices.

Example 109. *Let $\mathbf{M}$ be a monoid; the sets $\boldsymbol{I}_l(\mathbf{M})$, $\boldsymbol{I}_r(\mathbf{M})$ and $\boldsymbol{I}(\mathbf{M})$ of all left, right and two-sided ideals in $\mathbf{M}$, respectively (Notation 64), are complete lattices.*

Consider for example $\boldsymbol{I}_l(\mathbf{M})$. If $(\mathfrak{a}_i)$ be a family of left ideals in the monoid $\mathbf{M}$, then $\bigcap \mathfrak{a}_i$ is the largest left ideal included in all $\mathfrak{a}_i$, i.e., $\bigcap_i \mathfrak{a}_i = \bigwedge_i \mathfrak{a}_i$. The union $\bigcup_i \mathfrak{a}_i$ is not a left ideal but the left ideal generated by $\bigcup_i \mathfrak{a}_i$ is the smallest left ideal containing all $\mathfrak{a}_i$, i.e., is equal to $\bigvee_i \mathfrak{a}_i$.

1.5.4.2 Modular Lattices, and Distributive Lattices

Definition 110. *(i) A lattice L is* modular *if the first inequality of (1.18) is an equality.*
(ii) A lattice L is distributive *if (1.19) (or equivalently (1.20)) is an equality.*

From the above, a distributive lattice is modular, but the converse is not true in general. A chain of an ordered set is a distributive lattice. A sublattice of a modular (resp., distributive) lattice is modular (resp., distributive). The dual $\breve{L}$ of a lattice L is a lattice; it is modular (resp., distributive) if L is.

Example 111. *In Examples 70 & 94, $\mathbb{N}^{\times}$ with the usual divisibility is a commutative divisibility monoid; $a \wedge b$ is the greatest common divisor (a,b)*

of a *and* b *and* $a \vee b$ *is the least common multiple* $[a, b]$ *of* a *and* b. *We have* $(a \vee b) = (a) \cap (b)$.[2] *Obviously,* b *covers* a *if, and only if* $a \mid b$ *and* b/a *is a prime; therefore, a subset* $\{x_0, ..., x_m\}$ *is a connected chain if, and only if for* $1 \leq i \leq m$, $x_i = p_i \, x_{i-1}$ *where* p_i *is a prime.*

Let $a = \prod_i p_i^{\alpha_i}$, $b = \prod_i p_i^{\beta_i}$ *and* $c = \prod_i p_i^{\gamma_i}$ $(\alpha_i, \beta_i, \gamma_i \in \mathbb{N})$ *be the decompositions of* a, b *and* c *into prime factors. Then* $b \wedge c = \prod_i p_i^{\inf(\beta_i, \gamma_i)}$, $a \vee b = \prod_i p_i^{\sup(\alpha_i, \beta_i)}$ *and* $a \vee (b \wedge c) = \prod_i p_i^{\delta_i}$ *where* $\delta_i = \sup(\alpha_i, \inf(\beta_i, \gamma_i))$; *similarly,* $(a \vee b) \wedge (a \vee c) = \prod_i p_i^{\varepsilon_i}$ *where* $\varepsilon_i = \inf(\sup(\alpha_i, \beta_i), \sup(\alpha_i, \gamma_i))$. *One easily checks that* $\delta_i = \varepsilon_i$ *is the median of the three integers* $\alpha_i, \beta_i, \gamma_i$. *As a result,* $\mathbb{N}^{\times}$, *with the usual divisibility, is a distributive lattice.*

Take a natural number, for example 18; *the interval* $[1, 18]$ *(in the sense of divisibility) is the set of all divisors of* 18, *i.e.,* $1, 2, 3, 6, 9, 18$. *The maximal (i.e., connected) chains with end-points* 1, 18 *are* $\{1, 2, 6, 18\}$, $\{1, 3, 6, 18\}$ *and* $\{1, 3, 9, 18\}$; *notice that they have the same length (a fact which is explained in a more general context by Theorem 117 and clarified by Example 118 below) which is* 3, *the number of prime factors of* 18 *(counted according to their multiplicities). Note that* $(a \vee b)\,(a \wedge b) = \prod_i p_i^{\sup(\alpha_i, \beta_i) + \inf(\alpha_i, \beta_i)} = a\,b$.

1.5.4.3 Lattice-Homomorphisms

Let L, M be two lattices and $\varphi : L \to M$.

Definition 112. *(i)* φ *is a* lattice-homomorphism *if* $\varphi(a \vee b) = \varphi(a) \vee \varphi(b)$ *and* $\varphi(a \wedge b) = \varphi(a) \wedge \varphi(b)$ *for any* $a, b \in L$. *(ii)* φ *is a lattice-monomorphism (resp., -epimorphism, -isomorphism) if it is a lattice-homomorphism and if it is injective (resp., surjective, bijective). (iii)* φ *is a lattice-antiisomorphism if* φ *is a bijection and* $\varphi(a \vee b) = \varphi(a) \wedge \varphi(b)$, $\varphi(a \wedge b) = \varphi(a) \vee \varphi(b)$ *for any* $a, b \in L$.

A lattice-homomorphism (i.e., a morphism in the category of lattices) is isotone but the converse is not true in general (in other words, the category of lattices is not a full subcategory of the category of posets). However, one can prove the following ([18], Sect. II.3):

Lemma 113. *A map* $\varphi : L \to M$ *is an isotone* bijection *with isotone inverse if, and only if it is a lattice-isomorphism.*

Corollary 114. *Let* M *be a modular lattice and* $a, b \in M$. *The map* $x \mapsto a \wedge x$ *is a lattice-isomorphism* $\varphi_a : [b, a \vee b] \mapsto [a \wedge b, a]$ *with inverse* $\psi_b : y \mapsto b \vee y$.

Proof. The map φ_a is obviously isotone. Let $x \in [b, a \vee b]$; then $\psi_b(\varphi_a(x)) = b \vee (a \wedge x) = (b \vee a) \wedge x$ (since M is modular) and

[2] In the dual lattice, $a \vee b$ is changed to $a \wedge b$ and then this equality takes the more appealing form $(a \wedge b) = (a) \cap (b)$. The reader should have in mind that the symbols $\vee$ and $\wedge$ are interchanged when passing from a lattice to its dual.

$(b \vee a) \wedge x = x$ since $x \preceq b \vee a$. Therefore, φ_a is bijective, and by Lemma 113 it is a lattice-isomorphism. ■

Definition 115. *(1) In a modular lattice, two intervals related as in Corollary 114 are* perspective *(or relatively transposed).*
(2) Two intervals $[a, b]$ *and* $[a^*, b^*]$ *are* projective *if there exists a finite sequence of intervals* $[a_k, b_k]$, $0 \leq k \leq n$, *such that (i)* $[a_0, b_0] = [a, b]$, *(ii)* $[a_n, b_n] = [a^*, b^*]$, *(iii)* $[a_{k-1}, b_{k-1}]$ *and* $[a_k, b_k]$ *are perspective,* $1 \leq k \leq n$.
(3) Two finite chains with the same end-points are isomorphic *if their intervals can be paired off in such a way that the corresponding intervals are projective.*

Perspectivity and projectivity are equivalence relations in the set of intervals of a modular lattice and isomorphy is an equivalence relation in the set of its chains.

1.5.5 Four Theorems in Modular Lattices

1.5.5.1 Refinements of Chains and Maximal Chains

By Corollary 114 and Definition 115, in a modular lattice two perspective intervals have the same length, as well as two projective intervals or two isomorphic chains. The following result is proved in [76]:

Theorem 116. *(*Schreier refinement theorem*). In a modular lattice, two finite chains between the same end-points have refinements which are isomorphic.*

In a modular lattice L any two maximal chains between the same end-points have the same length, as specified below:

Theorem 117. *(*Jordan-Hölder-Dedekind theorem*). In a modular lattice* of finite length, *any chain can be refined to a maximal chain, and any two maximal chains between the same end-points are isomorphic.*

Proof. See ([225], Sect. XIV.5). ■

Example 118. *Consider in Example 111 the two maximal chains* $\{1, 3, 6, 18\}$ *and* $\{1, 3, 9, 18\}$*; the sublattice consisting of these two chains is represented below.*

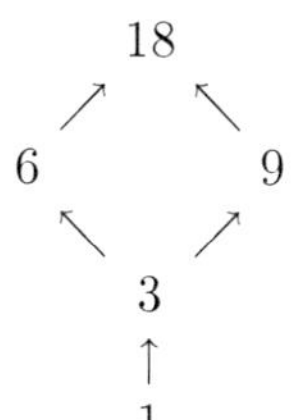

Projective (and, more specifically in this case, perspective) intervals are parallel arrows: $[6,18]$ *in the first chain and* $[3,9]$ *in the second are perspective since* $[6,18] = \psi_6([[3,9]])$ *(with* $a = 9$, $b = 6$*);* $[3,6]$ *in the first chain and* $[9,18]$ *in the second are perspective since* $[9,18] = \psi_9([[3,6]])$ *(with* $a = 6$, $b = 9$*). Finally,* $[1,3]$ *in the first chain and* $[1,3]$ *in the second are perspective (with* $a = 1$, $b = 3$*). Therefore, these two maximal chains are isomorphic. This rationale still holds in the general case and can be used to prove by induction the Jordan-Hölder-Dedekind theorem.*

1.5.5.2 Join- and Meet-Irreducibility, Independence, and Irredundancy

A lattice L of finite length has a least and a greatest element respectively denoted by $\dot{0}$ and $\dot{1}$.

Definition 119. *Let* L *be a modular lattice having a least element* $\dot{0}$.
(i) The length *of an element* c *(written* $|c|$*) is* $\left|\left[\dot{0}, c\right]\right|$.
(ii) $c \in L$ *is said to be* join-irreducible *if* $|c| = 1$.

Dualizing this definition, we get:

Definition 120. *Let* L *be a modular lattice having a greatest element* $\dot{1}$. *An element* $a \in L$ *is said to be* meet-irreducible *if* $\left|\left[a, \dot{1}\right]\right| = 1$.

Thus, an element $c \in L \backslash \{\dot{1}\}$ is meet-irreducible if, and only if whenever $a, b \in L$ are such that $a \succeq c$, $b \succeq c$ and $a \wedge b = c$, then necessarily $a = c$ or $b = c$.

If $\left|\left[\dot{0}, c\right]\right|$ is infinite, so is $|c|$; otherwise, $|c|$ is the length of any maximal chain between $\dot{0}$ and c by the Jordan-Hölder-Dedekind theorem (Theorem 117).

Example 121. *(i) The length of any element* $x > 0$ *in the totally ordered set* $\mathbb{Q}^+ = \{x \in \mathbb{Q} : x \geq 0\}$ *(with the usual ordering) is* $\aleph_0$ *(where* $\aleph_0$ *denotes the cardinal of any infinite countable set).*
(ii) The length of any element $x > 0$ *in the totally ordered set* $\mathbb{R}^+ = \{x \in \mathbb{R} : x \geq 0\}$ *(with the usual ordering) is* $\aleph = 2^{\aleph_0}$ *(i.e., the cardinal of the real line* $\mathbb{R}$*) by Cantor's theorem ([79], Prop. 1.1.5).*
(iii) Let n *be a natural integer (Examples 70 & 111); its length is the number of prime factors of* n *(counted according to their multiplicities) and* n *is join-irreducible if, and only if it is prime.*

Definition 122. *Let* L *be a modular lattice having a least element* $\dot{0}$.
(i) Let $a_1, ..., a_n$ *be elements* $\neq \dot{0}$ *and of finite length; they are said to be* independent *if*

$$\left| \bigvee_{1 \leq i \leq n} a_i \right| = \sum_{1 \leq i \leq n} |a_i| .$$

(ii) A finite decomposition $a = \bigvee_{1 \leq i \leq n} a_i$ *is said to be* irredundant *if no* a_i *can be omitted.*

Some of the results in Example 111 can be generalized. The following is clear:

Proposition 123. *Let* $\mathbf{M}$ *be an invariant cancellation monoid,* $\mathbf{M}/\mathbf{U}(M)$ *the conical monoid associated with* $\mathbf{M}$ *and* $\pi : \mathbf{M} \twoheadrightarrow \mathbf{M}/\mathbf{U}(M)$ *the canonical epimorphism (Definition 93). By Lemma 97,* $\mathbf{M}/\mathbf{U}(M)$ *is a divisibility monoid. Assuming that* $\mathbf{M}/\mathbf{U}(M)$ *is a lattice, for any* $a, b \in \mathbf{M}$*,* $\varnothing \neq \pi^{-1}(\bar{a} \wedge \bar{b}) = \operatorname{gcd}(a,b)$ *and* $\varnothing \neq \pi^{-1}(\bar{a} \vee \bar{b}) = \operatorname{lcm}(a,b)$ *(where* $\operatorname{gcd}(a,b)$ *and* $\operatorname{lcm}(a,b)$ *respectively denote the set of all* $\operatorname{gcd}'s$ *and of all* $\operatorname{lcm}'s$ *of* (a,b)*).*

If the monoid $\mathbf{M}/\mathbf{U}(M)$ is a *modular lattice*, the notions of *atom* and of *join-irreducible element* are closely connected: the least element of $\mathbf{M}/\mathbf{U}(M)$ is $\dot{0} = 1$. An element $a \in \mathbf{M}$ is an atom if, and only if $a = bc$ implies that b or c is a unit; this is equivalent to the following property: $\bar{a} = \bar{b}\,\bar{c}$ implies that $\bar{b} = 1$ or $\bar{c} = 1$, i.e., $\bar{a}$ is join-irreducible (denoting by $\bar{x}$ the canonical image of x in $\mathbf{M}/\mathbf{U}(M)$). To summarize:

Proposition 124. *An element* a *of an invariant monoid* $\mathbf{M}$ *such that* $\mathbf{M}/\mathbf{U}(M)$ *is a modular lattice is an atom if, and only if its canonical image* $\bar{a}$ *is join-irreducible.*

1.5.5.3 Decompositions of an Element

The two theorems below are proved in [18], [76].

Theorem 125. *(*Krull-Schmidt theorem*). In a modular lattice* L *with least element* $\dot{0}$*, let* $a \neq \dot{0}$ *be an element of finite length and*

$$a = \bigvee_{1 \leq i \leq n} a_i = \bigvee_{1 \leq i \leq m} b_i$$

be two representations of a *as a join of independent elements. Then* $n = m$ *and for some permutation* $i \mapsto i'$ *of* $\{1, 2, ..., n\}$*,* $[\dot{0}, a_i]$ *is projective with* $[\dot{0}, b_{i'}]$*. If* L *is distributive, projectivity can be replaced by equality:* $a_i = b_{i'}$*.*

Theorem 126. *(*Kurosh-Ore theorem*). In a modular lattice* L *with least element* $\dot{0}$*, let* $c \neq \dot{0}$ *be an element of finite length and*

$$c = \bigvee_{1 \leq i \leq n} p_i = \bigvee_{1 \leq i \leq m} q_i$$

be two irredundant decompositions of c *into join-irreducible elements. Then* $n = m = |c|$ *and for some permutation* $i \mapsto i'$ *of* $\{1, 2, ..., n\}$ *one can substitute in the first decomposition any* p_i *by* $q_{i'}$*; in addition, if* L *is distributive, then* $p_i = q_{i'}$*.*

Example 127. *Consider the distributive lattice* $\mathbb{N}^{\times}$ *with the usual divisibility (Examples 70, 111 & 121(iii)). The numbers* $a_1, ..., a_n$ *are* independent *if, and only if they are* pairwise coprime[3]. *A number* a *is* join-irreducible *if, and only if it is prime. By Theorem 125, if a natural number can be expressed as the product of* n *pairwise coprime factors, this representation is unique; for example,* $4 \times 9 \times 5$ *is the only decomposition of* 180 *as the product of pairwise coprime nonunit factors. Theorem 126 expresses the uniqueness of the decomposition of a number* ≥ 2 *into prime factors, e.g.,* $180 = 2^2 \times 3^2 \times 5$ *and the length of* 180 *is* $2 + 2 + 1 = 5$.

1.5.6 Unique Factorization Monoids

1.5.6.1 General UF-Monoids

Let $\mathbf{M}$ be a cancellation monoid.

Definition 128. *A* prime *in* $\mathbf{M}$ *is an invariant nonunit* p *such that* $p \mid a\,b$ $(a, b \in \mathbf{M})$ *implies* $p \mid a$ *or* $p \mid b$.

Lemma 129. *Every prime in* $\mathbf{M}$ *is an atom.*

Proof. If $p = a\,b$ is prime, then $p \mid a$ or $p \mid b$, e.g., $a = p\,q$ (Lemma 86), therefore $p = p\,q\,b$, thus $1 = q\,b$ since $\mathbf{M}$ is a cancellation monoid, and b is a unit. ∎

The converse of Lemma 129 is false in general ([77], Sect. 3.1); it holds true in *unique factorization monoids.*

Let $\mathcal{S}$, $\mathcal{A}$ be equivalence relations in $\mathbf{M}$ where $\mathcal{A}$ is defined as in Lemma 92, is finer than $\mathcal{S}$ in the general case, and is equal to $\mathcal{S}$ if $\mathbf{M}$ is invariant (and in particular if $\mathbf{M}$ is commutative). The following definition is taken from [74].

Definition 130. *A cancellation monoid* $\mathbf{M}$ *is an* $\mathcal{S}$-unique factorization monoid *(an* $\mathcal{S}$*-UF-monoid, for short, and a* UF-monoid *if* $\mathcal{S} = \mathcal{A}$*) if it is atomic and such that any two* complete factorizations *(i.e., factorizations into atoms) of an element* c *are isomorphic in the sense that (i) they have the same length* r*, say*

$$c = a_1 ... a_r = b_1 ... b_r$$

and (ii) there is a permutation $i \mapsto i'$ *such that* $a_i \equiv b_{i'} \,(\operatorname{mod} \mathcal{S})$.

1.5.6.2 Invariant UF-Monoids

If $\mathbf{M}$ is an invariant cancellation monoid, $\mathcal{S} = \mathcal{A}$ is compatible with the monoid structure of $\mathbf{M}$ according to Lemma 92 and the quotient monoid $\mathbf{M}/\mathbf{U}(\mathbf{M})$ is conical. The following is proved in ([77], Sect. 3.1):

[3] *Coprime* is synonymous with *relatively prime*.

Theorem 131. *For any invariant cancellation monoid* $\mathbf{M}$ *the following conditions are equivalent:*

(i) $\mathbf{M}$ *is a UF-monoid;*
(ii) $\mathbf{M}/\mathbf{U}(\mathbf{M})$ *is free commutative (Definition 95);*
(iii) $\mathbf{M}$ *satisfies* ACC_1 *and any two elements have a* gcd*;*
(iv) $\mathbf{M}$ *satisfies* ACC_1 *and any two elements have an* lcm*;*
(v) $\mathbf{M}$ *satisfies* ACC_1 *and the intersection of any two principal ideals is principal;*
(vi) $\mathbf{M}$ *is atomic and every atom of* $\mathbf{M}$ *is prime.*

Remark 132. *(a) Let* $\mathbf{M}$ *be an invariant UF-monoid; the* gcd *and the* lcm *of two elements of the conical monoid* $\mathbf{M}/\mathbf{U}(\mathbf{M})$ *can be calculated as in Example 111. Therefore:*
(i) $\mathbf{M}/\mathbf{U}(\mathbf{M})$ *is a distributive lattice, and Theorems 125 & 126 apply to that case; by Theorem 126 and Definition 130, for any* $a \in \mathbf{M}$, $|a| = |\bar{a}|$.
(ii) Setting $\bar{d} = \gcd(\bar{a}, \bar{b})$ *and* $\bar{m} = \operatorname{lcm}(\bar{a}, \bar{b})$, *we have* $\bar{d}\,\bar{m} = \bar{a}\,\bar{b}$ *and for any* $m \in \bar{m}$, $M\,m = M\,a \cap M\,b$.
(b) One can prove the following ([77], Sect. 3.1, Exercises 7 & 8): if two elements of an invariant cancellation monoid have an lcm*, they have a* gcd *but the converse is false in general. However, any pair of elements has an* lcm *if, and only if any pair of elements has a* gcd.

By Remark 132 one is led to the following

Definition 133. *An invariant cancellation monoid in which any pair of elements has a* gcd *is called a* GCD invariant cancellation monoid.

Corollary 134. *Any invariant UF-monoid is a GCD invariant cancellation monoid.*

1.5.6.3 Unique Factorization of Invariant Elements

Let $\mathbf{M}$ be a monoid and let $I(\mathbf{M})$ be the set of all invariant elements of $\mathbf{M}$; I is obviously an invariant monoid and, assuming that it is cancellable, the notion of prime can be defined in I using Definition 128. Let us call such a prime an I-prime. Since a prime in $\mathbf{M}$ belongs to I, every prime is an I-prime but not conversely in general. Similarly, an I-atom can be defined and every atom is an I-atom but not conversely in general. Last, $\mathbf{M}$ is said to have *unique factorization of invariant elements* (or I-unique factorization) if $I(\mathbf{M})$ is a UF-monoid.

1.6 Groups

1.6.1 Groups and Their Generators

1.6.1.1 Basic Notions

A group G is a monoid in which each element is invertible, i.e., is a unit. As already seen (**1.3.2.1**), the inverse of an element x is unique and written x^{-1}. In the remainder of this section, G is a group. Its cardinal is called the *order* of G, written $(G:1)$.

Example 135. *The set of permutations of a set E (i.e., of bijections $E \to E$), endowed with the composition of maps, is a group, called the* symmetric group *of E, and denoted by $\mathfrak{S}(E)$. If $E = \{1, ..., n\}$, $\mathfrak{S}(E)$ is denoted by $\mathfrak{S}_n$ and is called the symmetric group of degree n; as easily shown by induction, $(\mathfrak{S}_n : 1) = n!$.*

Example 136. *As already mentioned in Example 71, the set $\mathrm{GL}_n(\mathbf{R})$ of all invertible $n \times n$ matrices over a ring $\mathbf{R}$, endowed with the multiplication of matrices, is the* general linear group *of the matrices of order n over $\mathbf{R}$. If $n > 1$, this group is noncommutative, even if $\mathbf{R}$ is commutative.*

If G is commutative and additively noted, the inverse of x is called its opposite and is written $-x$. A commutative group is called an *abelian group*.

1.6.1.2 Generators of a Group

One defines a generator X of a group G in the same way as a generator of a monoid (Definition 69). Let $X \subseteq G$; then the subgroup of G generated by X is denoted by $\langle X \rangle$. If X is finite, the group $\langle X \rangle$ is said to be *finitely generated*. For any nonempty part Y of G, let $Y^{-1} \triangleq \{y^{-1} : y \in Y\}$. The following lemma is proved in ([27], n°I.4.3):

Lemma 137. *(i) If X is nonempty, $\langle X \rangle = [[X] \cup [X^{-1}]]$ (where $[X]$ is the monoid generated by X).*
(ii) Let $f : G \to G'$ be a group-homomorphism and $X \subseteq G$; then $f(\langle X \rangle) = \langle f(X) \rangle$.

Definition 138. *A group G is said to be* cyclic *if it is generated by a single element, i.e., if there exists an element x such that $G = \langle x \rangle$.*

The proposition below is obvious:

Proposition 139. *A group G is cyclic if, and only if there exists an element $x \in G$ such that $G = \{x^n : n \in \mathbb{Z}\}$. A cyclic group is abelian.*

1.6.2 Subgroups

1.6.2.1 Basic Notions

A subgroup H of G is a submonoid of G such that if $x \in H$, then $x^{-1} \in H$. We write $H \leq G$ to mean that H is a subgroup of the group G.

The kernel of a group-homomorphism $f : G \rightarrow G'$ is $\ker f \triangleq f^{-1}(\{1\}) = \{x \in G : f(x) = 1\}$. Its image is $\operatorname{im} f \triangleq f(G) = \{f(x) : x \in G\}$.

The proof of the following is straightforward:

Proposition 140. *Let $f : G \rightarrow G'$ be a group-homomorphism;*
(i) $\operatorname{im} f \leq G'$ *and* $\ker f \leq G$*;*
(ii) f is an epimorphism if, and only if $\operatorname{im} f = G'$*;*
(iii) f is a monomorphism if, and only if $\ker f = \{1\}$.

Let $H \leq G$ and $x \in G$; we set $x\,H \triangleq \{x\,y : y \in H\}$ and $H\,x \triangleq \{y\,x : y \in H\}$.

Example 141. *An even (resp., odd) permutation of $E = \{1, ..., n\}$ is one that can be produced by an even (resp., odd) number of exchanges of two elements (these exchanges are called* transpositions*). The subset* Alt_n *(also denoted by A_n) of $\mathfrak{S}_n$ consisting of all even permutations of E is a subgroup of $\mathfrak{S}_n$, called the* alternating group *of degree n. Let* $\operatorname{sgn} : \mathfrak{S}_n \rightarrow \{-1, 1\}$ *be defined by* $\operatorname{sgn}(\sigma) = 1$ *if σ is even and* $\operatorname{sgn}(\sigma) = -1$ *if σ is odd. The map* sgn *is called the* signature *and it is a group-epimorphism; its kernel is* Alt_n. *Obviously,* $\operatorname{sgn}(\sigma) = \operatorname{sgn}(\sigma^{-1})$.

The centre $\mathbf{Z}(G)$ is the set of all elements of $g \in G$ such that $x\,g = g\,x$ for any $x \in G$. The centralizer $\mathbf{Z}_G(A)$ and the normalizer $\mathbf{N}_G(A)$ of a part A of G are defined as in Definition 68.

1.6.2.2 Index of a Subgroup

Let $H \leq G$ and consider the relation $\mathcal{R}_{l,H}$ defined in G as follows: $\mathcal{R}_{l,H}(x, y)$ if $y^{-1}x \in H$. As easily seen, $\mathcal{R}_{l,H}$ is an equivalence relation and $x \equiv y \pmod{\mathcal{R}_{l,H}}$ if, and only if $x \in y\,H$. The subset $y\,H$ of G is called a *left coset* of H in G and y is called a representative of that coset; $y\,H$ is the equivalence class of $y \pmod{\mathcal{R}_{l,H}}$ and $y \mapsto y\,H$ is the canonical surjection. The map $x \mapsto y\,x$ is a bijection $H \xrightarrow{\sim} y\,H$, thus $\operatorname{card}(y\,H) = \operatorname{card}(H)$.

Similarly, consider the relation $\mathcal{R}_{r,H}$ defined in G as: $\mathcal{R}_{r,H}(x, y)$ if $x\,y^{-1} \in H$, i.e., $x \in H\,y$. The subset $H\,y$ of G is called a *right coset* of H in G and, by the same rationale as above, $\operatorname{card}(H\,y) = \operatorname{card}(H)$.

The set of left (resp., right) cosets $x\,H$ (resp., $H\,x$) ($x \in H$) is denoted by G/H (resp., $H\backslash G$).

Since two different left (resp., right) cosets have no element in common, G is the *disjoint union* of all left (resp., right) cosets of H in G. From the above, the number of left cosets of H in G is equal to the number of right cosets in G. This number is called the *index* of H in G and is denoted by $(G : H)$.

Let $(G:1)$ and $(H:1)$ be the orders of G and H, respectively (**1.6.1.1**); we have the equality

$$(G:1) = (G:H)\,(H:1)\,, \tag{1.21}$$

called *Lagrange's theorem*. More generally, if H, K are subgroups of G such that $K \subseteq H$, then

$$\boxed{(G:K) = (G:H)\,(H:K)\,.}$$

By (1.21), $(G:H) \mid (G:1)$ and $(H:1) \mid (G:1)$, if all these cardinals are finite.

1.6.3 Normal Subgroups and Quotient Groups

Let $\mathcal{R}$ be an equivalence relation in G. According to Lemma 76, $G/\mathcal{R}$ is a monoid if, and only if $\mathcal{R}$ is compatible with the monoid structure of G (Definition 75). More specifically, we have the following ([27], n°I.4.4):

Lemma and Definition 142. *(1) The following conditions are equivalent:*
(i) $\mathcal{R}$ is compatible with the monoid structure of G;
(ii) $G/\mathcal{R}$ is a group;
(iii) $N \triangleq \{y \in G : y \equiv 1 \,(\mathrm{mod}\,\mathcal{R})\}$ is a subgroup of G such that $x\,N = N\,x$ for any $x \in G$.
(2) The subgroup N of G is said to be normal *(written $N \lhd G$) if the above equivalent conditions hold.*
If $N \lhd G$, the quotient group $G/\mathcal{R}$ (where $\mathcal{R}$ is the equivalence relation in G defined by $x \equiv x' \,(\mathrm{mod}\,\mathcal{R})$ if $x\,x'^{-1} \in N$) is denoted by G/N. This equivalence relation is also written $x \equiv x' \,(\mathrm{mod}\,N)$.

If G is an abelian group, any subgroup of G is normal.

Remark 143. *(i) If $H \lhd G$ and $K \lhd H$, then K is a subgroup of G but is not necessarily normal ([27], §I.5, Exercise 10).*
(ii) If $H \lhd G$, $K \lhd G$ and $K \subseteq H$, then $K \lhd H$ ([206], Sect. I.3).
(iii) Let G, G' be groups and $f : G \to G'$; then–as easily shown–$\ker f \lhd G$ but the subgroup $\operatorname{im} f$ of G' is generally not normal in G'.
(iv) $\{1\} \lhd G$ and $G/\{1\} = G$; $G \lhd G$ and $G/G = \{1\}$.
(v) A group G, every subgroup of which is normal, is called a Dedekind group *[155]. For example, the quaternion group is a non-abelian Dedekind group ([27], §I.6, Exercise 4), i.e., a* Hamiltonian group.

Example 144. *Let* $\mathbf{R}$ *be a* commutative ring *and consider the monoid* $\mathrm{Mat}_n(\mathbf{R})$ *(Example 71). The map* $\det : \mathrm{Mat}_n(\mathbf{R}) \to \mathbf{R}$ *is a monoid-homomorphism since* $\det(X\,Y) = \det(X)\,\det(Y)$*. The general linear group* $\mathrm{GL}_n(\mathbf{R})$ *(Example 136) is the set of all matrices* $A \in \mathrm{Mat}_n(\mathbf{R})$ *such that* $\det(A)$ *is a unit (see Subsect. 2.11.1 below). Let* $\mathrm{SL}_n(\mathbf{R})$ *be the subset of* $\mathrm{GL}_n(\mathbf{R})$ *consisting of all matrices A such that* $\det(A) = 1$*. Obviously,*

$\mathrm{SL}_n(\mathbf{R}) \lhd \mathrm{GL}_n(\mathbf{R})$; $\mathrm{SL}_n(\mathbf{R})$ *is called the* special linear group *of the square matrices of order* n *over* $\mathbf{R}$. *A matrix* $X \in \mathrm{SL}_n(\mathbf{R})$ *is said to be* unimodular *([27], n° III.8.3). (Note that according to many authors, especially in applied mathematics, the set of unimodular matrices of order* n *over* $\mathbf{R}$ *is* $\mathrm{GL}_n(\mathbf{R})$.*)*

1.6.4 Induced Homomorphism, and Isomorphism Theorems

1.6.4.1 Induced Homomorphism

Theorem and Definition 145. *Let* G, G' *be two groups,* $N \lhd G$, $N' \lhd G'$ *and* $f : G \to G'$.

(1) The following properties are equivalent:
(a) $f(N) \leq N'$;
(b) there exists a unique group-homomorphism $\bar{f} : G/N \to G'/N'$ *such that the diagram below is commutative*

$$\begin{array}{ccc} G & \xrightarrow{f} & G' \\ \downarrow \pi & & \downarrow \pi' \\ G/N & \xrightarrow{\bar{f}} & G'/N' \end{array}$$

where $\pi : G \twoheadrightarrow G/N$ *and* $\pi' : G' \twoheadrightarrow G'/N'$ *are the canonical epimorphisms.*
(2) When $\bar{f} : G/N \to G'/N'$ *exists,* $\ker \bar{f} = \pi\left(f^{-1}(N')\right)$ *and* $\operatorname{im} \bar{f} = \pi'(f(G))$.
(3) Therefore, $\bar{f}$ *is a monomorphism (resp., an epimorphism) if, and only if* $f^{-1}(N') \leq N$ *(resp.,* $G' = N'.\operatorname{im} f$*). If* $N' \leq \operatorname{im} f$*, then* f *is a monomorphism if, and only if* $f(N) = N'$.
(4) The homomorphism $\bar{f}$ *is said to be* induced *by* f *(with respect to* N *and* N'*–in this order).*

Proof. (1) (a)⇒(b): Assuming that (a) holds true, let $x, x' \in G$ be such that $x{x'}^{-1} \in N$. Then, $f(x)\, f(x')^{-1} = f\left(x x'^{-1}\right) \in N'$, thus $\pi'(f(x))$ only depends on $\pi(x)$ for any $x \in G$. Let $\bar{f} : \pi(x) \mapsto \pi'(f(x))$; it is easy to check that $\bar{f}$ is a group-homomorphism. (b)⇒(a): If (b) holds true, let $x \in N$, i.e., such that $\pi(x) = 1$; then $\pi'(f(x)) = \bar{f}(1) = 1$, i.e., $f(x) \in N'$.
(2) $\bar{x} \in \ker \bar{f}$ if, and only if $\pi'(f(x)) = 1$ for all $x \in G$ such that $\pi(x) = \bar{x}$, and $\pi'(f(x)) = 1$ if, and only if $f(x) \in N'$, i.e., $x \in f^{-1}(N')$; therefore, $\bar{x} \in \ker \bar{f}$ if, and only if $\bar{x} \in \pi\left(f^{-1}(N')\right)$. The equality $\operatorname{im} \bar{f} = \pi'(f(G))$ is obvious.
(3) (A) By (2), $\ker \bar{f} = \{1\}$ if, and only if $f^{-1}(N') \leq N$.
(B) If $N' \leq \operatorname{im} f$, then $f\left(f^{-1}(N')\right) = N'$ and the above condition is equivalent to $N' \leq f(N)$. By Property (a) this is equivalent to $N' = f(N)$.
(C) Last, by (2), $\bar{f}$ is an epimorphism if, and only if $\pi'(\operatorname{im} f) = \pi'(G')$, and this happens if, and only if for every $x' \in G'$, there exist $x \in G$ and $y' \in N'$ such that $x'(f(g))^{-1} = y'$, i.e., $x' = y' f(g)$. This means that $G' = N'.\operatorname{im} f$. ∎

1.6.4.2 Noether's Isomorphism Theorems

The category **Grp** is not abelian, thus Theorem 52 cannot be applied. Noether's isomorphism theorems for groups take the following form ([27], n°I.4.6):

Theorem 146. *(*Noether's isomorphism theorems for groups*).*
(1) Let G, G' be groups and $f : G \to G'$ a group-homomorphism. Then

$$\boxed{G/\ker f \cong \operatorname{im} f} \tag{1.22}$$

(first isomorphism).
(2) Let G be a group and $N \lhd G$. For any subgroup G' of G, $G'N$ is a subgroup of G and $N \lhd G'N$, $G' \cap N \lhd G'$; in addition,

$$\boxed{(G'N)/N \cong G'/(G' \cap N)} \tag{1.23}$$

(second isomorphism).
(3) Let $G' \leq G$ and $N \lhd G$ be such that $G' \geq N$; then $G'/N \lhd G/N$ if, and only if $G' \lhd G$ and then

$$\boxed{G/G' \cong (G/N)/(G'/N)} \tag{1.24}$$

(third isomorphism).

1.6.4.3 Other Isomorphisms

Theorem 147. *(i)* (Correspondence theorem). *Let G be a group and $N \lhd G$. The map $G' \to G'/N$ is a one-to-one correspondence between the subgroups of G which contain N and the subgroups of G/N, given by $G' \to \pi(G')$ where $\pi : G \twoheadrightarrow G/N$ is the canonical epimorphism.*
(ii) Let $(G_i)_{i \in I}$ be a family of groups and, for each $i \in I$, $N_i \lhd G_i$. Then $\prod_{i \in I} N_i \lhd \prod_{i \in I} G_i$ and

$$\boxed{\left(\prod_{i \in I} G_i\right) / \left(\prod_{i \in I} N_i\right) \cong \prod_{i \in I} (G_i/N_i).} \tag{1.25}$$

Proof. The proof of (i) is easy. Let us prove (ii): Let $\pi_i : G_i \twoheadrightarrow G_i/N_i$ be the canonical epimorphism and $\psi : \prod_{i \in I} G_i \to \prod_{i \in I} (G_i/N_i)$ be the map defined by $\psi\left((x_i)_{i \in I}\right) = (\pi_i(x_i))_{i \in I}$; obviously, ψ is an epimorphism and $\ker \psi = \prod_{i \in I} N_i$. Thus (1.25) is a consequence of (1.22). ■

The proof of Item (3) below is similar to that of Theorem 147(ii).

Lemma and Definition 148. *(1) Let $(G_i)_{i \in I}$ be a family of* abelian groups *and let $x = (x_i)_{i \in I}$ be an element of $\prod_{i \in I} G_i$. The set $J \subseteq I$ consisting of all indices i for which $x_i \neq 0$ is called the* support *of x and is denoted by* $\operatorname{supp} x$. *If* $\operatorname{supp} x$ *is finite, the family x is said to be* finitely supported.

(2) The coproduct $\coprod_{i\in I} G_i$ *of the family* $(G_i)_{i\in I}$ *is the subgroup of* $\prod_{i\in I} G_i$ *consisting of all finitely supported elements*[4].
(3) If $N_i \lhd G_i$ *for each* $i \in I$, *then* $\coprod_{i\in I} N_i \lhd \coprod_{i\in I} G_i$ *and*

$$\boxed{\left(\coprod_{i\in I} G_i\right) / \left(\coprod_{i\in I} N_i\right) \cong \coprod_{i\in I} (G_i/N_i)\,.} \tag{1.26}$$

1.6.5 Simple Groups, and Cyclic Groups

Definition 149. *A group* G *is said to be* simple *if the only normal subgroups of* G *are* G *and* $\{1\}$.

Lemma 150. *(i) A cyclic group of finite order* a $(a \neq 0)$ *is isomorphic to* $\mathbb{Z}/a\mathbb{Z}$ *and a cyclic group of infinite order is isomorphic to* $\mathbb{Z}$.
(ii) The group $\mathbb{Z}$ *is not simple.*

Proof. (i): Let x be a generator of G. According to Proposition 139, $G = \{x^n : n \in \mathbb{Z}\}$. The map $f : \mathbb{Z} \ni n \mapsto x^n \in G$ is a group-epimorphism, thus $G \cong \mathbb{Z}/\ker f$ by Theorem 146(1), and $N \cong \ker f$ is a subgroup of $\mathbb{Z}$. If $N = \{0\}$, $G \cong \mathbb{Z}$. If $N \neq \{0\}$, let a be the smallest element > 0 of N. As easily seen, $N = a\,\mathbb{Z}$.
(ii): $2\mathbb{Z}$ is a normal subgroup of $\mathbb{Z}$. ∎

Proposition 151. *Every quotient of* $\mathbb{Z}/a\mathbb{Z}$ $(a \neq 0)$ *is isomorphic to* $\mathbb{Z}/b\mathbb{Z}$ *where* $b \mid a$*; conversely, if* $b \mid a$, *then* $\mathbb{Z}/b\mathbb{Z}$ *is isomorphic to a quotient of* $\mathbb{Z}/a\mathbb{Z}$.

Proof. By Theorem 147(i), a subgroup of $\mathbb{Z}/a\mathbb{Z}$ is of the form $N/a\mathbb{Z}$ where $N \lhd \mathbb{Z}$ and $N \supseteq a\mathbb{Z}$, thus $N = c\mathbb{Z}$ where $c \mid a$. Let $a = c\,b$ and consider the left multiplication by c, written $c\bullet : x + b\mathbb{Z} \mapsto c\,x + c\,b\mathbb{Z}$. This map is an isomorphism $\mathbb{Z}/b\mathbb{Z} \xrightarrow{\sim} c\mathbb{Z}/a\mathbb{Z}$. Conversely, if $b \mid a$, $\mathbb{Z}/b\mathbb{Z} \cong (\mathbb{Z}/a\mathbb{Z}) / (b\mathbb{Z}/a\mathbb{Z})$ according to Theorem 146(3). ∎

Corollary 152. *(i) The group* $\mathbb{Z}/p\mathbb{Z}$ *is simple if, and only if* p *is prime.*
(ii) Every abelian simple group is cyclic of prime order.

Proof. (i) is a consequence of Proposition 151.
(ii): If G is abelian and simple, let $1 \neq g \in G$. Then $\langle g \rangle$ is normal (since G is abelian) and $\langle g \rangle \neq \{1\}$, thus $\langle g \rangle = G$ and G is cyclic. Therefore, there exists $n \in \mathbb{N}$ such that $G \cong \mathbb{Z}/n\mathbb{Z}$ by Lemma 150 and $n > 0$ since $\mathbb{Z}$ is not simple. By (i), n is a prime. ∎

---

[4] The coproduct of a family of noncommutative groups can also be defined. This coproduct is a *free product* – a notion which is not needed in this book and which is explained in, e.g., ([27], n°*I*.7.3).

1.6.6 *The Lattice of Normal Subgroups*

1.6.6.1 The Complete Modular Lattice of Normal Subgroups

The following is proved in ([225], Sect. XIV.4):

Lemma 153. *The set* Lat (G) *of all normal subgroups of* G*, ordered by inclusion, is a complete modular lattice where for any* $H, K \in$ Lat (G), $HK = H \vee K$ *and* $H \cap K = H \wedge K$*. The set of all subgroups of* G*, ordered by inclusion, is a complete lattice with the same definition of the meet and the join.*

Let $S, T \in$ Lat (G) and consider the diamond below, where $S \vee T = ST$ and $S \wedge T = S \cap T$:

$$\begin{array}{ccccc} & & S \vee T & & \\ & \nearrow & & \nwarrow & \\ S & & & & T \\ & \nwarrow & & \nearrow & \\ & & S \wedge T & & \end{array} \tag{1.27}$$

The sides of the diamond are the parallel intervals $[S, S \vee T]$ and $[S \wedge T, T]$, as well as $[S \wedge T, S]$ and $[T, S \vee T]$. According to Definition 115, parallel intervals are perspective, and according to Noether's second isomorphism theorem (Theorem 146(2)) they yield isomorphisms $(ST)/S \cong T/(S \cap T)$ and $(ST)/T \cong S/(S \cap T)$. Thus, perspectivity yields isomorphism, and by induction so does projectivity. We have obtained the following result:

Lemma 154. *(*Diamond lemma*). In the modular lattice* Lat (G), *projective intervals* $[K, H]$ *and* $[K', H']$ $(K \lhd H,\ K' \lhd H')$*, correspond to isomorphic quotients* $H/K \cong H'/K'$.

Based on the two above lemmas, let us translate the terminology of lattices into that of groups:

Definition 155. *(1) A chain (of length* r*) in* Lat (G) *with end-points* G *and* $\{1\}$ *is of the form*

$$G = G_0 \supsetneq G_1 \supsetneq \ldots \supsetneq G_r = \{1\} \tag{1.28}$$

and is called an invariant chain *of* G.
(2) A normal chain *(of length* r*) of* G *is a chain of the form (1.28) such that* $G_{i+1} \lhd G_i$ *for any* $i \in \{1, \ldots, r-1\}$ *and it is written* $(G_i)_{0 \leq i \leq r}$*; then, each quotient* G_i/G_{i+1} *is called a* factor *of* G.

According to Remark 143, every invariant chain is a normal chain but the converse does not hold.

Consider a second normal chain of G :

$$G = H_0 \supsetneq H_1 \supsetneq \ldots \supsetneq H_s = \{1\}. \tag{1.29}$$

Definition 156. *The normal chains* (*1.28*) *and* (*1.29*) *are said to be* isomorphic *if there exists a permutation* $i \mapsto i'$ *of* $\{0, ..., r-1\}$ *such that* $G_i/G_{i+1} \cong H_{i'}/H_{i'+1}$.

1.6.6.2 Schreier-Zassenhaus Theorem and Jordan-Hölder Theorem

The theorem below is a consequence of the Schreier refinement theorem (Theorem 116) and of the Diamond Lemma (Lemma 154):

Theorem 157. *(*Schreier-Zassenhaus theorem*). Any two normal chains of* G *have isomorphic refinements.*

Remark 158. *Strictly speaking, Theorem 157 is a consequence of the* proof *(not given above) of Theorem 116, not of its statement, since the subgroups* G_i *and* H_i *in the normal towers* (*1.28*) *and* (*1.29*) *do not all belong to* Lat (G) *when these towers are not invariant. For further details, see, e.g., ([155], Theorems 8.4.2 & 8.4.3).*

Let Σ be a normal chain of G.

Definition 159. Σ *is called a* Jordan-Hölder series *of* G *if every refinement of* Σ *is equal to* Σ.

As easily shown ([27], §I.4, Prop. 9), the normal chain (1.28) is a Jordan-Hölder series of G if, and only if every quotient G_i/G_{i+1} is a simple group. The following result is a consequence of the Jordan-Hölder-Dedekind theorem (Theorem 117) or more specifically of its proof: see Remark 158.

Lemma and Definition 160. *(1) (*Jordan-Hölder theorem*). Any two Jordan-Hölder series of* G *are isomorphic.*
(2) If (*1.28*) *is a Jordan-Hölder series of* G*, then the* length *of* G *is* $|G| = r$*; if* G *has no Jordan-Hölder series, then* $|G| = +\infty$. *A group* G *is simple if, and only if* $|G| = 1$.

1.6.6.3 Polycyclic by Finite Groups

Definition 161. *(i) A group* G *is called* polycyclic *(resp.,* polycyclic by finite*) if there exists a finite normal chain* (*1.28*) *such that each factor* G_i/G_{i+1}, $i \in \{0, ..., r-1\}$, *is* infinite cyclic *(resp.,* either infinite cyclic or finite*).*
(ii) The Hirsch number $h(G)$ *of the polycyclic by finite group* G *is equal to the number of infinite cyclic factors of the normal chain* (*1.28*).

If G is polycyclic by finite, it is easy to show that, by choosing a different chain if necessary, one can arrange that the only finite factor, if any, is the first, G_0/G_1. Then, $h(G) = r$ if G is polycyclic and $h(G) = r - 1$ otherwise.

Example 162. *(1) An infinite cyclic subgroup of the additive group* $\mathbb{R}^n$ *is isomorphic to* $\mathbb{Z}$.
(2) A polycyclic subgroup of the additive group $\mathbb{R}^n$ *is a finitely generated additive subgroup of* $\mathbb{R}^n$, *i.e., a subgroup of the form* $\sum_{1\leq i\leq r} a_i\mathbb{Z}$ *where* $a_i \in \mathbb{R}^n$ *and the* a_i $(1 \leq i \leq r)$ *are* $\mathbb{R}$*-linearly independent. See ([313], Sect. 4.1) for more details.*
(3) A subgroup of a Dedekind group (Remark 143(v)) is finitely generated if, and only if it is polycyclic by finite.

1.6.7 Commutators, and Derived Group

1.6.7.1 Commutators

Definition 163. *Let* $x, y \in G$. *The* commutator *of* x *and* y, *written* (x, y), *is the element* $x^{-1}\, y^{-1}\, x\, y$ *of* G.

Lemma 164. *(i) For any* $x, y \in G$, $x\, y = y\, x\ (x, y)$.
(ii) The elements x, y commute *(i.e.,* $x\, y = y\, x$*) if, and only if* $(x, y) = 1$.
(iii) Let $f : G \rightarrow G'$ *be a group-homomorphism; then for any* $x, y \in G$, $f\left((x, y)\right) = \left(f\left(x\right), f\left(y\right)\right)$.

Proof. (i) and (iii) are clear and (ii) is an obvious consequence of (i). ∎

Notation 165. *Let* H, K *be subgroups of* G; *the subgroup of* G *generated by the commutators of* h *and* k $(h \in H,\ k \in K)$ *is denoted by* (H, K).

1.6.7.2 Derived Group

Definition 166. *The* derived group *of* G *(also called the* commutator group *of* G, *and written* $\mathbf{D}\left(G\right)$ *or* G^d*) is the subgroup generated by the commutators of all elements of* G.

According to Lemma 164(ii), $\mathbf{D}\left(G\right) = \{1\}$ if, and only if G is abelian.

Lemma 167. *Let* $f : G \rightarrow G'$ *be a group-homomorphism. Then* $f\left(\mathbf{D}\left(G\right)\right) \subseteq \mathbf{D}\left(G'\right)$ *and if* f *is surjective, so is its restriction* $f|_{\mathbf{D}(G)} \rightarrow \mathbf{D}\left(G'\right)$.

Proof. $f\left(\mathbf{D}\left(G\right)\right) \subseteq \mathbf{D}\left(G'\right)$ by lemma 164(iii) and if f is surjective, the image by f of the set of commutators of G is the set of commutators of G'. The result is now a consequence of Lemma 137. ∎

Theorem 168. *(i)* $\mathbf{D}\left(G\right) \lhd G$ *and* $G/\mathbf{D}\left(G\right)$ *is abelian.*
(ii) Let $\pi : G \rightarrow G/\mathbf{D}\left(G\right)$ *be the canonical epimorphism; for every abelian group* G' *and every group-homomorphism* $f : G \rightarrow G'$, *there exists a unique homomorphism* $\bar{f} : G/\mathbf{D}\left(G\right) \rightarrow G'$ *such that* $f = \bar{f} \circ \pi$.
(iii) Let H *be a subgroup of* G; *the following conditions are equivalent: (a)* $\mathbf{D}\left(G\right) \subseteq H$; *(b)* $H \lhd G$ *and* G/H *is abelian.*

Proof. (i): Consider $\pi : G \twoheadrightarrow \pi(\mathbf{D}(G)) = \{1\}$; since $\mathbf{D}(G) = \ker\pi$, $\mathbf{D}(G) \lhd G$ by Remark 143(iii). In addition, by Lemma 167, $\operatorname{im} \pi|_{\mathbf{D}(G)} = \mathbf{D}(\pi(\mathbf{D}(G))) = \mathbf{D}(G/\mathbf{D}(G))$; since $\operatorname{im} \pi|_{\mathbf{D}(G)} \subseteq \operatorname{im}\pi = \{1\}$, $\mathbf{D}(G/\mathbf{D}(G)) = \{1\}$ and $G/\mathbf{D}(G)$ is abelian.

(ii) is a consequence of Theorem 145, and $\bar{f}$ is the homomorphism induced by $\mathbf{D}(G)$ and $\mathbf{D}(G') = \{1\}$ (Theorem and Definition 145).

(iii): (a)⇒(b) by Theorem 146(2) and (b)⇒(a) by Theorem 145. ∎

Corollary 169. *Let* $\mathbf{R}$ *be a* commutative ring; *consider the general linear group* $\mathrm{GL}_n(\mathbf{R})$ *and the special linear group* $\mathrm{SL}_n(\mathbf{R})$ *(Example 144). Then* $\mathbf{D}(\mathrm{GL}_n(\mathbf{R})) \subseteq \mathrm{SL}_n(\mathbf{R})$.

Proof. Denoting by $\mathbf{U}(\mathbf{R})$ is the group of units of $\mathbf{R}$, the map $\det : \mathrm{GL}_n(\mathbf{R}) \to \mathbf{U}(\mathbf{R})$ is a group-homomorphism. Since $\det(\mathrm{diag}(1, ..., 1, u)) = u$, det is a group-epimorphism and $\mathrm{SL}_n(\mathbf{R}) = \ker\det$. By Theorem 146(1), $\mathrm{GL}_n(\mathbf{R})/\mathrm{SL}_n(\mathbf{R}) \cong \mathbf{U}(\mathbf{R})$ and $\mathbf{U}(\mathbf{R})$ is commutative since so is $\mathbf{R}$, thus $\mathbf{D}(\mathrm{GL}_n(\mathbf{R})) \subseteq \mathrm{SL}_n(\mathbf{R})$ by Theorem 168. ∎

Definition 170. *Let* G *be a group; then the abelian group* $G^{ab} \triangleq G/\mathbf{D}(G)$ *is called the* abelianization *of* G.

1.6.8 Solvable Groups

Let G be a group, $G^{(1)} \triangleq \mathbf{D}(G)$ and, for any $k \geq 1$, $G^{(k+1)} = \mathbf{D}\left(G^{(k)}\right)$. We can form the so-called *derived series* of G:

$$G \rhd G^{(1)} \rhd ... \rhd G^{(k)} \rhd ...$$

which terminates when we reach a group $G^{(s)}$ which is *perfect*, i.e., such that $G^{(s)} = \mathbf{D}\left(G^{(s)}\right)$.

Definition 171. *The group* G *is said to be* solvable *(or* soluble*) if* $G^{(s)} = \{1\}$.

The following result is proved in ([27], n°I.6.4):

Theorem 172. *Let* G *be a group of finite length. The following conditions are equivalent:*

(i) G *is solvable;*
(ii) G *has a normal chain* $(H_i)_{0 \leq i \leq s}$ *whose all factors* H_i/H_{i+1} $(0 \leq i \leq s-1)$ *are abelian.*
(iii) Considering a Jordan-Hölder series $(G_i)_{0 \leq i \leq n}$ *of* G, *all factors* G_i/G_{i+1} $(0 \leq i \leq n-1)$ *are abelian cyclic of prime order.*

Example 173. *The symmetric group* $\mathfrak{S}_n$ *is solvable for* $n \leq 4$ *and insolvable for* $n \geq 5$ *since for those values of* n, Alt_n *is simple ([79], Sect. 7.11).*

1.6.9 *Action of a Group on a Set*

1.6.9.1 Action, Orbit and Stabilizer

Let G be a group and let S be a set. An action of G on S is a group-homomorphism $\pi : G \rightarrow \mathfrak{S}(S)$ (Example 135). Let $g \in G$ and $x \in S$; the element $y = \pi(g)(x) \in S$ is denoted by $g\,.\,x$ in what follows.

The *orbit* of $x \in S$ under G (also called the G-orbit of x) is $G\,.\,x \triangleq \{g\,.\,x : g \in G\}$. Two different orbits are necessarily disjoint, thus $S = \bigcup_{i \in I} G\,.\,x_i$ where I is some indexing set and the x_i are the elements of distinct orbits.

The *stabilizer* of $x \in S$ is the subset G_x of G such that $G_x\,.\,x = \{x\}$; G_x is a subgroup of G. Let $(G : G_x)$ be the index of G_x. As easily seen, $(G : G_x) = \operatorname{card}(G\,.\,x)$. Therefore, if S is finite, we have the *orbit decomposition formula*

$$\boxed{\operatorname{card}(S) = \sum_{i \in I} (G : G_{x_i})\,.} \tag{1.30}$$

1.6.9.2 Homogeneous Spaces

Let $x \in S$ and consider the map $h_x : G \rightarrow G\,.\,x : g \mapsto g\,.\,x$. This map can be decomposed as follows:

$$G \xrightarrow{\pi_x} G/G_x \xrightarrow{\varphi_x} G\,.\,x$$

where G/G_x is the set of left cosets $g\,G_x$ of G_x in G (1.6.2.**2**), π_x is the canonical surjection $g \mapsto g\,G_x$ and φ_x is the canonical bijection $g\,G_x \mapsto g\,x$.

The action of G on S is said to be *transitive* (resp., *simply transitive*) if for any $x \in S$, the map $G \rightarrow S : x \mapsto g\,.\,x$ is surjective (resp., bijective). If the action of G on S is transitive, the set S consists of only one orbit (i.e., $S = G\,.\,x$ for any $x \in S$) and is said to be *homogeneous*. Then φ_x is a bijection $G/G_x \xrightarrow{\sim} S$. Conversely, if H is any subgroup of G, then G transitively acts on G/H in an obvious way, thus the set of left cosets G/H is homogeneous. A similar rationale holds for right cosets.

1.7 Digression 1: Topological Vector Spaces

1.7.1 *Topological Vector Spaces*

We assume that the reader has a basic knowledge of general topology and of topological vector spaces (see, e.g., [339], Part I, Chap. 1-6). Let us recall some basic facts.

1.7.1.1 Filters and Nets

Let S be a nonempty set. A *filter* Φ on S is a set of parts of S with the following properties: (a) $\varnothing \notin \Phi$, (b) if two parts P, Q of S are such that $P \in \Phi$ and $Q \supseteq P$, then $Q \in \Phi$, (c) if $P, Q \in \Phi$, then $P \cap Q \in \Phi$. If Φ, Φ' are two filters on S, Φ' is said to be *finer* than Φ if $\Phi \subseteq \Phi'$. Let $\mathfrak{B}$ be a set of parts of S; $\mathfrak{B}$ is called a *base* of the filter Φ if Φ is the set of all parts of S which contain a set belonging to $\mathfrak{B}$. A set $\mathfrak{B}$ of parts of S is a base of filter on S if, and only if (a') if $P, Q \in \mathfrak{B}$, then there exists $R \in \mathfrak{B}$ such that $P \cap Q \supseteq R$, (b') $\mathfrak{B} \neq \varnothing$, (c') $\varnothing \notin \mathfrak{B}$.

To a filter Φ on S one can associate a *net* (also called a *generalized sequence*) in S as follows: let I be a filtering index set (**1.2.3.4**) and let $(\mathfrak{B}_i)_{i \in I}$ be a direct system of elements of Φ (for inclusion), such that $\mathfrak{B} = (\mathfrak{B}_i)_{i \in I}$ is a base of Φ. Choose in each $\mathfrak{B}_i$ an element x_i using the axiom of choice (Axiom 1). Then $(x_i)_{i \in I}$ is a net in S, said to be associated to $\mathfrak{B}$ (or to Φ). If $\mathfrak{B}$ is countable, then one can assume that $I = \mathbb{N}$ without loss of generality, and every net associated to $\mathfrak{B}$ is a sequence.

Conversely, let $(x_i)_{i \in I}$ be a net in S, and for every $i \in I$, let $\mathfrak{B}_i = \{x_k : k \succcurlyeq i\}$. Then $\mathfrak{B} = (\mathfrak{B}_i)_{i \in I}$ is a base of filter on S (called the elementary base of filter associated with the net $(x_i)_{i \in I}$).

1.7.1.2 Topological Spaces

Consider a topology $\mathfrak{T}$ on a set S. The set S, endowed with the topology $\mathfrak{T}$, is often denoted by $S[\mathfrak{T}]$ for the sake of clarity. When speaking of the topological space S, the topology $\mathfrak{T}$ is understood. A topological space $S[\mathfrak{T}]$ (or the topology $\mathfrak{T}$) is *Hausdorff* if whenever $x_1, x_2 \in S$ are such that $x_1 \neq x_2$, there exist neighborhoods N_1, N_2 of x_1 and x_2 respectively, such that $N_1 \cap N_2 = \varnothing$. The set of all neighborhoods of $x \in S$ is a filter $\mathfrak{N}_S(x)$.

A filter Φ on $S[\mathfrak{T}]$ is said to converge to $x \in S$ (written $\lim \Phi = x$) if Φ is finer than $\mathfrak{N}_S(x)$. Let $\mathfrak{B}$ be a base of the filter Φ; $\mathfrak{B}$ is said to converge to x (written $\lim \mathfrak{B} = x$) if so does Φ, i.e., if for every neighborhood N of x there exists a part $P \in \mathfrak{B}$ such that $P \subseteq N$. A net $(x_i)_{i \in I}$ in S is said to converge to x (written $\lim_I x_i = x$) if so does the elementary base of filter $(\mathfrak{B}_i)_{i \in I}$ associated with the net $(x_i)_{i \in I}$, i.e., for every neighborhood N of x, there exists $i \in I$ such that $x_k \in N$ for all $k \succcurlyeq i$.

The proof of the following is easy [96].

Proposition 174. *Let $S[\mathfrak{T}]$ be a topological space, let $x \in S$, and let Φ be a filter on S.*

(1) The following conditions are equivalent:
(i) $\lim \Phi = x$;
(ii) $\lim \mathfrak{B} = x$, *where $\mathfrak{B}$ is any base of the filter Φ;*
(iii) there exists a base $\mathfrak{B}$ of the filter Φ such that $\lim \mathfrak{B} = x$;

(iv) every net $(x_i)_{i\in I}$ *associated to* Φ *converges to* x.
(2) Let $\mathfrak{T}$ *be Hausdorff. If* $\lim \Phi = x$ *and* $\lim \Phi = y$, *then* $x = y$.

For the use of filters and nets to characterize limits of functions, see Exercise 209. Let $S[\mathfrak{T}]$ be a topological space, let $M \subseteq S$ and let $\iota : M \hookrightarrow X$ be the inclusion; the topology $\mathfrak{T}_M$ induced by $\mathfrak{T}$ on M is the coarsest topology which makes ι continuous. The open sets in the topology $\mathfrak{T}_M$ are the sets $A \cap M$ where A is open in $\mathfrak{T}$.

Let us give three simple examples of topologies.

The *discrete topology* on S is the topology $\mathfrak{T}_d[S]$ in which each subset of S is open. Assuming that S is nonempty, let $x \in S$; a net $(x_i)_{i\in I}$ of elements of S converges toward x in $\mathfrak{T}_d[S]$ if, and only if there exists $i_0 \in I$ such that $x_i = x$ whenever $i \succeq i_0$.

The *finite topology* will be used in the sequel (Remark 728). Let X, Y be nonempty sets, let Y^X be the set of all mappings $X \to Y$, and let $S \subseteq Y^X$ be nonempty. The *finite topology* on S is the topology $\mathfrak{T}_f[S]$ defined as follows: let $f \in S$; a base of the filter of neighborhoods of f consists of all subsets of the form

$$\mathcal{N}(f; g, x_1, ..., x_n) = \{g \in S : g(x_i) = f(x_i), \quad 1 \le i \le n\}$$

where $\{x_i : 1 \le i \le n\}$ is a finite set of elements of X. Assuming that S is finite, the topologies $\mathfrak{T}_f[S]$ and $\mathfrak{T}_d[S]$ coincide, and they are the unique Hausdorff topology on S; in this topology, each part of S is both open and closed, and S is compact.

The *trivial topology* on S is the topology $\mathfrak{T}_t[S]$ in which the only open subsets are S and $\varnothing$.

Let $\mathfrak{T}, \mathfrak{T}'$ be topologies on a set S. For any $x \in S$, let $\mathcal{N}_S(x)$ (resp., $\mathcal{N}'_S(x)$) be the filter of neighborhoods of x in the topology $\mathfrak{T}$ (resp., $\mathfrak{T}'$). Then $\mathfrak{T}$ is said to be *finer* than $\mathfrak{T}'$, and $\mathfrak{T}'$ is said to be *coarser* than $\mathfrak{T}$, if for any $x \in S$, $\mathcal{N}'_S(x) \subseteq \mathcal{N}_S(x)$. The finest (resp., coarsest) topology on S is $\mathfrak{T}_d[S]$ (resp., $\mathfrak{T}_t[S]$). The topology $\mathfrak{T}_d[S]$ is Hausdorff, whereas $\mathfrak{T}_t[S]$ is not if S has more that one element.

1.7.1.3 Topological Vector Spaces

Let $\mathbf{K}$ be a division ring (Definition 226 below). A vector space over $\mathbf{K}$ (**2.2.4.4**) is called a $\mathbf{K}$-space, for short. The topology $\mathfrak{T}$ of a topological left vector space X over a valued non-discrete division ring $\mathbf{K}$ is defined by its filter $\mathfrak{N}_X$ of neighborhoods of zero.

Let $X[\mathfrak{T}]$ be a topological vector space (TVS), let $A \subseteq X$ and let Φ be a filter on A. The filter Φ on A is called a *Cauchy filter* if for every $N \in \mathfrak{N}_X$, there exists $M \in \Phi$ such that $M - M \subseteq N$. To a Cauchy filter Φ on $A \subseteq X$ one can associate as in (**1.7.1.1**) a net $(x_i)_{i\in I}$, called a *Cauchy net*, i.e., a net $(x_i)_{i\in I}$ such that for any $U \in \mathfrak{N}_X$, there exists $i_0 \in I$ for which $x_{i_1} - x_{i_2} \in U$ whenever $i_1, i_2 \succeq i_0$.

Let M be a subspace of the TVS X and let $\varphi : X \twoheadrightarrow X/M$ be the canonical surjection; the *quotient topology* on X/M is the finest topology on X/M which makes φ continuous.

Let A, B be two subsets of X; A is said to *absorb* B if there exists $\alpha > 0$ such that $\lambda A \supseteq B$ for every $\lambda \in \mathbf{K}$ such that $|\lambda| \geq \alpha$. A subset B of X is called *bounded* if it is absorbed by every neighborhood of zero. A set $A \subseteq X$ is called *absorbing* (resp., *bornivorous*) if it absorbs all finite (resp., bounded) subsets of X.

A precompact set in a TVS (in particular, a relatively compact set in a Hausdorff TVS) is bounded but the converse does not hold in general.

A TVS $X[\mathfrak{T}]$ is *semicomplete* (resp., *complete*) if every Cauchy sequence in (resp., every Cauchy filter on) $X[\mathfrak{T}]$ converges. Thus, every complete TVS is semicomplete. A Cauchy sequence is bounded, but a Cauchy filter is not so in general ([316], Subsect. I.5.1, Corol. 3).

A TVS $X[\mathfrak{T}]$ is *metrizable* if, and only if it is Hausdorff and $\mathfrak{N}_X$ has a countable base. A metrizable TVS is semicomplete if, and only if it is complete.

1.7.2 Locally Convex Topological Vector Spaces

1.7.2.1 Seminorms

A TVS X over $\mathbf{k} = \mathbb{R}$ or $\mathbb{C}$ is called a *locally convex topological vector space* (or a locally convex space, or an LCS, for short) if its topology $\mathfrak{T}$ is locally convex, i.e., if $\mathfrak{T}$ is defined by a family of seminorms $(p_i)_{i \in I}$; this means that the sets $p_i^{-1}(\{\lambda\})$, $i \in I$, $\lambda > 0$, form a base of the filter $\mathfrak{N}_X$ of neighborhoods of zero. The topology $\mathfrak{T}$ is called the *initial topology* of X.

In everything that follows, all TVSs are LCSs.

1.7.2.2 Duality

(I) Consider two $\mathbf{k}$-spaces X, Y and a bilinear form

$$\langle -, - \rangle : X \times Y \to \mathbf{k}.$$

The spaces X, Y are said to be in duality (or to form a *dual system*) with respect to $\langle -, - \rangle$. The *weak topology* $\sigma(Y, X)$ on Y defined by this duality is the coarsest topology which makes continuous all linear forms $\langle x, \bullet \rangle : y \mapsto \langle x, y \rangle$ $(x \in X)$. Therefore, a net $(y_i)_{i \in I}$ in Y converges to $0 \in Y$ in the topology $\sigma(Y, X)$ if $(\langle x, y_i \rangle)_{i \in I}$ converges to $0 \in \mathbf{k}$ for all $x \in X$. The weak topology $\sigma(Y, X)$ is locally convex. The weak topology $\sigma(X, Y)$ on X is defined similarly. The LCS $Y[\sigma(Y, X)]$ is Hausdorff if, and only if the bilinear form $\langle -,- \rangle$ is *separated at* Y, i.e., when the following condition holds true: $\langle x,y \rangle = 0$ for all $x \in X$ implies $y = 0$. The dual system (X,Y) (where $\langle -,- \rangle$ is understood) is called *separated* if $\langle -,- \rangle$ is separated at both X and Y.

(II) Consider an LCS $X[\mathfrak{T}]$. Its *topological dual* X' consists of all *continuous* linear forms $x' : X \to \mathbf{k}$, and $x'(x)$ is denoted by $\langle x, x' \rangle$. A subset A of X' is called *equicontinuous* if for every $V \in \mathfrak{N}_{\mathbf{k}}$ (where $\mathfrak{N}_{\mathbf{k}}$ is the filter of neighborhoods of zero in $\mathbf{k}$), there exists $U \in \mathfrak{N}_X$ such that $x'(U) \subseteq V$ for all $x' \in A$.

The canonical bilinear form

$$\langle -,- \rangle : X \times X' \to \mathbf{k} \tag{1.31}$$

is called the *duality bracket*. The two $\mathbf{k}$-spaces X, X' form a dual system with respect to the duality bracket, thus (I) applies; this dual system (X, X') is always separated at X'. The topology $\sigma(X', X)$ is called the *weak* topology* of X' and $X'[\sigma(X', X)]$ is called the *weak* dual* of $X[\mathfrak{T}]$; $\sigma(X', X)$ is the topology of uniform convergence on all finite parts of X. If $X[\mathfrak{T}]$ is Hausdorff, then the dual system (X, X') is separated ([36], n°II.6.1).

One can also define on X' the topology of uniform convergence on all bounded parts of $X[\mathfrak{T}]$; this topology is denoted by $\beta(X', X)$, and is called the *strong topology* of X'. The space $X'[\beta(X', X)]$ is called the *strong dual* of $X[\mathfrak{T}]$ and is denoted by X'_β. This LCS is Hausdorff.

The *weakened topology* on X is the topology $\sigma(X, X')$. Note that $\langle x, \bullet \rangle$ belongs to the *algebraic dual* X'^* of X' (see Subsect. 2.2.5 below). The $\mathbf{k}$-space X (resp., X') endowed with the topology $\sigma(X, X')$ (resp., $\sigma(X', X)$) is denoted by X_σ (resp., X'_σ). Every bounded set in X_σ is bounded in $X[\mathfrak{T}]$ ([36], §III.5, Corol. 3 of Theorem 2).

Assume that $X[\mathfrak{T}]$ is Hausdorff. Then, as a consequence of the *Hahn-Banach theorem* ([36], §II.4, Corol. 1 of Prop. 2) the map $x \mapsto \langle x, \bullet \rangle$ is injective, so that X can be embedded in X'^*, identifying x with $\langle x, \bullet \rangle$.

1.7.2.3 Transpose of a Continuous Linear Map

Let $X[\mathfrak{T}_1], Y[\mathfrak{T}_2]$ be two LCSs and $f : X \to Y$ be a $\mathbf{k}$-linear map; f is called *weakly continuous* if it continuous from X_σ to Y_σ. This happens if, and only if there exists a map (necessarily unique, $\mathbf{k}$-linear and continuous) ${}^t f : Y'_\sigma \to X'_\sigma$ such that $\langle f(x), y' \rangle = \langle x, {}^t f(y') \rangle$ for all $x \in X$, $y' \in Y'$. This map ${}^t f$ is called the *transpose*, or the *adjoint*, of f . If $f : X[\mathfrak{T}_1] \to Y[\mathfrak{T}_2]$ is continuous, then f is weakly continuous.

If $X[\mathfrak{T}_1]$ and $Y[\mathfrak{T}_2]$ are Hausdorff, then $(X'_\sigma)' = X$, $(Y'_\sigma)' = Y$ and ${}^{tt} f = f$.

1.7.3 Special Classes of Topological Vector Spaces

1.7.3.1 Fréchet Spaces, Banach Spaces, and Hilbert Spaces

Consider an LCS $X[\mathfrak{T}]$, and assume that $\mathfrak{T}$ is Hausdorff. The following conditions are equivalent:

(a) the filter $\mathfrak{N}_X$ has a countable base;

(b) $\mathfrak{T}$ is defined by a *sequence* $(p_n)_{n\in\mathbb{N}}$ of seminorms;

(c) $X[\mathfrak{T}]$ is metrizable.

An LCS is called a *Fréchet space* if it is metrizable and complete. Let X and Y be two Fréchet spaces and $f : X \to Y$ be linear, continuous and bijective; according to the *Banach inverse mapping theorem*, f is an isomorphism.

A *Banach space* is a complete normed vector space. Therefore, every Banach space is a Fréchet space.

Consider a $\mathbf{k}$-space X and a non-negative Hermitian form $\langle \bullet, \bullet \rangle_X : X \times X \to \mathbf{k}$. As a consequence of the Cauchy-Schwarz inequality, this form is nondegenerate if, and only if it is positive definite, i.e., $\langle x, x \rangle_X > 0$ whenever $x \neq 0$ ([36], n°V.1.2). The $\mathbf{k}$-space X is called a *pre-Hilbert space* if it is endowed with a nondegenerate Hermitian form $\langle \bullet, \bullet \rangle_X$. Such a space is canonically endowed with the norm $\|x\|_X = \sqrt{\langle x, x \rangle_X}$. A complete pre-Hilbert space is called a *Hilbert space*. Therefore, every Hilbert space is a Banach space. Note that $\langle \bullet, \bullet \rangle_X$ is $\mathbf{k}$-linear with respect to the first variable (i.e., $\langle \lambda x, y \rangle = \lambda \langle x, y \rangle$ for all $\lambda \in \mathbf{k}$) and $\mathbf{k}$-antilinear with respect to the second variable (i.e., $\langle x, \lambda y \rangle = \bar{\lambda} \langle x, y \rangle$, where $\bar{\lambda}$ is the conjugate of λ). When $\mathbf{k} = \mathbb{R}$, $\bar{\lambda} = \lambda$ and $\langle \bullet, \bullet \rangle_X$ is $\mathbb{R}$-bilinear.

1.7.3.2 Bidual, and Reflexive LCSs

The *bidual* of an LCS $X[\mathfrak{T}]$ is the dual of the strong dual of $X[\mathfrak{T}]$, i.e., it is the space $X'' = X'[\beta(X', X)]'$. When X is Hausdorff, there exists a $\mathbf{k}$-linear monomorphism $X \hookrightarrow X''$. The bidual X'' can be endowed with the strong topology $\beta(X'', X')$. The LCS $X[\mathfrak{T}]$ is called *semireflexive* if X coincides with X'', and $X[\mathfrak{T}]$ is called *reflexive* if it coincides with $X''[\beta(X'', X')]$, i.e., with its *strong bidual.* A Hilbert space is reflexive.

The *weak dual* of $X[\mathfrak{T}]$ is the LCS $X'[\sigma(X', X'')]$. If X is Hausdorff, $\sigma(X', X'')$ is finer than $\sigma(X', X)$, and these topologies coincides if, and only if $X = X''$, i.e., if $X[\mathfrak{T}]$ is semireflexive. Then, the weak* dual $X'[\sigma(X', X)] = X'_\sigma$ of $X[\mathfrak{T}]$ coincides with the weak dual $X'[\sigma(X', X'')]$.

Remark 175. *Let us emphasize that two situations must be distinguished:*

(I) Two $\mathbf{k}$*-spaces* X, Y *are given (without topology) along with a bilinear form*

$$\langle -, - \rangle : X \times Y \to \mathbf{k}$$

with respect to which (X, Y) *is a dual system. Then* $\sigma(X, Y)$ *(resp.,* $\sigma(Y, X)$*) is the weak topology on* X *(resp.,* Y*) defined by the duality.*

(II) An LCS $X[\mathfrak{T}]$ is given. Then X and the topological dual X' of $X[\mathfrak{T}]$, along with the duality bracket, form a dual system separated at X'. The topology $\sigma(X, X')$ is coarser than $\mathfrak{T}$ and is called the weakened topology on X. The topology $\sigma(X', X)$ is coarser than the strong topology $\beta(X', X)$ and is called the weak topology on X'. The dual of X'_β is the bidual X'', and the topology $\sigma(X', X'')$ is called the weak topology on X'.*
Therefore, the expression weak topology *on X' may have two different meanings, depending on the context.*

1.7.3.3 Barreled Spaces, Bornological Spaces, and Montel Spaces

Consider an LCS $X[\mathfrak{T}]$. A set $A \subseteq X$ is called *balanced* if for any $\lambda \in \mathbf{k}$ such that $|\lambda| \leq 1$, $\lambda A \subseteq A$. A *disk* in $X[\mathfrak{T}]$ is a part A which is convex and balanced. A disk is a *barrel* if it is absorbing. Let $(p_i)_{i \in I}$ be a family of seminorms which defines the topology $\mathfrak{T}$. Each set $p_i^{-1}(\{\lambda\})$, $i \in I$, $\lambda > 0$, is a barrel, but this does not imply that each barrel is a neighborhood of zero in general. We have the following ([36], n°III.4.1 and §III.2; [190], 21.2(2)):

Lemma and Definition 176. *(1) The following conditions are equivalent:*
(i) every barrel of $X[\mathfrak{T}]$ is a neighborhood of zero;
(ii) every bounded part of X'_σ is equicontinuous;
(iii) the topology $\mathfrak{T}$ coincides with the strong topology $\beta(X, X')$.
(2) The LCS $X[\mathfrak{T}]$ is called barreled *if the above equivalent conditions are satisfied.*
(3) The LCS $X[\mathfrak{T}]$ is called bornological *if every bornivorous disk is a neighborhood of zero.*

A Fréchet space is barreled. Every product of barreled spaces is barreled ([36], §IV.2, Corol. of Prop. 15). Obviously, an LCS is reflexive if, and only if it is both semireflexive and barreled.

A metrizable LCS is bornological ([36], §III.2, Prop. 2). A countable product of bornological spaces is bornological ([190], 28.4(4)). The strong dual of a bornological space is complete ([36], §III.3, Corol. 1 of Prop. 12). The strong dual $X[\mathfrak{T}]'_\beta$ of a metrizable LCS $X[\mathfrak{T}]$ is bornological if, and only if $X[\mathfrak{T}]$ is barreled ([149], Chap. 4, Part 3, Sect. 4, Corol. 2 of Theorem 6), and this happens if, and only if $X[\mathfrak{T}]'_\beta$ is barreled ([316], Chap. IV, 6.6).

A bornological semicomplete LCS is barreled ([36], §III.4, Corol. 2 of Theorem 2).

An LCS $X[\mathfrak{T}]$ is called a *Montel space* if it is Hausdorff, barreled, and if every bounded part is relatively compact. A Montel space is reflexive and the strong dual of a Montel space is again a Montel space ([36], n°IV.2.5). Every product of Montel spaces is a Montel space ([190], 27.2(4)). Montel spaces enjoy the following important property:

Theorem 177. *Let $X[\mathfrak{T}]$ be a Montel space and let $(x_i)_{i\in I}$ be a net which converges to x in X_σ.*

(i) Assume that there exists $i_0 \in I$ such that the set $\{x_i : i \succeq i_0\}$ is bounded in $X[\mathfrak{T}]$. Then $(x_i)_{i\in I}$ converges to x in $X[\mathfrak{T}]$.
(ii) If $I = \mathbb{N}$, then the set $\{x_n : n \in \mathbb{N}\}$ is bounded.
(iii) If Y is a closed subspace of $X[\mathfrak{T}]$, then Y, endowed with the topology $\mathfrak{T}_Y$ induced by $\mathfrak{T}$, is again a Montel space.

Proof. (i): See ([36], §IV.2, Prop. 8).

(ii): Since (x_n) converges to x in X_σ, it is a Cauchy sequence in that space, therefore the set $\{x_n : n \in \mathbb{N}\}$ is bounded in X_σ (Subsect. 1.7.1), thus in $X[\mathfrak{T}]$ (**1.7.2.2**).

(iii): Let B be a closed bounded part of $Y[\mathfrak{T}_Y]$. Then B is again closed and bounded in $X[\mathfrak{T}]$, thus it is compact. ∎

An LCS which is both a Fréchet space and a Montel space is called *Fréchet-Montel.*

1.7.3.4 Inductive Limits

Let I be a filtering set and let $\{X_i[\mathfrak{T}_i], \varphi_j^i\}$ be a direct system with index set I in the category **LCS** (Definition 15 and Example 36(2)). The inductive limit $\varinjlim X_i[\mathfrak{T}_i]$ is defined as usual (Definition 16) and is an LCS $X[\mathfrak{T}_\rightarrow]$.

Every inductive limit of barreled (resp., bornological) spaces is barreled (resp., bornological) ([36], §III.4, Corol. 3 of Prop. 3; Sect. III.2, Example 2).

Definition 178. *Let (X_n) be strictly increasing sequence of* **k**-*spaces and $X = \bigcup_n X_n$. Assume that each X_n is endowed with a locally convex topology $\mathfrak{T}_n$ such $\mathfrak{T}_{n+1}$ induces $\mathfrak{T}_n$ on X_n Then $X[\mathfrak{T}_\rightarrow] = \varinjlim X_n[\mathfrak{T}_n]$ is called the* strict inductive limit *of the sequence of LCSs $(X_n[\mathfrak{T}_n])$.*

Theorem 179. *Let $X[\mathfrak{T}_\rightarrow]$ be the strict inductive limit of the sequence $(X_n[\mathfrak{T}_n])$.*

(1) The following properties hold:
(i) for each n, $\mathfrak{T}_\rightarrow$ induces $\mathfrak{T}_n$ on X_n;
(ii) if each $\mathfrak{T}_n$ is Hausdorff (resp., complete), then $\mathfrak{T}_\rightarrow$ is Hausdorff (resp., complete).
(2) Assume that for each n, X_n is closed in $X_{n+1}[\mathfrak{T}_{n+1}]$. Then:
(i') each X_n is closed in $X[\mathfrak{T}_\rightarrow]$;
(ii') a subset B of $X[\mathfrak{T}_\rightarrow]$ is bounded if, and only if B lies in some $X_n[\mathfrak{T}_n]$ and is bounded there;
(iii') if each $X_n[\mathfrak{T}_n]$ is a Montel space, so is $X[\mathfrak{T}_\rightarrow]$.

Proof. See ([36], §II.4, Prop. 9), ([190], 19.4(4)) and ([36], n°IV.2.5, Example 3). ■

Corollary 180. *Let $X[\mathfrak{T}_{\rightarrow}]$ be an $(\mathcal{LF})$ (resp., $(\mathcal{LFM})$) space, i.e., the strict inductive limit of a sequence $(X_n[\mathfrak{T}_n])$ of Fréchet (resp., Fréchet-Montel) spaces. Then $X[\mathfrak{T}_{\rightarrow}]$ is Hausdorff, complete, bornological and barreled (resp., both barreled and Montel).*

Proof. Since $X_n[\mathfrak{T}_n] \hookrightarrow X_{n+1}[\mathfrak{T}_{n+1}]$ and $X_n[\mathfrak{T}_n]$ is complete, $X_n[\mathfrak{T}_n]$ is closed in $X_{n+1}[\mathfrak{T}_{n+1}]$ ([34], §II.3, Prop. 8). The result follows from the above. ■

Fréchet-Schwartz spaces and *Dual Fréchet-Schwartz spaces* (also called, respectively, $(\mathcal{FS})$ spaces and $(\mathcal{DFS})$ spaces, for short) can be defined as follows ([250], Def. A.4.1):

Definition 181. *(1) An LCS space E is $(\mathcal{FS})$ if there exists a decreasing sequence of Banach spaces*

$$E_1 \overset{\rho_1^2}{\longleftarrow} E_2 \longleftarrow \ldots \longleftarrow E_j \overset{\rho_j^{j+1}}{\longleftarrow} E_{j+1} \longleftarrow \ldots$$

such that all linear maps $\rho_j^{j+1} : E_{j+1} \to E_j$ are compact *(i.e., there exists a neighborhood V_{j+1} of 0 in E_{j+1} whose image by ρ_j^{j+1} is relatively compact in E_j; this implies that ρ_j^{j+1} is continuous) and*

$$E = \varprojlim E_j = \bigcap_{j \geq 1} E_j.$$

(2) An LCS space E is $(\mathcal{DFS})$ if there exists an increasing sequence of Banach spaces

$$E_1 \overset{\rho_2^1}{\longrightarrow} E_2 \longrightarrow \ldots \longrightarrow E_j \overset{\rho_{j+1}^j}{\longrightarrow} E_{j+1} \longrightarrow \ldots$$

such that all maps ρ_j^{j+1} are compact *and*

$$E = \varinjlim E_j = \bigcup_{j \geq 1} E_j.$$

An $(\mathcal{FS})$ space can also defined to be an LCS which is both Fréchet and Schwartz, but will not detail the notion of Schwartz space. This notion and the following result are due to Grothendieck [147] (see, also, [149], Chap. 4, Part 4):

Theorem 182. *(1) An LCS space is* $(\mathcal{DFS})$ *if, and only if it is the strong dual of an* $(\mathcal{FS})$*-space. Conversely, an LCS space is* $(\mathcal{FS})$ *if, and only if it is the strong dual of a* $(\mathcal{DFS})$*-space.*
(2) $(\mathcal{FS})$ *spaces and* $(\mathcal{DFS})$ *spaces are Montel, bornological and complete. An inductive limit of a* sequence *of* $(\mathcal{DFS})$ *spaces is a* $(\mathcal{DFS})$ *space; dually, a projective limit of a* sequence *of* $(\mathcal{FS})$ *spaces is an* $(\mathcal{FS})$ *space.*

1.7.3.5 Examples

Let Ω be an open subset of $\mathbf{k}^n$ ($\mathbf{k} = \mathbb{R}$ or $\mathbb{C}$). In what follows, all functions are $\mathbb{C}$-valued, except when otherwise specified.

(1) Let $\mathcal{E}(\Omega) = C^\infty(\Omega)$ be the $\mathbb{C}$-vector space of all indefinitely differentiable functions on Ω. For any multi-index $\alpha = (\alpha_1, ..., \alpha_n) \in \mathbb{N}^n$, any $\varphi \in \mathcal{E}(\Omega)$ and any $x \in \Omega$, we set

$$|\alpha| = \sum_{1 \leq i \leq n} \alpha_i, \quad D^\alpha \varphi = \frac{\partial^{|\alpha|} \varphi}{\partial_1^{\alpha_1} ... \partial_n^{\alpha_n}}. \tag{1.32}$$

For every integer $k > 0$, let K_k be the compact set $\|x\| \leq k$ and let

$$p_{k,m}(\varphi) = \sup_{\substack{x \in K_k \cap \Omega \\ |\alpha| \leq m}} |D^\alpha \varphi(x)|.$$

The topology $\mathfrak{T}_c^\infty$ of $\mathcal{E}(\Omega)$ is defined by the seminorms $p_{k,m}$ and is easily seen to be Hausdorff. Since $\{p_{k,m}\}$ is countable, $\mathfrak{T}_c^\infty$ is metrizable (**1.7.3.1**). One can check that $\mathcal{E}(\Omega)[\mathfrak{T}_c^\infty]$ is complete ([36], n°III.1.7, Example b)), thus is a Fréchet space. More specifically, $\mathcal{E}(\Omega)[\mathfrak{T}_c^\infty]$ is a Fréchet-Montel space ([36], n°IV.2.5, Example 4) and a . Its dual $\mathcal{E}'(\Omega)$ is the space of distributions with compact support included in Ω. The strong dual $\mathcal{E}'_\beta(\Omega)$ is a complete bornological Montel space.

(2) Let $\mathcal{D}(\Omega) = C_0^\infty(\Omega)$ be the $\mathbb{C}$-vector space of all indefinitely differentiable functions with compact support included in Ω. Let $K \subseteq \Omega$ be compact and let $\mathcal{D}_K(\Omega)$ be the subspace of $\mathcal{D}(\Omega)$ consisting of those functions, the support of which is included in K; $\mathcal{D}_K(\Omega)$ is endowed with the topology $\mathfrak{T}_K^\infty$ induced by the above topology $\mathfrak{T}_c^\infty$. As easily seen, $\mathcal{D}_K(\Omega)$ is closed in $\mathcal{E}(\Omega)[\mathfrak{T}_c^\infty]$, thus is a Fréchet space. If $K \subsetneq K'$, then $\mathcal{D}_K(\Omega) \subsetneq \mathcal{D}_{K'}(\Omega)$. Let $(K_k)_{k>0}$ be a strictly increasing sequence of compact sets, the interiors of which form a covering of Ω. Since $\bigcup_{k>0} \mathcal{D}_{K_k}(\Omega) = \mathcal{D}(\Omega)$, one can define on $\mathcal{D}(\Omega)$ the topology $\mathfrak{T}_\rightarrow^\infty$, strict inductive limit of the topologies $\mathfrak{T}_{K_k}^\infty$. Then $\mathcal{D}(\Omega)[\mathfrak{T}_\rightarrow^\infty]$ is the strict inductive limit of a sequence of Fréchet-Montel spaces. By Corollary 180, $\mathcal{D}(\Omega)[\mathfrak{T}_\rightarrow^\infty]$ is a complete bornological Montel space. The topology $\mathfrak{T}_\rightarrow^\infty$ is not metrizable. The dual of $\mathcal{D}(\Omega)[\mathfrak{T}_\rightarrow^\infty]$ is the space $\mathcal{D}'(\Omega)$ of all *distributions on* Ω. The strong dual $\mathcal{D}'_\beta(\Omega)$ is a complete Montel space.

(3) Let $\alpha = (\alpha_1, ..., \alpha_n) \in \mathbb{N}^n$ be a multi-index and $z = (z_1, ..., z_n) \in \mathbf{k}^n$. We set

$$\alpha! = \prod_{1 \leq i \leq n} \alpha_i!, \quad z^\alpha = \prod\nolimits_{1 \leq j \leq n} z_j^{\alpha_j}.$$

Consider a function $f : \Omega \to \mathbb{C}$. This function is called *analytic in* Ω if each point $w \in U$ has an open neighborhood $U \subseteq \Omega$ such that f has a *power series expansion*

$$f(z) = \sum_\alpha a_\alpha (z - w)^\alpha \tag{1.33}$$

($a_\nu \in \mathbb{C}$) which converges for all $z \in U$. Then $f \in \mathcal{E}(\Omega)$ and (1.33) is the Taylor expansion of f at the point w, i.e.,

$$a_\alpha = \frac{1}{\alpha!} D^\alpha f(w).$$

The function f is said to have a zero at w if $f(w) = a_{|0|} = 0$, and

$$\omega(f; w) = \min\{|\alpha| : a_\alpha \neq 0\}$$

is called the *total order* of f at w. (If $n = 1$, the *total order* is called the *order*.) Obviously, the analytic function f has a zero at w if, and only if $\omega(f; w) > 0$.

When $\mathbf{k} = \mathbb{R}$, the space of analytic functions in Ω is denoted by $C^\omega(\Omega)$, and a function $f \in C^\omega(\Omega)$ is said to be of class C^ω, or to be *real analytic*. For any nonnegative integer p, $C^p(\Omega) \subsetneq C^\infty(\Omega) \subsetneq C^\omega(\Omega)$, and this leads to the following convention:

$$p < \infty < \omega, \quad \forall p \in \mathbb{N}.$$

When $\mathbf{k} = \mathbb{C}$, an analytic function in Ω called *complex analytic* or *holomorphic*. The space of holomorphic functions in Ω is denoted by $\mathcal{O}(\Omega)$. According to the *Cauchy-Riemann criterion*, $\mathcal{O}(\Omega) = C^\infty(\Omega) = C^1(\Omega)$.

On the space $\mathcal{O}(\Omega)$, the topology $\mathfrak{T}_c^\infty$ and the topology $\mathfrak{T}_c^0 = \mathfrak{T}_c$ of uniform convergence on compact subsets of Ω coincide ([339], Chap. 10, Example II). Thus, let K_k be the compact set $\|x\| \leq k$ (where $k \in \mathbb{N}$) and for any $\varphi \in \mathcal{O}(\Omega)$, let

$$p_k(\varphi) = \sup_{x \in K_k \cap \Omega} |\varphi(x)|.$$

The topology $\mathfrak{T}_c$ of $\mathcal{O}(\Omega)$ is defined by the seminorms $p_k(\varphi)$. The LCS $\mathcal{O}(\Omega)[\mathfrak{T}_c] = \mathcal{O}(\Omega)[\mathfrak{T}_c^\infty]$ is a closed subspace of $\mathcal{E}(\Omega)[\mathfrak{T}_c^\infty]$, as a consequence of the Cauchy-Riemann equations. Since $\mathcal{E}(\Omega)[\mathfrak{T}_c^\infty]$ is a Fréchet space, so is $\mathcal{O}(\Omega)[\mathfrak{T}_c^\infty]$. Likewise, since $\mathcal{E}(\Omega)[\mathfrak{T}_c^\infty]$ is a Montel space, so is $\mathcal{O}(\Omega)[\mathfrak{T}_c]$. (This is *Montel's theorem* when $n = 1$ and *Vitali's theorem* when $n > 1$.) Therefore, $\mathcal{O}(\Omega)[\mathfrak{T}_c]$ is a Fréchet-Montel space. (This is not valid in the case of *real analytic functions*.)

Let K be a compact subset of $\mathbb{C}^n$ and let $\mathcal{O}(K)$ be the space of germs of analytic functions defined in an open neighborhood of K in $\mathbb{C}^n$. Let $(U_n)_{n\in\mathbb{N}}$ be a base of the filter consisting of all open neighborhoods of K in $\mathbb{C}$ (with $U_{n+1} \subsetneq U_n$ for each n). Then

$$\mathcal{O}(K) = \varinjlim \mathcal{O}(U_n)$$

The topology $\mathfrak{T}_n$ is not induced on $\mathcal{O}(U_n)$ by $\mathfrak{T}_{n+1}$ ([339], Exercise 13.7) thus the above limit is not strict. Nevertheless, $\mathcal{O}(K)$ is a $(\mathcal{DFS})$ space ([250], Theorem 1.5.5).

Interesting complements on Fréchet-Schwartz (FS) spaces and their duals ((DFS) spaces) can be found in the appendix of [250].

1.8 Digression 2: Manifolds and Lie Groups

The short presentation given below without proofs can be completed by [315]. Recall that $\mathbf{k} = \mathbb{R}$ or $\mathbb{C}$.

1.8.1 Lie Groups

1.8.1.1 Manifolds

A locally finite-dimensional $\mathbf{k}$-analytic manifold X is a topological space together with a collection of pairs (U_i, φ_i) where each U_i is an open subset of X, φ_i is an homeomorphism $U_i \to \varphi_i(U_i) \subseteq \mathbf{k}^{n(i)}$ (where $n(i)$ is a nonnegative integer depending on i), $\{U_i\}$ is a covering of X and $\varphi_j \circ \varphi_i^{-1} : \varphi_i(U_i \cap U_j) \to \varphi_j(U_i \cap U_j)$ is analytic for each pair of indices i, j. Each pair (U_i, φ_i) is called a chart; if $a \in U_i$, (U_i, φ_i) is called a chart at a, and if in addition $\varphi_i(a) = 0$, then the chart (U_i, φ_i) is said to be centered at a. If the integer $n(i)$ is independent of i, the manifold X is called *pure* and $n(i) = n$ is called its *dimension*. In what follows, every manifold X is $\mathbf{k}$-analytic, locally finite-dimentional and *countable at infinity*, i.e., X is the directed union of a sequence (K_j) of compact subsets (Example 19)[5]. Let X, Y be two manifolds. Then a map $f : X \to Y$ is said to be of class C^r ($r \in \mathbb{N} \cup \{\infty, \omega\}$) if for any charts (U, φ) of X and (V, ψ) of Y such that $f(U) \subseteq V$, the map

$$\psi \circ (f|_U) \circ \varphi^{-1} : \varphi(U) \to \psi(V)$$

is of class C^r.

A *path* in X is a continuous map $p : [a, b] \to X$ where $[a, b]$ is an interval of $\mathbb{R}$. Such a path is said to be *closed* if $p(a) = p(b)$. Two paths $p, q : [a, b] \to X$ are *homotopic* if there exists a continuous map $F : [a, b] \times [\alpha, \beta] \to X$

[5] It is sufficient to assume that X is $\mathbf{k}$-analytic and *paracompact* ([34], n°*I*.9.10).

such that $F(t,\alpha) = p(t)$ and $F(\beta,t) = q(t)$ for any $t \in [a,b]$. A path-connected manifold is said to be *simply connected* if every closed path p is homotopic with one single point (i.e., with the constant path $[a,b] \mapsto p(a)$) ([92], (16.27.5)).

Let (V,ξ) be a chart of X, let $\Omega = \xi(V) \subseteq \mathbf{k}^n$, and let $f: V \to \mathbb{C}$ be a function of class C^r. There exists a uniquely defined C^r map $\phi: \Omega \to \mathbf{k}$ such that $f(x) = \phi(\xi_1(x), ..., \xi_n(x))$ where the analytic maps ξ_i are the *local coordinates* (which can also be considered as variables). Generalizing (1.32), for any multi-index $\alpha = (\alpha_1, ..., \alpha_n) \in \mathbb{N}^n$ such that $|\alpha| \leq r$ and any $x \in V$, we set

$$\frac{\partial^{|\alpha|}}{\partial \xi^{\alpha}} = \frac{\partial^{|\alpha|}}{\partial \xi_1^{\alpha_1} ... \partial \xi_n^{\alpha_n}}, \quad D_{\xi}^{\alpha} f(x) = \frac{\partial^{|\alpha|} \phi}{\partial \xi^{\alpha}} (\xi(x)).$$

1.8.1.2 Lie Groups

A $\mathbf{k}$-Lie group $\mathbf{G}$ is a $\mathbf{k}$-manifold which is a group (multiplicatively noted in general, with unit element denoted by 1 or e, and additively noted if $\mathbf{G}$ is abelian, with zero element 0) such that the map $\mathbf{G} \times \mathbf{G} \ni (x,y) \mapsto x^{-1}y \in \mathbf{G}$ is analytic. A Lie group is a pure manifold.

According to *E. Cartan's theorem* ([92], (19.10.1)), a subgroup $\mathbf{H}$ of a Lie group $\mathbf{G}$ is again a Lie group if, and only if $\mathbf{H}$ is closed in $\mathbf{G}$.

The *universal covering* of a connected $\mathbf{k}$-Lie group $\mathbf{G}$ is a simply connected $\mathbf{k}$-Lie group $\tilde{\mathbf{G}}$ along with an analytic epimorphism $\rho: \tilde{\mathbf{G}} \twoheadrightarrow \mathbf{G}$ which is locally injective; $\tilde{\mathbf{G}}$ is unique up to isomorphism and ρ is called the *canonical projection* ([92], (16.30.1)). Therefore, $\mathbf{G} \cong \tilde{\mathbf{G}} / \ker \rho$; $\ker \rho$ is a discrete commutative subgroup of $\tilde{\mathbf{G}}$, denoted by $\pi_1(\mathbf{G})$ and called the *fundamental group* of $\mathbf{G}$ ([92], (16.30.2)). The Lie group $\mathbf{G}$ is simply connected if, and only if $\pi_1(\mathbf{G}) = \{1\}$ ([92], (16.27.5)).

In particular, let $\mathbf{G}$ be a *connected commutative* $\mathbf{k}$-Lie group. Then $\tilde{\mathbf{G}}$ can be identified with $\mathbf{k}^n$ and $\mathbf{G} \cong \mathbf{k}^n / D$ where D is a discrete subgroup of $\mathbf{k}^n$ ([38], n°III.6, Prop. 11; [92], (19.17.1)). If $\mathbf{k} = \mathbb{R}$, then $D \cong 0 \times \mathbb{Z}^q$ and $\mathbf{G} \cong \mathbb{R}^p \times \mathbb{T}^q$ where $\mathbb{T}$ is the torus $\mathbb{R}/\mathbb{Z}$. The commutative groups of the form

$$\mathbb{R}^m \times \mathbb{T}^q \times \mathbb{Z}^s \times \Phi,$$

where Φ is a finite group, are real Lie groups which are called *elementary*. Every real commutative Lie group generated by a compact neighborhood of 0 is isomorphic to a group of this form ([51], n°5.1). Every simply connected real Lie group is isomorphic to $\mathbb{R}^m$ for some integer $m \geq 0$.

1.8.1.3 Distributions

Let X be a $\mathbf{k}$-manifold and let $\mathcal{E}(X)$ be the space of all C^∞ functions $X \to \mathbb{C}$, endowed with the topology $\mathfrak{T}_c^\infty$ of compact convergence of a function and of all its derivatives ([92], (17.3.1)). When X is an open subset of $\mathbf{k}^n$, $\mathcal{E}(X)[\mathfrak{T}_c^\infty]$ is the LCS $\mathcal{E}(\Omega)[\mathfrak{T}_c^\infty]$ already studied in (**1.7.3.5**). By definition, a sequence (φ_n) of elements of $\mathcal{E}(X)$ converges to 0 in the topology $\mathfrak{T}_c^\infty$ if for any compact set $K \subseteq X$, any chart (V, ξ) of X such that $K \cap V \neq \varnothing$, any integer $p > 0$ and any multi-index $\alpha \in \mathbb{N}^n$ such that $|\alpha| \leq p$, the restrictions $D_\xi^\alpha \varphi|_{K \cap V}$ converge uniformly to zero. The topological vector space $\mathcal{E}(X)[\mathfrak{T}_c^\infty]$ is, like $\mathcal{E}(\Omega)[\mathfrak{T}_c^\infty]$, a Fréchet-Montel space. The space $\mathcal{E}'(X)$ of all continuous linear forms $\mathcal{E}(X) \to \mathbb{C}$ is the space of distributions with compact support $\subseteq X$. The strong dual $\mathcal{E}'_\beta(X)$ is Hausdorff, complete, and is a bornological Montel space.

The space $\mathcal{D}(X)$ is the subspace of $\mathcal{E}(X)$ consisting of those functions which have a compact support. This space is endowed with the topology $\mathfrak{T}_\rightarrow^\infty$, strict inductive limit of the topologies $\mathfrak{T}_{K_k}^\infty$ of Fréchet-Montel spaces defined as in (1.7.3.**5**). Therefore, $\mathcal{D}(X)[\mathfrak{T}_\rightarrow^\infty]$ is a complete bornological Montel space, and its dual $\mathcal{D}'(X)$ is the space of all distributions on X ([92], (17.3.1)); the *strong dual* $\mathcal{D}'_\beta(X)$ of $\mathcal{D}(X)[\mathfrak{T}_\rightarrow]$ is a complete Montel space.

For any $T \in \mathcal{E}'(X)$ (resp., $T \in \mathcal{D}'(X)$) and $\varphi \in \mathcal{E}(X)$ (resp., $\varphi \in \mathcal{D}(X)$), $T(\varphi)$ is written $\langle \varphi, T \rangle$ where $\langle -, - \rangle$ is the duality bracket (see (1.31)). With a slight abuse of notation ([92], (17.3.8)), $\langle \varphi, T \rangle$ is also written

$$\int_X dT(x)\,\varphi(x) = \int_X \varphi(x)\,dT(x). \tag{1.34}$$

Let (V, ξ) be a chart centered at $a \in X$, and let $T \in \mathcal{E}'(X)$ (resp., $T \in \mathcal{D}'(X)$); the restriction $T|_V$ is the linear form $\varphi \mapsto T(\varphi)$, $\varphi \in \mathcal{E}(V)$ (resp., $\varphi \in \mathcal{D}(V)$), and its derivative with multi-index α is expressed in function of (V, ξ) by $\left\langle \varphi, D_\xi^\alpha T \right\rangle = (-1)^{|\alpha|} \left\langle D_\xi^\alpha \varphi, T \right\rangle$.

The above holds when X is a $\mathbf{k}$-Lie group $\mathbf{G}$. Let $T, S \in \mathcal{E}'(\mathbf{G})$; their convolution product $T \star S \in \mathcal{E}'(\mathbf{G})$ is the linear form $\mathcal{E}(\mathbf{G}) \ni \varphi \mapsto \langle \varphi m, T \otimes S \rangle \in \mathbb{C}$ where m is the map $(x, y) \mapsto xy$. By (1.34), this can also be written ([92], (14.5.4), (17.10.3))

$$\begin{aligned}\int_{\mathbf{G}} \varphi d(T \star S) &= \int_{\mathbf{G}} \left(\int_{\mathbf{G}} \varphi(xy)\,dT(x) \right) dS(y) \\ &= \int_{\mathbf{G}} dS(y) \int_{\mathbf{G}} dT(x)\,\varphi(xy).\end{aligned}$$

Let $a \in \mathbf{G}$ and let $\delta_a : \varphi \mapsto \varphi(a)$ be the Dirac distribution at point a. From the above,

$$D_\xi^\alpha \varphi(a) = (-1)^{|\alpha|} \left\langle \varphi, D_\xi^\alpha \delta_a \right\rangle, \tag{1.35}$$

$$\langle \varphi, \delta_a \star S \rangle = \langle \varphi(a\bullet), S \rangle, \quad \langle \varphi, T \star \delta_a \rangle = \langle \varphi(\bullet a), T \rangle \tag{1.36}$$

where $\varphi(a\bullet)$ and $\varphi(\bullet a)$ respectively denote the maps $x \mapsto \varphi(ax)$ and $x \mapsto \varphi(xa)$.

A $\mathbb{C}$-algebra $\mathbf{A}$ is a $\mathbb{C}$-vector space along with a bilinear map $\mathbf{A} \times \mathbf{A} \to \mathbf{A} : (a, b) \mapsto a\, b$ (called the *internal law* of $\mathbf{A}$); the general definition of an algebra over a commutative ring is given in (**2.2.2.3**) below. A $\mathbb{C}$-algebra $\mathbf{A}$ is called a *convolution algebra* when its internal law is the convolution product.

The pair $(\mathcal{E}'(\mathbf{G}), \star)$ is an associative convolution algebra with unit element δ_e, and is commutative if, and only if $\mathbf{G}$ is commutative too. This ring is an integral domain in the classical case $\mathbf{G} = \mathbb{R}^n$ according to the Paley-Wiener-Schwartz theorem ([92], 22.18.7), but not in general. For example, let $\mathbb{T}$ be the *torus* $\mathbb{R}/\mathbb{Z}$ and let C be a nonzero constant; then $0 \neq C \in \mathcal{E}'(\mathbb{T})$, $\dot{\delta}_0 \neq 0$ and $\dot{\delta}_0 \star C = 0$.

1.8.1.4 Convolution of Functions and Distributions

Let $\mu = dx$ be a *left Haar measure* on $\mathbf{G}$ ([92], Sect. 14.1). Every left Haar measure on the Lie group $\mathbf{G}$ is of the form $a\mu$, $a \in \mathbb{C}$. The Haar measure $\mu = dx$ is chosen once for all in the sequel, and when $\mathbf{G} = \mathbb{R}^n$ or $\mathbb{C}^n \cong \mathbb{R}^{2n}$, dx is the Lebesgue measure. Let f be a function defined μ-almost everywhere on $\mathbf{G}$; we will abuse the language as in ([339], p. 102) and ([308], p. 67), and make no distinction between a function f, defined μ-almost everywhere on $\mathbf{G}$, and its class $(\operatorname{mod} \mu)$. This class consists of all functions g, defined μ-almost everywhere on $\mathbf{G}$, and such that ([92], Sect. 13.6)

$$\mu\left(\mathbf{G} \backslash \{x \in \mathbf{G} : f(x) = g(x)\}\right) = 0.$$

For any function $f \in \mathcal{E}(\mathbf{G})$ consider the distribution $T = f dx$. The support of T is not compact in general but for any $S \in \mathcal{E}'(\mathbf{G})$, the convolution products $T \star S$ and $S \star T$ exist and are absolutely continuous with respect to dx, with density function g (resp., h) belonging to $\mathcal{E}(\mathbf{G})$. Identifying $f \star S$ with g and $S \star f$ with h one obtains ([92], (14.8.2), (14.8.4))

$$(S \star f)(a) = \int_{\mathbf{G}} f\left(s^{-1}a\right) dS(s), \quad (f \star S)(a) = \int_{\mathbf{G}} f\left(as^{-1}\right) \Delta\left(s^{-1}\right) dS(s) \tag{1.37}$$

where Δ is the *modular function* which is positive ([92], Sect. XIV.3) and analytic ([38], §III.3, Prop. 56). The convolution product $\star : \mathcal{E}'(\mathbf{G}) \times \mathcal{E}(\mathbf{G}) \to \mathcal{E}(\mathbf{G})$ is separately weakly* continuous ([51], Chap. I, Prop. 1;2). In particular,

$$\delta_a \star f = f\left(a^{-1}\bullet\right), \quad f \star \delta_a = \Delta\left(a^{-1}\right) f\left(\bullet a^{-1}\right)$$

The operator $f(\bullet) \mapsto f(a\bullet)$ (resp., $f(\bullet) \mapsto f(\bullet a)$ is the *left* (resp., *right*) *shift* denoted by $\gamma\left(a^{-1}\right)$ (resp., $\boldsymbol{\delta}(a)$) ([92], (14.1.1). Let $a \in \mathbf{G}$, let (V, ξ) be a chart centered at e, and let $f \in \mathcal{E}(\mathbf{G})$; as easily shown, for any $x \in aV$,

$$D_\xi^\alpha \delta_a \star f(x) = D_\xi^\alpha f\left(a^{-1}x\right).$$

Denoting by $\mathcal{T}_x^\infty(\mathbf{G})$ the $\mathbb{C}$-vector space consisting of all distributions with support included in $\{x\}$ $(x \in \mathbf{G})$ and by $\mathcal{T}^\infty(\mathbf{G})$ the $\mathbb{C}$-vector space consisting of all punctual distributions (i.e., all distributions with finite support), $\mathcal{T}_e^\infty(\mathbf{G})$ and $\mathcal{T}^\infty(\mathbf{G})$ are subalgebras of $(\mathcal{E}'(\mathbf{G}), \star)$, and $\mathcal{T}^\infty(\mathbf{G})$ is commutative if, and only if $\mathbf{G}$ is commutative too. In addition,

$$\mathcal{T}^\infty(\mathbf{G}) = \bigoplus_{x \in \mathbf{G}} \mathcal{T}_x^\infty(\mathbf{G}) \tag{1.38}$$

and $\mathcal{T}_x^\infty(\mathbf{G}) = \delta_x \star \mathcal{T}_e^\infty(\mathbf{G}) = \mathcal{T}_e^\infty(\mathbf{G}) \star \delta_x$ ([38], n°III.3.1).

1.8.2 Lie Algebra of a Lie Group

1.8.2.1 General Case

Let $\mathfrak{g} \triangleq T_e(\mathbf{G})$ be the tangent space of $\mathbf{G}$ at e and let n be the dimension of the pure $\mathbf{k}$-manifold $\mathbf{G}$. Consider the n first order *elementary differential operators*

$$X_i = \partial_i|_e : f \to D_\xi^\alpha f(e), \quad \alpha_j = \delta_{ij} \quad i, j \in \{1, ..., n\}$$

(where δ_{ij} is the Kronecker symbol and (V, ξ) is a chart centered at e) and set $X = (X_1, ..., X_n)$; X is a $\mathbf{k}$-basis of $\mathfrak{g}$. By (1.35), for any $i \in \{1, ..., n\}$, X_i can be identified with the distribution $(-1)^{|\alpha|} D_\xi^\alpha \delta_e$ where $\alpha \in \mathbb{N}^n$ is the above multi-index; then $\mathfrak{g}$ is identified with the $\mathbf{k}$-vector space consisting of all distributions of order 1 with support included in $\{e\}$. By ([92], (17.14.2)), for any $Y, Z \in \mathfrak{g}$,

$$[Y, Z] \triangleq Y \star Z - Z \star Y \in \mathfrak{g}$$

and the bilinear map $[\bullet, \bullet] : \mathfrak{g} \times \mathfrak{g} \to \mathfrak{g}$ is a *Lie bracket*, i.e., for any $X, Y, Z \in \mathfrak{g}$,

$$[X, Y] = -[Y, X]$$

(*skew-symmetry*) and the following *Jacobi identity* is satisfied:

$$[X, [Y, Z]] + [Y, [Z, X]] + [Z, [X, Y]] = 0;$$

$\mathfrak{g}$, endowed with this bracket, is a $\mathbf{k}$-algebra, called the *Lie algebra* of $\mathbf{G}$ ([92], (19.3.3)).

All elements of the form e^A, $A \in \mathfrak{g}$ (where e^A is defined by its usual power series expansion) belong to $\mathbf{G}$; in addition, there exist an open neighborhood V of 0 in $\mathfrak{g}$ and an open neighborhood W of $\{e\}$ in $\mathbf{G}$ such that $\exp_{\mathbf{G}} : A \mapsto e^A$ is a diffeomorphism of V onto W.

Two Lie groups $\mathbf{G}, \mathbf{H}$ are locally isomorphic if, and only if their Lie algebras $\mathfrak{g}, \mathfrak{h}$ are isomorphic. The latter condition is necessary and sufficient for $\mathbf{G}, \mathbf{H}$ to be isomorphic if these Lie groups are simply connected ([92], (19.7.7)). In particular, since a Lie group $\mathbf{G}$ and its universal covering group $\tilde{\mathbf{G}}$ are locally isomorphic, they have the same Lie algebra.

As a consequence of *Levi's theorem*, every finite-dimensional $\mathbf{k}$-Lie algebra is isomorphic to the Lie algebra of a $\mathbf{k}$-Lie group ([92], (21.23.4)). If $\mathbf{H}$ is a Lie subgroup of a Lie group $\mathbf{G}$, then the Lie algebra $\mathfrak{h}$ of $\mathbf{H}$ is a subalgebra of the Lie algebra $\mathfrak{g}$ of $\mathbf{G}$ ([92], (19.4.3)). Conversely, to every subalgebra $\mathfrak{h}$ of the Lie algebra $\mathfrak{g}$ of a Lie group $\mathbf{G}$, there canonically corresponds one connected Lie group $\mathbf{H}$ "embedded in $\mathbf{G}$", having $\mathfrak{h}$ for its Lie algebra ([92], (19.7.4)).

1.8.2.2 Classical Groups

The general linear group $\mathrm{GL}_n(\mathbb{R})$ is endowed with the topology induced by the canonical topology of $\mathrm{Mat}_n(\mathbb{R})$. The *classical groups* are the closed subgroups of $\mathrm{GL}_n(\mathbb{R})$. According to Cartan's theorem, a classical group is a real Lie group. In particular, a real linear algebraic group, i.e., a subgroup of $\mathrm{GL}_n(\mathbb{R})$ defined by algebraic equations (see Subsect. 2.10.1 below for more details) is a classical group, thus a Lie group.

For example, the group $O(n,\mathbb{R})$ of all real orthogonal matrices of order n is a real linear algebraic group since $S \in \mathrm{GL}_n(\mathbb{R})$ belongs to $O(n,\mathbb{R})$ if, and only if S satisfies the algebraic equation $S^T S = I_n$, where S^T is the classical transpose of S (Definition 271); therefore, $O(n,\mathbb{R})$ is a Lie group, called the *orthogonal group*.

Let $\mathbf{G}$ be a classical group; then, $\mathfrak{g}$ is a subalgebra of $\mathrm{Mat}_n(\mathbb{R})$. The Lie bracket $[A,B] \triangleq AB - BA$ $(A,B \in \mathfrak{g})$ is sometimes called the *commutator* of A and B (not to be confused with the notion introduced in Definition 163): see (**2.2.2.3**) below. The Lie algebra $\mathfrak{g}$ can be calculated as follows: it consists of those elements $A \in \mathrm{Mat}_n(\mathbb{R})$ such that $I_n + \varepsilon A \in \mathbf{G}$ when $\varepsilon \to 0^+$, neglecting the terms of second order in ε.

1.8.2.3 Classification of Lie Algebras and Lie Groups

An *ideal* of $\mathfrak{g}$ is a $\mathbf{k}$-vector space $\mathfrak{a} \subseteq \mathfrak{g}$ such that $[\mathfrak{g},\mathfrak{a}] \subseteq \mathfrak{a}$. The Lie algebra $\mathfrak{g}$ is called

- *simple* if it is noncommutative and if its only proper ideal is $\{0\}$,
- *semisimple* if its only abelian ideal is $\{0\}$,
- *nilpotent* if there exists a finite decreasing sequence of ideals $(\mathfrak{g}_i)_{0\leq i\leq p}$ such that $\mathfrak{g}_0 = \mathfrak{g}$, $\mathfrak{g}_p = \{0\}$ with $[\mathfrak{g},\mathfrak{g}_i] \subseteq \mathfrak{g}_{i+1}$ $(0 \leq i \leq p-1)$,
- *solvable* if there exists a sequence

$$\mathfrak{g}_0 = \mathfrak{g} \supsetneqq \mathfrak{g}_1 \supsetneqq \dots \supsetneqq \mathfrak{g}_n = \{0\} \tag{1.39}$$

of Lie subalgebras such that $\mathfrak{g}_i$ is a one-codimensional ideal in $\mathfrak{g}_{i-1}$, $1 \leq i \leq n$ ([38], §I.5, Prop. 2).

All nilpotent Lie algebras are solvable. See in Exercise 216 another characterization of nilpotent Lie algebras and solvable Lie algebras.

Assuming that the Lie group $\mathbf{G}$ is connected, it is called semisimple (resp., nilpotent, solvable) if its Lie algebra is semisimple (resp., nilpotent, solvable). A Lie group is solvable if, and only if the underlying group is solvable (Subsect. 1.6.8): see ([92], (19.12.3)).

1.8.3 The Universal Enveloping Algebra, and Its Centre

1.8.3.1 General Universal Enveloping Algebra

Given any associative $\mathbf{k}$-algebra A, one can endow A with a $\mathbf{k}$-Lie algebra structure by defining the Lie bracket $[a,b] = ab - ba$ $(a,b \in A)$ (i.e., the *commutator* of a and b). A *representation* of $\mathfrak{g}$ is defined to be a $\mathbf{k}$-Lie algebra morphism from $\mathfrak{g}$ to such A. There exists an associative $\mathbf{k}$-algebra $\mathrm{U}_{\mathbf{k}}(\mathfrak{g})$ together with a representation $\theta : \mathfrak{g} \rightarrow \mathrm{U}_{\mathbf{k}}(\mathfrak{g})$ with the following universal property: for any associative $\mathbf{k}$-algebra representation $\varphi : \mathfrak{g} \rightarrow A$, there exists a unique $\mathbf{k}$-algebra morphism $\psi : \mathrm{U}_{\mathbf{k}}(\mathfrak{g}) \rightarrow A$ such that $\psi\theta = \varphi$. Thus $\mathrm{U}_{\mathbf{k}}(\mathfrak{g})$ is unique up to isomorphism and is called the *universal enveloping algebra* of $\mathfrak{g}$. Consider the *standard monomials*

$$\{X_{i_1} X_{i_2} ... X_{i_n} : i_j \in \{1,...,n\}, \quad i_1 \leq ... \leq i_n\}.$$

According to the *Poincaré-Birkhoff-Witt theorem*, these standard monomials form a $\mathbf{k}$-basis of $\mathrm{U}_{\mathbf{k}}(\mathfrak{g})$; as proved in ([247], §1.7), $\mathrm{U}_{\mathbf{k}}(\mathfrak{g})$ is a skew polynomial ring (Subsect. 2.7.4 below) and a noncommutative Noetherian domain (Subsect. 2.4.4 below).

The centre $\mathrm{Z}_{\mathbf{k}}(\mathfrak{g})$ of $\mathrm{U}_{\mathbf{k}}(\mathfrak{g})$ is a commutative domain. If $\mathfrak{g}$ is semisimple, then $\mathrm{Z}_{\mathbf{k}}(\mathfrak{g})$ is the polynomial ring $\mathbf{k}[y_1,...,y_r]$ where $r < \infty$ is the "rank" of $\mathfrak{g}$ ([94], 1.9.8, 7.3.8), thus $\mathrm{Z}_{\mathbf{k}}(\mathfrak{g})$ is Noetherian by Theorem 352 below. The case of solvable Lie algebras is quite different since there exists a nilpotent Lie algebra $\mathfrak{g}$ of dimension 45 such that $\mathrm{Z}_{\mathbf{k}}(\mathfrak{g})$ is not Noetherian ([94], §4.9.20).

1.8.3.2 The Borel Epimorphism

If $\mathfrak{g}$ is the Lie algebra of a Lie group $\mathbf{G}$, then $\mathrm{U}_{\mathbf{k}}(\mathfrak{g})$ is isomorphic to, and identified with, $\mathcal{T}_e^{\infty}(\mathbf{G})$. The dual $\mathrm{U}_{\mathbf{k}}^{*}(\mathfrak{g}) \triangleq \mathrm{Hom}_{\mathbf{k}}(\mathrm{U}_{\mathbf{k}}(\mathfrak{g}), \mathbf{k})$ is isomorphic to the power series ring $\mathbf{k}[[x]]$, $x = (x_i)_{1 \leq i \leq n}$, and according to E. Borel's theorem ([339], Theorem 38.1), there exists a canonical epimorphism

$$\mathcal{E}(\mathbf{G}) \twoheadrightarrow \mathrm{U}_{\mathbf{k}}^{*}(\mathfrak{g}) : f \mapsto \sum_{\alpha \in \mathbb{N}^n} \frac{1}{\alpha!} (X^{\alpha} f) x^{\alpha},$$

called the *Borel epimorphism.*

1.8.3.3 Invariant and Bi-invariant Differential Operators

Assume that the $\mathbf{k}$-Lie group $\mathbf{G}$ is connected and let $P : \mathcal{E}(\mathbf{G}) \to \mathcal{E}(\mathbf{G})$ be a differential operator ([92], Sect. XVII.13). Such an operator is called *left-* (resp., *right-*) *invariant* if for any $a \in \mathbf{G}$ and $\varphi \in \mathcal{E}(\mathbf{G})$,

$$P\varphi = \gamma(a)\left(P\gamma\left(a^{-1}\right)\varphi\right) \quad \text{(resp., } P\varphi = \delta(a)\left(P\delta\left(a^{-1}\right)\varphi\right)\text{)}.$$

For short, a differential operator is called *invariant* when it is left-invariant, and *bi-invariant* when it is both left- and right-invariant. The algebra of all invariant (resp., bi-invariant) differential operators is identified with $\mathrm{U}_{\mathbf{k}}(\mathfrak{g})$ (resp., with $\mathrm{Z}_{\mathbf{k}}(\mathfrak{g})$) ([92], (19.3.1), (23.36.6)).

1.9 Digression 3: Hyperfunctions

1.9.1 Hyperfunctions on an Open Subset of the Real Line

We consider hyperfunctions on a nonempty open subset Ω of the real line. Let U be an open subset of $\mathbb{C}$. The set of all analytic functions in U (resp., Ω) is denoted by $\mathcal{O}(U)$ (resp., $\mathcal{O}(\Omega)$); this is a $\mathbb{C}$-space. Thus, in this section, the space of *real analytic functions* on Ω is denoted by $\mathcal{O}(\Omega)$ instead of $C^{\omega}(\Omega)$.

A *complex neighborhood* of Ω is defined as an open subset U of $\mathbb{C}$ which contains Ω as a relatively closed subset. The set of all complex neighborhoods of Ω is denoted by $\mathfrak{U}(\Omega)$; this is a filtering set for the inclusion (**1.2.3.4**).

The set $U \backslash \Omega$ is open in $\mathbb{C}$ for $U \backslash \Omega$ is open in U and U is open in $\mathbb{C}$. In addition, $U \backslash \Omega = U_+ \cup U_-$ where $U_\pm = \{z \in U : \pm \operatorname{Im} z > 0\}$. The open sets U_+ and U_- are disjoint, thus every function $\varphi \in \mathcal{O}(U \backslash \Omega)$ can be uniquely written into the form $\varphi_+ - \varphi_-$ where $\varphi_\pm \in \mathcal{O}(U_\pm)$.

Let $\varphi \in \mathcal{O}(U)$; then $\varphi\left|_{U \backslash \Omega}\right. \in \mathcal{O}(U \backslash \Omega)$ and according to the analytic continuation principle ([208], Sect. III.1, Theorem 1.2), φ is the *unique* analytic continuation of $\varphi\left|_{U \backslash \Omega}\right.$ to U. Thus $\mathcal{O}(U)$ can be embedded in $\mathcal{O}(U \backslash \Omega)$ and the quotient $\mathcal{O}(U \backslash \Omega) / \mathcal{O}(U)$ is a $\mathbb{C}$-space.

Let $U_1, U_2 \in \mathfrak{U}(\Omega)$, $U_1 \subseteq U_2$. The embedding $\mathcal{O}(U_2 \backslash \Omega) \hookrightarrow \mathcal{O}(U_1 \backslash \Omega)$ is the restriction $\iota : f \longmapsto f\left|_{\mathcal{O}(U_1 \backslash \Omega)}\right.$. Then $\iota(\mathcal{O}(U_2)) = \mathcal{O}(U_1)$ and by Theorem and Definition 145 the induced map $\bar{\iota}$:

$\mathcal{O}(U_2 \backslash \Omega)/\mathcal{O}(U_2) \to \mathcal{O}(U_1 \backslash \Omega)/\mathcal{O}(U_1)$ exists and is injective. Therefore, both $\{\mathcal{O}(U) : U \in \mathfrak{U}(\Omega)\}$ and $\{\mathcal{O}(U \backslash \Omega)/\mathcal{O}(U) : U \in \mathfrak{U}(\Omega)\}$ are direct systems of $\mathbb{C}$-vector spaces with index set $\mathfrak{U}(\Omega)$ (Definition 15).

By Definition 16,

$$\mathcal{O}(\Omega) = \varinjlim_{U \in \mathfrak{U}(\Omega)} \mathcal{O}(U),$$

and we are led to the following.

Definition 183. *The $\mathbb{C}$-space of hyperfunctions on Ω is*

$$\boxed{\mathcal{B}(\Omega) = \varinjlim_{U \in \mathfrak{U}(\Omega)} \mathcal{O}(U \backslash \Omega)/\mathcal{O}(U)}.$$

Let $U \in \mathfrak{U}(\Omega)$ and $\varphi \in \mathcal{O}(U \backslash \Omega)$. The canonical image of φ in $\mathcal{B}(\Omega)$ is denoted by $[\varphi]$. By construction, the map $\mathcal{O}(U \backslash \Omega) \ni \varphi \mapsto [\varphi] \in \mathcal{B}(\Omega)$ is $\mathbb{C}$-linear. The *derivative* of order k of $[\varphi]$ is defined by $[\varphi]^{(k)} = [\varphi^{(k)}]$. Let $\varphi \in \mathcal{O}(U \backslash \Omega)$; then $[\varphi] = 0$ if, and only if φ has an analytic continuation to an open set $U' \in \mathfrak{U}(\Omega)$.

Abusing the language, one can write $[\varphi](x) = \varphi(x + i0) - \varphi(x - i0) = [\varphi(z)]_{z=x}$ where the "boundary values" $\varphi(x \pm i0) \in \mathcal{B}(\Omega)$ are defined by

$$\varphi(x + i0) = [\varepsilon(z)\varphi(z)]_{z=x}, \quad \varphi(x - i0) = -[\bar{\varepsilon}(z)\varphi(z)]_{z=x},$$
$$\varepsilon(z) = \begin{cases} 1 \\ 0 \end{cases}, \quad \bar{\varepsilon}(z) = \varepsilon(-z) = \begin{cases} 0 & (\operatorname{Im} z > 0) \\ 1 & (\operatorname{Im} z < 0). \end{cases}$$

Let $\psi \in \mathcal{O}(\Omega)$. Then there exists $U \in \mathfrak{U}(\Omega)$ such that ψ has an analytic continuation to U ([92], (9.4.5)), denoted by $\psi(z)$; this continuation of ψ to U is unique. Let $[\varphi] \in \mathcal{B}(\Omega)$; then the product $[\varphi]\psi$ is defined by: $[\varphi]\psi = [(\varphi\psi)(z)]_{z=x} \in \mathcal{B}(\Omega)$. Therefore, $\mathcal{B}(\Omega)$ is endowed with a structure of $\mathcal{O}(\Omega)$-module (see Subsect. 2.2.2 below).

Define $1 \in \mathcal{B}(\Omega)$ by $1 = [\varepsilon(z)]_{z=x}$. The map $\mathcal{O}(\Omega) \ni \psi \mapsto 1\psi \in \mathcal{B}(\Omega)$ is easily seen to be injective (using the analytic continuation principle), thus determines an embedding $\mathcal{O}(\Omega) \hookrightarrow \mathcal{B}(\Omega)$.

If Ω' is an open subset of Ω and $[\varphi] \in \mathcal{B}(\Omega)$, we define the restriction $[\varphi]|_{\Omega'}$ as follows:

$$[\varphi]|_{\Omega'} = [\varphi|_{\Omega'}].$$

In addition, one can prove that every $[\varphi'] \in \mathcal{B}(\Omega')$ can be expressed in that way, i.e., the following holds:

Theorem 184. *The canonical "restriction map" $\rho_{\Omega'}^{\Omega} : \mathcal{B}(\Omega) \to \mathcal{B}(\Omega')$ is surjective.*

In other words, the hyperfunctions on Ω form a *flabby sheaf* of vector spaces ([184], Theorem 1.3.1; [140], Subsect. II.1.3)); this is a key property of hyperfunctions, which is not shared by distributions. The *support* of the hyperfunction $[\varphi] \in \mathcal{B}(\Omega)$, denoted by $\operatorname{supp}[\varphi]$, is defined as the smallest closed set $F \subseteq \Omega$ such that $[\varphi]\,|_{\Omega \backslash F} = 0$.

The following will play an important role in the sequel:

Theorem and Definition 185. *Let*

$$P\left(x, \frac{d}{dx}\right) = \sum_{i=0}^{m} a_i(x) \frac{d^i}{dx^i}, \quad a_m \neq 0$$

be a nonzero linear differential operator (see Subsect. 2.7.2 below) with real analytic coefficients on a nonempty open interval Ω of $\mathbb{R}$.

*(i) This operator $P\left(x, \frac{d}{dx}\right) : \mathcal{B}(\Omega) \to \mathcal{B}(\Omega)$ (where $\mathcal{B}(\Omega)$ is the space of hyperfunctions on Ω) is surjective (*Sato's theorem*).*
(ii) Let

$$\ker_{\mathcal{B}(\Omega)}(P\bullet) \triangleq \left\{ f \in \mathcal{B}(\Omega) : P\left(x, \frac{d}{dx}\right) f = 0 \right\}.$$

Then

$$\dim_{\mathbb{C}} \ker_{\mathcal{B}(\Omega)}(P\bullet) = m + \sum_{x \in \Omega} ord_x a_m$$

where $ord_x a_m$ is the order of the zero at x of the leading coefficient a_m (**1.7.3.5**). *If $ord_x a_m > 0$, the point x is called* singular.

In addition, the following conditions are equivalent:

(a) $a_m(x) \neq 0$ for all $x \in \Omega$;
(b) $\ker_{\mathcal{B}(\Omega)}(P\bullet) \subseteq \mathcal{O}(\Omega)$;
(c) $P\left(x, \frac{d}{dx}\right) f \in \mathcal{O}(\Omega)$ implies $f \in \mathcal{O}(\Omega)$.

Assume that the point x is singular and consider the highest convex polyhedron under the $m+1$ points

$$(j, ord_x a_j), \quad 0 \leq j \leq m,$$

called the Newton polygon *([286], Sect. 3.3). The largest slope σ_x of this polyhedron is called the* nonregularity *of the singular point x. The singular point x is called* determined *if $\sigma_x \leq 1$. The following conditions are equivalent:*

(a') $\sigma_x \leq 1$ for all $x \in \Omega$;
(b') $\ker_{\mathcal{B}(\Omega)}(P\bullet) \subseteq \mathcal{D}'(\Omega)$;
(c') $P\left(x, \frac{d}{dx}\right) f \in \mathcal{D}'(\Omega)$ implies $f \in \mathcal{D}'(\Omega)$.
*(*Komatsu's theorem*).*

Proof. See ([184], Theorems 1.3.2, 1.3.6 and 1.3.7; [250], Theorem 3.9.5). ■

1.9.2 Examples of Hyperfunctions

In the examples below, Ω is a bounded open interval of $\mathbb{R}$.

1.9.2.1 The Dirac Hyperfunction

Let $x \in \Omega$ and consider the hyperfunction $\breve{\delta}_x \triangleq \left[\frac{1}{2\pi i(x-z)}\right]$. Let $\psi \in \mathcal{O}(\Omega)$ and let V be a simply connected open subset of U such that $\Omega \subseteq V$ (**1.8.1.1**). The hyperfunction $\breve{\delta}_x$ acts on ψ in the following way:

$$\left\langle \psi, \breve{\delta}_x \right\rangle \triangleq -\oint_{\partial V} \frac{1}{2\pi i\,(x-z)}\,\psi(z)\,dz$$

where ∂V is the boundary of V; ∂V is assumed to be canonically oriented and to be a closed path of class C^1. By the Cauchy theorem ([92], (9.9.1)), $\left\langle \psi, \breve{\delta}_x \right\rangle = \psi(x) = \langle \psi, \delta_x \rangle$ where δ_x is the Dirac distribution at x. Thus $\breve{\delta}_x$ is called the *Dirac hyperfunction* at x.

1.9.2.2 Hyperfunction Defined by a Distribution

More generally, let $T \in \mathcal{E}'(\Omega)$, $U \in \mathfrak{U}(\Omega)$, and let $\varphi_T \in \mathcal{O}(U \backslash \Omega)$ be defined by

$$\varphi_T(z) = \frac{1}{2\pi i} \int_\Omega \frac{1}{x-z}\,dT(x)$$

with the notation of (1.34). Then $\breve{T} = [\varphi_T]$ is called the hyperfunction associated with T. As above, let $\psi \in \mathcal{O}(\Omega)$, extended to U (assuming that U is sufficiently small) and let $\breve{T}$ act on ψ according to

$$\left\langle \psi, \breve{T} \right\rangle = -\oint_{\partial V} \varphi_T(z)\,\psi(z)\,dz. \tag{1.40}$$

We obtain

$$\begin{aligned} \left\langle \psi, \breve{T} \right\rangle &= -\int_\Omega dT(x)\,\frac{1}{2\pi i} \oint_{\partial V} \frac{1}{x-z}\,\psi(z)\,dz \\ &= \int_\Omega \psi(x)\,dT(x) = \langle \psi, T \rangle. \end{aligned} \tag{1.41}$$

We have the following:

Theorem 186. *There exists an embedding*

$$\mathcal{E}'(\Omega) \to \mathcal{B}(\Omega) : T \mapsto \breve{T} \tag{1.42}$$

such that $\operatorname{supp} \breve{T} = \operatorname{supp} T$. *Therefore,* $\mathcal{E}'(\Omega)$ *is identified with a subspace of* $\mathcal{B}(\Omega)$.

Proof. The map $\mathcal{E}'(\Omega) \ni T \mapsto \breve{T} \in \mathcal{B}(\Omega)$ is obviously $\mathbb{C}$-linear. Let $\mathcal{P}(\Omega)$ be the space of all polynomial functions in Ω; $\mathcal{P}(\Omega)$ is dense in $\mathcal{E}(\Omega)$ ([339], Chap. 15, Corollary 4 of Theorem 15.3). Since $\mathcal{P}(\Omega) \subseteq \mathcal{O}(\Omega) \subseteq \mathcal{E}(\Omega)$, $\mathcal{O}(\Omega)$ is dense in $\mathcal{E}(\Omega)$, therefore the map (1.42) is injective. See ([250], Prop. 3.9.1) for the statement about supports. ∎

1.9.2.3 Hadamard's Finite Parts

Let f be a meromorphic function in a complex neighborhood U of an open subset Ω of $\mathbb{R}$, and assume that the poles of f all lie in Ω. Then *Hadamard's finite part* of f is defined by

$$\mathfrak{f}.\mathfrak{p}.\left(f(x)\right) = \frac{1}{2}\left\{f(x+i0) + f(x-i0)\right\}$$

([184], Example 1.2.9). This is a distribution ([322], Sect. 2.2, Example 2).

1.9.3 General Hyperfunctions

1.9.3.1 Compactly Supported Hyperfunctions

Let K be a compact subset of $\mathbb{R}$, let $\mathcal{O}(K)$ be the space of germs of analytic functions defined in an open neighborhood of K in $\mathbb{C}$, and let $\mathcal{B}_K(\Omega)$ be the space of all hyperfunctions with support $\subseteq K$ and defined in a real open neighborhood Ω of K. The expression (1.40) can be generalized, setting

$$\langle \psi, u \rangle = -\oint_{\partial V} \varphi(z)\, \psi(z)\, dz.$$

where $u = [\varphi]$. With the notation in (1.41) we obtain

$$u(\psi) = \langle \psi, u \rangle = \int_{\Omega} f(x)\, \psi(x)\, dx \tag{1.43}$$

where $u = [\varphi]$ and $\psi \in \mathcal{O}(\Omega)$. This leads to the following:

Theorem 187. (Köthe's duality theorem.) *The spaces $\mathcal{B}_K(\Omega)$ and $\mathcal{O}'(K)$ are topologically isomorphic – thus* identified. *They are $(\mathcal{FS})$ spaces, thus Fréchet-Montel spaces* (**1.7.3.4**).

Proof. The topological isomorphism $\mathcal{B}_K(\Omega) \cong \mathcal{O}(K)'_\beta$ can be viewed as a consequence of (1.43) [188]. Since $\mathcal{O}(K)$ is a $(\mathcal{DFS})$ space (**1.7.3.5**), $\mathcal{O}(K)'_\beta$ is an $(\mathcal{FS})$ space, thus a Fréchet-Montel space (Theorem 182). ∎

1.9.3.2 Generalization

Let $\mathcal{O}(\mathbb{C}^n)$ be the space of entire functions in $\mathbb{C}^n$, endowed with the topology of uniform convergence on the compact subsets of $\mathbb{C}^n$. We know that $\mathcal{O}(\mathbb{C}^n)$ is a Fréchet-Montel space (**1.7.3.5**).

Definition 188. *(i) The elements of the dual space $\mathcal{O}'(\mathbb{C}^n)$ are called* analytic functionals.
(ii) An analytic functional u is said to be carried by the compact set $K \subsetneq \mathbb{C}^n$ if for every neighborhood ω of K, there is a real constant $C_\omega \geq 0$ such that

$$|u(\varphi)| \leq C_\omega \sup_\omega |\varphi|. \tag{1.44}$$

Let $u \in \mathcal{O}'(\mathbb{C}^n)$; since u is continuous on $\mathcal{O}(\mathbb{C}^n)$, for every compact set $\omega \subsetneq \mathbb{C}^n$, there exists $C_\omega \geq 0$ such that (1.44) holds, which implies that u has some compact carrier K. In general, there does not exist a carrier which is contained in all others and that we could call the *support* of u. The case of a *real analytic functionals* (i.e., of continuous linear forms on a space of *real analytic functions*) is quite different, since we have the following ([162], Theorems 9.1.2, 9.1.6 and 9.1.8):

Theorem 189. *(i) If $u \in \mathcal{O}'(\mathbb{R}^n)$ is carried by the compact set $K \subsetneq \mathbb{R}^n$, then $u(\psi)$ can be defined for every ψ which is analytic just in a neighborhood of K, so that the space of all such real analytic functionals can be identified with $\mathcal{O}'(K)$.*
(ii) If $u \in \mathcal{O}'(\mathbb{R}^n)$, then there is a smallest compact set $K \subsetneq \mathbb{R}^n$ such that $u \in \mathcal{O}'(K)$, and K is called the support *of u.*
(iii) If $K_1, ..., K_r$ are compact subsets of $\mathbb{R}^n$, $K = \bigcup_{1 \leq i \leq r} K_i$ and $u \in \mathcal{O}'(K)$, then one can find $u_j \in \mathcal{O}'(K_j)$ $(1 \leq j \leq r)$ so that

$$u = \sum_{1 \leq i \leq r} u_j.$$

One can now define hyperfunctions in $\mathbb{R}^n$ in such a way that they are locally equivalent to real analytic functionals with compact support in $\mathbb{R}^n$.

Lemma and Definition 190. *(1) If $X \subsetneq \mathbb{R}^n$ is open and bounded, let*

$$\mathcal{B}(X) = \mathcal{O}'\left(\bar{X}\right) / \mathcal{O}'(\partial X)$$

where $\bar{X}$ is the closure of X and $\partial X = \bar{X} \backslash X$ is the boundary of X.

(i) If $u, v \in \mathcal{O}'(\bar{X})$ and $u - v \in \mathcal{O}'(\partial X)$, then $X \cap \operatorname{supp} u = X \cap \operatorname{supp} v$. Therefore, $X \cap \operatorname{supp} u$ only depends of the class $u^\bullet$ of u in $B(X)$.
(ii) If $Y \subseteq X \subsetneq \mathbb{R}^n$ are open and bounded, then for every $u \in \mathcal{O}'(\bar{X})$ one can find $v \in \mathcal{O}'(\bar{Y})$ such that $Y \cap \operatorname{supp}(u - v) = \varnothing$. The class $u^\bullet$ of u in $\mathcal{B}(X)$ is uniquely determined by the class $v^\bullet$ of y in $\mathcal{B}(Y)$.
(2) The above $\mathbb{C}$-space $\mathcal{B}(X)$ is called the space of hyperfunctions on X; the set $X \cap \operatorname{supp} u$ (which is relatively closed in X) is called the support of $u^\bullet \in \mathcal{B}(X)$ and is denoted by $\operatorname{supp} u^\bullet$. The hyperfunction $v^\bullet \in \mathcal{B}(Y)$ (as defined in (ii) above) is called the restriction of $u^\bullet$ to Y *and is denoted by $u^\bullet|_Y$.*

Proof. (i): We have $\operatorname{supp}(u - v) \subseteq \partial X$, thus

$$\operatorname{supp} u \subseteq \operatorname{supp} v \cup \operatorname{supp}(u - v) \subseteq \operatorname{supp} v \cup \partial X.$$

Likewise, $\operatorname{supp} v \subseteq \operatorname{supp} u \cup \partial X$. It follows that $X \cap \operatorname{supp} u = X \cap \operatorname{supp} v$.
(ii): We have $\bar{X} = \bar{Y} \cup \overline{X \backslash Y}$. By Theorem 189(iii), there are $v \in \mathcal{O}'(\bar{Y})$ and $w \in \mathcal{O}'\left(\overline{X \backslash Y}\right)$ such that $u = v + w$. Thus $w = u - v$ and $\operatorname{supp} w \subseteq \overline{X \backslash Y}$, therefore $\operatorname{supp} w \cap Y = \varnothing$. ■

Theorem 191. *(i) The elements with compact support in $\mathcal{B}(X)$ can be identified with the elements of $\mathcal{O}'(\mathbb{R}^n)$ having support in X.*
(ii) There exists an embedding $\mathcal{D}'(X) \hookrightarrow \mathcal{B}(X)$ which preserves supports, thus $\mathcal{D}'(X)$ is identified with a subspace of $\mathcal{B}(X)$.

Proof. (i): Let $u \in \mathcal{O}'(\bar{X})$ and assume that the class $u^\bullet$ of u in $\mathcal{B}(X)$ has a compact support $K \subsetneq X$. Then $\operatorname{supp} u \subseteq K \cup \partial X$, and by Theorem 189(iii) there are $u_1 \in \mathcal{O}'(K)$, $u_2 \in \mathcal{O}'(\partial X)$ such that $u = u_1 + u_2$. Since K and ∂X are disjoint, this decomposition is unique, and we have $u^\bullet = u_1^\bullet$. The map $\mathcal{O}'(K) \ni u_1 \to u^\bullet \in B(X)$ (where $u^\bullet$ is compactly supported) is injective.
(ii): This generalization of Theorem 186 is proved in ([162], p. 337) and ([351], Theorem 4.2). ■

Let now X be a real analytic manifold. Using analytic charts, the space $C^\omega(X) = \mathcal{O}(X)$ has been defined (1.8.1.1). The space $\mathcal{B}(X)$ of hyperfunctions on X is defined similarly, and we have the following, which generalizes Theorem 184:

Theorem 192. *If X is a real analytic manifold and Y is an open subset of X, then every $u \in \mathcal{B}(Y)$ is the restriction to Y of a hyperfunction $v \in \mathcal{B}(X)$ with support included in $\bar{Y}$.*

1.9.4 Hyperfunctions Which Are Not Distributions

Let Ω be an open bounded subset of the real line. From the above, $\mathcal{D}'(\Omega) \subseteq \mathcal{B}(\Omega)$, and this inclusion is strict (i.e., $\mathcal{D}'(\Omega) \subsetneq \mathcal{B}(\Omega)$) since, if $\varphi \in \mathcal{O}(U \backslash \Omega)$

has an essential singularity at a point of Ω (e.g., $\varphi(z) = e^{1/z}$ when $0 \in \Omega$), then $[\varphi] \in \mathcal{B}(\Omega) \backslash \mathcal{D}'(\Omega)$. For example, $e^{1/z} = \sum_{n \geq 0} z^{-n}/n!$ can be viewed as defining a "distribution of infinite order" with support $\{0\}$ (Exercise 217), whereas all Schwartz's distributions are of finite order at each point ([322], Sect. III.7, Theorem XXVI, and Sect. III.8, Theorem XXXII).

1.10 Exercises

Exercise 193. *Let $(X_i)_{i \in I}$ be a family of sets and $\mathfrak{F} = \prod_{i \in I} (\bullet)^{X_i}$ the functor* **Set** $\rightarrow$ **Set** *defined as follows: for any sets S, T*

$$\mathfrak{F}(S) = \prod_{i \in I} S^{X_i} = \{(f_i)_{i \in I} ; f_i : X_i \rightarrow S\};$$

for any $\tau : S \rightarrow T$, $\mathfrak{F}(\tau) : \mathfrak{F}(S) \rightarrow \mathfrak{F}(T)$ is the family of maps

$$(\tau \circ f_i)_{i \in I} \in \prod_{i \in I} T^{X_i}.$$

(i) Prove that $\coprod_{i \in I} X_i$ is a universal element for $\mathfrak{F}$. (ii) Prove that for any set S,

$$S^{\coprod X_i} \cong \prod S^{X_i}$$

under the bijection $h \mapsto (h \circ \iota_i)_{i \in I}$. (Hint: for (ii), use Theorem 13(i). The solution can be found in ([224], Sect. III.3).)

Exercise 194. *Let $\mathfrak{F} : \mathcal{C} \rightarrow \mathcal{D}$ be a functor. $\mathfrak{F}$ is called* essentially surjective *(or* dense*) if for each $Y \in \mathcal{D}$ there exists $X \in \mathcal{C}$ and an isomorphism $\mathfrak{F}(X) \cong Y$. Prove that $\mathfrak{F}$ is an equivalence of categories (Definition 11(iv)) if, and only if it is fully faithful and essentially surjective. (Hint: the solution can be found in ([224], Sect. IV.4).)*

Exercise 195. *In the category* **HTVS** *(Example 37), let $f : X \rightarrow Y$ be a morphism. Prove that* $\ker f = f^{-1}(\{0\})$, $\operatorname{im} f = \overline{f(X)}$, $\operatorname{coker} f = Y/\operatorname{im} f$ *and* $\operatorname{coim} f = X/\ker f$.

Exercise 196. *Prove the following:*

(i) The category **Grp** *is not preadditive.*
(ii) The categories **TVS** *and* **HTVS** *(Examples 37 and 36) are semiabelian categories but are not abelian. The same holds for the categories* **LCS**, **Frt** *and* **Ban**.
(iii) The category **Ab** *of Abelian groups is abelian.*
(Hint: to prove that **Ban** *is not an abelian category, use the following fact: there exist Banach spaces X, Y and a continuous injective linear map $f : X \hookrightarrow Y$, the image of which is dense and non-closed; for example, $X = C^0([0,1])$, the space of continuous real-valued functions on $[0,1]$, $Y = L^1([0,1])$, the space of Lebesgue classes of integrable real-valued functions on $[0,1]$–each space endowed with its usual norm– and f is the inclusion ([92], (13.11.6)).)*

Exercise 197. *(i) Let $\mathcal{C}$ be a category and let f be a morphism of $\mathcal{C}$. Verify that f is a monomorphism in $\mathcal{C}$ if, and only if it is an epimorphism in $\mathcal{C}^{op}$.*
(ii) Let $\mathcal{C}$ be a category and let $\mathcal{D}$ be a subcategory of $\mathcal{C}$. Show that if a morphism f of $\mathcal{D}$ is a monomorphism (resp., an epimorphism) in $\mathcal{C}$, then it is a monomorphism (resp., an epimorphism) in $\mathcal{D}$.
(iii) Show that in the categories **Set**, ${}_{\mathbf{R}}\mathbf{Mod}$ *and* **Grp**, *a morphism is monic (resp., epic) if, and only if it is injective (resp., surjective) as a function on sets.*
(iv) Show that in the category **Ban** *there exist epimorphisms which are not surjective. (Hint: use the hint of Exercise 196.)*

Exercise 198. *In a preabelian category $\mathcal{C}$, let $f : X \to Y$.*

(i) Prove the statement in Remark 40.
(ii) Assuming that f is a strict bimorphism, prove that f is an isomorphism.

Exercise 199. *(1) Consider the following exact sequences in an abelian category $\mathcal{C}$:*

$$0 \longrightarrow N \xrightarrow{\iota} M, \tag{1.45}$$

$$N \xrightarrow{\varphi} M \longrightarrow 0, \tag{1.46}$$

$$0 \longrightarrow N \xrightarrow{\psi} M \longrightarrow 0. \tag{1.47}$$

Show that in (1.45), ι is a monomorphism, in (1.46), φ is an epimorphism, and in (1.47), ψ is an isomorphism.
(2) Let $f : M \to N$ be morphism, $(K, \iota) = \ker f$ and $(Q, \varphi) = \operatorname{coker} f$. Verify that the following sequence is exact:

$$0 \longrightarrow K \xrightarrow{\iota} M \xrightarrow{f} N \xrightarrow{\varphi} Q \longrightarrow 0.$$

Consider the case when $M \subseteq N$ and f is the inclusion $M \hookrightarrow N$.

Exercise 200. *(i) Prove Theorem and Definition 43(2) when $\mathcal{C}$ is abelian. (Hint: use Proposition 549 below.)*
(ii) Assuming that the additive category $\mathcal{C}$ has products and coproducts indexed by the set I (possibly filtering), prove that there exist canonical isomorphisms

$$\operatorname{Hom}_{\mathcal{C}}\left(\coprod_{i \in I} X_i, Y\right) \cong \prod_{i \in I} \operatorname{Hom}_{\mathcal{C}}\left(X_i, Y\right),$$
$$\operatorname{Hom}_{\mathcal{C}}\left(X, \prod_{i \in I} Y_i\right) \cong \prod_{i \in I} \operatorname{Hom}_{\mathcal{C}}\left(X, Y_i\right).$$

(iii) When I is arbitrary, prove that the functor $\mathfrak{A}_X = \operatorname{Hom}_{\mathcal{C}}(X, \bullet)$ (resp., $\mathfrak{B}_Y = \operatorname{Hom}_{\mathcal{C}}(\bullet, Y)$) is faithful if, and only if X a generator (resp., Y is a cogenerator) of $\mathcal{C}$.

Exercise 201. *(i) Let $\mathcal{C}$ be a small abelian category. Prove that there exists a ring* $\mathbf{R}$ *such that $\mathcal{C}$ is (isomorphic to) an exact full subcategory of* ${}_{\mathbf{R}}\mathbf{Mod}$. *(Hint: use Theorem 50, Lemma 32 and Definition 28(ii).)*

(ii) The Five lemma, the Nine lemma and the Snake lemma in the category $_{\mathbf{R}}\mathbf{Mod}$ *can be found in many classical books, e.g., [206], [306], [141] and [31]. Using Theorem 50, state and prove these lemmas in abelian categories.*

Exercise 202. *(1) In an abelian category, prove that a morphism* α *is a monomorphism (resp., an epimorphism) if, and only if* $\alpha = \ker \operatorname{coker} \alpha$ *(resp.,* $\alpha = \operatorname{coker} \ker \alpha$*).*
(2) Is this correct in any semiabelian category? (Hint: give counterexamples.)

Exercise 203. *Let* X *be an indeterminate and* $\mathbf{M} = \{X^n : n \in \mathbb{N}\}$*, endowed with the multiplication. Prove that* $\mathbf{M}$ *is a rigid cancellation monoid.*

Exercise 204. *Prove that any well-ordered poset is totally ordered. Is the converse true? (Hint: consider the interval* $[0, +\infty)$ *of the real line endowed with its usual order.)*

Exercise 205. *Let* G *be a group and* $A \subseteq G$*. (i) Prove that* $\mathbf{Z}_G(A)$ *and* $\mathbf{N}_G(A)$ *are subgroups of* G *and that* $\mathbf{Z}_G(A) \lhd \mathbf{N}_G(A)$*. (ii) If* A *is a subgroup of* G*, prove that* $\mathbf{N}_G(A)$ *is the largest subgroup of* G *containing* A *in which* A *is normal.*

Exercise 206. *Let* G *be a group. For any* $x, y \in G$*, set* $x^y = y^{-1} x y$*. The map* $\operatorname{Int}(y) : x \mapsto x^y$ *is called the* inner automorphism *induced by* y*. Consider the map* $\operatorname{Int} : G \to \operatorname{Aut}(G)$ *(where* $\operatorname{Aut}(G)$ *is the group of automorphisms of* G*) and prove that (i)* $\ker \operatorname{Int} = \mathbf{Z}(G)$ *and (ii)* $\operatorname{im} \operatorname{Int}$ *is a normal subgroup of* $\operatorname{Aut}(G)$*.*

Exercise 207. *(i) Prove that* $\mathrm{Alt}_n \lhd \mathfrak{S}_n$*, where* Alt_n *is the alternating group of degree* n *(Example 141), and that the order of* Alt_n *is* $n!/2$*. (Hint: use the fact that* $\mathrm{Alt}_n = \ker \operatorname{sgn}$*, Lagrange's theorem, and the appropriate isomorphism theorem.) (ii) Prove that the odd permutations of* $\mathfrak{S}_n$ *form a coset* Ω_n *of* Alt_n *in* $\mathfrak{S}_n$*, not a group, and that* $\operatorname{card} \Omega_n = n!/2$*.*

Exercise 208. *Let* S *be a set and let* G *be a group which* transitively *acts on* S*.*

(i) Prove that $\operatorname{card}(S) \mid (G : 1)$ *if these cardinals are finite.*
(ii) Assuming that G *is a topological group ([34], n° III.1.1),* S *is a topological space and the map* $G \times S \to S$ *is continuous, show that the bijection* $G/G_x \to S$ **(1.6.9.1)** *is continuous when* G/G_x *is endowed with the quotient topology ([34], n° III.2.5).*
(Hint: for (i), use the orbit decomposition formula and Lagrange's theorem.)

Remark: *If* G *is a Lie group,* X *is a manifold and the map* $G \times X \to X$ *is* C^∞*, one can prove that the bijection* $G/G_x \to S$ *is a diffeomorphism ([92], (16.10.8)).*

Exercise 209. *Let $S_1\,[\mathfrak{T}_1]$ and $S_2\,[\mathfrak{T}_2]$ be topological spaces and let $f : S_1 \to S_2$ be a map.*

(i) Let Φ be a filter on S_1. Check that $f(\Phi)$ is a base of filter on S_2.
(ii) Let $x \in S_1$, let Φ be the filter of neighborhoods of x, and let $y \in S_2$. Check that $\lim_{t \to x} f(t) = y$ if, and only if $\lim f(\Phi) = y$. Express this condition using nets.
(iii) Assume that f is continuous at $a \in S_1$ and let Φ be a filter on S_1 such that $\lim \Phi = a$. Show that $\lim f(\Phi) = f(a)$.
(iv) Conversely, assume that for every filter Φ on S_1 such that $\lim \Phi = a$, one has $\lim f(\Phi) = f(a)$. Show that f is continuous at a.

Exercise 210. *(i) Consider the set $A(n, \mathbb{R})$ of all matrices of the form*

$$\begin{bmatrix} U & \tau \\ 0 & 1 \end{bmatrix} \tag{1.48}$$

where $U \in \mathrm{GL}_n(\mathbb{R})$ and $\tau \in \mathbb{R}^n$. Show that $A(n, \mathbb{R})$ is a closed subgroup of $\mathrm{GL}_{n+1}(\mathbb{R})$ (called the affine group *of $\mathbb{R}^n$), thus a Lie group.*
(ii) Let $\mathrm{SO}(n, \mathbb{R})$ be the kernel in the orthogonal group $O(n, \mathbb{R})$ of the group-homomorphism det, *i.e.*

$$\mathrm{SO}(n, \mathbb{R}) = \{A \in O(n, \mathbb{R}) : \det(A) = 1\}.$$

Show that $\mathrm{SO}(n, \mathbb{R})$ is a normal subgroup of $O(n, \mathbb{R})$ and a closed subgroup of $\mathrm{SL}_n(\mathbb{R})$ (it is called the special orthogonal group, *or the* group of rotations, *of $\mathbb{R}^n$). Is $\mathrm{SO}(n, \mathbb{R})$ normal in $\mathrm{SL}_n(\mathbb{R})$? (Hint: give a counterexample when $n = 2$.)*
(iii) Let $\mathrm{SE}(n, \mathbb{R})$ be the subset of $A(n, \mathbb{R})$ consisting of all matrices of the form (1.48) where $U \in \mathrm{SO}(n, \mathbb{R})$ and $\tau \in \mathbb{R}^n$. Prove that $\mathrm{SE}(n, \mathbb{R})$ is a closed subgroup of $\mathrm{SL}_{n+1}(\mathbb{R})$ (it is called the special Euclidean group *of $\mathbb{R}^n$).*
(iv) The action of $A(n, \mathbb{R})$ on the points of $\mathbb{R}^n$ is given by

$$\begin{bmatrix} U & \tau \\ 0 & 1 \end{bmatrix} x = U\,x + \tau.$$

How can the induced action of $\mathrm{SE}(n, \mathbb{R})$ on the points of $\mathbb{R}^n$ be physically interpreted when $n = 2$ or 3?

Exercise 211. *(i) Verify that $\mathrm{Mat}_n(\mathbb{R})$, equipped with the bracket $[\bullet, \bullet]$ where $[A, B]$ is the commutator of A and B, is a Lie algebra, i.e., check that the Jacobi identity holds. This Lie algebra is denoted by $\mathfrak{gl}_n(\mathbb{R})$.*
(ii) Prove that $\mathfrak{gl}_n(\mathbb{R})$ is the Lie algebra of $\mathrm{GL}_n(\mathbb{R})$.

Exercise 212. *(i) Prove that the Lie algebra $\mathfrak{so}(n, \mathbb{R})$ of $\mathrm{SO}(n, \mathbb{R})$ consists of all skew-symmetric matrices belonging to $\mathrm{Mat}_n(\mathbb{R})$ (i.e., of those matrices A such that $A^T + A = 0$). (Exercises 210 and 211.)*

(ii) Conversely, verify that for any skew-symmetric matrix $A \in \mathrm{Mat}_n(\mathbb{R})$, $e^A \in \mathrm{SO}(n,\mathbb{R})$.
(iii) Show that any matrix belonging to $\mathrm{SO}(2,\mathbb{R})$ *is of the form* e^A, $A \in \mathfrak{so}(2,\mathbb{R})$. *(Hint: first calculate by induction the powers of* $A = \begin{bmatrix} 0 & \theta \\ -\theta & 0 \end{bmatrix}$.*)*
(iv) Show that for any $n \geq 2$, *any matrix* B *belonging to* $\mathrm{SO}(n,\mathbb{R})$ *is of the form* e^A, $A \in \mathfrak{so}(n,\mathbb{R})$. *(Hint: first prove that there exists a matrix* $P \in \mathrm{GL}_n(\mathbb{R})$ *such that* $P^{-1}BP = \mathrm{diag}(B_1,...,B_k)$, $B_j \in \mathrm{SO}(2,\mathbb{R})$ *for all* $j \in \{1,...,k\}$.*)*

Exercise 213. *(i) Prove that for any* $M \in \mathrm{Mat}_n(\mathbb{R})$, $\det\left(e^A\right) = e^{\mathrm{Tr}(A)}$. *(Hint: reduce* A *to its Jordan canonical form.)*
(ii) Prove that the Lie algebra $\mathfrak{sl}_n(\mathbb{R})$ *of* $\mathrm{SL}_n(\mathbb{R})$ *consists of all matrices* $A \in \mathrm{Mat}_n(\mathbb{R})$ *such that* $\mathrm{Tr}(A) = 0$.
(iii) Check that for any matrix $A \in \mathfrak{sl}_n(\mathbb{R})$, $e^A \in \mathrm{SL}_n(\mathbb{R})$.
(iv) Show that $\mathrm{diag}(\lambda, 1/\lambda) \in \mathrm{SL}_2(\mathbb{R})$ *cannot be represented in the form* e^A, $A \in \mathfrak{sl}_2(\mathbb{R})$, *if* $\lambda < 0$ *and* $\lambda \neq -1$.
(v) Set $A = \begin{bmatrix} 0 & 2\pi \\ 2\pi & 0 \end{bmatrix} \in \mathfrak{sl}_2(\mathbb{R})$ *and prove that* $e^A = I_2 = e^0$. *The Lie group* $\mathrm{SL}_n(\mathbb{R})$ *can be proved to be connected; deduce from (iv) and (v) that* $\exp_G : \mathfrak{g} \to G$, *where* G *is a general connected Lie group, is neither injective nor surjective in general.*
(vi) As easily shown (see, e.g., [34], §III.2, Prop. 6), a connected topological group is generated by any open neighborhood of its unit element e. *Deduce from this result that a connected Lie group* G *is generated by* $\mathrm{im}(\exp_G)$.
(vii) Prove that $\mathbf{Z}(\mathrm{SL}_2(\mathbb{R})) = \{I_2, -I_2\}$ *and that* $\mathrm{SL}_2(\mathbb{R}) / \pm I_2$ *is simple. (The last point is proved in ([206], Sect. XIII.8.)*
(viii) Poincaré's half plane $\mathbf{P}_+$ *is defined by* $\mathbf{P}_+ = \{z \in \mathbb{C} : \mathrm{Im}\, z > 0\}$. *The group* $\mathrm{SL}_2(\mathbb{R})$ *acts on* $\mathbf{P}_+$ *as follows:*

$$\begin{bmatrix} a & b \\ c & d \end{bmatrix} . z = \frac{a\,z + b}{c\,z + d} = f(z).$$

(a) Prove that $\mathbf{P}_+$ *is invariant by* $\mathrm{SL}_2(\mathbb{R})$, *and that the action of* $\mathrm{SL}_2(\mathbb{R})$ *on* $\mathbf{P}_+$ *is transitive (Subsect. 1.6.9) and continuous. (b) Prove that the set consisting of all above functions* f, *endowed with the composition of functions, is a group-isomorphic to* $\mathrm{SL}_2(\mathbb{R}) / \pm I_2$. *(c) Prove that the stabilizer of* i *is* $\mathrm{SO}(2,\mathbb{R})$ *(Exercise 212). (d) Show using Exercise 208(ii) that the homogeneous spaces* $\mathbf{P}_+$ *and* $\mathrm{SL}_2(\mathbb{R}) / \mathrm{SO}(2,\mathbb{R})$ *can be algebraically and topologically identified .*

Exercise 214. *Prove that the Lie algebra* $\mathfrak{se}(n,\mathbb{R})$ *of* $\mathrm{SE}(n,\mathbb{R})$ *consists of all matrices* $\begin{bmatrix} \Omega & \nu \\ 0 & 0 \end{bmatrix}$, $\Omega \in \mathfrak{so}(n,\mathbb{R})$, $\nu \in \mathbb{R}^n$ *(Exercises 210, 211 and 212). Physical interpretation of those matrices when* $n = 2$ *or* 3*?*

Exercise 215. *Let*

$$J=\begin{bmatrix}0 & I_n\\ -I_n & 0\end{bmatrix}.$$

A matrix $S\in\mathrm{Mat}_n(\mathbf{k})$ *(*$\mathbf{k}=\mathbb{R}$ *or* $\mathbb{C}$*) is said to be* symplectic *if* $S^T J S = J$*; the set of all symplectic matrices belonging to* $\mathrm{Mat}_{2n}(\mathbf{k})$ *is denoted by* $\mathrm{Sp}(2n,\mathbf{k})$*.*

(i) Prove that $\mathrm{Sp}(2n,\mathbf{k})$ *is a linear algebraic group over* $\mathbf{k}$ *(Subsect. 2.10.1);* $\mathrm{Sp}(2n,\mathbf{k})$ *is called the* symplectic group *of degree* $2n$ *over* $\mathbf{k}$.
(ii) Prove that if $\lambda\in\mathbb{C}$ *is an eigenvalue of* $S\in\mathrm{Sp}(2n,\mathbf{k})$*, then so are* $1/\lambda$, $\bar{\lambda}$ *and* $1/\bar{\lambda}$ *too.*
(iii) Deduce from the above that $\mathrm{Sp}(2n,\mathbf{k})$ *is a subgroup of* $\mathrm{SL}_{2n}(\mathbf{k})$*. (Hint: recall that* $\det S$ *is the product of all eigenvalues of* S*, repeated according to their algebraic multiplicities ([28], n° VII.5.5).)*
(iv) A matrix $H\in\mathrm{Mat}_{2n}(\mathbb{R})$ *is said to be* Hamiltonian *(resp.,* skew-Hamiltonian*) if* $(J H)^T = J H$ *(resp.,* $(J H)^T = -J H$*). Show that the square of a Hamiltonian matrix is skew-Hamiltonian and that if* $S\in\mathrm{Sp}(2n,\mathbb{R})$ *and* H *is Hamiltonian (resp., skew-Hamiltonian), then* $S^{-1} H S$ *is Hamiltonian (resp., skew-Hamiltonian).*
(v) Show that if H *is Hamiltonian, then* e^H *is symplectic. (Hint: calculate by induction the* n *first terms the power series expansion of* $S^T J S$ *where* $S=e^H$*.)*
(vi) The set of all Hamiltonian matrices belonging to $\mathrm{Mat}_{2n}(\mathbb{R})$ *is denoted by* $\mathfrak{S}_p(2n)$*. Show that* $\mathfrak{S}_p(2n)$ *is the Lie algebra of* $\mathrm{Sp}(2n,\mathbb{R})$ *(Exercise 211).*

Exercise 216. *(1) Let* $\mathfrak{a},\mathfrak{b}$ *be ideals in a Lie algebra* $\mathfrak{g}$*. Check that* $\mathfrak{g}/\mathfrak{a}$ *is a Lie algebra and that* $\mathfrak{a}+\mathfrak{b}$ *and* $\mathfrak{a}\cap\mathfrak{b}$ *are ideals in* $\mathfrak{g}$*.*
(2) The ideal $[\mathfrak{g},\mathfrak{g}]$ *in a Lie algebra* $\mathfrak{g}$ *is denoted by* $\mathfrak{D}(\mathfrak{g})$ *and is called the* derived algebra *of* $\mathfrak{g}$*. By induction, one defines* $\mathfrak{D}^1(\mathfrak{g})=\mathfrak{D}(\mathfrak{g})$ *and* $\mathfrak{D}^n(\mathfrak{g})=\mathfrak{D}\left(\mathfrak{D}^{n-1}(\mathfrak{g})\right)$ *for* $n>1$*. Check the following:*
(i) $\mathfrak{D}(\mathfrak{g})$ *is the largest ideal* $\mathfrak{a}$ *in* $\mathfrak{g}$ *such that* $\mathfrak{g}/\mathfrak{a}$ *is commutative.*
(ii) The Lie algebra $\mathfrak{g}$ *is solvable if, and only if there exists* $n>1$ *such that* $\mathfrak{D}^n(\mathfrak{g})=0$*.*
(3) Let $\mathfrak{g}^1=[\mathfrak{g},\mathfrak{g}]$ *and* $\mathfrak{g}^n=\left[\mathfrak{g},\mathfrak{g}^{n-1}\right]$ *for* $n>1$*. Check that for any* $n>1$, $\mathfrak{g}^n$ *is an ideal in* $\mathfrak{g}$ *and that* $\mathfrak{g}$ *is nilpotent if, and only if there exists* $n>1$ *such that* $\mathfrak{g}^n=0$*.*
(4) For any $X\in\mathfrak{g}$*, consider the map* $\mathrm{ad}(X):Y\mapsto[X,Y]$*.*
(i) Verify that $\mathrm{ad}(X)$ *is the inner derivation induced by* $-X$ *(see Definition 287 below).*
(ii) Let $\mathrm{Der}(\mathfrak{g})$ *be the set of all derivations of* $\mathfrak{g}$*, i.e., of all linear transformations* δ *which satisfy* $\delta[X,Y]=[\delta X,Y]+[X,\delta Y]$*. Prove that* $\mathrm{Der}(\mathfrak{g})$ *is a Lie algebra, that* $\mathrm{ad}:\mathfrak{g}\to\mathrm{Der}(\mathfrak{g})$ *is a Lie group homomorphism, and that for any derivation* $\delta\in\mathrm{Der}(\mathfrak{g})$, $[\delta,\mathrm{ad}(X)]=\mathrm{ad}(\delta X)$*. The map* ad *is called the* adjoint map *of* $\mathfrak{g}$*.*

Exercise 217. *(i) Show that the derivative of order $n \geq 0$ of the Dirac hyperfunction is given by*

$$\delta_0^{(n)} = \left[\frac{(-1)^{n+1} \, n!}{2\pi i} z^{-(n+1)} \right]$$

and that the "Heaviside hyperfunction" (corresponding to the classical Heaviside function [43]) is given by

$$\Upsilon(z) = \left[\frac{-1}{2\pi i} \ln(-z) \right].$$

(ii) Show that $\varphi(z) = e^{1/z}$ determines a hyperfunction $[\varphi]$ which is the sum of a series involving the Dirac hyperfunction and all its derivatives, and that $\operatorname{supp}[\varphi] = \{0\}$.
(iii) Let $\psi \in \mathcal{O}(\mathbb{R})$; calculate $\langle \psi, [\varphi] \rangle$ when $\varphi(z) = e^{1/z}$.

1.11 Notes

This chapter gathers, for the reader's convenience, classical notions and results which can be found in several references, especially [18], [27] [76], [77], [80], [183], [224] [225] and [249].

For an (almost) exhaustive presentation of categories we refer the reader to MacLane's classical book [224]. The evolution of the theory of categories (founded by Eilenberg and MacLane in 1942) is well described in the appendix of Freyd's book [127] and in the historical notes of MacLane's monographs [223], [224]. MacLane told Freyd that the word *category* goes back to Kant, but Kant himself took this term from Aristotle (Organon, Book I). The notion of *generator* and the dual notion of *cogenerator* (**1.2.2.4**) are due to Grothendieck ([148], Sect. I.9). Abelian categories were introduced by Buchsbaum ([55]; [59], Appendix), who called them *exact categories*, and Grothendieck [148] gave them their definitive name. *Preabelian categories* and *semiabelian categories* were defined by Popescu ([282], Sect. 2.1) and by Rump [311], respectively. Lemma 51 is due to Freyd ([127], Sect. 2.1) and Lemma 27 to Mitchell ([249], Sect. II.10). Theorem 54 in abelian categories is due to Buchsbaum ([55], Part III, Proposition 2.2). Direct proofs of the Nine lemma and of the Snake lemma in abelian categories (Exercise 201) can be found in ([224], Sect. VIII.4).

Divisibility, ordering and lattices are detailed in various references, among which Birkhoff's monograph [18]. The notion of unique factorization monoid is due to Cohn [74].

Our short presentation of groups is based on classical textbooks. Polycyclic by finite groups and are well described in [247], and Dedekind groups in Hall's monograph [155].

General topology, topological vector spaces, Lie groups and Lie algebras are detailed in the classical Bourbaki's and Dieudonné's treatises [34], [36], [38], [92], and universal enveloping algebras in Dixmier's monograph [94].

Hyperfunctions were introduced by K. Sato [314] as both the boundary values of analytic functions and as cohomology classes (the last point is not treated in this book). They play a key role in the theory of $\mathcal{D}$-modules and in Microlocal Analysis. The theory of hyperfunctions with application to systems of partial differential equations is nicely expounded in several books, among which [250], [184], and in Hörmander's treatise [162]. Detailed historical notes can be found in [184]. The space of hyperfunctions with compact support is endowed with a good topological structure, as shown by Theorem 187. The case of noncompactly supported hyperfunctions is quite different [182]. The definition of general hyperfunctions in Subsect. 1.9.3 is due to Martineau [245] (see also [351]) and our account is mainly based on ([162], Chap. IX).

2 Rings and Modules

2.1 Introduction

Section 2.2 contains present basic facts on rings and modules. Bimodules are also introduced, for they are very useful to study the properties of modules of homomorphisms and of tensor products.

Differential rings are introduced in Section 2.3.1. General derivatives are considered to cover the case of continous-time systems and that of discrete-time systems. Consider for example a system with polynomial coefficients with respect to time, i.e., with coefficients belonging to $\mathbf{K} = \mathbb{C}[t]$, the ring of polynomials with complex coefficients and indeterminate t. In case of "continuous-time", this ring is equipped with the usual derivative $\boldsymbol{\delta} = d/dt$; $\boldsymbol{\delta}$ is called an "outer derivation", and $(\mathbf{K}, \boldsymbol{\delta})$ is a *differential ring.* In case of "discrete-time", the same ring $\mathbf{K} = \mathbb{C}[t]$ is equipped with the shift $\boldsymbol{\alpha} : a(t) \mapsto a(t+1)$ and $(\mathbf{K}, \boldsymbol{\alpha})$ is a *difference ring.* In the latter case, one can slightly change the formulation and replace the shift $\boldsymbol{\alpha}$ by the "discrete derivative" $\boldsymbol{\alpha} - 1$, which is called an "inner derivation". In both cases, $\mathbf{K}$, endowed with the derivative under consideration, is a *generalized differential ring.*

Most properties of rings are inherited from the properties of their ideals. Therefore, ideals in rings are studied in Section 2.4. In particular, the notion of Noetherian ring is introduced.

The study of skew polynomial rings is divided into two parts. The first one is the subject of Section 2.7; the second one, which needs a wider background, is given in Section 4.3 (Chapter 4). Consider the continuous-time single-input-single-output (SISO) system

$$a(\partial)\, y = b(\partial)\, u$$

where u and y denote, e.g., the input and the output, respectively; ∂ is the derivative and $a(\partial), b(\partial)$ are differential operators which act on the variables y and u. One can write

$$a(\partial) = \sum_{0 \leq i \leq n} a_i \partial^{n-i}, \quad a_0 \neq 0, \quad a_i \in \mathbf{K} \ (1 \leq i \leq n).$$

H. Bourlès and B. Marinescu: Linear Time-Varying Systems, LNCIS 410, pp. 91–187.
springerlink.com

Consider a coefficient $a \in \mathbf{K}$ and a variable x. According to the Leibniz rule,

$$\partial (ax) = a (\partial x) + \left(a^{\boldsymbol{\delta}}\right) x \tag{2.1}$$

where $a^{\boldsymbol{\delta}} = \boldsymbol{\delta}(a) = da/dt$. Since (2.1) is valid for any variable x, one can write $\partial a = a\partial + a^{\boldsymbol{\delta}}$, i.e.,

$$\partial a - a\partial = a^{\boldsymbol{\delta}}.$$

The above differential operators form a ring which is isomorphic to the "skew polynomial ring" consisting of all polynomials with coefficients in $\mathbf{K}$, with indeterminate X, and endowed with the "commutation rule"

$$Xa - aX = a^{\boldsymbol{\delta}}; \tag{2.2}$$

this ring is denoted by $\mathbf{K}[X, \boldsymbol{\delta}]$. The commutation rule (2.2) shows that $Xa \neq aX$, except if $a^{\boldsymbol{\delta}} = 0$, i.e., a is a *constant.* Skew polynomials are of a paramount importance for the study of LTV systems.

Rings and fields of fractions are studied in Section 2.5. In the commutative case, these notions are classical [43], but the noncommutative case involves the so-called Ore condition.

The theory of rings of fractions makes it possible to pass from skew polynomials to skew Laurent polynomials in Section 2.8. This is very important for the study of discrete-time LTV systems.

Local rings and power series rings are studied Sections 2.6 and 2.9, respectively. They are needed for the study of the structure at infinity of LTV systems (Chapter 7).

The three following sections present notions which are necessary for the study of matrices with entries in a noncommutative ring or in a skew field. In particular, determinants, which are classical for square matrices with entries in a commutative ring, cannot be defined for square matrices with entries in a noncommutative ring. This is connected with the fact that there does not exist a unique notion of rank for a matrix with entries in a noncommutative ring (Section 2.12). Nevertheless, J. Dieudonné defined a notion of determinant for a square matrix with entries in a skew field (Section 2.11).

The chapter ends with Section 2.13, where specific (noncommutative) rings, such as GCD domains, unique factorization domains, Bézout domains, Dedekind domains, principal ideal domains, and Euclidean domains, are presented. The relations between most (not all) kinds of rings studied is this book are summarized in the table below, where $\rightarrow$ means "is more restrictive than" and where *ED*, *PID*, *UFD*, *GCD*, *EDR*, *GCD*, *EDR*, *BD*, *CSD*, *SD*, *NR*, *CR* are used for *Euclidean domain, principal ideal domain, unique factorization domain, GCD domain, elementary divisor ring* (without zerodivisor), *Bézout domain, coherent Sylvester domain, Sylvester domain, Noetherian ring,* and *coherent ring*, respectively.

$$
\begin{array}{ccccccccccc}
 & & & & UFD & \rightarrow & GCD & & & & \\
 & & & \nearrow & & & \uparrow & \nwarrow & & & \\
ED & \rightarrow & PID & \rightarrow & EDR & \rightarrow & BD & \rightarrow & CSD & \rightarrow & SD \\
 & & & \searrow & & & \downarrow & \swarrow & & & \\
 & & & & NR & \rightarrow & CR & & & &
\end{array}
$$

2.2 Rings and Modules

2.2.1 Rings, Ideals and Division Rings

2.2.1.1 Rings and Ideals

Definition 218. *A ring is a nonempty set* $\mathbf{R}$ *endowed with an addition* $+$ *and a multiplication* $.$ *with the following properties: (i)* $(\mathbf{R},+)$ *is an abelian group, additively noted; (ii)* $(\mathbf{R},.)$ *is a monoid; (iii) the following* distributivity law *holds: for any* $x,y,z\in\mathbf{R}$,

$$(x+y)\,z = x\,z + y\,z, \qquad z\,(x+y) = z\,x + z\,y. \tag{2.3}$$

The first identity is called *left-distributivity* (of multiplication with respect to addition), the second one *right-distributivity.* As easily shown, $0\,x = x\,0 = 0$ for any $x\in\mathbf{R}$.

The ring $(\mathbf{R},+,.)$ (written $\mathbf{R}$, for short) is said to be *commutative* if $(\mathbf{R},.)$ is a commutative monoid.

Definition 219. *The* opposite *of a ring* $\mathbf{R}$*, written* $\mathbf{R}^{\mathrm{op}}$*, is the ring such that (i) the abelian group* $(\mathbf{R}^{\mathrm{op}},+)$ *coincides with* $(\mathbf{R},+)$*; (ii) the monoid* $(\mathbf{R}^{\mathrm{op}},.)$ *is the opposite of* $(\mathbf{R},.)$ *(Definition 59).*

Remark 220. *(a) According to Definition 218, any ring has a unit element* 1 *(and* $1=0$ *if* $\mathbf{R}=\{0\}$*).*
(b) If $(\mathbf{R},+,.)$ *satisfies properties (ii) and (i'):* $(\mathbf{R},+)$ *is a commutative monoid (with unit element* 0*), then* $(\mathbf{R},+,.)$ *is called a* dioid*; this dioid is left- (resp., right-) distributive if the first (resp., the second) identity of (2.3) holds, and distributive if it is both left- and right-distributive [144]. If the monoid* $(\mathbf{R},.)$ *is commutative, the dioid* $(\mathbf{R},+,.)$ *is said to be* commutative *too.*

Definition 221. *Let* $\mathbf{R}$ *be a nonzero ring.*

(i) The characteristic *of* $\mathbf{R}$*, denoted by* $\operatorname{char}\mathbf{R}$*, is the least integer* $n\geq 0$ *such that* $n\,1=0$.
(ii) If the ring $\mathbf{R}$ *is commutative, it is called* perfect *if it is of characteristic zero, or if it is of characteristic* $p>0$ *and the map* $a\mapsto a^{p}$ *is bijective.*

The notions introduced in Definitions 63, 67 and 68 can be extended to the case of a ring $\mathbf{R}$, and Notation 64 as well. The details are left to the reader. Let us recall that *an ideal* means *a two-sided ideal*, except when otherwise stated. The product of a family $(\mathbf{R}_i)_{i \in I}$ of rings and the ring $\mathbf{R}^I$ are defined as in Subsect. 1.3.4 with obvious changes. The following is obvious:

Proposition 222. *Let $(x_\lambda)_{\lambda \in \Lambda}$ be a family of elements of the ring $\mathbf{R}$, let $X = \{x_\lambda : \lambda \in \Lambda\}$, and let $\mathfrak{a}$ (resp., $\mathfrak{b}$) be the set of all elements of the form $\sum_{\lambda \in \Lambda} a_\lambda x_\lambda$ where $a_\lambda \in \mathbf{R}$ and $a_\lambda = 0$ for all but a finite number of indices (resp., $\sum_{\lambda \in \Lambda} a_\lambda x_\lambda b_\lambda$ where $a_\lambda, b_\lambda \in \mathbf{R}$ and $a_\lambda b_\lambda = 0$ for all but a finite number of indices). Then $\mathfrak{a}$ is the left ideal generated by the set X (i.e., $\mathfrak{a} = (X)_l$) and $\mathfrak{b}$ is the ideal generated by the set X (i.e., $\mathfrak{b} = (X)$).*

The zero ideal and the zero ring are written 0. In the context of rings, Definition 149 is changed to the following one:

Definition 223. *A ring $\mathbf{R}$ is* simple *if it has no proper nonzero ideal.*

The notions below are very useful:

Definition 224. *(i) A ring $\mathbf{R}$ is a* domain *(also called an* integral ring*, or an* entire ring*) if $\mathbf{R} \neq \{0\}$ and $(\mathbf{R}^\times, .)$ is a cancellation monoid.*
(ii) A ring $\mathbf{R}$ is said to be atomic *if $\mathbf{R} \neq \{0\}$ and the monoid $(\mathbf{R}^\times, \times)$ is atomic (Definition 85(i)).*

In many references, e.g., [306], a domain is always assumed to be commutative; in the present book we will specify in such a case: a *commutative domain.*

Lemma 225. *(i) A ring $\mathbf{R}$ is a domain if, and only if $\mathbf{R} \neq \{0\}$ and all elements of $\mathbf{R}^\times$ are regular.*
(ii) If a ring $\mathbf{R}$ is atomic, it is a domain.
(iii) If a ring $\mathbf{R}$ is a domain, $\operatorname{char} \mathbf{R}$ *is a prime.*

Proof. (i) and (ii) are clear. (iii): see ([79], Sect. 4.1 and Sect. 4.5, Exercise 1). ∎

2.2.1.2 Division Rings

Definition 226. *A* division ring *(also called a* skew field*) is a ring $\mathbf{K} \neq 0$ such that every element of $\mathbf{K}^\times$ is a unit. A* field *is a* commutative *division ring (Remark 58).*

Theorem 227. *Let $\mathbf{R}$ be a ring.*

(1) The following conditions are equivalent:
(i) $\mathbf{R}$ is a division ring;
(ii) $\mathbf{R} \neq 0$ and this ring has no proper nonzero left nor right ideal.
(2) A division ring is a domain.

Proof. (1) (i)⇒(ii): If $\mathbf{R}$ is a division ring, $\mathbf{R} \neq 0$. Let $0 \neq \mathfrak{a}$ be a left ideal in $\mathbf{R}$ and $0 \neq a \in \mathfrak{a}$; for any $x \in \mathbf{R}$, $x = \left(x\, a^{-1}\right) a \in \mathfrak{a}$, thus $\mathfrak{a} = \mathbf{R}$.

(ii)⇒(i): If (ii) holds, let $0 \neq x \in \mathbf{R}$; since $\mathbf{R}x \neq 0$, $\mathbf{R}x = \mathbf{R}$ and x is left-invertible, i.e., there exists $x' \in \mathbf{R}$ such that $x'\,x = 1$. The above can be applied to x' and there exists $x'' \in \mathbf{R}$ such that $x''\,x' = 1$. We get $x'' = x''\,1 = x''\,x'\,x = 1\,x = x$, thus $x\,x' = 1$ and x is a unit.

(2) is clear. ∎

By Theorem 227, any division ring is a simple nonzero ring. If a ring is *commutative*, it is a division ring if, and only if it is nonzero and simple; this does not hold true in the noncommutative case. The centre of a division ring is a field, as easily seen.

2.2.2 Modules and Algebras

2.2.2.1 Modules

Let $\mathbf{R}$ be a ring.

Definition 228. *A nonempty set M is called a left $\mathbf{R}$-module if the following conditions hold:*
(i) M is an abelian group (additively noted);
(ii) there exists an action $\mathbf{R} \times M \to M : (a, x) \mapsto a\,x$ (Subsect. 1.6.9) which is bilinear and such that

$$(a\,b)\;x = a\;(b\,x)\,, \quad 1\,x = x$$

for any $a, b \in \mathbf{R}$ and $x \in M$.

If the above action is denoted by $(a, x) \mapsto x\,a$, with $x\;(a\,b) = (x\,a)\;b$ for any $a, b \in \mathbf{R}$ and $x \in M$, then M is called a *right* $\mathbf{R}$-module.

Proposition 229. *A right $\mathbf{R}$-module has a canonical structure of left $\mathbf{R}^{\mathrm{op}}$-module, setting*

$$a \,.\, x = x\,a \quad (a \in \mathbf{R},\; x \in M)\,. \tag{2.4}$$

In other words, the category $\mathbf{Mod}_{\mathbf{R}}$ of right $\mathbf{R}$-modules coincides with the category ${}_{\mathbf{R}^{\mathrm{op}}}\mathbf{Mod}$ of left $\mathbf{R}^{\mathrm{op}}$-modules.

Proof. We then have by Definitions 59 and 219, for any $a, b \in \mathbf{R}$ and $x \in M$,

$$x\;(a\,b) = (x\,a)\;b = b\,.\;(x\,a) = b\,.\;(a\,.\,x) = (b\,.\,a)\;.\,x. \quad ∎$$

When a ring $\mathbf{R}$ is commutative, it is irrelevant to distinguished between left and right modules. In the sequel, "an $\mathbf{R}$-module", without any additional specification, means indifferently a left or a right $\mathbf{R}$-module.

A nonempty set M is an abelian group if, and only if it is a $\mathbb{Z}$-module.

Let M be an $\mathbf{R}$-module. A *submodule* N of M is an $\mathbf{R}$-module such that $N \subseteq M$. We write $N \lhd M$ to mean that N is a submodule of M. This submodule is said to be *proper* if it is different from M (Definition 63(i)).

A left (resp., right) ideal $\mathfrak{a}$ in $\mathbf{R}$ is nothing but a left (resp., right) $\mathbf{R}$-module $\mathfrak{a} \subseteq \mathbf{R}$. In particular, the ring $\mathbf{R}$ can be considered as a left (resp., right) module over itself and is then denoted by ${}_{\mathbf{R}}\mathbf{R}$ (resp., $\mathbf{R}_{\mathbf{R}}$).

Let $(M_i)_{i\in I}$ be a family of left (resp., right) $\mathbf{R}$-modules. The product $\prod_{i\in I} M_i$ of the sets M_i (**1.2.1.2**) has an obvious structure of left (resp. right) $\mathbf{R}$-module. The coproduct $\coprod_{i\in I} M_i$ is the submodule of $\prod_{i\in I} M_i$ consisting of all finitely supported elements (see Lemma and Definition 148(2)). If $M_i = {}_{\mathbf{R}}\mathbf{R}$ (resp., $\mathbf{R}_{\mathbf{R}}$) for all $i \in I$, then $\prod_{i\in I} M_i$ is denoted by ${}_{\mathbf{R}}\mathbf{R}^I$ (resp., $\mathbf{R}^I_{\mathbf{R}}$) and $\coprod_{i\in I} M_i$ is denoted by ${}_{\mathbf{R}}\mathbf{R}^{(I)}$ (resp., $\mathbf{R}^{(I)}_{\mathbf{R}}$).

Let $\bullet\pi_i : \prod_{i\in I} M_i \to M_i$ be the canonical projection and $\bullet\iota_i : M_i \to \coprod_{i\in I} M_i$ be the canonical injection (**1.2.1.1**). For any $x \in \coprod_{i\in I} M_i$, we have

$$x = \sum_{i\in I} x\pi_i\iota_i. \tag{2.5}$$

Let $\mathbf{T}$ be a ring, let $\mathbf{S}$ be a subring of $\mathbf{T}$, and let M be a $\mathbf{T}$-module. M can be considered as a module over $\mathbf{S}$; this module is written $M_{[\mathbf{S}]}$ and is said to be obtained from M by *restriction of the ring of scalars*.

2.2.2.2 Bimodules

Let $\mathbf{R}$ and $\mathbf{S}$ be two rings and let M be an abelian group, additively noted; M is called an $(\mathbf{R},\mathbf{S})$-bimodule (written ${}_{\mathbf{R}}M_{\mathbf{S}}$) if (i) it is a left $\mathbf{R}$-module and a right $\mathbf{S}$-module and (ii) the following compatibility condition holds: for any $r \in \mathbf{R}$, $s \in \mathbf{S}$ and $x \in M$,

$$r\ (x\, s) = (r\, x)\ s.$$

Since by Proposition 229 a right $\mathbf{S}$-module has a canonical structure of left $\mathbf{S}^{\mathrm{op}}$-module, both $r \in \mathbf{R}$ and $s \in \mathbf{S}$ can be written on the left of $x \in M$ when this is useful.

Example 230. *(i) The ring* $\mathbf{R}$ *can be considered as an* $(\mathbf{R},\mathbf{R})$*-bimodule, then denoted by* ${}_{\mathbf{R}}\mathbf{R}_{\mathbf{R}}$.
(ii) Let M *be a left* $\mathbf{R}$*-module, let* $\mathbf{E} = \mathrm{End}_{\mathbf{R}}(M)$ *be the endomorphism ring of* M*, i.e., the set of all* left $\mathbf{R}$*-linear maps* $\bullet e : M \to M$*, endowed with the addition and the composition of maps (*$\mathbf{E}$ *is obviously a ring). For any* $r \in \mathbf{R}$ *and* $\bullet e \in \mathbf{E}$*,* $(\lambda\, x)\, e = \lambda\,(xe)$*, thus the compatibility condition holds and* M *is an* $(\mathbf{R},\mathbf{E})$*-bimodule.*
(iii) If $\mathbf{R}$ *is* commutative *and* M *is a* left $\mathbf{R}$*-module,* M *is canonically endowed with a structure of* right $\mathbf{R}$*-module, and* M *is clearly an* $(\mathbf{R},\mathbf{R})$*-bimodule.*

2.2.2.3 Algebras

Let $\mathbf{R}$ be a commutative ring. An $\mathbf{R}$-module $\mathbf{A}$ is an $\mathbf{R}$-algebra if it is endowed with a bilinear map $\mathbf{A} \times \mathbf{A} \to \mathbf{A} : (a, b) \mapsto a\,b$. All algebras considered in the sequel are associative, i.e., $(a\,b)\,c = a\,(b\,c)$ for any $a, b, c \in \mathbf{A}$. The *commutator* of two elements $a, b \in \mathbf{A}$ is $[a, b] \triangleq a\,b - b\,a$. When $\mathbf{R}$ is a field, $[\bullet, \bullet]$ is a Lie bracket.

An algebra $\mathbf{A}$ is said to by *unitary* if it possesses a unit element (often denoted by 1); then, $\mathbf{A}$ is a ring. Conversely, let $\mathbf{A}$ be a ring and let $\rho : \mathbf{R} \to \mathbf{A}$ be a ring-homomorphism such that $\rho(\mathbf{R}) \subseteq \mathbf{Z}(\mathbf{A})$ (where $\mathbf{Z}(\mathbf{A})$ is the centre of $\mathbf{A}$). Then, $\mathbf{A}$ may be viewed as an $\mathbf{R}$-module, defining the action of $\mathbf{R}$ on $\mathbf{A}$ by the map $(r, a) \mapsto \rho(r)\,a$ $(r \in \mathbf{R}, a \in \mathbf{A})$; thus $\mathbf{A}$ is an associative and unitary $\mathbf{R}$-algebra. An algebra $\mathbf{A}$ is commutative if, and only if $[a, b] = 0$ for all $a, b \in \mathbf{A}$.

As easily seen, $\mathrm{Mat}_n(\mathbf{R})$ and $\mathrm{GL}_n(\mathbf{R})$ are unitary $\mathbf{R}$-algebras with unit element I_n (Example 71). Every ring $\mathbf{R}$ is both a $\mathbb{Z}$-algebra and a $\mathbf{Z}(\mathbf{R})$-algebra.

2.2.3 *Homomorphisms and Quotients*

2.2.3.1 Homomorphisms

A ring-homomorphism is a morphism in the category **Rgn** (**1.2.2.2**). More specifically:

Definition 231. *A ring-homomorphism $\mathbf{R} \to \mathbf{R}'$ (where $\mathbf{R}$ and $\mathbf{R}'$ are rings) is an additive group-homomorphism $(\mathbf{R}, +) \to (\mathbf{R}', +)$ which is a monoid-homomorphism $(\mathbf{R}, .) \to (\mathbf{R}', .)$.*

The kernel of a ring-homomorphism $f : \mathbf{R} \to \mathbf{R}'$ is $f^{-1}(\{0\})$; the following is clear:

Lemma 232. *(i)* $\ker f$ *is an ideal in* $\mathbf{R}$*; (ii)* $\operatorname{im} f$ *is a subring of* $\mathbf{R}'$.

Definition 233. *(i) Let M, N be right $\mathbf{R}$-modules. A map $f : M \to N$ is a module-homomorphism (or is said to be $\mathbf{R}$-linear) if it is an additive group-homomorphism $x \mapsto fx$ (thus a* right map (**1.2.1.3**)*) such that $f(x\,r) = (fx)\,r$ for all $r \in \mathbf{R}$ and $x \in M$.*
(ii) A homomorphism of left $\mathbf{R}$-modules $f : M \to N$ is likewise defined but for convenience such a homomorphism is a left map $x \mapsto xf$ (**1.2.1.3**). *The $\mathbf{R}$-linearity is then expressed by $r\,xf = (r\,x)\,f$ for all $r \in \mathbf{R}$ and $x \in M$.*

Let $\mathbf{A}$, $\mathbf{A}'$ be two $\mathbf{R}$-algebras where $\mathbf{R}$ is a commutative ring.

Definition 234. *A right map $f : \mathbf{A} \to \mathbf{A}'$ is an algebra-homomorphism if it is a module-homomorphism such that $f(x\,y) = f(x)\,f(y)$ for any $x, y \in \mathbf{A}$.*

It is also useful for the sequel to define the notion of $(\mathbf{R},\mathbf{S})$-bimodule-homomorphism:

Definition 235. *Let M and N be $(\mathbf{R},\mathbf{S})$-bimodules. A map $f : M \to N$ is an $(\mathbf{R},\mathbf{S})$-bimodule-homomorphism if it is additive and for any $r \in \mathbf{R}$, $x \in M$, $s \in \mathbf{S}$, $r\,xf = (r\,x)\,f$ and $f\,(x\,s) = fx\,s$ (where $xf = fx$).*

A module-homomorphism $f : M \to N$ is a monomorphism (Definition 72) if, and only if $\ker f = 0$ (where 0 denotes the submodule of M consisting of 0 alone).

Let M and N be two left (or right) $\mathbf{R}$-modules. The set of all module-homomorphisms $f : M \to N$ is denoted by $\mathrm{Hom}_{\mathbf{R}}(M,N)$. As easily seen, $\mathrm{Hom}_{\mathbf{R}}(M,N)$ is an abelian group, i.e., a $\mathbb{Z}$-module. To express the general structure of $\mathrm{Hom}_{\mathbf{R}}(M,N)$, it is convenient to consider a left $\mathbf{R}$-module as an $(\mathbf{R},\mathbb{Z})$-bimodule and a right $\mathbf{T}$-module as a $(\mathbb{Z},\mathbf{T})$-bimodule. Then all situations are gathered in the following theorem:

Theorem 236. *Let $\mathbf{R}$, $\mathbf{S}$, $\mathbf{T}$ and $\mathbf{U}$ be rings.*

(i) If M is an $(\mathbf{R},\mathbf{S})$-bimodule (written ${}_{\mathbf{R}}M_{\mathbf{S}}$) and N an $(\mathbf{R},\mathbf{T})$-bimodule (written ${}_{\mathbf{R}}N_{\mathbf{T}}$), then $\mathrm{Hom}_{\mathbf{R}}(M,N)$ has a canonical structure of $(\mathbf{S},\mathbf{T})$-bimodule (written ${}_{\mathbf{S}}\mathrm{Hom}_{\mathbf{R}}({}_{\mathbf{R}}M_{\mathbf{S}},{}_{\mathbf{R}}N_{\mathbf{T}})_{\mathbf{T}}$) by defining for any $\bullet f : M \to N$, $s \in \mathbf{S}$ and $t \in \mathbf{T}$

$$f\,t : x \mapsto (xf)\,t, \quad s\,f : x \mapsto (x\,s)\,f. \tag{2.6}$$

(ii) If M is an $(\mathbf{R},\mathbf{T})$-bimodule (written ${}_{\mathbf{R}}M_{\mathbf{T}}$) and N an $(\mathbf{U},\mathbf{T})$-bimodule (written ${}_{\mathbf{U}}N_{\mathbf{T}}$), then $\mathrm{Hom}_{\mathbf{T}}(M,N)$ has a canonical structure of $(\mathbf{U},\mathbf{R})$-bimodule (written ${}_{\mathbf{U}}\mathrm{Hom}_{\mathbf{R}}({}_{\mathbf{R}}M_{\mathbf{T}},{}_{\mathbf{U}}N_{\mathbf{T}})_{\mathbf{R}}$) by defining for any $f\bullet : M \to N$, $r \in \mathbf{R}$ and $u \in \mathbf{U}$

$$u\,f : x \mapsto u\,(fx), \quad f\,r : x \mapsto f\,(r\,x). \tag{2.7}$$

Proof. (i): The maps $f\,t$ and $s\,f$ are obviously left $\mathbf{R}$-module homomorphisms. In addition, $x\,(s\,(f\,t)) = (x\,s)\,(f\,t) = x\,(s\,f)\,t$, therefore $s\,(f\,t) = (s\,f)\,t$ and the bimodule property holds. The proof of (ii) is similar. ∎

Definition 237. *Let $\mathbf{R}$ be a ring, u a unit in $\mathbf{R}$. The automorphism $x \mapsto u^{-1}\,x\,u$ $(x \in \mathbf{R})$ is said to be* inner, *induced by u (Exercise 206). An automorphism which is not inner is said to be* outer.

2.2.3.2 Quotient Rings

According to Definition 75, a relation $\mathcal{R}$ is compatible with the ring structure of $\mathbf{R}$ if $x \equiv x'(\mathrm{mod}\,\mathcal{R})$ and $y \equiv y'(\mathrm{mod}\,\mathcal{R})$ implies the two following conditions: (i) $x + y \equiv y + y'(\mathrm{mod}\,\mathcal{R})$; (ii) $x\,y \equiv x'\,y'(\mathrm{mod}\,\mathcal{R})$. The following result is classical ([27], n°I.8.7) and its proof is easy:

Lemma 238. *A relation* $\mathcal{R}$ *is compatible with the ring structure of* $\mathbf{R}$ *if, and only if there exists an ideal* $\mathfrak{a}$ *in* $\mathbf{R}$ *such that* $x \equiv x' (\operatorname{mod} \mathcal{R})$ *is equivalent to* $x - x' \in \mathfrak{a}$.

Notation 239. *The equivalence relation* $x \equiv x' (\operatorname{mod} \mathcal{R})$ *is written* $x \equiv x' (\operatorname{mod} \mathfrak{a})$ *(Lemma and Definition 142(2)).*

By Lemma 76, we obtain the following.

Proposition 240. *A ring* $\mathbf{R}'$ *is a quotient of a ring* $\mathbf{R}$ *if, and only if there exists an ideal* $\mathfrak{a}$ *in* $\mathbf{R}$ *such that* $\mathbf{R}' = \mathbf{R}/\mathfrak{a}$.

Theorems 145, 146 and 147 can be extended to the case of rings, changing "subgroup" to "one-sided ideal" (i.e., left ideal or right ideal), "normal subgroup" to "ideal" and multiplication to addition.

Let $\mathfrak{a}$ and $\mathfrak{b}$ be two ideals in the ring $\mathbf{R}$. The product $\mathfrak{a}\,\mathfrak{b}$ is the ideal generated by all products $a\,b$, $a \in \mathfrak{a}$, $b \in \mathfrak{b}$ (see Definition 295 below for more details). We have the following.

Theorem 241. *Let* $\mathbf{R}$ *be a ring and* $\mathfrak{a}$ *be an ideal in* $\mathbf{R}$.

(i) Every left (resp., right, two-sided) ideal in $\mathbf{R}/\mathfrak{a}$ *can be written in a unique way in the form* $\mathfrak{b}/\mathfrak{a}$ *where* $\mathfrak{b}$ *is a left (resp., right, two-sided) ideal in* $\mathbf{R}$ *such that* $\mathfrak{b} \supseteq \mathfrak{a}$.
(ii) Let $\mathfrak{b}$ *be an ideal in* $\mathbf{R}$*; then*

$$\mathfrak{b}/\mathfrak{a}\,\mathfrak{b} \cong \mathbf{R}/\mathfrak{a}.$$

Proof. (i): Let $\bullet\pi : \mathbf{R} \twoheadrightarrow \mathbf{R}/\mathfrak{a}$ be the canonical epimorphism and consider the case of left ideals. Let $\mathfrak{J}$ be a left ideal in $\mathbf{R}/\mathfrak{a}$ and $\mathfrak{b} = \mathfrak{J}\pi^{-1}$. Then, $\mathfrak{b}$ is a left ideal in $\mathbf{R}$, $\mathfrak{b} \supseteq \mathfrak{a}$ and $\mathfrak{b}\pi = \mathfrak{b}/\mathfrak{a}$, thus $\mathfrak{J} = \mathfrak{b}/\mathfrak{a}$. The converse is obvious.

(ii): The map $\bullet\varphi : \mathbf{R}/\mathfrak{a} \to \mathfrak{b}/\mathfrak{a}\,\mathfrak{b}$ given by $(x + \mathfrak{a})\,\varphi = x\,\mathfrak{b} + \mathfrak{a}\,\mathfrak{b}$ is an isomorphism. ∎

The following result is a straightforward consequence of Theorems 227 & 241(i):

Corollary 242. *Let* $\mathbf{R}$ *be a ring and* $\mathfrak{a}$ *be a two-sided ideal in* $\mathbf{R}$. *The quotient ring* $\mathbf{R}/\mathfrak{a}$ *is a division ring if, and only if* $\mathfrak{a}$ *is a maximal left (or right) ideal in* $\mathbf{R}$.

2.2.3.3 Quotient Modules

Let $\mathbf{R}$ be a ring, let M be a left $\mathbf{R}$-module and let $\mathcal{R}$ be an equivalence relation in M. According to Definition 75, $\mathcal{R}$ is compatible with the module structure of M if, and only if whenever $x \equiv x' (\operatorname{mod} \mathcal{R})$ and $y \equiv y' (\operatorname{mod} \mathcal{R})$, the two following conditions hold: (i) $x + y \equiv y + y' (\operatorname{mod} \mathcal{R})$; (ii) for any $\lambda \in \mathbf{R}$, $\lambda\,x \equiv \lambda\,x' (\operatorname{mod} \mathcal{R})$. The proof of the following result is similar to that of Lemma 238:

Lemma 243. *A relation $\mathcal{R}$ is compatible with the module structure of the left* $\mathbf{R}$*-module M if, and only if there exists a submodule N of M (written $N \lhd M$) such that $x \equiv x' (\bmod \mathcal{R})$ is equivalent to $x - x' \in N$.*

The proof of the following result is straightforward:

Theorem 244. *(i) The set* $\mathrm{Lat_l}(M)$ *of all submodules of a left* $\mathbf{R}$*-module M, ordered by inclusion, is a complete modular lattice where for any $N, N' \in$* $\mathrm{Lat_l}(M)$*, $N + N' = N \vee N'$ and $N \cap N' = N \wedge N'$.*
(ii) The statements of the isomorphism theorems in Subsect. 1.6.4 still hold, replacing throughout "group" by "module", "normal subgroup" by "submodule", and the multiplicative notation by the additive one.

Without changing the notation, Noether's isomorphism theorems in abelian categories (Theorem 52) are valid (**1.2.4.3**). With the notation in Theorem 244, the second one takes the following form:
Let M be a left $\mathbf{R}$-module and let $S, T \in \mathrm{Lat_l}(M)$. Then

$$(S+T)/T \cong S/(S \cap T), \quad (S+T)/S \cong T/(S \cap T).$$

Let S, T be as above and consider the diamond (1.27), where $S \vee T = S + T$ and $S \wedge T = S \cap T$. As in (**1.6.6.1**), projectivity yields isomorphism and the "diamond lemma" (Lemma 154) can be stated as follows: in the modular lattice $\mathrm{Lat_l}(M)$, projective intervals $[K, H]$ and $[K', H']$ ($K \lhd H$, $K' \lhd H'$) correspond to isomorphic quotients $H/K \cong H'/K'$. In particular:

Corollary 245. *Let $N \lhd M$ and $N' \lhd M$; then projective intervals $[N, M]$ and $[N', M]$ in* $\mathrm{Lat_l}(M)$ *correspond to isomorphic quotients $M/N \cong M/N'$.*

2.2.4 Free Modules and Vector Spaces

2.2.4.1 Direct Sum of Submodules

Lemma 246. *Let $(M_i)_{i \in I}$ be a family of left* $\mathbf{R}$*-modules, let M be a left* $\mathbf{R}$*-module, and let $\bullet f_i : M_i \to M$ be a module-homomorphism for each $i \in I$. There exists a unique* $\mathbf{R}$*-module-homomorphism $\bullet f : \coprod_{i \in I} M_i \to M$ such that for all $i \in I$,*

$$\iota_i \circ f = f_i \tag{2.8}$$

where $\bullet \iota_i : M_i \to \coprod_{i \in I} M_i$ is the canonical injection.

Proof. If f exists, we have by (2.5) $xf = \sum_{i \in I} x\pi_i f_i$ for all $x \in \coprod_{i \in I} M_i$, thus f is unique. Conversely, the above module-homomorphism satisfies (2.8). ∎

Definition 247. *Let M be an* $\mathbf{R}$*-module, let* $(M_i)_{i\in I}$ *be a family of submodules of* M *and let* $f_i : M_i \hookrightarrow M$ *be the inclusion. The module-homomorphism* f *given by (2.8) is called the* canonical homomorphism.

Let $(M_i)_{i\in I}$ be a family of submodules of an $\mathbf{R}$-module E. The *sum* Σ of this family, written $\sum_{i\in I} M_i$, consists of all elements of the form $\sum_{i\in I} x_i$ where $x_i \in M_i$ and $x_i = 0$ for all but a finite set of indices; Σ is a submodule of E.

Definition 248. *The sum* Σ *is said to be* direct *if the canonical homomorphism* $f : \coprod_{i\in I} M_i \to \Sigma$ *is an isomorphism.*

The above direct sum is denoted by $\bigoplus_{i\in I} M_i$. The proof of the following result is easy:

Proposition 249. *Let* $(M_i)_{i\in I}$ *be a family of submodules of an* $\mathbf{R}$*-module* E*. The following conditions are equivalent:*

(i) The sum $\sum_{i\in I} M_i$ *is direct;*
(ii) The relation $\sum_{i\in I} x_i = 0$*, where* $x_i \in M_i$ *for all* $i \in I$*, implies* $x_i = 0$ *for all* i*;*
(iii) For any $j \in I$*,* $M_j \cap \left(\sum_{i\neq j} M_i\right) = 0$*;*
(iv) Any element $x \in \sum_{i\in I} M_i$ *can be expressed in a unique way in the form* $x = \sum_{i\in I} x_i$ *where* $x_i \in M_i$ *for all* $i \in I$*.*

The element $x_i \in M_i$ of Condition (iv) above is called the *component* of x in M_i.

Remark 250. *Let* $(M_i)_{1\leq i\leq n}$ *be a* finite family *of* $\mathbf{R}$*-modules. From Proposition 4, the product* $\prod_{1\leq i\leq n} M_i$ *and the coproduct* $\coprod_{1\leq i\leq n} M_i$ *are easily seen to be isomorphic, thus they are a biproduct* (**1.2.4.1**) *denoted by* $\bigoplus_{1\leq i\leq n} M_i$*. This biproduct is also isomorphic to, thus is identified with, the sum* $S = \sum_{1\leq i\leq n} M_i$ *when the* M_i $(1 \leq i \leq n)$ *are submodules of a module* M*; the sum* S *is direct. Therefore, the ambiguity of the notation* $\bigoplus_{1\leq i\leq n} M_i$ *does not cause confusions in practice.*

Lemma and Definition 148(3) takes the following form: let $(M_i)_{i\in I}$ be a family of $\mathbf{R}$-modules and for each $i \in I$, let $N_i \lhd M_i$. Then $\left(\bigoplus_{i\in I} N_i\right) \lhd \left(\bigoplus_{i\in I} M_i\right)$ and there exists a canonical isomorphism

$$\boxed{\bigoplus_{i\in I} (M_i/N_i) \xrightarrow{\sim} \left(\bigoplus_{i\in I} M_i\right) / \left(\bigoplus_{i\in I} N_i\right)}. \tag{2.9}$$

2.2.4.2 Direct Summands

Definition 251. *Let* E *be an* $\mathbf{R}$*-module and let* $M \lhd E$*;* M *is called a* direct summand *of* E *if there exists* $N \lhd E$ *such that* $E = M \oplus N$*.*

Assuming that M is a direct summand of E, i.e., $E = M \oplus N$, we have by (2.9) $E/M \cong (M/M) \oplus N$, and since $M/M = 0$,

$$\boxed{(M \oplus N)/M \cong N.} \tag{2.10}$$

Definition 252. *An* $\mathbf{R}$*-module* E *is* decomposable *if it has proper direct summands (i.e., direct summands different from* E *and* 0*) and* indecomposable *otherwise.*

2.2.4.3 Bases and Free Modules

Definition 253. *Let* $(e_i)_{i\in I}$ *be a family of elements of a left* $\mathbf{R}$*-module* M*.*

(i) $(e_i)_{i\in I}$ *(or the set* $E = \{e_i : i \in I\}$*) is* free *or* $\mathbf{R}$-linearly independent *if the equality* $\sum_{i\in I} \lambda_i\, e_i = 0$ *(where* $(\lambda_i)_{i\in I}$ *is a family elements of* $\mathbf{R}$ *with finite support) implies* $\lambda_i = 0$ *for all* i*; it is* non-free *or* $\mathbf{R}$-linearly dependent *otherwise.*
(ii) $(e_i)_{i\in I}$ *is a* generating family *of* M *(and* $E = \{e_i : i \in I\}$ *is a* generating set *of* M*, written* $M = [E]_{\mathbf{R}}$*) if every element of* M *can be expressed as a left* $\mathbf{R}$*-linear combination of the* e_i*, i.e., for every* $m \in M$ *there exists a family* $(\lambda_i)_{i\in I}$ *of elements of* $\mathbf{R}$*, with finite support, such that* $m = \sum_{i\in I} \lambda_i\, e_i$*. Then,* M *is said to be generated by the family* $(e_i)_{i\in I}$*.*
(iii) $(e_i)_{i\in I}$ *is a* basis *of* M *if it is both a free and a generating family of all elements of* M*.*
(iv) An $\mathbf{R}$*-module* M *is* free *if it has a basis.*

If $M = [E]_{\mathbf{R}}$ and card(E) is finite, M is said to be *finitely generated* (*f.g.*, for short). The proof of the following is easy and left to the reader:

Lemma 254. *Let* M *be a left* $\mathbf{R}$*-module. The following conditions are equivalent:*

(i) M *is free;*
(ii) there exists a family $(e_i)_{i\in I}$ *such that every element of* M *can be expressed in a unique way as a left* $\mathbf{R}$*-linear combination of the* e_i*;*
(iii) there exist a set of indices I *and a module-isomorphism* ${}_{\mathbf{R}}\mathbf{R}^{(I)} \tilde{\to} M$*.*

Lemma and Definition 255. *Let* $(\varepsilon_i)_{i\in I}$ *be the family of elements of* ${}_{\mathbf{R}}\mathbf{R}^{(I)}$ *defined by* $\varepsilon_k = (1)\,\iota_k$ *where* $\bullet\iota_k : {}_{\mathbf{R}}\mathbf{R} \hookrightarrow {}_{\mathbf{R}}\mathbf{R}^{(I)}$ *is the canonical injection with index* $k \in I$ (**1.2.1.3**)*. The family* $(\varepsilon_i)_{i\in I}$ *is a basis of* ${}_{\mathbf{R}}\mathbf{R}^{(I)}$*, called the* canonical basis.

When this does not generate confusion, the free left module ${}_{\mathbf{R}}\mathbf{R}^{(I)}$ is written $\mathbf{R}^{(I)}$. If a free module M has a finite basis, M is said to be *finite free*. The *canonical basis* of the finite free module $\mathbf{R}^n$ is the basis $(\varepsilon_i)_{1\leq i\leq n}$ such that

$$\varepsilon_i = (0, ..., 0, 1, 0, ..., 0)$$

where the 1 is the ith component.

Let $(e_i)_{i\in I}$ be a basis of a free left $\mathbf{R}$-module M and $m = \sum_{i\in I} \lambda_i e_i$; the element λ_i is called the *component* of $m \in M$ in the basis $(e_i)_{i\in I}$.

Theorem 256. *Let M be an $\mathbf{R}$-module, let Γ be a generating set of M, and let $F \subseteq \Gamma$ be a free subset of M. The set $\mathcal{F}$ of all free sets S such that $F \subseteq S \subseteq \Gamma$ has a maximal element Φ.*

Proof. First, a set $S \subseteq M$ is free if, and only if every finite subset of S is free. Let $(L_i)_{i\in I}$ be a chain of $\mathcal{F}$ ordered by inclusion, let $L = \bigcup_{i\in I} L_i$, and let $a_1, ..., a_n$ be elements of L. There exists $i \in I$ such that $a_1, ..., a_n$ belong to L_i, thus $\{a_1, ..., a_n\}$ is free and L is free, i.e., L is the greatest element element of $(L_i)_{i\in I}$. Therefore, $\mathcal{F}$ is inductively ordered (1.5.3.**2**) and, according to Zorn's lemma (Lemma 107), $\mathcal{F}$ has a maximal element Φ. ∎

Theorem 257. *Let $\mathbf{R}$ be a ring, let M and N be left $\mathbf{R}$-modules, and let $\bullet f : M \to N$ be a homomorphism.*

(i) If $\ker f$ and $\operatorname{im} f$ are finitely generated, so is M.
(ii) If $\ker f \cong \mathbf{R}^n$ and $\operatorname{im} f \cong \mathbf{R}^m$, then $M \cong \mathbf{R}^{n+m}$.

Proof. (i): Let $(a_i)_{1\leq i\leq n}$ be a generating family of $\ker f$ and let $(b_j)_{1\leq j\leq m}$ be a generating family of $\operatorname{im} f$. Let $a_{n+1}, ..., a_{n+m}$ be elements of M such that $a_{n+j} f = b_j$ $(1 \leq j \leq m)$. For any element $x \in M$, there exists a finite family $(\lambda_{n+j})_{1\leq j\leq m}$ of elements of $\mathbf{R}$ such that $xf = \sum_{1\leq j\leq m} \lambda_{n+j} b_j$, thus $xf = \left(\sum_{1\leq j\leq n} \lambda_j a_{n+j}\right) f$. Therefore, $x = \sum_{1\leq j\leq n} \lambda_j a_{n+j} + y$ where $y \in \ker f$. In addition, there exists a finite family $(\lambda_i)_{1\leq i\leq n}$ of elements of $\mathbf{R}$ such that $y = \sum_{1\leq i\leq n} \lambda_i a_i$, thus $x = \sum_{1\leq i\leq n+m} \lambda_i a_i$.

(ii): Let us assume that $(a_i)_{1\leq i\leq n}$ is free in $\ker f$ and $(b_j)_{1\leq j\leq m}$ is free in $\operatorname{im} f$. If $x = 0$, we have $xf = 0$, thus $\lambda_{n+j} = 0$ $(1 \leq j \leq m)$; in addition, we have $y = 0$, thus $\lambda_i = 0$ $(1 \leq i \leq n)$. Therefore, $(a_i)_{1\leq i\leq n+m}$ is a basis of M. ∎

2.2.4.4 Vector Spaces

Definition 258. *If $\mathbf{K}$ is a division ring, a left (resp., right) $\mathbf{K}$-module is called a left (resp., right) $\mathbf{K}$-vector space.*

Lemma 259. *Let $(a_i)_{i\in I}$ be a free family of elements of a $\mathbf{K}$-vector space V; if $b \in V$ does not belong to the subspace W generated by $(a_i)_{i\in I}$, the subset of V consisting of all a_i, $i \in I$, and of b, is free.*

Proof. Assuming that there exists a relation

$$\mu b + \textstyle\sum_{i\in I} \lambda_i a_i = 0, \quad \mu \in \mathbf{K},\ \lambda_i \in \mathbf{K},$$

where $(\lambda_i)_{i\in I}$ has a finite support, either $\mu \neq 0$ and $b \in W$ (by left-dividing the above equality by μ) or $\mu = 0$ and $\sum_{i\in I} \lambda_i a_i = 0$. Both cases are impossible. ■

Corollary and Definition 260. *(1) Let Γ be a generating set of a* $\mathbf{K}$*-vector space V and let $F \subseteq \Gamma$ be a free subset of V; there exists a basis B of V such that $F \subseteq B \subseteq \Gamma$.*
(2) The cardinal of B is called the dimension *of the* $\mathbf{K}$*-vector space V (written* $\dim_{\mathbf{K}}(V)$*, or* $\dim(V)$ *when there is no confusion).*

Proof. (1) By Theorem 256, the set $\mathcal{F}$ of all free sets S such that $F \subseteq S \subseteq \Gamma$ has a maximal element Φ. By Lemma 259, the subspace of V generated by Φ is V, thus Φ is a basis B.

(2) is justified by Theorem 405 below. ■

Taking $F = \varnothing$ and $\Gamma = M$ in Corollary and Definition 260 we obtain the following.

Theorem 261. *Let* $\mathbf{K}$ *be a division ring and let V be a* $\mathbf{K}$*-vector space. Then:*

(i) V is free;
(ii) if $N \lhd V$, N is a direct summand.

2.2.4.5 Matrix of a Homomorphism of Free Modules

Lemma 262. *Let* $\mathbf{R}$ *be a ring, let M be a free left* $\mathbf{R}$*-module with basis $(\alpha_i)_{i\in I}$, let N be a left* $\mathbf{R}$*-module and let $(\eta_i)_{i\in I}$ be a family of elements of N.*

(i) The exists a unique $\mathbf{R}$*-linear homomorphism $\bullet f : M \to N$ such that*

$$\alpha_i f = \eta_i, \quad i \in I.$$

(ii) f is a monomorphism (resp., an epimorphism) if, and only if $(\eta_i)_{i\in I}$ is a free (resp., generating) family of N.

Proof. (i): Let $\mathbf{x} \in M$; then by Lemma 254 the exists a unique family $(\lambda_i)_{i\in I}$ of elements of $\mathbf{R}$, with finite support, such that $\mathbf{x} = \sum_{i\in I} \lambda_i \alpha_i$, thus $\mathbf{x}f = \sum_{i\in I} \lambda_i \eta_i$.

(ii): f is a monomorphism if, and only if $\mathbf{x}f = 0$ implies $\mathbf{x} = 0$, i.e., $(\eta_i)_{i\in I}$ is a free family; f is an epimorphism if, and only if for any element $\mathbf{y}$ of N there exists $\mathbf{x} \in M$ such that $\mathbf{y} = \mathbf{x}f$, i.e., $(\eta_i)_{i\in I}$ is a generating family of N. ■

Item (1) below is an obvious consequence of Lemmas 254 and 262:

Lemma and Definition 263. *(1) Let* $\bullet f : M \rightarrow N$ *be* $\mathbf{R}$*-linear. Assuming that* N *is free with basis* $(\beta_j)_{j\in J}$*, for any* $i \in I$ *there exists a unique family* $(a_{ij})_{j\in J}$ *of* ${}^{I}\mathbf{R}$*, with finite support, such that*

$$\boxed{\alpha_i f = \sum_{j\in J} a_{ij}\,\beta_j .} \tag{2.11}$$

(2) The above matrix $A = (a_{ij})_{(i,j)\in I\times J} \in {}^{I}\mathbf{R}^{(J)}$ *is the called the matrix of the homomorphism* $\bullet f$ *in the bases* $(\alpha_i)_{i\in I}$ *and* $(\beta_j)_{j\in J}$*. (Since the entries* a_{ij} *of* $\mathbf{A}$ *belong to* $\mathbf{R}$*,* A *is called a* matrix over $\mathbf{R}$*.)*

Remark 264. *(i) In the above definition,* M *and* N *are* left *free* $\mathbf{R}$*-modules, thus their elements are represented by* rows *in the bases* $(\alpha_i)_{i\in I}$ *and* $(\beta_j)_{j\in J}$*; in particular,* α_i *and* β_j *are represented by the rows* $\left[(\delta_{ij})_{j\in I}\right]$ *and* $\left[(\delta_{ji})_{i\in J}\right]$*, respectively, where* δ_{ij} *is the Kronecker index. The homomorphism* $\bullet f$ *is represented by the* right-multiplication *by the matrix* $A \in {}^{I}\mathbf{R}^{(J)}$*, written* $\bullet A$*, i.e.,*

$$y = x\,A \tag{2.12}$$

where x *and* y *are the* rows *representing* $\mathbf{x}$ *and* $\mathbf{x}f$*, respectively (see the proof of Lemma 262). Assuming that* I *and* J *are finite, say* $I = \{1, ..., n\}$ *and* $J = \{1, ..., m\}$*, we have* $x \in \mathbf{R}^n$*,* $y \in \mathbf{R}^m$*,* $A \in {}^{n}\mathbf{R}^m$*, and (2.12) can be explicitly written*

$$y_i = \sum_{1\leq j\leq n} x_j\, a_{ij} \quad (1 \leq i \leq m) . \tag{2.13}$$

(ii) If M *and* N *are* right *free* $\mathbf{R}$*-modules (or if* $\mathbf{R}$ *is commutative), as in most textbooks on Algebra, the elements of these modules are represented by* columns *in bases of these modules ([141], Sect. 12.3). Our notation is the one chosen by most (if not all) people working on "*$\mathcal{D}$*-modules" (see, e.g., [320], [181]) for reasons which will become clear in the sequel.*

Let M, N, Q be free left $\mathbf{R}$-modules with bases $(\alpha_i)_{i\in I}$, $(\beta_j)_{j\in J}$ and $(\gamma_k)_{k\in K}$, respectively. Let $\bullet f : M \rightarrow N$ and $\bullet g : N \rightarrow Q$ be $\mathbf{R}$-linear maps. Let $\mathbf{x}$ be an element of M, represented by the row $x \in \mathbf{R}^{(I)}$ in the basis $(\alpha_i)_{i\in I}$, let $\mathbf{y} = \mathbf{x}f \in N$ be represented by the row $y \in \mathbf{R}^{(J)}$ in the basis $(\beta_j)_{j\in J}$, and let $\mathbf{z} = \mathbf{y}g$ be represented by the row $z \in \mathbf{R}^{(K)}$ in the basis $(\gamma_k)_{k\in K}$; in addition, let $A \in {}^{I}\mathbf{R}^{(J)}$ be the matrix representing f in the bases $(\alpha_i)_{i\in I}$ and $(\beta_j)_{j\in J}$, and let $B \in {}^{J}\mathbf{R}^{(K)}$ be the matrix representing g in the bases $(\beta_j)_{j\in J}$ and $(\gamma_k)_{k\in K}$. We have $z = y\,B$, $y = x\,A$, thus $z = x\,A\,B$ where $A\,B \in {}^{I}\mathbf{R}^{(K)}$. Therefore, we have the following.

Proposition 265. *With the above notation,* $f \circ g$ *is represented by the matrix* $A\,B$ *in the bases* $(\alpha_i)_{i\in I}$ *and* $(\gamma_k)_{k\in K}$*.*

For short:

$$\boxed{\mathrm{Mat}\,(f \circ g) = \mathrm{Mat}\,(f)\ \mathrm{Mat}\,(g)\,.} \tag{2.14}$$

2.2.5 Duality

2.2.5.1 Dual of a Module

Let $\mathbf{R}$ be a ring and let M be a left $\mathbf{R}$-module. Its *dual*, written M^*, is the *right* $\mathbf{R}$-module $\mathrm{Hom}_{\mathbf{R}}\,(M, {}_{\mathbf{R}}\mathbf{R}_{\mathbf{R}})$ (Theorem 236(i)). For any $x^* \in M^*$ and $x \in M$, the element $x^*\,(x)$ is denoted by $\langle x, x^* \rangle$. The $\mathbf{R}$-bilinear map $\langle \bullet, \bullet \rangle : M \times M^* \to {}_{\mathbf{R}}\mathbf{R}_{\mathbf{R}}$ is called the *duality bracket*. Do not confuse these algebraic notions with those in Example 36, where topology and continuity are involved.

If $\mathbf{R}$ is noncommutative, M^* cannot be endowed with a structure of left $\mathbf{R}$-module: if we define $\lambda\, x^*$ by $\langle x, \lambda\, x^* \rangle = \lambda\, \langle x, x^* \rangle$, then the reader can easily check that $\lambda\, x^*$ is *not* $\mathbf{R}$-linear, thus does not belong to M^*.

2.2.5.2 Transpose of a Homomorphism

Let $\mathbf{R}$ be a ring, let M and N be two left $\mathbf{R}$-modules, and let $\bullet f : M \to N$ be a homomorphism. For any $\bullet y^* \in N^*$, $f \circ y^* \in M^*$, thus one can define a homomorphism ${}^t f : N^* \to M^*$ by ${}^t f\,(y^*) = f \circ y^*$; ${}^t f$ is called the *transpose* of f (and should not be confused with the transpose, or adjoint, of a linear continuous map in Example 36(2)). Using the duality bracket, we can write $\langle (x)\, f, y^* \rangle = \langle x, {}^t f\,(y^*) \rangle$ or more simply

$$\boxed{\langle x f, y^* \rangle = \langle x, {}^t f y^* \rangle} \tag{2.15}$$

for all $x \in M$, $y^* \in N^*$.

The transposition $\bullet f \ \mapsto\ {}^t f \bullet$ is $\mathbf{R}$-linear, i.e., additive and such that ${}^t\,(f\,\lambda) = ({}^t f)\,\lambda$ for any $f \in \mathrm{Hom}_{\mathbf{R}}\,(M, N)$ and $\lambda \in \mathbf{R}$. In addition, if $\bullet f : M \to N$ and $\bullet g : N \to Q$ are homomorphisms, we have, for any $x \in M$ and $z \in Q^*$,

$$\langle x f \circ g, z^* \rangle = \left\langle x f, {}^t g z^* \right\rangle = \left\langle x, {}^t f \circ {}^t g z^* \right\rangle, \text{ therefore}$$

$$\boxed{{}^t\,(f \circ g) = {}^t f \circ {}^t g.} \tag{2.16}$$

The above formula may look unfamiliar. The skew formula ${}^t\,(f \circ g) = {}^t g \circ {}^t f$ holds true when both a linear map *and its transpose* are *right* (or left) ones. Here the identity (2.16) holds true because if $f : M \to N$ is a *left* (resp., *right*) linear map, i.e., if M and N are *left* (resp., *right*) $\mathbf{R}$-modules, then the transpose ${}^t f : N^* \to M^*$ is a *right* (resp., *left*) linear map since the duals N^* and M^* are *right* (resp., *left*) $\mathbf{R}$-modules.

2.2.5.3 Dual of a Free Module

Let M be a *finite free* left $\mathbf{R}$-module with basis $(\alpha_i)_{1\leq i\leq n}$. Define $\alpha_i^* \in M^*$ by $\langle \alpha_j, \alpha_i^* \rangle = \delta_{ij}$ $(1 \leq i \leq n, 1 \leq j \leq n)$.

Theorem 266. *M^* is a free right $\mathbf{R}$-module with basis $(\alpha_i^*)_{1\leq i\leq n}$.*

Proof. For any $x^* \in M^*$, let $\theta(x^*) \in {}^n\mathbf{R}$ be the column with entries $\langle \alpha_i, x^* \rangle$ $(1 \leq i \leq n)$. By Lemma 262, $\theta : M^* \to {}^n\mathbf{R}$ is a bijection. It is obviously $\mathbb{Z}$-linear and $\theta(x^* \lambda) = \theta(x^*)\lambda$ for any $\lambda \in \mathbf{R}$, thus θ is an isomorphism of right $\mathbf{R}$-modules. Finally, $\theta(\alpha_i^*) = \varepsilon_i^*$ where $(\varepsilon_i^*)_{1\leq i\leq n}$ is the canonical basis of ${}^n\mathbf{R}$, i.e., ε_i^* is the column whose all entries are zero except the jth one which is 1. ∎

Definition 267. *$(\alpha_i^*)_{1\leq i\leq n}$ is the dual basis of $(\alpha_i)_{1\leq i\leq n}$.*

Using this notion, we see by (2.11) that $A = (a_{ij})$ is the matrix of $f : M \to N$ if, and only if (2.2.4.**3**)

$$\boxed{\langle \alpha_i f, \beta_j^* \rangle = a_{ij} \quad (1 \leq i \leq n, 1 \leq j \leq n).} \tag{2.17}$$

The following result is an obvious consequence of Theorem 266:

Corollary 268. *Let $\mathbf{R}$ be a ring; $(\mathbf{R}^n)^* \cong {}^n\mathbf{R}$ and $({}^n\mathbf{R})^* \cong \mathbf{R}^n$.*

Note that the first (resp., second) isomorphism is an isomorphism of right (resp., left) $\mathbf{R}$-modules.

2.2.5.4 Transpose of a Matrix

One way to define the transpose of a matrix is the following: let M and N be finite free left $\mathbf{R}$-modules, the first one with basis $(\alpha_i)_{1\leq i\leq n}$ and the second one with basis $(\beta_j)_{1\leq j\leq m}$, let $\bullet f : M \to N$ be a homomorphism and let $A = (a_{ij}) \in {}^n\mathbf{R}^m$ be its matrix in the above bases, i.e., $\alpha_i f = \sum_{1\leq j\leq m} a_{ij}\, \beta_j$ according to Lemma and Definition 263. The following definition sounds classical and reasonable:

Definition 269. *The* abstract transpose *of the matrix A is the matrix representing ${}^t f$ in the dual bases $(\beta_j^*)_{1\leq j\leq m}$ and $(\alpha_i^*)_{1\leq i\leq n}$ of N^* and M^*.*

In accordance with Definition 263, the transpose of A is the matrix (a'_{ij}) belonging to ${}^n\mathbf{R}^m$ such that ${}^t f(\beta_j^*) = \sum_{1\leq i\leq n} a'_{ij}\alpha_i^*$, thus $\langle \alpha_i, {}^t f(\beta_j^*) \rangle = a'_{ij}$. By (2.17), $a'_{ij} = a_{ij}$ (due to the notation in Definition 233), which makes the "abstract transpose" of a matrix in Definition 269 an irrelevant notion.

Remark 270. *(i) Let M and N be free left $\mathbf{R}$-modules with bases $(\alpha_j)_{1\leq j\leq n}$ and $(\beta_i)_{1\leq i\leq m}$ respectively, and let $A \in {}^n\mathbf{R}^m$ be the matrix representing*

$f : M \rightarrow N$ in those bases when the elements of M and N are represented by rows *(so that $\bullet f$ is identified with $\bullet A$). Representing the elements of the dual free right $\mathbf{R}$-modules N^* and M^* by* columns *in the dual bases $(\beta_i^*)_{1 \leq i \leq m}$ and $\left(\alpha_j^*\right)_{1 \leq j \leq n}$, ${}^t f : N^* \rightarrow M^*$ is identified with the* left-multiplication *by the matrix A (written $A\bullet$). The equality (2.15) takes the very attractive form*

$$\boxed{\langle x\, A, y^* \rangle = \langle x, A\, y^* \rangle\,.} \tag{2.18}$$

(ii) In most references, e.g., [77], the transpose of a matrix $A = (a_{ij})$ is defined as the matrix (a_{ji}). Assume that A represents a homomorphism of right free $\mathbf{R}$-modules $f : N \rightarrow M$ in bases of N and M (Remark 264(ii)); then ${}^t f : M^* \rightarrow N^*$ is a homomorphism of left free $\mathbf{R}$-modules which can be viewed as a homomorphism of right free $\mathbf{R}^{\mathrm{op}}$-modules by Proposition 229, where $\mathbf{R}^{\mathrm{op}}$ is the opposite ring. *Then ${}^t f$ is represented in the dual bases by the matrix $A^T = (a_{ji})$, defined over $\mathbf{R}^{\mathrm{op}}$. In particular, a product $(A\,B)^T$ must be calculated using the rule (2.4) ([141], §16, Theorem 4).*

According to Remark 270(ii), the following definition is useful:

Definition 271. *The* classical transpose *(or the* transpose, *for short) of a matrix $A = (a_{ij}) \in {}^n\mathbf{R}^m$ is the matrix $A^T = (a_{ji}) \in {}^n\left(\mathbf{R}^{\mathrm{op}}\right)^m$.*

2.2.5.5 Bidual

Let $\mathbf{R}$ be a ring, let M be a left $\mathbf{R}$-module, and let M^* be its dual. The *bidual* of M, denoted by M^{**}, is the left $\mathbf{R}$-module $\mathrm{Hom}_{\mathbf{R}}\left(M^*, {}_{\mathbf{R}}\mathbf{R}_{\mathbf{R}}\right)$ (Theorem 236(i)). For any $x \in M$, let $\langle x, \bullet \rangle : M^* \rightarrow {}_{\mathbf{R}}\mathbf{R}_{\mathbf{R}}$ be given by $\langle x, \bullet \rangle = \langle x, x^* \rangle$. As easily seen, $\langle x, \bullet \rangle \in M^{**}$ and the map $M \rightarrow M^{**} : x \mapsto \langle x, \bullet \rangle$ is $\mathbf{R}$-linear.

Definition 272. *The above $\mathbf{R}$-linear map $x \mapsto \langle x, \bullet \rangle$ is the* canonical map *$M \rightarrow M^{**}$.*

Theorem 273. *If the module M is finite free, the canonical map $M \rightarrow M^{**}$ is an isomorphism of left $\mathbf{R}$-modules.*

Proof. Let $(\alpha_i)_{1 \leq i \leq n}$ be a basis of M and let $(\alpha_i^*)_{1 \leq i \leq n}$ be its dual basis in M^*. Set $\beta_i = \alpha_i \psi$ $(1 \leq i \leq n)$ where $\bullet\psi$ is the canonical map; it is easy to check that $(\beta_i)_{1 \leq i \leq n}$ is the dual basis of $(\alpha_i^*)_{1 \leq i \leq n}$ in M^{**}. ■

More specifically, let $\mathcal{C}$ be the category of finite free left $\mathbf{R}$-modules and let $t_M : M \rightarrow M^{**}$ be defined as $x \mapsto \langle x, \bullet \rangle$; consider the "identity functor" $\mathfrak{F} = I_d$ defined by $I_d(M) = M$ for any $M \in \mathrm{Ob}(\mathcal{C})$ and $I_d(f) = f$ for any morphism $f : M \rightarrow N$; let $\mathfrak{G}$ be the functor $(\bullet)^{**}$ defined by $\mathfrak{G}(M) = M^{**}$ for any $M \in \mathrm{Ob}(\mathcal{C})$ and $\mathfrak{G}(f) = {}^{tt}f$–the bitranspose of f–for any $f : M \rightarrow N$. Then t is a natural transformation (**1.2.3.2**). By Theorem 273, we have the following.

Corollary 274. *The natural transformation t is a natural isomorphism.*

2.2.5.6 Orthogonal

Let $\mathbf{R}$ be a ring, let M be a left $\mathbf{R}$-module and let S be a subset of M. The *orthogonal* of S, denoted by $S^{\mathbf{0}}$, is the subset of M^* consisting of all linear forms x^* such that $\langle x, x^* \rangle = 0$ for all $x \in S$; then $S^{\mathbf{0}}$ is a submodule of the right $\mathbf{R}$-module M^*.

Remark 275. *Let $N \lhd M$.*

(i) Obviously, $N \subseteq N^{\mathbf{00}}$, but one can have $N \neq N^{\mathbf{00}}$ even if M is finite free ([27], §II.2, Exercise 9(b)).
(ii) However, if $\mathbf{K}$ is a division ring, then $N = N^{\mathbf{00}}$ even if N is not finite-dimensional ([27], §II.7, Theorem 7).
(iii) The map $(\bullet)^{\mathbf{0}}$ from the submodules of M to those of M^* and conversely, is a Galois connection *(Definition 98), thus the properties in Proposition 99 hold true.*

Proposition 276. *Let M, N be left $\mathbf{R}$-modules and let $\bullet f : M \to N$ be $\mathbf{R}$-linear. Then $(\operatorname{im} f)^{\mathbf{0}} = \ker ({}^t f)$.*

Proof. By (2.15), $(\operatorname{im} f)^{\mathbf{0}}$ consists of all $y^* \in N^*$ such that $\langle x, {}^t f(y^*) \rangle = 0$ for all $x \in M$; thus $y^* \in (\operatorname{im} f)^{\mathbf{0}}$ if, and only if ${}^t f(y^*) = 0$, i.e., $y^* \in \ker ({}^t f)$. ■

Corollary 277. *Let $\bullet f : M \to N$ be $\mathbf{R}$-linear.*

(1) Of the following, (i)⇒(ii)⇒(iii):
(i) f is left-invertible;
(ii) f is an epimorphism;
(iii) ${}^t f$ is a monomorphism.
(2) In addition:
(i) f is left- (resp., right-) invertible if, and only if ${}^t f$ is left- (resp., right-) invertible;
(ii) if f an isomorphism with inverse $g = f^{-1}$, then ${}^t f$ is an isomorphism with inverse ${}^t g$.

Proof. (1) (i)⇒(ii): If there exists $\bullet g : N \to M$ such that $gf = Id_N$, let $y \in N$ and $x = yg$; then $y = xgf = xf$. (ii)⇒(iii): If $\operatorname{im} f = N$, then $\ker ({}^t f) = (\operatorname{im} f)^{\mathbf{0}} = 0$.

Both statements of (2) are obvious consequences of (2.16). ■

Remark 278. *Of the conditions in Corollary 277(1), (ii)⇏(i) in general, as shown by Proposition 572 below. Regarding Condition (2)(ii), there exists an $\mathbf{R}$-linear map $f : M \to N$ such that ${}^t f$ is an isomorphism and f is neither injective nor surjective ([27], §II.2, Exercise 10); this also proves that of the conditions in (1), (iii)⇏(ii).*

Corollary 279. *Let M and N be finite free left* $\mathbf{R}$*-modules, let* $\bullet f : M \to N$ *be* $\mathbf{R}$*-linear and let S be the matrix of f in bases of M and N, so that* $\bullet f = \bullet S$ *in these bases. Let* ${}^t f : N^* \to M^*$ *be identified with* $S\bullet$ *in the dual bases (Remark 270(i)). Of the following,*

$$(i) \Leftrightarrow (ii) \Leftrightarrow (iii) \Rightarrow (iv) \Rightarrow (v) \Leftrightarrow (vi) \Leftrightarrow (vii):$$

(i) S is left-invertible;
(ii) $S\bullet$ is left-invertible;
(iii) $\bullet S$ is left-invertible;
(iv) $\bullet S$ is surjective;
(v) $S\bullet$ is injective;
(vi) the columns of S are right $\mathbf{R}$*-linearly independent;*
(vii) S is right-regular.
If $\mathbf{R}$ *is a division ring, then all the above conditions are equivalent.*
If $\mathbf{R}$ *is an arbitrary ring, then (v)$\nRightarrow$(iv)$\nRightarrow$(iii).*
Considering right modules, the above holds, interchanging respectively left *and* right, columns *and* rows, $\bullet S$ *and* $S\bullet$.

Proof. (i)$\Leftrightarrow$(ii)$\Leftrightarrow$(iii) is clear from the equalities $(TS)\bullet = (T\bullet) \circ (S\bullet)$ and $\bullet(TS) = (\bullet T) \circ (\bullet S)$.

(iii)$\Rightarrow$(iv)$\Rightarrow$(v) by Corollary 277(1).

(vii) means that for any column y of appropriate length, $S\,y = 0$ implies $y = 0$, thus (vii)$\Leftrightarrow$(vi)$\Leftrightarrow$(v).

If $\mathbf{R}$ is a division ring, then (vii)$\Rightarrow$(i) (see Proposition 420 below).

If $\mathbf{R}$ is arbitrary, (iv)$\nRightarrow$(iii) by Proposition 572 below and (v)$\nRightarrow$(iv) by Remark 278. ∎

2.2.6 Change of Basis

2.2.6.1 The Change of Basis Matrix

Let M be a finite free left $\mathbf{R}$-module, and let $(\alpha_i)_{1 \le i \le n}$, $(\beta_i)_{1 \le i \le n}$ be two bases of M, with the same cardinal. Let us identify these bases with the *columns* α and β with entries α_i and β_i, respectively $(1 \le i \le n)$. Consider $f = Id_M : (M, \beta) \to (M, \alpha)$ and let $P = (p_{ij})$ be the matrix (necessarily invertible, i.e., belonging to $\mathrm{GL}_n(\mathbf{R})$) of f. For each index i, $\beta_i f = \alpha_i$, thus by Lemma and Definition 263, $\beta_i = \sum_{1 \le i \le n} p_{ij}\, \alpha_i$. This can also be written

$$\boxed{\beta = P\,\alpha.} \tag{2.19}$$

Let $\mathbf{x} \in M$, and let x_α, x_β be the *rows* representing $\mathbf{x}$ in the bases α and β, respectively. Then, $\mathbf{x} = x_\alpha\,\alpha = x_\beta\,\beta$, therefore by (2.19), $x_\beta\,P\,\alpha = x_\alpha\,\alpha$, and

$$\boxed{x_\alpha = x_\beta\,P.} \tag{2.20}$$

2.2.6.2 Change of Basis in the Dual

Let M^* be the dual of M, and let $(\alpha_i^*)_{1\leq i\leq n}$, $(\beta_i^*)_{1\leq i\leq n}$ be the dual bases of $(\alpha_i)_{1\leq i\leq n}$ and $(\beta_i)_{1\leq i\leq n}$, respectively. Since M^* is a right $\mathbf{R}$-module, let us identify $(\alpha_i^*)_{1\leq i\leq n}$ and $(\beta_i^*)_{1\leq i\leq n}$ with the *rows* α^* and β^* with entries α_i^* and β_i^*, respectively $(1\leq i\leq n)$. The rows α^* and β^* can also be considered as morphisms of left $\mathbf{R}$-modules $M\to M^n$. Let $\mathbf{x}=x_\alpha\,\alpha=x_\beta\,\beta\in M$; we have $(\mathbf{x})\,\alpha^*=x_\alpha$ and $(\mathbf{x})\,\beta^*=x_\beta$. By (2.20), $(\mathbf{x})\,\alpha^*=(\mathbf{x})\,\beta^*\,P$, therefore

$$\boxed{\alpha^*=\beta^*\,P.} \tag{2.21}$$

Let $\mathbf{x}^*\in M^*$, and let x_α^*, x_β^* be the *columns* representing $\mathbf{x}^*$ in the bases α^* and β^*, respectively. Then, $\mathbf{x}^*=\alpha^*\,x_\alpha^*=\beta^*\,x_\beta^*$, therefore by (2.21), $\beta^*\,P\,x_\alpha^*=\beta^*\,x_\beta^*$, and

$$\boxed{x_\beta^*=P\,x_\alpha^*.} \tag{2.22}$$

Note the similarity between (2.19) and (2.22), and between (2.20) and (2.21).

2.2.6.3 Effect on the Matrix of a Homomorphism

Let M,N be two finite free left $\mathbf{R}$-modules and $f:M\to N$. Assume that M has bases $\alpha=(\alpha_i)_{1\leq i\leq n}$, $\alpha'=(\alpha_i')_{1\leq i\leq n}$, and that N has bases $\beta=(\beta_i)_{1\leq i\leq m}$, $\beta'=(\beta_i')_{1\leq i\leq m}$. Let A (resp., A') be the matrix of f in the bases α, β (resp., α', β'). Let $\mathbf{x}\in M$, $\mathbf{y}=\mathbf{x}\,f\in N$, and let x, y (resp., x', y') be the rows representing $\mathbf{x}$, $\mathbf{y}$ in the bases α, β (resp., α', β'). We have $y=x\,A$, $y'=x'\,A'$. Let P (resp., Q) be the change of basis matrix from α to α' (resp., β to β'). By (2.20), $x=x'\,P$, $\quad y=y'\,Q$. Therefore, $y'=y\,Q^{-1}=x\,A\,Q^{-1}=x'\,P\,A\,Q^{-1}$, thus

$$\boxed{A'=P\,A\,Q^{-1}}. \tag{2.23}$$

Definition 280. *(i) The matrices A, $A'\in{}^n\mathbf{R}^m$ are said to be* equivalent *(written $A\equiv A'$) if there exist matrices $P\in\mathrm{GL}_n(\mathbf{R})$ and $Q\in\mathrm{GL}_m(\mathbf{R})$ such that (2.23) holds.*
(ii) If $Q=I_m$ (resp., $P=I_n$), then A and A' are said to be left- (resp., right-) equivalent (written $A\equiv_l A'$ (resp., $A\equiv_r A'$).
(iii) If $n=m$, A, $A'\in\mathrm{Mat}_n(\mathbf{R})$ are said to be conjugate *or* similar-square *(written $A\sim A'$) if there exists a matrix $P\in\mathrm{GL}_n(\mathbf{R})$ such that $A'=P\,A\,P^{-1}$, i.e., if A and A' represent the same* endomorphism $f:M\to M$.

2.2.7 Tensor Products

2.2.7.1 General Definition

Let $\mathbf{R}$ be a ring, let A be a right $\mathbf{R}$-module, let B be a left $\mathbf{R}$-module and let G be an abelian group.

Definition 281. *A right function $f : A \times B \to G$ is said to be $\mathbf{R}$-biadditive if (i) it is $\mathbb{Z}$-bilinear and (ii) for any $(a,b) \in A \times B$ and $r \in \mathbf{R}$, $f(a\,r,b) = f(a, r\,b)$.*

Consider the following diagram:

$$\begin{array}{ccc} A \times B & \xrightarrow{h} & C \\ f \searrow & & \swarrow f' \\ & G & \end{array}$$

The problem to be solved here is to find an abelian group C and an $\mathbf{R}$-biadditive map $h : A \times B \to C$ such that for any abelian G and every $\mathbf{R}$-biadditive map $f : A \times B \to G$, there exists a unique $\mathbb{Z}$-linear map $f' : C \to G$ making the diagram commute. This is a universal problem (**1.2.3.3**); therefore, assuming that two abelian groups C_1 and C_2 are solutions to this problem, they are isomorphic (Theorem 13(ii)) and are identified. One can prove that a solution to this problem actually exists ([306], Theorem 1.4); it is called the *tensor product* of A and B over $\mathbf{R}$ and denoted by $A \bigotimes_{\mathbf{R}} B$. The $\mathbf{R}$-biadditive map $h : A \times B \to A \bigotimes_{\mathbf{R}} B$ is denoted by $\otimes$. A typical element of $A \bigotimes_{\mathbf{R}} B$ has the form $\sum_i a_i \otimes b_i$, $a_i \in A$, $b_i \in B$, where the family $(a_i \otimes b_i)$ has a finite support.

The proof of Item (1) below is easy:

Lemma and Definition 282. *(1) Let $f\bullet : A \to A'$ be a homomorphism of* right $\mathbf{R}$*-modules and $\bullet g : B \to B'$ be a homomorphism of* left $\mathbf{R}$*-modules. There exists a unique additive right map $h : A \bigotimes_{\mathbf{R}} B \to A' \bigotimes_{\mathbf{R}} B'$ with $h(a \otimes b) \mapsto fa \otimes bg$.*
(2) The above map is denoted by $f \otimes g$ and is called the tensor product *of f and g. The tensor product $Id_A \otimes g$ is usually denoted by $1 \otimes g$ when there is no confusion.*

To express the general structure of a tensor product, it is convenient to consider a left $\mathbf{R}$-module as an $(\mathbf{R}, \mathbb{Z})$-bimodule and a right $\mathbf{T}$-module as a $(\mathbb{Z}, \mathbf{T})$-bimodule (**2.2.3.1**). Then all situations are gathered in the following result (the proof of which is straightforward):

Theorem 283. *If A is an $(\mathbf{R}, \mathbf{S})$-bimodule (written ${}_{\mathbf{R}}A_{\mathbf{S}}$) and B is an $(\mathbf{S}, \mathbf{T})$-bimodule (written ${}_{\mathbf{S}}B_{\mathbf{T}}$), then $A \bigotimes_{\mathbf{S}} B$ is an $(\mathbf{R}, \mathbf{T})$-bimodule and the following identities hold (with an obvious notation)*

$$r\,(a \otimes b)\,t = (r\,a) \otimes (b\,t)\,, \quad a \otimes (s\,b) = (a\,s) \otimes b,$$

the last equality resulting from the $\mathbf{S}$*-biadditivity. Considering also bimodules* ${}_{\mathbf{R}}A'_{\mathbf{S}}$, ${}_{\mathbf{S}}B'_{\mathbf{T}}$, *an* $(\mathbf{R},\mathbf{S})$*-bimodule homomorphism* $f : {}_{\mathbf{R}}A_{\mathbf{S}} \to {}_{\mathbf{R}}A'_{\mathbf{S}}$ *and an* $(\mathbf{S},\mathbf{T})$*-bimodule homomorphism* $g : {}_{\mathbf{S}}B_{\mathbf{T}} \to {}_{\mathbf{S}}B'_{\mathbf{T}}$, *then* $f \otimes g$ *is an* $(\mathbf{R},\mathbf{T})$*-bimodule homomorphism.*

Let us state the *Adjoint Isomorphism Theorem:*

Theorem 284. *Let* $\mathbf{R},\mathbf{S},\mathbf{T},\mathbf{U}$ *be rings.*

(i) In the situation ${}_{\mathbf{R}}A_{\mathbf{S}}$, ${}_{\mathbf{S}}B_{\mathbf{T}}$ *and* ${}_{\mathbf{R}}C_{\mathbf{U}}$, *there exists a canonical* $(\mathbf{T},\mathbf{U})$*-bimodule isomorphism*

$$\tau : \mathrm{Hom}_{\mathbf{R}}\left(A \bigotimes\nolimits_{\mathbf{S}} B, C\right) \xrightarrow{\sim} \mathrm{Hom}_{\mathbf{S}}\left(B, \mathrm{Hom}_{\mathbf{R}}\left(A, C\right)\right).$$

(ii) In the situation ${}_{\mathbf{R}}A_{\mathbf{S}}$, ${}_{\mathbf{S}}B_{\mathbf{T}}$ *and* ${}_{\mathbf{U}}C_{\mathbf{T}}$, *there exists a canonical* $(\mathbf{U},\mathbf{R})$*-bimodule isomorphism*

$$\tau' : \mathrm{Hom}_{\mathbf{T}}\left(A \bigotimes\nolimits_{\mathbf{S}} B, C\right) \xrightarrow{\sim} \mathrm{Hom}_{\mathbf{S}}\left(A, \mathrm{Hom}_{\mathbf{T}}\left(B, C\right)\right)$$

Proof. Let us prove (i) (the proof of (ii) is similar). We know that $A \bigotimes_{\mathbf{S}} B$ is an $(\mathbf{R},\mathbf{T})$-bimodule (Theorem 283) and that $\mathrm{Hom}_{\mathbf{R}}(A,C)$ is an $(\mathbf{S},\mathbf{U})$-bimodule (Theorem 236(i)), thus both sides make sense and are $(\mathbf{T},\mathbf{U})$-bimodules (Theorem 236(i)). If $\bullet f : A \bigotimes_{\mathbf{S}} B \to C$ is $\mathbf{R}$-linear, for each $b \in B$ define $\bullet f_b : A \to C$ by $a f_b = (a \otimes b) f$; f_b is $\mathbf{R}$-linear, thus consider $\bullet \bar{f} : B \to \mathrm{Hom}_{\mathbf{R}}(A,C)$ such that $b\bar{f} = f_b$. The map $\bar{f}$ is additive and for any $s \in \mathbf{S}$,

$$(a)(s\,b)\bar{f} = (a \otimes s\,b)\,f = (a\,s \otimes b)\,f = (a\,s)\,f_b = a\,(s f_b)$$

(the last equality by Theorem 236(i)), thus $(s\,b)\bar{f} = s\left(b\bar{f}\right)$, and $\bar{f}$ is $\mathbf{S}$-linear, i.e., $\bar{f} \in \mathrm{Hom}_{\mathbf{S}}(B, \mathrm{Hom}_{\mathbf{R}}(A,C))$. Since $(a)(b)\bar{f} = (a \otimes b) f$, the map $\tau : f \mapsto \bar{f}$ is easily seen to be a $(\mathbf{T},\mathbf{U})$-bimodule homomorphism. To prove that it is an isomorphism, it remains to show that it has an inverse. Let $\bullet g : B \to \mathrm{Hom}_{\mathbf{R}}(A,C)$ be $\mathbf{S}$-linear; the map $(a,b) \mapsto (a)(b)\,g$ is $\mathbf{S}$-biadditive since for any $s \in \mathbf{S}$, $(a)(s\,b)\,g = (a\,s)(b)\,g$, therefore there exists $\bullet \tilde{g} \in \mathrm{Hom}_{\mathbf{R}}\left(A \bigotimes_{\mathbf{S}} B, C\right)$ such that $(a)(b)\,g = (a \otimes b)\,\tilde{g}$. ∎

2.2.7.2 Extension of the Ring of Scalars

Item (1) below is a consequence of Theorem 283:

Lemma and Definition 285. *(1) Let* $\mathbf{R}$, $\mathbf{S}$ *be two rings such that* $\mathbf{S}$ *is a right* $\mathbf{R}$*-module, and let* M *be a left* $\mathbf{R}$*-module (i.e., consider the situation* $\mathbf{S}_{\mathbf{R}}$, ${}_{\mathbf{R}}M$*). Then* $\mathbf{S} \bigotimes_{\mathbf{R}} M$ *is a left* $\mathbf{S}$*-module.*
(2) The above left $\mathbf{S}$*-module* $\mathbf{S} \bigotimes_{\mathbf{R}} M$ *is said to be obtained from* M *by* extension of the ring of scalars *and is denoted by* $M_{(\mathbf{S})}$. *The* $\mathbf{R}$*-linear map* $M \to \mathbf{S} \bigotimes_{\mathbf{R}} M$ *given by* $x \mapsto 1 \otimes x$ *is called the* canonical map.

This definition is generalized in Subsect. 3.4.2 below.

2.3 Generalized Differential Rings

2.3.1 Differential Rings

A *differential ring* is a ring $\mathbf{R}$ endowed with a derivation $\boldsymbol{\delta} : a \mapsto a^{\boldsymbol{\delta}}$ $(a \in \mathbf{R})$, i.e., an additive map from $\mathbf{R}$ to $\mathbf{R}$ such that $(a\,b)^{\boldsymbol{\delta}} = a^{\boldsymbol{\delta}}\,b + a\,b^{\boldsymbol{\delta}}$ for all $a, b \in \mathbf{R}$.

The ring $\mathbf{k}[t]$ of all polynomials in the indeterminate t with coefficients in $\mathbf{k} = \mathbb{R}$ or $\mathbb{C}$, endowed with the usual derivative $\boldsymbol{\delta} = d/dt$, is a typical example of commutative differential ring. The ring $\mathrm{Mat}_n(\mathbf{k}[t])$ of all square matrices of order n with entries in $\mathbf{k}[t]$ is a noncommutative differential ring for $n > 1$.

A *constant* in a differential ring is an element a such that $a^{\boldsymbol{\delta}} = 0$. The constants of a differential ring form a subring, called the *subring of constants.* For example, the subring of constants of $\mathbf{k}[t]$ is $\mathbf{k}$.

A *differential division ring* (resp., a *differential field*) is a differential ring which is a division ring (resp., a field).

The following rings (or fields), equipped with the usual derivative, are typical examples of differential rings (or fields):

- The ring $\mathbf{k}[t]$.
- The field $\mathbf{k}(t)$ of all rational functions with coefficients in $\mathbf{k} = \mathbb{R}$ or $\mathbb{C}$.
- The ring $\mathbf{R} = \mathbf{k}\{t\}$ of *convergent power series* in the variable t with coefficients in $\mathbf{k}$ (2.9.1.1).
- The field $\mathbf{k}(\{t\})$ of fractions of $\mathbf{k}\{t\}$, i.e., the field consisting of all elements f/g, $f \in \mathbf{k}\{t\}$, $g \in \mathbf{k}\{t\}^{\times}$.
- The ring $\mathfrak{E}(\mathbf{k})$ of all *entire functions* in $\mathbf{k}$, i.e., of all power series $\sum_{n \geq 0} a_n t^n$ with infinite convergence radius. (Recall that when $\mathbf{k} = \mathbb{C}$, $\mathfrak{E}(\mathbf{k}) = \mathcal{O}(\mathbf{k})$; more generally, the space $\mathcal{O}(\mathbf{k}^n)$ of all analytic functions in $\mathbf{k}^n$ coincides with $\mathfrak{E}(\mathbf{k}^n)$ when $\mathbf{k} = \mathbb{C}$. This does not hold when $\mathbf{k} = \mathbb{R}$: see [92], (9.9.6).)
- The field $\mathcal{M}(\mathbf{k})$ of meromorphic functions on $\mathbf{k}$, i.e., the field of fractions of $\mathfrak{E}(\mathbf{k})$.

Note that $\mathfrak{E}(\mathbf{k}) \subsetneq \mathbf{k}\{t\}$ and $\mathcal{M}(\mathbf{k}) \subsetneq \mathbf{k}(\{t\})$.

2.3.2 Difference Rings

A *difference ring* is a ring $\mathbf{R}$ endowed with a *shift operator* $\boldsymbol{\alpha} : a \mapsto a^{\boldsymbol{\alpha}}$ $(a \in \mathbf{R})$ which is an endomorphism of $\mathbf{R}$. A constant in a difference ring is an element a such that $a^{\boldsymbol{\alpha}} = a$ and the set of all constants of $\mathbf{R}$ is a subring, called the *subring of constants.* A difference ring which is a field is called a *difference field*.

The following rings (or fields), equipped with the *shift-forward operator* $a(t) \mapsto a(t+1)$, are typical examples of difference rings (or fields), and $\boldsymbol{\alpha}$ is an automorphism of these difference rings (or fields), with inverse $a(t) \mapsto a(t-1)$.

- The ring $\mathbf{k}[t]$.
- The field $\mathbf{k}(t)$.
- The ring $\mathfrak{E}(\mathbf{k})$.
- The field $\mathcal{M}(\mathbf{k})$.

Let $\boldsymbol{\alpha}$ be the above shift and let $N \in \mathbb{N}^{\times}$. A function f is said to be N-periodic ($N \in \mathbb{N}^{\times}$) if $f^{\boldsymbol{\alpha}^N} = f$.

The subring of $\mathfrak{E}(\mathbf{k})$ consisting of all N-periodic entire functions is denoted by $\mathcal{P}\mathfrak{E}_N(\mathbf{k})$ and the subfield of $\mathcal{M}(\mathbf{k})$ consisting of all N-periodic meromorphic functions is denoted by $\mathcal{PM}_N(\mathbf{k})$. The rings $\mathcal{P}\mathfrak{E}_N(\mathbf{k})$ and $\mathcal{PM}_N(\mathbf{k})$, endowed with the shift $\boldsymbol{\alpha}$, are a difference ring and a difference field, respectively, and for $N = 1$ they are rings of constants.

Definition 286. *A difference ring consisting of N-periodic functions is called an N-periodic difference ring.*

2.3.3 General Derivations

Consider one of the above difference rings and let $a^{\boldsymbol{\delta}} = a^{\boldsymbol{\alpha}} - a$. Then $\boldsymbol{\delta}$ is the *differencing operator* $a(t) \mapsto a(t+1) - a(t)$. We have

$$\begin{aligned}(ab)^{\boldsymbol{\delta}} &= (ab)^{\boldsymbol{\alpha}} - ab = a^{\boldsymbol{\alpha}}(b^{\boldsymbol{\alpha}} - b) + (a^{\boldsymbol{\alpha}} - a)b \\ &= a^{\boldsymbol{\alpha}} b^{\boldsymbol{\delta}} + a^{\boldsymbol{\delta}} b.\end{aligned}$$

We are led to the following general definition:

Definition 287. *(i) Let $\mathbf{R}$ be a ring and let $\boldsymbol{\alpha}$, $\boldsymbol{\beta}$ be endomorphisms of $\mathbf{R}$. An $(\boldsymbol{\alpha}, \boldsymbol{\beta})$-derivation is an additive map $\boldsymbol{\delta} : \mathbf{R} \to \mathbf{R}$ such that*

$$\boxed{(a\,b)^{\boldsymbol{\delta}} = a^{\boldsymbol{\alpha}}\, b^{\boldsymbol{\delta}} + a^{\boldsymbol{\delta}}\, b^{\boldsymbol{\beta}}} \tag{2.24}$$

for any $a, b \in \mathbf{R}$. (An $(\boldsymbol{\alpha}, \boldsymbol{\beta})$-derivation is also called a pseudo-derivation *with respect to $(\boldsymbol{\alpha}, \boldsymbol{\beta})$ [54].)*
(ii) A $(1,1)$-derivation is called a derivation *and an $(\boldsymbol{\alpha},1)$-derivation is called an $\boldsymbol{\alpha}$-*derivation.
(iii) An $(\boldsymbol{\alpha}, \boldsymbol{\beta})$-derivation is said to be inner *(induced by $m \in \mathbf{R}$) if $a^{\boldsymbol{\delta}} = a^{\boldsymbol{\alpha}} m - m a^{\boldsymbol{\beta}}$ for all $a \in \mathbf{R}$; a derivation which is not inner is said to be* outer.
(iv) A ring (resp., a division ring, a field) endowed with an $\boldsymbol{\alpha}$-derivation $\boldsymbol{\delta}$ is a generalized differential ring *(resp.,* division ring, field*).*

The reader can check that an $(\boldsymbol{\alpha}, \boldsymbol{\beta})$-derivation and a $(\boldsymbol{\beta}, \boldsymbol{\alpha})$-derivation are related as follows:

Proposition 288. *An $(\boldsymbol{\alpha}, \boldsymbol{\beta})$-derivation of the ring $\mathbf{R}$ is a $(\boldsymbol{\beta}, \boldsymbol{\alpha})$-derivation of the opposite ring $\mathbf{R}^{\mathrm{op}}$.*

Note that

$$1^{\alpha} = 1^{\beta} = 1, \quad 1^{\delta} = 0, \tag{2.25}$$

the first equality because α and β are ring-homomorphisms, the second one by taking $a = b = 1$ in (2.24).

A *constant* of a generalized differential ring $\mathbf{R}$ is an element c such that $c^{\alpha} = c^{\beta} = c$ and $c^{\delta} = 0$ [54].[1] The set of all constants of $\mathbf{R}$ is a subring, called the *subring of constants* of $\mathbf{R}$. If $\mathbf{R}$ is a division ring (resp., a field), the subring of constants of $\mathbf{R}$ is a division subring (resp., a subfield) (Exercise 477), called the *division subring of constants* (resp., the *subfield of constants*).

From the above, a (usual) differential ring is a generalized differential ring whose derivation is a $(1,1)$-derivation (outer if it is nonzero) and a difference ring is isomorphic to a generalized differential ring whose derivation is an inner α-derivation (induced by 1). The following is clear:

Lemma 289. *Let* $\mathbf{R}$ *be a difference ring with shift* α*, assumed to be an automorphism, and consider the following conditions:*

(i) $\mathbf{R}$ *is* N*-periodic;*
(ii) $\alpha^N = 1$*;*
(iii) the automorphism α^N *is inner.*
Conditions (i) and (ii) are equivalent and imply (iii). These three conditions are equivalent if $\mathbf{R}$ *is commutative.*

In the sequel, all (α, β)-derivations δ considered on a ring $\mathbf{R}$ are such that $c^{\delta} = 0$ implies $c^{\alpha} = c^{\beta} = c$. For such derivations, the definition of a constant is simpler and coincides with that given in ([77], Sect. 0.8):

Definition 290. *The subring of constants of* $\mathbf{R}$ *is the subring of* $\mathbf{R}$*, denoted by* $\mathbf{R}_{\delta}$*, consisting of all elements* c *such that* $c^{\delta} = 0$*.*

2.4 Ideals in Rings

2.4.1 Maximal Ideals

2.4.1.1 Krull's Theorem

The following result ([27], §I.8, Theorem 1) is a consequence of Zorn's Lemma (Lemma 107) and is generalized in Proposition 533 below.

Theorem 291. (Krull's theorem). *Let* $\mathbf{R}$ *be a ring and* $\mathfrak{a} \neq \mathbf{R}$ *be a left (resp., right, two-sided) ideal in* $\mathbf{R}$*. There exists a maximal left (resp., right, two-sided) ideal* $\mathfrak{m}$ *in* $\mathbf{R}$ *containing* $\mathfrak{a}$*.*

---

[1] For some authors, a constant is just an element c such that $c^{\delta} = 0$ ([77], Sect. 0.8).

2.4.1.2 Jacobson Radical

Lemma 292. *Let* $\mathbf{R} \neq \{0\}$ *be a ring; an element* x *belongs to a maximal left (resp., right, two-sided) ideal if, and only if* x *is not left-invertible (resp., right-invertible, invertible).*

Proof. (1) If x belongs to a left ideal $\mathfrak{a} \neq \mathbf{R}$, then $\mathbf{R}x \subseteq \mathfrak{a}$ and x is not left-invertible since otherwise $\mathbf{R}x = \mathbf{R}$. Similarly, if x belongs to an ideal $\mathfrak{a} \neq \mathbf{R}$, then $(x) \subseteq \mathfrak{a}$ and x is not invertible since otherwise $(x) = \mathbf{R}$. (2) If x is not left-invertible, then $\mathbf{R}x \subsetneq \mathbf{R}$ and by Theorem 291 there exists a maximal left ideal $\mathfrak{m} \supseteq \mathbf{R}x$. Similarly, if x is not invertible, then $(x) \subsetneq \mathbf{R}$ and by Theorem 291 there exists a maximal ideal $\mathfrak{m} \supseteq (x)$. ∎

Item (1) below is proved in ([198], §4):

Theorem and Definition 293. *Let* $\mathbf{R}$ *be a ring and* $\varnothing \neq \mathfrak{J} \subseteq \mathbf{R}$.

(1) The following conditions are equivalent:
(i) $\mathfrak{J}$ *is the intersection of all maximal left ideals in* $\mathbf{R}$*;*
(ii) An element x *belongs to* $\mathfrak{J}$ *if, and only if for any* $y \in \mathbf{R}$, $1 - yx$ *is left-invertible;*
(iii) An element x *belongs to* $\mathfrak{J}$ *if, and only if for any* $y, z \in \mathbf{R}$, $1 - yxz$ *is a unit;*
(iv) $\mathfrak{J}$ *is the only maximal ideal* $\mathfrak{a}$ *such that* $1 - x$ *is a unit for any* $x \in \mathfrak{a}$*;*
(v) $\mathfrak{J}$ *is the intersection of all maximal right ideals in* $\mathbf{R}$.
(2) The ideal satisfying the above equivalent conditions is called the Jacobson radical *of* $\mathbf{R}$, *and is denoted by* $\operatorname{rad}\mathbf{R}$.
(3) If $\operatorname{rad}\mathbf{R} = \{0\}$, $\mathbf{R}$ *is called* radical-free.

2.4.2 Lattices, and Products of Ideals

2.4.2.1 Lattices of Ideals

The following lemma is consistent with Example 109 but gives a more precise description of the lattices of ideals in a ring.

Lemma 294. *Let* $(\mathfrak{a}_i)_{i \in I}$ *be a family of left (resp., right, two-sided) ideals in* $\mathbf{R}$.

(i) $\bigcap_{i \in I} \mathfrak{a}_i$ *is a left (resp., right, two-sided) ideal in* $\mathbf{R}$.
(ii) So is also the set consisting of all elements of the form $\sum_i a_i$ *where* $a_i \in \mathfrak{a}_i$ *for any* i *and* a_i *is equal to zero for all but a finite number of indices. This set is denoted by* $\sum_i \mathfrak{a}_i$ *and is called the* sum *of the left (resp., right, two-sided) ideals* $\mathfrak{a}_i$. $(\boldsymbol{I}_l(\mathbf{R}), +)$, $(\boldsymbol{I}_r(\mathbf{R}), +)$ *and* $(\boldsymbol{I}(\mathbf{R}), +)$ *are commutative monoids (additively noted) with unit element* (0).

(iii) The sets $\boldsymbol{I}_l(\mathbf{R})$, $\boldsymbol{I}_r(\mathbf{R})$, $\boldsymbol{I}(\mathbf{R})$, $\boldsymbol{I}_l^{fg}(\mathbf{R})$, $\boldsymbol{I}_r^{fg}(\mathbf{R})$ *and* $\boldsymbol{I}^{fg}(\mathbf{R})$, *ordered by inclusion, are lattices (Definition 108(iii)),* $\boldsymbol{I}_l(\mathbf{R})$, $\boldsymbol{I}_r(\mathbf{R})$, $\boldsymbol{I}(\mathbf{R})$ *are complete,* $\boldsymbol{I}(\mathbf{R})$ *and* $\boldsymbol{I}^{fg}(\mathbf{R})$ *are modular. The greatest lower bound of* $(\mathfrak{a}_i)_{i\in I}$ *is* $\bigwedge_{i\in I}\mathfrak{a}_i = \bigcap_{i\in I}\mathfrak{a}_i$ *and its least upper bound is* $\bigvee_{i\in I}\mathfrak{a}_i = \sum_i \mathfrak{a}_i$.

Proof. (i) and (ii) are obvious; (iii) as well, except the fact that $\mathbf{I}(\mathbf{R})$ and $\mathbf{I}^{fg}(\mathbf{R})$ are a modular lattices. This point is a consequence of Lemma 153. ■

2.4.2.2 Products of Ideals

Definition 295. *Let* $\mathbf{R}$ *be a ring, and let* $\mathfrak{a}, \mathfrak{b}$ *be left, right, or two-sided ideals in* $\mathbf{R}$. *The* product $\mathfrak{a}\,\mathfrak{b}$ *is the abelian subgroup of* $(\mathbf{R}, +)$ *consisting of all elements of the form* $\sum_k a_k\, b_k$ *where* $a_k \in \mathfrak{a}$, $b_k \in \mathfrak{b}$ *and* $a_k\, b_k = 0$ *for all but a finite number of indices.*

Lemma 296. *If* $\mathfrak{a} \lhd_l \mathbf{R}$ *(resp.,* $\mathfrak{b} \lhd_r \mathbf{R}$*) then* $\mathfrak{a}\,\mathfrak{b} \lhd_l \mathbf{R}$ *(resp.,* $\mathfrak{a}\,\mathfrak{b} \lhd_r \mathbf{R}$*).*

Proof. Let $\mathfrak{a} \lhd_l \mathbf{R}$, $x \in \mathbf{R}$ and $\sum_k a_k\, b_k \in \mathfrak{a}\,\mathfrak{b}$. Then $x \sum_k a_k\, b_k = \sum_k (x\, a_k)\, b_k \in \mathfrak{a}\,\mathfrak{b}$ since $x\, a_k \in \mathfrak{a}$. ■

Theorem 297. *(i)* $(\boldsymbol{I}(\mathbf{R}), +, \times)$ *is a distributive dioid with zero element* (0) *and unit element* $\mathbf{R}$ *(Remark 220(b)); it is commutative if so is* $\mathbf{R}$.
(ii) Let $\mathfrak{a}, \mathfrak{b} \lhd \mathbf{R}$; *then* $\mathfrak{a}\,\mathfrak{b} \subseteq \mathfrak{a} \cap \mathfrak{b}$.
(iii) Let $\mathfrak{a}_1, \ldots, \mathfrak{a}_n$ *be ideals in* $\mathbf{R}$ *such that* $\mathfrak{a}_i + \mathfrak{a}_j = \mathbf{R}$ *for* $i \neq j$; *then*

$$\bigcap_{1\leq i\leq n} \mathfrak{a}_i = \sum_{\sigma\in\mathfrak{S}_n} \mathfrak{a}_{\sigma(1)} \ldots \mathfrak{a}_{\sigma(n)};$$

in particular, if $\mathbf{R}$ *is commutative, then* $\bigcap_{1\leq i\leq n} \mathfrak{a}_i = \mathfrak{a}_1 \ldots \mathfrak{a}_n$.

Proof. (i): We already know that $(\mathbf{I}(\mathbf{R}), +)$ is a commutative monoid with unit element (0) (Lemma 294). It is easy to check that the multiplication of ideals is associative and distributive with respect to the addition of those ideals. Last, for any $\mathfrak{a} \lhd \mathbf{R}$, $\mathfrak{a}\,\mathbf{R} = \mathbf{R}\,\mathfrak{a} = \mathfrak{a}$. (ii) is obvious and (iii) is proved in ([27], §I.8, Prop. 7). ■

2.4.3 Prime Ideals, and Completely Prime Ideals

2.4.3.1 Prime Ideals

In noncommutative rings, prime ideals are defined as follows ([198], Definition (10.1)):

Definition 298. *An ideal* $\mathfrak{p} \lhd \mathbf{R}$ *is said to be* prime *if* $\mathfrak{p} \neq \mathbf{R}$ *and for* $\mathfrak{a} \lhd \mathbf{R}$, $\mathfrak{b} \lhd \mathbf{R}$,

$$\mathfrak{a}\,\mathfrak{b} \subseteq \mathfrak{p} \quad \textit{implies that } \mathfrak{a} \subseteq \mathfrak{p} \textit{ or } \mathfrak{b} \subseteq \mathfrak{p}.$$

Prime ideals can be characterized as follows:

Lemma 299. *For an ideal $\mathfrak{p} \subsetneq \mathbf{R}$, the following conditions are equivalent:*

(i) $\mathfrak{p}$ is prime;
(ii) For $a, b \in \mathbf{R}$, $(a)(b) \subseteq \mathfrak{p}$ implies that $a \in \mathfrak{p}$ or $b \in \mathfrak{p}$;
(iii) For $a, b \in \mathbf{R}$, $a\mathbf{R}b \subseteq \mathfrak{p}$ implies that $a \in \mathfrak{p}$ or $b \in \mathfrak{p}$;
(iv) For $\mathfrak{a} \lhd_l \mathbf{R}$, $\mathfrak{b} \lhd_l \mathbf{R}$, $\mathfrak{a}\mathfrak{b} \subseteq \mathfrak{p}$ implies that $\mathfrak{a} \subseteq \mathfrak{p}$ or $\mathfrak{b} \subseteq \mathfrak{p}$;
(iv') For $\mathfrak{a} \lhd_r \mathbf{R}$, $\mathfrak{b} \lhd_r \mathbf{R}$, $\mathfrak{a}\mathfrak{b} \subseteq \mathfrak{p}$ implies that $\mathfrak{a} \subseteq \mathfrak{p}$ or $\mathfrak{b} \subseteq \mathfrak{p}$.

Proof. Obviously, (i)⇒(ii)⇒(iii) and (iv)⇒(i). Let us prove that (iii)⇒(iv): Assume that $\mathfrak{a} \lhd_l \mathbf{R}$, $\mathfrak{b} \lhd_l \mathbf{R}$, $\mathfrak{a}\mathfrak{b} \subseteq \mathfrak{p}$ and $\mathfrak{a} \nsubseteq \mathfrak{p}$. Let $a \in \mathfrak{a}$, $a \notin \mathfrak{p}$ and $b \in \mathfrak{b}$; then $a\mathbf{R}b \subseteq \mathfrak{p}$, thus $b \in \mathfrak{p}$ and $\mathfrak{b} \subseteq \mathfrak{p}$. ∎

Definition 300. *The set of all prime ideals in $\mathbf{R}$ is denoted by* $\operatorname{Spec}\mathbf{R}$ *and is called the* prime spectrum *of $\mathbf{R}$.*

2.4.3.2 Completely Prime Ideals

Definition 301. *An ideal $\mathfrak{p} \lhd \mathbf{R}$ is said to be* completely prime *if $\mathbf{R}/\mathfrak{p}$ is a domain.*

Note that $(0) \lhd \mathbf{R}$ is a completely prime ideal if, and only if $\mathbf{R}$ is a domain.

Lemma 302. *For $\mathfrak{p} \lhd \mathbf{R}$, the following properties are equivalent:*

(i) $\mathfrak{p}$ is completely prime;
(ii) $\mathfrak{p} \neq \mathbf{R}$, and for $a, b \in \mathbf{R}$, $ab \in \mathfrak{p}$ implies that $a \in \mathfrak{p}$ or $b \in \mathfrak{p}$;
(iii) $\mathfrak{p} \neq \mathbf{R}$, and for $a, b \in \mathbf{R}$, $a \in \mathbf{R}\backslash\mathfrak{p}$ and $b \in \mathbf{R}\backslash\mathfrak{p}$ implies that $ab \in \mathbf{R}\backslash\mathfrak{p}$.

Proof. Obviously, (ii)⇔(iii). (i)⇒(ii): Let $\pi : \mathbf{R} \to \mathbf{R}\backslash\mathfrak{p}$ be the canonical epimorphism and assume that $\mathbf{R}/\mathfrak{p}$ is a domain. If $ab \in \mathfrak{p}$, then $\pi(a)\,\pi(b) = \pi(ab) = 0$, thus $\pi(a) = 0$ or $\pi(b) = 0$, i.e., $a \in \mathfrak{p}$ or $b \in \mathfrak{p}$ and (ii) holds. The converse is similar. ∎

The following is a consequence of Lemmas 299 and 302:

Proposition 303. *If $\mathfrak{p} \lhd \mathbf{R}$ is completely prime, it is prime. The converse holds true if $\mathbf{R}$ is commutative.*

If $(\mathbf{R}^{\times}, \times)$ is a cancellation monoid, i.e., if $\mathbf{R}$ is a domain, we have the following.

Definition 304. *An element $p \neq 0$ is said to be a* prime *in $\mathbf{R}$ if it is a prime in the monoid $(\mathbf{R}^{\times}, \times)$ (Definition 128).*

Theorem 305. *Let $\mathbf{R}$ be a domain and $0 \neq p \in \mathbf{R}$ be such that $p\mathbf{R} = \mathbf{R}p$. The following properties are equivalent:*

(i) p is a prime in $\mathbf{R}$;
(ii) The principal ideal (p) is completely prime.

Proof. We know that for any $c \in \mathbf{R}$, $p \mid c$ if, and only if $c \in (p)$; in other words, $c \notin (p)$ if, and only if $p \nmid c$. By Definition 128, the invariant element p is a prime if, and only if $p \nmid a$ and $p \nmid b$ implies $p \nmid a\,b$. By Lemma 302, this means that (p) is completely prime. ∎

Proposition 306. *Let* $\mathbf{R}$ *be a ring and let* $\mathfrak{m} \lhd \mathbf{R}$. *If* $\mathfrak{m}$ *is a maximal ideal, then* $\mathfrak{m}$ *is completely prime.*

Proof. Let $a \in \mathbf{R} \backslash \mathfrak{m}$ and $b \in \mathbf{R} \backslash \mathfrak{m}$; then $(a) + \mathfrak{m} = \mathbf{R} = (b) + \mathfrak{m}$, thus

$$\mathbf{R} = ((a) + \mathfrak{m})\,((b) + \mathfrak{m}) = (a)\,(b) + \mathfrak{m},$$

therefore $(a)\,(b) \nsubseteq \mathfrak{m}$ which proves that $a\,b \in \mathbf{R} \backslash \mathfrak{m}$, and $\mathfrak{m}$ is completely prime by Lemma 302. ∎

2.4.4 Artinian Rings, Noetherian Rings, and Semisimple Rings

Definition 307. *Let* $\mathbf{A}$ *be a ring. A left* $\mathbf{A}$*-module* M *is said to be Artinian (resp., Noetherian) if every nonempty set of submodules of* M*, ordered by inclusion, is Artinian (resp., Noetherian) (Definition 101). This definition is still valid with* left *changed to* right.

Definition 308. *(i) A ring* $\mathbf{R}$ *is said to be* left *(resp.,* right*)* Noetherian *if the lattice* $\boldsymbol{I}_l(\mathbf{R})$ *(resp.,* $\boldsymbol{I}_r(\mathbf{R})$*) is left (resp., right) Noetherian (Definition 102 and Example 109).*
(ii) A ring $\mathbf{R}$ *is said to be* Noetherian *if it is both left and right Noetherian. (Statement (i) can also be expressed as follows: a ring* $\mathbf{R}$ *is left (resp., right) Noetherian if it satisfies left (resp., right) ACC.)*
(iii) Left Artinian, right Artinian *and* Artinian rings *are likewise defined. A ring* $\mathbf{R}$ *is left (resp., right) Artinian if it satisfied left (resp., right) DCC (Definition 102).*

Notice that a ring $\mathbf{A}$ is left Artinian (resp., Noetherian) if, and only if the left $\mathbf{A}$-module ${}_{\mathbf{A}}\mathbf{A}$ is Artinian (resp., Noetherian).

The following result is proved in ([225], Sect. XI.1) and ([29], §VIII.2):

Lemma 309. *Let* $\mathbf{A}$ *be a ring.*

(i) Let M *be a left* $\mathbf{A}$*-module; then* M *is Noetherian if, and only if every submodule of* M *is f.g.*
(ii) Let M *be a left* $\mathbf{A}$*-module and let* $N \lhd M$*; then* M *is Artinian (resp., Noetherian) if, and only if both* N *and* M/N *are Artinian (resp., Noetherian).*
(iii) Every finite product of left Artinian (resp., Noetherian) modules is left Artinian (resp., Noetherian); in particular, if $\mathbf{A}$ *is a left Artinian (resp.,*

Noetherian) ring, then every finite free left **A***-module is Artinian (resp., Noetherian).*

In the above statements, left *can be changed to* right.

The following is a consequence of Lemma 309.

Proposition 310. *Let* **A** *be a ring. The following conditions are equivalent:*

(i) the ring **A** *is left Artinian (resp., Noetherian);*
(ii) every finitely generated left **A***-module is Artinian (resp., Noetherian).*
In the above statements, left *can be changed to* right.

Lemma 311. *Let* **R** *be a ring. The following conditions are equivalent:*

(i) **R** *is left Noetherian;*
(ii) The lattice $\boldsymbol{I}_l^{fg}(\mathbf{R})$ *is Noetherian;*
(iii) Every left ideal in **R** *is finitely generated.*

Proof. (i)⇒(ii) and (iii)⇒(i) are obvious. To prove that (ii)⇒(iii), assume that (iii) fails. There exists an infinite sequence $(x_i)_{i\in\mathbb{N}}$ of elements of **R** such that $(x_0)_l \subsetneq (x_0, x_1)_l \subsetneq ... \subsetneq (x_0, x_1, ..., x_n)_l \subsetneq ....$ Therefore, (ii) fails. ∎

Theorem 312. *(*Hopkins-Levitzki theorem*). A left (resp., right) Artinian ring is left (resp., right) Noetherian.*

Proof. See ([247], 0.1.13). ∎

The above statement is not correct if modules are considered in place of rings ([198], §1). The definition below is connected with the definition of a simple Lie algebra and of a semisimple Lie algebra (**1.8.2.3**).

Definition 313. *Let* **R** *be a ring and let M be a left* **R***-module.*

(i) M is called a simple *(or* irreducible*)* **R***-module if $M \neq 0$ and M has no* **R***-submodule other than 0 and M.*
(ii) M is called a semisimple *(or* completely reducible*)* **R***-module if every* **R***-submodule of M is a direct summand of M.*

Definition 314. *Let* **R** *be a ring. A left (or right)* **R***-module M is said to be of* finite length *if it has a Jordan-Hölder series (Definition 159), i.e., if there exists a chain*

$$M = N_0 \supsetneq N_1 \supsetneq ... \supsetneq N_s = \{0\}$$

of submodules, such that each quotient N_i/N_{i+1} $(0 \leq i \leq s-1)$ is simple (Definition 313(i)). Then the integer s – which is well-defined by the Jordan-Hölder-Dedekind theorem (Theorem 117) – is called the length *of M.*

The following is easily proved ([29], §VIII.1, Prop. 3):

Theorem 315. *Let* $\mathbf{R}$ *be a ring and let* M *be a left (or right)* $\mathbf{R}$*-module. Then* M *is of finite length if, and only if* M *is both Artinian and Noetherian.*

Proposition 316. *Let* $\mathbf{R}$ *be a ring and* M *a left* $\mathbf{R}$*-module. The following conditions are equivalent:*

(i) M *is semisimple;*
(ii) M *is the direct sum of a family of simple modules;*
(iii) M *is the sum of a family of simple submodules.*

Proof. See ([198], §2, Theorem (2.4)). ■

Items (1) and (2) below are proved in ([198], Theorem (2.5) and Corol. (2.6), (3.7)).

Lemma and Definition 317. *Let* $\mathbf{R}$ *be a ring.*

(1) The following conditions are equivalent:
(i) all left $\mathbf{R}$*-modules are semisimple;*
(ii) all finitely generated left $\mathbf{R}$*-modules are semisimple;*
(iii) the left $\mathbf{R}$*-module* ${}_{\mathbf{R}}\mathbf{R}$ *is semisimple;*
(iv) the right $\mathbf{R}$*-module* $\mathbf{R}_{\mathbf{R}}$ *is semisimple;*
(v) all short exact sequences of left $\mathbf{R}$*-modules split (Definition 53(3)).*
(2) The ring $\mathbf{R}$ *is called* semisimple *if the above equivalent conditions are satisfied.*
(3) A semisimple ring is Artinian (thus Noetherian), and a division ring is semisimple.

The notion of semisimplicity is left/right symmetric.

Remark 318. *(i) The zero module is semisimple but not simple. A simple module is semisimple.*
(ii) A commutative *ring* $\mathbf{R}$ *is simple (Definition 223) if, and only if the module* ${}_{\mathbf{R}}\mathbf{R}$ *is simple.* This does not hold true if $\mathbf{R}$ is noncommutative.
(iii) A simple ring is (left) semisimple if, and only if, it is left Artinian ([198], Theorem (3.10)). Note that in Bourbaki ([29], §5), all simple modules are assumed to be semisimple (or equivalenly Artinian).

2.4.5 Prime Rings, and Semiprime Rings

Item (1) below is proved in ([247], 0.2.3):

Lemma and Definition 319. *Let* $\mathbf{R}$ *be a ring.*

(1) The following conditions are equivalent:
(i) if $0 \neq a, b \in \mathbf{R}$*, then* $a\mathbf{R}b \neq 0$*;*
(ii) if $0 \neq \mathfrak{a}, \mathfrak{b} \triangleleft_l \mathbf{R}$*, then* $\mathfrak{a}\mathfrak{b} \neq 0$*;*

(iii) if $0 \neq \mathfrak{a}, \mathfrak{b} \lhd_r \mathbf{R}$*, then* $\mathfrak{a}\mathfrak{b} \neq 0$*;*
(iv) if $0 \neq \mathfrak{a}, \mathfrak{b} \lhd \mathbf{R}$*, then* $\mathfrak{a}\mathfrak{b} \neq 0$.
(2) The ring $\mathbf{R}$ *is said to be* prime *if the above equivalent conditions are satisfied.*

Definition 320. *Let* $\mathbf{R}$ *be a ring and let* $\mathfrak{a}$ *be a left, right or two-sided ideal in* $\mathbf{R}$*;* $\mathfrak{a}$ *is called* nilpotent *if there exists a positive integer* n *such that* $\mathfrak{a}^n = 0$.

Item (1) below is proved in ([247], 0.2.7):

Lemma and Definition 321. *Let* $\mathbf{R}$ *be a ring.*

(1) The following conditions are equivalent:
(i) $\mathbf{R}$ *has no nonzero nilpotent left ideal;*
(ii) $\mathbf{R}$ *has no nonzero nilpotent right ideal;*
(iii) $\mathbf{R}$ *has no nonzero nilpotent ideal.*
(2) The ring $\mathbf{R}$ *is said to be* semiprime *if the above equivalent conditions are satisfied.*

The following is clear:

Lemma 322. *A domain is a prime ring; conversely, a prime* commutative *ring is a domain. A prime ring is semiprime.*

2.5 Rings of Fractions

2.5.1 The Commutative Case

Let $\mathbf{R}$ be a *commutative ring* and let S be a nonempty subset of $\mathbf{R}$. Our aim is to construct a ring $S^{-1}\mathbf{R}$ of fractions r/s ($r \in \mathbf{R}$, $s \in S$) such that all elements of S are units of $S^{-1}\mathbf{R}$ and there exists a canonical ring-homomorphism $\lambda_S : \mathbf{R} \to S^{-1}\mathbf{R}$.

Suppose that we have succeeded to construct $S^{-1}\mathbf{R}$. Consider two fractions r/s and r'/s'. To compare them, they must be reduced to the same denominator $d \in S$, therefore there must exist $q, q' \in \mathbf{R}^\times$ such that $d = q\,s = q'\,s'$. Then, for the two above fractions to be equal one must have $q\,r = q'\,r'$. Thus, consider the following relation $\mathcal{R}_1$ in $\mathbf{R} \times S$:

$$(r,s)\,\mathcal{R}_1(r',s') \Leftrightarrow q\,r = q'\,r' \text{ and } q\,s = q'\,s' \in S \text{ for some } q, q' \in \mathbf{R}.$$

From the above,

$$r/s = r'/s' \Rightarrow (r,s)\,\mathcal{R}_1(r',s').$$

Now, choosing $q = t\,s'$ and $q' = t\,s$ for some $t \in S$, we have obviously $q\,s = q'\,s' \in S$ provided that S is a *multiplicative set*, i.e., $s\,s' \in S$ whenever $s, s' \in S$. Then, $q\,r = q'\,r'$ if, and only if $t\,(s'\,r - s\,r') = 0$. Consider the following relation $\mathcal{R}_2$ in $\mathbf{R} \times S$:

$$(r,s)\,\mathcal{R}_2(r',s') \Leftrightarrow t\,(s'\,r - s\,r') = 0 \text{ for some } t \in S.$$

If $0 \in S$, then $(r,s)\,\mathcal{R}_2(0,1)$ for any $r \in \mathbf{R}$ and $s \in S$, thus $(\mathbf{R} \times S)/\mathcal{R}_2 = \{0\}$ and this case is trivial. To avoid it, we suppose that $0 \notin S$ and, for convenience, that $1 \in S$. Then, S is a submonoid of $(\mathbf{R}^{\times},.)$.

Definition 323. *Let $\mathbf{R}'$ be a ring and let S be a submonoid of $(\mathbf{R}^{\times},.)$. A ring-homomorphism $\mu : \mathbf{R} \to \mathbf{R}'$ is S-inverting if it maps the elements of S to units of $\mathbf{R}'$.*

Theorem 324. *Let S be a submonoid of $(\mathbf{R}^{\times},\times)$.*

(i) The relation $\mathcal{R}_2$ defined above is an equivalence relation in $\mathbf{R} \times S$;
(ii) Denote by r/s the class of $(r,s)\,(\mathrm{mod}\,\mathcal{R}_2)$. The set $S^{-1}\mathbf{R} \triangleq (\mathbf{R} \times S)/\mathcal{R}_2$ is a ring with the addition and the multiplication of fractions defined as usual, with $0/1$ as zero and $1/1$ as unit element;
(iii) Consider the canonical ring-homomorphism $\lambda_S : \mathbf{R} \to S^{-1}\mathbf{R}$ given by $r \mapsto r/1$. Then

$$\ker \lambda_S = \{r \in \mathbf{R} : t\,r = 0 \text{ for some } t \in S\};$$

(iv) λ_S is S-inverting;
(v) For every S-inverting ring-homomorphism $\mu : \mathbf{R} \to \mathbf{R}'$, there exists a unique ring-homomorphism $\mu' : S^{-1}\mathbf{R} \to \mathbf{R}'$ such that $\mu = \mu' \circ \lambda_S$; moreover, this property determines $S^{-1}\mathbf{R}$ up to isomorphism.

Proof. The proof of (i) and (ii) is straightforward. (iii): $r/1 = 0$ if, and only if $(r,1)\,\mathcal{R}_2(0,1)$, i.e., $t\,r = 0$ for some $t \in S$. (iv): If $r \in S$, then r/s is invertible with inverse s/r. (v) is proved in ([79], Sect. 10.3). ■

Remark 325. *(i) A unit of $S^{-1}\mathbf{R}$ is not necessarily of the form r/s, $r,s \in S$. Consider for example $\mathbf{R} = \mathbb{Z}$ and $S = {}^2\mathbb{Z}^{\times}$, the set of nonzero squares of $\mathbb{Z}$. Then $2/4 \in S^{-1}\mathbf{R}$ and $(2/4)\,(2/1) = 4/4 = 1/1$, thus $2/4$ is a unit of $S^{-1}\mathbf{R}$.*
(ii) By Properties (iv) and (v) in Theorem 324, the homomorphism λ_S is said to be universal S-inverting. *Let $\mu : \mathbf{R} \to \mathbf{R}'$ be universal S-inverting. There exists a unique ring-homomorphism $\lambda'_S : \mathbf{R}' \to S^{-1}\mathbf{R}$ such that $\lambda_S = \lambda'_S \circ \mu$; therefore $\mu' \circ \lambda'_S = Id_{\mathbf{R}'}$, $\lambda'_S \circ \mu' = Id_{S^{-1}\mathbf{R}}$, and μ' is an isomorphism with inverse $\lambda'_S : \mathbf{R}' \xrightarrow{\sim} S^{-1}\mathbf{R}$; identifying $\mathbf{R}'$ with $S^{-1}\mathbf{R}$, λ_S and μ are identified. The homomorphism λ_S is a* universal morphism *and, as such, is "essentially unique" (Theorem 13(ii)).*
(iii) The ring $\mathbf{R}$ can be embedded in $S^{-1}\mathbf{R}$ (identifying $\mathbf{R}$ with $\lambda_S(\mathbf{R}) \subseteq S^{-1}\mathbf{R}$) if, and only if λ_S is a ring-monomorphism; this happens if, and only if all elements of S are regular according to Property (iii).
(iv) The set $\mathbf{R}^{\times}$ is multiplicative, and $\mathbf{F} = (\mathbf{R}^{\times})^{-1}\mathbf{R}$ is called the total ring of fractions *of $\mathbf{R}$. Every non-zero divisor in $\mathbf{F}$ is a unit. If all elements of $\mathbf{R}^{\times}$ are regular, i.e., if $\mathbf{R}$ is a commutative domain, then $\mathbf{F}$ is the* field of fractions *of $\mathbf{R}$.*

Example 326. *Consider the ring of polynomials* $\mathbf{R}[X]$ *(where* X *is a central indeterminate). The set* $S = \{X^k : k \geq 0\}$ *is a submonoid of* $\left(\mathbf{R}[X]^\times, .\right)$ *and* $S^{-1}\mathbf{R}[X] = \mathbf{R}\left[X, X^{-1}\right]$ *is called the ring of* Laurent polynomials *with central indeterminate* X *and coefficients in* $\mathbf{R}$*. The elements of* $\mathbf{R}\left[X, X^{-1}\right]$ *can be uniquely written in the form*

$$\sum_{i=n}^{m} f_i X^i, \quad n, m \in \mathbb{Z},\ m \geq n.$$

2.5.2 The Noncommutative Case

2.5.2.1 Ore's Construction

If $\mathbf{R}$ is a noncommutative ring, the above relation $\mathcal{R}_2$ is no longer an equivalence relation. In addition, left fractions (of the form $s^{-1}r$) must be distinguished from right ones (of the form $r s^{-1}$). In what follows, some hints are given about the construction of a ring $S^{-1}\mathbf{R}$ of left fractions (written r/s) with denominator in a set $S \subseteq \mathbf{R}^\times$. The construction of a ring $\mathbf{R}\,S^{-1}$ of right fractions is similar (with obvious changes).

Definition 327. *Let* S *be a submonoid of* $(\mathbf{R}^\times, \times)$*;*
(i) S *is a* left Ore set *if* $S\,r \cap \mathbf{R}\,s \neq \varnothing$ *for all* $r \in \mathbf{R}$*,* $s \in \mathbf{S}$.
(ii) S *is a* left denominator set *if it is a left Ore set and the following condition holds: for each* $r \in \mathbf{R}$*,* $s \in S$*, if* $r\,s = 0$*, then* $t\,r = 0$ *for some* $t \in S$*. (The latter condition is automatically satisfied if all elements of* S *are left-regular.)*

Theorem 328. *(1) Let* $S \subseteq \mathbf{R}^\times$ *be a left denominator set.*
(i) The relation $\mathcal{R}_1$ *defined in Subsect. 2.5.1 is an equivalence relation in* $\mathbf{R} \times S$*;*
(ii) Denote by r/s *the class of* $(r, s) \,(\mathrm{mod}\,\mathcal{R}_1)$*. Two fractions* r/s *and* r'/s' *can be brought to a common denominator. Then, the addition in the quotient set* $S^{-1}\mathbf{R} \triangleq (\mathbf{R} \times S)/\mathcal{R}_1$ *is defined by the rule*

$$r/s + r'/s = (r + r')/s.$$

The multiplication is defined as follows: let $r/s,\ r'/s' \in S^{-1}\mathbf{R}$*; since* $S\,r \cap \mathbf{R}\,s' \neq \varnothing$*, there exist* $s_1' \in S$*,* $r_1' \in \mathbf{R}$ *such that* $s_1'\,r = r_1'\,s'$*, and then*

$$(r/s)(r'/s') = (r_1'\,r')/(s_1'\,s). \tag{2.26}$$

Endowed with this addition and this multiplication, $S^{-1}\mathbf{R}$ *is a ring with* $0/1$ *as zero element and* $1/1$ *as unit element.*
(iii) Statements (iii)-(v) of Theorem 324 are still valid and Remark 325(iii) as well.
(2) Conversely, if a ring of left fractions $S^{-1}\mathbf{R}$ *exists,* S *is a left denominator set.*

Proof. The proof of (1) and (2) are similar to the proofs given in ([80], Sect. 7.1) and ([247], §2.1), respectively. Let us detail some points. Regarding (1)(ii), it has been shown above that if $(r,s)\,\mathcal{R}_1(r',s')$, the fractions r/s and r'/s' can be brought to a common denominator $d = q\,s = q'\,s'$. Consider two fractions $r/s = r\,s^{-1}$ and $r'/s' = r'\,s'^{-1}$; since S is a left Ore set, there exist $s_1' \in S$, $r_1' \in \mathbf{R}$ such that $s_1'\,r = r_1'\,s'$. Therefore,

$$\left(s^{-1}\,r\right)\left(s'^{-1}\,r'\right) = s^{-1}\,s_1'^{-1}\,s_1'\,r\,s'^{-1}\,r' = s^{-1}\,s_1'^{-1}\,r_1'\,s'\,s'^{-1}\,r' \\ = \left(s_1'\,s\right)^{-1}\left(r_1'\,r'\right)$$

and (2.26) is obtained. Let us determine $\ker\lambda_S$ where $\lambda_S : \mathbf{R} \to S^{-1}\mathbf{R}$ is the canonical ring-homomorphism; an element $r_1 \in \mathbf{R}$ is such that $\lambda_S(r) = 0$ if, and only if $((r,1)\,\mathcal{R}_1(0,1))$, i.e., $t\,r = 0$ for some $t \in S$, as claimed. ■

Statement (1)(ii) of Theorem 328 implies that if S is a left denominator set, two elements s, s' of S have a common left multiple $d \in S$. By induction one obtains the following.

Corollary 329. *Let $S \subseteq \mathbf{R}^{\times}$ be a left denominator set.*

(i) Let $s_1, \ldots, s_n$ be elements of S; these elements have a common left multiple belonging to S.
(ii) Any finite set of $S^{-1}\mathbf{R}$ can be brought to a common denominator.

Let $\mathfrak{a} \lhd_l \mathbf{R}$ and S be a left denominator set. Obviously, $S^{-1}\mathfrak{a} \triangleq \{x/s : x \in \mathfrak{a}, s \in S\}$ is a left ideal in $S^{-1}\mathbf{R}$. The proof of the following result is straightforward:

Proposition 330. *The map $\psi_S : \boldsymbol{I}_l(\mathbf{R}) \to \boldsymbol{I}_l\left(S^{-1}\mathbf{R}\right)$ given by $\psi_S(\mathfrak{a}) = S^{-1}\mathfrak{a}$ is both a lattice- and a dioid-homomorphism, i.e., if $\mathfrak{a}, \mathfrak{b} \lhd_l \mathbf{R}$,*

$$S^{-1}(\mathfrak{a}+\mathfrak{b}) = S^{-1}\mathfrak{a} + S^{-1}\mathfrak{b}, \quad S^{-1}(\mathfrak{a}\cap\mathfrak{b}) = S^{-1}\mathfrak{a}\cap S^{-1}\mathfrak{b}, \\ S^{-1}(\mathfrak{a}\,\mathfrak{b}) = S^{-1}\mathfrak{a}\,S^{-1}\mathfrak{b}.$$

Definition 331. *An ideal $\mathfrak{b} \lhd_l \mathbf{R}$ is said to be* saturated *for the left denominator set S if $r \in \mathfrak{b}$ whenever $s\,r \in \mathfrak{b}$, $s \in S$, $r \in \mathbf{R}$.*

Theorem 332. *Let $S \subseteq \mathbf{R}^{\times}$ be a left denominator set. For any $\mathfrak{b}' \lhd_l S^{-1}\mathbf{R}$, write $\varphi_S(\mathfrak{b}') = \lambda_S^{-1}(\mathfrak{b}')$. The following properties hold:*

(i) If $\mathfrak{b}' \lhd_l S^{-1}\mathbf{R}$, then $\varphi_S(\mathfrak{b}') \lhd_l \mathbf{R}$ and $S^{-1}\varphi_S(\mathfrak{b}') = \mathfrak{b}'$.
(ii) If $\mathfrak{b}' \lhd_r S^{-1}\mathbf{R}$, then $\varphi_S(\mathfrak{b}') \lhd_r \mathbf{R}$.
(iii) If $\mathfrak{b} \lhd_l \mathbf{R}$ is saturated for S, then $\varphi_S\left(S^{-1}\mathfrak{b}\right) = \mathfrak{b}$.
(iv) If $\mathbf{R}$ is left Noetherian, so is $S^{-1}\mathbf{R}$.
(v) If $\mathfrak{b} \lhd \mathbf{R}$ and either $\mathbf{R}$ is commutative or $S^{-1}\mathbf{R}$ is left Noetherian, then $S^{-1}\mathfrak{b} \lhd S^{-1}\mathbf{R}$.
(vi) If $\mathbf{R}$ is Noetherian or commutative, the map $\psi_S : \mathfrak{p} \mapsto S^{-1}\mathfrak{p}$ is an isotone (for inclusion) bijection from the set of all prime (resp., maximal) ideals

$\mathfrak{p} \lhd \mathbf{R}$ such that $\mathfrak{p} \cap S = \varnothing$ onto the set of all prime (resp., maximal) ideals $\mathfrak{p}' \lhd S^{-1}\mathbf{R}$, and the inverse of ψ_S is given by $\mathfrak{p}' \mapsto \varphi_S(\mathfrak{p}')$.
(vii) Let $\mathfrak{b}' \lhd S^{-1}\mathbf{R}$, $\mathfrak{b} = \varphi_S(\mathfrak{b}')$ and $\pi : \mathbf{R} \twoheadrightarrow \mathbf{R}/\mathfrak{b}$ be the canonical epimorphism. There exists a canonical isomorphism $(S^{-1}\mathbf{R})/\mathfrak{b}' \cong (\pi(S))^{-1}(\mathbf{R}/\mathfrak{b})$.
(viii) If $S \subseteq \mathbf{R}^{\times}$ is a left and right denominator set, the ring of left fractions $S^{-1}\mathbf{R}$ coincides with the ring of right fractions $\mathbf{R}S^{-1}$.
(ix) If $\mathbf{R}$ is a domain, then $S^{-1}\mathbf{R}$ is again a domain.

Proof. See ([247], 2.1.16) and ([31], §II.2, Prop. 11). ■

2.5.2.2 Goldie's Theorem

Theorem 333. *(*Goldie's theorem*). Let $\mathbf{R}$ be a semiprime left (resp., right) Noetherian ring (Subsect. 2.4.5 & 2.4.4) and let $\mathcal{C}_{\mathbf{R}}(0)$ be the set of all regular elements of $\mathbf{R}$.*
(i) $\mathbf{R}$ can be embedded in the ring $\mathbf{Q}$ of left (resp., right) fractions $\mathcal{C}_{\mathbf{R}}(0)^{-1}\mathbf{R}$ (resp., $\mathbf{R}\,\mathcal{C}_{\mathbf{R}}(0)^{-1}$) which is semisimple.
(ii) Furthermore, $\mathbf{R}$ is prime if, and only if $\mathbf{Q}$ is simple.

Proof. (i) Assume $\mathbf{R}$ is semiprime left Noetherian. By ([247], 2.3.6), $\mathcal{C}_{\mathbf{R}}(0)$ is a left Ore set, thus is a left denominator set (Definition 327(ii)). By Theorem 328, $\mathbf{R}$ can be embedded in the ring of left fractions $\mathbf{Q} = \mathcal{C}_{\mathbf{R}}(0)^{-1}\mathbf{R}$. The ring $\mathbf{Q}$ is semisimple by ([247], 2.3.6). The same holds with *left* changed to *right.*
(ii): see ([247], 2.3.6). ■

Remark 334. *(1) The above theorem can be improved, using the notion of* Goldie ring *([247], §2.3). We do not detail this notion in the general case, for the following statement will be sufficient:*
A ring $\mathbf{R}$ is semiprime left (resp., right) Goldie if, and only if Condition (i) in Theorem 333 holds.
For any ring $\mathbf{R}$,

$$\text{left (resp., right) Noetherian} \Longrightarrow \text{left (resp., right) Goldie}$$

and a Goldie ring is a ring which is both left and right Goldie.
(2) If a ring $\mathbf{R}$ is semiprime left (resp., right) Goldie, then the set of all left-regular (resp., right-regular) elements of $\mathbf{R}$ coincides with the set $\mathcal{C}_{\mathbf{R}}(0)$ of all regular elements of $\mathbf{R}$ ([199], Prop. (11.4)).

2.5.3 Ore Domains

Definition 335. *(i) A ring $\mathbf{R}$ is a* left Ore domain *if it is a domain and $\mathbf{R}^{\times}$ is a left Ore set. A* right Ore domain *is likewise defined.*
(ii) An Ore domain *is a left Ore domain which is a right Ore domain.*

The following result is an obvious consequence of Theorem 328 and of the definitions.

Theorem 336. *Let* $\mathbf{R}$ *be a domain.*

(1) The following conditions are equivalent:
(i) $\mathbf{R}$ *is a left Ore domain;*
(ii) $\mathbf{R}^{\times}$ *is a left denominator set;*
(iii) $\mathbf{R}\, a \cap \mathbf{R}\, b \neq 0$ *for all* $a, b \in \mathbf{R}^{\times}$.
(iv) $\mathbf{R}$ *can be embedded in the* division ring $\mathbf{R}^{\times -1}\mathbf{R}$, *called the* division ring of left fractions *of* $\mathbf{R}$.
(2) Assuming that the above conditions hold, the embedding $\lambda : \mathbf{R} \hookrightarrow \mathbf{R}^{\times -1}\mathbf{R}$ *is universal* $\mathbf{R}^{\times}$*-inverting. If* $\mathbf{R}$ *is an Ore domain, its division ring of left fractions and its division ring of right fractions coincide; they are called the* division ring of fractions *of* $\mathbf{R}$ *and are denoted by* $\mathrm{Q}(\mathbf{R})$.
(3) Any commutative domain is an Ore domain.
(4) Any Ore domain is semiprime Goldie (Remark 334(1)).

Theorem 337. *A domain either is a left Ore domain or it contains free left ideals of infinite rank (see Definition 407 below). In particular, any left Noetherian domain is a left Ore domain.*

Proof. Assuming that the domain $\mathbf{R}$ is not a left Ore domain, there exist $a, b \in \mathbf{R}^{\times}$ such that $\mathbf{R}\, a \cap \mathbf{R}\, b = 0$. Therefore, $\mathbf{R}\, a \cap \mathbf{R}\, a\, b^{i} \subseteq \mathbf{R}\, a \cap \mathbf{R}\, b = 0$ for any $i \geq 1$, and for any $n > m$, $\mathbf{R}\, a\, b^{m} \cap \mathbf{R}\, a\, b^{n} = 0$ since the latter equality is equivalent to $\mathbf{R}\, a \cap \mathbf{R}\, a\, b^{n-m} = 0$. Let $\mathfrak{a}_{n} = \sum_{0 \leq i \leq n} \mathbf{R}\, a\, b^{i} \lhd_{l} \mathbf{R}$. This sum is direct, i.e., $\mathfrak{a}_{n} = \bigoplus_{0 \leq i \leq n} \mathbf{R}\, a\, b^{i}$. We have $\mathbf{R} \cong \mathbf{R}\, a\, b^{i}$ under the isomorphism $x \overset{\sim}{\mapsto} x\, a\, b^{i}$, thus $\mathfrak{a}_{n}$ is free of rank n. Since $\mathfrak{a}_{n} \subsetneq \mathfrak{a}_{n+1}$, $\varinjlim \mathfrak{a}_{n} = \bigcup_{n \geq 0} \mathfrak{a}_{n}$ is free of infinite rank (Example 19). ∎

The following definition will be useful in the sequel (a more general framework is considered in [247], 3.1.11):

Definition 338. *Let* $\mathbf{R}$ *be a left Ore domain with division ring of left fractions* $\mathbf{K}$. *A* fractional left ideal $\mathfrak{b}$ *is a submodule of the left* $\mathbf{R}$*-module* $\mathbf{K}$ *such that there exists* $c \in \mathbf{R}^{\times}$ *for which* $c\,\mathfrak{b} \subseteq \mathbf{R}$. *In a similar fashion fractional right ideals and fractional ideals are defined.*

2.5.4 Modules over Rings of Fractions

Let $\mathbf{R}$ be a ring, let M be a left $\mathbf{R}$-module, let $S \subseteq \mathbf{R}^{\times}$ be a left denominator set and let $\mathbf{K} = S^{-1}\mathbf{R}$; $\mathbf{K}$ is a right $\mathbf{R}$-module, thus the tensor product $M_{(\mathbf{K})} \triangleq \mathbf{K} \bigotimes_{\mathbf{R}} M$ is a left $\mathbf{K}$-module (**2.2.7.2**).

Theorem and Definition 339. *(i) The left* $\mathbf{K}$*-module* $M_{(\mathbf{K})}$ *can be described as the set of formal products* $s^{-1}x$ $(s \in S,\ x \in M)$ *subject to the*

relations $s^{-1}x = s'^{-1}x'$ *if, and only if there exist* $q, q' \in \mathbf{R}^{\times}$ *such that* $q\,x = q'\,x'$ *and* $q\,s = q'\,s' \in S$.
(ii) The kernel of the $\mathbb{Z}$*-linear canonical map* $M \to M_{(\mathbf{K})}$ *given by* $x \mapsto 1 \otimes x$ *is the submodule* $\mathcal{T}_S(M)$ *of* M *consisting of all* $x \in M$ *for which there exists* $t \in S$ *such that* $t\,x = 0$ *(*$\mathcal{T}_S(M)$ *is called the* torsion submodule of M with respect to S*). If* $M = \mathcal{T}_S(M)$*, then* M *is called a* torsion module with respect to S.
(iii) Let M *be a torsion module with respect to* S*; then any quotient* M/N *(*$N \lhd M$*) is torsion with respect to* S.
(iv) Assuming that $\mathbf{R}$ *is a domain, let* $(a_i)_{i\in I}$ *be a family of elements of* M*. If* $(a_i)_{i\in I}$ *is free (over* $\mathbf{R}$*), then* $(1 \otimes a_i)_{i\in I}$ *is free (over* $\mathbf{K} = S^{-1}\mathbf{R}$*). In particular, if* M *is free with basis* $(a_i)_{i\in I}$*, then* $M_{(\mathbf{K})}$ *is free with basis* $(1 \otimes a_i)_{i\in I}$.
(v) Let $\mathbf{R}$ *be a left Ore domain and let* M *be a left* $\mathbf{R}$*-module. The torsion submodule of* M *with respect to* $\mathbf{R}^{\times}$ *is called the* torsion submodule *of* M*, and is denoted by* $\mathcal{T}(M)$*. If* $M = \mathcal{T}(M)$*, then* M *is called a* torsion module.

Proof. (i): Let $(a_i)_{i\in I}$ be a generating family of M. Any element of $M_{(\mathbf{K})}$ has the form $y = \sum_{i\in I} s_i^{-1}\,r_i \otimes a_i$ where the families $(s_i)_{i\in I}$, $(r_i)_{i\in I}$ of S and $\mathbf{R}$, respectively, have a finite support, and $x_j \in M$. Let $s \in S$ be a left multiple of the elements s_i (Corollary 329), i.e., $s = q_i\,s_i$. Then

$$y = \sum_{i\in I} s^{-1}\,q_i\,r_i \otimes a_i = s^{-1}\sum_{i\in I} q_i\,r_i\,a_i = s^{-1}\,x$$

where $x \in M$. Let $p = s^{-1}\,x$ and $p' = s'^{-1}\,x'$ be two elements of $M_{(\mathbf{K})}$; there exist $q, q' \in \mathbf{R}^{\times}$ such that $q\,s = q\,s' = d \in S$ (Corollary 329), thus $d\,p = q\,x$, $d\,p' = q\,x'$; therefore, $p = p'$ if, and only if $q\,x = q\,x'$ and $q\,s = q\,s' \in S$.

(ii): Since s is a unit of $\mathbf{K}$, $p = 0$ if, and only if $s\,p = 0 = 1^{-1}0$; this equality holds if, and only if there exists $q \in \mathbf{R}^{\times}$ such that $q\,s\,x = 0$ and $q\,s \in S$; setting $t = q\,s$, we see that $p = 0$ if, and only if $t\,x = 0$ for some $t \in S$.

(iii): Let M be such that for any $x \in M$ there exists $t \in S$ such that $t\,x = 0$, and let $\bullet\varphi : M \twoheadrightarrow M/N$ be the canonical epimorphism. Then $t\,(x)\,\varphi = (t\,x)\,\varphi = 0$.

(iv): If $(a_i)_{i\in I}$ is a free family of M and the above element y of $M_{(\mathbf{K})}$ is zero, there exists $t \in S \subseteq \mathbf{R}^{\times}$ such that $t\,x = 0$, thus $\sum_{i\in I} t\,q_i\,r_i\,a_i = 0$ and $t\,q_i\,r_i = 0$ for all $i \in I$; therefore, $r_i = 0$ for all $i \in I$ and $(1 \otimes a_i)_{i\in I}$ is a free family of $M_{(\mathbf{K})}$. ■

Corollary and Definition 340. *340(1) Let* $\mathbf{R}$ *be a ring, let* M *be a left* $\mathbf{R}$*-module, let* $S \subseteq \mathbf{R}^{\times}$ *be a left denominator set, let* $\mathbf{K} = S^{-1}\mathbf{R}$ *and let* $M_{(\mathbf{K})} \triangleq \mathbf{K} \bigotimes_{\mathbf{R}} M$*; then* $M/\mathcal{T}_S(M) \cong M\tau$ *where* $\bullet\tau : M \to M_{(\mathbf{K})}$ *is the canonical map.*
(2) In particular, if $\mathcal{T}_S(M) = 0$*,* τ *is an embedding. In that case,* M *is said to be* S-torsion-free, *and* torsion-free *if* $\mathbf{R}$ *is a left Ore domain and* $S = \mathbf{R}^{\times}$.

Item (1) below is a consequence of Theorem and Definition 339:

Corollary and Definition 341. *Let* $\mathbf{R}$ *be a left Ore domain and let* M *be a left* $\mathbf{R}$*-module.*

(1) Then:
(i) if M *is free, it is torsion-free;*
(ii) the dimension of the left $\mathbf{K}$*-vector space* $\mathbf{K} \otimes_{\mathbf{R}} M$ *(see Definition 407 below) equals the cardinal of a maximal* $\mathbf{R}$*-linearly independent subset of* M *(Theorem 256).*
(2) The above cardinal is called the rank *of the module* M*.*

To complement the above result, let us state the following, due to Gentile ([136], Prop. 4.1):

Proposition 342. *Let* $\mathbf{R}$ *be a left Ore domain. Every finitely generated torsion-free left* $\mathbf{R}$*-module is embeddable in a finite free left* $\mathbf{R}$*-module if, and only if* $\mathbf{R}$ *is an Ore domain.*

2.6 Local Rings

2.6.1 Definition of a Local Ring

Lemma and Definition 343. *Let* $\mathbf{R} \neq \{0\}$ *be a ring.*

(1) The following conditions are equivalent:
(i) $\mathbf{R}$ *has a unique maximal left ideal;*
(ii) $\mathbf{R}$ *has a unique maximal right ideal;*
(iii) $\mathbf{R}/\operatorname{rad}\mathbf{R}$ *is a division ring;*
(iv) $\mathbf{R}\backslash\mathbf{U}(\mathbf{R})$ *is an ideal.*
(2) A ring $\mathbf{R} \neq \{0\}$ *is* local *if the above equivalent conditions hold. Then the division ring* $\mathbf{R}/\operatorname{rad}\mathbf{R}$ *is called the* residue class division ring *of* $\mathbf{R}$ *(and its* residue class field *if* $\mathbf{R}$ *is commutative).*

Proof. (1) (i)⇒(iii) by Theorem 227(1) and (iii)⇒(i) by Corollary 242. Therefore, (ii)⇔(iii) by symmetry. (iii)⇒(iv): By Lemma 292, any $x \notin \operatorname{rad}\mathbf{R}$ is a unit, thus $\mathbf{R}\backslash\mathbf{U}(\mathbf{R}) = \operatorname{rad}\mathbf{R}$ which is an ideal. ■

Example 344. *A division ring* $\mathbf{K}$ *is a local ring (and its residue class division ring is* $\mathbf{K}$*).*

2.6.2 Characterization of a Local Ring; Local Homomorphism

Proposition 345. *Let* $\mathbf{R} \neq \{0\}$ *be a ring. Then* $\mathbf{R}$ *is local if, and only if* $\operatorname{rad}\mathbf{R} = \mathbf{R}\backslash\mathbf{U}(\mathbf{R})$*.*

Proof. If $\mathbf{R}$ is local, then $\operatorname{rad}\mathbf{R} = \mathbf{R}\backslash\mathbf{U}(\mathbf{R})$, as shown in the proof of Lemma 343(1). Conversely, if $\mathbf{R} \neq \{0\}$ is a ring such that $\operatorname{rad}\mathbf{R} = \mathbf{R}\backslash\mathbf{U}(\mathbf{R})$, $\operatorname{rad}\mathbf{R}$ is the unique maximal left ideal of $\mathbf{R}$ and $\mathbf{R}$ is local. ∎

Definition 346. *Let $\mathfrak{o}, \mathfrak{o}'$ be two local rings; a ring-homomorphism $f : \mathfrak{o} \to \mathfrak{o}'$ is said to be* local *if $f(\operatorname{rad}\mathfrak{o}) \subseteq \operatorname{rad}\mathfrak{o}'$.*

2.6.3 $\mathfrak{a}$-Adic Completion

Let $\mathbf{R}$ be a ring and $\mathfrak{a}$ be an ideal in $\mathbf{R}$. By Theorem 241(ii),

$$\mathbf{R}/\mathfrak{a}^n \cong \mathfrak{a}/\mathfrak{a}^{n+1} \subseteq \mathbf{R}/\mathfrak{a}^{n+1} \quad (n \geq 1)$$

therefore,

$$\mathbf{R}/\mathfrak{a} \hookrightarrow \mathbf{R}/\mathfrak{a}^2 \hookrightarrow \ldots \hookrightarrow \mathbf{R}/\mathfrak{a}^n \hookrightarrow \ldots \tag{2.27}$$

Let $\mathcal{A} = \{\mathfrak{a}^n : n \geq 1\}$; $\mathcal{A}$ is a base of a filter $\mathcal{F}$ which is a fundamental system of neighborhoods of zero in a topology $\mathcal{T}$ such that $(\mathbf{R}, \mathcal{T})$ is a topological ring ([34], n°III.6.3, Example 3).

Definition 347. *The above topology $\mathcal{T}$ is called the $\mathfrak{a}$-adic topology, and $\mathfrak{a}$ is called a* defining ideal *of that topology.*

A sequence (a_i) of $\mathbf{R}$ is Cauchy in the $\mathfrak{a}$-adic topology if for any natural integer n, there exists a natural integer k such that $a_i - a_j \in \mathfrak{a}^n$ for $i, j \geq k$. The *completion* $\hat{\mathbf{R}}$ of $\mathbf{R}$ can be defined with respect to the $\mathfrak{a}$-adic topology if the latter is Hausdorff, i.e.,

$$\bigcap_{n \geq 1} \mathfrak{a}^n = 0. \tag{2.28}$$

Then, there exists a canonical embedding $\iota : \mathbf{R} \hookrightarrow \hat{\mathbf{R}}$.

Remark 348. *Consider the inverse system $\left\{F_i, \psi_i^j\right\}$ (Definition 21) where $F_i = \mathbf{R}/\mathfrak{a}^i$ and $\psi_i^j : F_j \to F_i$ is the map $x + \mathfrak{a}^j \mapsto x + \mathfrak{a}^i$ for $i \leq j$ (this map is that induced by $Id_{\mathbf{R}}$ with respect to $\mathfrak{a}^j$ and $\mathfrak{a}^i$ (Theorem and Definition 145)). Assuming that (2.28) holds, $\hat{\mathbf{R}} = \varprojlim F_i$.*

2.7 Skew Polynomials: Elementary Properties

2.7.1 Polynomials in a Central Indeterminate

The definition of the ring $\mathbf{R}[X]$ of all polynomials with indeterminate X and coefficients in a commutative ring $\mathbf{R}$ is classical. The same definition holds when $\mathbf{R}$ is any ring (not necessarily commutative) and X is a *central indeterminate*, i.e., an indeterminate which commutes with all elements of $\mathbf{R}$: then any element $f(X)$ of $\mathbf{R}[X]$ can be uniquely written in the form

$$\boxed{f(X) = \sum_{0 \le i \le n} f_i \, X^{n-i}, \quad f_i \in \mathbf{R}} \tag{2.29}$$

where $f_0 \neq 0$ if $f(X) \neq 0$.

Definition 349. *(i) The term $f_n X^0$ is called the* constant term *of $f = f(X)$ and it is identified with $f_n \in \mathbf{R}$. The term of degree i of f $(i \in \mathbb{N})$ is $f_{n-i} X^i$. The* degree *of $f \neq 0$ is n and is denoted by $d^\circ(f)$; if $f = 0$, $d^\circ(f) \triangleq -\infty$. The term $f_0 X^n$ is called the* leading term *of f and f_0 is called the* leading coefficient *of f. A polynomial whose leading coefficient is equal to 1 is said to be* monic.
(ii) Assuming that $\mathbf{R}$ is commutative, let $\mathbf{L}$ be a unitary $\mathbf{R}$-algebra; an element $a \in \mathbf{L}$ is called a root *of the polynomial $f(X)$ if $f(a) = 0$.*
(iii) Let $\mathbf{K}$ be a division ring; a polynomial $f(X) \in \mathbf{K}[X]$ (with central indeterminate X) is said to be irreducible *if it has degree ≥ 1 and it cannot be written as a product $g(X)\,h(X)$ with $g, h \in \mathbf{K}[X]$ and both $g, h \notin \mathbf{K}$.*

Remark 350. *(i) A polynomial $f(X) \in \mathbf{K}[X]$ (where $\mathbf{K}$ is a field) is irreducible if, and only if the ideal (f) is maximal in the ring $\mathbf{K}[X]$; by Proposition 459 below, this holds true if, and only if $f(X)$ is a prime in the principal ideal domain $\mathbf{K}[X]$.*
(ii) If the ring $\mathbf{R}$ is a domain, so is $\mathbf{R}[X]$.
(iii) The definition of a root of a polynomial can be extended in an obvious way to the case of a multivariate polynomial *(see Exercise 493 for the definition of this expression).*

Let $\mathbf{R}$ be a commutative ring. The *derivative* of a polynomial

$$f(X) = f_0 \, X^n + f_1 \, X^{n-1} + ... + f_n \in \mathbf{R}[X] \tag{2.30}$$

$(f_0 \neq 0)$ is defined as

$$f'(X) = n \, f_0 \, X^{n-1} + (n-1) \, f_1 \, X^{n-2} + ... + f_{n-1}. \tag{2.31}$$

Lemma 351. *Let $\mathbf{R}$ be a commutative ring, $f(X) \in \mathbf{R}[X]$ a polynomial of positive degree and α a root of f. The following conditions are equivalent:*
(i) the root α is simple;
(ii) $f'(\alpha) \neq 0$.

Proof. Let α be a root of f of order n. We have $f(X) = (X - \alpha)^n g(X)$ where $g \in \mathbf{R}[X]$ is such that $g(\alpha) \neq 0$, thus

$$f'(X) = n\,(X - \alpha)^{n-1} g(X) + (X - \alpha)^n g'(X).$$

Therefore, $f'(\alpha) = n\,0^{n-1} g(\alpha)$. If $n = 1$, then $f'(\alpha) = g(\alpha) \neq 0$; if $n > 1$, then $f'(\alpha) = 0$. ∎

The following result is well-known and is generalized below (Theorem 358).

Theorem 352. (Hilbert's basis theorem). *Let* $\mathbf{R}$ *be a left (resp., right) Noetherian ring. Then the polynomial ring* $\mathbf{R}[X]$ *in the central indeterminate* X *is again left (resp., right) Noetherian.*

Let $\mathbf{K}$ be a field. The field of fractions of $\mathbf{K}[X]$, i.e., the field consisting of all elements of the form $g(X)/f(X)$, $g(X) \in \mathbf{K}[X]$, $f(X) = \mathbf{K}[X]^{\times}$, is called the *field of rational functions* in the indeterminate X with coefficients in $\mathbf{K}$ and denoted by $\mathbf{K}(X)$.

Since $g(X)/f(X)$ remains unchanged when both $g(X)$ and $f(X)$ are divided by the leading coefficient of $f(X)$, the denominator $f(X)$ of a rational function $g(X)/f(X)$ can be assumed to be monic without loss of generality.

2.7.2 Differential Operators

Let us detail the hints given about skew polynomials in Section 2.1. Consider a differential ring $\mathbf{R}$, e.g., $\mathbb{C}[t]$ endowed with the usual derivative $\boldsymbol{\delta} = d/dt$, and the ring of polynomials $\mathbf{R}[X]$. The underlying additive group of this ring is denoted by $\mathbf{R}[X]_+$. Let this group be acting on, e.g., the space of functions $\mathcal{E}(\mathbb{R})$ (**1.7.3.5**) as follows:

(i) To $a \in \mathbf{R}$, we associate the left multiplication by a, i.e., the operator $a\bullet : f \mapsto a\,f$;

(ii) To the indeterminate X we associate the derivation $\partial : f \mapsto \partial f \triangleq df/dt$.

The derivation ∂ is the extension to $\mathcal{E}(\mathbb{R})$ of the derivation $\boldsymbol{\delta}$ defined in $\mathbf{R}$. We have according to the Leibniz rule: $\partial(a\,f) = a\,\partial f + a^{\boldsymbol{\delta}}\,f$.

Coming back to the indeterminate X, (ii) yields $(X\,a)\,f = \left(a\,X + a^{\delta}\right)f$ and we obtain the commutation rule (2.2) which determines the left multiplication of an element of $\mathbf{R}$ by X as a function of the right multiplication; by induction this rule determines a ring structure for $\mathbf{R}[X]_+$. The ring obtained in this way is denoted by $\mathbf{R}[X;\boldsymbol{\delta}]$ and is called a *ring of skew polynomials.* By (2.2), the ring $\mathbf{R}[X;\boldsymbol{\delta}]$ is noncommutative except if $\mathbf{R}$ is a commutative ring of constants.

The space $\mathcal{E}(\mathbb{R})$ has now an obvious structure of $\mathbf{R}[X,\boldsymbol{\delta}]$-module, and so do as well the space of polynomial functions $\mathbb{C}[t]$, the space of distributions $\mathcal{D}'(\mathbb{R})$ (**1.7.3.5**) and the space of hyperfunctions $\mathcal{B}(\mathbb{R})$ (Sect. 1.9).

In what follows, the coefficients of $\mathbf{R}[X;\boldsymbol{\delta}]$ are written on the left, i.e., an element of this ring is of the form (2.29).

By (i), (ii) we associate to the *skew polynomial ring* $\mathbf{R}[X,\boldsymbol{\delta}]$ the *ring of linear differential operators* $\mathbf{R}[\partial,\boldsymbol{\delta}]$. These two rings are isomorphic, thus can be identified when this is useful.

If $\mathbf{R}$ is a ring of constants, $\mathbf{R}[\partial,\boldsymbol{\delta}]$ is a ring of linear differential operators *with constant coefficients*; then, one can set $\boldsymbol{\delta} = 0$ and $\mathbf{R}[\partial, 0]$ is written $\mathbf{R}[\partial]$ (note that $\boldsymbol{\delta} = 0$ does not imply $\partial = 0$!).

2.7.3 Difference Operators

The statements in Subsect. 2.7.2 can be repeated almost word-for-word in the case of difference operators. Recall that $\mathbf{k} = \mathbb{R}$ or $\mathbb{C}$.

Let $\mathbf{k}^{\mathbb{Z}}$ be the $\mathbf{k}$-vector space consisting of all sequences $f : \mathbb{Z} \to \mathbf{k}$. Consider a difference ring $\mathbf{R}$, e.g., $\mathbf{k}[t]$ endowed with the shift-forward operator $\boldsymbol{\alpha} : a(t) \mapsto a(t+1)$. Let the additive group $\mathbf{R}[Y]_+$ be acting on $\mathbf{k}^{\mathbb{Z}}$ as follows:

(i) To $a \in \mathbf{k}$ we associate the left multiplication by a, i.e., the operator $a\bullet : f \mapsto a\, f$;

(ii) To the indeterminate X we associate the shift-forward operator $\mathbf{q} : f \mapsto \mathbf{q}f$ where $\mathbf{q}f(t) \triangleq f(t+1)$.

Let $a \in \mathbf{R}$ and $f \in \mathbf{k}^{\mathbb{Z}}$; we have $\mathbf{q}(a\, f) = a^{\boldsymbol{\alpha}}\, \mathbf{q}\, f$ and $\mathbf{q}$ is the extension to $\mathbf{k}^{\mathbb{Z}}$ of the shift $\boldsymbol{\alpha}$ defined in $\mathbf{R}$. Coming back to the indeterminate Y, (ii) yields $(Y\, a)\, f = (a^{\boldsymbol{\alpha}}\, Y)\, f$ and we obtain the "commutation rule"

$$Y\, a = a^{\boldsymbol{\alpha}}\, Y. \tag{2.32}$$

This rule determines a ring structure for $\mathbf{R}[Y]_+$ and the ring obtained in this way is denoted by $\mathbf{R}[Y; \boldsymbol{\alpha}]$; it is noncommutative except if $\mathbf{R}$ is a commutative ring of constants.

By (i), (ii) we associate to the skew polynomial ring $\mathbf{R}[Y; \boldsymbol{\alpha}]$ the *ring of linear difference operators* $\mathbf{R}[\mathbf{q}; \boldsymbol{\alpha}]$. These two rings are isomorphic, thus can be identified.

The space $\mathbf{k}^{\mathbb{Z}}$ has an obvious structure of left $\mathbf{R}[\mathbf{q}; \boldsymbol{\alpha}]$-module, and so does as well, e.g., the subspace $\mathbf{k}^{(\mathbb{Z})}$ of $\mathbf{k}^{\mathbb{Z}}$ consisting of those sequences which have a finite support.

2.7.4 General Skew Polynomials

2.7.4.1 Introduction

Let $\mathbf{R}[\mathbf{q}; \boldsymbol{\alpha}]$ be a ring of difference operators and let $\mathbf{R}[Y; \boldsymbol{\alpha}]$ be the associated skew polynomial ring. Set $X = Y - 1$. By (2.32), $Y\, a = a^{\boldsymbol{\alpha}}\, Y$, therefore $X\, a = a^{\boldsymbol{\alpha}}\, X - a = a^{\boldsymbol{\alpha}}\, X - a^{\boldsymbol{\alpha}} + a^{\boldsymbol{\alpha}} - a$, i.e.,

$$\boxed{X\, a = a^{\boldsymbol{\alpha}}\, X + a^{\boldsymbol{\delta}}.} \tag{2.33}$$

When $\boldsymbol{\alpha} = 1$, (2.33) reduces to the commutation rule (2.2).

2.7.4.2 Skew Polynomials

The general commutation rule (2.33) is considered in the sequel.

Definition 353. *A* left skew polynomial *(over a differential ring* $\mathbf{R}$*, endowed with an* $\boldsymbol{\alpha}$*-derivation* $\boldsymbol{\delta}$*) is an element written in a unique way in the form (2.29) where* a_n *is nonzero if* $a(X) \neq 0$*. Definition 349 is still valid.*

A *right skew polynomial* and the degree of such a polynomial can be defined similarly, writing the coefficients on the right of the powers of the indeterminate.

Denote by $\mathbf{R}[X;\boldsymbol{\alpha},\boldsymbol{\delta}]_l$ (resp., $\mathbf{R}[X;\boldsymbol{\alpha},\boldsymbol{\delta}]_r$) the set of all left (resp., right) skew polynomials as defined above. Obviously, $\mathbf{R}[X;\boldsymbol{\alpha},\boldsymbol{\delta}]_l$ and $\mathbf{R}[X;\boldsymbol{\alpha},\boldsymbol{\delta}]_r$ are additive groups.

Proposition 354. *(i)* $\mathbf{R}[X;\boldsymbol{\alpha},\boldsymbol{\delta}]_l$ *is a ring and if* $f,g \in \mathbf{R}[X;\boldsymbol{\alpha},\boldsymbol{\delta}]_l$*, then* $d^\circ(f-g) \leq \max\{d^\circ(f), d^\circ(g)\}$ *with equality if* $d^\circ(f) \neq d^\circ(g)$*. In addition,* $\mathbf{R}[X;\boldsymbol{\alpha},\boldsymbol{\delta}]_r \subseteq \mathbf{R}[X;\boldsymbol{\alpha},\boldsymbol{\delta}]_l$*.*
(ii) Let $\boldsymbol{\alpha}$ *be a monomorphism. (a) The degree of* $f \in \mathbf{R}[X;\boldsymbol{\alpha},\boldsymbol{\delta}]_r$ *is the same as that of* f *considered as an element of* $\mathbf{R}[X;\boldsymbol{\alpha},\boldsymbol{\delta}]_l$*. (b) If in addition* $\mathbf{R}$ *is a domain,* $\mathbf{R}[X;\boldsymbol{\alpha},\boldsymbol{\delta}]_l$ *is again a domain and for any* $f,g \in \mathbf{R}[X;\boldsymbol{\alpha},\boldsymbol{\delta}]_l^\times$*,* $d^\circ(fg) = d^\circ(f) + d^\circ(g)$*. (c) If* $\mathbf{R}$ *is of characteristic zero,* $\mathbf{R}[X;\boldsymbol{\alpha},\boldsymbol{\delta}]_l$ *is again of characteristic zero.*
(iii) If $\boldsymbol{\alpha}$ *is an automorphism,* $\mathbf{R}[X;\boldsymbol{\alpha},\boldsymbol{\delta}]_r = \mathbf{R}[X;\boldsymbol{\alpha},\boldsymbol{\delta}]_l$*.*

Proof. (i): Let $f(X)$ be given by (2.29) and $g(X) = \sum_{i=0}^{m} g_i X^{m-i}$ where $f_0 \neq 0$ and $g_0 \neq 0$. By the distributivity law, the product $f(X)\,g(X)$ is a sum of products $f_i X^{n-i} g_j X^{m-j}$ which can expressed as elements of $\mathbf{R}[X;\boldsymbol{\alpha},\boldsymbol{\delta}]_l$ using (2.33). The relation $d^\circ(f-g) \leq \max\{d^\circ(f), d^\circ(g)\}$ is obvious and the corresponding equality as well when $d^\circ(f) \neq d^\circ(g)$. Using (2.33), one can put an element of $\mathbf{R}[X;\boldsymbol{\alpha},\boldsymbol{\delta}]_r$ in the form of an element of $\mathbf{R}[X;\boldsymbol{\alpha},\boldsymbol{\delta}]_l$.

(ii): (a) and (c) are clear. Let us prove (b). Let $f(X)$ and $g(X)$ be as above. The leading term of $f(X)\,g(X)$ is $f_0\, g_0^{\boldsymbol{\alpha}^{n+m}} X^{n+m}$ since $g_0^{\boldsymbol{\alpha}^{n+m}} \neq 0$ (because $\boldsymbol{\alpha}$ is a monomorphism) which implies that $f_0\, g_0^{\boldsymbol{\alpha}^{n+m}} \neq 0$ (because $\mathbf{R}$ is a domain).

(iii) is easy: see ([77], Sect. 0.10, Prop. 10.2). ∎

In everything that follows, $\boldsymbol{\alpha}$ is an automorphism of the ring $\mathbf{R}$, thus $\mathbf{R}[X;\boldsymbol{\alpha},\boldsymbol{\delta}]_r$ is a ring which coincides with $\mathbf{R}[X;\boldsymbol{\alpha},\boldsymbol{\delta}]_l$; these two rings are denoted by $\mathbf{R}[X;\boldsymbol{\alpha},\boldsymbol{\delta}]$.

If $\boldsymbol{\alpha} = 1$, $\mathbf{R}[X;\boldsymbol{\alpha},\boldsymbol{\delta}]$ is denoted by $\mathbf{R}[X;\boldsymbol{\delta}]$. If $\boldsymbol{\delta} = 0$, $\mathbf{R}[X;\boldsymbol{\alpha},\boldsymbol{\delta}]$ is denoted by $\mathbf{R}[X;\boldsymbol{\alpha}]$.

When considering skew polynomials associated with (usual) differential operators, $\boldsymbol{\alpha} = 1$ and $\boldsymbol{\delta}$ is an outer derivation except if it is zero; identifying the indeterminate X with the derivation ∂, the ring of skew polynomials $\mathbf{R}[X;\boldsymbol{\delta}]$ is identified with the *ring of linear differential operators* $\mathbf{R}[\partial;\boldsymbol{\delta}]$ (with $\boldsymbol{\delta} = \partial|_{\mathbf{R}}$).

When considering skew polynomials associated with difference operators one can proceed in two equivalent ways: (i) Associating the indeterminate Y with the shift operator $\mathbf{q}$, $\boldsymbol{\delta} = 0$ and the ring of skew polynomials $\mathbf{R}[Y;\boldsymbol{\alpha}]$ is

identified with the *ring of linear difference operators* $\mathbf{R}[\mathbf{q};\boldsymbol{\alpha}]$ (with $\boldsymbol{\alpha}=\mathbf{q}|_{\mathbf{R}}$).
(ii) On changing the indeterminate Y to $X=Y-1$, X is associated with the *differencing operator* $\partial=\mathbf{q}-1$ and $\boldsymbol{\delta}$ is the inner $\boldsymbol{\alpha}$-derivation induced by 1; the ring of skew polynomials $\mathbf{R}[X;\boldsymbol{\alpha},\boldsymbol{\delta}]$ is identified with $\mathbf{R}[\partial;\boldsymbol{\alpha},\boldsymbol{\delta}]$, called the *ring of linear differencing operators*. The above is summarized in the following definition:

Definition 355. *Let* $\mathbf{R}$ *be a ring,* $\boldsymbol{\alpha}$ *an automorphism and* $\boldsymbol{\delta}$ *an* $\boldsymbol{\alpha}$*-derivation of* $\mathbf{R}$*. Consider the skew polynomial ring* $\mathbf{R}[X;\boldsymbol{\alpha},\boldsymbol{\delta}]$ *and let* ∂ *be the operator defined as the action of* X *on any* $\mathbf{R}[X;\boldsymbol{\alpha},\boldsymbol{\delta}]$*-module. Then* $\mathbf{R}[\partial;\boldsymbol{\alpha},\boldsymbol{\delta}]$ *is called a* ring of linear generalized differential operators *and* $\boldsymbol{\delta}=\partial|_{\mathbf{R}}$*. Consider the following cases: (i)* $\boldsymbol{\alpha}=1$ *or (ii)* $\boldsymbol{\delta}$ *is the inner* $\boldsymbol{\alpha}$*-derivation induced by* 1*.*

In case (i), $\mathbf{R}[\partial;1,\boldsymbol{\delta}]=\mathbf{R}[\partial;\boldsymbol{\delta}]$ *is a* ring of linear differential operators.
In case (ii), $\mathbf{R}[\partial;\boldsymbol{\alpha},\boldsymbol{\delta}]$ *is a* ring of linear differencing operators *and* $\mathbf{R}[\mathbf{q};\boldsymbol{\alpha}]$*, where* $\mathbf{q}=\partial+1$*, is a* ring of linear difference operators*; in addition,* $\boldsymbol{\alpha}=\mathbf{q}|_{\mathbf{R}}$*.*

When $\mathbf{R}$ is a field $\mathbf{K}$, cases (i) and (ii) of the above definition are essentially the only ones to be considered. Indeed:

Lemma 356. *Let* $\mathbf{K}$ *be a field. If* $\boldsymbol{\alpha}\neq 1$*, every* $\boldsymbol{\alpha}$*-derivation* $\boldsymbol{\delta}$ *of* $\mathbf{K}$ *is inner; in addition,* $\mathbf{K}[X;\boldsymbol{\alpha},\boldsymbol{\delta}]$ *is isomorphic to* $\mathbf{K}[Y;\boldsymbol{\alpha}]$ *where* $Y=(a^{\boldsymbol{\alpha}}-a)\,X+a^{\boldsymbol{\delta}}$ *and* a *is any element of* $\mathbf{K}$ *such that* $a\neq a^{\boldsymbol{\alpha}}$*. Therefore, if* $\boldsymbol{\delta}$ *is an outer* $\boldsymbol{\alpha}$*-derivation of* $\mathbf{K}$*, then* $\boldsymbol{\alpha}=1$*.*

Proof. Let $Y=X\,a-a\,X$ and let $c\in\mathbf{K}$. Then

$$Y\,c=X\,c\,a-a\,X\,c=\left(c^{\boldsymbol{\alpha}}\,X+c^{\boldsymbol{\delta}}\right)a-a\left(c^{\boldsymbol{\alpha}}\,X+c^{\boldsymbol{\delta}}\right)=c^{\boldsymbol{\alpha}}Y$$

and $Y=(a^{\boldsymbol{\alpha}}-a)\,X+a^{\boldsymbol{\delta}}$. If $\boldsymbol{\alpha}\neq 1$, one can prove that $\boldsymbol{\delta}$ is inner using the Skolem-Noether theorem ([78], Sect. 2.1, Exercise 3). ∎

Let $\mathbf{R}$ be a left Ore domain, $\mathbf{K}$ its division ring of left fractions, $\boldsymbol{\alpha}$ an automorphism and $\boldsymbol{\delta}$ an $\boldsymbol{\alpha}$-derivation of $\mathbf{R}$. Setting $\mathbf{F}=\mathbf{K}(X;\boldsymbol{\alpha},\boldsymbol{\delta})$ we have

$$\mathbf{R}[X;\boldsymbol{\alpha},\boldsymbol{\delta}]\subseteq\mathbf{K}[X;\boldsymbol{\alpha},\boldsymbol{\delta}]\subseteq\mathbf{F}.$$

Any element $u\in\mathbf{F}$ is of the form $g^{-1}f$ where $f,g\in\mathbf{K}[X;\boldsymbol{\alpha},\boldsymbol{\delta}]$, $g\neq 0$. On bringing the coefficients of f and g to a common left denominator (Corollary 329), we can write $f=c^{-1}f_1$, $g=c^{-1}g_1$ where $f_1,g_1\in\mathbf{R}[X;\boldsymbol{\alpha},\boldsymbol{\delta}]$. Therefore, $u=g_1^{-1}f_1$ which proves that $\mathbf{F}$ is the division ring of left fractions of $\mathbf{R}[X;\boldsymbol{\alpha},\boldsymbol{\delta}]$. In addition,

$$\mathbf{R}[X;\boldsymbol{\alpha},\boldsymbol{\delta}]\subseteq\mathbf{F}.$$

The following is now a consequence of Theorem 336:

Theorem 357. *Let* $\mathbf{R}$ *be a left Ore domain,* $\boldsymbol{\alpha}$ *an automorphism and* $\boldsymbol{\delta}$ *an* $\boldsymbol{\alpha}$*-derivation of* $\mathbf{R}$*, and let* $\mathbf{K}$ *be the division ring of left fractions of* $\mathbf{R}$*. The* *skew polynomial ring* $\mathbf{R}[X;\boldsymbol{\alpha},\boldsymbol{\delta}]$ *is again a left Ore domain and the set of nonzero elements forms a left denominator set of this ring. The division ring of left fractions of* $\mathbf{R}[X;\boldsymbol{\alpha},\boldsymbol{\delta}]$ *is* $\mathbf{K}(X;\boldsymbol{\alpha},\boldsymbol{\delta})$. *If* $\mathbf{R}$ *is of characteristic zero, then* $\mathbf{R}[X;\boldsymbol{\alpha},\boldsymbol{\delta}]$ *is again of characteristic zero.*

2.7.4.3 Extension of Hilbert's Basis Theorem

Theorem 358. *Let* $\mathbf{R}$ *be a left (resp., right) Noetherian ring,* $\boldsymbol{\alpha}$ *an automorphism and* $\boldsymbol{\delta}$ *an* $\boldsymbol{\alpha}$*-derivation of* $\mathbf{R}$. *Then the skew polynomial ring* $\mathbf{A}=\mathbf{R}[X;\boldsymbol{\alpha},\boldsymbol{\delta}]$ *is again left (resp., right) Noetherian.*

Proof. Assume that $\mathbf{R}$ is right Noetherian and consider $\mathbf{A}$ as a left skew polynomial ring. If $\mathbf{A}$ is not right Noetherian there exists a right ideal $\mathfrak{a}$ in $\mathbf{A}$ which is not f.g. by Lemma 311. Let $f_1\in\mathfrak{a}$ be a skew polynomial of least degree; given $f_1,...,f_k\in\mathfrak{a}$ we take f_{k+1} in $\mathfrak{a}\backslash(f_1,...,f_k)_r$ of least degree. Since $\mathfrak{a}$ in $\mathbf{A}$ is not f.g. we obtain an infinite sequence $f_1,f_2,...$ Let n_i and a_i be the degree and the leading coefficient of f_i, respectively; then $n_i\leq n_{i+1}$. We claim that $a_1\mathbf{R}\subsetneq a_1\mathbf{R}+a_2\mathbf{R}\subsetneq ...$ is an infinite chain, which contradicts the fact that $\mathbf{R}$ is right Noetherian. Indeed, assume that the above chain is not infinite; then there exists $k\geq 1$ such that $a_{k+1}\mathbf{R}\subseteq a_1\mathbf{R}+...+a_k\mathbf{R}$ and there exist elements $b_j\in\mathbf{R}$ $(1\leq j\leq k)$ such that $a_{k+1}=\sum_{1\leq j\leq k}a_j\,b_j$. Therefore,

$$g=f_{k+1}-\sum_{1\leq j\leq k}f_j\,b_j^{\alpha^{-n_j}}\,X^{n_{k+1}-n_j}\in\mathfrak{a}\backslash(f_1,...,f_k)_r$$

and $d^{\circ}(g)<d^{\circ}(f_{k+1})$, a contradiction. ∎

When $\mathbf{R}=\mathbf{k}[t]$, the ring $\mathbf{R}[\partial;\boldsymbol{\delta}]\cong\mathbf{k}[t][X;\boldsymbol{\delta}]=\mathbf{k}[t][X;1,\boldsymbol{\delta}]$ $(\boldsymbol{\delta}=d/dt)$ is called the *first Weyl algebra* over $\mathbf{k}$, denoted by $A_1(\mathbf{k})$; when $\mathbf{R}=\mathbf{k}(t)$, the ring $\mathbf{R}[\partial;\boldsymbol{\delta}]\cong\mathbf{k}(t)[X;\boldsymbol{\delta}]=\mathbf{k}(t)[X;1,\boldsymbol{\delta}]$ $(\boldsymbol{\delta}=d/dt)$ is denoted by $B_1(\mathbf{k})$. By Proposition 354(ii) and Theorem 358 we have the following.

Corollary 359. *The rings* $A_1(\mathbf{k})$ *and* $B_1(\mathbf{k})$ *are Noetherian domains.*

2.7.4.4 Transpose (or adjoint) of a Generalized Linear Differential Operator

(I) Consider a $q\times p$ matrix P, the entries of which belong to a ring of partial differential operators $\mathbf{A}=\mathbf{K}[\partial_1,...,\partial_n;\boldsymbol{\delta}_1,...,\boldsymbol{\delta}_n]$ where $\mathbf{K}$ is both a commutative $\mathbf{k}$-algebra and a differential ring, such as in Exercise 493, where $\boldsymbol{\alpha}_k=1$, ∂_k the usual derivative with respect to the kth variable $(1\leq k\leq n)$ and $\boldsymbol{\delta}_k=\partial_k|_{\mathbf{K}}$. Let Ω be an open subset of $\mathbb{R}^n$ and let $\mathcal{T}(\Omega)$ be a space of test functions in Ω. Assume that these test functions are C^{∞} and that $\mathcal{T}(\Omega)$ is a Hausdorff LCS. Let $\mathcal{T}'(\Omega)$ be the dual of $\mathcal{T}(\Omega)$.

Assume $P\bullet$ is a weakly* continuous linear map ${}^{p}\mathcal{T}'(\Omega) \to {}^{q}\mathcal{T}'(\Omega)$. The transpose ${}^{t}P\bullet$ of $P\bullet$ is the weakly continuous linear map ${}^{q}\mathcal{T}(\Omega) \to {}^{p}\mathcal{T}(\Omega)$ defined by

$$\langle \psi, PT \rangle = \langle {}^{t}P\,\psi, T \rangle \tag{2.34}$$

for all $T \in {}^{p}\mathcal{T}'(\Omega)$, $\psi \in {}^{q}\mathcal{T}(\Omega)$ (**1.7.2.3**). In (2.34) all maps are *right ones* (**1.2.1.1**). As in Subsect. 2.2.5, this equality becomes simpler when denoting the transpose of $(P\bullet)$ by the *left map* $(\bullet P)$. Then, indeed, (2.34) becomes

$$\langle \psi, PT \rangle = \langle \psi P, T \rangle \tag{2.35}$$

for all $T \in {}^{p}\mathcal{T}'(\Omega)$, $\psi \in \mathcal{T}(\Omega)^{q}$. Recall that ${}^{t}({}^{t}P) = P$.

(II) A similar rationale makes sense with spaces of multi-indexed sequences $\mathfrak{s}$ and $\mathfrak{s}'$ in duality with respect to the duality bracket

$$\langle s, s' \rangle = \sum_{m \in \mathbb{Z}^{n}} s(m)\, s'(m)\,.$$

(III) To treat the general case, one must change the above ring of differential operators $\mathbf{A}$ to the ring of *generalized* differential operators $\mathbf{K}\left[(\partial_k; \boldsymbol{\alpha}_k, \boldsymbol{\delta}_k)_{1\leq k\leq n}\right]$ (Exercise 493). Assuming that $\boldsymbol{\alpha}_k$ is an automorphism of $\mathbf{K}$ with inverse $\boldsymbol{\beta}_k$, the commutation rule

$$\partial_k a = a^{\boldsymbol{\alpha}_k} \partial_k + a^{\boldsymbol{\delta}_k} \tag{2.36}$$

yields

$$a(-\partial_k) = (-\partial_k)\, a^{\boldsymbol{\beta}_k} + a^{\boldsymbol{\beta}_k \boldsymbol{\delta}_k}. \tag{2.37}$$

Lemma and Definition 360. *(1) The* adjoint map *is* $\mathrm{ad} : (P\bullet) \mapsto (\bullet P)$.
(2) The differential operator $({}^{t}P\bullet) = (\bullet P)$ *is called* transpose, *or, by abuse of language, the* adjoint *of the linear differential operator* $(P\bullet)$ *and is also denoted by* $(P^{*}\bullet) = \mathrm{ad}(P\bullet)$.
(3) The adjoint map $\mathrm{ad} : (P\bullet) \mapsto (P^{*}\bullet)$ *is determined by the following conditions:*
(i) $\mathrm{ad}(P) = Q^{T}$ *where* $Q_{ij} = \mathrm{ad}(P_{ij})$*;*
(ii) for any $L, M \in \mathbf{A}$, $\mathrm{ad}(L\,M) = \mathrm{ad}(M)\,\mathrm{ad}(L)$.
(iii) for any $a \in \mathbf{K}$, $\mathrm{ad}(a) = a$*;*
(iv) $\mathrm{ad}(\partial_k) = -\partial_k$ $(1 \leq k \leq n)$*;*
(v) $\mathrm{ad}(\boldsymbol{\alpha}_k) = \boldsymbol{\beta}_k$ *(thus* $\mathrm{ad}(\boldsymbol{\beta}_k) = \boldsymbol{\alpha}_k$*)* $(1 \leq k \leq n)$.
(vi) $\mathrm{ad}(\boldsymbol{\delta}_k) = \boldsymbol{\beta}_k \boldsymbol{\delta}_k$ $(1 \leq k \leq n)$.

Proof. (3) is an obvious consequence of (1). ∎

Remark 361. *Consider a ring of difference operators* $\mathbf{K}\left[(\mathbf{q}_k; \boldsymbol{\alpha}_k)_{1\leq k\leq n}\right]$. *The commutation rule reduces to*

$$\mathbf{q}_k\, a = a^{\boldsymbol{\alpha}_k}\, \mathbf{q}_k,$$

therefore Item (iv) of Lemma and Definition 360(3) must be changed to
(iv') ad $(\mathbf{q}_k) = \mathbf{q}_k$ $(1 \leq k \leq n)$,
whereas (v) must be left unchanged and (vi) must be removed.

2.7.5 Simplicity of Some Skew Polynomial Rings

Let $\mathbf{R}$ be a ring, $\boldsymbol{\alpha}$ an endomorphism and $\boldsymbol{\delta}$ an $\boldsymbol{\alpha}$-derivation of $\mathbf{R}$. A subset S of $\mathbf{R}$ is said to be $\boldsymbol{\alpha}$-invariant (resp., $\boldsymbol{\delta}$-invariant) if $a^{\boldsymbol{\alpha}} \in S$ (resp., $a^{\boldsymbol{\delta}} \in S$) for any $a \in S$. We study below the simplicity of a skew polynomial ring in the case of an outer derivation (for the case of an inner derivation, see Subsect. 2.8.3).

Theorem 362. *Let* $\mathbf{R}$ *be a ring and let* $\mathbf{A} = \mathbf{R}[X, \boldsymbol{\delta}]$. *(i) If* $\mathbf{A}$ *is simple, then* $\mathbf{R}$ *has no proper nonzero* $\boldsymbol{\delta}$*-invariant ideal and* $\boldsymbol{\delta}$ *is an outer derivation of* $\mathbf{R}$. *(ii) If* $\mathbf{R}$ *is a* $\mathbb{Q}$*-algebra, the converse holds.*

Proof. (i): Consider the inner $\boldsymbol{\alpha}$-derivation $\boldsymbol{\delta}$ induced by m, i.e., $\boldsymbol{\delta} : a \mapsto a^{\boldsymbol{\alpha}} m - m\, a$; then for any $a \in \mathbf{R}$, $X\, a = a^{\boldsymbol{\alpha}}\, X + a^{\boldsymbol{\delta}} = a^{\boldsymbol{\alpha}}\, (X + m) - m\, a$ and

$$Y\, a = a^{\boldsymbol{\alpha}}\, Y \quad \text{where } Y = X + m. \tag{2.38}$$

Here, $\boldsymbol{\alpha} = 1$, thus $\mathbf{A} \cong \mathbf{R}[Y]$ is not simple. If $\mathbf{R}$ has a proper nonzero $\boldsymbol{\delta}$-invariant ideal $\mathfrak{a}$, then $\mathbf{A}\, \mathfrak{a} = \mathfrak{a}\, \mathbf{A} \subsetneq \mathbf{A}$ is an ideal in $\mathbf{A}$ and this ring is not simple.
(ii): Suppose the necessary condition holds and that $0 \neq \mathfrak{a} \lhd \mathbf{A}$. Let $\mathfrak{a}_n$ be the set of leading coefficients of all *left polynomials* belonging to $\mathfrak{a}$ and whose degree is $\leq n$; $\{\mathfrak{a}_n : n \in \mathbb{N}\}$ is called the set of *leading ideals* of $\mathfrak{a}$. Obviously, $\mathfrak{a}_n \lhd \mathbf{R}$ is $\boldsymbol{\delta}$-invariant. Choose the least n with $\mathfrak{a}_n \neq 0$; by assumption, $\mathfrak{a}_n = \mathbf{R}$, thus $1 \in \mathfrak{a}_n$. If $n = 0$, then $\mathfrak{a} = \mathbf{A}$ as required. If $n > 0$, there exists a monic left polynomial

$$f = X^n + a_{n-1}\, X^{n-1} + \ldots + a_0 \in \mathfrak{a}.$$

For any $a \in \mathbf{R}$,

$$f\, a - a\, f = \left(a_{n-1}\, a - a\, a_{n-1} - n\, a^{\boldsymbol{\delta}}\right)\, X^{n-1} + \ldots$$

and n is a unit since $\mathbf{R}$ is a $\mathbb{Q}$-algebra (so that any element of $\mathbf{R}$ can be multiplied by $1/n$). The minimality of n shows that

$$a_{n-1}\, a - a\, a_{n-1} - n\, a^{\boldsymbol{\delta}} = 0$$

thus

$$\boldsymbol{\delta} : a \mapsto \left(a_{n-1}\, a - a\, a_{n-1}\right)/n$$

is an inner derivation of $\mathbf{R}$, a contradiction. Therefore, $\mathfrak{a} = \mathbf{A}$ and $\mathbf{A}$ is simple. ∎

The following corollary is obvious:

Corollary 363. *Let* $\mathbf{R}$ *be a simple ring (e.g., a division ring) and let* $\mathbf{A} = \mathbf{R}[X, \boldsymbol{\delta}]$.

(i) If $\mathbf{A}$ *is simple, then* $\boldsymbol{\delta}$ *is an outer derivation of* $\mathbf{R}$.
(ii) If $\mathbf{R}$ *is a* $\mathbb{Q}$*-algebra, the converse holds.*

Corollary 364. *The rings* $A_1(\mathbf{k})$ *and* $B_1(\mathbf{k})$ *are simple.*

Proof. Let $\mathbf{R} = \mathbf{k}[t]$ or $\mathbf{k}(t)$. Since $\boldsymbol{\delta} = d/dt$ is an outer derivation of $\mathbf{R}$, it remains to prove that $\mathbf{R}$ has no proper nonzero $\boldsymbol{\delta}$-invariant ideal. This is obvious when $\mathbf{R} = \mathbf{k}(t)$ which is a field. Consider the case $\mathbf{R} = \mathbf{k}[t]$ and assume that there exists a nonzero proper ideal $\mathfrak{a} \lhd \mathbf{R}$ which is $\boldsymbol{\delta}$-invariant. Let $f \in \mathfrak{a}^{\times}$ with leading term $f_0\, t^n$, $f_0 \in \mathbf{k}$. We have $f^{\boldsymbol{\delta}^n} = n!\, f_0$. Therefore, $n!\, f_0 \in \mathfrak{a}$ and since $n!\, f_0$ is a unit of $\mathbf{R}$, $\mathfrak{a} = \mathbf{R}$, a contradiction. ∎

Let $\mathbf{R} = \mathbf{k}\{t\}$ be the ring of convergent power series in the variable t with coefficients in $\mathbf{k}$ (2.9.1.**1**). Then $\mathbf{R}[X, \boldsymbol{\delta}]$ is denoted by $A_{1c}(\mathbf{k})$. The proof of the following is left to the reader:

Corollary 365. *The ring* $A_{1c}(\mathbf{k}) = \mathbf{R}[X, \boldsymbol{\delta}]$ *is (like* $A_1(\mathbf{k})$ *and* $B_1(\mathbf{k})$*) a simple Noetherian domain.*

See complements in Theorem 734.

2.8 Skew Laurent Polynomials, and Rational Functions

2.8.1 Skew Laurent Polynomials

Let $\mathbf{R}$ be a ring, $\boldsymbol{\alpha}$ an automorphism and $\boldsymbol{\delta}$ an $\boldsymbol{\alpha}$-derivation of $\mathbf{R}$ (Subsect. 2.3.3). Consider the skew polynomial ring $\mathbf{R}[X; \boldsymbol{\alpha}, \boldsymbol{\delta}]$ (2.7.4.**2**). The monoid $S = \{X^n : n \geq 0\}$ is a left and right Ore set since for any $r \in \mathbf{R}$ and $s \in S$, $0\, r = 0\, s = 0$ and $r\, 0 = s\, 0 = 0$; moreover, S is a left and right denominator set since all elements of S are regular. Therefore, one can form the ring of fractions $S^{-1}\mathbf{R} = \mathbf{R}\, S^{-1}$, called the *ring of skew Laurent polynomials* denoted by $\mathbf{R}\left[X, X^{-1}; \boldsymbol{\alpha}, \boldsymbol{\delta}\right]$ (and by $\mathbf{R}\left[X, X^{-1}; \boldsymbol{\delta}\right]$ if $\boldsymbol{\alpha} = 1$).

Similarly, if $\boldsymbol{\delta}$ is inner, induced by $m \in \mathbf{R}$, i.e., $a^{\boldsymbol{\delta}} = a^{\boldsymbol{\alpha}} m - m\, a$, the monoid $T = \{Y^n : n \geq 0\}$ where $Y = X + m$ is a left and right denominator set by (2.38), and one can form the ring of skew Laurent polynomials $T^{-1}\mathbf{R}[Y; \boldsymbol{\alpha}]$ denoted by $\mathbf{R}\left[Y, Y^{-1}; \boldsymbol{\alpha}\right]$.

The theorem below is obtained by the same rationale as Theorem 357:

Theorem 366. *Let* $\mathbf{R}$ *be a left Ore domain,* $\boldsymbol{\alpha}$ *an automorphism and* $\boldsymbol{\delta}$ *an* $\boldsymbol{\alpha}$*-derivation of* $\mathbf{R}$*, and let* $\mathbf{K}$ *be the division ring of left fractions of* $\mathbf{R}$*. The ring of skew Laurent polynomials* $\mathbf{R}\left[X, X^{-1}; \boldsymbol{\alpha}, \boldsymbol{\delta}\right]$ *is again a left Ore domain and the set of nonzero elements forms a left denominator set of this ring. The division ring of left fractions of* $\mathbf{R}\left[X, X^{-1}; \boldsymbol{\alpha}, \boldsymbol{\delta}\right]$ *is* $\mathbf{K}(X; \boldsymbol{\alpha}, \boldsymbol{\delta})$*. If* $\mathbf{R}$ *is of characteristic zero, then* $\mathbf{R}\left[X, X^{-1}; \boldsymbol{\alpha}, \boldsymbol{\delta}\right]$ *is again of characteristic zero.*

2.8.2 Division Ring of Skew Rational Functions

(i) Let $\mathbf{K}$ be a division ring, $\boldsymbol{\alpha}$ an automorphism and $\boldsymbol{\delta}$ an $\boldsymbol{\alpha}$-derivation of $\mathbf{K}$. By Theorem 358, $\mathbf{K}[X;\boldsymbol{\alpha},\boldsymbol{\delta}]$ is a Noetherian domain, thus it is an Ore domain by Theorem 337. Its quotient division ring is denoted by $\mathbf{K}(X;\boldsymbol{\alpha},\boldsymbol{\delta})$ and is called the *division ring of rational functions* (generated by $\mathbf{K}$ and the indeterminate X).

2.8.3 Simplicity of Some Skew Laurent Polynomial Rings

The proof of the following result is similar to that of Theorem 362; see ([247], 1.8.5) for the details.

Theorem 367. *Let $\mathbf{R}$ be a ring, $\boldsymbol{\alpha}$ an automorphism of $\mathbf{R}$ and $\mathbf{T} = \mathbf{R}\left[X, X^{-1}; \boldsymbol{\alpha}\right]$. Then $\mathbf{T}$ is simple if, and only if $\mathbf{R}$ has no nonzero proper $\boldsymbol{\alpha}$-invariant ideal (Subsect. 2.7.5) and no power of $\boldsymbol{\alpha}$ is an inner automorphism of $\mathbf{R}$.*

Corollary 368. *Let $\mathbf{R}$ be a simple ring (e.g., a division ring), $\boldsymbol{\alpha}$ an automorphism of $\mathbf{R}$ and $\mathbf{T} = \mathbf{R}\left[X, X^{-1}; \boldsymbol{\alpha}\right]$. Then $\mathbf{T}$ is simple if, and only if no power of $\boldsymbol{\alpha}$ is an inner automorphism of $\mathbf{R}$.*

Example 369. *Let $\mathbf{R} \subseteq \mathcal{M}(\mathbf{k})$ and let $\boldsymbol{\alpha}$ be the usual shift. By Lemma 289, $\boldsymbol{\alpha}^N$ is inner if, and only if $\mathbf{R}$ is N-periodic. Therefore:*

(i) the ring $A_1'(\mathbf{k}) \triangleq \mathbf{k}[t]\left[X, X^{-1}; \boldsymbol{\alpha}\right]$ is simple;
(ii) the ring $\mathbf{R}\left[X, X^{-1}; \boldsymbol{\alpha}\right]$, where $\mathbf{R} = \mathcal{PM}_N(\mathbf{k})$ (Subsect. 2.3.2), is not *simple.*

2.8.4 Crossed Products

2.8.4.1 The General Case

The general notion of *crossed product* is defined as follows ([247], 1.5.8):

Definition 370. *Let $\mathbf{T}$ be a ring, let $\mathbf{S}$ be a subring of $\mathbf{T}$ and let G be a group; let $\bar{G} = \{\varphi(g); g \in G\}$ be a set of units of $\mathbf{T}$ such that $\varphi : G \to \bar{G}$ is a bijection and*

(i) $\mathbf{T}$ is a free right $\mathbf{S}$-module with basis $\bar{G}$;
(ii) for all $g_1, g_2 \in G$, $\bar{g}_1 \mathbf{S} = \mathbf{S}\,\bar{g}_1$ and $\bar{g}_1\,\bar{g}_2\,\mathbf{S} = \overline{g_1\,g_2}\,\mathbf{S}$.
Then $\mathbf{T}$ is called a crossed product *of $\mathbf{S}$ by G, and is denoted by $\mathbf{S} * G$.*

For all $g \in G$, $\bar{g}\,\mathbf{S} = \mathbf{S}\,\bar{g}$, thus each $\bar{g}$ normalizes $\mathbf{S}$ (Definition 68) and induces an endomorphism σ_g of $\mathbf{S}$, such that $s\,\bar{g} = \bar{g}\,\sigma_g(s)$.

Lemma 371. *The endomorphism σ_g is a group-automorphism.*

Proof. Assuming that $\sigma_g(s_1) = \sigma_g(s_2)$, we have $s_1\,\bar{g} = s_2\,\bar{g}$, thus $s_1 = s_2$ by right-multiplying this equality by $\bar{g}^{-1}$; this proves that σ_g is a monomorphism. In addition, for any $s \in \mathbf{S}$, $s = \sigma_g(s_1)$ where $s_1 = \bar{g}\,s\,\bar{g}^{-1}$, thus σ_g is an epimorphism. ∎

Theorem 372. *Let $\mathbf{S}$ be a left (resp., right) Noetherian domain of characteristic zero and G a polycyclic by finite group (Definition 161). Then $\mathbf{S} * G$ is a semiprime left (resp., right) Noetherian ring, and a left (resp., right) Noetherian domain if G is polycyclic.*

Proof. Since $\mathbf{S}$ is left (resp., right) Noetherian and G polycyclic by finite, $\mathbf{S} * G$ is left (resp., right) Noetherian by ([247], 1.5.12). Consider the normal chain (1.28) where G_i/G_{i+1} is infinite cyclic (thus isomorphic to $\mathbb{Z}$) for $i \in \{1, ..., r-1\}$, and infinite cyclic (resp., either infinite cyclic or finite) if G is polycyclic (resp., polycyclic by finite).

(a) The ring $\mathbf{S} * \mathbb{Z}$ is isomorphic to a skew Laurent polynomial ring $\mathbf{S}\left[X, X^{-1}; \boldsymbol{\sigma}\right]$ where $\boldsymbol{\sigma}$ is an automorphism of $\mathbf{S}$ ([247], Proposition 1.5.11), thus it is an integral domain of characteristic zero by Theorem 366.

(b) For $i \in \{1, ..., r-1\}$, $\mathbf{S} * G_i \cong (\mathbf{S} * G_{i+1}) * \mathbb{Z}$ is an integral domain of characteristic zero by induction. This is still true for $i \in \{0, ..., r-1\}$ if G_0/G_1 is infinite cyclic, i.e., if G is polycyclic.

(c) If G_0/G_1 is finite, then $\mathbf{S}*G = \mathbf{S}*G_0 \cong (\mathbf{S} * G_1)*(G_0/G_1)$ is semiprime by ([247], Proposition 10.5.11). ∎

Let M be a left $\mathbf{T}$-module, consider the left $\mathbf{S}$-module $M_{[\mathbf{S}]}$ obtained by restriction of the ring of scalars and consider the left $\mathbf{S}$-module

$$M_{[\mathbf{S}]} * G \triangleq \mathbf{T} \bigotimes\nolimits_{\mathbf{S}} M_{[\mathbf{S}]} = \textstyle\sum_{g \in G} \bar{g}\, M_{[\mathbf{S}]}.$$

(Lemma and Definition 285).

Theorem 373. *The functor $\mathfrak{C} = (\bullet)_{[\mathbf{S}]} * G : {}_{\mathbf{T}}\mathbf{Mod} \to {}_{\mathbf{S}}\mathbf{Mod}$ is faithful exact.*

Proof. The functor $\mathfrak{F}$ can be decomposed as

$${}_{\mathbf{T}}\mathbf{Mod} \xrightarrow{\rho_*} {}_{\mathbf{S}}\mathbf{Mod} \overset{\mathbf{T} \bigotimes_{\mathbf{S}} -}{\longrightarrow} {}_{\mathbf{S}}\mathbf{Mod}.$$

where $\rho_* : {}_{\mathbf{T}}\mathbf{Mod} \xrightarrow{\rho_*} {}_{\mathbf{S}}\mathbf{Mod}$ is the restriction of the ring of scalars. This functor is faithful exact (see Subsect. 3.4.2 below) and so is the functor $\mathbf{T} \bigotimes_{\mathbf{S}} -$ too since $\mathbf{T}$ is a free, thus faithfully flat, $\mathbf{S}$-module (**3.5.3.3**). ∎

Corollary 374. *(i) Of the sequences below, the first is exact in ${}_{\mathbf{T}}\mathbf{Mod}$ if, and only if so is the second in ${}_{\mathbf{S}}\mathbf{Mod}$:*

$$N \xrightarrow{f} M \xrightarrow{g} Q, \quad \mathfrak{F}(N) \xrightarrow{\mathfrak{F}(f)} \mathfrak{F}(M) \xrightarrow{\mathfrak{F}(f)} \mathfrak{F}(M).$$

(ii) If $(a_i)_{i \in I}$ is a generating (resp., free) family of a left $\mathbf{T}$-module M, then $(\bar{g}\, a_i)_{(g,i) \in G \times I}$ is a generating (resp., free) family of the left $\mathbf{S}$-module $\mathfrak{F}(M)$.
(iii) If $(M_i)_{1 \leq i \leq n}$ is a finite family of left $\mathbf{T}$-modules and $M = \bigoplus_{1 \leq i \leq n} M_i$, then $\mathfrak{F}(M) = \bigoplus_{1 \leq i \leq n} \mathfrak{F}(M_i)$.
(iv) A left $\mathbf{T}$-module M is free if, and only if so is the left $\mathbf{S}$-module $\mathfrak{F}(M)$. In addition, $\mathfrak{F}(M)$ is finite free if, and only if M is finite free and G is finite.

Proof. (i): If the first sequence is exact, so is the second by exactness of the functor $\mathfrak{F}$. Conversely, if the first sequence is inexact, so is the second one since $\mathfrak{F}$ if faithful (Proposition 49).

(ii): Consider an element $x = \sum_{g \in G} \bar{g}\, x_g$ of $M_{[\mathbf{S}]} * G$ (where all elements of this sum are zero except a finite number of them). Assume that $(a_i)_{i \in I}$ is generating in M; since $x_g \in M$, there exists for each $g \in G$ a family $(\lambda_{gi})_{i \in I}$ of elements of $\mathbf{T}$, with finite support, such that $x_g = \sum_i \lambda_{gi}\, a_i$. Each λ_{gi} can be expressed as $\sum_{g'} \mu_{gig'}\, \bar{g}'$, where $\left(\mu_{gig'}\right)_{g' \in G}$ is a family of elements of $\mathbf{S}$ with finite support, therefore

$$x = \sum_{g,i,g'} \bar{g}\, \mu_{gig'}\, \bar{g}'\, a_i = \sum_{g,i,g'} \sigma_g\left(\mu_{gig'}\right) \overline{g\, g'}\, a_i, \tag{2.39}$$

and $g\, g' = g'' \in G$, therefore $\left(\overline{g''}\, a_i\right)_{(g'',i) \in G \times I}$ is generating in $M_{[\mathbf{S}]} * G$.

If $(a_i)_{i \in I}$ is free in M and the expression (2.39) is zero, we have $\sum_{g,i,g'} \sigma_g\left(\mu_{gig'}\right) \overline{g\, g'} = 0$ for all $i \in I$, thus $\mu_{gig'} = 0$ for all $(g, i, g') \in G \times I \times G$ and – with the above notation – $\left(\overline{g''}\, a_i\right)_{(\overline{g''},i) \in G \times I}$ is free in $M_{[\mathbf{S}]} * G$. See also [27], §II.1, Prop. 25.

(iii) is a straightforward consequence of Theorem 373 (see Definition 28).

(iv) Consider the two sequences below:

$$0 \longrightarrow \mathbf{T}^{(I)} \longrightarrow M \longrightarrow 0, \quad 0 \longrightarrow \mathfrak{F}\left(\mathbf{T}^{(I)}\right) \longrightarrow \mathfrak{F}(M) \longrightarrow 0.$$

By (i), the first sequence is exact (i.e., $M \cong \mathbf{T}^{(I)}$) if, and only if so is the second (i.e., $\mathfrak{F}(M) \cong \mathfrak{F}\left(\mathbf{T}^{(I)}\right)$), and by (ii) $\mathfrak{F}\left(\mathbf{T}^{(I)}\right) \cong \mathbf{S}^{(G \times I)}$. ■

In the sequel, the group G is finite with order N, i.e., $G = \left\{1, g, ..., g^{N-1}\right\}$; then $\mathbf{T}$ is called a *finite normalizing extension* of $\mathbf{S}$.

2.8.4.2 Periodic Skew Laurent Polynomials

Our aim is now to study $\mathbf{T} = \mathbf{R}\left[X, X^{-1}; \boldsymbol{\alpha}\right]$ when there exists a natural integer N such that $\boldsymbol{\alpha}^N$ is an inner automorphism of $\mathbf{R}$. Then for any $r \in \mathbf{R}$, $r^{\boldsymbol{\alpha}^N} = u^{-1}\, r\, u$ for some unit u of $\mathbf{R}$.

Proposition 375. *Let* $\boldsymbol{\alpha}$ *be as above. Then* $\mathbf{T} = \mathbf{R}\left[X, X^{-1}; \boldsymbol{\alpha}\right]$ *is a crossed product* $\mathbf{S} * G$ *where* $\mathbf{S} = \mathbf{R}\left[Y, Y^{-1}\right]$*, the Laurent polynomial ring in the central indeterminate* $Y = u\,X^N$*, and* G *is cyclic of order* N*.*

Proof. For any $r \in \mathbf{R}$, $Y\,r = u\,X^N r = u\,r^{\boldsymbol{\alpha}^N} X^N = u\,u^{-1}\,r\,u\,X^N = r\,Y$, thus the indeterminate Y is central. Let G be a cyclic group of order n and let g be a generator of G; then $G = \left\{1, g, ..., g^{N-1}\right\}$. Set $\varphi\left(g^i\right) = X^i$; then $\bar{G}$ is a set of units of $\mathbf{T}$ which is a free left and right $\mathbf{S}$-module with basis $\bar{G}$. ■

Let $f(X) = \sum_{0 \leq i \leq N-1} f_j(Y)\, X^j \in \mathbf{T}$, let $\mathbf{e}_N(X)$ be the column with ith entry X^{i-1} $(1 \leq i \leq N)$ and write

$$f(X) = \left[\, f_0(Y)\ f_1(Y) \ldots f_{N-1}(Y) \right] \mathbf{e}_N(X) \tag{2.40}$$

where $(f_i(Y))_{0 \leq i \leq N-1} \in \mathbf{S}^N$. Since X is a unit of $\mathbf{T}$ and $X^N = u^{-1}Y$, (2.40) is equivalent to

$$X\,f(X) = \left[u^{-1}Y\, f_{N-1}^{\boldsymbol{\alpha}}(Y)\ f_0^{\boldsymbol{\alpha}}(Y) \ldots f_{N-2}^{\boldsymbol{\alpha}}(Y)\right] \mathbf{e}_N(X).$$

This equality can be again left-multiplied by X, etc., and by induction, it is easily shown that (2.40) is equivalent to

$$\mathbf{e}_N(X) \otimes f(X) = \tilde{f}(Y)\, \mathbf{e}_N(X) \tag{2.41}$$

where $\otimes$ is the Kronecker (or tensor) product, i.e., $\mathbf{e}_N(X) \otimes f(X)$ is the column with ith entry $X^{i-1} f(X)$ $(1 \leq i \leq N)$ and where $\tilde{f}(Y) \in \mathrm{Mat}_N(\mathbf{S})$ is given by

$$\tilde{f}(Y) = \begin{bmatrix} f_0 & f_1 & \cdots & f_{N-2} & f_{N-1} \\ u^{-1}Y\, f_{N-1}^{\boldsymbol{\alpha}} & f_0^{\boldsymbol{\alpha}} & \cdots & f_{N-3}^{\boldsymbol{\alpha}} & f_{N-2}^{\boldsymbol{\alpha}} \\ & \ddots & \ddots & \ddots & \\ u^{-1}Y\, f_2^{\boldsymbol{\alpha}^{N-2}} & u^{-1}Y\, f_3^{\boldsymbol{\alpha}^{N-2}} & & f_0^{\boldsymbol{\alpha}^{N-2}} & f_1^{\boldsymbol{\alpha}^{N-2}} \\ u^{-1}Y\, f_1^{\boldsymbol{\alpha}^{N-1}} & u^{-1}Y\, f_2^{\boldsymbol{\alpha}^{N-1}} & \cdots & u^{-1}Y\, f_{N-1}^{\boldsymbol{\alpha}^{N-1}} & f_0^{\boldsymbol{\alpha}^{N-1}} \end{bmatrix}.$$

By inspection we see that

$$\tilde{f}(Y)^{\boldsymbol{\alpha}}\, P_N(Y) = P_N(Y)\, \tilde{f}(Y) \tag{2.42}$$

where $P(Y) = \begin{bmatrix} 0 & I_{N-1} \\ u^{-1}Y & 0 \end{bmatrix}$. Conversely, let $\tilde{f}(Y) \in \mathrm{Mat}_N(\mathbf{S})$ be a matrix satisfying (2.42), let $\left[\, f_0 \cdots f_{N-1}\right]$ be its first row and let $f(X) = \sum_{0 \leq i \leq N-1} f_j(Y)\, X^j$. Then, the entries of the jth row of $\tilde{f}(Y)$ are the components of $X^j f(X)$ in the basis $\left(X^i\right)_{0 \leq i \leq N-1}$.

Let $\mathcal{M}_{\mathbf{T}}$ be the subring of $\mathrm{Mat}_N(\mathbf{S})$ consisting of all matrices satisfying (2.42).

Proposition 376. *The map $\tau : \mathbf{T} \to \mathcal{M}_{\mathbf{T}}$ defined by $\tau(f(X)) = \tilde{f}(Y)$ is a ring-isomorphism.*

Proof. From the above, $\tau : \mathbf{T} \to \mathcal{M}_{\mathbf{T}}$ is surjective; in addition, it is obviously $\mathbb{Z}$-linear and injective. Let $f, g \in \mathbf{T}$; we have

$$\mathbf{e}_N(X) \otimes f(X)\, g(X) = \tilde{f}(Y)\, \mathbf{e}_N(X)\, g(X) = \tilde{f}(Y)\, \tilde{g}(Y)\, \mathbf{e}_N(X),$$

therefore $\tau(f\,g) = \tau(f)\,\tau(g)$ which proves that τ is a ring-homomorphism. ∎

The $\mathbf{S}$-module $\mathfrak{C}(M) = M_{[\mathbf{S}]} * G$ is generated by the entries of the column $\mathbf{e}_N(X) \otimes x$, $x \in M_{[\mathbf{S}]}$. Consider a $\mathbf{T}$-linear relation $x \mapsto f(X)\,x$ in M; then we have

$$\mathbf{e}_N(X) \otimes f(X)\,x = \tilde{f}(Y)\, \mathbf{e}_N(X) \otimes x.$$

This is expressed in the following proposition:

Proposition 377. *The diagram below is commutative.*

$$\begin{array}{ccc} M & \overset{f(X)\bullet}{\longrightarrow} & M \\ e_N \otimes \bullet \downarrow & & \downarrow e_N \otimes \bullet \\ M_{[\mathbf{S}]} * G & \overset{\tilde{f}(Y)\bullet}{\longrightarrow} & M_{[\mathbf{S}]} * G \end{array}$$

where $e_N = e_N(X)$.

Note also the following.

Corollary 378. *(i) Assuming that $\mathbf{R}$ is a commutative integral domain, so is again $\mathbf{S}$; for any skew Laurent polynomial $0 \neq f(X) \in \mathbf{T}$, the $\mathbf{S}$-module spanned by the entries of $\mathbf{e}_N(X) \otimes f(X)$ is of rank N, thus the rank of $\tilde{f}(Y)$ over $\mathbf{S}$ is equal to N, and*

$$\det\left(\tilde{f}(Y)\right) \neq 0$$

(see Theorem 400 in Subsect. 2.11.1 below).
(ii) By Proposition 376,

$$f(X) = (X - a_1)\ldots(X - a_n) \text{ implies } \tilde{f}(X) = (\widetilde{X - a_1})\ldots(\widetilde{X - a_n}).$$

(iii) Consider the polynomial $f(X) = X - a \in \mathbf{T}$. Assuming that $\mathbf{R}$ is commutative,

$$\tilde{f}(Y) = \begin{bmatrix} -a & 1 & 0 & \cdots & 0 \\ 0 & -a^{\boldsymbol{\alpha}} & \ddots & \ddots & \vdots \\ \vdots & \ddots & \ddots & \ddots & 0 \\ 0 & & \ddots & \ddots & 1 \\ Y & 0 & \cdots & 0 & -a^{\boldsymbol{\alpha}^{N-1}} \end{bmatrix}, \textit{ thus}$$

$$\det\left(\tilde{f}(Y)\right) = Y - \prod_{0 \leq i \leq N-1} a^{\boldsymbol{\alpha}^i}.$$

Theorem 379. *Assume that* $\mathbf{R}$ *is a commutative domain, let* $M \in {}_{\mathbf{T}}\mathbf{Mod}$ *and let* $\tilde{M} = \mathfrak{C}(M)$ *(where* $\mathfrak{C} : {}_{\mathbf{T}}\mathbf{Mod} \to {}_{\mathbf{S}}\mathbf{Mod}$ *is the functor in Theorem 373). Then,* M *is torsion if, and only if* $\tilde{M}$ *is torsion.*

Proof. Let $M \in {}_{\mathbf{T}}\mathbf{Mod}$ be a nonzero torsion module. Thus every nonzero element m of M is torsion over $\mathbf{T}$, and there exists $0 \neq f(X) \in \mathbf{T}$ such that $f(X)m = 0$. Then, $\mathfrak{C}([m]_{\mathbf{T}}) = [\tilde{m}]_{\mathbf{S}}$ where $\tilde{m} = \mathbf{e}_N(X) \otimes m$ is such that $\tilde{f}(Y)\tilde{m} = 0$. Let $\mathbf{F}$ be the field of fractions of the commutative domain $\mathbf{S}$. By Corollary 378, $\tilde{f}(Y)$ is invertible over $\mathbf{F}$, therefore $\mathbf{F} \bigotimes_{\mathbf{S}} [\tilde{m}]_{\mathbf{S}} = 0$, and $[\tilde{m}]_{\mathbf{S}}$ is torsion by Theorem and Definition 339(ii). Since $\tilde{M} = \mathfrak{C}(M)$ is generated by the entries of the elements $\tilde{m}$ $(m \in M)$, we have $\tilde{M} = \sum_{m \in M} [\tilde{m}]_{\mathbf{S}}$; theferore,

$$\mathbf{F} \bigotimes\nolimits_{\mathbf{S}} \tilde{M} = \sum\nolimits_{m \in M} \mathbf{F} \bigotimes\nolimits_{\mathbf{S}} [\tilde{m}]_{\mathbf{S}} = 0$$

and $\tilde{M}$ is torsion.

Conversely, if m is not torsion, i.e. if m is free, then $[m]_{\mathbf{T}}$ is free (with basis $\{m\}$), and so is $[\tilde{m}]_{\mathbf{S}}$ by Corollary 374(ii). Therefore, if $M \in {}_{\mathbf{T}}\mathbf{Mod}$ is not torsion, neither is $\tilde{M} = \mathfrak{C}(M)$. ■

2.9 Rings of Power Series

2.9.1 Formal Power Series

2.9.1.1 The Commutative Case

Let $\mathbf{R}$ be a ring and consider the ring of polynomials $\mathbf{R}[X]$. The ideal $\mathfrak{a} = (X)$ in $\mathbf{R}[X]$ is two-sided and condition (2.28) holds. The completion of $\mathbf{R}[X]$ with respect to the (X)-adic topology is the ring $\mathbf{R}[[X]]$ of all formal power series in the central indeterminate X with coefficients in $\mathbf{R}$; an element of this ring is of the form

$$a = \sum_{i \geq 0} a_i X^i, \quad a_i \in \mathbf{R} \tag{2.43}$$

without condition of convergence, i.e., $\sum_{i \geq 0} a_i X^i$ is nothing but a notation for the sequence $(a_i)_{i \in \mathbb{N}}$. The set of all these formal power series is denoted by $\mathbf{R}[[X]]$. This set has an obvious structure of abelian group additively noted. If $b = \sum_{i \geq 0} b_i X^i$, the product $c = a\,b$ is defined as $c = \sum_{i \geq 0} c_i X^i$ where $c_i = \sum_{j+k=i} a_j\, b_k$. Then, $\mathbf{R}[[X]]$ is endowed with a ring structure.

When $\mathbf{R} = \mathbf{k}$ ($\mathbf{k} = \mathbb{R}$ or $\mathbb{C}$) one can also define the subring $\mathbf{k}\{X\}$ of $\mathbf{k}[[X]]$ consisting of all *convergent power series*, i.e., of those power series which have a positive convergence radius. Specifically, an element $a(X) \in \mathbf{k}[[X]]$ belongs to $\mathbf{k}\{X\}$ if there exists $\rho > 0$ such that $a(x)$ converges for all $x \in \mathbf{k}$ such that $|x| < \rho$ (and the *convergence radius* of $a(X)$ is $\inf \rho$ among those ρ for which this property holds).

2.9.1.2 Skew Power Series

Let $\mathbf{R}$ be a ring, $\boldsymbol{\alpha}$ an automorphism and $\boldsymbol{\delta}$ be an $\boldsymbol{\alpha}$-derivation of $\mathbf{R}$. One can consider the skew polynomial ring $\mathbf{R}[X;\boldsymbol{\alpha},\boldsymbol{\delta}]_l$ (2.7.4.**2**), but its (X)-completion cannot be defined if $\boldsymbol{\delta} \neq 0$ since, due to the commutation rule (2.33), the left ideal $(X)_l$ and the right ideal $(X)_r$ in $\mathbf{R}[X;\boldsymbol{\alpha},\boldsymbol{\delta}]_l$ do not coincide (in the case $\boldsymbol{\delta} = 0$, see [80], Sect. 7.3). However, setting $Z = X^{-1}$, (2.33) yields (left and right multiplying this equality by Z)

$$\boxed{a\,Z = Z\,a^{\boldsymbol{\alpha}} + Z\,a^{\boldsymbol{\delta}}\,Z.} \tag{2.44}$$

Therefore,

$$a\,Z = Z\,a^{\boldsymbol{\alpha}} + Z^2\,a^{\boldsymbol{\alpha}\,\boldsymbol{\delta}} + Z^2\,a^{\boldsymbol{\delta}^2}\,Z = Z\,a^{\boldsymbol{\alpha}} + Z^2\,a^{\boldsymbol{\alpha}\,\boldsymbol{\delta}} + Z^3\,a^{\boldsymbol{\alpha}\,\boldsymbol{\delta}^2} + Z^3\,a^{\boldsymbol{\delta}^3}\,Z$$

and by induction we obtain

$$a\,Z = \sum_{i \geq 1} Z^i\,a^{\boldsymbol{\alpha}\,\boldsymbol{\delta}^{i-1}}; \tag{2.45}$$

in addition, for any $n \geq 0$,

$$a\,Z^n = \sum_{i \geq n} Z^i\,a^{\boldsymbol{\gamma}_i}, \quad a^{\boldsymbol{\gamma}_n} = a^{\boldsymbol{\alpha}^n}. \tag{2.46}$$

Consider the abelian group $\mathbf{R}[[Z;\boldsymbol{\alpha},\boldsymbol{\delta}]]_+$ consisting of all elements of the form

$$a(Z) = \sum_{i \geq 0} a_i\,Z^i. \tag{2.47}$$

Using (2.45), two elements of $\mathbf{R}[[Z;\boldsymbol{\alpha},\boldsymbol{\delta}]]_+$ can be multiplied and the result belongs to $\mathbf{R}[[Z;\boldsymbol{\alpha},\boldsymbol{\delta}]]_+$; in addition, the above element $a(Z)$ can be put into the form

$$a(Z) = \sum_{i \geq 0} Z^i\,a'_i \tag{2.48}$$

where $a'_0 = a_0$, $a'_1 = a_1^{\boldsymbol{\alpha}}$, etc. Therefore, $\mathbf{R}\left[\left[Z;\boldsymbol{\alpha},\boldsymbol{\delta}\right]\right]_+$ is endowed with a ring structure. This ring, denoted by $\mathbf{R}\left[\left[X^{-1};\boldsymbol{\alpha},\boldsymbol{\delta}\right]\right]$, is called the *ring of formal skew power series* in the indeterminate X^{-1}.

Consider the element $a = a(Z)$ defined by (2.47) (with the coefficients on the left of the powers of the indeterminate $Z = X^{-1}$) and, assuming that $a \neq 0$, let

$$\omega(a) = \inf\{i \in \mathbb{N} : a_i \neq 0\}.$$

Consider now $a = a(Z)$ in the form (2.48) (with the coefficients on the right of the powers of the indeterminate). By (2.46), the coefficient of lowest degree is $a'_n = a_n^{\boldsymbol{\alpha}^n}$. Conversely, if a'_n is the coefficient of lowest degree of a defined by (2.48), the coefficient of lowest degree of a in the form (2.47) is $a_n = a'^{\boldsymbol{\alpha}^{-n}}_n$. Therefore, we also have

$$\omega(a) = \inf\{i \in \mathbb{N} : a'_i \neq 0\}.$$

Definition 380. *If $a \neq 0$, the* order *of this formal power series is $\omega(a)$, whereas $\omega(0) \triangleq +\infty$. The* constant term *of a is a_0.*

From the above,

$$\begin{aligned}\omega(a+b) &\geq \inf\{\omega(a), \omega(b)\} \text{ with equality if } \omega(a) \neq \omega(b),\\ \omega(a\,b) &\geq \omega(a) + \omega(b) \text{ with equality if } \mathbf{R} \text{ is a domain.}\end{aligned}$$

Theorem 381. *(i) The ring $\mathbf{S} = \mathbf{R}\left[\left[X^{-1};\boldsymbol{\alpha},\boldsymbol{\delta}\right]\right]$ is Hausdorff and complete in the $\left(X^{-1}\right)$-adic topology.*
(ii) The units in $\mathbf{S}$ are the formal power series whose constant term is a unit in $\mathbf{R}$.
(iii) If $\mathbf{R}$ is a domain, so is $\mathbf{S}$.
(iv) If $\mathbf{R}$ is left (resp., right) Noetherian, then $\mathbf{S}$ is again left (resp., right) Noetherian.

Proof. (i): The topological ring $\mathbf{S}$ is obviously Hausdorff. Let (a_n) be a Cauchy sequence in $\mathbf{S}$, i.e., given a power Z^k there exists $N \in \mathbb{N}$ such that for all $n, m \geq N$, $a_n - a_m \in \left(Z^k\right)$. Let $\sum_{0 \leq i \leq k-1} b_i\,Z^i$ be a representative of a_n and a_m in $\mathbf{S}/\left(Z^k\right)$. Obviously, (a_n) converges to $\sum_{i \geq 0} b_i\,Z^i$ in the (Z)-adic topology, thus $\mathbf{S}$ is complete.

(ii): Consider a given by (2.47). If a_0 is a unit in $\mathbf{R}$, then a can be written in the form $a_0\,(1 - Y)$, the inverse of which is $\sum_{i \geq 0} Y^i\,a_0^{-1} \in \mathbf{S}$, thus a is a unit of $\mathbf{S}$. Conversely, if a is a unit of $\mathbf{S}$ with inverse b, the constant terms a_0 and b_0 of a and b are such that $a_0\,b_0 = 1$, thus a_0 is a unit.

(iii): Let $a, b \in \mathbf{S}^{\times}$, with the coefficients written on the right and with term of lowest degree $Z^n\,a_n$ and $Z^m\,b_m$ respectively. By (2.46), the term of lowest degree of $a\,b$ (with the coefficients written on the right) is $Z^{n+m}\,a_n^{\boldsymbol{\alpha}^m}\,b_m \neq 0$, thus $\mathbf{S}$ is a domain.

The proof of (iv) is the quite similar to that of Theorem 358: one has only to replace everywhere in that proof the polynomials of lowest degree by the power series of highest order (for more details when $\boldsymbol{\alpha} = 1$ and $\boldsymbol{\delta} = 0$, see [206], Theorem 9.4; the proof of that theorem is easily adapted to the general case). ■

Remark 382. *The power series rings* $\mathbf{k}[[X]]$ *and* $\mathbf{k}\{X\}$ *are local Noetherian domains, and even local* principal ideal domains *(see Subsect. 2.13.5). Indeed, let* $\mathbf{S}$ *be any of these rings; any nonzero element of* $\mathbf{S}$ *is of the form* $X^n v$ *where* v *is a unit, thus every nonzero ideal in* $\mathbf{S}$ *is of the form* (X^n) *which is a principal ideal, thus* $\mathbf{S}$ *is a principal ideal domain. In addition,* (X) *is the only maximal ideal in* $\mathbf{S}$*, thus this ring is local.*

Theorem 383. *Let* $\mathfrak{o}$ *be a local ring,* $\boldsymbol{\alpha}$ *an automorphism and* $\boldsymbol{\delta}$ *an* $\boldsymbol{\alpha}$*-derivation of* $\mathfrak{o}$*; then the ring* $\mathbf{S} = \mathfrak{o}\left[\left[X^{-1}; \boldsymbol{\alpha}, \boldsymbol{\delta}\right]\right]$ *is local, the canonical injection* $\mathfrak{o} \hookrightarrow \mathbf{S}$ *is a local homomorphism and the corresponding injection of the residue class division rings is an isomorphism. The maximal ideal in* $\mathbf{S}$ *is* $(\mathfrak{m}, \mathfrak{a})$ *where* $\mathfrak{m}$ *is the maximal ideal in* $\mathfrak{o}$ *and* $\mathfrak{a} = \left(X^{-1}\right)$.

Proof. This is an easy generalization of ([31], n°II.3.1, Example 2. ■

Corollary 384. *Let* $\mathbf{K}$ *be a division ring,* $\boldsymbol{\alpha}$ *an automorphism and* $\boldsymbol{\delta}$ *an* $\boldsymbol{\alpha}$*-derivation of* $\mathbf{K}$*; then* $\mathbf{S} = \mathbf{K}\left[\left[X^{-1}; \boldsymbol{\alpha}, \boldsymbol{\delta}\right]\right]$ *is a local Noetherian domain with residue class division ring* $\mathbf{K}$*. All ideals in* $\mathbf{S}$ *are two-sided and of the form* (X^{-n}) *and every element of* $\mathbf{S}$ *is of the form* $v\, X^{-n} = X^{-n} v'$ *where* v, v' *are units.*

2.9.2 Formal Skew Laurent Series

Let $\mathbf{R}$ be a ring, $\boldsymbol{\alpha}$ an automorphism and $\boldsymbol{\delta}$ an $\boldsymbol{\alpha}$-derivation of $\mathbf{R}$; consider the ring of formal skew power series $\mathbf{S} = \mathbf{R}\left[\left[X^{-1}; \boldsymbol{\alpha}, \boldsymbol{\delta}\right]\right]$ (**2.9.1.2**). As easily seen, the monoid $D = \{X^{-n}; n \geq 1\}$ is a left and right denominator set (Subsect. 2.8.1). Therefore, the rings of fractions $D^{-1}\mathbf{S}$ and $\mathbf{S}\, D^{-1}$ exist an coincide by Theorem 332(viii). This ring is denoted by $\mathbf{R}\left(\left(X^{-1}; \boldsymbol{\alpha}, \boldsymbol{\delta}\right)\right)$ and is called the *ring of formal skew Laurent series* in the indeterminate X^{-1}.

Any element of $\mathbf{R}\left(\left(X^{-1}; \boldsymbol{\alpha}, \boldsymbol{\delta}\right)\right)$ is of the form

$$a = \sum_{i \geq \nu} a_i\, Z^i$$

where $Z = X^{-1}$ and $a_\nu \neq 0$ if $a \neq 0$. The *order* $\omega(a)$ of a is defined as in Subsect. 2.9.1 for a formal power series (except that $\omega(a) \in \mathbb{Z} \cup \{-\infty\}$) and has the same properties.

Theorem 385. *(i) If* $\mathbf{R}$ *is left (resp., right) Noetherian, then* $\mathbf{L} = \mathbf{R}\left(\left(X^{-1};\boldsymbol{\alpha},\boldsymbol{\delta}\right)\right)$ *is again left (resp., right) Noetherian.*
(ii) If $\mathbf{R}$ *is a domain,* $\mathbf{L}$ *is again a domain.*
(iii) Let $\mathbf{A} = \mathbf{R}[X;\boldsymbol{\alpha},\boldsymbol{\delta}]$ *and* $\mathbf{B} = \mathbf{R}\left[X,X^{-1};\boldsymbol{\alpha},\boldsymbol{\delta}\right]$ *(Subsect. 2.8.1); there exist embeddings* $\mathbf{A} \hookrightarrow \mathbf{B} \hookrightarrow \mathbf{L}$*. If* $\mathbf{R}$ *is a Noetherian domain, all these rings are again Noetherian domains.*
(iv) If $\mathbf{R}$ *is a division ring, then* $\mathbf{L}$ *is the* division ring of fractions $\mathrm{Q}(\mathbf{S})$ *of* $\mathbf{S}$.
(v) Let $\mathbf{A} = \mathbf{R}[X;\boldsymbol{\alpha},\boldsymbol{\delta}]$*. If* $\mathbf{R}$ *is an Ore domain with division ring of fractions* $\mathbf{K}$*, let* $\mathbf{Q} = \mathbf{K}(X;\boldsymbol{\alpha},\boldsymbol{\delta})$*,* $\mathbf{S} = \mathbf{R}\left[\left[X^{-1};\boldsymbol{\alpha},\boldsymbol{\delta}\right]\right]$*,* $\hat{\mathbf{S}} = \mathbf{K}\left[\left[X^{-1};\boldsymbol{\alpha},\boldsymbol{\delta}\right]\right]$ *and* $\hat{\mathbf{L}} = \mathbf{K}\left(\left(X^{-1};\boldsymbol{\alpha},\boldsymbol{\delta}\right)\right)$*; there exist embeddings* $\mathbf{A} \hookrightarrow \mathbf{Q} \hookrightarrow \hat{\mathbf{L}}$ *and* $\mathbf{S} \hookrightarrow \hat{\mathbf{S}} \hookrightarrow \hat{\mathbf{L}}$.

Proof. (i), (ii): Assume that $\mathbf{R}$ is left Noetherian (resp., is a domain); then so is $\mathbf{L}$ by Theorem 332(iv)&(ix) and Theorem 381.

(iii) is clear from Subsect. 2.8.1; indeed, $\mathbf{A}$ is the additive subgroup of $\mathbf{B}$ consisting of those Laurent polynomials, all terms of which only contain non-negative powers of X. The additive group monomorphism $\mathbf{A} \hookrightarrow \mathbf{B}$ is also a ring monomorphism. In addition, the embedding $\mathbf{B} \hookrightarrow \mathbf{L}$ is obvious.

(iv): If $\mathbf{R}$ is a division ring and $\mathbf{S} \ni a \neq 0$, one can write $a = a_\nu\, Z^\nu\, (1-b)$ where $\omega(b) > 0$, thus a is invertible over $\mathbf{L}$ with inverse $\sum_{j\geq 0} b^j\, Z^{-\nu}\, a_\nu^{-1}$.

(v): We know that $\mathbf{A}$ can be embedded in its division ring of fractions $\mathbf{Q} = \mathrm{Q}(\mathbf{A}) = \mathbf{K}(X;\boldsymbol{\alpha},\boldsymbol{\delta})$ (Theorem 357). Consider a left fraction

$$f(X) = D^{-1}(X)\, N(X) \in \mathbf{K}(X;\boldsymbol{\alpha},\boldsymbol{\delta})$$

where

$$\begin{aligned} D(X) &= X^n + \textstyle\sum_{i=1}^n X^i\, d_i = X^n\left(1 - E\left(X^{-1}\right)\right), \\ E\left(X^{-1}\right) &= -\textstyle\sum_{i=1}^n X^{-i}\, d_i. \end{aligned}$$

The polynomial $N(X)$ is embedded in $\hat{\mathbf{L}}$ in an obvious way, and so is also

$$D^{-1}(X) = \left(\textstyle\sum_{0\leq i\leq\infty} E\left(X^{-1}\right)^i\right) X^{-n}.$$

The rest is clear. ∎

2.10 Some Linear Groups

2.10.1 Linear Algebraic Groups

2.10.1.1 Linear Groups

A *matrix group* over a ring $\mathbf{R}$ is a subgroup of the general linear group $\mathrm{GL}_n(\mathbf{R})$ (Example 71). A *linear group* is a group that is isomorphic to a matrix group. In the sequel, the terms *linear group* and *matrix group* are considered synonymous.

Let us give some examples: if $\mathbf{R}$ is commutative, the special linear group $\mathrm{SL}_n(\mathbf{R})$ (Example 144) is a linear group over $\mathbf{R}$. The affine group $A(n, \mathbb{R})$, the special Euclidean group $\mathrm{SE}(n, \mathbb{R})$, the orthogonal group $O(n, \mathbb{R})$ and the special orthogonal group $\mathrm{SO}(n, \mathbb{R})$ are real linear groups (Exercise 210).

2.10.1.2 Algebraic Sets, Varieties, and Linear Algebraic Groups

When speaking about linear algebraic group, some elementary facts of Algebraic Geometry cannot be completely avoided. We begin with the traditional presentation. Let $\mathbf{K}$ be an algebraically closed field (see Definition 703(i) below). In the present context, the space $\mathbf{K}^n$ is called the *affine n-space over* $\mathbf{K}$ and is denoted by $\mathbf{A}^n_{\mathbf{K}}$, or $\mathbf{A}^n$ when $\mathbf{K}$ is understood. Let S be a subset of $\mathbf{R} = \mathbf{K}[X]$ $\left(X = (X_i)_{1 \leq i \leq n}\right)$ and let $\mathfrak{a}$ be the ideal in $\mathbf{R}$ generated by S. By a zero of S in $\mathbf{A}^n$ we mean a common root in $\mathbf{A}^n$ of all $P \in S$ (Remark 350(iii)). Let $\mathcal{Z}_S$ be the set of all these zeros; as easily seen, $\mathcal{Z}_S = \mathcal{Z}_{\mathfrak{a}}$. The following is taken from ([156], Chap. I.):

Definition 386. *(i) The set* $\mathcal{Z}_{\mathfrak{a}} \subseteq \mathbf{A}^n$ *is called an* affine algebraic set *(or an algebraic set, for short).*
(ii) We define the Zariski topology *on* $\mathbf{A}^n$ *taking the open subsets to be the complements of the algebraic sets. (The reader may check that this is indeed a topology, which is* not *Hausdorff.)*
(iii) An (affine) algebraic variety V *is an irreducible Zariski-closed set in* $\mathbf{A}^n$ *endowed with the induced topology. (A nonempty subset* Y *of a topological space* X *is* irreducible *if it cannot be expressed as the union of two proper closed subsets of* X*.)*
(iv) Let $f = g/h \in \mathbf{K}(X)$ *and let* U *be an open subset of an algebraic variety* $V \subseteq \mathbf{A}^n$*. If* $h(x) \neq 0$ *for all* $x \in U$*, then the function* $f : U \to \mathbf{K}$ *is called regular on* U*.*
(v) Let V, W *be two algebraic varieties and let* $\varphi : V \to W$ *be continuous (for the Zariski topology);* φ *is called a* morphism of varieties *if for every open subset* $U \subseteq W$ *and every regular function* $f : W \to \mathbf{K}$*, the function* $f \circ \varphi : \varphi^{-1}(U) \to \mathbf{K}$ *is regular.*

The following is useful for the sequel:

Theorem and Definition 387. *Let* $\mathbf{R}$ *be a commutative ring.*

(i) Let $\mathfrak{a}$ *be an ideal in* $\mathbf{R}$*. The* radical *of* $\mathfrak{a}$*, denoted by* $\sqrt{\mathfrak{a}}$*, is the set of all* $x \in \mathbf{R}$ *such that* $x^n \in \mathfrak{a}$ *for some natural integer* n*.*
(ii) $\sqrt{\mathfrak{a}}$ *is the intersection of all prime ideals in* $\mathbf{R}$ *containing* $\mathfrak{a}$ *(Definition 298), thus is an ideal in* $\mathbf{R}$*. In particular,* $\mathfrak{N}(\mathbf{R}) = \sqrt{0}$ *(the intersection of all prime ideals in* $\mathbf{R}$*) is an ideal, called the* nilradical *of* $\mathbf{R}$*, and consisting of all nilpotent elements of* $\mathbf{R}$ *(i.e., of those elements* x *for which there exists a*

natural integer m *such that* $x^m = 0$*);* $\mathfrak{N}(\mathbf{R})$ *is a proper ideal, for* $1 \notin \mathfrak{N}(\mathbf{R})$*. The ring* $\mathbf{R}$ *is called* reduced *if* $\mathfrak{N}(\mathbf{R}) = \{0\}$*.*
(iii) An ideal $\mathfrak{a}$ *in* $\mathbf{R}$ *is called a* radical ideal *if* $\mathfrak{a} = \sqrt{\mathfrak{a}}$*. By (ii), a prime ideal is a radical ideal.*
(iv) Let $\mathbf{K}$ *be an algebraically closed field and* $\mathbf{R} = \mathbf{K}[X_1, ..., X_n]$*.*
(a) Let $\mathfrak{a}$ *be an ideal in* $\mathbf{R}$ *and* $f \in \mathbf{R}$ *a polynomial vanishing at all points of* $\mathcal{Z}_{\mathfrak{a}}$*. Then* $f^m \in \mathfrak{a}$ *for some integer* $m > 0$ *(*Hilbert's Nullstellensatz*).*
(b) $\sqrt{\mathfrak{a}}$ *coincides with the intersection of all maximal ideals in* $\mathbf{R}$ *containing* $\mathfrak{a}$*. (c) In particular, the Jacobson ideal* $\operatorname{rad}\mathbf{R}$ *(Theorem and Definition 293) coincides with* $\mathfrak{N}(\mathbf{R})$*.*

Proof. (i): See ([79], p. 353). (ii): The intersection of all prime ideals in $\mathbf{R}$ is the nilradical of $\mathbf{R}$ by ([79], Prop. 10.2.9). (iv): (a) is proved in ([206], Chap. IX, Theorem 1.5). (b) is proved in ([198], (5.4)) and (c) is an obvious consequence of (b) and (ii). ■

Remark 388. *Let* $\mathbf{R}$ *be a reduced ring. Then the total ring of fractions of* $\mathbf{R}$ *(Remark 325(iv)) is again reduced.*

For any subset $Y \subseteq \mathbf{A}^n$, define the ideal $\mathfrak{J}(Y)$ of Y in $\mathbf{R} = \mathbf{K}[X]$ $\left(X = (X_i)_{1 \leq i \leq n}\right)$ according to

$$\mathfrak{J}(Y) = \{P \in \mathbf{K}[X] : P(x) = 0 \text{ for all } x \in Y\}.$$

Recall that the field $\mathbf{K}$ is algebraically closed.

Corollary and Definition 389. *(1) There exists a bijective antitone (for inclusion) Galois connection between the algebraic sets in* $\mathbf{A}^n$ *and the radical ideals in* $\mathbf{R}$*, given by*

$$Y \mapsto \mathfrak{J}(Y), \quad \mathfrak{a} \mapsto \mathcal{Z}_{\mathfrak{a}}.$$

Furthermore, an algebraic set is irreducible (i.e., is an affine algebraic variety) if, and only if its ideal is a prime ideal.
(2) If $\mathfrak{a}$ *is any ideal in* $\mathbf{R}$*, then* $\mathfrak{J}(\mathcal{Z}_{\mathfrak{a}}) = \sqrt{\mathfrak{a}}$*.*
(3) Let Y *be an algebraic set. The ring* $\mathbf{R}(Y) \triangleq \mathbf{R}/\mathfrak{J}(Y)$ *is called the* (affine) coordinate ring *of* Y*. Consider the polynomial functions* $P : \mathbf{A}^n \to \mathbf{K} : x \mapsto P(x)$ $(P(X) \in \mathbf{K}[X])$ *and their restrictions* $P|_Y$*; identify two such restrictions if they agree at all points* $x \in Y$*. The set of functions obtained in this way is a ring isomorphic to* $\mathbf{R}(Y)$*. The ring* $\mathbf{R}(Y)$ *is reduced.*
(4) If Y *is an algebraic variety, then* $\mathbf{R}(Y)$ *is an integral domain.*

Proof. (1): See ([156], Sect. I.1, Corol. 1.4).
(2) is a reformulation of the Nullstellensatz.
(3) is easy ([105], Sect. 1.6).
(4) is a consequence of Proposition 303. ■

Let G be an algebraic set over an algebraically closed field $\mathbf{K}$. It is an *affine algebraic group* (or simply an *algebraic group*) if it is endowed with a group structure. Not all algebraic groups are linear group ([187], Sect. VI.1). Nevertheless, all algebraic groups are linear groups in the sequel, thus are *linear algebraic groups.* A *morphism of linear algebraic groups* is a morphism of algebraic varieties which is a morphism of groups. This completely defines the *category of linear algebraic groups.*

2.10.1.3 Functorial Approach

Some hints must be given about the generalization of the above notions in the context of modern Algebraic Geometry [150]. Let $\mathbf{R}$ be a commutative ring and let I be a set of indices; for any commutative $\mathbf{R}$-algebra A, consider the space $\mathbf{A}^I(A) \triangleq A^I$. Define the covariant functor $\mathbf{A}^I$ with object map $A \mapsto \mathbf{A}^I(A)$ and arrow map $\varphi \mapsto \varphi^I$ where $\varphi : A \to A'$ is any $\mathbf{R}$-linear homomorphism and φ^I is the map $A \ni a \mapsto \varphi^I(a) \in A'^I$; this functor $\mathbf{A}^I$ is called the *affine space of dimension* $\operatorname{card}(I)$ over $\mathbf{R}$.

Let S be a subset of $\mathbf{R}[X]$ $\left(X = (X_i)_{i\in I}\right)$ and let $\mathfrak{a}$ be the ideal in $\mathbf{R}[X]$ generated by S. For any commutative $\mathbf{R}$-algebra A , there exists a homomorphism ρ of commutative rings $\mathbf{R} \to A$ and to each $P \in S$ there corresponds a polynomial P_A, the coefficients of which are the images under ρ of the coefficients of P. Let S_A be the set of all P_A $(P \in S)$. By a zero of S in A^I we now mean a common root in A^I of all $P_A \in S_A$. Let $\mathcal{Z}_S(A)$ be the set of all these zeros; again, $\mathcal{Z}_S(A) = \mathcal{Z}_{\mathfrak{a}}(A)$. The functor $\mathcal{Z}_{\mathfrak{a}}$ from the category of all commutative rings to the category of sets is a subfunctor of $\mathbf{A}^I$ (**1.2.3.1**), called an *algebraic set* over $\mathbf{R}$ in $\mathbf{A}^I$.

Likewise, let $(P_\beta)_{\beta\in B}$ be a family of elements of $\mathbf{R}\left[(X_{ij})_{1\leq i\leq n, 1\leq j\leq n}\right]$. Given a commutative $\mathbf{R}$-algebra A, let $\mathcal{G}(A)$ be the subset of $\operatorname{Mat}_n(A)$ consisting of all matrices $M = (m_{ij})$ such that $P_\beta\left((m_{ij})_{1\leq i\leq n, 1\leq j\leq n}\right) = 0$ for all $\beta \in B$. If for any $\mathbf{R}$-algebra A, $\mathcal{G}(A)$ is a subgroup of the general linear group $\mathrm{GL}_n(A)$, the family $(P_\beta)_{\beta\in B}$ is said to define the *linear algebraic group* $\mathcal{G}$ over $\mathbf{R}$ ([327], Chap. I, Def. 4), ([206], Chap. XIII, Exercise 30).

If $\mathcal{G}$ is an algebraic group over $\mathbf{k}$, then $\mathcal{G}(\mathbf{k})$ is closed in $\mathrm{GL}_n(\mathbf{k})$ endowed with the topology induced by the canonical topology of $\operatorname{Mat}_n(\mathbf{k})$, thus it is a Lie group by Cartan's theorem (**1.8.1.2**). More general, let $\mathcal{G}$ be an algebraic group over a commutative ring $\mathbf{R}$ and let $\mathcal{A}$ be the set of all $\mathbf{R}$-algebras A. Let $(\mathcal{H}(A))_{A\in\mathcal{A}}$ be a family such that each $\mathcal{H}(A)$ is a subgroup of $\mathcal{G}(A)$. Then $\mathcal{H}$ is an *algebraic subgroup* of $\mathcal{G}$ (written $\mathcal{H} \leq \mathcal{G}$) if for all $A \in \mathcal{A}$, $\mathcal{H}(A)$ is Zariski-closed in $\mathcal{G}(A)$ (and in that case, $\mathcal{H}$ is said to be Zariski-closed in $\mathcal{G}$). Let $\mathcal{H} \leq \mathcal{G}$; $\mathcal{H}$ is said to be *normal* in $\mathcal{G}$ (written $\mathcal{H} \lhd \mathcal{G}$) if for all $A \in \mathcal{A}$, $\mathcal{H}(A) \lhd \mathcal{G}(A)$.

2.10.1.4 The Linear Group $\mathrm{D}_n(\mathbf{R})$

Let $\mathbf{R}$ be a ring. The set of all diagonal matrices of the form $D(u) = \mathrm{diag}(1, 1, ..., 1, u)$ where u is a unit of $\mathbf{R}$ is a subgroup $\mathrm{D}_n(\mathbf{R})$ of $\mathrm{GL}_n(\mathbf{R})$ since $D(u)\,D(v) = D(u\,v)$, and

$$\mathrm{D}_n(\mathbf{R}) \cong \mathbf{U}(\mathbf{R}).$$

As easily shown, the linear group $\mathrm{D}_n(\mathbf{R})$ is not a normal subgroup of $\mathrm{GL}_n(\mathbf{R})$.

2.10.2 Elementary Matrices

Let $n > 1$ and let $E_{ij} \in \mathrm{GL}_n(\mathbf{R})$ be the matrix defined for $i \neq j$ as follows: all entries of E_{ij} are zero except the (i, j) entry which is equal to 1. In addition, for any $\lambda \in \mathbf{R}$, let $B_{ij}(\lambda) = I_n + \lambda\,E_{ij}$. As easily shown,

$$B_{ij}(\lambda)\,B_{ij}(\mu) = B_{ij}(\lambda + \mu), \quad B_{ij}^{-1}(\lambda) = B_{ij}(-\lambda) \tag{2.49}$$

Definition 390. *The above matrix $B_{ij}(\lambda)$ is called an* elementary matrix. *The multiplicative group generated by all elementary matrices belonging to $\mathrm{GL}_n(\mathbf{R})$ is denoted by $\mathrm{E}_n(\mathbf{R})$ and is called the* group of elementary matrices *of order n.*

Obviously, $\mathrm{E}_n(\mathbf{R}) \leq \mathrm{GL}_n(\mathbf{R})$, and moreover $\mathrm{E}_n(\mathbf{R}) \leq \mathrm{SL}_n(\mathbf{R})$ if $\mathbf{R}$ is commutative.

2.10.3 Elementary Operations

(i) The right multiplication of $A \in {}^m\mathbf{R}^p$ by an elementary matrix $B_{ij}(\lambda) \in \mathrm{E}_p(\mathbf{R})$ corresponds to the operation of replacing the jth column A_j of A by $A_j + A_i\,\lambda$ (first kind of elementary operations on the columns).

(ii) The right multiplication of A by $\mathrm{diag}(1, ..., 1, u, 1, ...1)$ where the unit u occurs at the ith place corresponds to the operation of right-multiplying the column A_i by u (second kind of elementary operations on the columns).

(iii) The right multiplication of A by a permutation matrix corresponds to the operation of interchanging two columns (third kind of elementary operations on the columns).

Three kinds of elementary operations on the rows are likewise defined (changing *right* to *left* everywhere).

Remark 391. *Elementary operations of kind (i),* but not those of kind (ii) or (iii), *correspond to multiplications by elementary matrices. Elementary operations of kind (iii) are combinations of elementary operations of kind (i) and of elementary operations of kind (ii).*

2.10.4 Reduction of a Matrix to an Equivalent Diagonal Form

Let $\mathbf{K}$ be a *division ring* and $A \in {}^{m}\mathbf{K}^{p}$. The following proposition is proved by induction ([27], §II.10, Prop. 14):

Theorem and Definition 392. *(i) There exist matrices $P \in \mathrm{E}_m(\mathbf{K})$ and $Q \in \mathrm{E}_p(\mathbf{K})$ such that*

$$P\,A\,Q^{-1} = \mathrm{diag}\,(1, ..., 1, \delta_r, 0, ...0)\,, \quad \delta_r \neq 0$$

and $r \in \mathbb{N}$ is determined by A.
(ii) The above integer r is the rank *of A, written $r = \mathrm{rk}\,(A)$.*

2.10.5 Structure of $\mathrm{GL}_n\,(\mathrm{K})$ and of $\mathrm{E}_n\,(\mathrm{K})$

For $n = 1$, we set $\mathrm{E}_1(\mathbf{K}) = \mathbf{D}(\mathbf{K}^{\times})$, i.e., the derived group of the multiplicative group $\mathbf{K}^{\times}$ (Definition 166).

Proposition 393. *Let n be any positive integer and $\mathbf{K}$ be a division ring. The following properties hold:*

(i) $\mathrm{GL}_n(\mathbf{K}) = \mathrm{D}_n(\mathbf{K})\,\mathrm{E}_n(\mathbf{K}) = \mathrm{E}_n(\mathbf{K})\,\mathrm{D}_n(\mathbf{K})$*;*
(ii) if $\mathbf{K}$ is commutative, $\mathrm{SL}_n(\mathbf{K}) = \mathrm{E}_n(\mathbf{K})$*;*
(iii) $\mathrm{E}_n(\mathbf{K}) \lhd \mathrm{GL}_n(\mathbf{K})$ *and* $\mathrm{GL}_n(\mathbf{K}) / \mathrm{E}_n(\mathbf{K})$ *is abelian.*

Proof. The case $n = 1$ is trivial, thus assume $n > 1$.

(i) is a straightforward consequence of Theorem and Definition 392 and (ii) is readily obtained from (i).

(iii): The first statement is proved if we show that for every $B_{ij}(\lambda) \in \mathrm{E}_n(\mathbf{K})$ and every $P \in \mathrm{GL}_n(\mathbf{K})$, $P^{-1}B_{ij}(\lambda)\,P \in \mathrm{E}_n(\mathbf{K})$. By Proposition 393(i), it is sufficient to prove this property for every $P \in \mathrm{D}_n(\mathbf{K})$, i.e., of the form $D(u)$. This is easy and left to the reader.

By (i), the second statement is proved if we show that $D(u)\,D(v) \equiv D(v)\,D(u) \pmod{\mathrm{E}_n(\mathbf{K})}$, i.e., using a finite number of operations of the first kind on the rows and the columns, $D(u)\,D(v)$ can be transformed into $D(v)\,D(u)$. It is sufficient to consider the case $n = 2$. As easily seen, $\mathrm{diag}\,(u, v) \equiv \mathrm{diag}\,(1, v\,u) \pmod{\mathrm{E}_2(\mathbf{K})}$ and $\mathrm{diag}\,(1, u\,v) = D(u)\ D(v)$. ■

Remark 394. *(a) Proposition 393(i) means that for any $n \geq 1$, any matrix $Q \in \mathrm{GL}_n(\mathbf{K})$ can be written as a finite product E of matrices belonging to $\mathrm{E}_n(\mathbf{K})$ left- or right-multiplied by a matrix $D(u) \in \mathrm{D}_n(\mathbf{K})$.*
(b) Statements (i) and (ii) of Proposition 393 are still valid when $\mathbf{K}$ is an Euclidean domain (see Remark 660 below).

Lemma 395. *Let $n > 1$ and let $B_{ij}(\lambda) \in \mathrm{E}_n(\mathbf{K})$, $\lambda \neq 0$; there exists a matrix $P \in \mathrm{GL}_n(\mathbf{K})$ such that*

$$P\, B_{ij}(\lambda)\, P^{-1} = B_{21}(1)\,. \tag{2.50}$$

Proof. Consider the automorphism $\mathbf{u}$ of $\mathbf{K}^n$ defined as follows: let $\mathbf{x} \in \mathbf{K}^n$, $\mathbf{x} = (x_1, ..., x_n)$. In the canonical basis of $\mathbf{K}^n$, $\mathbf{x}$ is represented by the row $x = \begin{bmatrix} x_1 & ... & x_n \end{bmatrix}$ and $\mathbf{u}$ is the map $x \mapsto x\, B_{ij}(\lambda)$. Consider the new basis of $\mathbf{K}^n$ defined as follows: in this basis, $\mathbf{x}$ is represented by the row $y = \begin{bmatrix} y_1 & ... & y_n \end{bmatrix}$ where $y_1 = x_j + x_i\,\lambda$, $y_2 = x_i$ and for $k \in \{3, ..., n\}$, $y_k = x_l$, $l \in \{1, ..., n\} \setminus \{i, j\}$ where the map $k \mapsto l$ is isotone. In this basis, $\mathbf{u}$ is the map

$$y \mapsto y\, B_{21}(1)\,. \tag{2.51}$$

This yields (2.50) where $P \in \mathrm{GL}_n(\mathbf{K})$ is the change of basis matrix (Subsect. 2.2.6). ■

Lemma 396. *For any $n \geq 1$, $\mathbf{D}(\mathrm{GL}_n(\mathbf{K})) = \mathrm{E}_n(\mathbf{K})$ if $\mathbf{K} \neq \mathbb{Z}/2\mathbb{Z}$.*

Proof. See, e.g., ([80], Prop. 9.2.4). ■

2.11 Determinants

2.11.1 *Commutative Case*

Let $\mathbf{R}$ be a commutative ring. All facts below are classical ([27], §III.8), ([206], Sect. XIII.4) and are stated without proof. Recall that A^T denotes the classical transpose of the matrix A (Definition 271).

2.11.1.1 Definition of a Determinant

Lemma and Definition 397. *(i) There exists a unique monoid-homomorphism* $\det : (\mathrm{Mat}_n(\mathbf{R}), \times) \to (\mathbf{R}, \times)$ *such that* $\mathrm{Mat}_n(\mathbf{R}) \ni A \mapsto \det(A)$ *is an n-linear alternating form with respect to the columns of the matrix A.*
(ii) The map $\mathrm{Mat}_n(\mathbf{R}) \ni A \mapsto \det(A)$ *is an n-linear alternating form with respect to the rows of A, and* $\det(A) = \det(A^T)$.
(iii) The map det *is the* determinant.
(iv) The following equalities hold, setting $A = (a_{ij})$:

$$\det(A) = \sum_{\sigma \in \mathfrak{S}_n} \operatorname{sgn}(\sigma) \prod_{1 \leq i \leq n} a_{\sigma(i),i} = \sum_{\sigma \in \mathfrak{S}_n} \operatorname{sgn}(\sigma) \prod_{1 \leq i \leq n} a_{i,\sigma(i)}.$$

2.11.1.2 Expansions

Let H, K be subsets of $I \triangleq \{1, ..., n\}$ having the same cardinal p. Then $A_{H,K}$ denotes the submatrix of A obtained by deleting those rows and columns whose indices do not belong to H and K, respectively; $\det(A_{H,K})$ is called the *minor* with set of indices H, K of the matrix A and $\operatorname{card}(H) = \operatorname{card}(K)$ is called the *order* of this minor.

For any subset K of I, let $K' \triangleq I \backslash K$ and let $\rho_{K,K'} \triangleq (-1)^{\nu}$ where ν is the number of pairs $(k, k') \in K \times K'$ such that $k > k'$. In addition, let $H \subseteq I$, let $p = \operatorname{card}(H)$ and let $\mathcal{F}_p(I)$ be the set of all subsets of I with p elements. The determinant of A can be calculated using the *Laplace expansion*:

$$\boxed{\det(A) = \sum_{R \in \mathcal{F}_p(I)} \rho_{H,H'}\, \rho_{R,R'} \det(A_{R,H}) \det(A_{R',H'}).} \tag{2.52}$$

With $H = \{j\}$, we have $p = 1$; $A_{R,H}$ is an entry a_{ij} of A, $A_{R',H'}$ is the minor m_{ij} of order $n-1$ obtained by deleting the row of index i and the column of index j, and $\rho_{H,H'}\, \rho_{R,R'} = (-1)^{i+j}$. Let

$$\alpha_{ij} \triangleq (-1)^{i+j}\, m_{ij},$$

called the *cofactor* of a_{ij}; since $\det(A) = \det(A^T)$, (2.52) yields

$$\boxed{\det(A) = \sum_{1 \leq i \leq n} a_{ij}\, \alpha_{ij} = \sum_{1 \leq j \leq n} a_{ij}\, \alpha_{ij}}$$

(expansion of $\det(A)$ according to the jth column, then according to the ith row). Let $\boldsymbol{\alpha}$ be the matrix with entries α_{ij}. Its classical transpose $\boldsymbol{\alpha}^T$ is called the *classical adjoint* or the *adjugate* of A ([225], Sect. IX.2) and

$$\boxed{\boldsymbol{\alpha}^T A = A\, \boldsymbol{\alpha}^T = \det(A).} \tag{2.53}$$

2.11.1.3 Cramer's Rule

Theorem 398. *Let $A_1, ..., A_n, B$ and $x = \begin{bmatrix} x_1 & ... & x_n \end{bmatrix}^T$ be elements of ${}^n\mathbf{R}$ such that*

$$A x = B \tag{2.54}$$

where A is the matrix with columns A_j $(1 \leq j \leq n)$. For each i we have

$$\boxed{\det(A)\, x_i = \det(A_1, ..., A_{i-1}, B, A_{i+1}, ..., A_n)} \tag{2.55}$$

and the system of n equalities (2.55) $(1 \leq i \leq n)$ is equivalent *to (2.54) if $\det(A)$ is a regular element of $\mathbf{R}$ ("Cramer's rule").*

2.11.1.4 Linear Independence and Invertibility

The definition below is valid when $\mathbf{R}$ is a noncommutative ring:

Definition 399. *The rows* $A_{i\bullet}$ $(1 \leq i \leq n)$ *of a matrix* $A \in \mathrm{Mat}_n(\mathbf{R})$ *are said to be* left $\mathbf{R}$-linearly independent *if*

$$\sum_{1 \leq i \leq n} r_i A_{i\bullet} = 0 \ (r_i \in \mathbf{R}) \text{ implies that } r_i = 0 \text{ for all } i.$$

The columns A_j $(1 \leq j \leq n)$ *of* A *are said to be* right $\mathbf{R}$-linearly independent *if*

$$\sum_{1 \leq j \leq n} A_j r_j = 0 \ (r_j \in \mathbf{R}) \text{ implies that } r_j = 0 \text{ for all } j.$$

In the following theorem, the ring $\mathbf{R}$ is commutative.

Theorem 400. *Let* $A \in \mathrm{Mat}_n(\mathbf{R})$.

(1) The following properties are equivalent:
(i) the rows of A *are left* $\mathbf{R}$*-linearly independent;*
(ii) the columns of A *are right* $\mathbf{R}$*-linearly independent;*
(iii) A *is* regular *(in the monoid* $\mathrm{Mat}_n(\mathbf{R})$*);*
(iv) $\det(A)$ *is a* regular element *of* $\mathbf{R}$.
(2) The following properties are equivalent:
(i') A is left-invertible in $\mathrm{Mat}_n(\mathbf{R})$*;*
(ii') A is right-invertible in $\mathrm{Mat}_n(\mathbf{R})$*;*
(iii') $A \in \mathrm{GL}_n(\mathbf{R})$*, i.e., A is invertible;*
(iv') $\det(A)$ *is a* unit *of* $\mathbf{R}$.

2.11.1.5 Determinant of a Composed Matrix

Proposition 401. *Let*

$$A = \begin{bmatrix} X & Y \\ Z & T \end{bmatrix}$$

where X *and* T *are square.*
(i) If $Z = 0$ *or* $Y = 0$*, then*

$$\det(A) = \det(X)\det(T).$$

(ii) If X *is invertible, then*

$$\boxed{\det(A) = \det(X)\det\left(T - Z X^{-1} Y\right).} \tag{2.56}$$

2.11.2 Noncommutative Case

One cannot properly define the determinant of a square matrix over a noncommutative ring. The definition the determinant of a square matrix over a noncommutative division ring $\mathbf{K}$ is due to Dieudonné [91]. It is the solution of an universal problem:

2.11.2.1 Definition of the Dieudonné Determinant

Recall that $\mathbf{K}^{\times} = \mathbf{K} \backslash \{0\}$ is a multiplicative group and that for any group G, G^{ab} denotes the abelianization of $G^{\times}$ (Definition 170), i.e., $G^{ab} = G/\mathbf{D}(G)$.

Lemma and Definition 402. *(1) There exists a unique group-homomorphism* $\Delta_n : \mathrm{GL}_n(\mathbf{K}) \to \mathbf{K}^{\times ab}$ *which is universal for homomorphisms into abelian groups (Definition 12). Moreover, if* $\mathbf{K} \neq \mathbb{Z}/2\mathbb{Z}$, $\mathrm{GL}_n(\mathbf{K})^{ab} \cong \mathrm{D}_n(\mathbf{K})^{ab} \cong \mathbf{K}^{\times ab}$.
(2) Let $\bar{\mathbf{K}} \triangleq \mathbf{K}_0^{\times ab} = \mathbf{K}^{\times ab} \cup \{0\}$. *The* Dieudonné determinant *of a matrix* $A \in \mathrm{Mat}_n(\mathbf{K})$, *denoted by* $\det(A)$, *is the element of* $\bar{\mathbf{K}}$ *defined as follows: if* $A \in \mathrm{GL}_n(\mathbf{K})$, $\det(A) = \Delta_n(\mathbf{K})$; *if* $A \notin \mathrm{GL}_n(\mathbf{K})$ *(i.e., if* $\mathbf{A}$ *is* singular, *or equivalently* non-regular*),* $\det(A) = 0$.

Proof. (1). (a) If $\mathbf{K} = \mathbb{Z}/2\mathbb{Z}$, $\mathbf{K}$ is commutative and this case has already been treated.

(b) Let us assume that $\mathbf{K} \neq \mathbb{Z}/2\mathbb{Z}$. Since $\mathbf{K}^{\times} \cong \mathrm{D}_n(\mathbf{K})$ under the isomorphism

$$u \tilde{\mapsto} \mathrm{diag}(1, ..., 1, u), \quad u \in \mathbf{K}^{\times},$$

$D(\bar{u}) = \mathrm{diag}(1, ..., 1, \bar{u})$ is the canonical image of $D(u) = \mathrm{diag}(1, ..., 1, u)$ in $\mathrm{D}_n(\mathbf{K})^{ab}$, and $\mathbf{K}^{\times ab} \cong \mathrm{D}_n(\mathbf{K})^{ab}$. By Proposition 393(i), for any matrix $A \in \mathrm{GL}_n(\mathbf{K})$, there exist a matrix $D(u) \in \mathrm{D}_n(\mathbf{K})$ and a matrix $E \in \mathrm{E}_n(\mathbf{K})$ such that $A = D(u)\, E$. Therefore, denoting by $\overline{A}$ and $\overline{D(u)}$ the classes $(\mathrm{mod}\, \mathrm{E}_n(\mathbf{K}))$ of A and $D(u)$ respectively, $\overline{A} = \overline{D(u)}$. In addition, by Lemma 396,

$$\mathrm{GL}_n(\mathbf{K})^{ab} = \mathrm{GL}_n(\mathbf{K}) / \mathrm{E}_n(\mathbf{K}), \tag{2.57}$$

thus $\overline{D(u)} = D(\bar{u})$ and the group-homomorphism we are looking for is

$$\Delta_n : \mathrm{GL}_n(\mathbf{K}) \ni A \mapsto \bar{u} \in \mathbf{K}^{\times ab}.$$

This is an epimorphism since $\Delta_n(\mathrm{diag}(1, ..., 1, u)) = \bar{u}$. Therefore, according to Theorem 168(ii) there exists a unique group-epimorphism $\bar{\Delta}_n$: $\mathrm{GL}_n(\mathbf{K})^{ab} \to \mathbf{K}^{\times ab}$ such that $\Delta_n = \bar{\Delta}_n \circ \pi$ where

$$\pi : \mathrm{GL}_n(\mathbf{K}) \twoheadrightarrow \mathrm{GL}_n(\mathbf{K})^{ab}$$

is the canonical epimorphism. By Proposition 393 and Noether's second isomorphism theorem (Theorem 146(2)),

$$\mathrm{GL}_n(\mathbf{K})^{ab} = \mathrm{D}_n(\mathbf{K})\, \mathrm{E}_n(\mathbf{K}) / \mathrm{E}_n(\mathbf{K}) \cong \mathrm{D}_n(\mathbf{K}) / (\mathrm{D}_n(\mathbf{K}) \cap \mathrm{E}_n(\mathbf{K})),$$

$$\mathrm{D}_n(\mathbf{K}) / (\mathrm{D}_n(\mathbf{K}) \cap \mathrm{E}_n(\mathbf{K})) = \mathrm{D}_n(\mathbf{K})^{ab}.$$

Therefore, $\bar{\Delta}_n$ is the isomorphism

$$\mathrm{GL}_n(\mathbf{K})^{ab} \ni \overline{A} = \overline{D(u)} = D(\bar{u}) \mapsto \bar{u} \in \mathbf{K}^{\times ab} \quad (u \in \mathbf{K}^{\times}).$$

(c) Let Γ be an abelian group (multiplicatively noted) and let $f : \mathrm{GL}_n(\mathbf{K}) \to \Gamma$ be a group-homomorphism. By Theorem 168(ii) there exists a unique homomorphism $\bar{f} : \mathrm{GL}_n(\mathbf{K})^{ab} \to \Gamma$ such that $f = \bar{f} \circ \pi$. Since $\mathrm{GL}_n(\mathbf{K})^{ab} \cong \mathbf{K}^{\times ab}$, the group-homomorphism $\Delta_n : \mathrm{GL}_n(\mathbf{K}) \to \mathbf{K}^{\times ab}$ is universal for homomorphisms into abelian groups. The universality of Δ_n proves its uniqueness. ∎

2.11.2.2 Properties of the Dieudonné Determinant

In what follows, the canonical image of any element $a \in \mathbf{K}^{\times}$ in $\mathbf{K}^{\times ab}$ is denoted by $\bar{a}$; the canonical image $\bar{0}$ of $0 \in \mathbf{K}$ in $\bar{\mathbf{K}}$ is identified with 0, thus $0\,\bar{a} = \bar{a}\,0 = 0$. When the division ring $\mathbf{K}$ is a field, $\bar{\mathbf{K}} = \mathbf{K}$, thus the Dieudonné determinant and the usual determinant coincide.

Theorem 403. *Let $A \in \mathrm{Mat}_n(\mathbf{K})$ where $\mathbf{K}$ is a division ring.*

(i) $\det(A\,B) = \det(A)\,\det(B)$.
(ii) If $\mathbf{K} \neq \mathbb{Z}/2\mathbb{Z}$,[2] *then for any integer $n \geq 1$ the sequence*

$$1 \longrightarrow \mathrm{E}_n(\mathbf{K}) \longrightarrow \mathrm{GL}_n(\mathbf{K}) \xrightarrow{\det} \mathbf{K}^{\times ab} \longrightarrow 1 \tag{2.58}$$

is exact. In particular, the determinant is unchanged by elementary row or column operations of the first kind.
(iii) If the row $A_{i\bullet}$ is changed to $\lambda\,A_{i\bullet}$, $\det(A)$ *is multiplied by $\bar{\lambda}$.*
(iv) If two rows $A_{i\bullet}$ and $A_{j\bullet}$ $(i \neq j)$ are interchanged, $\det(A)$ *is multiplied by $\overline{-1}$.*
(v) The statement of Proposition 401 is still valid.
(vi) Statement (iii) is valid if columns are considered in place of rows, provided that the multiplications by the elements of $\mathbf{K}$ or $\bar{\mathbf{K}}$ are done on the right.
(vii) The statement of Theorem 400 is still valid (with $\mathbf{R}$ changed to $\mathbf{K}$ and noticing that the regular elements, the units and the nonzero elements of $\mathbf{K}$ coincide).

Proof. (i) is a consequence of the definition.

(ii): By Lemma and Definition 402(1), $\mathrm{GL}_n(\mathbf{K})^{ab} \cong \mathbf{K}^{\times ab}$, thus the exactness of the sequence (2.58) is a consequence of (2.57).

(iii): Let $A = D(u)\,E$, $E \in \mathrm{E}_n(\mathbf{K})$; changing the row $A_{i\bullet}$ to $\lambda\,A_{i\bullet}$, one obtains a matrix A' which can be put into the form $D(\lambda\,u)\,E'$, $E' \in \mathrm{E}_n(\mathbf{K})$.

(iv): See Remark 391.

---

[2] This assumption can be removed ([80], Theorem 9.2.6).

(v): Obviously, $\det(X \oplus I) = \det(X)$, therefore by (i),

$$\det(X \oplus T) = \det((X \oplus I)(I \oplus T)) = \det(X)\det(T).$$

Let

$$A = \begin{bmatrix} X & Y \\ 0 & T \end{bmatrix} \text{ or } A = \begin{bmatrix} X & 0 \\ Z & T \end{bmatrix}.$$

If X and Y are regular, $A \equiv X \oplus T \pmod{\mathrm{E}_n(\mathbf{K})}$, thus $\det(A) = \det(X)\det(T)$; this is still true if X or T is singular since then so is A. Last, if X is invertible,

$$\begin{bmatrix} I & 0 \\ -Z\,X^{-1} & I \end{bmatrix} \begin{bmatrix} X & Y \\ Z & T \end{bmatrix} = \begin{bmatrix} X & Y \\ 0 & T - Z\,X^{-1}\,Y \end{bmatrix}$$

and one obtains (2.56).

(vi) is a straightforward consequence of Proposition 393(i).

(vii): (1) If the rows of $\mathbf{A}$ are left $\mathbf{R}$-linearly dependent, there exists a relation $\sum_{1 \leq i \leq n} \lambda_i A_{i\bullet} = 0$ ($\lambda_i \in \mathbf{K}$) and at least one coefficient λ_i is nonzero, say $\lambda_1 \neq 0$; then $A_{1\bullet} + \sum_{2 \leq i \leq n} \mu_i A_{i\bullet} = 0$ where $\mu_i = \lambda_1^{-1} \lambda_i$. Since $\det(A)$ is unchanged when $A_{1\bullet}$ is changed to $A_{1\bullet} + \mu_i A_{i\bullet}$ ($2 \leq i \leq n$), $\det(A) = \det(A')$ where A' is a matrix whose first row is zero, thus $\det(A) = 0$ and $A \notin \mathrm{GL}_n(\mathbf{K})$. Conversely, if the rows of A are left $\mathbf{R}$-linearly independent, $\operatorname{rk} A = n$ (Theorem and Definition 392) and $A \in \mathrm{GL}_n(\mathbf{K})$. A similar rationale can be done for the columns. (2) If $A, B \in \mathrm{Mat}_n(\mathbf{K})$ are such that $AB = I_n$, then $\det(A)\det(B) = \bar{1}$, thus $\det(A) \neq 0$ and $\det(B) \neq 0$, therefore A and B belong to $\mathrm{GL}_n(\mathbf{K})$. ∎

2.11.2.3 Cramer's Rule

Let $A_1, ..., A_n, B$ and x be as in Theorem 398 but where $\mathbf{R}$ is a division ring $\mathbf{K}$, and consider equation (2.54). According to this equation, the matrix

$$A' = (A_1, ..., A_{i-1}, B, A_{i+1}, ..., A_n)$$

is obtained by right-multiplying the column A_i by x_i and then adding to this column all columns of the form $A_j\,x_j$ ($1 \leq j \leq n$, $j \neq i$). By Theorem 403,

$$\boxed{\det(A)\,\bar{x}_i = \det(A_1, ..., A_{i-1}, B, A_{i+1}, ..., A_n)} \tag{2.59}$$

which is identical to Cramer's rule (2.55) when $\mathbf{K}$ is commutative. When $\mathbf{K}$ is noncommutative, (2.59) gives those indices i for which $x_i = 0$ and, for the other indices, the class $\bar{x}_i$ of x_i in $\mathbf{K}^{ab}$.

2.11.2.4 Specificity of the Dieudonné Determinant

The map $\mathrm{Mat}_n(\mathbf{K}) \ni A \mapsto \det(A) \in \bar{\mathbf{K}}$ is *not* n-linear with respect to the rows nor of the columns of the matrix A if the division ring $\mathbf{K}$ is noncommutative. This is due to the fact that $\bar{\mathbf{K}}$ is not endowed with an addition $+$ such that (i) $(\bar{\mathbf{K}}, +)$ is an abelian group and (ii) the distributivity property holds true ([6], Sect. IV.1). In addition, $\det(A)$ is a *rational function*–not a polynomial function–of the entries of A: see Exercise 480.

2.11.3 Invariant Basis Number

Let $\mathbf{R}$ be a ring and let $\bullet f : M \to N$ be a module-isomorphism where M and N are left free modules such that $M \cong \mathbf{R}^n$, $N \cong \mathbf{R}^m$; assume that $n \geq m$, without loss of generality since if $m \geq n$ we replace f by f^{-1}. Let $(\varepsilon_i)_{1 \leq i \leq n}$ be a basis of M and $(\omega_j)_{1 \leq j \leq m}$ be a basis of N (Lemma 254). Let $A \in {}^n\mathbf{R}^m$ be the matrix of f in those bases. Considering the equation $y = xf$, let Y be the row representing y in the basis $(\omega_j)_{1 \leq j \leq m}$, and let X be the row representing x in the basis $(\varepsilon_i)_{1 \leq i \leq n}$. The "matrix form" of the above equation is

$$Y = X\,A.$$

which is equivalent to

$$\begin{bmatrix} Y & 0 \end{bmatrix} = X\,\tilde{A} \tag{2.60}$$

where $\tilde{A} = \begin{bmatrix} A & 0 \end{bmatrix} \in \mathrm{Mat}_m(\mathbf{R})$ has $n-m$ zero columns. Since f is a monomorphism, given any $Y \in \mathbf{R}^m$, (2.60) uniquely determines $X \in \mathbf{R}^n$. Let us assume that the ring $\mathbf{R}$ is commutative or that it is a division ring. According to Theorems 400 & 403(vii), $\det\left(\tilde{A}\right) \neq 0$, but this is impossible if $n \neq m$.

Definition 404. *A ring* $\mathbf{R}$ *is said to have* invariant basis number (IBN) *if* $\mathbf{R}^n$ *is not isomorphic to* $\mathbf{R}^m$ *when* $n \neq m$.

We have obtained the following.

Theorem 405. *A commutative ring and a division ring have IBN.*

Remark 406. *(i) Let* $\mathbf{R}$ *be a ring (not necessarily commutative) and* M *a free module with* infinite *basis* B. *One can prove that for any basis* B' *of* M, $\mathrm{card}(B') = \mathrm{card}(B)$ *([27], §II.1, Corol. 2 of Prop. 23).*
(ii) There exist noncommutative rings which do not have IBN ([79], Sect. 4.6, Exercise 2; [111]).

Definition 407. *Let* $\mathbf{R}$ *be a ring having IBN, let* M *be a free* $\mathbf{R}$*-module and let* B *be a basis of* M.

(i) $\mathrm{card}(B)$ *is called the* rank *of the free module* M *(written* $\mathrm{rk}_{\mathbf{R}}(M)$, *or* $\mathrm{rk}(M)$ *when there is no confusion).*

(ii) Let $\mathbf{K}$ *be a division ring, let* V *and* W *be left* $\mathbf{K}$*-vector spaces and let* $f : V \to W$ *be a* $\mathbf{K}$*-linear map. The rank of* f*, denoted by* $\mathrm{rk}(f)$*, is the dimension of* $\mathrm{im}\, f$*.*

Let f be as in Definition 407(ii); as a consequence of Theorem 257, we have the classical identity

$$\mathrm{rk}(f) + \dim(\ker f) = \dim(V). \tag{2.61}$$

In addition, choosing bases in V and W, let A be the matrix representing f in these bases; then $\mathrm{rk}(f) = \mathrm{rk}(A)$, as is well-known (see, e.g., [141], Sect. 19.8, Theorem 14).

Item (1) below is a consequence of Theorem 339(iii) and Remark 406(i):

Theorem and Definition 408. *Let* $\mathbf{R}$ *be a left Ore domain and let* $\mathbf{K}$ *be its division ring of left fractions.*

(1) Then $\mathbf{R}$ *has IBN and for any free left* $\mathbf{R}$*-module* M*,* $\mathrm{rk}(M) = \dim\left(M_{(\mathbf{K})}\right)$*.*
(2) More generally, let M *be a left* $\mathbf{R}$*-module. Then* $\dim\left(M_{(\mathbf{K})}\right)$ *is called the* rank *of* M *(over* $\mathbf{R}$*) and is denoted by* $\mathrm{rk}_{\mathbf{R}}(M)$*, or by* $\mathrm{rk}(M)$ *when there is no confusion.*

2.12 Rank of a Matrix

2.12.1 Outer Rank

Let $\mathbf{R}$ be an Ore domain and let $A \in {}^{n}\mathbf{R}^{m}$. Embedding $\mathbf{R}$ into $\mathbf{K} = \mathrm{Q}(\mathbf{R})$, A can be viewed as an element of ${}^{n}\mathbf{K}^{m}$. The rank of this matrix (i.e., the rank of A over $\mathbf{K}$) is called the *outer rank* (or the *rank*, for short) of A, and is denoted by $\mathrm{rk}(A)$.

Theorem 409. *(i)* $\mathrm{rk}(A)$ *is the dimension of the left* $\mathbf{K}$*-vector space generated by the rows of* A*.*
(ii) $\mathrm{rk}(A)$ *is the dimension of the right* $\mathbf{K}$*-vector space generated by the columns of* A*.*
(iii) $\mathrm{rk}(A)$ *is equal to the maximum of the orders of the invertible submatrices of* A*.*
(iv) $\mathrm{rk}(A)$ *is equal to the maximum of the orders of the square submatrices with* $\mathbf{R}$*-linearly independent rows (or columns).*
(v) The following property, called Sylvester's law of nullity*, holds: if* $A \in {}^{m}\mathbf{R}^{n}$ *and* $B \in {}^{n}\mathbf{R}^{p}$ *are such that* $AB = 0$*, then* $n \geq \mathrm{rk}(A) + \mathrm{rk}(B)$*.*
(vi) For any $A \in {}^{m}\mathbf{R}^{n}$*,* $B \in {}^{n}\mathbf{R}^{p}$*,*

$$\mathrm{rk}(A) + \mathrm{rk}(B) - n \leq \mathrm{rk}(AB) \leq \min\{\mathrm{rk}(A), \mathrm{rk}(B)\}$$

and the left inequality is still called Sylvester's law of nullity.

Proof. (i)-(iii) and (v), (vi) are classical: see ([141], Subsect. 19.9) and Exercise 482. (iv) is a consequence of (iii) and of Proposition 341(ii). ■

2.12.2 Inner Rank

2.12.2.1 Semifirs

There exists a class of rings, larger than that of division rings, for which an identity similar to (2.61) holds.

Lemma and Definition 410. *Let* $\mathbf{R}$ *be a nonzero ring and* $n \geq 1$.

(1) The following conditions are equivalent:
(i) every n*-generated (i.e., generated by at most* n *elements) left ideal in* $\mathbf{R}$ *is free, as left* $\mathbf{R}$*-module, of unique rank;*
(ii) every n*-generated submodule of a free left* $\mathbf{R}$*-module is free and* $\mathbf{R}^n$ *has unique rank;*
(iii) every n*-generated right ideal in* $\mathbf{R}$ *is free, as right* $\mathbf{R}$*-module, of unique rank.*
(iv) every n*-generated submodule of a free right* $\mathbf{R}$*-module is free and* ${}^n\mathbf{R}$ *has unique rank;*
(v) for any matrix $A \in {}^q\mathbf{R}^k$ $(q \leq n)$*, there exists an invertible matrix* U *such that* $U\,A = \begin{bmatrix} T \\ 0 \end{bmatrix}$ *where the rows of* T *are left* $\mathbf{R}$*-linearly independent.*
(2) The ring $\mathbf{R}$ *is called an* n-fir *if the above equivalent conditions are satisfied.*
(3) A ring which is an n*-fir for all* $n \geq 1$ *is called a* semifir.

Proof. (1). See ([77], Sect. 1.1, Theorem 1.1) and ([80], Sect. 8.7, Theorem 8.7.1). ■

Note that an n-fir is an n'-fir for all $n' \leq n$; a 1-fir is just a domain. A ring $\mathbf{R}$ is a semifir if, and only if $\mathbf{R}$ has IBN and every finitely generated left ideal is free (equivalently, every finitely generated right ideal is free). A division ring is a semifir, thus the following result, which is a consequence of Theorem 257 and of Lemma and Definition 410(1)(ii), generalizes (2.61) (see Proposition 413(3) below):

Proposition 411. *Let* $\mathbf{R}$ *be a semifir, let* M *be a finite free left* $\mathbf{R}$*-module, let* N *be a free left* $\mathbf{R}$*-module and let* $f : M \to N$ *be an* $\mathbf{R}$*-linear map. Then*

$$\boxed{\operatorname{rk}(\operatorname{im} f) + \operatorname{rk}(\ker f) = \operatorname{rk}(M)\,.}$$

Assuming that N is finite free, let $A \in {}^n\mathbf{R}^m$ be the matrix representing f in bases of M and N. As easily seen, $\operatorname{rk}(\operatorname{im} f)$ is equal to the rank of the free left $\mathbf{R}$-module generated by the rows of A, thus it is called the row-rank of

A and is denoted by $\rho_r(A)$. Similarly, $\operatorname{rk}({}^t f)$ is equal to the rank of the free right $\mathbf{R}$-module generated by the columns of A, thus it is called the column-rank of A and is denoted by $\rho_c(A)$. However, $\rho_r(A)$ and $\rho_c(A)$ are unrelated over an arbitrary semifir.

2.12.2.2 Inner Rank

The following notion is left-right symmetric and valid over an arbitrary ring:

Definition 412. *Let* $\mathbf{R}$ *be a ring and* $A \in {}^n\mathbf{R}^m$.

(i) The inner rank of A*, written* $\rho(A)$*, is the least integer* r *such that*

$$A = B\,C \quad \text{where } B \in {}^n\mathbf{R}^r \text{ and } C \in {}^r\mathbf{R}^m \tag{2.62}$$

(and $\rho(0) = 0$*).*
(ii) Any factorization such as (2.62) where $r = \rho(A)$ *is called a* minimal factorization *of* A*.*

The following properties of the inner rank are easily proved:

Proposition 413. *Let* $\mathbf{R}$ *be a ring and let* $A \in {}^n\mathbf{R}^m$.

(1) The following inequalities hold:
(i)

$$\rho(A) \leq \min\{n, m\}, \quad \rho(A \oplus B) \leq \rho(A) + \rho(B),$$
$$\rho(A\,B) \leq \min\{\rho(A), \rho(B)\}.$$

(ii) if A' *and* A'' *have the same number of rows,*

$$\rho([\,A' \; A''\,]) \geq \max\{\rho(A'), \rho(A'')\}.$$

(2) If $\mathbf{R}$ *is a semifir, then*

$$\rho(A) \leq \min\{\rho_r(A), \rho_c(A)\}.$$

(3) If $\mathbf{R}$ *is an Ore domain, then*

$$\operatorname{rk}(A) \leq \rho(A)$$

with equality if $\mathbf{R}$ *is a division ring.*

See Theorem 418 and Corollary 442 below.

Definition 414. *(i) Let* $\mathbf{R}$ *be a ring; a matrix* $A \in {}^n\mathbf{R}^m$ *is left- (resp., right-) full if* $\rho(A) = n$ *(resp.,* $\rho(A) = m$*); a matrix* $A \in \operatorname{Mat}_n(\mathbf{R})$ *is said to be* full *if* $\rho(A) = n$*.*

(ii) A ring **R** *is said to be* rank-stable *if the set of full matrices over* **R** *is closed under products (when defined) and diagonal sums.*

Theorem 415. *Let* **R** *be a rank-stable ring. Then the following properties hold:*

(i) **R** *is a domain;*
(ii) $\rho(I_n) = n$ *and* **R** *has IBN;*
(iii) the inner rank of a matrix over **R** *is the maximum of the orders of its full submatrices.*

Proof. (i) Let $0 \neq a, b \in \mathbf{R}$. Then a and b are full, thus so is $a\,b$, hence $a\,b \neq 0$.

(ii) Since $I_n = 1 \oplus ... \oplus 1$ (n terms) and $\rho(1) = 1$, $\rho(I_n) = n$. Let $\bullet f : \mathbf{R}^n \xrightarrow{\sim} \mathbf{R}^m$ be represented by the matrix A in the canonical bases and $\bullet g = f^{-1}$ be represented by the matrix B in the same bases. Then $f \circ g = Id_{\mathbf{R}^n}$ and $g \circ f = Id_{\mathbf{R}^m}$ are represented by I_n and I_m, respectively. Therefore, $AB = I_n$ and $BA = I_m$; by the first equality, $n = \rho(AB) \leq \min\{\rho(A), \rho(B)\} \leq \min\{n, m\}$ (Proposition 413(1)), thus $n \leq m$; by the second equality, $m \leq n$, thus $n = m$.

(iii) is proved by induction in [89] (this statement is consistent with that of Theorem 409(iii) when **R** is a division ring). ■

2.12.2.3 Sylvester Domains

Definition 416. *A ring* **R** *is called a* Sylvester domain *if Sylvester's law of nullity holds for the inner rank, i.e., whenever two matrices* $A \in {}^m\mathbf{R}^n$ *and* $B \in {}^n\mathbf{R}^p$ *are such that* $A\,B = 0$*, then* $n \geq \rho(A) + \rho(B)$*.*

Proposition 417. *(i) Sylvester's law of nullity is equivalent to the following: for any* $A \in {}^m\mathbf{R}^n$ *and* $B \in {}^n\mathbf{R}^p$*,*

$$\rho(A) + \rho(B) - n \leq \rho(A\,B).$$

(ii) A Sylvester domain is a rank-stable ring.
(iii) A semifir is a Sylvester domain.

Proof. (i) Assuming that Sylvester's law of nullity holds in the form given in Definition 416, let $A\,B = P\,Q$ be a minimal factorization, so that Q has $\rho(A\,B)$ rows and

$$\begin{bmatrix} A & P \end{bmatrix} \begin{bmatrix} B \\ -Q \end{bmatrix} = 0.$$

Therefore, $n + \rho(A\,B) \geq \rho(\begin{bmatrix} A & P \end{bmatrix}) + \rho\left(\begin{bmatrix} B \\ -Q \end{bmatrix}\right) \geq \rho(A) + \rho(B)$. The converse is obvious.

(ii) Let **R** be a Sylvester domain and $A, B \in \mathrm{Mat}_n(\mathbf{R})$ be full matrices. Then $\rho(A\,B) = n$ by by (i) and Proposition 413(1), therefore the set of full matrices over $\mathbf{R}$ is closed under products when defined. Let $A \in \mathrm{Mat}_n(\mathbf{R})$ and $B \in \mathrm{Mat}_m(\mathbf{R})$; by suitably partitioning a minimal factorization of $A \oplus B$ we have

$$\begin{bmatrix} A & 0 \\ 0 & B \end{bmatrix} = \begin{bmatrix} P \\ P' \end{bmatrix} \begin{bmatrix} Q & Q' \end{bmatrix}$$

where the number of columns of P is $\rho(A \oplus B)$. Since $P Q' = 0$, we have by Sylvester's law of nullity $\rho(A \oplus B) \geq \rho(P) + \rho(Q') \geq \rho(P Q) + \rho(P' Q')$ (Proposition 413(1)) and $P Q = A$, $P' Q' = B$, thus $\rho(A \oplus B) \geq \rho(A) + \rho(B) \geq \rho(A \oplus B)$ (Proposition 413(1)).

(iii) is proved in ([77], Sect. 5.5, Prop. 5.1). ■

Theorem 418. *Let* $\mathbf{R}$ *be an Ore domain. The following properties are equivalent:*

(i) $\mathbf{R}$ *is a Sylvester domain;*
(ii) every full matrix is left-regular;
(iii) the outer rank and the inner rank of a matrix over $\mathbf{R}$ *coincide.*
In statement (ii), left *can be changed to* right.

Proof. (i)⇒(ii): Let $A \in \mathrm{Mat}_n(\mathbf{R})$ be a full matrix (i.e., $\rho(A) = n$) and assume that A is not left-regular, i.e., there exists a nonzero matrix $X \in \mathrm{Mat}_n(\mathbf{R})$ such that $X A = 0$. Then $\rho(A) + \rho(X) > n$ and $\mathbf{R}$ is not a Sylvester domain.

(ii)⇒(iii): Let us assume that every full matrix is left-regular. Let $A \in \mathrm{Mat}_n(\mathbf{R})$ be full; then it is left-regular over $\mathbf{R}$ and it remains left-regular over $\mathbf{K} = \mathrm{Q}(\mathbf{R})$, thus invertible over $\mathbf{K}$ (Theorem 403(vii)). Conversely, if $A \in \mathrm{Mat}_n(\mathbf{R})$ is invertible over $\mathbf{K}$, then $\mathrm{rk}(A) = n$ and A is full by Proposition 413(3). Since the set of all invertible matrices over $\mathbf{K}$ is closed under products (when defined) and diagonal sums, $\mathbf{R}$ is a rank-stable ring. Therefore, for any matrix $B \in {}^n\mathbf{R}^p$, $\rho(B)$ is the maximum of the orders of the full submatrices of B. This number is $\mathrm{rk}(B)$ by Theorem 409(iii).

(iii)⇒(i) is an obvious consequence of Theorem 409(iv). ■

2.12.3 Invertibility, Primeness, and Fullness

Let $\mathbf{R}$ be a ring.

Definition 419. *A matrix* $A \in {}^m\mathbf{R}^n$ *is* left-prime *if in any equation* $A = P Q$, *where* P *is square,* P *necessarily has a right inverse. A* right-prime *matrix is defined correspondingly. A square matrix* $A \in \mathrm{Mat}_n(\mathbf{R})$ *is* prime *if it is both left- and right-prime.*

Proposition 420. *(i) For a matrix* $A \in {}^m\mathbf{R}^n$, *we have the implication*

right-invertible ⇒ left-prime

which is an equivalence for $A \in \mathrm{Mat}_n(\mathbf{R})$.

(ii) If $\mathbf{R}$ *is weakly finite (Exercise 481(2)) we have for a matrix* $A \in {}^m\mathbf{R}^n$ *the implication*

$$\textit{left-prime} \Rightarrow \textit{left-full}.$$

(iii) If $\mathbf{R}$ *is a Sylvester domain, we have for a matrix* $A \in {}^m\mathbf{R}^n$ *the implication*

$$\textit{left-full} \Rightarrow \textit{left-regular}.$$

(iv) If $\mathbf{R}$ *is a division ring, we have for a matrix* $A \in {}^m\mathbf{R}^n$ *the implication*

$$\textit{left-regular} \Rightarrow \textit{right-invertible}.$$

The above still holds true with "left" and "right" interchanged.

Proof. (i) If there exists $B \in {}^n\mathbf{R}^m$ such that $A\,B = I_m$ and $A = P\,Q$, where $P \in \mathrm{Mat}_m(\mathbf{R})$, then $P\,(Q\,B) = I_m$ and P is right-invertible, which proves the implication of (i). If $A \in \mathrm{Mat}_n(\mathbf{R})$ is left-prime, we have the equation $A = P\,Q$ with $Q = I_n$, thus $P = A$ is right-invertible.

(ii) Assuming that $\mathbf{R}$ is weakly finite, if $A \in {}^m\mathbf{R}^n$ is not left-full, we can write $A = P_1\,Q_1$, $P_1 \in {}^m\mathbf{R}^r$, $Q_1 \in {}^r\mathbf{R}^n$, $r < m$. Then A can be written in the form $P\,Q$ where $P = \begin{bmatrix} P_1 & 0 \end{bmatrix}$ is square and has $m - r$ zero columns, thus P is not invertible and not right-invertible (Exercise 481(2)). Hence A is not left-prime.

(iii) Assuming that $\mathbf{R}$ is a Sylvester domain and that $A \in {}^m\mathbf{R}^n$ is not left-regular, let $C \in {}^p\mathbf{R}^m$ be a nonzero matrix such that $C\,A = 0$. By Sylvester's law of nullity (Definition 416), $m \geq \rho(C) + \rho(A)$ with $\rho(C) > 0$, thus $\rho(A) < m$ and A is not left-full, which proves (iii).

(iv) is an easy consequence of Theorem and Definition 392. ■

Note that Sylvester domains are very special rings (see Theorem 634 and Exercise 697 below).

2.13 Specific Rings

2.13.1 GCD Domains

2.13.1.1 The Commutative Case

Definition 421. *A commutative domain* $\mathbf{R}$ *is called a* GCD domain *if the monoid* $(\mathbf{R}^\times, \times)$ *is a GCD cancellation monoid (Definition 133).*

Note that $(\mathbf{R}^\times, \times)$ is an invariant cancellation monoid. A commutative domain $\mathbf{R}$ is a GCD domain if, and only if the following equivalent conditions are satisfied (Remark 132(b) and Definition 133):

(i) Any pair of nonzero elements has a gcd.
(ii) Any pair of nonzero elements has an lcm.
(iii) The set of all principal ideals in $\mathbf{R}$ is a lattice.

Let $c \in \mathbf{R}^{\times}$ and let $\mathbf{L}((c), \mathbf{R})$ be the set of all principal ideals containing (c). If (iii) holds, $\mathbf{L}((c), \mathbf{R})$ is a lattice, included in the lattice $\mathrm{Lat}(\mathbf{R})$ of all submodules of the $\mathbf{R}$-module $\mathbf{R}_{\mathbf{R}}$ which is modular (Theorem 244), thus $\mathbf{L}((c), \mathbf{R})$ is a modular lattice too. Therefore, (iii) is equivalent to

(iv) For any $c \in \mathbf{R}^{\times}$, $\mathbf{L}((c), \mathbf{R})$ is a modular lattice.

Example 422. *The Hardy space $\mathcal{H}_{\infty}$, widely used in robustness theory, was proved to be both a GCD domain [334] and a Sylvester domain [288].*

2.13.1.2 The Noncommutative Case

In the noncommutative case, the above is generalized as follows:

Definition 423. *A domain $\mathbf{R}$ is a left (resp., right) GCD domain if for any $c \in \mathbf{R}^{\times}$ the set $\mathbf{L}(\mathbf{R}c, \mathbf{R})$ (resp., $\mathbf{L}(c\mathbf{R}, \mathbf{R})$) of all principal left (resp., right) ideals in $\mathbf{R}$ containing $\mathbf{R}c$ (resp., $c\mathbf{R}$) is a modular lattice for the inclusion. A GCD domain is a left GCD domain which is a right GCD domain.*

The following result is proved in Subsect. 3.2.4 below (see Corollary 521):

Lemma and Definition 424. *Let $\mathbf{R}$ be a ring and let a, b be regular elements of $\mathbf{R}$.*

(1) The following conditions are equivalent:
(i) $\mathbf{R}/a\mathbf{R} \cong \mathbf{R}/b\mathbf{R}$;
(ii) $\mathbf{R}/\mathbf{R}a \cong \mathbf{R}/\mathbf{R}b$.
(2) The regular elements a and b are said to be similar *if they satisfy the above equivalent conditions.*

Obviously, if two regular elements $a, b \in \mathbf{R}$ are associates, they are similar, and the converse holds true when the ring $\mathbf{R}$ is commutative (Exercise 485).

Consider two factorizations of $c \in \mathbf{R}^{\times}$:

$$c = q_1 ... q_r, \tag{2.63}$$

$$c = q_1' ... q_s'. \tag{2.64}$$

Definition 425. *The factorizations (2.63), (2.64) are said to be* isomorphic *if the corresponding chains are isomorphic in the modular lattice $\mathbf{L}(c\mathbf{R}, \mathbf{R})$ (Definition 115(3)).*

By Corollary 245, we have the following.

Corollary 426. *The factorizations (2.63), (2.64) are isomorphic if, and only if $r = s$ and there is a permutation $i \to i'$ of $\{1, ..., r\}$ such that q_i is similar to $q_{i'}'$.*

Note that if a factor q_i is a unit, q_i and $q_{i'}'$ are similar to 1, therefore one can assume that all factors in the factorizations (2.63), (2.64) are nonunits. Such factorizations are said to be *proper*.

Definition 427. *A* Schreier ring *is a domain in which any two factorizations of a nonzero element have isomorphic proper refinements.*

In a right GCD domain, $\mathbf{L}(c\mathbf{R}, \mathbf{R})$ is a modular lattice, thus the following is a consequence of the Schreier refinement theorem (Theorem 116):

Corollary 428. *A right GCD domain is a Schreier ring. (The converse does not hold [72].)*

2.13.2 Unique Factorization Domains

2.13.2.1 The Commutative Case

Definition 429. *A commutative domain* $\mathbf{R}$ *is said to be a* unique factorization domain *(*UFD*, for short) if the monoid* $(\mathbf{R}^{\times}, .)$ *is a UF-monoid (Definition 130).*

By Theorem 131 and Lemma 129, a commutative unique factorization domain is a GCD domain and an element p of a unique factorization domain is a prime (Definition 304) if, and only if it is an atom.

Definition 430. *Let* $\mathbf{R}$ *be a domain. A* system of representatives of atoms *in* $\mathbf{R}$ *is a family* $(p_\alpha)_{\alpha \in A}$ *of atoms in* $\mathbf{R}$ *such that every atom in* $\mathbf{R}$ *is associated with one (and only one)* p_α.

Let $(p_\alpha)_{\alpha \in A}$ be a system of representatives of atoms in a commutative domain $\mathbf{R}$. According to Definition 130, $\mathbf{R}$ is a unique factorization domain if, and only if every nonzero element $x \in \mathbf{R}$ can be written, in a unique way, in the form

$$x = u \prod_{\alpha} p_\alpha^{n_\alpha} \tag{2.65}$$

where u is a unit and $n_\alpha \in \mathbb{N}$ is zero for all but a finite number of indices.

Let $y = v \prod_{\alpha} p_\alpha^{m_\alpha}$ be another element of $\mathbf{R}$. Then

$$\gcd(x, y) = \mathbf{U}(\mathbf{R}) \prod_{\alpha} p_\alpha^{\inf(n_\alpha, m_\alpha)}, \quad \operatorname{lcm}(x, y) = \mathbf{U}(\mathbf{R}) \prod_{\alpha} p_\alpha^{\sup(n_\alpha, m_\alpha)}.$$

The Hardy space $\mathcal{H}_\infty$ (Example 422) is *not* a unique factorization domain.

An important result about commutative UFDs is due to Gauss ([206], Sect. IV.2):

Theorem 431. *(*Gauss' theorem*). Let* $\mathbf{R}$ *be a commutative UFD; the ring* $\mathbf{R}[X]$ *is again a UFD.*

Let us also state without proof the following result about the irreducibility of a polynomial in $\mathbf{R}[X]$ ([206], Sect. IV.3):

Proposition 432. (Eisenstein's Criterion). *Let* $\mathbf{R}$ *be a commutative UFD, let* $\mathbf{K}$ *be its field of fractions and let* $f(X) = f_0 X^n + ... + f_n$ *be a polynomial of degree* $n \geq 1$ *in* $\mathbf{R}[X]$*. Let* p *be a prime in* $\mathbf{R}$ *and assume:*

$$f_0 \not\equiv 0 \pmod{p}, \quad f_i \equiv 0 \pmod{p} \quad \text{for all } i \in \{1, ..., n\},$$
$$f_n \not\equiv 0 \pmod{p^2}.$$

Then, $f(X)$ *is irreducible in* $\mathbf{K}[X]$*.*

2.13.2.2 The Noncommutative Case

Lemma and Definition 433. *Let* $\mathbf{R}$ *be a domain (not necessarily commutative).*

(1) The following conditions are equivalent:
(i) $(\mathbf{R}^\times, .)$ *is an* $\mathcal{S}$*-unique factorization monoid (Definition 130) where the equivalence relation* $\mathcal{S}$ *is similarity.*
(ii) $\mathbf{R}$ *is a GCD domain and is atomic (Definition 224(ii)).*
(iii) For any $c \in \mathbf{R}^\times$*, the set* $\mathbf{L}(c\mathbf{R}, \mathbf{R})$ *of all principal right ideals containing* $c\mathbf{R}$ *is a modular lattice* of finite length.
(2) The domain $\mathbf{R}$ *is called a* unique factorization domain *if the above equivalent conditions are satisfied.*

Proof. (1). (i)⇔(ii) and (ii)⇒(iii) are clear. (iii)⇒(ii) is a consequence of the Jordan-Hölder-Dedekind theorem (Theorem 117). ■

Corollary 434. *A GCD domain which satisfies left and right* ACC_1 *is a unique factorization domain. In particular, a GCD domain which is Noetherian is a unique factorization domain.*

Proof. This is a consequence of Lemma and Definition 433, Proposition 103 and Definition 308. ■

2.13.2.3 Rigid Unique Factorization Domains

A domain $\mathbf{R}$ is said to be rigid if the cancellation monoid $(\mathbf{R}^\times, .)$ is rigid (Definition 60). The following is clear:

Lemma 435. *For a domain* $\mathbf{R}$*, the following conditions are equivalent:*

(i) $\mathbf{R}$ *is rigid;*
(ii) whenever $c = a\,b' = b\,a' \in \mathbf{R}^\times$*,* $a\mathbf{R} \subseteq b\mathbf{R}$ *or* $b\mathbf{R} \subseteq a\mathbf{R}$*;*
(iii) for any $c \in \mathbf{R}^\times$*, the lattice* $\mathbf{L}(c\mathbf{R}, \mathbf{R})$ *is a chain.*

Lemma and Definition 436. *Let* $\mathbf{R}$ *be a ring.*

(1) The following conditions are equivalent:
(i) $\mathbf{R}$ *is an atomic rigid domain.*

(ii) $\mathbf{R}$ *is a unique factorization domain such that for any* $c \in \mathbf{R}^{\times}$, *the set* $\mathbf{L}(c\mathbf{R}, \mathbf{R})$ *is a chain.*
(iii) For any $c \in \mathbf{R}^{\times}$, *the set* $\mathbf{L}(c\mathbf{R}, \mathbf{R})$ *is a chain* of finite length.
(2) The ring $\mathbf{R}$ *is called a* rigid unique factorization domain *if the above equivalent conditions are satisfied.*

In a rigid UFD, let us consider two complete factorizations (2.63), (2.64) of an element $c \neq 0$ (Definition 130). We have $r = s$ and for all $i \in \{1, ...r\}$, q_i is associated with q_i' (whereas in a general UFD, we have $r = s$ and there exists a permutation $i \to i'$ of $\{1, ...r\}$ such that q_i is similar to $q_{i'}'$).

A rigid unique factorization domain can be characterized by the following result ([77], Sect. 3.4, Corol. 4.8):

Proposition 437. *A ring is a rigid UFD if, and only if it is an atomic 2-fir (Lemma and Definition 410(2)) and a local ring.*

2.13.3 Bézout Domains

2.13.3.1 Definition and Example

Definition 438. *A ring* $\mathbf{R}$ *is called a* left *(resp.,* right*)* Bézout domain *if it is a domain and every* finitely generated *left (resp., right) ideal in* $\mathbf{R}$ *is principal. A* Bézout domain *is a left Bézout domain which is a right Bézout domain.*

Example 439. *Let* Ω *be an open connected subset of the complex plane and let* $\mathcal{O}(\Omega)$ *be the ring of all analytic (or, equivalently, holomorphic) functions in* Ω. *The ring* $\mathcal{O}(\Omega)$ *is a Bézout domain ([308], Theorem 15.15).*
This can be easily proved when $\Omega = \mathbb{C}$ *(*$\mathcal{O}(\mathbb{C})$ *is the ring of all* entire functions *in* $\mathbb{C}$*). Indeed, consider an ideal* $\mathfrak{a} = (f_1, ..., f_n)$ *of* $\mathcal{O}(\mathbb{C})$ *generated by a finite number of functions* $f_1, ...f_n$. *By* Weierstrass' factorization theorem *([308], Theorem 15.10), there exists a sequence* $(a_j)_{j \in \mathbb{N}}$ *of isolated points of* $\mathbb{C}$, n *functions* $g_1, ..., g_n \in \mathcal{O}(\mathbb{C})$ *and non-negative integers* n_{ij} $(1 \leq i \leq n,\ j \in \mathbb{N})$ *such that*

$$f_i(z) = \prod_{j=0}^{+\infty} (z - a_j)^{n_{ij}}\, e^{g_i(z)}$$

where $n_{ij} > 0$ *if* a_j *is a zero of order* n_{ij} *of* f_i. *Therefore setting* $n_i = \inf\{n_{ij}, 1 \leq j \leq n\}$, $\mathfrak{a} = (f)$ *where* $f(z) = \prod_{j=0}^{+\infty} (z - a_j)^{n_i}$. *Note that an element of* $\mathcal{O}(\mathbb{C})$ *is a unit if, and only if it has no zero.*

Example 440. *In [137], the ring* $\mathcal{H} = \mathbf{Q}(\mathbb{R}[s, e^s, e^{-s}]) \cap \mathcal{O}(\mathbb{C})$ *was introduced for the study of linear delay-differential systems and proved to be a commutative Bézout domain.*

2.13.3.2 Properties

Proposition 441. *For a ring* $\mathbf{R}$*, the following conditions are equivalent:*

(i) $\mathbf{R}$ *is a left Bézout domain.*
(ii) $\mathbf{R}$ *is both a semifir and a left Ore domain.*
(iii) $\mathbf{R}$ *is both a 2-fir and a left Ore domain.*
The above still holds with left *everywhere changed to* right.

Proof. (i)⇒(ii): Let $\mathbf{R}$ be left Bézout domain and let $\mathfrak{a} \neq 0$ be a f.g. left ideal in $\mathbf{R}$. Then $\mathfrak{a}$ is a left principal ideal, i.e., there exists $0 \neq a \in \mathbf{R}$ such that $\mathfrak{a} = \mathbf{R}\,a$; thus $x \mapsto x\,a$ is an $\mathbf{R}$-linear isomorphism ${}_{\mathbf{R}}\mathbf{R} \xrightarrow{\sim} \mathfrak{a}$ and $\mathfrak{a}$ is free of rank 1; therefore $\mathbf{R}$ is a semifir. Let us assume that $\mathbf{R}$ is not an Ore domain. Then the f.g. ideal $\mathfrak{a}_n$ in the proof of Theorem 337 is free of rank n, a contradiction.

(ii)⇒(i): Let $\mathbf{R}$ be a semifir and let $a, b \in \mathbf{R}^{\times}$; the left ideal $\mathbf{R}\,a \cap \mathbf{R}\,b$ is free of finite rank by Lemma and Definition 410(1) and $\mathbf{R}\,a \cap \mathbf{R}\,b \subseteq \mathbf{R}\,a$ which is of rank 1. If $\mathbf{R}$ is a left Ore domain, this implies that $\mathbf{R}\,a \cap \mathbf{R}\,b$ is of rank 1 by Theorem and Definition 408, thus there exists $c \in \mathbf{R}^{\times}$ such that $\mathbf{R}\,a \cap \mathbf{R}\,b = \mathbf{R}\,c$.

(ii)⇔(iii): See ([77], Sect. 1.1, Prop. 1.7). ■

Corollary 442. *Let* $\mathbf{R}$ *be a Bézout domain and let* A *be a matrix over* $\mathbf{R}$*; then* $\operatorname{rk}(A) = \rho(A) = \rho_r(A) = \rho_c(A)$.

Proof. The first equality is a consequence of Theorem 418, Proposition 417(iii) and Proposition 441. For the two last equalities, see ([77], Sect. 5.4, Prop. 4.4). ■

2.13.3.3 Multiples and Divisors in Bézout Domains

Theorem 443. *Let* $\mathbf{R}$ *be a right Bézout domain and let* $a, b \in \mathbf{R}^{\times}$*; then* a, b *have a* lcrm c *and a* gcld d *(Definition 81) characterized up to right associates by the relations* $c\,\mathbf{R} = a\,\mathbf{R} \cap b\,\mathbf{R}$ *and* $d\,\mathbf{R} = a\,\mathbf{R} + b\,\mathbf{R}$.

Proof. (a) By Proposition 441 and its proof, there exists $c \in \mathbf{R}^{\times}$ such that $c\,\mathbf{R} = a\,\mathbf{R} \cap b\,\mathbf{R}$. We have $c \in a\,\mathbf{R}$ and $c \in b\,\mathbf{R}$, thus c is a common right multiple of (a, b). If c' is a common right multiple of (a, b), then there exist $x, y \in \mathbf{R}^{\times}$ such that $c' = a\,x = b\,y$, thus $c'\,\mathbf{R} \subseteq a\,x\,\mathbf{R} \subseteq a\,\mathbf{R}$ and similarly $c\,\mathbf{R} \subseteq b\,\mathbf{R}$, thus $c'\,\mathbf{R} \subseteq c\,\mathbf{R}$ and c' is a right multiple of c, thus c and its right-associates are the lcrms of (a, b).

(b) There exists $d \in \mathbf{R}^{\times}$ such that $d\,\mathbf{R} = a\,\mathbf{R} + b\,\mathbf{R}$. We have $a\,\mathbf{R} \subseteq d\,\mathbf{R}$ and $b\,\mathbf{R} \subseteq d\,\mathbf{R}$, thus d is a common left divisor of (a, b). If d' is a common left divisor of (a, b), there exist $x', y' \in \mathbf{R}^{\times}$ such that $a = d'\,x'$ and $b = d'\,y'$, thus for any $u, v \in \mathbf{R}$, $a\,u + b\,v = d'\,(x'\,u + y'\,v) \in d'\,\mathbf{R}$; therefore, $a\,\mathbf{R} + b\,\mathbf{R} \subseteq d'\,\mathbf{R}$, hence $d \in d'\,\mathbf{R}$ and d' is a left-divisor of d. It follows that d and its right-associates are gclds of (a, b). ■

Corollary 444. *Let* $\mathbf{R}$ *be a right Bézout domain and* $c \in \mathbf{R}^{\times}$*; then the set* $\mathbf{L}(c\mathbf{R}, \mathbf{R})$ *of all right principal ideals containing* $c\mathbf{R}$ *is a modular lattice.*

Proof. Let $a, b \in \mathbf{R}$ be such that $c\mathbf{R} \subseteq a\mathbf{R}$ and $c\mathbf{R} \subseteq b\mathbf{R}$. Then $c\mathbf{R} \subseteq a\mathbf{R} \cap b\mathbf{R}$, $c\mathbf{R} \subseteq a\mathbf{R} + b\mathbf{R}$, thus both $a\mathbf{R} \cap b\mathbf{R}$ and $a\mathbf{R} + b\mathbf{R}$ belong to $\mathbf{L}(c\mathbf{R}, \mathbf{R})$ by Theorem 443 which proves that $\mathbf{L}(c\mathbf{R}, \mathbf{R})$ is a lattice. In addition, $\mathbf{L}(c\mathbf{R}, \mathbf{R})$ is included in the lattice $\mathrm{Lat}_{\mathbf{r}}(\mathbf{R}_{\mathbf{R}})$ of all submodules of the right $\mathbf{R}$-module $\mathbf{R}_{\mathbf{R}}$ which is modular (Theorem 244), thus $\mathbf{L}(c\mathbf{R}, \mathbf{R})$ is a modular lattice. ∎

Corollary 445. *A right Bézout domain is a right GCD domain.*

Proof. This is a consequence of Corollary 444 and of Definition 423. ∎

Let us state without proof the following result about (two-sided) Bézout domains ([70], Theorem 4.2):

Proposition 446. *A domain* $\mathbf{R}$ *is a Bézout domain if, and only if the sum of any two principal right ideals or of any two principal left ideals is again principal.*

The invariant elements (Definition 85(ii)) play an important role in the study of Bézout domains due to the following result:

Lemma 447. *Let* $\mathbf{R}$ *be weakly finite ring (Exercise 481) and let* $\mathfrak{a} \neq 0$ *be a principal left ideal which is a principal right ideal. Then* $\mathfrak{a} = \mathbf{R}\, a\, \mathbf{R}$ *where* a *is an invariant element of* $\mathbf{R}$.

Proof. We have $\mathfrak{a} = \mathbf{R}\, a = a'\, \mathbf{R}$; let $u, v \in \mathbf{R}^{\times}$ be such that $a = u\, a'$ and $a' = a\, v$. We have $a' \in \mathbf{R}\, a' = a\, \mathbf{R}$, thus $a'\, v \in a\, \mathbf{R} = \mathbf{R}\, a'$ and there exists $u' \in \mathbf{R}^{\times}$ such that $a'\, v = u'\, a'$, hence $a' = u\, a'\, v = u\, u'\, c'$ and $u\, u' = 1$. Therefore, u is right-invertible over $\mathbf{R}$, thus it is invertible since $\mathbf{R}$ is weakly finite (Exercise 481(iv)), i.e., u is a unit. By symmetry, v is a unit too. ∎

2.13.4 Dedekind Domains

2.13.4.1 Commutative Case

Lemma and Definition 448. *Let* $\mathbf{R}$ *be a commutative domain and let* $\mathbf{K}$ *be its field of fractions.*

(1) Let $x \in \mathbf{K}$*. The following conditions are equivalent:*
(i) There exists a f.g. nonzero $\mathbf{R}$*-module* M *such that* $x\, M \subseteq M$.
(ii) The element x *satisfies an equation*

$$x^{n} + a_{n-1}\, x^{n-1} + \ldots + a_0 = 0 \tag{2.66}$$

where $a_i \in \mathbf{R}$ $(0 \leq i \leq n-1)$ *and* $n \geq 1$. *(Such an equation is called an* integral equation *over* $\mathbf{R}$.*)*
(2) An element $x \in \mathbf{K}$ *satisfying the above equivalent conditions is said to be* integral *over* $\mathbf{R}$.
(3) The domain $\mathbf{R}$ *is said to be* integrally closed *if every element of* $\mathbf{K}$*, integral over* $\mathbf{R}$*, belongs to* $\mathbf{R}$.

Proof. (1) (ii)$\Rightarrow$(i): Let M be the $\mathbf{R}$-module generated $1, x, \ldots, x^{n-1}$. If (ii) holds, then (i) holds too.

(i)$\Rightarrow$(ii): Assume there exists $M = [v_1, \ldots, v_n]_{\mathbf{R}}$ such that $x\,M \subseteq M$ and $M \neq 0$. Then each $x\,v_i$ $(1 \leq i \leq n)$ is an $\mathbf{R}$-linear combination of the v_j $(1 \leq j \leq n)$, i.e., there exists a matrix $A = (a_{ij})$ such that $v\,x = v\,A$ where $v = \begin{bmatrix} v_1 & \cdots & v_n \end{bmatrix}$. Therefore, $v\,(x\,I_n - A) = 0$ and $v \neq 0$, thus $\det(x\,I_n - A) = 0$ by Theorem 400(3), and x satisfies an integral equation over $\mathbf{R}$. ∎

The following is easily proved ([207], Sect. 1.2, Prop. 7):

Proposition 449. *Every commutative unique factorization domain is integrally closed.*

Definition 450. *A* commutative Dedekind domain *is a Noetherian domain, integrally closed, in which every non-zero prime ideal is maximal.*

The main property of commutative Dedekind domains is expressed in the theorem below (see [306], Theorem 4.24, and [207], Sect. 1.6):

Theorem 451. *Let* $\mathbf{R}$ *be a commutative domain. The following conditions are equivalent:*

(i) $\mathbf{R}$ *is a Dedekind domain.*
(ii) Every nonzero ideal $\mathfrak{a}$ *in* $\mathbf{R}$ *is invertible, i.e., there exists a fractional ideal* $\mathfrak{b}$ *(Definition 338) such that* $\mathfrak{a}\,\mathfrak{b} = \mathfrak{b}\,\mathfrak{a} = \mathbf{R}$.
(iii) Every ideal in $\mathbf{R}$ *can be factored into prime ideals, and the nonzero fractional ideals form a group under multiplication.*

Let $\mathfrak{P}$ be the set of all nonzero prime ideals in a commutative Dedekind domain $\mathbf{R}$. According to Condition (iii) of the above theorem, every fractional ideal $\mathfrak{b}$ in $\mathbf{R}$ can be uniquely written in the form

$$\mathfrak{b} = \prod_{\mathfrak{p} \in \mathfrak{P}} \mathfrak{p}^{n(\mathfrak{p};\mathfrak{b})} \tag{2.67}$$

where all but a finite number of the integers $n(\mathfrak{p};\mathfrak{b})$ are zero. The above fractional ideal $\mathfrak{b}$ is an ideal if, and only if all numbers $n(\mathfrak{p};\mathfrak{b})$ are ≥ 0.

2.13.4.2 Noncommutative Dedekind Domains

Let $\mathbf{R}$ be a left Ore domain and let $\mathbf{K}$ be its division ring of left fractions. The following is consistent with the definition given in Theorem 451(ii):

Definition 452. *A* left ideal $\mathfrak{a}$ *in* $\mathbf{R}$ *is called* invertible *if there exist elements* $q_1, ..., q_n$ *in* $\mathbf{K}$ *and* $a_1, ..., a_n$ *in* $\mathfrak{a}$ *such that* $\mathfrak{a}\, q_i \subseteq \mathbf{R}$ *and* $\sum_{1 \leq i \leq n} q_i\, a_i = 1$. *An* invertible right ideal *is likewise defined.*

One can prove the following ([136], Corol. 3.1):

Lemma 453. *Let* $\mathbf{R}$ *be a left Ore domain. If a left ideal* $\mathfrak{a}$ *in* $\mathbf{R}$ *is invertible, it is finitely generated. Therefore, if all nonzero left ideals in* $\mathbf{R}$ *are invertible,* $\mathbf{R}$ *is left Noetherian.*

We are now led to the following.

Definition 454. *A* left Dedekind domain *is a left Ore domain in which all nonzero left ideals are invertible. A right Dedekind domain is likewise defined and a* noncommutative Dedekind domain *is a left Dedekind domain which is a right Dedekind domain.*

Noncommutative Dedekind domains play an important role in Systems Theory due to the following result, stated here without proof (see [247], 7.11.2):

Theorem 455. *Let* $\mathbf{R}$ *be a commutative Dedekind domain and let* $\mathbf{S} = \mathbf{R}[X; \delta]$ *or* $\mathbf{S} = \mathbf{R}[X, X^{-1}; \boldsymbol{\alpha}]$ *(Subsect. 2.7.2 & 2.8.1). Then* $\mathbf{S}$ *is a noncommutative Dedekind domain if, and only if* $\mathbf{S}$ *is simple.*

The following consequence is clear:

Corollary 456. *The rings* $A_1(\mathbf{k})$, $B_1(\mathbf{k})$ *and* $\mathbf{A}_{1c}(\mathbf{k})$ *(Subsect. 2.7.5) are noncommutative Dedekind domains. So is also the ring* $A_1'(\mathbf{k})$ *(Example 369).*

Last, let us state the following without proof (see [247], 5.7.7):

Proposition 457. *Let* $\mathbf{R}$ *be a Dedekind domain. Then, every left (or right) ideal in* $\mathbf{R}$ *is generated by two elements.*

More specifically, let $\mathfrak{a}$ *be a left (resp., right) ideal in* $\mathbf{R}$. *If* $\mathfrak{a}$ *is essential (see, below, Exercise 692), then there exists a left (resp., right) essential ideal* $\mathfrak{b}$ *such that* $\mathfrak{a} \oplus \mathfrak{b} \cong \mathbf{R}^2$ *and* $\mathfrak{a}$ *has two generators. If* $\mathfrak{a}$ *is non-essential, then there exists a left (resp., right) non-essential ideal* $\mathfrak{b}$ *such that* $\mathfrak{a} \oplus \mathfrak{b} \cong \mathbf{R}$ *and* $\mathfrak{a}$ *is principal.*

2.13.5 Principal Ideal Domains

Definition 458. *A ring* $\mathbf{R}$ *is a principal left ideal ring if every left ideal in* $\mathbf{R}$ *is principal. A principal ideal ring is a principal left ideal ring which is a principal right ideal ring.*

We know that in any domain $\mathbf{R}$, an invariant element p is a prime if, and only if the principal ideal (p) is completely prime (Theorem 305) and this holds true if (p) is a maximal ideal (Proposition 306). Conversely, we have the following (which is generalized in Theorem 586(3) below).

Proposition 459. *Let* $\mathbf{R}$ *be a principal ideal domain and let* p *be an invariant element of* $\mathbf{R}$*. Then* p *is a prime in* $\mathbf{R}$ *if, and only if* (p) *is a maximal ideal.*

Proof. Let p be a prime and assume that there exists a proper ideal $\mathfrak{a}$ such that $(p) \subsetneq \mathfrak{a}$. There exist an invariant element a of $\mathbf{R}$ which is a nonunit and such that $\mathfrak{a} = (a) = \mathbf{R}\, a = \mathbf{R}\, a\, \mathbf{R}$, and a nonzero nonunit b, such that $p = a\, b$ (Lemma 86). Thus, $\mathbf{R}\, a\, b = \mathbf{R}\, a\, \mathbf{R}\, b = (a)\,(b) = (p)$ whereas $a \notin (p)$ and $b \notin (p)$, thus (p) is not a prime ideal (Lemma 299(ii)), a contradiction. ∎

The following is clear:

Proposition 460. *Let* $\mathbf{R}$ *be a ring. Then* $\mathbf{R}$ *is a principal ideal domain if, and only if* $\mathbf{R}$ *is both a Bézout domain and a Noetherian ring.*

Corollary 461. *A principal ideal domain is atomic (Definition 224(ii)).*

Proof. Let $\mathbf{R}$ be a principal ideal domain. By Proposition 460, $\mathbf{R}$ is Noetherian, thus it satisfies left and right ACC (Definition 308) and since every left or right ideal in $\mathbf{R}$ is principal, $\mathbf{R}$ satisfies left and right ACC_1 (Definition 102 and Notation 64(ii)). Since $\mathbf{R}$ is a domain, the monoid $(\mathbf{R}^{\times}, \times)$ is cancellable, thus it is atomic (Proposition 103) and so is $\mathbf{R}$ too. ∎

Theorem 462. *Let* $\mathbf{R}$ *be a ring. Of the following, (i)⇔(ii)⇔(iii)⇒(iv); if* $\mathbf{R}$ *is commutative, (iv)⇒(i):*

(i) $\mathbf{R}$ *is a principal ideal domain;*
(ii) $\mathbf{R}$ *is both a Bézout domain and a UFD.*
(iii) $\mathbf{R}$ *is an atomic Bézout domain.*
(iv) $\mathbf{R}$ *is is both a Dedekind domain and a UFD.*

Proof. (i)⇒(ii)⇒(iii)⇒(iv): We know that a principal ideal domain is a Bézout domain, thus it is a GCD domain (Corollary 445). Since it is atomic by Corollary 461, it is a unique factorization domain (Lemma and Definition 433). In addition, it is obviously a Dedekind domain.

(iii)⇒(ii)⇒(i): A unique factorization domain is atomic, and one can easily prove that a ring is a principal ideal domain if, and only if it is an atomic Bézout domain ([77], Sect. 3.3, Exercise 14).

Assuming that $\mathbf{R}$ is commutative and (iv) holds, let $\mathfrak{p}$ be a nonzero prime ideal in $\mathbf{R}$ and let $0 \neq x \in \mathfrak{p}$ be of the form (2.65). By Lemma 299, there exists an index α such that $p_\alpha \in \mathfrak{p}$, thus $(p_\alpha) \subseteq \mathfrak{p}$ and (p_α) is a prime ideal by Theorem 305, thus is maximal by Definition 450; therefore, $(p_\alpha) = \mathfrak{p}$ and $\mathfrak{p}$ is a principal ideal. So is every ideal in $\mathbf{R}$ by Theorem 451. ∎

Lemma 463. *Let* $\mathbf{R}$ *be a left principal ideal ring and let* S *be a left denominator set; then* $\mathbf{T} = S^{-1}\mathbf{R}$ *is again a left principal ideal ring.*

Proof. Let $\mathfrak{b}' \vartriangleleft_l S^{-1}\mathbf{R}$; by Theorem 332(i), $\varphi_S(\mathfrak{b}') \vartriangleleft_l \mathbf{R}$ and $S^{-1}\varphi_S(\mathfrak{b}') = \mathfrak{b}'$; $\varphi_S(\mathfrak{b}')$ is a left principal ideal in $\mathbf{R}$, i.e., there exists $r \in \mathbf{R}$ such that $\varphi_S(\mathfrak{b}') = \mathbf{R}a$. Therefore, $\mathfrak{b}' = S^{-1}\varphi_S(\mathfrak{b}') = \mathbf{T}(a/1)$ is a left principal ideal in $\mathbf{T}$. ■

Theorem 464. *Let* $\mathbf{K}$ *be a division ring,* $\boldsymbol{\alpha}$ *an automorphism and* $\boldsymbol{\delta}$ *an* $\boldsymbol{\alpha}$*-derivation of* $\mathbf{K}$*. The ring of skew power series* $\mathbf{S} = \mathbf{K}\left[\left[X^{-1}; \boldsymbol{\alpha}, \boldsymbol{\delta}\right]\right]$ *is a principal ideal domain and is local with Jacobson radical* $\left(X^{-1}\right)$*; every element of* $\mathbf{S}$ *is of the form* $v\,X^{-n} = X^{-n}\,v'$ *where* v*,* v' *are units; the (two-sided) ideals, left ideals and right ideals of* $\mathbf{S}$ *coincide and are all* $\left(X^{-n}\right)$*,* $n \geq 0$*. The ring* $\mathbf{S}$ *is a rigid UFD.*

Proof. This is a consequence of Corollary 384, of Proposition 437 and of Theorem 462. ■

Remark 465. *Let* $\mathbf{R} = \mathbb{Z}\left[\sqrt{-5}\right]$ *be the ring consisting of all elements of the form* $a + \sqrt{-5}b$*,* $a, b \in \mathbb{Z}$*. One can prove that (i)* $\mathbf{R}$ *is a Dedekind domain, (ii)* $1 + \sqrt{-5}$*,* $1 - \sqrt{-5}$*,* 2 *and* 3 *are primes in* $\mathbf{R}$ *([313], Sect. 3.4). Since* $\left(1 + \sqrt{-5}\right)\left(1 - \sqrt{-5}\right) = 2 \times 3$*,* $\mathbf{R}$ *is not a UFD, thus it is not a principal ideal domain.*

Let $\mathbf{R}$ be a commutative principal ideal domain and let $p \in \mathbf{R}$ be a prime. Set $\mathbf{R}\left(p^i\right) = \mathbf{R}/\left(p^i\right)$ $(i \geq 1)$. We have by Theorem 241(iii) $\mathbf{R}\left(p^i\right) \cong (p)/\left(p^{i+1}\right) \subseteq \mathbf{R}\left(p^{n+1}\right)$. Therefore, the sequence $\left(\mathbf{R}\left(p^i\right)\right)$ is a direct system (Example 19) and

$$\boxed{\mathbf{R}\left(p^{\infty}\right) \triangleq \varinjlim \mathbf{R}\left(p^i\right).}$$

This construction is *not* the same as that in Remark 348.

2.13.6 Euclidean Domains

2.13.6.1 Definitions and Properties

(i) A domain $\mathbf{R}$ is said to be *left Euclidean* if there exists a function $\theta : \mathbf{R} \to \mathbb{N} \cup \{-\infty\}$, called a *left Euclidean function*, such that:

(E1) $\theta(0) = -\infty$.
(E2) For any $a, b \in \mathbf{R}^{\times}$, $\theta(a\,b) \geq \theta(a) > -\infty$.

(E3) For any $a \in \mathbf{R}$ and $b \in \mathbf{R}^{\times}$, there exists $q, r \in \mathbf{R}$ such that

$$a = q\,b + r, \quad \theta(r) < \theta(b), \tag{2.68}$$

called the *left division algorithm*.

(ii) The above holds with *left* changed throughout to *right*, a and b interchanged in (E2), and (2.68) replaced by the *right division algorithm*

$$a = b\,q + r, \quad \theta(r) < \theta(b). \tag{2.69}$$

The elements q and r of (2.68) (resp., (2.69)) are called a quotient and a remainder of the *right* (resp., *left*) division of a by b (notice that using the *left* division algorithm, one *right* divides a by b!).

(iii) An Euclidean domain is a left Euclidean domain which is a right Euclidean domain.

(iv) A left Euclidean domain $\mathbf{R}$ is said to be *strongly Euclidean* if the left Euclidean function θ satisfies the following condition:

(E2') For any $a, b \in \mathbf{R}^{\times}$, $\theta(a-b) \leq \max\{\theta(a), \theta(b)\}$ and $\theta(a\,b) = \theta(a) + \theta(b)$.

Obviously, (E2') implies (E2). In addition, we have the following.

Lemma 466. *(i) In an Euclidean domain, $\theta(b) \geq \theta(1)$ if $b \in \mathbf{R}^{\times}$, and $\theta(b) = \theta(1)$ if, and only if b is a unit.*
(ii) Condition (E3) is equivalent to the following:
(E3') For any $a, b \in \mathbf{R}$ such that $b \neq 0$ and $\theta(a) \geq \theta(b)$, there exists $c \in \mathbf{R}$ such that

$$\theta(a - c\,b) < \theta(a).$$

(iii) In a left Euclidean domain $\mathbf{R}$, the remainder r in the division (2.68) is unique if, and only if $\mathbf{R}$ is strongly Euclidean, and then the quotient q is unique too.

Proof. (i) Let $b \in \mathbf{R}^{\times}$; then $\theta(b) = \theta(1\,b) \geq \theta(1)$ by (E1) and (E2). If u is a unit, there exists u' such that $1 = u\,u'$, thus $\theta(1) \geq \theta(u)$ by (E1) and (E2), thus $\theta(u) = \theta(1)$. Conversely, let b be such that $\theta(b) = \theta(1)$; right-dividing 1 by b we obtain $1 = q\,b + r$ with $\theta(r) < \theta(1)$, thus $r = 0$, and $1 = q\,b$; left-dividing 1 by b we obtain by the same rationale $q' \in \mathbf{R}$ such that $1 = b\,q'$, thus b is both left- and right-invertible, i.e., is a unit (**1.3.2.1**).

(ii) If (E3) holds, then (E3') follows by taking $c = q$. Conversely, if (E3') holds and $a, b \in \mathbf{R}$, $b \neq 0$, choose $q \in \mathbf{R}$ such that $\theta(a - q\,b)$ takes its least value. If $\theta(a - q\,b) \geq \theta(b)$, then by (E3') there exists $c \in \mathbf{R}$ such that $\theta(a - q\,b - c\,b) < \theta(a - q\,b)$, a contradiction. Therefore, $\theta(a - q\,b) < \theta(b)$ and (E3) is satisfied with $r = a - q\,b$.

The proof of (iii) is easy ([77], Sect. 2.1, Prop. 1.3). ■

Using the above lemma, it is easy to prove the following (see, e.g., [80], Theorem 7.3.3):

Theorem 467. *Let* $\mathbf{K}$ *be a division ring,* $\boldsymbol{\alpha}$ *an automorphism and* $\boldsymbol{\delta}$ *an* $\boldsymbol{\alpha}$*-derivation of* $\mathbf{K}$*. Then* $\mathbf{K}[X;\boldsymbol{\alpha},\boldsymbol{\delta}]$ *is a strongly Euclidean domain with Euclidean function the* degree (**2.7.4.2**).

In a general left (or right) Euclidean domain, $\theta(a)$ is often called the *degree* of a, abusing the language.

Theorem 468. *A left (resp., right) Euclidean domain is a principal left (resp., right) ideal domain.*

Proof. Let $\mathfrak{a} \neq 0$ be a left ideal in a left Euclidean domain $\mathbf{R}$ with left Euclidean function θ. Let b be an element of $\mathfrak{a}$ such that $\theta(b)$ is the smallest element of the set $\{\theta(x) : x \in \mathfrak{a}, \theta(x) \geq \theta(1)\} \subseteq \mathbb{N}$. Let $a \neq 0$ be an element of $\mathfrak{a}$. By the left division algorithm (E3), we find $q, r \in \mathbf{R}$ such that $a = q\,b + r$ and $\theta(r) < \theta(b)$. Since $r = a - q\,b$, necessarily $r \in \mathfrak{a}$. Therefore, $\theta(r) = -\infty$ and $r = 0$, thus $\mathfrak{a} = \mathbf{R}\,b$ and $\mathbf{R}$ is a principal left ideal domain. ■

2.13.6.2 Calculation of a gcrd and of an lclm

Euclid's algorithm takes the following form in a left Euclidean domain:

Theorem 469. *In a left Euclidean domain* $\mathbf{R}$*, let* a, b *be nonzero elements and consider the following algorithm where* $a_1 = a$ *and* $a_2 = b$ *:*

$$a_1 = q_1\,a_2 + a_3$$

$$\cdots$$

$$a_i = q_i\,a_{i+1} + a_{i+2} \tag{2.70}$$

$$\cdots$$

where (2.70) is the right Euclidean division of a_i *by* a_{i+1}*. There exists a natural integer* n *such that* $a_n \neq 0$ *and the remainder* a_{n+1} *is zero;* a_n *is a gcrd of* a, b*.*

Proof. (1) Since $\theta(a_{i+2}) < \theta(a_{i+1})$ and $\theta(a_2)$ is finite, there exists n such that $a_{n+1} = 0$.

(2) Thus, $a_{n-1} = q_{n-1}\,a_n$ and a_n right-divides a_{n-1}. Since

$$a_{n-2} = q_{n-2}\,a_{n-1} + a_n,$$

a_n right-divides a_{n-2} too and successively one sees than a_n right-divides all a_{n-i} $(1 \leq i \leq n-1)$, thus a_n right-divides a and b.

(3) If an element c right-divides $a = a_1$ and $b = a_2$, it right-divides $a_3 = a_1 - q_1\,a_2$. Continuing this process, one sees that c right-divides all a_i $(1 \leq i \leq n)$, thus a_n is a gcrd of a, b. ■

Our aim is now to calculate an lclm in a left Euclidean domain. We know that an lclm of a nonzero finite family $(a_i)_{1 \leq i \leq n}$ is an element $a \in \mathbf{R}^{\times}$, uniquely determined up to left associates, such that

$$\mathbf{R}\, a = \bigcap_{1 \leq i \leq n} \mathbf{R}\, a_i.$$

(Definition 81 and Theorem 443). The following recursive formula is obvious:

$$[a_1, ..., a_n]_l = \upsilon \left[[a_1, ..., a_{n-1}]_l \,, a_n \right]_l$$

where υ is a unit. Write

$$a \equiv c \,(\mathrm{mod}\, b) \tag{2.71}$$

$(a, b, c \in \mathbf{R})$ when b right-divides $a - c$.

Lemma 470. *When (2.71) holds, then there exists a unit υ such that*

$$[a, b]_l = \upsilon \, [c, b]_l \, c^{-1} \, a. \tag{2.72}$$

Proof. Let $m = [a, b]_l$. Then one must have $m = b_1\, a$ and b_1 is an element of lowest degree such that $b_1\, a$ is right-divisible by b. By (2.71), there exists $q \in \mathbf{R}$ such that $a = q\, b + c$, thus $m = b_1\, (q\, b + c)$ and the product $b_1\, c$ must be right-divisible by b. The lowest degree of b_1 is therefore obtained when there exists a unit υ such that $b_1\, c = \upsilon \, [c, b]_l$, which yields (2.72). ■

Theorem 471. *Let $a, b \in \mathbf{R}$ (where $\mathbf{R}$ is a left Euclidean domain) and consider the algorithm in Theorem 469. Then $[a, b]_l = [a_1, a_2]_l$ is given by*

$$[a_1, a_2]_l = \upsilon\, a_{n-1}\, a_n^{-1}\, a_{n-2}\, a_{n-1}^{-1} ... a_{n-i}\, a_i^{-1} ... a_2\, a_3^{-1}\, a_1 \tag{2.73}$$

where υ is a unit.

Proof. By (2.72),

$$[a_{i+1}, a_i]_l = \upsilon_i \, [a_{i+2}, a_{i+1}]_l \, a_{i+2}^{-1}\, a_i, \quad 1 \leq i \leq n-2.$$

Therefore,

$$\begin{aligned} [a_{n-1}, a_{n-2}]_l &= \upsilon_{n-2}\, a_{n-1}\, a_n^{-1}\, a_{n-2}, \\ [a_{n-2}, a_{n-3}]_l &= \upsilon_{n-3}\, \upsilon_{n-2}\, a_{n-1}\, a_n^{-1}\, a_{n-2}\, a_{n-1}^{-1}\, a_{n-3} \end{aligned}$$

and one finally obtains (2.73). ■

2.14 Exercises

Exercise 472. *Let $\mathbf{K}$ be a division ring and V a left $\mathbf{K}$-vector space. Prove that any subspace of V is a direct summand.*

Exercise 473. *Let* $\mathbf{R}$ *be a ring and let* $(M_i)_{i\in I}$ *be a family of left* $\mathbf{R}$*-modules. Prove that* $\left(\bigoplus_{i\in I} M_i\right)^* \cong \prod_{i\in I} M_i^*$ *and* $\left(\prod_{i\in I} M_i\right)^* \cong \bigoplus_{i\in I} M_i^*$.

Exercise 474. *Prove that* $(\mathbb{Z}/p\mathbb{Z}) \bigotimes_{\mathbb{Z}} (\mathbb{Z}/q\mathbb{Z}) = 0$ *for any primes* $p \neq q$. *(Hint: write the Bézout identity.) Prove that* $(\mathbb{Q}/\mathbb{Z}) \bigotimes_{\mathbb{Z}} (\mathbb{Q}/\mathbb{Z}) = 0$.

Exercise 475. *(i) Let* $\mathbf{R}$ *be a commutative ring,* E_i $(1 \leq i \leq n)$ *be* $\mathbf{R}$*-modules and* $x_i^* \in E_i^*$. *Prove that*

$$x_1^* \otimes ... \otimes x_n^*$$

coincides with the map

$$\prod_{1\leq i\leq n} E_i \ni (x_1, ..., x_n) \mapsto \prod_{1\leq i\leq n} \langle x_i^*, x_i\rangle .$$

(ii) Assuming that each module E_i *is finite free, let* $(e_{ji})_{1\leq j\leq n_i}$ *be the canonical basis of* E_i $(1 \leq i \leq n)$ *and* $\left(e_{ji}^*\right)_{1\leq j\leq n_i}$ *be the dual basis. Prove that*

$$\left(e_{j_1 1}^* \otimes ... \otimes e_{j_n n}^*\right)_{1\leq j_i\leq n_i}$$

is a basis of $E_1^* \otimes ... \otimes E_n^*$ *which is, therefore, a finite free* $\mathbf{R}$*-module. Calculate its rank (Subsect. 2.11.3).*
(iii) Let E_1, E_2, F_1, F_2 *be finite free* $\mathbf{R}$*-modules. Check that the map*

$$\mathrm{Hom}_{\mathbf{R}}(E_1, F_1) \times \mathrm{Hom}_{\mathbf{R}}(E_2, F_2) \to \mathrm{Hom}_{\mathbf{R}}(E_1 \otimes E_2, F_1 \otimes F_2) :$$
$$(u_1, u_2) \mapsto u_1 \otimes u_2$$

is bilinear. Using bases, prove that this map is an isomorphism. Then, prove that this isomorphism is canonical.
(iv) Prove that there exist canonical isomorphisms

$$E_1^* \otimes E_2^* \cong (E_1 \otimes E_2)^* , \quad E^* \otimes F \cong \mathrm{Hom}_{\mathbf{R}}(E, F) ,$$

generalize the former isomorphism to the case of n *modules and make the latter isomorphism explicit.*

Exercise 476. *Let* M, N *be left* $\mathbf{A}$*-modules and assume that the ring* $\mathbf{A}$ *is an* $\mathbf{R}$*-algebra (where* $\mathbf{R}$ *is a commutative ring). Prove that* $\mathrm{Hom}_{\mathbf{A}}(M, N)$ *is an* $\mathbf{R}$*-module.*

Exercise 477. *Let* $\mathbf{K}$ *be a division ring,* $\boldsymbol{\alpha}$ *an automorphism and* $\boldsymbol{\delta}$ *an* $\boldsymbol{\alpha}$*-derivation of* $\mathbf{K}$.

(i) Prove that for any $a \in \mathbf{K}^\times$, $a^{-\boldsymbol{\delta}} = -a^{-\boldsymbol{\alpha}}\, a^{\boldsymbol{\delta}}\, a^{-1}$ *(where* $a^{-\boldsymbol{\delta}} \triangleq \left(a^{-1}\right)^{\boldsymbol{\delta}}$, $a^{-\boldsymbol{\alpha}} \triangleq \left(a^{-1}\right)^{\boldsymbol{\alpha}}$*).*
(ii) Prove that the subring of constants of $\mathbf{K}$ *is a division subring.*

Exercise 478. *Let* $\mathbf{R}$ *be a commutative ring, let* $\mathfrak{p}$ *be a prime ideal in* $\mathbf{R}$ *and let* $S = \mathbf{R}\backslash\mathfrak{p}$. *(i) Prove that* $S = \mathbf{R}\backslash\mathfrak{p}$ *is a submonoid of* $(\mathbf{R}^\times, \times)$ *and* *that the ring* $\mathbf{R}_{\mathfrak{p}} = S^{-1}\mathbf{R}$ *is local with* $\operatorname{rad}\mathbf{R}_{\mathfrak{p}} = \mathfrak{p}\,\mathbf{R}_{\mathfrak{p}}$*, the ideal generated by the canonical image of* $\mathfrak{p}$ *in* $\mathbf{R}_{\mathfrak{p}}$*. (ii) Prove that the residue class field of* $\mathbf{R}_{\mathfrak{p}}$ *is canonically isomorphic to the field of fractions of* $\mathbf{R}/\mathfrak{p}$*. (Hint: use Proposition 303, Lemma 302 and Theorem 332.) The ring* $\mathbf{R}_{\mathfrak{p}}$ *is called the local ring of* $\mathbf{R}$ *at* $\mathfrak{p}$*.*

Exercise 479. *Let* $\mathbf{R}$ *be a ring and let* $\mathfrak{p}$ *be an ideal in* $\mathbf{R}$*. Prove that* $\mathfrak{p} \in \operatorname{Spec}\mathbf{R}$ *if, and only if* $\mathbf{R}/\mathfrak{p}$ *is a prime ring.*

Exercise 480. *Let* $\mathbf{K}$ *be a division ring.*

(1) Let $A \in \operatorname{Mat}_2(\mathbf{K})$ *be the matrix* $\begin{bmatrix} a & b \\ c & d \end{bmatrix}$*. (i) If* $a \neq 0$*, show that* $\det(A) = \overline{ad - aca^{-1}b}$*. (ii) If* $a = 0$*, show that* $\det(A) = \overline{-cb}$*.*
(2) Show that $\det(A)$ *is not a polynomial function of the entries of* A *if* $\mathbf{K}$ *is noncommutative.*

Exercise 481. *Let* $\mathbf{R}$ *be a ring.*

(1) Prove that $\mathbf{R}$ *has IBN if, and only if for any integers* n, m*, if* $A \in {}^{n}\mathbf{R}^{m}$*,* $B \in {}^{m}\mathbf{R}^{n}$ *and* $A\,B = I_n$*,* $B\,A = I_m$*, then* $n = m$*.*
(2) A ring $\mathbf{R}$ *is said to be* weakly finite *[79] (or stably finite [199]) if for any* $n \geq 1$*, each generating set of* n *elements of* $\mathbf{R}^n$ *is free. The following conditions are equivalent ([79], Prop. 4.6.6; [199], Prop. 1.7): (i)* $\mathbf{R}$ *is weakly finite; (ii) for any* n*, if an* $\mathbf{R}$*-module* H *is such that* $\mathbf{R}^n \cong H \oplus \mathbf{R}^n$*, then* $H = 0$*; (iii) for any* n*, any epimorphism* $\mathbf{R}^n \twoheadrightarrow \mathbf{R}^n$ *is an isomorphism; (iv) for any* n*, if* $A, B \in \operatorname{Mat}_n(\mathbf{R})$ *are such that* $A\,B = I_n$*, then* $B\,A = I_n$ *(thus a square matrix which is left- or right-invertible is invertible). Prove that any non-trivial weakly finite ring has IBN. (Hint: if IBN fails, then* $\mathbf{R}^n \cong \mathbf{R}^m$ *for some* n, m *such that* $m > n$ *and* $\mathbf{R}^n \cong \mathbf{R}^{m-n} \oplus \mathbf{R}^n$*.)*
(3) Prove that a semifir is a weakly finite ring. (Hint: use Proposition 411.)
(4) Prove that a (left or right) Ore domain is weakly finite. (Hint: use (2)(iv).)
(5) Prove that a commutative ring is weakly finite. (Hint: use determinants.)

Exercise 482. *Let* $\mathbf{K}$ *be a division ring, and let* $A \in {}^{m}\mathbf{K}^{n}$*,* $B \in {}^{n}\mathbf{K}^{p}$*. (i) Prove that* $\operatorname{rk}(A\,B) \leq \min\{\operatorname{rk}(A), \operatorname{rk}(B)\}$*. (ii) Prove the following, called* Sylvester's law of nullity*: if* $A\,B = 0$*, then* $n \geq \operatorname{rk}(A) + \operatorname{rk}(B)$*. (iii) Show that Sylvester's law of nullity holds in the following form: for any* $A \in {}^{m}\mathbf{K}^{n}$*,* $B \in {}^{n}\mathbf{K}^{p}$*,* $\operatorname{rk}(A) + \operatorname{rk}(B) - n \leq \operatorname{rk}(A\,B)$*. (Hint: use Theorem and Definition 392.)*

Exercise 483. *(i) Let* $\mathbf{R}$ *be a semifir, let* $M \cong \mathbf{R}^n$ *and let* N *be a submodule of* M*. Prove that if* N *is a direct summand of* M *(2.2.4.1), then* $\operatorname{rk}(N^{\mathbf{0}}) = n - \operatorname{rk}(N)$*. (ii) Let* $\mathbf{K}$ *be a division ring, let* V *and* W *be left* $\mathbf{K}$*-vector spaces such that* $\dim(V) < +\infty$*, and let* $f : V \to W$ *be a* $\mathbf{K}$*-linear map. Using (i), (2.61), Proposition 276 and Exercise 472, prove that* $\dim(\operatorname{im} f) = \dim(\operatorname{im}{}^{t}f)$*.*

Exercise 484. *Show that a domain with left and right ACC_1 is atomic and, conversely, that a unique factorization domain has left and right ACC_1.*

Exercise 485. *Let* $\mathbf{R}$ *be a commutative domain. Prove that two regular elements $a, b \in \mathbf{R}$ are similar (Lemma and Definition 424) if, and only if they are associates.*

Exercise 486. *A left* free ideal ring *(or left* fir*, for short), is a ring* $\mathbf{R}$ *in which all left ideals are free of unique rank, as left* $\mathbf{R}$*-modules.*
Prove that the following conditions are equivalent: (i) $\mathbf{R}$ *is a left fir and a left Ore domain; (ii)* $\mathbf{R}$ *is a principal left ideal domain; (iii)* $\mathbf{R}$ *is a left Noetherian left fir; (iv)* $\mathbf{R}$ *is a left Bézout domain with left ACC_1.*

Exercise 487. *An integral domain* $\mathbf{R}$ *is said to have* distributive factor lattice *if for any $c \in \mathbf{R}^{\times}$, the lattice $\mathbf{L}(c\mathbf{R}, \mathbf{R})$ is distributive ([77], Chap. IV). Prove that (i) this notion is left-right symmetric and that (ii) any ring with distributive factor lattice is a 2-fir (Lemma and Definition 410). (Hint: for (i), use Lemma and Definition 424.)*

Exercise 488. *Let* $\mathbf{R}$ *be a left GCD domain which is a left Ore domain, and let $\mathbf{Q} = \mathrm{Q}(\mathbf{R})$ be the division ring of left fractions of* $\mathbf{R}$*. Define (and prove the existence of) a least common left denominator of two elements $a, b \in \mathbf{Q}$. To what extent is that element unique?*

Exercise 489. *(i) Prove that the ring $\mathbb{Z}$ is an Euclidean domain with Euclidean function $n \mapsto |n|$ if $n \neq 0$, but that it is not strongly Euclidean.*
(ii) Let $\mathbf{K}$ *be a field. Prove that the polynomial ring $\mathbf{K}[X]$ is a commutative strongly Euclidean with Euclidean function $f(X) \mapsto d^{\circ}(f(X))$.*

Exercise 490. *Let* $\mathbf{R}$ *be a commutative domain and let* $\mathbf{Q}$ *be its field of fractions.*

(i) Prove that the following conditions are equivalent: (a) Given two elements, one divides the other; (b) $x \in \mathbf{Q} \backslash \mathbf{R} \Rightarrow x^{-1} \in \mathbf{R}$; (c) the set of all principal ideals in $\mathbf{R}$ *is totally ordered by inclusion; (d) the set of all ideals in* $\mathbf{R}$ *is totally ordered by inclusion. (Hint: (c)⇒(d) is proved in ([31], § VI.1, Theorem 1).)*
A commutative domain for which these equivalent conditions hold is called a valuation domain.
(ii) Let $\mathbf{K}$ *be a field and let T be an indeterminate. Prove that $\mathbf{K}[[T]]$ is a valuation domain.*
(iii) Prove that a valuation domain is local and integrally closed.

Exercise 491. *Let* $\mathbf{R}$ *be a left Dedekind domain.*

(i) Let S be a left denominator set; prove that $S^{-1}\mathbf{R}$ is again a left Dedekind domain.
(ii) Assuming that $\mathbf{R}$ *is commutative, let $\mathfrak{a} = \prod \mathfrak{p}^{m(\mathfrak{p})}$, $\mathfrak{b} = \prod \mathfrak{p}^{n(\mathfrak{p})}$ be two ideals in* $\mathbf{R}$*, where each $\mathfrak{p}$ is a prime ideal (see (2.67)). Prove that*

$$\mathfrak{a}\,\mathfrak{b}=\prod\mathfrak{p}^{m(\mathfrak{p})+n(\mathfrak{p})},\quad \mathfrak{a}:\mathfrak{b}\triangleq\mathfrak{a}\,\mathfrak{b}^{-1}=\prod\mathfrak{p}^{m(\mathfrak{p})-n(\mathfrak{p})},$$
$$\mathfrak{a}+\mathfrak{b}=\prod\mathfrak{p}^{\min(m(\mathfrak{p}),n(\mathfrak{p}))},\quad \mathfrak{a}\cap\mathfrak{b}=\prod\mathfrak{p}^{\max(m(\mathfrak{p}),n(\mathfrak{p}))}.$$

Exercise 492. *Let Ω be a nonempty interval of $\mathbb{R}$ and let $\mathcal{R}(\Omega)$ be the largest ring of rational functions analytic in Ω, i.e., $\mathcal{R}(\Omega)=\mathbb{C}(t)\cap\mathcal{O}(\Omega)$ (Example 439).*

(i) Prove that $\mathcal{R}(\Omega)$ is a commutative principal ideal domain and that its primes are (up to associates) the $X-\lambda$, $\lambda\in\Omega$.
(ii) Prove that the ring $A_o(\mathbb{C})=\mathcal{R}(\Omega)[X;\boldsymbol{\delta}]\supsetneq A_1(\mathbb{C})$ is a simple noncommutative Dedekind domain.

Exercise 493. *Let $\mathbf{R}$ be a ring and let $X=(X_1,...,X_n)$ be a finite sequence of indeterminates. When these indeterminates are independent but commute with the elements of $\mathbf{R}$, one can form the* free $\mathbf{R}$-ring $\mathbf{S}$ *generated by X over $\mathbf{R}$ and denoted by $\mathbf{R}\langle X\rangle$ ([199], (1.2)). When in addition the indeterminates X_i pairwise commute, $\mathbf{S}$ is called the* multivariate *(*univariate *when $n=1$)* polynomial ring *generated by X over $\mathbf{R}$ and is denoted by $\mathbf{R}[X]$. This construction is generalized below: for $1\leq i\leq n$, let $\boldsymbol{\alpha}_i$ be an automorphism and $\boldsymbol{\delta}_i$ an $\boldsymbol{\alpha}_i$-derivation of $\mathbf{R}$ with commutation rule $X_i\,a=a^{\boldsymbol{\alpha}_i}X_i+a^{\boldsymbol{\delta}_i}$.*

(i) Show that the indeterminates X_i and X_j commute if, and only if $X_j^{\boldsymbol{\alpha}_i}=X_j$, $X_j^{\boldsymbol{\delta}_i}=0$, $i\neq j$, and that this implies $\boldsymbol{\alpha}_i\,\boldsymbol{\alpha}_j=\boldsymbol{\alpha}_j\,\boldsymbol{\alpha}_i$, $\boldsymbol{\delta}_i\,\boldsymbol{\delta}_j=\boldsymbol{\delta}_j\,\boldsymbol{\delta}_i$ and $\boldsymbol{\alpha}_j\,\boldsymbol{\delta}_i=\boldsymbol{\delta}_i\,\boldsymbol{\alpha}_j$. These conditions are assumed to be satisfied in the sequel.
(ii) Prove that $\boldsymbol{\delta}_i$ is an $\boldsymbol{\alpha}_i$-derivation of $\mathbf{R}[X_j;\boldsymbol{\alpha}_j,\boldsymbol{\delta}_j]$ $(i\neq j)$ and, by induction, define the iterated skew multivariate polynomial ring *$\mathbf{A}=\mathbf{R}[X_1,...,X_n;\boldsymbol{\alpha}_1,...\boldsymbol{\alpha}_n,\boldsymbol{\delta}_1,...,\boldsymbol{\delta}_n]$ (written $\mathbf{R}\left[(X_i;\boldsymbol{\alpha}_i,\boldsymbol{\delta}_i)_{1\leq i\leq n}\right]$, for short) using the relation (for $1\leq j\leq n-1$)*

$$\mathbf{R}\left[(X_i;\boldsymbol{\alpha}_i,\boldsymbol{\delta}_i)_{1\leq i\leq j+1}\right]=\mathbf{R}\left[(X_i;\boldsymbol{\alpha}_i,\boldsymbol{\delta}_i)_{1\leq i\leq j}\right][X_{j+1};\boldsymbol{\alpha}_{j+1},\boldsymbol{\delta}_{j+1}].$$

If $\boldsymbol{\alpha}_i=1$ and $\boldsymbol{\delta}_i=0$ $(1\leq i\leq n)$, the indeterminates X_i are called central. *Why?*
(iii) Let $a=\sum_{\nu\in\mathbb{N}^n}a_\nu X^\nu\in\mathbf{S}$ where $\nu=(\nu_1,...,\nu_n)\in\mathbb{N}^n$, $X^\nu=\prod_{0\in i\leq n}X_i^{\nu_i}$; setting $|\nu|=\sum_{1\leq i\leq n}\nu_i$, the terms of degree n of a $(n\in\mathbb{N})$ are the terms $a_\nu X^\nu$ such that $|\nu|=n$, the degree *of $a\neq 0$ is the highest $|\nu|$ such that $a_\nu\neq 0$ and is denoted by $d^\circ(a)$, and $d^\circ(0)\triangleq-\infty$. If $|\nu|=d^\circ(a)$, the term $a_\nu X^\nu$ is called a* leading term *of a and a_ν is called a* leading coefficient *of a.*

Prove that $d^\circ(a-b)\leq\max\{d^\circ(a),d^\circ(b)\}$ with equality if $d^\circ(a)\neq d^\circ(b)$ and that $d^\circ(a\,b)\leq d^\circ(a)+d^\circ(b)$ with equality if $\mathbf{R}$ is a domain.
(iv) If $\mathbf{R}$ be a left (resp., right) Noetherian domain, prove that the skew polynomial ring $\mathbf{A}$ is again a left (resp., right) Noetherian domain. (Hint: use (iii) and Theorem 358).

(v) If $\mathbf{R} = \mathbf{k}[x_1, ..., x_n]$*,* $\boldsymbol{\alpha}_i = 1$ *and* $\boldsymbol{\delta}_i = \partial/\partial x_i$*, then* $\mathbf{R}\left[(\partial_i; \boldsymbol{\alpha}_i, \boldsymbol{\delta}_i)_{1 \leq i \leq n}\right]$ *(where* ∂_i *extends* $\boldsymbol{\delta}_i$*) is denoted by* $A_n(\mathbf{k})$ *and is called the nth Weyl algebra over* $\mathbf{k}$*; if* $\mathbf{R} = \mathbf{k}(x_1, ..., x_n)$*, then* $\mathbf{R}\left[(\partial_i; \boldsymbol{\alpha}_i, \boldsymbol{\delta}_i)_{1 \leq i \leq n}\right]$ *is denoted by* $B_n(\mathbf{k})$*. Prove that the rings* $A_n(\mathbf{k})$ *and* $B_n(\mathbf{k})$ *are Noetherian domains (generalization of Corollary 359).*
(vi) If $\mathbf{R}$ *is a* $\mathbb{Q}$*-algebra, prove that the ring* $\mathbf{R}\left[(X_i; \boldsymbol{\delta}_i)_{1 \leq i \leq n}\right]$ *is simple if, and only if for any* $i \in \{1, ..., n\}$*, (a)* $\mathbf{R}$ *has no proper nonzero* $\boldsymbol{\delta}_i$*-invariant ideal, and (b)* $\boldsymbol{\delta}_i$ *is an outer derivation of* $\mathbf{A}_j = \mathbf{R}[X_j; \boldsymbol{\delta}_j]$*,* $j \in \{1, ..., n\}$*,* $j \neq i$*. (Hint: use Theorem 362.) Deduce that the rings* $A_n(\mathbf{k})$ *and* $B_n(\mathbf{k})$ *are simple.*
(vii) Prove that the monoid $S = \left\{X_1^{k_1} ... X_n^{k_n} : k_i \geq 0 \ (1 \leq i \leq n)\right\}$ *is a left and right denominator set of* $\mathbf{A}$*, and form the* skew multivariate Laurent polynomial ring $\mathbf{T} = S^{-1}\mathbf{R} = \mathbf{R}S^{-1}$*, denoted by* $\mathbf{R}\left[\left(X_i, X_i^{-1}; \boldsymbol{\alpha}_i, \boldsymbol{\delta}_i\right)_{1 \leq i \leq n}\right]$*. Write the general form of an element of this ring. If* $\mathbf{R}$ *is a left (resp., right) Noetherian ring, prove that* $\mathbf{T}$ *is left (resp., right) Noetherian, and that if* $\mathbf{R}$ *is a domain,* $\mathbf{T}$ *is again a domain.*
(viii) Let $\mathbf{R}$ *be an Ore domain, let* $\mathbf{K}$ *be its division ring of fractions, and let* $\mathbf{F} = \mathbf{K}\left((X_i; \boldsymbol{\alpha}_i, \boldsymbol{\delta}_i)_{1 \leq i \leq n}\right)$ *be the division ring of fractions of* $\mathbf{K}\left[(X_i; \boldsymbol{\alpha}_i, \boldsymbol{\delta}_i)_{1 \leq i \leq n}\right]$*. Prove that* $\mathbf{F}$ *is the division ring of fractions of* $\mathbf{A}$ *and of* $\mathbf{T}$*.*
(ix) Prove that the ring $\mathbf{R}\left[\left(X_i, X_i^{-1}; \boldsymbol{\alpha}_i\right)_{1 \leq i \leq n}\right]$ *is simple if, and only if for any* $i \in \{1, ..., n\}$*, (a)* $\mathbf{R}$ *has no proper nonzero* $\boldsymbol{\alpha}_i$*-invariant ideal, and (b) no power of* $\boldsymbol{\alpha}_i$ *is an inner derivation of* $\mathbf{R}$*. (Hint: use Theorem 367).*
(x) Show how one can form the iterated formal skew multivariate power series ring $\mathbf{S} = \mathbf{R}\left[\left[\left(X_i^{-1}; \boldsymbol{\alpha}_i, \boldsymbol{\delta}_i\right)_{1 \leq i \leq n}\right]\right]$*, and show that Theorem 383 can be generalized to this ring.*
(xi) Similarly, construct the formal skew multivariate Laurent series ring $\mathbf{L} = \mathbf{R}\left(\left(\left(X_i^{-1}; \boldsymbol{\alpha}_i, \boldsymbol{\delta}_i\right)_{1 \leq i \leq n}\right)\right)$ *and check that Theorem 385 can be generalized to this case.*

2.15 Notes

The main references for this chapter are [77] and [247] but other ones, especially [27], [31], [79], [80] and [198], have also been useful to establish the text. Skew polynomials were introduced by Segre, van der Waerden and, in their general form, by Ore [265]. The theory of rings of fractions (Sect. 2.5) was mainly established by Ore [264] and Goldie [142].

Theorem 372 is taken from Bourlès and Oberst [47]. Proposition 376 has been proved by El Mrabet and Bourlès [108] but the connection between this result and crossed products, as well as Proposition 377, are new. Our presentation of Dieudonné determinants (Subsect. 2.11.2) is essentially based on the original paper [91] and on ([6], Chapter IV). Linear algebraic groups over a field are extensively studied in Borel's monograph [23].

Everett [111] was the first to construct examples of finite free modules having bases of different sizes. Cohn introduced n-firs, semifirs and firs (Lemma and Definition 410, and Exercise 486) [71], and the notion of full matrix [73]. The term "rank-stable ring" has been introduced here to simplify the presentation. Our presentation of Sylvester domains is mainly based on the original paper of Dicks and Sontag [89] and on ([77], Sect. 5.5 & 5.6). Our presentation of GCD domains and unique factorization domains is based of Cohn's papers [70], [74]. Proposition 446 was independently obtained by Cohn [70] and Amitsur. Dedekind domains (and the notions of ideal, fractional ideal and module as well) were first defined by Dedekind shortly before the end of the 19th century to improve Kummer's theory of ideal factors (dating back to 1847 and created to partially prove Fermat's Last Theorem [98]); in 1927, E. Noether, who also discovered Noetherian rings and modules, gave the axiomatic characterization of Dedekind domains in Definition 450, which is still the most useful in Number Theory (see the historical notes of [31] and [165]); Definition 454 of a noncommutative Dedekind domain is based on that given by Goldie [143] (improving Robson's definition [297]) and on a result due to Gentile [136]. Goldie's original definition [143] is given in Subsect. 3.5.1 below (see the necessary and sufficient condition in Theorem 586). Definition 454 of a Dedekind domain, valid in the noncommutative case, is that given by Cartan and Eilenberg [59]. Theorem 469 can be traced back to Euclid (Elements, Book VII) whereas Theorem 471 is a slight generalization of a result due to Ore ([265], Sect. I.2, Theorem 8). The notion of valuation domain (Exercise 490) is due to Krull [195].

The Bézout domain $\mathcal{H}$ in Example 440 was introduced by Glüsing-Lüerssen [137] for the study of linear delay-differential systems, as said in the text; a similar ring, which is also a Bézout domain, was introduced with the same aim by Loiseau [221]. Exercise 492(i)&(ii) is taken from Fröhler and Oberst [129]. Exercise 493 proposes a quite natural extension (detailed in, e.g., [66]) of skew polynomial rings, skew Laurent polynomial rings, formal skew power series rings and formal skew Laurent series rings to the case of several indeterminates.

3 Homological Algebra

3.1 Introduction

In this chapter, we continue the study of modules. Let $\mathbf{A}$ be a ring. A left $\mathbf{A}$-module M is said to be *finitely presented* if it is generated by a set of variables $w_1, ..., w_k$ satisfying an equation of the form

$$\boxed{R\,w = 0} \tag{3.1}$$

where R is a $q \times k$ matrix with entries in $\mathbf{A}$ and $w = \begin{bmatrix} w_1 & \cdots & w_k \end{bmatrix}^T$. Such a module (denoted by $M = [w]_{\mathbf{A}}$) is said to be defined by generators – the variables w_i $(1 \leq i \leq k)$ – and relations – the q equalities $R_i\,w = 0$, where R_i is the ith row of the matrix R. If $\mathbf{A}$ is a ring of differential operators, e.g., $\mathbf{A} = \mathbf{K}[\partial, \boldsymbol{\delta}]$, $\mathbf{K} = \mathbb{C}[t]$, then (3.1) is the equation of an LTV system. If M' is a second left $\mathbf{A}$-module generated by a set of variables $w_1, ..., w'_{k'}$ satisfying an equation $R'\,w' = 0$, where $w' = \begin{bmatrix} w_1 & \cdots & w'_{k'} \end{bmatrix}^T$ and R' is a $q' \times k'$ matrix with entries in $\mathbf{A}$, and if $M \cong M'$, then M and M' are identified and the matrices M, M' are called left-similar. Presentations and similarity are studied in Section 3.2.

A module is said to be *cyclic* if it is generated by one element. As will be seen in Subsection 3.6.2, any finitely generated module over a principal ideal domain is isomorphic to a direct sum of cyclic modules. Cyclic modules are studied in Section 3.3.

In Section 3.4, the most important *functors* in the categories of modules are studied. They include the tensor product, the functor Hom, and related functors. The importance of the functor Hom for the study of linear systems is further detailed in the Introduction of Chapter 5 (and, more technically, in Subsection 3.4.3).

The homological properties of certain modules are studied in Section 3.5. At this stage, it is probably useful to indicate the etymology of the word "homology". As already said (**1.3.3.1**), *όμο* means *the same*; *λόγοσ* can be translated by *word* or *views*. In algebraic topology, integration along two

H. Bourlès and B. Marinescu: Linear Time-Varying Systems, LNCIS 410, pp. 189–268.
springerlink.com

paths in the complex plane yield the same result if these paths belong to the same *homology class*; in that case, $\acute{o}\mu o+\lambda\acute{o}\gamma o\sigma$ means *agreeing*, as emphasized by Rotman [306]. Homological algebra is the branch of mathematics which studies homology in a general algebraic setting (following H. Cartan and S. Eilenberg [59]). Several fundamental notions for the study of linear systems are detailed in Section 3.5, especially that of *injective cogenerator.*

Last, Section 3.6 is devoted to study of the *structure of finitely presented modules* over specific noncommutative rings, especially Bézout domains, principal ideal domains, and Dedekind domains.

The table in Section 2.1 can be completed by that below, where *PID*, *BD*, and *SD* have already been defined, and where *PFR*, *HR*, *SFID*, *SSFID*, *OD*, *DD*, and *ND* mean *projective free ring, Hermite ring, stably-free ideal domain, semistably-free ideal domain, Ore domain, Dedekind domain*, and *Noetherian domain*, respectively.

$$\begin{array}{ccccc} PID & \rightarrow & BD & \rightarrow SD \rightarrow PFR \rightarrow HR \\ \downarrow & & \downarrow & \\ SFID & \rightarrow & SSFID & \rightarrow OD \\ & \searrow & & \uparrow \\ & & DD & \rightarrow ND \end{array}$$

3.2 Presentations and Similarity

3.2.1 Presentation of a Module

3.2.1.1 Quotient Modules

Let $\mathbf{A}$ be a ring.

Theorem 494. *Every left* $\mathbf{A}$*-module* M *is isomorphic to the quotient of a free left* $\mathbf{A}$*-module, which is finite if* M *is finitely generated. Therefore,* $\mathbf{A}$ *is a projective generator in the category* ${}_{\mathbf{A}}\mathbf{Mod}$.

Proof. Let M be a left $\mathbf{A}$-module M and let $(\eta_i)_{i\in K}$ be a generating family of M (M is f.g. if, and only if one can take $\operatorname{card}(K) < +\infty$). By Lemma 262 there exists a unique $\mathbf{A}$-linear epimorphism $\bullet\alpha : \mathbf{A}^{(K)} \rightarrow M$ such that $\varepsilon_i\alpha = \eta_i$ where $(\varepsilon_i)_{i\in K}$ is the canonical basis of $\mathbf{A}^{(K)}$ (Lemma and Definition 255). By Noether's first isomorphism theorem (Theorem 52(1)), the following diagram is commutative

$$\begin{array}{ccc} \mathbf{A}^{(K)} & \xrightarrow{\alpha} & M \longrightarrow 0 \\ \downarrow \varphi & \nearrow \psi & \\ \mathbf{A}^{(K)}/\ker\alpha & & \end{array}$$

where φ is the canonical epimorphism $\mathbf{A}^{(K)} \twoheadrightarrow \mathbf{A}^{(K)}/\ker\alpha$ and ψ is an isomorphism. Therefore, $\mathbf{A}$ is a generator in ${}_{\mathbf{A}}\mathbf{Mod}$ (**1.2.2.4**). The functor $\mathrm{Hom}_{\mathbf{A}}(\mathbf{A}, \bullet)$ is exact, thus $\mathbf{A}$ is projective in ${}_{\mathbf{A}}\mathbf{Mod}$ (Proposition 49(2)). ∎

The above theorem can be depicted by the exact sequence

$$\ker\alpha \xrightarrow{\iota} \mathbf{A}^{(K)} \xrightarrow{\alpha} M \longrightarrow 0 \tag{3.2}$$

where ι is the inclusion.

3.2.1.2 Finitely Generated and Finitely Presented Modules

Let $(\mu_j)_{j\in Q}$ be a generating family of $\ker\alpha$; there exists by Lemma 262 a unique $\mathbf{A}$-linear epimorphism $\bullet\beta : \mathbf{A}^{(Q)} \to \ker\alpha$ such that $\zeta_j\beta = \mu_j$ where $(\zeta_j)_{j\in Q}$ is the canonical basis of $\mathbf{A}^{(Q)}$. Let $\bullet f = \beta \circ i : \mathbf{A}^{(Q)} \to \mathbf{A}^{(K)}$ where $\bullet i : \ker\alpha \hookrightarrow \mathbf{A}^{(K)}$ is the inclusion. This yields the exact sequence

$$\mathbf{A}^{(Q)} \xrightarrow{\bullet f} \mathbf{A}^{(K)} \xrightarrow{\bullet\alpha} M \longrightarrow 0. \tag{3.3}$$

Definition 495. *The exact sequence (3.3) is called a* presentation *of the module M.*

Notice that $M = \mathrm{coker}\, f$ (**1.2.4.1**). By Theorem 494, the module M is finitely generated if, and only if there exists a presentation of M such as (3.3) with $\mathrm{card}\,(K) = k$, a natural number. If there exists a presentation of M such as (3.3) with $\mathrm{card}\,(Q) = q$, a natural number, M is said to be *finitely related*.

Definition 496. *The left $\mathbf{A}$-module M is said to be* finitely presented *(*f.p.*, for short) if it is both finitely generated and finitely related, i.e., it has a presentation of the form*

$$\mathbf{A}^{q} \xrightarrow{\bullet f} \mathbf{A}^{k} \xrightarrow{\bullet\alpha} M \longrightarrow 0. \tag{3.4}$$

Notation 497. *In what follows, ${}_{\mathbf{A}}\mathbf{Mod}^{fg}$ (resp., ${}_{\mathbf{A}}\mathbf{Mod}^{fp}$) denotes the full subcategory of ${}_{\mathbf{A}}\mathbf{Mod}$, the objects of which are finitely generated (resp., finitely presented).*

Lemma 498. *Let $\mathbf{A}$ be a ring and consider the exact sequence of left $\mathbf{A}$-modules*

$$0 \longrightarrow N \longrightarrow M \longrightarrow P \longrightarrow 0.$$

(i) If N and P are finitely generated, so is M.
(ii) If P is finitely presented and M is finitely generated, N is finitely generated
(iii) If M is finitely presented, N is finitely generated if, and only if P is finitely presented.

Proof. (i): See ([27], §II.1, Corol. 5 of Prop. 9). (ii): See ([31], §I.2, Lemma 9).

(iii): Identifying P with M/N, let $M = F/K$ where $F = \mathbf{A}^k$ and $K = (\mathbf{A}^q) f$ is f.g., and let $N = G/K$ where $K \lhd G \lhd F$ (Theorem 147(i)). Obviously, N is f.g. if, and only if G is f.g. too; since $M/N = (F/K)/(G/K) \cong F/G$ according to Noether's third isomorphism theorem (Theorem 52(3)), this is the condition for $P = M/N$ to be f.p. ∎

Theorem 499. *Let* $\mathbf{A}$ *be a left Noetherian ring or a semifir; then every finitely generated left* $\mathbf{A}$*-module is finitely presented.*

Proof. Let $\mathbf{A}$ be a ring and let M be a f.g. left $\mathbf{A}$-module. We have the exact sequence (3.2) where $\mathrm{card}(K) = k$, a natural number, and $\ker \alpha \lhd \mathbf{A}^k$. (1) If $\mathbf{A}$ is left Noetherian, $\mathbf{A}^k$ is a left Noetherian $\mathbf{A}$-module by Lemma 309(iii), thus $\ker \alpha$ is f.g. and M is finitely presented. (2) If $\mathbf{A}$ is a semifir, $\ker \alpha$ is finite free by Lemma and Definition 410(iv), thus M is finitely presented. ∎

3.2.1.3 Generalization

Let $\mathcal{C}$ be an additive category (**1.2.4.1**), let U be an object of $\mathcal{C}$ and let $\mathbf{A} \triangleq \mathrm{Hom}_{\mathcal{C}}(U, U)$ be the ring of all left endomorphisms of U. Then, for any object N of $\mathcal{C}$, the abelian group $\mathrm{Hom}_{\mathcal{C}}(U, N)$ of all left morphisms $\bullet g : U \to N$ is a left $\mathbf{A}$-module via composition of morphisms. Let I be a set of indices, arbitrary if $\mathcal{C}$ has arbitrary coproducts and finite otherwise.

Definition 500. *An object* M *of* $\mathcal{C}$ *is said to be* U-generated *if* M *is a factor of* $U^{(I)}$, *and* U-finitely generated *(*U*-f.g.) if* I *is finite.*

Note that all objects of $\mathcal{C}$ are U-generated if, and only if U is a generator (**1.2.2.4**).

Example 501. *Let* $\mathbf{R}$ *be a ring and let* $\mathcal{C}$ *be the category* ${}_{\mathbf{R}}\mathbf{Mod}$ *of left* $\mathbf{R}$*-modules with its left morphisms* $x \mapsto (x) f = xf$. *Then* $U \triangleq \mathbf{R}$ *is a projective generator and the map*

$$\mathbf{A} \triangleq \mathrm{Hom}_{\mathbf{R}}(\mathbf{R}, \mathbf{R}) \to \mathbf{R}, \quad f \mapsto (1) f,$$

is a ring-isomorphism with inverse $r \mapsto \bullet r$. *We identify* $\mathbf{A}$ *and* $\mathbf{R}$, *and with this identification*

$$\mathrm{Hom}_{\mathbf{R}}(\mathbf{R}, M) \to M : g \mapsto (1) g$$

is a canonical isomorphism of left $\mathbf{A}$*-modules which is again an identification.*

Let us assume that the category $\mathcal{C}$ is preabelian (**1.2.4.2**) and consider a subobject N of $U^{(I)}$ (viewed as an object of $\mathcal{C}$ such that $N \subseteq U^{(I)}$ according to Remark 6); let $\iota : N \hookrightarrow U^{(I)}$ be the inclusion, and consider the associated factor object $M = \operatorname{coker} \iota$ (also written $M = U^{(I)}/N$) defined by the canonical exact sequence of left morphisms

$$N \xrightarrow{\bullet\iota} U^{(I)} \xrightarrow{\bullet\varphi} M \longrightarrow 0. \tag{3.5}$$

Definition 502. *The above object $M = \operatorname{coker} \iota = U^{(I)}/N$ is called a U-cokernel.*

If the category $\mathcal{C}$ is not abelian, a U-generated object need not be a U-cokernel (see Example 503(i) below).

Example 503. *(i) Let $\mathcal{C}$ be the category* **LCS** *(Example 36), and let U and M be objects of $\mathcal{C}$ such that there exists an integer $k > 0$ and a continuous linear surjection $\varphi : U^k \twoheadrightarrow M$. Let $\varphi_{ind} : U^k/N \to M$ be the induced linear bijection (Theorem 52(1)). Then φ_{ind} is continuous (Subsect. 1.7.1) but φ_{ind}^{-1} is not continuous in general (since φ need not be a strict morphism), thus M and U^k/N cannot be identified. In other words, M is U-finitely generated but is not a U-cokernel.*
(ii) Let φ and φ_{ind} be as above when $\mathcal{C}$ is the category **Frt** *(Example 37). According to the Banach inverse mapping theorem* (**1.7.3.1**), *φ_{ind} is an isomorphism, thus every U-finitely generated object is a U-cokernel.*

Lemma and Definition 504. *Let $\mathcal{C}$ be a preabelian category and U be an object of $\mathcal{C}$.*

(i) An object $M \in \mathcal{C}$ is called U-finitely presented *(U-f.p.) if there exists an exact sequence*

$$U^q \xrightarrow{\bullet f} U^k \xrightarrow{\bullet\varphi} M \longrightarrow 0. \tag{3.6}$$

(ii) In that case, $M = \operatorname{coker} \iota$ where $\iota \triangleq \operatorname{im} f$, and $\iota = \operatorname{im} \iota = \ker \varphi$.

Proof. (ii): In this proof, all morphisms are right ones. We have $\varphi = \operatorname{coker} f$; therefore, setting $\iota = \operatorname{im} f$ we obtain $\operatorname{im} \iota = \operatorname{im} f$ by Lemma 34(iii). Therefore, by (1.6), (1.7), $\operatorname{im} \iota = \ker \operatorname{coker} \iota$, and

$$\operatorname{coker} \iota = \operatorname{coker} \operatorname{im} f = \operatorname{coker} \ker \operatorname{coker} f = \operatorname{coker} f.$$

This yields $\operatorname{im} \iota = \ker \operatorname{coker} f = \ker \varphi$. ∎

3.2.2 Coherent Modules, and Coherent Rings

3.2.2.1 Annihilators

The following will be useful:

Definition 505. *Let* **A** *be a ring and let N be a left (resp., right)* **A***-module.*
(i) The annihilator *of a nonempty part S of N in* **A** *is the subset A of*

$\mathbf{A}$ *consisting of all elements* r *such that* $r\,S = 0$ *(resp.,* $S\,r = 0$*). This set* A *is denoted by* $\mathrm{Ann}_l^{\mathbf{A}}(S)$ *(resp.,* $\mathrm{Ann}_r^{\mathbf{A}}(S)$*), and by* $\mathrm{Ann}^{\mathbf{A}}(S)$ *if* $\mathbf{A}$ *is commutative.*
(ii) Conversely, let S' *be a nonempty part of* $\mathbf{A}$*; the annihilator of* S' *in* N *is the subset* A' *of* N *consisting of all elements* $n \in N$ *such that* $S'\,n = 0$ *(resp.,* $n\,S' = 0$*). This set is denoted by* $\mathrm{Ann}_r^N(S')$ *(resp.,* $\mathrm{Ann}_l^N(S')$*).*

As easily seen, $\mathrm{Ann}_l^{\mathbf{A}}(S)$ (resp., $\mathrm{Ann}_r^{\mathbf{A}}(S)$) is a left (resp., right) ideal in $\mathbf{A}$ and $\mathrm{Ann}_r^N(S')$ (resp., $\mathrm{Ann}_l^N(S')$) is a submodule of the left (resp., right) $\mathbf{A}$-module N. In addition, we have the following.

Lemma 506. *Let* $\mathbf{A}$ *be a ring and let* M *be a left (resp., right)* $\mathbf{A}$*-module; then* $\mathrm{Ann}_l^{\mathbf{A}}(M)$ *(resp.,* $\mathrm{Ann}_r^{\mathbf{A}}(M)$*) is an* ideal *in* $\mathbf{A}$.

Proof. Let $\lambda \in \mathrm{Ann}_l^{\mathbf{A}}(M)$; this means that $\lambda\,m = 0$ for any $m \in M$. Now, $\mu\,m \in M$ for any $\mu \in \mathbf{A}$ whenever $m \in M$, which implies that $\lambda\,(\mu\,m) = (\lambda\,\mu)\,m = 0$, thus $\lambda\,\mu \in \mathrm{Ann}_l^{\mathbf{A}}(M)$. Therefore, $\mathrm{Ann}_l^{\mathbf{A}}(M)$ is a right ideal in $\mathbf{A}$; since it is a left ideal (as already said), it is an ideal in $\mathbf{A}$. ∎

3.2.2.2 Coherent Modules, and Coherent Rings

Definition 507. *(i) Let* $\mathbf{A}$ *be a ring. A left (resp., right)* $\mathbf{A}$*-module* M *is* coherent *if it is f.g. and every f.g. submodule of* M *is finitely presented.*
(ii) A ring $\mathbf{A}$ *is* left *(resp.,* right*)* coherent *if every finitely generated left (resp., right) ideal in* $\mathbf{A}$ *is finitely presented. A* coherent ring *is a ring which is both left and right coherent.*

Let us gather in the two following lemmas some results about coherent modules and rings.

Lemma 508. *Let* $\mathbf{A}$ *be a ring and let* M *be a left* $\mathbf{A}$*-module.*

(1) The following conditions are equivalent:
(i) M *is coherent;*
(ii) M *is f.g., and for any integer* $n \geq 0$*, the kernel of any homomorphism of left modules* $\mathbf{A}^n \to M$ *is f.g.;*
(iii) M *is f.g., and for any f.g. left module* N *and any* $f : N \to M$*,* $\ker f$ *is f.g.*
(2) If M *is coherent and* $N \lhd M$ *is f.g., then* N *is again coherent.*
(3) The ring $\mathbf{A}$ *is left coherent if, and only if the left* $\mathbf{A}$*-module* ${}_{\mathbf{A}}\mathbf{A}$ *is coherent.*
(4) Let

$$0 \longrightarrow M_1 \longrightarrow M_2 \longrightarrow M_3 \longrightarrow 0$$

be an exact sequence of left $\mathbf{A}$*-modules. If two of these modules are coherent, so is the third. In addition, if* M_2 *is coherent and* M_1 *is f.g., then* M_3 *is coherent.*
(5) Any finite direct sum of left coherent modules is left coherent.

Proof. We only prove (1) since the proof of the other parts is given in ([77], Theorem A.6), ([199], §4G), and ([31], §II.2, Exercise 11).

(i)⇒(ii): Let M be left coherent, thus f.g., and let $\bullet f : \mathbf{A}^n \to M$. Then $\operatorname{im} f$ is f.g., thus $\operatorname{im} f$ is f.p. Let $\bullet g : \mathbf{A}^n \twoheadrightarrow \operatorname{im} f$ be the epimorphism such that $(x)\,g = (x)\,f$ for all $x \in \mathbf{A}^n$. We obtain the short exact sequence of left morphisms

$$0 \longrightarrow \ker f \xrightarrow{\bullet\iota} \mathbf{A}^n \xrightarrow{\bullet g} \operatorname{im} f \longrightarrow 0$$

where $\operatorname{im} f$ is f.p. and $\mathbf{A}^n$ is f.g., thus $\ker f$ is f.g.

(ii)⇒(iii): Assume that (ii) holds and let N be f.g. Let $f : N \to M$. There exist a positive integer n and an epimorphism $\bullet\varphi : \mathbf{A}^n \twoheadrightarrow N$. Consider $\varphi f : \mathbf{A}^n \to M$. Then $\ker \varphi f$ is f.g., say generated by $\kappa_1, ..., \kappa_m$, and $\ker f$ is generated by $(\kappa_1)\,\varphi, ..., (\kappa_m)\,\varphi$.

(iii)⇒(ii) is clear.

(ii)⇒(i): Assume that (ii) holds and let $N \lhd M$ be f.g. Let $\bullet\varphi : \mathbf{A}^n \twoheadrightarrow N$ and let $f : \mathbf{A}^n \twoheadrightarrow M$ be the map such that $(x)\,\varphi = (x)\,f$ for all $x \in \mathbf{A}^n$. Then $\ker \varphi = \ker f$ is f.g., therefore there exist a positive integer q and an epimorphism $\bullet\psi : \mathbf{A}^q \twoheadrightarrow \ker \varphi$. Let $\bullet g : \mathbf{A}^q \to \mathbf{A}^n$ be such that $(y)\,g = (y)\,\psi$ for all $y \in \mathbf{A}^q$. The sequence of left morphisms

$$\mathbf{A}^q \xrightarrow{\bullet g} \mathbf{A}^n \xrightarrow{\bullet\varphi} N \longrightarrow 0$$

is exact and is a finite presentation of N. ■

Lemma 509. *Let* $\mathbf{A}$ *be a ring; the following conditions are equivalent:*

(i) $\mathbf{A}$ *is right coherent;*
(ii) every finitely generated submodule of a free right $\mathbf{A}$*-module is finitely presented;*
(iii) the right annihilator of any row vector over $\mathbf{A}$ *is finitely generated, i.e., given* $u \in \mathbf{A}^n$, *if* $u\,B = 0$ *for some* $B \in {}^n\mathbf{A}^I$, *then there exists* $C \in {}^n\mathbf{A}^r$, $D \in {}^r\mathbf{A}^I$ *such that* $B = C\,D$, $u\,C = 0$;
(iv) every finitely presented right $\mathbf{A}$*-module is coherent;*
(v) every f.g. right ideal $\mathfrak{a}$ *in* $\mathbf{A}$ *is finitely related (thus finitely presented);*
(vi) for any $a \in \mathbf{A}$, *the right annihilator of* a *is f.g., and the intersection of any two f.g. right ideals in* $\mathbf{A}$ *is f.g.;*
(vii) for any integer $n \geq 0$, *the kernel of any homomorphism of right modules* $\mathbf{A}^n \to \mathbf{A}$ *is finitely generated.*

Proof. (i)⇔(ii)⇔(iii)⇔(iv)⇔(v)⇔(vi): see ([77], Theorem A.6), ([199], §4G), and ([31], §II.2, Exercise 11).

(i)⇔(vii) is easily deduced from Lemma 508(1). ■

The reason why coherent modules are important is clear from the following theorem (see also Theorem 612 below).

Theorem 510. *Let* $\mathbf{A}$ *be a left (resp., right) coherent ring. Then the category* ${}_{\mathbf{A}}\mathbf{Mod}^{fp}$ *(resp.,* $\mathbf{Mod}_{\mathbf{A}}^{fp}$*) of finitely presented left (resp., right)* $\mathbf{A}$*-modules is an abelian full subcategory of all left (resp., right)* $\mathbf{A}$*-modules.*

Proof. (1). a) Let $\mathbf{A}$ be any ring. The category ${}_{\mathbf{A}}\mathbf{Mod}^{fp}$ of finitely presented left $\mathbf{A}$-modules is obviously preadditive. If M and N are finitely presented $\mathbf{A}$-modules, so is $M \oplus N \cong M \times N$, thus ${}_{\mathbf{A}}\mathbf{Mod}^{fp}$ is additive (since the module 0 is an object of ${}_{\mathbf{A}}\mathbf{Mod}^{fp}$).

b) Assuming that $\mathbf{A}$ is left coherent, let $M, Q \in {}_{\mathbf{A}}\mathbf{Mod}^{fp}$ and $\bullet f : M \to Q$; then $\operatorname{im} f$ is a f.g. submodule of Q, thus $\operatorname{im} f$ is finitely presented by Lemma 509(iv). By Lemma 498(iii) with $N = \ker f$ and $P = \operatorname{im} f \cong M/\ker f$, $\ker f$ is finitely presented. Therefore ${}_{\mathbf{A}}\mathbf{Mod}^{fp}$ has kernels. We have the short exact sequence

$$0 \longrightarrow \operatorname{im} f \longrightarrow N \longrightarrow N/\operatorname{im} f \longrightarrow 0$$

where N is finitely presented and $\operatorname{im} f$ is f.g., thus $\operatorname{coker} f \cong N/\operatorname{im} f$ is finitely presented by Lemma 498(iii), and ${}_{\mathbf{A}}\mathbf{Mod}^{fp}$ has cokernels. This proves that ${}_{\mathbf{A}}\mathbf{Mod}^{fp}$ is preabelian. Since ${}_{\mathbf{A}}\mathbf{Mod}^{fp}$ is a subcategory of the abelian category ${}_{\mathbf{A}}\mathbf{Mod}$, ${}_{\mathbf{A}}\mathbf{Mod}^{fp}$ is abelian. ∎

Proposition 511. *A left Noetherian ring and a semifir are left coherent.*

Proof. Let $\mathbf{A}$ be a ring.

(1) Assuming that $\mathbf{A}$ is left Noetherian, let M be a finitely presented left $\mathbf{A}$-module and let $N \vartriangleleft M$; since M is Noetherian by Proposition 310, N is f.g., therefore N is finitely presented by Theorem 499, thus M is coherent, and $\mathbf{A}$ is left coherent by Lemma 509.

(2) Assuming that $\mathbf{A}$ is a semifir, consider a f.g. left ideal $\mathfrak{a}$; by Lemma 410(i), $\mathfrak{a}$ is finite free, say of rank k, and has a presentation

$$0 \longrightarrow \mathbf{A}^k \longrightarrow \mathfrak{a} \longrightarrow 0,$$

thus $\mathbf{A}$ is left coherent by Definition 507(i). ∎

3.2.3 The Matrix of a Module

3.2.3.1 Definition

Let $\mathbf{A}$ be a ring, let $\bullet f : \mathbf{A}^{(Q)} \to \mathbf{A}^{(K)}$ be an $\mathbf{A}$-linear homomorphism, where Q and K are sets of indices, and let $R = (r_{ij})$ be the matrix of f in the canonical bases $(\zeta_i)_{i \in Q}$ and $(\varepsilon_j)_{j \in K}$ of $\mathbf{A}^{(Q)}$ and $\mathbf{A}^{(K)}$, respectively (Lemma and Definition 263); in addition, let $\bullet\varphi : \mathbf{A}^{(K)} \twoheadrightarrow \operatorname{coker} f$ be the canonical epimorphism, let $w_j = \varepsilon_j \varphi$ and let $M = \operatorname{im} \varphi$. Then $M = [w]_{\mathbf{A}}$ where $w \triangleq (w_j)_{j \in K}$ and $\ker \varphi = \operatorname{im} f$, thus the sequence (3.3) is exact. Every module over $\mathbf{A}$ is isomorphic to a module constructed in that way (Definition 495).

Identifying $\bullet f$ with $\bullet R$, the right-multiplication by R (Remark 264), $\operatorname{im} f$ is the left $\mathbf{A}$-module generated by the rows of R. The row of R with index i is $\varsigma_i R = \sum_{j \in K} r_{ij}\, \varepsilon_j$; its canonical image in $\operatorname{coker} f$ is zero, thus $\sum_{j \in I} r_{ij}\, w_j = 0$, and since this holds for all $i \in Q$ we obtain the equation (3.1). This equation gathers all relations existing between the generators w_j of M, thus it *characterizes* the module $M = [w]_{\mathbf{A}}$; therefore it is an equation of M in the generators w_j $(j \in K)$. This equation is *not unique*; indeed, assume that $Q = \{1, ..., q\}$ and choose another basis in $\mathbf{A}^q$; then we obtain another equation

$$R'\, w = 0$$

in the same generators, where R' is *left-associated* with R (Definition 63), i.e., $R' = U\, R$ for some $U \in \mathrm{GL}_q(\mathbf{A})$. Nevertheless, M is determined up to isomorphism by the matrix R in equation (3.1) as $M = \operatorname{coker}_{\mathbf{A}}(\bullet R)$.

Definition 512. *(1) The above matrix* $R \in {}^{Q}\mathbf{A}^{(K)}$ *is called* a matrix of definition *of the module* $M = \operatorname{coker}_{\mathbf{A}}(\bullet R)$.
(2) The module $M = [w]_{\mathbf{A}}$ *characterized by (3.1) is said to be* defined by generators and relations.

Recall that (3.1) defines a cokernel, not a kernel (however, see Subsect. 3.4.3 below). In usual cases, the module M under consideration is *finitely presented*, i.e., $Q = \{1, ..., q\}$, $K = \{1, ..., k\}$ and $R \in {}^{q}\mathbf{A}^{k}$.

3.2.3.2 Generalization

The situation is that in (**3.2.1.3**) when $\mathcal{C}$ is a preabelian category. Let $\bullet\iota_j^q : U \hookrightarrow U^q$ be the jth injection and $\bullet\pi_i^k : U^k \twoheadrightarrow U$ the ith projection. There exists a canonical isomorphism $\mathbf{A}$-linear isomorphism $\operatorname{Hom}_{\mathcal{C}}(U^q, U^k) \xrightarrow{\sim} {}^{q}\mathbf{A}^{k}$ given by

$$\bullet f \mapsto (R_{ij})_{i,j}\,, \quad R_{ij} \triangleq \iota_i^q f \pi_j^k, \quad f = \sum_{i,j} \pi_i^q R_{ij} \iota_j^k.$$

The matrix $R = (R_{ij})_{i,j} \in {}^{q}\mathbf{A}^{k}$ is written $\operatorname{Mat}(f)$. Let $\bullet f : U^q \to U^k$ and $\bullet g : U^k \to U^l$. Then, $\operatorname{Mat}(fg) = \operatorname{Mat}(f)\operatorname{Mat}(g)$. Let M be an object of $\mathcal{C}$ and $x \in \operatorname{Hom}_{\mathcal{C}}(M, U^q) = \operatorname{Hom}_{\mathcal{C}}(M, U)^q$; then $x = (x_i)_{1 \leq i \leq q}$ is denoted by a row with entries x_i $(1 \leq i \leq q)$ and

$$xf = \left(\sum_{1 \leq i \leq q} x_i R_{ij} \right)_{1 \leq j \leq k} \in \operatorname{Hom}_{\mathcal{C}}(M, U^k) = \operatorname{Hom}_{\mathcal{C}}(M, U)^k$$

is the row xR, thus the left morphism $\bullet f$ is identified with the right multiplication $\bullet R$ by the matrix $R = \operatorname{Mat}(f)$.

3.2.3.3 Some Properties

Item (1) of Definition 513 below generalizes that given in Corollary 340 and coincides with it when $\mathbf{A}$ is a left Ore domain:

Definition 513. *Let* $\mathbf{A}$ *be a ring and* M *a left* $\mathbf{A}$*-module.*

(1) M *is* torsion-free *if* $\mathrm{Ann}_r^M(a) = 0$ *whenever* $a \in \mathbf{A}$ *is right regular, i.e.,* $\mathrm{Ann}_r^{\mathbf{A}}(a) = 0$.
(2) M *is* torsionless *if it can be embedded in some direct product* $\mathbf{A}^I$.

We have the following.

Theorem 514. *Let* $\mathbf{A}$ *be a ring, consider the sequence (3.7) below where* $R_1 \in {}^q\mathbf{A}^k, R_2 \in {}^k\mathbf{A}^r$*, and set* $M_1 = \mathrm{coker}_{\mathbf{A}}(\bullet R_1)$.

$$\mathbf{A}^q \xrightarrow{\bullet R_1} \mathbf{A}^k \xrightarrow{\bullet R_2} \mathbf{A}^r \to 0. \tag{3.7}$$

(i) The following conditions are equivalent: (a) the sequence (3.7) is exact at $\mathbf{A}^r$*; (b)* $\bullet R_2$ *is left-invertible; (c) the algebraic transpose* $R_2 \bullet$ *of* $\bullet R_2$ *is left-invertible; (d) the matrix* R_2 *is left-invertible over* $\mathbf{A}$.
(ii) Given $R_2 \in {}^k\mathbf{A}^r$*, if* $\mathbf{A}$ *is a left coherent ring, then there exist a natural integer* q *and a matrix* $R_1 \in {}^q\mathbf{A}^k$ *such that the sequence (3.7) is exact at* $\mathbf{A}^k$.
(iii) Given $R_1 \in {}^q\mathbf{A}^k$*, there exists a matrix* $R_2 \in {}^k\mathbf{A}^r$ *such that the sequence (3.7) is exact (resp., exact at* $\mathbf{A}^k$*) if, and only if* $M_1 \cong \mathbf{A}^r$ *(resp., there exists an embedding* $M_1 \hookrightarrow \mathbf{A}^r$*, i.e., Condition* ***(F)*** *in Proposition 516(iv) below is satisfied).*

Proof. (i) Obviously, (b)⇔(c)⇔(d)⇒(a). If (a) holds, $\mathbf{A}^k / \ker_{\mathbf{A}}(\bullet R_2) \cong \mathbf{A}^r$ by Noether's first isomorphism theorem, thus $\ker_{\mathbf{A}}(\bullet R_2)$ is a direct summand of $\mathbf{A}^k$ and since $\bullet R_2$ is surjective, it is left-invertible (see Proposition 572 below).

(ii): If $\mathbf{A}$ is left coherent, then $\ker_{\mathbf{A}}(\bullet R_2) \subseteq \mathbf{A}^k$ is finitely generated by Theorem 510. Consider a matrix R_1, the rows of which form a finite generator of $\ker_{\mathbf{A}}(\bullet R_2)$; then the sequence (3.7) is exact at $\mathbf{A}^k$.

(iii): (a) If the sequence (3.7) is exact at $\mathbf{A}^k$, there exists an induced map $(\bullet R_2)_{ind} : M_1 \to \mathbf{A}^r$ which is an embedding.

(b) Conversely, consider the exact sequence

$$\mathbf{A}^q \xrightarrow{\bullet R_1} \mathbf{A}^k \xrightarrow{\varphi_1} M_1 \to 0. \tag{3.8}$$

If there exists an embedding $\bullet\iota : M_1 \hookrightarrow \mathbf{A}^r$, set $\bullet R_2 = \bullet\varphi_1\iota : \mathbf{A}^k \to \mathbf{A}^r$. Then the sequence (3.7) is exact at $\mathbf{A}^k$.

(c) Assume $M_1 \cong \mathbf{A}^r$. Then the above monomorphism $\bullet\iota : M_1 \hookrightarrow \mathbf{A}^r$ is an isomorphism, thus $\bullet R_2 = \bullet\varphi_1\iota : \mathbf{A}^k \to \mathbf{A}^r$ is an epimorphism and the sequence (3.7) is exact.

(d) Conversely, if the sequence (3.7) is exact, the monomorphism $(\bullet R_2)_{ind}$ is an isomorphism, thus $M_1 \cong \mathbf{A}^r$. ∎

Remark 515. *(a) The case when* $\mathbf{A}^q$, $\mathbf{A}^k$ *and* $\mathbf{A}^r$ *are changed to* $\mathbf{A}^{(Q)}$, $\mathbf{A}^{(K)}$ *and* $\mathbf{A}^{(R)}$*, respectively, where* Q, K *and* R *are arbitrary sets of indices, can also be considered; it is simpler and left to the reader.*
(b) The matrix R_1 *(resp., the map* $\bullet R_1$*) in Theorem 514(ii) is called a* syzygy *of the matrix* R_2 *(resp., the map* $\bullet R_2$*).*

For a module, torsion-freeness, torsionlessness and being embeddable in a free module are closely connected, as shown by Proposition 342, and as proved below.

Proposition 516. *Let* $\mathbf{A}$ *be a ring and* M *a left* $\mathbf{A}$*-module.*

(i) If M *is free, it is both torsion-free and torsionless.*
(ii) If M *torsion-free (resp., torsionless) and* $N \lhd M$*, then* N *is torsion-free (resp., torsionless).*
(iii) If M *is torsionless, then* M *is torsion-free.*
(iv) Denote by ***(F)*** *the following condition:*
(F) *There exists an embedding* $M \hookrightarrow \mathbf{A}^r$ *for some integer* $r \geq 1$.
(a) Condition ***(F)*** *holds if, and only if* M *is torsionless and has finite uniform dimension* (**3.5.5.1**).
(b) If the ring $\mathbf{A}$ *is semiprime Goldie and* M *is f.g., Condition* ***(F)*** *holds if, and only if* M *is torsion-free (see Proposition 342).*
(c) If the ring $\mathbf{A}$ *is right-coherent and* M *is f.p., Condition* ***(F)*** *holds if, and only if* M *is torsionless.*
(v) In particular, if $\mathbf{A}$ *is semiprime Goldie and* M *is f.g. torsion-free, then* M *is torsionless.*

Proof. (i) and (ii) are clear.

(iii): Assume that $M \lhd \mathbf{A}^I$ and let $w = (b_i)_{i \in I} \in M$. If $a\,w = 0$, then $a\,b_i = 0$ for all $i \in I$. If $\operatorname{Ann}_r^{\mathbf{A}}(a) = 0$, this implies $b_i = 0$ for all $i \in I$, thus $w = 0$; therefore, M is torsion-free.

(iv)(a),(b): see ([247], 3.4.3, 3.4.7).

(iv)(c): If $M_1 = M$ is f.p., there exists an exact sequence such as (3.8). If M_1 is torsionless, there exists an embedding $\varepsilon : M_1 \hookrightarrow \mathbf{A}^I$ for some set of indices I. Set $\bullet\tau = \bullet\varphi_1\,\varepsilon : \mathbf{A}^k \to \mathbf{A}^I$; for any $x \in \mathbf{A}^k$, $x\tau$ can be written xT where the matrix $T \in {}^k\mathbf{A}^I$ has possibly infinitely many columns. The sequence

$$\mathbf{A}^q \xrightarrow{\bullet R_1} \mathbf{A}^k \xrightarrow{\bullet T} \mathbf{A}^I \tag{3.9}$$

is exact, thus $R_1 T = 0$. If $\mathbf{A}$ is right coherent, there exists a natural integer r and matrices $R_2 \in {}^k\mathbf{A}^r$, $D \in {}^r\mathbf{A}^I$ such that $R_1 R_2 = 0$, $T = R_2 D$ by Lemma 509(iii), and the exactness of the sequence (3.9) at $\mathbf{A}^k$ implies that of the sequence (3.7) at $\mathbf{A}^k$, thus there exists an embedding $M_1 \hookrightarrow \mathbf{A}^r$ by Theorem 514(iii).

(v) is an obvious consequence of (iv). ∎

3.2.4 Similar Matrices

Given a finitely presented module M over a ring $\mathbf{A}$, it is important to characterize all matrices determining M up to isomorphism.

Let $M = \operatorname{coker}_{\mathbf{A}}(\bullet R)$; all matrices R' we are looking for are those for which $\operatorname{coker}_{\mathbf{A}}(\bullet R) \cong \operatorname{coker}_{\mathbf{A}}(\bullet R')$. The definition below is a generalization of Definition 424(2):

Definition 517. *Two matrices R, R' are* left-similar *if the* left $\mathbf{A}$*-modules they define are isomorphic.* Right-similar *matrices are defined correspondingly, and two matrices R, R' are called* similar *(written $R \simeq R'$) if they are left- and right-similar.*

In many textbooks, two square matrices are called *similar* when they are *conjugate* (or *similar-square*): see Definition 280(iii) and Exercise 679. Let us further study the conditions under which two matrices are left-similar. To obtain explicit conditions, the ring $\mathbf{A}$ is assumed to be *weakly finite* (Exercise 481(2)); this is not a strong assumption. Consider a relation

$$\begin{bmatrix} R' & -T' \end{bmatrix} \begin{bmatrix} T \\ R \end{bmatrix} = 0 \tag{3.10}$$

between matrices.

Definition 518. *(i) The relation (3.10) is called* right-comaximal *if $\begin{bmatrix} R' & -T' \end{bmatrix}$ is right-invertible, i.e., if there is a pair (X, Y) such that the following Bézout identity holds:*

$$R' X - T' Y = I. \tag{3.11}$$

Then the pair $(R', T'))$ is called right-comaximal, *or* strongly left-coprime.
(ii) The pair (R', T') is said to be left-coprime *if the matrix $A' = \begin{bmatrix} R' & -T' \end{bmatrix}$ is left-prime (Definition 419).*
(iii) The relation (3.10) is called left-comaximal *if $\begin{bmatrix} T \\ R \end{bmatrix}$ is left-invertible, i.e., there exists a pair (X', Y') such that*

$$X' R - Y' T = I. \tag{3.12}$$

Then the pair $(R, T))$ is called left-comaximal, *or* strongly right coprime.
(iv) The pair (R, T) is right-coprime *pair if the matrix $\begin{bmatrix} T \\ R \end{bmatrix}$ is right-prime.*
(v) The relation (3.10) is comaximal *if it is both right- and left-comaximal. This relation is* coprime *if (R', T') is left-coprime and (R, T) is right-coprime.*

The *index* of a matrix $R \in {}^{q}\mathbf{A}^{k}$ is $i(R) = k - q$. A relation (3.10) is said to be *proper* if R, R' (or equivalently T, T') have the same index.

Definition 519. *Let* $\mathbf{A}$ *be a weakly finite ring and let* $R \in {}^{q}\mathbf{A}^{k}$, $R' \in {}^{q'}\mathbf{A}^{k'}$ *be two matrices having the same index; R and R' are said to be* stably associated *if there exists a* $(q'+k) \times (q+k')$ *matrix* $\begin{bmatrix} R' & -T' \\ * & * \end{bmatrix}$ *with inverse of the form* $\begin{bmatrix} * & T \\ * & R \end{bmatrix}$.

Theorem 520. *Let* $\mathbf{A}$ *be a weakly finite ring and let* $R \in {}^{q}\mathbf{A}^{k}$, $R' \in {}^{q'}\mathbf{A}^{k'}$.

(1) Of the following, (a) and (b) are equivalent and imply (c) and (d):
(a) R and R' are stably associated;
(b) R and R' satisfy a proper comaximal relation (3.10);
(c) R and R' are left-similar;
(d) setting $S' = \begin{bmatrix} R' & -T' \end{bmatrix}$ *and* $S = \begin{bmatrix} T \\ R \end{bmatrix}$ *the sequence*

$$\xrightarrow{\bullet S'} \mathbf{A}^{k'+q} \xrightarrow{\bullet S} \tag{3.13}$$

is exact and the pair (R, T) *is strongly right-coprime.*
(2) If R and R' are left-regular, then Conditions (a), (b) and (c) are equivalent.

Proof. (b)⇒(a): Assuming that (b) holds, we have (3.11) and (3.12); setting

$$M = \begin{bmatrix} R' & -T' \\ -Y' & X' \end{bmatrix}, \quad N = \begin{bmatrix} X & T \\ Y & R \end{bmatrix}$$

we obtain $M\,N = \begin{bmatrix} I & 0 \\ * & I \end{bmatrix} \triangleq J$; the matrix J is invertible, thus M and N are left- and right-invertible (Exercise 481(2)(iv)), therefore those matrices are invertible (**1.3.2.1**) and (a) holds.

(a)⇒(b) is obvious.

For the other statements regarding Conditions (a), (b) and (c), see ([77], Sect. 0.6, Th. 6.2).

(b)⇒(d): Assume that (b) holds; then we have obviously $\mathrm{im}_{\mathbf{A}}(\bullet S') \vartriangleleft \ker_{\mathbf{A}}(\bullet S)$, thus it remains to prove that $\ker_{\mathbf{A}}(\bullet S) \vartriangleleft \mathrm{im}_{\mathbf{A}}(\bullet S')$. We have $M\,N\,J^{-1} = I$, thus $N\,J^{-1}\,M = I$ (since $\mathbf{A}$ is weakly finite) and J^{-1} is of the form $\begin{bmatrix} I & 0 \\ Z & I \end{bmatrix}$. Let $\begin{bmatrix} x & y \end{bmatrix} \in \ker_{\mathbf{A}}(\bullet S)$; then an easy calculation shows that

$$\begin{bmatrix} x & y \end{bmatrix} N\,J^{-1} M = z\,S'$$

with $z = x\,X + y\,Y$. Therefore, $\begin{bmatrix} x & y \end{bmatrix} \in \mathrm{im}_{\mathbf{A}}(\bullet S')$. ∎

See complements in Corollary 847 below. Due to the symmetry of Condition (a) above, we have the following.

Corollary 521. *If* $R \in \mathrm{Mat}_{q}(\mathbf{A})$ *and* $R' \in \mathrm{Mat}_{q'}(\mathbf{A})$ *are regular matrices, then R and R' are left- (or right-) similar if, and only if they are similar.*

Definition 522. *Over a ring* $\mathbf{A}$*, a pair* (A, B) *of matrices is said to be* comaximally transposable *if there exists matrices* A'*,* B' *over* $\mathbf{A}$ *such that* $A\,B = B'\,A'$ *is a proper comaximal relation.*

The following result is proved in ([77], Sect. 3.4, Lemma 3.4):

Proposition 523. *Let* $\mathbf{A}$ *be a weakly finite ring. If a pair* (A, B) *of matrices over* $\mathbf{A}$ *is comaximally transposable, then there exist matrices* X, Y *over* $\mathbf{A}$ *such that the following Bézout identity holds:*

$$X\,A - B\,Y = I. \tag{3.14}$$

Conversely, if $\mathbf{A}$ *is a semifir (or more general, an Hermite ring: see Definition 579 below) and there exist matrices* X, Y *such that (3.14) holds, then the pair* (A, B) *is comaximally transposable.*

3.2.5 Similarity over Bézout Domains

Let $\mathbf{A}$ be a right Bézout domain and a, a' be nonzero elements of $\mathbf{A}$. Since $\mathbf{A}$ is weakly finite (Exercise 481), a and a' are similar if, and only if there exists a proper comaximal relation

$$\begin{pmatrix} a' & -b' \end{pmatrix} \begin{pmatrix} b \\ a \end{pmatrix} = 0 \tag{3.15}$$

(Theorem 520 and Corollary 521).

Theorem 524. *The elements* a *and* a' *are similar if, and only if there exists* $b \in \mathbf{A}^{\times}$ *such that* (a, b) *is right-coprime and* $a' = u\,[a, b]_l\,b^{-1}$ *where* u *is a unit.*

Proof. Recall that $[a, b]_l$ is a least common left multiple of a and b, uniquely determined up to left associates (Proposition 82). Since $\mathbf{A}$ is a right Bézout domain, a pair (a', b') of elements of $\mathbf{A}$ is strongly left-coprime if, and only if it is left-coprime by Theorem 443 (see also Lemma 650 below). Assuming that the relation (3.15) holds with (a, b) right-coprime, then

$$a'b = b'\,a, \quad a' = (b'\,a)\,b^{-1}. \tag{3.16}$$

The element $b'\,a$ is a left-multiple of a and by the first equality of (3.16) it is a left-multiple of b. Conversely, if $b'\,a$ is a left-multiple of b, then (3.16) holds. Let $c \in \mathbf{A}^{\times}$ and let b' be such that $b'\,a = u\,[a, b]_l$; let $b'' = c\,b'$ and let a'' be such that $a''b = b''a$. From the above, $a'\,b = b'\,a$ and $a'' = b''\,a\,b^{-1} = c\,b'\,a\,b^{-1} = c\,a'$, thus

$$\begin{bmatrix} a'' & -b'' \end{bmatrix} = c \begin{bmatrix} a' & -b' \end{bmatrix}.$$

Therefore, the relation (3.15) is comaximal if, and only if c is a unit. ∎

3.3 Cyclic Modules

3.3.1 Cyclic Modules over a General Ring

3.3.1.1 Cyclic Modules

Definition 525. *A left* $\mathbf{A}$*-module* C *is said to be* cyclic *if it is generated by one element.*

Proposition 526. *(i) A left* $\mathbf{A}$*-module* C *is cyclic, generated by* m*, if, and only* $C \cong \mathbf{A}/\mathfrak{a}$ *where* $\mathfrak{a} = \operatorname{Ann}_l^{\mathbf{A}}(m)$*.*
(ii) $C \cong \mathbf{A}$ *if, and only if* $\mathfrak{a} = \{0\}$*.*
(iii) Let $S \subseteq \mathbf{A}^{\times}$ *be a left denominator set; then* $S^{-1}C = 0$ *if* $S \cap \mathfrak{a} \neq \varnothing$ *(i.e.,* S meets $\mathfrak{a}$*).*

Proof. (i): Let C be a cyclic module with generator m, i.e., $C = [m]_{\mathbf{A}}$, and let $\mathfrak{a} = \operatorname{Ann}_l^{\mathbf{A}}(m)$. Let $\bullet\alpha : \mathbf{A} \twoheadrightarrow C$ be the epimorphism defined by $(1)\,\alpha = m$. Since $\ker\alpha = \mathfrak{a}$, there exists by Noether's first isomorphism theorem (Theorem 52(1)) an isomorphism $C \cong \mathbf{A}/\mathfrak{a}$. Conversely, let $\mathfrak{a}$ be a left ideal in $\mathbf{A}$; then the quotient $\mathbf{A}/\mathfrak{a}$ is cyclic, generated by $(1)\,\varphi$ where $\varphi : \mathbf{A} \twoheadrightarrow \mathbf{A}/\mathfrak{a}$ is the canonical epimorphism. (ii) is obvious.

(iii): If S meets $\mathfrak{a}$, there exists $s \in S \cap \mathfrak{a}$; since s is a unit of $S^{-1}\mathbf{A}$, $1/1 \in S^{-1}\mathfrak{a}$, thus $S^{-1}\mathfrak{a} = S^{-1}\mathbf{A}$ and $\left(S^{-1}\mathbf{A}\right) / \left(S^{-1}\mathfrak{a}\right) = S^{-1}\left(\mathbf{A}/\mathfrak{a}\right) = 0$. ∎

Corollary 527. *Let* $\mathbf{A}$ *be a left Ore domain and let* $C \cong \mathbf{A}/\mathfrak{a}$ *be a cyclic left* $\mathbf{A}$*-module.*

(1) The following conditions are equivalent:
(i) $\mathfrak{a} \neq \{0\}$*;*
(ii) the module C *is torsion;*
(iii) $C \ncong \mathbf{A}$*.*
(2) If $\mathbf{A}$ *is a left Bézout domain, then there exists* $a \in \mathbf{A}$ *such that* $C \cong \mathbf{A}/\mathbf{A}\,a$*, and* $a \neq 0$ *if, and only if* C *is torsion.*

Proof. It remains to prove (2). Let $\mathfrak{a}$ be a left ideal in $\mathbf{A}$ such that $C \cong \mathbf{A}/\mathfrak{a}$. Since $\mathbf{A}$ is a semifir, the f.g. module C is finitely presented by Theorem 499. We have the short exact sequence

$$0 \longrightarrow \mathfrak{a} \longrightarrow \mathbf{A} \longrightarrow C \longrightarrow 0$$

thus $\mathfrak{a}$ is f.g. by Lemma 498(ii), and there exists $a \in \mathbf{A}$ such that $\mathfrak{a} = \mathbf{A}\,a$ by Definition 438. The rest is clear. ∎

Regarding quotients of cyclic modules, we have the following.

Proposition 528. *Let* $\mathbf{A}$ *be a ring.*

(i) Every quotient of a cyclic module is cyclic.
(ii) If $a, b \in \mathbf{A}$ *and* b *is left-regular (Definition 83), then* $\mathbf{A}/\mathbf{A}\,a \cong \mathbf{A}\,b/\mathbf{A}\,a\,b$, *and this is a direct summand of* $\mathbf{A}/\mathbf{A}\,a\,b$ *if, and only if there exist* $x, y \in \mathbf{A}$ *such that*

$$x\,a - b\,y = 1. \tag{3.17}$$

Moreover, in that case we have

$$\mathbf{A}\,/\mathbf{A}\,a\,b \cong \mathbf{A}/\mathbf{A}\,a \oplus \mathbf{A}/\mathbf{A}\,b.$$

(iii) Every submodule M *of a cyclic left* $\mathbf{A}$*-module* $C \cong \mathbf{A}/\mathfrak{c}$ *(where* $\mathfrak{c}$ *is a left ideal) is cyclic and isomorphic to a module* $\mathfrak{b}/\mathfrak{c}$ *where* $\mathfrak{b}$ *is a left ideal such that* $\mathfrak{c} \subseteq \mathfrak{b}$*. If* $\mathbf{A}$ *is a left Bézout domain and* $M \neq 0$, $\mathfrak{c} = \mathbf{A}\,c$, $\mathfrak{b} = \mathbf{A}\,b$ *and* $\mathfrak{b}/\mathfrak{c} \cong \mathbf{A}/\mathbf{A}\,a$ *where* $c = a\,b$*.*
(iv) A left $\mathbf{A}$*-module* M *is simple if, and only if* $M \cong \mathbf{A}/\mathfrak{m}$ *where* $\mathfrak{m}$ *is a maximal left ideal.*

Proof. (i): Let C be a cyclic left $\mathbf{A}$-module with generator m and let D be a quotient of C. Let $\bullet\varphi : C \twoheadrightarrow D$ be the canonical epimorphism. Then D is generated by $m\varphi$, thus is cyclic.

(ii): See ([77], Sect. 3.6, Lemma 6.7).

(iii): Let $\mathfrak{c} = \mathrm{Ann}_l^{\mathbf{A}}(m)$ and let M be a submodule of $C = [m]_{\mathbf{A}}$. By Proposition 526, there exists an isomorphism $\bullet\psi : C \xrightarrow{\sim} \mathbf{A}/\mathfrak{c}$, thus $(M)\,\psi$ is a submodule of $\mathbf{A}/\mathfrak{c}$. By Theorem 147(i) there exists a left ideal $\mathfrak{b}$ such that $\mathfrak{c} \subseteq \mathfrak{b}$ and $(M)\,\psi = \mathfrak{b}/\mathfrak{c}$. If $\mathbf{A}$ is a left Bézout domain and $M \neq 0$, $\mathfrak{b}$ and $\mathfrak{c}$ are left principal ideals, i.e., $\mathfrak{c} = \mathbf{A}\,c$ and $\mathfrak{b} = \mathbf{A}\,b$ where $b \neq 0$ (Corollary 527(2)); since $\mathfrak{c} \subseteq \mathfrak{b}$, $c \in \mathbf{A}\,b$ and there exists a such that $c = a\,b$. Then $\mathbf{A}\,b/\mathbf{A}\,a\,b \cong \mathbf{A}/\mathbf{A}\,a$ by (ii).

(iv): A simple left $\mathbf{A}$-module is obviously cyclic, thus isomorphic to $\mathbf{A}/\mathfrak{m}$ for some left ideal $\mathfrak{m}$. By (iii), $\mathbf{A}/\mathfrak{m}$ is simple if, and only if $\mathfrak{m}$ is a maximal left ideal. ∎

Corollary 529. *Let* $\mathbf{A}$ *be a semifir and let* $a, b \in \mathbf{A}^{\times}$*. The following conditions are equivalent:*

(i) $\mathbf{A}/\mathbf{A}\,a \oplus \mathbf{A}/\mathbf{A}\,b$ *is cyclic;*
(ii) $\mathbf{A}/\mathbf{A}\,a \oplus \mathbf{A}/\mathbf{A}\,b \cong \mathbf{A}/\mathbf{A}\,a\,b$*;*
(iii) the pair (a, b) *is strongly right-coprime.*

Proof. Let a, a' be similar. Then $\mathbf{A}/\mathbf{A}\,a' \oplus \mathbf{A}/\mathbf{A}\,b \cong \mathbf{A}/\mathbf{A}\,a \oplus \mathbf{A}/\mathbf{A}\,b$. By Proposition 528(ii), $\mathbf{A}/\mathbf{A}\,a \cong \mathbf{A}\,b/\mathbf{A}\,a\,b$ and the latter module is a direct summand of $\mathbf{A}/\mathbf{A}\,a\,b$ if, and only if there exist x, y such that (3.17) holds, and then $\mathbf{A}/\mathbf{A}\,a\,b \cong \mathbf{A}/\mathbf{A}\,a \oplus \mathbf{A}/\mathbf{A}\,b$. By Proposition 523, there exist x, y such that (3.17) holds if, and only if (a, b) is comaximally transposable, say $a\,b = b'\,a'$ is a proper comaximal relation and (a', b) is strongly right-coprime. Conversely, if (a', b) is strongly right-coprime, there exists a proper comaximal relation $a\,b = b'\,a'$. In the above a and a' can be anywhere interchanged. ∎

3.3.1.2 Decomposable Elements

Theorem and Definition 530. *Let* $\mathbf{A}$ *be a left Bézout domain and let* $c \in \mathbf{A}^{\times}$.

(i) The element c *is said to be* left-decomposable *if there exist* $a_1, a_2 \in \mathbf{A}$ *such that*

$$\mathbf{A}\,c = \mathbf{A}\,a_1 \cap \mathbf{A}\,a_2, \quad \mathbf{A}\,c \neq \mathbf{A}\,a_1, \mathbf{A}\,a_2, \tag{3.18}$$

i.e., $c = b_2\,a_1 = b_1\,a_2$ *where* a_1, a_2, b_1, b_2 *are nonunits and* c *is an* lclm *of the pair* (a_1, a_2) *(Theorem 443);* c *is left-decomposable if, and only if* $\mathbf{A}/\mathbf{A}\,c$ *is an irredundant subdirect sum of two torsion cyclic modules:*

$$\mathbf{A}/\mathbf{A}\,c \hookrightarrow \mathbf{A}/\mathbf{A}\,a_1 \oplus \mathbf{A}/\mathbf{A}\,a_2. \tag{3.19}$$

A right-decomposable *element is likewise defined and an element which is not left- (resp., right-) decomposable is said to be left- (resp., right-)* indecomposable.
(ii) The element c *is said to be* decomposable *if (3.18) holds with* (a_1, a_2) *right-coprime, i.e.,*

$$\mathbf{A}\,a_1 + \mathbf{A}\,a_2 = \mathbf{A} \tag{3.20}$$

(and indecomposable *otherwise);* c *is decomposable if, and only if*

$$\mathbf{A}/\mathbf{A}\,c \cong \mathbf{A}/\mathbf{A}\,a_1 \oplus \mathbf{A}/\mathbf{A}\,a_2,$$

where both $\mathbf{A}/\mathbf{A}\,a_1, \mathbf{A}/\mathbf{A}\,a_2 \neq 0$ *or* $\mathbf{A}/\mathbf{A}\,c = 0$.
(iii) Therefore, $c \in \mathbf{A}^{\times}$ *is (left, right) decomposable if, and only if every element similar to it is (left, right) decomposable too.*

Proof. (i): All $\mathbf{A}$-linear left maps $\mathbf{A} \to \mathbf{A} \times \mathbf{A}$ are of the form $\bullet f_\lambda : x \mapsto (\lambda x, \lambda x)$ $(\lambda \in \mathbf{A})$. By Theorem and Definition 145, there exists an $\mathbf{A}$-linear map $\bullet \bar{f}_\lambda : \mathbf{A}/\mathbf{A}\,c \to (\mathbf{A} \times \mathbf{A}) / (\mathbf{A}\,a_1 \times \mathbf{A}\,a_2)$ induced by f_λ with respect to $\mathbf{A}\,c$ and $\mathbf{A}\,a_1 \times \mathbf{A}\,a_2$ if, and only if $(\lambda \mathbf{A}\,c) \times (\lambda \mathbf{A}\,c) \subseteq \mathbf{A}\,a_1 \times \mathbf{A}\,a_2$, i.e., $\lambda \mathbf{A}\,c \subseteq \mathbf{A}\,a_1 \cap \mathbf{A}\,a_2$. This implies $\mathbf{A}\,c \subseteq \mathbf{A}\,a_1 \cap \mathbf{A}\,a_2$, thus the weakest condition is obtained with $\lambda = 1$. The left $\mathbf{A}$-module $(\mathbf{A} \times \mathbf{A}) / (\mathbf{A}\,a_1 \times \mathbf{A}\,a_2)$ can be identified with $\mathbf{A}/\mathbf{A}\,a_1 \oplus \mathbf{A}/\mathbf{A}\,a_2$ by (2.9). The induced map $\bullet \bar{f}_1$ is injective if, and only if $\left((\mathbf{A}\,a_1 \times \mathbf{A}\,a_2)\, f_1^{-1}\right) \pi = 0$ where $\bullet \pi : \mathbf{A} \twoheadrightarrow \mathbf{A}/\mathbf{A}\,c$ is the canonical epimorphism; this condition is equivalent to $\mathbf{A}\,a_1 \cap \mathbf{A}\,a_2 \subseteq \mathbf{A}\,c$. To summarize, $\mathbf{A}\,c = \mathbf{A}\,a_1 \cap \mathbf{A}\,a_2$ if, and only if there exists an embedding (3.19). This embedding induces an *irredundant* subdirect sum if, and only if $\mathbf{A}\,c \neq \mathbf{A}\,a_1, \mathbf{A}\,a_2$.

(ii): The map $\bullet\pi$ is surjective if, and only if for any element $\bar{y} \in \mathbf{A}/\mathbf{A}\,a_1 \oplus \mathbf{A}/\mathbf{A}\,a_2$, there exists $x \in \mathbf{A}$ such that $\bar{y} = (x)\,\pi$. Set $\bar{y} = (y_1 + \mathbf{A}\,a_1, y_2 + \mathbf{A}\,a_2)$; we have $\bar{y} = (x)\,\pi$ if, and only if $x \in (y_1 + \mathbf{A}\,a_1) \cap (y_2 + \mathbf{A}\,a_2)$. This intersection is nonempty if, and only if there exist $\alpha_1, \alpha_2 \in \mathbf{A}$ such that $y_1 + \alpha_1\,a_1 = y_2 + \alpha_2\,a_2$, or equivalently $\alpha_1\,a_1 - \alpha_2\,a_2 = y_2 - y_1$. This Bézout equation has a solution (α_1, α_2) for arbitrary y_1, y_2 if, and only if (3.20) holds.

(iii) is clear. ∎

3.3.1.3 Simple Modules

Let us further study simple modules (Definition 313(i)).

Definition 531. *Let* $\mathbf{A}$ *be a ring. A submodule of a left* $\mathbf{A}$*-module* M *is* maximal *if it is maximal in the set of all proper submodules of* M*.*

Proposition 532. *Let* $\mathbf{A}$ *be a left principal ideal domain.*

(i) A left $\mathbf{A}$*-module* S *is simple if, and only if* $S \cong \mathbf{A}/\mathbf{A}\,c$ *where* c *is an atom (Definition 83(ii)).*
(ii) If $\mathbf{A}$ *is commutative, an* $\mathbf{A}$*-module* S *is simple if, and only if* $S \cong \mathbf{A}/\mathbf{A}\,p$ *where* p *is a prime (Definitions 128 and 304).*

Proof. This is a consequence of Lemma 84, and of Propositions 459 & 528(iv). ∎

Proposition 533. *Let* M *be a nonzero finitely generated left* $\mathbf{A}$*-module.*

(i) Let L *be a proper submodule of* M*. There exists a maximal submodule of* M *containing* L*.*
(ii) M *has a simple quotient.*

Proof. (i): Let $\mathfrak{M}$ be the set of all proper submodules of M containing L and let us show that $\mathfrak{M}$ is inductively ordered by inclusion (**1.5.3.2**). Let $\mathfrak{C}$ be a chain of $\mathfrak{M}$, N be the union of all modules belonging to $\mathfrak{C}$ and G a finite generator of M; if $N = M$, then each element g of G belongs to a module $P_g \in \mathfrak{C}$, thus G belongs to $P = \sum_{g \in G} P_g$, a contradiction since $P \in \mathfrak{C}$. Therefore, according to Zorn's lemma (Lemma 107), $\mathfrak{M}$ has a maximal element N.

(ii): By (i) with $L = 0$, M has a maximal submodule N. Then $S = M/N$ is simple by Theorem 147(i). ∎

Definition 534. *Let* $(S_i)_{i \in I}$ *be a nonempty family of simple left* $\mathbf{A}$*-modules. This family is called a* representative system of simple left A-modules *if (i)* $S_i \neq S_j$ *for* $i \neq j$*, and (ii) for any simple left* $\mathbf{A}$*-module* S*, there exists one index* $i \in I$ *and an isomorphism* $S \cong S_i$*.*

3.3.2 Cyclic Modules over Principal Ideal Domains

3.3.2.1 Bounded Elements and Bounded Modules

Definition 535. *(i) A left* $\mathbf{A}$*-module* M *is said to be* bounded *if there exists a regular element* c *of* $\mathbf{A}$ *such that* $c \in \mathrm{Ann}_l^{\mathbf{A}}(M)$*; then, the two-sided ideal* $\mathrm{Ann}_l^{\mathbf{A}}(M)$ *is called the* bound *of* M*.*
(ii) An element a *in a ring* $\mathbf{A}$ *is said to be* left *(resp.,* right*)* bounded *if the left (resp., right) cyclic* $\mathbf{A}$*-module* $\mathbf{A}/\mathbf{A}\,a$ *(resp.,* $\mathbf{A}/a\,\mathbf{A}$*) is bounded;* $a \in \mathbf{A}$ *is said to be* bounded *if it is both left and right bounded.*

First note that if a is a unit in any ring $\mathbf{A}$, then $\mathbf{A}/\mathbf{A}\,a = \mathbf{A}/a\,\mathbf{A} = 0$, thus a is bounded and $\mathrm{Ann}_l^{\mathbf{A}}(\mathbf{A}/\mathbf{A}\,a) = \mathrm{Ann}_r^{\mathbf{A}}(\mathbf{A}/a\,\mathbf{A}) = \mathbf{A}$. If $a = 0$, $\mathbf{A}/\mathbf{A}\,a = \mathbf{A}/a\,\mathbf{A} = \mathbf{A}$ and a is unbounded.

Assuming that the ring $\mathbf{A}$ is a principal ideal domain and that $a \in \mathbf{A}^\times$ is left (resp., right) bounded, $\mathrm{Ann}_l^{\mathbf{A}}(\mathbf{A}/\mathbf{A}\,a) = \mathbf{A}\,a_l^*\mathbf{A}$ (resp., $\mathrm{Ann}_r^{\mathbf{A}}(\mathbf{A}/a\,\mathbf{A}) = \mathbf{A}\,a_r^*\mathbf{A}$) where a_l^* (resp., a_r^*) is an invariant element (Lemma 447). We have the following.

Proposition 536. *Let* $\mathbf{A}$ *be a principal ideal domain and* $a \in \mathbf{A}^\times$*.*

(i) $\mathrm{Ann}_l^{\mathbf{A}}(\mathbf{A}/\mathbf{A}\,a) = \mathrm{Ann}_r^{\mathbf{A}}(\mathbf{A}/a\,\mathbf{A})$*, thus* a *is left (or right) bounded if, and only if* a *is bounded; then* a_l^* *and* a_r^* *are associates.*
(ii) If a *is bounded, let* $\mathfrak{n} = \mathrm{Ann}_l^{\mathbf{A}}(\mathbf{A}/\mathbf{A}\,a) \neq 0$*; one has* $\mathfrak{n} = \mathbf{A}\,a^*$ *where* a^* *is an invariant element in* $\mathbf{A}$*, unique up to associates and called the* least bound *of* a*;* a^* *is also the least bound of any element similar to* a*, and it is both a left multiple and a right multiple of* a*.*
(iii) If $\mathbf{A}$ *is commutative,* a *is bounded.*

Proof. (i) and (ii): see ([77], Sect. 6.4). (iii): If $\mathbf{A}$ is commutative, then $a \in \mathrm{Ann}_l^{\mathbf{A}}(\mathbf{A}/a\,\mathbf{A})$. ■

Definition 537. *Let* a *and* b *be nonzero elements of a principal ideal domain* $\mathbf{A}$*;* a *and* b *are said to be* totally coprime *if no nonunit factor of* a *is similar to a factor of* b*.*

Lemma 538. *Let* a *and* u *be nonzero elements of a principal ideal domain* $\mathbf{A}$*; if* a *is bounded and totally coprime to* u*, there is a comaximal relation*

$$a\,u = u_1\,a_1 \tag{3.21}$$

and hence

$$\mathbf{A}/\mathbf{A}\,a\,u \cong \mathbf{A}/\mathbf{A}\,a \oplus \mathbf{A}/\mathbf{A}\,u. \tag{3.22}$$

Proof. If we can find a relation (3.21) where a_1 is totally coprime to u and u_1 is totally coprime to a, then the relation $\begin{bmatrix} a & -u_1 \end{bmatrix} \begin{bmatrix} u \\ a_1 \end{bmatrix}$ is comaximal (**3.2.4.1**) by Theorem 443 and the decomposition (3.22) holds by Corollary 529.

Let a^* be the least bound of a; by Proposition 536(ii), there exists a' such that $a^* = a' a$. Then $a' a u = a^* u = u_0 a^*$ since a^* is invariant. Let $a' u_1 = u_0 b$ be an lcrm of a' and u_0. Since $a' a u$ and $u_0 a^*$ are common right multiples of a' and u_0, there exists a_1 such that $a^* = b a_1$ and $a u = u_1 a_1$ (Definition 81(1)). By Theorem 462 and Lemma 433, u_1 is similar to a right factor of u_0 and so totally coprime to a, while a_1 is a factor of a^* and so totally coprime to u. Therefore we have obtained the required relation and (3.22) holds. ∎

3.3.2.2 Totally Unbounded Elements

Definition 539. *Let* $\mathbf{A}$ *be a ring; an element* $a \in \mathbf{A}$ *is totally unbounded if it has no bounded factors apart from units.*

Note that a unit is totally unbounded.

Lemma 540. *Let* $\mathbf{A}$ *be a simple ring. Then every nonunit in* $\mathbf{A}$ *is totally unbounded.*

Proof. Let $a \in \mathbf{A}$ be a nonunit. Since $\mathfrak{n} = \mathrm{Ann}_l^{\mathbf{A}}(\mathbf{A}/\mathbf{A}\,a)$ is a two-sided ideal (Lemma 506), $\mathfrak{n} = 0$ or $\mathfrak{n} = \mathbf{A}$. In the second case, $\mathbf{A}/\mathbf{A}\,a = 0$, thus a is a unit, a contradiction. Therefore, any nonunit in $\mathbf{A}$ is unbounded, hence any element of $\mathbf{A}$ is totally unbounded. ∎

3.3.2.3 Primary Decomposition

First, let us consider the primary decomposition in a commutative Noetherian ring. The following is classical and can be found in, e.g., ([31], §§IV.1, IV.2).

Lemma and Definition 541. *Let* $\mathbf{A}$ *be commutative ring.*

(1) An ideal $\mathfrak{q}$ *in* $\mathbf{A}$ *is said to be* primary *if whenever* $a, b \in \mathbf{A}$ *are such that* $ab \in \mathfrak{q}$ *and* $a \notin \mathfrak{q}$*, then* $b^n \in \mathfrak{q}$ *for some* $n \geq 1$.
(2) If $\mathfrak{q}$ *is a primary ideal, its radical* $\sqrt{\mathfrak{q}}$ *(Theorem and Definition 387(i)) is a prime ideal* $\mathfrak{p}$*, and* $\mathfrak{q}$ *(denoted by* $\mathfrak{q}(\mathfrak{p})$*) is said to be* $\mathfrak{p}$*-primary.*
(3) Let M *be an* $\mathbf{A}$*-module and let* $\mathfrak{p}$ *be a prime ideal in* $\mathbf{A}$*. The following conditions are equivalent:*
(i) There exists $x \in M$ *such that* $\mathfrak{p} = \mathrm{Ann}^{\mathbf{A}}(x)$.
(ii) There exists a submodule N *of* M *such that* $N \cong \mathbf{A}/\mathfrak{p}$.
When the above equivalent conditions hold, the prime ideal $\mathfrak{p}$ *is said to be* associated with M*. The set of all prime ideals associated with* M *is denoted by* $\mathrm{Ass}(M)$.
(4) If the ring $\mathbf{A}$ *is Noetherian and the* $\mathbf{A}$*-module* M *is f.g., then* $\mathrm{Ass}(M)$ *is finite.*

Theorem 542. (Lasker-Noether theorem.) *Let* $\mathbf{A}$ *be commutative Noetherian ring and.*

(i) Every ideal $\mathfrak{a}$ *in* $\mathbf{A}$ *can be written as an intersection*

$$\mathfrak{a}=\bigcap_{\mathfrak{p}\in \operatorname{Ass}(\mathbf{A}/\mathfrak{a})}\mathfrak{q}(\mathfrak{p});$$

which is the "primary decomposition" of $\mathfrak{a}$*, or more specifically the "reduced primary decomposition" of* $\mathfrak{a}$*, since that decomposition is irredundant.*
(ii) Let $\operatorname{Ass}(\mathbf{A}/\mathfrak{a})=\{\mathfrak{p}_1,...,\mathfrak{p}_r\}$ *and let* $\mathfrak{q}_i=\mathfrak{q}(\mathfrak{p}_i)$ $(1\leq i\leq r)$*. There exists an embedding*

$$\mathbf{A}/\mathfrak{a}\hookrightarrow\bigoplus_{1\leq i\leq r}\mathbf{A}/\mathfrak{q}_i.$$

Proof. (i): See ([31], §IV.2, Theorem 1).
The proof of (ii) is similar to that of Theorem and Definition 530(i). Let

$$f\bullet:\mathbf{A}\rightarrow\prod_{1\leq i\leq r}\mathbf{A}:x\mapsto(x,...,x).$$

Since $\mathfrak{a}\subseteq\bigcap_{1\leq i\leq r}\mathfrak{q}_i$, there exists an $\mathbf{A}$-linear map

$$\bar{f}\bullet:\mathbf{A}/\mathfrak{a}\rightarrow\left(\prod_{1\leq i\leq r}\mathbf{A}\right)/\left(\prod_{1\leq i\leq r}\mathfrak{q}_i\right)$$

induced by f with respect to $\mathfrak{a}$ and $\prod_{1\leq i\leq r}\mathfrak{q}_i$ (Theorem 145). Then, $\ker\bar{f}=\pi\left(f^{-1}\left(\prod_{1\leq i\leq r}\mathfrak{q}_i\right)\right)$ where $\pi:\mathbf{A}\rightarrow\mathbf{A}/\mathfrak{a}$ is the canonical epimorphism, i.e.,

$$\ker\bar{f}=\left(\bigcap_{1\leq\leq r}\mathfrak{q}_i\right)/\mathfrak{a}=0,$$

thus $\bar{f}$ is injective. Last,

$$\left(\prod_{1\leq i\leq r}\mathbf{A}\right)/\left(\prod_{1\leq i\leq r}\mathfrak{q}_i\right)\cong\bigoplus_{1\leq i\leq r}\mathbf{A}/\mathfrak{q}_i$$

by (2.9). ∎

Although the above primary decomposition is reduced, it is in general non-unique ([31], §IV.2, Exercise 24(c)). With additional conditions on the primary ideals, a canonical primary decomposition can be obtained, which is unique ([31], §IV.2, Exercise 4).

Let $\mathbf{A}$ be a principal ideal domain (not necessarily commutative). The following is obtained from Lemma 538 by induction ([77], Sect. 6.4, Theorem 4.5; [78], Sect. 1.5, Theorem 1.5.5):

Lemma and Definition 543. *(1) An element* a *of* $\mathbf{A}$ *is such that the left ideal* $\mathbf{A}\,a$ *has a decomposition*

$$\mathbf{A}\,a = \mathbf{A}\,q_1 \cap ... \cap \mathbf{A}\,q_k \cap \mathbf{A}\,u \tag{3.23}$$

where each q_i is a product of similar bounded atoms, while atoms in different q's are dissimilar and u is totally unbounded. Moreover, the q_i and u are unique up to left associates, (3.23) gives rise to a direct sum decomposition

$$\mathbf{A}/\mathbf{A}\,a \cong \mathbf{A}/\mathbf{A}\,q_1 \oplus ... \oplus \mathbf{A}/\mathbf{A}\,q_k \oplus \mathbf{A}/\mathbf{A}\,u, \tag{3.24}$$

and $k = 0$ (i.e., $\mathbf{A}/\mathbf{A}\,a \cong \mathbf{A}/\mathbf{A}\,u$) if $\mathbf{A}$ is simple.
(2) As in Lemma and Definition 541, the ideals $\mathbf{A}\,q_i$ $(1 \leq i \leq k)$ are called primary. *As is Theorem 542, (3.23) (resp., (3.24)) is called the* primary decomposition *of $\mathbf{A}\,a$ (resp., $\mathbf{A}/\mathbf{A}\,a$).*

Remark 544. *(i) If $\mathbf{A}$ is any simple ring (not necessarily a principal ideal domain), the statement of the above theorem is still valid by Lemma 540.*
(ii) If $\mathbf{A}$ is commutative, then u is a unit by Proposition 536(iii) and for each $i \in \{1, ..., k\}$, their exists a prime p_i and an integer $n_i \geq 1$ such that $\mathbf{A}\,q_i = \mathbf{A}\,p_i^{n_i}$ (Exercise 485). Thus $\mathbf{A}\,q_i$ is a primary ideal and (3.23) (resp., (3.24)) is the primary decomposition of $\mathbf{A}\,a$ (resp., $\mathbf{A}/\mathbf{A}\,a$) given by the Lasker-Noether theorem. In addition, each q_i $(1 \leq i \leq k)$ is left- and right-indecomposable in the sense of Definition 530.
(iii) When $\mathbf{A}$ is noncommutative, neither u nor the q_i are in general indecomposable in the sense of Definition 530. Thus we will be led to study irredundant decompositions.

Remark 544(ii) can be generalized as follows ([78], Proposition 1.5.6):

Corollary 545. *Let $\mathbf{A}$ be a principal ideal domain. Then a bounded indecomposable element is a product of similar atoms and has a bound of the form p^n, where p is an I-atom* (**1.5.6.3**). *Two bounded indecomposable elements have the same bound if, and only if they are similar.*

Proof. This is a consequence of Lemma and Definition 543 when $\mathbf{A}/\mathbf{A}\,u = 0$. ∎

3.3.2.4 Irredundant Decomposition and Complete Direct Decomposition

Let $\mathbf{A}$ be a principal ideal domain and $c \in \mathbf{A}^{\times}$.

(1) Since $\mathbf{L}\,(c\,\mathbf{A}, \mathbf{A})$ is a modular lattice of finite length, one may consider by the Kurosh-Ore Theorem (Theorem 126) an irredundant decomposition of $c\,\mathbf{A}$ (Definition 122)

$$c\,\mathbf{A} = a_1\,\mathbf{A} \cap ... \cap a_r\,\mathbf{A} \tag{3.25}$$

where each a_i is indecomposable (Definition 530). Considering a second irredundant decomposition of $c\,\mathbf{A}$

$$c\,\mathbf{A} = b_1\,\mathbf{A} \cap ... \cap b_s\,\mathbf{A} \tag{3.26}$$

where each b_i is indecomposable, then (by the Kurosh-Ore theorem) $r = s$ and there exists a permutation $i \mapsto i'$ of $\{1, ...r\}$ such that one can substitute in the decomposition (3.25) any a_i by $b_{i'}$.

(2) Consider a decomposition such as (3.25) where for each index i,

$$a_i\,\mathbf{A} \neq \mathbf{A}, \quad a_i\,\mathbf{A} + \left(\bigcap_{j \neq i} a_j\,\mathbf{A}\right) = \mathbf{A}.$$

Such a decomposition is called a *complete direct decomposition* of $c\,\mathbf{A}$. By Theorem and Definition 530, the complete direct decomposition (3.25) of $c\,\mathbf{A}$ corresponds to the direct sum representation

$$\mathbf{A}/c\,\mathbf{A} \cong \mathbf{A}/a_1\,\mathbf{A} \oplus ... \oplus \mathbf{A}/a_r\,\mathbf{A}.$$

Consider another complete direct decompositions (3.26) of $c\,\mathbf{A}$; by the Krull-Schmidt theorem (Theorem 125), $r = s$ and there exists a permutation $i \mapsto i'$ of $\{1, ...r\}$ such that a_i and $b_{i'}$ are similar (since projectivity corresponds to similarity by Corollary 245). Any complete direct decomposition can be refined to an irredundant decomposition but an irredundant decomposition is not a complete direct decomposition in general.

Example 546. *(Example 127 cont'd). The following decompositions of* $180\mathbb{Z}$ *are irredundant and complete direct:* $180\mathbb{Z} = \pm 4\mathbb{Z} \pm 9\mathbb{Z} \pm 5\mathbb{Z}$. *They correspond to the direct sum decompositions*

$$\mathbb{Z}/180\mathbb{Z} \cong \mathbb{Z}/\left(\pm 4\mathbb{Z}\right) \oplus \mathbb{Z}/\left(\pm 9\mathbb{Z}\right) \oplus \mathbb{Z}/\left(\pm 5\mathbb{Z}\right).$$

3.4 Functors in Categories of Modules

All functors below are additive functors between categories of modules. Left exact, right exact and exact functors have already been defined in general categories (**1.2.3.6**) and in preabelian ones (Theorem 42). Recall that the category ${}_{\mathbf{A}}\mathbf{Mod}$ is abelian.

3.4.1 $\bigotimes$ *and* Hom

3.4.1.1 $A \bigotimes_{\mathbf{S}} -$ and $\mathbf{Hom}_{\mathbf{A}}(A, \bullet)$

Consider an $(\mathbf{A}, \mathbf{S})$-bimodule A; the functor $\mathfrak{F} = A \bigotimes_{\mathbf{S}} - : {}_{\mathbf{S}}\mathbf{Mod} \to {}_{\mathbf{A}}\mathbf{Mod}$ is defined as follows: for any left $\mathbf{S}$-module B, $\mathfrak{F}(B)$ is the left $\mathbf{A}$-module $A \bigotimes_{\mathbf{S}} B$ (Theorem 283); for any homomorphism of left $\mathbf{S}$-module $\bullet\beta : B \to B'$, $\mathfrak{F}(\beta)$ is the left $\mathbf{A}$-module homomorphism $\mathbf{1} \otimes \beta$ where $\mathbf{1} = Id_A$ (Lemma and Definition 282). As easily seen, $\mathfrak{F}(\beta \circ \beta') = \mathfrak{F}(\beta) \circ \mathfrak{F}(\beta')$, thus the functor $A \bigotimes_{\mathbf{S}} -$ is covariant.

The functor $\mathfrak{G} = \mathrm{Hom}_{\mathbf{A}}(A, \bullet) : {}_{\mathbf{A}}\mathbf{Mod} \to {}_{\mathbf{S}}\mathbf{Mod}$ is the covariant functor defined in Theorem and Definition 43 (see also Theorem 236). For any left $\mathbf{A}$-module C, $\mathfrak{G}(C)$ is the left $\mathbf{S}$-module $\mathrm{Hom}_{\mathbf{A}}(A, C)$ and for any homomorphism of left $\mathbf{A}$-module $\bullet\gamma : C \to C'$, $\mathfrak{G}(\gamma)$ is the right-composition by γ, as depicted by the commutative diagram below:

$$\begin{array}{ccc} A & \xrightarrow{\bullet f} & C \\ \bullet\mathfrak{G}(\gamma)(f) & \searrow & \downarrow \bullet\gamma \\ & & C' \end{array} \tag{3.27}$$

By the Adjoint Isomorphism Theorem (Theorem 284(i)),

$$\boxed{\left(A \bigotimes_{\mathbf{S}} -, \mathrm{Hom}_{\mathbf{A}}(A, \bullet)\right) \text{ is an adjoint pair.}}$$

Therefore, the following is a consequence of Theorem 30:

Proposition 547. *(1) The functor $A \bigotimes_{\mathbf{S}} -$ has the following properties:*
(i) $A \bigotimes_{\mathbf{S}} -$ is right exact and commutes with direct limits:

$$A \bigotimes_{\mathbf{S}} \varinjlim B_i \cong \varinjlim A \bigotimes_{\mathbf{S}} B_i;$$

(ii) in particular,

$$A \bigotimes_{\mathbf{S}} \left(\coprod_i B_i\right) \cong \coprod_i \left(A \bigotimes_{\mathbf{S}} B_i\right)$$

(Example 18).
(2) The functor $\mathrm{Hom}_{\mathbf{A}}(A, \bullet)$ has the following properties:
(i') $\mathrm{Hom}_{\mathbf{A}}(A, \bullet)$ is left exact and commutes with inverse limits:

$$\mathrm{Hom}_{\mathbf{A}}\left(A, \left(\varprojlim C_i\right)\right) \cong \varprojlim \left(\mathrm{Hom}_{\mathbf{A}}(A, C_i)\right)$$

and if (C_i) is a family of $(\mathbf{A}, \mathbf{T})$-bimodules, this is an isomorphism of $(\mathbf{S}, \mathbf{T})$-bimodules;
(ii') in particular,

$$\mathrm{Hom}_{\mathbf{A}}\left(A, \left(\prod_i C_i\right)\right) \cong \prod_i \left(\mathrm{Hom}_{\mathbf{A}}(A, C_i)\right)$$

(Example 24).

Remark 548. *(i) $A \bigotimes_{\mathbf{S}} -$ is not left exact (Exercise 680).*
(ii) $A \bigotimes_{\mathbf{S}} \prod_i B_i \ncong \prod_i A \bigotimes_{\mathbf{S}} B_i$ in general ([306], Exercise 2.25).
(iii) $\mathrm{Hom}_{\mathbf{A}}(A, \coprod_i C_i) \ncong \coprod_i \mathrm{Hom}_{\mathbf{A}}(A, C_i)$ and $\mathrm{Hom}_{\mathbf{A}}(A, \coprod_i C_i) \ncong \prod_i \mathrm{Hom}_{\mathbf{A}}(A, C_i)$ in general ([306], Exercises 2.18 & 2.19).

3.4.1.2 $-\bigotimes_{\mathbf{S}} B$ and $\mathbf{Hom}_{\mathbf{T}}(B, \bullet)$

Assuming that B is an $(\mathbf{S}, \mathbf{T})$-bimodule, the functor $\mathfrak{F}' = -\bigotimes_{\mathbf{S}} B : \mathbf{Mod}_{\mathbf{S}} \rightarrow \mathbf{Mod}_{\mathbf{T}}$ is defined as follows: for any right $\mathbf{S}$-module A, $\mathfrak{F}'(A)$ is the right $\mathbf{T}$-module $A \bigotimes_{\mathbf{S}} B$ (Theorem 283); for any homomorphism of right $\mathbf{S}$-modules $\alpha : A \rightarrow A'$, $\mathfrak{F}'(\alpha)$ is the homomorphism of right $\mathbf{T}$-modules $\alpha \otimes \mathbf{1}$ where $\mathbf{1} = Id_B$ (Lemma and Definition 282). This functor is covariant.

Therefore, two functors can be defined from $\bigotimes_{\mathbf{S}}$: $A \bigotimes_{\mathbf{S}} -$ (**3.4.1.1**) and $-\bigotimes_{\mathbf{S}} B$ that we are now considering; $\bigotimes_{\mathbf{S}}$ is a *bifunctor* (**1.2.3.1**).

The functor $\mathfrak{G}' = \mathrm{Hom}_{\mathbf{T}}(B, \bullet) : \mathbf{Mod}_{\mathbf{T}} \rightarrow \mathbf{Mod}_{\mathbf{S}}$ is defined as follows: for any right $\mathbf{T}$-module C, $\mathfrak{G}'(C)$ is the right $\mathbf{S}$-module $\mathrm{Hom}_{\mathbf{T}}(B, C)$ (Theorem 236(ii)); for any homomorphism of right $\mathbf{T}$-modules $\gamma\bullet : C \rightarrow C'$, $\mathfrak{G}'(\gamma)\bullet$ is the left-composition by γ. This functor is covariant.

By the Adjoint Isomorphism Theorem (Theorem 284(ii)),

$$\boxed{(-\bigotimes_{\mathbf{S}} B, \mathrm{Hom}_{\mathbf{T}}(B, \bullet)) \text{ is an adjoint pair.}}$$

The above has consequences similar to those in Proposition 547 and are not repeated here.

3.4.1.3 $\mathbf{Hom}_{\mathbf{A}}(\bullet, C)$

Let C be a left $\mathbf{A}$-module and consider the functor $\bar{\mathfrak{G}} = \mathrm{Hom}_{\mathbf{A}}(\bullet, C)$ defined as follows: for any left $\mathbf{A}$-module A, $\bar{\mathfrak{G}}(A)$ is the abelian group $\mathrm{Hom}_{\mathbf{A}}(A, C)$; for any for any homomorphism of left $\mathbf{A}$-module $\bullet\alpha : A' \rightarrow A$, $\bar{\mathfrak{G}}(\alpha)$ is the left-composition by α, as depicted by the commutative diagram below (obtained from (3.27) by reversing the arrows).

$$\begin{array}{ccc} C & \overset{\bullet g}{\longleftarrow} & A \\ \bullet\bar{\mathfrak{G}}(\alpha)(g) & \nwarrow & \uparrow \bullet\alpha \\ & & A' \end{array}$$

Since the functor $\mathfrak{G}$ is covariant, the functor $\bar{\mathfrak{G}}$ is contravariant (Theorem and Definition 43 and Exercise 200). To detail this point, consider the commutative diagram below:

$$\begin{array}{ccccc} & & C & & \\ & \overset{\varphi}{\nearrow} & \uparrow & \overset{g}{\nwarrow} & \\ A & \overset{\alpha}{\longrightarrow} & A' & \overset{\alpha'}{\longrightarrow} & A" \end{array}$$

The vertical arrow is $\varphi' = \alpha' \circ g = \bar{\mathfrak{G}}(\alpha')(g)$, therefore

$$\varphi = \alpha \circ \varphi' = \bar{\mathfrak{G}}(\alpha)(\varphi') = \bar{\mathfrak{G}}(\alpha) \circ \bar{\mathfrak{G}}(\alpha')(g).$$

In addition, $\varphi = (\alpha \circ \alpha') \circ g = \bar{\mathfrak{G}}(\alpha \circ \alpha')(g)$, hence

$$\bar{\mathfrak{G}}(\alpha \circ \alpha') = \bar{\mathfrak{G}}(\alpha) \circ \bar{\mathfrak{G}}(\alpha')$$

which proves that the functor $\bar{\mathfrak{G}} = \mathrm{Hom}_{\mathbf{A}}(A, \bullet)$ is contravariant since it transforms *left morphisms* $\bullet\alpha$ into *right morphisms* $\bar{\mathfrak{G}}(\alpha)\bullet$ (Remark 10).

Let $\mathcal{C} = {}_{\mathbf{A}}\mathbf{Mod}$. From the above, the contravariant functor $\bar{\mathfrak{G}} = \mathrm{Hom}_{\mathbf{A}}(\bullet, C) : \mathcal{C} \to \mathbf{Ab}$ uniquely defines a covariant functor $\mathfrak{G} = \mathrm{Hom}_{\mathbf{A}}(C, \bullet) : \mathcal{C}^{\mathrm{op}} \to \mathbf{Ab}$, which has been already studied (**3.4.1.1**). We know that $\mathfrak{G}$ is left exact, therefore so is $\bar{\mathfrak{G}}$. Let us summarize the obtained results:

Proposition 549. *The functor* $\mathrm{Hom}_{\mathbf{A}}(\bullet, C)$ *has the following properties:*

(i)

$$\mathrm{Hom}_{\mathbf{A}}\left(\varinjlim A_i,\, C\right) \cong \varprojlim \mathrm{Hom}_{\mathbf{A}}(A_i, C),$$

and if (A_i) *is a family of* $(\mathbf{A}, \mathbf{S})$*-bimodules and* C *is an* $(\mathbf{A}, \mathbf{T})$*-bimodule, this is an isomorphism of* $(\mathbf{S}, \mathbf{T})$*-bimodules;*
(ii) in particular,

$$\mathrm{Hom}_{\mathbf{A}}\left(\left(\coprod_i A_i\right), C\right) \cong \prod_i \left(\mathrm{Hom}_{\mathbf{A}}(A_i, C)\right);$$

(iii) $\mathrm{Hom}_{\mathbf{A}}(\bullet, C)$ *is contravariant left exact from* ${}_{\mathbf{A}}\mathbf{Mod}$ *to* $\mathbf{Mod}_{\mathbf{E}}$ *where* $\mathbf{E} = \mathrm{End}_{\mathbf{A}}(C)$*, the endomorphism ring of* C*.*

Proof. Due to the importance of (iii) for the sequel, let us give a direct proof of this result:

(1) Consider the exact sequence (1.9) and let us show that the sequence (1.11), where $\mathfrak{F} = \mathrm{Hom}_{\mathbf{A}}(\bullet, C)$, is exact. First recall that for every left morphism $\bullet f : X \to Y$, $\mathfrak{F}(f)$ is a right morphism.

(i) $\mathfrak{F}(\bullet f_2)$ is a monomorphism: if $\bullet\varepsilon : X_3 \to C$ is such that $(\mathfrak{F}(f_2))(\varepsilon) = 0$, one has $(y)(f_2 \circ \varepsilon) = 0$ for any $y \in X_2$; as f_2 is an epimorphism, this implies $\varepsilon = 0$.

(ii) $\mathrm{im}\,\mathfrak{F}(\bullet f_2) \subseteq \ker \mathfrak{F}(\bullet f_1)$: let $\bullet\gamma : X_2 \to C$ be such that $\gamma \in \mathrm{im}\,\mathfrak{F}(f_2)$. There exists $\varepsilon : X_3 \to C$ such that $\gamma = (\mathfrak{F}(f_2))(\varepsilon) = f_2 \circ \varepsilon$; therefore, $(\mathfrak{F}(f_1))(\gamma) = f_1 \circ f_2 \circ \varepsilon = 0$ since $f_1 \circ f_2 = 0$.

(iii) $\ker \mathfrak{F}(\bullet f_1) \subseteq \mathrm{im}\,\mathfrak{F}(\bullet f_2)$: let $\bullet\gamma : X_2 \to C$ be such that $\gamma \in \ker \mathfrak{F}(f_1)$, i.e., $(\mathfrak{F}(f_1))(\gamma) = f_1 \circ \gamma = 0$. Let $y \in \ker f_2$; as $\ker f_2 = \mathrm{im}\, f_1$, there exists

$x \in X_1$ such that $y = (x) f_1$, thus $(y) \gamma = (x) (f_1 \circ \gamma) = 0$, hence $\ker f_2 \subseteq \ker \gamma$. Let us identify X_3 with $X_2/\ker f_2$ and f_2 with the canonical epimorphism. Let $\bar{\gamma} : X_2/\ker f_2 \to C$ be the induced homomorphism with respect to $\ker f_2$ and 0 (Theorem 145). One obtains $\gamma = (\mathfrak{F}(f_2))(\bar{\gamma}) \in \operatorname{im} \mathfrak{F}(f_2)$. This proves that the covariant functor $\operatorname{Hom}_{\mathbf{A}}(\bullet, C) : {}_{\mathbf{A}}\mathbf{Mod} \to \mathbf{Ab}$ is left exact.

(2) We know that C is an $(\mathbf{A}, \mathbf{E})$-bimodule (Example 230(ii)), thus for any left $\mathbf{A}$-module M, $\operatorname{Hom}_{\mathbf{A}}(M, C)$ has a canonical structure of $(\mathbb{Z}, \mathbf{E})$-bimodule (Theorem 236(i)), i.e., of right $\mathbf{E}$-module. Therefore, $\operatorname{Hom}_{\mathbf{A}}(\bullet, C)$ is a functor ${}_{\mathbf{A}}\mathbf{Mod} \to \mathbf{Mod}_{\mathbf{E}}$. ∎

Remark 550. *If* $\mathbf{A}$ *is commutative, then for any* $\mathbf{A}$*-modules* A *and* C, $\operatorname{Hom}_{\mathbf{A}}(A, C)$ *is an* $\mathbf{A}$*-module by Example 230(iii).*

3.4.2 Restriction and Extension of the Ring of Scalars

3.4.2.1 Change-of-Rings Functor

Let $\mathbf{A}$ and $\mathbf{S}$ be two rings and let $\rho : \mathbf{A} \to \mathbf{S}$ be a ring-homomorphism. The *change-of-ring functor* is the functor $\rho_* : {}_{\mathbf{S}}\mathbf{Mod} \to {}_{\mathbf{A}}\mathbf{Mod}$ defined as follows: every left $\mathbf{S}$-module M is endowed with a structure of left $\mathbf{A}$-module by setting

$$r \,.\, m = \rho(r)\, m \quad \text{for any } m \in M \text{ and } r \in \mathbf{A};$$

and then every $\mathbf{S}$-linear map $\bullet f : M \to N$ becomes an $\mathbf{A}$-linear map since

$$(r \,.\, m) f = (\rho(r)\, m) f = \rho(r)\, (m) f = r \,.\, (m) f.$$

The functor ρ_* is obviously covariant and the following holds:

Theorem 551. *(i) The functor* ρ_* *is faithful exact and every generating family of* $\rho_*(M)$ $(M \in {}_{\mathbf{S}}\mathbf{Mod})$ *is a generating family of* M.
(ii) If $(M_i)_{i \in I}$ *is a family of left* $\mathbf{S}$*-modules, then*

$$\rho_*\left(\prod_i M_i\right) = \prod_i \rho_*(M_i), \quad \rho_*\left(\bigoplus_i M_i\right) = \bigoplus_i \rho_*(M_i).$$

(iii) If ρ *is surjective, then the functor* ρ_* *is full and every generating family of* M $(M \in {}_{\mathbf{S}}\mathbf{Mod})$ *is a generating family of* $\rho_*(M)$.
(iv) If ρ *is injective, then every free family in* M $(M \in {}_{\mathbf{S}}\mathbf{Mod})$ *is free in* $\rho_*(M)$.

Proof. (i): For any $\mathbf{S}$-linear map $f : M \to N$, f and $\rho_*(f)$ are equal *as functions*, thus the arrow map of ρ_* is injective and ρ_* is faithful. In addition, $\rho_*(f)$ has the same kernel and image as does f, thus ρ_* is exact.

(ii)-(iv) are clear (see [27], §II.1, Prop. 24). ■

In particular, if $\mathbf{A} \subseteq \mathbf{S}$ and ρ is the inclusion, ρ_* is the *restriction of the ring of scalars* (2.2.2.1).

3.4.2.2 Extension of the Ring of Scalars

Let $\rho : \mathbf{A} \to \mathbf{S}$ be a ring-homomorphism and $\rho_* : \mathbf{Mod_S} \to \mathbf{Mod_A}$ the change-of-ring functor. Then, setting

$$\boxed{s\,\rho(r) = s\,.\,r \text{ for any } s \in \mathbf{S} \text{ and } r \in \mathbf{A}} \tag{3.28}$$

$\rho_*(_{\mathbf{S}}\mathbf{S}_{\mathbf{S}})$ is an $(\mathbf{S}, \mathbf{A})$-bimodule with the following compatibility condition (**2.2.2.2**) holding true:

$$s'\,(s\,.\,r) = s'\,(s\,\rho(r)) = (s'\,s)\,\rho(r) = (s'\,s)\,.\,r$$

for any $s, s' \in \mathbf{S}$, $r \in \mathbf{A}$. Let M be a left $\mathbf{A}$-module; then $\rho_*(_{\mathbf{S}}\mathbf{S}_{\mathbf{S}}) \bigotimes_{\mathbf{A}} M$ has a canonical structure of left $\mathbf{S}$-module (Theorem 283) and is denoted by $\rho^*(M) = {}_{(\mathbf{S})}M$. The definition below is a generalization of Lemma and Definition 285(2):

Definition 552. *The left* $\mathbf{S}$*-module* $\rho^*(M) = {}_{(\mathbf{S})}M$ *is called the induced* extension *of* M *along* ρ.

Item (1) below is a consequence of Theorem 284(i); its proof is detailed in ([27], §II.5, Prop. 1):

Theorem and Definition 553. *(1) Let* M *be a left* $\mathbf{A}$*-module; the left map* $\bullet\varphi_M : M \to \rho_*(\rho^*(M))$ *given by* $x \mapsto 1 \otimes x$ *is* $\mathbf{A}$*-linear and the set* $(M)\,\varphi_M$ *is a generator of the* $\mathbf{S}$*-module* $\rho^*(M)$. *In addition, for any left* $\mathbf{S}$*-module* N *and any* $\mathbf{A}$*-linear map* $\bullet f : M \to \rho_*(M)$, *there exists one* $\mathbf{S}$*-linear map* $\bullet\bar{f} : \rho^*(M) \to N$ *such that* $(1 \otimes x)\,\bar{f} = xf$ *for any* $x \in M$.
(2) The $\mathbf{A}$*-linear map* $\varphi_M : M \to \rho_*(\rho^*(M))$ *is called the* canonical map *(see Lemma and Definition 285(2)).*

By (3.28), for any $x \in M$ and $r \in \mathbf{A}$,

$$(r\,x)\,\varphi_M = 1 \otimes (r\,x) = (1\,.\,r) \otimes x = \rho(r)\,(1 \otimes x) = \rho(r)\;(x)\,\varphi_M = r\,.\;(x)\,\varphi_M.$$

Corollary 554. *Let M, M' be left* $\mathbf{A}$*-modules. For any* $\mathbf{A}$*-linear map* $\bullet u : M \to M'$, $v = 1_S \otimes u$ *is the unique* $\mathbf{S}$*-linear map making the following diagram commute:*

$$\begin{array}{ccc} M & \xrightarrow{\varphi_M} & \rho^*(M) \\ \downarrow u & & \downarrow v \\ M' & \xrightarrow{\varphi_{M'}} & \rho^*(M') \end{array}$$

Proof. Apply Theorem and Definition 553 to the $\mathbf{A}$-linear map $u \circ \varphi_{M'} : M \to \rho^*(M')$. ■

Let us denote by $\rho^*(u)$ the above map v. We see that ρ^* is a functor ${}_{\mathbf{A}}\mathbf{Mod} \to {}_{\mathbf{S}}\mathbf{Mod}$, called the "extension of the ring of scalars induced along ρ" . By Proposition 547, we have the following.

Proposition 555. *The functor ρ^* is covariant and*
(i) ρ^* is right exact and commutes with direct limits: $\rho^*\left(\varinjlim M_i\right) \cong \varinjlim \rho^*(M_i)$;
(ii) in particular, $\rho^*\left(\coprod_i M_i\right) \cong \coprod_i \rho^*(M_i)$.

Corollary 556. *Let M be a free left* $\mathbf{A}$*-module with basis* $(a_i)_{i\in I}$*. Then:*

(i) $\rho^*(M)$ *is a free left* $\mathbf{S}$*-module with basis* $((a_i)\varphi_M)_{i\in I}$ *(where φ_M is the canonical map);*
(ii) if ρ is a monomorphism, so is φ_M.

Proof. (i): We have $M = \coprod_i \mathbf{A}\, a_i$, thus by Proposition 555(ii), $\rho^*(M) \cong \coprod_i \rho^*(\mathbf{A}\, a_i)$ and $\rho^*(\mathbf{A}\, a_i) = \mathbf{S}\,(a_i)\,\varphi_M$.
(ii): Let $(r_i)_{i\in I}$ be a family of elements of $\mathbf{A}$ with finite support; then

$$\left(\sum_{i\in I} r_i\, a_i\right)\varphi_M = \sum_{i\in I} \rho(r_i)\ a_i.$$

Let $\left(\sum_{i\in I} r_i\, a_i\right)\varphi_M = 0$; then $\rho(r_i) = 0$ for all i, thus $r_i = 0$ if ρ is a monomorphism. ■

3.4.2.3 Relation with the Restriction of the Ring of Scalars

The reader should not believe that ρ_* is the inverse of ρ^*. Rather,

Proposition 557. *(ρ_*, ρ^*) is an adjoint pair of functors.*

Proof. Consider two modules ${}_{\mathbf{A}}M$, ${}_{\mathbf{S}}N$. Then by the Adjoint Isomorphism Theorem (Theorem 284(i)),

$$\mathrm{Hom}_{\mathbf{S}}(\rho^*(M), N) \cong \mathrm{Hom}_{\mathbf{A}}(M, \rho_*(N)).$$ ■

Since the functor ρ_* is exact, it is not surprising–according to Freyd's Adjoint Functor Theorem (Remark 31)–that it has both a right and a left adjoint: see Exercise 686.

Let N be a left $\mathbf{S}$-module, let $M = \rho_*(N)$ and let $f = Id_M$. By Theorem and Definition 553 there exists one $\mathbf{S}$-linear map $\bar{f} : \rho^*(\rho_*(N)) \to N$ such that $(1 \otimes x)\bar{f} = x$ for any $x \in N$ (since the *sets* N and M coincide). Setting $\bar{f} = \psi_N$, this yields $(1 \otimes x)\psi_N = x$ for any $x \in N$; therefore, $(s \otimes x)\psi_N = s\,x$ for any $s \in \mathbf{S}$ (ψ_N, which is $\mathbf{S}$-linear, should not confused with φ_M which is $\mathbf{A}$-linear). The following is proved in ([27], §II.5, Prop. 5).

Proposition 558. *Let M be a left $\mathbf{A}$-module and N a left $\mathbf{S}$-module. The composed maps*

$$\rho^*(M) \overset{\rho^*(\varphi_M)}{\longrightarrow} \rho^*(\rho_*(\rho^*(M))) \overset{\psi_{\rho^*(M)}}{\longrightarrow} \rho^*(M) \tag{3.29}$$

$$\rho_*(N) \overset{\varphi_{\rho^*(N)}}{\longrightarrow} \rho_*(\rho^*(\rho_*(N))) \overset{\rho_*(\psi_N)}{\longrightarrow} \rho_*(N) \tag{3.30}$$

respectively coincide with $Id_{\rho^*(M)}$ and $Id_{\rho_*(N)}$.

Corollary 559. *The maps $\rho^*(\varphi_M) : \rho^*(M) \to \rho^*(\rho_*(\rho^*(M)))$ and $\varphi_{\rho^*(N)} : \rho_*(N) \to \rho_*(\rho^*(\rho_*(N)))$ are respectively an $\mathbf{S}$-linear monomorphism and an $\mathbf{A}$-linear monomorphism; they identify $\rho^*(M)$ with a direct summand of $\rho^*(\rho_*(\rho^*(M)))$ and $\rho_*(N)$ with a direct summand of $\rho_*(\rho^*(\rho_*(N)))$, respectively.*

Proof. This is an obvious consequence of Lemma 571 below. ∎

Remark 560. *Let M be a left $\mathbf{A}$-module and let $\rho : \mathbf{A} \to \mathbf{S}$ a ring-homomorphism.*

(i) In general, the canonical map $\varphi_M : M \to \rho_*(\rho^*(M))$ is not an embedding. For example, let $\mathbf{A}$ be a left Ore domain, let $M \neq 0$ be a torsion $\mathbf{A}$-module and let $\rho : \mathbf{A} \to \mathrm{Q}(\mathbf{A})$; then $\rho^*(M) = 0$ (Theorem 339), thus $\rho_*(\rho^*(M)) = 0$ and $M \nsubseteq \rho_*(\rho^*(M))$.
(ii) φ_M is an embedding in the following cases: (a) there exists a left $\mathbf{S}$-module N such that $M = \rho_*(N)$ (Corollary 559); (b) ρ is a monomorphism and M is torsionless (see Exercise 685).
(iii) Let $M \cong \bigoplus_{i \in I} M_i$. By Proposition 555(ii) and Theorem 551(ii),

$$\rho_*(\rho^*(M)) \cong \bigoplus_{i \in I} \rho_*(\rho^*(M_i)).$$

3.4.3 Relation between Kernels and Cokernels

3.4.3.1 Kernels

Let $\mathbf{A}$ be a ring, let $R \in {}^{Q}\mathbf{A}^{(K)}$ be a matrix with entries in $\mathbf{A}$ (where Q and K are sets of indices) and let W be a left $\mathbf{A}$-module. Let $R\bullet$ denote the left-multiplication by R and consider the kernel

$$\ker_W(R\bullet) \triangleq \left\{\mathbf{w} \in {}^{K}W : R\,\mathbf{w} = 0\right\}. \tag{3.31}$$

Example 561. *To clarify ideas, consider the case when* $\mathbf{A} = A_1(\mathbb{C})$, *the first Weyl algebra over* $\mathbb{C}$ (**2.7.4.3**) *and* $K = \{1, ..., k\}$*; then* $W = \mathcal{E}(\mathbb{R})$ *– as defined in* (**1.7.3.5**) *– is an* $\mathbf{A}$*-module,* $R\,w = 0$ *is a system of* k *differential equations, and the kernel* $\ker_W(R\bullet)$ *is the set of all columns* $\mathbf{w} \in {}^{k}W$ *satisfying this system of differential equations.*

Consider the covariant functor $\mathfrak{C}_K : {}_{\mathbf{A}}\mathbf{Mod} \rightarrow \mathbf{Ens}$ defined as follows: (a) for any $W_1 \in {}_{\mathbf{A}}\mathbf{Mod}$, $\mathfrak{C}_K(W_1)$ is the set ${}^{K}W_1 = \prod_{i \in K} W_1$; (b) for any $\mathbf{A}$-linear homomorphism $\bullet g : W_1 \rightarrow W_2$, $\bullet\mathfrak{C}_K(g) : {}^{K}W_1 \rightarrow {}^{K}W_2$ is the function which to the column $\mathbf{w} \in {}^{K}W_1$ associates the column $\mathbf{w}\mathfrak{C}_K(g) \in {}^{K}W_2$ with entries $\mathbf{w}_i g$ $(i \in K)$.

Lemma 562. *There exists a functor*

$$\mathfrak{K}_R : W \rightarrow \ker_W(R\bullet)$$

which is a subfunctor of $\mathfrak{C}_K$ (**1.2.3.6**).

Proof. We have $\mathfrak{K}_R(W) \subseteq \mathfrak{C}_K(W)$ for any $W \in {}_{\mathbf{A}}\mathbf{Mod}$; in addition, for any $\mathbf{A}$-linear left homomorphism $\bullet g : W_1 \rightarrow W_2$ and any $\mathbf{w} \in {}^{K}W_1$ such that $R\,\mathbf{w} = 0$, we have $R\,\mathbf{w}\mathfrak{C}_K(g) = 0$. ∎

The study of the kernel (3.31) can be divided into two parts:

(1) The study of the functor $\mathfrak{K}_R$, independently of a specific embedding $\ker_W(R\bullet) \hookrightarrow {}^{K}W$.

(2) The study of the properties of a specific embedding $\ker_W(R\bullet) \hookrightarrow {}^{K}W$ once the properties of the functor $\mathfrak{K}_R$ are known.

Example 563. *(Example 561 cont'd). Part (1) only depends on the matrix* R*. Part (2) is related to the space in which the solutions are sought: depending on whether they are calculated in the space of functions* $\mathcal{E}(\mathbb{R})$ (**1.7.3.5**)*, or in the space of distributions* $\mathcal{D}'(\mathbb{R})$*, different results can be obtained. All these spaces have an obvious structure of* $A_1(\mathbb{C})$*-module, as already seen (Subsect. 2.7.2), and of* $A_o(\mathbb{C})$*-module as well, where* $A_o(\mathbb{C})$ *is the ring of differential operators in Exercise 492.*

Consider the equation (3.1) with $R = t - a$, where $a \in \mathbb{R}$ and $t - a$ is considered as an element of $A_1(\mathbb{C})$ or of $A_o(\mathbb{C})$.

(a) If $W = \mathcal{E}(\mathbb{R})$, then the set of solutions is $\mathfrak{B} = \ker_W((t-a)\bullet) = \{0\}$.
(b) If $W = \mathcal{D}'(\mathbb{R})$ then the set of solutions is $\mathfrak{B} = \{\lambda\,\delta_a : \lambda \in \mathbb{C}\}$, where δ_a is the Dirac distribution.

3.4.3.2 The Functor $\mathfrak{K}_N$

Regarding the above Part (1), two matrices R_1 and R_2 are considered "equivalent" if the associated functors $\mathfrak{K}_{R_1}$ and $\mathfrak{K}_{R_2}$ are isomorphic (**1.2.3.2**). A functor $\mathfrak{K}_R$ (where $R \in {}^Q\mathbf{A}^{(K)}$) is left unchanged if R is changed into $\begin{bmatrix} R \\ R' \end{bmatrix}$ where the rows of R' are left $\mathbf{A}$-linear combinations of the rows of R. Therefore, $\mathfrak{K}_R$ only depends on the left $\mathbf{A}$-module $N = \left(\mathbf{A}^Q\right) R \triangleleft \mathbf{A}^{(K)}$; this functor $\mathfrak{K}_R$ and this module N are respectively denoted by $\mathfrak{K}_N$ and $\operatorname{im}_{\mathbf{A}}(\bullet R)$ in the sequel; therefore,

$$\boxed{\ker_W(R\bullet) = \mathfrak{K}_N(W)}$$

which means that

$$\boxed{\ker_W(R\bullet) = \left\{\mathbf{w} \in {}^K W : r\,\mathbf{w} = 0, \forall r \in N\right\}} \tag{3.32}$$

where $N = \operatorname{im}_{\mathbf{A}}(\bullet R)$.

For any $\mathbf{w} \in {}^K W$, consider the left map $\bullet t_W^K(\mathbf{w}) : \mathbf{A}^{(K)} \ni r \mapsto r\,\mathbf{w} \in W$; the elements of the endomorphism ring $\mathbf{E}$ of W can be extended to ${}^K W$, acting component-wise, and then we have the following.

Lemma 564. *The right map $t_W^K\bullet : \mathbf{w} \mapsto t_W^K(\mathbf{w})$ is a canonical $\mathbf{E}$-linear isomorphism ${}^K W \xrightarrow{\sim} \operatorname{Hom}_{\mathbf{A}}\left(\mathbf{A}^{(K)}, W\right)$.*

Proof. For any $\mathbf{w} \in {}^K W$, $t_W^K(\mathbf{w}) \in \operatorname{Hom}_{\mathbf{A}}\left(\mathbf{A}^{(K)}, W\right)$ and the map t_W^K is $\mathbf{E}$-linear. Let $(\varepsilon_j)_{j\in K}$ be the canonical basis of $\mathbf{A}^{(K)}$; then $\varepsilon_j\,\mathbf{w} = \mathbf{w}_j$, where $\mathbf{w}_j$ is the entry with index j of the column $\mathbf{w}$. Therefore, the $\mathbf{E}$-linear map t_W^K is a canonical isomorphism. ■

Consider the left $\mathbf{A}$-module $M = \mathbf{A}^{(K)}/N$ where $N = \operatorname{im}_{\mathbf{A}}(\bullet R)$ (i.e., $M = \operatorname{coker}_{\mathbf{A}}(\bullet R)$). Let $\tau_W^{K,N}$ be the restriction of t_W^K to $\ker_W(R\bullet)$.

Theorem 565. *The map*

$$\tau_W^{K,N} : \mathfrak{K}_N(W) \to \operatorname{Hom}_{\mathbf{A}}\left(\mathbf{A}^{(K)}/N, W\right)$$

is an $\mathbf{E}$-linear isomorphism; furthermore,

$$\tau^{K,N} : \mathfrak{K}_N \to \operatorname{Hom}_{\mathbf{A}}\left(\mathbf{A}^{(K)}/N, \bullet\right)$$

is a natural isomorphism.

Proof. By (3.32), $t_W^K(\mathfrak{K}_N(W))$ is the set of all $\mathbf{A}$-linear homomorphisms $g : \mathbf{A}^{(K)} \to W$ which vanish on N. This set is a right $\mathbf{E}$-module and by Noether's first isomorphism theorem (Theorem 52(1)), $t_W^K(\mathfrak{K}_N(W)) \cong \mathrm{Hom}_{\mathbf{A}}\left(\mathbf{A}^{(K)}/N, W\right)$. The naturality of $\tau^{K,N}$ is obvious (**1.2.3.2**). ■

Remark 566. *(1) By Theorem 565, the functors* $\mathfrak{K}_N$ *and* $\mathrm{Hom}_{\mathbf{A}}\left(\mathbf{A}^{(K)}/N, \bullet\right)$ *are identified.*
(2) Once again, recall the following:
(i) $M = \mathrm{coker}_{\mathbf{A}}(\bullet R)$ *is generated by the components* w_i *of* w *such that (3.1) holds, i.e.,* $M = [w]_{\mathbf{A}}$ *(Definition 253(ii)).*
(ii) $\mathfrak{B} = \ker_W(R\bullet)$ *is defined by (3.31).*
We use bold letters *for the elements of* $\mathfrak{B}$ *to distinguish them from the "abstract variable"* w *which generates* M*.* Formally, *w and* $\mathbf{w}$ *satisfy the same equation.*

3.4.3.3 From Cokernels to Kernels

Conversely, let M be a left $\mathbf{A}$-module and let $R \in {}^{Q}\mathbf{A}^{(K)}$ be a matrix of definition of M (Definition 512), i.e., $M = \mathrm{coker}_{\mathbf{A}}(\bullet R) = \mathbf{A}^{(K)}/N$ where $N = \mathrm{im}_{\mathbf{A}}(\bullet R)$. The sequence

$$0 \longleftarrow M \overset{\varphi}{\longleftarrow} \mathbf{A}^{(K)} \overset{\bullet R}{\longleftarrow} \mathbf{A}^{(Q)} \tag{3.33}$$

is exact, where $\varphi : \mathbf{A}^{(K)} \twoheadrightarrow \mathbf{A}^{(K)}/N$ is the canonical epimorphism. Let W be a left $\mathbf{A}$-module; as shown by Lemma 564, there exists a canonical $\mathbf{E}$-linear isomorphism

$$t_W^K : {}^{K}W \overset{\sim}{\longrightarrow} \mathrm{Hom}_{\mathbf{A}}\left(\mathbf{A}^{(K)}, W\right), \quad t_W^K : \mathbf{w} \mapsto t_W^K(\mathbf{w})$$

$$\text{where } t_W^K(\mathbf{w}) : \mathbf{A}^{(K)} \ni r \mapsto r\,\mathbf{w} \in W \quad \text{for any } \mathbf{w} \in {}^{K}W.$$

The contravariant functor $\mathfrak{B}_W = \mathrm{Hom}_{\mathbf{A}}(\bullet, W) : {}_{\mathbf{A}}\mathbf{Mod} \to {}_{\mathbf{E}}\mathbf{Mod}$ is left exact (Proposition 549), thus the exact sequence (3.33) yields the exact sequence

$$0 \longrightarrow \mathfrak{B}_W(M) \overset{\mathfrak{B}_W(\varphi)}{\longrightarrow} \mathfrak{B}_W\left(\mathbf{A}^{(K)}\right) \overset{\mathfrak{B}_W(\bullet R)}{\longrightarrow} \mathfrak{B}_W\left(\mathbf{A}^{(Q)}\right).$$

Let $u_W^K : \mathrm{Hom}_{\mathbf{A}}\left(\mathbf{A}^{(K)}, W\right) \to {}^{K}W$ and $v_W^{K,N} : \mathrm{Hom}_{\mathbf{A}}(M, W) \to \ker_W(R\bullet)$ be the inverse of t_W^K and $\tau_W^{K,N}$, respectively. By Lemma 564 and Theorem 565, u_W^K and $v_W^{K,N}$ are canonical $\mathbf{E}$-linear isomorphisms which determine natural isomorphisms $u^K : \mathrm{Hom}_{\mathbf{A}}\left(\mathbf{A}^{(K)}, \bullet\right) \overset{\sim}{\longrightarrow} (\bullet)^{(K)}$ and $\tau^{K,N}$: $\mathrm{Hom}_{\mathbf{A}}(M, \bullet) \overset{\sim}{\longrightarrow} \mathfrak{K}_N$, and we have the following commutative diagram

$$\begin{array}{ccccccc} 0 \longrightarrow & \mathfrak{B}_W(M) & \overset{\mathfrak{B}_W(\varphi)}{\longrightarrow} & \mathfrak{B}_W\left(\mathbf{A}^{(K)}\right) & \overset{\mathfrak{B}_W(\bullet R)}{\longrightarrow} & \mathfrak{B}_W\left(\mathbf{A}^{(Q)}\right) \\ & \downarrow v_W^{K,N} & & \downarrow u_W^K & & \downarrow u_W^Q \\ 0 \longrightarrow & \ker_W(R\bullet) & \overset{\iota_W^K}{\longrightarrow} & {}^K W & \overset{R\bullet}{\longrightarrow} & {}^Q W \end{array}$$

where both rows are exact sequences, ι_W^K is the canonical injection, and $R\bullet$ is the left multiplication by R (**3.4.1.3**). Therefore, we have the following.

Corollary 567. *(1) Applying the functor* $\mathfrak{B}_W = \mathrm{Hom}_{\mathbf{A}}(\bullet, W)$ *to the exact sequence (3.33) one obtains the exact sequence*

$$0 \longrightarrow \ker_W(R\bullet) \overset{\iota_W^K}{\longrightarrow} {}^K W \overset{R\bullet}{\longrightarrow} {}^Q W$$

up to canonical **E***-linear isomorphisms.*
(2) The two right **E***-modules* $\mathfrak{B}_W(\mathrm{coker}_{\mathbf{A}}(\bullet R)) \triangleq \mathrm{Hom}_{\mathbf{A}}(\mathrm{coker}_{\mathbf{A}}(\bullet R), W)$ *and* $\ker_W(R\bullet)$ *are canonically isomorphic and identified.*
(3) Let M_1 *and* M_2 *be two left* **A***-modules such that there exists an* **A***-linear isomorphism* $\varphi : M_1 \xrightarrow{\sim} M_2$*. Then* $\mathfrak{B}_W(\varphi)$ *is an* **E***-linear isomorphism* $\mathfrak{B}_W(M_2) \xrightarrow{\sim} \mathfrak{B}_W(M_1)$*.*

Proof. Only (3) remains to be proved. Let $\bullet\varphi : M_1 \xrightarrow{\sim} M_2$ be an **A**-linear isomorphism and $\tilde{\varphi} = \mathfrak{B}_W(\varphi) : \mathrm{Hom}_{\mathbf{A}}(M_2, W) \to \mathrm{Hom}_{\mathbf{A}}(M_1, W)$; $\tilde{\varphi}$ is defined by $\tilde{\varphi}(f) = \varphi f$ for any $\bullet f : M_2 \to W$. The map $\tilde{\varphi}\bullet$ is **E**-linear since it is additive and for any $\tilde{\alpha} : W \to W$, $\tilde{\varphi}(f\tilde{\alpha}) = \varphi f \tilde{\alpha} = \tilde{\varphi}(f)\tilde{\alpha}$. In addition, $\tilde{\varphi}\bullet$ is bijective with inverse $g \mapsto \varphi^{-1} g$ (where $\bullet g : M_1 \to W$). ■

3.4.3.4 Generalization

The situation is now that in (**3.2.1.3**) when $\mathcal{C}$ is a preabelian category. Let W be an object of $\mathcal{C}$ and let $\mathbf{E} = \mathrm{End}_{\mathcal{C}}(W)$ (where $\mathrm{End}_{\mathcal{C}}(W) = \mathrm{Hom}_{\mathcal{C}}(W, W)$). Then for any object M of $\mathcal{C}$, the abelian group $\mathrm{Hom}_{\mathcal{C}}(M, W)$ of all left morphisms $\bullet f : M \to W$ is a right **E**-module via composition of morphisms. In addition, the abelian group $\mathrm{Hom}_{\mathcal{C}}(U, W)$ is an $(\mathbf{A}, \mathbf{E})$-bimodule (**2.2.2.2**).

let M be the U-cokernel defined by the exact sequence of left morphisms (3.5). Let $\mathfrak{B}_W$ be the functor $\mathrm{Hom}_{\mathcal{C}}(\bullet, W)$–a notation which is consistent with that in Example 43 and in (**3.4.3.3**). The functor $\mathfrak{B}_W$ is contravariant left exact (Example 43(2)), thus one obtains from (3.5) the exact sequence of right morphisms

$$0 \longrightarrow \mathfrak{B}_W(M) \overset{\mathfrak{B}_W(\varphi)}{\longrightarrow} \mathfrak{B}_W\left(U^{(I)}\right) \overset{\mathfrak{B}_W(\iota)}{\longrightarrow} \mathfrak{B}_W(N). \tag{3.34}$$

Set

$$N^{\perp} \triangleq \left\{\mathbf{w} \in \mathrm{Hom}_{\mathcal{C}}\left(U^{(I)}, W\right) : \iota\mathbf{w} = 0\right\}. \tag{3.35}$$

Item (1) below generalizes Corollary 567(1),(2).

Lemma and Definition 568. *(1) There exists a canonical* $\mathbf{E}$*-linear isomorphism*

$$\sigma : \mathrm{Hom}_{\mathcal{C}}(M, W) \xrightarrow{\sim} N^{\perp} : \hat{\mathbf{w}} \mapsto \mathbf{w}$$

given by $\mathbf{w} = \varphi\hat{\mathbf{w}}$*: see below.*

$$\begin{array}{ccccc} N & \xrightarrow{\iota} & U^{(I)} & \xrightarrow{\varphi} & M & \longrightarrow 0 \\ & & & \searrow^{\mathbf{w}} & \downarrow \hat{\mathbf{w}} & \\ & & & & W & \end{array}$$

(2) The right $\mathbf{E}$*-module* $N^{\perp}$ *is called the* W-kernel *associated with the* U*-cokernel* M*.*

Proof. First note that there exists a canonical isomorphism $\mathrm{Hom}_{\mathcal{C}}\left(U^{(I)}, W\right) \cong {}^{I}\mathrm{Hom}_{\mathcal{C}}(U, W)$ (Exercise 200), thus these two right $\mathbf{E}$-modules are identified. The kernel $\ker \mathfrak{B}_W(\iota)$ is the largest subobject of $\mathfrak{B}_W\left(U^{(I)}\right) = {}^{I}\mathrm{Hom}_{\mathcal{C}}(U, W)$ annihilated by $\mathfrak{B}_W(\iota) : \mathbf{w} \mapsto \iota\mathbf{w}$ (**1.2.4.1**), thus $N^{\perp}$ is a representative of $\ker \mathfrak{B}_W(\iota)$. So is also $\mathfrak{B}_W(M)$ since (3.34) is exact. The isomorphism $\sigma : \hat{\mathbf{w}} \mapsto \varphi\hat{\mathbf{w}}$ is canonical and $\mathbf{E}$-linear. ■

Conversely, let $\mathfrak{B} \subseteq \mathrm{Hom}_{\mathcal{C}}\left(U^{(I)}, W\right)$ and consider the object

$$\mathfrak{B}^{\perp} \triangleq \bigcap_{\mathbf{w} \in \mathfrak{B}} \ker \mathbf{w} = \ker\left((\mathbf{w})_{w \in \mathfrak{B}} : U^{(I)} \longrightarrow W^{\mathfrak{B}}\right) \subseteq U^{(I)}. \tag{3.36}$$

For any U-cokernel M (i.e., for any M such as in (3.5)) and $\mathfrak{B} = N^{\perp}$ defined by (3.35) we define the *canonical (left) morphism*

$$\bullet\phi_M \triangleq (\hat{\mathbf{w}})_{\mathbf{w} \in \mathfrak{B}} : M \longrightarrow W^{\mathfrak{B}}; \tag{3.37}$$

we have by Lemma and Definition 568 $\varphi\phi_M = (\mathbf{w})_{\mathbf{w} \in \mathfrak{B}}$, thus by (3.36)

$$\ker(\bullet\varphi\phi_M) = \mathfrak{B}^{\perp}. \tag{3.38}$$

Lemma 569. *The two maps* $(\bullet)^{\perp} : N \mapsto N^{\perp}$ *and* $\mathfrak{B} \mapsto \mathfrak{B}^{\perp}$ *constitute a Galois connection (for the inclusion).*

Proof. (i) Let $N_1 \subseteq N_2$ and let $\iota_1 : N_1 \hookrightarrow U^{(I)}$ be the inclusion. For any $w \in N_2^{\perp}$, $\iota_1 w = 0$, thus $N_2^{\perp} \subseteq N_1^{\perp}$. Similarly, $\mathfrak{B}_2^{\perp} \subseteq \mathfrak{B}_1^{\perp}$ whenever $\mathfrak{B}_1 \subseteq \mathfrak{B}_2$.

(ii) We have also to prove that

$$N \subseteq N^{\perp\perp}, \quad \mathfrak{B} \subseteq \mathfrak{B}^{\perp\perp}.$$

We have $N \subseteq \ker\varphi \subseteq \ker\varphi\phi_M = N^{\perp\perp}$ and the proof of the second inclusion is similar. ■

Remark 570. *Let* $\mathcal{D}$ *be a full subcategory of* $\mathcal{C}$*.*

(i) For any $W \in \mathcal{C}$*, consider the functor* $\mathfrak{B}_W : \mathcal{D}^{op} \to \mathbf{Ab}$:

$$M \mapsto \mathfrak{B}_W(M), \quad \bullet\varphi \mapsto \mathfrak{B}_W(\varphi) : \bullet f \mapsto \bullet\varphi f$$

where $\bullet\varphi \in \operatorname{Hom}_{\mathcal{D}}(M_1, M_2)$ *and* $\bullet f \in \operatorname{Hom}_{\mathcal{C}}(M_2, W)$. *This functor is covariant, and all functors* $\mathfrak{B}_W$ $(W \in \mathcal{C})$ *are called* representable *([150], section 0.1.1). The notion of representable functor (not fully detailed here) is fundamental is the theory of categories [183].*
(ii) Let $\bullet g : W_1 \to W_2$ *be a morphism of* $\mathcal{C}$. *For any* $M \in \mathcal{D}$ *and any* $\bullet\varphi \in \operatorname{Hom}_{\mathcal{C}}(M, W_1) = \mathfrak{B}_{W_1}(M)$, *we have* $\bullet g\varphi \in \operatorname{Hom}_{\mathcal{C}}(M, W_2) = \mathfrak{B}_{W_2}(M)$. *Let* $\mathfrak{B}_g : \operatorname{Hom}_{\mathcal{C}}(M, W_1) \to \operatorname{Hom}_{\mathcal{C}}(M, W_2)$ *be the map* $\bullet\varphi \mapsto \bullet g\varphi$. *Let* $\operatorname{Hom}(\mathcal{D}, \mathbf{Ab})^{rep}$ *be the category whose objects are the representable functors* $\mathfrak{B}_W$ *and whose morphisms are the* morphisms of functors $\mathfrak{B}_g$ (**1.2.3.2**). *Doing so, we get a canonical covariant functor*

$$\mathfrak{B} : \mathcal{D}^{op} \to \operatorname{Hom}(\mathcal{D}, \mathbf{Ab})^{rep} :$$

$$M \mapsto \operatorname{Hom}_{\mathcal{C}}(M, \bullet) : W \mapsto \operatorname{Hom}_{\mathcal{C}}(M, W), \quad \bullet\varphi \mapsto \mathfrak{B}_{\bullet}(\bullet\varphi) : (\bullet g \mapsto \bullet g\varphi).$$

This functor is an isomorphism ([150], (0.1.1.7), (0.1.1.8)) which makes it possible to identify $\operatorname{Hom}(\mathcal{D}, \mathbf{Ab})^{rep}$ *with* $\mathcal{D}^{op}$. *Roughly speaking, this means that the study of all behaviors* $\mathfrak{B}_W(M)$ *(when* M *varies among all objects of* $\mathcal{C}$*, i.e., all signal spaces) and all relations between them, is equivalent to the study of all objects* $M \in \mathcal{D}^{op}$ *and all relations between them.*
(iii) Let $W \in \mathcal{C}$ *be given. The functor* $\mathfrak{B}_W : \mathcal{D}^{op} \to \mathbf{Ab}$ *is not faithful in general. See Corollary and Definition 822 below.*

3.5 Homological Properties of Certain Modules

3.5.1 Projective Modules

3.5.1.1 Split Exact Sequences

Lemma and Definition 53 can be completed as follows: ([27], §II.1, Prop. 15):

Lemma 571. *Let* $\mathbf{A}$ *be a ring and consider the short exact sequence of* $\mathbf{A}$*-modules*

$$0 \longrightarrow M \xrightarrow{\bullet f} E \xrightarrow{\bullet\varphi} N \longrightarrow 0. \tag{3.39}$$

(1) The following conditions are equivalent:
(i) the short exact sequence (*3.39*) *is split;*
(ii) the submodule $\operatorname{im} f$ *of* E *is a direct summand;*
(iii) there exists a linear retraction $\bullet r : E \to M$ *associated with* $\bullet f$ *(such that* $\bullet fr = Id_M$*);*
(iv) there exists a linear section $\bullet s : N \to E$ *associated with* $\bullet\varphi$ *(such that* $\bullet s\varphi = Id_E$*).*
(2) Then, $f + s : M \oplus N \to E$ *(given by*

$$(m, n)(f + s) = mf + ns,$$

$m \in M, n \in N$*) is an isomorphism.*

We have the following ([27], §II.1, Corol. 1 & 2 of Prop. 15):

Proposition 572. *Let M, N be left* **A***-modules and* $\bullet f : M \rightarrow N$.

(i) f is left-invertible (or a retraction, *i.e., there exists a section $\bullet g : N \rightarrow M$ such that $\bullet gf = Id_M$) if, and only if f is surjective and* $\ker f$ *is a direct summand. Then,* $\operatorname{im} g \oplus \ker f = M$.
(ii) f is right-invertible (or a section, *i.e., there exists a retraction $\bullet h : M \rightarrow N$ such that $\bullet fh = Id_N$) if, and only if f is injective and* $\operatorname{im} f$ *is a direct summand. Then,* $\operatorname{im} f \oplus \ker h = N$.

Proof. (i) For f to be left-invertible, f must be surjective (Corollary 277); (i) is a consequence of Lemma 571 with φ and r changed to f and g, respectively.
(ii) For f to be right-invertible, f must be injective; (ii) is a consequence of Lemma 571 with r changed to h. ∎

Note that a linear *left* map is a retraction (resp., a section) if, and only if it is left- (resp., right-) invertible. A linear *right* map is a retraction (resp., a section) if, and only if it is right- (resp., left-) invertible.

3.5.1.2 Projective Modules and Torsion-Free Modules

A left **A**-module P is called *projective* if it is a projective object in the category ${}_{\mathbf{A}}\mathbf{Mod}$ (Definition 44), and *finite projective* if it is both finitely generated and projective. Theorem 54 holds true as well as the following.

Theorem 573. *The following conditions are equivalent:*

(i) P is projective;
(ii) any short exact sequence

$$0 \longrightarrow M \xrightarrow{f} E \xrightarrow{\varphi} P \longrightarrow 0 \tag{3.40}$$

is split (thus P is isomorphic to a direct summand of E);
(iii) P is a direct summand of a free module.

Proof. (i)⇒(ii): Consider the diagram below

$$\begin{array}{ccccccccc} & & & & & & P & & \\ & & & & & \varepsilon \swarrow & \downarrow \nu & & \\ 0 & \longrightarrow & M & \xrightarrow{f} & E & \xrightarrow{u} & P & \longrightarrow & 0 \end{array}$$

where the vertical arrow is $\nu = Id_P$, the horizontal row is a short exact sequence, $M = \ker u$ and f is the inclusion.

If (i) holds, there exists $\varepsilon : P \rightarrow E$ such that $u \circ \varepsilon = Id_P$ (assuming that all modules and all maps are right ones). Therefore, the exact sequence is split by Lemma and Definition 53.

(ii)⇒(iii): Let P be any module. By Theorem 494, there exists a free module E such that the sequence (3.40) is exact, where φ is the canonical epimorphism and M, f have the above meaning. If (ii) holds, P is isomorphic to a direct summand P' of E by Lemma 571.

(iii)⇒(i) by Theorem 54. ■

Proposition 574. *(i) A projective module P is finitely generated if, and only if it is the direct summand of a finite free module F; then $F = P \oplus Q$ where Q is finitely generated and projective.*
(ii) A free module is projective.
(iii) Every module is isomorphic to a quotient of a projective module.

Proof. (i): See the proof of the implication (ii)⇒(iii) of Theorem 573. (ii) is a consequence of Theorem 573(iii), whereas (iii) is a consequence of (ii) and of Theorem 494. ■

It is easy to find examples showing that the converse of Statement (ii) above does not hold (Exercise 681). The following is a consequence of Proposition 516 and Theorem 573(iii):

Corollary 575. *A projective module is both torsion-free and torsionless.*

3.5.1.3 Presentation of a Projective Module

Lemma and Definition 576. *(1) A left $\mathbf{A}$-module P is projective if, and only if there exists a family (p_i) of elements of P and a family $(\bullet\varphi_i)$ of $\mathbf{A}$-linear morphisms $P \rightarrow \mathbf{A}$ such that for any $p \in P$,*

$$p = \sum_i (p\varphi_i)\, p_i \tag{3.41}$$

where $p\varphi_i = 0$ for all but a finite number of indices.
(2) A family (p_i) as above is called a projective basis *of P.*

Proof. (1): Let $\bullet\psi : \mathbf{A}^{(I)} \twoheadrightarrow P$. By Theorem 573, P is projective if, and only if there exists a section $\bullet\varphi : P \rightarrow \mathbf{A}^{(I)}$ associated with ψ, i.e., such that $\bullet\varphi\psi = Id_P$. Let (ε_i) be the canonical basis of $\mathbf{A}^{(I)}$ and $p_i = (\varepsilon_i)\,\psi$; as easily shown, (3.41) holds for any $p \in P$ and $p\varphi_i = 0$ for all but a finite number of indices. ■

Note that a projective basis is not a basis in the sense of Definition 253(iii).

Proposition 577. *(i) Every finitely generated projective module is finitely presented.*
(ii) Let $P = \operatorname{coker}_{\mathbf{A}}(\bullet R)$ where $R \in {}^q\mathbf{A}^k$. Then P is projective if, and only if there exists $S \in {}^k\mathbf{A}^q$ such that $R = R\,S\,R$.
(iii) Let $P = \operatorname{coker}_{\mathbf{A}}(\bullet R)$ where $R \in {}^q\mathbf{A}^k$ is left-regular. Then P is projective if, and only if R is right-invertible.

Proof. (i): Let P be a f.g. projective module. There exists by Theorem 494 a finite free module F such that the following sequence is exact:

$$F \xrightarrow{g} P \longrightarrow 0.$$

Therefore we have the short exact sequence

$$0 \longrightarrow K \xrightarrow{\iota} F \xrightarrow{g} P \longrightarrow 0$$

where $K = \ker g$ and ι is the inclusion. This exact sequence is split by Theorem 573, thus $F = K \oplus P'$ where $P' \cong F/K \cong P$. This proves that $K \cong F/P'$ is f.g., thus P is finitely related, hence finitely presented.

(ii): See ([30], §X.2, Prop. 6). (iii): Since R is left-regular, P is presented by a short exact sequence

$$0 \longrightarrow \mathbf{A}^q \xrightarrow{\bullet R} \mathbf{A}^k \xrightarrow{\varphi} P \longrightarrow 0. \tag{3.42}$$

By Theorem 573, if P is projective, then (3.42) is split. Therefore, by Lemma 571, $\operatorname{im}_{\mathbf{A}}(\bullet R)$ is a direct summand of $\mathbf{A}^k$, thus $\bullet R$ is right-invertible by Proposition 572, and so is the matrix R. Conversely, if R is right-invertible, $\bullet R$ is right-invertible too, i.e., is a section. By Lemma and Definition 53(2), there exists an isomorphism $\bullet\psi : \mathbf{A}^k \xrightarrow{\sim} P \oplus \mathbf{A}^q$, therefore P is a direct summand of the finite free module $\left(\mathbf{A}^k\right)\psi$, and P is projective by Proposition 574(i). ∎

3.5.1.4 Stably-Free Modules

Definition 578. *Let* $\mathbf{A}$ *be a ring with IBN and let* P *be a left* $\mathbf{A}$*-module;* P *is said to be* stably-free *of rank* $r \geq 0$ *if* $P \oplus \mathbf{A}^q \cong \mathbf{A}^{q+r}$ *for some non-negative integer* q.

Obviously, every finite free module is stably-free and every stably-free module is finitely generated projective. These notions do not coincide in general.

Definition 579. *Let* $\mathbf{A}$ *ring be a ring with IBN.*

(i) $\mathbf{A}$ *is called* projective-free *if every finitely generated projective module is free.*
(ii) $\mathbf{A}$ *is called an* Hermite ring *if any stably-free module is free.*

Note that these two notions are left/right symmetric.

Theorem 580. *(1) The following implications hold true for a ring* $\mathbf{A}$*:*

$$\textit{semifir} \Rightarrow \textit{projective-free} \Rightarrow \textit{Hermite}$$

(2) A local ring is projective-free (see also Theorem 634 below).

Proof. (1) Recall that a semifir has IBN (**2.12.2.1**). Let $\mathbf{A}$ be a ring with IBN and F a finite free left $\mathbf{A}$-module.

If $M \vartriangleleft F$ is projective, it is f.g. (Proposition 574(ii)), thus M is free if $\mathbf{A}$ is a semifir (Lemma 410) which proves the first implication.

If $M \vartriangleleft F$ is stably-free, it is projective, thus it is free if $\mathbf{A}$ is projective-free, which proves the second implication.

(2): See [176] or ([77], Sect. 0.5, Corol. 5.5). ∎

Other important examples of projective-free rings are given in the theorem below which gives positive answers to the so-called *Serre's conjecture* and some of its generalizations. Items (i) and (ii) are proved in ([200], Sect. V.2, Theorem 2.9; Sect. VIII.7, p. 334-335; Sect. V.4, Corol. 4.10); Item (iii) is proved in ([81], Theorem 7).

Theorem 581. *Let $X = (X_1, ..., X_n)$ and $Y = (Y_1, ..., Y_m)$ be finite sequences (possibly empty) of indeterminates and $Y^{\pm 1}$ denote the sequence $\left(Y_1, ..., Y_m, Y_1^{-1}, ..., Y_m^{-1}\right)$. The following rings are projective-free:*

*(i) $\mathbf{K}[X]$ if $\mathbf{K}$ is a field (*Quillen-Suslin theorem*) or, more generally, if $\mathbf{K}$ is a commutative Bézout domain or a valuation domain of finite Krull dimension (Exercise 490 and Subsect. 3.5.5 below);*
(ii) $\mathbf{K}\left[X, Y^{\pm 1}\right]$ if $\mathbf{K}$ is a field;
(iii) $\mathbf{A}[[T]]$ if $\mathbf{A}$ is a projective-free ring and T is a finite sequence of central indeterminates (Exercise 493).

Remark 582. *Let $\mathbf{K}$ be a division ring, $n \geq 2$ and $\mathbf{S} = \mathbf{K}[X_1, ..., X_n]$ where the indeterminates $X_1, ..., X_n$ are central. Then $\mathbf{S}$ has a stably-free non-free right ideal if, and only if $\mathbf{K}$ is not a field ([247], 11.2.9).*

Theorem 583. *Let $\mathbf{A}$ be a ring with IBN and let P be a left $\mathbf{A}$-module. The following conditions are equivalent:*

(i) P is stably-free with $P \oplus \mathbf{A}^q \cong \mathbf{A}^k$;
(ii) $P \cong \operatorname{coker} f$ for some split monomorphism $f : \mathbf{A}^q \hookrightarrow \mathbf{A}^k$;
(iii) the matrix Bézout equation

$$\boxed{RX = I_q} \tag{3.43}$$

has a solution X, where R is the matrix of f in the canonical bases of $\mathbf{A}^q$ and $\mathbf{A}^k$, i.e., R is right-invertible.

Proof. (i)⇒(ii): Let $F = P \oplus \mathbf{A}^q$ be a free module of rank n and let $\bullet\psi : F \xrightarrow{\sim} \mathbf{A}^k$. Let $\bullet\iota : \mathbf{A}^q \hookrightarrow F$ be the inclusion, let $\bullet\pi : P \oplus \mathbf{A}^q \twoheadrightarrow P$ be the projection and consider the commutative diagram below

$$\begin{array}{ccccccccc} 0 & \longrightarrow & \mathbf{A}^q & \xrightarrow{\iota} & F & \xrightarrow{\pi} & P & \longrightarrow & 0 \\ & & & f \searrow & \downarrow \psi & \nearrow \varphi & & & \\ & & & & \mathbf{A}^k & & & & \end{array}$$

where the row is a split exact sequence. Since $\bullet f = \bullet\iota\psi$ and ι is split, so is f too and we get the following split exact sequence:

$$0 \longrightarrow \mathbf{A}^q \xrightarrow{f} \mathbf{A}^k \xrightarrow{\varphi} P \longrightarrow 0. \tag{3.44}$$

(ii)⇒(i): Assuming that P is presented by the split exact sequence (3.44), we have $\mathbf{A}^k \cong (\mathbf{A}^q) f \oplus P$ and $(\mathbf{A}^q) f \cong \mathbf{A}^q$, thus (i) holds.

(ii)⇔(iii): We know from Lemma 571(1) that $\bullet f : \mathbf{A}^q \hookrightarrow \mathbf{A}^k$ is split if, and only if f is right-invertible, i.e., there exists $\bullet g : \mathbf{A}^k \to \mathbf{A}^q$ such that $f \circ g = Id_{\mathbf{A}^q}$. By (2.14), this is equivalent to (3.43) where X is the matrix of g in the canonical bases of $\mathbf{A}^q$ and $\mathbf{A}^k$. ■

3.5.1.5 Projective Ideals

The following is easily deduced from Lemma 576(1); see ([59], Sect. 7.3, Prop. 3.2) and ([136], Prop. 3.1 and Corol. 3.1) for more details:

Proposition 584. *Let* $\mathbf{A}$ *be a left Ore domain.*

(i) A nonzero left ideal in $\mathbf{A}$ *is projective if, and only if it is invertible (Definition 452).*
(ii) Such an ideal is finitely generated.

Definition 585. *A ring* $\mathbf{A}$ *is* left hereditary *if every left ideal in* $\mathbf{A}$ *is projective. A right hereditary ring is likewise defined and a ring is called* hereditary *if it is both left and right hereditary.*

Theorem 586. *(1) A ring* $\mathbf{A}$ *is a left Dedekind domain if, and only if it is a left Ore domain which is left hereditary.*
(2) Let $\mathbf{A}$ *be a left hereditary ring. If* M *is a submodule of a free left* $\mathbf{A}$*-module* F*, then* M *is a direct sum of modules isomorphic to left ideals in* $\mathbf{A}$*.*
(3) Let $\mathbf{A}$ *be a hereditary Noetherian prime ring; then any nonzero prime ideal in* $\mathbf{A}$ *is maximal.*

Proof. (1) is a consequence of Proposition 584(i). (2): See ([28], §VII.3, Theorem 1). (3): See ([247], 5.6.2). ■

See Exercise 688.

3.5.2 Injective Modules

3.5.2.1 Definition and Consequences

An $\mathbf{A}$-module I is *injective* in the category $\mathcal{C} = {}_{\mathbf{A}}\mathbf{Mod}$ if, and only if it is *projective* in the category $\mathcal{C}^{\mathrm{op}}$ (Lemma 45). Theorem 54 holds and most (but not all) properties of injective modules can be obtained by dualizing those of projective modules. The following is a consequence of Theorem 573:

Theorem 587. *The following conditions are equivalent:*

(i) I is injective;
(ii) any short exact sequence

$$0 \longleftarrow M \overset{g}{\longleftarrow} E \overset{i}{\longleftarrow} I \longleftarrow 0 \tag{3.45}$$

is split.

Note that Theorem 573(iii) has no analogous. The "dual" of Proposition 574(i) is given below:

Proposition 588. *Let $I = \prod_{j \in J} I_j$; then I is injective if, and only if every I_j $(j \in J)$ is injective. In particular, every summand of an injective module is injective.*

Theorem 589. *Let $\mathbf{A}$ be a ring. The following conditions are equivalent:*

(i) $\mathbf{A}$ is semisimple;
(ii) all left $\mathbf{A}$-modules are projective;
(iii) all right $\mathbf{A}$-modules are injective.
In the above statements, left *can be changed to* right.

Proof. This is a straightforward consequence of Definitions 313 & 317(2), and of Theorems 573 & 588. ■

The following, which is connected with Proposition 588, involves the notion of indecomposable submodule (Definition 252) and is proved in ([199], Theorem (3.48)).

Theorem 590. *Let $\mathbf{A}$ be a ring. The following conditions are equivalent:*

(i) $\mathbf{A}$ is left Noetherian;
(ii) any injective left module I is a direct sum of indecomposable (injective) submodules I_j $(j \in J)$.

3.5.2.2 Baer Criterion and Fundamental Principle

The following result, called the *Baer criterion*, is very useful to check if a module is injective. It is a consequence of Zorn's lemma: see ([30], §X.1, Prop. 10).

Criterion 591 *A left $\mathbf{A}$-module I is injective if, and only if every homomorphism $f : \mathfrak{a} \to I$, where $\mathfrak{a}$ is a left ideal of $\mathbf{A}$, can be extended to ${}_{\mathbf{A}}\mathbf{A}$.*

Corollary and Definition 592. *(1) Let $\mathbf{A}$ be a ring and W a left $\mathbf{A}$-module. The following conditions are equivalent:*
(i) W is injective;
(ii) the module W satisfies the "fundamental principle" *defined as follows: if a sequence*

$$\mathbf{A}^{(N)} \xrightarrow{\bullet R_1} \mathbf{A}^{(Q)} \xrightarrow{\bullet R_2} \mathbf{A}^{(K)} \tag{3.46}$$

is exact $\left(R_2 \in {}^{Q}\mathbf{A}^{(K)},\ R_1 \in {}^{N}\mathbf{A}^{(Q)}\right)$*, where* N*,* Q *and* K *are arbitrary sets of indices then so are the transformed sequences*

$$\begin{array}{ccccc} \mathrm{Hom}_{\mathbf{A}}\left(\mathbf{A}^{(K)}, W\right) & \overset{\mathfrak{B}_W(\bullet R_2)}{\longrightarrow} & \mathrm{Hom}_{\mathbf{A}}\left(\mathbf{A}^{(Q)}, W\right) & \overset{\mathfrak{B}_W(\bullet R_1)}{\longrightarrow} & \mathrm{Hom}_{\mathbf{A}}\left(\mathbf{A}^{(N)}, W\right) \\ \updownarrow & & \updownarrow & & \updownarrow \\ {}^{K}W & \overset{R_2\bullet}{\longrightarrow} & {}^{Q}W & \overset{R_1\bullet}{\longrightarrow} & {}^{N}W \end{array}$$

where $\mathfrak{B}_W = \mathrm{Hom}_{\mathbf{A}}(\bullet, W)$ *and the vertical arrows are canonical isomorphisms.*
(2) If $\mathbf{A}$ *is left Noetherian, the above still holds with the arbitrary sets of indices* N*,* Q *and* K *replaced by* finite *arbitrary sets of indices* $\{1, ..., n\}$*,* $\{1, ..., q\}$ *and* $\{1, ..., k\}$*, respectively.*

Proof. (1). (i)$\Rightarrow$(ii) is obvious since (i) means that the functor $\mathfrak{B}_W$ is exact (Definition 44); the isomorphisms between the first row and the second one have been established in Lemma 564.

(ii)$\Rightarrow$(i): Let $\mathfrak{a}$ be a left ideal in $\mathbf{A}$. According to the Baer criterion, W is injective if, and only if the restriction map $W \cong \mathrm{Hom}_{\mathbf{A}}(\mathbf{A}, W) \rightarrow \mathrm{Hom}_{\mathbf{A}}(\mathfrak{a}, W) : w \mapsto (a \mapsto a\,w)$ is surjective. Let $(a_j)_{j\in Q}$ be a generating family of $\mathfrak{a}$ and let R_2 be the column with entries a_j $(j \in Q)$. Let $(r_l)_{l\in N}$ be a generating family of $\ker_{\mathbf{A}}(\bullet R_2) \lhd \mathbf{A}^{(Q)}$ and let R_1 be the matrix with rows r_l $(l \in N)$. Then $R_1 \in {}^{N}\mathbf{A}^{(Q)}$ and the sequence (3.46) is exact (with $K = \{1\}$); since $\mathrm{im}_{\mathbf{A}}(\bullet R_2) = \mathfrak{a}$, the sequence

$$\mathbf{A}^{(N)} \overset{\bullet R_1}{\longrightarrow} \mathbf{A}^{(Q)} \overset{(\bullet R_2)_{ind}}{\longrightarrow} \mathfrak{a} \longrightarrow 0 \tag{3.47}$$

is also exact, where $(\bullet R_2)_{ind}$ is the map from $\mathbf{A}^{(Q)}$ to $\mathfrak{a}$ induced by $\bullet R_2$. The fundamental principle implies the exact sequence

$$W \overset{R_2\bullet}{\longrightarrow} {}^{Q}W \overset{R_1\bullet}{\longrightarrow} {}^{N}W. \tag{3.48}$$

The exact sequence (3.47) yields $\mathrm{coker}_{\mathbf{A}}(\bullet R_1) = \mathfrak{a}$, thus $\ker_W(R_1\bullet) = \mathrm{Hom}_{\mathbf{A}}(\mathfrak{a}, W)$ by Corollary 567; therefore, the exact sequence (3.48) yields $\mathrm{im}_W R_2\bullet = \mathrm{Hom}_{\mathbf{A}}(\mathfrak{a}, W)$ and $(R_2\bullet)_{ind} : W \twoheadrightarrow \mathrm{Hom}(\mathfrak{a}, W)$ is the desired epimorphism.

(2). If $\mathbf{A}$ is left Noetherian, $\mathfrak{a}$ is f.g. (Lemma 311), thus Q is finite, say $Q = \{1, ..., q\}$; $\ker_{\mathbf{A}}(\bullet R_2)$ is also f.g. since $\mathrm{coker}_{\mathbf{A}}(\bullet R_2)$ is f.p. (Lemma 498(ii)), thus N is finite, say $N = \{1, ..., n\}$. ■

Remark 593. *Let us clarify the meaning of the fundamental principle: let* $\mathbf{y} \in {}^{Q}W$ *and consider the equation*

$$R_2\,\mathbf{x} = \mathbf{y} \tag{3.49}$$

$\left(R_2 \in {}^{(Q)}\mathbf{A}^{K},\ \mathbf{x} \in {}^{K}W\right)$.

(i) Let $R_1 \in {}^{(N)}\mathbf{A}^Q$ *be such that the sequence (3.46) is exact. This implies* $R_1 R_2 = 0$, *thus (3.49) implies the* compatibility condition

$$R_1 \, \mathbf{y} = 0. \tag{3.50}$$

(ii) Conversely, assume that the fundamental principle holds and that $\mathbf{y} \in {}^Q W$ *satisfies the compatibility condition (3.50), i.e.,* $\mathbf{y} \in \ker_W (R_1 \bullet)$. *Then,* $\mathbf{y} \in \operatorname{im}_W (R_2 \bullet)$ *and there exists a solution* $\mathbf{x} \in {}^K W$ *to (3.49).*

A coproduct of injective $\mathbf{A}$-modules is not injective in general, as shown by the following ([306], Theorem 4.9):

Theorem 594. *The conditions below are equivalent for a ring* $\mathbf{A}$*:*

(i) $\mathbf{A}$ *is left Noetherian;*
(ii) every direct limit of injective left $\mathbf{A}$*-modules is injective.*
(iii) every coproduct of injective left $\mathbf{A}$*-modules is injective.*

3.5.2.3 Divisible Modules

Let $\mathbf{A}$ be a ring, let M be a left $\mathbf{A}$-module, let $m \in M$ and let $a \in \mathbf{A}$.

Definition 595. *The element* m *is said to be* divisible *by* a *if* $m \in a\,M$, *i.e., there exists* $m' \in M$ *such that* $m = a\,m'$.

If $m \in M$ is divisible by $a \in \mathbf{A}$, then $\operatorname{Ann}_l^{\mathbf{A}}(a) \subseteq \operatorname{Ann}_l^{\mathbf{A}}(m)$ (where a is considered as an element of the left $\mathbf{A}$-module ${}_{\mathbf{A}}\mathbf{A}$), but the converse does not hold true in general. Note that if $r\,a = 0$ and $m \in a\,M$, then $r\,m \in r\,a\,M = 0$, hence $a\,M \subseteq \operatorname{Ann}_r^M \left(\operatorname{Ann}_l^{\mathbf{A}}(a)\right)$.

Definition 596. *The left* $\mathbf{A}$*-module* M *is said to be* divisible *if* $m \in M$ *is divisible by* $a \in \mathbf{A}$ *whenever* $\operatorname{Ann}_l^{\mathbf{A}}(a) \subseteq \operatorname{Ann}_l^{\mathbf{A}}(m)$.

Theorem 597. *For a left* $\mathbf{A}$*-module* M*, the following properties are equivalent:*

(i) M *is divisible;*
(ii) for any $a \in \mathbf{A}$, $\operatorname{Ann}_r^M \left(\operatorname{Ann}_l^{\mathbf{A}}(a)\right) = a\,M$*;*
(iii) for any principal left ideal $\mathfrak{a} = \mathbf{A}\,a$, *any homomorphism* $f : \mathfrak{a} \to M$ *can be extended to* ${}_{\mathbf{A}}\mathbf{A}$.

Proof. (i)⇔(ii): Assuming that (i) holds, we have to prove that $\operatorname{Ann}_r^M \left(\operatorname{Ann}_l^{\mathbf{A}}(a)\right) \subseteq a\,M$. Let $m \in \operatorname{Ann}_r^M \left(\operatorname{Ann}_l^{\mathbf{A}}(a)\right)$; then $\operatorname{Ann}_l^{\mathbf{A}}(a) \subseteq \operatorname{Ann}_l^{\mathbf{A}}(m)$, thus m is divisible by a and $m \in a\,M$. The converse is obvious.

(i)⇒(iii): Let $\bullet f : \mathbf{A}a \to M$ and $m = (a) f$. If $r \in \mathrm{Ann}_l^{\mathbf{A}}(a)$, then $(r\,a) f = r\,(a) f = r\,m = 0$, thus $r \in \mathrm{Ann}_l^{\mathbf{A}}(m)$ and we have $\mathrm{Ann}_l^{\mathbf{A}}(a) \subseteq \mathrm{Ann}_l^{\mathbf{A}}(m)$. If (i) holds, there exists $m' \in M$ such that $m = a\,m'$. Let $\tilde{f} : {}_{\mathbf{A}}\mathbf{A} \to M$ be given by $(1)\tilde{f} = m'$; then $(a)\tilde{f} = m$ and $\tilde{f}$ extends f to ${}_{\mathbf{A}}\mathbf{A}$.

(iii)⇒(ii): Assuming that (iii) holds, we have to prove that $\mathrm{Ann}_r^M\left(\mathrm{Ann}_l^{\mathbf{A}}(a)\right) \subseteq a\,M$. Let $m \in \mathrm{Ann}_r^M\left(\mathrm{Ann}_l^{\mathbf{A}}(a)\right)$ and let $f : \mathbf{A}\,a \to M$ be given by $(r\,a) f = r\,m$; f is a homomorphism of left $\mathbf{A}$-module. By (iii), there exists $\tilde{f} : {}_{\mathbf{A}}\mathbf{A} \to M$ which extends f. Let $m' = (1)\tilde{f}$; then $m = (a)\tilde{f} = a\,m'$. ■

Proposition 598. *(i) If a left* $\mathbf{A}$*-module* I *is injective, it is divisible.*
(ii) All divisible left $\mathbf{A}$*-modules are injective, if, and only if* $\mathbf{A}$ *is left-hereditary (Definition 585). In particular, if* $\mathbf{A}$ *is a left Dedekind domain, then a left* $\mathbf{A}$*-module is divisible if, and only if it is injective.*

Proof. (i) is an obvious consequence of Theorem 597.
(ii) is proved in ([199], Corollary (3.23)). If $\mathbf{A}$ is a left Dedekind domain, we know from Theorem 586 that it is left hereditary. ■

See Exercise 690.

3.5.2.4 Injective Hull

Dualizing Proposition 574(iii) one obtains the following.

Theorem 599. *Let* $\mathbf{A}$ *be a ring; every left* $\mathbf{A}$*-module* M *can be embedded in an injective left* $\mathbf{A}$*-module.*

The injective hull we are now looking for is, roughly speaking, the smallest injective module in which a given module can be embedded. More specifically, it is (if it exists) a solution to the universal problem described below:

Theorem and Definition 600. *Let* $\mathbf{A}$ *be a ring and let* M *be a left* $\mathbf{A}$*-module.*

(1) There exists a left $\mathbf{A}$*-module* $E(M)$ *with the following properties:*
(i) $E(M)$ *is injective and there is an embedding* $M \hookrightarrow E(M)$;
(ii) for any injective module I *such that there is an embedding* $M \hookrightarrow I$*, there is an embedding* $E(M) \hookrightarrow I$ *as shown in the commutative diagram below where all lines are exact sequences.*

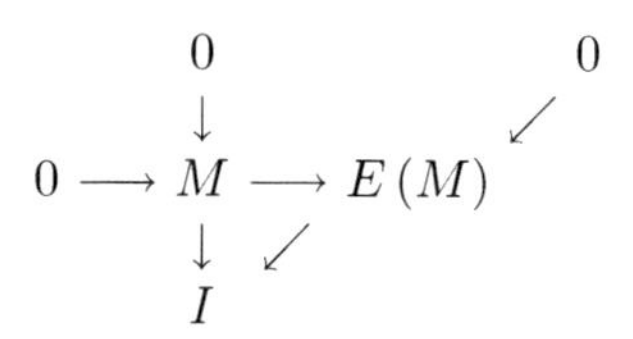

(2) Such a module $E(M)$ is unique up to isomorphism fixing M pointwise, and is called the injective hull *of M.*

Proof. (1): See Exercise 692. (2): By Theorem 13, a solution to the above universal problem is unique up to isomorphism. ∎

Let $\mathbf{A}$ be a ring. By Lemma 153, the set $\mathbf{I}_l(\mathbf{A})$ of all left ideals in $\mathbf{A}$ is a complete modular lattice. Therefore, a left ideal $\mathfrak{c} \neq \mathbf{A}$ in $\mathbf{A}$ is *meet-irreducible* if, and only if whenever two left ideals $\mathfrak{a}, \mathfrak{b}$ in $\mathbf{A}$ are such that $\mathfrak{a} \supseteq \mathfrak{c}$, $\mathfrak{b} \supseteq \mathfrak{c}$ and $\mathfrak{a} \cap \mathfrak{b} = \mathfrak{c}$, then necessarily $\mathfrak{a} = \mathfrak{c}$ or $\mathfrak{b} = \mathfrak{c}$ (**1.5.5.2**).

Lemma 601. *Let $\mathbf{A}$ be a ring and let $\mathfrak{c}$ be a left ideal in $\mathbf{A}$.*

(i) If $\mathfrak{c}$ is a maximal left ideal, then $\mathfrak{c}$ is a meet-irreducible left ideal.
(ii) If $\mathbf{A}$ is commutative and $\mathfrak{p}$ is a prime ideal (2.4.3.1), then $\mathfrak{p}$ is meet-irreducible.
(iii) If $\mathbf{A}$ is a commutative Noetherian ring, and if $\mathfrak{q}$ is a meet-irreducible ideal in $\mathbf{A}$, then $\mathfrak{q}$ is primary (E. Noether's theorem). In addition, $E(\mathbf{A}/\mathfrak{q}) \cong E\left(\mathbf{A}/\sqrt{\mathfrak{q}}\right)$.
(iv) If $\mathbf{A}$ is left principal ideal domain, then the following conditions are equivalent:
(a) $\mathfrak{c} = \mathbf{R}c$ is meet-irreducible;
*(b) if $a, b \in \mathbf{A}$ are such that $c = [a,b]_l$ (with the notation from (**1.4.1.2**)), then $c = u\,a$ or $c = u'\,b$ where u and u' are units.*

Proof. (i) and (iv) are clear.

(ii): Assume that $\mathfrak{p}$ is a prime ideal and let $\mathfrak{a}, \mathfrak{b}$ be ideals such that $\mathfrak{a}, \mathfrak{b} \supseteq \mathfrak{p}$ and $\mathfrak{a} \cap \mathfrak{b} = \mathfrak{p}$. Then $\mathfrak{a}\mathfrak{b} \subseteq \mathfrak{p}$ by Theorem 297(i), therefore $\mathfrak{a} \subseteq \mathfrak{p}$ or $\mathfrak{b} \subseteq \mathfrak{p}$, and $\mathfrak{p}$ is a meet-irreducible ideal.

(iii): For a proof of E. Noether's theorem, see, e.g., ([269], Lemma 28.6). For the other statement, see ([199], §3, Exercise 40A). ∎

The following is proved in, e.g., ([199], Theorem (3.52)).

Lemma 602. *For a left injective left $\mathbf{A}$-module I, the following conditions are equivalent:*

(i) I is indecomposable;
(ii) $I \cong E(\mathbf{A}/\mathfrak{c})$ for some meet-irreducible left-ideal $\mathfrak{c}$ in $\mathbf{A}$.

Let us consider two important examples of injective hulls:

Proposition 603. *Let $\mathbf{A}$ be a commutative principal ideal domain, let $p \in \mathbf{A}^{\times}$ be a prime, let $\mathbf{A}\left(p^i\right) \triangleq \mathbf{A}/\mathbf{A}\,p^i$ and let $\mathbf{A}\left(p^{\infty}\right) = \varinjlim \mathbf{A}\left(p^i\right)$ (Subsect. 2.13.5); then $\mathbf{A}\left(p^{\infty}\right)$ is the injective hull of $\mathbf{A}\left(p^i\right)$ for any $i \geq 1$.*

Proof. (1) By definition of $\mathbf{A}\left(p^{\infty}\right)$, $\mathbf{A}\left(p^i\right) \subseteq \mathbf{A}\left(p^{\infty}\right)$ for any $i \geq 1$.

(2) Let us show that $\mathbf{A}\left(p^{\infty}\right)$ is divisible: let $0 \neq x \in \mathbf{A}\left(p^{\infty}\right)$ and $0 \neq r \in \mathbf{A}$; let i be the least positive integer for which $x \in \mathbf{A}\left(p^i\right)$ and write $r = p^k q$

where k is a natural integer and $q \in \mathbf{A}$ is not divisible by p. There exists $y \in \mathbf{A}\left(p^{i+k}\right)$ such that $x = p^k y$ and then $p^m y = 0$ if $m = i + k$. Since p^m and q are coprime, there exist elements $u, v \in \mathbf{A}$ such that $u\,p^m + v\,q = 1$ (Bézout identity). Therefore, $x = v\,q\,x$, thus $x = p^k v\,q\,y = r\,z$ with $z = v\,y$.

(3) Therefore, $\mathbf{A}\left(p^{\infty}\right)$ is injective by Proposition 598(ii).

(4) The class $1 + \left(p^i\right) \in \mathbf{A}\left(p^i\right)$ is identified with $p + \left(p^{i+1}\right) \in \mathbf{A}\left(p^{i+1}\right)$ for any $i \geq 1$ under multiplication by p (Theorem 241(ii)), thus a divisible module containing $1 + \left(p^i\right)$ must contain $1 + \left(p^{i+1}\right)$. Therefore, $E\left(\mathbf{A}\left(p^i\right)\right) \supseteq \mathbf{A}\left(p^{i+1}\right)$, and by induction $E\left(\mathbf{A}\left(p^i\right)\right) \supseteq \mathbf{A}\left(p^{\infty}\right)$. Thus $E\left(\mathbf{A}\left(p^i\right)\right) = \mathbf{A}\left(p^{\infty}\right)$. ■

Example 604. *Let* $\mathbf{A}$ *be a commutative domain and let* $\mathbf{K}$ *be its field of fractions; then* $E\left({}_{\mathbf{A}}\mathbf{A}\right) = \mathbf{K}$. *Indeed, (i)* $\mathbf{K}$ *is a torsion-free divisible* $\mathbf{A}$-*module, thus it is injective (Exercise 690); (ii) if* $\mathbf{A} \subseteq S \lhd \mathbf{K}$ *and* S *is injective,* S *is divisible, thus for any* $n \in S$, $d \in \mathbf{A}^{\times}$, *there exists* $m \in S$ *such that* $n = d\,m$, *therefore the* $\mathbf{A}^{\times -1} S \subseteq S$; *since* $\mathbf{K} \subseteq \mathbf{A}^{\times -1} S$, $\mathbf{K} = S$.

3.5.3 Flat Modules

3.5.3.1 Flat Modules

Definition 605. *A right* $\mathbf{A}$-*module* U *is* flat *if the functor* $U \bigotimes_{\mathbf{A}} -$ *is exact (in the category* ${}_{\mathbf{A}}\mathbf{Mod}$).

Since the functor $U \bigotimes_{\mathbf{A}} -$ is always right exact (Proposition 547(1)), the right $\mathbf{A}$-module U is flat if, and only if whenever $A \hookrightarrow B$ is an embedding in ${}_{\mathbf{A}}\mathbf{Mod}$, so is $U \bigotimes_{\mathbf{A}} A \rightarrow U \bigotimes_{\mathbf{A}} B$ in the category $\mathbf{Ab}$ of Abelian groups.

The following is often useful ([80], theorem 4.6.2; [77], Appendix):

Lemma 606. *For a right* $\mathbf{A}$-*module* U, *the following conditions are equivalent:*

(i) U *is flat;*
(ii) whenever

$$u\,c = 0 \quad \text{where } u \in U^n, \quad c \in {}^n\mathbf{A},$$

there exists $B \in {}^m\mathbf{A}^n$ *and* $x \in U^m$ *such that*

$$u = x\,B, \quad B\,c = 0;$$

(iii) for any f.g. left ideal $\mathfrak{a}$ *of* $\mathbf{A}$ *the map* $U \otimes \mathfrak{a} \rightarrow U$, *induced by the inclusion* $\mathfrak{a} \subseteq \mathbf{A}$, *is injective, so that* $U \otimes \mathfrak{a} \cong U\mathfrak{a}$;
(iv) as (iii) but for any left ideal (not necessarily f.g.);
(v) every f.g. submodule of U *is contained in a flat submodule.*

Let us clarify the above condition (ii): it means that solutions in U of a linear equation with coefficients in $\mathbf{A}$ ($u\,c = 0$) can be expressed as linear combinations with coefficients in U ($u = x\,B$) of solutions in $\mathbf{A}$ ($B\,c = 0$).

The right **A**-module $\mathbf{A}_{\mathbf{A}}$ is obviously flat. As easily seen, a direct sum $U = \bigoplus_i U_i$ is flat if, and only if each U_i is flat. Therefore, any free right **A**-module is flat and more generally:

Corollary 607. *(i) For any right (or left)* **A***-module,*

$$\textit{projective} \Longrightarrow \textit{flat} \Longrightarrow \textit{torsion-free.}$$

(ii) A finitely related *right* **A***-module is flat if, and only if it is projective.*
(iii) A right **A***-module U is flat if, and only if it is a direct limit of f.g. free modules.*
(iv) If $\{M_i, \varphi_j^i\}$ is a direct system of right **A***-modules with filtering index set I (Definition 15) and each module M_i is flat, then the right* **A***-module $\varinjlim M_i$ is flat.*
(v) Assuming that **A** *is semiprime Goldie, a right* **A***-module M is torsion-free if, and only if it is embeddable in a flat module.*

Proof. (i) is clear; see ([199], (4.3), (4.18)) for more details. For (ii), see ([80], Corol. 4.6.4); for (iii) and (iv), see ([199], Theorem 4.34 and Proposition 4.4, respectively).

(v): If there is an embedding $M \hookrightarrow F$ where F is flat, then M is torsion-free by (i) and Proposition 516(ii). Conversely, a torsion-free module M is the directed union of a family $(M_i)_{i \in I}$ of f.g. torsion-free modules M_i, i.e., $M = \bigcup_{i \in I} M_i$. By Proposition 516(iv), for each index $i \in I$ there exists a finite free right module F_i such that $M_i \subseteq F_i$. Therefore, $M \subseteq F$ where $F = \bigcup_{i \in I} F_i$ is flat by (iii). ∎

The following result, proved in ([30], §X.1, Prop. 11) involves flatness and injectivity.

Proposition 608. *Let* **A**, **S** *be rings, let I be a left* **A***-module and let U be an* $(\mathbf{A}, \mathbf{S})$*-bimodule. If I is injective and U is a flat right* **S***-module, then the left* **S***-module* $\mathrm{Hom}_{\mathbf{A}}(U, I)$ *(Theorem 236(i)) is injective.*

Theorem 609. *Let* **A** *be a ring and let $S \subseteq \mathbf{A}^{\times}$ be a left denominator set (Definition 327); then the right* **A***-module $\mathbf{K} = S^{-1}\mathbf{A}$ is flat.*

Proof. We will prove that Condition (ii) in Lemma 606 holds true. Suppose $0 = u\,c = \sum u_i\,c_i$ where $c_i \in \mathbf{A}$ and $u_i \in S^{-1}\mathbf{A}$. Since there are finitely many u_i, they can be brought to a common denominator (Corollary 329), i.e., $u_i = s^{-1}b_i$, $s \in S$, $b_i \in \mathbf{A}$. Then $\sum b_i\,c_i = 0$, i.e., $b\,c = 0$, and $u = s^{-1}b$ as desired. ∎

Corollary 610. *Let* **A** *be a right Noetherian domain which is a left Ore domain, and let $R \in {}^q\mathbf{A}^k$. Let $N \lhd \mathbf{A}^k$ be the left* **A***-module* $\mathrm{im}_{\mathbf{A}}(R\bullet)$ *and let $N^{\mathbf{0}} \lhd {}^k\mathbf{A}$ be the right* **A***-module orthogonal to N* (**2.2.5.6**)*. Let $\{s_1, ..., s_r\}$ be a generator of $N^{\mathbf{0}}$ and let $S \in {}^k\mathbf{A}^r$ be the matrix with columns $s_1, ..., s_r$.*

(i) The sequence

$$0 \longrightarrow \mathrm{im}_{\mathbf{A}}(\bullet R) \xrightarrow{\iota} \ker_{\mathbf{A}}(\bullet S) \longrightarrow \ker_{\mathbf{A}}(\bullet S) / \mathrm{im}_{\mathbf{A}}(\bullet R) \longrightarrow 0 \tag{3.51}$$

is exact, where ι is the inclusion, and the module $\ker_{\mathbf{A}}(\bullet S) / \mathrm{im}_{\mathbf{A}}(\bullet R)$ is torsion.
(ii) Let $M = \mathrm{coker}_{\mathbf{A}}(R\bullet)$ and let $\mathcal{T}(M)$ be the torsion submodule of M; then $\ker_{\mathbf{A}}(\bullet S) / \mathrm{im}_{\mathbf{A}}(\bullet R) = \mathcal{T}(M)$. In particular, $\mathrm{im}_{\mathbf{A}}(\bullet R) = \ker_{\mathbf{A}}(\bullet S)$ if, and only if, M is torsion-free.

Proof. (i): The right $\mathbf{A}$-module $N^{\mathbf{0}}$ is f.g. since $\mathbf{A}$ is right Noetherian. The sequence (3.51) is exact, for $N^{\mathbf{00}} = \ker_{\mathbf{A}}(\bullet S)$. Let $\mathbf{K}$ be the division ring of left fractions of $\mathbf{A}$, and let $\hat{N} = \mathbf{K} \bigotimes_{\mathbf{A}} N$, $\hat{N}^{\mathbf{00}} = \mathbf{K} \bigotimes_{\mathbf{A}} N^{\mathbf{00}}$. We know that $\hat{N}^{\mathbf{00}} = \hat{N}$ (Remark 275), therefore

$$\mathbf{K} \bigotimes_{\mathbf{A}} (\ker_{\mathbf{A}}(\bullet S) / \mathrm{im}_{\mathbf{A}}(\bullet R)) = 0,$$

thus $\ker_{\mathbf{A}}(\bullet S) / \mathrm{im}_{\mathbf{A}}(\bullet R)$ is torsion by Theorem and Definition 339(ii).

(ii) From the above equality, $\ker_{\mathbf{A}}(\bullet S) / \mathrm{im}_{\mathbf{A}}(\bullet R) \subseteq \mathcal{T}(M)$. Conversely, let $r \in \mathbf{A}^k$ be such that $a\, r \in N$ for some $a \in \mathbf{A}^{\times}$. This means that $a\, \bar{r} = 0$ where $\bar{r}$ is the canonical image of r in $M = \mathbf{A}^k / \mathrm{im}_{\mathbf{A}}(\bullet R)$, i.e., $\bar{r} \in \mathcal{T}(M)$. Then, for any $r' \in N^{\mathbf{0}}$, $\langle a\, r, r' \rangle = a \langle r, r' \rangle = 0$, thus $\langle r, r' \rangle = 0$ since $\mathbf{A}$ is a domain. Therefore, $r \in N^{\mathbf{00}}$, and $\bar{r} \in N^{\mathbf{00}} / N$. ■

The following is similar to Corollary and Definition 592:

Proposition 611. *(1) Let $\mathbf{A}$ be a ring and let U be a right $\mathbf{A}$-module. The following conditions are equivalent:*
(i) U is flat;
(ii) For an arbitrary exact sequence of the form

$$\mathbf{A}^{(N)} \xrightarrow{\bullet f} \mathbf{A}^{(Q)} \xrightarrow{\bullet g} \mathbf{A}^{(K)},$$

the transformed sequence (by the functor $U \bigotimes_{\mathbf{A}} -$)

$$U^{(N)} \xrightarrow{Id_U \otimes f} U^{(Q)} \xrightarrow{Id_U \otimes g} U^{(K)}$$

is again exact.
(2) If $\mathbf{A}$ is left Noetherian, the above still holds with the arbitrary sets of indices N, Q and K replaced by finite *arbitrary sets of indices $\{1, ..., n\}$, $\{1, ..., q\}$ and $\{1, ..., k\}$, respectively.*

Proof. This is a generalization, based on ([59], Chap. I, Prop. 2.5), of ([266], Part I, Sect. I.3, Prop. 5), where $\mathbf{A}$ is assumed to be commutative and Noetherian. This generalization is a good exercise. ■

Theorem 612. *Let $\{\mathbf{A}_i, \varphi_j^i\}$ be a direct system of right coherent rings with filtering index set I, such that $\mathbf{A}_j$ is a flat right $\mathbf{A}_i$-module whenever $i \preccurlyeq j$. Then $\mathbf{A} = \varinjlim \mathbf{A}_i$ is right coherent.*

Proof. The ring $\mathbf{A}$ is a flat right $\mathbf{A}_i$-module for each $i \in I$ by Corollary 607(iv). Let $\mathfrak{a}$ be a f.g. right ideal in $\mathbf{A}$. By definition of the inductive limit (Definition 16), there exists an index $i \in I$ and a f.g. right ideal $\mathfrak{a}_i$ in $\mathbf{A}_i$ such that

$$\mathfrak{a} = \mathbf{A}\,\mathfrak{a}_i \cong \mathbf{A} \bigotimes_{\mathbf{A}_i} \mathfrak{a}_i.$$

Since $\mathbf{A}_i$ is right coherent, $\mathfrak{a}_i$ is a f.p. right ideal in $\mathbf{A}_i$ (Lemma 509), thus $\mathfrak{a}$ is a f.p. right ideal in $\mathbf{A}$ by Proposition 611, therefore the ring $\mathbf{A}$ is right coherent. ∎

See Exercises 688 and 696.

3.5.3.2 Faithful Modules

With the notation in Definition 505 we have the following.

Definition 613. *(i) A left $\mathbf{A}$-module U is* faithful *if* $\mathrm{Ann}_l^{\mathbf{A}}(U) = 0$.
(ii) U is completely faithful *if every nonzero quotient of U is faithful.*

For example, the $\mathbb{R}[\partial]$-module $\mathcal{E}(\mathbb{R})$ (where $\partial = d/dt$) is easily seen to be faithful.

3.5.3.3 Faithfully Flat Modules

Lemma and Definition 614. *(1) For a right $\mathbf{A}$-module U, the following conditions are equivalent:*

(i) a sequence

$$N' \xrightarrow{v} N \xrightarrow{w} N''$$

of left $\mathbf{A}$-modules is exact if, and only if the sequence

$$U \bigotimes_{\mathbf{A}} N' \xrightarrow{1 \otimes v} U \bigotimes_{\mathbf{A}} N \xrightarrow{1 \otimes w} U \bigotimes_{\mathbf{A}} N''$$

is exact;
(ii) U is flat, and for any left $\mathbf{A}$-module N, $U \bigotimes_{\mathbf{A}} N = 0 \Longrightarrow N = 0$;
(iii) U is flat, and for any homomorphism of left $\mathbf{A}$-modules $v : N' \rightarrow N$, $Id_U \otimes v = 0 \Longrightarrow v = 0$;
(iv) U is flat, and for any maximal left ideal $\mathfrak{m}$ of $\mathbf{A}$, $U\mathfrak{m} \neq U$;
(v) the functor $U \bigotimes_{\mathbf{A}} -$ is faithful exact.
(2) A right $\mathbf{A}$-module U is said to be faithfully flat *if the above equivalent conditions hold.*

Proof. The equivalences (i)⇔(ii)⇔(iii)⇔(iv) are proved in ([31], n°I.3.1). (i)⇔(v): (i) holds if, and only if the functor $U \bigotimes_{\mathbf{A}} -$ is exact (by Theorem 42) and faithful (by Proposition 49). ∎

Proposition 615. *A faithfully flat module is both faithful and flat.*

Proof. Let U be faithfully flat and $a \in \mathbf{A}$ be such that $x\,a = 0$ for all $x \in U$. The map $v : \mathbf{A} \to \mathbf{A}$ defined by $v(b) = ba$ is such that for any $y \in U$, $b \in \mathbf{A}$, $(Id_U \otimes v)(y \otimes b) = y \otimes ba = yb \otimes a = x \otimes a$ where $x = yb$, thus $Id_U \otimes v = 0$. Therefore, $v = 0$ by Lemma and Definition 614(iii), thus $v(1) = a = 0$, and U is faithful. ■

The converse of the above statement does not hold: there exist faithful projective modules which are not faithfully flat ([199], p. 150).

Corollary 616. *Let U, V be two right $\mathbf{A}$-modules.*

(i) If U is flat and V is faithfully flat, then $U \oplus V$ is faithfully flat.
(ii) If $U \neq 0$ is free, it is faithfully flat.

Proof. (i): As already said, $U \oplus V$ is flat, and it is faithfully flat by Lemma and Definition 614(iv). Since $\mathbf{A}_\mathbf{A}$ is obviously faithfully flat, (ii) is a consequence of (i). ■

3.5.4 Resolutions

3.5.4.1 Left Resolutions and Right Resolutions

Let M be an $\mathbf{A}$-module and consider exact sequences

$$\longrightarrow \ldots \longrightarrow E_n \longrightarrow \ldots \longrightarrow E_0 \longrightarrow M \longrightarrow 0, \tag{3.52}$$

$$\longleftarrow \ldots \longleftarrow E^n \longleftarrow \ldots \longleftarrow E^0 \longleftarrow M \longleftarrow 0. \tag{3.53}$$

Definition 617. *(i) The exact sequence (3.52) (resp., (3.53)) is called a* left *(resp.,* right*) resolution of M.*
(ii) If $E_n \neq 0$ and $E_i = 0$ for all $i \geq n$, then the left resolution (3.52) is said to be of length n. *Likewise, if $E^n \neq 0$ and $E^i = 0$ for all $i \geq n$, then the right resolution (3.53) is said to be of* length n.
(iii) If in the left resolution (3.52), each E_i is free (resp., projective, flat), then this resolution is called free *(resp.,* projective, flat*).*
(iv) If in the right resolution (3.53), each E^i is injective, *then this resolution is called* injective.
(v) If the left resolution (3.52) is free of finite length *and if each E_i is of* finite rank, *then this resolution is called* finite free.

Theorem 618. *Let M be an $\mathbf{A}$-module.*

(i) The module M has free, projective, flat and injective resolutions.
(ii) If the ring $\mathbf{A}$ is left (resp. right) Noetherian and M is f.g. left (resp. right) $\mathbf{A}$-module, then M has a finite free resolution.

Proof. (i): (1) We know that M is a quotient of a free module E_0, i.e. the sequence

$$E_0 \longrightarrow M \longrightarrow 0$$

is exact (Theorem 494). Then, there exists a presentation of M, thus there exists a free module F_1 such that the sequence

$$E_1 \longrightarrow E_0 \longrightarrow M \longrightarrow 0$$

is exact (see (3.3)). This rationale can be continued, and by induction M is shown to have a free presentation.

(2) By Proposition 574(ii) and Corollary 607(i), a free presentation is both a projective and a flat presentation.

(3) The projective resolution (3.52) of M in the category $\mathbf{Mod}_{\mathbf{A}^{\mathrm{op}}}$ is an injective resolution of M in the category ${}_{\mathbf{A}}\mathbf{Mod}$ by Proposition 229 and Lemma 45.

(ii): See the proof of Theorem 499. ∎

More generally, let $\mathcal{C}$ be an abelian category.

Lemma and Definition 619. *(1) The following conditions are equivalent:*
(i) For every object M there exists a projective object P_0 and an epimorphism

$$P_0 \longrightarrow M \longrightarrow 0.$$

(ii) Every object M has a projective resolution (3.52).
When the above equivalent conditions hold, the category $\mathcal{C}$ is said to have enough projectives.
(2) Dually, the following conditions are equivalent:
(i') For every object M there exists an injective object I^0 and a monomorphism

$$I^0 \longleftarrow M \longleftarrow 0.$$

(ii') Every object M has an injective resolution (3.53).
When the above equivalent conditions hold, the category $\mathcal{C}$ is said to have enough injectives.

Therefore, according to Theorem 618, the category ${}_{\mathbf{A}}\mathbf{Mod}$ has both enough projectives and enough injectives. More general, the following can be proved ([148], Theorem 1.10.1):

Theorem 620. *A Grothendieck category* (**1.2.4.4**) *has enough injectives.*

See Exercise 676.

3.5.4.2 Characteristic of a Module

Let M be an $\mathbf{A}$-module which has a finite free resolution

$$0 \longrightarrow F_n \longrightarrow ... \longrightarrow F_0 \longrightarrow M \longrightarrow 0 \tag{3.54}$$

Definition 621. *If* $\mathbf{A}$ *has IBN, the* Euler characteristic *of* M *is*

$$\chi(M) = \sum\nolimits_{0 \leq i \leq n} (-1)^i \operatorname{rk}(F_i).$$

The Euler characteristic $\chi(M)$ is well-defined (i.e., independent of the finite free resolution of M) according to Schanuel's lemma ([199], §5A).

Consider a module with finite free resolution of length ≤ 1, i.e., having a presentation of the form

$$0 \longrightarrow \mathbf{A}^q \xrightarrow{\bullet R} \mathbf{A}^k \longrightarrow M \longrightarrow 0. \tag{3.55}$$

Then $\chi(M) = k - q = i(R)$, the index of R (**3.2.4.1**). This happens for f.g. modules over semifirs:

Lemma 622. *Let* $\mathbf{A}$ *be a semifir and* M *a f.g. left* $\mathbf{A}$*-module. Then* M *has a finite free resolution of length* ≤ 1.

Proof. The f.g. module M is finitely presented by Theorem 499, thus there exists a matrix $A \in {}^p\mathbf{A}^k$ such that $M \cong \operatorname{coker}_{\mathbf{A}}(\bullet A)$. There exists by Lemma and Definition 410 an invertible matrix U such that $U A = \begin{bmatrix} R \\ 0 \end{bmatrix}$ where the q rows of R ($q \leq p$) are left $\mathbf{A}$-linearly independent, i.e., $\bullet R$ is injective (Corollary 279). Then (3.55) is a presentation of M. ∎

3.5.4.3 Presentation of a Stably-Free Module

Corollary 623. *Let* $\mathbf{A}$ *be a ring with IBN and let* P *be a left* $\mathbf{A}$*-module. The following conditions are equivalent:*

(i) P *is stably-free;*
(ii) $P \cong \operatorname{coker}_{\mathbf{A}}(\bullet R)$ *where* R *is a right-invertible;*
(iii) P *is projective and has finite free resolution of length* ≤ 1 *(Definition 621);*
(iv) P *is projective and has finite free resolution.*

Proof. (i)⇔(ii)⇔(iii) is an obvious consequence of Theorem 583.

(iii)⇒(iv) and the converse is easily shown by induction (see, e.g., [278], Chap. II, Prop. 2.36). ∎

3.5.4.4 Presentation of a Free Module

Let $\mathbf{A}$ be a weakly finite ring (Exercise 481) and let $R \in {}^{q}\mathbf{A}^{k}$, $k \geq q$.

Definition 624. *The matrix R is said to be* completable *there exists a matrix $R' \in {}^{k-q}\mathbf{A}^{k}$ such that $\begin{bmatrix} R \\ R' \end{bmatrix}$ is invertible.*

If a left $\mathbf{A}$-module F is free of rank $k-q$, it is stably-free of rank $k-q$, thus by Corollary 623 it has a matrix of definition which is right-invertible.

Remark 625. *(i) The above does not mean that any matrix of definition of a finite free (or stably-free) module is right-invertible. For example, the $\mathbb{Z}$-module with matrix of definition*

$$\begin{bmatrix} 1 & 0 \\ 0 & 0 \end{bmatrix}$$

is free of rank 1.
(ii) Any completable matrix is right-invertible.

Theorem 626. *Let F be a left $\mathbf{A}$-module and let $R \in {}^{q}\mathbf{A}^{k}$ be a* right-invertible *matrix of definition of F. The module F is free of rank $k-q$ if, and only if R is completable.*

Proof. (i) Assuming that R is completable, there exists $\begin{bmatrix} X & X' \end{bmatrix} \in \mathrm{GL}_k(\mathbf{A})$ such that

$$\begin{bmatrix} R \\ R' \end{bmatrix} \begin{bmatrix} X & X' \end{bmatrix} = \begin{bmatrix} I_q & 0 \\ 0 & I_{k-q} \end{bmatrix} \tag{3.56}$$

thus $R \begin{bmatrix} X & X' \end{bmatrix} = \begin{bmatrix} I_q & 0_{k-q} \end{bmatrix}$. We have $F \cong [w]_{\mathbf{A}}$ where w is only submitted to the relation $R\,w = 0$ (Subsect. 3.2.3). Let $v = \begin{bmatrix} X & X' \end{bmatrix} w$; then we have again $F \cong [v]_{\mathbf{A}}$ and v is only submitted to $\begin{bmatrix} I_q & 0_{k-q} \end{bmatrix} v = 0$, i.e., $v_1 = \dots = v_q = 0$. Setting $v' = (v_i)_{q+1 \leq i \leq k}$ we obtain $F \cong [v']_{\mathbf{A}}$ and since the components of v' are submitted to no relation, $F \cong \mathbf{A}^{k-q}$.

(ii) Conversely, let $F \cong \mathbf{A}^{k-q}$ with right-invertible matrix of definition R. We have a split exact sequence

$$0 \longrightarrow \mathbf{A}^{q} \xrightarrow{\bullet R} \mathbf{A}^{k} \xrightarrow{\bullet \varphi} F \longrightarrow 0;$$

where $\bullet\varphi$ has a left-inverse $\bullet\rho$ (Lemma 571(1)). Choosing a basis ξ in F, let X' and R' be the matrices of φ and ρ, respectively, in the canonical basis of $\mathbf{A}^{k}$ and in ξ. We have the equalities (3.43), $R'\,X' = I_{k-q}$ (since $\rho \circ \varphi = I_F$) and $R\,X' = 0$ since $(\bullet R, \varphi)$ is exact. Therefore,

$$\begin{bmatrix} R \\ R' \end{bmatrix} \begin{bmatrix} X & X' \end{bmatrix} = \begin{bmatrix} I_q & 0 \\ * & I_{k-q} \end{bmatrix}$$

and $\begin{bmatrix} R \\ R' \end{bmatrix}$ is invertible (see the proof of the part (b)⇒(a) of Theorem 520). ∎

Notice that the relation (3.56) contains two Bézout equations.

3.5.5 Dimension

3.5.5.1 Introduction

Several kinds of dimensions are defined for a ring (and also for a module), among which the *uniform dimension* (also called the Goldie dimension), the *Krull dimension* and the *global dimension*. A Goldie ring–and in particular a Noetherian ring–has finite uniform dimension ([247], 2.3.1); Goldie showed that a domain $\mathbf{A}$ has uniform dimension 1 if, and only if it is an Ore domain, and has infinite uniform dimension otherwise ([199], (10.22)). The Krull dimension and the global dimension can be viewed as measures how far a ring is from being Artinian and semisimple, respectively. Due to the limitation of space, only the global dimension (both the usual one and the *weak* one), and the Krull dimension *in the commutative case*, are detailed in the sequel; some hints are given about the right Krull dimension of noncommutative rings (when it exists).

3.5.5.2 Projective, Injective and Flat Dimensions

Definition 627. *Let M be a left $\mathbf{A}$-module. The projective (resp., flat, injective) dimension of M, denoted by* $\mathrm{pd}\,({}_{\mathbf{A}}M)$ *(resp.,* $\mathrm{fd}\,({}_{\mathbf{A}}M)$, $\mathrm{id}\,({}_{\mathbf{A}}M)$*), is defined as the shortest length of a projective (resp., flat, injective) resolution of M and as $+\infty$ if M has no projective (resp., flat, injective) resolution of finite length.*

A left $\mathbf{A}$-module M is such that $\mathrm{pd}\,({}_{\mathbf{A}}M) = 0$ (resp., $\mathrm{fd}\,({}_{\mathbf{A}}M) = 0$, $\mathrm{id}\,({}_{\mathbf{A}}M) = 0$) if, and only if M is projective (resp., flat, injective). If M has a finite free resolution of length $\leq n$, then $\mathrm{pd}\,({}_{\mathbf{A}}M) \leq n$ since a projective module is free. The converse holds if M is f.g. and the ring $\mathbf{A}$ is both Noetherian and projective-free (Definition 579). In addition, as shown in ([306], Lemma 9.40):

Lemma 628. *If a module M has a projective resolution (3.52) in which each E_i is stably-free, then M has a finite free resolution of length $\leq n+1$.*

3.5.5.3 Global Dimension and Krull Dimension

Theorem and Definition 629. *(1) The following quantities are equal ([306], Chap. 9) and are called the* left global dimension *of* $\mathbf{A}$*, which is denoted by* $\mathrm{lgld}\,\mathbf{A}$*:*

$$(i) \qquad \sup\{\mathrm{pd}({}_{\mathbf{A}}M) : M \in {}_{\mathbf{A}}\mathbf{Mod}\}$$
$$(ii) \qquad \sup\{\mathrm{id}({}_{\mathbf{A}}M) : M \in {}_{\mathbf{A}}\mathbf{Mod}\}$$

(2) The right global dimension *of* $\mathbf{A}$, *denoted by* $\mathrm{rgld}\,\mathbf{A}$, *is likewise defined.*
(3) The following quantities are equal ([306], Chap. 9) and are called the weak *(or* flat*)* global dimension *of* $\mathbf{A}$, *written* $\mathrm{wgld}\,\mathbf{A}$*:*

$$(i) \qquad \sup\{\mathrm{fd}({}_{\mathbf{A}}M) : M \in {}_{\mathbf{A}}\mathbf{Mod}\}$$
$$(ii) \qquad \sup\{\mathrm{fd}(M_{\mathbf{A}}) : M \in \mathbf{Mod}_{\mathbf{A}}\}$$

(4) A ring $\mathbf{A}$ *is called* left (resp., right) regular *if* $\mathrm{pd}(M) < \infty$ *for all finitely generated modules* $M \in {}_{\mathbf{A}}\mathbf{Mod}$ *(resp.,* $M \in \mathbf{Mod}_{\mathbf{A}}$*). In particular, if* $\mathrm{lgld}\,\mathbf{A} < \infty$ *(resp.,* $\mathrm{rgld}\,\mathbf{A} < \infty$*), then* $\mathbf{A}$ *is left (resp., right) regular.*
(5) If $\mathbf{A} \neq 0$ *is commutative ring, its* Krull dimension *(or simply its* dimension*), denoted by* $\mathcal{K}(\mathbf{A})$ *(or by* $\dim(\mathbf{A})$*) is the supremum of the lengths of all chains of prime ideals* $\mathfrak{p}_0 \supsetneq \ldots \supsetneq \mathfrak{p}_d$ *in* $\mathbf{A}$ *(Subsect. 1.5.2 & 2.4.3) and is equal to a non-negative integer or* $+\infty$.

Remark 630. *(i) We have given in (5) above the classical definition of the Krull dimension of a commutative ring ([32], n° VIII.1.3). This definition can be extended (in a non obvious way) to noncommutative rings, for which the* left Krull dimension *and the* right *one (when they exist) must be distinguished; the latter is always denoted by* $\mathcal{K}(\mathbf{A})$. *See ([247], 6.2.2) for more details.*
(ii) If X *is a topological space, its dimension* $\dim X$ *is defined to be* -1 *if* $X = \varnothing$ *and otherwise to be the supremum of all integers* n *such that there exists a chain* $Z_0 \subsetneq Z_1 \subsetneq \ldots \subsetneq Z_n$ *of distinct irreducible closed subsets of* X *(Definition 386(iii)). One can prove ([156], Sect. I.1, Prop 1.7) that if* Y *is an algebraic set, then* $\dim Y$ *coincides with the Krull dimension of the affine coordinate ring* $\mathbf{A}(Y)$ *(Corollary and Definition 389(3)).*

The following result involves faithfully flat ring extensions (**3.5.3.3**) and the Krull dimension. It is proved in, e.g., ([32], §VIII.2, Prop. 2).

Proposition 631. *Let* $\mathbf{A}, \mathbf{B}$ *be commutative rings such that* $\mathbf{A} \subseteq \mathbf{B}$ *and the* $\mathbf{A}$*-module* $\mathbf{B}$ *is faithfully flat. Then*

$$\dim(\mathbf{B}) \geq \dim(\mathbf{A}) + \inf_{\mathfrak{m} \in S} \dim(\mathbf{B}/\mathfrak{m}\mathbf{B}) \tag{3.57}$$

where S *is the set of all maximal ideals in* $\mathbf{A}$. *Furthermore, if the rings* $\mathbf{A}, \mathbf{B}$ *are local Noetherian, then (3.57) is an equality.*

If $\mathbf{A} \neq 0$ is a commutative Noetherian ring, any chain of prime ideals in $\mathbf{A}$ has finite length, but this does not imply $\mathcal{K}(\mathbf{A}) < \infty$ ([32], §VIII.1, Exercise 13). The following, is classical ([247], Sect. 7.1):

Theorem and Definition 632. *(1) For any ring* $\mathbf{A}$*,*

$$\operatorname{wgld}\mathbf{A} \leq \sup\{\operatorname{lgld}\mathbf{A},\ \operatorname{rgld}\mathbf{A}\}.$$

(2) If $\mathbf{A}$ *is left Noetherian, then*

$$\operatorname{wgld}\mathbf{A} = \operatorname{lgld}\mathbf{A}.$$

(3) If $\mathbf{A}$ *is Noetherian, then*

$$\operatorname{wgld}\mathbf{A} = \operatorname{lgld}\mathbf{A} = \operatorname{rgld}\mathbf{A},$$

and this common value, called the global dimension *of* $\mathbf{A}$*, is denoted by* $\operatorname{gld}\mathbf{A}$.

Lemma 633. *Let* $\mathbf{A}$ *be a nonzero ring.*

(i) $\operatorname{gld}\mathbf{A} = 0$ *if, and only if* $\mathbf{A}$ *is semisimple.*
(ii) $\operatorname{rgld}\mathbf{A} = 1$ *if, and only if* $\mathbf{A}$ *is right-hereditary.*
(iii) If $\mathbf{A}$ *is commutative,* $\mathcal{K}(\mathbf{A}) = 0$ *if, and only if every prime ideal is maximal. In particular, if* $\mathbf{A}$ *is commutative and Noetherian,* $\mathcal{K}(\mathbf{A}) = 0$ *if, and only if* $\mathbf{A}$ *is Artinian (Subsect. 2.4.4); if* $\mathbf{A}$ *is a commutative domain,* $\mathcal{K}(\mathbf{A}) = 0$ *if, and only if* $\mathbf{A}$ *is a field.*
(iv) If $\mathbf{A}$ *is a hereditary Noetherian prime ring (Definitions 319 & 585), then* $\mathcal{K}(\mathbf{A}) \leq 1$*. In particular, if* $\mathbf{A}$ *is a Dedekind domain, then* $\mathcal{K}(\mathbf{A}) \leq 1$*; more specifically, a* commutative domain $\mathbf{A}$ *is a Dedekind domain if, and only if* $\mathcal{K}(\mathbf{A}) \leq 1$ *and* $\mathbf{A}$ *is integrally closed (Lemma and Definition 448).*
(v) If $\mathbf{A}$ *is commutative, Noetherian and regular, then* $\mathcal{K}(\mathbf{A}) = \operatorname{gld}\mathbf{A}$.
(vi) If $\mathbf{A}$ *is commutative and local, then* $\operatorname{gld}\mathbf{A} < \infty$ *if, and only if* $\mathbf{A}$ *is regular.*

Proof. (i) is a consequence of Theorem 589. (ii): See ([30], §X.8, Prop. 7). (iii): See ([32], n°VIII.1.3, Example 1). (iv): See ([247], 6.2.8) and ([32], n°VIII.1.3, Example 2). (v): See ([199], (5.94)). (vi): See ([105], Theorem 19.12). ∎

Sylvester domains (Definition 416) can be characterized using the global dimension, as shown by the following, stated without proof (see [14], [89]):

Theorem 634. *Let* $\mathbf{A}$ *be a ring.*
(i) Assume that $\mathbf{A}$ *is Noetherian domain. Then,* $\mathbf{A}$ *is a Sylvester domain if, and only if* $\operatorname{gld}\mathbf{A} \leq 2$ *and* $\mathbf{A}$ *is projective-free.*
(ii) Assume that $\mathbf{A}$ *is an Ore domain. Then* $\mathbf{A}$ *is a Sylvester domain if, and only if* $\operatorname{gld}\mathbf{A} \leq 2$ *and every flat module is a directed union of free submodules.*

The following gathers useful results (see [247], 7.4.3, 7.5.3, 7.5.5, 6.5.4, 6.9.24).

Theorem 635. *(1) Let* $\mathbf{A}$ *be a ring, let* S *be a right denominator set (Definition 327) and let* $\mathbf{Q} = \mathbf{A}\,S^{-1}$*. Then* $\operatorname{rgld}\mathbf{Q} \leq \operatorname{rgld}\mathbf{A}$ *and* $\operatorname{wgld}\mathbf{Q} \leq \operatorname{wgld}\mathbf{A}$.

(2) Let $\mathbf{A}$ *be a ring,* $\boldsymbol{\alpha}$ *an automorphism and* $\boldsymbol{\delta}$ *a derivation or an* $\boldsymbol{\alpha}$*-derivation.*
(i) $\mathrm{rgld}\,\mathbf{A} \leq \mathrm{rgld}\,\mathbf{A}[X;\boldsymbol{\alpha},\boldsymbol{\delta}] \leq \mathrm{rgld}\,\mathbf{A}+1$ *if* $\mathrm{rgld}\,\mathbf{A} < +\infty$;
(ii) $\mathrm{rgld}\,\mathbf{A} \leq \mathrm{rgld}\,\mathbf{A}\left[X,X^{-1};\boldsymbol{\alpha}\right] \leq \mathrm{rgld}\,\mathbf{A}+1$;
(iii) $\mathrm{rgld}\,\mathbf{A}[X;\boldsymbol{\alpha}] = \mathrm{rgld}\,\mathbf{A}+1$;
(iv) $\mathrm{rgld}\,\mathbf{A}\left[X,X^{-1}\right] = \mathrm{rgld}\,\mathbf{A}+1$;
(v) $\mathrm{rgld}\,\mathbf{A}[[X;\boldsymbol{\alpha}]] = \mathrm{rgld}\,\mathbf{A}+1$ *if* $\mathbf{A}$ *is Noetherian;*
(vi) $\mathrm{rgld}\,\mathbf{A}[X;\boldsymbol{\alpha},\boldsymbol{\delta}] = \mathrm{rgld}\,\mathbf{A}\left[X,X^{-1};\boldsymbol{\alpha}\right] = 1$ *if* $\mathbf{A}$ *is semisimple;*
(vii) if $\mathbf{A}$ *is Noetherian commutative with* $\mathrm{gld}\,\mathbf{A} = d < \infty$*, let* $\mathbf{S}$ *be either* $\mathbf{A}[X;\boldsymbol{\delta}]$ *or* $\mathbf{A}\left[X,X^{-1};\boldsymbol{\alpha}\right]$ *and suppose* $\mathbf{S}$ *is a simple ring. Then* $\mathrm{rgld}\,\mathbf{S} = \sup\{1,d\}$.
(3) If $\mathbf{A}$ *is a right Noetherian ring, then its right Krull dimension* $\mathcal{K}(\mathbf{A})$ *exists and (i)-(v) of Item (2) hold true with the right global dimension changed to the right Krull dimension. In addition,*
(vi') $\mathcal{K}(\mathbf{A}[X;\boldsymbol{\alpha},\boldsymbol{\delta}]) = \mathcal{K}\left(\mathbf{A}\left[X,X^{-1};\boldsymbol{\alpha}\right]\right) = 1$ *if* $\mathbf{A}$ *is right Artinian;*
(vii') if $\mathbf{A}$ *is Noetherian commutative, not Artinian, let* $\mathbf{S}$ *be a simple ring as in (vii) above; then* $\mathcal{K}(\mathbf{S}) = \mathcal{K}(\mathbf{A})$.

Theorem 636. *(i) (*Hilbert's theorem on syzygies*). Let* $\mathbf{A}$ *be a commutative ring; then*

$$\mathrm{gld}\,\mathbf{A}[X_1,...,X_n] = \mathrm{gld}\,\mathbf{A}+n$$

and the same holds with the Krull dimension if $\mathbf{A}$ *is Noetherian.*
(ii) Let $A_n(\mathbf{k})$ *(*$k=\mathbb{R}$ *or* $\mathbb{C}$*) be the* n*th Weyl algebra (Exercise 493(v)); then*

$$\mathrm{gld}\,A_n(\mathbf{k}) = \mathcal{K}(A_n(\mathbf{k})) = n.$$

(iii) Let $\mathbf{A}$ *be a* $\mathbb{Q}$*-algebra (resp., a right Noetherian ring) and* $\mathbf{S} = \mathbf{A} * G$ *be the crossed product of* $\mathbf{A}$ *by a polycyclic by cyclic group* G *with Hirsch number* h *(Definition 161). Then* $\mathrm{rgld}\,\mathbf{S} \leq \mathrm{rgld}\,\mathbf{A}+h$ *(resp.,* $\mathcal{K}(\mathbf{S}) \leq \mathcal{K}(\mathbf{A})+h$*).*

Proof. (i) is a consequence of Theorem 635. (ii) and (iii): see ([247], 7.5.8, 7.5.6, 6.6.15, 6.5.5). ■

In connection with Hilbert's theorem on syzygies, and using the notions of filtration and grading (not presented in this book), one obtains Item (1) below, proved in ([247], 12.3.3).

Theorem 637. *(1) Let* $\mathbf{R}$ *be a right Noetherian ring such that every cyclic module has finite projective dimension, and such that f.g. projective right* $\mathbf{R}$*-modules are all stably-free. Let* $\mathbf{S}$ *be one of the following rings:*
(i) $\mathbf{R} * G$ *where the group* G *is polycyclic;*
(ii) $\mathbf{R}[X;\boldsymbol{\alpha},\boldsymbol{\delta}]$ *where* $\boldsymbol{\alpha}$ *is an automorphism;*
(iii) $\mathbf{R}\left[X,X^{-1};\boldsymbol{\alpha}\right]$ *where* $\boldsymbol{\alpha}$ *is an automorphism.*
Then every f.g. projective right $\mathbf{S}$*-module is stably-free.*
(2) The above also holds true when $\mathbf{S}$ *is one of the following rings:*
(iv) $A_1(\mathbf{k})$*, the first Weyl algebra;*
(v) $A_{1c}(\mathbf{k})$ *(Corollary 365);*

(vi) $A'_1(\mathbf{k})$ *(Example 369);*
(vii) $A_o(\mathbb{C})$ *(Exercise 492);*
(viii) $A_n(\mathbf{k})$*, the nth Weyl algebra (Exercise 493(v)).*

Proof. (2): Over $A_1(\mathbf{k})$, $A_{1c}(\mathbf{k})$ and $A_o(\mathbb{C})$, every f.g. projective right $\mathbf{S}$-module is stably-free by (1)(ii) since $\mathbf{k}[t]$, $\mathbf{k}\{t\}$ (2.9.1.1) and $\mathcal{R}(\Omega)$ (Exercise 492) are principal ideal domains. The rings $A'_1(\mathbf{k})$ and $A_n(\mathbf{k})$ enjoy the same property, the former by (iii), the latter by induction from (iv) and (ii). ∎

3.5.6 Cogenerators

3.5.6.1 Basic Properties

Generators (resp., cogenerators) have been defined in a category $\mathcal{C}$ having arbitrary products (resp., coproducts) (**1.2.2.4**), and are characterized by Theorem 43(3) when $\mathcal{C}$ is additive: for any objects U and W of $\mathcal{C}$, $\mathfrak{A}_U = \mathrm{Hom}_{\mathcal{C}}(U, \bullet)$ (resp., $\mathfrak{B}_W(\bullet) = \mathrm{Hom}_{\mathcal{C}}(\bullet, W)$) is a covariant (resp., contravariant) functor $\mathcal{C} \to \mathbf{Ab}$, and U is a generator (resp., W is a cogenerator) in $\mathcal{C}$ if, and only if $\mathfrak{A}_U$ (resp., $\mathfrak{B}_W(\bullet)$) is faithful. This applies to the category ${}_{\mathbf{A}}\mathbf{Mod}$ (where $\mathbf{A}$ is any ring) since ${}_{\mathbf{A}}\mathbf{Mod}$ has arbitrary products and coproducts and is abelian. See Example 501.

Theorem 638. *(1) For a left* $\mathbf{A}$*-module* W*, the following properties are equivalent:*
(i) W *is a cogenerator;*
(ii) for any nonzero homomorphism of left $\mathbf{A}$*-modules* $\bullet f : X_1 \to X_2$*, there exists* $\bullet g : X_2 \to W$ *such that* $\bullet fg \neq 0$*;*
(iii) for any $X \in {}_{\mathbf{A}}\mathbf{Mod}$ *and* $0 \neq x \in X$*, there exists* $\bullet g : X \to W$ *such that* $(x)\, g \neq 0$*;*
(iv) any $X \in {}_{\mathbf{A}}\mathbf{Mod}$ *can be embedded in some product* $W^I = \prod_i W$*;*
(v) for every simple left $\mathbf{A}$*-module* S*,* W *contains a copy of* $E(S)$ *(the injective hull of* S*).*
(2) If W *is a cogenerator and* $W \lhd V$*, then* V *is a cogenerator.*

Proof. (1) (i)⇔(ii) by Definition 8 and (i)⇔(iv) by Lemma 9.
(ii)⇒(iii): Let $\bullet f : \mathbf{A} \to X$ be defined by $(1)\, f = x \neq 0$ and take $\bullet g : X \to W$ such that $\bullet fg \neq 0$. Then $(x)\, g = (1)(fg) \neq 0$.
(iii)⇒(iv): For any $0 \neq x \in X$, let $\pi_x : X \to W$ such that $(x)\, \pi_x \neq 0$. Then $\pi = (\pi_x)_{x \neq 0}$ is an embedding $X \hookrightarrow \prod_{x \neq 0} W$.
(2) is a consequence of (1). ∎

Corollary 639. *Any cogenerator* $W \in {}_{\mathbf{A}}\mathbf{Mod}$ *is faithful.*

Proof. Let $\lambda \in \mathbf{A}$ be such that $\lambda W = 0$ and consider (by the above property (iv)) an embedding ${}_{\mathbf{A}}\mathbf{A} \hookrightarrow W^I$. Then $\lambda\, {}_{\mathbf{A}}\mathbf{A} = 0$ and in particular $\lambda 1 = \lambda = 0$. ∎

Definition 640. *Let* $\mathbf{A}$ *be a ring and* $X \in {}_{\mathbf{A}}\mathbf{Mod}$. *The module* X *is said to be* finitely cogenerated *if whenever there exist a family* $(W_i)_{i \in I}$ *of left* $\mathbf{A}$*-modules and an embedding* $X \hookrightarrow \prod_{i \in I} W_i$, *there exists already an embedding* $X \hookrightarrow \prod_{i \in J} W_i$ *for some* finite *subset* $J \subseteq I$.

The following is proved in ([199], (19.1), (19.4)), and can be viewed as the "dual" of Lemma 311.

Proposition 641. *A module* $X \in {}_{\mathbf{A}}\mathbf{Mod}$ *is Artinian if, and only if every quotient of* X *is finitely cogenerated.*

Theorem 642. *Let* W *be an injective left* $\mathbf{A}$*-module. The following conditions are equivalent:*

(i) W *is cogenerator;*
(ii) $\mathrm{Hom}_{\mathbf{A}}(S, W) \neq 0$ *for every simple left* $\mathbf{A}$*-module* S*;*
(iii) for every simple left $\mathbf{A}$*-module* S, *there exists a set of indices* I *and an embedding* $S \hookrightarrow W^I$.

Proof. (i)⇒(ii) and (i)⇒(iii) are obvious.

(ii)⇒(i): Let X be a left $\mathbf{A}$-module and $0 \neq x \in X$. The submodule $\mathbf{A}\,x$ of X has a simple quotient S (Proposition 533(ii)); if $\mathrm{Hom}_{\mathbf{A}}(S, W) \neq 0$, then $\mathrm{Hom}_{\mathbf{A}}(\mathbf{A}\,x, W) \neq 0$ and there exists $f : \mathbf{A}\,x \to W$ which is nonzero. Since W is injective, f can be extended to a nonzero homomorphism $g : X \to W$.

(iii)⇒(i) is a consequence of Theorem 638(1). Indeed, if S embeds in W^I, so does $E(S)$ since W^I is injective by Proposition 588. Therefore, W^I is a cogenerator, and so is W. ∎

Corollary and Definition 643. *(i) Let* $(S_i)_{i \in I}$ *be a representative system of simple left* $\mathbf{A}$*-modules (Definition 534). Then* $W_0 = \bigoplus_{i \in I} E(S_i)$ *is a cogenerator, called the* canonical cogenerator *of* ${}_{\mathbf{A}}\mathbf{Mod}$, *and* W_0 *is injective if* $\mathbf{A}$ *is left-Noetherian.*
(ii) A left $\mathbf{A}$*-module* W *is a cogenerator if, and only if there exists an embedding* $W_0 \hookrightarrow W$.
(iii) In particular, $E(W_0)$ *is an* injective cogenerator.

Proof. (i) is an obvious consequence of Theorems 642(ii) & 594(iii).

(ii): If there exists an embedding $W_0 \hookrightarrow W$, then W is a cogenerator by Theorem 638(2). Conversely, let W be a cogenerator. By Theorem 642(ii), we may assume that $E(S_i) \subseteq W$ for all $i \in I$ (recall that $E(S_i)$ is defined up to isomorphism). Since $S_i \ncong S_j$ for $i \neq j$, the sum $\sum_{i \in I} S_i$ is direct and so is the sum $\sum_{i\in I} E(S_i)$ by Exercise 692(v), therefore W contains a copy of W_0.

(iii): The module W_0 is not necessarily injective, but by Theorem and Definition 600 there exists an embedding $W_0 \hookrightarrow E(W_0)$. By (ii), $E(W_0)$ is an injective cogenerator. ∎

The reason why injective cogenerators are important is explained by the following.

Theorem 644. *(i) Let* $\mathbf{A}$ *be any ring and let* M_i $(1 \leq i \leq 3)$ *and* W *be left* $\mathbf{A}$*-modules. Assume that* W *is an* injective cogenerator*. Then the sequence*

$$M_1 \xrightarrow{\bullet\alpha} M_2 \xrightarrow{\bullet\beta} M_3 \qquad (3.58)$$

is exact in ${}_{\mathbf{A}}\mathbf{Mod}$ *if, and only if the sequence*

$$\mathfrak{B}_W(M_1) \xleftarrow{\mathfrak{B}_W(\alpha)\bullet} \mathfrak{B}_W(M_2) \xleftarrow{\mathfrak{B}_W(\beta)\bullet} \mathfrak{B}_W(M_3) \qquad (3.59)$$

is exact in $\mathbf{Ab}$ *(and in the category* $\mathbf{Mod}_{\mathbf{E}}$ *as well, where* $\mathbf{E} = \mathrm{End}_{\mathbf{A}}(W)$*).*
(ii) Let $\mathcal{C}$ *be an abelian category with arbitrary products and let* $\mathbf{E} \triangleq \mathrm{Hom}_{\mathcal{C}}(W,W)$*. Then for any object* M *of* $\mathcal{C}$*, the abelian group* $\mathrm{Hom}_{\mathcal{C}}(M,W)$ *of all left morphisms* $\bullet f : M \to W$ *is a right* $\mathbf{E}$*-module via composition of morphisms, and the above correspondence is still valid between* $\mathcal{C}$ *and* $\mathbf{Mod}_{\mathbf{E}}$*. Therefore, the object* W *is an injective cogenerator if, and only if the functor* $\mathfrak{B}_W$ *is faithfully exact.*

Proof. (i) The "only if" part is obvious since W is injective. Let us prove the "if" part and assume that (3.59) is exact.

(1) Suppose that $\mathrm{im}\,\alpha \nsubseteq \ker\beta$ and let $b \in \mathrm{im}\,\alpha \backslash \ker\beta$. Then $(b)\,\beta \neq 0$, and since W is a cogenerator there exists (by Theorem 638(iii)) a map $\bullet f : M_3 \to W$ such that $(b)\,\beta f \neq 0$. In addition there exists $a \in M_1$ such that $b = (a)\,\alpha$, therefore $(a)\,\alpha\beta f \neq 0$. This implies $\alpha\beta f \neq 0$, i.e., $\mathfrak{B}_W(\alpha\beta)(f) \neq 0$. Therefore, $\mathfrak{B}_W(\alpha\beta) = \mathfrak{B}_W(\alpha) \circ \mathfrak{B}_W(\beta) \neq 0$, a contradiction.

(2) Suppose that $\mathrm{im}\,\alpha \subsetneq \ker\beta$ and let $b \in \ker\beta \backslash \mathrm{im}\,\alpha$. Then, $\bar{b} \neq 0$, where $\bar{b}$ is the canonical image of $\bar{b}$ in $M_2/\,\mathrm{im}\,\alpha$. Since W is a cogenerator, there exists $\bullet f : M_2/\,\mathrm{im}\,\alpha \to W$ such that $(\bar{b})\, f \neq 0$. Let $\bullet g : M_2 \to W$ be defined by the composition below

$$M_2 \xrightarrow{\bullet\varphi} M_2/\,\mathrm{im}\,\alpha \xrightarrow{\bullet f} W$$

where $\bullet\varphi$ is the canonical epimorphism. Then $(b)\,g \neq 0$. But $\mathfrak{B}_W(\alpha)(g) = \bullet\alpha g = 0$ since $(\mathrm{im}\,\alpha)\,g = 0$, therefore $g \in \ker\mathfrak{B}_W(\alpha) = \mathrm{im}\,\mathfrak{B}_W(\beta)$, and there exists $\bullet h \in \mathfrak{B}_W(M_3)$ such that $g = \mathfrak{B}_W(\beta)(h) = \beta h$. Since $b \in \ker\beta$, $(b)\,g = (b)\,\beta h = 0$, a contradiction.

(ii) is deduced from (i) using the Freyd-Mitchell theorem (Theorem 50). ∎

Remark 645. *Consider in* $\mathbf{Mod}_{\mathbf{E}}$ *an exact sequence*

$$\mathfrak{B}_W(M_1) \xleftarrow{\lambda\bullet} \mathfrak{B}_W(M_2) \xleftarrow{\mu\bullet} \mathfrak{B}_W(M_3)\,.$$

There does not necessarily exist in ${}_{\mathbf{A}}\mathbf{Mod}$ *an exact sequence (3.58), even if* W *is an injective cogenerator, for* $\mathfrak{B}_W\left({}_{\mathbf{A}}\mathbf{Mod}\right)$ *is not necessarily a full subcategory of* $\mathbf{Mod}_{\mathbf{E}}$. *See the notion of* large injective cogenerator *in* (**5.3.2.2**) *below.*

3.5.6.2 Construction of Cogenerators

The general result given below is often useful ([30], §X.1, Prop. 13):

Theorem 646. *Let* $\mathbf{A}$, $\mathbf{S}$ *be rings,* U *a left* $\mathbf{A}$*-module which is injective cogenerator and let* P *be an* $(\mathbf{A},\mathbf{S})$*-bimodule. If* P *is faithfully flat over* $\mathbf{S}$*, then the left* $\mathbf{S}$*-module* $\mathrm{Hom}_{\mathbf{A}}(P,U)$ *is an injective cogenerator.*

Proof. We already know that $\mathrm{Hom}_{\mathbf{A}}(P,U)$ is an injective left $\mathbf{S}$-module (Proposition 608). Let S be a simple left $\mathbf{S}$-module; then, according to the Adjoint Isomorphism theorem (Theorem 284), we have the isomorphism of Abelian groups

$$\mathrm{Hom}_{\mathbf{A}}\left(S,\mathrm{Hom}_{\mathbf{A}}(P,U)\right)\cong\mathrm{Hom}_{\mathbf{S}}\left(P\bigotimes\nolimits_{\mathbf{A}}S,U\right).$$

Since P is faithfully flat and $S\neq 0$, $P\bigotimes_{\mathbf{A}}S\neq 0$ by Lemma and Definition 614(ii), thus $\mathrm{Hom}_{\mathbf{S}}\left(P\bigotimes_{\mathbf{A}}S,U\right)\neq 0$ since U is cogenerator. Thus $\mathrm{Hom}_{\mathbf{A}}\left(S,\mathrm{Hom}_{\mathbf{A}}(P,U)\right)\neq 0$ and $\mathrm{Hom}_{\mathbf{A}}(P,U)$ is cogenerator by Theorem 642(i). ■

Corollary 647. *Let* $\mathbf{A}$, $\mathbf{S}$ *be rings, let* U *be a left* $\mathbf{A}$*-module which is an injective cogenerator and let* $\rho:\mathbf{A}\to\mathbf{S}$ *be a ring-homomorphism. Using the change-of-ring functor* (**3.4.2.1**)*, the* $(\mathbf{S},\mathbf{S})$*-bimodule* ${}_{\mathbf{S}}\mathbf{S}_{\mathbf{S}}$ *is endowed with an obvious structure of* $(\mathbf{A},\mathbf{S})$*-bimodule, setting* $r\,.\,s\,s'=\rho(r)\,s\,s'$ *for any* $r\in\mathbf{A}$ *and* $s,s'\in\mathbf{S}$*. Then the left* $\mathbf{S}$*-module* $\mathrm{Hom}_{\mathbf{A}}(\mathbf{S},U)$ *is an injective cogenerator.*

Proof. This is a consequence of Theorem 646 since the right $\mathbf{S}$-module $\mathbf{S}_{\mathbf{S}}$ is faithfully flat (Corollary 616). ■

Corollary 648. *Let* $\mathbf{K}$ *be a field and let* $\mathbf{S}$ *be a unitary* $\mathbf{K}$*-algebra* (**2.2.2.3**)*. Then the* $\mathbf{S}$*-module* $\mathbf{S}^{*}\triangleq\mathrm{Hom}_{\mathbf{K}}(\mathbf{S},\mathbf{K})$ *is a left and right injective cogenerator.*

Proof. The $\mathbf{K}$-vector space $\mathbf{K}$ is an injective $\mathbf{K}$-module by the Baer Criterion (Criterion 591). The only maximal ideal of $\mathbf{K}$ is 0, thus ${}_{\mathbf{K}}\mathbf{K}_{\mathbf{K}}=\mathbf{K}/0$ is the only simple $\mathbf{K}$-module, and $\mathrm{Hom}_{\mathbf{K}}(\mathbf{K},\mathbf{K})\cong\mathbf{K}\neq 0$, thus ${}_{\mathbf{K}}\mathbf{K}_{\mathbf{K}}$ is an injective cogenerator. Therefore the $\mathbf{S}$-module $\mathbf{S}^{*}$ is a left and right injective cogenerator by Corollary 647 (with $\mathbf{A}=\mathbf{K}$ and $U={}_{\mathbf{K}}\mathbf{K}_{\mathbf{K}}$). ■

3.6 Modules over Specific Rings

3.6.1 Modules over Bézout Domains

3.6.1.1 Matrices over Bézout Domains

In what follows, $\operatorname{diag}(a_1, ..., a_n)$ denotes any $q \times k$ matrix with $\min(q, k) = n$, $a_1, ..., a_n$ on the main diagonal (from top to down) and 0s elsewhere. If $\mathbf{A}$ is a Bézout domain, the notion of rank of a matrix over $\mathbf{A}$ is not ambiguous by Corollary 442.

By Lemma and Definition 410, a matrix A over a semifir is left-equivalent (Definition 280) to a matrix $\begin{bmatrix} T \\ 0 \end{bmatrix}$ where the rows of T are left $\mathbf{A}$-linearly independent. By symmetry, A is right-equivalent to a matrix $\begin{bmatrix} T & 0 \end{bmatrix}$ where the columns of T are right $\mathbf{A}$-linearly independent.

More precision can be given over a Bézout domain. For this, *elementary operations on the rows and on the columns* are necessary (Subsect. 2.10.3) and *secondary operations on the rows and on the columns* as well (see below).

Definition 649. *A* secondary operation *on the rows (resp., on the columns) consists in left- (resp., right-) multiplying two rows (resp., two columns) by an invertible* 2×2 *matrix.*

Lemma 650. *Let* $\mathbf{A}$ *be a domain. An arbitrary row* $\begin{pmatrix} a & b \end{pmatrix}$ $(a, b \in \mathbf{A})$ *is right-equivalent to a diagonal form* $\begin{pmatrix} d & 0 \end{pmatrix}$ *if, and only if* $\mathbf{A}$ *is a right Bézout domain. Then,* d *is a* gcld *of* a *and* b.

Proof. (i) Let $\mathbf{A}$ be a right Bézout domain, let $\begin{pmatrix} a & b \end{pmatrix}$ be a row over $\mathbf{A}$ and let d be a gcld of a and b (Theorem 443). Then $a = d\,a'$, $b = d\,b'$ where (a', b') is strongly left-coprime, i.e., $\begin{pmatrix} a' & b' \end{pmatrix}$ is right-invertible. Therefore, $\begin{pmatrix} a' & b' \end{pmatrix}$ can be completed to an invertible matrix Q^{-1}. This yields $\begin{pmatrix} a' & b' \end{pmatrix} Q = \begin{pmatrix} 1 & 0 \end{pmatrix}$. Left-multiplying this expression by d, one obtains

$$\begin{pmatrix} a & b \end{pmatrix} Q = \begin{pmatrix} d & 0 \end{pmatrix}. \tag{3.60}$$

(ii) Conversely, assume that for any row $\begin{pmatrix} a & b \end{pmatrix}$, there exists Q such that (3.60) holds. Then $d \in a\,\mathbf{A} + b\,\mathbf{A}$, and $\begin{pmatrix} a & b \end{pmatrix} = \begin{pmatrix} d & 0 \end{pmatrix} Q^{-1}$, therefore $a, b \in d\,\mathbf{A}$. It follows that $a\,\mathbf{A} + b\,\mathbf{A} = d\,\mathbf{A}$. By induction, every f.g. right ideal in $\mathbf{A}$ is principal, thus $\mathbf{A}$ is a right Bézout domain. ■

Corollary and Definition 651. *(1) Let* $\mathbf{A}$ *be a left (resp., right) Bézout domain and* $A \in {}^{q}\mathbf{A}^{k}$. *Then* A *is left (resp., right) equivalent to an upper (resp., a lower) triangular matrix* T, *called a left (resp., right)* Hermite form *of* A.
(2) If $\mathbf{A}$ *is a Bézout domain, there exists a permutation matrix* P *such that the left (resp., right) Hermite form of* AP *(resp.,* PA*) is* $\begin{bmatrix} T_1 & 0 \\ 0 & 0 \end{bmatrix}$ *where* $T_1 \in \operatorname{Mat}_r(A)$ *is upper (resp., lower) triangular of rank* r.

Proof. (1) Assume that $\mathbf{A}$ is a right Bézout domain. If the ith row of A is nonzero, one can bring a nonzero element a_i in the (i,i) position in A using elementary operations of the third kind on the columns (Subsect. 2.10.3). Then, using secondary operations on the columns, one can replace successively $a_{i,i}$ by a gcld of the pair $(a_{i,i}, a_{1,i+1})$ and $a_{1,i+1}$ by 0, then the new $a_{i,i}$ by a gcld of the pair $(a_{i,i}, a_{i,i+2})$ and $a_{i,i+2}$ by 0, etc. Doing this for all nonzero rows, a lower triangular matrix is obtained.

(2) There exists a permutation matrix P such that $PA = \begin{bmatrix} A_1 \\ 0 \end{bmatrix}$ where no row of A_1 is zero. The left Hermite form of PA is $\begin{bmatrix} T_1 & 0 \\ 0 & 0 \end{bmatrix}$ where $T_1 \in \operatorname{Mat}_r(A)$ is upper triangular and no diagonal element of T_1 is zero. Since $\mathbf{A}$ is an Ore domain, $\operatorname{rk}(T_1) = r$. ∎

Definition 652. *(1) Let $\mathbf{A}$ be a ring, $A \in {}^n\mathbf{A}^m$ and $B \in {}^n\mathbf{A}^p$ two matrices over $\mathbf{A}$ with the same number of rows.*
(i) A matrix $L \in \operatorname{Mat}_n(\mathbf{A})$ is a common left divisor *of A, B if L is a left divisor of $\begin{bmatrix} A & B \end{bmatrix}$, i.e., there exists a matrix $\begin{bmatrix} A' & B' \end{bmatrix}$ over $\mathbf{A}$ such that*

$$\begin{bmatrix} A & B \end{bmatrix} = L \begin{bmatrix} A' & B' \end{bmatrix}.$$

(ii) A matrix $L \in \operatorname{Mat}_n(\mathbf{A})$ is a greatest common left divisor (gcld) *of A, B if (a) L is a common divisor of A, B, (b) every common left divisor of A, B is a left divisor of L.*
(2) A common right divisor and a greatest common right divisor (gcrd) *of two matrices with the same number of columns are likewise defined (see Definition 81).*

Theorem 653. *Let $\mathbf{A}$ be a Bézout domain, let $A \in {}^n\mathbf{A}^m$, let $B \in {}^n\mathbf{A}^p$ and let $r = \operatorname{rk}\begin{bmatrix} A & B \end{bmatrix}$.*

(i) A and B have a gcld $L \in {}^n\mathbf{A}^r$, of rank r, such that $\begin{bmatrix} A & B \end{bmatrix} \equiv_r \begin{bmatrix} L & 0 \end{bmatrix}$.
(ii) If $r = n$, then L is unique up to right-equivalence.
(iii) A and B are left-coprime if, and only if they are strongly left-coprime. Similar properties hold for two matrices $A \in {}^m\mathbf{A}^n$ and $B \in {}^p\mathbf{A}^n$, and left *and* right *interchanged.*

Proof. (i) Let $r = \operatorname{rk}\begin{bmatrix} A & B \end{bmatrix}$. By Corollary and Definition 651 there exist a permutation matrix $P \in \operatorname{GL}_n(\mathbf{A})$, a matrix $Q \in \operatorname{GL}_k(\mathbf{A})$ $(k = m + p)$ and an upper triangular matrix $T \in {}^n\mathbf{A}^r$ of rank r, such that

$$P \begin{bmatrix} A & B \end{bmatrix} Q^{-1} = \begin{bmatrix} T & 0 \end{bmatrix}.$$

Therefore,

$$\begin{bmatrix} A & B \end{bmatrix} = P^{-1} T \begin{bmatrix} I_n & 0 \end{bmatrix} Q$$

i.e.,

$$\begin{bmatrix} A \ B \end{bmatrix} \equiv_r \begin{bmatrix} L \ 0 \end{bmatrix}, \quad L = P^{-1} T,$$

thus $L \in {}^n\mathbf{A}^r$ is a gcld of A, B and $\mathrm{rk}\,(L) = r$.

(ii) If L, L' are two gclds of A, B, L' left-divides L and there exists a matrix $C \in \mathrm{Mat}_n\,(\mathbf{A})$ such that $L = L'C$. Likewise there exists a matrix $C' \in \mathrm{Mat}_n\,(\mathbf{A})$ such that $L' = L\,C'$. Therefore, $L\,(I_n - C'\,C) = 0$ and since $\mathrm{rk}\,(L) = n$, this implies $C'\,C = I_n$. Since $\mathbf{A}$ is weakly finite (Exercise 481), C' and C are invertible, thus $L \equiv_r L'$.

(iii) is a consequence of (i), taking $L = I_n$. ∎

3.6.1.2 Modules over Bézout Domains

Theorem 654. *Let* $\mathbf{A}$ *be a Bézout domain and let* M *be a f.g. left* $\mathbf{A}$*-module. Then there exists a finite free left* $\mathbf{A}$*-module* Φ *such that*

$$M = \mathcal{T}\,(M) \oplus \Phi \tag{3.61}$$

where $\mathcal{T}\,(M)$ *is the torsion submodule of* M *(Definition 339(v)). The module module* Φ *is uniquely defined up to isomorphism, since* $\Phi \cong M/\mathcal{T}\,(M)$*, and* $\mathrm{rk}\,(\Phi) = \mathrm{rk}\,(M)$ *(Definition 408).*
In particular, if the f.g. left $\mathbf{A}$*-module is torsion-free, then* $M = \Phi$ *is free.*

Proof. Since $\mathbf{A}$ is a semifir, the f.g. module M is finitely presented (Theorem 499). Let $R \in {}^q\mathbf{A}^k$ be a nonzero matrix of definition of the k-generated module M. Then there exist by Corollary 651 invertible matrices P and Q such that

$$P^{-1}\,R\,Q = \begin{bmatrix} T_1 & 0 \\ 0 & 0 \end{bmatrix}, \quad T_1 \in \mathrm{Mat}_r\,(\mathbf{A})\,, \quad \mathrm{rk}\,(T_1) = r \tag{3.62}$$

and $M \cong \mathrm{coker}_{\mathbf{A}} \begin{bmatrix} T_1 & 0 \\ 0 & 0 \end{bmatrix} \triangleq M'$. Let $m = \begin{bmatrix} m' \\ m'' \end{bmatrix}$ be a column of generators of M' where m' and m'' have respectively r and $k - r$ elements. An equation of the module M' is (see (3.1))

$$\begin{bmatrix} T_1 \ 0 \end{bmatrix} m = 0, \tag{3.63}$$

thus a presentation of M is

$$0 \longrightarrow \mathbf{A}^r \xrightarrow{\bullet R'} \mathbf{A}^k \longrightarrow M \longrightarrow 0$$

where $R' = \begin{bmatrix} T_1 \ 0 \end{bmatrix}$; this is a finite free resolution of length 1. In addition, (3.63) is equivalent to $T_1 m' = 0$. Therefore, $M' = [m]_{\mathbf{A}} = [m']_{\mathbf{A}} \oplus [m'']_{\mathbf{A}}$ where $[m']_{\mathbf{A}} = \mathrm{coker}_{\mathbf{A}}\,(\bullet T_1)$ and $[m'']_{\mathbf{A}} = \mathbf{A}^{k-r}$. By Theorem and Definition 339, the module $\mathrm{coker}_{\mathbf{A}}\,(\bullet T_1)$ is torsion. Therefore, (3.61) holds and by (2.10), $\Phi \cong M/\mathcal{T}\,(M)$. ∎

The above shows that a f.g. left $\mathbf{A}$-module M (where $\mathbf{A}$ is a Bézout domain) is free if, and only if $M \cong \operatorname{coker}_{\mathbf{A}}(\bullet R)$ where R is right-equivalent to a matrix $\begin{bmatrix} I_r & 0 \end{bmatrix} \in {}^r\mathbf{A}^k$; then, $M \cong \mathbf{A}^{k-r}$.

3.6.2 Modules over Principal Ideal Domains

3.6.2.1 Diagonal Reduction and Elementary Divisor Rings

Lemma and Definition 655. *(i) A matrix $R \in {}^q\mathbf{A}^k$ admits* diagonal reduction *if*

$$R \equiv \operatorname{diag}(e_1, ..., e_r, 0, ..., 0), \quad e_i \parallel e_{i+1},\ e_r \neq 0$$

(see Definition 89). Let $M = \operatorname{coker}_{\mathbf{A}}(\bullet R)$; then $M = \mathcal{T}(M) \oplus \Phi$ where Φ is finite free with $\operatorname{rk} \Phi = \operatorname{rk} M$ *and*

$$\mathcal{T}(M) \cong \bigoplus\nolimits_{1 \leq i \leq r} \mathbf{A}/\mathbf{A}\, e_i. \tag{3.64}$$

(ii) If the elements e_i are uniquely determined up to similarity, then $\operatorname{diag}(e_1, ..., e_r, 0, ..., 0)$ *is called the* Smith normal form *of R, and the elements e_i $(1 \leq i \leq r)$ are called the* nonzero invariant factors *of R or of M; R and M are said to have* $\operatorname{rk}(M)$ *zero invariant factors. (When $\mathbf{A}$ is noncommutative, the above normal form is often called the* Jacobson-Teichmüller normal form *of R, but we will keep the name* Smith form, *for the sake of simplicity.) When the ring $\mathbf{A}$ is simple, then $e_i = 1$ for all $i \in \{1, ..., r-1\}$ and the Smith normal form of R is* $\operatorname{diag}(1, ..., 1, e_r, 0, ..., 0)$.
(iii) A domain $\mathbf{A}$ is called an elementary divisor ring *if every finite matrix over $\mathbf{A}$ admits diagonal reduction.*

More general elementary divisor rings, which can have zerodivisors, were defined by Kaplansky [175].

Theorem 656. *A domain $\mathbf{A}$ is an elementary divisor ring if, and only if (i) it is a Bézout domain and (ii) all 2×2 matrices over $\mathbf{A}$ admit diagonal reduction.*

Proof. The necessary condition is a consequence of Lemma 650. Let us prove the sufficient condition. Let $A \in {}^q\mathbf{A}^k$, $3 \geq q \geq k$ (without loss of generality) and let us proceed by induction. The induction assumption is that diagonal reduction is possible for smaller q, and for a given q for smaller k.

Let $A = \begin{bmatrix} A_1 \\ A_2 \end{bmatrix}$ where A_1 is the first row of A. We can find invertible matrices P_1, Q_1 such that $P_1 A_2 Q_1 = \operatorname{diag}(b_1, ..., b_{q-1}) = B$, $b_i \parallel b_{i+1}$ if $b_{i+1} \neq 0$, hence

$$C = \begin{bmatrix} 1 & 0 \\ 0 & P_1 \end{bmatrix} \begin{bmatrix} A_1 \\ A_2 \end{bmatrix} Q_1 = \begin{bmatrix} A_1 Q_1 \\ B \end{bmatrix}.$$

Let $C = \begin{bmatrix} C_1 \\ C_2 \end{bmatrix}$ where C_1 consists of the first two rows of C. We can find invertible matrices P_2, Q_2 such that $P_2\, C_1\, Q_2 = \mathrm{diag}\,(e_1, e_1') = C$, $e_1 \parallel e_1'$ if $e_1' \neq 0$, hence

$$F = \begin{bmatrix} P_2 & 0 \\ 0 & I_{q-2} \end{bmatrix} \begin{bmatrix} E_1 \\ E_2 \end{bmatrix} Q_2 = \begin{bmatrix} C \\ D \end{bmatrix}$$

where $D = E_2\, Q_2$. Now, e_1 is a total divisor of all entries of E_1, thus $e_1 \parallel b_1$; likewise b_1 is a total divisor of all entries of D, thus e_1 is a total divisor of all entries of F. Using elementary operations to sweep out the first column of F we reach

$$\begin{bmatrix} e_1 & 0 \\ 0 & G \end{bmatrix}$$

where e_1 is a total divisor of all entries of G. Applying induction to G we obtain the complete reduction. ∎

Corollary 657. *Let* $\mathbf{A}$ *be a commutative domain;* $\mathbf{A}$ *is an elementary divisor ring if, and only if (i)* $\mathbf{A}$ *is a commutative Bézout domain, and (ii) if for any* $a, b, c \in \mathbf{A}$ *such that* $(a, b, c) = 1$*, there exist* $p, q \in \mathbf{A}$ *such that* $(p\,a, p\,b + q\,c) = 1$ *(with the notation after Proposition 82).*

Proof. (1) We know that (i) is necessary. (2) Let us prove the necessity of (ii). Let

$$A = \begin{bmatrix} a & b \\ 0 & c \end{bmatrix} \tag{3.65}$$

where $(a, b, c) = 1$ and assume $P\,A\,Q$ effects the diagonal reduction of A. Then $P\,A\,Q = \mathrm{diag}\,(v, e)$ where v is a unit. Setting $P = \begin{bmatrix} p & * \\ q & * \end{bmatrix}$, $Q = \begin{bmatrix} x & y \\ * & * \end{bmatrix}$, we obtain $p\,a\,x + p\,b\,y + q\,c\,y = v$, whence $(p\,a, p\,b + q\,c) = 1$.

(3) Let us prove the sufficiency of (i), (ii). By (i) and Lemma 650, we may assume A of the form (3.65). Assuming $(p\,a, p\,b + q\,c) = 1$, then $(p, q) = 1$ and there exists an invertible matrix $P = \begin{bmatrix} p & q \\ * & * \end{bmatrix}$ which yields

$$P\,A = \begin{bmatrix} p\,a & p\,b + q\,c \\ * & * \end{bmatrix} \equiv \begin{bmatrix} 1 & 0 \\ 0 & * \end{bmatrix}.$$

∎

The Bézout domain in Example 440 was proved in [137] to be an elementary divisor ring which is not a principal ideal domain. So is also the ring in Example 439 (see Exercise 698). No example of commutative domain Bézout which is not an elementary divisor ring is presently known.

3.6.2.2 The Case of Principal Ideal Domains

Theorem 658. *A principal ideal domain* $\mathbf{A}$ *is an elementary divisor ring and every nonzero matrix* A *over* $\mathbf{A}$ *has diagonal reduction* $\operatorname{diag}(e_1, ..., e_r, 0, ..., 0)$, $e_r \neq 0$, *where the elements* e_i $(1 \leq i \leq r)$ *are uniquely determined up to similarity.*

Proof. By Theorem 656, it is sufficient to prove that any 2×2 matrix A over $\mathbf{A}$ has diagonal reduction.

(1) If $A = 0$, there is nothing to prove. If not, bring a nonzero element in position (1,1). Since $\mathbf{A}$ is atomic (Corollary 461), the length $|a_{11}|$ (Definition 119) of a_{11} is finite. By Lemma 650, A can be transformed, using a secondary column operation, into a form where the first row is $\begin{bmatrix} a_{11} & 0 \end{bmatrix}$ with a new a_{11}, the length of which is not greater than the length of the old a_{11}. The same process can be applied to the first column of A, using a secondary row operation, and the element in position (2,1) becomes zero. Doing this, the element in position (1,2) may again become nonzero, but, as easily seen, this can happen only if the old a_{21} is not a right-multiple of the old a_{11}, and then $|a_{11}|$ is reduced. Therefore, after a finite number of steps, A becomes diagonal, say $\operatorname{diag}(a_1, a_2)$.

(2) For any $x, y \in \mathbf{A}$,

$$\begin{bmatrix} 1 & x \\ 0 & 1 \end{bmatrix} \begin{bmatrix} a_1 & 0 \\ 0 & a_2 \end{bmatrix} \begin{bmatrix} y & 1 \\ 1 & 0 \end{bmatrix} = \begin{bmatrix} a_1' & a_1 \\ a_2 & 0 \end{bmatrix}.$$

where $a_1' = a_1 y + x\, a_2 \in a_1\mathbf{A} + \mathbf{A}a_2$. We have $a_1\mathbf{A} + \mathbf{A}a_2 \supsetneq a_1\mathbf{A}$ except if $\mathbf{A}a_2 \subseteq a_1\mathbf{A}$, i.e., if the following condition (C) holds: $\mathbf{A}\, a_2 \mathbf{A} \subseteq a_1 \mathbf{A}$.

(a) Condition (C) holds if, and only if $a_1 \mid c \mid a_2$ where c is an invariant generator of the ideal $\mathbf{A}\, a_2 \mathbf{A}$ (Lemma 447), and this is equivalent to $a_1 \parallel a_2$.

(b) If Condition (C) does not hold, one can choose x, y such that $a_1 = a_1' q_1$, for some *nonunit* q_1. Then, $|a_1'| < |a_1|$.

Therefore, by iterating the whole process, $|a_1|$ can be diminished until $a_1 \parallel a_2$ and, as a result, any 2×2 matrix over $\mathbf{A}$ has diagonal reduction.

The proof of the uniqueness of the diagonal elements e_i up to similarity ([77], Sect. 8.2, Theorem 2.4) is based on the lemma below which is proved in ([77], Sect. 8.2, Corol. 2.5). ∎

Lemma 659. *Over a principal ideal domain* $\mathbf{A}$, *finitely generated modules can be cancelled, i.e., if*

$$M \oplus N \cong M \oplus N',$$

where M, N, N' *are f.g., then* $N \cong N'$.

Remark 660. *(i) To our knowledge, the statement of Lemma 659 is no longer valid if* $\mathbf{A}$ *is only assumed to be a Bézout domain or even an elementary divisor ring. Indeed, for the proof of ([77], Sect. 8.2, Corol. 2.5) to be correct,* $\mathbf{A}$ *must be atomic (see [77], Sect. 3.6, Theorem 6.3).*

(ii) The diagonal reduction is obtained using elementary and secondary operations on the rows and the columns. If $\mathbf{A}$ *is an Euclidean domain, the diminution the length* $|a_{11}|$ *can be replaced by the diminution of the degree* $\theta(a_{11})$ *using the Euclidean algorithm, and for this secondary operations are unnecessary. Therefore, if* $\mathbf{A}$ *is a commutative Euclidean domain,* $\mathrm{SL}_n(\mathbf{A})$ *(Example 144) is generated by all elementary matrices (Subsect. 2.10.2), i.e.,* $\mathrm{SL}_n(\mathbf{A}) = \mathrm{E}_n(\mathbf{A})$, *and* $\mathrm{GL}_n(\mathbf{A}) = \mathrm{D}_n(\mathbf{A})\,\mathrm{E}_n(\mathbf{A}) = \mathrm{E}_n(\mathbf{A})\,\mathrm{D}_n(\mathbf{A})$ *(see Proposition 393(i)).*

Based on the above lemma, the following is proved in ([43], Lemma 560 and Proposition 561):

Corollary 661. *Let* $\mathbf{A}$ *be a principal ideal domain, let* M *be a f.g. left* $\mathbf{A}$*-module and let* F *be a free submodule of* M.

(i) There exists a maximal free submodule Φ_F *of* M *containing* F *and* $M = \mathcal{T}(M) \oplus \Phi_F$.
(ii) Let $N \lhd M$*; there exists a free submodule* Φ_N *of* M *such that* $M = \mathcal{T}(M) \oplus \Phi_N$ *and* $N = \mathcal{T}(N) \oplus (\Phi_N \cap N)$. *If* Φ'_N *is another free submodule of* M *such that* $M = \mathcal{T}(M) \oplus \Phi'_N$ *and* $N = \mathcal{T}(N) \oplus (\Phi'_N \cap N)$, *then* $\Phi'_N \cong \Phi_N$ *and* $\Phi'_N / (\Phi'_N \cap N) \cong \Phi_N / (\Phi_N \cap N)$.

The diagonal reduction in Theorem 658 yields the following decomposition of $M = \operatorname{coker}_{\mathbf{A}}(\bullet R)$ into cyclic modules:

$$M \cong \bigoplus\nolimits_{1 \leq i \leq r} \mathbf{A}/\mathbf{A}\, e_i \bigoplus\nolimits^{k-r} \mathbf{A} \tag{3.66}$$

and the condition that $e_i \parallel e_{i+1}$ $(1 \leq i \leq r-1)$ ensures that this decomposition has as few terms as possible provided that the zero modules $\mathbf{A}/\mathbf{A}\, e_i$ where e_i is a unit are removed ([77], Sect. 8.2, Exercise 3). Now, considering the primary decomposition of each nonzero cyclic torsion module (Lemma and Definition 543) we obtain the following.

Theorem and Definition 662. *Let* $\mathbf{A}$ *be a principal ideal domain and let* T *be a finitely generated torsion left module over* $\mathbf{A}$.

(1) Then

$$T \cong \bigoplus\nolimits_{1 \leq i \leq m} \mathbf{A}/\mathbf{A}\, q_i \bigoplus \mathbf{A}/\mathbf{A}\, u \tag{3.67}$$

where each q_i *is a product of pairwise similar bounded atoms, while* u *is totally unbounded. If* $\mathbf{A}$ *is simple,* $T \cong \mathbf{A}/\mathbf{A}\, u$*; if* M *is bounded,* $\mathbf{A}/\mathbf{A}\, u = 0$.
(2) The similarity classes of the elements q_i *(or, abusing the language, the ideals* $\mathbf{A}\, q_i$*, or the elements of these ideals, one for each ideal, taking into account multiplicities) are called the* elementary divisors *of* T *(and of any matrix of definition of* T*).*

Proof. (1) is easily deduced from Theorem 658 when T is bounded, and one is led to consider the case $m = 0$. Then by (3.66) and (3.24), $T \cong \bigoplus_{1 \leq i \leq r} \mathbf{A}/\mathbf{A}\, u_i$ where $u_i \parallel u_{i+1}$ and each u_i $(1 \leq i \leq r)$ is a unit or is totally unbounded; thus u_i is a unit for $1 \leq i \leq r-1$ and $T \cong \mathbf{A}/\mathbf{A}\, u_r$. ∎

The integer $k - r$ in (3.66) is called the *multiplicity of the elementary divisor* 0 of M ([28], §VII.4, Definition 4).

Corollary 663. *Let T_1, T_2 be two f.g. torsion* **A***-modules.*

(i) Let $\varepsilon(T_i)$ be the set of elementary divisors of T_i $(i = 1, 2)$. Then,

$$\varepsilon(T_1 \oplus T_2) = \varepsilon(T_1)\, \dot{\cup} \varepsilon(T_2)$$

where $\dot{\cup}$ is the disjoint union (**1.2.1.3**).
(ii) Let $T_2 \lhd T_1$, and let e_{ij} $(1 \leq i \leq r)$ be the invariant factors of T_j $(j = 1, 2)$. There exist elements e'_{i2} such that $e_{i1} = e_{i2}\, e'_{i2}$ for each $i \in \{1, ..., r\}$, and

$$T_1/T_2 \cong \bigoplus\nolimits_{1 \leq i \leq r} \mathbf{A}/\mathbf{A}\, e'_{i2}.$$

Proof. (i) is clear by (3.67).
(ii): We have by (3.64), for $j = 1, 2$,

$$T_j = \bigoplus\nolimits_{1 \leq i \leq r} C_{ij}, \quad C_{ij} \cong \mathbf{A}/\mathbf{A}\, e_{ij}.$$

Since $T_2 \lhd T_1$, we have $C_{i2} \lhd C_{i1}$ $(1 \leq i \leq r)$. Therefore, by Proposition 528, there exists for each i an element e'_{i2} such that $e_{i1} = e_{i2}\, e'_{i2}$. In addition, by (2.9),

$$T_1/T_2 \cong \bigoplus\nolimits_{1 \leq i \leq r} C_{i1}/C_{i2}$$

and by Noether's third isomorphism theorem (Theorem 52(3)),

$$C_{i1}/C_{i2} \cong \mathbf{A}/\mathbf{A}\, e'_{i2}.$$ ∎

3.6.3 Modules over Dedekind Domains

Finitely generated modules over Dedekind domains have properties similar to those of modules over principal ideal domains. First, such a module is the direct sum of a torsion module of finite length (Theorem and Definition 160) and of a torsion-free module. Second, this torsion-free module is the direct sum of a finite free module and of an ideal. More specifically (see [247], 5.7.4, 5.7.8, 5.7.12, 5.7.17, [31], §6.4, Prop. 23, and Proposition 457):

Theorem 664. *(1) Let* **A** *be a* hereditary Noetherian prime ring *and let M be a* f.g. left *A-module. Then*

$$M = \mathcal{T}(M) \oplus P$$

where $\mathcal{T}(M)$ has finite length and $P \cong M/\mathcal{T}(M)$ is projective.

(2) If $\mathbf{A}$ *is a* Dedekind domain *(not necessarily commutative), then*

$$P \cong \mathbf{A}^n \oplus \mathfrak{I}$$

for some n *and some left ideal* $\mathfrak{I}$ *in* $\mathbf{A}$ *(generated by two elements according to Proposition 457) and* $\mathcal{T}(M)$ *is a direct sum of cyclic submodules, i.e.*

$$\mathcal{T}(M) \cong \bigoplus_{1 \leq i \leq r} \mathbf{A}/\mathfrak{a}_i$$

where each $\mathfrak{a}_i$ $(1 \leq i \leq r)$ *is a nonzero left ideal generated by two elements. If* $\mathbf{A}$ *is simple non-Artinian, then* $\mathcal{T}(M)$ *is cyclic (i.e.,* $r = 1$*) and completely faithful.*
(3) If $\mathbf{A}$ *is a* commutative Dedekind domain, *then there exist two finite families* $(n_i)_{i \in I}$ *and* $(\mathfrak{p}_i)_{i \in I}$ *where the* n_i'*s are positive integers and the* $\mathfrak{p}_i'$*s are nonzero prime ideals in* $\mathbf{A}$, *such that*

$$\mathcal{T}(M) \cong \bigoplus_{i \in I} \mathbf{A}/\mathfrak{p}_i^{n_i};$$

these families are unique up to a permutation of I.

Remark 665. *(i) Over the Dedekind domains* $A_1(\mathbf{k})$, $A_{1c}(\mathbf{k})$, $A_1'(\mathbf{k})$ *(where* $\mathbf{k} = \mathbb{R}$ *or* $\mathbb{C}$*) and* $A_o(\mathbb{C})$, *every f.g. torsion-free module is projective by Theorem 664, thus is stably-free by Theorem 637(i).*
(ii) Let $\mathbf{A}$ *be the* n*th Weyl Algebra* $A_n(\mathbf{k}), n \geq 1$ *(Exercise 493(v)). As shown by Stafford [336], every f.g. left* $\mathbf{A}$*-module is isomorphic to* $N \oplus \mathbf{A}^s$ *for some integer* $s \geq 0$, *where* N *is a module generated by two elements, and every stably-free* $A_n(\mathbf{k})$*-module of rank* ≥ 2 *is free.*
(iii) Over a commutative Dedekind domain, all stably-free modules are free ([247], 11.1.5); in other words, all commutative Dedekind domains are Hermite rings (Definition 579(ii)).

Item (i) of the above remark leads to the following.

Lemma and Definition 666. *Let* $\mathbf{A}$ *be a ring and consider the following conditions.*

(i) Every left or right ideal in $\mathbf{A}$ *is stably-free.*
(ii) Every f.g. torsion-free $\mathbf{A}$*-module is stably-free.*
(iii) Every f.g. left or right ideal in $\mathbf{A}$ *is stably-free.*
(1) If $\mathbf{A}$ *is a Noetherian domain, then (i)⇔(ii)⇔(iii). If these equivalent conditions hold,* $\mathbf{A}$ *is called a* stably-free ideal domain.
(2) If $\mathbf{A}$ *is an Ore domain, then (ii)⇔(iii). If these equivalent conditions hold,* $\mathbf{A}$ *is called a* semistably-free ideal domain.

Proof. (1) (ii)⇒(i): Assume that (ii) holds and let $\mathfrak{I}$ be a left ideal in $\mathbf{A}$. Then $\mathfrak{I}$ is a f.g. torsion-free module, therefore it is stably-free.
(i)⇒(ii): Assume that (i) holds and let P be a f.g. torsion-free $\mathbf{A}$-module. Since every left or right ideal is projective, $\mathbf{A}$ is a Dedekind domain by Theorem 586. Therefore, by Theorem 664, $P \cong \mathbf{A}^n \oplus \mathfrak{I}$ where $\mathfrak{I}$ is a left ideal.

Since $\mathfrak{I}$ is stably-free, say of rank $r \geq 0$, there exists an integer $q \geq 0$ such that $\mathfrak{I} \oplus \mathbf{A}^q \cong \mathbf{A}^{q+r}$. Therefore, $P \oplus \mathbf{A}^q \cong \mathbf{A}^{n+q+r}$ and P is stably-free of rank $n+r$. (i)$\Leftrightarrow$(iii) is clear.

(2) (ii)$\Rightarrow$(iii) is clear.

(iii)$\Rightarrow$(ii): If (iii) holds, $\mathbf{A}$ is semihereditary (Exercise 688). Let P be a torsion-free left $\mathbf{A}$-module. Since $\mathbf{A}$ is an Ore domain, there exists an integer $n > 0$ and an embedding $P \hookrightarrow \mathbf{A}^n$ (Proposition 342). Therefore, there exists a finite sequence of f.g. left ideals $(\mathfrak{I}_i)_{1 \leq i \leq k}$ such that $P \cong \bigoplus_{i=1}^k \mathfrak{I}_i$ ([199], Theorem (2.29)). For every index $i \in \{1, ..., k\}$, $\mathfrak{I}_i$ is stably-free, therefore there exist non-negative integers q_i and r_i such that $\mathfrak{I}_i \oplus \mathbf{A}^{q_i} \cong \mathbf{A}^{q_i+r_i}$. As a consequence,

$$P \oplus \mathbf{A}^q \cong \mathbf{A}^{q+r}$$

where $q = \sum_{1 \leq i \leq k} q_i$ and $r = \sum_{1 \leq i \leq k} r_i$, and P is stably-free. ■

All stably-free ideal domains are Dedekind domains. The Dedekind domains $A_1(\mathbf{k})$, $A_{1c}(\mathbf{k})$, $A_1'(\mathbf{k})$ and $A_o(\mathbb{C})$ are all stably-free ideal domains. All Bézout domains are semistably-free ideal domains. A Bézout domain is a stably-free ideal domain if, and only if it is a principal ideal domain.

3.7 Coprime Factorizations of Matrices

In what follows, $\mathbf{R}$ is an Ore domain and $\mathbf{F} = \mathrm{Q}(\mathbf{R})$ is its division ring of fractions.

3.7.1 Left-Coprime Factorizations and Right-Coprime Factorizations

Definition 667. *Let $G \in {}^p\mathbf{F}^m$.*

(i) The matrix G is said to have a left-coprime factorization *(D_l, N_l) over $\mathbf{R}$ if $D_l \in \mathrm{Mat}_p(\mathbf{R})$ is invertible over $\mathbf{F}$, $N_l \in {}^p\mathbf{R}^m$, the pair (D_l, N_l) is strongly left-coprime over $\mathbf{R}$ (Definition 518(i)), and $G = D_l^{-1} N_l$.*
(ii) A right-coprime factorization *(D_r, N_r) of G over $\mathbf{R}$ is likewise defined.*

Lemma 668. *Let $f \in \mathbf{F}$, $f = a^{-1}b$. Then f has a left-coprime factorization over $\mathbf{A}$ if, and only if the right ideal generated by a and b is principal.*

Proof. "if" Assume $a\mathbf{R} + b\mathbf{R} = c\mathbf{R}$. Then there exist d, n, x, y in $\mathbf{R}$ such that $a = cd$, $b = cn$, $c = ax + by$. Since $a \neq 0$, we have $c, d \neq 0$. Then $f = (cd)^{-1} cn = d^{-1}n$. In addition, $c = cdx + cny$, therefore $dx + ny = 1$, and (d, n) is a left-coprime factorization of f.

"only if" Suppose (d, n) is a left-coprime factorization of f and let $x, y \in \mathbf{R}$ be such that $nx + dy = 1$. Let $r = ad^{-1}$. We have

$$r = ad^{-1}(nx + dy) = a\left(d^{-1}nx + y\right) = a\left(a^{-1}bx + y\right) = bx + ay \in \mathbf{R}.$$

Therefore, $a = rd$, $b = rn$ and $ax + by = r$. As a result, $a\mathbf{R} + b\mathbf{R} = c\mathbf{R}$ is a principal right ideal. ■

Lemma 669. *The following conditions are equivalent:*

(i) Every $f \in \mathbf{F}$ has a left-coprime factorization over $\mathbf{R}$.
(ii) Every $f \in \mathbf{F}$ has a right-coprime factorization over $\mathbf{R}$.
(iii) $\mathbf{R}$ is a Bézout domain.

Proof. The ring $\mathbf{R}$ is an Ore domain, therefore, according to Proposition 441, $\mathbf{R}$ is a Bézout domain if, and only if every pair of elements in $\mathbf{R}$ generates a principal right ideal. By Lemma 668, (i)⇔(iii). Therefore, (ii)⇔(iii) by symmetry. ■

Let $\mathbf{M}(\mathbf{F})$ be the set of all matrices of finite size with entries in $\mathbf{F}$.

Theorem 670. *The following conditions are equivalent:*

(i) Every $G \in \mathbf{M}(\mathbf{F})$ has a left-coprime factorization over $\mathbf{R}$.
(ii) Every $G \in \mathbf{M}(\mathbf{F})$ has a right-coprime factorization over $\mathbf{R}$.
(iii) $\mathbf{R}$ is a Bézout domain.

Proof. By Lemma 669, (i)⇒(iiii) and (ii)⇒(iii).

(iii)⇒(i): Assume that (iii) holds and let $G \in {}^{p}\mathbf{F}^{m}$. Put $G = (g_{ij})$. By Corollary 329, there exist $d \in \mathbf{R}^{\times}$ and $\tilde{N} \in {}^{p}\mathbf{R}^{m}$ such that $G = d^{-1}\tilde{N}$. By Theorem 653, the matrix $\begin{bmatrix} dI_p \; \tilde{N} \end{bmatrix}$ has a gcld $L \in \mathrm{Mat}_p(\mathbf{R})$ such that

$$\begin{bmatrix} dI_p \; \tilde{N} \end{bmatrix} \equiv_r \begin{bmatrix} L \; 0 \end{bmatrix}$$

and since the matrix in the left-hand side is of rank p over $\mathbf{R}$, we have $L \in \mathrm{GL}_p(\mathbf{F})$. Let $D_l = L^{-1}dI_p$ and $N_l = L^{-1}\tilde{N}$. Then $G = D_l^{-1}N_l$ and (D_l, N_l) is strongly left-coprime over $\mathbf{R}$, thus (D_l, N_l) is a left-coprime factorization of G, and (i) holds.

By symmetry, (iii)⇒(ii). ■

3.7.2 Doubly Coprime Factorizations

Lemma 671. *Let $G \in {}^{p}\mathbf{F}^{m}$.*

(1) The following conditions are equivalent:
(i) G admits both a left-coprime factorization (D_l, N_l) and a right-coprime factorization (D_r, N_r) where $D_l \in \mathrm{Mat}_p(\mathbf{R})$, $D_r \in \mathrm{Mat}_m(\mathbf{R})$, $N_l, N_r \in {}^{p}\mathbf{R}^{m}$.
(ii) $G = D_l^{-1}N_l = N_r D_r^{-1}$ and

$$\begin{bmatrix} * & * \\ D_l & N_l \end{bmatrix} \begin{bmatrix} -N_r & * \\ D_r & * \end{bmatrix} = I$$

where the asterisks denote submatrices with entries in $\mathbf{R}$.

(2) When (ii) holds, then the pairs (D_l, N_l), (D_r, N_r) *(in this order) are called a* doubly coprime factorization *of* G *over* $\mathbf{R}$.

Proof. (1): see ([354], Sect. 4.1, Theorem 60), the proof of which is still valid without any change. ■

Proposition 672. *If* (D_l, N_l), (D_r, N_r) *is a doubly coprime factorization of* $G \in {}^p\mathbf{F}^m$, *then* $\operatorname{coker}_{\mathbf{R}}(\bullet D_l) \cong \operatorname{coker}_{\mathbf{R}}(\bullet D_r)$ *and* $\operatorname{coker}_{\mathbf{R}}(\bullet N_l) \cong \operatorname{coker}_{\mathbf{R}}(\bullet N_r)$.

Proof. If (D_l, N_l), (D_r, N_r) is a doubly coprime factorization of G, then D_l and D_r are stably associated (Definition 519), therefore they are left-similar by Theorem 520. Likewise, N_l and N_r are stably associated, thus left-similar. ■

The following is proved in ([354], Sect. 8.1, Theorem 66) when $\mathbf{R}$ is commutative, but the same rationale as in proof of the cited theorem is still valid in the general case (see Exercise 683).

Theorem 673. *The following conditions are equivalent:*

(i) $\mathbf{R}$ *is an Hermite ring.*
(ii) If $G \in \mathbf{M}(\mathbf{F})$ *admits a left-coprime factorization, then it admits a right-coprime factorization, thus a doubly coprime factorization.*
(iii) Same statement as (ii) with left *and* right *interchanged.*

3.8 Exercises

Exercise 674. *In an abelian category* $\mathcal{C}$, *consider the diagram below*

$$\begin{array}{ccccc} 0 & \longrightarrow & N & \xrightarrow{\alpha} & M \\ & & & & \uparrow \beta \\ & & & & P \end{array}$$

where the first row is an exact sequence and $\operatorname{im}\beta \subseteq \operatorname{im}\alpha$. *Assuming that all morphisms are right ones* (**1.2.2.1**), *prove that there exists* $\gamma : P \to N$ *such that* $\beta = \alpha\gamma$. *(Hint: assume that* $\mathcal{C} = {}_{\mathbf{A}}\mathbf{Mod}$ *and then use the Freyd-Mitchell Full Embedding Theorem (Theorem 50). In* ${}_{\mathbf{A}}\mathbf{Mod}$, *show that* $\gamma = \bar{\alpha}^{-1}\bar{\beta}$ *where* $\bar{\alpha} : N \to \operatorname{im}\alpha$ *and* $\bar{\beta} : M \to \operatorname{im}\alpha$ *are the maps induced by* α *and* β, *respectively.)*

Exercise 675. *Let* $\mathbf{A}$ *be a ring. Prove the following:*

(i) A submodule N *of* $\mathbf{A}^k$ *is of the form* $\mathbf{A}^q R$ *for some matrix* $R \in {}^q\mathbf{A}^k$ *if, and only if the f.g.* $M = \mathbf{A}^k / N$ *is f.p.*

(ii) Every submodule N of $\mathbf{A}^k$ is of the form $\mathbf{A}^q\, R$ for some matrix $R \in {}^q\mathbf{A}^k$ if, and only if $\mathbf{A}$ is Noetherian.

Exercise 676. *Let $\mathbf{A}$ be a ring.*

*(i) Show that the category $\mathcal{C} = {}_{\mathbf{A}}\mathbf{Mod}$ is a Grothendieck category (**1.2.4.4**).*
(ii) Show that a left $\mathbf{A}$-module is f.g. (resp., coherent) if, and only if it is of finite type (resp., coherent) in $\mathcal{C}$ (Definition 55).

Exercise 677. *Let $\mathbf{R}$ be an Ore domain.*

(i) Prove that $\mathbf{R}$ is a left coherent Sylvester domain if, and only if for any integer $q \geq 1$ and any column $\Gamma \in {}^q\mathbf{R}$, the left annihilator of Γ is free.
(ii) Let $S \subseteq \mathbf{R}^\times$ be a denominator set. Prove that if $\mathbf{R}$ is a left coherent Sylvester domain, so is again $S^{-1}\mathbf{R}$.

Exercise 678. *Prove the following:*

(i) M is torsionless if, and only if the canonical map $\psi : M \to M^{**}$ (Definition 272) is injective.
(ii) Any product or any direct sum of torsionless modules is again torsionless.

Exercise 679. *Let $\mathbf{A}$ be a ring and let $A, A' \in \mathrm{Mat}_n(\mathbf{A})$. Prove that if A, A' are conjugate (Subsect. 2.2.6), they are similar (Definition 517).*

Exercise 680. *Let $A = \mathbb{Z}/2\mathbb{Z}$, let $B = B' = \mathbb{Z}^2$, let $(\varepsilon_1, \varepsilon_2)$ be the canonical basis of the free $\mathbb{Z}$-module $\mathbb{Z}^2$ and let $\beta : B \to B'$ be the map defined by $\beta(\varepsilon_1) = \varepsilon_1 + \varepsilon_2$ and $\beta(\varepsilon_2) = \varepsilon_1 - \varepsilon_2$. (i) Prove that $\ker \beta = 0$. (ii) Prove that $(\mathbf{1} \otimes \beta)(\varepsilon_1 - \varepsilon_2) = 0$ where $\mathbf{1} = Id_A$.*

Exercise 681. *(i) Using the Bézout identity, prove that $\mathbb{Z} = 2\mathbb{Z} + 3\mathbb{Z}$ and using Theorem 241 show that $\mathbb{Z}/(6) \cong \mathbb{Z}/(2) \oplus \mathbb{Z}/(3)$. (ii) Set $\mathbf{A} = \mathbb{Z}/(6)$; prove that any nonzero free $\mathbf{A}$-module has at least 6 elements. (iii) Show that $\mathbb{Z}/(2)$ is a projective $\mathbf{A}$-module but not a free one.*

Exercise 682. *Let $\mathbf{A}$ be a fir (Exercise 486), let F be a free left $\mathbf{A}$-module and let M a submodule of F. Prove that M is free. (Hint: use Theorem 586(2).)*

Exercise 683. *Let $\mathbf{A}$ be a ring with IBN. A row $r = (r_i) \in \mathbf{A}^k$ is said to be* unimodular *if there exist $b_i \in \mathbf{A}$ $(1 \leq i \leq k)$ such that $\sum_{1 \leq i \leq r} r_i\, b_i = 1$; $\mathbf{A}$ is said to have the* unimodular extension property *if every unimodular row is the first row of a matrix $A \in \mathrm{GL}_k(\mathbf{A})$.*

(1) Prove that of the following, (a) and (b) are equivalent and imply (c): (a) $\mathbf{A}$ *is Hermite; (b) every right-invertible matrix* $R \in {}^{q}\mathbf{A}^{k}$ $(q \leq k)$ *is completable; (c)* $\mathbf{A}$ *has the unimodular extension property.*
(2) Moreover, prove that if $\mathbf{A}$ *is commutative, (c) is equivalent to (a) and (b). (Hint: for (1), first show that any Hermite ring is weakly finite; then apply Theorem 499. For (2), proceed by induction using determinants ([206], Sect. XXI.3, Th. 3.6).)*
(3) Detail the proof of Theorem 673.

Exercise 684. *Let* $\rho : \mathbf{A} \rightarrow \mathbf{S}$ *be a ring-homomorphism and let* M *be a left* $\mathbf{S}$*-module. Let* $(\alpha_i)_{i\in I}$ *be a generating family (resp., a free family, a basis) of the left* $\mathbf{A}$*-module* ${}_{\mathbf{A}}\mathbf{S}$ *and let* $(a_j)_{j\in J}$ *be a generating family (resp., a free family, a basis) of the* $\mathbf{S}$*-module* M*. Prove that* $(\alpha_i\, a_j)_{(i,j)\in I\times J}$ *is a generating family (resp.,–when* ρ *is injective–a free family, a basis) of the left* $\mathbf{A}$*-module* $\rho_*(M)$*.*

Exercise 685. *Let* $r = (r_i)_{i\in I}$ *be a family of elements of a ring* $\mathbf{A}$*, let* $\rho : \mathbf{A} \rightarrow \mathbf{S}$ *be a ring-homomorphism and let* $\varphi_{\mathbf{A}^I} : \mathbf{A}^I \rightarrow \rho_*\left(\rho^*\left(\mathbf{A}^I\right)\right)$ *be the canonical map. Check that* $\varphi_{\mathbf{A}^I}(r) = (\rho(r_i))_{i\in I}$ *and deduce from this equality that Remark 560(ii)(b) holds true.*

Exercise 686. *Let* $\rho : \mathbf{A} \rightarrow \mathbf{S}$ *be a ring-homomorphism, and for any left* $\mathbf{A}$*-module* M *set* ${}^*\rho(M) \triangleq \mathrm{Hom}_R(\rho_*(S), M)$*.*
Prove that ${}^*\rho$ *is a functor* ${}_{\mathbf{A}}\mathbf{Mod} \rightarrow {}_{\mathbf{S}}\mathbf{Mod}$ *and that* $({}^*\rho, \rho_*)$ *is an adjoint pair of functors. (The module* ${}^*\rho(M)$ *is called the* coinduced extension *of* M *along* ρ*. For more details regarding the functor* ${}^*\rho$*, see ([80], Prop. 2.3.5).)*

Exercise 687. *Let* $\rho : \mathbf{A} \rightarrow \mathbf{S}$ *be a ring-homomorphism and let* P *be a projective left* $\mathbf{A}$*-module. Prove that the left* $\mathbf{S}$*-module* $\rho^*(P)$ *is projective and that if* ρ *is a monomorphism, so is the canonical map* $\varphi_P : P \rightarrow \rho_*(\rho^*(P))$*. (Hint: use Corollary 556.)*

Exercise 688. *A ring* $\mathbf{A}$ *is called* left semihereditary *if every f.g. left ideal in* $\mathbf{A}$ *is projective. A semihereditary commutative domain is called a* Prüfer domain*.*

(i) Prove that a ring $\mathbf{A}$ *is left semihereditary if, and only if all torsionless right* $\mathbf{A}$*-modules are flat. If such a ring* $\mathbf{A}$ *is a left Ore domain, then a right* $\mathbf{A}$*-module is flat if, and only if it is torsion-free.*
Prove that a left semihereditary ring is left coherent.
(ii) Show that a ring $\mathbf{A}$ *is a left Dedekind domain if, and only if* $\mathbf{A}$ *is left Noetherian domain which is left semihereditary.*
(iii) Let $\mathbf{A}$ *be a left Dedekind domain. Deduce from (i) and Proposition 516(v) that for a f.g. right* $\mathbf{A}$*-module* M*, torsion-free* $\Longleftrightarrow$ *torsionless* $\Longleftrightarrow$ *flat.*
(iv) Prove that every valuation domain (Exercise 490) is a Prüfer domain.

(v) Let $\mathbf{A}$ *be a Prüfer domain. Prove that an* $\mathbf{A}$*-module is flat if, and only if it is torsion-free.*
(vi) Let $\mathbf{A}$ *be a commutative domain and assume that all torsion-free* $\mathbf{A}$*-modules are flat; show that* $\mathbf{A}$ *is a Prüfer domain. Prove that the same holds if* torsion-free *is changed to* torsionless.
Hint: for (i), (v) and (vi), see ([136], Note after Prop. 3.2) and ([199], (4.67), (4.20), (4.69)).

Exercise 689. *Let* $\mathbf{A} \subseteq \mathbf{S}$ *be left Ore domains, let* $\rho : \mathbf{A} \hookrightarrow \mathbf{S}$ *be the inclusion and let* T *be a torsion left* $\mathbf{A}$*-module.*

(i) Prove that $\rho^* (T)$ *is again torsion.*
(ii) If $\mathbf{A}$ *and* $\mathbf{S}$ *are commutative, prove that* $\rho_* (\rho^* (T))$ *is torsion too.*
(Hint: for (i), use Theorem 339(ii); for (ii), write any $x \in \rho^* (T)$ *in the form* $\sum_{1 \leq i \leq n} \lambda_i \otimes x_i$, $\lambda_i \in \mathbf{S}$, $x_i \in T$*, and left-multiply this expression by* $\prod_{1 \leq i \leq n} \mu_i$ *where for each* i, $0 \neq \mu_i \in \mathbf{A}$ *is such that* $\mu_i x_i = 0$.*)*

Exercise 690. *Let* $\mathbf{A}$ *be a left Ore domain and let* M *be a torsion-free* $\mathbf{A}$*-module (Definition 513). Prove that* M *is injective if, and only if it is divisible. (Hint: let* $\mathfrak{a}$ *be a nonzero left ideal in* $\mathbf{A}$ *and let* $f : \mathfrak{a} \to M$*; for any* $0 \neq a \in \mathfrak{a}$*, there exists* $m_a \in M$ *such* $(a) f = a\, m_a$*. Let* $0 \neq b \in \mathfrak{a}$ *and let* $u, v \in \mathbf{A}$, $u \neq 0$*, be such that* $u\, a = v\, b$*; form the product* $u\, a\, m_a$ *and show that* $m_a = m_b = m$*. Deduce that* f *can be extended to* ${}_{\mathbf{A}}\mathbf{A}$ *and use the Baer Criterion.)*

Exercise 691. *Let* $\mathbf{A}$ *be a commutative domain. Prove that (i) every quotient of a divisible* $\mathbf{A}$*-module is divisible; (ii) all products and all coproducts of divisible* $\mathbf{A}$*-modules are divisible. (Hint: since* $\mathbf{A}$ *is a commutative domain, the condition for an* $\mathbf{A}$*-module* M *to be divisible reduces to* $M = a\, M$ *for any* $0 \neq a \in \mathbf{A}$.*)*

Exercise 692. *Let* $\mathbf{A}$ *be a ring and let* M *be a left* $\mathbf{A}$*-module. A left* $\mathbf{A}$*-module* $E \supseteq M$ *is said to be an* essential extension *of* M *(or* M *is said to be an* essential submodule *of* E*, written* $M \subseteq_e E$*) if every nonzero submodule of* E *intersects* M *nontrivially.*

(i) Prove that $M \subseteq_e E$ *if, and only if for any* $0 \neq a \in E$*, there exits* $r \in \mathbf{A}$ *such that* $0 \neq r\, a \in M$*; deduce from this necessary and sufficient condition that the relation* $\subseteq_e$ *is transitive.*
(ii) Prove that a left $\mathbf{A}$*-module* M *is injective if, and only if it has no proper essential extension.*
(Hint: Assuming that M *is injective and that* E *is a proper extension of* M*, show using Theorem 587(iii) that* $E = M \oplus N$ *where* $N \neq 0$ *and conclude that* $M \nsubseteq_e E$*. Conversely, if* M *has no proper essential extension, embed* M *in an injective module* I *(Theorem 599) and using Zorn's lemma (Lemma 107) show that there exists a submodule* $S \subseteq I$ *maximal with respect to the property that* $S \cap M = 0$*; denoting by* $\varphi : I \twoheadrightarrow I/S$ *the canonical epimorphism, show that* $\varphi(M) \subseteq_e I/S$, $I = M \oplus S$ *and that* M *is injective by Proposition 588.)*

(iii) Show any module M has a maximal essential extension.
(Hint: Fix an injective module $I \supseteq M$ and consider any family $\mathfrak{F}$ of essential extensions of M in I ordered by inclusion; show using (i) that $\mathfrak{F}$ is inductively ordered and that any maximal element E of $\mathfrak{F}$ is a maximal essential extension of M.)
(iv) Show that the following are equivalent: (a) E is maximal essential over M; (b) E is injective, and is essential over M; (c) E is an injective hull of M.
(v) Let U be a module that contains a direct sum $\bigoplus_\alpha V_\alpha$. (a) Let $V_\alpha \subseteq_e E_\alpha \subseteq U$ for every α. Show that the sum $\sum_\alpha E_\alpha$ is direct. (b) Deduce from the above that the sum $\sum_\alpha E(V_\alpha)$ is direct.

Exercise 693. *The notion of* projective cover *is obtained from that of injective hull by "dualizing" it, i.e., reversing the arrows. (Note that, contrary to injective hulls, projective covers do not always exist: see ([198], §24). A ring over which all modules have projective covers is said to be* perfect*; this is* not *the same notion as in Definition 221(ii).)*

(i) Detail the notion of projective cover and specify in what sense a projective cover, when it exists, is uniquely defined.
(ii) Let M be a left $\mathbf{A}$-module and let $N \lhd M$. The submodule N is said to be superfluous, *or* small *(written $N \subseteq_s M$) if, whenever $L \lhd M$ with $L + N = M$, then $L = M$. Let P be a projective module and let $\varepsilon : P \twoheadrightarrow M$ be an epimorphism; prove that P is a projective cover of M if, and only if* $\ker \varepsilon$ *is superfluous.*

Exercise 694. *Let $\mathbf{K}$ be a field and let $X = (X_i)$ be an* infinite *family of indeterminates. Prove that $\mathbf{A} = \mathbf{K}[X]$ is a coherent domain but is not Noetherian. (Hint: use Theorem 612.)*

Exercise 695. *Let $\mathbf{A}$ be a simple ring and let U be a nonzero left $\mathbf{A}$-module. Prove that U is completely faithful (Definition 613(ii)).*

Exercise 696. *Let $\mathbf{A}$ be a ring, let $S \subseteq \mathbf{A}^\times$ be a left denominator set and let (3.52) be a free (resp., projective, flat) resolution of length $\leq n$ of the left $\mathbf{A}$-module M. Prove that*

$$0 \longrightarrow S^{-1}F_n \longrightarrow ... \longrightarrow S^{-1}F_0 \longrightarrow S^{-1}M \longrightarrow 0$$

is a free (resp., projective, flat) resolution of length $\leq n$ of the left $S^{-1}\mathbf{A}$-module $S^{-1}M$.

Exercise 697. *(i) Prove that $\mathbf{k}[X_1, ..., X_n]$ ($\mathbf{k} = \mathbb{R}$ or $\mathbb{C}$) is a Sylvester domain if, and only if $n \leq 2$.*
(ii) Prove that $B_1(\mathbf{k})$ is a Sylvester domain but that the first Weyl algebra $A_1(\mathbf{k})$ is not a Sylvester domain.

Exercise 698. *A commutative ring* $\mathbf{A}$ *is called* adequate *if every f.g. ideal in* $\mathbf{A}$ *is principal and the following condition holds: for any* $a, c \in \mathbf{A}$ *with* $a \neq 0$*, there exist* r, s *such that* $a = r\,s$ *with* $(r, c) = 1$ *and* $(s', c) \neq 1$ *for any nonunit divisor* s' *of* s*.*

(i) Prove that a commutative domain which is adequate is an elementary divisor ring.
(ii) Prove that the commutative domain in Example 439 is adequate, thus is an elementary divisor ring, but is not a principal ideal domain. (Hint: for (i), use Corollary 657. Considering a, b, c *such that* $(a, b, c) = 1$ *and* $a \neq 0$*, show that there exists* q *such that* $(a, b + q\,c = 1)$*.)*

3.9 Notes

The functors $\bigotimes$ and Hom are also denoted by Tor_0 and Ext^0, respectively. Tor_n and Ext^n $(n \geq 1)$ are their "derived functors". According to MacLane [223], Tor_n and Ext^n $(n \geq 0)$ are the subject of Homological Algebra, thus our account, where the use of derived functors is avoided, is very basic. Those functors, as well as $\varinjlim$ and $\varprojlim$, were first introduced in the context of Algebraic Topology [104].

Coherent rings and coherent sheaves of rings were introduced by Serre [324] and further studied by Chase [62]. Proposition 342 and Exercise 690 are due to Gentile [136]. Theorem 514 and Proposition 516 are taken from Bourlès and Oberst [47]. The part of Theorem 520 involving Conditions (a), (b) and (c) goes back to Fitting [113] (see also [77], Sect. 0.6, Th. 6.2); that involving Condition (d) is new, to our knowledge. Proposition 536 is due to Bowtell and Cohn [53]; the proof of Lemma 538 is taken from [78].

The key role played by the functor $\mathfrak{K}_N$ (**3.4.3.2**) was pointed out by Grothendieck and Dieudonné [150] in the context of Algebraic Geometry. Corollary 567 and the identification of $\mathrm{Hom}_{\mathbf{A}}\left(\mathrm{coker}_{\mathbf{A}}\left(\bullet R\right), W\right)$ with $\ker_W\left(R\bullet\right)$, which play a crucial role in Algebraic Systems Theory, are due to Malgrange [232].

Hermite rings were introduced by Lissner [219] (this term has a quite different meaning in Kaplansky's classical paper [175] where "left Hermite domain" means "left Bézout domain" in the present terminology). In 1955, Serre wondered whether $\mathbf{K}\left[X_1, ..., X_n\right]$ is projective-free when $\mathbf{K}$ is a field. A positive answer to that question was given independently by Quillen and Suslin in January 1976 [290], [337]. A nice presentation of the main results related to this question is proposed in [200]. The notion of injective module (as well as Theorem 599) is due to Baer [9], that of injective hull to Eckmann and Schopf [97] (but the term "injective hull" to Rosenberg and Zelinsky [300]), and the dual notion of projective cover (Exercise 693) essentially to Bass [12] who also introduced the notion of torsionless module (Definition 513(2)). Injective and projective resolutions play a key role in Homological Algebra (see [59] in the context of modules and [148], [134] in the context of abelian categories). An extensive study of finite free resolutions can be found in Northcott [256].

The first book where the notions of hereditary ring (Definition 585) and of semi-hereditary ring (Exercise 688) were presented is that of Cartan and Eilenberg [59]; the result in Exercise 688(i) is due to Chase [62]. Definitions 513 and 595 of a torsion-free module and of a divisible module, respectively, are more general than Levy's usual definitions [215] and are due to Lam [199]. Flat modules were introduced by Serre [325]. Corollary 610 is extracted from Pillai and Shankar [275]. The global dimension and the weak global dimension were defined and studied by Cartan and Eilenberg [59]. The Krull dimension of Noetherian commutative domains was introduced by Krull (in 1938) in the case of local rings and shortly after by Zariski without this restriction; this notion was extended to noncommutative rings by Gabriel in 1962. The characterization of Sylvester domains in Theorem 634 is due to Bedoya and Lewin [14] and to Dicks and Sontag [89].

The equivalence between injectivity and the fundamental principle (Corollary and Definition 592) was established in the Noetherian case by Malgrange [232]. Ehrenpreis [101] announced that the fundamental principle holds true when **A** is the ring of linear partial differential operators with constant coefficients $\mathbb{C}[\partial_1, ..., \partial_n]$ and W is a space of entire functions in $\mathbb{C}^n$ that satisfy a condition of bounded growth at infinity; see also Ehrenpreis' and Palomodov's monographs ([102], Sect. IV.2), ([266], Sect. IV.5). The validity of the fundamental principle was established by Malgrange [231] when $\mathbf{A} = \mathbb{C}[\partial_1, ..., \partial_n]$ and $W = \mathcal{E}(\mathbb{C}^n)$, the space of indefinitely differentiable functions in $\mathbb{C}^n$. Theorems 638 and 642 on cogenerators are classical ([199], §19).

The theory of finitely generated modules over principal ideal domains goes back essentially to Frobenius (around 1879) in the commutative case and to van der Waerden, Wedderburn, Jacobson and Teichmüller (in the years 1932-1937) in the noncommutative case. Nakayama [254] established in 1938 the uniqueness result in Theorem 658. Corollary 663 is taken from Bourlès and Fliess [44], with slight modifications, and Corollary 661 is new. The structure of finitely generated modules over Bézout domains and elementary divisor rings was clarified essentially by Kaplansky [175], and by Larsen *et al.* [209]. That of finitely generated modules over Dedekind domains was made explicit by Steinitz (in 1912) in the commutative case, and by Eisenbud and Robson [106], [107] (in 1970) in the noncommutative case. Stafford's theorem (Remark 665(ii)) has been generalized by Quadrat and Roberts [289]. Lemma and Definition 666 is new.

Or account on coprime factorization of matrices is an easy generalization of Vidyasagar's presentation [354] where commutative rings are considered.

4
Galois Theory and Skew Polynomials

4.1 Introduction

Let $\mathbf{K}$ be a field, and let $f(X) \in \mathbf{K}[X]$ be a polynomial with coefficients in $\mathbf{K}$. In general, the equation $f(x) = 0$ has solutions in an extension of $\mathbf{K}$, not in $\mathbf{K}$ itself. For example, if $\mathbf{K} = \mathbb{Q}$, the equation $x^2 + 2 = 0$ has no solution in $\mathbb{Q}$, it has in $\mathbb{C}$ its full set of solutions $\{i\sqrt{2}, -i\sqrt{2}\}$, and these solutions are already in the extension $\mathbb{Q}(i, \sqrt{2}) \subsetneq \mathbb{C}$ of $\mathbb{Q}$. The study of algebraic field extensions is the Galois theory, recalled in Section 4.2.

Let $\boldsymbol{\delta} = d/dt$ and consider the differential field $(\mathbf{K}, \boldsymbol{\delta})$. Let $f(X) \in \mathbf{K}[X; \boldsymbol{\delta}]$. The solutions of the differential equation $f(\partial) y = 0$ are closely connected with the roots of the skew polynomial $f(X)$. These roots must be properly defined. All these points are treated in Section 4.3.

So, given a skew polynomial $f(X) \in \mathbf{K}[X; \boldsymbol{\delta}]$, we want to determine a field extension of $\mathbf{K}$ in which the differential equation $f(\partial) y = 0$ has a full set of solutions. Equivalently, we want to determine a field extension of $\mathbf{K}$ in which the skew polynomial $f(X)$ has a full set of roots. This problem can be solved, and the method to do it constitutes the Picard-Vessiot theory, also called the differential Galois theory. It is expounded in Section 4.4. The case when $\boldsymbol{\delta}$ is an inner derivation (corresponding to discrete-time systems) is more difficult, for a *ring extension* (not a *field extension*) is then needed. This is explained in Subsect. 4.4.3.

4.2 Field Extensions

4.2.1 Algebraic Field Extensions

4.2.1.1 Extensions of Rings and Fields

Definition 699. *(i) Let* $\mathbf{K}$ *be a ring. A* ring extension *of* $\mathbf{K}$ *is a ring* $\mathbf{L}$ *such that* $\mathbf{K} \subseteq \mathbf{L}$*; such an extension is written* $\mathbf{L}/\mathbf{K}$.

H. Bourlès and B. Marinescu: Linear Time-Varying Systems, LNCIS 410, pp. 269–308.
springerlink.com

(ii) Let $\mathbf{K} \subseteq \mathbf{L}$ *be fields; then* $\mathbf{L}/\mathbf{K}$ *is called a* field extension, *and it is said to be* simple *if* $\mathbf{L} = \mathbf{K}(a)$ *where* $a \in \mathbf{L}$, *i.e., if* $\mathbf{L}$ *is generated by a single element* a *over* $\mathbf{K}$.

For example, let $\mathbb{Q}$ be the field of rational numbers; then $\mathbb{Q}(\pi)$ is the smallest field containing both $\mathbb{Q}$ and $\pi = 3.141592...$, i.e., $\mathbb{Q}(\pi) = \{N(\pi)/D(\pi) : N, D \in \mathbb{Q}(X)\}$.[1] The extension $\mathbb{Q}(\pi)/\mathbb{Q}$ is simple.

Definition 700. *(i) Let* $\mathbf{L}/\mathbf{K}$ *and* $\mathbf{L}'/\mathbf{K}$ *be two ring extensions. A* $\mathbf{K}$-homomorphism $\mathbf{L} \to \mathbf{L}'$ *is a homomorphism* $\mathbf{L} \to \mathbf{M}$ *which leaves fixed all elements of* $\mathbf{K}$.
(ii) Let $\mathbf{L}/\mathbf{K}$ *be a field extension. The dimension of the* $\mathbf{K}$*-vector space* $\mathbf{L}$ *is denoted by* $[\mathbf{L} : \mathbf{K}]$ *and called the* degree *of the extension* $\mathbf{L}/\mathbf{K}$. *This extension is* finite *if* $[\mathbf{L} : \mathbf{K}]$ *is finite.*

As easily shown, we have the following:

Proposition 701. *(i) Let* $\mathbf{K} \subseteq \mathbf{L} \subseteq \mathbf{M}$ *be fields such that both* $[\mathbf{L} : \mathbf{K}]$ *and* $[\mathbf{M} : \mathbf{L}]$ *are finite. Then* $[\mathbf{M} : \mathbf{K}] = [\mathbf{M} : \mathbf{L}][\mathbf{L} : \mathbf{K}]$.
(ii) More specifially, let $(u_i)_{1 \leq i \leq n}$ *be a basis of the* $\mathbf{K}$*-vector space* $\mathbf{L}$ *and let* $(v_j)_{1 \leq j \leq m}$ *be a basis of the* $\mathbf{L}$*-vector space* $\mathbf{M}$. *Then* $(u_i v_j)_{1 \leq i \leq n, 1 \leq j \leq m}$ *is a basis of the* $\mathbf{K}$*-vector space* $\mathbf{M}$.

The following is proved in, e.g., ([28], §V.6, Corol. 2 of Theorem 1):

Theorem 702. *(*Dedekind's theorem*). Let* $\mathbf{L}/\mathbf{K}$, $\mathbf{L}'/\mathbf{K}$ *be two field extensions. If* $[\mathbf{L} : \mathbf{K}]$ *is finite, the number* n *of distinct* $\mathbf{K}$*-homomorphisms* $\mathbf{L} \to \mathbf{L}'$ *is upper bounded by* $[\mathbf{L} : \mathbf{K}]$.

As shown below (Theorem 725), if $\mathbf{L} = \mathbf{L}'$ and $\mathbf{L}/\mathbf{K}$ is a *Galois extension*, then $n = [\mathbf{L} : \mathbf{K}]$.

4.2.1.2 Algebraic and Transcendental Elements

Definition 703. *(i) A field* $\mathbf{K}$ *is said to be* algebraically closed *when every non constant polynomial in* $\mathbf{K}[X]$ *has a root in* $\mathbf{K}$.
(ii) Let $\mathbf{L}/\mathbf{K}$ *be a field extension. A family* $(a_i)_{i \in I}$ *of elements of* $\mathbf{L}$ *is called* algebraically free over $\mathbf{K}$ *if for a polynomial* $f \in \mathbf{K}[(X_i)_{i \in I}]$, $f((a_i)) = 0$ *is equivalent to* $f = 0$. *An element of* $\mathbf{L}$ *which is algebraically free over* $\mathbf{K}$ *is called* transcendental. *If* $a \in \mathbf{L}$ *is not transcendental over* $\mathbf{K}$, *it is called* algebraic.
(iii) A field extension $\mathbf{L}/\mathbf{K}$ *is said to be* algebraic *if every element of* $\mathbf{L}$ *is algebraic over* $\mathbf{K}$*; a field extension which is not algebraic is called* transcendental.

[1] $D(\pi)$ cannot be zero, for π is a transcendental number, as proved by Lindemann in 1882.

(iv) A field extension $\mathbf{L}/\mathbf{K}$ *is called an* algebraic closure *of* $\mathbf{K}$ *if* $\mathbf{L}/\mathbf{K}$ *is both algebraic and algebraically closed. If the field extension* $\mathbf{L}/\mathbf{K}$ *is such that every element of* $\mathbf{L}$ *which is not in* $\mathbf{K}$ *is transcendental over* $\mathbf{K}$*, then* $\mathbf{K}$ *is called* algebraically closed in $\mathbf{L}$ *([82], p. 4).*
(v) Let $\mathbf{L}/\mathbf{K}$ *be a field extension. A part* B *of* $\mathbf{L}$ *is called a* transcendence basis *of* $\mathbf{L}$ *(over* $\mathbf{K}$*) if* B *is algebraically free over* $\mathbf{K}$ *and* $\mathbf{L}$ *is algebraic over* $\mathbf{K}(B)$*. A field extension* $\mathbf{L}/\mathbf{K}$ *is called* purely transcendental *or* pure *if it is of the form* $\mathbf{L} = \mathbf{K}(B)$ *where* B *is a transcendence basis of* $\mathbf{L}$ *over* $\mathbf{K}$*.*

The following can be found in ([28], §V.14, Theorems 1 & 3, Definition 4 and Prop. 14).

Theorem and Definition 704. *(1) (*Steinitz' theorem*). Every field extension* $\mathbf{L}/\mathbf{K}$ *has a transcendence basis over* $\mathbf{K}$*. In other words, every field extension is an algebraic extension of a pure extension.*
(2) Any two transcendence bases over $\mathbf{K}$ *of a field extension* $\mathbf{L}/\mathbf{K}$ *have the same cardinal, called the* transcendence degree *of* $\mathbf{L}/\mathbf{K}$ *and denoted by* $\deg.\mathrm{tr}_{\mathbf{K}}\mathbf{L}$.
(3) Let $B = (x_i)_{i\in I}$ *be a transcendence basis of the pure extension* $\mathbf{L}/\mathbf{K}$*. Then* $\mathbf{L} \cong \mathbf{K}\left((X_i)_{i\in I}\right)$.

For example, $\sqrt{2}$ is algebraic over $\mathbb{Q}$ (since it is a root of the polynomial $X^2 - 2$) whereas π is not ([206], Appendix 1). Therefore, $\mathbb{Q}\left(\sqrt{2}\right)$ is an algebraic (simple) field extension of $\mathbb{Q}$ whereas the field extension $\mathbb{Q}(\pi)/\mathbb{Q}$ is pure (with $\deg.\mathrm{tr}_{\mathbb{Q}}\mathbb{Q}(\pi) = 1$). The field extension $\mathbb{Q}\left(\sqrt{2},\pi\right)/\mathbb{Q}$ is transcendental, so is also the field extension $\mathbb{Q}\left(\sqrt{2},\pi\right)/\mathbb{Q}\left(\sqrt{2}\right)$, but the field extension $\mathbb{Q}\left(\sqrt{2},\pi\right)/\mathbb{Q}(\pi)$ is algebraic. Let us detail this point: we have $\mathbb{Q}\left(\sqrt{2},\pi\right) = \mathbb{Q}\left(\sqrt{2}\right)(\pi)$ and π is transcendental over $\mathbb{Q}$, thus it is transcendental over any algebraic field extension of $\mathbb{Q}$ and in particular over $\mathbb{Q}\left(\sqrt{2}\right)$. We also have $\mathbb{Q}\left(\sqrt{2},\pi\right) = \mathbb{Q}(\pi)\left(\sqrt{2}\right)$ and $\sqrt{2}$ is algebraic over $\mathbb{Q}$, thus over any field extension of $\mathbb{Q}$ and in particular over $\mathbb{Q}(\pi)$.

4.2.1.3 Splitting Fields

Definition 705. *Let* $f[X] \in \mathbf{K}[X]$ *be a polynomial of degree* $n \geq 1$*;* $f(X)$ *is said to* split into linear factors *if there exist* $a_1, \ldots, a_n \in \mathbf{K}$ *(not necessarily distinct) such that*

$$f(X) = c(X - a_1)\ldots(X - a_n)$$

where $c = f_0 \in \mathbf{K}$.

Lemma 706. *Let* $\mathbf{K}$ *be a field and let* $f[X] \in \mathbf{K}[X]$ *be a polynomial of degree* $n \geq 0$*. Then* f *has at most* n *distinct roots in* $\mathbf{K}$*, and* a *is a root of* f *in* $\mathbf{K}$ *if, and only if* $X - a$ *divides* $f(X)$.

Proof. The degree is an Euclidean function over $\mathbf{K}[X]$ (Exercise 489), thus one can find $q, r \in \mathbf{K}[X]$ such that

$$f(X) = q(X)(X-a) + r(X)$$

with $d^\circ(r) < 1$, i.e., $r \in \mathbf{K}$. Thus $f(a) = 0$ if, and only if $r = 0$, i.e., $X - a$ divides $f(X)$. If $a_1, ..., a_m$ are distinct roots of $f(X)$ in $\mathbf{K}$, then by induction we see that the product

$$p(X) = (X-a_1)...(X-a_m)$$

divides $f(X)$ and since $d^\circ(p) = m$ we have $m \leq n$. ■

An irreducible polynomial (Definition 349(iii)) does not necessary reduce to one linear factor. For example, $f(X) = X^2 - 2$ is irreducible in $\mathbb{Q}[X]$. Embedding $\mathbf{K} = \mathbb{Q}$ in the algebraic extension $\mathbf{L} = \mathbb{Q}(\sqrt{2})$, $f(X)$ splits into linear factors in $\mathbf{L}[X]$, since $f(X) = (X-\sqrt{2})(X+\sqrt{2})$.

Theorem 707. (Kronecker's theorem). *Let $f[X] \in \mathbf{K}[X]$ be a polynomial of positive degree. There is a simple extension $\mathbf{L}/\mathbf{K}$ such that f has a root in $\mathbf{L}$.*

Proof. The polynomial f is a product of irreducible polynomials, thus one can assume that f is irreducible without loss of generality. Then the ideal (f) is maximal in $\mathbf{K}[X]$ by Remark 350(i), and $\mathbf{L} \triangleq \mathbf{K}[X]/(f)$ is a field by Proposition 459. Let $\iota : \mathbf{K} \hookrightarrow \mathbf{R}$ be the inclusion, let $\varphi : \mathbf{R} \twoheadrightarrow \mathbf{R}/(f)$ be the canonical epimorphism and let $a = \varphi(X)$. We have $f(a) = 0$. Identifying $\mathbf{K}$ and $\iota(\mathbf{K})$, let $y \in \mathbf{K}^\times$; since φ is a $\mathbf{K}$-algebra homomorphism, we have $\varphi(y)\varphi(y^{-1}) = 1$, thus $\varphi(y) \neq 0$. Therefore, $\varphi|_{\mathbf{K}}$ is injective and $\mathbf{K}$ can be identified with $\varphi(\mathbf{K})$. Thus $\mathbf{L}$ is a field extension of $\mathbf{K}$ and there exists $a \in \mathbf{L}$ such that $f(a) = 0$. ■

By Theorem 707, one can split any polynomial of degree $n > 2$ into one linear factor and a polynomial of degree $n-1$. Therefore, by induction one can split any polynomial of positive degree into linear factors; thus we are led to define a *splitting field* of a polynomial or, more generally, of a family of polynomials:

Definition 708. *Let $(f_i)_{i\in I}$ be a family of polynomials over* $\mathbf{K}$. *A* splitting field *of $(f_i)_{i\in I}$ over $\mathbf{K}$ is an extension field $\mathbf{L}/\mathbf{K}$ such that every f_i splits into linear factors in $\mathbf{L}[X]$ and $\mathbf{L}$ is generated over $\mathbf{K}$ by $\bigcup_{i\in I} R_i$ where R_i is the set of roots in $\mathbf{L}$ of the polynomial f_i.*

Consider the following example: let $f(X) = (X^2+1)(X^2-2)^2$ over $\mathbb{Q}$. This polynomial is not irreducible since it is the product of the three irreducible polynomials X^2+1, X^2-2, X^2-2. A splitting field of X^2-2 is $\mathbf{L}_1 = \mathbb{Q}(\sqrt{2})$, as already seen, and $\mathbf{L}_1 \cong \mathbb{Q}/(X^2-2)$ (where $\cong$ is a $\mathbf{K}$-isomorphism, as all isomorphisms below). The polynomial X^2+1 does not split into linear factors over $\mathbf{L}_1$, but it does over $\mathbf{L} = \mathbf{L}_1/\left(X^2+1\right) \cong \mathbf{L}_1(i)$ ($i \triangleq \sqrt{-1}$) which is a field extension of $\mathbf{L}_1$. We have $\mathbf{L} \cong \mathbb{Q}\left(\sqrt{2}\right)(i) = \mathbb{Q}\left(\sqrt{2}, i\right) = \mathbb{Q}(i)\left(\sqrt{2}\right)$, which shows that the field extension $\mathbf{L}/\mathbb{Q}$ can be constructed in a different order: one can first construct the splitting field $\mathbf{L}_2 = \mathbb{Q}(i)$ of X^2+1 and then the field extension $\mathbf{L}_2/\left(X^2-2\right) \cong \mathbf{L}_2\left(\sqrt{2}\right) = \mathbf{L}$, which is a splitting field of X^2-2. The field $\mathbf{L}$ is a splitting field of $f(X)$ since over $\mathbf{L}$,

$$f(X) = (X-i)(X+i)\left(X-\sqrt{2}\right)^2\left(X+\sqrt{2}\right)^2.$$

Remark 709. *The polynomial $f(X)$ also splits into linear factors over, say, $\mathbb{Q}\left(\sqrt{2}, i, \sqrt{3}\right)$ which is called a splitting field of $f(X)$ by certain authors who specify that $\mathbb{Q}\left(\sqrt{2}, i\right)$ is a* minimal *splitting field of $f(X)$. Our terminology is that of Bourbaki [28], where a splitting field of a polynomial is a* minimal *field extension over which this polynomial splits into linear factors.*

Generalizing the rationale in the above example, we obtain the following result ([28], §V.4.2, Corol. of Prop. 5):

Proposition 710. *Let $\mathbf{K}$ be a field and let $(f_i)_{i\in I}$ be a family of polynomials of positive degree over $\mathbf{K}$. Then $(f_i)_{i\in I}$ admits a splitting field over $\mathbf{K}$ and any two splitting fields of $(f_i)_{i\in I}$ are $\mathbf{K}$-isomorphic.*

Since all splitting fields of a given family of polynomials of positive degree are $\mathbf{K}$-isomorphic, they can be identified. Applying Proposition 710 to the family of all polynomials of positive degree over $\mathbf{K}$, we obtain :

Theorem 711. *(*Steinitz' theorem*). Let $\mathbf{K}$ be a field. There exists an algebraic closure of $\mathbf{K}$; if $\mathbf{\Omega}$ and $\mathbf{\Omega}'$ are two algebraic closures of $\mathbf{K}$, there exists a $\mathbf{K}$-isomorphism $\mathbf{\Omega} \cong \mathbf{\Omega}'$.*

4.2.1.4 Minimal Polynomials and Normal Extensions

Let $\mathbf{K}$ be a field and let $A = \{a_i : 1 \leq i \leq n\}$ be a finite set of algebraic elements over $\mathbf{K}$ (Definition 703(ii)). The set $\mathfrak{I}$ of all polynomials $g \in \mathbf{K}[X]$ such that each a_i $(1 \leq i \leq n)$ is a root of g is a nonzero ideal of $\mathbf{K}[X]$, and since $\mathbf{K}[X]$ is a principal ideal domain (Theorem 468), there exists a unique monic polynomial f_A such that $\mathfrak{I} = (f_A)$. This polynomial is obviously irreducible.

Definition 712. *The above polynomial f_A is the* minimal polynomial *of the set A. If $A = \{a\}$, then f_A is denoted by f_a and $d^\circ(f_a)$ is called the* degree *of a (written $[a : \mathbf{K}]$).*

Note that $[a : \mathbf{K}] = [\mathbf{K}(a) : \mathbf{K}]$, as shown by the proof of Kronecker's theorem (Theorem 707).

Let a be algebraic over $\mathbf{K}$, let f_a be its minimal polynomial and let $a_1, ..., a_k$ be the distinct roots of f_a in an algebraic closure $\mathbf{\Omega}$ of $\mathbf{K}$; there exists one index i for which $a_i = a$. By Lemma 706, $k \leq [a : \mathbf{K}]$, and as shown below (Corollary 721(ii)), $k = [a : \mathbf{K}]$ if $\mathbf{K}$ is perfect (Definition 221(ii)). Setting $\sigma_i(x) = x$ for any $x \in \mathbf{K}$ and $\sigma_i(a) = a_i$, $i \in \{1, ..., k\}$, the k maps σ_i are distinct $\mathbf{K}$-monomorphisms $\mathbf{K}(a) \hookrightarrow \mathbf{\Omega}$. Considering the field $\mathbf{L} = \mathbf{K}(a_1, ..., a_k)$, we see that each σ_i is a $\mathbf{K}$-automorphism of $\mathbf{L}$.

Definition 713. *The elements $a_i \in \mathbf{\Omega}$ $(1 \leq i \leq k)$ are the* conjugates *of a over $\mathbf{K}$ in $\mathbf{\Omega}$.*

We have proved the following.

Proposition 714. *Let a be algebraic over $\mathbf{K}$ and let k be the number of distinct roots of its minimal polynomial. There exist k distinct $\mathbf{K}$-monomorphisms $\sigma_i : \mathbf{K}(a) \hookrightarrow \mathbf{\Omega}$ and the $\sigma_i(a)$ $(1 \leq i \leq k)$ are the conjugates of a over $\mathbf{K}$; in addition, there are k $\mathbf{K}$-automorphisms $\mathbf{L} \xrightarrow{\sim} \mathbf{L}$, where $\mathbf{L}$ is the field generated over $\mathbf{K}$ by all conjugates of a, and these automorphisms are the above σ_i.*

Item (1) below is proved in ([206], Sect. V.3, Theorem 3.3).

Lemma and Definition 715. *Let $\mathbf{K}$ be a field, $\mathbf{\Omega}$ an algebraic closure of $\mathbf{K}$ and $\mathbf{L}/\mathbf{K}$ be an algebraic extension such that $\mathbf{L} \subseteq \mathbf{\Omega}$.*

(1) The three conditions below are equivalent:
(i) every $\mathbf{K}$-monomorphism $\mathbf{L} \hookrightarrow \mathbf{\Omega}$ induces a $\mathbf{K}$-automorphism of $\mathbf{L}$;
(ii) $\mathbf{L}$ is the splitting field of a family of polynomials in $\mathbf{K}[X]$;
(iii) every irreducible polynomial of $\mathbf{K}[X]$ which has a root in $\mathbf{L}$ splits into linear factors in $\mathbf{K}$.
(2) An algebraic field extension $\mathbf{L}/\mathbf{K}$ is said to be normal *if it satisfies the above equivalent conditions.*

Example 716. *(i) Both algebraic field extensions $\mathbb{Q}\left(\sqrt{2}\right)/\mathbb{Q}$ and $\mathbb{Q}(i)/\mathbb{Q}$ are normal of degree 2, the former because it is the splitting field of $X^2 - 2$, the latter for a similar reason.*
(ii) The algebraic extension $\mathbb{Q}\left(\sqrt{2}, i\right)/\mathbb{Q}$ is normal, since it is the splitting field of $\left(X^2 - 2\right)\left(X^2 + 1\right)$. In addition,

$$\left[\mathbb{Q}\left(\sqrt{2}, i\right) : \mathbb{Q}\right] = \left[\mathbb{Q}\left(\sqrt{2}, i\right) : \mathbb{Q}\left(\sqrt{2}\right)\right]\left[\mathbb{Q}\left(\sqrt{2}\right) : \mathbb{Q}\right]$$

(Proposition 701(i)), thus the extension $\mathbb{Q}\left(\sqrt{2}, i\right)/\mathbb{Q}$ is of degree 4.
(iii) The algebraic field extension $\mathbb{Q}(\alpha)/\mathbb{Q}$, where $\alpha = \sqrt[3]{2}$, is not *normal since the complex roots of $X^3 - 2$ do not all belong to it. We have indeed*

$$X^3 - 2 = (X - \alpha)(X - \omega\alpha)\left(X - \omega^2\alpha\right)$$

where $\omega = \left(-1 + i\sqrt{3}\right)/2$. *The extension* $\mathbb{Q}(\alpha, \omega\alpha)/\mathbb{Q}$ *is the splitting field of* $X^3 - 2$, *by Proposition 701(ii) we have* $\mathbb{Q}(\alpha, \omega\alpha) = \mathbb{Q}(\alpha, \omega)$, *thus the extension* $\mathbb{Q}(\alpha, \omega)/\mathbb{Q}$ *is normal. We have by Proposition 701(i)*

$$[\mathbb{Q}(\alpha, \omega) : \mathbb{Q}] = [\mathbb{Q}(\alpha, \omega) : \mathbb{Q}(\alpha)] [\mathbb{Q}(\alpha) : \mathbb{Q}]$$

and $[\mathbb{Q}(\alpha, \omega) : \mathbb{Q}(\alpha)] = 2$, $[\mathbb{Q}(\alpha) : \mathbb{Q}] = 3$. *See Exercise 796.*

4.2.1.5 Separable Extensions

Definition 717. *Let* $\mathbf{K}$ *be a field and* $\mathbf{\Omega}/\mathbf{K}$ *an algebraically closed extension.*

(i) A polynomial $f(X) \in \mathbf{K}[X]$ *is* separable *if it has no multiple roots in* $\mathbf{\Omega}$.
(ii) An element $a \in \mathbf{\Omega}$ *is* separable *if it is algebraic over* $\mathbf{K}$ *and its minimal polynomial over* $\mathbf{K}$ *is separable.*
(iii) A finite field extension $\mathbf{L}/\mathbf{K}$ *is* separable *if every element of* $\mathbf{L}$ *is separable over* $\mathbf{K}$. *An arbitrary field extension* $\mathbf{L}/\mathbf{K}$ *is separable if every finitely generated subextension* $\mathbf{L}'/\mathbf{K}$, $\mathbf{K} \subseteq \mathbf{L}' \subseteq \mathbf{L}$, *is separable.*
(iv) Let $\mathbf{A}$ *be a* $\mathbf{K}$*-algebra;* $\mathbf{A}$ *is called* separable *(over* $\mathbf{K}$*) if for every field extension* $\mathbf{L}/\mathbf{K}$, *the algebra* $\mathbf{L} \bigotimes_{\mathbf{K}} \mathbf{A}$ *is radical-free (Theorem and Definition 293(3)).*

Proposition 718. *Let* $\mathbf{K}$ *be a perfect field (Definition 221(ii)). Then every algebraic field extension of* $\mathbf{K}$ *is separable over* $\mathbf{K}$.

Proof. In the present proof, we assume for simplicity that $\operatorname{char} \mathbf{K} = 0$. Let $\mathbf{L}/\mathbf{K}$ be an algebraic field extension, let $a \in \mathbf{L}$ and let $f(X)$ be the minimal polynomial of a. If a is a multiple root of $f(X)$, then $f'(a) = 0$ by Lemma 351. Therefore, according to Definition 712, either $f(X) \mid f'(X)$ or $f'(X) = 0$. The former case is impossible since $d^\circ(f'(X)) < d^\circ(f(X))$. Assuming that $f(X)$ is given by (2.30) with $f_0 = 1$ and $n \geq 1$, then $f'(X)$ is given by (2.31). If $f'(X) = 0$, then $n\, X^{n-1} = 0$, thus $n\,1 = 0$ and this is impossible. ∎

The *Theorem of the primitive element* below is proved in ([206], Sect. V.4, Theorem 4.6):

Theorem 719. *If* $\mathbf{L}/\mathbf{K}$ *is a finite separable extension, then there exists an element* $\beta \in \mathbf{L}$, *called a primitive element, such that* $\mathbf{L} = \mathbf{K}(\beta)$.

For example, let $\mathbf{K} = \mathbb{Q}$ and let $\mathbf{L}$ be the field generated over $\mathbf{K}$ by the elements $e^{2\pi k i/8}$, $1 \leq k \leq 7$. As easily seen, $e^{2\pi k i/8}$ is a primitive element of $\mathbf{L}/\mathbf{K}$ if, and only if k is odd. More general, one can show that $e^{2\pi k i/p}$ is a primitive element of the field generated over $\mathbb{Q}$ by the elements $e^{2\pi k i/p}$ $(1 \leq k \leq p-1)$ if, and only if k and p are coprime ([28], §V.11, Prop. 3).

4.2.2 Galois Field Extensions

4.2.2.1 Galois Theory

Definition 720. *Let* $\mathbf{K}$ *be a field. A* Galois field extension $\mathbf{L}/\mathbf{K}$ *is an algebraic extension which is both normal and separable. The Galois group of such an extension* $\mathbf{L}/\mathbf{K}$*, denoted by* $\mathrm{Gal}(\mathbf{L}/\mathbf{K})$*, is the group of all* $\mathbf{K}$*-automorphisms of* $\mathbf{L}$.

In the sequel, *all fields are perfect* (but the proofs are given only for fields of characteristic zero) and except when otherwise stated *all Galois extensions are finite* (Definition 700(ii)). By Theorem 702, $\mathrm{Gal}(\mathbf{L}/\mathbf{K})$ is finite and we have the following.

Corollary 721. *(i) An algebraic extension* $\mathbf{L}/\mathbf{K}$ *is Galois if, and only if it is normal.*
(ii) An irreducible polynomial is separable.

Proof. (i) is a consequence of Proposition 718. To prove (ii), let $g \in \mathbf{K}[X]$ be a non-separable polynomial. Then, g has a multiple root a in an algebraically closed field extension $\boldsymbol{\Omega}/\mathbf{K}$ and g is a multiple of the minimal polynomial f_a of a. The proof of Proposition 718 shows that a is a simple root of f_a, thus there exists a polynomial h of positive degree such that $g = f_a\, h$. Since both g and f_a belong to $\mathbf{K}[X]$, so does h, and g is not irreducible. ∎

The converse of Statement (ii) of Corollary 721 does not hold. For example, $(X^2-1)(X^2-2)$ is separable but not irreducible over $\mathbb{Q}[X]$. Let $\mathbf{K}$ be a field and $f \in \mathbf{K}[X]$ a separable polynomial over $\mathbf{K}$.

Definition 722. *Let* R *be the set of roots of* f *in an algebraically closed field extension* $\boldsymbol{\Omega}/\mathbf{K}$*. Then* $\mathrm{Gal}(\mathbf{K}(R)/\mathbf{K})$ *is called a Galois group of* f *over* $\mathbf{K}$.

As stated in Proposition 710, every two splitting fields $\mathbf{L}$ and $\mathbf{M}$ of f are $\mathbf{K}$-isomorphic. Let $u : \mathbf{L} \xrightarrow{\sim} \mathbf{M}$ be a $\mathbf{K}$-isomorphism. The map $s \mapsto u \circ s \circ u^{-1}$ from $\mathrm{Gal}(\mathbf{L}/\mathbf{K})$ to $\mathrm{Gal}(\mathbf{M}/\mathbf{K})$ is an isomorphism. The splitting fields as well as the Galois groups of a given polynomial are thus unique up to isomorphism and one can speak about "the" splitting field and "the" Galois group of f over $\mathbf{K}$. The latter is a group of permutations of the roots of f by Proposition 714, and is denoted by $\mathrm{Gal}_{\mathbf{K}}(f)$. With the notation in Definition 722, let $\mathfrak{S}(R)$ be the symmetric group of R (Example 135). There exists by Lemma 715 and Corollary 721(i) an embedding $\iota : \mathrm{Gal}_{\mathbf{K}}(f) \hookrightarrow \mathfrak{S}(R)$. Thus $\mathrm{Gal}_{\mathbf{K}}(f)$ can be identified with $\iota(\mathrm{Gal}_{\mathbf{K}}(f)) \triangleq \Gamma \subseteq \mathfrak{S}(R)$.

Remark 723. Γ *is a subgroup of* $\mathfrak{S}(R)$ *which does not coincide with* $\mathfrak{S}(R)$ *in general. For example, if* f *has a root* $r \in \mathbf{K}$*, then* r *is a fixed point of all* $u \in \mathrm{Gal}_{\mathbf{K}}(f)$*, thus* $\Gamma \subsetneq \mathfrak{S}(R)$*. This is not the only case when the inclusion* $\Gamma \subset \mathfrak{S}(R)$ *is strict (Exercise 799).*

The proof of the following is easy ([28], n°V.10.2):

Proposition 724. *Let $x, y \in R$, where R is as in Definition 722.*

(a) The following conditions are equivalent:
(i) x and y are conjugate over $\mathbf{K}$;
*(ii) x and y belong to the same orbit of Γ (1.6.9.**1**);*
(iii) x and y are roots of the same irreducible factor of f.
*(b) In particular, f is irreducible if, and only if $R \neq \varnothing$ and Γ acts transitively on R (1.6.9.**2**).*

Let $\mathbf{L}/\mathbf{K}$ be a (finite) Galois extension. For any subgroup H of $\mathrm{Gal}(\mathbf{L}/\mathbf{K})$, let $\mathbf{L}^H$ be the subfield of $\mathbf{L}$ consisting of all elements x such that $u(x) = x$ for all $u \in H$. For any field $\mathbf{M}$ such that $\mathbf{K} \subseteq \mathbf{M} \subseteq \mathbf{L}$, let $G(\mathbf{L}/\mathbf{M})$ be the set of all automorphisms $u \in \mathrm{Gal}(\mathbf{L}/\mathbf{K})$ such that $u(x) = x$ for all $x \in \mathbf{M}$. Let $\mathfrak{F}(\mathbf{L})$ be the set of all subfields $\mathbf{M}$ of $\mathbf{L}$ as considered above and let $\mathfrak{G}(\mathrm{Gal}(\mathbf{L}/\mathbf{K}))$ the set of all subgroups of $\mathrm{Gal}(\mathbf{L}/\mathbf{K})$; $\mathfrak{F}(\mathbf{L})$ and $\mathfrak{G}(\mathrm{Gal}(\mathbf{L}/\mathbf{K}))$ are lattices ordered by inclusion (Exercise 798 and Lemma 153). The following is proved in ([28], n°V.10.7).

Theorem 725. *(*Fundamental theorem of the Galois theory*).*
(i) There exists a lattice-antiisomorphism $\mathfrak{G}(\mathrm{Gal}(\mathbf{L}/\mathbf{K})) \to \mathfrak{F}(\mathbf{L})$ (Definition 112) given by

$$H \mapsto \mathbf{L}^H$$

with inverse

$$\mathbf{M} \mapsto G(\mathbf{L}/\mathbf{M}).$$

In addition

$$(G(\mathbf{L}/\mathbf{M}) : 1) = [\mathbf{L} : \mathbf{M}]; \quad (\mathrm{Gal}(\mathbf{L}/\mathbf{K}) : G(\mathbf{L}/\mathbf{M})) = [\mathbf{M} : \mathbf{K}] \tag{4.1}$$

*(with the notation from (1.6.2.**2**)).*
(ii) Let $\mathbf{M} \in \mathfrak{F}(\mathbf{L})$. The extension $\mathbf{M}/\mathbf{K}$ is Galois if, and only if $G(\mathbf{L}/\mathbf{M})$ is a normal subgroup *of* $\mathrm{Gal}(\mathbf{L}/\mathbf{K})$*, and then*

$$G(\mathbf{L}/\mathbf{M}) = \mathrm{Gal}(\mathbf{L}/\mathbf{M}). \tag{4.2}$$

When $\mathbf{M}/\mathbf{K}$ is Galois, the restriction to $\mathbf{M}$ of an automorphism $u \in \mathrm{Gal}(\mathbf{L}/\mathbf{K})$ is an automorphism $u_{\mathbf{M}}$ of $\mathbf{M}$ and the restriction $\rho_{\mathbf{M}} : u \mapsto u_{\mathbf{M}}$ is such that $\ker \rho_{\mathbf{M}} = \mathrm{Gal}(\mathbf{L}/\mathbf{M})$, $\mathrm{im}\, \rho_{\mathbf{M}} = \mathrm{Gal}(\mathbf{M}/\mathbf{K})$, thus it induces an isomorphism

$$\mathrm{Gal}(\mathbf{L}/\mathbf{K}) / \mathrm{Gal}(\mathbf{L}/\mathbf{M}) \xrightarrow{\sim} \mathrm{Gal}(\mathbf{M}/\mathbf{K}). \tag{4.3}$$

Remark 726. *(i) The above lattice-antiisomorphism is a special case of Galois connection (Definition 98).*
(ii) The equalities (4.1) are called Artin's theorem.

Let us consider an illustrating example:

Example 727. *The polynomial* $f(X) = X^4 - 3$ *is irreducible over* $\mathbf{K} = \mathbb{Q}$*, as shown by Eisenstein's Criterion (Proposition 432). The four roots of* $f(X)$ *over* $\mathbb{C}$ *are* $r, i\,r, -r, -i\,r$ *where* $r = \sqrt[4]{3}$ *and* $i = \sqrt{-1}$*. The splitting field of* $f(X)$ *is* $\mathbf{N} = \mathbb{Q}(r,i) = \mathbb{Q}(r)(i)$*. We have* $[\mathbb{Q}(r) : \mathbb{Q}] = 4$ *since a basis of the* $\mathbb{Q}$*-vector space* $\mathbb{Q}(r)$ *is* $\{1, r, r^3, r^3\}$*; and* $[\mathbb{Q}(r,i) : \mathbb{Q}(r)] = 2$ *since a basis of the* $\mathbb{Q}(r)$*-vector space* $\mathbb{Q}(r,i)$ *is* $\{1, i\}$*. Therefore,* $[\mathbb{Q}(r,i) : \mathbb{Q}] = 8$ *(Proposition 701) and by (4.2) and the first equality of (4.1) with* $\mathbf{M} = \mathbf{K} = \mathbb{Q}$*,* $\mathbf{N} = \mathbb{Q}(r,i)$*, we have* $(\mathrm{Gal}(\mathbb{Q}(r,i)/\mathbb{Q}) : 1) = 8$*. A* $\mathbb{Q}$*-automorphism* τ *of* $\mathbf{N}$ *is determined by* $\tau(r)$ *and* $\tau(i)$*. Consider the extension* $\mathbb{Q}(r,i)/\mathbb{Q}(r)$ *of degree 2. The two* $\mathbb{Q}(r)$*-automorphisms of* $\mathbb{Q}(r,i)$ *are* $\tau : i \mapsto -i$ *and* $\tau^0 = Id$*, thus* $(\langle\tau\rangle : 1) = 2$*. The four* $\mathbb{Q}(i)$*-automorphisms of* $\mathbb{Q}(r,i)$ *are* $\sigma : r \mapsto i\,r$ *and its powers* $\sigma^0 = Id$*,* σ^2*,* σ^3*, thus* $(\langle\sigma\rangle : 1) = 4$*. Let us check that we have found all elements of* $\mathbb{Q}(r,i)$*: according to Noether's second isomorphism theorem (Theorem 146(2)), we have* $\langle\tau\rangle \cong \langle\sigma,\tau\rangle / \langle\sigma\rangle$ *thus by Lagrange's theorem (1.21),* $(\langle\sigma,\tau\rangle : 1) = (\langle\tau\rangle : 1)(\langle\sigma\rangle : 1) = 8 = (G : 1)$*, therefore* $G = \langle\sigma,\tau\rangle$*. Combining all these* $\mathbb{Q}$*-automorphisms, the eight elements of* $\mathrm{Gal}(\mathbb{Q}(r,i)/\mathbb{Q})$ *are obtained, as shown in the table below where* $1 = Id_G$ *:*

	1	σ	σ^2	σ^3	τ	$\sigma\circ\tau$	$\sigma^2\circ\tau$	$\sigma^3\circ\tau$
image of r	r	$i\,r$	$-r$	$-i\,r$	r	$i\,r$	$-r$	$-i\,r$
image of i	i	i	i	i	$-i$	$-i$	$-i$	$-i$

Consider now the subgroups of $G = \mathrm{Gal}(\mathbb{Q}(r,i)/\mathbb{Q})$*. Let* $H_1 = \{1, \sigma, \sigma^2, \sigma^3\} = \langle\sigma\rangle$ *and* $H_2 = \{1, \sigma^2\} = \langle\sigma^2\rangle$*. The automorphisms belonging to* H_1 *and* H_2 *are* $\mathbb{Q}(i)$*-automorphisms and we have the decreasing chain of subgroups*

$$G \supseteq H_1 \supseteq H_2 \supseteq 1$$

which corresponds through the Galois connection to the increasing chain of field extensions

$$\mathbb{Q} \subseteq \mathbb{Q}(i) \subseteq \mathbb{Q}\left(i, \sqrt{3}\right) \subseteq \mathbb{Q}(i,r)$$

obtained by successively adding the roots of the polynomials $X^2 + 1$*,* $X^2 - 3$*,* $X^2 - \sqrt{3}$*.*

Remark 728. *Consider an* infinite Galois extension $\mathbf{L}/\mathbf{K}$*.*

(i) This extension can be put into the form of a directed union of finite Galois extensions $\mathbf{M}_i/\mathbf{K}$ *(Example 19), i.e.,* $\mathbf{L} = \varinjlim \mathbf{M}_i$*.*

(ii) Let us endow each finite Galois group $\mathrm{Gal}(\mathbf{M}_i/\mathbf{K})$ *with the finite topology (1.7.1.***2***). Then* $\mathrm{Gal}(\mathbf{M}_i/\mathbf{K})$ *becomes a topological group which is compact and* totally discontinuous, *i.e., the connected component of the unit element* 1 *is* $\{1\}$ *([34], n°I.11.5).*

(iv) The Galois group $\mathrm{Gal}(\mathbf{L}/\mathbf{K})$ *is identified with the projective limit* $\varprojlim \mathrm{Gal}(\mathbf{M}_i/\mathbf{K})$ *of the topological groups* $\mathrm{Gal}(\mathbf{M}_i/\mathbf{K})$ *([28], § V.10, Prop. 4); the topology of* $\mathrm{Gal}(\mathbf{L}/\mathbf{K})$ *is called the* profinite topology *(also called the* Krull topology*). The topological group* $\mathrm{Gal}(\mathbf{L}/\mathbf{K})$ *is again compact and totally discontinuous. In the statement of Theorem 725,* $\mathfrak{G}(\mathrm{Gal}(\mathbf{L}/\mathbf{K}))$ *must be replaced by the set of all* closed *subgroups of* $\mathrm{Gal}(\mathbf{L}/\mathbf{K})$*, and (4.3) is an isomorphism of topological groups ([28], n° V.10.7).*

4.2.2.2 Solvability by Radicals

The explicit formula to calculate the roots of a general polynomial of degree 2 using a square root is classical. Similar (although more complicated) formulas are also existing to calculate the roots of general polynomials of degree 3 and 4 using cubic and quartic roots. Abel showed in 1826 that such a formula (using also quintic roots) cannot exist to calculate the roots of a general polynomial of degree 5. But to make Abel's theorem clear, Galois' theory is needed.

Definition 729. *Let* $\mathbf{K}$ *be a field of characteristic zero.*

(i) A radical extension *of* $\mathbf{K}$ *is a simple extension* $\mathbf{K}(\alpha)/\mathbf{K}$ *where* α *is a root of a* binomial equation *irreducible over* $\mathbf{K}$

$$X^n - a = 0, \quad a \in \mathbf{K}.$$

(ii) An algebraic equation $f(X) = 0$ *is solvable by radicals over* $\mathbf{K}$ *if there exists a finite tower of radical extensions*

$$\mathbf{K} = \mathbf{K}_0 \subsetneq ... \subsetneq \mathbf{K}_r$$

(called a root tower *for* f *over* $\mathbf{K}$*) such that* $f(X)$ *splits into linear factors in* $\mathbf{K}_r$*.*

Theorem 730. *(*Galois' theorem*). Let* $\mathbf{K}$ *be a field of characteristic zero. An equation* $f(X) = 0$ *is solvable by radicals over* $\mathbf{K}$ *if, and only if the Galois group of* $f(X)$ *over* $\mathbf{K}$ *is solvable (Definition 171).*

A proof of the above theorem is given in, e.g., ([79], Theorem 7.11.1). For any $n \geq 1$ there exists an irreducible polynomial $f_n(X) \in \mathbb{Q}[X]$ of degree n, the Galois group of which is $\mathfrak{S}_n$ ([79], Theorem 7.11.4). By Example 173 and Theorem 730 we obtain *Abel's theorem*: for any $n \geq 5$, the algebraic equation $f_n(X) = 0$ is *not* solvable by radicals over $\mathbb{Q}$.

4.3 Skew Polynomials and Related Notions

4.3.1 Skew Polynomial Rings

4.3.1.1 Further Properties of Skew Polynomial Rings

The rings of skew polynomial rings in one indeterminate and the associated rings of generalized differential operators (Definition 355) have been introduced in Subsect. 2.7. Let us consider a skew polynomial ring $\mathbf{R} = \mathbf{K}[X; \boldsymbol{\alpha}, \boldsymbol{\delta}]$ where $\mathbf{K} \supseteq \mathbb{Q}$ is a division ring, $\boldsymbol{\alpha}$ is an automorphism and $\boldsymbol{\delta}$ is an $\boldsymbol{\alpha}$-derivation of $\mathbf{K}$. The following is a generalization of the result to be proved in Exercise 489(ii):

Theorem 731. *The skew polynomial ring* $\mathbf{R} = \mathbf{K}[X; \boldsymbol{\alpha}, \boldsymbol{\delta}]$ *is a strongly Euclidean domain.*

Proof. We know that $\mathbf{R}$ is a domain and that the degree d° of a skew polynomial (Definitions 349 and 353) is a function θ which satisfies conditions (E1) and (E2') in (2.13.6.1) (Proposition 354). Let us show that d° satisfies (E3'). Consider the nonzero left differential polynomials $f(X)$ defined by (2.29) (with $\mathbf{R}$ changed to $\mathbf{K}$) and $g(X) = \sum_{i=0}^{m} g_i X^{m-i}$, where $f_0 \neq 0, g_0 \neq 0$ and $m \leq n$. Then $d^\circ\left(f - f_0\, g_0^{-1}\, g\, X^{n-m}\right) < d^\circ(f)$. Thus (E3') is satisfied and $\mathbf{R}$ is a left strongly Euclidean domain. By symmetry, $\mathbf{R}$ is right strongly Euclidean too. ∎

Corollary 732. *The ring of skew Laurent polynomials* $\mathbf{T} = \mathbf{K}\left[X, X^{-1}; \boldsymbol{\alpha}, \boldsymbol{\delta}\right]$ *(where* $\boldsymbol{\alpha}$ *is an automorphism and* $\boldsymbol{\delta}$ *is an* $\boldsymbol{\alpha}$*-derivation of* $\mathbf{K}$*) is a principal ideal domain.*

Proof. By Theorems 468 and 731, $\mathbf{R} = \mathbf{K}[X; \boldsymbol{\alpha}, \boldsymbol{\delta}]$ is a principal ideal domain. In addition, $\mathbf{T} = S^{-1}\mathbf{R}$ where S is the left and right denominator set $\{X^n; n \geq 1\}$ (Subsect. 2.8.1). Therefore, $\mathbf{T}$ is a principal ideal ring by Lemma 463 and is a domain by Theorem 332(ix). ∎

The proof of the following result is easy ([77], Sect. 8.3, Prop. 3.2 and Corol. 3.3):

Lemma 733. *Let* $f = \sum_{i=0}^{n} a_i X^{n-i} \in \mathbf{R}$, $a_0 = 1$. *This element* f *is invariant in* $\mathbf{R}$ *(Definition 85) if, and only if Conditions (i) and (ii) below are satisfied:*

(i) $f\, c = c^{\boldsymbol{\alpha}^n} f$ *for all* $c \in \mathbf{K}$,
(ii) $f\, X = (X + a_1 - a_1^{\boldsymbol{\alpha}})\, f$.

Theorem 734. *(i) If* $\boldsymbol{\delta}$ *is outer and* $\boldsymbol{\alpha} = 1$, *the invariant elements of* $\mathbf{R} = \mathbf{K}[X; \boldsymbol{\alpha}, \boldsymbol{\delta}]$ *are the units.*
(ii) If $\boldsymbol{\delta}$ *is inner, induced by* 1 *and no power of* $\boldsymbol{\alpha}$ *is an inner automorphism of* $\mathbf{R}$, *then the invariant elements of* $\mathbf{R}$ *are* 1, *the powers of* $Y = X + 1$ *and their associates.*

(iii) If $\boldsymbol{\delta}$ is inner, induced by 1 and $N \geq 1$ is the least integer such that $\boldsymbol{\alpha}^N$ is an inner automorphism of $\mathbf{R}$ *(induced by the unit u), then the invariant elements are the polynomials of the form*

$$f = Y^n \sum_{0 \leq i \leq k} b_i \left(u\, Y^N\right)^{k-i} \tag{4.4}$$

and their associates, where $Y = X + 1$ and the coefficients b_i are constant and central.

Proof. (i) is an obvious consequence of Corollary 363 since by hypothesis $\mathbf{K} \supseteq \mathbb{Q}$.

(ii): By Corollary 368, if f is an invariant element of $\mathbf{R}$, necessarily it is associated with 1 or a power of $Y = X + 1$. Conversely, if $f = Y^n$ $(n \geq 0)$, then Conditions (i) and (ii) in Lemma 733 are obviously satisfied, thus f is an invariant element of $\mathbf{R}$.

(iii): We know that $u\, Y^N$ is central in $\mathbf{K}\,[Y; \boldsymbol{\alpha}]$ (Proposition 375); thus, if f given by (4.4) where the coefficients b_i are constant and central, Conditions (i) and (ii) in Lemma 733 are satisfied. Conversely, it is easy to check that if these two conditions are satisfied, then f is necessarily of the form (4.4) with b_i constant and central. (Note that the indeterminate Y has not the same meaning here as in Proposition 375.) ■

In particular, the ring $B_1\,(\mathbf{k})$ (2.7.4.**3**) is a strongly Euclidean domain (thus a principal ideal domain) and is simple.

4.3.1.2 Companion Matrix

Recall that the action of the indeterminate X on any $\mathbf{R}$-module is the operator denoted by ∂, that ∂ is an extension of the $\boldsymbol{\alpha}$-derivation $\boldsymbol{\delta}$, and that $\mathbf{K}\,[\partial; \boldsymbol{\alpha}, \boldsymbol{\delta}]$ is a called a ring of generalized differential operators (Definition 355). Let $f\,(X) = X^n + a_1\, X^{n-1} + ... + a_n$ be an element of $\mathbf{R}^{\times}$. The cyclic left $\mathbf{R}$-module $M = \mathbf{R}/\mathbf{R}\,f$ is generated by an element y (i.e., $M = [y]_{\mathbf{R}}$) submitted to the only relation $\left(\partial^n + a_1\, \partial^{n-1} + ... + a_n\right)\, y = 0$.

Set $x_1 = \partial^{n-1}\, y, ..., x_n = y$, let x be the column with entries x_i $(1 \leq i \leq n)$ and $[x]_{\mathbf{R}}$ be the left $\mathbf{R}$-module generated by $(x_i)_{1 \leq i \leq n}$. We have obviously $M = [x]_{\mathbf{R}}$ where x is submitted to the only relation

$$\partial\, x = C_f\, x, \quad C_f = \begin{bmatrix} -a_1 & -a_2 & \cdots & \cdots & -a_n \\ 1 & 0 & \cdots & \cdots & 0 \\ 0 & \ddots & \ddots & & \vdots \\ \vdots & & \ddots & \ddots & \vdots \\ 0 & \cdots & 0 & 1 & 0 \end{bmatrix}; \tag{4.5}$$

$C_f \in \mathrm{Mat}_n(\mathbf{K})$ is called a *companion matrix* of the polynomial $f(X)$. Therefore,

$$M \cong \mathrm{coker}_{\mathbf{R}}(\bullet(\partial I_n - C_f));$$

by Definition 517, Theorem 520 and Corollary 521 we have the following.

Proposition 735. *The polynomial $f(X)$ and the matrix $X I_n - C_f$ are similar or, equivalently, stably associated.*

Let V be the left $\mathbf{R}$-module $M = [y]_{\mathbf{R}}$ considered as a left $\mathbf{K}$-vector space. Obviously, $(x_i)_{1 \leq i \leq n}$ is a basis of V and it is generated over $\mathbf{R}$ by the unique element y as follows: $x_n = y$, $x_{n-1} = \partial y$, ..., $x_1 = \partial^{n-1} y$. This leads to the following

Lemma and Definition 736. *(1) The above basis $(x_i)_{1 \leq i \leq n}$ of V is said to be* cyclic.
(2) Let $A \in \mathrm{Mat}_n(\mathbf{K})$. The following conditions are equivalent:
(i) A is similar to a companion matrix;
(ii) $\mathrm{coker}_{\mathbf{R}}(\bullet(\partial I_n - A))$ *is a cyclic* $\mathbf{R}$*-module.*
When these equivalent conditions hold, the matrix A is said to be cyclic.

4.3.2 Pseudo-linear Transformations

4.3.2.1 A General Definition

Let $\mathbf{R} = \mathbf{K}[\partial; \boldsymbol{\alpha}, \boldsymbol{\delta}]$, let M be a left $\mathbf{R}$-module and let $V = M_{[\mathbf{K}]}$ be the left $\mathbf{K}$-vector space obtained by restriction of the ring of scalars (3.4.2.1). For any $v_1, v_2 \in M$ and any $\lambda \in \mathbf{K}$, we have by (2.33)

$$\partial(v_1 + v_2) = \partial v_1 + \partial v_2,$$
$$\partial(\lambda v_1) = \lambda^{\boldsymbol{\alpha}} \partial v_1 + \lambda^{\boldsymbol{\delta}} v_1.$$

Let $\theta\bullet$ be the restriction of ∂ to the $\mathbf{K}$-vector space V. We obtain from the above equations

$$\theta(v_1 + v_2) = \theta(v_1) + \theta(v_2), \tag{4.6}$$
$$\theta(\lambda v_1) = \lambda^{\boldsymbol{\alpha}} \theta(v_1) + \lambda^{\boldsymbol{\delta}} v_1. \tag{4.7}$$

By (4.6), the map $\theta : V \to V$ is additive but by (4.7) *it is not* $\mathbf{K}$*-linear*, except if $\boldsymbol{\alpha} = 1$ and $\boldsymbol{\delta} = 0$, i.e., $\mathbf{R} = \mathbf{K}[X]$ (therefore, it would be misleading to denote θ as left map). We are led to the following

Definition 737. *The above map θ is called the* pseudo-linear transformation *(*PLT*, for short) induced by ∂ on V.*

A left eigenvalue and an eigenvector of a PLT $\theta : V \to V$ are defined as follows [214], [204]:

Definition 738. *Let* $\lambda \in \mathbf{K}$ *and let* $v \in V^{\times}$*;* λ *is a* left eigenvalue *of* θ *and* v *is an* eigenvector *associated with* λ *if* $\theta(v) = \lambda\, v$.

Remark 739. *A right eigenvalue* λ *and an associated eigenvector can also be defined; the latter is a nonzero element* v *of a right* $\mathbf{K}$*-vector space, such that* $\theta(v) = v\,\lambda$.

4.3.2.2 PLTs in the Finite-Dimensional Case

Theorem 740. *The finitely generated left* $\mathbf{R}$*-module* M *is torsion if, and only if the left* $\mathbf{K}$*-vector space* $V = M_{[\mathbf{K}]}$ *is finite-dimensional. Then, considering the decomposition (3.66) of* M *into a direct sum of cyclic modules (with* $r = k$*),* $\dim_{\mathbf{K}} V = \sum_{1 \leq i \leq k} d^{\circ}(e_i)$.

Proof. If M is not torsion, there exists an $\mathbf{R}$-independent element $m \in M$, i.e., an element for which there does not exist a nonzero skew polynomial $f(X) \in \mathbf{R}$ such that $f(\partial)\, m = 0$ (Theorem and Definition 339). Therefore, $\left(\partial^i m\right)_{i \geq 0}$ is an infinite sequence of left $\mathbf{K}$-independent elements, and V is infinite-dimensional.

Conversely, if M is torsion, then by (3.66), $M = \bigoplus_{1 \leq i \leq r} M_i$, $M_i \cong \mathbf{R}/\mathbf{R}e_i$, $e_i \neq 0$. Therefore by Theorem 551(ii),

$$M_{[\mathbf{K}]} = \bigoplus\nolimits_{1 \leq i \leq r} M_{i[\mathbf{K}]}. \tag{4.8}$$

Let $N = M_i$, let $f = e_i$ and let $n = d^{\circ}(f)$; then $N = [x]_{\mathbf{R}}$ where x satisfies (4.5). Therefore, the components of $\partial\, x$ belong to $[x]_{\mathbf{K}}$ (i.e., to the left $\mathbf{K}$-vector space generated by the components of x) and by induction so do the components of $\partial^j x$ for all $j \geq 0$. Thus $N_{[\mathbf{K}]} = [x]_{\mathbf{K}}$ is finite-dimensional. Since the components of x are left $\mathbf{K}$-linearly independent, $(x_j)_{1 \leq i \leq n}$ is a basis of $N_{[\mathbf{K}]}$ and $\dim_{\mathbf{K}} N_{[\mathbf{K}]} = d^{\circ}(f)$. With the original notation, $\dim_{\mathbf{K}} M_{i[\mathbf{K}]} = d^{\circ}(e_i)$. By (4.8), $V = M_{[\mathbf{K}]}$ is finite-dimensional and $\dim_{\mathbf{K}} V = \sum_{1 \leq i \leq k} d^{\circ}(e_i)$. ∎

Let $(x_i)_{1 \leq i \leq n}$ be a basis of the left $\mathbf{K}$-vector space $V = M_{[\mathbf{K}]}$. There exist elements $a_{ij} \in \mathbf{K}$ $(1 \leq i \leq n, 1 \leq j \leq n)$ such that

$$\theta(x_i) = \sum_{1 \leq j \leq n} a_{ij}\, x_j.$$

Let $y_i = \theta(x_i)$. Denoting by x and y the columns with entries x_i and y_i, respectively, and by A the matrix (a_{ij}), we obtain

$$y = A\, x. \tag{4.9}$$

The following definition is consistent with Definition 263, although θ is not $\mathbf{K}$-linear:

Definition 741. *The matrix $A \in \mathrm{Mat}_n(\mathbf{K})$ is called the matrix of the PLT θ in the basis x.*

Let $\mathbf{v} \in V$, let $v_i \in \mathbf{K}$ $(1 \leq i \leq n)$ be the components of $\mathbf{v}$ in the basis x and let v be the row $\begin{bmatrix} v_1 \cdots v_n \end{bmatrix}$, so that $\mathbf{v} = v\,x$. Identifying $\mathbf{v}$ and v in the basis x, V is identified with $\mathbf{K}^n$. Let $\mathbf{w} = \theta(\mathbf{v})$. We have

$$\begin{aligned}\mathbf{w} = \theta\left(\sum_{1\leq i\leq n} v_i\,x_i\right) &= \sum_{1\leq i\leq n} \theta(v_i\,x_i) = \sum_{1\leq i\leq n} \left(v_i^{\boldsymbol{\alpha}}\,\theta(x_i) + v_i^{\boldsymbol{\delta}}\,x_i\right) \\ &= \left[v_1^{\boldsymbol{\alpha}} \ldots v_n^{\boldsymbol{\alpha}}\right] A\,x + \left[v_1^{\boldsymbol{\delta}} \ldots v_n^{\boldsymbol{\delta}}\right] x.\end{aligned}$$

We have obtained the following

Lemma 742. *In the basis x, θ is the map $v \mapsto v^{\boldsymbol{\alpha}}\,A + v^{\boldsymbol{\delta}}$.*

4.3.2.3 Change of Basis

Let $y = (y_i)_{1\leq i\leq n}$ be a second basis of V and let $P \in \mathrm{GL}_n(\mathbf{K})$ the change of basis matrix from x to y, i.e., the matrix of the map $Id_V : (V, y) \to (V, x)$. By (4.9) (which is valid since Id_V is linear, thus pseudo-linear), we obtain

$$y = P\,x.$$

Let $\mathbf{v} \in V$ and let v_x, v_y be the rows representing $\mathbf{v}$ in the bases x and y, respectively. We have $\mathbf{v} = v_x\,x = v_y\,y = v_y\,P\,x$, therefore

$$v_x = v_y\,P \tag{4.10}$$

which is consistent with (2.20).

Let A and B be the matrices of the PLT θ in the bases x and y, respectively, and let $\mathbf{w} = \theta(\mathbf{v})$. By Lemma 742 and by (4.10) we have

$$w_x = v_x^{\boldsymbol{\alpha}}\,A + v_x^{\boldsymbol{\delta}}, \quad w_y = v_y^{\boldsymbol{\alpha}}\,B + v_y^{\boldsymbol{\delta}}, \quad w_x = w_y\,P \tag{4.11}$$

therefore (using the expression in Lemma 742)

$$w_y\,P = (v_y\,P)^{\boldsymbol{\alpha}}\,A + (v_y\,P)^{\boldsymbol{\delta}} = v_y^{\boldsymbol{\alpha}}\left(P^{\boldsymbol{\alpha}}\,A + P^{\boldsymbol{\delta}}\right) + v_y^{\boldsymbol{\delta}}\,P.$$

Right-multiplying this expression by P^{-1} and using the second equality of (4.11) we obtain

$$\boxed{B = \left(P^{\boldsymbol{\alpha}}\,A + P^{\boldsymbol{\delta}}\right)P^{-1}.} \tag{4.12}$$

The following is a generalization of Definition 280(iii):

Definition 743. *Two matrices $A, B \in \mathrm{Mat}_n(\mathbf{K})$ are said to be $(\boldsymbol{\alpha}, \boldsymbol{\delta})$-conjugate if there exists a matrix $P \in \mathrm{GL}_n(\mathbf{K})$ such that A and B are related by (4.12). More specifically, B is called the $(\boldsymbol{\alpha}, \boldsymbol{\delta})$-conjugate of A by P.*

Indeed, conjugacy in Definition 280(iii) corresponds to $(1,0)$-conjugacy in Definition 743. The following is now clear:

Theorem 744. *(i) Two matrices* $A, B \in \mathrm{Mat}_n(\mathbf{K})$ *are* $(\boldsymbol{\alpha}, \boldsymbol{\delta})$*-conjugate if, and only if they represent (in different bases) the same PLT.*
(ii) $(\boldsymbol{\alpha}, \boldsymbol{\delta})$*-conjugacy is an equivalence relation.*

Let f, g be two nonzero monic polynomials in $\mathbf{R} = \mathbf{K}[X; \boldsymbol{\alpha}, \boldsymbol{\delta}]$, and let $M = \mathbf{R}/\mathbf{R}f$, $N = \mathbf{R}/\mathbf{R}g$. If $M \cong N$, then $d^\circ(f) = d^\circ(g)$ by Theorem 740. As already seen (Proposition 735 and Definition 737), the companion matrix C_f represents the PLT θ induced by ∂ in M considered as a $\mathbf{K}$-vector space, and by definition $M \cong N$ if, and only if f and g are similar (Definition 517). We have therefore the following by Theorem 744:

Lemma 745. *Two monic polynomials* f, g *are similar if, and only if their companion matrices are* $(\boldsymbol{\alpha}, \boldsymbol{\delta})$*-conjugate.*

The above can be applied to two polynomials $p(X)$ and $X - a$ in $\mathbf{R} = \mathbf{K}[X; \boldsymbol{\alpha}, \boldsymbol{\delta}]$. The companion matrices of $X - a$ and $X - b$ are a and b, respectively, thus by Lemma 745 we obtain:

Corollary 746. *Two polynomials* $p(X)$ *and* $X - a$ *in* $\mathbf{R} = \mathbf{K}[X; \boldsymbol{\alpha}, \boldsymbol{\delta}]$ *are similar if, and only if* $p(X) = c(X - b)$ *where* $c \in \mathbf{K}^\times$ *and* a, b *are* $(\boldsymbol{\alpha}, \boldsymbol{\delta})$*-conjugate (written* $a \overset{(\boldsymbol{\alpha},\boldsymbol{\delta})}{\sim} b$*).*

Definition 747. *Let* $\mathbf{F}$ *be a division ring such that* $\mathbf{F} \subseteq \mathbf{K}$*.*

(i) A PLT θ *is said to be* algebraic *over* $\mathbf{F}$ *if there exist an integer* $n > 1$ *and elements* $a_0, a_1, ..., a_n \in \mathbf{F}^\times$ *such that* $a_0 \theta^n + a_1 \theta^{n-1} + ... + a_n I = 0$*. This expression is denoted by* $f(\theta)$ *where* $f(X) = \sum_{0 \le i \le n} a_i X^{n-i}$*. A square matrix* A *over* $\mathbf{K}$ *is called* algebraic *over* $\mathbf{F}$ *if it is the matrix of an algebraic PLT over* $\mathbf{F}$*.*
(ii) A PLT θ *is said to be* totally transcendental *over* $\mathbf{F}$ *if* $f(\theta)$ *is non-singular for any polynomial* $f(X) \in \mathbf{F}[X]^\times$*. A square matrix* A *over* $\mathbf{K}$ *is called* totally transcendental *over* $\mathbf{F}$ *if it is the matrix of a totally transcendental PLT over* $\mathbf{F}$*.*

4.3.3 Jordan Canonical Form

4.3.3.1 Diagonal Decomposition

Consider the "classical case" $\boldsymbol{\alpha} = 1$, $\boldsymbol{\delta} = 0$; then $\mathbf{R} = \mathbf{K}[X]$ where X is a central indeterminate, and $\mathbf{F}$ is assumed to be the centre of $\mathbf{K}$. Identifying $\mathbf{R}$ with $\mathbf{K}[\partial]$, let M be a left $\mathbf{R}$-module and let $V = M_{[\mathbf{K}]}$; the PLT $\theta = \partial|_V$ is $\mathbf{K}$-linear (**4.3.2.1**), and V is finite-dimensional if, and only if $M = T$ is torsion (Theorem 740). Then (3.67) holds and in the suitable basis, the matrix A of θ is of the form

$$A = A_1 \oplus ... \oplus A_m \oplus U$$

where each submatrix is cyclic (Lemma and Definition 736(2)). One can prove that for each $i \in \{1, ..., m\}$, A_i is algebraic over $\mathbf{F}$ (with a single elementary divisor $q_i \in \mathbf{F}[X]$, by construction), and that U is totally transcendental over $\mathbf{F}$ ([78], Sect. 8.3).

4.3.3.2 Case of a Totally Transcendental Matrix

Assume that A is totally transcendental, i.e., $A = U$. Item (2) below is proved in ([78], Theorem 5.5.5).

Corollary and Definition 748. *(1) Let* $\mathbf{K}$ *be a division ring with central subfield* $\mathbf{F}$*;* $\mathbf{K}$ *is said to be* matrix-homogeneous *over* $\mathbf{F}$ *if* $\mathbf{K}$ *has a skew field extension* $\mathbf{L}/\mathbf{K}$ *(Definition 699(i)) such that any two matrices over* $\mathbf{K}$*, of the same size and totally transcendental over* $\mathbf{F}$*, are conjugate.*
(2) Every skew field $\mathbf{K}$ *with central subfield* $\mathbf{F}$ *has a skew field extension* $\mathbf{L}$ *which is matrix-homogeneous over* $\mathbf{F}$*.*

Corollary 749. *Let* $A \in \mathrm{Mat}_n(\mathbf{K})$ *be totally transcendental over* $\mathbf{F}$ *and let* $\lambda \in \mathbf{L}$ *be a transcendental element over* $\mathbf{F}$*. Then* A *and* λI_n *are conjugate, i.e., there exists a matrix* $P \in \mathrm{GL}_n(\mathbf{L})$ *such that* $P\,A\,P^{-1} = \lambda I_n$*.*

4.3.3.3 Case of an Algebraic Matrix

Consider an algebraic cyclic matrix A with a single elementary divisor q. The following can be deduced from Corollary 545 (see [78], Prop. 8.3.5):

Lemma 750. *Assume that the centre* $\mathbf{F}$ *of* $\mathbf{K}$ *is perfect and that* $\mathbf{K}$ *contains an algebraic closure* $\mathbf{\Omega}$ *of* $\mathbf{F}$*. Then every indecomposable bounded polynomial over* $\mathbf{K}$ *is similar to a polynomial of the form* $(X-\alpha)^n$*,* $\alpha \in \mathbf{\Omega}$*.*

If the assumptions of the above lemma are satisfied, $q(X)$ is of the form $(X-\alpha)^n$. Let w be a generator of $\mathbf{R}/\mathbf{R}q \cong \mathrm{coker}_{\mathbf{R}}(\bullet(\partial I_n - A))$, so that $q\,w = (\partial-\alpha)^n w = 0$, set $x_i = (\partial-\alpha)^i w$ $(1 \leq i \leq n)$ and let x be the column with entries x_i. Then $(\partial I_n - J_n(\alpha))\,x = 0$ where $J_n(\alpha)$ is the Jordan block of order n

$$J_n(\alpha) = \begin{bmatrix} \alpha & 1 & 0 & 0 & \dots & 0 \\ 0 & \alpha & 1 & & \dots & 0 \\ & 0 & & & & \\ & & & & 1 & 0 \\ 0 & 0 & \dots & & \alpha & 1 \\ 0 & 0 & \dots & & 0 & \alpha \end{bmatrix}$$

and we obtain the following by Corollary 521 and Exercise 679:

Corollary 751. *(i) The matrix* A *is conjugate to* $J_n(\alpha)$*. (ii)* $q(A) = 0$ *(*Cayley-Hamilton theorem*).*

From the above corollary, one can deduce the "classical form" of the Cayley-Hamilton theorem when $\mathbf{K}$ is a field. The statement below shows that it is sufficient to assume that $\mathbf{K}$ is a commutative ring ([141], §35, Theorem 1):

Theorem 752. (Cayley-Hamilton theorem). *Let $\mathbf{K}$ be a commutative ring, let $A \in \mathrm{Mat}_n(\mathbf{K})$, and let*

$$p(X) = \det(X I_n - A) = X^n + p_1 X^{n-1} + \ldots + p_n$$

be the characteristic polynomial of A. Then $p_1 = \mathrm{Tr}(A)$, the trace of A, $p_n = \det(A)$, and for each $i \in \{1, \ldots, n\}$, p_i is a polynomial function of the entries of A, with coefficients belonging to $\mathbb{Z}$. In addition,

$$p(A) = 0$$

where $p(A) \triangleq A^n + p_1 A^{n-1} + \ldots + p_n I_n$.

4.3.3.4 General Case

When $\mathbf{K}$ is a skew field with perfect centre $\mathbf{F}$, we can extend $\mathbf{K}$ so as to contain an algebraic closure $\boldsymbol{\Omega}$ of $\mathbf{F}$ as well as an element λ transcendental over $\mathbf{F}$. Let $\mathbf{L}/\mathbf{K}$ be such a skew field extension. The following is now obvious:

Theorem 753. (Jordan theorem). *Every square matrix of order n over $\mathbf{L}$ is conjugate to a matrix with the following* Jordan canonical form*:*

$$J_{r_1}(\alpha_1) \oplus \ldots \oplus J_{r_m}(\alpha_m) \oplus \lambda I_s$$

where $\alpha_i \in \boldsymbol{\Omega}$ and $\sum_{1 \leq i \leq m} r_i + s = n$.

4.3.4 Roots of a Skew Polynomial

4.3.4.1 Evaluations of a Skew Polynomial

Let $\mathbf{K}$ be a division ring, $\boldsymbol{\alpha}$ an automorphism and $\boldsymbol{\delta}$ an $\boldsymbol{\alpha}$-derivation of $\mathbf{K}$. When necessary to avoid confusions, the automorphism $\boldsymbol{\alpha}$ and the derivation $\boldsymbol{\delta}$ are respectively written $\boldsymbol{\alpha} : a \mapsto \boldsymbol{\alpha}(a)$ and $\boldsymbol{\delta} : a \mapsto \boldsymbol{\delta}(a)$ in the sequel. Following the notation in ([201], end of Sect 2), the $(\boldsymbol{\alpha}, \boldsymbol{\delta})$-conjugate of $a \in \mathbf{K}$ by $c \in \mathbf{K}^{\times}$ is denoted by ${}^{c}a$. By (4.12),

$$\boxed{{}^{c}a = \boldsymbol{\alpha}(c)\, a\, c^{-1} + \boldsymbol{\delta}(c)\, c^{-1}.} \tag{4.13}$$

Definition 754. *(i) The set $\{{}^{c}a : c \in \mathbf{K}^{\times}\}$ is called the $(\boldsymbol{\alpha}, \boldsymbol{\delta})$-conjugacy class of a in $\mathbf{K}$ and is denoted by $\Delta^{\boldsymbol{\alpha}, \boldsymbol{\delta}}(a)$.*
(ii) Let $c \in \mathbf{K}^{\times}$; then ${}^{c}0 = \boldsymbol{\delta}(c)\, c^{-1}$ is called the logarithmic derivative *of $c \in \mathbf{K}^{\times}$.*

If $(\boldsymbol{\alpha}, \boldsymbol{\delta}) = (Id_{\mathbf{K}}, 0)$, then

$$\Delta^{\boldsymbol{\alpha},\boldsymbol{\delta}}(a) = \left\{c\,a\,c^{-1} : c \in \mathbf{K}^{\times}\right\} = \left\{\mathrm{Int}(c)(a) : c \in \mathbf{K}^{\times}\right\}$$

(Exercise 206); if in addition $\mathbf{K}$ is a field, then $\Delta^{\boldsymbol{\alpha},\boldsymbol{\delta}}(a) = \{a\}$. The following will be useful in the sequel ([201], Prop. 2.9), ([88], Example 2.2):

Lemma 755. *For any $a, b \in \mathbf{K}$ such that $b - a \neq 0$ and $c, d \in \mathbf{K}^{\times}$*

$$^{d}\left(^{c}a\right) = {}^{dc}a, \tag{4.14}$$

$$\left(X - {}^{c}a\right)c = \boldsymbol{\alpha}(c)\left(X - a\right), \tag{4.15}$$

$$\left(X - {}^{b-a}b\right)\left(X - a\right) = \left(X - {}^{a-b}a\right)\left(X - b\right). \tag{4.16}$$

Proof. Apply the commutation rule (2.33); all expressions are readily obtained after some algebra. ■

We know that the skew polynomial ring $\mathbf{R} = \mathbf{K}[X; \boldsymbol{\alpha}, \boldsymbol{\delta}]$ is a strongly Euclidean domain with Euclidean function $\theta(f) = d^{\circ}(f)$ (Theorem 731). Let $f \in \mathbf{R}^{\times}$, let $a \in \mathbf{K}$ and let us right-divide $f(X)$ by $X - a$. We obtain

$$f(X) = q(X)\,(X - a) + r$$

where $r \in \mathbf{K}$ since $d^{\circ}(r) < 1$. If $\mathbf{R} = \mathbf{K}[X]$ where $\mathbf{K}$ is a field, we have $r = f(a)$, the "evaluation" of f at a. In the general case we are led to the following

Definition 756. *The* right-evaluation *of $f(X) \in \mathbf{R}$ at $a \in \mathbf{K}$ is the above element $r \in \mathbf{K}$, and is denoted by $f_r(a)$. The* left-evaluation *of $f(X)$ at $a \in \mathbf{K}$ is likewise defined and denoted by $f_l(a)$.*

The right evaluation $f_r(a)$ is made explicit below ([201], [88]).

Lemma 757. *(1) Assume that $f(X)$ is written in the form a* left *skew polynomial*

$$f(X) = \sum_{0 \leq i \leq n} f_{n-i}\,X^{i}, \quad f_0 \neq 0$$

(Definition 353) and define

$$N_0(a) = 1,$$
$$N_{k+1}(a) = \boldsymbol{\alpha}(N_k(a))\,a + \boldsymbol{\delta}(N_k(a)).$$

Then,

$$f_r(a) = \sum_{0 \leq i \leq n} f_{n-i}\,N_i(a). \tag{4.17}$$

(2) The terms $N_i(a)$ can be expressed as

$$N_i(a) = X_r^{i}(a), \quad i \geq 0, \quad a \in \mathbf{K}, \tag{4.18}$$

so that (4.17) takes the following appealing form:

$$\boxed{f_r(a) = \sum_{0 \le i \le n} f_{n-i} X_r^i(a).} \tag{4.19}$$

Note that if $\mathbf{K}$ is a field and $\boldsymbol{\alpha} = Id_{\mathbf{K}}$, $\boldsymbol{\delta} = 0$, then $N_i(a) = a^i$ and (4.17) is the usual expression of $f(a)$. If $f(X)$ is given in the form of a *right* skew polynomial, it is possible to convert this form into that of a left skew polynomial (Proposition 354(iii)) and then to apply formula (4.17).

The right evaluation of the product of two skew polynomials at a point $a \in \mathbf{K}$ is given by the following

Lemma 758. *For any $f, g \in \mathbf{K}[X; \boldsymbol{\alpha}, \boldsymbol{\delta}]$ and $a \in \mathbf{K}$,*

$$(f\,g)_r(a) = \begin{cases} 0 & \text{if } g_r(a) = 0, \\ f_r\left({}^{g_r(a)}a\right) g_r(a) & \text{if } g_r(a) \neq 0. \end{cases}$$

Proof. (1) If $g_r(a) = 0$, $X - a$ right-divides $g(X)$, thus it right-divides $f(X)\,g(X)$ and $(f\,g)_r(a) = 0$.

(2) Assuming that $c \triangleq g_r(a) \neq 0$, write $g(X) = q_1(X)(X - a) + c$, $b = {}^c a$ and $f(X) = q_2(X)(X - b) + f_r(b)$. We obtain

$$f(X)\,g(X) = f(X)\,q_1(X)(X - a) + q_2(X)\,\boldsymbol{\alpha}(c)(X - a) + f_r(b)\,c. \quad \blacksquare$$

4.3.4.2 Roots of a Skew Polynomial

Definition 759. *An element $a \in \mathbf{K}$ is said to be a* left *(resp.,* right*)* root *of a nonzero polynomial $f \in \mathbf{K}[X; \boldsymbol{\alpha}, \boldsymbol{\delta}]$ if $f_l(a) = 0$ (resp., $f_r(a) = 0$). The set of left (resp., right) roots of f in $\mathbf{K}$ is denoted by $V_l(f)$ (resp., $V_r(f)$).*

Lemma 760. *Let $a \in \mathbf{K}$ and let $f \in \mathbf{R}^\times$ where $\mathbf{R} = \mathbf{K}[X; \boldsymbol{\alpha}, \boldsymbol{\delta}]$. If a is a left (resp., right) root of f, so is any element of $\Delta^{\boldsymbol{\alpha}, \boldsymbol{\delta}}(a)$.*

Proof. We have $f_r(a) = 0$ if, and only if the canonical image $(f)\,\varphi_a$ of f in the left $\mathbf{R}$-module $\mathbf{R}/\mathbf{R}(X - a)$ is zero. For any element $b \in \Delta^{\boldsymbol{\alpha}, \boldsymbol{\delta}}(a)$ there exists an isomorphism $\bullet\psi : \mathbf{R}/\mathbf{R}(X - a) \xrightarrow{\sim} \mathbf{R}/\mathbf{R}(X - b)$. Therefore, $(f)\,\varphi_b = 0$ where $\varphi_b = \varphi_a \circ \psi$ is the canonical epimorphism $\mathbf{R} \twoheadrightarrow \mathbf{R}/\mathbf{R}(X - b)$. For a left root, the proof is similar. $\blacksquare$

Consider a matrix $A \in \mathrm{Mat}_n(\mathbf{K})$. The following definition is classical ([77], Sect. 8.5):

Definition 761. *Let $\lambda \in \mathbf{K}$ and let $v \in {}^n\mathbf{K}$ (resp., $v \in \mathbf{K}^n$); λ is a* left *(resp.,* right*)* eigenvalue *of A and v is a left (resp., right) eigenvector associated with λ if $v\,A = \lambda\,v$ (resp., $A\,v = v\,\lambda$).*

Proposition 762. *The left (resp., right) roots of a monic skew polynomial $f \in \mathbf{R}^\times$ coincide with the $(\boldsymbol{\alpha}, \boldsymbol{\delta})$-conjugates of the left (resp., right) eigenvalues of its companion matrix C_f.*

Proof. (1): Let us consider the case of left roots and left eigenvalues. By Proposition 735, $\operatorname{coker}_{\mathbf{R}}(\bullet(\partial I_n - C_f)) \cong \mathbf{R}/\mathbf{R}f$. Let V be the left $\mathbf{R}$-module $M = \mathbf{R}/\mathbf{R}f$ considered as a left $\mathbf{K}$-vector space of dimension $n = d^\circ(f)$ (Theorem 740). We know that C_f represents the PLT $\theta = \partial|_V$ in a cyclic basis of V (Definition 736); the explicit relation between θ and C_f is given by Lemma 742. Let $\bar{v} \in V^\times$ be such that $\theta(\bar{v}) = a\,\bar{v}$. We have $(\partial - a)\,\bar{v} = 0$ in $\mathbf{R}/\mathbf{R}f$ and $\bar{v}$ is the canonical image of an element $v \in \mathbf{R}$; this polynomial $v = v(X)$ can be chosen of degree $\leq n-1$ by the Euclidean algorithm, and the above equality yields $(\partial - a)\,v \in \mathbf{R}f$. Therefore, there exists $\mu \in \mathbf{K}^\times$ such that $(X-a)\,v(X) = \mu\, f(X)$, i.e., $\mu^{-1}(X-a)\,v(X) = f(X)$. Therefore, $\left(X - {}^{\boldsymbol{\alpha}^{-1}(\mu^{-1})}a\right)$ left-divides $f(X)$ by (4.15). Conversely, if $\left(X - {}^{\boldsymbol{\alpha}^{-1}(\mu^{-1})}a\right)$ left-divides $f(X)$, then $(\partial - a)\,v \in \mathbf{R}f$ and $(\partial - a)\,\bar{v} = 0$, i.e., $\theta(\bar{v}) = a\,\bar{v}$.

(2): In the case of right roots and right eigenvalues, one must consider the PLT associated with the right multiplication by ∂; a right eigenvalue and a right eigenvector are defined in a obvious way for such a PLT [204]. ■

Theorem 763. *Let $f \in \mathbf{K}[X;\boldsymbol{\alpha},\boldsymbol{\delta}]$ be a skew polynomial of degree $n \geq 1$.*

(i) The right roots of f lie in at most n conjugacy classes.
(ii) Assuming that f splits into linear factors, i.e.,

$$f(X) = b\,(X - c_1)\ldots(X - c_n)\,, \quad b \neq 0, \tag{4.20}$$

then any right root of f is $(\boldsymbol{\alpha},\boldsymbol{\delta})$-conjugate to one of the c_i.

Proof. (i): Let us proceed by induction: the case $n = 1$ is trivial.

Assume that statement (i) holds true for some $n \geq 1$, let f be a polynomial of degree $n+1$ and let a_n, a_{n+1} be two right roots of f. Then $f(X) = g(X)(X - a_{n+1})$ where $d^\circ(g) = n$. By Lemma 758, a_n is $(\boldsymbol{\alpha},\boldsymbol{\delta})$-conjugate to a right root of g, thus there exist at most n conjugacy classes $\Delta_1, \ldots, \Delta_n$ and an index $k \in \{1, \ldots, n\}$ such that $a_n \in \Delta_k$, and (i) is proved.

(ii): The ring $\mathbf{R} = \mathbf{K}[X;\boldsymbol{\alpha},\boldsymbol{\delta}]$ is a unique factorization domain (Theorems 731, 468 and 462), thus $\mathbf{L}(f\,\mathbf{R},\mathbf{R})$ is a modular lattice of finite length (Lemma and Definition 433). By the factorization (4.20),

$$f\,\mathbf{R} \subsetneq b(X - c_1)\ldots(X - c_{n-1})\,\mathbf{R} \subsetneq \ldots \subsetneq b(X - c_1)\,\mathbf{R}$$

is a maximal chain in the lattice $\mathbf{L}(f\,\mathbf{R},\mathbf{R})$. Let a be a right root of f; then $X - a$ is an atomic factor of f and by the Jordan-Hölder-Dedekind theorem (Theorem 117), $X - a$ is similar to one of the factors $X - c_i$ (Corollary 426). By Corollary 746, $a \overset{(\boldsymbol{\alpha},\boldsymbol{\delta})}{\sim} c_i$. ■

Corollary 764. *Let $f \in \mathbf{R}$ ($\mathbf{R} = \mathbf{K}[X;\boldsymbol{\alpha},\boldsymbol{\delta}]$) be a skew polynomial of degree $n \geq 1$ and assume that*

$$f(X) = (X - c_1)^{d_1} \dots (X - c_m)^{d_m}$$

where the d_i's are positive integers and the c_i's are elements of $\mathbf{K}$ which are pairwise non-conjugated. Then,

$$\frac{\mathbf{R}}{\mathbf{R}f} \cong \bigoplus_{1 \leq i \leq m} \frac{\mathbf{R}}{\mathbf{R}(X - c_i)^{d_i}}.$$

Proof. If c_m is not conjugated with c_j for $1 \leq j \leq m-1$, then c_m is not a left root of

$$f_{m-1}(X) = (X - c_1)^{d_1} \dots (X - c_{m-1})^{d_{m-1}}$$

by Theorem 763, therefore $(X - c_m)^{d_m}$ and $f_{m-1}(X)$ are left-coprime. By Corollary 529,

$$\frac{\mathbf{R}}{\mathbf{R}f} \cong \frac{\mathbf{R}}{\mathbf{R}f_{m-1}} \oplus \frac{\mathbf{R}}{\mathbf{R}(X - c_m)^{d_m}}.$$

One can continue this rationale in an obvious way, and establish the result by induction. ■

4.3.4.3 gcrd and lclm of Skew Polynomials

Let $\mathbf{R} = \mathbf{K}[X;\boldsymbol{\alpha},\boldsymbol{\delta}]$ and let $(f_i)_{1 \leq i \leq n}$ be a finite family of elements of $\mathbf{R}^{\times}$. There exists a unique *monic* skew polynomial which is a gcrd (resp., an lclm) of $(f_i)_{1 \leq i \leq n}$. This skew polynomial is denoted by $(f_1, \dots, f_n)_r$ (resp., $[f_1, \dots, f_n]_l$) and is called *the* gcrd (resp., *the* lclm) of $(f_i)_{1 \leq i \leq n}$. The gcrd and the lclm of two nonzero skew polynomials are given by Theorems 469 and 471. One can easily check the following ([204], Remark 4.5):

$$[f, X - a]_l = \begin{cases} f & \text{if } f_r(a) = 0, \\ \left(X - {}^{f_r(a)}a\right) f & \text{if } f_r(a) \neq 0. \end{cases} \tag{4.21}$$

In particular, (4.16) is the lclm of $X - a$ and $X - b$.

4.3.4.4 Wedderburn Polynomials

The classical definition of an algebraic set (Definition 386) has been generalized in [203] as follows:

Definition 765. *A set $\Delta \subseteq \mathbf{K}$ is $(\boldsymbol{\alpha},\boldsymbol{\delta})$-algebraic if there exists a nonzero skew polynomial $f \in \mathbf{K}[X;\boldsymbol{\alpha},\boldsymbol{\delta}]$ such that $\Delta \subseteq V_r(f)$.*

Let Δ be an $(\boldsymbol{\alpha},\boldsymbol{\delta})$-algebraic subset of $\mathbf{K}$. The set S of all skew polynomials vanishing on Δ is a nonzero left ideal of $\mathbf{R} = \mathbf{K}[X;\boldsymbol{\alpha},\boldsymbol{\delta}]$; since this ring is a principal ideal domain, there exists a unique monic polynomial f_Δ such that $S = \mathbf{R}\, f_\Delta$.

Definition 766. *The above monic polynomial f_Δ is called the* minimal polynomial *of Δ; $d^\circ(f_\Delta)$ is called the* rank *of Δ and is denoted by* $\mathrm{rk}(\Delta)$.

Let $\Delta = \{a_i : 1 \leq i \leq n\}$. Obviously, f_Δ is characterized by

$$f_\Delta = [X - a_i : 1 \leq i \leq n]_l \, .$$

Definition 767. *(i) The algebraic set $\Delta = \{a_i : 1 \leq i \leq n\}$ is* P-independent *if* $\mathrm{rk}(\Delta) = n$.
(ii) A skew polynomial $f \in \mathbf{R}$ is said to be a Wedderburn polynomial *(or a W-polynomial, for short) if it is the minimal polynomial of an $(\boldsymbol{\alpha}, \boldsymbol{\delta})$-algebraic subset of* $\mathbf{K}$.

According to the definition, a polynomial $f \in \mathbf{R}$ of degree n has at most n P-independent right roots. If $f \in \mathbf{R}$ is the minimal polynomial of an $(\boldsymbol{\alpha}, \boldsymbol{\delta})$-algebraic set $\Delta = \{a_1, ..., a_n\} \subseteq \mathbf{K}$, then we have the complete direct decomposition (3.3.2.**2**)

$$\mathbf{R}\, f = \bigcap_{1 \leq i \leq n} \mathbf{R}\,(X - a_i)\, ,$$

which yields the direct sum representation

$$\mathbf{R}/\mathbf{R}\, f \cong \bigoplus_{1 \leq i \leq n} \mathbf{R}/\mathbf{R}\,(X - a_i)\, .$$

The following is proved in [203] and is clear from the above:

Proposition 768. *Let $f \in \mathbf{R}$ be a monic polynomial of degree $n \geq 1$. The following conditions are equivalent:*

(i) f is a W-polynomial;
(ii) $\mathrm{rk}(V_r(f)) = n$;
(iii) for any $g \in \mathbf{R}$, $V_r(f) \subseteq V_r(g) \Longrightarrow g \in \mathbf{R}\, f$;
(iv) the companion matrix C_f of f is $(\boldsymbol{\alpha}, \boldsymbol{\delta})$-conjugate (Definition 743) to a diagonal matrix.

4.3.5 Vandermonde, Wronskian, and Casoratian Matrices

Vandermonde matrices over commutative rings and Wronskian matrices over commutative differential rings are classical, respectively in linear algebra ([27], n°III.8.6, Example 1) and in the theory of linear differential equations ([37], n°IV.2.7). There exists a connection between these matrices when they are defined over a generalized differential division ring ([197], Epilogue). Let

$\mathbf{K}$ be a division ring, let $\mathbf{A} \supseteq \mathbf{K}$ be a commutative ring or an Ore domain endowed with an automorphism $\boldsymbol{\alpha}$ and an $\boldsymbol{\alpha}$-derivation $\boldsymbol{\delta}$, let $a_1, ..., a_n$ be elements of $\mathbf{K}$, let $y_1, ..., y_n$ be elements of $\mathbf{A}$, let $a = (a_i)_{1 \leq i \leq n}$, $y = (y_i)_{1 \leq i \leq n}$ and $\mathbf{R} = \mathbf{K}[X; \boldsymbol{\alpha}, \boldsymbol{\delta}]$. The indeterminate X is identified with the derivation operator ∂ and the right evaluation $X_r^i(a) = N_i(a)$ (Lemma 757) is denoted by $\partial_r^i(a)$ for any $a \in \mathbf{K}$ and any integer $i \geq 0$.

Lemma and Definition 769. *(1) The* Vandermonde matrix $V(a)$ *and the* Wronskian matrix $W(y)$ *are respectively*

$$V(a) = \begin{bmatrix} 1 & 1 & \dots & 1 \\ \partial_r(a_1) & \partial_r(a_2) & \dots & \partial_r(a_n) \\ \vdots & \vdots & & \vdots \\ \partial_r^{n-1}(a_1) & \partial_r^{n-1}(a_2) & \dots & \partial_r^{n-1}(a_n) \end{bmatrix},$$

$$W(y) = \begin{bmatrix} y_1 & y_2 & \dots & y_n \\ y_1^{\boldsymbol{\delta}} & y_2^{\boldsymbol{\delta}} & \dots & y_n^{\boldsymbol{\delta}} \\ \vdots & \vdots & & \vdots \\ y_1^{\boldsymbol{\delta}^{n-1}} & y_2^{\boldsymbol{\delta}^{n-1}} & \dots & y_n^{\boldsymbol{\delta}^{n-1}} \end{bmatrix}.$$

If $\boldsymbol{\delta}$ is an inner derivation induced by 1, *then the Wronskian matrix $W(y)$ is left-equivalent to the* Casoratian matrix

$$C(y) = \begin{bmatrix} y_1 & y_2 & \dots & y_n \\ y_1^{\boldsymbol{\alpha}} & y_2^{\boldsymbol{\alpha}} & \dots & y_n^{\boldsymbol{\alpha}} \\ \vdots & \vdots & & \vdots \\ y_1^{\boldsymbol{\alpha}^{n-1}} & y_2^{\boldsymbol{\alpha}^{n-1}} & \dots & y_n^{\boldsymbol{\alpha}^{n-1}} \end{bmatrix}.$$

(the words Wronskian *and* Casoratian *refer to the mathematicians Wronski and Casorati, respectively).*
(2) If $\mathbf{A}$ *is a division ring, the following equality holds:*

$$W(y) = V({}^{y}0)\,\mathrm{diag}(y_1, ..., y_n) \tag{4.22}$$

where ${}^{y}0$ is the finite sequence of logarithmic derivatives $({}^{y_i}0)_{1 \leq i \leq n}$ (Definition 754(ii)).

Proof. (1) If $\boldsymbol{\delta} = \boldsymbol{\alpha} - 1$, then $W(y)$ is easily shown to be left-equivalent to $C(y)$ using elementary row operations of the first kind.

(2) By definition of ${}^{y_i}0$, $y_i^{\boldsymbol{\delta}} = {}^{y_i}0\, y_i = a_i\, y_i$ where $a_i = {}^{y_i}0$. By Lemma 757, this implies $y_i^{\boldsymbol{\delta}^j} = N_j(a_i)\, y_i$ for any natural integer j, and this can be written $y_i^{\boldsymbol{\delta}^j} = \partial_r^j(a_i)\, y_i$ with the above notation. ■

Theorem 770. *(1) Let $\Delta = \{a_1, ..., a_n\} \subseteq \mathbf{K}$. The set Δ is P-independent if, and only if $V(a)$ is invertible, where $a = (a_i)_{1 \leq i \leq n}$.*

(2) Let $f \in \mathbf{R}$. The following conditions are equivalent:
(i) f is a W-polynomial (Definition 767) and $f = f_\Delta$ where $\Delta = \{a_1, ..., a_n\}$;
(ii) there exists a P-independent set $\Delta = \{a_1, ..., a_n\} \subseteq \mathbf{K}$ such that $f = f_\Delta$;
(iii) the companion matrix C_f of f is $(\boldsymbol{\alpha}, \boldsymbol{\delta})$-conjugate to the diagonal matrix $\operatorname{diag}(a) \triangleq \operatorname{diag}(a_1, ..., a_n)$, *and more specifically*

$$C_f\, V(a) = V(a)^{\boldsymbol{\alpha}} \operatorname{diag}(a) + V(a)^{\boldsymbol{\delta}}.$$

(3) Of the following, (iv) implies (v); in addition, if $\mathbf{A}$ is a division ring, then (iv) and (v) are equivalent:
(iv) the matrix $W(y)$ is regular (Theorem 400, Lemma and Definition 402(2));
(v) the elements $y_1, ..., y_n$ are right linearly independent over the subring of constants $\mathbf{A}_{\boldsymbol{\delta}}$ of $\mathbf{A}$ (Definition 290).
(4) Suppose (ii) holds, $\mathbf{A}$ is a division ring and for each $i \in \{1, ..., n\}$ there exists $y_i \in \mathbf{A}^\times$ such that ${}^{y_i}0 = a_i$. Then (iv) holds.

Proof. (1) is proved in ([204], Lemma 5.6).

(2): (i)⇔(ii) by Definition 767(ii) and (ii)⇔(iii) by ([204], Theorem 5.7).

(3): (iv)⇒(v): Assume $y_1, ..., y_n$ are right linearly dependent over $\mathbf{A}_{\boldsymbol{\delta}}$. Then there exist $c_1, ..., c_n$ in $\mathbf{A}_{\boldsymbol{\delta}}$, not all zero, and such that $\sum_{1 \leq i \leq n} y_i\, c_i = 0$. This implies $\sum_{1 \leq i \leq n} \partial^j y_i\, c_i = 0$ for all $j \in \{0, ..., n-1\}$. Therefore, $W(y)\, c = 0$ where c is the column with entries c_i, and $W(y)$ is singular.

(v)⇒(iv) when $\mathbf{A}$ is a division ring: see ([201], Theorem 4.9), replacing $\mathbf{K}$ by $\mathbf{A}$.

(4) is a consequence of (4.22). ■

4.3.6 Solutions of Differential Operators

Let M be a left $\mathbf{R}$-module, let $V = M_{[\mathbf{K}]}$ be the left $\mathbf{K}$-vector space obtained from M by restriction of the ring of scalars, and let $f \in \mathbf{R}$. Let $\theta : V \to V$ be the PLT defined by ∂ (4.3.2.1) and for simplicity write $\theta = \partial$; then it is no longer necessary to distinguish between M and V. When M is a ring $\mathbf{A} \supseteq \mathbf{K}$ such as in Subsect. 4.3.5, and then $\partial = \boldsymbol{\delta}$.

Lemma and Definition 771. *(i) An element $y \in M$ is called a* solution *of the differential operator $f(\partial)$ if $f(\partial)\, y = 0$. Then, f is called a left annihilator of y, i.e., $f \in \operatorname{Ann}_l^{\mathbf{R}}(y)$.*
(ii) The set of all solutions of $f \in \mathbf{R}$ is a right $\mathbf{K}_{\boldsymbol{\delta}}$-vector space (Definition 290).
(iii) If $f, g \in \mathbf{R}$ and y is a solution of g, then y is a solution of fg. Therefore, if $y \in M$ is such that $\operatorname{Ann}_l^{\mathbf{R}}(y) \neq 0$, there exists a unique monic polynomial of least degree $f \in \mathbf{R}^\times$ such that $f(\partial)\, y = 0$. This polynomial is called the minimal annihilator *of y over $\mathbf{K}$.*
(iv) Let $y \in M$ be such that $\operatorname{Ann}_l^{\mathbf{R}}(y) \neq 0$ and let f_y be its minimal annihilator; then $\operatorname{Ann}_l^{\mathbf{R}}(y) = \mathbf{R}\, f_y$.

Proof. (ii) and (iii) are clear.

(iv): Let $g \in \mathrm{Ann}_l^{\mathbf{R}}(y)$ and let $g = a\, f_y + r$ be the right Euclidean division of g by f_y, where $d^\circ(r) < d^\circ(f_y)$. Then $g(\partial)\, v = a(\partial)\, f_y(\partial)\, y + r(\partial)\, y = r(\partial)\, y$, thus $r(\partial)\, y = 0$, and $r \in \mathrm{Ann}_l^{\mathbf{R}}(y)$. Since f_y is of minimal degree among all polynomials belonging to $\mathrm{Ann}_l^{\mathbf{R}}(y)$, this proves that $r = 0$; therefore, $\mathrm{Ann}_l^{\mathbf{R}}(y) \subseteq \mathbf{R}\, f_y$. The converse is obvious. ∎

Consider now a left eigenvalue λ of the PLT θ and an associated eigenvector y (Definition 738). Since $\theta = \partial\,|_V$, the equality $\theta(v) = \lambda\, y$ is equivalent to $(\partial - \lambda)\, y = 0$, i.e., $\partial - \lambda \in \mathrm{Ann}_l^{\mathbf{R}}(y)$.

Lemma and Definition 772. *(1) An element $y \in M$ for which there exists $\lambda \in \mathbf{K}$ such that $(\partial - \lambda)\, y = 0$ is said to be $\boldsymbol{\delta}$*-hyperexponential *over* $\mathbf{K}$.
(2) If $f \in \mathbf{R}^\times$ is such that $f(\partial)$ has a nonzero $\boldsymbol{\delta}$-hyperexponential solution y in a left $\mathbf{R}$-module M, then f has a right root $\lambda \in \mathbf{K}$ which is uniquely determined by y.
(3) If M is a ring $\mathbf{A}$, the above element λ is denoted by ${}^{y}0$, and this extends Definition 754(ii).
(4) The $\boldsymbol{\delta}$-hyperexponential elements of the left $\mathbf{R}$-module M form a right $\mathbf{K}_{\boldsymbol{\delta}}$-vector space. Suppose $M = \mathbf{A}$ as in (3) and $\mathbf{K}$ is a field; then the $\boldsymbol{\delta}$-hyperexponential elements of $\mathbf{A}$ over $\mathbf{K}$ form a subring of $\mathbf{A}$, called the $\boldsymbol{\delta}$-hyperexponential subring of $\mathbf{A}$.

Proof. (2): Let $y \in M^\times$ be a $\boldsymbol{\delta}$-hyperexponential solution of $f(\partial)$. There exists $\lambda \in \mathbf{K}$ such that $(\partial - \lambda)\, y = 0$. Clearly, $\partial - \lambda$ is the minimal annihilator of y, thus $f \in \mathbf{R}(\partial - \lambda)$, and the element λ is uniquely determined by y, by Lemma and Definition 771.

(3): If M is a ring $\mathbf{A}$, then $\partial = \boldsymbol{\delta}$ as already said, and λ is such that $y^{\boldsymbol{\delta}} = \lambda\, y$.

The first part of (4) is clear. For the second one, let $y, z \in \mathbf{A}$ be $\boldsymbol{\delta}$-hyperexponential elements over $\mathbf{K}$. There exist $a, b \in \mathbf{K}$ such that $y^{\boldsymbol{\delta}} = a\, y$, $z^{\boldsymbol{\delta}} = b\, z$, thus $(yz)^{\boldsymbol{\delta}} = y^{\boldsymbol{\alpha}} z^{\boldsymbol{\delta}} + y^{\boldsymbol{\delta}} z$. If $\boldsymbol{\delta}$ is outer, then $\boldsymbol{\alpha} = 1$ by Lemma 356, thus $(yz)^{\boldsymbol{\delta}} = (a + b)\, yz$ and yz is $\boldsymbol{\delta}$-hyperexponential since $a + b \in \mathbf{K}$. If $\boldsymbol{\delta}$ is inner, say induced by $m \in \mathbf{K}$, then $y^{\boldsymbol{\delta}} = (y^{\alpha} - y)\, m$. If $m = 0$, then $\boldsymbol{\delta} = 0$ and yz is 0-hyperexponential since so are y and z. If $m \neq 0$, $y^{\boldsymbol{\alpha}} = y^{\delta}/m + y$ and $(yz)^{\boldsymbol{\delta}} = (ab/m + a + b)\, yz$, thus yz is $\boldsymbol{\delta}$-hyperexponential since $ab/m + a + b \in \mathbf{K}$. ∎

Example 773. *(1) Let $\mathbf{K} = \mathbb{C}$ and $\boldsymbol{\delta}$ be the usual derivation d/dt. Consider the ring of differential operators $\mathbf{R} = \mathbb{C}[d/dt]$ and the $\mathbb{C}$-vector space $\mathbf{A}$ generated by all exponential $t \mapsto e^{\lambda t}$ $(\lambda \in \mathbb{C})$. Obviously, $\mathbf{A}$ is both an $\mathbf{R}$-module and a ring, and it is d/dt-hyperexponential. Thus the exponentials are the d/dt-hyperexponential elements over $\mathbb{C}$.*
(2) Let $\mathbf{K}$ be as above and let $\boldsymbol{\delta}$ be the inner derivation $\mathbf{q} - 1$ where $\mathbf{q} = \boldsymbol{\alpha}$ is the shift-forward operator (Subsect. 2.7.3). Consider the ring of differencing operators $\mathbf{R} = \mathbb{C}[\mathbf{q} - 1]$ and the $\mathbb{C}$-vector space $\mathbf{A}$ generated by all

geometric sequences $n \mapsto \lambda^n$ $(\lambda \in \mathbb{C})$. *Likewise,* $\mathbf{A}$ *is both an* $\mathbf{R}$*-module and a ring, and it is* $(\mathbf{q}-1)$*-hyperexponential. Thus the geometric sequences are the* $(\mathbf{q}-1)$*-hyperexponential elements over* $\mathbb{C}$.
(3) Let $\mathbf{K} = \mathbb{C}(n)$ *and let* $\boldsymbol{\delta} = \mathbf{q} - 1$. *Consider the sequences* (y_n) *such that* (y_{n+1}/y_n) *belongs to* $\mathbb{C}(n)$. *These sequences, called* hypergeometric, *are the* $(\mathbf{q}-1)$*-hyperexponential elements over* $\mathbb{C}(n)$.
(4) Similarly, let $\mathbf{K} = \mathbb{C}(t)$, *let* $\boldsymbol{\delta} = d/dt$, *and consider all complex functions* f *such that* $\dot{f}/f \in \mathbb{C}(t)$ *(where* $\dot{f} = df/dt$*). These functions, called* hyper-exponential *([85], Sect. 20.2), are the* d/dt*-hyperexponential elements over* $\mathbb{C}(t)$.

Corollary 774. *(i) With the notation in Lemma and Definition 772(3), the Wronskian in Theorem and Definition 769 still makes sense and the equality (4.22) holds.*
(ii) Assuming that the ring $\mathbf{A}$ *is commutative and* $\mathbf{A}_{\boldsymbol{\delta}} = \mathbf{K}_{\boldsymbol{\delta}}$, *let* $\mathbf{H}$ *be the set of all* $\boldsymbol{\delta}$*-hyperexponential elements of* $\mathbf{A}$, *let* $y_i \in \mathbf{H}$, *let* $a_i = {}^{y_i}0$ $(1 \leq i \leq n)$, *let* $\Delta = \{a_1, ..., a_n\} \subseteq \mathbf{K}$ *and let* $f = f_\Delta \in \mathbf{R}$. *Then* $\mathbf{K} \subseteq \mathbf{H} \subseteq \mathbf{A}$ *are* $\mathbf{K}_{\boldsymbol{\delta}}$*-algebras and each of the conditions (i)-(iv) of Theorem 770 is equivalent to:*
(v') the elements $y_1, ..., y_n$ *are linearly independent over* $\mathbf{K}_{\boldsymbol{\delta}}$.

Proof. (i) is clear. (ii): Since $\mathbf{A}$ is commutative, $\mathbf{K}$ is a field, and $\mathbf{K}_{\boldsymbol{\delta}}$ is again a field (Exercise 477). Therefore, $\mathbf{K} \subseteq \mathbf{H} \subseteq \mathbf{A}$ are $\mathbf{K}_{\boldsymbol{\delta}}$-algebras by Lemma and Definition 772(4). Furthermore, of the conditions of Theorem 770, (i)$\Leftrightarrow$(ii)$\Leftrightarrow$(iii)$\Rightarrow$(iv)$\Leftrightarrow$(v') by the proof of that theorem. If (v') holds, then $V(a)$ is regular by (4.22), thus $V(a)$ is invertible, and (i) holds. ■

4.4 Picard-Vessiot Extensions

4.4.1 General Notions

Let $\mathbf{K}$ be a field of characteristic zero and assume that $\mathbf{K}$ is either a differential field with (outer) derivation $\boldsymbol{\delta}$ (Subsect. 2.3) or a difference field with shift operator $\boldsymbol{\alpha}$ (Subsect. 2.3.2); recall that $\boldsymbol{\alpha}$ is an automorphism. Consider an equation

$$\partial x = A x, \quad A \in \mathrm{Mat}_n(\mathbf{K}) \tag{4.23}$$

where ∂ is an extension of $\boldsymbol{\delta}$ in the differential case. In the difference case, ∂ is an extension of $\boldsymbol{\alpha} - 1$; setting $\partial = \mathbf{q} - 1$ where $\mathbf{q}$ is an extension of $\boldsymbol{\alpha}$, and (4.23) is equivalent to

$$\mathbf{q} x = B x, \quad B = A - I_n. \tag{4.24}$$

Let $\mathbf{C}$ be the subfield of constants of $\mathbf{K}$ (Exercise 477) and assume that $\mathbf{C}$ is algebraically closed. In general, the equation (4.23) does not have a full set of solutions in a power of $\mathbf{K}$, i.e., there does not exist a matrix $Y \in \mathrm{GL}_n(\mathbf{K})$ such that $\partial Y = A Y$.

Example 775. *(i) In the differential case, let* $\mathbf{K} = \mathbb{C}$, $n = 1$ *and let* $A = a \neq 0$. *The solutions of (4.23) are of the form* $t \mapsto \lambda\, e^{at}$ $(\lambda \in \mathbb{C})$ *and the only solution in* $\mathbb{C}$ *is zero.*
(ii) In the difference case, let $\mathbf{K} = \mathbb{C}$, $n = 1$ *and let* $B = b \neq 0$. *The solutions of (4.24) are of the form* $k \mapsto \lambda\, b^k$ *and the only solution in* $\mathbb{C}$ *is zero.*

Therefore, it is necessary to properly extend $\mathbf{K}$ in order to find all the solutions of (4.23) or of (4.24). The proof of Item (iii) below is easy and left to the reader.

Lemma and Definition 776. *(i) Let* $(\mathbf{A}, \boldsymbol{\delta}_A)$ *and* $(\mathbf{B}, \boldsymbol{\delta}_\mathbf{B})$ *be differential rings (Subsect. 2.3.1). A ring-homomorphism* $\sigma : \mathbf{A} \to \mathbf{B}$ *is called a morphism of differential rings if for any* $x \in \mathbf{A}$, $\sigma(x)^{\boldsymbol{\delta}_\mathbf{B}} = \sigma\left(x^{\boldsymbol{\delta}_\mathbf{A}}\right)$. *The category of differential rings is defined in this way.*

The category of difference rings is likewise defined: if $(\mathbf{A}, \boldsymbol{\alpha}_A)$ *and* $(\mathbf{B}, \boldsymbol{\alpha}_\mathbf{B})$ *are difference rings (Subsect. 2.3.2) where* $\boldsymbol{\alpha}_A$ *is a shift of* $\mathbf{A}$ *and* $\boldsymbol{\alpha}_\mathbf{B}$ *is a shift of* $\mathbf{B}$, *a ring-homomorphism* $\sigma : \mathbf{A} \to \mathbf{B}$ *is a morphism of difference rings if for any* $x \in \mathbf{A}$, $\sigma(x)^{\boldsymbol{\alpha}_\mathbf{B}} = \sigma(x^{\boldsymbol{\alpha}_\mathbf{A}})$.
(ii) A differential ideal *in a differential ring* $(\mathbf{A}, \boldsymbol{\delta}_\mathbf{A})$ *is an ideal* $\mathfrak{a}$ *such that* $a^{\boldsymbol{\delta}_\mathbf{A}} \in \mathfrak{a}$ *whenever* $a \in \mathfrak{a}$.

A difference ideal *in a difference ring* $(\mathbf{A}, \boldsymbol{\alpha}_A)$ *is an ideal* $\mathfrak{a}$ *such that* $a^{\boldsymbol{\alpha}_\mathbf{A}} \in \mathfrak{a}$ *whenever* $a \in \mathfrak{a}$.
(iii) If $(\mathbf{A}, \boldsymbol{\delta}_A)$ *and* $(\mathbf{B}, \boldsymbol{\delta}_\mathbf{B})$ *are two differential rings such that* $\mathbf{B} \subseteq \mathbf{A}$ *and* $\boldsymbol{\delta}_A|_\mathbf{B} = \boldsymbol{\delta}_\mathbf{B}$, *then* $(\mathbf{A}, \boldsymbol{\delta}_A)$ *is called an* extension *of* $(\mathbf{B}, \boldsymbol{\delta}_\mathbf{B})$, *written* $(\mathbf{A}, \boldsymbol{\delta}_A) \geq (\mathbf{B}, \boldsymbol{\delta}_\mathbf{B})$.

An extension of difference ring is likewise defined (and denoted).
(iii) Let $\mathbf{A}$ *be a differential (resp., difference) ring and let* $\mathfrak{a}$ *be a differential (resp., difference) ideal in* $\mathbf{A}$. *Then* $\mathbf{A}/\mathfrak{a}$ *is again a differential (resp., difference) ring.*
(iv) A differential (resp., difference) ring $\mathbf{A}$ *is called* simple *if it has no nonzero proper differential (resp., difference) ideal.*

Given a differential (resp., difference) field $\mathbf{K}$, the first question is whether there exists a differential (resp., difference) field or ring $\mathbf{L}$ extending $\mathbf{K}$.

Lemma 777. *(A) Let* $\mathbf{L}/\mathbf{K}$ *be a field extension, let* $(x_i)_{i \in I}$ *be a transcendence basis of* $\mathbf{L}$ *over* $\mathbf{K}$ *(Definition 703(v)) and let* $(u_i)_{i \in I}$ *be a family of elements of* $\mathbf{L}$.
(i) Let $\boldsymbol{\delta}$ *be a derivation of* $\mathbf{K}$; *there exists a unique derivation* $\bar{\boldsymbol{\delta}}$ *of* $\mathbf{L}$ *extending* $\boldsymbol{\delta}$ *and such that* $x_i^{\bar{\boldsymbol{\delta}}} = u_i$ *for all* $i \in I$.
(ii) Let $\boldsymbol{\alpha}$ *be a shift of* $\mathbf{K}$ *and assume that the family* $(u_i)_{i \in I}$ *is algebraically free over* $\mathbf{K}$ *(Definition 703(i)). There exists a shift* $\bar{\boldsymbol{\alpha}}$ *of* $\mathbf{L}$ *extending* $\boldsymbol{\alpha}$ *and such that* $x_i^{\bar{\boldsymbol{\alpha}}} = u_i$ *for all* $i \in I$ *in the two following cases: (a) the extension* $\mathbf{L}/\mathbf{K}$ *is purely transcendental (then,* $\bar{\boldsymbol{\alpha}}$ *is unique);* $\mathbf{L}/\mathbf{K}$ *is a Galois extension of a purely transcendental extension. (Recall that, according to Steinitz' theorem in Theorem and Definition 704(1), a field extension of* $\mathbf{K}$ *is an algebraic extension of a purely transcendental extension.)*

(B) Let $X = (X_{ij})_{1 \leq i \leq n, 1 \leq j \leq n}$ be a family of n^2 indeterminates, let $\mathbf{L}_1 = \mathbf{K}[X]$, let $S \subseteq \mathbf{L}_1$ be a multiplicative set, let $\mathbf{L} = \mathbf{L}_1 S^{-1}$ and let $A, B \in \mathrm{GL}_n(\mathbf{K})$. There exists a unique derivation $\bar{\boldsymbol{\delta}}$ (resp., a shift $\bar{\boldsymbol{\alpha}}$) of $\mathbf{L}$ which extends $\boldsymbol{\delta}$ (resp., $\boldsymbol{\alpha}$) and is such that $X^{\bar{\boldsymbol{\delta}}} = AX$ (resp., $X^{\bar{\boldsymbol{\alpha}}} = BX$).

Proof. (A) (i) and (ii)(a): It is sufficient to consider simple field extensions (Definition 699(ii)) since finite extensions are then treated by induction, and general extensions using Zorn's lemma ([28], §V.16, Prop. 4). We proceed in two steps, denoted below by (1) and (2).

(1) Consider a simple purely transcendental extension $\mathbf{L} = \mathbf{K}(x)$. Then $\mathbf{L} \cong \mathbf{K}(X)$ (Theorem and Definition 704(3)) and the elements of $\mathbf{L}$ are of the form $y = f(x)/g(x)$, $f, g \in \mathbf{K}[X]$, $g(x) \neq 0$.

In the differential case, let $x^{\bar{\boldsymbol{\delta}}} = u$; then $y^{\bar{\boldsymbol{\delta}}}$ is uniquely determined by

$$y^{\bar{\boldsymbol{\delta}}} = (f'(x)\, g(x) - f(x)\, g'(x))\, u / g(x)^2$$

where $f'(X)$ and $g'(X)$ are the derivatives of $f(X)$ and $g(X)$, defined according to (2.31).

In the difference case, let $x^{\bar{\boldsymbol{\alpha}}} = u$ and, for any polynomial $h(X) = \sum_{i=0}^{n} h_i X^i$, set $h^{\boldsymbol{\alpha}}(X) = \sum_{i=0}^{n} h_i^{\boldsymbol{\alpha}} X^i$. If u is transcendent, $g^{\boldsymbol{\alpha}}(u) \neq 0$, $y^{\bar{\boldsymbol{\alpha}}}$ is uniquely determined by

$$y^{\bar{\boldsymbol{\alpha}}} = f^{\boldsymbol{\alpha}}(u) / g^{\boldsymbol{\alpha}}(u)$$

and $\bar{\boldsymbol{\alpha}}$ is obviously an automorphism of $\mathbf{L}$.

(2) Consider a simple algebraic field extension $\mathbf{L}/\mathbf{K}$, say $\mathbf{L} = \mathbf{K}(y)$, and let $g(X) \in \mathbf{K}[X]$ be the minimal polynomial of y (Definition 712). Set $g(X) = \sum_{0 \leq i \leq n} a_i X^i$, $g^{\boldsymbol{\delta}}(X) = \sum_{0 \leq i \leq n} a_i^{\boldsymbol{\delta}} X^i$ and, as above, let $g'(X)$ be the derivative of $g(X)$. Since $g(y) = 0$ and $\boldsymbol{\delta}$ is a derivation (Definition 287(ii)), one must have $g^{\boldsymbol{\delta}}(y) + g'(y)\, y^{\bar{\boldsymbol{\delta}}} = 0$. The extension $\mathbf{L}/\mathbf{K}$ is separable (Proposition 718), thus $g'(y) \neq 0$ by Lemma 351, therefore $y^{\bar{\boldsymbol{\delta}}} = -g^{\boldsymbol{\delta}}(y) / g'(y)$. This proves that $\bar{\boldsymbol{\delta}}$ exists and is uniquely determined.

(ii)(b): Consider a Galois extension $\mathbf{L}/\mathbf{K}$, $\mathbf{K} \subsetneq \mathbf{L}$. Let $\boldsymbol{\Omega}$ be an algebraic closure of $\mathbf{K}$, $g(X) \in \mathbf{K}(X)$ an irreducible polynomial and $y_1, ..., y_n$ the roots of $g(X)$ in $\boldsymbol{\Omega}$. These roots are distinct (Corollary 721), thus so are the roots $\check{y}_1, ..., \check{y}_n$ of $g^{\boldsymbol{\alpha}}(X)$ in $\boldsymbol{\Omega}$. The extension $\mathbf{L}_1 = \mathbf{K}(y_1, ..., y_n)$ is the splitting field of $g(X)$ and $\mathbf{L} \supseteq \mathbf{L}_1 \supseteq \mathbf{K}$ by Lemma and Definition 715. Let $\sigma \in \mathrm{Gal}(\mathbf{L}_1/\mathbf{K})$ and $\bar{\boldsymbol{\alpha}}$ be the field morphism such that $\bar{\boldsymbol{\alpha}}|_{\mathbf{K}} = \boldsymbol{\alpha}$ and $y_i^{\bar{\boldsymbol{\alpha}}} = \check{y}_{\sigma(i)}$; then $\bar{\boldsymbol{\alpha}}$ is an extension of $\boldsymbol{\alpha}$ to $\mathbf{L}_1$ and is an automorphism of $\mathbf{L}_1$. As above, $\boldsymbol{\alpha}$ can be extended to $\mathbf{L}$, first by induction and then by applying Zorn's lemma.

(B) Set $X^{\bar{\boldsymbol{\delta}}} = AX$ (resp., $X^{\bar{\boldsymbol{\alpha}}} = BX$). Then, $\bar{\boldsymbol{\delta}}$ (resp., $\bar{\boldsymbol{\alpha}}$) is an extension of $\boldsymbol{\delta}$ (resp., $\boldsymbol{\alpha}$) to $\mathbf{L}_1$; $\bar{\boldsymbol{\delta}}$ can be extended to $\mathbf{L}$ using the expression in Exercise 477(i), and $\bar{\boldsymbol{\alpha}}$ can be extended to $\mathbf{L}$ according to $(x/s)^{\bar{\boldsymbol{\alpha}}} = x^{\bar{\boldsymbol{\alpha}}}/s^{\bar{\boldsymbol{\alpha}}}$ $(x \in \mathbf{L}_1, s \in S)$. As easily seen, $\bar{\boldsymbol{\alpha}}$ is an automorphism of $\mathbf{L}$. ∎

A crucial point when constructing extensions of differential (or difference) fields is to avoid introducing new constants.

Example 778. *Consider the differential equation in Example 775(i), let* $e_a : t \mapsto e^{a\,t}$ *and* $\mathbf{K} = \mathbb{C}(e_a)$. *Consider the ring* $\mathbf{A} = \mathbf{K}[Y]$ *and the differential ideal* $\mathfrak{a}$ *in* $\mathbf{A}$ *generated by* $\partial Y - Y$. *Let* y *be the canonical image of* Y *in* $\mathbf{A}/\mathfrak{a}$ *and let* $\mathbf{L}$ *be the quotient field of* $\mathbf{A}/\mathfrak{a}$. *Denoting by* $\dot{a}$ *the derivative of any element* $a \in \mathbf{L}$, *one has* $\dot{y} - y = 0$. *Let* $c = y/e_a$. *Then* $c \neq \mathbf{K}$ *(for* $0 \neq y \notin \mathbf{K}$*) and* $\dot{c} = (\dot{y}\,e_a - y\,\dot{e}_a)/e_a^2 = 0$, *thus* c *is a "new constant".*

Lemma 779. *Let* $\mathbf{A}$ *be a commutative ring and let* $\mathfrak{N}(\mathbf{A})$ *be the nilradical of* $\mathbf{A}$ *(Theorem and Definition 387(ii)).*

(i) Let $\boldsymbol{\alpha}$ *be an automorphism of* $\mathbf{A}$ *and assume that* $\mathbf{A}$ *is a simple difference ring (Lemma and Definition 776(iii)); then the ring* $\mathbf{A}$ *is reduced and its subring of constants is a field* $\mathbf{C}$.
(ii) Let $\boldsymbol{\delta}$ *be a derivation of* $\mathbf{A}$ *and assume that* $\mathbf{A}$ *is a simple differential ring of characteristic zero. Then* $\mathbf{A}$ *is an integral domain and its quotient field* $\mathrm{Q}(\mathbf{A})$ *is a differential field.*

Proof. (i): (a) Let $\mathfrak{N}(\mathbf{A})$ be the nilradical of $\mathbf{A}$, let $x \in \mathfrak{N}(\mathbf{A})$ and let n be the least positive integer such that $x^n = 0$. Then, $0 = (x^n)^{\boldsymbol{\sigma}} = (x^{\boldsymbol{\sigma}})^n$, thus $x^{\boldsymbol{\sigma}} \in \mathfrak{N}(\mathbf{A})$ and $\mathfrak{N}(\mathbf{A})$ is a difference ideal. The difference ring $\mathbf{A}$ is simple and $\mathfrak{N}(\mathbf{A}) \neq \mathbf{A}$, therefore $\mathfrak{N}(\mathbf{A}) = 0$ and $\mathbf{A}$ is reduced.

(b) Let c be a constant. Then $c\,\mathbf{A}$ is a nonzero difference ideal, therefore $c\,\mathbf{A} = \mathbf{A}$ and there exists $d \in \mathbf{A}$ such that $c\,d = 1$. So, both c and d are units in $\mathbf{A}$, and $d = 1/c$ is a constant for $d^{\boldsymbol{\alpha}} = 1/c^{\boldsymbol{\alpha}} = 1/c = d$. This proves that the subring of constants is a field.

(ii): (a) First, let us prove that every zerodivisor is nilpotent. Let $a \in \mathbf{A} \backslash \mathfrak{N}(\mathbf{A})$; the set

$$\mathfrak{I} = \{b \in \mathbf{A} : a^n\,b = 0 \text{ for some } n > 0\}$$

is an ideal in $\mathbf{A}$. Let $b \in \mathfrak{I}$ and let $n > 0$ be such that $a^n\,b = 0$; therefore $a^{n+1}\,b = 0$ and $\left(a^{n+1}\,b\right)^{\boldsymbol{\delta}} = 0$, thus $(n+1)\,a^{\boldsymbol{\delta}}\,a^n\,b + a^{n+1}\,b^{\boldsymbol{\delta}} = 0$. Therefore, $a^{n+1}\,b^{\boldsymbol{\delta}} = 0$ and $b^{\boldsymbol{\delta}} \in \mathfrak{I}$. As a result, $\mathfrak{I}$ is a differential ideal. Since $1 \notin \mathfrak{I}$, $\mathfrak{I}$ is a proper ideal. Since $\mathbf{A}$ is simple, $\mathfrak{I} = 0$, thus a is a nonzerodivisor.

(b) As above, let $x \in \mathfrak{N}(\mathbf{R})$ and let n be the least positive integer such that $x^n = 0$. This implies $n\,x^{n-1}\,x^{\boldsymbol{\delta}} = 0$, therefore $x^{n-1}\,x^{\boldsymbol{\delta}} = 0$ since $\operatorname{char}\mathbf{A} = 0$. Therefore, $x^{\boldsymbol{\delta}}$ is a zerodivisor, thus $x^{\boldsymbol{\delta}} \in \mathfrak{N}(\mathbf{A})$ by (a). This proves that $\mathfrak{N}(\mathbf{A})$ is a differential ideal, and it is proper, thus $\mathfrak{N}(\mathbf{A}) = 0$ since $\mathbf{A}$ is a simple differential ring. By (a), $\mathbf{A}$ has no nonzero zerodivisor, i.e., $\mathbf{A}$ is an integral domain. ■

Remark 780. *As shown below, a "suitable" field extension always exists for a differential field, but not for a difference field. Consider the difference equation*

$$\boldsymbol{\alpha}(y) = -y \tag{4.25}$$

(where $\boldsymbol{\alpha}(y)$ is written for $y^{\boldsymbol{\alpha}}$, for the sake of clarity), and let us look for a full set of solutions in a difference field $\mathbf{K}$ *with algebraically closed subfield of constants* $\mathbf{C}$*. This set is necessarily a* $\mathbf{C}$*-vector space of dimension* 1*. Let x be a nonzero solution. Then $\boldsymbol{\alpha}(x^2) = \boldsymbol{\alpha}(x)^2 = (-x)^2 = x^2$, therefore $x^2 \in \mathbf{C}$. Since $\mathbf{C}$ is algebraically closed, $x \in \mathbf{C}$, thus $\boldsymbol{\alpha}(x) = x$ and by (4.25) $x = -x$, a contradiction.*

4.4.2 Differential Galois Theory

4.4.2.1 Picard-Vessiot Differential Fields

In what follows, $\mathbf{K}$ is a differential field of characteristic zero with derivation δ, the subfield of constants $\mathbf{K}_{\delta} = \mathbf{C}$ of $\mathbf{K}$ is algebraically closed and $\mathbf{R} = \mathbf{K}[X;\delta]$. Let $(\mathbf{A}, \bar{\delta})$ be a differential ring such that $(\mathbf{A}, \bar{\delta}) \geq (\mathbf{K}, \delta)$.

Lemma and Definition 781. *(1) Let $f \in \mathbf{R}^{\times}$ be such that $d^{\circ} f = n$.*
(i) The equation $f(\partial)y = 0$ is equivalent to (4.23) where $A = C_f \in \mathrm{Mat}_n(\mathbf{K})$ is the companion matrix (4.5) of the polynomial f.
(ii) Let $V = \{y \in \mathbf{A} : f(\partial)y = 0\}$, i.e., the set of solutions of the differential operator $f(\partial)$ where $\partial = \bar{\delta}$; V is a $\mathbf{C}$*-vector space.*
(iii) f is said to have a full set of solutions in $\mathbf{A}$ *if* $\dim_{\mathbf{C}}(V) = n$.
(2) Let $A \in \mathrm{Mat}_n(\mathbf{K})$.
(iv) A fundamental matrix *of A over* $\mathbf{A}$ *(if any) is a matrix $Y \in \mathrm{GL}_n(\mathbf{A})$ such that $Y^{\bar{\delta}} = AY$.*
(v) There exists $f \in \mathbf{R}^{\times}$ which is similar to $X I_n - A$, i.e., $\mathbf{R}/\mathbf{R}f \cong T$ where $T = \mathrm{coker}_{\mathbf{A}}(\bullet(X I_n - A))$.
(vi) The $\mathbf{C}$*-vector space V defined in (ii) is isomorphic to* $\mathfrak{B}_{\mathbf{L}}(T) = \mathrm{Hom}_{\mathbf{R}}(T, \mathbf{K})$ *(Corollary 567(2)).*
(vii) The following conditions are equivalent:
(a) f has a full set of solutions V in $\mathbf{A}$*;*
(b) any matrix A which is $\boldsymbol{\delta}$-conjugate to the companion matrix C_f has a fundamental matrix Y over $\mathbf{A}$*.*
(c) $\dim_{\mathbf{C}} \mathfrak{B}_{\mathbf{L}}(T) = n$.

Proof. (i) is clear and (ii) is a consequence of Lemma and Definition 771(ii).

(v): Since $\mathrm{char}\,\mathbf{K} = 0$, $\mathbf{K}$ is easily seen to be a $\mathbb{Q}$-algebra ([141], Sect. 30.6, Remark 1); by Corollary 363 the ring $\mathbf{R}$ is simple, therefore $\partial I_n - A$ is cyclic and similar to a companion matrix C_f (Lemma and Definition 736(2)).

(vi): Let $\mathbf{E}$ be the endomorphism ring of the differential ring $\mathbf{L}$. The field $\mathbf{C}$ can be identified with a subring of $\mathbf{E}$ since for any $c \in \mathbf{C}$, the map $\bullet c : \mathbf{K} \to \mathbf{K} : x \mapsto x\,c$ belongs to $\mathbf{E}$, and $\bullet c = \bullet c'$ implies $c = c'$. Since the right $\mathbf{E}$-modules V and $\mathfrak{B}_{\mathbf{L}}(T)$ are isomorphic (Corollary 567(3)), the $\mathbf{C}$-vector spaces V and $\mathfrak{B}_{\mathbf{L}}(T)$ are again isomorphic (by restriction of the ring of scalars). (vii) is a consequence of (vi). ∎

Theorem and Definition 782. *Let $f \in \mathbf{R}^{\times}$.*

(1) The differential ring $\left(\mathbf{A}, \bar{\boldsymbol{\delta}}\right) \geq (\mathbf{K}, \boldsymbol{\delta})$ is called Picard-Vessiot *for f (or for any matrix $A \in \mathrm{Mat}_n(\mathbf{K})$ conjugate to the companion matrix C_f of f) if:*
(i) $\left(\mathbf{A}, \bar{\boldsymbol{\delta}}\right)$ is a simple differential ring;
(ii) there exists for the matrix A a fundamental solution matrix Y over $\mathbf{A}$;
(iii) $\mathbf{A}$ is generated over $\mathbf{K}$ by the entries y_{ij} of Y and $(\det Y)^{-1}$ (written $\mathbf{A} = \mathbf{K}\left[y_{ij}, (\det Y)^{-1}\right]$).
(2) If the differential ring $\left(\mathbf{A}, \bar{\boldsymbol{\delta}}\right) \geq (\mathbf{K}, \boldsymbol{\delta})$ is Picard-Vessiot, then $\mathbf{A}$ is a commutative domain; in addition, the subring of constants of both $\mathbf{A}$ and its field of fractions $\mathbf{L} = \mathrm{Q}(\mathbf{A})$ coincide with $\mathbf{C}$ ("no new constants").
(3) There exists a Picard-Vessiot ring $\mathbf{A}$ for f or for any matrix $A \in \mathrm{Mat}_n(\mathbf{K})$, and any two Picard-Vessiot rings for f or for A are isomorphic differential rings.
(4) Let $\mathbf{A}$ be as in (3). The field $\mathbf{L} = \mathrm{Q}(\mathbf{A})$ is called the Picard-Vessiot differential field *of f or A, and is unique up to differential isomorphism (Lemma and Definition 776(i)). The derivation of $\mathbf{L}$ induced by $\bar{\boldsymbol{\delta}}$ (see Exercise 477) is again denoted by $\boldsymbol{\delta}$.*
(5) The above Picard-Vessiot field $\mathbf{L}$ can be characterized as follows:
(i) the subfields of constants $\mathbf{L}_{\boldsymbol{\delta}}$ and $\mathbf{K}_{\boldsymbol{\delta}} = \mathbf{C}$ coincide;
(ii) there exists $Y \in \mathrm{GL}_n(\mathbf{L})$ such that $Y^{\boldsymbol{\delta}} = AY$;
(iii) $\mathbf{L} = \mathbf{K}(y_{ij})$.

Proof. (2): If $\left(\mathbf{A}, \bar{\boldsymbol{\delta}}\right)$ is Picard-Vessiot over $(\mathbf{K}, \boldsymbol{\delta})$, it is a simple differential ring finitely generated over $\mathbf{K}$, thus its is a domain (Lemma 779(ii)) and its subring of constants is $\mathbf{C}$ ([286], Sect. 1.3, Lemma 1.15). Therefore, $\mathbf{A}$ has a field of fractions $\mathbf{L} = \mathrm{Q}(\mathbf{A})$ and the subfield of fractions of $\mathbf{L}$ is again $\mathbf{C}$.

(3): Let $Z = (z_{ij})$ be an $n \times n$ matrix with entries z_{ij} algebraically independent over $\mathbf{K}$. By Lemma 777(B), we can extend uniquely $\boldsymbol{\delta}$ to $\mathbf{K}[z_{ij}]$ by $Z^{\bar{\boldsymbol{\delta}}} = AZ$, and then to $\mathbf{A}_0 = \mathbf{K}\left[z_{ij}, (\det Z)^{-1}\right]$. By Zorn's lemma, there exists a maximal differential ideal $\mathfrak{m}$ in $\mathbf{A}_0$. The quotient $\mathbf{A} = \mathbf{A}_0/\mathfrak{m}$ is a simple differential ring. For each pair (i, j) of indices, let y_{ij} be the canonical image of z_{ij} in $\mathbf{A}$; then $Y^{\bar{\boldsymbol{\delta}}} = AY$ and $\mathbf{A} = \mathbf{K}\left[y_{ij}, (\det Y)^{-1}\right]$, thus $\mathbf{A}$ is a Picard-Vessiot ring for A. The existence of a differential isomorphism between two Picard-Vessiot rings for the same equation is proved in [226] and [286].

(5): See ([226], Definition 3.2). ■

Let $(f_i)_{i \in I}$ be a family of elements of $\mathbf{R}^{\times}$. One can define the Picard-Vessiot differential ring of that family; this is the smallest differential ring which contains the Picard-Vessiot ring of each f_i. When $(f_i)_{i \in I}$ is the family of all elements of $\mathbf{R}^{\times}$, the corresponding Picard-Vessiot ring is called *universal*. One can prove that a universal Picard-Vessiot differential ring exists and is unique up to $\mathbf{K}$-linear differential isomorphism ([286], Chap. 10); it is denoted by UnivR. It has no zero divisor and the field of constants of its field of fractions is again $\mathbf{K}_{\delta} = \mathbf{C}$.

Corollary and Definition 783. *The field of fractions* UnivF *of* UnivR *exists and is unique up to* $\mathbf{K}$*-linear differential isomorphism. Its subfield of constants is* $\mathbf{K}_{\delta} = \mathbf{C}$. *The differential field* UnivF *is called* universal.

The notion of universal differential field corresponds, in the context of Differential Algebra, to the notion of algebraically closed field in the context of (usual) Algebra.

4.4.2.2 Galois Correspondence

Lemma and Definition 784. *(i) Let* $\mathbf{L} \supseteq \mathbf{K}$ *and* $\mathbf{L}' \supseteq \mathbf{K}$ *be differential fields (with the same derivation* $\boldsymbol{\delta}$ *to simplify the notation). A morphism of differential fields* $\mathbf{L} \to \mathbf{L}'$ *which leaves fixed all elements of* $\mathbf{K}$ *is called a* $(\mathbf{K}, \boldsymbol{\delta})$–homomorphism, *in accordance with Definition 700.*
(ii) The set of all $(\mathbf{K}, \boldsymbol{\delta})$*-automorphisms of the differential field* $(\mathbf{L}, \boldsymbol{\delta})$ *is a group denoted by* $\mathrm{Aut}_{\mathbf{K},\boldsymbol{\delta}}(\mathbf{L}) = G_{\boldsymbol{\delta}}(\mathbf{L}/\mathbf{K})$.
(iii) If $\mathbf{L}/\mathbf{K}$ *is a Picard-Vessiot field extension,* $G_{\boldsymbol{\delta}}(\mathbf{L}/\mathbf{K})$ *is called the* differential Galois group *of* $(\mathbf{L}, \boldsymbol{\delta})$ *over* $(\mathbf{K}, \boldsymbol{\delta})$ *and is denoted by* $\mathrm{Gal}_{\boldsymbol{\delta}}(\mathbf{L}/\mathbf{K})$.
(iv) $\mathrm{Gal}_{\boldsymbol{\delta}}(\mathbf{L}/\mathbf{K})$ *has a canonical structure of linear algebraic group.*

Proof. (iv): Let $\mathbf{L}/\mathbf{K}$ and $Y \in \mathrm{GL}_n(\mathbf{L})$ be respectively a Picard-Vessiot field extension and a fundamental solution matrix for $A \in \mathrm{Mat}_n(\mathbf{K})$. Then

$$\mathrm{Gal}_{\boldsymbol{\delta}}(\mathbf{L}/\mathbf{K}) = \{M \in \mathrm{GL}_n(\mathbf{C}) : q(Y\,M) = 0 \text{ for all } q \in \mathfrak{p}\} \tag{4.26}$$

where $\mathfrak{p}$ is the annihilator ideal in $\mathbf{K}[x_{ij}]$ defined by

$$\mathfrak{p} = \{q \in \mathbf{K}[x_{ij}] : q(Y) = 0\}$$

([226], Corol. 4.10). Since $\mathbf{K}[x_{ij}]$ is Noetherian (Exercise 493(iv)), the ideal $\mathfrak{p}$ is finitely generated, and $\mathrm{Gal}_{\boldsymbol{\delta}}(\mathbf{L}/\mathbf{K})$ consists of a Zariski-closed subgroup of $\mathrm{GL}_n(\mathbf{C})$, thus is a linear algebraic group $\mathcal{G}$ (Subsect. 2.10.1). ■

The following, proved in ([226], Theorem 6.5), is an extension of the fundamental theorem of the finite Galois theory (Theorem 725) in the context of differential algebra:

Theorem 785. *(*Fundamental theorem of differential Galois theory*). Let* $\mathbf{L}/\mathbf{K}$ *be a finite Picard-Vessiot field extension (with derivative* $\boldsymbol{\delta}$*). Then and there is a lattice-antiisomorphism between differential fields* $(\mathbf{E}, \boldsymbol{\delta})$ *such that* $\mathbf{K} \subseteq \mathbf{E} \subseteq \mathbf{L}$ *and Zariski-closed subgroups* H *of* $\mathrm{Gal}_{\boldsymbol{\delta}}(\mathbf{L}/\mathbf{K})$ *given by*

$$H \mapsto \mathbf{L}^H$$

with inverse

$$\mathbf{M} \mapsto G_{\boldsymbol{\delta}}\left(\mathbf{L}/\mathbf{M}\right)$$

(with the same meaning as in Theorem 725). The extension $\mathbf{M}/\mathbf{K}$ *is Picard-Vessiot if, and only if* $G_{\boldsymbol{\delta}}\left(\mathbf{L}/\mathbf{M}\right)$ *is a* normal subgroup *of* $\mathrm{Gal}_{\boldsymbol{\delta}}\left(\mathbf{L}/\mathbf{K}\right)$ *and then*

$$G_{\boldsymbol{\delta}}\left(\mathbf{L}/\mathbf{M}\right) = \mathrm{Gal}_{\boldsymbol{\delta}}(\mathbf{L}/\mathbf{M}).$$

When $\mathbf{M}/\mathbf{K}$ *is Galois, the restriction to* $\mathbf{M}$ *of a differential automorphism* $u \in \mathrm{Gal}_{\boldsymbol{\delta}}(\mathbf{L}/\mathbf{K})$ *is a differential automorphism* $u_{\mathbf{M}}$ *of* $\mathbf{M}$ *and the restriction* $\rho_{\mathbf{M}} : u \mapsto u_{\mathbf{M}}$ *is such that* $\ker \rho_{\mathbf{M}} = \mathrm{Gal}_{\boldsymbol{\delta}}(\mathbf{L}/\mathbf{M})$, $\operatorname{im} \rho_{\mathbf{M}} = \mathrm{Gal}_{\boldsymbol{\delta}}(\mathbf{M}/\mathbf{K})$, *thus it induces an isomorphism of algebraic groups*

$$\mathrm{Gal}_{\boldsymbol{\delta}}(\mathbf{L}/\mathbf{K})/\,\mathrm{Gal}_{\boldsymbol{\delta}}(\mathbf{L}/\mathbf{M}) \xrightarrow{\sim} \mathrm{Gal}_{\boldsymbol{\delta}}(\mathbf{M}/\mathbf{K}).$$

Example 786. *In what follows,* $\mathbf{K} = \mathbb{C}\left(t\right)$ *and* $\boldsymbol{\delta} = d/dt$.

(1) Let $f\left(\partial\right) = \partial - a \in \mathbf{K}\left[\partial\right]$ *and let* $a \in \mathbb{C}^{\times}$. *The solution space is* $V = \mathbb{C}\,y$, $y = e^{a\,t}$, *thus the Picard-Vessiot field for* f *is* $\mathbf{K}\left(y\right)$. *By (4.26),*

$$\mathrm{Gal}_{\boldsymbol{\delta}}(\mathbf{L}/\mathbf{K}) = \left\{m \in \mathrm{GL}_1\left(\mathbb{C}\right) : \left(d/dt - a\right)\left(e^{a\,t}m = 0\right)\right\} = \mathrm{GL}_1\left(\mathbb{C}\right) = \mathbb{C}^{\times}.$$

(2) Let $f\left(\partial\right) = \partial^2 + \frac{1}{t}\partial$. *The solution space is* $V = \mathbb{C} \oplus \mathbb{C}\,y$, $y = \ln\left(t\right)$. *Let us regard* $\mathrm{Gal}_{\boldsymbol{\delta}}(\mathbf{L}/\mathbf{K})$ *as a subgroup of* $\mathrm{GL}\left(V\right)$. *A basis of the* $\mathbb{C}$*-vector space* V *is* $B = \{1, y\}$. *Let* $\sigma \in \mathrm{Gal}_{\boldsymbol{\delta}}(\mathbf{L}/\mathbf{K}) \subseteq \mathrm{GL}\left(V\right)$ *and let us determine the action of* σ *on* V. *Since* $\mathbf{K}$ *is fixed,* $\sigma\left(1\right) = 1$ *and* $\sigma\left(y^{\boldsymbol{\delta}}\right) = \sigma\left(1/t\right) = 1/t = y^{\boldsymbol{\delta}}$. *In addition,* $\sigma\left(y^{\boldsymbol{\delta}}\right) = \sigma\left(y\right)^{\boldsymbol{\delta}}$, *thus* $\sigma\left(y\right)^{\boldsymbol{\delta}} = y^{\boldsymbol{\delta}}$. *Therefore,* $d/dt\left(\sigma\left(y\right) - y\right) = 0$, *hence* $\sigma\left(y\right) - y = c \in \mathbb{C}$. *Therefore,* $\mathrm{Gal}_{\boldsymbol{\delta}}(\mathbf{L}/\mathbf{K})$ *is isomorphic to the translation group of* $\mathbb{C}$, *i.e., to* $\mathbb{C}$ *itself. The matrix of the above element* σ *in the basis* B *is* $\begin{bmatrix} 1 & c \\ 0 & 1 \end{bmatrix} \in \mathrm{GL}_2\left(\mathbb{C}\right)$. *The group consisting of all matrices* $\begin{bmatrix} 1 & c \\ 0 & 1 \end{bmatrix}$, $c \in \mathbb{C}$, *is denoted by* $\mathrm{T}\left(2,1;\mathbb{C}\right)$*; it defines the algebraic group* $\mathrm{T}\left(2,1\right)$ *(2.10.1.***3***).*

The following is closely connected with Remark 728:

Remark 787. *Let* $\mathbf{L} \supseteq \mathbf{K}$ *be a directed union of finite Picard-Vessiot extensions. Then the group* $\mathrm{Gal}_{\boldsymbol{\delta}}\left(\mathbf{L}/\mathbf{K}\right)$ *of all differential* $\mathbf{K}$*-automorphisms of* $\mathbf{L}$ *has a canonical structure of linear* proalgebraic group *(i.e., of projective limit of linear algebraic groups). Changing* finite *to* infinite *and* algebraic *to* proalgebraic *in the statement of Theorem 785, this statement is again valid.*

4.4.3 Difference Galois Theory

4.4.3.1 Picard-Vessiot Difference Rings

As shown through Remark 780, the Picard-Vessiot theory of differential fields cannot be simply translated into the frame of difference fields since there exist difference equations for which a Picard-Vessiot *field* cannot be obtained. In the remainder of this subsection, $\mathbf{K}$ is a difference field (with automorphism $\boldsymbol{\alpha}$) with algebraically closed subfield of constants $\mathbf{C}$.

Theorem and Definition 788. *Consider the difference equation (4.24) where $B \in \mathrm{Mat}_n(\mathbf{K})$.*

(i) There exist a ring $\mathbf{A} \supseteq \mathbf{K}$ and an automorphism $\bar{\boldsymbol{\alpha}}$ of $\mathbf{A}$, extending $\boldsymbol{\alpha}$, with the following properties:
(a) $\mathbf{A}$, endowed with the automorphism $\bar{\boldsymbol{\alpha}}$, is a simple difference ring;
(b) there exists a fundamental matrix $X \in \mathrm{GL}_n(\mathbf{A})$ *for (4.24) (i.e., $\mathbf{q}X = BX$), such that* $\mathbf{A} = \mathbf{K}\left[X, \det(X)^{-1}\right]$.
(ii) The difference ring $(\mathbf{A}, \bar{\boldsymbol{\alpha}})$ is called a Picard-Vessiot ring *for (4.24). For the sake of simplicity, $\bar{\boldsymbol{\alpha}}$ is again denoted by $\boldsymbol{\alpha}$ in the sequel.*
(iii) A Picard-Vessiot ring for (4.24) is unique up to difference isomorphism (Lemma and Definition 776(i)).
(iv) The total ring of fractions (Remark 325(iv)) $\mathbf{L}$ of the above Picard-Vessiot ring is called the total Picard-Vessiot ring *of the equation (4.24).*
(v) The above total Picard-Vessiot ring $\mathbf{L} \supseteq \mathbf{K}$ is characterized up to difference isomorphism by the following conditions:
(a) $\mathbf{L}$ is reduced and every every regular element of $\mathbf{L}$ is invertible;
(b) the subring of constants of $\mathbf{L}$ is $\mathbf{C}$;
(c) there exists a fundamental matrix $X \in \mathrm{GL}_n(\mathbf{L})$ for (4.24);
(d) $\mathbf{L}$ is minimal with respect to (a), (b) and (c).

Proof. The proof of (i) is quite similar to that of Theorem and Definition 782(3), using Lemma 777(B). (iii) and (v) are proved in ([285], Prop.1.9 and 1.23). ∎

Remark 789. *For certain kinds of difference equations, a total Picard-Vessiot difference* field *can be obtained [126]. Especially:*

(i) for the first order difference equation $\boldsymbol{\alpha}(y) = a\,y$ when there is no nonzero solution to $\boldsymbol{\alpha}(y) = a^n\,y$, $n > 1$ ([126], Prop. 6). (The example in Remark 780 is not of this kind.)
(ii) for the second order difference equation $\boldsymbol{\alpha}^2(y) = a\,\boldsymbol{\alpha}(y) + b\,y$ when the only nonzero solution to $\boldsymbol{\alpha}(y) = b^n\,y$ is obtained for $n = 0$ and is a constant ([126], Theorem 11).
(iii) for the second order difference equation $\boldsymbol{\alpha}^2(y) = -\boldsymbol{\alpha}(y) - b(t)\,y$ when $\mathbf{K} = \mathbb{C}(t)$ and $b(t) \in \mathbb{C}[t]$ is monic of degree one ([126], Theorem 12).

4.4.3.2 Galois Correspondence

The following is analogous to Lemma and Definition 784 ([285], Sect. 1.2).

Lemma and Definition 790. *(i) Let* $\mathbf{L} \supseteq \mathbf{K}$ *be difference rings (endowed with an automorphism* $\bar{\boldsymbol{\alpha}}$ *which extends* $\boldsymbol{\alpha}$*, and which is still denoted by* $\boldsymbol{\alpha}$ *to simplify the notation). A morphism of difference rings* $\mathbf{L} \to \mathbf{L}$ *which leaves fixed all elements of* $\mathbf{K}$ *is called a* $(\mathbf{K}, \boldsymbol{\alpha})$-endomorphism.
(ii) The set of all $(\mathbf{K}, \boldsymbol{\alpha})$*-automorphisms of the difference ring* $(\mathbf{L}, \boldsymbol{\alpha})$ *is a group denoted by* $\mathrm{Aut}_{\mathbf{K},\boldsymbol{\alpha}}(\mathbf{L}) = G_{\boldsymbol{\alpha}}(\mathbf{L}/\mathbf{K})$.
(iii) If $\mathbf{L}/\mathbf{K}$ *is a total Picard-Vessiot ring over* $\mathbf{K}$*,* $G_{\boldsymbol{\alpha}}(\mathbf{L}/\mathbf{K})$ *is called the* difference Galois group *of* $(\mathbf{L}, \boldsymbol{\alpha})$ *over* $(\mathbf{K}, \boldsymbol{\alpha})$ *and is denoted by* $\mathrm{Gal}_{\boldsymbol{\alpha}}(\mathbf{L}/\mathbf{K})$.
(iv) $\mathrm{Gal}_{\boldsymbol{\alpha}}(\mathbf{L}/\mathbf{K})$ *has a canonical structure of linear algebraic group* $\mathcal{G}$.

The following is proved in ([285], Theorem 1.29 and Corol. 1.30).

Theorem 791. *(*Fundamental theorem of difference Galois theory*). Let* $\mathbf{L}/\mathbf{K}$ *be a total Picard-Vessiot ring over* $\mathbf{K}$*, and let* $G_{\boldsymbol{\alpha}}(\mathbf{L}/\mathbf{K})$ *be its difference Galois group. Let* $\mathcal{F}$ *be the set* $\mathbf{F}$ *of difference rings with* $\mathbf{K} \subseteq \mathbf{F} \subseteq \mathbf{L}$*, and such that every regular element of* $\mathbf{F}$ *is a unit of* $\mathbf{F}$*. Let* $\mathfrak{G}$ *denote the set of algebraic subgroups of* $\mathrm{Gal}_{\boldsymbol{\alpha}}(\mathbf{L}/\mathbf{K})$.

(i) For any $\mathbf{F} \in \mathcal{F}$*, the subgroup* $G_{\boldsymbol{\alpha}}(\mathbf{L}/\mathbf{F}) \leq \mathrm{Gal}_{\boldsymbol{\alpha}}(\mathbf{L}/\mathbf{K})$ *consisting of the elements of* $\mathrm{Gal}_{\boldsymbol{\alpha}}(\mathbf{L}/\mathbf{K})$ *which fix all elements of* $\mathbf{F}$ *pointwise, belongs to* $\mathfrak{G}$.
(ii) For any $H \in \mathfrak{G}$*, the subring* $\mathbf{L}^H$ *of* $\mathbf{L}$ *consisting of all elements* $x \in \mathbf{L}$ *such that* $u(x) = x$ *for all* $u \in H$*, belongs to* $\mathcal{F}$.
(iii) There exists a lattice-antiisomorphism $\mathfrak{G} \to \mathcal{F}$ *given by*

$$H \mapsto \mathbf{L}^H$$

with inverse

$$\mathbf{F} \mapsto G_{\boldsymbol{\alpha}}(\mathbf{L}/\mathbf{F}).$$

(iv) Let $H \in \mathfrak{G}$*; then* $H \lhd \mathrm{Gal}_{\boldsymbol{\alpha}}(\mathbf{L}/\mathbf{K})$ *if, and only if the difference ring* $\mathbf{F} = \mathbf{L}^H$ *has the property that for every* $z \in \mathbf{F}\backslash\mathbf{K}$ *there exists a* $(\mathbf{K}, \boldsymbol{\alpha})$*–automorphism* σ *such that* $\sigma(z) \neq z$*. When this happens,*

$$G_{\boldsymbol{\alpha}}(\mathbf{L}/\mathbf{F}) \cong \mathrm{Gal}_{\boldsymbol{\alpha}}(\mathbf{L}/\mathbf{K})/H.$$

4.5 Exercises

Exercise 792. *Prove that every algebraically closed field* $\mathbf{K}$ *is infinite. (Hint: assuming that* $\mathbf{K}$ *is finite, consider the polynomial* $f(X) = 1 + \prod_{a \in \mathbf{K}} (X - a)$*.)*

Exercise 793. *Find the minimum polynomial over* $\mathbb{Q}$ *of (i)* $(1+i)/\sqrt{2}$*; (ii)* $i+\sqrt{2}$*; (iii)* $e^{2\pi i/3}+2$.

Exercise 794. *Prove that if* α *is transcendental over* $\mathbf{K}$*, then* $[\mathbf{K}(\alpha):\mathbf{K}] = +\infty$ *and that* $\mathbf{K}(\alpha) \cong \mathbf{K}(X)$.

Exercise 795. *Find the degrees of the algebraic field extensions:* $\mathbb{Q}\left(\sqrt{7}\right)/\mathbb{Q}$ *and* $\mathbb{Q}\left(\sqrt{5},\sqrt{7},\sqrt{35}\right)/\mathbb{Q}$.

Exercise 796. *Show that extension* $\mathbb{Q}\left(\sqrt[4]{2}\right)$ *is not normal. Let* $\alpha=\sqrt{2}$, $\mathbf{L}=\mathbb{Q}(\alpha)$ *and* $\mathbf{M}=\mathbf{L}(\sqrt{\alpha})$. *Check that both extensions* $\mathbf{M}/\mathbf{L}$ *and* $\mathbf{L}/\mathbb{Q}$ *are normal and deduce that a tower of normal extensions need not be normal.*

Exercise 797. *Prove that the subgroups* H_1 *and* H_2 *of* G *in Example 727 are normal. What can you conclude?*

Exercise 798. *Prove that the set denoted by* $\mathfrak{F}(\mathbf{L})$ *in Theorem 725 is a lattice; specify the meet and the join of two elements of* $\mathfrak{F}(\mathbf{L})$.

Exercise 799. *Let* $f \in \mathbf{K}[X]$ *be a polynomial of degree* $n \geq 1$.

(i) Prove that $(\mathrm{Gal}_{\mathbf{K}}(f):1) \mid k! \mid n!$ *where* k *is the number of distinct roots of* f.
(ii) If f *is irreducible, prove that* $n \mid (\mathrm{Gal}_{\mathbf{K}}(f):1)$.
(iii) Show that in Example 727, $\Gamma \subsetneq \mathfrak{S}(R)$ *(with the notation in Remark 723).*
(Hint: for (i), use Remark 723, Example 135 and Lagrange's theorem (1.21); for (ii), use Proposition 724(b) and Exercise 208.)

Exercise 800. *Determine the lattice of all subgroups of* G *in Example 727. (This exercise is solved in [206], Sect. VI.2, Example 3.)*

Exercise 801. *Let*

$$A=\begin{bmatrix} 0 & 0 & 0 & 0 & 1 \\ 1 & 0 & 0 & 1 & 1 \\ -1 & 1 & 0 & -1 & 0 \\ 0 & 0 & 0 & 0 & -1 \\ 0 & 0 & 0 & 0 & 0 \end{bmatrix}.$$

Prove that the set of elementary divisors of $X\,I_5 - A$ *is* $\left\{X^2, X^3\right\}$ *and determine the Jordan canonical form of* A.

Exercise 802. *Prove the following for any* $a \in \mathbf{K}$ *and* $b,c \in \mathbf{K}^{\times}$ *(with the notation in (4.3.4.1)):*

(i) $N_n\left({}^{c}a\right)c = \sum_{1\leq i\leq n} f_{n-i}(c)\, N_i(a)$;
(ii) $N_{n+1}(c) = N_n\left({}^{c}c\right)c$;
(iii) ${}^{b}0 - {}^{c}0 = \boldsymbol{\alpha}(c)\left({}^{c^{-1}b}0\right)c^{-1}$.

Exercise 803. *Let* $(\mathbf{K}, \boldsymbol{\delta})$ *be a differential field, let* $(\mathbf{L}, \boldsymbol{\delta}) \geq (\mathbf{K}, \boldsymbol{\delta})$ *be a universal differential field (Corollary and Definition 783), and let* $\alpha \in \mathbf{L}^{\times}$. *Recall that* α *is called an* exponential *(resp., a* primitive*) over* $\mathbf{K}$ *if there exists* $a \in \mathbf{K}$ *such that* $\boldsymbol{\delta}\alpha = a\,\alpha$ *(resp.,* $\boldsymbol{\delta}\alpha = a$*). If* α *is an exponential or a primitive over* $\mathbf{K}$*, prove that* $\mathbf{K}(\alpha)$ *is a Picard-Vessiot extension of* $\mathbf{K}$.

Exercise 804. *Let* $(\mathbf{K}, \boldsymbol{\delta})$ *be a differential field with algebraically closed field of constants* $\mathbf{C}$*, and let* $\mathbf{L} \supseteq \mathbf{K}$ *be a finite Galois extension. Prove that* $\mathbf{L}$ *is a Picard-Vessiot extension of* $\mathbf{K}$*. (Hint: first use Lemma 777(A) to extend* $\boldsymbol{\delta}$*; then show that the subfield of constants of* $(\mathbf{L}, \boldsymbol{\delta})$ *is again* $\mathbf{C}$*. See [226], Lemma 3.19 and Prop. 3.20 for more details.)*

4.6 Notes

The modern presentation of the finite Galois theory is due to Artin [7] who showed the advantage of going down from a field to its subfields, instead of going up from a field to its extensions as in the classical approach. Nevertheless, the so-called Artin's theorem (4.1) goes back to Dedekind who established this result before 1894. The infinite Galois theory (Remark 728) is due to Krull [194] who introduced for this purpose the profinite topology ([28], Chap. V), ([79], Chap. 11), ([206], Chap. VI). The Galois theory was extended to division rings by E. Noether, N. Jacobson, and H. Cartan [58] (see [78] for an up-to-date presentation).

Theorem 734 is due to Jacobson [168] who also introduced the notion of pseudo-linear transformation [167]. Definitions 747(i) & 761 of an algebraic PLT and of left eigenvalues and eigenvectors are due to Leroy [214]; they generalize Cohn's definitions given in [75] when $\boldsymbol{\alpha} = 1$, $\boldsymbol{\delta} = 0$. The definition and the existence of a matrix-homogeneous division ring (Theorem and Definition 748) were stated and established in [75]. Theorem 753 was obtained by Sizer in his Ph.D. thesis [332]. The definitions of the left (resp., right) evaluation and of the left (resp., right) root of a skew polynomial (Definitions 756 and 759), as well as Lemma 758, are due to Lam and Leroy [201], [202]. Lemma 745 and Proposition 762 were established by Lam *et al.* [204]. The equalities in Exercise 802 are proved in [201]. Theorem 763 is classical (see [197], Theorem 4); our proof is that given by Cohn ([77], Chap. 8, Theorem 6.1). The notion of *P-independence* (Definition 767(i)) was introduced by Lam [197] in the case of a zero derivation $\boldsymbol{\delta}$, and then generalized by Lam and Leroy [203]. Definition 767(ii) is a generalization, given in [203], of Wedderburn's original definition [357] of a polynomial $f \in \mathbf{K}[X]$ (where $\mathbf{K}$ is a division ring with centre $\mathbf{F}$ and X is a central indeterminate) which is the minimal polynomial over $\mathbf{F}$ of an element a algebraic over $\mathbf{F}$. Lemma and Definition 769 was established (in collaboration with Leroy) in the Epilogue of Lam's paper [197]. General hyperexponential solutions have been considered by Abramov [2] (see also [54]).

The modern theory of Picard-Vessiot differential fields is essentially due to Kolchin [187], one of the fathers–with Ritt [294]–of Differential Algebra. In particular, Kolchin emphasized the importance of extending differential fields *without adding new constants.* Picard-Vessiot differential field extensions are also nicely expounded in Magid [226] and in van der Put and Singer [286]. The latter account is an important contribution for its clarity and the algorithms presented. R.M. Cohn [82] noticed that Difference Algebra is more complicated than Differential Algebra. This was confirmed by Franke [126], who showed that severe restrictions must be made on a linear difference equation for a Picard-Vessiot difference *field* to exist for that equation. Van der Put and Singer [285] eventually established a theory of Picard-Vessiot difference *rings.*

Part II
Algebraic Theory of Linear Systems

5 Systems and Behaviors: A General Setting

5.1 Introduction

Historically, a linear system was first considered as a "black box", excited by inputs, generating outputs, and characterized by a transfer matrix [22]. In 1960, Kalman founded the theory of state-space systems [171], inaugurating the era of "modern control theory". This approach is however limited by several facts.

1° Multidimensional systems do not have a state-space representation, except in very peculiar cases. Regarding this point, Rosenbrock representations [301] are more general.

2° As emphasized by Willems [360], [361], in many applications it is not appropriate to *a priori* distinguish inputs and outputs among all system variables.

Therefore, the general representation of a linear system is of the form

$$Rw = 0 \tag{5.1}$$

where R is a $q \times k$ matrix with entries in a ring of operators $\mathbf{A}$ and

$$w = \begin{bmatrix} w_1 & \cdots & w_k \end{bmatrix}^T. \tag{5.2}$$

Equation (5.1) is identical to (3.1) and, therefore, can be viewed as the equation of a finitely presented (f.p.) left $\mathbf{A}$-module M. As shown in (3.4.3.**2**), the matrix R is not an intrinsic object, as opposed to the module M. Therefore, following Kashiwara ([181], Definition 1.1.1) and – in the context of Automatic Control – Fliess [115], *one is led to define a (linear) system as a finitely presented A-module.*

In practice, the variables $w_1, ..., w_k$ must belong to a well-defined space of functions (or distributions, hyperfunctions, etc.). Let W be such a space which is also a left $\mathbf{A}$-module. The object $\ker_W(R\bullet)$ defined by (3.32) is called the *W-behavior* associated with M, and is denoted by $\mathfrak{B}_W(M)$. In the

H. Bourlès and B. Marinescu: Linear Time-Varying Systems, LNCIS 410, pp. 311–401.
springerlink.com

behavioral theory, first developed by Willems, $\mathfrak{B}_W(M)$ plays a preeminent role.

We already know that there are connections between M and $\mathfrak{B}_W(M)$. First, as shown in (**3.4.3.3**), $\mathfrak{B}_W(M)$ is deduced from M using the functor $\mathfrak{B}_W = \mathrm{Hom}_{\mathbf{A}}(\bullet, W)$. Conversely, M is uniquely determined by $\mathfrak{B}_W(M)$ if the functor $\mathfrak{B}_W$ is faithful, i.e., if W is a cogenerator (Theorem and Definition 43(3)). If W is an *injective cogenerator*, then the correspondence between M and $\mathfrak{B}_W(M)$ is even stronger (Theorem 644). This shows that in the "post-modern control theory" expounded in the present book, homological algebra cannot be avoided (at least at the basic level of Chapter 3).

For certain problems, such as those in Section 5.7 below, the above algebraic considerations are not sufficient. Topological properties must also be taken into account; then, the calculations can no longer be made in an abelian category such as ${}_{\mathbf{A}}\mathbf{Mod}$, but in a category which is only semiabelian (Definition 38). In addition, for the above-mentioned problems, the classical notion of cogenerator is too restrictive. This motivates the generalizations in Section 5.3 below.

5.2 The Module-Theoretic Setting

In what follows, $\mathbf{A}$ is a *ring of operators*. Let us detail the introduction in Subsect. 5.1 and review some important points of Chapter 3 in the context of systems theory.

5.2.1 Linear Systems

Definition 805. *A* linear system M *over* $\mathbf{A}$ *is a finitely presented (f.p.) left* $\mathbf{A}$*-module.*

By Definition 496, there exists a matrix $R \in {}^{q}\mathbf{A}^{k}$ such that the sequence

$$\mathbf{A}^{q} \xrightarrow{\bullet R} \mathbf{A}^{k} \xrightarrow{\bullet \alpha} M \longrightarrow 0 \tag{5.3}$$

is exact, where $\bullet\alpha$ is the canonical epimorphism. Therefore,

$$M = \mathrm{im}_{\mathbf{A}}(\bullet\alpha) \cong \mathbf{A}^{k} / \ker_{\mathbf{A}}(\bullet\alpha)$$

according to Noether's first isomorphism theorem (Theorem 52(1)). Since $\ker_{\mathbf{A}}(\bullet\alpha) = \mathrm{im}_{\mathbf{A}}(\bullet R)$, we obtain

$$M \cong \mathbf{A}^{k} / \mathrm{im}_{\mathbf{A}}(\bullet R) \triangleq \mathrm{coker}_{\mathbf{A}}(\bullet R).$$

Consider another exact sequence

$$\mathbf{A}^{q'} \xrightarrow{\bullet R'} \mathbf{A}^{k} \xrightarrow{\bullet \alpha'} M' \longrightarrow 0.$$

Then $M' \cong \operatorname{coker}_{\mathbf{A}} (\bullet R')$, therefore, according to Definition 517, the following conditions are equivalent:

(i) $M \cong M'$;
(ii) $\operatorname{coker}_{\mathbf{A}} (\bullet R) \cong \operatorname{coker}_{\mathbf{A}} (\bullet R')$;
(iii) R and R' are left-similar.

Let $(\varepsilon_i)_{1 \leq i \leq k}$ be the canonical basis of $\mathbf{A}^k$ and w_i the canonical image of ε_i in M, i.e., $w_i = (\varepsilon_i)\, \alpha = \varepsilon_i + \operatorname{im}_{\mathbf{A}} (\bullet R)$. Consider the finite sequence of generators w defined by (5.2). As shown in (**3.2.3.1**), $M = [w]_{\mathbf{A}}$ where the finite sequence of generators w is only subject to the relation (5.1). According to Definition 512, this module $M = [w]_{\mathbf{A}}$ is said to be *defined by generators and relations*: the generators are the elements w_i; the relations are the q rows of equation (5.1)). The free module $\mathbf{A}^k$ (resp., $\mathbf{A}^q$) is sometimes called the module of generators (resp., of relations).

Example 806. *Let* $\mathbf{A}$ *be the first Weyl algebra* $A_1 (\mathbb{C})$ (**2.7.4.3**) *and consider the equation over* $\mathbf{A}$

$$\left(t\partial^2 + 1\right) y = (1 + t)\, u. \tag{5.4}$$

This equation represents a linear system with time-varying coefficients, and more specifically with coefficients in $\mathbb{C}[t]$. *To see this, consider the exact sequence*

$$\mathbf{A} \xrightarrow{\bullet R} \mathbf{A}^2 \longrightarrow M \longrightarrow 0$$

where $R = \left[\, \left(t\partial^2 + 1\right) \;\; -(1+t) \,\right]$. *Let* (e) *be the canonical basis of* $\mathbf{A}$ *(i.e.,* $e = 1$*) and let* $(\varepsilon_1, \varepsilon_2)$ *be the canonical basis of* $\mathbf{A}^2$. *The left* $\mathbf{A}$*-module* M *is generated by the canonical images* $\bar{\varepsilon}_1 = y$ *and* $\bar{\varepsilon}_2 = u$ *in* $\mathbf{A}^2 / \operatorname{im}_{\mathbf{A}} (\bullet R)$. *Since* $eR = \left(t\partial^2 + 1\right) \varepsilon_1 - (1+t)\, \varepsilon_2 \in \operatorname{im}_{\mathbf{A}} (\bullet R)$, *the generators* y *and* u *satisfy (5.4), i.e.,* $M = [y, u]_{\mathbf{A}}$ *where* y *and* u *are only submitted to the relation (5.4). This is a typical example of the above construction.*

5.2.2 Behaviors

Let W be a left $\mathbf{A}$-module and consider the functor $\mathfrak{B}_W = \operatorname{Hom}_{\mathbf{A}} (\bullet, W)$ from the category ${}_{\mathbf{A}}\mathbf{Mod}$ to the category $\mathbf{Mod}_{\mathbf{E}}$ where $\mathbf{E} = \operatorname{End}_{\mathbf{A}} (W)$ is the endomorphism ring of W. The functor $\mathfrak{B}_W$ is contravariant left exact (Proposition 549(iii)), therefore the exact sequence (5.3) yields the exact sequence

$$\mathfrak{B}_W (\mathbf{A}^q) \overset{\mathfrak{B}_W(\bullet R)}{\longleftarrow} \mathfrak{B}_W \left(\mathbf{A}^k\right) \overset{\mathfrak{B}_W(\bullet \alpha)}{\longleftarrow} \mathfrak{B}_W (M) \longleftarrow 0.$$

We have the canonical isomorphisms $\mathfrak{B}_W \left(\mathbf{A}^k\right) \cong {}^k W$ and $\mathfrak{B}_W (\mathbf{A}^q) \cong {}^q W$. Therefore, the following identifications are made (Corollary 567):

$$\begin{aligned} \mathfrak{B}_W \left(\mathbf{A}^k\right) &= {}^k W, \qquad & \mathfrak{B}_W (\mathbf{A}^q) &= {}^q W, \\ \mathfrak{B}_W (\bullet \alpha) &= \iota \bullet, \qquad & \mathfrak{B}_W (\bullet R) &= R \bullet, \\ \mathfrak{B}_W (M) &= \ker_W (R \bullet). \end{aligned}$$

Summing up, using the functor $\mathfrak{B}_W$, one passes from $M = \operatorname{coker}_{\mathbf{A}}(\bullet R)$ to $\mathfrak{B}_W(M) = \ker_W(R\bullet)$.

This can also be expressed as follows: let $N = \operatorname{im}_{\mathbf{A}}(\bullet R) \lhd \mathbf{A}^k$. Then:

(i) $M = \mathbf{A}^k / N$;
(ii) $\mathfrak{B}_W(M) = N^{\perp} \triangleq \{\mathbf{w} \in {}^kW : r\,\mathbf{w} = 0,\ \forall r \in N\}$.

Recall that:

(a) the elements of $\mathbf{A}^k$ (thus those of N) are rows with k entries in $\mathbf{A}$;
(b) the elements of ${}^kW$ (thus those of $N^{\perp}$) are columns of k entries in W.

Definition 807. *The right* $\mathbf{E}$*-module* $\mathfrak{B}_W(M) = N^{\perp}$ *is the* W-behavior *associated with the linear system* M*. The elements of* $\mathfrak{B}$ *are called* trajectories.

Setting $\mathfrak{B}_W(M) = \mathfrak{B}$ we obtain (in accordance with (3.32))

$$\boxed{\mathfrak{B} = \{\mathbf{w} \in {}^kW : R\,\mathbf{w} = 0\} \triangleq \ker_W(R\bullet).} \tag{5.5}$$

Note that the equation

$$R\,\mathbf{w} = 0 \tag{5.6}$$

which defines $\mathfrak{B}_W(M)$ looks like (5.1) but must not be confused with it (see Remark 566). The W-behavior $\mathfrak{B}_W(M)$ is the set of all solutions in ${}^kW$ of (5.6). The set $\mathfrak{B}_W(M)$ is a right $\mathbf{E}$-module, thus an abelian group, and if $\mathbf{A}$ is an $\mathbf{R}$-algebra (where $\mathbf{R}$ is a commutative ring), then $\mathfrak{B}_W(M)$ is an $\mathbf{R}$-module (Exercise 476).

Example 808. *Let* W *be, e.g., the space* $\mathcal{E}(\mathbb{R})$ (**1.7.3.5**). *The* W*-behavior associated with the linear system in Example 806 is*

$$\mathfrak{B}_W(M) = \left\{ \begin{bmatrix} \mathbf{y} \\ \mathbf{u} \end{bmatrix} \in {}^2W : (t\partial^2 + 1)\,\mathbf{y} = (1+t)\,\mathbf{u} \right\}.$$

Example 809. *Let* $\mathbf{A} = \mathbb{C}[\partial]$ *where* $\partial = d/dt$ *is the usual derivative and consider the system* M *defined by* $\dot{y} = y$ *(i.e.,* $M = \operatorname{coker}_{\mathbf{A}}(\partial - 1)$*).*

(a) Let $W = \mathcal{E}(\mathbb{R})$*. Then*

$$\mathfrak{B}_W(M) = \{\mathbf{y} : t \mapsto Ae^t, \quad A \in \mathbb{C}\}.$$

(b) Let $W = \mathbb{C}[t]$*. Then*

$$\mathfrak{B}_W(M) = \{0\}.$$

5.2.3 The Connection between Systems and Behaviors

5.2.3.1 Passage from M to $\mathfrak{B}_W(M)$

Let M and W be given. We have $M = \operatorname{coker}_{\mathbf{A}}(\bullet R) = \mathbf{A}^k/N$, $N = \operatorname{im}_{\mathbf{A}}(\bullet R)$. Then we can calculate $\mathfrak{B}_W(M) = \operatorname{Hom}_{\mathbf{A}}(M, W)$. We have indeed $\mathfrak{B}_W(M) = N^{\perp} = \ker_W(R\bullet)$.

5.2.3.2 The Fundamental Galois Connection

Let $\mathfrak{B} = \mathfrak{B}_W(M) \subseteq {}^k W$ be given. Consider

$$\mathfrak{B}^\perp \triangleq \left\{ r \in \mathbf{A}^k : r\,\mathbf{w} = 0, \quad \forall \mathbf{w} \in \mathfrak{B} \right\}.$$

Proposition 810. *The Galois connection in Lemma 569 induces a Galois connection*

$$\mathcal{S}_k \longleftrightarrow \mathcal{S}_k^\perp : N \mapsto N^\perp, \quad \mathfrak{B} \mapsto \mathfrak{B}^\perp$$

where $\mathcal{S}_k$ is the set of all f.g. submodules of $\mathbf{A}^k$ and $\mathcal{S}_k^\perp = \left\{ N^\perp : N \in \mathcal{S} \right\}$.

Proof. By Lemma 569, the above correspondence is a Galois connection with

$$\mathcal{S}_k = \left\{ N \lhd \mathbf{A}^k : \mathbf{A}^k / N \text{ is f.p.} \right\}.$$

We have the short exact sequence

$$0 \longrightarrow N \longrightarrow \mathbf{A}^k \longrightarrow \mathbf{A}^k/N \longrightarrow 0,$$

therefore, by Lemma 498, $\mathbf{A}^k/N$ is f.p. if, and only if N is f.g. ■

By Proposition 810 and Proposition 99(ii), we have for any $N \in \mathcal{S}_k$ and any $\mathfrak{B} \in \mathcal{S}_k^\perp$

$$N \lhd N^{\perp\perp} \text{ and } \mathfrak{B} \lhd \mathfrak{B}^{\perp\perp}.$$

These inclusions need not be equalities. In Example 809,

$$\begin{aligned} N &= \mathbb{C}[\partial](\partial - 1), \quad & \mathfrak{B} &= N^\perp = 0, \\ N^{\perp\perp} &= \mathbb{C}[\partial], \quad & \mathfrak{B}^{\perp\perp} &= N^{\perp\perp\perp} = 0. \end{aligned}$$

Remark 811. *A submodule N of $\mathbf{A}^k$ is said to be* W-closed *if $N = N^{\perp\perp}$.*

We will now give a *sufficient condition* for any f.g. submodule of $\mathbf{A}^k$ to be W-closed.

5.2.3.3 Passage from $\mathfrak{B}_W(M)$ to M

Theorem 812. *Let W be a* cogenerator *(Subsect. 3.5.6). Then:*

(i) The map $M \mapsto \mathfrak{B}_W(M)$ $\left(M \in {}_{\mathbf{A}}\mathbf{Mod}^{fp}\right)$ is one-to-one.

(ii) For any $N \in \mathcal{S}_k$, $N = N^{\perp\perp}$, i.e., the Galois connection in Proposition 810 is bijective.

(iii) Let $R \in {}^q\mathbf{A}^k$ and $R' \in {}^{q'}\mathbf{A}^k$. Then $\ker_W(R'\bullet) \subseteq \ker_W(R\bullet)$ *if, and only if there exists a matrix $X \in {}^q\mathbf{A}^{q'}$ such that $R = X\,R'$. In particular,* $\ker_W(R'\bullet) = \ker_W(R\bullet)$*, if, and only if $q = q'$ and there exists $X \in \mathrm{GL}_q(\mathbf{R})$ such that $R = X\,R'$ ("quasi-uniqueness of the matrix of definition").*

Proof. (i): If W is a cogenerator, the functor $\mathfrak{B}_W$ is faithful (Theorem and Definition 43(3)), thus it is injective (Lemma 32).

(ii): We know that $N \lhd N^{\perp\perp}$. Let us prove the reverse inclusion by contradiction. If $m \in \mathbf{A}^k$ does not belong to N, let $\bar{m} = m + N \in \mathbf{A}^k/N$. Then $\bar{m} \neq 0$ and by Theorem 638, there exists $\bar{\eta} : \mathbf{A}^k/N \to W$ such that $\bar{\eta}(m) \neq 0$. By Theorem and Definition 145, the $\mathbf{A}$-homomorphism $\bar{\eta}$ is induced by an $\mathbf{A}$-homomorphism $\eta : \mathbf{A}^k \to W$ such that $\eta(N) = 0$, i.e. $\eta \in N^\perp$. As $\eta(m) \neq 0$, $m \notin N^{\perp\perp}$.

(iii): Let $N = \operatorname{im}_{\mathbf{A}}(\bullet R)$ and $N' = \operatorname{im}_{\mathbf{A}}(\bullet R') \subseteq \mathbf{A}^k$. If $N'^\perp \lhd N^\perp$, then $N^{\perp\perp} \lhd N'^{\perp\perp}$. By (i), $N = N^{\perp\perp}$ and $N' = N'^{\perp\perp}$, thus $N \lhd N'$, i.e. $\operatorname{im}_{\mathbf{A}}(\bullet R) \lhd \operatorname{im}_{\mathbf{A}}(\bullet R')$. Now, $\bullet R'$ is an epimorphism $\mathbf{A}^{q'} \to \operatorname{im}_{\mathbf{A}}(\bullet R')$ and the free module $\mathbf{A}^{q'}$ is projective (Proposition 574(ii)), therefore there exists an $\mathbf{A}$-homomorphism $\bullet X : \mathbf{A}^{q'} \to \mathbf{A}^q$ such that $R = X\, R'$ (Lemma 45(2)).

The rest is clear. ■

5.2.4 The Categorical Point of View

Recall that $\mathbf{A}$ is a ring of operators and that $\mathbf{E} = \operatorname{End}_{\mathbf{A}}(W)$.

Lemma and Definition 813. *(i) The category* $\mathfrak{Sys}$ *of systems over the ring of operators* $\mathbf{A}$ *is the category* ${}_{\mathbf{A}}\mathbf{Mod}^{fp}$ *of finitely presented left* $\mathbf{A}$*-modules.*[1]
(ii) The class of W*-behaviors is* $\mathfrak{Beh} = \mathfrak{B}_W(\mathfrak{Sys})$*.*
(iii) If W *is a cogenerator, then* $\mathfrak{Beh}$ *is a subcategory of* $\mathbf{Mod}_{\mathbf{E}}$ *and* $\mathfrak{Sys}^{op} \cong \mathfrak{Beh}$*.*
(iv) Let $M_1, M_2 \in \mathfrak{Sys}$*. Then* $\mathfrak{B}_W(M_1) \oplus \mathfrak{B}_W(M_2) \in \mathfrak{Beh}$*. If* $M_1 \lhd M_2$*, then* $\mathfrak{B}_W(M_2/M_1) \lhd \mathfrak{B}_W(M_2)$ *is an element of* $\mathfrak{Beh}$*, called a* subbehavior *of* $\mathfrak{B}_W(M_2)$*.*
(v) Assume that $\mathbf{A}$ *is left coherent ring; then* $\mathfrak{Sys}$ *is an abelian category. If in addition* $W \in {}_{\mathbf{A}}\mathbf{Mod}$ *is a cogenerator, then* $\mathfrak{Beh}$ *is an abelian category too.*

Proof. (iii) is a consequence of Lemma 32.

(iv): The functor $\mathfrak{B}_W$ is left exact (Theorem and Definition 43), therefore

$$\mathfrak{B}_W(M_1 \oplus M_2) \cong \mathfrak{B}_W(M_1) \oplus \mathfrak{B}_W(M_2)$$

where the isomorphism is canonical, thus is an identification.

In addition, if $M_1 \lhd M_2$, then canonical exact sequence

$$0 \longrightarrow M_1 \xrightarrow{\bullet\iota} M_2 \xrightarrow{\bullet\varphi} M_2/M_1 \longrightarrow 0$$

where ι is the inclusion and φ is the canonical epimorphism, yields the exact sequence of right morphisms

[1] $\mathfrak{Sys}$ reads *Sys* (Gothic).

$$\mathfrak{B}_W(M_1) \overset{\mathfrak{B}_W(\iota)}{\longleftarrow} \mathfrak{B}_W(M_2) \overset{\mathfrak{B}_W(\varphi)}{\longleftarrow} \mathfrak{B}_W(M_2/M_1) \longleftarrow 0$$

where the canonical monomorphism $\mathfrak{B}_W(\varphi)$ is identified with the inclusion.
(v) is a consequence of Theorem 510 and of (ii). ∎

5.2.5 The Use of Large Cogenerators

Neither Theorem 644 nor Lemma and Definition 813 allow us to specify when the category $\mathfrak{Beh}$ of behaviors is a full subcategory of $\mathbf{Mod}_{\mathbf{E}}$ (Remark 645). This leads us to consider the notion of *large cogenerator*:

Definition 814. *A module $W \in {}_{\mathbf{A}}\mathbf{Mod}$ is called a* large cogenerator *if it is a cogenerator and for every $M \in {}_{\mathbf{A}}\mathbf{Mod}^{fg}$, there exists a* finite *set of indices J and an embedding $M \hookrightarrow W^J$.*

The following is a useful characterization of *injective* large cogenerators in the Noetherian case:

Theorem 815. *Let $\mathbf{A}$ be a left Noetherian ring and $W \in {}_{\mathbf{A}}\mathbf{Mod}$ be an injective module.*

(1) Of the following, (i)⇔(ii)⇒(iii):
(i) W is a large cogenerator.
(ii) for any meet-irreducible left ideal $\mathfrak{c}$ in $\mathbf{A}$ (3.5.2.4), there exist a positive integer k and an embedding $\mathbf{A}/\mathfrak{c} \hookrightarrow W^k$.
(iii) W is a cogenerator.
(2) If $\mathbf{A}$ is commutative, then (i)⇔(ii)⇔(iv) where (iv) is the following:
(iv) for any prime ideal $\mathfrak{c} \in \operatorname{Spec} \mathbf{A}$ (2.4.3.2), there exist a positive integer k and an embedding $\mathbf{A}/\mathfrak{c} \hookrightarrow W^k$.
(3) If $\mathbf{A}$ is a Dedekind domain (not necessarily commutative) for which there exist a positive integer n and an embedding $\mathbf{A} \hookrightarrow W^n$, then (i)⇔(ii)⇔(iii).

Proof. (1). (i)⇒(ii) is clear since $\mathfrak{c}$ is f.g. by the Noetherian assumption (Lemma 311), and (ii)⇒(iii) is obvious.

(ii)⇒(i): Let M be a f.g. left $\mathbf{A}$-module with injective envelope $E(M)$, and consider the decomposition of $E(M)$ into a direct sum of indecomposable injective modules E_i (Theorem 590):

$$E(M) = \bigoplus_{i \in I} E_i.$$

For every $i \in I$, $E_i \cap M \neq 0$ since $M \subseteq_e E(M)$ (Exercise 692), therefore I is finite since M is f.g. By Lemma 602, for every $i \in I$, there exists a meet-irreducible left ideal $\mathfrak{c}_i$ such that $E_i \cong \mathbf{A}/\mathfrak{c}_i$. If (ii) holds, there exists a positive integer k_i and an embedding $\bullet\iota_i : \mathbf{A}/\mathfrak{c}_i \hookrightarrow W^{k_i}$. Since W is injective, so is also W^{k_i}, and ι_i can be extended to an $\mathbf{A}$-linear map

$$\bullet f_i : E(\mathbf{A}/\mathfrak{c}_i) \to W^{k_i}.$$

Then, $\ker f_i \cap \mathbf{A}/\mathfrak{c}_i = \ker \iota_i = 0$. Therefore, $\ker f_i = 0$ since $\mathbf{A}/\mathfrak{c}_i \subseteq_e E(\mathbf{A}/\mathfrak{c}_i)$ (Exercise 692), and $E(\mathbf{A}/\mathfrak{c}_i) \hookrightarrow W^{k_i}$. As a result,

$$M \subseteq E(M) \cong \bigoplus_{i \in I} E_i \hookrightarrow W^k, \quad k = \sum_{i \in I} k_i,$$

and (i) holds.

(2) is an obvious consequence of the above and of Lemma 601(iii).

(3). (iii)⇒(i): Assume that (iii) holds, and let $\mathfrak{c}$ be a nonzero left ideal in $\mathbf{A}$. The quotient module $\mathbf{A}/\mathfrak{c}$ is torsion, thus it is of finite length by Theorem 664; therefore, the module $\mathbf{A}/\mathfrak{c}$ is Artinian by Theorem 315, thus it is finitely cogenerated by Proposition 641, and (i) holds. ■

Example 816. *Let* $\mathbf{K}$ *be a field and let* $\mathbf{S}$ *be a unitary* $\mathbf{K}$*-algebra,* finitely generated *over* $\mathbf{K}$*. Then the* $\mathbf{S}$*-module* $\mathbf{S}^* \triangleq \mathrm{Hom}_{\mathbf{K}}(\mathbf{S}, \mathbf{K})$ *is a left and right large injective cogenerator. See ([258], (3.15)) and ([199], (19.31)). Compare with Corollary 648.*

Theorem 817. *(i) Let* W *be a* large injective cogenerator*. Then* $\mathfrak{Beh} \cong \mathfrak{Sys}^{op}$ *is a full subcategory of* $\mathbf{Mod}_{\mathbf{E}}$*.*
(ii) In addition, assume that the ring $\mathbf{A}$ *is left Noetherian. Then* $\mathfrak{Beh}$ *is the full subcategory of* $\mathbf{Mod}_{\mathbf{E}}$*, the objects of which are f.g. submodules of powers* ${}^{n}W$ $(n \in \mathbb{N})$*. The minimal number of* $\mathbf{E}$*-generators of* $\mathfrak{B}_W(M)$ $(M \in \mathfrak{Sys})$ *coincides with the minimal number* m *for which there exists an embedding* $M \hookrightarrow {}^{m}W$*.*

Proof. (i): (1) Since W is a cogenerator, $\mathfrak{Beh} \cong \mathfrak{Sys}^{\mathrm{op}}$ is a subcategory of $\mathbf{Mod}_{\mathbf{E}}$ by Lemma and Definition 813(ii).

(2) Let us prove that when W is a *large injective cogenerator*, the functor $\mathfrak{B}_W : \mathfrak{Sys} \to \mathbf{Mod}_{\mathbf{E}}$ is full. Let $N, M \in \mathfrak{Sys}$ and let $f\bullet : \mathfrak{B}_W(N) \to \mathfrak{B}_W(M)$ be an $\mathbf{E}$-linear right morphism. Since W is a large cogenerator, there exists a finite set of indices I and a monomorphism $\bullet\alpha : N \hookrightarrow W^I$. Consider the short exact sequence

$$0 \longrightarrow N \xrightarrow{\bullet\alpha} W^I \xrightarrow{\bullet\varphi} W^I / \mathrm{im}(\bullet\alpha) \longrightarrow 0$$

where $\bullet\varphi$ is the canonical epimorphism. The module $W^I / \mathrm{im}\,\alpha \in {}_{\mathbf{A}}\mathbf{Mod}$ is not f.g. in general, but since W is a cogenerator, there exists a set of indices J (possibly infinite) and an embedding $\bullet\iota : W^I / \mathrm{im}\,\alpha \hookrightarrow W^J$. Let $\bullet\beta = \bullet\varphi\iota : W^I \to W^J$; then $\bullet\alpha\beta = (\bullet\alpha\varphi) \bullet \iota = 0$, therefore $\ker \beta = \ker \varphi = \mathrm{im}\,\alpha$, and the sequence

$$0 \longrightarrow N \xrightarrow{\bullet\alpha} W^I \xrightarrow{\bullet\beta} W^J \qquad (5.7)$$

is exact. Let $\bullet\pi_i : W^I \twoheadrightarrow W$ be the canonical projection with index $i \in I$ (**1.2.1.2**), and let $\bullet\alpha_i = \bullet\pi_i\alpha$. By Exercise 200(i), since I is finite,

$$\mathrm{Hom}_{\mathbf{A}}\left(\prod_{i\in I} W_i, \prod_{j\in J} W_j\right) \cong \prod_{i,j} \mathrm{Hom}_{\mathbf{A}}\left(W_i, W_j\right)$$

where $W_i = W_j = W$. Let $\bullet\sigma_i : W_i \hookrightarrow W^I$ be the canonical injection with index i (**1.2.1.3**) and let $\bullet\lambda_j : W^J \twoheadrightarrow W$ be the canonical projection with index j; in addition, let

$$\bullet\beta_{ij} = \bullet\sigma_i\beta\lambda_j : W_i \to W_j.$$

The equality $\bullet\alpha\beta = 0$ is equivalent to

$$\sum_{i\in I} \alpha_i\beta_{ij} = 0, \quad j \in J.$$

Since $f\bullet : \mathfrak{B}_W(N) \to \mathfrak{B}_W(M)$ is a homomorphism of right $\mathbf{E}$-modules, $\alpha_i \in \mathfrak{B}_W(N)$ and $\beta_{ij} \in \mathbf{E}$, the above equality yields

$$0 = f\left(\sum_{i\in I} \alpha_i\beta_{ij}\right) = \sum_{i\in I} f(\alpha_i)\,\beta_{ij}, \quad j \in J.$$

Let

$$\bullet\alpha' = \sum_{i\in I} f(\alpha_i)\,\sigma_i : M \to W^I.$$

From the above, for all $j \in J$,

$$\alpha'\beta\lambda_j = \sum_{i\in I} f(\alpha_i)\,\sigma_i\beta\lambda_j = \sum_{i\in I} f(\alpha_i)\,\beta_{ij} = 0,$$

therefore $\alpha'\beta = 0$, i.e., $\mathrm{im}\,\alpha' \subseteq \ker\beta = \mathrm{im}\,\alpha$. Since the sequence (5.7) is exact, there exists $\gamma : M \to N$ such that

$$\gamma\alpha = \alpha' \tag{5.8}$$

(Exercise 674). Let us show that $f = \mathfrak{B}_W(\gamma)$. Since $\alpha = \sum_{i\in I} \alpha_i\sigma_i$, (5.8) is equivalent to

$$\sum_{i\in I} \gamma\alpha_i\sigma_i = \sum_{i\in I} f(\alpha_i)\,\sigma_i,$$

i.e., to $\gamma\alpha_i = f(\alpha_i)$ for all $i \in I$. Let $\bullet\delta : N \to W$. We obtain the diagram below, the row of which is an exact sequence:

$$\begin{array}{ccccc} 0 & \longrightarrow & N & \xrightarrow{\alpha} & W^I \\ & & \delta\downarrow & & \\ & & W & & \end{array}$$

Since W is injective, there exists a morphism $\bullet\varepsilon : W^I \to W$ such that $\alpha\varepsilon = \delta$. Let $\varepsilon_i = \sigma_i \varepsilon$ $(i \in I)$. Then

$$f(\delta) = f(\alpha\varepsilon) = f\left(\sum_{i\in I} \alpha_i \sigma_i \varepsilon\right) = f\left(\sum_{i\in I} \alpha_i \varepsilon_i\right) = \sum_{i\in I} f(\alpha_i)\,\varepsilon_i = \sum_{i\in I} \gamma\alpha_i\varepsilon_i$$
$$= \gamma\delta = \mathfrak{B}_W(\gamma)(\delta),$$

therefore $f = \mathfrak{B}_W(\gamma)$ as desired. This means that $\mathfrak{B}_W : \mathfrak{Sys} \to \mathbf{Mod}_{\mathbf{E}}$ is full.

(ii): If $\mathbf{A}$ is left Noetherian, then $\mathfrak{Sys} = {}_{\mathbf{A}}\mathbf{Mod}^{fg}$ and one can conclude as in ([257], Prop. 3.3) and ([259], Lemma 2.8). ∎

Remark 818. *Statement (iii) in Lemma and Definition 813 and Statement (i) in Theorem 817 still hold if* $\mathfrak{Sys}$ *is changed to any full subcategory* $\mathcal{D}$ *of* ${}_{\mathbf{A}}\mathbf{Mod}^{fg}$ *and* $\mathfrak{Beh}$ *is changed to* $\mathfrak{B}_W(\mathcal{D})$.

5.3 Generalization of the Module-Theoretic Setting

The results in Section 5.2 are now generalized using the theory of categories. This section can be skipped in a first reading, especially by the readers who are not interested in the application in Section 5.7.

5.3.1 Generalized Systems and Associated Behaviors

Let $\mathcal{C}$ a preabelian category and U, W be objects of $\mathcal{C}$. The two definitions below are generalizations of Definitions 805 and 807.

Definition 819. *(i) A U-*system *M is a f.g. U-cokernel (Definitions 500 and 502), i.e., and object M defined by a canonical exact sequence*

$$N \xrightarrow{\bullet\iota} U^k \xrightarrow{\bullet\varphi} M \longrightarrow 0 \tag{5.9}$$

where $N \subseteq U^k$. This means that $M = \operatorname{coker}\iota = U^k/N$.
(ii) The above U-system M is called finitely presented *(f.p.) if the object M is U-finitely presented, i.e., if there exists an exact sequence (3.6):*

$$U^q \xrightarrow{\bullet R} U^k \xrightarrow{\bullet\varphi} M \longrightarrow 0. \tag{5.10}$$

Definition 820. *The W-*behavior *$\mathfrak{B}$ associated with the U-system M defined by (5.9) is the W-kernel associated with the U-cokernel M, i.e., the right $\mathbf{E}$-module (where $\mathbf{E} = \operatorname{End}_{\mathcal{C}}(W)$) defined as*

$$\mathfrak{B} = N^{\perp} \triangleq \left\{\mathbf{w} \in \operatorname{Hom}_{\mathcal{C}}\left(U^k, W\right) : \iota\mathbf{w} = 0\right\} \tag{5.11}$$

where $\iota : N \hookrightarrow U^k$ is the inclusion.

5.3.2 Further Study of Cogenerators

5.3.2.1 Cogenerators for a Subcategory

We will define the notion of a cogenerator for a subcategory of an additive category.

Lemma 821. *Let $\mathcal{C}$ be an additive category with arbitrary products, let M and W be objects of $\mathcal{C}$, and consider the canonical morphism*

$$\bullet\phi_M \triangleq (\hat{w})_{\hat{w}\in \mathrm{Hom}_{\mathcal{C}}(M,W)} : M \to W^{\mathrm{Hom}_{\mathcal{C}}(M,W)}.$$

(1) The following conditions are equivalent:
(i) ϕ_M is a monomorphism;
(ii) There exists a monomorphism $M \hookrightarrow W^J$ for some set of indices J;
(iii) For every object M_1 of $\mathcal{C}$, the arrow map

$$\mathfrak{B}_W : \mathrm{Hom}_{\mathcal{C}}(M_1, M) \mapsto \mathrm{Hom}_{\mathbf{E}}(\mathfrak{B}_W(M), \mathfrak{B}_W(M_1))$$

is injective.
(2) If U is a generator, the above conditions are equivalent to:
(iv) For any nonzero morphism $\bullet h : U \to M$, there exists $\bullet\beta : M \to W$ with $\bullet h\beta \neq 0$.

Proof. (1). (i)$\Rightarrow$(ii): This is clear (take $J = \mathrm{Hom}_{\mathcal{C}}(M,W)$). (ii)$\Rightarrow$(iii): Let $\bullet f : M_1 \to M$ and $\bullet\eta = (\bullet\eta_j)_{i\in J} : M \hookrightarrow W^J$. If $\mathfrak{B}_W(f) = 0$, then $\mathfrak{B}_W(f)(\eta_j) = \bullet f\eta_j = 0$ for all $j \in J$, thus $\bullet f\eta = 0$ and since $\bullet\eta$ is a monomorphism, this implies $f = 0$. (iii)$\Rightarrow$(i): By (3.37), $\mathfrak{B}_W(f) = 0$ if, and only if $f\phi_M = 0$; thus, if $\mathfrak{B}_W$ is injective, then ϕ_M is a monomorphism.

(2). Assume that U is a generator. (ii)$\Rightarrow$(iv): Let $\bullet\eta = (\bullet\eta_j)_{j\in J} : M \hookrightarrow W^J$ and $\bullet h : U \to M$, $h \neq 0$. Then $\bullet h\eta \neq 0$, thus there exists $j \in J$ such that $h\eta_j \neq 0$ and we have found $\beta = \eta_j : M \to W$ such that $\bullet h\beta \neq 0$. (iv)$\Rightarrow$(iii): Let $\bullet f : M_1 \to M$ be such that $\mathfrak{B}_W(f) = 0$, thus $\mathfrak{B}_W(f)(\beta) = f\beta = 0$ for any $\bullet\beta : M \to W$. If $f \neq 0$, there exists $\bullet h_1 : U \to M_1$ such that $h_1 f \neq 0$ whereas $h_1 f\beta = 0$, i.e., $h\beta = 0$ with $h = h_1 f \neq 0$, and (iv) does not hold. ∎

Item (ii) below is a consequence of Lemma 32:

Corollary and Definition 822. *(i) Let $\mathcal{C}$ be an additive category with arbitrary products. An object $W \in \mathcal{C}$ is called a* cogenerator *for a full subcategory $\mathcal{D}$ of $\mathcal{C}$ if the conditions of Lemma 821(1) are satisfied for all objects $M \in \mathcal{D}$.*
(ii) Then $\mathfrak{B}_W(\mathcal{D})$ is a subcategory of $\mathbf{Mod}_{\mathbf{E}}$ *and the faithful functor $\mathfrak{B}_W$ induces an isomorphism $\mathfrak{B}_W : \mathcal{D}^{op} \xrightarrow{\sim} \mathfrak{B}_W(\mathcal{D})$.*

When taking $\mathcal{D} = \mathcal{C}$, one meets the classical definition of a cogenerator again. Item (ii) above generalizes Lemma and Definition 813. The following is a generalization of Corollary 639.

Lemma 823. *Let $\mathcal{D}$ be a full subcategory of the additive category $\mathcal{C}$, let $W \in \mathcal{C}$ be a cogenerator for $\mathcal{D}$, let $U \in \mathcal{D}$ and let $\mathbf{A} \triangleq \mathrm{Hom}_{\mathcal{C}}(U, U)$. Then $\mathfrak{B}_W(U)$ is a faithful left $\mathbf{A}$-module (Definition 613(i)).*

Proof. By Lemma 821(1)(iii), the map

$$\mathfrak{B}_W : \mathbf{A} \to \mathrm{Hom}_{\mathbf{E}}(\mathfrak{B}_W(U), \mathfrak{B}_W(U)), \quad a \mapsto a\bullet$$

(where $a\bullet$ is the left composition by a) is injective, thus, if a is nonzero, $a\bullet$ and $a\mathfrak{B}_W(U)$ are again nonzero, which implies that the left $\mathbf{A}$-module $\mathfrak{B}_W(U)$ is faithful. ∎

Remark 824. *In the situation of Corollary and Definition 822, an object $W \in \mathcal{C}$ is called a* weak injective cogenerator *for $\mathcal{D}$ if the functor $\mathfrak{B}_W : \mathcal{D}^{op} \to \mathbf{Mod}_{\mathbf{E}}$ is faithfully exact [49].*

The notation of (**3.4.3.4**) is used in the following.

Theorem 825. *Let $\mathcal{C}$ be a semiabelian category (Definition 38) and let $U, W \in \mathcal{C}$.*

(1) Let I be a set of indices, let $N \subseteq U^{(I)}$, let $\iota : N \hookrightarrow U^{(I)}$ be the inclusion and let $M = U^{(I)}/N$ be a U-cokernel. The conditions in Lemma 821(1) are equivalent to:
(v) $\mathrm{im}\,\iota = N^{\perp\perp}$.
(2) Let $\mathcal{D}$ be a full subcategory of $\mathcal{C}$, and let $\mathcal{S}_{\mathcal{D}}\left(U^{(I)}\right)$ be the set of all $N \subseteq U^{(I)}$ such that $U^{(I)}/N \in \mathcal{D}$ and $\iota : N \hookrightarrow U^{(I)}$ is a strict morphism (Definition 39); in addition, let $\mathcal{S}_{\mathcal{D}}\left(U^{(I)}\right)^{\perp}$ be the set of all $N^{\perp}$, $N \in \mathcal{S}_{\mathcal{D}}\left(U^{(I)}\right)$. If $W \in \mathcal{C}$ is a cogenerator for $\mathcal{D}$, the Galois connection in Lemma 569 induces a Galois connection

$$\mathcal{S}_{\mathcal{D}}\left(U^{(I)}\right) \longleftrightarrow \mathcal{S}_{\mathcal{D}}\left(U^{(I)}\right)^{\perp} : N \mapsto N^{\perp}, \quad \mathfrak{B} \mapsto \mathfrak{B}^{\perp}$$

which is bijective.

Proof. (1). (i)⇒(v): By (3.38), $\ker(\bullet\varphi\phi_M) = N^{\perp\perp}$. If (i) holds, then $N^{\perp\perp} = \ker\varphi = \ker\mathrm{coker}\,\iota = \mathrm{im}\,\iota$. (v)⇒(i): If (v) holds, then $\ker\varphi = \ker(\varphi\phi_M)$, hence also by (1.6), (1.7)

$$\mathrm{coim}(\varphi\phi_M) = \mathrm{coim}\,\varphi = \mathrm{coker}\ker\mathrm{coker}\,\iota = \mathrm{coker}\,\iota = \varphi.$$

The canonical decomposition of $\varphi\phi_M$ yields $\varphi\phi_M = \mathrm{coim}(\varphi\phi_M)\,g = \varphi g$, thus $\phi_M = g$ which is a monomorphism since $\mathcal{C}$ is semiabelian.

(2). If $\iota : N \hookrightarrow U^{(I)}$ is strict, then $\iota = \mathrm{im}\,\iota$ by Remark 40 and Lemma 34, and by (v), $N = N^{\perp\perp}$ since W is a cogenerator for $\mathcal{D}$, hence $N = \mathfrak{B}^{\perp}$ where $\mathfrak{B} = N^{\perp}$; thus the map $\mathcal{S}_{\mathcal{D}}\left(U^{(I)}\right) \ni N \mapsto N^{\perp} \in \mathcal{S}_{\mathcal{D}}\left(U^{(I)}\right)^{\perp}$ is bijective with inverse $\mathfrak{B} \mapsto \mathfrak{B}^{\perp}$. ∎

Item (2) of Theorem 825 is a generalization of Proposition 810.

In everything that follows, we make the following

Assumption 826. *$\mathcal{C}$ is a semiabelian category and $\mathcal{D}$ is a full subcategory of $\mathcal{C}$, all objects of which are f.g. U-cokernels. The two* $\mathbf{E}$*-modules in Lemma and Definition 568(1) are identified under σ (this is a generalization of the identification in Corollary 567(2)).*

Consider a finitely presented U-system M, defined by the exact sequence (5.10). The W-behavior $\mathfrak{B} \triangleq N^{\perp}$, identified with $\mathfrak{B}_W(M)$, is obtained by applying the functor $\mathfrak{B}_W$ to (5.10) which yields the exact sequence of right morphisms

$$0 \longrightarrow \mathfrak{B}_W(M) \overset{\mathfrak{B}_W(\varphi)}{\longrightarrow} \mathfrak{B}_W\left(U^k\right) \overset{\mathfrak{B}_W(\bullet R)}{\longrightarrow} \mathfrak{B}_W(U^q)$$

with the canonical $\mathbf{E}$-linear isomorphisms $\mathfrak{B}_W\left(U^k\right) \cong {}^kW$ and $\mathfrak{B}_W(U^q) \cong {}^qW$; $\mathfrak{B}_W(\bullet R)$ is identified with the left multiplication by R, written $R\bullet$, and $\mathfrak{B}_W(\varphi)$ with the inclusion $\mathfrak{B}_W(M) \hookrightarrow {}^kW$. We obtain the following, which is a generalization of Theorem 812:

Theorem 827. *Let $M_i = \operatorname{coker}(\bullet R_i)$ be finitely presented U-systems $\left(R_i \in {}^{q_i}\mathbf{A}^k\right)$ and $N_i = \operatorname{im}(\bullet R_i)$ $(i = 1, 2)$, and let $\mathcal{D}$ be a full subcategory of $\mathcal{C}$, all objects of which are finitely presented U-systems.*

(i) If there exists a matrix $X \in {}^{q_1}\mathbf{A}^{q_2}$ such that $R_1 = X R_2$, then $\ker_W(R_2\bullet) \subseteq \ker_W(R_1\bullet)$.
(ii) The converse of (i) holds true if $M_2 \in \mathcal{D}$, U is projective *in $\mathcal{C}$ and $W \in \mathcal{C}$ is a* cogenerator *for $\mathcal{D}$.*
(iii) In particular, if U is projective in $\mathcal{C}$, $\ker_W(R_2\bullet) = \ker_W(R_1\bullet)$, *$M_2 \in \mathcal{D}$ and $W \in \mathcal{C}$ is a cogenerator for $\mathcal{D}$, then there exists an* invertible *matrix X over* $\mathbf{A}$ *such that $R_1 = X R_2$ ("quasi-uniqueness of the matrix of definition").*
(iv) If U is a generator and the converse of (i) holds true whenever $M_2 \in \mathcal{D}$, then $W \in \mathcal{C}$ is a cogenerator for $\mathcal{D}$.

Proof. (i) is obvious. (ii): Since $\mathcal{C}$ is semiabelian, the morphism $\bullet R_i$ can be factored as $\bullet R_i = \beta_i \omega_i$ where $\bullet \beta_i : U^{q_i} \twoheadrightarrow N_i$ is an epimorphism and $\omega_i : N_i \hookrightarrow U^k$ is the inclusion. We have $N_2^{\perp} \subseteq N_1^{\perp}$, thus $N_1 \subseteq N_1^{\perp\perp} \subseteq N_2^{\perp\perp}$ by Lemma 569, and $N_2^{\perp\perp} = N_2$ by Theorem 638(1) and Lemma and Definition 504(ii). Since $N_1 \subseteq N_2$, let $\iota : N_1 \hookrightarrow N_2$ be the inclusion. We have $\bullet R_1 = \gamma \omega_2$ with $\gamma = \beta_1 \iota : U^{q_1} \to N_2$. Since U is projective, so is U^{q_1} too and there exists $\bullet \xi : U^{q_1} \to U^{q_2}$ such that $\gamma = \xi \beta_2$, thus $\gamma \omega_2 = \xi \beta_2 \omega_2$, i.e., $\bullet R_1 = \xi R_2$. Therefore, $R_1 = X R_2$ where $X = \operatorname{Mat}(\xi)$.

(iii) is an obvious consequence of (ii). (iv): Since U is a generator there is an epimorphism $f = (\bullet R_i)_{i \in I} : U^{(I)} \twoheadrightarrow N_2^{\perp\perp}$ where

$$\bullet R_i : U \overset{\iota_i}{\hookrightarrow} U^{(I)} \overset{f}{\twoheadrightarrow} N_2^{\perp\perp} \hookrightarrow U^k$$

i.e., $R_i \in \mathbf{A}^k$, $\operatorname{im}(\bullet R_i) \subseteq N_2^{\perp\perp}$. Then,

$$\ker_W(R_2\bullet) = N_2^{\perp} = N_2^{\perp\perp\perp} \subseteq \operatorname{im}(\bullet R_i)^{\perp} = \ker_W(R_i\bullet).$$

If the converse of (i) holds, there are rows $X_i \in \mathbf{A}^{q_2}$ with $R_i = X_i R_2$ and therefore $\operatorname{im}(\bullet R_i) \subseteq \operatorname{im}(\bullet R_2) = N_2$, hence $f = (\bullet R_i)_{i\in I} : U^{(I)} \to N_2$ and $\operatorname{im} f \subseteq N_2$. Since $N_2^{\perp\perp}$ is a kernel and $f : U^{(I)} \twoheadrightarrow N_2^{\perp\perp}$, we have $N_2^{\perp\perp} = \operatorname{im} f$ and thus also $N_2^{\perp\perp} \subseteq N_2$; finally, $N_2^{\perp\perp} = N_2$, thus W is a cogenerator for $\mathcal{D}$ by Theorem 638(1). ■

We are also led to the following generalization of Lemma and Definition 813:

Lemma and Definition 828. *(i) The category* $\mathfrak{Sys}$ *of* U*-systems is the above category* $\mathcal{D}$*.*
(ii) The class of associated W*-behaviors is* $\mathfrak{Beh} = \mathfrak{B}_W(\mathfrak{Sys})$*. If* $W \in \mathcal{C}$ *is a cogenerator for* $\mathfrak{Sys}$*, then* $\mathfrak{Beh}$ *is a subcategory of* $\mathbf{Mod}_{\mathbf{E}}$ *and* $\mathfrak{Sys}^{op} \cong \mathfrak{Beh}$*.*
(iii) Assume that the category $\mathfrak{Sys}$ *is additive and let* $M_1, M_2 \in \mathfrak{Sys}$*. Then* $\mathfrak{B}_W(M_1) \oplus \mathfrak{B}_W(M_2) \in \mathfrak{Beh}$*. If in addition* $\mathfrak{Sys}$ *is abelian and* $M_1 \subseteq M_2$*, then* $\mathfrak{B}_W(M_2/M_1) \subseteq \mathfrak{B}_W(M_2)$ *is an element of* $\mathfrak{Beh}$*, called a* subbehavior *of* $\mathfrak{B}_W(M_2)$*.*

5.3.2.2 Large Cogenerators

The notion of *large cogenerator* can be generalized as follows:

Definition 829. *Let* $\mathcal{C}$ *be an additive category with arbitrary products and coproducts* (**1.2.4.1**) *and let* U, W *be objects of* $\mathcal{C}$*. Assume that* U *is a generator and that* W *is a cogenerator* (**1.2.2.4**)*. The object* W *is called a* large cogenerator *in* $\mathcal{C}$ *if for every (*U*-)*finitely generated *object* M*, there exists a* finite *set of indices* J *and an embedding* $M \hookrightarrow W^J$*.*

Theorem 817(i) is then generalized as follows (with the same proof):

Theorem 830. *Let* $\mathcal{C}$ *and* U *be as in Definition 829, let* $\mathcal{D}$ *be a full subcategory of* $\mathcal{C}$*, all objects of which are (*U*-)finitely generated, and let* $\mathbf{A} \triangleq \operatorname{Hom}_{\mathcal{C}}(U, U)$*. Let* W *be a* large injective cogenerator *in* $\mathcal{C}$ *and let* $\mathbf{E} \triangleq \operatorname{Hom}_{\mathcal{C}}(W, W)$*. Then* $\mathfrak{B}_W(\mathcal{D}) \cong \mathcal{D}^{\mathrm{op}}$ *is a full subcategory of* $\mathbf{Mod}_{\mathbf{E}}$*.*

5.4 Examples of Linear Systems

5.4.1 LTI Systems

The most simple linear systems are the linear time-invariant (LTI) systems.

5.4.1.1 Continuous-Time Case

In the continuous-time case, the appropriate ring of operators for the definition of LTI systems is $\mathbf{A} = \mathbb{C}[\partial]$ ($\partial = d/dt$), i.e., the ring of linear differential operators with constant coefficients (Subsect. 2.7.2). We know that $\mathbf{A}$ is a commutative strongly Euclidean domain (Exercise 489(ii)), thus $\mathbf{A}$ is a commutative principal ideal domain (Theorem 468).

The primes in $\mathbf{A}$ are the linear polynomials $p_a = \partial - a$ ($a \in \mathbb{C}$), thus every simple $\mathbf{A}$-module is of the form $S_a = \mathbf{A}(p_a)$ whose injective hull is $E(S_a) = \mathbf{A}(p_a^\infty)$ by Proposition 603. Therefore the canonical cogenerator of ${}_{\mathbf{A}}\mathbf{Mod}$ is $W_0 = \bigoplus_{a \in \mathbb{C}} \mathbf{A}(p_a^\infty)$ (Corollary and Definition 643), and W_0 is injective by Corollary and Definition 643(i) since $\mathbf{A}$ is Noetherian. By definition of $E(S_a)$, there exists an embedding $S_a \hookrightarrow E(S_a)$, thus there exists an embedding

$$S_a \hookrightarrow W_0. \tag{5.12}$$

For any $a \in \mathbb{C}$ and $n \geq 1$, let $C_{n,a}$ be the $\mathbb{C}$-vector space generated by the n functions

$$t \mapsto t^{k-1} e^{a\,t} \quad (1 \leq k \leq n)$$

defined in a nonempty open interval I of the real line; $C_{n,a}$ is an $\mathbf{A}$-module since ∂ is an endomorphism of $C_{n,a}$. Consider the map

$$\psi : \mathbf{A} \to C_{n,a} : r(\partial) \mapsto r(\partial)\, t^{n-1} e^{a\,t}$$

(with a slight abuse of notation); it is an $\mathbf{A}$-linear epimorphism and $\ker \psi = \mathbf{A}\, p_a^n$ (i.e., $\operatorname{Ann}^{\mathbf{A}}(C_{n,a}) = \mathbf{A}\, p_a^n$: see Definition 505). Therefore, $C_{n,a} \cong \mathbf{A}(p_a^n)$ by Noether's first isomorphism theorem (Theorem 52(1)), thus

$$W_0 \cong \sum_{n \geq 1,\, a \in \mathbb{C}} C_{n,a} = \bigoplus_{a \in \mathbb{C}} \mathbb{C}[t]\, e^{a\,t} \triangleq \mathcal{PE}(I).$$

The space $\mathcal{PE}(I)$, called the space of *polynomial-exponential functions*, is identified with the canonical (injective) cogenerator W_0.

Let $\mathcal{O}(I)$ (resp., $C^\infty(I)$, $\mathcal{D}'(I)$) be the space of complex-valued analytic functions (resp., indefinitely differentiable functions, distributions) on the interval I. We have with the usual identifications

$$\mathcal{PE}(I) \subsetneq \mathcal{O}(I) \subsetneq C^\infty(I) \subsetneq \mathcal{D}'(I)$$

and by Corollary and Definition 643(ii), all the above spaces (which are $\mathbf{A}$-modules) are cogenerators. Let W be any of the three spaces $\mathcal{O}(I), C^\infty(I)$ and $\mathcal{D}'(I)$.

Let $p(\partial) \in \mathbf{A}$ be a nonzero differential operator and consider the equation

$$p(\partial)\,\mathbf{y} = \mathbf{u}(t), \quad t \in I \tag{5.13}$$

where $\mathbf{u}$ is given. As is well known ([37], n°IV.2.3 and IV.2.8; [322], p. 213), for any $\mathbf{u} \in W$, there exists $\mathbf{y} \in W$ for which (5.13) holds. In other words, W is divisible. Therefore, W is injective by Proposition 598(ii).

Consider the differential equation (5.13) with $\mathbf{u} = 0$. Every solution $\mathbf{y}$ belongs to $\mathcal{PE}(I)$ ([37], n°IV.2.8). Let $\mathbf{f} \in W \backslash \mathcal{PE}(I)$. Then, for any $p(\partial) \in \mathbf{A}$, $p(\partial)\,\mathbf{f}$ is nonzero. The map

$$\iota_{\mathbf{f}} : \mathbf{A} \to W : p(\partial) \longmapsto p(\partial)\,\mathbf{f} \tag{5.14}$$

is $\mathbf{A}$-linear and $\ker \iota_{\mathbf{f}} = \{0\}$. Therefore, $\iota_{\mathbf{f}}$ is an embedding. In addition, by (5.12), for any $a \in \mathbb{C}$ there exists an embedding $S_a \hookrightarrow W$. By Theorem 815(3) we obtain the following (with the above notation).

Theorem 831. *W is a large injective cogenerator.*

Remark 832. *The statement of the above theorem is still valid when "continuous multidimensional systems", also called "continuous nD systems" are considered, i.e., when $\mathbf{A} = \mathbb{C}[\partial_1, ..., \partial_n]$ ($\partial_i = \partial/\partial x_i$), and when I is changed to a nonempty open subset Ω of $\mathbb{R}^n$ [258]. The proof of this result is however much more difficult.*

5.4.1.2 Discrete-Time Case

In the discrete-time case, the suitable ring of operators for the definition of LTI systems is $\mathbf{A} = \mathbb{C}[\mathbf{q}]$ where $\mathbf{q}$ is the shift-forward operator, i.e., the ring of linear difference operators with constant coefficients (Subsect. 2.7.3). As above, $\mathbf{A}$ is a commutative strongly Euclidean domain, thus is a commutative principal ideal domain.

The primes in $\mathbf{A}$ are the linear polynomials $p_a = \mathbf{q} - a$ ($a \in \mathbb{C}$) and the rationale in (**5.4.1.1**) can be repeated.

For any $a \in \mathbb{C}$ and $n \geq 1$, let $C_{n,a}$ be the $\mathbb{C}$-vector space generated by the n sequences $k \mapsto \binom{j+k-1}{j} a^k$ ($1 \leq j \leq n$) where $\binom{n}{p}$ is the binomial coefficient. We have again $\mathrm{Ann}^{\mathbf{A}}(C_{n,a}) = \mathbf{A}\, p_a^n$, thus the canonical cogenerator of ${}_{\mathbf{A}}\mathbf{Mod}$ is

$$W_0 \cong \sum_{n \geq 1,\, a \in \mathbb{C}} C_{n,a} \triangleq \mathcal{AG}$$

where $\mathcal{AG}$ is the space of *arithmetic-geometric sequences*. This is a *injective cogenerator*. So is $\mathbb{C}^{\mathbb{N}}$ too since $\mathcal{AG} \subsetneq \mathbb{C}^{\mathbb{N}}$ and $\mathbb{C}^{\mathbb{N}}$ is divisible. In addition, $\mathbb{C}^{\mathbb{N}}$ can be embedded in $\mathbb{C}^{\mathbb{Z}}$ which is again divisible.

Let $p(\mathbf{q}) \in \mathbf{A}$. Then, every solution to $p(\mathbf{q})\,\mathbf{y} = 0$ belongs to $\mathcal{AG}$. We can deduce as above the following.

Theorem 833. *$\mathbb{C}^{\mathbb{N}}$ and $\mathbb{C}^{\mathbb{Z}}$ are large injective cogenerators.*

Remark 834. *The statement of the above theorem is still valid when considering "discrete multidimensional systems", also called "discrete nD systems" are considered, i.e., when* $\mathbf{A} = \mathbb{C}[\mathbf{q}_1, ..., \mathbf{q}_n]$

$$\mathbf{q}_i : a(x) = a(x_1, ..., x_i, ..., x_n) \mapsto a(x_1, ..., x_{i-1}, x_i + 1, x_{i+1}, ..., x_n)$$

and when $\mathbb{C}^{\mathbb{N}}$ *(resp.,* $\mathbb{C}^{\mathbb{Z}}$*) is changed to* $(\mathbb{C}^n)^{\mathbb{N}}$ *(resp.,* $(\mathbb{C}^n)^{\mathbb{Z}}$*) [258].*

5.4.2 LTV Systems (1)

In this subsection, we consider linear time-varying (LTV) systems with coefficients in a field. LTV systems with coefficients in a ring are studied in the next subsection.

5.4.2.1 Ore Fields

The following will be useful:

Lemma and Definition 835. *(1) Let* $A \in \mathbb{R}$ *and let* $g : [A, +\infty)$ *be a complex-valued function. The function* g *is called* moderate *(or* hypoexponential*) if for every* $\alpha > 0$,

$$\lim_{t \to +\infty} e^{-\alpha t} g(t) = 0.$$

(2) Let f *be a real-valued analytic function defined in a neighborhood of* $+\infty$ *in* $\mathbb{R}$*, for which the following conditions hold*

$$\lim_{t \to +\infty} f(t) = +\infty, \quad \dot{f}(t) > 0 \text{ in a neighborhood of } +\infty, \quad \lim_{t \to +\infty} \frac{\dot{f}(t)}{f(t)} = 0.$$

Then f *is moderate.*
(3) A function f *as above is called an* Ore function*. The set of all Ore functions, endowed with the multiplication, is a commutative cancellable monoid* $\mathfrak{O}$ *with the following properties:*
(i) If $f \in \mathfrak{O}$*, then for any real* $a > 0$, $f^a \in \mathfrak{O}$.
(ii) If $f, g \in \mathfrak{O}$ *and the function* $t \mapsto \dot{g}$ *is bounded in a neighborhood of* $+\infty$*, then* $f \circ g \in \mathfrak{O}$.
(iii) If $f \in \mathfrak{O}$ *and* $g \in \mathbb{C}(f)$*, then* g *is moderate and* $\lim_{t \to +\infty} \frac{\dot{g}(t)}{g(t)} = 0$*. If* $g \in \mathbb{R}(f)$*, then* $g(t)$ *has a constant sign as* $t \to +\infty$.
(4) (i) A complex (resp., real) Ore field *is a field* $\mathbb{C}(f)$ *(resp.,* $\mathbb{R}(f)$*) where* $f \in \mathfrak{O}$.
(ii) The set of all complex (resp., real) Ore fields is denoted by $\mathbb{C}(\mathfrak{O})$ *(resp.,* $\mathbb{R}(\mathfrak{O})$*), and* $\mathfrak{O} \subsetneq \mathbb{R}(\mathfrak{O}) \subsetneq \mathbb{C}(\mathfrak{O})$.
(iii) Let $f \in \mathfrak{O}$ *and* $n \geq 1$ *be an integer. Then* $\mathbb{C}(f^{1/n}) \supseteq \mathbb{C}(f)$.
(5) If f *belongs to an Ore field, there exists* $A \in \mathbb{R}$ *such that* $f \in \mathcal{O}((A, +\infty))$*, thus* f *is a representative of a germ* $\overline{f} \in \mathcal{O}_\infty$.

Proof. (2) We have $\dot{f}(t)/f(t) \to 0^+$ as $t \to +\infty$. Therefore, for any $\varepsilon > 0$, there exists a real A such that for any $t \geq A$, $\dot{f}(t)/f(t) \leq \varepsilon$. Therefore, there exists a constant $c > 0$ such that $\ln(f(t)) \leq \varepsilon t + c$, whenever $t \geq A$, i.e.,

$$f(t) \leq \gamma e^{\varepsilon t}.$$

where $\gamma = e^c$. Given $\alpha > 0$, choose $\varepsilon \in (0, \alpha)$. The above inequality implies

$$e^{-\alpha t} f(t) \leq \gamma e^{(\varepsilon - \alpha)t}$$

and $e^{(\varepsilon-\alpha)t} \to 0$ as $t \to +\infty$.

(3) If $f, g \in \mathfrak{O}$, then $fg \in \mathfrak{O}$, as easily seen, thus $(\mathfrak{O}, .)$ is obviously a commutative cancellable monoid.

(i) is easily proved.

(ii): Let $f, g \in \mathfrak{O}$ where $t \mapsto \dot{g}(t)$ is bounded in a neighborhood of $+\infty$. Then $\lim_{t \to +\infty} (f \circ g)(t) = +\infty$ and $d/dt\,(f \circ g)(t) = \dot{f}(g(t))\,\dot{g}(t) > 0$ in a neighborhood of $+\infty$. Last,

$$\frac{d/dt\,(f \circ g)(t)}{(f \circ g)(t)} = \frac{\dot{f}(g(t))}{f(g(t))} \dot{g}(t) \to 0.$$

(iii): Let $f \in \mathfrak{O}$ and $g \in \mathbb{C}(f)$. The case where g is a constant is trivial, thus let us assume that g is not a constant. There exist finite set of indices I and J and complex numbers a_i, b_j $(i \in I, j \in J)$, not necessarily distinct, such that

$$g = \frac{\prod_{j \in J} (f - b_j)}{\prod_{i \in I} (f - a_i)}$$

and the above rational function is irreducible. Taking the logarithmic derivative,

$$\frac{\dot{g}}{g} = \sum\nolimits_{j \in J} \frac{\dot{f}}{f - b_j} - \sum\nolimits_{i \in I} \frac{\dot{f}}{f - a_i}.$$

Therefore, $\lim_{t \to +\infty} \dot{g}(t)/g(t) = 0$. In addition,

$$\frac{\dot{g}}{g} = \frac{\dot{f}}{f} \left(\sum_{j \in J} \sum_{m=0}^{+\infty} \left(\frac{b_j}{f} \right)^m - \sum_{i \in I} \sum_{n=0}^{+\infty} \left(\frac{a_i}{f} \right)^n \right).$$

The right-hand member of this equality is nonzero (for g is not a constant), therefore there exist numbers $c, d \in \mathbb{R}^\times$ and integers $k, l \geq 0$ such that

$$\frac{\dot{g}}{g} = \frac{\dot{f}}{f} \left(\left(\frac{c}{f^k} + o\left(\frac{1}{f^k} \right) \right) + i \left(\frac{d}{f^l} + o\left(\frac{1}{f^l} \right) \right) \right).$$

where $o\left(\frac{1}{f^k}\right)$ and $o\left(\frac{1}{f^l}\right)$ are real-valued functions such that $f^k o\left(\frac{1}{f^k}\right)$ and $f^l o\left(\frac{1}{f^l}\right)$ tend to zero as $t \to +\infty$. Thus therefore, $\operatorname{Re}\left(\dot{g}(t)/g(t)\right) \to 0$ and $\operatorname{Im}\left(\dot{g}(t)/g(t)\right) \to 0$ as $t \to +\infty$, therefore $\dot{g}(t)/g(t) \to 0$ as $t \to +\infty$. In addition, the two functions $\operatorname{Re}(\dot{g}/g)$ and $\operatorname{Im}(\dot{g}/g)$ have a constant sign in a neighborhood of $+\infty$. By ([37], n°V.3.4, Prop. 7), $\int \operatorname{Re}\left(\dot{g}(t)/g(t)\right) dt = o(t)$ and $\int \operatorname{Im}\left(\dot{g}(t)/g(t)\right) dt = o(t)$ as $t \to +\infty$, thus

$$\ln|g(t)| = \int \dot{g}(t)/g(t)\, dt = o(t)$$

and one concludes as in (2).

The second assertion is clear.

(4) (ii): The inclusion $\mathfrak{O} \subseteq \mathbb{C}(\mathfrak{O})$ is obvious. Let $f \in \mathfrak{O}$; then $1/f \notin \mathfrak{O}$.

(iii) is clear. ∎

Example 836. *(a) The function* $t \mapsto t$ *belongs to* $\mathfrak{O}$.
(b) The function $t \mapsto \ln(t)$ *belongs to* $\mathfrak{O}$.
(c) By (a), (b) and property (i), for any reals $a > 0$ *and* $b > 0$*, the function* $t \mapsto t^a (\ln(t))^b$ *belongs to* $\mathfrak{O}$.
(d) By (b) and property (ii), the nth iterate $\ln^{(n)} = \ln \circ \ldots \circ \ln$ *(n terms,* $n \geq 1$*) belongs to* $\mathfrak{O}$.

See Exercise 934.

5.4.2.2 Continuous-Time Case (1)

Consider a differential field $\mathbf{K}$ equipped with the usual derivative $\boldsymbol{\delta} = d/dt$ (Subsect. 2.3.1). Then the ring of linear differential operators $\mathbf{A} = \mathbf{K}[\partial; \boldsymbol{\delta}]$ (where ∂ is an extension of $\boldsymbol{\delta}$) is a principal ideal domain by Theorem 731. The linear systems defined over $\mathbf{A}$ are LTV continuous-time systems with coefficients in $\mathbf{K}$. Consider the following cases:

(a) $\mathbf{K} = \mathbb{C}(t)$.
(b) $\mathbf{K}$ is an Ore field (Lemma and Definition 835(3)).
(c) $\mathbf{K}$ is the field of meromorphic functions on the real line.

Cases (b) and (c) both include case (a), but there is no connection between cases (b) and (c).

In cases (a) or (c), consider the left $\mathbf{A}$-module $\mathcal{E}_d$ consisting of all complex-valued functions $\mathbf{f}$ for which there exists a discrete subset $S_{\mathbf{f}}$ of $\mathbb{R}$ (depending on $\mathbf{f}$) such that $\mathbf{f}$ is defined and indefinitely differentiable in $\mathbb{R} \backslash S_{\mathbf{f}}$. This set is a the union of a countable family of open nonempty intervals I_i.

(1) Let us prove that $\mathcal{E}_d$ is injective. By Proposition 598(ii), we have only to prove that $\mathcal{E}_d$ is divisible. Let $p(\partial) \in \mathbf{A}^{\times}$ be a linear differential

operator. We have to show that for any $\mathbf{u} \in \mathcal{E}_d$, there exists $\mathbf{y} \in \mathcal{E}_d$ such that $p(\partial)\,\mathbf{y} = \mathbf{u}$. Write

$$p(\partial) = \partial^n + a_1\,\partial^{n-1} + ... + a_n \tag{5.15}$$

and let $S_t = S_{\mathbf{u}} \cup S_a$ where S_a is the set of all poles of the coefficients $a_1, ..., a_n$ which belong to the real line; S_t is still a discrete set. It remains to show that the differential equation $p(\partial)\,\mathbf{y} = \mathbf{u}$ has a solution $\mathbf{y} \in \mathcal{E}_d$ in any open nonempty interval $I \subseteq \mathbb{R} \backslash S_t$; but since $\mathbf{u}$ and all coefficients a_i $(1 \leq i \leq n)$ are C^∞ in I, this differential equation has a C^∞ solution in I ([92], Sect. X.8).

(2) Let us prove that $\mathcal{E}_d$ is a cogenerator. By Theorem 642(i), we have to prove that $\mathrm{Hom}_{\mathbf{A}}(S, \mathcal{E}_d) \neq 0$ for every simple left $\mathbf{A}$-module S. By Proposition 532(i), any simple left $\mathbf{A}$-module S is of the form $S \cong \mathbf{A}/\mathbf{A}\,c$ where $c = c(\partial)$ is an atom in $\mathbf{A}$ (thus $c \neq 0$). By Corollary 567, $\mathrm{Hom}_{\mathbf{A}}(S, \mathcal{E}_d) = \mathrm{Hom}_{\mathbf{A}}(\mathrm{coker}_{\mathbf{A}}(\bullet c), \mathcal{E}_d)$ is identified with

$$\ker_{\mathcal{E}_d}(c\bullet) = \{\mathbf{y} \in \mathcal{E}_d : c(\partial)\,\mathbf{y} = 0\}.$$

Let S_a be the discrete set as defined above, and let be I any open nonempty interval such that $I \subseteq \mathbb{R} \backslash S_a$. Since the coefficients of the differential operator $c(\partial)$ are C^∞ in I, the equation $c(\partial)\,y = 0$ has a nonzero C^∞ solution in that interval, and $\mathcal{E}_d$ is a cogenerator. Therefore, we have obtained the following.

Theorem 837. *$\mathcal{E}_d$ is an injective cogenerator.*

Whether $\mathcal{E}_d$ is a *large* cogenerator is an open question.

5.4.2.3 Continuous-Time Case (2)

We consider again the problem in (**5.4.2.2**) but with a different signal space. Assume that $\mathbf{K}$ is an Ore field (case (b)) and that $\mathbf{A} = \mathbf{K}[\partial; \delta]$. Let $\mathcal{O}_\infty$ be the space of germs of analytic functions defined in an open connected neighborhood of $+\infty$ in $\mathbb{R}$, i.e., in an interval I of the form $(A, +\infty)$ for A large enough (Example 20).

(1) Let us prove that $\mathcal{O}_\infty$ is injective, or equivalently that $\mathcal{O}_\infty$ is divisible. Let $p(\partial) \in \mathbf{A}^\times$ be a linear differential operator. We have to show that for any $\mathbf{u} \in \mathcal{O}_\infty$, there exists $\mathbf{y} \in \mathcal{O}_\infty$ such that $p(\partial)\,\mathbf{y} = \mathbf{u}$. Let $A \in \mathbb{R}$; if A is large enough, all coefficients of $p(\partial)$ (which are functions $\mathbf{K} \ni p_i : t \mapsto p_i(t)$) are defined and analytic in $I = (A, +\infty)$. Let $\mathbf{u} \in W$, i.e., $\mathbf{u} \in \mathcal{O}(J)$ where $J = (B, +\infty)$ and $B \geq A$ is "large enough"; consider the differential equation $p(\partial)\,\mathbf{y} = \mathbf{u}(t)$, $t \in J$. This equation has a solution $\mathbf{y} \in \mathcal{O}(J)$ (Theorem and Definition 185(ii)), thus $\mathcal{O}_\infty$ is divisible.

(2) By the same rationale as in (**5.4.2.2**), $\mathcal{O}_\infty$ is a cogenerator.

(3) Consider the analytic function $\mathbf{f} \in \mathcal{O}(\mathbb{R})$ defined (up a multiplicative constant) by

$$\mathbf{f}(t) = e^{\int e^t dt}.$$

As easily shown, as $t \to +\infty$,

$$\frac{\mathbf{f}^{(n)}(t)}{\mathbf{f}(t)} \sim e^{n\,t}.$$

Assume that there exists a nonzero linear differential operator $p(\partial)$, of the form (5.15) with $a_i \in \mathbf{K}$, such that $p(\partial)\,\mathbf{f}(t) = 0$, $t \in (A, +\infty)$, where A is large enough for all coefficients a_i to be analytic in $I = (A, +\infty)$. Then,

$$\frac{\mathbf{f}^{(n)}(t)}{\mathbf{f}(t)} + a_1(t)\,\frac{\mathbf{f}^{(n-1)}(t)}{\mathbf{f}(t)} + ... + a_n(t) = 0, \quad t \in I.$$

By Lemma and Definition 835(2, iii), the left-hand member of the above equality is equivalent to $\frac{\mathbf{f}^{(n)}(t)}{\mathbf{f}(t)} \sim e^{n\,t}$ as $t \to +\infty$, a contradiction. Therefore, the map $\iota_{\mathbf{f}}$ defined as in (5.14) is an embedding $\mathbf{A} \hookrightarrow \mathcal{O}_\infty$, and by Theorem 815(3) we obtain the following.

Theorem 838. *$\mathcal{O}_\infty$ is a large injective cogenerator.*

5.4.2.4 Discrete-Time Case

Let $\mathbf{K} = \mathbb{R}(t)$. Equipped with the usual shift-forward operator $\boldsymbol{\alpha}$, $\mathbf{K}$ is a difference field (Subsect. 2.3.2), thus the ring of linear difference operators $\mathbf{A} = \mathbf{K}[\mathbf{q};\boldsymbol{\alpha}]$ (where $\mathbf{q}$ is an extension of $\boldsymbol{\alpha}$) is a principal ideal domain by Theorem 731. Let us study linear systems defined over $\mathbf{A}$, i.e., LTV discrete-time systems with coefficients in $\mathbf{K}$.

Let $\mathcal{S}_\infty$ be the space of germs of all complex-valued sequences defined in a neighborhood of $+\infty$ in $\mathbb{Z}$, i.e., of all complex-valued sequences $(\mathbf{x}_n)_{n\geq N}$ defined for N large enough. This space $\mathcal{S}_\infty$ is easily seen to be a left $\mathbf{A}$-module.

(1) Let us prove that $\mathcal{S}_\infty$ is divisible, thus injective. Let $p(\mathbf{q}) \in \mathbf{A}^\times$ be a linear difference operator. We have to show that for any $\mathbf{u} \in \mathcal{S}_\infty$, there exists $\mathbf{y} \in \mathcal{S}_\infty$ such that $p(\mathbf{q})\,\mathbf{y} = \mathbf{u}$. Let N be a natural integer; if N is large enough, all coefficients of $p(X)$ (which are functions $\mathbf{K} \ni p_i : t \mapsto p_i(t)$) are defined for any $t \geq N$. Let $\mathbf{u} \in \mathcal{S}_\infty$, i.e., $\mathbf{u} = (\mathbf{u}_n)_{n\geq N}$ where N is "large enough"; consider the difference equation $(p(\mathbf{q})\,\mathbf{y})_n = \mathbf{u}_n$, $n \geq N$. By induction, this equation has a solution $(\mathbf{y}_n)_{n\geq N}$. Therefore, $\mathcal{S}_\infty$ is divisible.

(2) By the same rationale as in (**5.4.2.2**), $\mathcal{S}_\infty$ is a cogenerator.

(3) Let $a > 1$ and consider the sequence $(\mathbf{z}_n)_{n\geq N}$ defined by

$$\mathbf{z}_n = a^{\frac{n(n+1)}{2}}.$$

For any $k \geq 0$,

$$\frac{\mathbf{z}_{N+k}}{\mathbf{z}_N} = a^{k\,N + \frac{k(k-1)}{2}}$$

and by the same rationale as in (**5.4.2.3**), the only difference operator $p(\mathbf{q}) \in \mathbf{A}$ such that $p(\mathbf{q})\,\mathbf{z}(n) = 0$ $(n \geq N)$, is zero. This implies that the **A**-linear map

$$\iota_{\mathbf{z}} : \mathbf{A} \to \mathcal{S}_\infty : p(\mathbf{q}) \mapsto p(\mathbf{q})\,\mathbf{z}$$

is an embedding. By Theorem 815(3) we obtain the following.

Theorem 839. *$\mathcal{S}_\infty$ is a large injective cogenerator.*

5.4.3 LTV Systems (2)

Let **A** be the first Weyl algebra $A_1(\mathbb{C})$ (**2.7.4.3**), or the ring $A_o(\mathbb{C})$ in Exercise 492 with $\Omega = \mathbb{R}$. In both cases, the ring of linear differential operators **A** is a noncommutative Dedekind domain, and is a good candidate for the definition of an LTV system with coefficients in a ring (in the continuous-time case).

5.4.3.1 Difficulties with Classical Signal Spaces

(i) Let $a \in \mathbb{R}$ and consider the module $M = \mathbf{A}/\mathbf{A}(t-a) \neq 0$. The only solution of (5.11) with $R = t - a$ is zero in $\mathcal{E}(\mathbb{R})$ (Example 563(a)), therefore the left **A**-module $\mathcal{E}(\mathbb{R})$ is *not* a cogenerator by Theorem and Definition 43(3) and Lemma 32.

Consider the module $M = \mathbf{A}/\mathbf{A}\left(t^3\partial + 2\right) \neq 0$ $(\partial = d/dt)$. The only solution of (5.6) with $R = t^3\partial + 2$ can be shown to be zero in $\mathcal{D}'(\mathbb{R})$ ([322], (V,6;15)), therefore the left **A**-module $\mathcal{D}'(\mathbb{R})$ is *not* a cogenerator.

(ii) Consider the differential equation

$$P(\partial)\,\mathbf{y} = \varphi, \quad \partial = d/dt \tag{5.16}$$

where $P(\partial) = -\left(1-t^2\right)^2\partial + 2t \in \mathbf{A}$ and $\varphi \in \mathcal{E}(\mathbb{R}) \hookrightarrow \mathcal{D}'(\mathbb{R})$ is the function defined by

$$\varphi(t) = \begin{cases} e^{1/(t^2-1)}, & |t| < 1 \\ 0, & |t| \geq 1. \end{cases}$$

The differential operator P is a continuous linear map $\mathcal{D}'(\mathbb{R}) \longrightarrow \mathcal{D}'(\mathbb{R})$, thus its adjoint P^* is a continuous linear map $\mathcal{D}(\mathbb{R}) \longrightarrow \mathcal{D}(\mathbb{R})$. By Lemma and Definition 360, $P^* = \left(1-t^2\right)^2\partial + 2t$. As easily checked, $P^*\varphi = 0$. Thus, if (5.16) has a solution $\mathbf{T} \in \mathcal{D}'(\mathbb{R})$, i.e., if $P\mathbf{T} = \varphi$, then

$$0 = \langle P^*\varphi, \mathbf{T} \rangle = \langle \varphi, P\mathbf{T} \rangle = \langle \varphi, \varphi \rangle = \int |\varphi(t)|^2\, dt > 0,$$

a contradiction. This proves that (5.16) has no solution in $\mathcal{D}'(\mathbb{R})$, hence the **A**-module $\mathcal{D}'(\mathbb{R})$ is not divisible, and hence not injective. Neither is the **A**-module $\mathcal{E}(\mathbb{R})$.

The above shows that the classical signal spaces $\mathcal{E}(\mathbb{R})$ and $\mathcal{D}'(\mathbb{R})$ are not appropriate, and that a larger space is necessary.

5.4.3.2 The Use of Hyperfunctions

The $\mathbf{A}$-module $\mathcal{B}(\mathbb{R})$ is divisible by Sato's theorem (Theorem and Definition 185(i)), hence injective by Proposition 598. The following shows the importance of hyperfunctions for the study of linear systems with time-varying coefficients.

Theorem 840. *The left $A_o(\mathbb{C})$-module $\mathcal{B}(\mathbb{R})$ is a large injective cogenerator.*

Proof. (1) We have already seen that the left $\mathbf{A}$-module $\mathcal{B}(\mathbb{R})$ is injective.

(2) We have to show that $\mathcal{B}(\mathbb{R})$ is a cogenerator. By Theorem 642, $\mathcal{B}(\mathbb{R})$ has this property if, and only if for any simple left $A_o(\mathbb{C})$-module S, $\mathrm{Hom}_{\mathbf{A}}(S, \mathcal{B}(\mathbb{R})) \neq 0$. By Proposition 528, $S \cong A_o(\mathbb{C})/\mathfrak{m}$ where $\mathfrak{m}$ is a maximal left ideal. By Proposition 457, $\mathfrak{m}$ is a principal left ideal, i.e., $\mathfrak{m} = A_o(\mathbb{C})\, c$ where, by Lemma 84, c is an atom. Two cases must be considered: (a) $d^\circ(c) > 0$, (b) $d^\circ(c) = 0$. In the latter case, $c(\partial)$ must have a real singular point, for if $d^\circ(c) = 0$ and c has no real singular point, c is a unit in $A_o(\mathbb{C})$, thus is not an atom. Therefore, according to Komatzu's theorem (Theorem and Definition 185(ii)),

$$\ker_{\mathcal{B}(\mathbb{R})}(c\bullet) = \{\mathbf{f} \in \mathcal{B}(\mathbb{R}) : c(\partial)\,\mathbf{f}\} \neq 0,$$

and $\mathcal{B}(\mathbb{R})$ is an injective cogenerator.

(3) According to Theorem 815(3), to prove that $\mathcal{B}(\mathbb{R})$ is a *large* cogenerator, it remains to show that there exist a positive integer n and an embedding $A_o(\mathbb{C}) \hookrightarrow \mathcal{B}(\mathbb{R})^n$. Let $a \in \mathbb{R}$ and let $\varphi \in \mathcal{O}(\mathbb{C}\backslash\mathbb{R})$ be the analytic function defined by

$$\varphi(z) = \frac{1}{\sin \frac{1}{z-a}}.$$

Consider the hyperfunction $\mathbf{f} = [\varphi]$ (Subsect. 1.9.1) and let $p(\partial) \in A_o(\mathbb{C})$. If $p(\partial) \neq 0$, one can choose $b > a$ such that the leading coefficient of $p(\partial)$ has no zero in the interval (a, b). Then, any solution to the equation $p(\partial)\,\mathbf{y} = 0$ is analytic on (a, b) (Theorem and Definition 185(ii)). Since $\mathbf{f}$ is singular at $a + \frac{1}{n\pi}$ for any positive integer n, the only differential operator $p(\partial) \in A_o(\mathbb{C})$ such that $p(\partial)\,\mathbf{f} = 0$ is zero, therefore the map $A_o(\mathbb{C}) \to \mathcal{B}(\mathbb{R}) : p(\partial) \mapsto p(\partial)\,\mathbf{f}$ is an embedding. ∎

Part (2) of the above proof is not valid if $A_o(\mathbb{C})$ is replaced by $A_1(\mathbb{C})$. See Exercise 939.

5.5 Linear Systems, Behaviors, and Their Relations

5.5.1 Relations between Behaviors

Let $\mathbf{A}$ be a ring of operators. Consider a relation of the form (3.10), i.e.,

$$\begin{bmatrix} R' & -T' \end{bmatrix} \begin{bmatrix} T \\ R \end{bmatrix} = 0 \tag{5.17}$$

between finite matrices over $\mathbf{A}$, let $W \in {}_{\mathbf{A}}\mathbf{Mod}$ be an *injective module*, and let $\mathbf{E} = \mathrm{End}_{\mathbf{A}}(W)$.

5.5.1.1 Subbehaviors and Quotient Behaviors

Assume that T is square and invertible. Let $M' = [w']_{\mathbf{A}} = [T\,w']_{\mathbf{A}}$ be the left $\mathbf{A}$-module defined up to isomorphism by the equation $R'\,T\,w' = 0$, i.e., $T'\,R\,w' = 0$ which is equivalent to

$$\begin{cases} T'\,v' = 0 \\ v' = R\,w'. \end{cases} \tag{5.18}$$

We have $M' \cong \mathrm{coker}_{\mathbf{A}}(\bullet R')$; setting $N' = [v']_{\mathbf{A}}$ we have $N' \lhd M'$ and w' is only submitted to the relation $T'\,w' = 0$, thus $N' \cong \mathrm{coker}\bullet T'$. Let $M = M'/N'$. Denoting by w_i the canonical image of w'_i in M'/N', we have $M = [w]_{\mathbf{A}}$ and (5.18) shows that w is only submitted to the relation $R\,w = 0$, therefore $M \cong \mathrm{coker}_{\mathbf{A}}(\bullet R)$. This proves the following.

Proposition 841. *(i) Under the above assumptions, the relation (5.17) yields the short exact sequence*

$$0 \longrightarrow \mathrm{coker}_{\mathbf{A}}(\bullet T') \longrightarrow \mathrm{coker}_{\mathbf{A}}(\bullet R') \longrightarrow \mathrm{coker}_{\mathbf{A}}(\bullet R) \longrightarrow 0. \tag{5.19}$$

involving linear systems over $\mathbf{A}$*; therefore, there exists an isomorphism of left* $\mathbf{A}$*-modules*

$$\mathrm{coker}_{\mathbf{A}}(\bullet R) \cong \mathrm{coker}_{\mathbf{A}}(\bullet R') \,/\, \mathrm{coker}_{\mathbf{A}}(\bullet T').$$

(ii) The sequence

$$0 \longrightarrow \ker_W(R\bullet) \longrightarrow \ker_W(R'\bullet) \longrightarrow \ker_W(T'\bullet) \longrightarrow 0 \tag{5.20}$$

is exact; therefore, the exists an isomorphism of right $\mathbf{E}$*-modules*

$$\ker_W(T'\bullet) \cong \ker_W(R'\bullet) \,/\, \ker_W(R\bullet).$$

Corollary 842. *Assume that the ring* $\mathbf{A}$ *is weakly finite, that (5.17) holds and that both matrices* R *and* R' *belong to* ${}^{q}\mathbf{A}^{k}$ *and are left-regular. Then the following conditions are equivalent:*

(i) R and R' are equivalent (Definition 280(i));
(ii) $\operatorname{coker}_{\mathbf{A}}(\bullet R) \cong \operatorname{coker}_{\mathbf{A}}(\bullet R')$*;*
(iii) $\ker_W(R\bullet) \cong \ker_W(R'\bullet)$.

Proof. (i)⇔(ii): By Proposition 841, $\operatorname{coker}_{\mathbf{A}}(\bullet R) \cong \operatorname{coker}_{\mathbf{A}}(\bullet R')$ if, and only if $\operatorname{coker}_{\mathbf{A}}(\bullet T') = 0$, i.e., T' is left-invertible. Since T' is square, T' is left-invertible if, and only if $T' \in \mathrm{GL}_q(\mathbf{A})$ (Exercise 481(2)).
(ii)⇔(iii) is a consequence of the injectivity of W. ∎

5.5.1.2 General Relations

By Proposition 841, there exist an embedding $\operatorname{coker}_{\mathbf{A}}(\bullet R') \hookrightarrow \operatorname{coker}_{\mathbf{A}}(\bullet T' R)$ and an epimorphism $\operatorname{coker}_{\mathbf{A}}(\bullet T' R) \twoheadrightarrow \operatorname{coker}_{\mathbf{A}}(\bullet R)$. The composition of these two maps is studied in the sequel.

Proposition 843. *(i) There exists a left $\mathbf{A}$-linear homomorphism*

$$\tau : \operatorname{coker}_{\mathbf{A}}(\bullet R') \to \operatorname{coker}_{\mathbf{A}}(\bullet R).$$

(ii) Therefore there exists a right $\mathbf{E}$-linear homomorphism

$$\mathfrak{B}_W(\tau) : \ker_W(R\bullet) \to \ker_W(R'\bullet).$$

Proof. (i): Consider the following commutative diagram with exact rows

$$\begin{array}{ccccccc}
\mathbf{A}^{q'} & \xrightarrow{\bullet R'} & \mathbf{A}^{k'} & \xrightarrow{\varphi'} & M' & \longrightarrow & 0 \\
\bullet T' \downarrow & & \bullet T \downarrow & & \tau \downarrow & & \\
\mathbf{A}^{q} & \xrightarrow{\bullet R} & \mathbf{A}^{k} & \xrightarrow{\varphi} & M & \longrightarrow & 0
\end{array}$$

where $M' = \operatorname{coker}_{\mathbf{A}}(\bullet R')$ and $M = \operatorname{coker}_{\mathbf{A}}(\bullet R)$. By (5.17), $\operatorname{im}_{\mathbf{A}}(\bullet R'T) = \operatorname{im}_{\mathbf{A}}(\bullet T'R) \subseteq \operatorname{im}_{\mathbf{A}}(\bullet R)$. Therefore there exists a map $\bullet\tau : M' \to M$ induced by $\bullet T$ with respect to $\operatorname{im}_{\mathbf{A}}(\bullet R')$ and $\operatorname{im}_{\mathbf{A}}(\bullet R)$ (Theorem 145).
(ii): Applying the functor $\mathfrak{B}_W$ to the above diagram, one obtains the commutative diagram with exact rows

$$\begin{array}{ccccccc}
{}^{q'}W & \xleftarrow{R'\bullet} & {}^{k'}W & \xleftarrow{\mathfrak{B}_W(\varphi')} & \mathfrak{B}_W(M') & \longleftarrow & 0 \\
T'\bullet \uparrow & & T\bullet \uparrow & & \mathfrak{B}_W(\tau) \uparrow & & \\
{}^{q}W & \xleftarrow{R\bullet} & {}^{k}W & \xleftarrow{\mathfrak{B}_W(\varphi)} & \mathfrak{B}_W(M) & \longleftarrow & 0
\end{array}$$

∎

Remark 844. *The homomorphism τ can be made more explicit. We have*

$$\bullet\varphi'\tau = \bullet T\varphi.$$

Let $(\varepsilon_i')_{1 \le i \le k'}$ and $(\varepsilon_i)_{1 \le i \le k}$ be the canonical bases of $\mathbf{A}^{k'}$ and $\mathbf{A}^{k}$, respectively, let $w_i' = (\varepsilon_i')\varphi'$ $(1 \le i \le k')$, and let $w_i = (\varepsilon_i)\varphi$ $(1 \le i \le k)$. We have

$$\varepsilon_i'\varphi'\tau = \varepsilon_i' T\varphi \quad (1 \le i \le k')$$

and $\bullet T$ *is the map* $\varepsilon'_i \mapsto \sum_{1 \leq j \leq k} t_{ij}\, \varepsilon_j$*, where* T *is the matrix with entries* t_{ij} *(Lemma and Definition 263). Therefore,*

$$(w'_i)\,\tau = \sum_{1 \leq j \leq k} t_{ij}\, w_j \quad (1 \leq i \leq k')\,.$$

This can be written

$$(w')\,\tau = T\,w \tag{5.21}$$

where w' *and* w *are the columns with entries* w'_i *and* w_i*, respectively, and where* τ *acts on* w' *component-wise.*

Proposition 845. *The following conditions are equivalent:*

(i) (R, T) *is strongly right-coprime;*
(ii) the induced map τ *is an epimorphism* $\operatorname{coker}_{\mathbf{A}}(\bullet R') \twoheadrightarrow \operatorname{coker}_{\mathbf{A}}(\bullet R)$*;*
(iii) $\mathfrak{B}_W(\tau)$ *is an embedding* $\ker_W(R\bullet) \hookrightarrow \ker_W(R'\bullet)$*, and*

$$\mathfrak{X} \triangleq T\left\{\mathbf{w} \in {}^kW : R\,\mathbf{w} = 0\right\} \subseteq \left\{\mathbf{w}' \in {}^{k'}W : R'\,\mathbf{w}' = 0\right\}.$$

If $\mathbf{A}$ *is left Noetherian and* ${}_{\mathbf{A}}W$ *is a large injective cogenerator, then* $\mathfrak{X}$ *is a* W*-behavior.*

Proof. (i)$\Leftrightarrow$(ii): By Theorem 145, $\operatorname{im}\tau = (\operatorname{im}_{\mathbf{A}}(\bullet T))\,\varphi$; therefore, τ is an epimorphism if, and only if for any $y \in \mathbf{A}^k$, there exists $y' \in \mathbf{A}^{k'}$ such that $y'\,T - y \in \operatorname{im}_{\mathbf{A}}(\bullet R)$, i.e., $y'\,T - x\,R = y$ for some $x \in \mathbf{A}^q$. This equality can be written

$$\begin{bmatrix} y' & -x \end{bmatrix} \begin{bmatrix} T \\ R \end{bmatrix} = y$$

and this equality can be satisfied for any y and some y', x if, and only if (R, T) is strongly right-coprime.

(ii)$\Leftrightarrow$(iii): Clearly, $\mathfrak{B}_W(\tau)$ is an embedding $\iota : \ker_W(R\bullet) \hookrightarrow \ker_W(R'\bullet)$. In addition, $(T\bullet) \circ \mathfrak{B}_W(\varphi) = \mathfrak{B}_W(\varphi') \circ \iota$, thus

$$(T\bullet) \circ \mathfrak{B}_W(\varphi)\,(\ker_W(R\bullet)) = \mathfrak{B}_W(\varphi') \circ \iota\,(\ker_W(R\bullet))\,.$$

We have

$$(T\bullet) \circ \mathfrak{B}_W(\varphi)\,(\ker_W(R\bullet)) = \mathfrak{X}$$

and $\iota\,(\ker_W(R\bullet)) \subseteq \ker_W(R'\bullet)$. Last, $\mathfrak{B}_W(\varphi')$ is the inclusion $\ker_W(R'\bullet) \hookrightarrow {}^{k'}W$.

The last statement is a consequence of Theorem 817(ii). Indeed, $\ker_W(R\bullet)$ is a f.g. $\mathbf{E}$-submodule of ${}^kW$, thus so is $T\ker_W(R\bullet)$. ∎

Assume that the ring $\mathbf{A}$ is left coherent. Then by Theorem 514(ii), given two matrices $R \in {}^q\mathbf{A}^k$ and $T \in {}^{k'}\mathbf{A}^k$, there exist matrices $R' \in {}^{q'}\mathbf{A}^{k\prime}$ and $T' \in {}^{q'}\mathbf{A}^q$ such that the sequence

$$\xrightarrow{\bullet S'} \mathbf{A}^{k'+q} \xrightarrow{\bullet S} \tag{5.22}$$

is exact with
$$S = \begin{bmatrix} T \\ R \end{bmatrix}, \quad S' = \begin{bmatrix} R' & -T' \end{bmatrix}. \tag{5.23}$$

Proposition 846. *Assume that the sequence (5.22) is exact. Then:*

(i) the induced map τ is a monomorphism $\operatorname{coker}_{\mathbf{A}}(\bullet R') \hookrightarrow \operatorname{coker}_{\mathbf{A}}(\bullet R)$.
(ii) therefore, $\mathfrak{B}_W(\tau)$ is an epimorphism $\ker_W(R\bullet) \twoheadrightarrow \ker_W(R'\bullet)$, *and*

$$T\left\{\mathbf{w} \in {}^kW : R\,\mathbf{w} = 0\right\} = \left\{\mathbf{w}' \in {}^{k'}W : R'\,\mathbf{w}' = 0\right\}.$$

Proof. (i): By Theorem 145, $\ker \tau = \left((\operatorname{im}_{\mathbf{A}}(\bullet R))\left(\bullet T^{-1}\right)\right)\varphi'$ where $\bullet\varphi'$ is the canonical epimorphism $\mathbf{A}^{k'} \twoheadrightarrow \mathbf{A}^{k'}/\operatorname{im}_{\mathbf{A}}(\bullet R')$. An element $\bar{y}'$ belongs to $\left((\operatorname{im}_{\mathbf{A}}(\bullet R))\left(\bullet T^{-1}\right)\right)\varphi'$ if, and only if $\bar{y}' = y'\varphi'$ where $y'T = xR$ for some $x \in \mathbf{A}^q$; this equality can be written $\begin{bmatrix} y' & -x \end{bmatrix} S = 0$, i.e., $\begin{bmatrix} y' & -x \end{bmatrix} \in \ker_{\mathbf{A}}(\bullet S)$. If $\ker_{\mathbf{A}}(\bullet S) = \operatorname{im}_{\mathbf{A}}(\bullet S')$, there exists $x' \in \mathbf{A}^{q'}$ such that $y' = x'\,R'$ and $x = x'\,T'$; then $y' \in \operatorname{im}_{\mathbf{A}}(\bullet R')$, thus $\bar{y}' = 0$ and $\ker \tau = 0$ as desired.
(ii) is now clear. ∎

The following is now obvious from Theorem 520 and the above.

Corollary 847. *(i) If the sequence (5.22) is exact and (R, T) is strongly right-coprime, then R and R' are left-similar and the induced map τ in Proposition 843(i) is an isomorphism* $\operatorname{coker}_{\mathbf{A}}(\bullet R') \xrightarrow{\sim} \operatorname{coker}_{\mathbf{A}}(\bullet R)$ *(or equivalently, $\mathfrak{B}_W(\tau)$ is an isomorphism* $\ker_W(R\bullet) \xrightarrow{\sim} \ker_W(R'\bullet)$*).*
(ii) Conversely, if R and R' are left-regular and left-similar, then they satisfy a proper comaximal relation (5.17); the sequence (5.22) is exact and (R, T) is strongly right-coprime (which implies that τ and $\mathfrak{B}_W(\tau)$ are isomorphisms).
(iii) To summarize, over a weakly finite ring, *conditions (a), (b), (c) and (d) of Theorem 520 are related as follows:*

$$\begin{array}{ccc} (a) & \Longleftrightarrow & (b) \\ \Downarrow & & \Downarrow \\ (c) & \Longleftrightarrow & (d) \end{array}$$

and all these conditions are equivalent if both R and R' are left-regular. The latter condition can be assumed to be satisfied without loss of generality over a Bézout domain *by Corollary and Definition 651.*

Remark 848. *Assume that τ is an isomorphism. By (5.17), $R'\,T = T'\,R$ and by (5.1), $R'\,T\,w = 0$, therefore by (5.21), $R'\,(w')\,\tau = 0$. Since* $M' = \operatorname{coker}_{\mathbf{A}}(\bullet R')$ *is defined by the equation $R'\,w' = 0$, w' and $(w')\,\tau$ can be identified. Then by (5.21),*
$$w' = T\,w.$$

5.5.1.3 Elimination Problem

Proposition 846 can be applied to the elimination problem. Let $R_1 \in {}^q\mathbf{A}^k, R_2 \in {}^q\mathbf{A}^l$, and consider

$$\mathfrak{Y} = \left\{ \mathbf{w} \in {}^{k}W, \exists \mathbf{z} \in {}^{l}W : R_1 \mathbf{w} = R_2 \mathbf{z} \right\}$$

where $W \in {}_{\mathbf{A}}\mathbf{Mod}$ is injective. As recalled above, if the ring $\mathbf{A}$ is left coherent, there exists a finite matrix T' over $\mathbf{A}$ such that the sequence (5.24) below is exact.

Corollary 849. *Assume that there exists a finite matrix T' over* $\mathbf{A}$ *such that the sequence*

$$\xrightarrow{\bullet T'} A^q \xrightarrow{\bullet R_2} \tag{5.24}$$

is exact. Then $\mathfrak{Y}$ is a W-behavior, and more specifically

$$\mathfrak{Y} = \left\{ \mathbf{w} \in {}^{k}W : T' R_1 \mathbf{w} = 0 \right\}.$$

Proof. Set $R = \begin{bmatrix} R_1 & -R_2 \end{bmatrix} \in {}^{q}\mathbf{A}^{k+l}$ and let $T = \begin{bmatrix} I_k & 0 \end{bmatrix} \in {}^{k}\mathbf{A}^{k+l}$. Then $\mathfrak{Y} = T \ker_W (R\bullet)$. Let S and S' be given by (5.23). The relation (5.10) is equivalent to

$$R' = T\, R_1, \quad T'\, R_2 = 0,$$

and the sequence (5.22) is exact if, and only if $R' = T\, R_1$ and the sequence (5.24) is exact. ∎

5.5.1.4 Autonomy

Let $\mathbf{A}$ be a left Ore domain, let $W \in {}_{\mathbf{A}}\mathbf{Mod}$ be a "signal space", assumed to be *injective*, and let $\mathbf{E} = \mathrm{End}_{\mathbf{A}}(W)$. Let M be a linear system over $\mathbf{A}$ (i.e., $M \in \mathfrak{Sys}$ where $\mathfrak{Sys} = {}_{\mathbf{A}}\mathbf{Mod}^{fp}$) and let $\mathfrak{B} = \mathfrak{B}_W(M)$ be the associated W-behavior (thus, $\mathfrak{B} \in \mathfrak{Beh}$ where $\mathfrak{Beh} = \mathfrak{B}_W(\mathfrak{Sys})$ is a subcategory of $\mathbf{Mod}_{\mathbf{E}}$).

Lemma and Definition 850. *(1) Of the following, (i)⇒(ii)⇒(iii). If W is an* injective cogenerator, *then (ii)⇒(i). If W is a* large injective cogenerator, *then (iii)⇒(ii).*

(i) M is not a torsion module.
(ii) There exists in the category $\mathfrak{Beh}$ an exact sequence.

$$\mathfrak{B} \xrightarrow{\pi} W \longrightarrow 0 \tag{5.25}$$

and then $\pi(\mathfrak{B}) \subseteq W$ is called a free variable *generated by $\mathfrak{B}$ (do not confuse with a free element in a module).*
(iii) There exists in $\mathbf{Mod}_{\mathbf{E}}$ *an exact sequence (5.25).*
(2) The behavior $\mathfrak{B}$ is called nonautonomous *if (ii) holds, and* autonomous *otherwise, i.e., when $\mathfrak{B}$ does not generate any free variable.*
(3) Assume that W is an injective cogenerator.
(iv) $\mathfrak{B}$ is autonomous if, and only if M is a torsion module.
(v) There exists a short exact sequence

$$0 \longrightarrow \mathfrak{B}_c \longrightarrow \mathfrak{B} \longrightarrow \mathfrak{B}_a \longrightarrow 0 \tag{5.26}$$

where $\mathfrak{B}_a = \mathfrak{B}_W(\mathcal{T}(M)) \cong \mathfrak{B}/\mathfrak{B}_c$ *is autonomous. If* $\mathcal{T}(M)$ *is f.g. (in particular, if the module* M *is Noetherian, or if* $\mathbf{A}$ *is a Bézout domain), then* $M/\mathcal{T}(M)$ *is finitely presented –thus is a linear system, i.e. an object of* $\mathfrak{Sys}$*–, and* $\mathfrak{B}_c = \mathfrak{B}_W(M/\mathcal{T}(M))$ *is the* largest nonautonomous subbehavior *of* $\mathfrak{B}$*.*

Proof. (1) (i)⇒(ii): Assume that M is not torsion, and let $x \in M$ be a free element (Definition 253(i); Theorem and Definition 339). Define $\bullet\alpha : \mathbf{A} \to M : a \mapsto a\,x$. Then $\ker\alpha = 0$, i.e., the sequence

$$0 \longrightarrow \mathbf{A} \stackrel{\bullet\alpha}{\longrightarrow} M \tag{5.27}$$

is exact. Since W is injective, the functor $\mathfrak{B}_W(\bullet)$ is exact, therefore the sequence (5.25) is exact with $\pi = \mathfrak{B}_W(\alpha)$, and $\mathfrak{B}$ is nonautonomous.

(ii)⇒(iii) is clear. Assume that (ii) holds and that W is an injective cogenerator. There exists an $\mathbf{E}$-linear map $\pi = \mathfrak{B}_W(\alpha) : \mathfrak{B} \to W$ such that the sequence (5.25) is exact. Therefore, the sequence (5.27) is exact by Theorem 644, and (i) holds.

Assume that (iii) holds and that W is a large injective cogenerator. Then $\mathfrak{Beh}$ is a full subcategory of $\mathbf{Mod}_{\mathbf{E}}$ by Theorem 817(i), thus there exists an $\mathbf{A}$-linear map $\bullet\alpha : \mathbf{A} \to M$ such that $\pi = \mathfrak{B}_W(\alpha)$, and (ii) holds.

(3) (iv) is a consequence of (1). (v): Consider the exact sequence

$$0 \longrightarrow \mathcal{T}(M) \longrightarrow M \longrightarrow M/\mathcal{T}(M) \longrightarrow 0.$$

If M is Noetherian, then $\mathcal{T}(M) \lhd M$ is f.g., and the same holds if $\mathbf{A}$ is a Bézout domain by Theorem 654. If $\mathcal{T}(M) \lhd M$ is f.g., then $M/\mathcal{T}(M)$ is f.p. by Lemma 498(iii). The rest is clear. ∎

5.5.2 Control Systems

5.5.2.1 The Definition of a Control System

Consider a linear system M over a ring $\mathbf{A}$ (Definition 805). To control that system, it is necessary to distinguish its input variables u_i $(1 \leq i \leq m)$ and its output variables y_i $(1 \leq i \leq p)$ among all its variables w_j $(1 \leq j \leq k)$. The finite sequences $u = (u_i)_{1 \leq i \leq m}$ and $y = (y_i)_{1 \leq i \leq p}$ are identified with the columns with entries u_i and y_i, respectively.

(1) Our first requirement is that the input variables can be freely chosen. To clarify this, consider a signal space $W \in {}_{\mathbf{A}}\mathbf{Mod}$ which is an *injective cogenerator* (such a signal space is not necessarily "classical", but by Corollary and Definition 643(iii), we know that it always exists). That the input variables can be freely chosen means that there exists an isomorphism of right $\mathbf{E}$-modules (where $\mathbf{E} = \mathrm{End}_{\mathbf{A}}(W)$) $\mathfrak{B}_W([u]_{\mathbf{A}}) \cong {}^m W$, or in other words that there exists an exact sequence of right $\mathbf{E}$-modules

$$0 \longrightarrow \mathfrak{B}_W([u]_{\mathbf{A}}) \longrightarrow {}^m W \longrightarrow 0.$$

Since W is an injective cogenerator, by Theorem 644(i) this is equivalent to the exactness of the sequence

$$0 \longrightarrow \mathbf{A}^m \longrightarrow [u]_{\mathbf{A}} \longrightarrow 0.$$

This means that $[u]_{\mathbf{A}} \cong \mathbf{A}^m$, i.e., the module $[u]_{\mathbf{A}}$ is free of rank m.

(2) Our second requirement is that, once all input variables have been fixed, there does not remain any free variable. By Lemma and Definition 850, this happens if, and only if the quotient $M/[u]_{\mathbf{A}}$ is torsion (assuming that $\mathbf{A}$ is a left Ore domain and that W is an injective cogenerator). Thus we are led to the following.

Definition 851. *Let $\mathbf{A}$ be a left Ore domain with division ring of left fractions $\mathbf{F}$. Let M be a linear system over $\mathbf{A}$, and let u_i $(1 \leq i \leq m)$ and y_i $(1 \leq i \leq p)$ be elements of M such that (a) the module $[u]_{\mathbf{A}}$ is free of rank m, (b) the module $M/[u]_{\mathbf{A}}$ is torsion. Then the triple (M, u, y) is called a* control (linear) system.

Remark 852. *Fliess [115] called the pair (M, u) a* dynamics. *In the above definition, we assume that both the input- and output-variables have been chosen when we speak of a control system.*

In everything that follows, $\mathbf{A}$ is a left Ore domain with division ring of left fractions $\mathbf{F}$.

5.5.2.2 Transfer Matrix

Consider the functor $\mathfrak{L} = \mathbf{F} \bigotimes_{\mathbf{A}} -$. Let (M, u, y) be a control system over $\mathbf{A}$, let $\hat{M} = \mathfrak{L}(M)$ and for any element x of M, set $\hat{x} = 1 \otimes x \in \hat{M}$. The sequence

$$0 \longrightarrow [u]_{\mathbf{A}} \longrightarrow M \longrightarrow M/[u]_{\mathbf{A}} \longrightarrow 0$$

is exact, and since the right $\mathbf{A}$-module $\mathbf{F}$ is flat (Theorem 609), the covariant functor $\mathfrak{L}$ is exact. Since $M/[u]_{\mathbf{A}}$ is torsion, $\mathfrak{L}(M/[u]_{\mathbf{A}}) = 0$ (Theorem and Definition 339(ii)), therefore the sequence below is exact:

$$0 \longrightarrow [\hat{u}]_{\mathbf{F}} \longrightarrow \hat{M} \longrightarrow 0.$$

This means that $\hat{M} = [\hat{u}]_{\mathbf{F}}$. By Theorem and Definition 339(iv), $\hat{u}$ is a basis of the left $\mathbf{F}$-vector space $\hat{M}$.

Corollary and Definition 853. *(i) There exists a uniquely defined matrix $G \in {}^p\mathbf{F}^m$ such that $\hat{y} = G\,\hat{u}$, and G is called the* transfer matrix *(or the* transfer function *if $m = p = 1$) of the control system (M, u, y).*
(ii) The functor $\mathfrak{L}$ is called the Laplace functor.
(iii) The left $\mathbf{F}$-vector space $\hat{M} = \mathfrak{L}(M)$ is called the transfer vector space *of the system M.*

Example 854. *The transfer function of the control system (5.4) is*

$$G = \left(t\partial^2 + 1\right)^{-1}(1+t).$$

5.5.2.3 Rosenbrock Descriptions

Since $\mathbf{A}$ is a left Ore domain, it is weakly finite (Exercise 481(4)).

Definition 855. *(1) A* Rosenbrock description *over* $\mathbf{A}$ *is a set of two equations of the form*

$$\begin{cases} D\,\xi = N\,u, \\ y = Q\,\xi + W\,u \end{cases} \tag{5.28}$$

where $\xi = (\xi_i)_{1\leq i\leq r}$ *is called the* partial state, $D \in {}^{p}\mathbf{A}^{r}$ $N \in {}^{p}\mathbf{A}^{m}$, $Q \in {}^{p}\mathbf{A}^{r}$, $W \in {}^{p}\mathbf{A}^{m}$. *The matrix*

$$P = \begin{bmatrix} D & -N \\ Q & W \end{bmatrix}$$

is called the system matrix.
(2) The Rosenbrock description (5.28) is said to be in input-output form *if* $y = \xi$, *i.e.,* $Q = I_p$ *and* $W = 0$.

Lemma and Definition 856. *(1) The Rosenbrock description (5.28) represents a control system if, and only if (a) D is right-regular, and (b)* $\ker_{\mathbf{A}}(\bullet D) \subseteq \ker_{\mathbf{A}}(\bullet N)$.
If D is square and regular, then conditions (a) and (b) are both satisfied.
(2) The Rosenbrock description is called admissible *if both conditions (a) and (b) are satisfied.*

Proof. (i) The second equality of (5.28) means that the elements y_i $(1 \leq i \leq p)$ belong to $M = [\xi, u]_{\mathbf{A}}$. Therefore, we have only to consider the first equality.

(ii) With the notation in (**5.5.2.2**), the left $\mathbf{F}$-vector space $\mathfrak{L}(M/[u]_{\mathbf{A}})$ is equal to $[\bar{\xi}]_{\mathbf{F}}$ which is defined by $D\bar{\xi} = 0$. As recalled above, $M/[u]_{\mathbf{A}}$ is torsion if, and only if $\mathfrak{L}(M/[u]_{\mathbf{A}}) = 0$, i.e., $D\bar{\xi} = 0 \Rightarrow \bar{\xi} = 0$. This holds true if, and only if D is left-invertible over $\mathbf{F}$, i.e., D is right-regular (Proposition 420).

(iii) The remaining condition is that one must have $[u]_{\mathbf{A}} \cong \mathbf{A}^m$. Choose an injective cogenerator W. Then this condition is equivalent to $\mathfrak{B}_W([u]_{\mathbf{A}}) \cong {}^{m}W$. By the first equation of (5.28),

$$\mathfrak{B}_W([u]_{\mathbf{A}}) = \left\{\mathbf{u} \in {}^{m}W; \exists \boldsymbol{\xi} \in {}^{r}W : N\,\mathbf{u} = D\boldsymbol{\xi}\right\}.$$

The determination of $\mathfrak{B}_W([u]_{\mathbf{A}})$ from this equality is an elimination problem (**5.5.1.3**). Let $R = \begin{bmatrix} N & -D \end{bmatrix}$. The sequence (5.24) is exact if, and only if

$$\operatorname{im}_{\mathbf{A}}(\bullet T') = \ker_{\mathbf{A}}(\bullet D), \tag{5.29}$$

without assuming that $\ker_{\mathbf{A}}(\bullet D)$ is f.g. (see Remark 515), i.e., that the matrix T' has a finite number of rows. Then

$$\mathfrak{B}_W([u]_{\mathbf{A}}) = \{\mathbf{u} \in {}^m W : T' N \, \mathbf{u} = 0\}.$$

Thus, $\mathfrak{B}_W([u]_{\mathbf{A}}) \cong {}^m W$ if, and only if $T'N = 0$, i.e.,

$$\operatorname{im}_{\mathbf{A}}(\bullet T') \subseteq \ker_{\mathbf{A}}(\bullet N). \tag{5.30}$$

(5.29) and (5.30) yield condition (b).

The last statement is clear. ∎

Corollary 857. *Let (5.28) be an admissible Rosenbrock description. The transfer matrix of the corresponding control system is*

$$\boxed{G = Q\, D^{\dagger}\, N + W} \tag{5.31}$$

where $D^{\dagger}$ is any left-inverse of D over $\mathbf{F}$.

Proof. By condition (a) of Lemma and Definition 856, D has a left-inverse $D^{\dagger}$ over $\mathbf{F}$ (Proposition 420). Therefore, $\hat{\xi} = D^{\dagger} \hat{u}$ and $\hat{y} = G\, \hat{u}$ where G is given by (5.31). ∎

Remark 858. *(1) Our definition of admissibility of a Rosenbrock description does not correspond to that given in, e.g., [258], [372]. Let $W \in {}_{\mathbf{A}}\mathbf{Mod}$ be an injective module. In those references, the Rosenbrock description (5.28) is called admissible when*

$$\mathfrak{B}_{io} = \left\{ \begin{bmatrix} \mathbf{u} \\ \mathbf{y} \end{bmatrix} \in {}^{m+p}W, \exists \boldsymbol{\xi} \in {}^r W : \left\{ \begin{array}{c} D\,\boldsymbol{\xi} = N\,\mathbf{u}, \\ \mathbf{y} = Q\,\boldsymbol{\xi} + W\,\mathbf{u} \end{array} \right\} \right\}$$

is a W-behavior such that the input variable $\mathbf{u}$ *is free. Only the quotient module $[y, u]_{\mathbf{A}} / [u]_{\mathbf{A}}$ (not $M / [u]_{\mathbf{A}}$) is required to be torsion. In Lemma and Definition 856, the Rosenbrock description (5.28) is admissible when*

$$\mathfrak{B} = \left\{ \begin{bmatrix} \mathbf{u} \\ \mathbf{y} \\ \boldsymbol{\xi} \end{bmatrix} \in {}^{m+p+r}W, \left\{ \begin{array}{c} D\,\boldsymbol{\xi} = N\,\mathbf{u}, \\ \mathbf{y} = Q\,\boldsymbol{\xi} + W\,\mathbf{u} \end{array} \right\} \right\}$$

is a W-behavior such that the input $\mathbf{u}$ *is free. See another equivalent formulation in Exercise 940.*

(2) In most references, it is assumed that the matrix $\begin{bmatrix} D & -N \end{bmatrix}$ is left-regular. This happens when the module $M = [\xi, u]_{\mathbf{A}}$ has finite free resolution of length ≤ 1 (**3.5.4.1**). *Every f.g. left* $\mathbf{A}$*-module has this property when the ring* $\mathbf{A}$ *is Noetherian, projective-free, and has global dimension ≤ 1 (Definition 579, and Theorem and Definition 632). When* $\mathbf{A}$ *is a ring of linear differential operators $\mathbf{K}[\partial; \boldsymbol{\alpha}, \boldsymbol{\delta}]$, this happens when* $\mathbf{K}$ *is a* division ring *(Lemma 622, and Theorems 635, 731).*

(3) A Rosenbrock description over a ring of linear differential operators $\mathbf{K}[\partial;\boldsymbol{\alpha},\boldsymbol{\delta}]$ *(where* $\mathbf{K}$ *is, e.g., a Noetherian domain) is called a* polynomial matrix description (PMD).

Consider two control systems (S,u,y) and (S',u,y), the former with admissible Rosenbrock description with partial state ξ and system matrix $P=\begin{bmatrix} D & -N \\ Q & W \end{bmatrix}$, the latter with admissible Rosenbrock description with partial state ξ' and system matrix $P'=\begin{bmatrix} D' & -N' \\ Q' & W' \end{bmatrix}$. Note that $S=[\xi,u]_{\mathbf{A}}$ and $S'=[\xi',u]_{\mathbf{A}}$.

Definition 859. *The system matrices P and P' are said to be* strictly equivalent *if there is an isomorphism $[\xi',u]_{\mathbf{A}}\cong[\xi,u]_{\mathbf{A}}$ fixing $[y,u]_{\mathbf{A}}$ pointwise.*

In other words, two system matrices are strictly equivalent when they are associated with the same control system.

Theorem 860. *Consider two system matrices P and P' with $\rho+r'=\rho'+r$.*

(i) The matrices P and P' are strictly equivalent if there exist matrices M,X,M',X' over $\mathbf{A}$*, of appropriate size, such that*

$$\begin{bmatrix} M' & 0 \\ -X' & I_p \end{bmatrix}\begin{bmatrix} D & -N \\ Q & W \end{bmatrix}=\begin{bmatrix} D' & -N' \\ Q' & W' \end{bmatrix}\begin{bmatrix} M & X \\ 0 & I_m \end{bmatrix}, \tag{5.32}$$

where

(D',M') is strongly left-coprime, (D,M) is strongly right-coprime.

(See Definition 518.)
(ii) If the matrices D and D' are square and regular, then this sufficient condition is necessary as well.

Proof. (1) The matrices $R=\begin{bmatrix} D & -N \end{bmatrix}$ and $R'=\begin{bmatrix} D' & -N' \end{bmatrix}$ are have the same index (Subsect. 3.2.4), thus by Theorem 520 and Corollary 847, there exists an isomorphism $\tau:[\xi',u]_{\mathbf{A}}\xrightarrow{\sim}[\xi,u]_{\mathbf{A}}$ if the matrices R and R' satisfy a comaximal relation, i.e., there exist two matrices $T'=M'$ and $T=\begin{bmatrix} M & X \\ Y & Z \end{bmatrix}$ such that $R'T=T'R$ with the suitable comaximality properties. The converse holds true if D and D' are regular since then R and R' are left-regular.

(2) For $[u]_{\mathbf{A}}$ to be fixed pointwise, we must have $\tau([u]_{\mathbf{A}})=[u]_{\mathbf{A}}$; then there exists an isomorphism $\bar{\tau}:[\xi',u]_{\mathbf{A}}/[u]_{\mathbf{A}}\xrightarrow{\sim}[\xi,u]_{\mathbf{A}}/[u]_{\mathbf{A}}$ induced by τ (Theorem 145). Therefore, denoting by $\bar{\xi}_i'$ and $\bar{\xi}_j$ the canonical images of ξ_i and ξ_j in $[\xi',u]_{\mathbf{A}}/[u]_{\mathbf{A}}$ and $[\xi,u]_{\mathbf{A}}/[u]_{\mathbf{A}}$ respectively, we must have $\left[\bar{\xi}'\right]_{\mathbf{A}}\cong\left[\bar{\xi}\right]_{\mathbf{A}}$ under $\bar{\tau}$. The columns $\bar{\xi}'$ and $\bar{\xi}$ are characterized by the

relations $D'\bar{\xi}' = 0$, $D\bar{\xi} = 0$. By Theorem 520, $\left[\bar{\xi}'\right]_{\mathbf{A}} \cong \left[\bar{\xi}\right]_{\mathbf{A}}$ if D and D' satisfy a comaximal relation

$$\begin{bmatrix} D' & -M' \end{bmatrix} \begin{bmatrix} M \\ D \end{bmatrix} = 0 \tag{5.33}$$

and the converse holds true when D and D' are square and regular. The relation (5.33) is comaximal if, and only if (D, M) is strongly right-coprime (i.e., left-comaximal) and (D', M') is strongly left-coprime (i.e., right-comaximal). These comaximality conditions are assumed to be satisfied in what follows.

(3) By Remark 848, with the suitable identification we have

$$\begin{bmatrix} \xi' \\ u \end{bmatrix} = \begin{bmatrix} M & X \\ Y & Z \end{bmatrix} \begin{bmatrix} \xi \\ u \end{bmatrix}$$

thus $Y = 0$ and $Z = I_m$; in addition, $y = Q\,\xi + W\,u = Q'\,\xi' + W'\,u$. The latter quantity can be expressed as: $Q'\,\xi' + W'\,u = Q'\,(M\,\xi + X\,u) + W'\,u + X'\,(D\,\xi - N\,u)$, thus $Q = Q'\,M + X'\,D$, $W = Q'\,X + W' - X'\,N$, i.e., the equality (5.32) must hold.

(4) Conversely, if (5.32) holds, then (3.10) holds too with T' and T as specified above, and (3.10) is a comaximal relation since (D, M) is strongly right-coprime and (D', M') is strongly left-coprime. In addition, $[u, y]_{\mathbf{A}}$ is fixed pointwise by the isomorphism τ. ■

5.5.3 State-Space Systems

Consider a control system (M, u, y) over a ring of linear generalized differential operators $\mathbf{A} = \mathbf{K}\,[\partial; \boldsymbol{\alpha}, \boldsymbol{\delta}]$ (Definition 355), where $\boldsymbol{\alpha}$ is an automorphism and $\boldsymbol{\delta}$ is an $\boldsymbol{\alpha}$-derivation of the ring $\mathbf{K}$. If $\mathbf{K}$ is a division ring, we know that the left $\mathbf{K}$-vector space $M/\,[u]_{\mathbf{A}}$ (obtained from the left $\mathbf{A}$-module $M/\,[u]_{\mathbf{A}}$ by restriction of the ring of scalars) is finite-dimensional (Theorem 740). The condition that $\mathbf{K}$ be a division ring is however restrictive.

Theorem and Definition 861. *(1) If the left* $\mathbf{K}$*-module* $M/\,[u]_{\mathbf{A}}$ *is finite free, there exist matrices* $F \in \mathrm{Mat}_n(\mathbf{K})$, $G \in {}^{n}\mathbf{K}^{m}$, $H \in {}^{p}\mathbf{K}^{n}$ *and* $J_i \in {}^{p}\mathbf{K}^{m}$ $(0 \leq i \leq \alpha)$ *such that* (M, u, y) *is defined by the equations*

$$\partial x = F\,x + G\,u \tag{5.34}$$

$$y = H\,x + \sum_{i=0}^{\alpha} J_i\,\partial^i\,u. \tag{5.35}$$

(2) Then the control system (M, u, y) *is called a* state-space system, *and (5.34), (5.35) is called a* state-space realization *of that system; (5.34) and (5.35) are called the* state equation *and the output equation, respectively.*

The matrix F (resp., G, H) is called the state *(resp.,* input, output*) matrix, and the matrices J_i $(0 \leq i \leq \alpha)$ are called the* matrices of the direct term. *For short, the state-space system (5.34), (5.35) is denoted by $\left(F,\ G,\ H,\ (J_i)_{0 \leq i \leq \alpha}\right)$.*
(3) This state-space system is said to be proper *(resp.,* strictly proper*) if $J_i = 0$ for $i \geq 1$ (resp., $i \geq 0$).*
(4) The integer $n = \mathrm{rk}_{\mathbf{K}} \left(M / [u]_{\mathbf{A}}\right)$ is called the order *of the control system (M, u, y).*

Proof. Let $\bar{\eta} = (\bar{\eta}_i)_{1 \leq i \leq n}$ be a basis of the $\mathbf{K}$-module $M / [u]_{\mathbf{A}}$. For all $i \in \{1, ..., n\}$, $\partial \bar{\eta}_i \in M / [u]_{\mathbf{A}}$; therefore, there exists a matrix $F \in \mathrm{Mat}_n(\mathbf{K})$ such that $\partial \bar{\eta} = F \bar{\eta}$. There exist n elements $\eta_i \in M$ such that $\bar{\eta}_i = \eta_i + [u]_{\mathbf{A}}$ $(1 \leq i \leq n)$. Therefore, there exist matrices $G_j \in {}^n\mathbf{K}^m$ $(0 \leq j \leq s)$ such that

$$\partial \eta = F \eta + \sum_{j=0}^{s} G_j \partial^j u.$$

Assume that $s \geq 1$ and $G_s \neq 0$, and set

$$\eta^* = \eta - G_s^{\boldsymbol{\beta}} \partial^{s-1} u$$

(where $\boldsymbol{\beta} = \boldsymbol{\alpha}^{-1}$). Using the commutation rule (2.33), which can be written

$$\partial a = a^{\boldsymbol{\alpha}} \partial + a^{\boldsymbol{\delta}}$$

it is easily seen that

$$\partial \eta^* = F \eta^* + \left(\sum_{j=0}^{s-2} G_j \partial^j + G'_{s-1} \partial^{s-1} \right) u,$$

where G'_{s-1} is expressed in function of G_{s-1}, $G_s^{\boldsymbol{\beta}}$ and $G_s^{\boldsymbol{\delta}}$. By induction, this elimination procedure yields the state equation (5.34) which no longer contains derivatives of the input variables. Let $\bar{y}_i = y_i + [u]_{\mathbf{R}}$ $(1 \leq i \leq p)$. As $\bar{y}_i \in M / [u]_{\mathbf{A}}$, there exist matrices $H \in {}^p\mathbf{K}^n$ and $J_i \in {}^p\mathbf{K}^m$ $(0 \leq i \leq \alpha)$ such that (5.35) holds. ∎

5.6 Structural Properties of Linear Systems

5.6.1 Controllability

5.6.1.1 Controllability and Image Representation

Let $M = \mathrm{coker}_{\mathbf{A}} (\bullet R)$ be the linear system defined by the exact sequence (5.3), let $W \in {}_{\mathbf{A}}\mathbf{Mod}$ be a signal space, and let $\mathfrak{B}_W(M) = \ker_W (R \bullet)$ be the W-behavior associated with M. As usual, $\mathbf{E} = \mathrm{End}_{\mathbf{A}}(W)$. Recall that

${}_{\mathbf{A}}\mathbf{Mod}^{fp}$ denotes the full subcategory of ${}_{\mathbf{A}}\mathbf{Mod}$, the objects of which are finitely presented (Notation 497).

Definition 862. *(i) The behavior* $\mathfrak{B}_W(M)$ *is said to have an* image representation *if there exists a matrix* $S \in {}^k\mathbf{A}^r$ *such that*

$$\mathfrak{B}_W(M) = \operatorname{im}_W(S\bullet) \triangleq \left\{\mathbf{w} \in {}^kW : \exists \mathbf{v} \in {}^rW, \mathbf{w} = S\,\mathbf{v}\right\}. \tag{5.36}$$

Then, the variable $\mathbf{v}$ *is called a* potential *of* $\mathfrak{B}_W(M)$ *(see Exercise 949).*
(ii) This image representation is called observable *if* $S\bullet : {}^rW \to {}^kW$ *is injective. Then, the variable* $\mathbf{v}$ *is called a* free output.

Theorem and Definition 863. *Let* M *be a linear system over* $\mathbf{A}$ *and let* $W \in {}_{\mathbf{A}}\mathbf{Mod}$.

(1) If $\bullet S : \mathbf{A}^k \to \mathbf{A}^r$ *is left-invertible, then the image representation (5.36) is observable, and the converse holds true if* W *is a cogenerator for* $\mathcal{D} = {}_{\mathbf{A}}\mathbf{Mod}^{fp}$.
(2) Of the following, (i)⇒(ii)⇒(iii); if W *is a* cogenerator *for* $\mathcal{D} = {}_{\mathbf{A}}\mathbf{Mod}^{fp}$, *then (ii)⇒(i); last, if* W *is a* large injective cogenerator, *then (iii)⇒(i).*
(i) M *is free;*
(ii) $\mathfrak{B}_W(M)$ *has an observable image representation;*
(iii) There exist a positive integer r *and an* $\mathbf{E}$*-linear isomorphism* $\mathfrak{B}_W(M) \cong {}^rW$.
(3) The linear system M *(resp., the behavior* $\mathfrak{B}_W(M)$*) is said to be* freely controllable, *or* flat, *if (i) (resp., (ii)) holds.*

Proof. (1) If $\bullet S : \mathbf{A}^k \to \mathbf{A}^r$ is left-invertible, then the image representation (5.36) is obviously observable. Conversely, consider the exact sequence

$$\mathbf{A}^k \xrightarrow{\bullet S} \mathbf{A}^r \xrightarrow{\psi} N \to 0$$

which yields the exact sequence

$$0 \longrightarrow \mathfrak{B}_W(N) \longrightarrow {}^rW \xrightarrow{S\bullet} {}^kW.$$

If $S\bullet : {}^rW \to \mathfrak{B}_W(M)$ is injective, this implies $\mathfrak{B}_W(N) = 0$, and finally $N = 0$ if in addition W is a cogenerator for $\mathcal{D}$. Therefore, $\bullet S$ is surjective, thus left-invertible by Theorem 514(i).

(2) (a) (i)⇒(ii): If M is free, there exists an isomorphism $\bullet S : \mathbf{A}^r \tilde{\longrightarrow} M$ for some positive integer r, thus by Corollary 567(3), $S\bullet$ is an isomorphism ${}^rW \tilde{\longrightarrow} \mathfrak{B}_W(M)$, and $\mathfrak{B}_W(M) = S({}^rW)$ has an observable image representation.

(ii)⇒(iii): If $\mathfrak{B}_W(M) = \operatorname{im}_W(S\bullet)$ and $S\bullet : {}^rW \to {}^kW$ is injective, then $\mathfrak{B}_W(M) \cong {}^rW$ by Noether's first isomorphism theorem.

(b) If W is a cogenerator for $\mathcal{D} = {}_{\mathbf{A}}\mathbf{Mod}^{fp}$, let us show that (ii)⇒(i):

Assuming that $\mathfrak{B}_W(M)$ has an observable image representation, (5.36) holds with $\bullet S$ left-invertible by (1), thus the sequence

$$0 \longrightarrow {}^rW \xrightarrow{S\bullet} {}^kW \xrightarrow{R\bullet} {}^qW \tag{5.37}$$

is exact in $\mathbf{Mod_E}$ and there exists a matrix $T \in {}^r\mathbf{A}^k$ such that $TS = I_r$. This equality proves that $\bullet S$ is surjective and the exactness of (5.37) proves that $S\bullet$ is injective and $(RS\bullet) = 0$, thus $RS = 0$ since W is a cogenerator for $\mathcal{D}$. Therefore, $\mathbf{A}^qR \subseteq \ker_{\mathbf{A}}(\bullet S)$ and there exists a map $\bullet f$ such that the diagram below is commutative

$$\begin{array}{lll} \mathbf{A}^k & \xrightarrow{\bullet S} & \mathbf{A}^r \\ \downarrow \varphi & f \nearrow & \\ M & & \end{array}$$

where $\bullet\varphi : \mathbf{A}^k \twoheadrightarrow M = \mathbf{A}^k/\mathbf{A}^qR \in {}_{\mathbf{A}}\mathbf{Mod}^{fp}$ is the canonical epimorphism and $\bullet f : M \rightarrow \mathbf{A}^r$ is an epimorphism too since so is $\bullet S$. Thus $\mathfrak{B}_W(M) = \ker_W(R\bullet) = \operatorname{im}_W(S\bullet) \cong {}^rW$ (by exactness of (5.37)) and since the functor $\mathfrak{B}_W$ is left exact, the surjectivity of $\bullet f$ implies the exactness of

$$0 \longrightarrow {}^rW \xrightarrow{\mathfrak{B}_W(\bullet f)} \mathfrak{B}_W(M),$$

thus $\mathfrak{B}_W(\bullet f) : {}^rW \rightarrow \mathfrak{B}_W(M)$ is an embedding. Therefore, $\mathfrak{B}_W(\bullet f)$ is an isomorphism ${}^rW \xrightarrow{\sim} \mathfrak{B}_W(M)$. Now we have

$$Id_{\mathbf{A}^r} = \bullet TS = \bullet T\varphi f = gf, \quad g \triangleq \bullet T\varphi : \mathbf{A}^r \rightarrow M.$$

Thus $\bullet f = \bullet fgf$, hence $\mathfrak{B}_W(\bullet f) = \mathfrak{B}_W(\bullet fg)\,\mathfrak{B}_W(\bullet f)$ which implies $\mathfrak{B}_W(\bullet fg) = Id_{{}^rW}$ for $\mathfrak{B}_W(\bullet f)$ is invertible. Since W is a cogenerator for $\mathcal{D}$, this implies $\bullet fg = Id_{\mathbf{A}^r}$, thus $\bullet f$ is injective; finally, $\bullet f$ an isomorphism $M \xrightarrow{\sim} \mathbf{A}^r$, and M is free.

(c) If W is a large injective cogenerator, let us show that (iii)⇒(i).

If $\mathfrak{B}_W(M) \cong {}^rW$, there exists in $\mathbf{Mod_E}$ an exact sequence

$$0 \longrightarrow \mathfrak{B}_W(M) \xrightarrow{\psi} {}^rW \longrightarrow 0.$$

By Theorem 817(i), $\mathfrak{B}_W(\mathcal{D})$ is a full subcategory of $\mathbf{Mod_E}$, thus there exists an $\mathbf{A}$-linear map $\varphi : \mathbf{A}^r \rightarrow M$ such that $\psi = \mathfrak{B}_W(\varphi)$. By Theorem 644(i), φ is an isomorphism, and (i) holds. ∎

So, freely controllable linear systems, and W-behaviors having an observable image representation, are closely connected, and are clear notions. We still have to give an interpretation of behaviors having an image representation when this representation is nonobservable.

To clarify ideas, let us consider a *continuous multidimensional system* (Remark 832). So, assume that $\mathbf{A} = \mathbb{C}[\partial_1, ..., \partial_n]$ $(\partial_i = \partial/\partial x_i)$ and $W = \mathcal{E}(\mathbb{R}^n)$

(**1.7.3.5**). According to Remark 350(ii) and to Hilbert's basis theorem (Theorem 352), $\mathbf{A}$ is a Noetherian domain. First recall the following.

Lemma 864. *Let $0 \neq a \in \mathbb{C}[\partial_1, ..., \partial_n]$. Then the equation $a(\partial_1, ..., \partial_n)\mathbf{u} = 0$ has no solution, other than zero, in the space $\mathcal{D}(\mathbb{R}^n)$ of indefinitely differentiable functions with compact support.*

Proof. Let $\mathbf{u} \in \mathcal{D}(\mathbb{R}^n) \subseteq \mathcal{E}'(\mathbb{R}^n)$ be nonzero and such that $a(\partial_1, ..., \partial_n)\mathbf{u} = 0$. Then $a(s)\hat{\mathbf{u}}(s) = 0$ where $s = (s_1, ..., s_n)$ and $\hat{\mathbf{u}}$ is the Laplace transform of $\mathbf{u}$. By the Paley-Wiener-Schwartz theorem ([92], (22.18.7)), $\hat{\mathbf{u}}$ is an entire function, thus $\hat{\mathbf{u}} = 0$, which implies $\mathbf{u} = 0$, a contradiction. ■

Definition 865. *Let U be a nonempty open subset of $\mathbb{R}^n$, and let V be a closed subset of $\mathbb{R}^n$, the interior of which contains $\bar{U}$. Let $\mathbf{w} \in {}^k W$. An element $\mathbf{w}' \in {}^k W$ is a* cut-off *of $\mathbf{w}$ with respect to U and V if $\mathbf{w}'$ coincides with $\mathbf{w}$ on U and with zero in V^c, the complementary of V.*

Lemma and Definition 866. *Let $\mathfrak{B} \subseteq {}^k W$ be a right $\mathbf{E}$-module.*

(1) The following conditions are equivalent:
(i) Given any two trajectories $\mathbf{w}_1, \mathbf{w}_2 \in \mathfrak{B}$, and any two nonempty open subsets U_1, U_2 of $\mathbb{R}^n$ with closures $\bar{U}_1$ and $\bar{U}_2$ respectively, such that $\bar{U}_1 \cap \bar{U}_2 = \varnothing$, there exists a trajectory $\mathbf{w} \in \mathfrak{B}$ which coincides with $\mathbf{w}_1$ on U_1 and with $\mathbf{w}_1$ on U_2.
(ii) Given any trajectory $\mathbf{w} \in \mathfrak{B}$, and any sets U and V as in Definition 865, there exists in $\mathfrak{B}$ a cut-off $\mathbf{w}'$ of $\mathbf{w}$ with respect to U and V.
(2) The right $\mathbf{E}$-module $\mathfrak{B}$ is called concatenable *if it satisfies the above equivalent conditions.*

Proof. (1). (i)⇒(ii) is clear, taking $U = U_1$ and $V = U_2^c$.
(ii)⇒(i): Let $\mathbf{w}_1, \mathbf{w}_2 \in \mathfrak{B}$, and consider two nonempty open subsets U_1, U_2 of $\mathbb{R}^n$ with closures $\bar{U}_1$ and $\bar{U}_2$ respectively, such that $\bar{U}_1 \cap \bar{U}_2 = \varnothing$. As $\mathbb{R}^n$ is a metric space, it is a normal topological space ([35], §IX.4), and there exist two closed sets V_1 and V_2, the interiors of which contain U_1 and U_2, respectively. Let $\mathbf{w}_1' \in \mathfrak{B}$ be a cut-off of $\mathbf{w}_1$ with respect to U_1 and V_1, and let $\mathbf{w}_2' \in \mathfrak{B}$ be a cut-off of $\mathbf{w}_2$ with respect to U_2 and V_2. Then $\mathbf{w} = \mathbf{w}_1' + \mathbf{w}_2' \in \mathfrak{B}$; in addition, $\mathbf{w}$ coincides with $\mathbf{w}_1$ on U_1 and with $\mathbf{w}_2$ on U_2. ■

Consider a W-behavior $\mathfrak{B}_W(M)$.

Theorem 867. *The following conditions are equivalent:*

(i) $\mathfrak{B}_W(M)$ has an image representation;
(ii) $\mathfrak{B}_W(M)$ is concatenable;
(iii) $M = \operatorname{coker}_{\mathbf{A}}(\bullet R)$ is torsion-free.

Proof. (i)⇒(ii): Assume that (i) holds. Let $\mathbf{w}_1, \mathbf{w}_2 \in \mathfrak{B}$, and let $\mathbf{v}_1, \mathbf{v}_2 \in {}^r W$ be such that $\mathbf{w}_i = S\,\mathbf{v}_i$ $(i = 1, 2)$. Let U_1, U_2 be two subsets of $\mathbb{R}^n$ with closures $\bar{U}_1$ and $\bar{U}_2$ respectively, such that $\bar{U}_1 \cap \bar{U}_2 = \varnothing$. There exists $\mathbf{v} \in^r W$ such that $\mathbf{v}|_{U_1} = \mathbf{v}_1|_{U1}$ and $\mathbf{v}|_{U_2} = \mathbf{v}_2|_{U_2}$ ([322], Sect. I.2, Theorem II). Therefore, $\mathbf{w} = S\,\mathbf{v}$ is such that $\mathbf{w}|_{U_1} = \mathbf{w}_1|_{U_1}$ and $\mathbf{w}|_{U_2} = \mathbf{w}_2|_{U_2}$. Therefore, (ii) holds.

(ii)⇒(iii): Assume that (iii) does not hold, and consider the matrix $S \in {}^k\mathbf{A}^r$ in Corollary 610. According to that corollary, $\ker_{\mathbf{A}}(\bullet S)/\operatorname{im}_{\mathbf{A}}(\bullet R) = \mathcal{T}(M)$. Let $\lambda \in \ker_{\mathbf{A}}(\bullet S) \setminus \operatorname{im}_{\mathbf{A}}(\bullet R)$. The canonical image $\bar{\lambda}$ of λ in $\ker_{\mathbf{A}}(\bullet S)/\operatorname{im}_{\mathbf{A}}(\bullet R)$ is torsion, thus there exists $a \in \mathbf{A}^{\times}$ such that $a\,\bar{\lambda} = 0$, i.e., $a\,\lambda \in \operatorname{im}_{\mathbf{A}}(\bullet R)$. If $\lambda\,\mathbf{w} = 0$ for all $\mathbf{w} \in \mathfrak{B}_W(M)$, the sequence

$$0 \longrightarrow \mathfrak{B}_W(M) \longrightarrow {}^kW \xrightarrow{\lambda\bullet} W$$

is exact, and since the signal space W is an injective cogenerator (Remark 832), the sequence

$$\mathbf{A} \xrightarrow{\bullet\lambda} \mathbf{A}^k \longrightarrow \mathbf{A}^k/\operatorname{im}_{\mathbf{A}}(\bullet R) \longrightarrow 0$$

is exact by Theorem 644(i); but this means that $\lambda \in \operatorname{im}_{\mathbf{A}}(\bullet R)$, a contradiction. Therefore, $\lambda\,\mathbf{w}$ is not zero for all $\mathbf{w} \in \mathfrak{B}_W(M)$, in other words there exists a nonzero $\mathbf{E}$-linear map

$$\Lambda : \mathfrak{B}_W(M) \to W : \mathbf{w} \mapsto \lambda\,\mathbf{w}.$$

Define

$$\alpha : W \to W : \mathbf{u} \mapsto a\,\mathbf{u}.$$

Let $\mathbf{w} \in \mathfrak{B}$ be a trajectory such that $\Lambda(\mathbf{w}) \neq 0$. Let U be a bounded open subset of $\mathbb{R}^n$ on which $\Lambda(\mathbf{w})$ is nonzero, and let V be a compact subset of $\mathbb{R}^n$, the interior of which contains $\bar{U}$. If $\mathfrak{B}_W(M)$ is concatenable, there exists a cut-off $\mathbf{w}' \in \mathfrak{B}_W(M)$ of $\mathbf{w}$ with respect to U and V. Then $\Lambda(\mathbf{w}')$ is nonzero and has a compact support. In addition,

$$\Lambda(\mathbf{w}') = a\lambda\mathbf{w}' = 0$$

since $\lambda \in \operatorname{im}_{\mathbf{A}}(\bullet R)$ and $\mathbf{w}' \in \ker_W(R\bullet)$, thus $\Lambda(\mathbf{w}') \in \ker\alpha$. This is impossible by Lemma 864. Therefore, (ii) does not hold.

(iii)⇒(i): If (iii) holds, there exist by Proposition 342 a positive integer r and an embedding $M \hookrightarrow \mathbf{A}^r$; thus, by Theorem 514(iii), there exists an exact sequence

$$\mathbf{A}^q \xrightarrow{\bullet R} \mathbf{A}^k \xrightarrow{\bullet S} \mathbf{A}^r.$$

Since W is injective, this yields the exact sequence

$${}^rW \xrightarrow{S\bullet} {}^kW \xrightarrow{R\bullet} {}^qW,$$

i.e., (i) holds. ■

To close the discussion on multidimensional systems, let us give some additional definitions and results. Let $\mathbf{A} = \mathbb{C}[X_1, ..., X_n]$ and let $R \in {}^q\mathbf{A}^k$. In the case of a continuous (resp., discrete) system, the action of the indeterminate X_i on a variable is ∂_i (resp., $\mathbf{q}_i$).

Definition 868. *(i) The matrix R is called* zero left prime *(ZLP) if there is no common zero of all minors* (**2.11.1.2**) *of R in $\mathbb{C}^n$.*
(ii) The matrix R is called weakly zero left prime *(WLP) if the number of common zeros of all minors of R is finite.*
(iii) The matrix R is called minor left prime *(MLP) if the minors of R have no nonconstant common factor.*
(iv) If $\mathrm{rk}_{\mathbf{A}}(R) = q$, the matrix R is called factor left prime *(FLP) if in any factorization $R = L\,R'$ where $L \in \mathrm{Mat}_q(\mathbf{A})$, necessarily $L \in \mathrm{GL}_q(\mathbf{A})$.*
(v) The matrix R is called factor left prime in the generalized sense *(GFLP) if the existence of a factorization $R = D\,R'$ (D not necessarily square) with $\mathrm{rk}_{\mathbf{A}}(R) = \mathrm{rk}_{\mathbf{A}}(R')$ implies the existence of a matrix E such that $R' = E\,R$. (Definitions (iv) & (v) are still valid when $\mathbf{A}$ is any left Ore domain.)*

The proof of the following is easy:

Lemma 869. *Let R be left-regular.*

(1) R is FLP if, and only if it is GLFP.
(2) Let $\mathcal{V}(R)$ be the algebraic set (Definition 386(i)) defined as follows:

$$\mathcal{V}(R) = \{\xi \in \mathbb{C}^n : \mathrm{rk}_{\mathbb{C}}(R(\xi)) < \mathrm{rk}_{\mathbf{A}}(R)\}$$

and let $\dim \mathcal{V}(R)$ be its dimension (Remark 630(ii)). (i) The matrix R is ZLP (resp., WLP, MLP) if, and only if $\dim \mathcal{V}(R) = -1$ (resp., $\dim \mathcal{V}(R) \leq 0$, $\dim \mathcal{V}(R) \leq n-2$).
(ii) Therefore, when $n = 1$ (univariate case), ZLP $\Leftrightarrow$ MLP whereas WLP is always true. When $n = 2$ (bivariate case), ZLP $\Rightarrow$ WLP $\Leftrightarrow$ MLP. When $n \geq 3$, ZLP $\Rightarrow$ WLP $\Rightarrow$ MLP and these conditions are mutually non-equivalent.

Theorem 870. *Let $M = \mathrm{coker}(\bullet R)$ where $R \in {}^q\mathbf{A}^k$ is* left-regular.

(1) The following conditions are equivalent:
(i) The matrix R is ZLP.
(ii) There exists a matrix $X \in {}^k\mathbf{A}^q$ which solves the matrix Bézout equation $R\,X = I_q$.
(iii) $M = \mathrm{coker}_{\mathbf{A}}(\bullet R)$ is a freely controllable system.
(2) The following conditions are equivalent:
(iv) The matrix R is MLP.
(v) The matrix R is FLP.
(vi) $M = \mathrm{coker}_{\mathbf{A}}(\bullet R)$ is torsion-free.

Proof. (1) (i)$\Leftrightarrow$(ii): see ([368], Theorem 2).

(ii)$\Leftrightarrow$(iii): The matrix Bézout equation $RX = I_q$ has a solution if, and only if M is stably-free by Theorem 583. By the Quillen-Suslin theorem (Theorem 581(i)), this holds if, and only if M is free, i.e., freely controllable.

(2) (iv)$\Leftrightarrow$(vi): See, e.g., ([278], Chap. V, Theorem 1.39). (v)$\Leftrightarrow$(vi): See Theorem and Definition 874(1) below. ■

Remark 871. *(1) According to Hilbert's theorem on syzygies (Theorem 636(i)),* $\operatorname{gld} \mathbf{A} = n$ *, and by the Quillen-Suslin theorem (Theorem 581(i)),* $\mathbf{A}$ *is projective-free, thus all f.g.* $\mathbf{A}$*-modules have finite free resolution of length* $\leq n$ (**3.5.5.2**). *Let* $M = \operatorname{coker}_{\mathbf{A}} (\bullet R)$*; the matrix* R *is left-regular when* M *has finite free resolution of length* ≤ 1. *Therefore, the assumption that* R *be left-regular is restrictive when* $n > 1$.
(2) Without assuming that R *is left-regular,* R *is GFLP if, and only if* $M = \operatorname{coker}_{\mathbf{A}} (\bullet R)$ *is torsion-free (Exercise 951).*

Let us consider again the general case. Let $\mathbf{A}$ be an Ore domain (it is not sufficient to assume that $\mathbf{A}$ is a *left* Ore domain for the sequel), let M be a linear system over $\mathbf{A}$, and let $W \in {}_{\mathbf{A}}\mathbf{Mod}$.

Lemma and Definition 872. *(1) Of the following, (i)$\Rightarrow$(ii)$\Rightarrow$(iii). If* W *is an* injective cogenerator, *(ii)$\Rightarrow$(i). If* W *is a* large injective cogenerator, *(iii)$\Rightarrow$(i).*
(i) M *is torsion-free.*
(ii) There exist a positive integer r *and, in the category* $\mathfrak{B}_W(\mathcal{D})$ $(\mathcal{D} = {}_{\mathbf{A}}\mathbf{Mod}^{fp})$*, an exact sequence*

$$^{r}W \xrightarrow{\varphi} \mathfrak{B}_W(M) \longrightarrow 0, \tag{5.38}$$

i.e., there exists an $\mathbf{E}$*-linear epimorphism* $\varphi : {}^{r}W \twoheadrightarrow \mathfrak{B}_W(M)$ *where* $\varphi = \mathfrak{B}_W(\iota)$ *for some* $\mathbf{A}$*-linear monomorphism* ι.
(iii) There exist a positive integer r *and, in* $\mathbf{Mod}_{\mathbf{E}}$*, an exact sequence (5.38), i.e., there exists an* $\mathbf{E}$*-linear epimorphism* $\varphi : {}^{r}W \twoheadrightarrow \mathfrak{B}_W(M)$.
(2) The linear system M *(resp., the behavior* $\mathfrak{B}_W(M)$*) is said to be controllable if (i) (resp., (ii)) holds.*

Proof. (1) By Proposition 342, (i) holds if, and only if there exist a positive integer r and an embedding $\iota : M \hookrightarrow \mathbf{A}^r$, i.e., an exact sequence

$$0 \longrightarrow M \xrightarrow{\iota} \mathbf{A}^r.$$

The rest of the proof is similar to that of, e.g., Lemma and Definition 850(1). ■

By the above proof and by Theorem 514(iii), $M = \operatorname{coker}_{\mathbf{A}} (\bullet R)$ is controllable if, and only if there exists an exact sequence

$$\mathbf{A}^q \xrightarrow{\bullet R} \mathbf{A}^k \xrightarrow{\bullet S} \mathbf{A}^r.$$

Therefore, assuming that W is an injective cogenerator, $\mathfrak{B}_W(M) = \ker_W(R\bullet)$ is controllable if, and only if there exists an exact sequence

$$^rW \xrightarrow{S\bullet} {}^kW \xrightarrow{R\bullet} {}^qW,$$

i.e., $\mathfrak{B}_W(M)$ has an image representation. In addition, if $\mathcal{T}(M)$ is f.g., consider the short exact sequence (5.26) where $\mathfrak{B} = \mathfrak{B}_W(M)$; then the largest nonautonomous subbehavior $\mathfrak{B}_c = \mathfrak{B}_W(M/\mathcal{T}(M)) \subseteq \mathfrak{B}$ is the *controllable part* of $\mathfrak{B}$ and $M/\mathcal{T}(M)$ is the *controllable quotient* of M. The exact sequence (3.51) yields the exact sequence

$$0 \longrightarrow \mathfrak{B}_a \longrightarrow \operatorname{im}_W(S\bullet) \longrightarrow \mathfrak{B} \longrightarrow 0$$

where $\mathfrak{B}_a = \mathfrak{B}_W(\mathcal{T}(M)) \cong \mathfrak{B}/\mathfrak{B}_c$ and $\operatorname{im}_W(S\bullet)$ is controllable, therefore $\mathfrak{B}$ is $\mathfrak{B}_W(\mathcal{D})$-isomorphic (i.e., isomorphic in the category $\mathfrak{B}_W(\mathcal{D})$) to a quotient of the controllable behavior $\operatorname{im}_W(S\bullet)$.

Corollary 873. *Let M be a linear system over an Ore domain $\mathbf{A}$.*

(i) M is freely controllable if, and only if $M \cong \operatorname{coker}_{\mathbf{A}}(\bullet R)$ where the matrix R is completable (Theorem 626 and Corollary 623).
(ii) Assume that $\mathbf{A}$ is a semistably-free ideal domain (Lemma and Definition 666). Then M is controllable if, and only if $M \cong \operatorname{coker}_{\mathbf{A}}(\bullet R)$ where the matrix R is right-invertible (Corollary 623).

Let us add some important points.

Theorem and Definition 874. *Let $M \cong \operatorname{coker}_{\mathbf{A}}(\bullet R)$ be a linear system over an Ore domain $\mathbf{A}$ and assume that $R \in {}^q\mathbf{A}^k$ is left-regular.*

(1) If M is controllable, then R is FLP, and the converse holds true if the torsion submodule $\mathcal{T}(M)$ is f.g.
(2) Of the following, (i)⇔(ii)⇔(iii):
(i) R is right-invertible over $\mathbf{A}$;
(ii) M is stably-free;
(iii) Let U be a right *$\mathbf{A}$-module which is an injective cogenerator (in $\mathbf{Mod}_{\mathbf{A}}$), and let $\boldsymbol{\lambda} \in U^q$. The equality $\boldsymbol{\lambda} R = 0$ implies $\boldsymbol{\lambda} = 0$ ("*generalized Popov-Belevitch-Hautus (PBH) test*").*
(3) The linear system M is said to be strongly controllable *if the equivalent conditions of (2) hold.*
(4) The following conditions are equivalent:
(iv) M is freely controllable;
(v) There exists a matrix $V \in \mathrm{GL}_k(\mathbf{A})$ such that $RV^{-1} = \begin{bmatrix} I_q & 0 \end{bmatrix}$.

Proof. (1) (A) Write $R = LR'$, and set $M = \operatorname{coker}_{\mathbf{A}}(\bullet R)$, $M' = \operatorname{coker}_{\mathbf{A}}(\bullet R')$ and $\Lambda = \operatorname{coker}_{\mathbf{A}}(L)$. Since R is left-regular, so is L too, therefore L is invertible over the field of fractions $\mathbf{F}$ of $\mathbf{A}$ (Proposition 420(iv) and Exercise 481). By Proposition 841(i) one has the short exact sequence

$$0 \longrightarrow \Lambda \longrightarrow M \longrightarrow M' \longrightarrow 0 \tag{5.39}$$

which yields the short exact sequence

$$0 \longrightarrow \hat{\Lambda} \longrightarrow \hat{M} \longrightarrow \hat{M}' \longrightarrow 0$$

since the right **A**-module **F** is flat (Theorem 609). Since L is invertible over **F**, $\hat{\Lambda} = 0$ and $\hat{M} = \hat{M}'$, i.e., M and M' have the same transfer vector space (Corollary and Definition 853(iii)). By Theorem and Definition 339(ii), Λ is torsion, and $\Lambda = 0$ if, and only if $L \in \mathrm{GL}_q(\mathbf{A})$. Since (5.39) is exact, $M' \cong M/\Lambda$. Therefore, if $\Lambda \neq 0$ (i.e., if R is not FLP), M is not torsion-free.

(B) Conversely, assume that $M = \mathrm{coker}_{\mathbf{A}}(\bullet R)$ is not torsion-free. We have the short exact sequence

$$0 \longrightarrow T \longrightarrow M \longrightarrow M' \longrightarrow 0$$

where $T = \mathcal{T}(M) \neq 0$ and $M' \cong M/T$. If T is f.g., then M' is f.p. by Lemma 498(iii), and there exists $R' \in {}^{q}\mathbf{A}^{k}$ such that $M' = \mathrm{coker}_{\mathbf{A}}(\bullet R')$. Let W be an injective cogenerator. Since W is injective, we obtain the short exact sequence

$$0 \longrightarrow \mathfrak{B}_W(M') \longrightarrow \mathfrak{B}_W(M) \longrightarrow \mathfrak{B}_W(T) \longrightarrow 0$$

where $\mathfrak{B}_W(T) \cong \mathfrak{B}_W(M)/\mathfrak{B}_W(M')$; since $T \neq 0$ and W is a cogenerator, $\mathfrak{B}_W(T) \neq 0$, therefore $\mathfrak{B}_W(M') \subsetneq \mathfrak{B}_W(M)$. By Theorem 827(ii), there exists $L \in \mathrm{Mat}_q(\mathbf{A})$ such that $R = L\,R'$. Since $\mathfrak{B}_W(M') \subsetneq \mathfrak{B}_W(M)$, $L \notin \mathrm{GL}_q(\mathbf{A})$, thus R is not FLP.

(2) (i)⇔(ii) is clear by Theorem 583.

(i)⇔(iii): By Theorem 514(i), (i) means that the sequence

$${}^{k}\mathbf{A} \xrightarrow{R\bullet} {}^{q}\mathbf{A} \longrightarrow 0$$

is exact. By Theorem 644(i), this holds true if, and only if $\ker_U(\bullet R) = 0$, which is equivalent to (iii).

(4) (iv)⇒(v): If (iv) holds, then R is right-invertible by Proposition 577(ii) (since M is projective). Therefore, R is completable by Theorem 626, and there exists a matrix $R' \in {}^{k-q}\mathbf{A}^{k}$ such that $V = \begin{bmatrix} R \\ R' \end{bmatrix}$ is invertible. Therefore, $R\,V^{-1} = \begin{bmatrix} I_q & 0 \end{bmatrix}$.

(v)⇒(iv): If (v) holds, let $R' = \begin{bmatrix} 0 & I_{k-q} \end{bmatrix} V$. Then,

$$\begin{bmatrix} R \\ R' \end{bmatrix} V^{-1} = I_k$$

which proves that R is completable. Therefore, M is freely controllable by Corollary 873. ∎

For a behavioral interpretation of strong controllability, we refer to [289]. For a linear system over an Ore domain $\mathbf{A}$,

$$\text{freely controllable} \Rightarrow \text{strongly controllable} \Rightarrow \text{controllable.}$$

If $\mathbf{A}$ is Hermite,

$$\text{freely controllable} \Leftrightarrow \text{strongly controllable.}$$

If $\mathbf{A}$ is a semistably-free ideal domain,

$$\text{strongly controllable} \Leftrightarrow \text{controllable.}$$

If $\mathbf{A}$ is a Bézout domain,

$$\text{freely controllable} \Leftrightarrow \text{controllable.}$$

The problem of seeking torsion elements is solved by the following. Let $\mathbf{A}$ be a ring of operators. Consider a sequence of matrices $R_i \in {}^{k_{i-1}}\mathbf{A}^{k_i}$ $(i \in \mathbb{Z})$ and the associated operators $(R_i \bullet) = \mathcal{D}_i : E_{i-1} \to E_i$ where $E_i = \operatorname{Hom}\left(\mathbf{A}^{k_i}, {}_{\mathbf{A}}\mathbf{A}\right) \cong {}^{k_i}\mathbf{A}_{\mathbf{A}}$.

Definition 875. *The operator $\mathcal{D}_{i+1}$ said to generate all* compatibility conditions *of $\mathcal{D}_i$ (and $\mathcal{D}_{i+1}$ is said to be* parametrized *by $\mathcal{D}_i$) if the sequence*

$$E_{i-1} \xrightarrow{\mathcal{D}_i} E_i \xrightarrow{\mathcal{D}_{i+1}} E_{i+1}$$

is exact.

For any operator $\mathcal{D}_i$, let $\mathcal{D}_i^*$ denote its adjoint $\operatorname{ad}(\mathcal{D}_i)$ (Lemma and Definition 360). Assuming that $\mathbf{A}$ is semiprime Goldie, consider an operator $\mathcal{D}_{i+1}$. By Theorem 514(iii) and Proposition 516(iv), the following conditions are equivalent:

(a) $\mathcal{D}_{i+1}$ is parametrizable by an operator $\mathcal{D}_i$;
(b) the left $\mathbf{A}$-module $M_{i+1} = \operatorname{coker}_{\mathbf{A}} R_{i+1}$ is torsion-free;
(c) the linear system M_{i+1} is controllable.

Assume that the ring $\mathbf{A}$ is coherent. Consider an operator $\mathcal{D}_i : E_{i-1} \to E_i$. By Theorem 514(ii), $\mathcal{D}_i$ parametrizes some operator $\mathcal{D}_{i+1} : E_i \to E_{i+1}$, i.e., one can find $\mathcal{D}_{i+1}$ which generates all compatibility conditions of $\mathcal{D}_i$. Similarly, an operator $\mathcal{D}_{i+1}^*$ parametrizes some operator $\mathcal{D}_i^*$, i.e., one can find $\mathcal{D}_i^*$ which generates all compatibility conditions of $\mathcal{D}_{i+1}^*$.

Theorem 876. *Let $\mathbf{A}$ be a semiprime Goldie coherent ring, consider an operator $\mathcal{D}_{i+1}$, and the following algorithm.*

(I) Construct $\mathcal{D}_{i+1}^* : \mathbf{A}^{k_{i+1}} \to \mathbf{A}^{k_i}$.
(II) Construct $\mathcal{D}_i^* : \mathbf{A}^{k_i} \to \mathbf{A}^{k_{i-1}}$ generating all compatibility conditions of $\mathcal{D}_{i+1}^*$, so that the sequence

$$\mathbf{A}^{k_{i-1}} \xleftarrow{\mathcal{D}_i^*} \mathbf{A}^{k_i} \xleftarrow{\mathcal{D}_{i+1}^*} \mathbf{A}^{k_{i+1}} \tag{5.40}$$

is exact.
(III) Construct $\mathcal{D}_i = \mathrm{ad}\,(\mathcal{D}_i^*) : E_{i-1} \to E_i$.
(IV) Construct an operator $\mathcal{D}'_{i+1} : E_i \to E'_{i+1}$ *generating all compatibility conditions of* $\mathcal{D}_i$.
Then $\mathcal{D}'_{i+1} \circ \mathcal{D}_i = 0$, *and likewise* $\mathcal{D}_{i+1} \circ \mathcal{D}_i = 0$ *by exactness of (5.40), and by Lemma and Definition 360. However,* $\mathcal{D}_{i+1}$ *does not generate all compatibility conditions of* $\mathcal{D}_i$ *in general.*
Let R_{i+1} *(resp.,* R'_{i+1}*) be the matrix of* $\mathcal{D}_{i+1}$ *(resp.,* $\mathcal{D}'_{i+1}$*) and set*

$$R_{i+1}\eta = \zeta, \quad R'_{i+1}\eta = \zeta'.$$

Then, denoting by $\bar{\varsigma}'$ *the canonical image of* ζ' *in* $M_{i+1} = [\eta]_{\mathbf{A}} / [\varsigma]_{\mathbf{A}}$, $[\bar{\varsigma}']_{\mathbf{A}}$ *is the torsion submodule of* M_{i+1}.

Example 877. *Let* $\mathbf{A} = \mathbb{C}[\partial]$ *and consider the control system* M_1 *given by*

$$\begin{bmatrix} \partial^2 & -\partial \end{bmatrix} \begin{bmatrix} w_1 \\ w_2 \end{bmatrix} = 0.$$

(To clarify ideas, one can think of w_1 *as the output and of* w_2 *as the input). Then, with* $i = 0$,

$$\mathcal{D}_1 = \begin{bmatrix} \partial^2 & -\partial \end{bmatrix}, \quad \mathcal{D}_1^* = \begin{bmatrix} \partial^2 \\ \partial \end{bmatrix},$$
$$\mathcal{D}_0^* = \begin{bmatrix} 1 & -\partial \end{bmatrix}, \quad \mathcal{D}_0 = \begin{bmatrix} 1 \\ \partial \end{bmatrix},$$
$$\mathcal{D}'_1 = \begin{bmatrix} \partial & -1 \end{bmatrix}.$$

Let

$$R_1\eta = \begin{bmatrix} \partial^2 & -\partial \end{bmatrix} \begin{bmatrix} \eta_1 \\ \eta_2 \end{bmatrix} = \varsigma,$$
$$R'_1\eta = \begin{bmatrix} \partial & -1 \end{bmatrix} \begin{bmatrix} \eta_1 \\ \eta_2 \end{bmatrix} = \varsigma'$$

i.e., $\zeta = \partial^2\eta_1 - \partial\eta_2 = \partial\varsigma'$. *Since* $P = \partial$ *is not a unit,* $\bar{\varsigma}'$ *is a nonzero torsion element of* M_1, *and the system* M_1 *is noncontrollable.*

Example 878. *Let* $\mathbf{A} = \mathbb{C}[\partial_1, \partial_2, \partial_3]$ *and consider the linear system* $M_1 = \mathrm{coker}_{\mathbf{A}}\,(\bullet R_1)$ *where*

$$R_1 = \begin{bmatrix} 0 & -\partial_3 & \partial_2 \\ \partial_3 & 0 & \partial_1 \end{bmatrix}.$$

Thus, $\mathcal{D}_1 = R \bullet$ *and*

$$\mathcal{D}_1^* = \begin{bmatrix} 0 & -\partial_3 \\ \partial_3 & 0 \\ -\partial_2 & -\partial_1 \end{bmatrix}.$$

To determine $\mathcal{D}_0^*$, *consider a column* ς^* *of two elements, calculate* $\eta^* = \mathcal{D}_1^* \varsigma^*$ *and determine the corresponding compatibility conditions:*

$$\eta^* = \begin{bmatrix} -\partial_3 \varsigma_2^* \\ \partial_3 \varsigma_1^* \\ -\partial_2 \varsigma_1^* - \partial_1 \varsigma_2^* \end{bmatrix}$$

is subject to $\mathcal{D}_0^* \eta^* = 0$ *where* $\mathcal{D}_0^* = \begin{bmatrix} \partial_1 & -\partial_2 & -\partial_3 \end{bmatrix}$. *Therefore,*

$$\mathcal{D}_0 = \begin{bmatrix} -\partial_1 \\ \partial_2 \\ \partial_3 \end{bmatrix}.$$

To determine $\mathcal{D}_1'$, *consider an element* μ, *calculate* $\mathcal{D}_0 \mu = \nu$ *and determine the corresponding compatibility conditions. One obtains* $\mathcal{D}_1' \nu = 0$ *where*

$$\mathcal{D}_1' = \begin{bmatrix} \partial_2 & \partial_1 & 0 \\ 0 & \partial_3 & -\partial_2 \end{bmatrix}.$$

Observe that $\mathcal{D}_1 \circ \mathcal{D}_0 = \mathcal{D}_1' \circ \mathcal{D}_0 = 0$. *Set* $\varsigma = R_1 \eta$ *and* $\varsigma' = R_1' \eta$. *This yields*

$$\begin{cases} \varsigma_1 = -\partial_3 \eta_2 + \partial_2 \eta_3, \\ \varsigma_2 = \partial_3 \eta_1 + \partial_1 \eta_3, \end{cases}$$

$$\begin{cases} \varsigma_1' = \partial_2 \eta_1 + \partial_1 \eta_2, \\ \varsigma_2' = \partial_3 \eta_2 - \partial_2 \eta_3. \end{cases}$$

Therefore, $\varsigma_2' = -\varsigma_1$, *and* $\bar{\varsigma}_2' = 0$. *On the other hand, one can easily check that* $\bar{\varsigma}_1'$ *is only submitted to* $\partial_3 \bar{\varsigma}_1' = 0$. *Therefore,* $\bar{\varsigma}_1' = \partial_2 \bar{\eta}_1 + \partial_1 \bar{\eta}_2$ *is a nonzero torsion element of* M_1 *(with equation* $R_1 \bar{\eta} = 0$*), and* M_1 *is a noncontrollable system.*

5.6.1.2 Controllability of State-Space Systems

Let $\mathbf{K}$ be an Ore domain, $\boldsymbol{\alpha}$ an automorphism and $\boldsymbol{\delta}$ an $\boldsymbol{\alpha}$-derivation of $\mathbf{K}$, and consider the state-space system (M, u, y) with state equation (5.34) over $\mathbf{A} = \mathbf{K}[\partial; \boldsymbol{\alpha}, \boldsymbol{\delta}]$. Assume that $\boldsymbol{\alpha} \boldsymbol{\delta} = \boldsymbol{\delta} \boldsymbol{\alpha}$ (as this always holds in concrete situations), let $\boldsymbol{\beta} = \boldsymbol{\alpha}^{-1}$, and assume that be the subring of constants of $\mathbf{K}$ (Subsect. 2.3.3) is an Ore domain denoted by $\mathbf{C}$.

Theorem and Definition 879. *(1) Consider the matrices*

$$\Gamma_s = \left[P_0 \;\vdots\; P_1 \;\vdots\; \cdots \;\vdots\; P_{s-1} \right] \tag{5.41}$$

defined for $s \geq 1$*, where*

$$P_0 = G, \quad P_{i+1} = (F - \boldsymbol{\delta}\, I_n)\, P_i^{\boldsymbol{\beta}}, \quad i \geq 0, \tag{5.42}$$

and

$$\Gamma_\infty = \left[P_0 \;\vdots\; P_1 \;\vdots\; \cdots \;\vdots\; P_{s-1} \vdots \cdots \right].$$

(i) The matrix Γ_∞ *is right-invertible over* $\mathbf{K}$ *if, and only if there exists an integer* $s \geq 1$ *such that* Γ_s *is already right-invertible over* $\mathbf{K}$*.*
(ii) If $\mathbf{A} = \mathbf{K}[\partial]$ *and* $\mathbf{K}$ *is a* commutative domain*, then* Γ_∞ *is right-invertible over* $\mathbf{K}$ *if, and only if* Γ_n *is right-invertible over* $\mathbf{K}$*.*
(2) If there exists $s \geq 1$ *such that* Γ_s *is right-invertible over* $\mathbf{K}$*, then* M *is strongly controllable.*
(3) Assume that there exists a fundamental matrix $X \in \mathrm{GL}_n(\mathbf{K})$ *such that* $X^{\boldsymbol{\delta}} = F\, X$*. If* M *is strongly controllable, then there exists* $s \geq 1$ *such that* Γ_s *is right-invertible over* $\mathbf{K}$*. If* M *is controllable, then* $\mathrm{rk}_{\mathbf{C}}\, X^{\beta}\, \Gamma_1 = n$*. The state-space system is said to be* Kalman-controllable *when the latter condition holds. Therefore, under the above assumption,*

$$\textit{controllable} \Rightarrow \textit{Kalman-controllable.}$$

(4) If $\mathbf{K}$ *is a field and* $\mathbf{A} = \mathbf{K}[\partial]$*, then the following conditions are equivalent (*Kalman test for controllability*):*
(iii) $\mathrm{rk}_{\mathbf{K}}\, \Gamma_n = n$*;*
(iv) M *is Kalman-controllable;*
(v) M *is freely controllable.*

Proof. (1)(i) If the matrix $\Gamma_\infty \in {}^n\mathbf{K}^{\mathbb{N}}$ is right-invertible, there exists a matrix $\Upsilon_\infty \in {}^{(\mathbb{N})}\mathbf{K}^n$ such that $\Gamma_\infty \Upsilon_\infty = I_n$. The rows of Υ_∞ are all zero except for a finite number of them, therefore the product $\Gamma_\infty \Upsilon_\infty$ involves only a finite number of columns of Γ_∞. As a result, there exists $s \geq 1$ such that Γ_s is right-invertible. The converse is obvious.

(ii): If $\mathbf{A} = \mathbf{K}[\partial]$ where $\mathbf{K}$ is a commutative, then (since $\boldsymbol{\delta} = 0$ and $\boldsymbol{\beta} = 1$)

$$\Gamma_s = \left[G \;\vdots\; F\,G \;\vdots\; \cdots \;\vdots\; F^{s-1}\,G \right];$$

by the Cayley-Hamilton theorem (Theorem 752) and by induction, for any $i \geq 0$, the columns of $F^{n+i}\,G$ are $\mathbf{K}$-linear combinations of those of Γ_{n-1}. It follows that Γ_s is right-invertible for some $s > n$ if, and only if so is already Γ_n.

(2) Equation (5.34) can be put into the form (5.1), setting

$$R = \begin{bmatrix} \partial I_n - F & \vdots & G \end{bmatrix}, \quad w = \begin{bmatrix} x \\ -u \end{bmatrix}.$$

By Theorem and Definition 874, M is strongly controllable if, and only if the two equalities

$$(\alpha) \quad \boldsymbol{\lambda}\,(\partial I_n - F) = 0, \quad (\beta) \quad \boldsymbol{\lambda}\, G = 0,$$

imply $\boldsymbol{\lambda} = 0$. (β) implies

$$0 = \boldsymbol{\lambda}\, G = \boldsymbol{\lambda}\, G\, \partial = \boldsymbol{\lambda}\, G\, \partial^2 = ... = \boldsymbol{\lambda}\, G\, \partial^{r-1} = ...$$

The first equality is equivalent to $\boldsymbol{\lambda}\, P_0 = 0$.

Assume that $\boldsymbol{\lambda}\, P_i = 0$, $i \geq 0$. The commutation rule (2.33) can be written $a\,\partial = \partial\, a^{\boldsymbol{\beta}} - a^{\boldsymbol{\beta\delta}}$, thus the equality $\boldsymbol{\lambda}\, P_i\, \partial = 0$ is equivalent to $\boldsymbol{\lambda}\left(\partial\, P_i^{\boldsymbol{\beta}} - P_i^{\boldsymbol{\beta\delta}}\right) = 0$. Using (α), this yields

$$0 = \boldsymbol{\lambda}\left(F\, P_i^{\boldsymbol{\beta}} - P_i^{\boldsymbol{\beta\delta}}\right) = \boldsymbol{\lambda}\,(F - \boldsymbol{\delta}\, I_n)\, P_i^{\boldsymbol{\beta}} = \boldsymbol{\lambda}\, P_{i+1}.$$

By induction, $\boldsymbol{\lambda}\, \Gamma_s = 0$; if Γ_s is right-invertible, this implies $\boldsymbol{\lambda} = 0$.

(3) Set $V = X^{-1}$. Since $V\,X = I_n$, $(V\,X)^{\boldsymbol{\delta}} = 0$, therefore $V^{\boldsymbol{\alpha}}\, X^{\boldsymbol{\delta}} + V^{\boldsymbol{\delta}}\, X = 0$, and

$$V^{\boldsymbol{\delta}} = -V^{\boldsymbol{\alpha}}\, F.$$

Consider the state equation (5.34) and the new state $\eta = V\,x$. We obtain

$$\begin{aligned} \partial\eta &= V^{\boldsymbol{\alpha}}\, \partial x + V^{\boldsymbol{\delta}} x = V^{\boldsymbol{\alpha}}\,(F\,x + G\,u) + V^{\boldsymbol{\delta}} x \\ &= \left(V^{\boldsymbol{\alpha}}\, F + V^{\boldsymbol{\delta}}\right) x + V^{\boldsymbol{\alpha}}\, G\, u \\ &= \tilde{G}\, u \end{aligned}$$

where $\tilde{G} = V^{\boldsymbol{\alpha}}\, G$. The corresponding "controllability matrix" is

$$\tilde{\Gamma}_s = \begin{bmatrix} \tilde{G} & \vdots & -\tilde{G}^{(\boldsymbol{\delta\beta})} & \vdots & \cdots & \vdots\, (-1)^{s-1}\, \tilde{G}^{(\boldsymbol{\delta\beta})^{s-1}} \end{bmatrix}.$$

(A) If Condition (i) of Theorem and Definition 874 holds, there exist matrices $\tilde{Y} \in {}^m\mathbf{A}^n$ and $\tilde{Z} \in \mathrm{Mat}_n(\mathbf{A})$ such that

$$(\partial I_n)\, \tilde{Z} - \tilde{G}\, \tilde{Y} = -I_n.$$

Write

$$\tilde{Y} = \sum_{i=0}^{s-1} \partial^i\, Y_i, \quad Y_i \in {}^m\mathbf{K}^n$$

and let

$$\tilde{\Upsilon}_s = \begin{bmatrix} Y_0 \\ \vdots \\ Y_{s-1} \end{bmatrix} \in {}^{ms}\mathbf{K}^n$$

By induction, it is easy to prove that

$$\tilde{G}\,\partial^i = (-1)^i\,\tilde{G}^{(\boldsymbol{\beta\delta})^i} - \partial\,M_i(\partial)$$

for some matrix $M_i(\partial) \in {}^n\mathbf{A}^m$. Therefore,

$$(\partial I_n)\,\tilde{Z} = \tilde{G}\,\tilde{Y} - I_n = \sum\nolimits_{i=0}^{s-1} (-1)^i\,\tilde{G}^{(\boldsymbol{\beta\delta})^i}\,Y_i - \partial \sum\nolimits_{i=0}^{s-1} M_i(\partial)\,Y_i - I_n,$$

or equivalently

$$\partial\left(\tilde{Z} - \sum\nolimits_{i=0}^{s-1} M_i(\partial)\,Y\right) = \tilde{\Gamma}_s\,\tilde{\Upsilon}_s - I_n.$$

Therefore, $\tilde{\Gamma}_s\,\tilde{\Upsilon}_s = I_n$, and $\tilde{\Gamma}_s$ is right-invertible over $\mathbf{K}$.

(B) By Theorem 409, $\mathrm{rk}_{\mathbf{C}}\,X^{\beta}\,\Gamma_1 = \mathrm{rk}_{\mathbf{C}}\,\tilde{\Gamma}_1 < n$ if, and only if there exists a nonzero row $\mathbf{v}^* \in \mathbf{C}^n$ such that $\mathbf{v}^*\,\tilde{\Gamma}_1 = 0$; this is equivalent to $\mathbf{v}^*\,\partial\eta = \partial(\mathbf{v}^*\,\eta) = 0$. The element $\mathbf{v}^*\,\eta$ of M is only submitted to the relation $\partial(\mathbf{v}^*\,\eta) = 0$, thus it is a nonzero torsion element of M, and M is not controllable.

(4) (iv)⇔(v) by Theorem 654 since $\mathbf{A}$ is a Bézout domain, and (iii)⇔(iv) is classical (see, e.g., [43]). ■

Remark 880. *For discrete-time systems, $\boldsymbol{\delta} = \boldsymbol{\alpha} - 1$, $\partial = \mathbf{q} - 1$ (**2.7.4.2**), therefore the state equation (5.34) takes the form*

$$\mathbf{q}\,x = \check{F}\,x + G\,u \tag{5.43}$$

where $\check{F} = F + I_n$, and (5.42) becomes

$$P_0 = G, \quad P_{i+1} = \left(\check{F} - \boldsymbol{\alpha}\,I_n\right)\,P_i^{\boldsymbol{\beta}} = \check{F}\,P_i^{\boldsymbol{\beta}} - P_i$$

and Γ_s can be changed to

$$\Gamma'_s = \left[P'_0 \;\vdots\; P'_1 \;\vdots\; \cdots \;\vdots\; P'_{s-1}\right],$$
$$P'_0 = G, \quad P'_{i+1} = \check{F}\,P_i'^{\boldsymbol{\beta}}, \quad i \geq 0.$$

Recall that $\mathbf{k} = \mathbb{R}$ or $\mathbb{C}$.

Proposition 881. *Let Ω be a nonempty open interval of $\mathbb{R}$, let $\mathbf{K}$ be a subring of the ring $\mathcal{O}(\Omega)$ of all analytic $\mathbf{k}$-valued functions in Ω (Subsect. 1.9.1), and let $\mathbf{A} = \mathbf{K}[\partial, \boldsymbol{\delta}]$ where $\boldsymbol{\delta}$ is the usual derivative. Let (M, u, y) be a* *state-space system with state equation (5.34) over* $\mathbf{A}$*, let* $t_0 \in \Omega$*, and assume that there exists a fundamental matrix* $X \in \mathrm{GL}_n(\mathbf{K})$ *such that* $X^{\delta} = F X$ *(this is guaranteed if* $\mathbf{K} = \mathcal{O}(\Omega)$*).*

(i) Then M *is Kalman-controllable if, and only if there exists an integer* $s \geq 1$ *such that* $\mathrm{rk}_{\mathbf{k}}\, \Gamma_s(t_0) = n$*.*
(ii) The latter condition holds if, and only if there exists a subset S *of* Ω*, dense in* Ω*, such that* $\mathrm{rk}_{\mathbf{k}}\, \Gamma_n(t) = n$ *for all* $t \in S$ *(see Exercise 956).*

Proof. If $\mathbf{K} = \mathcal{O}(\Omega)$, there exists a fundamental matrix $X \in \mathrm{GL}_n(\mathbf{K})$ such that $X^{\delta} = F X$ ([92], (10.7.5)). In addition, $\mathbf{C} = \mathbf{k}$.

(i) (a) Assume that for all $s \geq 1$, $\mathrm{rk}_{\mathbf{k}}\, \Gamma_s(t_0) < n$. For any $s \geq 1$, let

$$\begin{aligned} V_s &= \{\mathbf{u}^* \in \mathbf{k}^n \backslash \{0\}, \mathbf{u}^* P_i(t_0) = 0 : 0 \leq i \leq s-1\} \\ &= \{\mathbf{u}^* \in \mathbf{k}^n \backslash \{0\} : \mathbf{v}\, \Gamma_s = 0\} \neq \varnothing, \\ V_\infty &= \textstyle\bigcap_{r \geq 1} V_r = \{\mathbf{u}^* \in \mathbf{k}^n \backslash \{0\} : \mathbf{u}^*\, \Gamma_\infty = 0\}. \end{aligned}$$

Since

$$\mathbf{k}^n \supseteq V_1 \supseteq V_2 \supseteq \ldots$$

and $\mathbf{k}$ is Artinian (Lemma and Definition 317(3)), there exists $s \geq 1$ such that $V_{s+1} = V_s$. Therefore, $V_\infty = V_s$ is nonempty. Let $\mathbf{u}^* \in V_\infty$ and let $\mathbf{v}^* = \mathbf{u}^* X^{-1}(t_0)$ (which makes sense since $X(t_0) \in \mathrm{GL}_n(\mathbf{k})$). The row $\mathbf{v}^*$ is nonzero, and $0 = \mathbf{u}^*\, \Gamma_\infty(t_0) = \mathbf{v}^*\, \tilde{\Gamma}_\infty(t_0)$. Therefore, $\mathbf{v}^*\, \tilde{G}^{\delta^i}(t_0) = 0$ for all $i \geq 0$, and there exists an open interval $\mathcal{I} \ni t_0$ such that the analytic function $\mathbf{v}^*\, \tilde{G}$ is zero in $\mathcal{I}$. By analytic continuation, $\mathbf{v}^*\, \tilde{G}$ is zero in Ω, and by Theorem and Definition 879(3), M is not controllable.

(b) Conversely, if the system is not Kalman-controllable, there exists $\mathbf{u}^* \in \mathbf{k}^n \backslash \{0\}$ such that $\mathbf{v}^*\, \tilde{G} = 0$. This implies $\mathbf{v}^*\, \tilde{G}^{\delta^i} = 0$ for all $i \geq 0$, thus $\mathbf{v}^*\, \tilde{\Gamma}_s(t_0)$ for all $s \geq 1$. Therefore, with the above notation, $\mathbf{u}^*\, \Gamma_s(t_0) = 0$, thus $\mathrm{rk}_{\mathbf{k}}\, \Gamma_s(t_0) < n$, for all $s \geq 1$.

For (ii), see [329] or ([335], Sect. 3.5). ■

Example 882. *Let* $\mathbf{A} = \mathcal{O}(\Omega)[\partial, \delta]$ *where* $\delta = d/dt$ *and consider the first order state-space system* $\partial x = t^r u$*. Then, for* $1 \leq s \leq r+1$

$$\Gamma_s = \left[t^r \vdots \cdots \vdots (-1)^{s-i} \binom{r}{r-s+i} t^{r-s+i} \vdots \cdots \vdots (-1)^{s-1} \binom{r}{r-s+1} t^{r-s+1} \right]$$

($1 \leq i \leq s$*). Since* $\mathrm{rk}_{\mathbb{C}}\, \Gamma_1(t) = 1$ *for all* $t \neq 0$*, this state-space system is Kalman-controllable by Proposition 881(ii). For* $s \leq r$*,* Γ_s *is not right-invertible, but* Γ_{r+1} *is right-invertible; therefore this system is strongly controllable by Theorem and Definition 879(2).*

Let us give an interpretation of Kalman controllability. Let K be any compact interval $\subseteq \Omega$ and let W_K be the space of distributions on Ω with support included in K, or the space of hyperfunctions on Ω with support included in K (Subsect. 1.9.3). Consider the operator

$$L_K : {}^m W_K(\Omega) \to {}^n\mathbf{k} : \mathbf{u} \mapsto \int_\Omega \tilde{G}(\tau)\, \mathbf{u}(\tau)\, d\tau.$$

The operator L_K is surjective if, and only if when $\mathbf{v}^* L_K = 0$ ($\mathbf{v}^* \in \mathbf{k}^n$), necessarily $\mathbf{v}^* = 0$. But $\mathbf{v}^* L_K = 0$ if, and only if for all $\mathbf{u} \in {}^m W_K(\Omega)$,

$$0 = \mathbf{v}^* L\, \mathbf{u} = \int_\Omega \mathbf{v}^*\, \tilde{G}(\tau)\, \mathbf{u}(\tau)\, d\tau.$$

Let $(\varepsilon_i)_{1 \le i \le m}$ be the canonical basis of ${}^m\mathbb{C}$, let $t \in K$, let δ_t be the Dirac distribution or hyperfunction at t (**1.9.2.1**) and take $\mathbf{u} = \delta_t\, \varepsilon_i$. Then

$$\int_\Omega \mathbf{v}^*\, \tilde{G}(\tau)\, \mathbf{u}(\tau)\, d\tau = \mathbf{v}^*\, \tilde{g}_i(t)$$

where $\tilde{g}_i$ is the ith column of $\tilde{G}$. Therefore, all operators L_K are surjective if, and only if $\mathbf{v}^*\, \tilde{G} = 0 \Rightarrow \mathbf{v}^* = 0$, i.e., the state-space system is Kalman-controllable. Last, consider any two instants $t_1, t_2 \in \Omega$ such that $t_2 > t_1$; the surjectivity of $L_{[t_1,t_2]}$ means that for two arbitrary points $\boldsymbol{\eta}_1, \boldsymbol{\eta}_2 \in {}^n\mathbf{k}$, there exists an input $\mathbf{u} \in {}^m W_{[t_1,t_2]}(\Omega)$ such that $L_{[t_1,t_2]}\mathbf{u} = \boldsymbol{\eta}_2 - \boldsymbol{\eta}_1$. With this input, the system state $\boldsymbol{\eta} : t \mapsto \boldsymbol{\eta}(t)$ is transferred from $\boldsymbol{\eta}_1$ (assumed to be the value of $\boldsymbol{\eta}$ at time t_1) to the value $\boldsymbol{\eta}_2$ in the interval of time $[t_1, t_2]$. This is Kalman's original definition of controllability [172]. (Kalman used neither hyperfunctions nor distributions, but the reader may check that the above equivalence still holds if the signal space is changed to, e.g., $\mathcal{E}(\Omega)$: see Exercise 955.)

For discrete-time systems, "0-controllability" is an important notion: see Exercise 958.

Proposition 883. *Let* $\mathbf{A} = \mathbf{K}[\partial; \boldsymbol{\alpha}, \boldsymbol{\delta}]$ *be a stably-free ideal domain and let* (M, u, y) *be a control system over* $\mathbf{A}$. *Let* $\mathcal{T}(M)$ *be the torsion submodule of* M *and let* $P = M/\mathcal{T}(M)$. *Let* $\bar{u}_i$ *be the canonical image of* u_i *in the quotient* P $(1 \le i \le m)$. *If the left* $\mathbf{K}$*-modules* $\mathcal{T}(M)$ *and* $P/[\bar{u}]_{\mathbf{A}}$ *are finite free (this happens, in particular, when* $\mathbf{K}$ *is a division ring), then* (M, u, y) *admits a state-space realization of the form*

$$\partial \begin{bmatrix} x_c \\ x_{\bar{c}} \end{bmatrix} = \begin{bmatrix} F_c & * \\ 0 & F_{\bar{c}} \end{bmatrix} x_c + \begin{bmatrix} G_c \\ 0 \end{bmatrix} u$$

where x_c *and* $x_{\bar{c}}$ *are columns with* n_c *and* $n_{\bar{c}}$ *entries, respectively,* $*$ *denotes an* $n_c \times n_{\bar{c}}$ *submatrix with entries in* $\mathbf{K}$, *the quotient system* P *with equation*

$$\partial \bar{x}_c = F_c\, \bar{x}_c + G_c\, \bar{u}$$

is strongly controllable, and $n_{\bar{c}} > 0$ *if, and only if* M *is noncontrollable ("Kalman controllability decomposition").*

Proof. Since $\mathbf{A}$ is a stably-free ideal domain,

$$M = \mathcal{T}(M) \oplus \Phi$$

where $\Phi \cong P$ is stably-free. Let $x_{\bar{c}_1}, ..., x_{\bar{c}_\nu}$ be a $\mathbf{K}$-basis of $\mathcal{T}(M)$. There exists a matrix $F_{\bar{c}} \in \mathrm{Mat}_\nu(\mathbf{K})$ such that $\partial x_{\bar{c}} = F_{\bar{c}} x_{\bar{c}}$. Let $\bar{\eta}_{c_1}, ..., \bar{\eta}_{c_\gamma}$ be a $\mathbf{K}$-basis of $P/[\bar{u}]_{\mathbf{A}}$. There exists a matrix $F_c \in \mathrm{Mat}_\gamma(\mathbf{K})$ such that $\partial \bar{\eta}_c = F_c \bar{\eta}_c$. There exist γ elements $\eta_i \in P$ such that $\bar{\eta}_i = \eta_i + [\bar{u}]_{\mathbf{A}}$ $(1 \leq i \leq \gamma)$. Therefore, there exist matrices $G_{c_j} \in {}^\gamma\mathbf{K}^m$ such that

$$\partial \eta = F_c \eta + \sum_{j=0}^{s} G_{c_j} \partial^j \bar{u}.$$

By the same rationale as in the proof of Theorem and Definition 861(1), there exist a column $\bar{x}_c$ and a matrix $G_c \in {}^\gamma\mathbf{K}^m$ such that $P = [\bar{x}_c, \bar{u}]_A$ and $\partial \bar{x}_c = F_c \bar{x}_c + G_c \bar{u}$. Since P is stably-free, this quotient system is strongly controllable. Last, there exist elements $x_{c_i} \in M$ $(1 \leq i \leq \gamma)$ such that $\bar{x}_{c_i} = x_{c_i} + \mathcal{T}(M)$, and there exists a matrix $[*]$ such that

$$\partial x_c = F_c x_c + [*] x_{\bar{c}} + G_c \bar{u}.$$

The proposition is proved with $\nu = n_{\bar{c}}$ and $\gamma = n_c$. ■

5.6.2 *Duality*

The duality studied below associates to a control system another control system, the input of the former corresponds to the output of the latter and *vice-versa*. Consider a control system with input $u = (u_i)_{1 \leq i \leq m}$ and output $y = (y_i)_{1 \leq i \leq p}$. This system can be assumed to be defined by the admissible Rosenbrock description (5.28). This description can also be written

$$\begin{bmatrix} D & 0 & -N \\ Q & 0 & W \\ 0 & 0 & I_m \end{bmatrix} \begin{bmatrix} \xi \\ y \\ u \end{bmatrix} = \begin{bmatrix} 0 \\ y \\ u \end{bmatrix}.$$

Let R_a be the "augmented" matrix of definition in the left-hand side.

Let $W \in {}_{\mathbf{A}}\mathbf{Mod}$, and assume that W is a $\mathbf{k}$-space. Let W' be another $\mathbf{k}$-space and let $\langle -, - \rangle : W' \times W \to \mathbf{k}$ be a nondegenerate bilinear form. Then, W' is canonically endowed with a structure of right $\mathbf{A}$-module, setting $\langle \mathbf{w}' a, \mathbf{w} \rangle = \langle \mathbf{w}', a \mathbf{w} \rangle$ for all $\mathbf{w} \in W$, $\mathbf{w}' \in W'$ and $a \in \mathbf{A}$.

Consider the W-behavior

$$\mathfrak{B}_W(M) = \left\{ \begin{bmatrix} \boldsymbol{\xi} \\ \mathbf{y} \\ \mathbf{u} \end{bmatrix} \in {}^{r+2m}W : \begin{bmatrix} D & 0 & -N \\ Q & 0 & W \\ 0 & 0 & I_m \end{bmatrix} \begin{bmatrix} \boldsymbol{\xi} \\ \mathbf{y} \\ \mathbf{u} \end{bmatrix} = \begin{bmatrix} 0 \\ \mathbf{y} \\ \mathbf{u} \end{bmatrix} \right\}.$$

For simplicity, set $\mathfrak{B} = \mathfrak{B}_W(M)$. As already said, the behavior $\mathfrak{B}$ is (identified with) a right $\mathbf{E}$-module, where $\mathbf{E} = \mathrm{End}_{\mathbf{A}}(W)$. By restriction of the ring of scalars, $\mathfrak{B}$ is a $\mathbf{k}$-vector space. Let us extend the bilinear map $\langle -, - \rangle$ to $W'^k \times {}^kW$ in an obvious way, for any integer $k \geq 1$.

Lemma and Definition 884. *(i) The* dual behavior *of* $\mathfrak{B}$ *is*

$$\mathfrak{B}^{\star}=\left\{\begin{array}{c}\left[\boldsymbol{\xi}^{\star}\ \mathbf{u}^{\star}\ \mathbf{y}^{\star}\right]\in W'^{r+p+m}:\langle\mathbf{u}^{\star},\mathbf{y}\rangle=\langle\mathbf{y}^{\star},\mathbf{u}\rangle\\ \textit{whenever}\ \left[\boldsymbol{\xi}^{T}\ \mathbf{u}^{T}\ \mathbf{y}^{T}\right]^{T}\in\mathfrak{B}\ \textit{for some}\ \boldsymbol{\xi}\in{}^{r}W.\end{array}\right\}.$$

(ii) The $\mathbf{k}$*-space* $\mathfrak{B}^{\star}$ *is the* W'*-behavior given by*

$$\left[\boldsymbol{\xi}^{\star}\ \mathbf{u}^{\star}\ \mathbf{y}^{\star}\right]\in W'^{r+p+m}:\left[\boldsymbol{\xi}^{\star}\ \mathbf{u}^{\star}\ -\mathbf{y}^{\star}\right]R_{a}=0$$

i.e., $\mathfrak{B}^{\star}=\ker_{W'}(\bullet R_{a})=\mathfrak{B}_{W'}(M^{\star})$ *where* $M^{\star}=\operatorname{coker}_{\mathbf{A}}(R_{a}\bullet)$ *is the f.p. right* $\mathbf{A}$*-module* $[\xi^{\star},u^{\star},y^{*}]_{\mathbf{A}}$ *defined by*

$$[\xi^{\star},u^{\star},y^{*}]_{\mathbf{A}}\ R_{a}=0. \tag{5.44}$$

(iii) The presentation (5.44) is a Rosenbrock representation, and it is admissible (Lemma and Definition 856) if, and only if D is square and regular.

Proof. (ii): Let $\left[\boldsymbol{\xi}^{T}\ \mathbf{u}^{T}\ \mathbf{y}^{T}\right]^{T}\in{}^{r+m+p}W$; $\left[\boldsymbol{\xi}^{T}\ \mathbf{u}^{T}\ \mathbf{y}^{T}\right]^{T}\in\mathfrak{B}$ if, and only if

$$R_{a}\left[\boldsymbol{\xi}^{T}\ \mathbf{u}^{T}\ \mathbf{y}^{T}\right]^{T}=\left[0\ \mathbf{u}^{T}\ \mathbf{y}^{T}\right]^{T}.$$

On the other hand, $\langle\mathbf{u}^{\star},\mathbf{y}\rangle=\langle\mathbf{y}^{\star},\mathbf{u}\rangle$ if, and only if for some $\boldsymbol{\xi}^{\star}\in W'^{r}$,

$$\left\langle\left[\boldsymbol{\xi}^{\star}\ \mathbf{u}^{\star}\ -\mathbf{y}^{\star}\right],\left[0\ \mathbf{u}^{T}\ \mathbf{y}^{T}\right]^{T}\right\rangle=0.$$

The last equality is equivalent to

$$\left\langle\left[\boldsymbol{\xi}^{\star}\ \mathbf{u}^{\star}\ -\mathbf{y}^{\star}\right],R_{a}\left[\boldsymbol{\xi}^{T}\ \mathbf{u}^{T}\ \mathbf{y}^{T}\right]^{T}\right\rangle=0,$$

i.e., to

$$\left\langle\left[\boldsymbol{\xi}^{\star}\ \mathbf{u}^{\star}\ -\mathbf{y}^{\star}\right]R_{a},\left[\boldsymbol{\xi}^{T}\ \mathbf{u}^{T}\ \mathbf{y}^{T}\right]^{T}\right\rangle=0.$$

This hold true for all $\left[\boldsymbol{\xi}^{T}\ \mathbf{u}^{T}\ \mathbf{y}^{T}\right]^{T}\in{}^{r+m+p}W$ if, and only if

$$\left[\boldsymbol{\xi}^{\star}\ \mathbf{u}^{\star}\ -\mathbf{y}^{\star}\right]R_{a}=0,$$

i.e.,

$$\left[\boldsymbol{\xi}^{\star}\ \mathbf{u}^{\star}\ -\mathbf{y}^{\star}\right]\in\ker_{W'}(\bullet R_{a}).$$

(iii) Let us make the presentation of $M^{\star}=[\xi^{\star},u^{\star},y^{*}]_{\mathbf{A}}$ more explicit. We have

$$\left\{\begin{array}{c}\xi^{*}D+u^{*}Q=0,\\ -\xi^{*}N+u^{*}W-y^{*}=0.\end{array}\right. \tag{5.45}$$

By Lemma and Definition 856, a necessary condition for the Rosenbrock description (5.45) to be admissible is that D be left-regular. Since D is already right-regular, the Rosenbrock description (5.45) is admissible if, and only if D is square and regular. ∎

The above can be formalized in a more intrinsic way. First note that the necessary and sufficient condition in Lemma and Definition 884(iii) is equivalent to the following: the module M has finite free resolution of length ≤ 1 (see Exercise 960). Then, setting $\omega^\star = u^*$, we obtain

$$\left[\xi^\star \ \omega^\star \ -y^\star\right] R_a = \left[\xi^\star D + \omega^\star Q \ 0 \ -\xi^\star N + \omega^\star W - y^*\right].$$

On the other hand, the module M is given by

$$R \left[\xi^T \ y^T \ u^T\right]^T = 0, \quad R = \begin{bmatrix} D & 0 & -N \\ Q & -I_p & W \end{bmatrix}.$$

Therefore,

$$\left[\xi^\star \ \omega^\star \ -y^\star\right] R_a = \left[\xi^\star \ \omega^\star\right] R + \left[0 \ u^\star \ -y^\star\right].$$

We are led to the following definition, where Rosenbrock descriptions are no longer used:

Lemma and Definition 885. *(1) Consider a control system (M, u, y) with equation $R\left[\xi^T \ y^T \ u^T\right]^T = 0$ over $\mathbf{A}$ (where $M \in {}_{\mathbf{A}}\mathbf{Mod}^{fp}$), and assume that M has finite free resolution of length ≤ 1. The* dual *of (M, u, y) is the triple $(M^\star, u^\star, y^\star)$ (where $M^\star \in \mathbf{Mod}_{\mathbf{A}}^{fp}$) with equation*

$$\left[\xi^\star \ \omega^\star\right] R + \left[0 \ u^\star \ -y^\star\right] = 0. \tag{5.46}$$

The input of this dual system is $u^\star = (u_i^\star)_{1 \leq i \leq p}$, its output is $y^\star = (y_i^\star)_{1 \leq i \leq m}$.
(2) Let $W \in {}_{\mathbf{A}}\mathbf{Mod}$, and assume that W is a $\mathbf{k}$-space. Let W' be another $\mathbf{k}$-space and let $\langle -, - \rangle : W' \times W \to \mathbf{k}$ be a nondegenerate bilinear form. Let W' be canonically endowed with a structure of right $\mathbf{A}$-module as above, so that $W' = {}_{\mathbf{E}'}W'_{\mathbf{A}}$ where $\mathbf{E}' = \operatorname{End}_{\mathbf{A}}(W')$. The W'-behavior associated with $(M^\star, u^\star, y^\star)$ is $\mathfrak{B}^\star = \operatorname{Hom}_{\mathbf{A}}(M^\star, W')$; $\mathfrak{B}^\star$ is a left $\mathbf{E}'$-module, called the dual *of the W-behavior $\mathfrak{B} = \operatorname{Hom}_{\mathbf{A}}(M, W)$.*

Proof. It remains to prove that $\mathfrak{B}^\star$ is a left $\mathbf{E}'$-module, but this is clear by Theorem 236. ∎

Remark 886. *(i) The module $M^\star \in \mathbf{Mod}_{\mathbf{A}}^{fp}$ is* not *the dual M^* of M* (**2.2.5.1**).
(ii) It can be convenient to think of $M^\star$ as a left module $\breve{M}$ over the opposite ring $\mathbf{A}^{\mathrm{op}}$ (Definition 219 and Proposition 229). When $\mathbf{A}$ is a ring of linear generalized differential operators, one is led to use the adjoint of $(\bullet R)$ (Lemma and Definition 360) since $(R^* \bullet) = \operatorname{ad}(R \bullet)$. Then, (5.46) can be written

$$R^* \begin{bmatrix} \breve{\xi} \\ \breve{\omega} \end{bmatrix} + \begin{bmatrix} 0 \\ \breve{u} \\ -\breve{y} \end{bmatrix} = 0, \tag{5.47}$$

and (5.47) determines the dual control system $\left(\breve{M}, \breve{u}, \breve{y}\right)$ over $\mathbf{A}^{\mathrm{op}}$.

Corollary 887. *Consider the state-space system (5.34), (5.35). Its dual is*

$$-\partial \breve{x} = F^T \breve{x} + H^T \breve{u}, \tag{5.48}$$

$$\breve{y} = G^T \breve{x} + \sum_{i=0}^{\alpha} (-1)^i J_i^T \partial^i \breve{u}. \tag{5.49}$$

Proof. The state-space system (5.34), (5.35) can be written

$$R \begin{bmatrix} x^T & y^T & u^T \end{bmatrix}^T = 0, \quad R = \begin{bmatrix} \partial I_n - F & 0 & -G \\ H & -I_p & \sum_{i=0}^{\alpha} J_i \partial^i \end{bmatrix}.$$

Therefore, (5.47) yields by Lemma and Definition 360

$$\begin{bmatrix} -\partial I_n - F^T & H^T \\ 0 & -I_p \\ -G^T & \sum_{i=0}^{\alpha} J_i^T \partial^i \end{bmatrix} \begin{bmatrix} \breve{\xi} \\ \breve{\omega} \end{bmatrix} + \begin{bmatrix} 0 \\ \breve{u} \\ -\breve{y} \end{bmatrix} = 0.$$

Setting $\breve{x} = -\breve{\xi}$, one obtains (5.48), (5.49). ∎

5.6.3 Observability

Let $\mathbf{A}$ be an Ore domain which is a $\mathbf{k}$-algebra ($\mathbf{k} = \mathbb{R}$ or $\mathbb{C}$), and let (M, u, y) be a control system over $\mathbf{A}$ (Definition 851).

Definition 888. *The above control system is said to be* strongly observable *if $M = [u, y]_{\mathbf{A}}$, i.e., every element of M can be expressed as an $\mathbf{A}$-linear combination of the elements u_i $(1 \leq i \leq m)$ and y_j $(1 \leq j \leq p)$.*

Corollary 889. *(i) Let the control system (M, u, y) be given by an admissible Rosenbrock description (5.28) (Lemma and Definition 856). Then (M, u, y) is strongly controllable if, and only if $\begin{bmatrix} D \\ Q \end{bmatrix}$ is left-invertible over $\mathbf{A}$.*
(ii) Assuming that M has finite free resolution of length ≤ 1, (M, u, y) is strongly observable if, and only if its dual $\left(\breve{M}, \breve{u}, \breve{y}\right)$ is strongly controllable.

Proof. (i) We have $M = [u, y]_{\mathbf{A}}$ if, and only if $M / [u, y]_{\mathbf{A}} = 0$. Let ξ be the partial state with entries ξ_i $(1 \leq i \leq r)$ and let $\bar{\xi}_i$ be the canonical image of ξ_i in $M / [u, y]_{\mathbf{A}}$. Last, let $\bar{\xi}$ be the column with entries $\bar{\xi}_i$ $(1 \leq i \leq r)$. The column $\bar{\xi}$ is submitted to the only relation

$$\begin{bmatrix} D \\ Q \end{bmatrix} \bar{\xi} = 0.$$

Therefore, $M / [u, y]_{\mathbf{A}} = [\bar{\xi}]_{\mathbf{A}}$ is zero if, and only if $\begin{bmatrix} D \\ Q \end{bmatrix}$ is left-invertible over $\mathbf{A}$.

(ii) If M has finite free resolution of length ≤ 1, then its dual $\left(\breve{M}, \breve{u}, \breve{y}\right)$ exists (Lemma and Definition 885). It is given by the Rosenbrock description (5.45), the first equation of which can be written

$$\left[\xi^{\star}\ u^{\star}\right]\begin{bmatrix} D \\ Q \end{bmatrix} = 0.$$

By Theorem and Definition 874(2), this control system is strongly controllable if, and only if $\begin{bmatrix} D \\ Q \end{bmatrix}$ is left-invertible over $\mathbf{A}$. ∎

In what follows, all control systems (M, u, y) are assumed to be such that the left $\mathbf{A}$-module M has finite free resolution of length ≤ 1. Due to the above, we are led to the following.

Definition 890. *A control system (M, u, y) is said to be observable (resp., freely observable,...) if its dual is controllable (resp., freely controllable, ...). If (M, u, y) is a state-space system, it is said to be Kalman-observable if its dual is Kalman-controllable.*

The reader is requested to dualize Theorem and Definition 879, and Propositions 881, 883. By Corollary 887, and by Lemma and Definition 360, the matrix Γ_s is changed to Ω_s defined for $s \geq 1$ as follows:

$$\Omega_s = \left[Q_0 \ \vdots\ Q_1 \ \vdots\ \cdots \ \vdots\ Q_{s-1}\right], \tag{5.50}$$

$$Q_0 = \breve{H}^T, \quad Q_{i+1} = \left(F^T + \boldsymbol{\beta}\,\delta\, I_n\right) Q_i^{\boldsymbol{\alpha}}, i \geq 0. \tag{5.51}$$

For discrete-time systems (see Remark 880), Ω_s can be changed to

$$\Omega_s' = \left[Q_0' \ \vdots\ Q_1' \ \vdots\ \cdots \ \vdots\ Q_{s-1}'\right],$$

$$Q_0' = \breve{H}^T, \quad Q_{i+1}' = F^T\, Q_i'^{\boldsymbol{\alpha}}, i \geq 0.$$

5.6.4 Bicoprime Factorizations

Let $\mathbf{R}$ be an Ore domain and let $\mathbf{F}$ be its division ring of fractions.

Definition 891. *Let $G \in {}^p\mathbf{F}^m$, $N_r \in {}^p\mathbf{R}^\rho$, $D \in \operatorname{Mat}_\rho(\mathbf{R})$, $N_l \in {}^\rho\mathbf{R}^m$ and $W \in {}^p\mathbf{R}^m$, where D is invertible over $\mathbf{F}$. The quadruple (N_r, D, N_l, W) is called a* bicoprime factorization *of G over $\mathbf{F}$ if (i) $G = N_r D^{-1} N_l + W$, (ii) (D, N_l) is strongly left-coprime, and (iii) (D, N_r) is strongly right-coprime.*

Lemma 892. *If $\mathbf{R}$ is a Bézout domain, then every $G \in \mathbf{M}(\mathbf{F})$ (with the notation in Subsect. 3.7.1) admits a bicoprime factorization over $\mathbf{F}$.*

Proof. If $\mathbf{R}$ is a Bézout domain, then every $G \in \mathbf{M}(\mathbf{F})$ admits a left-coprime factorization (D, N_l) over $\mathbf{F}$ according to Theorem 670, and then $(I_p, D, N_l, 0)$ is a bicoprime factorization of G over $\mathbf{R}$. ■

Theorem 893. *Let (N_r, D, N_l, W) be a bicoprime factorization of $G \in \mathbf{M}(\mathbf{F})$ over $\mathbf{R}$.*

(i) If G admits a left-coprime factorization $\left(\tilde{D}_l, \tilde{N}_l\right)$ over $\mathbf{R}$, then $\operatorname{coker}_{\mathbf{R}}(\bullet D) \cong \operatorname{coker}_{\mathbf{R}}\left(\bullet \tilde{D}_l\right)$.

(ii) If G admits a right-coprime factorization $\left(\tilde{D}_r, \tilde{N}_r\right)$ over $\mathbf{R}$, then $\operatorname{coker}_{\mathbf{R}}(\bullet D) \cong \operatorname{coker}_{\mathbf{R}}\left(\bullet \tilde{D}_r\right)$.

Proof. (i) Consider the following Rosenbrock description over $\mathbf{R}$:

$$\begin{cases} D\,\xi = N_l\, u, \\ y = N_r\, \xi + W\, u. \end{cases}$$

Since D is square and regular, this Rosenbrock description is admissible according to Lemma and Definition 855. Therefore, it represents a control system (M, u, y) whose transfer matrix is G. Obviously, $M/[u] = \operatorname{coker}_{\mathbf{R}}(\bullet D)$. Since (D, N_r) is strongly right-coprime, the matrix $\begin{bmatrix} D \\ N_r \end{bmatrix}$ is left-invertible over $\mathbf{R}$, therefore the control system (M, u, y) is strongly observable by Corollary 889(i), which implies that $M = [\xi, u] = [y, u]$. The module M is defined by

$$\begin{bmatrix} D & -N_l \end{bmatrix} \begin{bmatrix} \xi \\ u \end{bmatrix} = 0$$

and since (D, N_l) is strongly left-coprime, $\begin{bmatrix} D & -N_l \end{bmatrix}$ is right-invertible, which implies that the system M is strongly controllable by Theorem and Definition 874(2). Let $\tilde{M} = \operatorname{coker}_{\mathbf{R}}\left(\bullet \begin{bmatrix} \tilde{D}_l & -\tilde{N}_l \end{bmatrix}\right)$. Since $G = \tilde{D}_l^{-1}\tilde{N}_l$, M and $\tilde{M}$ have the same transfer vector space (Corollary and Definition 853), i.e. $\mathbf{F} \bigotimes_{\mathbf{R}} M = \mathbf{F} \bigotimes_{\mathbf{R}} \tilde{M}$. Since both M and $\tilde{M}$ are torsion-free, they are isomorphic by Corollary and Definition 341, thus they can be identified. Therefore, the control system (M, u, y) can also be described by

$$\begin{bmatrix} \tilde{D}_l & -\tilde{N}_l \end{bmatrix} \begin{bmatrix} y \\ u \end{bmatrix} = 0$$

and $M/[u] \cong \operatorname{coker}_{\mathbf{R}}\left(\bullet \tilde{D}_l\right)$.

(ii) is proved similarly. ■

5.7 Differential-Difference Systems

As already said, for certain problems a purely algebraic approach is insufficient and topological notions must be added. Then, the notion of module is no longer appropriate, and one is led to define the notion of "topological module" (in a sense which does not coincide with the usual one, as defined in, e.g., [34], n°III.6.6).

Convolutions systems over Lie groups, which include delay-differential systems and differential-difference systems over Lie groups, can be viewed as special cases of topological systems. Convolution systems over Lie groups have applications in various fields such as robotics, image processing, vision and polymer science [56], [366], and [64].

Whereas the category ${}_{\mathbf{A}}\mathbf{Mod}$ of left modules over a ring $\mathbf{A}$ is abelian, the category ${}_{\mathbf{A}}\mathbf{TMod}$ of topological modules is only semiabelian (see Lemma 900 below). Therefore, all the notions introduced in Sections 1.2 and 5.3 are needed. Some results about topological vector spaces (Section 1.7) and about Lie groups and Lie algebras (Section 1.8) are also needed.

5.7.1 *Introductive Examples*

Example 894. *Consider the space* $\mathbf{A} = \mathcal{E}'(\mathbb{R})$ *of distributions with compact support included in* $\mathbb{R}$ (**1.7.3.5**). *The pair* $(\mathcal{E}'(\mathbb{R}), \star)$ *is a convolution algebra. A convolution system over* $\mathbb{R}$ *is defined by an equation of the form*

$$R \star w = 0, \tag{5.52}$$

$R \in {}^{q}\mathbf{A}^{k}$. *The Laplace transform* $\hat{R}$ *of* R *is well-defined as recalled in* (**5.7.5.3**) *below. Taking the Laplace transform of (5.52), one obtains*

$$\hat{R}\,\hat{w} = 0.$$

To simplify the notation, let us denote the convolution product as a usual product. Then, (5.52) takes the usual form (5.1).
The convolution algebra $\mathbf{A}$ *acts (via the convolution product) on the space* $W = \mathcal{E}(\mathbb{R})$ *of indefinitely differentiable functions in* $\mathbb{R}$, *thus* W *can be viewed as a left* $\mathbf{A}$*-module* (**1.8.1.4**). *Therefore,* W*-behaviors of linear systems over* $\mathbf{A}$ *can be defined as usual (Definition 807). The question which naturally arises is whether* W *is a cogenerator (for the category* ${}_{\mathbf{A}}\mathbf{Mod}$ *or for a subcategory of* ${}_{\mathbf{A}}\mathbf{Mod}$*).*
Let $R_i \in {}^{q_i}\mathbf{A}^{k}$, $i = 1, 2$, *and consider condition (C) and Property (P) below:*
(C): $\quad \mathrm{im}_{\mathbf{A}}(\bullet R_1)$ *is closed in* ${}^{q_1}\mathbf{A}$.

(P): $\quad \ker_W(R_1\bullet) \subseteq \ker_W(R_2\bullet) \Longleftrightarrow R_2 = X R_1$ *for some* $X \in {}^{q_2}\mathbf{A}^{q_1}$.

As shown in [103] (see also [139]), property (P) holds for aribitrary R_2 *if, and only if condition (C) holds. Property (P) coincides with the property* *obtained in Theorem 812(iii). By Theorem 827(iv), W is a cogenerator for the class of modules $M_1 = \mathrm{coker}_{\mathbf{A}}(\bullet R_1)$ which satisfy condition (C). The point is that condition (C) is not algebraic, thus those modules do not form a subcategory of* $_{\mathbf{A}}\mathbf{Mod}$. *However, as will be seen, they form a full subcategory of the category* $_{\mathbf{A}}\mathbf{TMod}$. *The above result is generalized in Subsect. 5.7.3.*

Example 895. *Consider the situation in Example 894. Let $\hat{R}_1(s)$ be the Laplace transform of the matrix of distributions $R_1 \in {}^q\mathbf{A}^k$. The system*

$$M_1 = \mathrm{coker}_{\mathbf{A}}(\bullet R_1) = \mathbf{A}^k / N_1 \quad (N = {}^q\mathbf{A}\, R_1)$$

is called spectrally controllable *if*

$$\mathrm{rk}_{\mathbb{C}}(R_1(s)) \text{ is constant as } s \text{ varies in } \mathbb{C}$$

(see, e.g., [352]).

As shown in the above-quoted reference, M_1 is spectrally controllable *if, and only if $\mathfrak{B}_1 = \ker_W(R_1 \bullet)$ admits a dense image representation, i.e., there exists a matrix $R_2 \in {}^k\mathbf{A}^r$ such that*

$$\mathfrak{B}_1 = \overline{\mathrm{im}_W(R_2 \bullet)}$$

where the closure is taken in W^k (recall that the multiplication is the convolution product).

Now assume that condition (C) in Example 894 holds. Then, as shown in ([353], Prop. 8 & Theorem 14), $\mathfrak{B}_1 = \ker_W(R_1 \bullet)$ admits a dense image representation if, and only if M_1 is torsion-free.
These results are connected with Theorem 867 and with Lemma and Definition 872 with the following differences:

(a) According to Lemma and Definition 872, M_1 is controllable if, and only if it is torsion-free (without assuming that condition (C) holds).
(b) According to Theorem 867, an nD system $M_1 = \mathrm{coker}_{\mathbf{A}}(\bullet R_1)$ is controllable if, and only if the associated behavior $\ker_W(R_1 \bullet)$ $(W = \mathcal{E}(\mathbb{R}^n))$ *admits an image representation – a condition which is more restrictive than to admit a* dense *image representation. In accordance with Gluesing-Luerssen* et al. *[139], we will say that a behavior which admits a dense image representaion is* weakly controllable.

Again, weak controllability is not a purely algebraic notion, thus we are faced with the same difficulty as in Example 894. The above result is generalized in Subsect. 5.7.4, where differential systems with lumped shifts are studied.

Example 896. *Let $\partial = d/dt$ and let σ_i $(i = 1, ..., m)$ be the shift-forward operator $\mathbf{w}(t) \mapsto \mathbf{w}(t + \tau_i)$ where the positive numbers τ_i are pairwise incommensurate, and set $\sigma = (\sigma_1, ..., \sigma_m)$, $\sigma^{-1} = \left(\sigma_1^{-1}, ..., \sigma_m^{-1}\right)$. Our aim is to study delay-differential delay systems (with incommensurate delays $\tau_1, ..., \tau_m$)*

as well as the associated behaviors in a power of W where W is the classic signal space $\mathcal{E}(\mathbb{R})$. We first have to choose a suitable ring of operators.

(i) At first glance, the polynomial rings $\mathbf{B}_1 = \mathbb{C}[\partial, \sigma]$ and $\mathbf{B} = \mathbb{C}[\partial, \sigma, \sigma^{-1}]$ are good candidates (see, e.g., [251], [298]). Notice that the indeterminates ∂ and σ commute. However, the following difficulty arises: assume, to simplify the discussion, that $m = 1$, and let $R_1, R_2 \in \mathbf{B}$ be given by $R_1 = \partial$, $R_2 = \sigma - 1$. The W-behavior $\ker_W(R_1\bullet)$ *consists of all constant functions, whereas* $\ker_W(R_2\bullet)$ *consists of all periodic functions of period* 1. *Therefore,* $\ker_W(R_2\bullet) \subsetneq \ker_W(R_1\bullet)$. *However, R_1 does not divide R_2 in $\mathbf{B}$, i.e. property (P) in Example 894 does not hold. Therefore, according to Theorem 827, W is* not *a cogenerator in* ${}_{\mathbf{B}}\mathbf{Mod}$. *Moreover, W is* not *a cogenerator for the full subcategory of* ${}_{\mathbf{B}}\mathbf{Mod}$ *whose objects are cyclic and torsion. Neither is the space $\mathcal{D}'(\mathbb{R})$ of distributions on the real line or the space $\mathcal{B}(\mathbb{R})$ of hyperfunctions. Indeed, the above rationale still holds when W is replaced by any of these two spaces. Therefore, one must choose another ring of operators.*
(ii) Assume again that $m = 1$ and, to simplify the notation, assume that σ is the shift-forward operator $\mathbf{w}(t) \mapsto \mathbf{w}(t+1)$. *The Laplace transform yields the ring-isomorphism*

$$\mathbf{B} \xrightarrow{\sim} \mathbb{C}\left[s, e^{s}, e^{-s}\right].$$

These two rings are identified in what follows. We know that the dual $W' = \mathcal{E}'(\mathbb{R})$ is the space of compactly supported distributions on the real line (**1.7.3.5**). *Consider the subring $PW(\mathbb{C})$ of $\mathcal{O}(\mathbb{C})$ consisting of those functions f for which there exist real numbers $a \geq 0$, $c \geq 0$ and an integer $N \geq 0$ such that for every $s \in \mathbb{C}$,*

$$|f(s)| \leq c\,(1 + |s|)^{N}\; e^{a|\mathrm{Re}\, s|}.$$

According to the Paley-Wiener-Schwartz theorem ([43], Section 12.3.4), the Laplace transform is an isomorphism

$$\mathcal{E}'(\mathbb{R}) \xrightarrow{\sim} PW(\mathbb{C}),$$

thus let us identify $\mathcal{E}'(\mathbb{R})$ with $PW(\mathbb{C})$. The set

$$\mathcal{H} = \left\{ \frac{b(s)}{a(s)} : b(s) \in \mathbb{C}\left[s, e^{s}, e^{-s}\right],\ a(s) \in \mathbb{C}[s]^{\times},\ \frac{b(s)}{a(s)} \in PW(\mathbb{C}) \right\}$$

is both a subring of $PW(\mathbb{C})$ and an over-ring of $\mathbf{B}$ (with the above identifications). The elements of $\mathcal{H}$ are Laplace transforms of delay-differential operators with distributed delays. *For example, $f(s) = \frac{1-e^{-s}}{s}$ is the transfer function of the operator*

$$\mathbf{w}(t) \mapsto \int_{t-1}^{t} \mathbf{w}(\zeta)\, d\zeta.$$

As shown in [137], although the commutative domain $\mathcal{H}$ is is neither Noetherian nor a UFD, $\mathcal{H}$ is an elementary divisor ring (Lemma and Definition 655(iii)), therefore linear systems over $\mathcal{H}$ have very nice algebraic properties which are very similar to those of linear systems over a commutative principal ideal domain (Lemma and Definition 655(i)). In addition, $W = \mathcal{E}(\mathbb{R})$ is a cogenerator for ${}_{\mathcal{H}}\mathbf{Mod}^{fp}$ by ([139], Theorem 5.3(1)).
(iii) The above construction of the ring $\mathcal{H}$ can be generalized to the case $m > 1$. However, the ring $\mathcal{H}$ then obtained is no longer an elementary divisor ring (nor a Bézout domain [153]).
Let $R_i \in {}^{q_i}\mathcal{H}^k$, $i = 1, 2$ and consider condition (C') and Property (P') below:
(C'): $\quad \mathrm{rk}_{\mathcal{H}}(\bullet R_1) = q_1.$

$$\textit{(P')}: \qquad \ker_W(R_1\bullet) \subseteq \ker_W(R_2\bullet) \iff R_2 = X\,R_1 \textit{ for some } X \in {}^{q_2}\mathcal{H}^{q_1}.$$

As shown in [153], property (P') holds if condition (C') holds. Therefore, W is a cogenerator for the full subcategory of ${}_{\mathcal{H}}\mathbf{Mod}$ consisting of all modules M_1 of the form $\mathrm{coker}_{\mathcal{H}}(\bullet R_1)$ *where R_1 satisfies condition (C'); these $\mathcal{H}$-modules M_1 are those which have finite free resolution of length ≤ 1* (**3.5.4.1**).
Although the result in (iii) seems to be purely algebraic, to our knowledge it cannot be established using only algebraic arguments: once again, the category ${}_{\mathbf{A}}\mathbf{TMod}$ must be used. The result in (iii) is generalized in Subsect. 5.7.5.

5.7.2 The Category of Topological Modules

In everything that follows, we make the following

Assumption 897. **k** *is the field of real or complex numbers and all algebras are associative, unitary and defined over* **k**.

Let **A** be an algebra, let W be a **k**-vector space and consider a nondegenerate bilinear form

$$\langle -, - \rangle : \mathbf{A} \times W \to \mathbf{k}.$$

We also consider the nondegenerate bilinear form

$$\langle -, - \rangle^{\tilde{}} : W \times \mathbf{A} \to \mathbf{k} \tag{5.53}$$

defined by $\langle \mathbf{w}, a \rangle^{\tilde{}} = \langle a, \mathbf{w} \rangle$ for any $a \in \mathbf{A}$ and $\mathbf{w} \in W$. The following facts are classical ([36], n°II.6.2): the space W is canonically isomorphic to, and identified with, the topological dual $\mathbf{A}'$ of $(\mathbf{A}, \sigma(\mathbf{A}, W))$, i.e., of **A** endowed with the topology $\sigma(\mathbf{A}, W)$; this statement is also valid with W and **A** interchanged; $(\mathbf{A}, \sigma(\mathbf{A}, W))$ and $(W, \sigma(W, \mathbf{A}))$ (respectively written **A** and W in what follows, for short). We know that **A** and W are *locally convex topological vector spaces* (LCSs) and *Hausdorff* (**1.7.2.2**).

A left **A**-module M is called *topological* if it is endowed with a structure of LCS and if all maps $\mathbf{A} \to M : a \mapsto am$, $m \in M$, are continuous; a topological right **A**-module M is similarly defined. (Note that the definition of a topological module over a topological ring is different ([34], n°III.6.6).) The algebra $\mathbf{A}$ may be itself a topological left or right $\mathbf{A}$-module. The algebraic dual $\mathbf{A}^* = \mathrm{Hom}_{\mathbf{k}}(\mathbf{A}, \mathbf{k})$ is a left and right $\mathbf{A}$-module via

$$(x)\, a\alpha = (xa)\, \alpha \quad \text{and} \quad (\alpha a)\, x = \alpha\,(ax)\,, \quad \alpha \in \mathbf{A}^*,\, a, x \in \mathbf{A}.$$

In general, $\mathbf{A}' = W$ is not an $\mathbf{A}$-submodule of $\mathbf{A}^*$ and when this happens is characterized below:

Lemma 898. *(1) The following conditions are equivalent:*
(i) The left module ${}_{\mathbf{A}}\mathbf{A}$ *is topological, i.e., all maps* $a \mapsto ab$ $(b \in \mathbf{A})$ *are continuous.*
(ii) W *is a left* $\mathbf{A}$*-module with* $\langle a, b\mathbf{w}\rangle = \langle ab, \mathbf{w}\rangle$ *for all* $a, b \in \mathbf{A}$ *and* $\mathbf{w} \in W$. *Then also all maps* $W \to W : w \mapsto b\mathbf{w}$ $(b \in \mathbf{A})$ *are continuous.*
(2) The following properties are equivalent:
(i') The right module $\mathbf{A}_{\mathbf{A}}$ *is topological.*
(ii') W *is a right* $\mathbf{A}$*-module with* $\widetilde{\langle \mathbf{w}a, b\rangle} = \widetilde{\langle \mathbf{w}, ab\rangle}$ *for all* $a, b \in \mathbf{A}$. *Then also all maps* $W \to W : w \mapsto \mathbf{w}b$ $(b \in \mathbf{A})$ *are continuous.*
(3) If the conditions of (1) hold, then those of (2) are equivalent to:
(iii') The left module ${}_{\mathbf{A}}W$ *is topological.*

Proof. (1). (i)$\Rightarrow$(ii): If $a \mapsto ab$ is continuous, then so is $b\alpha : a \mapsto a\,(b\alpha) = (ab)\,\alpha$ for any $\alpha \in \mathbf{A}' = W$ with the identification $W \ni \mathbf{w} \mapsto \langle -, \mathbf{w}\rangle \in \mathbf{A}'$. Hence $\mathbf{A}' = W$ is an $\mathbf{A}$-submodule of $\mathbf{A}^*$. (ii)$\Rightarrow$(i): If (ii) holds, then $b\mathbf{w} \in W$ for any $b \in \mathbf{A}$, $\mathbf{w} \in W$, therefore the map $\mathbf{A} \ni a \mapsto \langle a, b\mathbf{w}\rangle = \langle ab, \mathbf{w}\rangle \in \mathbf{k}$ is continuous which implies that so is the map $a \mapsto ab$. The maps $w \mapsto \langle a, b\mathbf{w}\rangle$ are continuous by the above equality. (2) is shown like (1).

(3). If (ii) holds, the following conditions are equivalent: all maps $b \mapsto b\mathbf{w}$ are continuous, i.e., (iii') holds; all maps $b \mapsto \langle a, b\mathbf{w}\rangle = \langle ab, \mathbf{w}\rangle$ are continuous; all maps $b \mapsto ab$ are continuous, i.e., (i') holds. ∎

Definition 899. *Assuming that all conditions of Lemma 898 are satisfied, the category of topological left* $\mathbf{A}$*-modules and continuous linear maps is called the* category of topological left $\mathbf{A}$-modules *and written* ${}_{\mathbf{A}}\mathbf{TMod}$.

Lemma 900. ${}_{\mathbf{A}}\mathbf{TMod}$ *is a semiabelian category with arbitrary products and coproducts and with projective generator* $\mathbf{A}$.

Proof. (a) The category $\mathcal{C} = {}_{\mathbf{A}}\mathbf{TMod}$ is a subcategory of ${}_{\mathbf{A}}\mathbf{Mod}$; $\mathcal{C}$ is obviously additive and has kernels and cokernels which are the algebraic ones endowed with the induced and coinduced topology, respectively, thus $\mathcal{C}$ is preabelian. Let $\bullet f : X \to Y$ be a morphism of $\mathcal{C}$ and f_{ind} be its induced morphism in ${}_{\mathbf{A}}\mathbf{Mod}$. Then f_{ind} is bijective (since ${}_{\mathbf{A}}\mathbf{Mod}$ is abelian) and continuous by definition of the topologies of $\operatorname{coim} f$ and of $\operatorname{im} f$. Therefore, f_{ind} is the induced morphism of f in $\mathcal{C}$. Since f_{ind} is both a monomorphism and an epimorphism, the preabelian category $\mathcal{C}$ is semiabelian. Last, $\mathcal{C}$ has arbitrary products and coproducts which are the algebraic ones endowed with the product topology and the locally convex direct sum topology, respectively ([36], §II.4, Definition 2).

(b) Let $M \in \mathcal{C}$. There exists an $\mathbf{A}$-linear surjection $\bullet\varphi : \mathbf{A}^{(I)} \twoheadrightarrow M$. Let $\bullet\iota : \mathbf{A} \hookrightarrow \mathbf{A}^{(I)}$ be the canonical injection. Let $\bullet f = \bullet\iota\varphi : \mathbf{A} \to M$; for any $a \in \mathbf{A}$, $(a) f = a\, m$ where $m = (1) f$, thus f is continuous since the module M is topological. Therefore, φ is continuous (if $\mathbf{A}^{(I)}$ is endowed with the locally convex direct sum topology) according to ([36], §II.4, Prop. 6). This proves that $\mathbf{A}$ is a generator of $\mathcal{C}$.

(c) Let $\bullet f : \mathbf{A} \to M$ and $\bullet g : M_1 \twoheadrightarrow M$ be respectively a morphism and an epimorphism of $\mathcal{C}$. Since $\bullet f$ and $\bullet g$ are respectively a morphism and an epimorphism of ${}_{\mathbf{A}}\mathbf{Mod}$, there exists a morphism $\bullet h : \mathbf{A} \to M_1$ of ${}_{\mathbf{A}}\mathbf{Mod}$ such that $\bullet f = \bullet hg$. Since M_1 is topological, $\bullet h$ is continuous, i.e., is a morphism of $\mathcal{C}$, thus $\mathbf{A}$ is projective in $\mathcal{C}$. ∎

5.7.3 Cogenerators for Topological Modules

For $X, Y \in {}_{\mathbf{A}}\mathbf{TMod}$, we denote by $\mathrm{Hom}_{\mathbf{A},c}(X, Y)$ the $\mathbf{k}$-vector space consisting of all continuous $\mathbf{A}$-linear maps $X \to Y$. Endowed with the topology of pointwise convergence, $\mathrm{Hom}_{\mathbf{A},c}(X, Y)$ is an LCS; in particular we have the additive left exact contravariant functor

$$\begin{aligned} \mathfrak{B}_W &: {}_{\mathbf{A}}\mathbf{TMod} \to \mathbf{Mod}_{\mathbf{E}} : M \mapsto \mathfrak{B}_W \triangleq \mathrm{Hom}_{\mathbf{A},c}(M, W), \\ \mathbf{E} &\triangleq \mathrm{Hom}_{\mathbf{A},c}(W, W). \end{aligned}$$

The image $\mathfrak{B}_W({}_{\mathbf{A}}\mathbf{TMod})$ is called the class of all (topological) W-behaviors. All results in Subsect. 5.3.2 are valid, in particular Theorems 638 and 827.

Remark 901. *Let $M \in \mathcal{C}$. By Lemma 900, there exists a set of indices I such that M is a factor of $\mathbf{A}^{(I)}$, i.e., there exists a continuous $\mathbf{A}$-linear surjection $\varphi : \mathbf{A}^{(I)} \twoheadrightarrow M$. There exists a continuous bijection $\varphi_{ind} : \mathbf{A}^{(I)}/N \to M$ where $N = \ker\varphi$. But φ_{ind}^{-1} is not continuous in general, i.e., is not an isomorphism of $\mathcal{C}$, thus $\mathbf{A}^{(I)}/N$ and M cannot be identified; then, M is not an $\mathbf{A}$-cokernel.*

Let I be a set of indices. The nondegenerate form $\langle -, - \rangle$ is extended to the nondegenerate form

$$\langle -, - \rangle : \mathbf{A}^{(I)} \times {}^{I}W : (x, \mathbf{w}) \mapsto \sum_{i \in I} \langle x_i, \mathbf{w}_i \rangle$$

where all terms of the sum are zero except a finite number of them. The locally convex topologies induced by this form in ${}^{I}W$ and $\mathbf{A}^{(I)}$ are respectively the product topology $\sigma\left({}^{I}W, \mathbf{A}^{(I)}\right) = \prod_{i \in I} \sigma(W, \mathbf{A})$ ([36], §II.6, Prop. 8) and the topology $\sigma\left(\mathbf{A}^{(I)}, {}^{I}W\right)$. The latter coincides with the locally convex direct sum topology $\bigoplus_{i \in I} \sigma(\mathbf{A}, W)$ if, and only if I is finite ([316], Chap. IV, Exercise 8). In the sequel, $\mathbf{A}^{(I)}$ is endowed with the topology $\sigma\left(\mathbf{A}^{(I)}, {}^{I}W\right)$

and I may be assumed to be finite, to avoid confusions. If $N \subseteq \mathbf{A}^{(I)}$ in ${}_{\mathbf{A}}\mathbf{TMod}$ (which means that N is a subspace of $\mathbf{A}^{(I)}$), its *polar* $N^0 \subseteq {}^I\mathbf{A}$ is defined as usual:

$$N^0 = \left\{ \mathbf{w} \in {}^I W : \langle N, \mathbf{w} \rangle = 0 \right\} = \bigcap_{x \in N} \ker \langle x, - \rangle$$

and N^0 is closed in ${}^I W$. According to the bipolar theorem ([36], n°II.6.3), $N^{00} = \overline{N}$ where $\overline{N}$ denotes the closure of N in $\mathbf{A}^{(I)}$.

Theorem 902. *(i) Let $N \subseteq \mathbf{A}^{(I)}$; then*

$$N^{\perp} = N^0 \quad \text{and} \quad N^{\perp\perp} = N^{00} = \overline{N}.$$

(ii) Let $M = \mathbf{A}^{(I)}/N$ be an $\mathbf{A}$-cokernel in $\mathcal{C} = {}_{\mathbf{A}}\mathbf{TMod}$; then M is Hausdorff if, and only if $N = N^{\perp\perp}$.
(iii) Let $\mathcal{D}$ be a full subcategory of $\mathcal{C}$, all objects of which are $\mathbf{A}$-cokernels. Then $W \in \mathcal{C}$ is a cogenerator for $\mathcal{D}$ if, and only if all objects $M = \mathbf{A}^{(I)}/N$ of $\mathcal{D}$ are Hausdorff.

Proof. (i): (a) The identity

$$\langle ax, \mathbf{w} \rangle = \langle a, x\mathbf{w} \rangle, \; a \in \mathbf{A}, \, x \in \mathbf{A}^{(I)}, \, \mathbf{w} \in {}^I W$$

holds. By definition, $\mathbf{w} \in N^{\perp}$ if, and only if for any $x = (x_i)_{i \in I} \in \mathbf{A}^{(I)}$, $x\mathbf{w} \triangleq \sum_{i \in I} x_i \mathbf{w}_i = 0$. Since $\langle -, - \rangle$ is nondegenerate, the last equality is equivalent to $0 = \langle a, x\mathbf{w} \rangle = \langle ax, \mathbf{w} \rangle$ for any $a \in \mathbf{A}$, thus to $\langle x, \mathbf{w} \rangle = 0$. Thus $N^{\perp} = N^0$. (b) Since the module $\mathbf{A}_{\mathbf{A}}$ is topological, the maps $\mathbf{A} \to \mathbf{A} : x \mapsto ax$ and $\mathbf{A}^{(I)} \to \mathbf{A}^{(I)} : x \to ax$ $(a \in \mathbf{A})$ are continuous ([36], n°II.4.5). This implies that $a\overline{N} \subseteq \overline{aN} \subseteq \overline{N}$ and that $\overline{N}$ is a left $\mathbf{A}$-submodule of $\mathbf{A}^{(I)}$. Now, $x \in N^{\perp\perp}$ if, and only if for any $\mathbf{w} \in N^{\perp} = N^0$, $x\mathbf{w} = 0$. The last equality holds if, and only if for all $a \in \mathbf{A}$, $0 = \langle a, x\mathbf{w} \rangle = \langle ax, \mathbf{w} \rangle$, i.e., $ax \in N^{00} = \overline{N}$. Since $\overline{N}$ is a left $\mathbf{A}$-submodule of $\mathbf{A}^{(I)}$, this is equivalent to $x \in \overline{N}$.

(i) implies (ii) which implies (iii) by Lemma and Definition 504(ii), and Theorem 638. ∎

Remark 903. *(i) Theorem 902 generalizes the result recalled in Example 894. See another nontrivial example in [49].*
(ii) Let W be an LCS and $\mathbf{A} \subseteq W'$. Then, the restriction to $\mathbf{A} \times W$ of the canonical bilinear form $\langle -, - \rangle : W' \times W \to \mathbf{k}$ is possibly degenerate; the topology $\sigma(\mathbf{A}, W)$ is Hausdorff (since it is induced from $\sigma(W', W)$) but $\sigma(W, \mathbf{A})$ is not in general ([36], §II.6, Prop. 1). Therefore, the statements of Theorem 902 are no longer valid. This situation is further investigated in the two next sections where convolution systems are considered.

5.7.4 Differential Systems with Lumped Shifts over Lie Groups

5.7.4.1 Actions over a Lie Group and Convolution Algebra

In everything that follows, we'll use the following

Notation 904. **G** *is a Lie group with Lie algebra* $\mathfrak{g}$, $W = \mathcal{E}(\mathbf{G})$ *is the space all* **k***-valued indefinitely differentiable functions on* **G***, and its topological dual* W' *is the space of all* **k***-valued distributions with compact support included in* **G**.

The group **G** continuously acts on W via a left and a right action, both of them denoted by $\circ$, and defined by

(i) $(y)(a \circ \varphi) = (ya)\varphi$ (*left action* of **G** on a *left* map, which is a *right shift*),

(ii) $(\varphi \circ a)(x) = \varphi(ax)$ (*right action* of **G** on a *right* map, which is a *left shift*),

for any $x, y, a \in \mathbf{G}$ and $\varphi \in W$. By (1.36), this yields for any $S, T \in W'$

$$\langle S, a \circ \varphi \rangle = \langle S \star \delta_a, \varphi \rangle, \quad \langle \varphi \circ a, T \rangle^{\tilde{}} = \langle \varphi, \delta_a \star T \rangle^{\tilde{}}$$

with Notation (5.53).

These actions induce respectively a right and a left action on W', both of them also denoted by $\circ$, according to

$$\langle S \circ a, \varphi \rangle = \langle S, a \circ \varphi \rangle, \quad \langle \varphi, a \circ T \rangle^{\tilde{}} = \langle \varphi \circ a, T \rangle^{\tilde{}} \tag{5.54}$$

and from the above

$$S \circ a = S \star \delta_a, \quad a \circ T = \delta_a \star T.$$

We also define a left and a right action of W' on W, both of them denoted by $\circ$, respectively by

$$S\circ : W \to W : (S \circ \varphi)(x) = \langle S, \varphi \circ x \rangle \tag{5.55}$$

$$\circ S : W \to W : (x)(\varphi \circ S) = \langle x \circ \varphi, S \rangle^{\tilde{}} \tag{5.56}$$

for any $\varphi \in W$ and $S \in W'$. As a consequence of (1.37), one easily obtains

$$\varphi \circ S = \check{S} \star \varphi, \quad S \circ \varphi = \varphi \star \check{\Delta}^{-1} \check{S}$$

where Δ is the modulus function, $\check{\varphi}(y) = \varphi\left(y^{-1}\right)$ and $\left\langle \check{S}, \varphi \right\rangle = \langle S, \check{\varphi} \rangle$ for any $y \in \mathbf{G}$, $\varphi \in W$ and $S \in W'$. We have the following:

Lemma 905. *(i) The equality*

$$\langle T, S \circ \varphi \rangle = \langle \varphi \circ T, S \rangle^{\sim}$$

holds for any $S, T \in W'$ *and* $\varphi \in W$.
(ii) Both actions $\circ$ *of* W' *on* W *induce one bilinear law of composition* $\circ :$ $W' \times W' \to W'$ *according to*

$$\langle T \circ S, \varphi \rangle = \langle T, S \circ \varphi \rangle, \quad \langle \varphi, T \circ S \rangle^{\sim} = \langle \varphi \circ T, S \rangle^{\sim} \tag{5.57}$$

and this law of composition is the usual convolution product $\star$.
(iii) For any $S, T \in W'$ *and* $\varphi \in W$,

$$(T \circ S) \circ \varphi = T \circ (S \circ \varphi), \quad \varphi \circ (T \circ S) = (\varphi \circ T) \circ S, \tag{5.58}$$

i.e., W *is a left and right* W'*-module.*

Proof. (i): We have

$$\langle T, S \circ \varphi \rangle = \int_{\mathbf{G}} dT(x) \int_{\mathbf{G}} dS(y) (\varphi \circ x)(y) = \int_{\mathbf{G}} dT(x) \int_{\mathbf{G}} dS(y) \varphi(xy)$$

thus

$$\langle T, S \circ \varphi \rangle = \langle T \star S, \varphi \rangle . \tag{5.59}$$

Similarly (denoting φ as a left map)

$$\langle \varphi \circ T, S \rangle^{\sim} = \int_{\mathbf{G}} dS(y) \int_{\mathbf{G}} dT(x)(x)(y \circ \varphi) = \int_{\mathbf{G}} dS(y) \int_{\mathbf{G}} dT(x)(xy) \varphi$$

thus

$$\langle \varphi \circ T, S \rangle^{\sim} = \langle T \star S, \varphi \rangle . \tag{5.60}$$

(ii): This is an obvious consequence of (5.59) and (5.60).
(iii): By (5.55), (5.57) we get

$$\langle U, T \circ (S \circ \varphi) \rangle = \langle U \star T * S, \varphi \rangle = \langle U, (T \star S) \circ \varphi \rangle = \langle U, (T \circ S) \circ \varphi \rangle$$

which proves the first equality of (5.58). The second one is likewise deduced from (5.56) and (5.57). ∎

Definition 906. *A convolution algebra is a subalgebra of* $(W', \circ)$.

5.7.4.2 Fundamental Principle and Divisibility

Let $\mathbf{B}$ be a convolution algebra. The restriction to $\mathbf{B} \times W$ of the action $\circ : W' \times W \to W$ and that of the canonical bilinear form $\langle -, - \rangle : W' \times W \to \mathbf{k}$ are still denoted by $\circ$ and $\langle -, - \rangle$, respectively. In general the latter is degenerate, then the topology $\sigma(\mathbf{B}, W)$ of $\mathbf{B}$ is Hausdorff (and induced by $\sigma(W', W)$) but not the topology $\sigma(W, \mathbf{B})$ of W (Remark 903). Therefore, the theory in Subsect. 5.7 cannot be directly applied.

Example 907. *Consider the case when* $\mathbf{B} = \mathrm{U}_{\mathbf{k}}(\mathfrak{g})$ (**1.8.3.1**)*; we have* $\mathbf{B} = \bigoplus_{\alpha \in \mathbb{N}^n} D_{\xi}^{\alpha} \delta_e$*, where* $\mathbb{N}$ *is the set of non-negative integers and the notation in* (**1.8.1.3**) *is used; the polar* $\mathbf{A}^o$ *of* $\mathbf{A}$ *is given by*

$$\mathbf{A}^o = \left\{ \mathbf{w} \in W : \forall \alpha \in \mathbb{N}^n, D_{\xi}^{\alpha} \mathbf{w}(e) = 0 \right\}$$

thus $\mathbf{A}^o \neq 0$ *(except in the trivial case* $\mathbf{G} = \{e\}$*). We have*

$$\mathbf{A}^{\perp} = \left\{ \mathbf{w} \in W : \forall \alpha \in \mathbb{N}^n, D_{\xi}^{\alpha} \delta_e \circ \mathbf{w} = 0 \right\} = 0$$

where $D_{\xi}^{\alpha} \delta_e \circ \mathbf{w} = D_{\xi}^{\alpha} \mathbf{w}$*, thus* $\mathbf{A}^o \neq \mathbf{A}^{\perp}$ *and Statement (ii) of Theorem 902 does not hold.*

Let I be a set of indices; the canonical bilinear form $\langle -, - \rangle : \mathbf{B} \times W \to \mathbf{k}$ and the action $\circ : \mathbf{B} \times W \to W$ are extended to $\mathbf{B}^{(I)} \times {}^{I}W$ according to

$$\langle -, - \rangle : \mathbf{B}^{(I)} \times {}^{I}W : (T, \mathbf{w}) \mapsto \sum_{i \in I} \langle T_i, \mathbf{w}_i \rangle$$

$$\circ : \mathbf{B}^{(I)} \times {}^{I}W : (T, \mathbf{w}) \mapsto \sum_{i \in I} T_i \circ \mathbf{w}_i.$$

As usual, the orthogonal space with respect to $\langle -, - \rangle$, i.e., the polar, is denoted by $(-)^0$, and that with respect to $\circ$ by $(-)^{\perp}$.

The following definition will also be needed in the sequel:

Definition 908. *Let* $\mathbf{G}$ *be a Lie group; then* $\mathfrak{P}^{cf}(\mathbf{G})$ *is the set of all subgroups* $\mathbf{H}$ *of* $\mathbf{G}$ *such that all finitely generated subgroups of* $\mathbf{H}$ *are polycyclic by finite (Definition 161).*

In everything that follows, we will use the following

Notation 909. $\mathbf{H}$ *is a subgroup of* $\mathbf{G}$ *and* $\mathbf{A}$ *is a* Noetherian subalgebra *of* $\mathrm{U}_{\mathbf{k}}(\mathfrak{g})$.

The two main cases are $\mathbf{A} = \mathrm{U}_{\mathbf{k}}(\mathfrak{g})$ –the algebra of invariant differential operators– and $\mathbf{A} = \mathrm{Z}_{\mathbf{k}}(\mathfrak{g})$ –the algebra of bi-invariant differential operators– when the latter is Noetherian (**1.8.3.3**). Let

$$\mathbf{B} = \sum\nolimits_{x \in \mathbf{H}} \delta_x \circ \mathbf{A}, \quad \mathbf{C} = \sum\nolimits_{x \in \mathbf{G}} \delta_x \circ \mathbf{A}.$$

Setting $\varepsilon(x) S \triangleq \delta_x \circ S$ ($x \in \mathbf{G}$, $S \in \mathbf{A}$), this yields by (1.38) $\mathbf{B} = \mathbf{A} * \mathbf{H}$ and $\mathbf{C} = \mathbf{A} * \mathbf{G}$ where $*$ denoted the crossed product (Definition 370). We have

$$\mathbf{C} = \sum\nolimits_{x \in \bar{\mathbf{G}}} \delta_x \circ \left(\sum\nolimits_{x \in \mathbf{H}} \delta_x \circ \mathbf{A} \right) = \sum\nolimits_{x \in \bar{\mathbf{G}}} \delta_x \circ \mathbf{B}$$

where $\bar{\mathbf{G}} \subseteq \mathbf{G}$ is a set of representatives of the cosets of $\mathbf{G}$ modulo $\mathbf{H}$. All distributions belonging to $\mathbf{A}$ have their support included in $\{e\}$, thus again by (1.38) all above sums are direct since $\mathbf{G}$ is the disjoint union of its distinct cosets modulo $\mathbf{H}$. In particular, $\mathbf{B}$ is a subalgebra of $\mathbf{C}$ and

$$\mathbf{C} = \bigoplus_{x \in \bar{\mathbf{G}}} \delta_x \circ \mathbf{B} = \mathbf{C} \bigotimes_{\mathbf{B}} \mathbf{B} \tag{5.61}$$

is a free right $\mathbf{B}$-module.

When $\mathbf{G} = \mathbb{R}^n$ and $\mathbf{H}$ is a *finitely generated* subgroup of $\mathbf{G}$, then $\mathbf{B}$ is a convolution algebra of differential-difference operators with lumped shifts, denoted by $\mathbf{A}^{shift}$.

Lemma 910. *Let* $\mathbf{H} \in \mathfrak{P}^{cf}(\mathbf{G})$*; then* $\mathbf{B} = \mathbf{A} * \mathbf{H}$ *is coherent.*

Proof. Let $\mathfrak{H}$ be the directed system of all finitely generated subgroups $\mathbf{H}$ of $\mathbf{G}$. The ring $\mathbf{A}$ is Noetherian , thus if $\mathbf{H}_1 \in \mathfrak{H}$, then the ring $\mathbf{A} * \mathbf{H}_1$ is Noetherian again (Theorem 372), thus it is coherent (Proposition 511). Moreover, let $\mathbf{H}_1, \mathbf{H}_2 \in \mathfrak{H}$ be such that $\mathbf{H}_1 \subseteq \mathbf{H}_2$, let $X_1 \subseteq \mathbf{H}_1$ be a set of representatives of the cosets of $\mathbf{H}_2$ modulo $\mathbf{H}_1$, and let $\varphi : \mathbf{H}_1 \to \bar{\mathbf{H}}_1$ be a bijection where $\bar{\mathbf{H}}_1$ is a set of units of $\mathbf{A} * \mathbf{H}_1$ (Definition 370). Let $\bar{X}_1 = \varphi(X_1)$. Then $\mathbf{A} * \mathbf{H}_2$ is easily seen to be a left and right $\mathbf{A} * \mathbf{H}_1$-module freely generated by $\bar{X}_1$; thus it is a flat left and right $\mathbf{A} * \mathbf{H}_1$-module (**3.5.3.1**), and therefore

$$\mathbf{B} = \bigcup_{\mathbf{I} \in \mathfrak{H}} \mathbf{A} * \mathbf{I} = \varinjlim_{\mathbf{I} \in \mathfrak{H}} \mathbf{A} * \mathbf{I}$$

is coherent (Theorem 612). ∎

Since the two rings $\mathbf{B} \subseteq \mathbf{C}$ are subrings of W', W is both a left and right module over $\mathbf{B}$ and over $\mathbf{C}$ via the actions $\circ$ induced by those defined in (5.55) and (5.56). The nondegenerate bilinear form $\langle -, - \rangle : W' \times W \to \mathbf{k}$ induces a bilinear form $\langle -, - \rangle : \mathbf{C} \times W \to \mathbf{k}$. We have the following:

Lemma 911. *The bilinear form* $\langle -, - \rangle : \mathbf{C} \times W \to \mathbf{k}$ *is nondegenerate.*

Proof. The ring $\mathbf{C}$ contains all distributions δ_x, $x \in \mathbf{G}$. If $w \in W$ is such that $\langle \mathbf{C}, w \rangle = 0$, then $\langle \delta_x, w \rangle = w(x) = 0$ for all $x \in \mathbf{G}$, thus $w = 0$. ∎

Let $R_2 \in {}^q\mathbf{B}^k$, let $N = \mathbf{B}^q \circ R_2$ and let $M = \mathbf{B}^k / \mathbf{B}^q \circ R_2$; then the sequence

$$\mathbf{B}^q \overset{\circ R_2}{\longrightarrow} \mathbf{B}^k \longrightarrow M \longrightarrow 0$$

is exact in ${}_{\mathbf{B}}\mathbf{Mod}$ and M is finitely presented in that category. The map $\circ R_2$ is weakly* continuous since the convolution product $\star : \mathcal{E}'(\mathbf{G}) \times \mathcal{E}(\mathbf{G}) \to \mathcal{E}(\mathbf{G})$ is separately weakly* continuous (**1.8.1.4**). This map $\circ R$ induces a weakly* continuous map $\circ R_2 : \mathbf{C}^q \to \mathbf{C}^k$.

Lemma 912. *(i) Assume that* $\mathbf{H} \in \mathfrak{P}^{cf}(\mathbf{G})$*. Then there exist* $r \in \mathbb{N}$ *and a matrix* $R_1 \in {}^r\mathbf{B}^q$ *such that the sequence below is exact in* ${}_{\mathbf{B}}\mathbf{Mod}$*:*

$$\mathbf{B}^r \xrightarrow{\circ R_1} \mathbf{B}^q \xrightarrow{\circ R_2} \mathbf{B}^k \tag{5.62}$$

(ii) The exactness of the sequence (5.62) in ${}_{\mathbf{B}}\mathbf{Mod}$ *is equivalent to the exactness of the sequence below in* ${}_{\mathbf{C}}\mathbf{Mod}$*:*

$$\mathbf{C}^r \xrightarrow{\circ R_1} \mathbf{C}^q \xrightarrow{\circ R_2} \mathbf{C}^k. \tag{5.63}$$

Proof. (i): The ring $\mathbf{B}$ is coherent by Lemma 910, thus there exists a matrix R_1 such that the sequence (5.62) is exact in ${}_{\mathbf{B}}\mathbf{Mod}$ by Theorem 514(ii).

(ii): As seen above, the right $\mathbf{B}$-module $\mathbf{C} = \mathbf{C} \otimes_{\mathbf{B}} \mathbf{B}$ is free, thus faithfully flat (Corollary 616(ii)), thus the exactness of the sequence (5.62) is equivalent to that of (5.63) (Lemma and Definition 614). ∎

The following expresses when a *fundamental principle* for W-behaviors is valid (compare with Corollary and Definition 592) using the above notation:

Corollary 913. *Let* $R_2 \in {}^q\mathbf{B}^k$ *be given and let* $R_1 \in {}^r\mathbf{B}^q$ *be as in Lemma 912(i) if it exists. The following properties are equivalent:*
(a) The fundamental principle holds, i.e.,

$$R_2 \circ {}^kW = \{\mathbf{u} \in {}^rW : R_1 \circ \mathbf{u} = 0\}$$

($R_2 \in {}^q\mathbf{B}^k$*,* $R_1 \in {}^r\mathbf{B}^q$*); i.e., the equation* $R_2 \circ \mathbf{w} = \mathbf{u}$ *(*$\mathbf{u} \in {}^rW$*) has a solution* $\mathbf{w} \in {}^kW$ *if, and only if* $\mathbf{u}$ *satisfies the compatibility (or integrability) condition* $R_1 \circ \mathbf{u} = 0$*.*
(b) The map $\circ R_2 : \mathbf{C}^q \to \mathbf{C}^k$ *is strict, i.e.,* $\mathrm{coim}_{\mathbf{C}} \circ R_2 \cong \mathrm{im}_{\mathbf{C}} \circ R_2 = \mathbf{C}^q \circ R_2$*.*

Proof. By Lemma 911 and (5.57), the transpose of $R_2 \circ : {}^kW \to {}^qW$ is $\circ R_2 : \mathbf{C}^q \to \mathbf{C}^k$ and that of $R_1 \circ : {}^qW \to {}^rW$ is $\circ R_1 : \mathbf{C}^r \to \mathbf{C}^q$. The fundamental principle holds if, and only if the sequence

$${}^kW \xrightarrow{R_2 \circ} {}^qW \xrightarrow{R_1 \circ} {}^rW$$

is exact in the category $\mathbf{Vct}_{\mathbf{k}}$ of right $\mathbf{k}$-vector spaces. By ([36], §II.6, Remark 1 after Corol. 4 of Prop. 7), this holds true if, and only if the map $\circ R_2 : \mathbf{C}^q \to \mathbf{C}^k$ is strict. ∎

Remark 914. *(i) If* $\mathbf{B} = \mathrm{U}_{\mathbf{k}}(\mathfrak{g})$*, then* R_1 *can be constructed via the Gröbner basis algorithm which is applicable to universal enveloping algebras of finite-dimensional Lie groups ([258], Sect. 5, Definition and Corollary (37)). (For a general presentation of Gröbner bases, we refer to, e.g., [105], Chap. 15.)*
(ii) Let $0 \neq \mu \in \mathcal{D}(\mathbb{R}^n) \subseteq W'$*, where* $\mathcal{D}(\mathbb{R}^n)$ *is the usual space of indefinitely differentiable* $\mathbf{k}$*-valued functions with compact support in* $\mathbb{R}^n$*; then* $\mu \star W \subsetneq W$ *([230], Subsect. II.3.1, Corollary 1), thus the module* ${}_{W'}W$ *is not divisible (Definition 596) and* a fortiori *not injective (Proposition 598).*

(iii) If $\mathbf{G} = \mathbb{R}^n$ *and* $\mathbf{B} = \mathbf{A}$*, the fundamental principle holds ([231], Theorem 3.2), thus Condition (b) in Corollary 913 is always satisfied.*
(iv) If $\mathbf{G}$ *is solvable simply connected* (**1.8.2.3**) *and* $\mathbf{A} = \mathrm{Z}_{\mathbf{k}}(\mathfrak{g})$ *then* ${}_{\mathbf{A}}W$ *is a divisible ([95], Prop. 2).*
(v) If $\mathbf{A} = \mathrm{U}_{\mathbf{k}}(\mathfrak{g})$*, the ring of all invariant differential operators over* $\mathbf{G}$*, and* ${}_{\mathbf{A}}W$ *is divisible, then either* $\mathbf{G}$ *is abelian or has an abelian normal subgroup of codimension 1 ([60], Prop. 2).*
(vi) If $\mathbf{G}$ *is the compact abelian torus* $\mathbb{R}/\mathbb{Z}$ *and* $\mathbf{A} = \mathrm{U}_{\mathbf{k}}(\mathfrak{g}) = \mathrm{Z}_{\mathbf{k}}(\mathfrak{g}) = \mathbf{k}[d/dt]$*, then* ${}_{\mathbf{A}}W$ *is not divisible ([322], p. 225), hence not injective.*
(vii) If $\mathbf{G} = \mathbb{R}^n$ *and* $\mathbf{B} = \mathbf{A}^{shift}$*, then* ${}_{\mathbf{B}}W$ *is divisible, as shown in Ehrenpreis [99].*

Items (iv) and (v) of the above remark suggest preferring the commutative algebra $\mathrm{Z}_{\mathbf{k}}(\mathfrak{g})$ of bi-invariant differential operators over the algebra $\mathrm{U}_{\mathbf{k}}(\mathfrak{g})$ of invariant differential operators regarding the existence of global C^{∞} solutions of non-homogeneous differential equations over Lie groups.

Corollary and Definition 915. *(1) Let* μ *be a left-regular element of* $\mathbf{B}$*. The following conditions are equivalent:*
(i) $\mu\circ : W \to W$ *is an epimorphism;*
(ii) $\circ\mu : \mathbf{C} \to \mathbf{C}$ *is a strict monomorphism, i.e.,* $\mathbf{C} \circ \mu \cong \mathbf{C}$.
(iii) $\circ\mu : W' \to W'$ *is a strict monomorphism for the weak* topology, i.e.,* $W' \circ \mu \cong W'$*;*
(iv) the principal ideal $W' \circ \mu$ *is weakly* closed in* W'*;*
(v) the principal ideal $W' \circ \mu$ *is strongly closed in* W'.
(2) A distribution $\mu \in \mathbf{B}$ *is called* invertible[2] *if it satisfies the conditions in (1).*
(3) Assuming that all nonzero elements of $\mathbf{B}$ *are left-regular, the left* $\mathbf{B}$*-module* W *is divisible if, and only if all nonzero elements of* $\mathbf{B}$ *are invertible distributions, and then the* $\mathbf{B}$*-module* W *is faithful.*

Proof. (1): Let μ be left-regular in $\mathbf{B}$. Then the sequence (5.62) is exact with $r = 0$, $q = k = 1$, $R_1 = 0$ and $R_2 = \mu$; therefore, by Lemma 912(ii), $\circ\mu : \mathbf{C} \to \mathbf{C}$ is a monomorphism, i.e., $\mathrm{coim}_{\mathbf{C}} \circ\mu = \mathbf{C}/\ker\circ\mu = \mathbf{C}/\{0\} = \mathbf{C}$. Thus, by Corollary 913, (i) holds if, and only if $\circ\mu : \mathbf{C} \to \mathbf{C}$ is strict , i.e., $\mathbf{C} = \mathrm{coim}_{\mathbf{C}} \circ\mu \cong \mathrm{im}_{\mathbf{C}} \circ\mu$. As a result, (i)$\Leftrightarrow$(ii).

(i)$\Leftrightarrow$(iii) by ([36], §IV.4, Corollary 2 of Theorem 1) since W is a Fréchet space, and (i)$\Leftrightarrow$(iv) by ([36], §IV.4, Theorem 1).

Since W is a Montel space (**1.8.1.3**), it is reflexive (**1.7.3.3**), thus the weak* closure of $W' \circ \mu$ is equal to its weak closure (**1.7.3.2**), and therefore also to its strong closure. Therefore, (iv)$\Leftrightarrow$(v).

---

[2] This denomination is in accordance with that of Ehrenpreis [100] and Hörmander ([162], Definition 16.3.12), but in a more general context. In Schwartz's terminology ([322], Sect. VI.10), such a distribution μ is called *completely invertible.*

(3): If all nonzero elements of $\mathbf{B}$ are left-regular, the divisibility of W over $\mathbf{B}$ is equivalent to condition (1)(i) for all $0 \neq \mu \in \mathbf{B}$ (Definition 596). This condition implies the faithfulness of W over $\mathbf{B}$, i.e., $\mu \circ W = 0$ implies $\mu = 0$. ■

5.7.4.3 A Cogenerator Property

Let $\mathbf{G}$ be a Lie group and consider W-behaviors ($W = \mathcal{E}(\mathbf{G})$) over the ring $\mathbf{B} = \mathbf{A} * \mathbf{H}$ where $\mathbf{H}$ is a polycyclic by finite subgroup of $\mathbf{G}$. Thus let $\mathbf{C} = \mathbf{A} * \mathbf{G}$, $R_1 \in {}^q\mathbf{B}^k$, $N = \mathbf{B}^q \circ R_1 \subseteq \mathbf{B}^k$, $M = \mathbf{B}^k / N$, $\mathfrak{B} = \mathfrak{B}_W(M) = N^{\perp}$ and $\mathfrak{B}^{\perp} = N^{\perp\perp} = \left\{T \in \mathbf{B}^k : T \circ \mathfrak{B}\right\} = 0$.

Theorem 916. *(1) If M is torsion-free, or equivalently torsionless, then the following equivalent properties hold:*
(i) The canonical map $\phi_M : M \to W^{\mathfrak{B}} : m \mapsto (m \circ w)_{w \in \mathcal{B}}$ is injective;
(ii) $N = N^{\perp\perp}$.
(2) W is a cogenerator for the full subcategory $\mathcal{D}$ of ${}_{\mathbf{B}}\mathbf{Mod}^{fp}$, all objects of which are torsion-free.

Proof. (1): By Theorem 372, $\mathbf{B}$ is semiprime Noetherian, thus semiprime Goldie (Remark 334(1)); therefore, by Proposition 516, M is torsion-free if, and only if it is torsionless, and then there exists an embedding $M \hookrightarrow \mathbf{A}^r$ for some integer $r \geq 1$. In that case, by Theorem 514(iii), there exists a matrix $R_2 \in {}^k\mathbf{B}^r$ such that the sequences (5.62), (5.63) are exact. In particular, with $N = \mathbf{B}^q \circ R_1$, the module $\mathbf{C} \bigotimes_{\mathbf{B}} N = \mathbf{C}^q \circ R_1 = \ker_{\mathbf{C}} \circ R_2$ is closed, hence by Lemma 911 and Theorem 902(i), $\mathbf{C} \bigotimes_{\mathbf{B}} N = (\mathbf{C} \bigotimes_{\mathbf{B}} N)^{\perp\perp}$ and the canonical map

$$\tilde{\phi}_{\tilde{M}} : \tilde{M} \to W^{\mathfrak{B}} : \tilde{m} \mapsto (\tilde{m} \circ \mathbf{w})_{\mathbf{w} \in \mathfrak{B}}$$

($\tilde{M} = \mathbf{C} \bigotimes_{\mathbf{B}} M$, $\tilde{m} = 1 \otimes m$) is injective. By (5.61), the map

$$\Psi : M \to \tilde{M} : m \mapsto 1 \otimes m = \delta_e \otimes m$$

is injective too. In addition, $\bullet\phi_M = \bullet\Psi\,\tilde{\phi}_{\tilde{M}}$ since $m \circ \mathbf{w} = (1 \otimes m) \circ \mathbf{w}$ for the given identification, hence $\bullet\phi_M$ is injective.

(2): This is a straightforward consequence of (1)(i) and of Corollary and Definition 822. ■

Remark 917. *(i) If $M \in {}_{\mathbf{B}}\mathbf{Mod}^{fp}$ is not torsion-free, W is not a cogenerator of the category consisting of M alone, as shown by ([137], Example 2.3).*
(ii) All $\mathbf{B}$-modules $M \in \mathcal{D}$ are Hausdorff by Proposition 516(iv) and Theorem 514(iv) since the map $\circ R_2$ in (5.62) induces a continuous injection $M = \mathbf{B}^k / \mathbf{B}^q R_1 \hookrightarrow \mathbf{B}^r$ where $\mathbf{B}^r \subseteq (W')^r$ is Hausdorff.
(iii) The standard case of delay-differential systems is that when $\mathbf{G} = \mathbb{R}$ and $\mathbf{H} = \bigoplus_{1 \leq i \leq m} \mathbb{Z}\tau_i$ with delays $\tau_1, ..., \tau_m$ (thus $\mathbf{B} = \mathbf{k}\left[d/dt, \sigma, \sigma^{-1}\right]$, $\sigma = (\sigma_1, ..., \sigma_m)$, $\sigma^{-1} \triangleq \left(\sigma_1^{-1}, ..., \sigma_m^{-1}\right)$, $\sigma_i : \mathbf{w}(t) \mapsto \mathbf{w}(t + \tau_i)$).

5.7.4.4 Dense Image Representation and Weak Controllability

Theorem 918. *Let* $(\mathbf{B}, \circ)$ *be a convolution algebra, let* $R_1 \in {}^q\mathbf{B}^k$, *let* $N = \mathbf{B}^q \circ R_1$, *let* $M = \mathbf{B}^k / N$, *and let* $\mathfrak{B} = \mathfrak{B}_W(M) = \{\mathbf{w} \in {}^kW : R_1 \circ \mathbf{w} = 0\}$.

(1) Assume that $\mathbf{G}$ *is a Lie group and that* $\mathbf{H} \in \mathfrak{P}^{cf}(\mathbf{G})$, *so that* $\mathbf{B} = \mathbf{A} * \mathbf{H}$ *is coherent according to Lemma 910. If* M *is torsionless, then it is Hausdorff and* $\mathfrak{B}$ *admits a* dense image representation, *i.e.,* $\mathfrak{B}$ *is* weakly controllable, *or equivalently, the following property* (**P**) *holds:*
*(**P**): There exist* $r \in \mathbb{N}$ *and* $R_2 \in {}^k\mathbf{B}^r$ *such that* $\mathfrak{B} = \overline{R_2 \circ ({}^rW)}$ *(where the closure is taken in the usual Fréchet topology of compact convergence in* ${}^rW$ (**1.8.1.3**)*).*
In addition, if $\mathbf{H}$ *is polycyclic by finite, then* M *is torsionless if, and only if it is torsion-free by Theorem 916(1).*
(2) Conversely, if $\mathbf{B}$ *is dense in* W', M *is Hausdorff and* $\mathfrak{B}$ *admits a dense image representation, then* M *is both torsion-free and torsionless.*
(3) If $\mathbf{B}$ *is semiprime Goldie (e.g., when* $\mathbf{G}$ *is a Lie group and* $\mathbf{B} = \mathbf{A} * \mathbf{H}$ *where* $\mathbf{H}$ *is polycyclic by finite, by Theorem 372) and* $\langle -, - \rangle : \mathbf{B} \times W \to \mathbf{k}$ *is nondegenerate, or in other words if* $\mathbf{B}$ *is wealkly* dense in* W', *then* M *is torsion-free if, and only if* M *is Hausdorff (i.e.,* N *is closed in* $\mathbf{B}^k$*) and* $\mathfrak{B}$ *admits a dense image representation.*

Proof. (1) By Proposition 516(iv) and Theorem 514(iii), there exists a matrix $R_2 \in {}^k\mathbf{B}^r$ such that both sequences (5.62), (5.63) are exact. The exactness of (5.62) implies that $N = \operatorname{im}_{\mathbf{B}}(\circ R_1) = \ker_{\mathbf{B}}(\circ R_2)$. Since $\circ R_2 : \mathbf{B}^k \to \mathbf{B}^r$ is continuous and $\mathbf{B}$ is Hausdorff, the kernel N of this map is closed and the factor $M = \mathbf{B}^k / N$ is Hausdorff. The exactness of (5.63) implies that $\operatorname{im}_{\mathbf{C}}(\circ R_1) = \ker_{\mathbf{C}}(\circ R_2)$. Therefore, by ([36], §II.6, Corol. 3 of Prop. 4, and Corol. 4 of Prop. 6),

$$\overline{R_2 \circ {}^rW} = (\ker_{\mathbf{C}} \circ R_2)^0 = (\operatorname{im}_{\mathbf{C}} \circ R_1)^0 = \ker_W(R_1 \circ) = \mathfrak{B}$$

where the closure is taken in the topology $\sigma({}^rW, \mathbf{C}^r)$ or equivalently in the initial Fréchet topology of ${}^rW$.

(2) Since $M = \mathbf{B}^k / N$ is Hausdorff, $N = \mathbf{B}^q \circ R_1$ is closed in $\mathbf{B}^k$. In addition, the bilinear form $\langle -, - \rangle : \mathbf{B} \times W \to \mathbf{k}$ is nondegenerate since $\mathbf{B}$ is dense –thus weakly* dense– in W' ([36], §II.6, Corol. 4 of Prop. 3). Therefore, $N = N^{\perp\perp} = \mathfrak{B}^{\perp}$ by 902(i), thus $N = \left(\overline{R_2 \circ {}^rW}\right)^{\perp} = (R_2 \circ {}^rW)^{\perp} = \ker_{\mathbf{B}} \circ R_2$. Thus the sequence (5.62) is exact and M is both torsion-free and torsionless by Proposition 516 and Theorem 514.

(3) (a) Necessary condition: Since $\mathbf{B}$ is an semiprime Goldie, the finitely generated torsion-free module M admits by Proposition 516 a continuous embedding in some $\mathbf{B}^r$ $(r \geq 1)$, and there exists an exact sequence (5.62). Since the form $\langle -, - \rangle : \mathbf{B} \times W \to \mathbf{k}$ is nondegenerate, the property $\overline{R_2 \circ {}^rW} = \mathfrak{B}$ is shown as in (1).

(b) The proof of the sufficient condition is the same as that of (2). ∎

Corollary 919. *Assume* $\mathbf{G} = \mathbb{R}^n$ *with the data of Theorem 918. Then* W' *is a commutative domain and so is any convolution algebra* $\mathbf{B}$. *A module* $M = \mathbf{B}^k / \mathbf{B}^q \circ R_1$ *is torsion-free if, and only if it admits an embedding in some* $\mathbf{B}^r$. *Therefore, Theorem 918 gets the following form:*

(1) If $\mathbf{H}$ *is any subgroup of* $\mathbb{R}^n$, $\mathbf{B} = \mathbf{A} * \mathbf{H}$ *and* M *is torsion-free, then* $N = \mathbf{B}^q \circ R_1$ *is closed in* $\mathbf{B}^k$, *and* $\mathfrak{B}$ *admits a dense image representation.*
(2) If $\langle -, - \rangle : \mathbf{B} \times W \to \mathbf{k}$ *is nondegenerate, i.e., if* $\mathbf{B}$ *is wealkly* dense in* W', *then* M *is torsion-free if, and only if* M *is Hausdorff and* $\mathfrak{B}$ *admits a dense image representation.*

Corollary 919(2) generalizes the result in Example 895.

The matrix R_2 in Theorem 918(1) can be constructed using the Gröbner basis algorithm (see Remark 914(i)).

5.7.5 *Differential Systems with Distributed Shifts over Lie Groups*

5.7.5.1 The Ring $\mathcal{H}(\mathbf{B})$ and Its Basic Properties

Recall that $\mathbf{A}$ is a Noetherian subalgebra of $\mathrm{U}_{\mathbf{k}}(\mathfrak{g})$ (Notations 904 and 909). In addition, in the sequel we make the following

Assumption 920. $\mathbf{H}$ *is a polycyclic subgroup of* $\mathbf{G}$ *such that all nonzero elements of* $\mathbf{B} = \mathbf{A} * \mathbf{H}$ *are invertible distributions (Corollary and Definition 915).*

Example 921. *(i) If* $\mathbf{G} = \mathbb{R}^n$, *Assumption 920 is satisfied if* $\mathbf{A} \triangleq \mathrm{U}_{\mathbf{k}}(\mathfrak{g}) = \mathbf{k}\left[\frac{\partial}{\partial x_1}, ..., \frac{\partial}{\partial x_n}\right]$ *and* $\mathbf{H}$ *is any finitely generated subgroup of* $\mathbf{G}$ *by Remark 914(vi).*
(ii) Let $\mathbf{G}$ *be a solvable simply connected Lie group such that* $\mathbf{A} \triangleq \mathrm{Z}_{\mathbf{k}}(\mathfrak{g})$ *is Noetherian (not all solvable Lie group have this property* (**1.8.3.3**)*). Then Assumption 920 is satisfied with* $\mathbf{H} = \{e\}$ *by Remark 914(iv).*

By Theorem 372, $\mathbf{B}$ is a Noetherian domain, thus an Ore domain, and it admits a quotient division ring $\mathrm{Q}(\mathbf{B})$ of *left fractions* $(a)^{\circ -1} \circ b$ and of *right fractions* $b \circ (a)^{\circ -1}$, where $c = (a)^{\circ -1} \circ b$ (resp., $c = b \circ (a)^{\circ -1}$) means by definition that $a \circ c = b$ (resp., $c \circ a = b$).

Recall that $\mu \circ \nu = \mu \star \nu$ for any $\mu, \nu \in W'$ by Lemma 905(ii).

Theorem and Definition 922. *(1) Let*

$$\mathcal{H}(\mathbf{B}) \triangleq \left\{ b \circ (a)^{\circ -1} \in \mathrm{Q}(\mathbf{B}) : \exists \mu \in W',\ b = \mu \circ a \right\}.$$

Then $\mathcal{H}(\mathbf{B})$ *is a subring of both* $\mathrm{Q}(\mathbf{B})$ *and* W', *and* $\mathcal{H}(\mathbf{B}) = \mathrm{Q}(\mathbf{B}) \cap W'$.

(2) The following conditions are equivalent for a right fraction $b \circ (a)^{\circ -1} \in \mathrm{Q}(\mathbf{B})$*:*
(i) $b \circ (a)^{\circ -1} \in \mathcal{H}(\mathbf{B})$*;*
(ii) $\ker_W (a\circ) \subseteq \ker_W (b\circ)$.
(3) $\mathcal{H}(\mathbf{B})$ *is an Ore domain, thus is weakly finite (Exercise 481(2)), and its quotient division ring is* $\mathrm{Q}(\mathbf{B})$.
(4) When $\mathbf{G} = \mathbb{R}$ *and* $\mathbf{H} \neq \{e\}$ *is cyclic,* $\mathcal{H}(\mathbf{B})$ *is not Noetherian.*

Proof. (1) is clear. (2) By definition, $b \circ (a)^{\circ -1} \in \mathrm{Q}(\mathbf{B})$ belongs to $\mathcal{H}(\mathbf{B})$ if, and only if there exists $\mu \in W'$ such that $b = \mu \circ a$. Let $w \in \ker_W a\circ$; then $a \circ w = 0$ and $b \circ w = (\mu \circ a) \circ w = \mu \circ (a \circ w) = 0$ by Lemma 905(iii), which implies that $\ker_W (a\circ) \subseteq \ker_W (b\circ)$.

Conversely, let $a, b \in \mathbf{B}$, $a \neq 0$, such that $\ker_W (a\circ) \subseteq \ker_W (b\circ)$. For any $\mu \in \mathbf{B}$, let $\mathfrak{B}_\mu = (W' \circ \mu)^\perp$. We have $\mathfrak{B}_\mu = \ker_W (\mu\circ)$ and $\overline{W' \circ \mu} = \mathfrak{B}_\mu^\perp$ by Theorem 902(i) since $\langle -, - \rangle : W' \times W \to \mathbf{k}$ is nondegenerate. By Corollary and Definition 915(iii), $\overline{W' \circ \mu} = W' \circ \mu$. Thus the assumption $\ker_W (a\circ) \subseteq \ker_W (b\circ)$ implies $W' \circ b \subseteq W' \circ a$, hence $b \in W' \circ a$ and $b \circ (a)^{\circ -1} \in \mathcal{H}(\mathbf{B})$.

(3) $\mathcal{H}(\mathbf{B})$ is an integral domain since $\mathcal{H}(\mathbf{B}) \subseteq \mathrm{Q}(\mathbf{B})$. Let us show that $\mathcal{H}(\mathbf{B})$ is right Ore, i.e., whenever $f_1, f_2 \in \mathcal{H}(\mathbf{B})^\times$, $f_1 \circ \mathcal{H}(\mathbf{B}) \cap f_2 \circ \mathcal{H}(\mathbf{B}) \neq 0$, in other words there exist $g_1, g_2 \in \mathcal{H}(\mathbf{B})^\times$ such that $f_1 \circ g_1 = f_2 \circ g_2$. This happens if, and only if $g_2 = \left(f_2^{\circ -1} \circ f_1\right) \circ g_1$ where $f_2^{\circ -1} \circ f_1 \in \mathrm{Q}(\mathbf{B})$. There exist $b, a \in \mathbf{B}^\times \subseteq W'^\times$ such that $f_2^{\circ -1} \circ f_1 = b \circ (a)^{\circ -1}$. Take $g_1 = a$, $g_2 = b$. Then $g_1, g_2 \in \mathcal{H}(\mathbf{B})^\times$ and $f_1 \circ g_1 = f_2 \circ g_2$ as desired. The definition of $\mathcal{H}(\mathbf{B})$ is left/right symmetric, thus $\mathcal{H}(\mathbf{B})$ is a left Ore domain too.

An Ore domain is weakly finite according to (Exercise 481(4)). We have $\mathbf{B} \subseteq \mathcal{H}(\mathbf{B}) \subseteq \mathrm{Q}(\mathbf{B})$, thus the quotient division ring of $\mathcal{H}(\mathbf{B})$ is $\mathrm{Q}(\mathbf{B})$.

(4) When $\mathbf{G} = \mathbb{R}$ and $\mathbf{H} \neq \{e\}$ is cyclic, then $\mathcal{H}(\mathbf{B})$ is not Noetherian by ([137], Prop. 3.1(c)). ∎

5.7.5.2 Cogenerator Properties

Since $\mathcal{H}(\mathbf{B}) \subseteq W'$, W has a canonical structure of $\mathcal{H}(\mathbf{B})$-module, by restriction of the ring of scalars.

Theorem 923. *(1) For* $i = 1, 2$*, let* $R_i \in {}^{q_i}\mathcal{H}(\mathbf{B})^k$*, let* $U_i = W'^{q_i} \circ R_i$ *and let* $\mathfrak{B}_i = U_i^\perp = \ker_W (R_i\circ)$*. Assume that* $\operatorname{rk} R_2 = q_2$*. Then:*
(i) The map $R_2\circ : {}^kW \to {}^{q_2}W$ *is surjective; in particular all nonzero distributions belonging to* $\mathcal{H}(\mathbf{B})$ *are invertible and the* $\mathcal{H}(\mathbf{B})$*-module* W *is both divisible and faithful.*
(ii) $\mathfrak{B}_2^\perp = U_2^{\perp\perp} = U_2$ *where* $\mathfrak{B}_2^\perp = \mathfrak{B}_2^{\perp(W')} \triangleq \left\{w' \in {}^kW' : w' \circ \mathfrak{B}_2\right\} = 0$.
(iii) $\ker_W (R_2\circ) \subseteq \ker_W (R_1\circ)$ *if, and only if there exists* $X \in {}^{q_i}\mathcal{H}(\mathbf{B})^{q_2}$ *such that* $R_1 = X \circ R_2$.
(2) The $\mathcal{H}(\mathbf{B})$*-module* W *is a cogenerator for the full subcategory* $\mathcal{D}$ *of* ${}_{\mathcal{H}(\mathbf{B})}\mathbf{Mod}$*, the objects* M *which have finite free resolution of length* ≤ 1.

Proof. (1): (i): The assumption means that R_2 is right-invertible over $\mathrm{Q}(\mathbf{B})$ and implies that R_2 is right-invertible over $\mathrm{Q}(\mathbf{B})$, i.e., there exists a matrix $X_2 \in {}^k\mathrm{Q}(\mathbf{B})^{q_2}$ such that $R_2 \circ X_2 = I_{q_2}$. Since $\mathbf{B}$ is an Ore domain, for each $j \in \{1, ..., q_2\}$, all entries of the j-th column of X_2 have a common right denominator $\mu_j \in \mathbf{B}^\times$ (Corollary 329), i.e., there exist elements $(Y_2)_{ij} \in \mathbf{B}$ $(1 \leq i \leq k)$ such that $(X_2)_{ij} = (Y_2)_{ij} \circ \mu_j^{\circ-1}$; then $(Y_2)_{ij} = (X_2)_{ij} \circ \mu_j \in \mathbf{B}$ and

$$R_2 \circ Y_2 = R_2 \circ \left(X_2 \circ \mathrm{diag}\left(\mu_j\right)_{1\leq j\leq q_2}\right) = \mathrm{diag}\left(\mu_j\right)_{1\leq j\leq q_2}.$$

Therefore, for any $w \in {}^{q_2}W$,

$$R_2 \circ (Y_2 \circ w) = \mathrm{diag}\left(\mu_j\right)_{1\leq j\leq q_2} \circ w = \left(\mu_j \circ w_j\right)_{1\leq j\leq q_2}.$$

By hypothesis, each distribution μ_j is invertible and $\mu_j \circ : W \to W$ is surjective. Therefore, $R_2 \circ : {}^kW \to {}^{q_2}W$ is surjective too.

(ii): The transpose of $R_2 \circ : {}^kW \to {}^{q_2}W$ is $\circ R_2 : W'^{q_2} \to W'^k$ and the spaces ${}^kW$, ${}^{q_2}W$ are Fréchet spaces. Therefore, by (i) and ([162], Lemma 16.5.8), $\circ R_2 : W'^{q_2} \to W'^k$ is injective and $W'^{q_2} \circ R_2$ is weakly* closed in W'^k. Therefore, $\mathfrak{B}_2^\perp = U_2^{\perp\perp} = U_2$ by Theorem 902(i) since $\langle -, - \rangle : {}^kW' \times W^k \to \mathbf{k}$ is nondegenerate.

(iii): The sufficient condition is obvious. Let us prove the necessary condition. If $\mathfrak{B}_2 \subseteq \mathfrak{B}_1$, then $W'^{q_1} \circ R_1 \subseteq \mathfrak{B}_1^\perp \subseteq \mathfrak{B}_2^\perp$ and $\mathfrak{B}_2^\perp = W'^{q_2} \circ R_2$ by (ii). Thus $W'^{q_1} \circ R_1 \subseteq W'^{q_2} \circ R_2$, hence every row of R_1 and therefore R_1 itself are left-multiples of R_2, i.e., there exists $X \in {}^{q_2}W'^{q_1}$ such that $R_1 = X \circ R_2$. Since $\mathrm{rk}\, R_2 = q_2$, after a possible permutation of columns we obtain

$$R_i = \left[\, R_i' \; R_i'' \,\right] \in {}^{q_i}\mathcal{H}(\mathbf{B})^{q_i+(k-q_i)} \quad (i = 1, 2)$$

where R_2' is invertible over $\mathrm{Q}(\mathbf{B})$. Then $R_1' = X \circ R_2'$ and $X = R_1' \circ R_2'^{\circ-1}$. As in (i) we obtain $Y \in {}^{q_1}\mathbf{B}^{q_2}$ and $\nu_i \in \mathbf{B}^\times$ $(1 \leq i \leq q_2)$ with $R_2' \circ Y = \mathrm{diag}\left(\nu_i\right)_{1\leq i\leq q_2} = \Delta$, hence

$$X = R_1' \circ R_2'^{\circ-1} = R_1' \circ Y \circ \Delta^{\circ-1} \in {}^{q_1}\left(\mathrm{Q}(\mathbf{B}) \bigcap W'\right)^{q_2} = {}^{q_1}\mathcal{H}(\mathbf{B})^{q_2}.$$

(2) is a consequence of (1)(iii) and of Theorem 827(iii). ∎

Remark 924. *Theorem 923(2) generalizes the result recalled in Example 896(iii).*

Corollary 925. *Let $R \in {}^q\mathcal{H}(\mathbf{B})^k$, let $N = \mathcal{H}(\mathbf{B})^q \circ R$, let $M = \mathcal{H}(\mathbf{B})^k / N$, and let $\mathfrak{B} = N^\perp = \ker_W (R\circ)$.*

(i) $\mathrm{rk}\, R = q$ *if, and only if* $R \circ {}^kW = {}^qW$.
(ii) If R has a left-inverse X with coefficients in W', i.e., $X \in {}^kW'^q$, $X \circ R = \delta_e I_k$, then $\mathfrak{B} = 0$.
(iii) Assuming that $\mathrm{rk}\, R = q$, $\ker_W (R\circ) = 0$ *if and only if* $R \in \mathrm{GL}_q(\mathcal{H}(\mathbf{B}))$.

(iv) In particular, for $R \in \mathcal{H}(\mathbf{B})$, $\ker_W(R\circ) = 0$ if, and only if R is a unit in $\mathcal{H}(\mathbf{B})$.

Proof. (i): The necessary condition follows from Theorem 923(i).

Conversely, assume $\operatorname{rk} R < q$. Since $\operatorname{rk} R$ is the dimension of the left $\mathrm{Q}(\mathbf{B})$-vector space generated by the rows of R (Theorem 409(i)), there exists a nonzero row $y \in \mathrm{Q}(\mathbf{B})^q$ such that $y \circ R = 0$. Let d be a common left denominator of all elements of y and $x = d \circ y \in \mathcal{H}(\mathbf{B})^q$; then $x \circ R = 0$ and $x \neq 0$. Thus $R \circ ({}^k W) \subseteq \ker_W(x\circ) \subseteq {}^q W$. Assuming that $\ker_W(x\circ) = {}^q W$, then $x \circ {}^q W = 0$ and $x_i \circ W = 0$ for all $i \in \{1, ..., q\}$, a contradiction since the $\mathcal{H}(\mathbf{B})$-module W is faithful by Theorem 923(i). Therefore $R \circ ({}^k W) \subsetneq {}^q W$.

(iii): Assume that $\operatorname{rk} R = q$. Clearly, $\ker_W(R\circ) = 0$ if, and only if $\ker_W(R\circ) \subseteq \ker_W(I_k\, \delta_e \circ)$, and by Theorem 923(iii) this is equivalent to the existence of a matrix $X \in {}^k\mathcal{H}(\mathbf{B})^{q_2}$ such that $I_k\, \delta_e = X \circ R$, thus R is left-invertible over $\mathcal{H}(\mathbf{B})$ and $\operatorname{rk} R = k$. Therefore, R is square, and since $\mathcal{H}(\mathbf{B})$ is weakly finite by Theorem and Definition 922(2), the left-invertibility of the square matrix R over $\mathcal{H}(\mathbf{B})$ is equivalent to its invertibility over that ring.

(ii) is obvious and (iv) is a particular case of (iii). ■

Remark 926. *Consider the data of Theorem 923 and Corollary 925 for the special case $\mathbf{G} = \mathbb{R}$.*

(a) Statement (iii) of Theorem 923 is proved in ([139], Theorem 5.3) without any assumption on $\operatorname{rk} R_2$ *when $\mathbf{H} \neq 0$ is cyclic. In that case, indeed, $\mathcal{H}(\mathbf{B})$ is a Bézout domain – and even an* elementary divisor ring, *as already said –, thus R_2 is left-equivalent to a matrix* $\begin{bmatrix} Q \\ 0 \end{bmatrix}$ *where Q is full row rank* (**3.6.1.1**), *thus the assumption on* $\operatorname{rk} R_2$ *becomes superfluous. We deduce that* when $\mathbf{H}$ is cyclic, *the $\mathcal{H}(\mathbf{B})$-module W is a cogenerator for* ${}_{\mathcal{H}(\mathbf{B})}\mathbf{Mod}^{fp}$.
When $\mathbf{H}$ *is not cyclic, the situation is quite different: see Exercise 962.*
(b) The units in $\mathcal{H}(\mathbf{B})$ are explicitly described in Corollary 931 below when $\mathbf{G} = \mathbb{R}^n$.

5.7.5.3 Differential Systems with Distributed Shifts over $\mathbb{R}^n$

In what follows, $\mathbf{G} = \mathbb{R}^n$ and $\mathbf{H}$ is a finitely generated subgroup of $\mathbf{G}$. The ring of all entire complex-valued functions on $\mathbb{C}^n$ is denoted by $\mathcal{O}(\mathbb{C}^n)$, as usual. For $s, z \in \mathbb{C}^n$, we set

$$s \bullet z \triangleq \sum_{1 \leq i \leq n} s_i z_i, \quad |s| = \left(\sum_{1 \leq i \leq n} |s_i|^2 \right)^{1/2}.$$

Let $\mathbf{A} \triangleq \mathrm{U}_{\mathbf{k}}(\mathfrak{g}) = \mathbf{k}\left[\frac{\partial}{\partial x_1}, ..., \frac{\partial}{\partial x_n}\right]$ and $\mathbf{B} = \mathbf{A} * \mathbf{H}$. The convolution algebra $\mathbf{B}$ is now a commutative domain, thus the left fractions $a^{\circ -1} \circ b \in \mathrm{Q}(\mathbf{B})$

coincide with the right fractions $b \circ a^{\circ -1} \in \mathrm{Q}(\mathbf{B})$ and are denoted by $b/^{\circ}a$ in the sequel. As is well known, the Laplace transform

$$\mathcal{L} : W' \ni \mu \mapsto \hat{\mu}(s) \triangleq \int_{\mathbb{R}^n} e^{-s \bullet x} d\mu(x), \quad s \in \mathbb{C}^n$$

yields the following isomorphisms:

$$\mathbf{A} \cong \mathbb{C}[s], \quad \mathbf{B} \cong \mathcal{PE}(\mathbf{H}) \triangleq \sum_{\tau \in \mathbf{H}} \mathbb{C}[s] e^{-\tau \bullet s}, \quad W' \cong PW(\mathbb{C}^n)$$

where $PW(\mathbb{C}^n)$ is the set of all functions $f \in \mathcal{O}(\mathbb{C}^n)$ for which there exist $c > 0$, $a > 0$ and $N \in \mathbb{N}$ such that

$$|f(s)| \leq c(1 + |s|)^N e^{a|\mathrm{Re}(s)|}.$$

The elements of $\mathcal{PE}(\mathbf{H})$ are called *complex exponential-polynomial functions* [17] and the third isomorphism is given by the Paley-Wiener-Schwartz theorem ([92], (22.18.7)). We have the following [1]:

Definition 927. *Let $\mu \in W'$ and let $\hat{\mu}$ be its Laplace transform. Then $\hat{\mu}$ is called* slowly decreasing *if there exist constants A, B and m such that*

$$\sup_{|s - s_0| \leq A \log(2 + |s_0|)} |\hat{\mu}(s)| \geq B(1 + |s_0|)^{-m}.$$

For other characterizations of slowly decreasing $\hat{\mu}$, see ([162], Theorem 16.3.10 and Definition 16.3.12). The following is proved in ([162], Theorems 16.3.10, 16.5.7 and 16.5.19), and in ([1], Appendix).

Lemma 928. *For $\mu \in W'$, consider the convolution equation*

$$\mu \star u = f \tag{5.64}$$

where the distribution μ is given and the distribution u is looked for. The following conditions are equivalent:

(i) If $f \in W'$ and $\hat{f}(s) / \hat{\mu}(s) \in \mathcal{O}(\mathbb{C}^n)$, then there exists a (unique) $u \in W'$ such that $\hat{f}(s) / \hat{\mu}(s) = \hat{u}(s)$ (which implies (5.64));
(ii) Equation (5.64) has a solution $u \in W$ for any $f \in W$, i.e., μ is invertible;
(iii) Equation (5.64) has a solution $u \in \mathcal{D}'(\mathbb{R}^n)$ for any $f \in \mathcal{D}'(\mathbb{R}^n)$;
(iv) $\hat{\mu}$ is slowly decreasing.

Malgrange established the equivalence between Corollary and Definition 915(1)(v) and Lemma 928(i) for $0 \neq \mu \in \mathcal{E}'(\mathbb{R}^n)$ ([230], Corollary on p. 310). Ehrenpreis [100] introduced the notion of slowly decreasing $\hat{\mu}$ and proved the equivalences (ii)⇔(iii)⇔(iv) of Lemma 928. The elements of $\mathcal{PE}(\mathbf{H})$ are easily seen to be slowly decreasing, and this proves Ehrenpreis' result in Remark 914(vii).

Lemma 929. *Let $F, G \in \mathcal{PE}(\mathbf{H})$, $G \neq 0$, be such that $F/G \in \mathcal{O}(\mathbb{C}^n)$. There exist $H \in \mathcal{PE}(\mathbf{H})$ and $0 \neq P \in \mathbb{C}[s]$ such that $F/G = H/P$.*

Proof. (a) By ([17], Main Theorem), there exist $H \in \mathcal{PE}(\mathbb{R}^n)$ and $0 \neq P \in \mathbb{C}[s]$ such that $F/G = H/P$.

(b) For any $0 \neq L(s) \in \mathcal{PE}(\mathbb{R}^n)$, write $L(s) = \sum_{\tau \in \mathbf{T}} l_\tau(s) e^{-\tau \bullet s}$ where $\mathbf{T}$ is a finite subset of $\mathbb{R}^n$ and $0 \neq l_\tau(s) \in \mathbb{C}[s]$ for any $\tau \in \mathbf{T}$. The elements $\tau \in \mathbf{T}$ are the exponents [153] (or, more loosely speaking, the "frequencies" [17]) of $L(s)$. The above element $H \in \mathcal{PE}(\mathbb{R}^n)$ is constructed by induction in ([17], Proof of Theorem 1). At each step, the exponents involved are $\mathbb{Z}$-linear combinations of those of F and G , i.e., belong to $\mathbf{H}$; see ([17], (2.20)-(2.26)) for more details. Since the procedure stops in a finite number of steps, $H \in \mathcal{PE}(\mathbf{H})$. (When $n = 1$, the fact that the exponents of H are $\mathbb{Z}$-linear combinations of those of F and G is used in the proof of ([153], Theorem 5.9).) ∎

Theorem 930. *The following rings coincide with the ring $\mathcal{H}(\mathbf{B})$:*

$$\mathcal{H}_1(\mathbf{B}) = \left\{ b/^\circ a \in \mathrm{Q}(\mathbf{B}) : \hat{b}(s)/\hat{a}(s) \in PW(\mathbb{C}^n) \right\},$$
$$\mathcal{H}_2(\mathbf{B}) = \left\{ b/^\circ a \in \mathrm{Q}(\mathbf{B}) : \hat{b}(s)/\hat{a}(s) \in \mathcal{O}(\mathbb{C}^n) \right\},$$
$$\mathcal{H}_3(\mathbf{B}) = \left\{ b/^\circ a \in \mathrm{Q}(\mathbf{B}) \cap W' : b \in \mathbf{B},\ a \in \mathbf{A}^\times \right\}.$$

Proof. (a) By definition, if $u = b/^\circ a \in \mathcal{H}(\mathbf{B})$, then $\hat{u}(s) \in PW(\mathbb{C}^n) \subseteq \mathcal{O}(\mathbb{C}^n)$ which proves that $\mathcal{H}(\mathbf{B}) \subseteq \mathcal{H}_1(\mathbf{B})$. The inclusion $\mathcal{H}_1(\mathbf{B}) \subseteq \mathcal{H}_2(\mathbf{B})$ is obvious.

(b) Let $b \in \mathbf{B}$, $a \in \mathbf{B}^\times$ be such that $\hat{b}(s)/\hat{a}(s) \in \mathcal{O}(\mathbb{C}^n)$. By Lemma 928, the distribution a is invertible and there exists $u \in W'$ such that $a \circ u = b$. Therefore, $b/^\circ a \in \mathcal{H}(\mathbf{B})$, hence $\mathcal{H}_2(\mathbf{B}) \subseteq \mathcal{H}(\mathbf{B})$.

(c) Last, $\mathcal{H}_3(\mathbf{B}) = \mathcal{H}_2(\mathbf{B})$ by Lemma 929. ∎

Corollary 931. *An element $T \in \mathcal{H}(\mathbf{B})$ is a unit if, and only if it has the form $T = \alpha\delta_\tau$, $\tau \in \mathbf{H}$, $\alpha \in \mathbf{k}^\times$, i.e., is also a unit in $\mathbf{B}$.*

Proof. The indicated elements are obviously units. The ring $\mathbf{B}$ can be written $\mathbf{A}\left[\sigma, \sigma^{-1}\right]$ with the notation in Remark 917(iii) and $\mathbf{A} = \mathbf{k}[\partial]$, $\partial = (\partial/\partial x_i)_{1 \leq i \leq n}$. Assume that $T_1 \in \mathcal{H}(\mathbf{B})$ is invertible and that

$$\delta_0 = T_1 \circ T_2,\ T_i = b_i/^\circ a_i,\ b_i \in \mathbf{B},\ a_i \in \mathbf{A}^\times\ (i = 1, 2)$$

therefore $b_1 \circ b_2 = a_1 \circ a_2$ in $\mathbf{B} = \mathbf{A}\left[\sigma, \sigma^{-1}\right]$, where we have used $\mathcal{H}(\mathbf{B}) = \mathcal{H}_3(\mathbf{B})$. We infer $b_1 = a_1' \circ \delta_{\tau_1}$ for some $a_1' \in \mathbf{A}$ and $\tau_1 \in \mathbf{H}$, thus $T_1 = b_1/^\circ a_1 = (a_1'/^\circ a_1) \circ \delta_{\tau_1}$. Since $\hat{T}_1(s)$ is entire, so is the rational function $\hat{a}_1'(s)/\hat{a}_1(s) = \hat{a}_1''(s)$. Thus $\hat{a}_1''(s)$ is a polynomial and $T_1 = a_1'' \circ \delta_{\tau_1}$ with $a_1'' \in \mathbf{A}$. The same argument applies to T_2 and furnishes

$$\delta_0 = T_1 \circ T_2 = (a_1'' \circ a_2'') \circ \delta_{\tau_1+\tau_2}$$

thus $a_1'' \circ a_2'' = \delta_0$ and $\tau_1 + \tau_2 = 0$. Since $\mathbf{A}$ is a polynomial algebra, $a_1'' = \alpha \in \mathbf{k}^\times$, hence $T = \alpha\delta_\tau$ as asserted. ∎

Remark 932. *(a) We consider the space $\mathcal{PE} \triangleq \mathcal{PE}(\mathbb{R}^n)$ of real polynomial-exponential functions ($\mathcal{PE} \subseteq \mathcal{E}(\mathbb{R}^n)$). Let $R_i \in {}^{q_i}\mathcal{E}'(\mathbb{R}^n)^k$ ($i = 1,2$). As shown by Malgrange ([230], Preliminaries, Sect. 4, Theorem 3), the following conditions are equivalent:*
(i) $\ker_{\mathcal{PE}}(R_1\circ) \subseteq \ker_{\mathcal{PE}}(R_2\circ)$;
(ii) there exists a matrix $X \in {}^{q_1}\mathcal{O}(\mathbb{C}^n)^{q_2}$ such that

$$\hat{R}_2 = X\,\hat{R}_1.$$

(b) Let $R_i \in {}^{q_i}\mathcal{E}'(\mathbb{R})^k$ ($i = 1,2$) (notice that $n = 1$). Vettori and Zampieri ([352], Theorem 3), proved that Condition (ii) above is equivalent to:
(i') $\ker_W(R_1\circ) \subseteq \ker_W(R_2\circ)$ ($W = \mathcal{E}'(\mathbb{R})$).
This result is based on Schwartz's theorem on mean periodic functions, extended to vector-valued functions [321], [185], [268], which implies that for any natural integer k, every shift-invariant closed subspace of ${}^kW$ is spanned by the ${}^k\mathbb{C}$-valued exponential-polynomial functions it contains (when $n = 1$).
(c) As proved by Gurevič ([152], Theorem 3.2), there exists a column $R \in {}^k\mathcal{E}'(\mathbb{R}^n)$, $n \geq 2$, $k = 6$, such that

$$0 = \ker_{\mathcal{PE}}(R\circ) = {}^k(\mathcal{PE}) \bigcap \ker_W(R\circ), \quad \ker_W(R\circ) \neq 0.$$

See Exercise 963.
(d) With the notation from Remark 917(iii), the ring $\mathcal{H}(\mathbf{B}) = \mathcal{H}_3(\mathbf{B})$ is isomorphic to $\mathbf{D} \cap \mathcal{O}(\mathbb{C}^n)$ where $\mathbf{D} = \mathbf{k}(s)[\sigma]$, $s = (s_1, ...s_n)$, $\sigma = (\sigma_1, ..., \sigma_m)$. The ring $\mathbf{D}$ is projective-free (Theorem 581(i)), thus a finitely presented $\mathbf{D}$-module M has a full row rank matrix of definition if, and only M has projective dimension ≤ 1. Assume $m > 1$. Since $\mathbf{D}$ has global dimension m (Theorem 635(2)), such a module M is very peculiar, and so are probably also the finitely presented $\mathcal{H}(\mathbf{B})$-modules in Theorem 923(iv), for we conjecture that $\mathcal{H}(\mathbf{B})$ has global dimension $\geq m$.

5.8 Exercises

Exercise 933. *Consider the Galois connection in Proposition 810. Assume that the ring $\mathbf{A}$ is left coherent and that W is a cogenerator.*
(i) Prove the following where $N_1, N_2 \in \mathcal{S}_k$ and $\mathfrak{B}_1, \mathfrak{B}_2 \in {}^kW$:

$$(N_1 + N_2)^\perp = N_1^\perp \cap N_2^\perp, \quad (\mathfrak{B}_1 \cap \mathfrak{B}_2)^\perp = \mathfrak{B}_1^\perp + \mathfrak{B}_2^\perp, \tag{5.65}$$

$$(N_1 \cap N_2)^\perp = N_1^\perp + N_2^\perp, \quad (\mathfrak{B}_1 + \mathfrak{B}_2)^\perp = \mathfrak{B}_1^\perp \cap \mathfrak{B}_2^\perp. \tag{5.66}$$

(ii) Deduce from the above that this Galois connection is a lattice-antiisomorphism (Definition 112).
(Hint: to prove (5.66), show that whenever N_1 and N_2 are f.p., $N_1 \cap N_2$ is again f.p. using Lemma 509.).

Exercise 934. *(1) Let* $g : t \mapsto t \arctan(t) - \frac{1}{2} \ln(t^2 + 1)$.
(i) Prove that $g \in \mathfrak{O}$ *(i.e., is an Ore function: see Lemma and Defintion 835).*
(ii) Let $f \in \mathfrak{O}$*; prove that* $f \circ g \in \mathfrak{O}$.
(2) Determine other examples of functions $g \in \mathfrak{O}$ *such that* $f \circ g \in \mathfrak{O}$ *whenever* $f \in \mathfrak{O}$.

Exercise 935. *Let* $\mathbf{A} = A_0(\mathbb{C})$ *and* $\mathfrak{m} = \mathbf{A}(t - x)$, $x \in \mathbb{R}$.

(i) Prove that $\mathfrak{m}$ *is a left maximal ideal.*
(ii) Prove that $\mathbf{A}/\mathfrak{m} \cong \bigoplus_{n \geq 0} \mathbb{C}\, \delta_x^{(n)} \subseteq \mathcal{B}(\mathbb{R})$.

Exercise 936. *Let* $R(\partial) = \partial t^2 \in A_1(\mathbb{C})$*, and consider the system* $R(\partial)\, y = 0$.

(i) Using Theorem and Definition 185(ii), calculate $\dim_{\mathbb{C}} \ker_{\mathcal{B}(\mathbb{R})}(R\bullet)$.
(ii) Using the commutation rule, and Theorem and Definition 185(ii), calculate σ_0 *and show that* $\ker_{\mathcal{B}(\mathbb{R})}(R\bullet) \subseteq \mathcal{D}'(\mathbb{R})$.
(iii) Prove that $\ker_{\mathcal{B}(\mathbb{R})}(R\bullet) = \mathbb{C}\,\delta \oplus \mathbb{C}\,\dot{\delta} \oplus \mathbb{C}\, \mathfrak{f.p.}\left(1/t^2\right)$.

Exercise 937. *Let* $R(\partial) = t(\partial t - 1) \in A_1(\mathbb{C})$.

(a) Answer the same questions as Questions (i) and (ii) of Exercise 936.
(b) Show that $\ker_{\mathcal{B}(\mathbb{R})}(R\bullet) = \mathbb{C}\,t \oplus \mathbb{C}\,t\,\Upsilon \oplus \mathbb{C}\,\delta$*, where* Υ *is the Heaviside function.*

Exercise 938. *Let* $R(\partial) = t^2\partial - 1 \in A_1(\mathbb{C})$.

(i) Calculate $\dim_{\mathbb{C}} \ker_{\mathcal{B}(\mathbb{R})}(R\bullet)$.
(ii) Prove that $\ker_{\mathcal{B}(\mathbb{R})}(R\bullet) \subsetneq \mathcal{D}'(\mathbb{R})$.
(iii) Prove that $\ker_{\mathcal{B}(\mathbb{R})}(R\bullet) = \mathbb{C}\, f \oplus \mathbb{C}\, e^{-(1/(t+i0))} \oplus \mathbb{C}\left[e^{1/z}\right]$ *where* $f \in \mathcal{E}(\mathbb{R})$ *is the function defined by* $f(t) = e^{-t}, t \neq 0$*, and* $f(0) = 0$.

Exercise 939. *Let* $R(\partial) = 1 + t^2 \in A_1(\mathbb{C})$.

(i) Prove that $\operatorname{coker}_{A_1(\mathbb{C})}(\bullet R) \neq 0$*, but that* $\ker_{\mathcal{B}(\mathbb{R})}(R\bullet) = 0$.
(ii) Conclude that the $A_1(\mathbb{C})$*-module* $\mathcal{B}(\mathbb{R})$ *is not a cogenerator.*

Exercise 940. *Consider a Rosenbrock description (5.28) over a left Ore domain* $\mathbf{A}$*, let* W *be an injective cogenerator, and consider the set*

$$\mathfrak{B}_c = \left\{ \begin{array}{c} \mathbf{u} \in {}^m W, \exists \begin{bmatrix} \boldsymbol{\xi} \\ \mathbf{y} \end{bmatrix} \in {}^{r+p} W : \\ \begin{bmatrix} D \\ W \end{bmatrix} \mathbf{u} = \begin{bmatrix} D & 0 \\ -Q & I_p \end{bmatrix} \begin{bmatrix} \boldsymbol{\xi} \\ \mathbf{y} \end{bmatrix} \end{array} \right\}$$

Using a rationale similar to that in the proof of Lemma and Definition 856, show that the Rosenbrock description (5.28) is admissible if, and only if $\mathfrak{B}_c = {}^m W$.

Exercise 941. *Consider the Rosenbrock description*

$$\begin{bmatrix} \partial & -1 \\ 0 & \partial \\ 0 & t \end{bmatrix} \begin{bmatrix} \xi_1 \\ \xi_2 \end{bmatrix} = \begin{bmatrix} 1 \\ 0 \\ 0 \end{bmatrix} u,$$
$$y = \begin{bmatrix} 1 & 0 \end{bmatrix} \begin{bmatrix} \xi_1 \\ \xi_2 \end{bmatrix}$$

over $\mathbf{A} = \mathbb{C}[\partial]$. *Show that this Rosenbrock description is admissible and calculate its transfer function.*

Exercise 942. *Consider the state-space system (5.34), (5.35). Let* $\zeta = V\,x$ *where* $V \in \mathrm{GL}_n(\mathbf{K})$ *is a change of basis matrix. Show that the new state* ζ *satisfies the state equation* $\partial\zeta = \check{F}\,\zeta + \check{G}\,u$ *where* $\check{F}\,V = V^{\boldsymbol{\alpha}}\,F + V^{\boldsymbol{\delta}}$ *(i.e.,* $\check{F}$ *is* $(\boldsymbol{\alpha}, \boldsymbol{\delta})$*-conjugate to* F *by* V*) and* $\check{G} = V^{\alpha}\,G$.

Exercise 943. *(1) Consider the control system*

$$D(\partial)\,y = N(\partial)\,u$$

where $D(\partial)$ *and* $N(\partial)$ *belong to the first Weyl algebra* $A_1(\mathbb{C})$, $D(\partial)$ *is monic (Definition 349(i)) and* $d^{\circ}(N) < d^{\circ}(D)$. *Put* $D(\partial)$ *and* $N(\partial)$ *into the form of* right *polynomials* (**2.7.4.2**), *i.e.,*

$$\begin{aligned} D(\partial) &= \partial^n + \partial^{n-1}a_1 + \ldots + a_n, \\ N(\partial) &= \partial^{n-1}b_1 + \ldots + b_n \end{aligned}$$

and show that this system has a state-space realization of the form

$$\dot{x} = \begin{bmatrix} -a_1 & 1 & 0 & \cdots & 0 \\ -a_2 & 0 & \vdots & \ddots & \vdots \\ \vdots & 0 & \vdots & \ddots & 0 \\ \vdots & \vdots & \vdots & 0 & 1 \\ -a_n & 0 & 0 & \cdots & 0 \end{bmatrix} x + \begin{bmatrix} b_1 \\ \vdots \\ \vdots \\ \vdots \\ b_n \end{bmatrix} u, \tag{5.67}$$

$$y = \begin{bmatrix} 1 & 0 & \cdots & \cdots & 0 \end{bmatrix} x. \tag{5.68}$$

(2) Consider the control system given by the Rosenbrock description

$$\begin{cases} y = N(\partial)\,\xi, \\ u = D(\partial)\,\xi, \end{cases} \tag{5.69}$$

where $D(\partial)$ and $N(\partial)$ belong to the first Weyl algebra $A_1(\mathbb{C})$, $D(\partial)$ is monic, and $d^\circ(N) < d^\circ(D)$. Put $D(\partial)$ and $N(\partial)$ into the form of left *polynomials, i.e.,*

$$D(\partial) = \partial^n + a_1 \partial^{n-1} + ... + a_n,$$
$$N(\partial) = b_1 \partial^{n-1} + ... + b_n$$

and show that this system has a state-space realization of the form

$$\dot{x} = \begin{bmatrix} -a_1 & -a_2 & \cdots & \cdots & -a_n \\ 1 & 0 & \cdots & 0 & 0 \\ 0 & 1 & \ddots & \vdots & \vdots \\ \vdots & \ddots & \ddots & 0 & \vdots \\ 0 & 0 & 0 & 1 & 0 \end{bmatrix} x + \begin{bmatrix} 1 \\ 0 \\ \vdots \\ \vdots \\ 0 \end{bmatrix} u, \tag{5.70}$$

$$y = \begin{bmatrix} b_1 \; b_2 \cdots \cdots b_n \end{bmatrix} x. \tag{5.71}$$

Exercise 944. *Consider the linear system M defined by $\dot{y} = 0$ over $A_1(\mathbb{C})$, and let $W = \mathcal{E}(\mathbb{R})$.*

(i) Determine $\mathfrak{B}_W(M)$.
(ii) Show that the function $t \mapsto t$ does not belong to $\mathfrak{B}_W(M)$. Deduce that $\mathfrak{B}_W(M)$ is not an $A_1(\mathbb{C})$-module.

Exercise 945. *Let $\mathbf{A} = \mathbb{R}[d/dt]$, let $R \in {}^q\mathbf{A}^k$ and let $M = \operatorname{coker}_{\mathbf{A}}(\bullet R)$. Let $I = (t_1, t_2)$ be a nonempty open interval of $\mathbb{R}$ and let $W = \mathcal{E}(I)$. A* trajectory *on I is an element of $\ker_W(R\bullet) = \operatorname{Hom}_{\mathbf{A}}(M, W)$, i.e., is an $\mathbf{A}$-linear map $\mathbf{w}: M \to W$. Let*

$$-\infty \leq t_1 \leq t_2 < t_3 \leq t_4 \leq +\infty.$$

A trajectory $\mathbf{w}' : M \to \mathcal{E}((t_2, t_3))$ is said to be a restriction *of $\mathbf{w} : M \to \mathcal{E}((t_1, t_4))$ if, and only if for every $m \in M$, the functions $\mathbf{w}'(m)$ and $\mathbf{w}(m)$ coincide on (t_2, t_3). Let*

$$-\infty \leq t_1' < t_2' < t_3' < t_4' \leq +\infty.$$

Two trajectories $\mathbf{w}' : M \to \mathcal{E}((t_1', t_2'))$ and $\mathbf{w}'' : M \to \mathcal{E}((t_3', t_4'))$ are said to be compatible *if, and only if there exists a trajectory $\mathbf{w} : M \to \mathcal{E}((t_1', t_4'))$ such that $\mathbf{w}'$ and $\mathbf{w}''$ are restrictions of $\mathbf{w}$. Then, $\mathbf{w}'$ (resp., $\mathbf{w}''$) is called the past (resp., future) trajectory.*

(i) The behavior $\mathfrak{B}_W(M)$ is said to be controllable "à la Willems" if, and only if past and future trajectories are always compatible. Show that this definition is equivalent to free controllability in Theorem and Definition 863, to concatenability in Lemma and Definition 866, and to controllability in Lemma and Definition 872.

(ii) Assume that M is free with basis $b_1, ..., b_k$. For each $i \in \{1, ..., k\}$, choose $f_i \in \mathcal{E}((t_1', t_4'))$ such that $f_i(t) = \mathbf{w}'(b_i)(t)$ for $t \in (t_1', t_2')$ and $f_i(t) = \mathbf{w}''(b_i)(t)$ for $t \in (t_3', t_4')$. Conclude that if M is free, then $\mathfrak{B}_W(M)$ is controllable "à la Willems".
(iii) Assume that M is not free and take $m \in \mathcal{T}(M)$. Let $0 \neq a(\partial) \in \mathbf{A}$ be such that $a(\partial)m = 0$. Check that any trajectory $\mathbf{w}$ is such that $a(\partial)\mathbf{w}(m) = 0$. Consider two trajectories $\mathbf{w}' : M \to \mathcal{E}((t_1', t_2'))$ and $\mathbf{w}'' : M \to \mathcal{E}((t_3', t_4'))$ such that $\mathbf{w}'(m)$ and $\mathbf{w}''(m)$ do not lie on the same integral curve of the differential equation $a(\partial)y = 0$, and show that $\mathbf{w}'$ and $\mathbf{w}''$ are not compatible. Conclude that if $\mathfrak{B}_W(M)$ is controllable "à la Willems", then M is free.

Exercise 946. *Let $\mathbf{A} = \mathbb{C}[\partial_1, \partial_2, \partial_3]$ and $W = \mathcal{E}(\mathbb{R}^3)$.*

(i) Let

$$R = \begin{bmatrix} 0 & -\partial_3 & \partial_2 \\ \partial_3 & 0 & \partial_1 \\ -\partial_2 & \partial_1 & 0 \end{bmatrix}.$$

(a) Check that $\ker_W(R\bullet)$ consist of all vector fields $\mathbf{w} \in {}^3W$ such that $\operatorname{curl}\mathbf{w} = 0$. *By* Poincaré's theorem *(see, e.g., [323], Theorem 6.7.19 and Corollary 6.7.21), this implies that there exists $\mathbf{f} \in W$ such that $\mathbf{w} = \nabla\mathbf{f}$ where ∇ is the gradient.*
(b) Show directly that $\ker_W(R\bullet) = \operatorname{im}_W(S\bullet)$ *where $S = \begin{bmatrix} \partial_1 & \partial_2 & \partial_3 \end{bmatrix}^T$.*
(c) Is $\ker_W(R\bullet)$ controllable?
(d) Is $\ker_W(R\bullet)$ freely controllable?
(Answer for (c): Yes. Answer for (d): No.)

Exercise 947. *Consider the same ring and the same signal space as in Exercise 946, and let*

$$R = \begin{bmatrix} \partial_1 & \partial_2 & \partial_3 \end{bmatrix}.$$

(a) Check that $\ker_W(R\bullet)$ consist of all vector fields $\mathbf{w} \in {}^3W$ such that $\operatorname{div}\mathbf{w} = 0$ *where* div *is the divergence. By Poincaré's theorem (see above), this implies that there exists $\mathbf{f} \in {}^3W$ such that $\mathbf{w} = \operatorname{curl}\mathbf{f}$.*
(b) Show directly that $\ker_W(R\bullet)$ has this image representation.
(c) Is $\ker_W(R\bullet)$ controllable? Is it freely controllable?

Exercise 948. *Consider $\mathfrak{B} = \mathfrak{B}_1 + \mathfrak{B}_2$ where $\mathfrak{B}_1$ is the kernel in Exercise 946 (i.e., $\mathfrak{B}_1 = \ker\operatorname{curl}$) and $\mathfrak{B}_2$ is the kernel in Exercise 947 (i.e., $\mathfrak{B}_2 = \ker\operatorname{div}$). Determine an image representation for $\mathfrak{B}$.*

(Hint: An image representation for $\mathfrak{B}_1 + \mathfrak{B}_2$ is given by $\mathfrak{I}_1 + \mathfrak{I}_2$ where $\mathfrak{I}_i$ is an image representation of $\mathfrak{B}_i$ $(i = 1, 2)$. This is known as the Helmholtz-Hodge decomposition.*)*

Exercise 949. *Let $\mathbf{A} = \mathbb{C}[\partial_1, \partial_2, \partial_3, \partial_4]$, where $x_4 = t$, and let $W = \mathcal{E}(\mathbb{R}^4)$. Recall that the Maxwell equations are given by*

$$\begin{aligned} \operatorname{div}(B) &= 0, \\ \partial_4 B + \operatorname{curl}(E) &= 0. \end{aligned}$$

(i) Put these equations into the classical form $R\,\mathbf{w} = 0$.
(ii) Using Exercise 948, determine an image representation for $\ker_{\mathcal{W}}(R\bullet)$, *and finally show that the Maxwell equations can be written*

$$\begin{aligned} B &= \operatorname{curl}(A), \\ E &= -\partial_4 A - \nabla\phi. \end{aligned}$$

A is the magnetic vector potential, and ϕ is the scalar electric potential.

Exercise 950. *Consider the classical operators in Exercises 947 and 946. We use Definition 875.*

(i) Let $\mathcal{D}_1 = \operatorname{div}$*; show that* $\mathcal{D}_1^* = -\nabla$.
(ii) Check that $\mathcal{D}_1^*$ *is parametrized by* $\mathcal{D}_0^* = \operatorname{curl}$ *and that* $\mathcal{D}_0 = \operatorname{ad}(\mathcal{D}_0^*) = \mathcal{D}_0^*$, *i.e.,* curl *is a* self-adjoint operator.
(iii) Check that curl *is parametrized by* $-\operatorname{div}$, *so that* $\mathcal{D}_0^*$ *is parametrized by* $\mathcal{D}_{-1}^* = -\operatorname{div}$, *and* $\operatorname{ad}\left(\mathcal{D}_{-1}^*\right) = \mathcal{D}_{-1} = \nabla$.
(iv) Summarize the above by two exact sequences: the first one involving ∇, curl *and* div, *the second one involving the adjoints of these operators.*

Exercise 951. *Generalize Theorem and Definition 874(1) to the case when R is GFLP and not necessarily left-regular.*

Exercise 952. *(i) Consider the linear system M given by (5.4) over the first Weyl algebra* $A_1(\mathbb{C})$. *(i) Prove that* $a = \left(t\partial^2 + 1\right)$ *and* $b = (1+t)$ *have no nonunit common left divisor, and deduce that M is controllable.*
(ii) (a) Show that there necessarily exists a pair (x, x') *of elements of* $A_1(\mathbb{C})$, *solution of the Bézout equation* $a\,x - b\,x' = 1$. *(b) Determine such elements* x, x'.
(iii) Consider the linear system $\check{M} = \mathbb{C}(t) \bigotimes_{A_1(\mathbb{C})} M$, *defined over* $B_1(\mathbb{C})$ (**2.7.4.3**). *Show that $\check{M}$ is freely controllable.*
(Hint for (ii)(b): $x = (1+t)^2/2, \quad x' = (1+t)\,\partial/2 + 2\partial + (1+t)/2.$*)*

Exercise 953. *Prove that the system M defined by (5.70), (5.71) over* $A_1(\mathbb{C})$ *is freely controllable.*

(Hint: define the partial state ξ such that $y = D(\partial)\,\xi$, $u = N(\partial)\,\xi$, *and check that* $\{\xi\}$ *is a basis of M.)*

Exercise 954. *Let* $\mathbf{K}$ *be a division ring, let* $\mathbf{A} = \mathbf{K}[\partial; \boldsymbol{\alpha}, \boldsymbol{\delta}]$ *where* $\boldsymbol{\alpha}$ *is an automorphism and* $\boldsymbol{\delta}$ *is an* $\boldsymbol{\alpha}$*-derivation of* $\mathbf{K}$, *let* $\mathbf{Q} = \mathrm{Q}(\mathbf{A}) = \mathbf{K}(\partial; \boldsymbol{\alpha}, \boldsymbol{\delta})$, *and consider a transfer matrix* $G \in \mathbf{Q}^{p \times m}$. *Denote by* g_{ij} *the entries of G* $(1 \le i \le p, 1 \le j \le m)$, *and write* $g_{ij} = a_{ij}^{-1} b_{ij}$, $0 \ne a_{ij} \in \mathbf{A}$, $b_{ij} \in \mathbf{A}$. *Let* $a \in \mathbf{A}$ *be an* lclm *of all elements* a_{ij} *(i.e., a least common left denominator* *of all elements* g_{ij}*: see Exercise 488). Let* $C \triangleq a\, G$*, let* $L \in \mathrm{Mat}_p(\mathbf{A})$ *be a* gcld *of* $a\, I_p$ *and* C*, set* $D_l = L^{-1} a\, I_p$ *and* $N_l = L^{-1} C$*.*

(i) Prove that $(D_l(\partial), N_l(\partial))$ *is a* left-coprime factorization *of* $G(\partial)$ *over* $\mathbf{A}$ *(Definition 667).*
(ii) Prove that the control system with equation

$$D_l(\partial)\, y = N_l(\partial)\, u$$

is freely controllable and freely observable.

(Hint: $\mathbf{A}$ *is a Bézout domain, thus use Theorems 653 and 654.)*

Exercise 955. *(1) Let* $\mathbf{E}$ *be a pre-Hilbert space* (**1.7.3.1**)*, let* $\mathbf{F} = {}^n\mathbf{k}$ *be endowed with its canonical inner product* $\langle \bullet, \bullet \rangle_{\mathbf{F}}$*, and let* $L : \mathbf{E} \to {}^n\mathbf{k}$ *be a continuous* $\mathbf{k}$*-linear map. Recall that the adjoint of* L*, when it exists (its existence is guaranteed when* $\mathbf{E}$ *is a Hilbert space), is the operator* $L^* : {}^n\mathbf{k} \to \mathbf{E}$ *such that* $\langle L^*\, \mathbf{z}, \mathbf{u} \rangle_{\mathbf{E}} = \langle \mathbf{z}, L\, \mathbf{u} \rangle_{\mathbf{F}}$ *for all* $\mathbf{u} \in \mathbf{E}$ *and all* $\mathbf{z} \in \mathbf{F}$ *([92], (11.5.1)). Show that the following conditions are equivalent:*
(a) L *is surjective.*
(b) L^* *is injective.*
(c) The Hermitian matrix operator $\mathbf{W} = L\, L^* : {}^n\mathbf{k} \to {}^n\mathbf{k}$ *is positive definite.*
(2) Let Ω *be a nonempty open interval of* $\mathbb{R}$ *and for any compact interval* $T \subseteq \Omega$*, let* W_T *be the space of the restrictions to* T *of the functions belonging to* $\mathcal{E}(\Omega)$*.*
(i) Show that the Hermitian form

$$\langle \mathbf{u}, \mathbf{v} \rangle_{W_T} = \int_T \mathbf{u}^*(t)\, \mathbf{v}(t)\, dt$$

is positive definite on ${}^m W_T$ *(denoting by* $\mathbf{u}^*(t)$ *the conjugate-transpose of* $\mathbf{u}(t)$*). We consider in what follows the pre-Hilbert space* W_T *endowed with the Hermitian form* $\langle \bullet, \bullet \rangle_{W_T}$*.*
(ii) Let $\mathbf{K} = \mathcal{O}(\Omega)$ *be the ring of real-valued analytic functions in* Ω*, let* $\tilde{G} \in {}^n\mathbf{K}^m$*, and let*

$$L_T : {}^m W_T \to {}^n\mathbf{k} : \mathbf{u} \mapsto \int_T \tilde{G}(t)\; \mathbf{u}(t)\; dt.$$

Prove that L_T *is continuous and that its adjoint is*

$$L_T^* : {}^n\mathbf{k} \to {}^m W_T : \mathbf{z} \mapsto \tilde{G}^* \left|_T \right. \mathbf{z}.$$

(iii) Define $\mathbf{W}_T = L_T\, L_T^*$ *and show that*

$$\mathbf{W}_T = \int_T \tilde{G}(t)\, \tilde{G}^* dt = \int_T X^{-1}(t)\, G(t)\, G^*(t)\, X^{*-1}(t)\, dt.$$

The matrix $\mathbf{W}_T$ *is called the* controllability gramian *over the interval* T. *Show that* $\mathbf{W}_T$ *is positive definite if, and only if the equality* $\mathbf{z}^* \tilde{G}|_T = 0$ *implies* $\mathbf{z}^* = 0$. *Prove that this holds for any compact interval* $T \subseteq \Omega$ *if, and only if the state-space system* $\dot{\eta} = \tilde{G}\, u$ *is Kalman-controllable.*
(iv) Deduce from the above that the state-space system (5.34) with coefficients in $\mathcal{O}(\Omega)$ *is Kalman-controllable if, and only if for any two points* $\mathbf{x}_1, \mathbf{x}_2 \in {}^n\mathbb{R}$ *and any two instants* $t_1, t_2 \in \Omega$ *such that* $t_2 > t_1$, *there exists an input* $\mathbf{u} \in {}^m\mathcal{E}(\Omega)$ *for which the system state* $\mathbf{x} : t \mapsto \mathbf{x}(t)$ *is transferred from* $\mathbf{x}_1$ *(assumed to be the value of* $\mathbf{x}$ *at time* t_1*) to* $\mathbf{x}_2$ *in the interval of time* $[t_1, t_2]$.

Exercise 956. *Consider the state-space system (5.34) with coefficients in* $\mathcal{O}(\Omega)$ *where*

$$F = \begin{bmatrix} t & 1 & 0 \\ 0 & t^3 & 0 \\ 0 & 0 & t^2 \end{bmatrix}, \quad G = \begin{bmatrix} 0 \\ 1 \\ 1 \end{bmatrix}.$$

Show that $\mathrm{rk}_{\mathbb{R}}\, \Gamma_3(0) = 2$ *and that* $\mathrm{rk}_{\mathbb{R}}\, \Gamma_4(0) = 3$. *Deduce the following: (i) This system is Kalman-controllable. (ii) The set* S *in the statement of Proposition 881(ii) cannot be reduced to* $\{t_0\}$.

Exercise 957. *Consider the state-space system (5.34) with coefficients in* $\mathcal{O}(\Omega)$ *where*

$$F(t) = \begin{bmatrix} 0 & 1 \\ -1 & 0 \end{bmatrix}, \quad G(t) = \begin{bmatrix} \cos t \\ -\sin t \end{bmatrix}.$$

Prove that this system is not Kalman-controllable, although the "frozen system" (i.e., that obtained when changing $G(t)$ *to* $G(\sigma)$ *where* σ *is any fixed parameter) is freely controllable.*

Exercise 958. *Consider a state-space system over* $\mathbf{A} = \mathbf{k}[\mathbf{q}]$ *where* $\mathbf{k} = \mathbb{R}$ *or* $\mathbb{C}$ *and* $\mathbf{q}$ *is the shift-forward operator* $\mathbf{f}(t) \mapsto \mathbf{f}(t+1)$. *Let (5.43) be the state equation of this system and let* M *the* $\mathbf{A}$*-module defined by this equation. Consider the multiplicative set*

$$S = \{\mathbf{q}^i : i \geq 0\}$$

and let $\mathbf{L} = S^{-1}\mathbf{A}$ *be the ring of Laurent polynomials with coefficients in* $\mathbf{k}$ *(Sect. 2.8).*

(1) Prove that the following conditions are equivalent:
(i) The system $\mathbf{L} \bigotimes_{\mathbf{A}} M$ *(over* $\mathbf{L}$*) is freely controllable.*
(ii) For any $0 \neq z \in \mathbb{C}$, $\mathrm{rk}_{\mathbb{C}} \left[z\, I_n - \check{F}\; G \right] = n$.
(iii) If $\mathbf{v}^* \in \mathbf{k}^n$ *is such that* $\mathbf{v}^* \check{F}^i G = 0 \ (0 \leq i \leq n-1)$, *then necessarily* $\mathbf{v}^* \check{F}^n = 0$.
(iii) $\mathrm{im}_{\mathbf{k}} \left(\check{F}^n \bullet \right) \subseteq \mathrm{im}_{\mathbf{k}} \left(\check{\Gamma}_n \right)$ *where*

$$\check{\Gamma}_n = \left[G \;\vdots\; \check{F} G \;\vdots\; \cdots \;\vdots\; \check{F}^{n-1} G \right].$$

(iv) For any initial state $\mathbf{x}_0 \in {}^n\mathbf{k}$ *at time* t_0*, there exists a time* $t_1 > t_0$ *and a finite sequence* $\{\mathbf{u}(t) : t_0 \leq t \leq t_1\}$ *for which the state of the system is transferred from* $\mathbf{x}_0$ to the origin *in the interval of time* $[t_0, t_1] \subseteq \mathbb{Z}$.
(2) The system is said to be 0-controllable *if the above equivalent conditions hold.*
(3) Generalize the above: for a system over $\mathbf{K}[\mathbf{q}, \boldsymbol{\alpha}]$*, where* $\mathbf{K}$ *is a commutative ring and* $\boldsymbol{\alpha}$ *is an automorphism of* $\mathbf{K}$*, define free* 0*-controllability, strong* 0*-controllability,* 0*-controllability, and for a state-space system, Kalman* 0*-controllability.*
(4) Consider the system M *with coefficients in* $\mathbb{C}[t]$ *and with equation*

$$q^2 y = q(t-1)u.$$

(i) Show that this system is not controllable, but that it is 0*-controllable.*
(ii) Set $\eta = \begin{bmatrix} y \\ \mathbf{q}y \end{bmatrix}$. *Using the construction in the proof of Theorem and Definition 861, show that a state-space realization of this system is*

$$q\,x = \begin{bmatrix} 0 & 1 \\ 0 & 0 \end{bmatrix} + \begin{bmatrix} t-1 \\ 0 \end{bmatrix} u,$$
$$y = \begin{bmatrix} 1 & 0 \end{bmatrix} x$$

and prove that M *is not Kalman-controllable.*
(Hint: for (1), see ([43], Sect. 10.4.2); for (3), see Proposition 1040 below.)

Exercise 959. *Prove that the bidual (i.e., the dual of the dual) of a control system* (M, u, y) *is again* (M, u, y).

Exercise 960. *Consider the control system in Exercise 941 and show that its dual cannot be properly defined.*

Exercise 961. *"Dualize" Exercises 955 and 958. Define the* observability gramian *and show that a state-space system with coefficients in* $\mathcal{O}(\Omega)$ *(where* Ω *is a nonempty open interval of* $\mathbb{R}$*) is Kalman-observable if, and only if for any instants* $t_1, t_2 \in \Omega$ *such that* $t_2 > t_1$*, the state* $\mathbf{x}(t_1)$ *can be determined from the knowledge of* $\mathbf{y}\big|_{(t_1,t_2)}$ *and* $\mathbf{u}\big|_{(t_1,t_2)}$ *(assuming that, e.g.,* $\mathbf{u} \in {}^m\mathcal{E}(\Omega)$*).*

Exercise 962. *(i) Let* $\tau_1 \neq \tau_2$ *be two positive incommensurable time-delays, let* $\mathbf{H}$ *be the subgroup of* $\mathbb{R}$ *generated by* τ_1 *and* τ_2*, and let* $\mathbf{B} = \mathbf{A} * \mathbf{H}$ *where* $\mathbf{A} = \mathbb{C}[d/dt]$ (**5.7.5.3**). *Show that* $\hat{f}(s) = s + e^{-\tau_1 s}$ *and* $\hat{g}(s) = e^{-\tau_1 s} - e^{-\tau_2 s}$ *have non common zeros. (For this, prove that all zeros of* $\hat{g}(s)$ *lie on the imaginary axis, and that* $\hat{f}(s)$ *cannot be zero when* s *is imaginary.)*

(ii) Recall that $\mathcal{O}(\mathbb{C})$ is a Bézout domain (Example 439). Deduce from (i) that the ideal in $\mathcal{O}(\mathbb{C})$ generated by $\hat{f}(s)$ and $\hat{g}(s)$ is $\mathcal{O}(\mathbb{C})$.
(iii) Using Malgrange's theorem (Remark 932(a)), show that

$$\ker_{PW(\mathbb{C})}\begin{bmatrix}\hat{f}(s)\\ \hat{g}(s)\end{bmatrix} = \ker_{PW(\mathbb{C})}[1] = 0$$

and deduce using Remark 932(b) that

$$\ker_{W}\left(\begin{bmatrix} f \\ g\end{bmatrix}\circ\right) = 0.$$

(iv) Let $\mathcal{L}(\mathcal{H}(\mathbf{B})) = \left\{\hat{h} : h \in \mathcal{H}(\mathbf{B})\right\}$. Consider the ideal $\mathfrak{I}$ in $\mathcal{L}(\mathcal{H}(\mathbf{B}))$ generated by $\hat{f}(s)$ and $\hat{g}(s)$ and assume that $\mathfrak{I}$ is principal, generated by $\hat{e}(s) = \hat{b}(s)/\hat{a}(s) \in PW(\mathbb{C})$. Show that $\hat{e}(s)$ has no zero in $\mathbb{C}$; using Weierstrass' factorization theorem [308] and Corollary 931, deduce the following: e is a unit in $\mathcal{H}(\mathbf{B})$, $\mathfrak{I} = \mathcal{L}(\mathcal{H}(\mathbf{B}))$, and there exist $\hat{c}, \hat{d} \in \mathcal{L}(\mathcal{H}(\mathbf{B}))$ such that the Bézout equality

$$\hat{c}\hat{f} + \hat{d}\hat{g} = 1$$

holds. Using the expression $\mathcal{H}(\mathbf{B}) = \mathcal{H}_3(\mathbf{B})$, show that the ideal $\mathfrak{J}$ in $\mathcal{L}(\mathbf{B})$ generated by $\hat{f}(s)$ and $\hat{g}(s)$ contains a nonzero polynomial $\hat{p}(s) \in \mathbb{C}[s]$.
(v) Let $\mathbf{D} = \mathbb{C}[s]^{-1}\mathcal{L}(\mathbf{B})$. Show that there exists an isomorphism

$$\psi : \mathbf{D} \xrightarrow{\sim} \mathbf{L} \triangleq \mathbb{C}(s)\left[z_1, z_1^{-1}, z_1, z_2^{-1}\right] : s \mapsto s, e^{-\tau_1 s} \mapsto z_1, e^{-\tau_2 s} \mapsto z_2$$

and that the ideal in the Laurent polynomial ring $\mathbf{L}$ generated by $s + z_1$ and $z_1 - z_2$ is equal to $\mathbf{L}$. Write the corresponding Bézout equality and prove that there is a contradiction. (For all details, see [153], Example 5.13.)
(vi) Deduce from the above that (a) $\mathcal{H}(\mathbf{B})$ is not a Bézout ring, and (b) statement (iii) of Theorem 916 is not valid if R_2 is not assumed to be full row rank.
(vii) Deduce that the $\mathcal{H}(\mathbf{B})$-module W is not *a cogenerator for ${}_{\mathcal{H}(\mathbf{B})}\mathbf{Mod}^{fp}$.*

Exercise 963. *Consider the column $R \in {}^{k}\mathcal{E}'(\mathbb{R}^n)$ of Gurevič's example (Remark 932(c)), which is such that $\ker_W(R\circ) \neq 0$, and show that by Malgrange's theorem (Remark 932(a)), there exists a row $X \in \mathcal{O}(\mathbb{C}^n)^k$ such that $X\hat{R} = 1$. Deduce that the equivalence in Remark 932(b) is false when $n > 1$ and, as a consequence, that Schwartz's theorem on mean periodic functions is not valid in that case.*

5.9 Notes

As already said in Section 5.1, Willems [360], [361], [362] was the first to emphasize that a linear system cannot be properly characterized by a transfer matrix or a state-space model. He developed the "behavioral approach" in the above references and in [276]. In this approach, a system is characterized by its behavior (5.5) in a power of a signal space W. As already said in the Notes of Chapter 3, the connection between a finitely presented $\mathbf{A}$-module $M = \operatorname{coker}_{\mathbf{A}} (\bullet R)$ and $\ker_W (R\bullet)$ (Theorem 565) was established by Malgrange [232]. Since M can be viewed as an intrinsic description of the system by generators and relations, it was called a linear system by Kashiwara [181], and, in the context of Automatic Control, by Fliess [115].

The results in Section 5.2 are essentially due to Oberst [258]. W-closedness (Remark 811) is a notion which was introduced by Pillai and Shankar [275]. The definition of a large cogenerator and its main properties (Definitions 814 and 829, Theorems 815 and 830) were established by Roos [299] and especially Oberst [257].

The generalizations in Section 5.3 (except those regarding large injective cogenerators) are due to Bourlès and Oberst [47].

In Section 5.4, Theorems 831 and 833 were obtained by Oberst [258] for multidimensional systems (Remarks 832 and 834); see also [260]. The proof we have given for these theorems is quite simple and taken from Bourlès [41]. Several authors, especially Blomberg and Ylinen [21], and Callier and Desoer [57], used these results in an *implicitly way*, because without referring to Homological Algebra. Theorem 837 is due to Zerz [375], whereas Theorems 838 and 833 are new. Theorem 840 is due to Fröhler and Oberst [128], [129]. The definitions of an Ore function and of an Ore field, and the corresponding properties (**5.4.2.1**), are new.

In Section 5.5, Propositions 843 and 846, as well as Corollary 849, are due to Oberst [258]. Autonomy is a classical notion which was emphasized by Willems [362]; Lemma and Definition 850, which characterizes autonomy in the general case, is related (taking into account the fact that the ring $\mathbf{A}$ is not necessarily commutative) to the formulations of Fornasini *et al.* [124], and of Wood *et al.* [363]. Control systems, as defined in (**5.5.2.1**), were introduced by Fliess [115] under the name of *dynamics*. Rosenbrock descriptions and the notion of strict equivalence go back to Rosenbrock [301]. Definition 859 is due to Bourlès and Fliess [44]. An equivalent "behavioral definition" (in the case of an injective signal space W) was given by Perbeno [270]; see also Fuhrmann [131] and, for a more recent account, Zerz [372]. In the present form, Lemma and Definition 856, as well as Theorem 860, are new (Remark 858).

The Laplace functor (Corollary and Definition 853) and the expression "transfer vector space", were introduced by Fliess [117], [115]; similar notions can be found in Oberst [258]. The state-space realization in Theorem and Definition 861 is a slight generalization of that due to Fliess [115].

The key notions in Section 5.6 were introduced by Kalman [171] in 1960 in the context of state-space systems (controllability, observability, and controllability/observability duality). Kalman's point of view has been dramatically questioned in the late eighties. In particular, controllability of a linear system is now considered as a system property which is independent of the choice of the input variables. This point was first emphasized by Willems [360], [361], [362] in the case of LTI systems; his original definition of controllability is recalled in Exercise 945 (concatenability is just a generalization of this notion). For LTI systems, controllability and strong controllability are equivalent notions which have not to be distinguished. The connection between the controllability of an LTI system and the freeness of the corresponding module was pointed out by Fliess [116] (see Exercise 945). The situation is completely different for multidimensional systems or for LTV systems with coefficients in a ring. The connection between concatenability of an nD behavior, the existence of an image representation, and the torsion-freeness of the corresponding system (Theorem 867), was established by Pillai and Shankar [275]. The notion of flat system and of flat output were introduced by Fliess *et al.* [121] in the context of nonlinear systems. Zero primeness, minor primeness and factor primeness of a polynomial matrix were defined by Youla and Gnavi [368], who also established the result in Theorem 870(1). That in Theorem 870(2) is due to Wood *et al.* [364]. Lemma 869 is due to Zerz [371], and Exercise 951 to Oberst [258] who introduced the notion of GFLP matrix. Theorem and Definition 863 is classical when the signal space W is assumed to be injective (see, e.g., [68]); in the present form, it is due to Bourlès and Oberst [47]. Exercises 946-949 are taken from Zerz [373], [374]. Theorem 876 and Exercise 950 are due to Pommaret [277], [278]. Theorem and Definition 879 is due to several authors, including Ilchmann *et al.* [163] for Statement (1), Popov, Belevitch and Hautus [157] for the Popov-Belevitch-Hautus test in Statement (2)[3], LaSalle [210] and Weiss [359] for Statement (3), Kalman [172] and Fliess [116] for Statement (4). However, in the present form, Theorem and Definition 879, which also improves the corresponding result in Bourlès [41], is more general than the above contributions. Proposition 881 is essentially due to Silverman and Meadows [329]. No example of Kalman-controllable but uncontrollable state-space system is presently known. The decomposition in Proposition 883 generalizes that discovered by Kalman for LTI systems [172]. The duality between observability and controllability of state-space systems was emphasized by Kalman in 1960 for LTI systems [171] (see also [172]), and by Kreindler and Sarachik [193] in 1964 for LTV systems. Our approach of duality is intrinsic, and is based on van der Schaft [317] and especially Rudolph [309]. The former contribution is not completely general, and the latter has not a clear interpretation. Our presentation is a synthesis of both of them.

---

[3] See Sontag's historical Note in ([335], Sect. 3.8).

Section 5.7 is essentially due to Bourlès and Oberst [47]. As already said, Theorem 902 and Corollary 919 generalize results obtained by Eindhoven and Habets [103], by Gluesing-Luerssen, Vettori and Zampieri [139], and by Vettori and Zampieri [352], [353], where it is assumed that $\mathbf{G} = \mathbb{R}$, $W = \mathcal{E}(\mathbb{R})$, and $\mathbf{B} = \mathbf{A} = (W', \star)$.

In Habets' terminology [153], the ring $\mathcal{H}(\mathbf{B})$ in Subsect. 5.7.5 is the ring of all *admissible right fractions* $b \circ (a)^{\circ -1} \in \mathrm{Q}(\mathbf{B})$ with respect to W. This is a generalization of the ring $\mathcal{H}$ introduced, in the case $\mathbf{G} = \mathbb{R}$, by Gluesing-Luerssen [137], [138] (see also Bréthé and Loiseau [50]) when the delays are commensurable (i.e., when $\mathbf{H} \cong \mathbb{Z}$), and by Habets [153] when the delays are noncommensurate (i.e., when $\mathbf{H}$ is not cyclic: see Exercise 962).

6
Finite Poles and Zeros of LTV Systems

6.1 Introduction

Stability of an LTV system can be evaluated from the stability of its autonomous part. This is shown in Chapter 12 where an analytic approach for stability of the LTV systems is given. However, stability can be studied using the *poles* of the system. This is the direction followed in the present section.

The approaches which establish for LTV systems a relation between poles and stability can be mainly split into two classes: the ones in the first class define the poles as the roots of the factors of decompositions of the differential operator associated with the system while for the approaches in the second class, the system is transformed, using operations which preserve stability, into another system which has a state matrix of upper triangular form; the poles are then defined as the diagonal terms. An approach which belongs to the first class is presented in this section, while the second class is discussed in Chapter 12.

The concept of poles of a system has received special attention as a mean to evaluate stability. Although this relation is clear and well known for linear time-invariant systems, the time-varying case is much more difficult since the frozen-time (pointwise-in-time) eigenanalysis may not contain information about stability as it is the case in the following example.

Example 964. *Consider the second order Euler equation*

$$\ddot{y} + 0.25t^{-1}\dot{y} + 0.01t^{-2}y = 0 \tag{6.1}$$

which can be written $P(\partial)y = 0$ with $P(\partial) = \partial^2 + 0.25t^{-1}\partial + 0.01t^{-2}$. The roots of polynomial $P(\partial)$ computed in a classical manner - called frozen *roots as they are parameterized by t, the variable time - are $-0.2t^{-1}$ and $-0.05t^{-1}$. Both roots have negative real parts for all $t \geq t_0 > 0$. However, the system given by (6.1) is unstable. Indeed, a fundamental set of solutions of (6.1) is $y_1(t) = t^{\frac{3}{8}+\frac{\sqrt{209}}{40}}$, $y_2(t) = t^{\frac{3}{8}-\frac{\sqrt{209}}{40}}$ and $lim_{t\to+\infty} y_1(t) = +\infty$.*

H. Bourlès and B. Marinescu: Linear Time-Varying Systems, LNCIS 410, pp. 403–452.
springerlink.com

Intrinsic definitions which overcome the difficulties pointed out above are given in this chapter. First, in Section 6.3 we clarify the connection between the roots of the factors of the skew polynomial which defines the autonomous part of the LTV system and the trajectories of the system. Next, in Section 6.6 the poles and the zeros of the system are defined using the module framework developed in Chapter 5. The usual relations between the poles and zeros are investigated in Section 6.6.5. Section 6.4 shows how field extensions which make it possible to factorize skew polynomials can be constructed (Those skew polynomials define the autonomous parts of the systems we are interested in).

6.2 Exponential Stability

Consider an autonomous LTV system Σ and let T be the associated torsion module. T can be given by a n-th order differential equation in *one* variable in the following polynomial form

$$P(\partial)y = 0, P(\partial) = \partial^n + a_{n-1}\partial^{n-1} + ... + a_1\partial + a_0, a_i \in \mathbf{K} \tag{6.2}$$

where P is a skew polynomial with coefficients in the differential field $\mathbf{K}$ (see Theorem and Definition 662(1), Lemma and Definition 781 (2)(v) and Example 964).

Let $\mathbf{R} = \mathbf{K}[\partial; \delta]$ $(\delta = d/dt)$ and consider the $\mathbb{C}$-space $W = \mathcal{O}_\infty$ introduced in (5.4.2.**2**). Recall that any element y of $\mathcal{O}_\infty$ can be viewed as an analytic function $(a, +\infty) \ni t \mapsto y(t) \in \mathbb{C}$ where a is large enough (Example 20). Assume that W is an injective cogenerator of $_{\mathbf{R}}\mathbf{Mod}$. This happens if $\mathbf{K}$ is one of the differential fields considered in (5.4.2.**2**) according to Theorem 838. A solution of (6.2) in $W = \mathcal{O}_\infty$ is an element $y \in Hom_{\mathbf{R}}(T, W)$ (Corollary 567; Lemma and Definition 771(i)).

Definition 965. *The autonomous LTV system Σ given by the torsion module T is said*

- exponentially stable *if any solution $y \in Hom_{\mathbf{R}}(T, W) : t \mapsto y(t)$ of (6.2) approaches zero exponentially for $t \to +\infty$,* i.e., *if there exist constants $\alpha > 0$, $\beta > 0$ and $t_0 > 0$ such that*

$$\|y(t)\| \leq \alpha e^{-\beta(t-t_0)}, t \geq t_0. \tag{6.3}$$

- exponentially unstable *if there exists a solution $y : t \mapsto y(t)$ of (6.2) which is exponentially unbounded,* i.e., *if there exist constants $\alpha > 0$, $\beta > 0$ and $t_0 > 0$ such that*

$$\|y(t)\| \geq \alpha e^{\beta(t-t_0)}, t \geq t_0. \tag{6.4}$$

Remark 966. *For short, the ring of differential operators $\mathbf{R} = \mathbf{K}[\partial; \delta]$ is denoted by $\mathbf{K}[\partial]$ in the sequel.*

6.3 Roots, Factors and Solutions

6.3.1 Linear Factors

Consider the polynomial description (6.2) of a general (n-th order) autonomous system Σ. Clearly, if $P(\partial)$ has a right factor, i.e., $P(\partial) = P'(\partial)(\partial - \gamma_1)$, then γ_1 is a right root of $P(\partial)$ and $y_1(t) = e^{\int \gamma_1(t)dt}$ is a solution of (6.2). One can thus conclude about the asymptotic behavior of $y_1(t)$ by investigating $\gamma_1(t)$. The right roots of $P(\partial)$ are thus good candidates for system poles. However, in general, $P(\partial)$ does not necessarily have roots over $\mathbf{K}$, the initial field of definition of the LTV system ($a_i \in \mathbf{K}$, $i = 0, ..., n-1$) in (6.2). A special class of polynomials which satisfy this constraint is the one of Wedderburn polynomials (W-polynomials for short). More specifically, if $P(\partial)$ of degree n is a W-polynomial (over $\mathbf{K}$), following Definition 767, there exists a set $\Delta = \{\gamma_1, ..., \gamma_n\}$, $\gamma_i \in \mathbf{K}$ such that

$$P(\partial) = P_\Delta, \tag{6.5}$$

where P_Δ denotes the minimal polynomial of Δ. Conversely, a set Δ which satisfies (6.5) is a set of n P-independent right roots of $P(\partial)$ according to Definition 767. W-polynomials are thus polynomials which have the maximum number of P-independent roots on the field of definition of their coefficients. As a matter of fact, as already said after Definition 767, a polynomial of degree n has at most n P-independent roots.

Assume that $\Delta = \{\gamma_1, ..., \gamma_n\} \subseteq \mathbf{K}$ is such that for each $i \in \{1, ..., n\}$, (a) $y_i(t) = e^{\int \gamma_i(t)dt}$ exists for t large enough, and (b) $y_i \in \mathcal{O}_\infty$ (this happens if $\mathbf{K}$ is one of the differential fields considered in (5.4.2.**2**)). By Lemma and Definition 772, each y_i is δ-hyperexponential over $\mathbf{K}$, thus the elements of the set Δ satisfy the following set of *elementary equations*

$$(\partial - \gamma_i)y_i = 0 \Leftrightarrow \dot{y}_i = \gamma_i y_i. \tag{6.6}$$

Moreover, from Theorem 770, $\{y_1, ..., y_n\}$ is a *fundamental set of solutions* of (6.2).

Definition 967. *Let P be a polynomial of degree n. A set of n P-independent roots is called a* fundamental set of roots of P.

Based on the elementary equations (6.6), if $P(\partial)$ is a W-polynomial, the stability of system Σ can be investigated from a fundamental set of roots of $P(\partial)$.

Proposition 968. *Consider the homogeneous differential equation (6.2) and let $\{\gamma_1, ..., \gamma_n\}$ be a fundamental set of roots of $P(\partial)$. If*

$$\limsup_{t \to +\infty} Re\{\gamma_i(t)\} < 0, i \in \{1, ..., n\}, \tag{6.7}$$

then any solution of (6.2) approaches exponentially zero when $t \to +\infty$*. If* $lim_{t\to+\infty}\gamma_i(t)$ *exists,* $i \in \{1, ..., n\}$*, then condition (6.7) is also necessary for all solutions of (6.2) to approach zero exponentially when* $t \to +\infty$*. At least one of the solutions of (6.2) is exponentially unbounded if at least one of the* $\gamma_i(t)$*'s satisfies the condition*

$$\liminf_{t\to+\infty} Re\{\gamma_i(t)\} > 0. \tag{6.8}$$

If $lim_{t\to+\infty}\gamma_i(t)$ *exists,* $i \in \{1, ..., n\}$*, then condition (6.8) is also necessary for at least one of the solutions of (6.2) to be exponentially unbounded.*

Proof. Any element $\gamma_i(t)$ of the considered fundamental set of roots of $P(\partial)$ is related to an element y_i of a fundamental set of solutions $\{y_1, ..., y_n\}$ of (6.2) by an elementary equation (6.6). If (6.7) holds, then $Re\{\gamma_i\} \leq -\beta_i$, for some $\beta_i > 0$, $\forall i \in \{1, ..., n\}$, when t is large enough. It follows that for $i = 1, ..., n$ there exist finite, positive constants α_i and β_i, such that $|e^{\int_{t_1}^{t} \gamma_i d\tau}| \leq \alpha_i e^{-\beta_i(t-t_1)}$, $\forall t \geq t_1$ and an instant $t_1 > t_0$, from which follows

$$|y_i| \leq \alpha_i e^{-\beta_i(t-t_1)}, \; t \geq t_1 > t_0 \tag{6.9}$$

and thus one concludes that any solution $y(t)$ satisfies (6.3).

Suppose now that $lim_{t\to+\infty}\gamma_i(t) = \overline{\gamma}_i$ and any solution of (6.2) approaches zero exponentially when $t \to +\infty$. Consider a solution y_i of (6.2) such that $\gamma_i = \frac{\dot{y}_i}{y_i}$. It follows that for any $\epsilon > 0$, there exists $t_1 > 0$ such that whenever $t \geq t_1$, $\mid y_i(t) \mid \geq \alpha_i \mid e^{\overline{\gamma}_i t} \mid -\epsilon$, where α_i is a positive constant. As $\mid y_i(t) \mid$ decreases exponentially to 0 when $t \to +\infty$, we must have $Re\{\overline{\gamma}_i\} < 0$, or, equivalently, $lim_{t\to+\infty} Re\{\gamma_i(t)\} < 0$.

The conclusion on the instability follows in the same manner. ∎

A subclass of W-polynomials are the polynomials with real coefficients and distinct real roots:

Example 969. $P(\partial) = \partial^2 - (a+b)\partial + ab$, $a, b \in \mathbb{R}$, $a \neq b$ *is a W-polynomial. A fundamental set of roots is* $\{a, b\}$*. The associated autonomous LTI system is exponentially stable if, and only if* $a, b < 0$*.*

Notice that all the polynomials with real coefficients are not necessarily W-polynomials as shown by the following example:

Example 970. $P(\partial) = (\partial - a)^2$, $a \in \mathbb{R}$*. A fundamental set of solutions of (6.2) is in this case* $y_1(t) = e^{at}$, $y_2(t) = te^{at}$ *from which follows, using (6.6), that a fundamental set of roots of* $P(\partial)$ *in* $\mathbb{R}(t)$ *is* $\gamma_1 = a$, $\gamma_2 = a + t^{-1}$*. Thus,* γ_2 *does not belong to* $\mathbb{R}$*. Notice however, that, following Proposition 968, the autonomous system is exponentially stable if, and only if* $a < 0$*.*

Let's now make the converse rationale: let $\{y_1, ..., y_n\}$ be a fundamental set of solutions of (6.2). The y_i's do not necessarily belong to $\mathbf{K}$, but may belong to a field extension $\overline{\mathbf{K}} \supseteq \mathbf{K}$ which is a Picard-Vessiot extension of (6.2). The behavior of these solutions in the vicinity of $+\infty$ makes sense if one can define the restriction of each element of $\overline{\mathbf{K}}$ to a neighborhood $(A, +\infty)$ of $+\infty$. Proceeding as in Example 20, one can then define the germs of the elements of $\overline{\mathbf{K}}$ in a neighborhood of $+\infty$. If the space of these germs can be embedded in $\mathcal{O}_\infty$, one is led back to the above formulation.

In Section 6.4 it is shown how such an extension can systematically be built for continuous-time LTV systems. The connections with W-polynomials are now further investigated.

Starting from a fundamental set of solutions of (6.2), say $\{y_1, ..., y_n\} \in \overline{\mathbf{K}}$, define $\Delta = \{\gamma_1, ..., \gamma_n\} \in \widetilde{\mathbf{K}} \subseteq \overline{\mathbf{K}}$ as solutions of the elementary equations (6.6). From Theorem 770, Δ is P-independent, thus a fundamental set of roots of $P(\partial)$. In other words, $P(\partial)$ is a W-polynomial over $\widetilde{\mathbf{K}}$. For the linear time-invariant systems with real coefficients, obviously, $\widetilde{\mathbf{K}} = \mathbb{C}$.

Each polynomial $P(\partial)$ with coefficients in the field $\mathbf{K}$ is thus a W-polynomial over a well-chosen extension $\widetilde{\mathbf{K}}$ of $\mathbf{K}$.

Let now α be a right root of $P(\partial)$. Then $y = e^{\int \alpha(t)dt}$ is a solution of (6.2): $P(\partial)y = 0$. Also, there exist $c_i \in \mathbf{C}$, $i = 1, ..., n$, where $\mathbf{C}$ is the subfield of constants of $\widetilde{\mathbf{K}}$, such that $y = \sum_i^n c_i y_i$. Let now $P'(\partial)$ be the minimal polynomial of Δ: $P'(\partial) = P_\Delta$. Obviously, $P'(\partial)y_i = 0$, $i = 1, ..., n$. As the c_i's are constants, $P'(\partial)y = \sum_i^n c_i P'(\partial)y_i = 0$ and thus α is also a right root of $P'(\partial)$. Thus $V(P) \subseteq V(P')$ where $V(.)$ denotes the set of zeros of a polynomial. As $P(\partial)$ is monic and it is a W-polynomial (over $\widetilde{\mathbf{K}}$), it follows from Proposition 768 (iii) that there exists a polynomial $P''(\partial)$ for which $P'(\partial) = P''(\partial)P(\partial)$. But, as $deg(P') = deg(P) = n$, one obtains $deg(P'') = 0$. Moreover, as $P'(\partial)$ is monic, $P''(\partial) = 1$ and thus $P(\partial) = P'(\partial)$. This leads to

Proposition 971. *If $P(\partial) \in \boldsymbol{K}[\partial]$ is a W-polynomial over $\widetilde{\boldsymbol{K}}$ then $P = P_\Delta$ where Δ is any fundamental set of roots of P.*

Example 972. *Let $P(\partial) = \partial^2 + (t^{-1} - 2a)\partial + a^2 - t^{-2} - at^{-1}$, $a \in \mathbb{R}$. Thus, $\boldsymbol{K} = \mathbb{R}(t)$, i.e., the field of rational functions in the indeterminate t and with real coefficients. A fundamental set of solutions of (6.2) with $P(\partial)$ as above is $y_1 = te^{at}$, $y_2 = e^{at}$ and a fundamental set of roots of $P(\partial)$ is $\gamma_1 = a + t^{-1}$, $\gamma_2 = a - t^{-1}$. Since $\gamma_1, \gamma_2 \in \mathbb{R}(t)$, $P(\partial)$ is a W-polynomial over the initial field of definition of the LTV system, i.e., $\widetilde{\boldsymbol{K}} = \boldsymbol{K}$, and no field extension is needed in this case.*

Using Definition 712 of the minimal polynomial and the computation formula (4.21), (6.5) leads to a factorization of P into n *linear* (i.e., of degree 1) *distinct* factors:

$$P(\partial) = (\partial - a_n)...(\partial - a_1). \tag{6.10}$$

As a matter of fact, using (4.21), $P(\partial)$ for Example 972 can be factorized as $P(\partial) = (\partial - {}^{\gamma_2 - \gamma_1}\gamma_2)(\partial - \gamma_1) = (\partial - a + 2t^{-1})(\partial - a - t^{-1})$ or $P(\partial) = (\partial - {}^{\gamma_1 - \gamma_2}\gamma_1)(\partial - \gamma_2) = (\partial - a)(\partial - a + t^{-1})$.

6.3.2 Multiple Factors

In the preceding section, a relation between a set of n P-independent (right) roots of polynomial $P(\partial)$ and the solutions of the differential equation (6.2) has been established using the elementary equations (6.6). This obviously allows one to conclude on the asymptotic behavior of the solutions y_i of (6.2) by investigating the roots γ_i as stated in Proposition 968. However, as shown below, this analysis can also be done in many cases with a factorization (6.10) of $P(\partial)$ where there exist indices $i, j \in \{1, ..., n\}$ such that $a_i = a_j$ and $i \neq j$. In this case, the polynomial $P(\partial)$ in (6.10) is not necessarily a W-polynomial over the field $\mathbf{K}$ of definition of the elements a_i.

To analyze this case, some preliminary results are needed.

Definition 973. *1. Let $P \in \mathbf{K}[\partial]^{\times}$ and let $\check{\mathbf{K}} \supseteq \mathbf{K}$ be a field extension of $\mathbf{K}$. If there exist polynomials P', $P'' \in \check{\mathbf{K}}[\partial]$ and $a \in \check{\mathbf{K}}$ such that $P(\partial) = P'(\partial)(\partial - a)P''(\partial)$, then a is called a* zero *of P.*
2. If P has a factorization (6.10) into n linear factors $\partial - a_i$ (not necessarily distinct), then $\{a_1, ..., a_n\} \subseteq \check{\mathbf{K}}$ is called a full set of zeros *of P.*

Let $f : I \times X \to \mathbb{R}^n$ be a function which satisfies the following assumptions:

(A1) $f(., x) : I \to \mathbb{R}^n$ is measurable for each fixed x.
(A2) $f(t, .) : \mathbb{R}^n \to \mathbb{R}^n$ is continuous for each fixed t.
(A3) f is locally Lipschitz on x; that is, there are for each $x^0 \in X$ a real number $\rho > 0$ and a locally integrable function $\alpha : I \to \mathbb{R}_+$ such that the ball $B_\rho(x^0)$ of radius ρ centered at x^0 is contained in X and $\| f(t,x) - f(t,y) \| \leq \alpha(t) \| x - y \|$ for each $t \in I$ and $x, y \in B_\rho(x^0)$.
(A4) f is locally integrable on t; that is, for each fixed x^0 there is a locally integrable function $\beta : I \to \mathbb{R}_+$ such that $\| f(t, x^0) \| \leq \beta(t)$ for almost all t.

Lemma 974. *Assume that f satisfies the assumptions A(1) to (A4) above. Let $\xi \in X \subseteq \mathbb{R}^n$ be a solution to the initial value problem*

$$\dot{\xi} = f(t, \xi(t)), \ \xi(\sigma^0) = x^0 \tag{6.11}$$

on the interval $[\sigma^0, \tau] \subseteq I$. Then, there exist numbers $c, \Delta > 0$ so that for each $\delta \in (0, \Delta]$ the following holds: if h is any mapping satisfying the assumptions A(1) to (A4) and in addition

$$\left\| \int_{\sigma^0}^{t} h(s, \xi(s)) ds \right\| \leq \delta \ for \ all \ t \in [\sigma^0, \tau], \tag{6.12}$$

and if $z^0 \in X$ is so that $\| z^0 - x^0 \| < \delta$, then the solution ζ of

$$\dot{\zeta} = f(t, \zeta) + h(t, \zeta), \ \zeta(\sigma^0) = z^0 \tag{6.13}$$

is defined on the entire interval $[\sigma^0, \tau]$, *and*

$$esssup\{\| \xi - \zeta \|\} < c\delta. \tag{6.14}$$

If f and h are analytical, then so is ξ.

See [335] for a proof (Theorem 37, Section C.4).

Consider the differential equation

$$(\partial - a)\, y = z, \tag{6.15}$$

$a, z \in \mathcal{O}_\infty$. Let $A \in \mathbb{R}$ be large enough, so that a, z can be viewed as analytic functions in $(A, +\infty)$. Let $B > A$. Then, for $t \geq B$, all solutions of (6.15) are given by

$$\begin{array}{c} y(t) = CU(t, B) + U(t, B) \int_B^t z(\tau)\, e^{-\int_B^\tau a(\varsigma) d\varsigma} d\tau, \\ U(t, B) = e^{\int_B^t a(\tau) d\tau}, \quad C \in \mathbb{C}, \end{array} \tag{6.16}$$

and they are analytic in $(A, +\infty) \supsetneqq [B, +\infty)$. Therefore, all solutions of (6.15) are the germs in a neighborhood of $+\infty$ of the analytic functions given by (6.16). These solutions belong to $\mathcal{O}_\infty$ and span an affine $\mathbb{C}$-vector space of dimension 1. Denote this vector space by $\mathcal{S}(z)$. From the above,

$$\mathcal{S}(z) \subsetneqq \mathcal{O}_\infty.$$

The above function $U(., B)$ is clearly the unique solution of the homogeneous differential equation

$$\partial v - av = 0 \tag{6.17}$$

such that $U(B, B) = 1$, i.e., U is the *fundamental solution* of (6.17).

Let $a = \alpha + i\omega$ where α, ω are real-valued (i.e., $\alpha = \operatorname{Re}(a)$, $\omega = \operatorname{Im}(a)$) and let

$$\overline{\alpha} = \limsup_{t \to +\infty} \frac{\int_B^t \alpha(\tau)\, dt}{t - B}, \quad \underline{\alpha} = \liminf_{t \to +\infty} \frac{\int_B^t \alpha(\tau)\, d\tau}{t - B}. \tag{6.18}$$

Lemma 975. *The following equalities hold:*

$$\overline{\alpha} = \limsup_{t \to +\infty} \frac{\ln |U(t, B)|}{t - B}, \quad \underline{\alpha} = \liminf_{t \to +\infty} \frac{\ln |U(t, B)|}{t - A},$$

i.e., the quantity $\overline{\alpha}$ (resp., $\underline{\alpha}$) (which belongs to $\overline{\mathbb{R}} \triangleq \mathbb{R} \cup \{-\infty, +\infty\}$) is the upper *(resp.,* lower*)* Lyapunov exponent *of the differential equation (6.17) (or, with an abuse of language, of the differential equation (6.15) as introduced by Definition 1198).*

Proof. We have

$$U(t,B) = e^{\int_B^t \alpha(\tau)d\tau} e^{\int_B^t i\omega(\tau)d\tau} \Longrightarrow |U(t,B)| = e^{\int_B^t \alpha(\tau)d\tau}$$
$$\Longrightarrow \ln|U(t,B)| = \int_B^t \alpha(\tau)\,d\tau.$$

The rest is clear. ∎

Lemma 976. *Consider the four conditions below:*

(i) There exists $B > A$ such that $U(t,B) \to 0$ exponentially as $t \to +\infty$.
(ii) $\overline{\alpha} < 0$.
(iii)

$$\limsup_{t\to+\infty} \alpha(t) < 0.$$

(iv) For every $y \in \mathcal{S}(z)$, $y(t) \to 0$ exponentially as $t \to +\infty$.
In general, (iii)⇒(i)⇔(ii)⇏(iii).
(2) If $\lim_{t\to+\infty} \alpha(t)$ exists (in particular, if a belongs to an Ore field), then (i)⇔(ii)⇔(iii).
(3) Assume that $z(t) \to 0$ exponentially as $t \to +\infty$. Then (iv)⇔(i).

Proof. (1). (i)⇒(ii): If (ii) holds, there exist $k > 0$ and $\varepsilon > 0$ such that for any $t \geq B$,

$$|U(t,B)| \leq ke^{-\varepsilon(t-B)}.$$

Since $|U(t,B)| = e^{\int_B^t \alpha(\tau)d\tau}$, this implies the following for any $t > B$:

$$e^{\int_B^t (\alpha(\tau)+\varepsilon)d\tau} \leq k \Longrightarrow \int_B^t (\alpha(\tau)+\varepsilon) \leq \ln k$$
$$\Longrightarrow \frac{1}{t-B}\int_B^t \alpha(\tau)\,d\tau \leq -\varepsilon + \frac{1}{t-B}\ln k,$$

therefore (ii) holds.

(ii)⇒(i): If (ii) holds there exists $\varepsilon > 0$ and $T > B$ such that for all $t \geq T$,

$$\frac{1}{t-B}\int_B^t \alpha(\tau)\,d\tau \leq -\varepsilon.$$

Therefore, whenever $t \geq T$,

$$\int_B^t \alpha(\tau)\,d\tau \leq -\varepsilon(t-B) \Longrightarrow |U(t,B)| \leq e^{-\varepsilon(t-B)},$$

and (i) holds.

(iii)$\Rightarrow$(ii): If (iii) holds, there exist $\varepsilon > 0$ and $T \geq B$ such that for all $t \geq B$,

$$\alpha(t) \leq -\varepsilon \Longrightarrow \int_B^t \alpha(\tau)\,d\tau \leq -\varepsilon(t-B) \Longrightarrow \frac{1}{t-B}\int_B^t \alpha(\tau)\,d\tau \leq -\varepsilon$$

and (ii) holds.

(ii)$\not\Rightarrow$(iii): Let $a(t) = \cos(t) - 1/2$. Then

$$\int_0^t a(\tau)\,d\tau = \sin(t) - t/2,$$

therefore $\overline{\alpha} = -1/2 < 0$, and (ii) hold. However, $\alpha(t) = a(t)$ and

$$\limsup_{t\to+\infty} \alpha(t) = 1/2 > 0,$$

thus (iii) does not hold.

(2). Assume that $l \triangleq \lim_{t\to+\infty} \alpha(t)$ exists. Then,

$$\limsup_{t\to+\infty} \alpha(t) = l.$$

For any $\varepsilon > 0$, there exists $B > 0$ such that whenever $t \geq B$, $|\alpha(t) - l| \leq \varepsilon$. This implies

$$\begin{aligned}\int_B^t |\alpha(\tau) - l|\,d\tau &\leq \varepsilon(t-B) \Longrightarrow \left|\int_B^t \alpha(\tau)\,d\tau - l(t-B)\right| \leq \varepsilon(t-B) \\ &\Longrightarrow \left|\frac{1}{t-B}\int_B^t \alpha(\tau)\,d\tau - l\right| \leq \varepsilon \Longrightarrow l - \varepsilon \leq \frac{1}{t-B}\int_B^t \alpha(\tau)\,d\tau.\end{aligned}$$

Taking $t \to +\infty$, we get

$$l \leq \overline{\alpha} + \varepsilon$$

If (ii) holds, we have $\overline{\alpha} < 0$. Choose $\varepsilon = -\overline{\alpha}/2$. We get $l \leq \overline{\alpha}/2 < 0$, and (iii) holds.

(3). Assume that $z(t) \to 0$ exponentially as $t \to +\infty$.

(i)$\Rightarrow$(iv): Let $\beta > 0$ be such that $|z| = O\left(e^{-\beta t}\right)$ as $t \to +\infty$ and let $\gamma \in (0, \min(-\bar{\alpha}, \beta))$. Set $\tilde{y} = e^{\gamma t} y$, $\tilde{z} = e^{\gamma t} z$, and $\tilde{U}(t,B) = e^{\gamma t} U(t,B)$. This yields

$$\partial \tilde{y} = (a + \gamma)\,\tilde{y} + \tilde{z}.$$

There exist $K > 0$ and $B > A$ such that $|z(t)| \leq K e^{-\beta t}$ for $t \geq B$, therefore

$$\left|\int_B^t \tilde{z}(\tau)\,d\tau\right| \leq K \int_B^t e^{(\gamma-\beta)\tau} d\tau = \delta_1 < +\infty$$

where δ_1 can be chosen arbitrarily small by taking B large enough. In addition,

$$\partial\tilde{U}(t,B) = (a+\gamma)\,\tilde{U}(t,B).$$

Let $\delta_2 = \left|\tilde{y}(B) - \tilde{U}(B,B)\right|$ and let $\delta = \sup(\delta_1,\delta_2)$. Without loss of generality for our purpose, we can choose C so that $\delta_2 = 0$. Then, $\delta = \delta_1$, and by Lemma 974, there exists $c > 0$ such that for all $t \geq B$, $\left|\tilde{y}(t) - \tilde{U}(t,B)\right| \leq c\delta$; the latter inequality implies

$$|y(t) - U(t,B)| < c\delta e^{-\gamma t}.$$

Thus, for all $t \geq B$,

$$|y(t)| \leq |U(t,B)| + c\delta e^{-\gamma t}.$$

Therefore, if (i) holds, then (iv) holds too.

(iv)$\Rightarrow$(i): This is clear by (6.16). ■

Lemma 977. *Consider the four conditions below:*

(i') There exists $B > A$ such that $|U(t,B)| \rightarrow +\infty$ exponentially as $t \rightarrow +\infty$.
(ii') $\underline{\alpha} > 0$.
(iii')

$$\liminf_{t\rightarrow+\infty} \alpha(t) > 0.$$

(iv') For any $z \in \mathcal{O}_\infty$, there exists $y \in \mathcal{S}(z)$ such that $|y(t)| \rightarrow +\infty$ exponentially as $t \rightarrow +\infty$.
In general, (iii')$\Rightarrow$(i')$\Leftrightarrow$(ii')$\Leftrightarrow$(iv')$\nRightarrow$(iii').

Proof. (1). The proof of the equivalence (i')$\Leftrightarrow$(ii') is similar to the proof of the equivalence (i)$\Leftrightarrow$(ii) in Item 1 of Lemma 976 (changing the sign of the inequalities). Likewise, (iii')$\Rightarrow$(i'). To see that (ii')$\nRightarrow$(iii'), consider the function $a : t \mapsto \cos(t) + t/2$.

(iv')$\Rightarrow$(i'): If (iv') holds, there exists $y \in \mathcal{S}(0)$ such that $|y(t)| \rightarrow +\infty$ exponentially as $t \rightarrow +\infty$, therefore (i') holds.

(i')$\Rightarrow$(iv'): Assume that (iv') does not hold, i.e., there exists $z \in \mathcal{O}_\infty$ such that for all $y \in \mathcal{S}(z)$, $|y(t)| \nrightarrow +\infty$ exponentially as $t \rightarrow +\infty$. Denote by $y(t;C,z)$ the solution (6.16). We know that $|y(t;1,z)| \nrightarrow +\infty$ and $|y(t;0,z)| \nrightarrow +\infty$ as $t \rightarrow +\infty$. Therefore, as $t \rightarrow +\infty$,

$$|U(t,B)| = |y(t;1,z) - y(t;0,z)| \nrightarrow +\infty$$

and (i') does not hold.

The proof of (2) is similar to the proof of Item 2 of Lemma 976. ■

Lemma 978. *Consider the differential equation (6.15) where $a \in \mathbb{C}$ and* $\operatorname{Re}(a) < 0$. *Of the following, (b)$\Rightarrow$(a) but the converse does not hold:*
(a) $y(t) \rightarrow 0$ exponentially as $t \rightarrow +\infty$.
(b) $z(t) \rightarrow 0$ exponentially as $t \rightarrow +\infty$.

Proof. (b)$\Rightarrow$(a): see Lemma 976(3).

(a)$\not\Rightarrow$(b): Assume $a = -1$ and $y(t) = e^{-t}\cos\left(e^{2t}\right)$. Then (a) holds. However, $z(t) = -2\sin\left(e^{2t}\right)e^{t}$, thus (b) does not hold. ■

Lemma 979. *Consider the differential equation*

$$(\partial - a)^n y = 0, \quad a \in \mathcal{O}_\infty, \tag{6.19}$$

where n is a positive integer. Let $P(\partial) = (\partial - a)^n$ and consider the set $\mathcal{S}_P$ consisting of all solutions $y \in \mathcal{O}_\infty$ of (6.19). Let $\overline{\alpha}$, $\underline{\alpha}$ be defined as in (6.18) with $\alpha = \mathrm{Re}(a)$.

(i) The set $\mathcal{S}_P$ is a $\mathbb{C}$-vector space of dimension n.
(ii) If $\overline{\alpha} < 0$, then all $y \in \mathcal{S}_P$ tend to zero exponentially as $t \to +\infty$.
(iii) If $\underline{\alpha} > 0$, then all solutions $y \in \mathcal{S}_P{}^{\times}$ are such that $|y(t)| \to +\infty$ exponentially as $t \to +\infty$.
(iv) If $\lim_{t\to+\infty} \alpha(t) = 0$, then all solutions $y \in \mathcal{S}_P{}^{\times}$ are bounded as $t \to +\infty$ if, and only if $n = 1$ and α is integrable on $[B, +\infty)$ for B large enough. These bounded solutions do not tend to zero as $t \to +\infty$.

Proof. (i) The germ a can be viewed as an analytic function in $(A, +\infty)$ for A large enough. Let $B > A$ and for any $i \in \{1, ..., n\}$, let y_n be the general solution of (6.19) in $[B, +\infty)$. We have for any $t \geq B$

$$y_1(t) = C_1 U(t, B)$$

where

$$U(t, B) = e^{\int_B^t a(\tau)d\tau}.$$

By induction, one can easily check that

$$y_n(t) = U(t, B) \sum_{0 \leq i \leq n-1} C_{n-i} \frac{(t - B)^i}{i!} \tag{6.20}$$

which proves (i).

(ii) If $\overline{\alpha} < 0$, then $U(t, B) \to 0$ exponentially as $t \to +\infty$ by Lemma 976, thus so does $|y_n(t)|$.

(iii) If $\underline{\alpha} > 0$, then $|U(t, B)| \to +\infty$ as $t \to +\infty$ by Lemma 977. Assuming that y_n is nonzero, let

$$j = \min\{i : 0 \leq i \leq n - 1 \text{ and } C_{n-i} \neq 0\}.$$

Then, as $t \to +\infty$,

$$|y_n(t)| \sim |U(t,B)| \frac{\left|C_{n-j}(t-B)^j\right|}{j!},$$

thus $|y_n(t)| \to +\infty$ exponentially as $t \to +\infty$.

(iv) is clear by the expression (6.20). ■

Notice that for any a which belongs to an Ore field (according to Lemma and Definition 835), there exists a constant A such that a is analytic on $(A, +\infty)$. Assume that $z \in \mathcal{O}_\infty$ in (6.15) and that the sign of $z(t)$ is constant as $t \to +\infty$. We have $a \in \mathcal{O}_\infty$, therefore $y(t) \in \mathcal{O}_\infty$.

Lemma 980. *Consider (6.15) where a belong to a* real *Ore field (see Lemma and Definition 835), $z(t)$ is real-valued and has a constant sign as $t \to +\infty$ and let $lim_{t\to+\infty}a(t) = \overline{a}$. If $\overline{a} < 0$ and $y(t)$ approaches zero exponentially when $t \to +\infty$ (i.e., satisfies (6.3)), then $z(t)$ also approaches zero exponentially when $t \to +\infty$.*

Proof. The solutions of (6.15) are

$$y(t) = cy_1(t) + y_2(t),\ y_1(t) = e^{\int adt},\ y_2(t) = y_1(t)\int ze^{-\int adt}dt \tag{6.21}$$

where c is any constant. As, by hypothesis, $\overline{a} < 0$, there exists $\beta > 0$ such that $\overline{a} \le -\beta$. Then, for a given time t_1, there exists $\alpha > 0$ such that $\mid y_1(t) \mid = \mid e^{\int_{t_1}^{t} ad\tau} \mid \le \alpha e^{-\beta(t-t_1)}$, $\forall t \ge t_1$. Thus, y_1 approaches zero exponentially for $t \to +\infty$. By hypothesis, $y(t)$ also approaches zero exponentially when $t \to +\infty$. From (6.21), it follows that $y_2(t) = y(t) - y_1(t)$ and thus y_2 has the same property as y_1 and y, i.e., it approaches zero exponentially when $t \to +\infty$. Moreover, y_2 can be written in the form

$$y_2(t) = \frac{f(t)}{g(t)},\ f(t) = \int ze^{-\int adt}dt,\ g(t) = e^{-\int adt} \tag{6.22}$$

where $e^{-\int adt} \to +\infty$ as $t \to +\infty$. Notice that $f(t)$ and $g(t)$ are real-valued functions. Notice first that $\frac{df/dt}{dg/dt} = -\frac{z(t)}{a(t)}$. Since a belongs to a real Ore field, it has a constant sign as $t \to +\infty$, f and g are comparable of order 1 near $+\infty$. Following Proposition 7, Chapter V in [37], $z(t)$ and $a(t)$ are also comparable. Thus, $lim_{t\to+\infty}y_2(t) = lim_{t\to+\infty}\frac{df/dt}{dg/dt} = -lim_{t\to+\infty}\frac{z(t)}{a(t)}$. As y_2 exponentially decreases with time, there exists $\alpha > 0$ such that $y_2(t) = O(e^{-\alpha t})(t \to +\infty)$. As $y_2(t) = \frac{f(t)}{g(t)} \sim \frac{df/dt}{dg/dt} = -\frac{z(t)}{a(t)}$, it follows that $\frac{z(t)}{a(t)} = O(e^{-\alpha t})(t \to +\infty)$. Moreover, as a belongs to an Ore field, there exists $\beta > 0$ such that $z(t) = O(e^{-\beta t})(t \to +\infty)$. ■

Notice however that Lemma 980 was proved under supplementary assumptions on a and z. Indeed, this result is not valid in general as shown by the following example:

Example 981. *Consider (6.15) with* $a(t) = -e^{2t}$ *and* $z(t) = e^{\int -\frac{1+e^{2t}}{1-e^{2t}}dt}$. *Then,* $\overline{a} = -\infty$ *and thus* $Re\{\overline{a}\} < 0$. *Moreover,* $y(t)$ *approaches zero exponentially when* $t \to +\infty$. *However,* $z(t) \to +\infty$ *when* $t \to +\infty$.

The asymptotic behavior of the solutions of (6.2) can be investigated from a general factorization (6.10) of P in which the a_i's are not necessarily distinct.

Definition 982. *Let* $a \in \mathbb{R}$. *The set of all* $f \in \mathcal{O}_\infty$ *such that* $f : [B, +\infty) \to \mathbb{R}$, $f(t) \geq 0$ $(t \geq B)$ *and* $\chi(f) = a$ *(Definition 1198) is denoted by* $Exp(a)$. *A germ* $f \in Exp(a)$ *is said to be quasi-exponential of type* a.

Remark 983. *In Definition 982, a germ* $\overline{f} \in \mathcal{O}_\infty$ *is not distinguished from one of its representatives* $f : [B, +\infty) \to \mathbb{R}$, *as usual.*

Lemma 984. *(a) If* $a \in \mathcal{O}_\infty$ *is real-valued and is such that* $lim_{t\to+\infty}a(t) = \bar{a} \in \bar{\mathbb{R}}$, *then* $e^{\int a(t)dt} \in Exp(\bar{a})$.
(b)If $f \in Exp(a_1)$ *and* $g \in Exp(a_2)$ $(a_1, a_2 \in \bar{\mathbb{R}})$, *then*

(i) $f + g \in Exp(a_1, a_2)$ *where* $a = sup(a_1, a_2)$*;*
(ii) $fg \in Exp(a_1 + a_2)$ *where* $(a_1, a_2) \neq (+\infty, -\infty)$.

(c) If $f \in Exp(a)$, *then* $\int f(t)dt \in Exp(a)$.

Proof. Points (a) and (b) are obvious. For point (c), $ln(y(t)) \sim \bar{a}t$ (where $\bar{a}$ is finite). Thus, $ln(y(t)) = \bar{a}t + O(t) = \bar{a}(t + O(1))$ from which follows

$$y(t) = e^{\bar{a}(t+O(1))}. \tag{6.23}$$

For any $\varepsilon > 0$, there exists $B > 0$ such that, for $t \geq B$, $-\varepsilon \leq O(1) \leq \varepsilon$. Thus, for $t \geq B$, (6.23) leads to

$$e^{(\bar{a}-\varepsilon)t} \leq y(t) \leq e^{(\bar{a}+\varepsilon)t}$$

and to

$$\frac{1}{\bar{a}-\varepsilon}e^{(\bar{a}-\varepsilon)t} \leq \int y(t)dt \leq \frac{1}{\bar{a}+\varepsilon}e^{(\bar{a}+\varepsilon)t}$$

and, finally, to $\chi(y) = \bar{a}$ (see Definition 1198). ∎

Theorem 985. *Consider the homogeneous differential equation (6.2).*

1. *Suppose* P *admits a factorization (6.10) where* $\{a_1, ..., a_n\}$ *is a full set of zeros of* P *(Definition 973). If*

$$\limsup_{t\to+\infty} Re\{a_i(t)\} < 0, \ i = 1, ..., n \tag{6.24}$$

then any solution of (6.2) approaches exponentially zero as $t \to +\infty$.

2. *If P admits a factorization (6.10) for which all a_i's belong to a real Ore field, then condition (6.24) is also necessary for all the solutions of (6.2) to approach exponentially zero as $t \to +\infty$. In the latter situation, there exists a solution of type $Exp(\bar{a})$ (according to Definition 982) where $\bar{a} = sup\{\bar{a}_i,\ i = 1, ..., n\}$. Therefore, condition*

$$lim_{t\to+\infty} a_i(t) > 0 \ for \ some \ i \in \{1, ..., n\} \tag{6.25}$$

is necessary and sufficient for the existence of an exponentially unbounded solution of (6.2).

3. *Assume that $P(\partial) \in \mathbf{K}[\partial]$ where $\mathbf{K}$ is an Ore field and*

$$P(\partial) \sim (\partial - b_1)^{d_1} ... (\partial - b_m)^{d_m}$$

where $\sim$ means "is similar to" (over $\mathbf{K}[\partial]$), $m \leq n$, the d_i's are positive integers and the b_i's are pairwise non-conjugate over $\mathbf{K}$. We know that $\lim_{t\to+\infty} b_i(t)$ exists in $\bar{\mathbb{R}}$. If there exists $i \in \{1, ..., n\}$ such that $\lim_{t\to+\infty} b_i(t) = 0$ (resp., $\lim_{t\to+\infty} b_i(t) > 0$), then there exists $y \in \mathcal{S}_P$ such that $|y(t)| \nrightarrow 0$ (resp., $|y(t)| \to +\infty$ exponentially) as $t \to +\infty$. Therefore, if there exists $i \in \{1, ..., n\}$ such that $\lim_{t\to+\infty} b_i(t) \geq 0$, then there exists $y \in \mathcal{S}_P$ which is exponentially unstable.

Proof.

1. Equation (6.2) with $P(\partial)$ given by (6.10) can equivalently be written in the form

$$\begin{cases} (\partial - a_1)y = z_1 \\ (\partial - a_2)z_1 = z_2 \\ \vdots \\ (\partial - a_{n-1})z_{n-2} = z_{n-1} \\ (\partial - a_n)z_{n-1} = 0. \end{cases} \tag{6.26}$$

Set $z_0 = z_1$ and suppose that (6.24) holds. The dynamics of z_{n-1} depends only on the last equation of (6.26) which is of elementary type (6.6):

$$z_{n-1}(t) = \lambda_n e^{\int a_n(t)dt} \tag{6.27}$$

where λ_n is any constant. As a_n satisfies (6.24), $Re\{a_n(t)\} \leq \beta_n$ for some $\beta_n > 0$ and for t large enough. It follows that for a time t_1 which is large enough, there exist a finite, positive constant α_n such that

$$\begin{gathered} |e^{\int_{t_1}^t a_n(\tau)d\tau}| \leq \frac{\alpha_n}{|\lambda_n|} e^{-\beta_n(t-t_1)},\ \forall t \geq t_1 \\ \Leftrightarrow |z_{n-1}(t)| \leq \alpha_n e^{-\beta_n(t-t_1)}, \forall t \geq t_1 \end{gathered} \tag{6.28}$$

from which one concludes that $z_{n-1}(t)$ approaches exponentially zero as $t \to +\infty$. The rest of the variables z_i, $i = n-2, ..., 1$ and $y(t)$ have the same property. Indeed, using Lemma 976 and the fact that a_{n-1} satisfies

(6.24), z_{n-2}, the solution of the last but one equation of (6.26), approaches exponentially zero as $t \to +\infty$. The conclusion follows by induction towards the first equation of (6.26).

2. Let now the a_i's belong to a real Ore field. Then $lim_{t\to+\infty}a_i(t)$ exists and let $lim_{t\to+\infty}a_i(t) = \overline{a}_i$, $i = 1, ..., n$. Suppose that all solutions of (6.2) approach exponentially zero when $t \to +\infty$. A particular solution of (6.2) or, equivalently, of (6.26), is $y(t) = y_1(t) = e^{\int a_1(t)dt}$, $z_1(t) = z_2(t) = ... = z_{n-1}(t) = 0$. More specifically, y_1 corresponds to the solution of the homogeneous part of the first equation of (6.26). Moreover, $lim_{t\to+\infty}y_1(t) = e^{\int \overline{a}_1 dt} = e^{\overline{a}_1 t}$. As $y_1(t)$ decreases exponentially to 0 when $t \to +\infty$, from the latter relation it follows that $\overline{a}_1 < 0$. Moreover, from (6.27) it follows that the sign of z_{n-1} is constant as $t \to +\infty$. Next, the solutions of (6.26) are

$$z_{k-1}(t) = ce^{\int a_k(t)dt} + e^{\int a_k(t)dt} \int z_k e^{-\int a_k(t)dt} dt \tag{6.29}$$

where c is any constant and $k = 2, ..., n$. It follows by induction that z_{n-2}, ..., z_1 have the same property. Using now Lemma 980 for the first equation in (6.26) one concludes that $z_1(t)$ decreases exponentially to 0 when $t \to +\infty$. Another particular solution of (6.26) is $y(t)$, $z_1(t) = e^{\int a_2(t)dt}$, $z_2(t) = z_3(t) = ... = z_{n-1}(t) = 0$. It follows, as before, that $\overline{a}_2 < 0$. Again, using Lemma 980 for the second equation in (6.26) one concludes that $z_2(t)$ decreases exponentially to 0 when $t \to +\infty$. The conclusion follows using the same rationale down to the last equation.
For the last part of this point, by (6.27), $\chi(z_{n-1}) = \bar{a}_n$. Also, by (6.29), $z_{n-2} = cz'_{n-2} + z''_{n-2}$ where $z'_{n-2} = e^{\int a_{n-1}(t)dt}$, $z''_{n-2} = e^{\int a_{n-1}(t)dt} \int z_{n-1}(t)e^{-\int a_{n-1}(t)dt}dt$. From Lemma 984 (a), $z'_{n-2} \in Exp(\bar{a}_{n-1})$. From parts (a) and (b) of the same lemma, $z_{n-1}(t)e^{-\int a_{n-1}(t)dt} \in Exp(\bar{a}_n - \bar{a}_{n-1})$. From part (c) of the same result, $\int z_{n-1}(t)e^{-\int a_{n-1}(t)dt}dt \in Exp(\bar{a}_n - \bar{a}_{n-1})$ and from part (b), $z''_{n-2} \in Exp(\bar{a}_n)$ and, choosing $c > 0$, $z_{n-2} \in Exp(sup(\bar{a}_{n-1}, \bar{a}_{n-2}))$. The result follows by induction.

3. Under the assumptions of this point, let $\mathbf{A} = \mathbf{K}[\partial]$ and $T = \mathbf{A}/\mathbf{A}P(\partial)$. We have by Corollary 764

$$T = \bigoplus_{1\leq i\leq m} T_i, \quad T_i \cong \frac{\mathbf{A}}{\mathbf{A}(\partial - b_i)^{d_i}},$$

and with the usual identifications,

$$\mathcal{S}_P = \mathrm{Hom}_{\mathbf{A}}\left(\bigoplus_{1\leq i\leq m} T_i, \mathcal{O}_\infty\right);$$

thus by Proposition 549, there exists a $\mathbb{C}$-linear isomorphism

$$\mathcal{S}_P \cong \bigoplus_{1\leq i\leq m} \mathrm{Hom}_{\mathbf{A}}\left(T_i, \mathcal{O}_\infty\right).$$

By Corollary 567(3), there exists for each $i \in \{1, ..., m\}$ a $\mathbb{C}$-linear isomorphism

$$\mathrm{Hom}_{\mathbf{A}}\left(T_i, \mathcal{O}_\infty\right) \cong \mathrm{Hom}_{\mathbf{A}}\left(\frac{\mathbf{A}}{\mathbf{A}\left(\partial - b_i\right)^{d_i}}, \mathcal{O}_\infty\right).$$

By Theorem 763(ii), each b_i $(1 \leq i \leq m)$ is conjugate to one of the a_j's $(1 \leq j \leq m)$, i.e., there exists $c_{ij} \in \mathbf{K}^\times$ such that

$$b_i - a_j = \frac{\dot{c}_{ij}}{c_{ij}}.$$

Let $\beta_i = \mathrm{Re}\left(b_i\right)$ and

$$\underline{\beta_i} = \liminf_{t\rightarrow+\infty} \frac{\int_B^t \beta_i\left(\tau\right) d\tau}{t - B}$$

$(1 \leq i \leq n)$. By Lemma and Definition 835(2), $b_i\left(t\right) - a_j\left(t\right) \rightarrow 0$ as $t \rightarrow +\infty$, thus

$$\min_{1\leq i\leq n} \underline{\beta_i} = \min_{1\leq j\leq m} \underline{\alpha_i} > 0.$$

The result is now a consequence of Lemma 979. ∎

Consider first the simple second order example:

Example 986. *Consider (6.2) with* $P(\partial) = (\partial - a(t))^2$. *It can be equivalently written*

$$\begin{cases} (\partial - a(t))y = z \\ (\partial - a(t))z = 0 \end{cases} \tag{6.30}$$

that is

$$\begin{bmatrix} \partial - a(t) & -1 \\ 0 & \partial - a(t) \end{bmatrix} \begin{bmatrix} y \\ z \end{bmatrix} = 0 \Leftrightarrow \partial w = A(t)w \tag{6.31}$$

with $A(t) = \begin{bmatrix} a(t) & 1 \\ 0 & a(t) \end{bmatrix}$, $w = \begin{bmatrix} y \\ z \end{bmatrix}$.

Remark 987. *The companion form in the example above can be generalized to factorizations of polynomial* P *of the form (6.51). The autonomous system (6.10) can be equivalently written*

$$\begin{cases} (\partial - a_1)y = z_1 \\ (\partial - a_2)z_1 = z_2 \\ \vdots \\ (\partial - a_n)z_{n-1} = 0. \end{cases} \tag{6.32}$$

In the matrix form this leads to

$$\partial w = A(t)w \tag{6.33}$$

with

$$A(t) = \begin{bmatrix} a_1(t) & 1 & 0 & \dots & 0 \\ 0 & \ddots & \ddots & & \vdots \\ \vdots & \ddots & \ddots & \ddots & 0 \\ \vdots & & \ddots & \ddots & 1 \\ 0 & \dots & \dots & 0 & a_n(t) \end{bmatrix} \tag{6.34}$$

and $w = \begin{bmatrix} y \; z_1 \; \dots \; z_{n-1} \end{bmatrix}^T$.

6.4 Field Extensions for the Continuous-Time Case

In this section only the continuous-time case is considered. Thus, ∂ exclusively stands for the usual derivation. The problem to solve is: given the field $\mathbf{K}$ to which the coefficients of a polynomial $P(\partial)$ belong, find a field $\check{\mathbf{K}}$ over which $P(\partial)$ has a factorization (6.10) and a field $\widetilde{\mathbf{K}}$ over which $P(\partial)$ has a fundamental set of roots.

6.4.1 The Field of Formal Laurent Series

We address the problem with the field $\mathbf{K} = \mathbb{C}((t))$ of formal Laurent series with coefficients in $\mathbb{C}$ equipped with the usual derivation d/dt. This field has the advantage to allow coherent theoretic bases for the developments and allows also direct calculations in many cases. However, the algorithms currently existing for factoring linear operators are in majority based on the calculation of the submodules of a given module from the solutions of the differential equations: see [10], [11], [330], [295], [296], [54] on this point. For those calculations we will use in the sequel the field $\mathbb{C}(t)$ of rational functions in t over $\mathbb{C}$ instead of $\mathbb{C}((t))$.

An element of $\mathbf{K} = \mathbb{C}((t))$ is of the form $a = \sum_{j \geq \nu} \alpha_j t^j, \quad \alpha_i \in \mathbb{C}$.

The field

$$\mathbf{K}_m = \mathbb{C}((z)), \; z^m = t, \; z = t^{1/m} \tag{6.35}$$

is a field extension of degree m over $\mathbf{K}$ and is a Galois extension of $\mathbf{K}$. The Galois automorphisms σ are given by $\sigma(t^{1/m}) = \xi t^{1/m}$ with $\xi \in \{e^{2\pi i k/m}, 0 \leq k \leq m\}$. The Galois group is $G \simeq \mathbf{Z}/m\mathbf{Z}$.

Consider the *valuation* v on $\mathbf{K}$ (and $\mathbf{K}_m$) defined as a map

$$v : \mathbf{K} \to \mathbf{Z} \cup \{\infty\} \tag{6.36}$$

with $v(0) = \infty$ and $v(a) = n$ if $a = \sum_{j\geq n} \alpha_j t^j$ and $\alpha_n \neq 0$. This valuation is extended to $\mathbf{K}_m$ as a map $v : \mathbf{K}_m \to (1/m)\mathbf{Z} \cup \{\infty\}$, $v(a) = \frac{n}{m}$ if $a = \sum_{j\geq n} \alpha_j t^{j/m}$ and $\alpha_n \neq 0$.

Consider the following rings of formal power series are considered:

Definition 988

- $\boldsymbol{O}_m = \mathbb{C}[[t^{1/m}]] = \{a \in \boldsymbol{K}_m, v(a) \geq 0\}$
- $\boldsymbol{O} = \mathbb{C}[[t]] = \{a \in \boldsymbol{K}, v(a) \geq 0\}$

They have the following properties:

Proposition 989

(i) The field of fractions of $\boldsymbol{O}_m$ (respectively of $\boldsymbol{O}$) is $\boldsymbol{K}_m$ (respectively $\boldsymbol{K}$).
(ii) $t^{1/m}\boldsymbol{O}_m$ is the unique maximal ideal of $\boldsymbol{O}_m$ and $\boldsymbol{O}_m/(t^{1/m}\boldsymbol{O}_m) \simeq \mathbb{C}$.

For a given skew polynomial $P(\partial)$ with coefficients in $\mathbf{K}$ we are looking for a extension $\mathbf{K}_m$ which allows a factorization of P.

6.4.2 Factorization of Skew Polynomials

A way to factorize $P(\partial)$ is to transform the initial differential equation into a form in which the only coefficients involved belong to the ring of formal power series $\mathbf{O}_m$ for some m. Indeed, as shown below, such equations can be reduced to equations with constant coefficients (over $\mathbb{C}$) which can be easily analyzed and from which the conclusions about the original system follow.

Definition 990. *A differential operator $P(\partial) = \partial^n + \sum_{i=0}^{n-1} a_i \partial^{n-i-1}$, $a_i \in$ $\mathbf{K}$ is said to be* regular singular *if $v(a_i) \geq 0$, $i = 0, ..., n-1$ where $v(.)$ is the valuation as introduced in (6.36).*

To bring $P(\partial)$ from (6.2) to a regular singular form one can consider the new variable $\xi = t^{-\lambda}\partial$ with $\lambda = min\{\frac{v(a_{n-i-1})}{i+1}\}$, $0 \leq i \leq n-1$. This leads to a monic polynomial with coefficients in $\boldsymbol{O}_m$ where m is the denominator of λ:

$$P(\xi) = \sum_{i=0}^{n} b_{n-i}\xi^{n-i}, \quad b_i \in \mathbf{K}_m,\ b_0 = 1. \tag{6.37}$$

Let $\overline{P}$ denote the polynomial obtained by the reduction of all coefficients of P given by (6.37) modulo $\pi = t^{1/m}$. $\overline{P}$ has coefficients in $\mathbb{C}$. Hensel's lemma can be used to factorize P using the factors of $\overline{P}$ obtained over the field of constants $\mathbb{C}$:

Lemma 991. *(Hensel's Lemma): If $\overline{P} = F_1 F_2$ with F_1, F_2 having coefficients in $\mathbb{C}$ and $gcd(F_1, F_2) = 1$, then there is a unique decomposition $P = P_1 P_2$ of P into monic polynomials such that $\overline{P}_i = F_i$ for $i = 1, 2$.*

The factors P_1 and P_2 of Hensel's lemma can be provided by an iterative process which is detailed, for example, in [286].

This is shown through the following example:

Example 992. *Let*

$$P(\partial)y(t) = 0, \; P(\partial) = \partial^2 + (\frac{1}{2}t^{-1} - 2\alpha)\partial - t^{-3} - \frac{1}{2}t^{-1}\alpha + \alpha^2 \tag{6.38}$$

with $\alpha \in \mathbb{C}$. $P(\partial)$ *is a skew polynomial which obeys to the commutation rule (2.2) with* $\delta = d/dt$. *Let* $\lambda = min\{-1, -\frac{3}{2}\} = -\frac{3}{2}$, $m = 2$. *The change of variable* $\xi = t^{\frac{3}{2}}\partial$ *transforms (6.38) into a differential equation with coefficients in* $\mathbf{O}_2$. *Setting* $P(\partial) = L(\xi)$, *(6.38) yields* $L(\xi)y = 0$, $L(\xi) = t^{-3}Q(\xi)$,

$$Q(\xi) = \xi^2 + (\frac{1}{2}t^{\frac{1}{2}} - 2\alpha t^{\frac{3}{2}})\xi - 1 - \frac{1}{2}t^2\alpha + \alpha^2 t^3$$

where the commutation rule for the new variable ξ *deduced from (2.2) with* $\delta = d/dt$ *is*

$$\xi a = a\xi + t^{\frac{3}{2}}\frac{da}{dt}, \; \forall a \in \boldsymbol{K}_2. \tag{6.39}$$

Therefore, the polynomial whose coefficients are the canonical images of those of $Q(\xi)$ *in* $\boldsymbol{O}_2/\pi\boldsymbol{O}_2$ $(\pi = t^{1/2})$ *is* $\overline{Q}(\xi) = \xi^2 - 1$ *which trivially factorizes on* $\mathbb{C}$: $\overline{Q}(\xi) = F_1F_2$ *with* $F_1 = \xi - 1$ *and* $F_2 = \xi + 1$. *The factorization in Lemma 991 can be obtained via an iterative process in which terms of the form* $\pi^k T_k$, $\pi^k S_k$ *are added to polynomials* F_1 *and* F_2 *respectively:*

$$\begin{aligned} Q_1(\xi) &= F_1 + \pi T_1 + \pi^2 T_2 + \ldots, \\ Q_2(\xi) &= F_2 + \pi S_1 + \pi^2 S_2 + \ldots. \end{aligned}$$

In the additional terms $\pi^i T_i$ *and* $\pi^i S_i$ *at step* i $(i \geq 1)$, $\pi = t^{\frac{1}{2}}$ *and* T_i *and* S_i *are polynomials with coefficients in* $\mathbb{C}$ *and such that* $d^0(T_i) < d^0(F_1)$ *and* $d^0(S_i) < d^0(F_2)$. *For this example,* $T_i, \; S_i \in \mathbb{C}$ *and the factors can be found in one step:*

$$\begin{aligned} Q_1(\xi) &= \xi - 1 - \tfrac{1}{2}t^{\frac{1}{2}} - \alpha t^{\frac{3}{2}}, \\ Q_2(\xi) &= \xi + 1 - \tfrac{1}{2}\tau^{\frac{1}{2}} - \alpha t^{\frac{3}{2}}. \end{aligned}$$

Obviously, $\overline{Q}_i = F_i$, $i = \{1, \; 2\}$ *and one can easily check using the new commutation rule (6.39) that* $Q(\xi) = Q_1(\xi)Q_2(\xi)$. *These factors can be transformed to find a factorization of* $P(\partial)$. *In the present case,*

$$\begin{aligned} P(\partial) &= t^{-3}Q_1(t^{3/2}\partial)Q_2(t^{3/2}\partial) = \\ &= t^{-3}(t^{\frac{3}{2}}\partial - 1 - \tfrac{1}{2}t^{\frac{1}{2}} - \alpha t^{\frac{3}{2}})(t^{\frac{3}{2}}\partial + 1 - \tfrac{1}{2}t^{\frac{1}{2}} - \alpha t^{\frac{3}{2}}) = \\ &= t^{-3}(t^{\frac{3}{2}}\partial t^{\frac{3}{2}} - t^{\frac{3}{2}} - \tfrac{1}{2}t^2 - \alpha t^3)(\partial + t^{-\frac{3}{2}} - \tfrac{1}{2}t^{-1} - \alpha) = \\ &= (\partial - t^{-\frac{3}{2}} + t^{-1} - \alpha)(\partial + t^{-\frac{3}{2}} - \tfrac{1}{2}t^{-1} - \alpha). \end{aligned}$$

This leads to a decomposition of P into two linear factors:

$$P(\partial) = (\partial - a_2)(\partial - a_1), \tag{6.40}$$

where $a_1(t) = -t^{-\frac{3}{2}} + \frac{1}{2}t^{-1} + \alpha$ *and* $a_2(t) = t^{-\frac{3}{2}} - t^{-1} + \alpha$.

Generally, a polynomial $P(\partial)$ can be decomposed as in (6.51) over a suitable extended field ($a_i \in \mathbf{K}_m$ for some m). In such a decomposition, it can happen that $a_i = a_j$ with $i \neq j$.

The above algorithm allows one to find for a given polynomial a root in a field extension $\mathbf{K}_m$ with $m \in \mathbf{N}$ sufficiently large. This leads to the following conclusions about the extensions of $\mathbf{K}$:

Proposition 993

(i) *The fields* $\boldsymbol{K}_m$ *are the only finite algebraic extensions of* $\boldsymbol{K}$.
(ii) $\overline{\boldsymbol{K}} = \cup_m \boldsymbol{K}_m$ *is the algebraic closure of* $\boldsymbol{K}$. $\overline{\boldsymbol{K}}$ *is the field of fractions of* $\overline{\boldsymbol{O}} = \{a \in \overline{\overline{K}}, v(a) \geq 0\}$.

6.4.3 Fundamental Sets of Roots and Picard-Vessiot Extensions

Note that in a factorization of type (6.10), only a_1 is a root (zero) of P in the sense of Definition 759. We are interested in obtaining a fundamental set of roots of $P(\partial)$. This set is obtained over the type of field denoted by $\widetilde{\mathbf{K}}$ in Section 6.3. A fundamental set of roots can be found from the set of factors of (6.10) in an iterative way: first, $\gamma_1 = a_1$. Next, γ_2 is solution of the equation

$$^{\gamma_2 - \gamma_1}\gamma_2 = \begin{cases} a_1 \; if \; a_1 = a_2 \\ a_2 \; if \; a_1 \neq a_2 \end{cases} \tag{6.41}$$

which leads to the following *Riccati equation* in γ_2

$$\frac{d\gamma_2}{dt} + \gamma_2^2 - (a_i + \gamma_1)\gamma_2 - \frac{d\gamma_1}{dt} + \gamma_1 a_1 = 0, \; i = 1 \; or \; 2. \tag{6.42}$$

Notice that the Riccati equation (6.42) has as a particular solution $\gamma_2 = \gamma_1$ and is thus solvable by reduction to a linear first order differential equation.

Proceeding in the same manner for the rest of the factors a_i (multiple or not), one obtains a set $\Delta = \{\gamma_1, ..., \gamma_n\}$ of P-independent roots of $P(\partial)$, or, equivalently, P is the minimal polynomial of Δ according to Definition 767.

For Example 992, the first root is $\gamma_1 = a_1 = -t^{-\frac{3}{2}} + \frac{1}{2}t^{-1} + \alpha$. The second one can be found as a solution of the equation

$$^{\gamma_2 - \gamma_1}\gamma_2 = a_2 \tag{6.43}$$

which leads to the Riccati equation

$$\frac{d\gamma_2}{dt} + \gamma_2^2 - (2\alpha - \frac{1}{2})\gamma_2 - t^{-3} - \frac{\alpha}{2}t^{-1} + \alpha^2 = 0 \tag{6.44}$$

which has γ_1 as a particular solution. One has thus to look for a general solution of the form $\gamma_2 = -t^{\frac{3}{2}} + \frac{1}{2}t^{-1} + \alpha + z$ which, substituted into (6.44), leads to the *Bernoulli equation* $\frac{dz}{dt} + z^2 + (\gamma_1 - a_2)z = 0$. Another change of variable $z = \frac{1}{u}$ transforms the Bernoulli equation into a linear one $\frac{du}{dt} + (2t^{-\frac{3}{2}} - \frac{3}{2}t^{-1})u = 1$, a solution of which is $u = \frac{1}{2}t^{\frac{3}{2}}$. It follows that $\gamma_2 = t^{-\frac{3}{2}} + \frac{1}{2}t^{-1} + \alpha$. $\Delta = \{\gamma_1, \gamma_2\}$ is a set of P-independent roots of P, or, equivalently, P is the minimal polynomial of Δ. Indeed, the solutions of the elementary equations (6.6) are

$$y_{1,2} = e^{(\pm 2t^{-\frac{1}{2}} + \frac{1}{2}lnt + \alpha t)} \tag{6.45}$$

which are $\mathbb{C}$-independent. As a matter of fact, the *Wronskian matrix* of y_1 and y_2 is $W(y_1, y_2) = \begin{bmatrix} y_1 & y_2 \\ \frac{dy_1}{dt} & \frac{dy_2}{dt} \end{bmatrix}$ and $det(W(y_1, y_2)) = 2y_1(t)y_2(t)t^{-\frac{3}{2}} \neq 0$ for $t > 0$, therefore condition (3-iv) of Theorem 770 is satisfied.

As shown in the above example, the fields in which a fundamental set of roots can be calculated provide a direct connection with the *solutions* of equation (6.2). Indeed, the elementary equations (6.6) can be solved over these fields. Notice that this kind of fields, denoted by $\widetilde{\mathbf{K}}$ in Section 6.3, may be larger that those over which a factorization with possibly multiple linear factors (6.10) can be obtained; the latter fields are denoted by $\check{\mathbf{K}}$ in Section 6.3.

Remark 994. *If we except periodic systems, which are treated separately (Sections 6.7.2 and 6.8), the coefficients of physical systems are not described by infinite series in practice (since they usually correspond to physical entities like,* e.g.*, time-constants, gains, parameters of actuators, etc.). Therefore, we will limit ourselves in the sequel to subfields of the Puiseux-type field* $\mathbf{K}^{\infty}$ *defined by*

$$\mathbf{K}^{\infty} = \bigcup_{m \geq 1} \mathbb{C}(t^{1/m}). \tag{6.46}$$

Notice that $\mathbf{K}^{\infty}$ is an Ore field (according to Lemma and Definition 835).

Remark 995. *From a theoretical point of view it may happen that* $\check{\mathbf{K}} \nsubseteq \mathbf{K}^{\infty}$. *However, in this case, from a practical point of view, the computation of the factors* $(\partial - a_i)$ *is not feasible since an infinite number of steps is needed with, for example, the iterative procedure described in Example 6.4.3.*

6.5 The Module Framework

6.5.1 Modules and Full Sets of Zeros

The results obtained in the preceding sections can be interpreted in the intrinsic framework of definition of an LTV system by a module. Let $\mathbf{R} = \mathbf{K}[\partial]$. The LTV system defined by (6.2) is, up to an $\mathbf{R}$-isomorphism, the left torsion $\mathbf{R}$-module T defined by (6.2).

Let again $P \in \mathbf{K}[\partial]$ be a skew polynomial (or, equivalently, a differential operator) of degree n. In Section 6.4 it was shown how a factorization of type (6.10) may be obtained for P over a field extension $\breve{\mathbf{K}} \supseteq \mathbf{K}$. This extension may be $\breve{\mathbf{K}} = \check{\mathbf{K}}$, in which case a full set of zeros of P is provided, or, $\breve{\mathbf{K}} = \widetilde{\mathbf{K}}$ which leads to a fundamental set of roots of P.

The general notion of conjugacy class (Definition 754) becomes simpler in the present case:

Definition 996. *The set*

$$\Delta_{\breve{K}}(a_i) = \{\overline{a}_i = a_i + \frac{\dot{c}}{c},\ c \in \breve{\boldsymbol{K}}\} \tag{6.47}$$

is called the *conjugacy class* of the zero $a_i \in \breve{\mathbf{K}}$ of P given by (6.10).

Following Theorem 763 (ii), there exist at most n distinct conjugacy classes of zeros of P.

Moreover, in the module-based framework, equation (6.2) is not the unique differential equation which defines the autonomous system. Indeed, if $\overline{T} \cong T$ is an $\mathbf{R}$-module defined by the equation

$$\overline{P}(\partial)\overline{y} = 0 \tag{6.48}$$

where $\overline{P}(\partial) \sim P(\partial)$ in the sense of Definition 517 and Lemma 745, then (6.2) and (6.48) define the same autonomous system. Let $\overline{P}(\partial) = (\partial - \overline{a}_{\overline{n}}) \ldots (\partial - \overline{a}_1)$ be a decomposition of type (6.10) of $\overline{P}(\partial)$ over $\breve{\mathbf{K}}$. As $\breve{\mathbf{K}}[\partial]$ is a principal ideal domain, thus, according to Theorem 462, a unique factorization domain, it follows from Lemma and Definition 433 that $n = \overline{n}$ and each a_i, $i = 1, \ldots, n$ is conjugated (in the sense of Definition 754) with some $\overline{a}_j$, i.e., there exists $c \in \breve{\mathbf{K}}$ such that

$$a_i - \overline{a}_j = \frac{\dot{c}}{c}. \tag{6.49}$$

If the ring of scalars is extended to $\breve{\mathbf{R}} = \breve{\mathbf{K}}[\partial]$, (6.10) leads to the *composition series*

$$\breve{\mathbf{R}}/\breve{\mathbf{R}}P(\partial) \supset \breve{\mathbf{R}}P_1(\partial)/\breve{\mathbf{R}}P(\partial) \supset ... \supset \breve{\mathbf{R}}P_n(\partial)/\breve{\mathbf{R}}P(\partial) \tag{6.50}$$

of the module $\breve{\mathbf{R}}/\breve{\mathbf{R}}P(\partial)$, where $P_1(\partial) = \partial - a_1$ and $P_i(\partial) = P_{i-1}(\partial)(\partial - a_i)$, $i = 2, ..., n$.

If the factors of P are pairwise non similar, (6.50) particularizes to a direct sum. This happens when the a_i's of (6.10) are pairwise nonconjugate but also when P has multiple zeros and can be written in the form

$$P(\partial) = (\partial - a_m)^{d_m} ...(\partial - a_1)^{d_1}, m \leq n \tag{6.51}$$

where the a_i's are pairwise nonconjugate.

Then,

$$\check{T} \cong \check{\mathbf{R}}/\check{\mathbf{R}}P(\partial) \cong \oplus_{i=1}^{m} \check{\mathbf{R}}/\check{\mathbf{R}}(\partial - a_i)^{d_i} \tag{6.52}$$

and

$$\check{\mathbf{R}}P = \bigcap_i \check{\mathbf{R}}(\partial - a_i)^{d_i}, \ i = 1, ..., m. \tag{6.53}$$

Obviously, when the a_i's of (6.10) are pairwise nonconjugate, then $m = n$ and $d_i = 1$, $i = 1, ..., n$ in (6.52) and (6.53).

Decomposition (6.52) is a *complete direct decomposition* of $\mathbf{R}$ according to the Definition given in **(3.3.2.4)**.

Particularly, P is a W-polynomial over $\check{\mathbf{K}}$; thus

$$\widetilde{T} \cong \widetilde{\mathbf{R}}/\widetilde{\mathbf{R}}P(\partial) = \oplus_{i=1}^{n} \widetilde{\mathbf{R}}/\widetilde{\mathbf{R}}(\partial - \gamma_i). \tag{6.54}$$

where $\{\gamma_1, ..., \gamma_n\} \in \check{\mathbf{K}}$ is a fundamental set of roots of $P(\partial)$. If, moreover, $\check{\mathbf{K}} = \widetilde{\mathbf{K}}$, then

$$\forall i, \exists j \textit{ such that } a_i \sim \gamma_j \textit{ over } \check{\mathbf{K}}. \tag{6.55}$$

The polynomial in Example 972 has a factorization (6.51) with $n = 2$, $a_1 = a + t^{-1}$, $a_2 = a - 2t^{-1}$, $d_1 = d_2 = 1$. It follows that $\check{T} \cong \check{\mathbf{R}}/\check{\mathbf{R}}(\partial - a - t^{-1}) \oplus \check{\mathbf{R}}/\check{\mathbf{R}}(\partial - a + 2t^{-1})$. Moreover $\gamma_1 = a + t^{-1}$, $\gamma_2 = a - t^{-1}$ and thus $a_2 \sim \gamma_2$.

6.5.2 Modules and Poles

The conjugacy classes (6.47) of the zeros of $P(\partial)$ have interesting properties, at least when the coefficients of P belong to an Ore field (according to Lemma and Definition 835). Indeed, following Lemma and Definition 835, $lim_{t \to +\infty}(\overline{a}_i(t) - a_i(t))$, $\forall \overline{a}_i \in \Delta_{\mathbf{K}}(a_i)$ where $\mathbf{K}$ is an Ore field.

The poles of an LTV system can now be defined in an intrinsic way:

Corollary and Definition 997. *Consider an autonomous LTV system given by a torsion $\boldsymbol{R}$-module $T \cong \boldsymbol{R}/\boldsymbol{R}P(\partial)$ where $\boldsymbol{R} = \boldsymbol{K}[\partial]$ and let $a_i \in \check{\boldsymbol{K}} \supseteq \boldsymbol{K}$, $i = 1, ..., n$ be such that $\{a_1, ..., a_n\}$ is a full set of zeros of $P(\partial)$ and let $\gamma_i \in \widetilde{\boldsymbol{K}} \supseteq \check{\boldsymbol{K}}$, $i = 1, ..., n$, be such that $\{\gamma_1, ..., \gamma_n\}$ is a fundamental set of roots of $P(\partial)$ (which is Wedderburn over $\widetilde{\boldsymbol{K}}$, $\widetilde{\boldsymbol{K}} \supseteq \check{\boldsymbol{K}} \supseteq \boldsymbol{K}$). Let $\pi_1, ..., \pi_p$ be the conjugacy classes of the a_i's. The multiplicity of π_j is the number ν_j of zeros a_i which belong to π_j. Therefore, $\sum_{i=1}^{p} \nu_i = n$. If $\check{\boldsymbol{K}}$ is an Ore field (according to Lemma and Definition 835) over which there exists a direct sum decomposition (6.52), the π_i's are called the* quasi-poles *of the*

LTV system T; $\{a_1, ..., a_n\}$ is a full set of quasi-poles *of T. If $\widetilde{\boldsymbol{K}}$ is an Ore field, the* poles *of the LTV system T are the conjugacy classes of the γ_i's; $\{\gamma_1, ..., \gamma_n\}$ is a* fundamental set of poles *of T.*

Notice that, in this context, the poles of a system have multiplicity one.

For Example 972, $P(\partial) = (\partial - a + 2t^{-1})(\partial - a - t^{-1})$, thus, the quasi-poles of the system are $\{\Delta_{\mathbb{C}(t)}(a - 2t^{-1}),\ \Delta_{\mathbb{C}(t)}(a + t^{-1})\}$. Since $a - 2t^{-1} \sim a + t^{-1}$ over $\mathbb{C}(t)$, finally $\Delta_{\mathbb{C}(t)}(a - 2t^{-1}) = \Delta_{\mathbb{C}(t)}(a + t^{-1})$ is the only quasi-pole with multiplicity 2. A fundamental set of poles of the same system is $\{a + t^{-1}, a - t^{-1}\}$.

6.5.3 Poles and Stability

Stability can be directly evaluated using the two types of poles of the system defined above.

Theorem 998. *Consider an autonomous system $\sum$ which is defined by the $\boldsymbol{R}$-torsion module $T \cong \boldsymbol{R}/\boldsymbol{R}P(\partial)$.*

- *Let $\{\gamma_1, ..., \gamma_n\}$ be a full set of quasi-poles or a fundamental set of poles of $\sum$ over an Ore field (assuming that such a set exists). Then, $\sum$ is* exponentially stable *if, and only if, for all $i \in \{1, ..., n\}$,*

$$\lim_{t \to +\infty} Re\{\gamma_i(t)\} < 0. \tag{6.56}$$

  $\sum$ is exponentially unstable *if, and only if at least one of the γ_i's satisfies the condition*

$$\lim_{t \to +\infty} Re\{\gamma_i(t)\} > 0. \tag{6.57}$$

- *If $P(\partial)$ has a full set of zeros $\{\gamma_1, ..., \gamma_n\}$ in an Ore field, then $\sum$ is* exponentially stable *if (6.56) holds for all $i \in \{1, ..., n\}$. If the γ_i's belong to a* real *Ore field, condition (6.56) is also necessary for exponential stability; furthermore, condition (6.57) is also necessary for exponential instability.*

The proof follows from Theorem 985.

In Example 972, $lim_{t \to +\infty} \gamma_1(t) = lim_{t \to +\infty} \gamma_2(t) = a$. The system is thus exponentially stable if, and only if $a < 0$; it is exponentially unstable if, and only if $a > 0$. Notice that one can conclude on exponential stability by investigating a_1 or a_2, i.e., $lim_{t \to +\infty} a_1(t) = lim_{t \to +\infty} a_2(t) = a$ as both of them belong to the same conjugacy class.

6.5.4 The Case of Linear Time-Invariant Systems

Theorem 998 covers also the case of LTI systems with the simplification that each conjugacy class of an element of a full set of zeros or a fundamental set of poles is a singleton. In this case, Theorem 998 leads to the well-known result for LTI systems:

Corollary 999. *An LTI autonomous system with poles* $\{a_1, ..., a_n\} \in \mathbb{C}$ *is exponentially stable if, and only if*

$$Re\{a_i\} < 0, i = 1, ..., n. \tag{6.58}$$

The system is exponentially unstable if, and only if there exists a pole a_i *such that*

$$Re\{a_i\} > 0. \tag{6.59}$$

Example 1000. *Consider the following LTI system:*

$$(\partial - 1)^2 x = (\partial - 1)(\partial + 1)u. \tag{6.60}$$

There is one pole at 1 *with multiplicity* 2 *and it satisfies condition (6.59). The system is thus exponentially unstable.*

6.6 Modules of Poles and Zeros

6.6.1 System Poles

The above developments make it possible to unify the different ways to characterize stability of an LTV system in the intrinsic framework of modules. With the notations of the preceding section, let $\breve{\mathbf{K}}$ be a field extension over which there exists a direct sum decomposition of type (6.54) or (6.52) of the torsion module which defines the autonomous system. This corresponds respectively to $\widetilde{\mathbf{K}}[\partial]$ in the case of the computation of a fundamental set of poles or to $\breve{\mathbf{K}}[\partial]$ in the case of the computation of quasi-poles which correspond to a factorization of type (6.52). All the usual poles and zeros entities (like, e.g., system poles, invariant zeros, decoupling zeros, etc.) are next defined using modules. Following the definition of those modules over $\widetilde{\mathbf{K}}$ or $\breve{\mathbf{K}}$ one will obtain entities or quasi-entities (i.e., poles or quasi-poles, invariant zeros or quasi-invariant zeros, etc.). To simplify the language, only one definition is given in each case. The formulation of the other one is left to the reader who has only to replace $\widetilde{\mathbf{K}}$ by $\breve{\mathbf{K}}$. One has the following

Lemma and Definition 1001. *Consider an LTV control system* $\sum = (M, u, y)$[1] *over* $\boldsymbol{R} = \boldsymbol{K}[\partial]$. *Let* $\breve{\boldsymbol{R}} = \breve{\boldsymbol{K}}[\partial]$, $\breve{M} = \breve{\boldsymbol{R}} \otimes_{\boldsymbol{R}} M$, $\breve{U} = \breve{\boldsymbol{R}} \otimes_{\boldsymbol{R}} [u]_{\breve{\boldsymbol{R}}}$. *The* $\breve{\boldsymbol{R}}$*-module* $M_{sp} = \breve{M}/\breve{U}$ *is torsion and is called the* module of system poles *of* $\sum$. M_{sp} *is the quotient module which defines the "autonomous part" (6.2) of the LTV system* $\breve{M}$.

Proof. Take, e.g., a Rosenbrock description (5.28) of $\sum$ where D is square and invertible over the field of fractions Q of R, thus over the field of fractions $\breve{Q}$ of $\breve{R}$. ∎

[1] The vector y here should not be confused with the scalar variable y in (6.2).

Let $B(\partial)$ be a matrix of definition of M_{sp} ($B(\partial)w = 0$) and consider its *Smith normal form* over $\breve{\mathbf{K}}$ as introduced in Lemma and Definition 655: there exist invertible matrices $U(\partial)$, $V(\partial)$ over $\breve{\mathbf{R}}$ such that

$$U(\partial)B(\partial)V^{-1}(\partial) = diag\{1, ..., 1, P_m(\partial)\}.$$

$P_m(\partial)$ is defined up to similarity and $P_m \cong P$, where P is the polynomial in (6.2) which defines M_{sp}. The poles of the LTV system associated with M thus correspond to the conjugacy classes of a full set of zeros of P_m.

As for polynomials, zeros of a torsion module can be defined:

Lemma and Definition 1002. *Let T be a finitely generated torsion module and let*

$$diag\{1, ..., 1, P_m(\partial)\}$$

be the Smith normal form of one of its matrices of definition B. Let $a \in \boldsymbol{K}$ be a zero of $P_m(\partial)$; a it is called a Smith zero *of T. Let $\{a_1, ..., a_m\}$ be a full set of zeros of $P_m(\partial)$; $\{a_1, ..., a_m\}$ it is called* a full set of Smith zeros *of T and noted $a(T)$ in the sequel.*

Remark 1003. *Notice that, in the case $\breve{\boldsymbol{K}} = \widetilde{\boldsymbol{K}}$, a full set of Smith zeros of T coincides with a fundamental set of roots of P_m.*

The matrices B and $diag\{1, ..., 1, P_m(\partial)\}$ define isomorphic $\mathbf{R}$-modules. Therefore, they are *left-similar* (Definition 517) and we write $B \sim diag\{1, ..., 1, P_m(\partial)\}$.

6.6.2 Invariant Zeros

The same construction can be made for the zeros if one considers an LTV control system (M, u, y). Let z be a generator of $\mathcal{T}(M/[y]_{\mathbf{R}})$ where $\mathcal{T}(.)$ denotes the torsion submodule. There exists $Z(\partial) \in \mathbf{K}[\partial]$ such that $Z(\partial)z = 0$ and let $\overline{\mathbf{K}}$ an extension of $\mathbf{K}$ over which a full set of zeros of Z can be provided. Let now $\widehat{\mathbf{K}}$ be such that $\widehat{\mathbf{K}} \supseteq \overline{\mathbf{K}}$ and $\widehat{\mathbf{K}} \supseteq \breve{\mathbf{K}}$ where $\breve{\mathbf{K}}$ is the field over which the poles of system M are defined. Let $\widehat{M} = \widehat{\mathbf{R}} \otimes_{\mathbf{R}} M$, where $\widehat{\mathbf{R}} = \widehat{\mathbf{K}}[\partial]$ and $\widehat{y} = (\widehat{y}_1, ..., \widehat{y}_p)$, $\widehat{y}_i = 1_{\widehat{\mathbf{R}}} y_i$.

Definition 1004. *The* module of the invariant zeros *of the LTV control system (M, u, y) is $M_{iz} = \mathcal{T}(\widehat{M}/[\widehat{y}]_{\widehat{\mathbf{R}}})$. The* invariant zeros *of the control system (M, u, y) are the conjugacy classes of the elements of a full set of Smith zeros of M_{iz}.*

The module of the invariant zeros of the system is generated by the variables of the system for which the output is zero ($y \equiv 0$). As an example, for the state-representation

$$\begin{cases} \dot{x} = Ax + Bu \\ y = Cx + Du \end{cases} \tag{6.61}$$

the module of the invariant zeros M_{iz} is defined by $R(\partial)\begin{bmatrix}\overline{x}\\ \overline{u}\end{bmatrix}$ where $R(\partial) = \begin{bmatrix}\partial I - A & -B\\ C & D\end{bmatrix}$ is the system's matrix (Definition 5.28). As M_{iz} is torsion, matrix $R(\partial)$ is singular for some $\partial = \alpha_i$. It is known (see, *e.g.*, [170]) that, in this case, for an input $u(t) = u_0 e^{\alpha t}$, $t \geq 0$ there exists an initial state x_0 such that the output is zero: $y(t) \equiv 0$, $t \geq 0$.

The same equivalent characterizations discussed above for the poles can be used to provide the invariant zeros of an LTV control system using a full set of Smith zeros of the module of the invariant zeros. From a computational point of view, this leads to the factorization of a skew polynomial.

Example 1005. *For the autonomous system of Example 992, consider the output* $\ddot{y}$. *The module of invariant zeros of the control system with the chosen output is defined by the equation*

$$(\frac{1}{2}t^{-1} - 2\alpha)\partial\overline{y} + (-t^{-3} - \frac{1}{2}t^{-1}\alpha + \alpha^2)\overline{y} = 0 \tag{6.62}$$

where $\overline{y}$ *denotes the image of* y *in* M_{iz}. *It follows that the invariant zero is* $\Delta_{\mathbb{C}(t)}(-\frac{-t^{-3}-\frac{1}{2}t^{-1}\alpha+\alpha^2}{\frac{1}{2}t^{-1}-2\alpha})$.

There exists an embedding $\mathbf{K} \hookrightarrow \hat{\mathbf{K}}$. Therefore, the system coefficients belong to $\hat{\mathbf{K}}$ and we assume without loss of generality in the sequel that $\mathbf{K} = \hat{\mathbf{K}}$.

6.6.3 Transmission Poles and Zeros

6.6.3.1 Modules of Transmission Poles and Zeros

For an LTV control system (M, u, y) the transfer (i.e., input-output) dynamics can be defined. Consider for that the submodule $[y, u]_{\mathbf{R}}$ of M. From Theorem 654, it exists a free submodule F of $[y, u]_{\mathbf{R}}$ such that $[y, u]_{\mathbf{R}} = \mathcal{T}([y, u]_{\mathbf{R}}) \oplus F$. Then $\mathcal{T}([y, u]_{\mathbf{R}}) \subseteq \mathcal{T}(M)$ and from Corollary 657, follows that there exists a free submodule $\Phi_{[y,u]_{\mathbf{R}}}$ of M such that $M = \mathcal{T}(M) \oplus \Phi_{[y,u]_{\mathbf{R}}}$ and $F \subseteq \Phi_{[y,u]_{\mathbf{R}}}$. In what follows $\Phi_{[y,u]_{\mathbf{R}}}$ is simply denoted by Φ and thus

$$M = \mathcal{T}(M) \oplus \Phi. \tag{6.63}$$

Definition 1006. *The module* $M_{td} = (\Phi \cap [y, u]_{\mathbf{R}})$ *is the* transfer module *of the LTV control system* (M, u, y).

Roughly speaking, the transfer module is the torsion-free module generated by the equations of the system after elimination of all variables with the exception of u and y. This is similar to the classical construction of the transfer matrix for a given system.

The modules of transmission poles and zeros can be defined from the transmission dynamics:

Definition 1007. *The* module of transmission zeros *of the LTV control system* (M, u, y) *is* $M_{tz} = \mathcal{T}((\Phi \cap [y,u]_{\boldsymbol{R}})/(\Phi \cap [y]_{\boldsymbol{R}}))$. *The* transmission zeros *are the conjugacy classes of the elements of a full set of Smith zeros of* M_{tz}.

Definition 1008. *The* module of transmission poles *of the LTV control system* (M, u, y) *is* $M_{tp} = (\Phi \cap [y,u]_{\boldsymbol{R}})/[u]_{\boldsymbol{R}}$. *The* transmission poles *are the conjugacy classes of the elements of a full set of Smith zeros of* M_{tp}.

Remark 1009. *As mentioned in Remark 6.6.1,* $M/[u]_{\boldsymbol{R}}$ *is torsion. From (6.63) follows that* $M/[u]_{\boldsymbol{R}} \cong \mathcal{T}(M) \oplus \Phi/[u]_{\boldsymbol{R}}$ *and thus* $\Phi/[u]_{\boldsymbol{R}}$ *is torsion.* $(\Phi \cap [y,u]_{\boldsymbol{R}})/[u]_{\boldsymbol{R}} \subseteq \Phi/[u]_{\boldsymbol{R}}$ *from which it follows that* M_{tp} *is torsion.*

To practically put into evidence the notions defined above, the following matrix computations are used.

6.6.3.2 Transfer Matrix Factorizations

The following notations are used in the sequel to denote matrices: H (or $H(t)$) is a matrix with entries in **K**, $H(\partial)$ a matrix with entries in **R** (a polynomial matrix) and $\mathcal{H}(\partial)$ a matrix with entries in **Q**, the field of fractions of **R** (a transfer matrix).

Using the gcld's and gcrd's (Definition 81) we are led to the following factorizations of a given LTV transfer matrix:

Proposition 1010. *Any LTV transfer matrix* $\mathcal{H}(\partial)$ *has the following factorizations:*

(i) left-*coprime factorizations:* $\mathcal{H}(\partial) = A_L^{-1}(\partial) B_L(\partial)$ *(i.e.,* $A_L(\partial) y = B_L(\partial) u$*), where* $A_L(\partial)$ *and* $B_L(\partial)$ *are left-coprime*
(ii) right-*coprime factorizations:* $\mathcal{H}(\partial) = B_R(\partial) A_R^{-1}(\partial)$ *(i.e.,* $y = B_R(\partial)\xi$ *and* $u = A_R(\partial)\xi$*), where* $A_R(\partial)$ *and* $B_R(\partial)$ *are right-coprime*

related by the diophantine equation

$$A_L(\partial) B_R(\partial) - B_L(\partial) A_R(\partial) = 0. \tag{6.64}$$

From a computational point of view, (6.64) can be solved either using elementary row and column operations or directly. In the latter case, when a left factorization is to be deduced from a right one, (6.64) is a system of *algebraic* equations and, in the opposite case, (6.64) leads to a system of *differential* equations.

Example 1011. $\mathcal{H}(\partial) = (\partial + t)(\partial - t)^{-1}$, *thus* $A_R(\partial) = \partial - t$, $B_R(\partial) = \partial + t$. *Let* $A_L(\partial) = a\partial + b$, $B_L(\partial) = c\partial + d$, $a, b, c, d \in \boldsymbol{K}$. *(6.64) leads to* $c = a$, $b = a(-t^{-1} - t)$, $d = a(-t^{-1} + t)$, *thus* $\mathcal{H}(\partial) = (\partial - t^{-1} - t)^{-1}(\partial - t^{-1} + t)$.

Matrices of definition of the modules of transmission poles and zeros defined above can be extracted from the left and right factorizations of the transfer matrix of the system: if a factorization (i) is used for $\mathcal{H}(\partial)$, the module of transmission poles is defined by $A_L(\partial)\overline{y} = 0$ and the module of transmission zeros is defined by $B_L(\partial)\overline{u} = 0$. Starting from (ii), the module of transmission poles is defined by $A_R(\partial)\overline{\xi} = 0$ and the module of transmission zeros is defined by $B_R(\partial)\overline{\xi} = 0$. It follows

Proposition 1012. *Consider two factorizations (i) and (ii) of a transfer matrix $\mathcal{H}$: $\mathcal{H}(\partial) = A_L^{-1}(\partial)B_L(\partial) = B_R(\partial)A_R^{-1}(\partial)$. Then $A_L(\partial) \sim A_R(\partial)$ and $B_L(\partial) \sim B_R(\partial)$. The transmission poles of the LTV transfer matrix $\mathcal{H}(\partial)$ are given by the full sets of zeros of polynomials $P(\partial)$ and $\overline{P}(\partial)$ where $A_L(\partial) \sim \{1, ..., 1, P(\partial)\}$, $A_R(\partial) \sim \{1, ..., 1, \overline{P}(\partial)\}$ and $P(\partial) \sim \overline{P}(\partial)$. The transmission zeros of the LTV transfer matrix $\mathcal{H}(\partial)$ are given by the full sets of zeros of polynomials $Q(\partial)$ and $\overline{Q}(\partial)$ where $B_L(\partial) \sim \{1, ..., 1, Q(\partial)\}$, $B_R(\partial) \sim \{1, ..., 1, \overline{Q}(\partial)\}$ and $Q(\partial) \sim \overline{Q}(\partial)$.*

Definition 1013. *A transfer matrix whose poles or quasi-poles belong to an Ore field and satisfy the stability condition of Theorem 998 is called exponentially stable.*

Example 1014. *Consider the LTV system given by*

$$\begin{cases} D(\partial)x_1 = N(\partial)x_2 \\ x_3 = Q(\partial)x_1 \end{cases} \tag{6.65}$$

where $D(\partial) = (\partial - t)^2$, $N(\partial) = (\partial - t)(\partial + 1)$ and $Q(\partial) = (\partial + t)$. We are looking for the transfer module with the choice $u = x_2$ and $y = x_3$. First, notice that $M/[x_2]_{\mathbf{R}}$ is torsion. Indeed, this module is generated by the relation: $(\partial - t)^2\overline{x}_1 = 0$, where $\overline{x}_1$ is the canonical image of x_1 in $M/[x_2]_{\mathbf{R}}$. So, $u = x_2$ is an input in the sense of the Definition 851. Next, notice that the first equation in (6.65) defines a torsion element $v = (\partial - t)x_1 - (\partial + 1)u$ since

$$(\partial - t)v = 0. \tag{6.66}$$

As shown in Section 5.6.1, this corresponds to a lack of controllability of the system. Left coprime factors $D^0(\partial) = (\partial - t)$, $N^0(\partial) = (\partial + 1)$ are obtained from $D(\partial)$ and $N(\partial)$ by skipping their gcld: $D(\partial) = L^0(\partial)D^0(\partial)$, $N(\partial) = L^0(\partial)N^0(\partial)$, $L^0(\partial) = \partial - t$. If $y = x_3$, the transfer module *can be defined using the equation*

$$\overline{\overline{y}} = Q(\partial)(D^0(\partial))^{-1}N^0(\partial)\overline{\overline{u}} \tag{6.67}$$

from which any nonobservable dynamics must be eliminated. As $Q(\partial)$ and $D^0(\partial)$ are right coprime, there is no lack of observability for this example. At this stage, it can be already concluded using Proposition 1012 that the transmission pole of (6.65) is $\Delta_{\mathbb{C}(t)}(t)$, i.e., it is given by the zeros of $D^0(\partial)$. To obtain the zeros, one should further look for a factorization (i) or (ii) of the transfer in (6.67). Let's start by computing a factorization (i) of $Q(\partial)D^0(\partial)^{-1}$, i.e., *computing $\overline{Q}(\partial)$ and $\overline{D}^0(\partial)$ left coprime such that*

$$Q(\partial)(D^0(\partial))^{-1} = (\overline{D}^0(\partial))^{-1}\overline{Q}(\partial) \tag{6.68}$$

which is (6.64) with $B_R(\partial) = Q(\partial)$, $A_R(\partial) = D^0(\partial)$. *Let the unknowns be* $\overline{D}^0(\partial) = a\partial + b$ *and* $\overline{Q}(\partial) = c\partial + d$. *Then, the coefficients* a, b, c, d *satisfy the identity:*

$$(a\partial + b)(\partial + t) \equiv (c\partial + d)(\partial - t), \; a, b, c, d \in \mathbb{C} \tag{6.69}$$

which leads to $a\partial^2 + (at+b)\partial + a + bt \equiv c\partial^2 + (d-ct)\partial - c - dt$, *i.e., to a system of algebraic equations with the following class of solutions parameterized by* a*:*

$$\begin{aligned} c &= a \\ b &= a(-t^{-1} - t) \\ d &= a(-t^{-1} + t) \end{aligned} \tag{6.70}$$

It follows that $Q(\partial)(D^0(\partial))^{-1} = (\partial+t)(\partial-t)^{-1} = (\partial-t^{-1}-t)^{-1}(\partial-t^{-1}+t)$ *which, along with (6.67) leads to*

$$(\partial - t^{-1} - t)\overline{\overline{\overline{y}}} = (\partial - t^{-1} + t)(\partial + 1)\overline{\overline{\overline{u}}} \tag{6.71}$$

which defines the module M_{td}. *The transfer function as defined in Corollary and Definition 853 is* $\mathcal{G}(\partial) = (\partial - t^{-1} - t)^{-1}(\partial - t^{-1} + t)(\partial + 1)$.

An equation of definition of M_{tz} *is* $(\partial - t^{-1} + t)(\partial + 1)\widetilde{u} = 0$. *The quasi-transmission zeros are thus* $\Delta_{\mathbb{C}(t)}(-1) \cup \Delta_{\mathbb{C}(t)}(-t + t^{-1})$. *An equation of definition of* M_{tp} *is* $(\partial - t^{-1} - t)y = 0$, *from which it follows that the transmission pole is* $\Delta_{\mathbb{C}(t)}(t^{-1} + t)$. *Notice that* $t^{-1} + t$ *is conjugated with* t *and thus the latter transmission pole is the same as the one found before:* $\Delta_{\mathbb{C}(t)}(t^{-1} + t) = \Delta_{\mathbb{C}(t)}(t)$.

The same conclusions can be reached if a right factorization (ii) is computed: consider (6.64) with $A_L(\partial) = N(\partial)$, $B_L(\partial) = D(\partial)$ *and the unknowns* $B_R(\partial) = \overline{\overline{D}}(\partial)$, $A_R(\partial) = \overline{\overline{N}}(\partial)$. *Obviously, a right coprime solution has the form* $\overline{\overline{D}}(\partial) = \partial + b$, $\overline{\overline{N}}(\partial) = c\partial + b$. *(6.64) leads in this case to the following system of* differential *equations:* $c = 1$, $1 + b = d - t$, $b + \dot{b} = \dot{d} - td$. *We have the following solution:* $d = 1 + \frac{1}{t+1}$, $b = -t + \frac{1}{t+1}$ $(c = 1)$. *Thus* $\mathcal{G}(\partial) = (\partial + t)(\partial + 1 + \frac{1}{t+1})(\partial - t + \frac{1}{t+1})^{-1}$. *From the latter factorization of* $\mathcal{G}(s)$ *it can be deduced that another equation of definition of* M_{tp} *is* $(\partial - t + \frac{1}{t+1})\xi = 0$, *thus the transmission pole is also given by* $\Delta_{\mathbb{C}(t)}(t - \frac{1}{t+1})$. *Indeed,* $t - \frac{1}{t+1} \sim t^{-1} + t \sim t$. *Similarly, the quasi-transmission zeros are also given by* $\Delta_{\mathbb{C}(t)}(-1 - \frac{1}{t+1}) \cup \Delta_{\mathbb{C}(t)}(-t)$ *as* $-1 \sim -1 - \frac{1}{t+1}$ *and* $-t \sim -t + t^{-1}$.

The transmission zeros have the usual significance: when a control system has a transmission zero, a zero output y can be obtained with a nonzero input u, as shown in the following example.

Example 1015. *Let* $\mathcal{G}(\partial) = \begin{bmatrix} 1 & (\partial+t)^{-1} & (\partial+1)^{-1}(\partial+2t) \\ 0 & 0 & (\partial+1)^{-1} \end{bmatrix}$. *A factorization of* $\mathcal{G}(\partial)$ *is provided by computing the lclm of* $(\partial+1)$ *and* $(\partial+t)$*:* $P(\partial) = [\partial+1, \partial+t]_l = (\partial - \frac{t-2}{1-t})(\partial+t) = (\partial - \frac{-t+t^2-1}{1-t})(\partial+1) = \partial^2 + \frac{2-t^2}{1-t}\partial + \frac{t^2-t-1}{t-1}$. *Then,* $\mathcal{G}(\partial) = D_l(\partial^{-1})N_l(\partial)$*, where* $D_l(\partial) = \begin{bmatrix} P(\partial) & 0 \\ 0 & \partial+1 \end{bmatrix}$, $N_l(\partial) = \begin{bmatrix} P(\partial) & \partial - \frac{t-2}{1-t} & (\partial - \frac{-t+t^2-1}{1-t})(\partial+2t) \\ 0 & 0 & 1 \end{bmatrix}$. *Notice that* $D_l(\partial)$, $N_l(\partial)$ *are left-coprime (when this does not happen, a left-coprime pair* $(D_l^0(\partial), N_l^0(\partial))$ *such that* $D_l^0(\partial)^{-1}N_l^0(\partial) = D_l^{-1}(\partial)N_l(\partial)$ *must be constructed from* $D_l(\partial)$, $N_l(\partial)$, i.e., *a gcld of* $D_l(\partial)$ *and* $N_l(\partial)$ *must be cancelled). Thus* $D_l(\partial)$ *and* $N_l(\partial)$ *are matrices of definition of the modules of poles and of transmission zeros respectively. The Smith normal form of* $N_l(\partial)$ *is* $\begin{bmatrix} 1 & 0 & 0 \\ 0 & \partial - \frac{t-2}{1-t} & 0 \end{bmatrix}$*, from which it follows that the transmission zero is* $\Delta_{\mathbb{C}(t)}(\frac{t-2}{1-t})$. *Therefore, there exists a nonzero* $\boldsymbol{R}$*-linear combination* $\overline{u}$ *of the inputs satisfying* $(\partial - \frac{t-2}{1-t})\overline{u} = 0$ *and a* $\boldsymbol{R}$*-linear combination* $\overline{y}$ *of the outputs, with nonzero coefficients, such that* $\overline{y} = 0$ *for the above inputs.*

In case of transmission zeros, the output is zero for some particular inputs of the form $u(t) = u_0\varphi(t)$, where u_0 is a *particular* matrix with constant entries. A particular case is when the output is zero for *any* matrix u_0 which is the so-called *blocking property* of a transfer matrix. This happens when a non-unit gcrd of all entries of $N_l(\partial)$ exists, where $\mathcal{G}(\partial) = D_l(\partial)^{-1}N_l(\partial)$ if a left-coprime factorization of the transfer matrix $\mathcal{G}(\partial)$. In the latter situation, blocking zeros can be defined from a non-unit gcrd of all entries of $N_l(\partial)$.

Definition 1016. *The* module of blocking zeros M_{bz} *of an LTV control system* (M, u, y) *is* $\operatorname{coker}_{\boldsymbol{R}}(\bullet P_N(\partial))$*, where* $P_N(\partial)$ *is the gcrd of all entries of* $N_l(\partial)$ *where* $(D_l(\partial), N_l(\partial))$ *is a left-coprime factorization of the transfer matrix* $\mathcal{G}(\partial)$ *of the control system:* $\mathcal{G}(\partial) = D_l(\partial)^{-1}N_l(\partial)$. *The* blocking zeros *are the conjugacy classes of the elements of a full set of Smith zeros of* M_{bz}.

The transfer matrix of Example 1015 has no blocking zero. If the transfer matrix is changed to

$$\mathcal{H}(\partial) = \begin{bmatrix} (\partial+2t) & (\partial+t)^{-1}(\partial+2t) & (\partial+1)^{-1}(\partial+2t) \\ 0 & 0 & (\partial+1)^{-1}(\partial+2t) \end{bmatrix}$$

then $N_l(\partial) = \begin{bmatrix} 1 & (\partial+t)^{-1} & (\partial+1) \\ 0 & 0 & 1 \end{bmatrix} P_N(\partial)$ with $P_N(\partial) = (\partial+2t)$. There exists a unique blocking zero which is $\Delta_{\mathbb{C}(t)}(-2t)$. Indeed, one can easily check that the Smith normal form of N_l is $\begin{bmatrix} 1 & 0 & 0 \\ 0 & (\partial+1)(\partial+2t)^2 & 0 \end{bmatrix}$.

6.6.4 Hidden Modes

Definition 1017. *The* module of the input-decoupling zeros *of the LTV system M is $M_{idz} = \mathcal{T}(M)$. The* input-decoupling zeros *are the conjugacy classes of the elements of a full set of Smith zeros of M_{idz}.*

For the system of Example 6.65, M_{idz} is defined by (6.66). The input-decoupling zero is $\Delta_{\mathbb{C}(t)}(t)$ and it corresponds to the noncontrollable dynamics of the system. Indeed, from (6.66), the trajectory of v is $v(t) = ce^{\frac{t^2}{2}}$, $c \in \mathbb{C}$ and it is thus independent of the choice of the input u. v is thus a torsion element of the system defined by M and (6.66) defines the noncontrollable quotient system.

Definition 1018. *The LTV system defined by M is said to be* exponentially stabilizable *if its noncontrollable quotient system is exponentially stable.*

Definition 1019. *The* module of output-decoupling zeros *of the LTV control system (M, u, y) is $M_{odz} = M/[y, u]_{\mathbf{R}}$. The* output-decoupling zeros *are the conjugacy classes of the elements of a full set of Smith zeros of M_{odz}.*

Remark 1020. *$M/[y, u]_{\mathbf{R}}$ is isomorphic to a quotient of $M/[u]_{\mathbf{R}}$ by Noether's third isomorphism theorem (Theorem 52). As $M/[u]_{\mathbf{R}}$ is torsion (see Definition 851), it follows that M_{odz} is also torsion.*

Example 1021. *Let*

$$\begin{cases} (\partial - \frac{t}{2})x = u \\ y = (\partial - \frac{t}{2})x. \end{cases} \tag{6.72}$$

The input-output relation is obviously $y = u$ which leads to $y = 0$ for $u = 0$. However, $M/[y, u]_{\mathbf{R}}$ is defined by

$$S(\partial)\overline{x} = 0, \; S(\partial) = \partial - \frac{t}{2} \tag{6.73}$$

from which it follows that the output-decoupling zero is $\Delta_{\mathbb{C}(t)}(\frac{t}{2})$. This corresponds to dynamics which cannot be observed from the chosen output y and input u. More precisely, if $u = 0$, from the first equation of (6.72), $x(t) = x(0)e^{\frac{t^2}{4}}$, $x(0) \in \mathbb{C}$ and thus $x(t) \neq 0$, $t \geq 0$ if $x(0) \neq 0$. This trajectory, which corresponds to the nonobservable dynamics (6.73) has no incidence on the output y which remains zero: $y(t) = 0$, $t \geq 0$. Thus, the state of the system x is nonobservable.

Definition 1022. *The LTV control system (M, u, y) is said to be* exponentially detectable *if its module of output-decoupling zeros is exponentially stable.*

Consider a direct sum decomposition of $[y, u]_{\mathbf{R}}$ into a torsion and free parts respectively: $[y, u]_{\mathbf{R}} = T \oplus F$. Obviously, $T \subseteq \mathcal{T}(M)$ and $F \subseteq \Phi$ and, using the form (2.9) of Theorem 147(ii) and (6.63), it follows that

$$M/[y,u]_{\mathbf{R}} \cong \mathcal{T}(M)/T \oplus \Phi/F \tag{6.74}$$

The module T is the noncontrollable subsystem of the observable subsystem $[y,u]_{\mathbf{R}}$ of the LTV control system, thus $\mathcal{T}(M)/T \cong \mathcal{T}(M)/\mathcal{T}(M) \cap [y,u]_{\mathbf{R}}$ and (6.74) provides a decomposition of the nonobservable quotient into noncontrollable subsystem and controllable quotient respectively. This leads us to the following definition of the input-output decoupling zeros which characterize the nonobservable quotient which is also noncontrollable:

Definition 1023. *The* module of input-output decoupling zeros *of the LTV control system* (M,u,y) *is* $M_{iodz} = \mathcal{T}(M)/(\mathcal{T}(M) \cap [y,u]_{\boldsymbol{R}})$. *The* input-output decoupling zeros *are the conjugacy classes of the elements of a full set of Smith zeros of* M_{iodz}.

The input-output decoupling zeros are thus given by the nonobservable quotient of the noncontrollable subsystem or, equivalently, by the noncontrollable subsystem of the nonobservable quotient.

Remark 1024. *The above terminology is explained as follows: we have* $[y,u]_{\boldsymbol{R}} \subseteq M$, *i.e.,* $[y,u]_{\boldsymbol{R}}$ *is a submodule of* M. *Since* $\boldsymbol{R}$*-modules are systems, we say that* $[y,u]_{\boldsymbol{R}}$ *is a* subsystem *of* M, *and, more specifically, is the* observable subsystem *of* M. *The quotient module* $M/[y,u]_{\boldsymbol{R}}$ *is the* nonobservable quotient *of the system* M. *Likewise,* $T \subseteq M$ *is a submodule of* M, *thus is the* noncontrollable subsystem *of the system* M. *The quotient module* M/T *is the* controllable quotient *of the system* M.

Example 1025. *Consider the LTV system given by*

$$\begin{cases} (\partial - t)x = (\partial - t)(\partial + 1)u \\ y = (\partial + 2)(\partial - t)x. \end{cases} \tag{6.75}$$

The module $\mathcal{T}(M)$ *of the input-decoupling zeros is defined by* $(\partial - t)v = 0$, *where* $v = x - (\partial + 1)u$, i.e.,

$$\mathcal{T}(M) \cong \boldsymbol{R}/\boldsymbol{R}(\partial - t) \tag{6.76}$$

and the module of the output-decoupling zeros $M/[y,u]_{\boldsymbol{R}}$ *is defined by*

$$\begin{bmatrix} (\partial - t) \\ (\partial + 2)(\partial - t) \end{bmatrix} \overline{x} = 0. \tag{6.77}$$

As $\begin{bmatrix} (\partial - t) \\ (\partial + 2)(\partial - t) \end{bmatrix} \cong \begin{bmatrix} (\partial - t) \\ 0 \end{bmatrix}$ *it follows that* $M/[y,u]_{\boldsymbol{R}}$ *is defined by* $(\partial - t)\overline{\overline{x}} = 0$, i.e.,

$$M/[y,u]_{\boldsymbol{R}} \cong \boldsymbol{R}/\boldsymbol{R}(\partial - t) \tag{6.78}$$

therefore (M,u,y) *has a unique output-decoupling zeros which is* $\Delta_{\mathbb{C}(t)}(t)$. *Consider now the controllable quotient* Φ *of the control system*

$$\begin{cases} \widetilde{x} = (\partial + 1)\widetilde{u} \\ \widetilde{y} = (\partial + 2)(\partial - t)\widetilde{x}. \end{cases} \tag{6.79}$$

Notice that the module of output-decoupling zeros of this quotient system is trivial: $\Phi/[\widetilde{y}, \widetilde{u}]_{\mathbf{R}} \cong 0$, *thus* $\Delta_{\mathbb{C}(t)}(t)$ *is again the (only) input-output decoupling zero.*

6.6.5 Relations between Poles and Zeros

In this section, relations between various poles and zeros are established in the module framework. They are connected to the well-known distinctions between the transmission modes and hidden modes and correspond to the separation of the system into controllable/noncontrollable, respectively observable/nonobservable subsystems and quotients.

6.6.5.1 Poles and Hidden Modes

First, let us distinguish among the system poles those which are controllable, respectively observable.

Definition 1026. *The* module of controllable poles *of the LTV control system* (M, u, y) *is* $M_{cp} = \Phi/[u]_{\mathbf{R}}$. *The* controllable poles *are the conjugacy classes of the elements of a full set of Smith zeros of* M_{cp}.

Following Remark 1009, M_{cp} is torsion.

For the system of Example 1021, the module of controllable poles is defined by $(\partial - \frac{t}{2})\overline{x} = 0$, so there exists a unique controllable pole, which is $\Delta_{\mathbb{C}(t)}(\frac{t}{2})$.

Definition 1027. *The* module of observable poles *of the LTV control system* (M, u, y) *is* $M_{op} = [y, u]_{\mathbf{R}}/[u]_{\mathbf{R}}$. *The* observable poles *are the conjugacy classes of the elements of a full set of Smith zeros of* M_{op}.

Remark 1028. *As mentioned in Remark 6.6.1,* $M/[u]_{\mathbf{R}}$ *is torsion. Since* $[y, u]_{\mathbf{R}} \subseteq M$, $[y, u]_{\mathbf{R}}/[u]_{\mathbf{R}}$ *is torsion too.*

For the system of Example 1025, the module of observable poles is given by $(\partial + 2)\overline{x} = 0$ and the only observable pole is $\Delta_{\mathbb{C}(t)}(-2)$.

Poles and zeros of LTV systems have been defined as torsion modules. To establish relations between them, one must investigate the relations between the Smith zeros of a given torsion module and its submodules. In comparison with the LTI case, some cautions must be taken in the LTV case since the non trivial invariant factor of a given module is a polynomial which is not unique but is defined up to similarity (Definition 1002). Let T be a finitely generated torsion module over $\mathbf{R} = \mathbf{K}[\partial]$ and let $T_1 \subseteq T$. If $T = coker \bullet P$ and $T_1 = coker \bullet P_1$, then there exists a polynomial P_2 such that

$$P(\partial) = P_1(\partial)P_2(\partial). \tag{6.80}$$

or

$$P(\partial) = P_2(\partial)P_1(\partial) \tag{6.81}$$

and $T/T_1 = coker \bullet P_2$.

Thus, roughly speaking, well-chosen relations of definition of the submodules lead to relations of skew polynomial division.

The following notations are used:

- $a(T) = \{a_1, ..., a_n\}$ is a full set of Smith zeros of the torsion module T as introduced by Definition 1002.
- $A(T) = \dot{\cup}_i \{\Delta_{\mathbf{K}}(a_1), ..., \Delta_{\mathbf{K}}(a_n)\}$ where $\Delta_{\mathbf{K}}(a_i)$ is the conjugacy class of a_i and $\dot{\cup}$ denotes the disjoint union.

Definition 1029. *Let $T_1 \subseteq T$ be finitely generated torsion modules over $\boldsymbol{R} = \boldsymbol{K}[\partial]$ defined by $P(\partial)y = 0$, $P_1(\partial)y_1 = 0$.*

(i) *The polynomials $P(\partial)$, $P_1(\partial)$ are said* compliant *if there exists P_2 to satisfy (6.80) or (6.81).*
(ii) *$a(T_1)$ and $a(T/T_1)$ are said* compliant *if $a(T_1)\dot{\cup}a(T/T_1)$ is a full set of Smith zeros of T.*

The following result is used in the sequel and is a direct consequence of Proposition 528:

Proposition 1030. *Let $T_1 \subseteq T$ be two finitely generated torsion modules over $\boldsymbol{R} = \boldsymbol{K}[\partial]$. Then $A(T) = A(T_1)\dot{\cup}A(T/T_1)$.*

Indeed, if compliant polynomials $P(\partial)$, $P_1(\partial)$ are chosen to define T and T_1 it follows that compliant full sets of Smith zeros can be obtained for T and T/T_1.

Corollary 1031. *If $T = T_1 \oplus T_2$, then $A(T) = A(T_1)\dot{\cup}A(T_2)$.*

Proof. $T_2 \cong T/T_1$ which leads to the conditions of Proposition 1030. ∎

It follows from (6.63) and the form (2.9) of Theorem 147(ii) that $M/[u]_{\mathbf{R}} \cong \mathcal{T}(M) \oplus \Phi/[u]_{\mathbf{R}}$, *i.e.*,

$$M_{sp} \cong M_{idz} \oplus M_{cp}. \tag{6.82}$$

Therefore, by Corollary 1031, one can chose $a(M_{cp})$ and $a(M_{idz})$ such that

$$a(M_{sp}) = a(M_{idz})\dot{\cup}a(M_{cp}) \tag{6.83}$$

and, generally,

$$A(M_{sp}) = A(M_{idz})\dot{\cup}A(M_{cp}). \tag{6.84}$$

Example 1032. *Consider the LTV system given by*

$$(\partial - t)^2 x = (\partial - t)(\partial + 1)u. \tag{6.85}$$

The system pole is thus $\Delta_{\boldsymbol{K}}(t)$ with multiplicity 2, i.e., *the module of the poles of the system is $M/[u]_{\boldsymbol{R}} \cong \boldsymbol{R}/\boldsymbol{R}(\partial - t)^2$. The module M_{idz} of input-decoupling zeros is defined by $(\partial - t)v = 0$,* i.e., *$\mathcal{T}(M) \cong \boldsymbol{R}/\boldsymbol{R}(\partial - t)$. The controllable dynamics is $(\partial - t)\overline{x} = (\partial + 1)\overline{u}$ and from this one directly obtains*

$$(\partial - t)\overline{\overline{x}} = 0 \tag{6.86}$$

as an equation of definition of the module of controllable poles M_{cp}. The controllable pole is thus $\Delta_{\mathbf{K}}(t)$ and (6.83) is thus satisfied. Notice also that, for example, the controlable poles are also given by $\Delta_{\mathbf{K}}(t+t^{-1})$ as $t \sim t+t^{-1}$ (if the change of variable $\widetilde{x} = t\overline{\overline{x}}$ is done, i.e., M_{cp} is defined by $(\partial - (t + t^{-1}))\widetilde{x} = 0$ instead of (11.22)). However, the corresponding $a(M_{cp}) = t + t^{-1}$ is not compliant with the set the definition of M_{sp} by $a(M_{sp})$ obtained from (6.85).

Obviously $M/[y,u]_{\mathbf{R}} \subseteq M/[u]_{\mathbf{R}}$ and, using Theorem 52(3), one has $M/[y,u]_{\mathbf{R}} \cong (M/[u]_{\mathbf{R}})/([y,u]_{\mathbf{R}}/[u]_{\mathbf{R}})$ from which follows using Proposition 1030 that $a(M_{op})$ and $a(M_{odz})$ can be chosen such that

$$a(M_{sp}) = a(M_{odz})\dot{\cup}a(M_{op}) \tag{6.87}$$

and, generally,

$$A(M_{sp}) = A(M_{odz})\dot{\cup}A(M_{op}). \tag{6.88}$$

This can be checked for the system of Example 1021 for which $a(M_{sp}) = \frac{t}{2}$, $a(M_{odz}) = \frac{t}{2}$ and $a(M_{op}) = \{\emptyset\}$.

From (6.63) and the form (2.9) of Theorem 147(ii) it follows that $[y,u]_{\mathbf{R}}/[u]_{\mathbf{R}} \cong (\mathcal{T}(M) \cap [y,u]_{\mathbf{R}}) \oplus (\Phi \cap [y,u]_{\mathbf{R}})/[u]_{\mathbf{R}}$, *i.e.*, $M_{op} \cong (\mathcal{T}(M) \cap [y,u]_{\mathbf{R}}) \oplus M_{tp}$ from which follows by Corollary 1031 that $a(M_{op})$ and $a(M_{tp})$ can be chosen such that

$$a(M_{op}) = a(\mathcal{T}(M) \cap [y,u]_{\mathbf{R}})\dot{\cup}a(M_{tp}) \tag{6.89}$$

thus,

$$A(M_{tp}) \subset A(M_{op}). \tag{6.90}$$

where $a(\mathcal{T}(M) \cap [y,u]_{\mathbf{R}})$ is a full set of Smith zeros of $\mathcal{T}(M) \cap [y,u]_{\mathbf{R}}$ compliant with $a(M_{tp})$. Moreover, as $\mathcal{T}(M) \cap [y,u]_{\mathbf{R}} \subseteq \mathcal{T}(M)$, it follows from Proposition 1030 that there exist $a(M_{idz})$ and $a(M_{iodz})$ such that

$$a(M_{idz}) = a(\mathcal{T}(M) \cap [y,u]_{\mathbf{R}})\dot{\cup}a(M_{iodz}) \tag{6.91}$$

thus,

$$A(M_{iodz}) \subset A(M_{idz}). \tag{6.92}$$

If, moreover, $a(\mathcal{T}(M) \cap [y,u]_{\mathbf{R}})$ is chosen simultaneously compliant with $a(M_{idz})$ and $a(M_{op})$, i.e., if $a(\mathcal{T}(M) \cap [y,u]_{\mathbf{R}})$ is the same in (6.117) and (6.89), then, from the latter relations one can conclude that

$$a(M_{op}) \cup a(M_{iodz}) = a(M_{tp})\dot{\cup}a(M_{idz}) \tag{6.93}$$

or, moreover

$$A(M_{tp})\dot{\cup}A(M_{idz}) = A(M_{op})\dot{\cup}A(M_{iodz}). \tag{6.94}$$

The hidden modes describe the unredundant union of the noncontrollable and nonobservable dynamics.

Definition 1033. *The* module of hidden modes *of the LTV control system* (M, u, y) *is* $M_{hm} = M/(\Phi \cap [y, u]_{\mathbf{R}})$.

From (6.63) and the form (2.9) of Theorem 147(ii) it follows that

$$M/(\Phi \cap [y,u]_{\mathbf{R}}) \cong \mathcal{T}(M) \oplus \Phi/(\Phi \cap [y,u]_{\mathbf{R}}) \tag{6.95}$$

and thus one can chose $a(M_{hm})$ and $a(M_{idz})$ such that

$$a(M_{hm}) = a(M_{idz}) \dot{\cup} a(\Phi/(\Phi \cap [y,u]_{\mathbf{R}})). \tag{6.96}$$

Obviously, it follows that

$$A(M_{idz}) \subset A(M_{hm}). \tag{6.97}$$

Now,

$$M/[y,u]_{\mathbf{R}} = (\mathcal{T}(M) \oplus \Phi)/((\mathcal{T}(M) \cap [y,u]_{\mathbf{R}}) \oplus (\Phi \cap [y,u]_{\mathbf{R}})$$

and, using again the form (2.9) of Theorem 147(ii),

$$M/[y,u]_{\mathbf{R}} \cong \mathcal{T}(M)/(\mathcal{T}(M \cap [y,u]_{\mathbf{R}}) \oplus \Phi/(\Phi \cap [y,u]_{\mathbf{R}})$$

or, equivalently,

$$M_{odz} \cong M_{iodz} \oplus \Phi/(\Phi \cap [y,u]_{\mathbf{R}}). \tag{6.98}$$

If one chooses $a(M_{iodz})$ compliant with $a(\Phi/\Phi \cap [y,u]_{\mathbf{R}})$, (6.98) leads to

$$a(M_{odz}) = a(M_{iodz}) \dot{\cup} a(\Phi/(\Phi \cap [y,u]_{\mathbf{R}})). \tag{6.99}$$

If, moreover, $a(\Phi/\Phi \cap [y,u]_{\mathbf{R}})$ chosen for (6.98) is also compliant with $a(M_{idz})$ in (6.96), (6.99) and (6.96) lead to

$$a(M_{hm}) = a(M_{idz}) \dot{\cup} a(M_{odz}) \backslash a(M_{iodz}) \tag{6.100}$$

and

$$A(M_{hm}) = A(M_{idz}) \dot{\cup} A(M_{odz}) \setminus A(M_{iodz}). \tag{6.101}$$

From (6.87), (6.93) and (6.87) follows that $a(M_{sp}) = a(M_{tp}) \dot{\cup} a(M_{hm})$ and

$$A(M_{sp}) = A(M_{tp}) \dot{\cup} A(M_{hm}). \tag{6.102}$$

The hidden modes for the system of Example 1025 are $\Delta_{\mathbb{C}(t)}(t)$, there are no transmission poles and the system poles are $\Delta_{\mathbb{C}(t)}(t)$. Note that $\Delta_{\mathbb{C}(t)}(t)$ give also the input-decoupling zeros as well as the output-decoupling zeros.

6.6.5.2 Zeros and Hidden Modes

The same kind of relations can be established among zeros and hidden modes using their modules of definition.

From the form (2.9) of Theorem 147(ii) one has

$$\begin{aligned} M/[y]_{\mathbf{R}} &\cong (\mathcal{T}(M) \oplus \Phi)/[y]_{\mathbf{R}} \cong \\ &\cong \mathcal{T}(M)/(\mathcal{T}(M) \cap [y]_{\mathbf{R}}) \oplus \Phi/(\Phi \cap [y]_{\mathbf{R}}) \end{aligned} \tag{6.103}$$

from which follows

$$\mathcal{T}(M/[y]_{\mathbf{R}}) \cong \mathcal{T}(M)/(\mathcal{T}(M) \cap [y]_{\mathbf{R}}) \oplus \mathcal{T}(\Phi/\Phi \cap [y]_{\mathbf{R}}). \tag{6.104}$$

As $(\Phi \cap [y,u]_{\mathbf{R}})/(\Phi \cap [y]_{\mathbf{R}}) \subseteq \Phi/(\Phi \cap [y]_{\mathbf{R}})$, from (6.104) results $\mathcal{T}(M)/(\mathcal{T}(M) \cap [y]_{\mathbf{R}}) \oplus (\Phi \cap [y,u]_{\mathbf{R}})/(\Phi \cap [y]_{\mathbf{R}}) \subseteq \mathcal{T}(M/[y]_{\mathbf{R}})$, *i.e.*,

$$M_{tz} \oplus M_{iodz} \subseteq M_{iz}. \tag{6.105}$$

From Proposition 1030 follows that $a(M_{tz})$ and $a(M_{iodz})$ can be chosen such that $a(M_{tz}) \dot{\cup} a(M_{iodz}) \in a(M_{iz})$

and

$$A(M_{tz}) \dot{\cup} A(M_{iodz}) \subset A(M_{iz}). \tag{6.106}$$

If the transfer matrix of the system is right-invertible, then $[y]_{\mathbf{R}}$ is free and $\mathcal{T}(M) \cap [y]_{\mathbf{R}} = 0$. Equation (6.103) can be written in this case

$$M/[y]_{\mathbf{R}} \cong \mathcal{T}(M) \oplus \Phi/(\Phi \cap [y]_{\mathbf{R}}) \tag{6.107}$$

from which follows as before $\mathcal{T}(M) \oplus (\Phi \cap [y,u]_{\mathbf{R}})/(\Phi \cap [y]_{\mathbf{R}}) \subseteq \mathcal{T}(M/[y]_{\mathbf{R}})$, *i.e.*,

$$M_{tz} \oplus M_{idz} \subseteq M_{iz} \tag{6.108}$$

or, furthermore, there exist $a(M_{tz})$ and $a(M_{idz})$ such that $a(M_{tz}) \dot{\cup} a(M_{idz}) \in a(M_{iz})$

and thus

$$A(M_{tz}) \dot{\cup} A(M_{idz}) \subset A(M_{iz}). \tag{6.109}$$

Note that in this case, the set of output-decoupling zeros is not generally included in the set of invariant zeros, as is shown by the following example:

Example 1034

$$\begin{cases} \partial x_1 = u_1 \\ \partial x_2 = x_2 + u_2 \\ y = x_1 \end{cases} \tag{6.110}$$

where the output-decoupling zero is $\{1\}$, but the set of invariant zeros is empty.

Consider now the case where the transfer matrix is left-invertible. Let $\mathcal{G}(\partial)$ be the $p \times m$ transfer matrix ($p > m$) and let $\mathcal{G}(\partial) = D_G^{-1}(\partial) N_G(\partial)$ be a left-coprime factorization of $G(\partial)$. The transfer module M_{td} is defined by

$D_G(\partial)\overline{y} = N_G(\partial)\overline{u}$. Obviously, $m = rank(\mathcal{G}) \leq rank(N_G)$, thus $N_G(\partial)$ is left-invertible. As $N_G(\partial)\overline{\overline{u}} = 0$ defines the module $M_{td}/[y]_{\mathbf{R}} \cong (\Phi \cap [y,u]_{\mathbf{R}})/[y]_{\mathbf{R}}$ it follows that $(\Phi \cap [y,u]_{\mathbf{R}})/[y]_{\mathbf{R}}$ is torsion. Now, using the form (2.9) of Theorem 147(ii),

$$M/[y]_{\mathbf{R}} \cong (\mathcal{T}(M) \cap [y,u]_{\mathbf{R}}) \oplus (\Phi \cap [y,u]_{\mathbf{R}}))/[y]_{\mathbf{R}}$$

so $M/[y]_{\mathbf{R}}$ is also torsion. Furthermore, its submodule $[y,u]_{\mathbf{R}}/[y]_{\mathbf{R}}$ is also torsion. Thus, from Theorem 52 (3) one obtains

$$(M/[y]_{\mathbf{R}})/([y,u]_{\mathbf{R}}/[y]_{\mathbf{R}}) \cong M_{odz},$$

i.e.,

$$a(M_{iz}) = a(M_{odz})\dot{\cup}a([y,u]_{\mathbf{R}}/[u]_{\mathbf{R}}) \tag{6.111}$$

where $a([y,u]_{\mathbf{R}}/[u]_{\mathbf{R}})$ is compliant with the chosen $a(M_{iz})$.

Next, from (6.63) and the form (2.9) of Theorem 147(ii) one obtains

$$\begin{aligned}&[y,u]_{\mathbf{R}}/[y]_{\mathbf{R}} \cong \\ &\cong (\mathcal{T}(M) \cap [y,u]_{\mathbf{R}})/(\mathcal{T}(M) \cap [y]_{\mathbf{R}}) \oplus (\Phi \cap [y,u]_{\mathbf{R}})/(\Phi \cap [y]_{\mathbf{R}}).\end{aligned}$$

or, moreover,

$$[y,u]_{\mathbf{R}}/[y]_{\mathbf{R}} \cong (\mathcal{T}(M) \cap [y,u]_{\mathbf{R}})/(\mathcal{T}(M) \cap [y]_{\mathbf{R}}) \oplus M_{tz} \tag{6.112}$$

from which

$$a([y,u]_{\mathbf{R}}/[u]_{\mathbf{R}}) = a((\mathcal{T}(M) \cap [y,u]_{\mathbf{R}})/(\mathcal{T}(M) \cap [y]_{\mathbf{R}}))\dot{\cup}a(M_{tz}) \tag{6.113}$$

where $a([y,u]_{\mathbf{R}}/[u]_{\mathbf{R}})$ is compliant with $a(M_{tz})$ and $a(M_{odz})$ chosen in (6.111). As a consequence, from (6.111) and (6.113) one can conclude that $a(M_{tz})\dot{\cup}a(M_{odz}) \in a(M_{iz})$ and

$$A(M_{tz})\dot{\cup}A(M_{odz}) \subseteq A(M_{iz}). \tag{6.114}$$

Note that in this case, the sets of input-decoupling zeros are generally not included in the sets of invariant zeros, as shown by the following example which is the "dual" of Example 1034:

Example 1035

$$\begin{cases} \partial x_1 = u \\ \partial x_2 = x_2 \\ y_1 = x_1 \\ y_2 = x_2 \end{cases} \tag{6.115}$$

In this case, $\{1\}$ is the input-decoupling zero, but the set of invariant zeros is empty.

Finally, consider the case where the transfer matrix is invertible. Then, both $[y]_{\mathbf{R}}$ is free and $M/[y]_{\mathbf{R}}$ is torsion. In these conditions, (6.112) can be written as $[y,u]_{\mathbf{R}}/[y]_{\mathbf{R}} \cong (\mathcal{T}(M) \cap [y,u]_{\mathbf{R}}) \oplus M_{tz}$ from which follows that $a(M_{tz})$ can be chosen such that

$$a([y,u]_{\mathbf{R}}/[y]_{\mathbf{R}}) = a(\mathcal{T}(M) \cap [y,u]_{\mathbf{R}}) \dot{\cup} a(M_{tz}) \tag{6.116}$$

where $a(\mathcal{T}(M \cap [y,u]_{\mathbf{R}})$ is compliant with $a([y,u]_{\mathbf{R}}/[y]_{\mathbf{R}})$.

From Definition 1023 and Proposition 1030 one can chose $a(M_{iodz})$ compliant with $a(\mathcal{T}(M \cap [y,u]_{\mathbf{R}})$ in (6.116) such that

$$a(M_{idz}) = a(M_{iodz}) \dot{\cup} a(\mathcal{T}(M \cap [y,u]_{\mathbf{R}})). \tag{6.117}$$

Moreover, from Theorem 52 (3)

$$M/[y,u]_{\mathbf{R}} \cong (M/[y]_{\mathbf{R}})/([y,u]_{\mathbf{R}}/[y]_{\mathbf{R}})$$

and, following Proposition 1030 one can chose $a(M_{odz})$ compliant with $a(\mathcal{T}(M \cap [y,u]_{\mathbf{R}})$ in (6.116) such that

$$a(M_{iz}) = a(M_{odz}) \dot{\cup} a(\mathcal{T}(M \cap [y,u]_{\mathbf{R}}). \tag{6.118}$$

From (6.118), (6.116), (6.117) and (6.100) one can conclude that

$$a(M_{tz}) \dot{\cup} a(M_{hm}) = a(M_{iz}) \tag{6.119}$$

and

$$A(M_{tz}) \dot{\cup} A(M_{hm}) = A(M_{iz}). \tag{6.120}$$

The main relations among the poles and zeros of an LTV control system can thus be summarized in the following

Theorem 1036.
(i)

$$\begin{gathered}\alpha(M_{sp}) = \alpha(M_{tp}) \dot{\cup} \alpha(M_{hm}) \\ \alpha(M_{tz}) \dot{\cup} \alpha(M_{iodz}) \subset \alpha(M_{iz})\end{gathered}$$

(ii) if the transfer matrix is left-invertible,

$$\alpha(M_{tz}) \dot{\cup} \alpha(M_{odz}) \subset \alpha(M_{iz})$$

(iii) if the transfer matrix is right-invertible,

$$\alpha(M_{tz}) \dot{\cup} \alpha(M_{idz}) \subset \alpha(M_{iz})$$

(iv) if the transfer matrix is invertible,

$$\alpha(M_{tz}) \dot{\cup} \alpha(M_{hm}) = \alpha(M_{iz})$$

where α stands for a or A.

6.6.5.3 The Case of Linear-Time Invariant Systems

The relations established in Section 6.6.5 are simpler in the case of LTI systems since the sets of poles and zeros are uniques: $A(M) = a(M) = \{a_1, ..., a_n\}$ where M is the module of any pole or zero defined in Section 6.6 and $a_i \in \overline{\mathbf{K}}$, $i = 1, ..., n$. The relations among various poles and zeros are thus the ones established in Section 6.6.5 for the A's and for which no compliance precautions are needed.

For Example 1000 for instance, the set of input-decoupling zeros is $\{1\}$ and relation (6.84) which must be red in the LTI case is

$$\{system\ poles\} = \{input - decoupling\ zeros\}\dot{\cup}\{controllable\ poles\} \quad (6.121)$$

and leads to $\{1, 1\} = \{1\}\dot{\cup}\{1\}$ for this example.

6.7 The Case of Discrete-Time Systems

6.7.1 Factors and Roots in the General Case

For the discrete-time case, ∂ stands for $q-1$ where q denotes the shift forward operator: $qy(t) = y(t+1)$. Similar definitions of poles and conditions for asymptotic stability can be given for discrete-time systems. However, we are faced with some supplementary difficulties in this case. First, there is no general method to provide a factorization (6.10) for $P(\partial)$, i.e., in so-called *hyperexponential* factors. The existing results (e.g., [54], [274] and related references) make it only possible to check if, over a given field $\breve{\mathbf{K}}$, $P(\partial)$ has a hyperexponential factor, but they do not provide the way to construct the field extension over which the polynomial factorizes. Next, condition

$$\frac{\dot{c}(t)}{c(t)} \to 0\ as\ t \to +\infty \quad (6.122)$$

satisfied by the elements of an Ore field transposes into a more restrictive form in the discrete-time case. Indeed, following (4.13), an element $\overline{a}$ of the conjugacy class of a root $a \in \breve{\mathbf{K}}$ of $P(\partial)$ is of the form $\overline{a} = \frac{c(t+1)}{c(t)}[a(t)+1]-1$ and thus $lim_{t\to+\infty}(\overline{a} - a(t)) = 0$ if, and only if $lim_{t\to+\infty}\frac{c(t+1)}{c(t)} = 1$. The well-suited fields in the discrete-time case are thus *asymptotically* constant fields.

Provided that a full set of factors or/and a fundamental set of roots of $P(\partial)$ can be provided over an asymptotically constant field $\breve{\mathbf{K}}$, the approach presented for the continuous-time case can be extended to the discrete-time one. A fundamental set of roots γ_i of $P(\partial)$ leads in the discrete-time case to the following *elementary equations*:

$$(\partial - \gamma_i(t))y_i(t) = 0 \Leftrightarrow y_i(t+1) - y_i(t) = \gamma_i(t)y_i(t). \quad (6.123)$$

Obviously,

$$y_i(t) = \prod_{k=m}^{t} (1 + \gamma_i(k)), \quad i = 1, ..., n \tag{6.124}$$

where $n = deg(P(\partial))$ and m is an integer which depends on the initial conditions of (6.2), form a fundamental set of solutions of (6.2). The γ_i's form a fundamental set of poles of the system according to Definition 997. Using the expression (6.124) for the solutions of (6.2), the condition (6.56) for exponential stability can be extended to the discrete-time case:

Theorem 1037. *The discrete-time autonomous system defined by the* $\mathbf{R}$-*torsion module* T *is* exponentially stable *if any of its poles* $\gamma_i(t)$ *satisfy the condition*

$$\limsup_{t\to+\infty} |1 + \gamma_i(t)| < 1. \tag{6.125}$$

If $lim_{t\to+\infty}|1 + \gamma_i(t)|$ *exists, then condition above is also necessary for exponential stability.*

Notice that polynomial P can be written and factorized using operator q instead of ∂ as in the example below:

Example 1038. $P(\partial) = \partial^2 - (2t - 1)\partial + t^2 - 2t$. *With* $\partial = q - 1$ *it follows that* $P(q) = q^2 - (2t + 1)q + t^2$. *One can easily check that*

$$P(q) = (q - t)(q - t). \tag{6.126}$$

If a fundamental set of roots is found working with q instead of ∂, (6.124) is replaced by

$$y_i(t) = \prod_{k=m}^{t} \gamma_i(k), \quad i = 1, ..., n \tag{6.127}$$

and condition (6.125) is replaced by

$$\limsup_{t\to+\infty} |\gamma_i(t)| < 1 \tag{6.128}$$

which, in the case of LTI systems, particularizes to the well-known stability condition: $|\gamma_i| < 1$.

For Example 1038, a fundamental set of solutions is $y_1(t) = (t - 1)!y_1(1)$, $y_2(t) = y_1(t)(\sum_{k=0}^{t-1} \frac{1}{k} + y_2(1))$ to which correspond the fundamental set of poles $\gamma_1(t) = t$, $\gamma_2(t) = t + \frac{1}{\sum_{k=0}^{t-1} \frac{1}{k} + y_2(1)}$. The system is exponentially unstable since $lim_{t\to+\infty}\gamma_1(t) = +\infty$ (as well as $lim_{t\to+\infty}\gamma_2(t) = +\infty$).

6.7.2 *Poles and Zeros of Periodic Discrete-Time Systems*

6.7.2.1 N-Piled Form

Let $\mathbf{K}$ be an algebraically closed field and let $\boldsymbol{\alpha}$ be an automorphism of $\mathbf{K}$ such that $\boldsymbol{\alpha}^N = 1$ for some integer $N > 1$. The noncommutative rings $\mathbf{A} = \mathbf{K}[X;\boldsymbol{\alpha}]$ and $\mathbf{T} = \mathbf{K}\left[X, X^{-1};\boldsymbol{\alpha}\right]$ are principal ideal domains (and $\mathbf{K}[X;\boldsymbol{\alpha}]$ is an Euclidean domain) by Theorem 467 and Lemma 463, but none of them is simple by Corollaries 363 & 368. Therefore, the case of "periodic discrete-time systems" is quite special, and the methods of the preceding sections cannot be used. Nevertheless, this case if almost as simple as that of systems with constant coefficients, due to the results obtained in (2.8.4.**2**).

Indeed, by Proposition 375 we know that $\mathbf{T}$ is is a crossed product $\mathbf{S} * G$ where $\mathbf{S} = \mathbf{K}\left[Y, Y^{-1}\right]$, the skew Laurent polynomial ring in the central indeterminate $Y = X^N$, and the group G is cyclic of order N.

Let M be a system over $\mathbf{A}$. Very little information is lost when replacing M by $\breve{M} = S^{-1}M$ where $S = \{X^n : n \geq 1\}$; $\breve{M}$ is a system (i.e., a f.g. module) over $\mathbf{T}$. Consider the functor $\mathfrak{C}$ in Theorem 373. The $\mathbf{T}$-module $\breve{M}$ and the $\mathbf{S}$-module $\tilde{M} = \mathfrak{C}\left(\breve{M}\right)$ contain the same information. Since the indeterminate Y of the Laurent polynomial ring $\mathbf{S}$ is central, the structure of $\tilde{M}$ can be studied using the same techniques as for a f.g. module over a commutative Laurent polynomial ring. In what follows, when this will be useful, we'll set $X = \mathbf{q}$ and $Y = \Delta$, considering these indeterminates as operators.

Now, let (M, u, y) be a control system over $\mathbf{A}$. Again, it is preferable to replace (M, u, y) by $\left(\breve{M}, \breve{u}, \breve{y}\right)$ where $\breve{u} = 1 \otimes u$ and $\breve{y} = 1 \otimes y$, extending the ring $\mathbf{A}$ to $\mathbf{T}$. Let $\tilde{u} = e_N(\mathbf{q}) \otimes \breve{u}$ and $\tilde{y} = e_N(\mathbf{q}) \otimes \breve{y}$.

Theorem and Definition 1039. *(1)* $\left(\breve{M}, \breve{u}, \breve{y}\right)$ *is a control system over* $\mathbf{T}$ *and* $\left(\tilde{M}, \tilde{u}, \tilde{y}\right)$ *is a control system over* $\mathbf{S}$.

(2) The control system $\left(\tilde{M}, \tilde{u}, \tilde{y}\right)$ *is called the* N*-piled form of* $\left(\breve{M}, \breve{u}, \breve{y}\right)$ *(and, abusing the language, of* (M, u, y)*).*

Proof. (1) Let $\mathbf{F} = \mathbf{K}(X;\boldsymbol{\alpha})$; $\mathbf{F}$ is the field of fractions of both $\mathbf{A}$ and $\mathbf{T}$. Since $M/[u]_{\mathbf{A}}$ is torsion, $\mathbf{F} \bigotimes_{\mathbf{A}} (M/[u]_{\mathbf{A}}) = 0$; therefore, $\mathbf{F} \bigotimes_{\mathbf{T}} \left(\breve{M}/[\breve{u}]_{\mathbf{T}}\right) = 0$ and $\breve{M}/[\breve{u}]_{\mathbf{T}}$ is torsion. In addition, since the m entries of u form a basis of $[u]_{\mathbf{A}}$, the m entries of $\breve{u}$ form a basis of $[\breve{u}]_{\mathbf{T}}$ by Theorem and Definition 339. This proves that $\left(\breve{M}, \breve{u}, \breve{y}\right)$ is a control system over $\mathbf{T}$.

By Corollary 374 and by (1), the Nm entries of the column $\tilde{u}$ form a basis of $[\tilde{u}]_{\mathbf{S}}$. Consider the following diagram with exact rows:

$$\begin{array}{ccccccccc}
0 & \longrightarrow & [u]_{\mathbf{A}} & \longrightarrow & M & \longrightarrow & M/[u]_{\mathbf{A}} & \longrightarrow & 0 \\
 & & \downarrow & & \downarrow & & \downarrow & & \\
0 & \longrightarrow & [\breve{u}]_{\mathbf{T}} & \longrightarrow & \breve{M} & \longrightarrow & \breve{M}/[\breve{u}]_{\mathbf{T}} & \longrightarrow & 0 \\
 & & \downarrow & & \downarrow & & \downarrow & & \\
0 & \longrightarrow & [\tilde{u}]_{\mathbf{S}} & \longrightarrow & \tilde{M} & \longrightarrow & \tilde{M}/[\tilde{u}]_{\mathbf{S}} & \longrightarrow & 0
\end{array}$$

Both functors $S^{-1}-$ and $\mathfrak{C}$ are exact by Theorem 609 and Corollary 374, therefore one passes from the first row to the second one using the functor $S^{-1}-$, and from the second row to the third one using the functor $\mathfrak{C}$. Therefore, by Theorem 379, $\tilde{M}/[\tilde{u}]_{\mathbf{S}} = \mathfrak{C}\left(\breve{M}/[\breve{u}]_{\mathbf{T}}\right)$ is torsion. This proves that $\left(\tilde{M}, \tilde{u}, \tilde{y}\right)$ is a control system over $\mathbf{S}$. ■

6.7.2.2 Controllability and Observability

Proposition 1040. *(1) The following conditions are equivalent:*
(i) M is 0-controllable (Exercise 958(3));
(ii) $\breve{M}$ is controllable;
(iii) $\tilde{M}$ is controllable.
(2) Likewise, the following conditions are equivalent:
(i') (M, u, y) is 0-observable [43];
(ii') $\left(\breve{M}, \breve{u}, \breve{y}\right)$ is observable;
(iii') $\left(\tilde{M}, \tilde{u}, \tilde{y}\right)$ is observable.

Proof. (1) By definition, (i)⇔(ii).

(ii)⇔(iii): $\breve{M}$ is controllable if, and only if the $\mathbf{T}$-module $\breve{M}$ is free (**5.6.1.1**). This holds if, and only if the $\mathbf{S}$-module $\tilde{M}$ is free by Corollary 374(iv).

(2) By definition, (i')⇔(ii').

(ii')⇔(iii'): $\left(\breve{M}, \breve{u}, \breve{y}\right)$ is observable if, and only if $\breve{M} = [\breve{u}, \breve{y}]_{\mathbf{T}}$, and this happens if, and only if the following sequence is exact:

$$0 \longrightarrow [\breve{u}, \breve{y}]_{\mathbf{T}} \longrightarrow \breve{M} \longrightarrow 0.$$

By Corollary 374, this happens if, and only if the following sequence is exact:

$$0 \longrightarrow [\tilde{u}, \tilde{y}]_{\mathbf{S}} \longrightarrow \tilde{M} \longrightarrow 0,$$

i.e., $\left(\tilde{M}, \tilde{u}, \tilde{y}\right)$ is observable. ■

6.7.2.3 Poles and Zeros

Recall that the module of i.d.z. of $\breve{M}$ is $\mathcal{T}\left(\breve{M}\right)$; that of $\tilde{M}$ is $\mathcal{T}\left(\tilde{M}\right)$; we have $\mathcal{T}\left(\tilde{M}\right) = \mathfrak{C}\left(\mathcal{T}\left(\breve{M}\right)\right)$ by Corollary 374 and Theorem 379.

Similarly, the module of system poles of $\left(\breve{M}, \breve{u}, \breve{y}\right)$ is $\breve{M}/\left[\breve{u}\right]_{\mathbf{T}}$, that of $\left(\tilde{M}, \tilde{u}, \tilde{y}\right)$ is $\tilde{M}/\left[\tilde{u}\right]_{\mathbf{S}}$, and as shown in the proof of Theorem and Definition 1039, $\tilde{M}/\left[\tilde{u}\right]_{\mathbf{S}} = \mathfrak{C}\left(\breve{M}/\left[\breve{u}\right]_{\mathbf{T}}\right)$.

The module of invariant zeros of $\left(\breve{M}, \breve{u}, \breve{y}\right)$ is $\mathcal{T}\left(\breve{M}/\left[\breve{y}\right]_{\mathbf{T}}\right)$, that of $\left(\tilde{M}, \tilde{u}, \tilde{y}\right)$ is $\mathcal{T}\left(\tilde{M}/\left[\tilde{y}\right]_{\mathbf{S}}\right)$, and the reader may easily check that $\mathcal{T}\left(\tilde{M}/\left[\tilde{y}\right]_{\mathbf{S}}\right) = \mathfrak{C}\left(\breve{M}/\left[\breve{y}\right]_{\mathbf{T}}\right)$.

This rationale can be generalized and one obtains the following result:

Proposition 1041. *Let $\breve{M}_p$ be a module of poles or zeros of $\left(\breve{M}, \breve{u}, \breve{y}\right)$ (e.g., i.d.z., o.d.z., i.o.d.z., system poles, transmission poles, or transmission zeros). The module of poles or zeros of $\left(\tilde{M}, \tilde{u}, \tilde{y}\right)$ of the same kind is $\tilde{M}_p = \mathfrak{C}\left(\breve{M}_p\right)$.*

The poles and zeros of $\left(\tilde{M}, \tilde{u}, \tilde{y}\right)$ are calculated in the same way, and have the same properties, as those of an LTI control system with coefficients in $\mathbb{C}$. In particular, the control system $\left(\tilde{M}, \tilde{u}, \tilde{y}\right)$ is exponentially stable if, and only if all its poles lie in the interior of the unit disc, and it is exponentially unstable if, and only if at least one of its poles lies in the exterior of the unit disc. By Corollary 378 we have the following.

Corollary 1042. *A element $a \in \mathbf{K}$ is a system pole of $\left(\breve{M}, \breve{u}, \breve{y}\right)$ if, and only if $\prod_{0 \leq i \leq N-1} a^{\boldsymbol{\alpha}^i}$ is a system pole of $\left(\tilde{M}, \tilde{u}, \tilde{y}\right)$.*

The same holds for all kinds of poles and zeros.

6.8 The Case of State-Space Continuous-Time Periodic Systems

Consider an LTV state-space autonomous system

$$\dot{x} = Ax \tag{6.129}$$

where the entries of matrix A are in the ring $\mathbf{K}_{t_0}^c$ defined as follows: let $C_{t_0} = C((t_0, +\infty))$ be the ring of continuous functions on $(t_0, +\infty)$. Consider a family $(\varphi_i)_{i \in I}$ of ω-periodic functions, $\omega > 0$, defined on the real line. Then, $\mathbf{K}_{t_0}^c = C_{t_0} \cap \mathbf{K}^\omega$, where $\mathbf{K}^\omega$ is the field $\mathbb{C}((\varphi_i)_{i \in I})$. Consider the rings of differential operators: $\mathbf{R}^\omega = \mathbf{K}^\omega[\partial]$ and $\mathbf{R}_{t_0}^c = \mathbf{K}_{t_0}^c[\partial]$. From the state-space system (6.129) we define the left $\mathbf{R}^\omega$-module

$$M = \mathbf{R}^\omega \otimes_{\mathbf{R}_{t_0}^c} [x]_{\mathbf{R}_{t_0}^c} \tag{6.130}$$

Since $\mathbf{K}^{\omega}$ is a differential field and the derivation is outer, $\mathbf{R}^{\omega}$ is a simple principal ideal domain, thus there exists $P(\partial) \in \mathbf{R}^{\omega}$ such that $M \cong coker \bullet P(\partial)$, *i.e.*, M is represented by the differential equation (6.2). Stability can be characterized by the roots of P as follows: each root γ_i of of P can be written as $\gamma_i = \gamma_i^{\omega} + \overline{\gamma}_i$ where $\overline{\gamma}_i \in \mathbb{C}$ is the mean of γ_i and γ_i^{ω} is ω-periodic with zero mean. Notice that the $\overline{\gamma}$'s are the Floquet exponents of (6.129) as introduced by Definition 1175. Thus, stability can be evaluated by investigating the real part of those coefficients as a direct consequence of Proposition 1187:

Corollary 1043. *The state-space system (6.129) is*

- *exponentially stable if, and only if all the γ_i's satisfy the condition $Re\{\overline{\gamma}_i\} < 0$.*
- *exponentially unstable if, and only if there exists at least one γ_i such that $Re\{\overline{\gamma}_i\} > 0$.*

For example, consider the first order system:

Example 1044

$$(\partial - a - sin(t))x = 0, \ a \in \mathbb{R}. \tag{6.131}$$

A fundamental set of roots is trivially found to be $\gamma(t) = a + sin(t)$. Thus, $\overline{\gamma} = a$ and $\gamma^{\omega} = sin(t)$. The system is exponentially stable if, and only if $a < 0$. Indeed, the solutions of (6.131) are $x(t) = ce^{at}e^{-cos(t)}$, $c \in \mathbb{R}$ and $lim_{t\to+\infty}x(t) = 0$ if, and only if $a < 0$.

The following second order system has been used several times as an example in the analysis of the stability of LTV systems (see, e.g., [335]):

Example 1045

$$\dot{x} = A(t)x, \ A(t) = \begin{bmatrix} -1 + acos^2(t) & 1 - asin(t)cos(t) \\ -1 - asin(t)cos(t) & -1 + asin^2(t) \end{bmatrix} \tag{6.132}$$

with $1 < a < 2$. If the variable x_1 is eliminated from the equations (6.132), one obtains

$$\alpha_2\ddot{x}_2 + \alpha_1\dot{x}_2 + \alpha_0 x_2 = 0 \Leftrightarrow P(\partial)x_2 = 0, \ P(\partial) = \alpha_2\partial^2 + \alpha_1\partial + \alpha_0 \tag{6.133}$$

where

$$\begin{aligned} \alpha_2 &= -1 - \tfrac{a}{2}sin(2t) \\ \alpha_1 &= a - 2 + \tfrac{a^2-2a}{2}sin(2t) + acos(2t) \\ \alpha_0 &= \tfrac{a^2}{2} + a - 2 + \tfrac{a^2}{2}sin(2t) + \tfrac{2a-a^2}{2}cos(2t). \end{aligned} \tag{6.134}$$

Notice that for $0 < a < 2$, α_2 does not vanish on any point of the real axis so the equation (6.133) has unique solution for given initial conditions (see, e.g., [164]). Also, it exists at least one periodic solution with the period equal to the one of the coefficients of the differential operator in (6.133) (see, e.g., [164]). *It follows that, in the case of periodic systems, no field extension is needed and at least one solution is of the form:*

$$x_2(t) = e^{bt}\varphi(t) \tag{6.135}$$

where $b \in \mathbb{C}$ *is a characteristic coefficient and* $\varphi(t)$ *is a function of period* 2π. *Replaced in (6.133), this leads to*

$$\alpha_2 \frac{d^2\varphi}{dt^2} + (\alpha_1 + 2\varphi\alpha_2)\frac{d\varphi}{dt} + (\alpha_2\varphi^2 + \alpha_1\varphi + \alpha_0)\varphi = 0 \tag{6.136}$$

One can assume $\varphi(t) = \alpha sin(t+\beta)$, *with* α *and* β *constants,* $\alpha \neq 0$ *and look for the solution which corresponds to* $x_2(0) = 0$ *and* $\frac{dx_2}{dt} = 1$, i.e., *to* $\beta = 2k\pi$. *For these values of* β *(6.136) can be written*

$$\alpha_2 b^2 + [\alpha_1 sin(t) + 2\alpha_2 cos(t)]b + \alpha_1 cos(t) + (\alpha_0 - \alpha_2)sin(t) = 0 \tag{6.137}$$

which for $t = 0$ *leads to* $b = a-1$ *which is one of the characteristic coefficients of (6.133) [164]. One solution to (6.133) is thus*

$$x_2(t) = e^{(a-1)t}sin(t) \tag{6.138}$$

from which one can chose the first member of a fundamental set of roots of (6.133):

$$\gamma_1 = \frac{dx_2}{dt} = a - 1 + ctg(t),\ \overline{\gamma}_1 = a-1, \gamma_1^{\omega} = ctg(t). \tag{6.139}$$

One can now factorize $P(\partial)$ *as* $P(\partial) = \widetilde{a}_2(\partial - a_2)(\partial - a_1)$ *with* $a_1 = \gamma_1$, $\widetilde{a}_2 = \alpha_2$ *and* $a_2 = -\gamma_1 - \frac{\alpha_1}{\alpha_2}$. *The second member* γ_2 *of a fundamental set of roots of (6.133) can be found as a solution to the equation* ${}^{\gamma_2-\gamma_1}\gamma_2 = a_2$ *which leads to the Riccati equation*

$$\gamma_2 + \frac{d(\gamma_2-\gamma_1)}{(\gamma_2-\gamma_1)dt} = -\gamma_1 - \frac{\alpha_1}{\alpha_2}. \tag{6.140}$$

It can be checked (rather by using a tool of formal computation) that (6.140) has a periodic solution with zero mean $\gamma_2(t) = a - 1 + \varphi(t)$ *where* φ *is periodic with zero mean. So,* $\overline{\gamma}_2 = \overline{\gamma}_1 = a-1$ *and the system is thus exponentially stable if, and only if* $a < 1$ *and exponentially unstable if, and only if* $a > 1$.

6.9 Exercises

Exercise 1046. *Let* K *be a field of characteristic zero. Show that every polynomial* $P \in \boldsymbol{K}[X]$ *has a splitting field. (Hint: the splitting field is constructed by extending* $\boldsymbol{K}$ *by the zeros of nonlinear factors of* P *over suitable field extensions.)*

Exercise 1047. *Compute the minimal polynomial of* $\{1, t\}$.

Exercise 1048. *Let*

$$P(\partial)y = 0 \text{ with } P(\partial) = t^2\partial^2 + (t - 2t^2)\partial - t + t^2 \tag{6.141}$$

Compute the system poles for the system defined by this differential equation and evaluate the stability. From the poles obtained, deduce the solutions of the differential equation. (Hint: the change of variable $\xi = t\partial$ *leads to* $P(\xi) = (\xi - t)^2$ *which, pulled-back to* ∂*, leads to* $P(\partial) = t^2(\partial - t^{-1} - 1)(\partial - 1)$*.)*

Exercise 1049. *Consider the 3rd order Euler equation:* $S(\partial)y = 0$, $S(\partial) = t^3\partial^3 - t^2\partial^2 - 2t\partial - 4$. *Compute a fundamental set of poles of the system and, next, a set of independent solutions of the differential equation.*

Exercise 1050. *Consider the system* $(\partial - t)(\partial + 1)y = (\partial - t)(\partial + 2t)u$. *Compute the transfer poles and zeros. Are there any hidden modes?*

Exercise 1051. *Consider a system given by the state-space form:*

$$\begin{cases} \partial x = \begin{bmatrix} 0 & 1 \\ 0 & t^{-1} \end{bmatrix} x + \begin{bmatrix} 1 \\ 0 \end{bmatrix} u \\ y = \begin{bmatrix} t & 1 \end{bmatrix} x \end{cases} \tag{6.142}$$

Give the different poles and zeros of this system and check the relations among them.

Exercise 1052. *Same questions as before for the system*

$$\begin{cases} \partial x = \begin{bmatrix} -1 & 0 & 0 \\ 1 & -3 & 0 \\ 1 & -4 & 1 \end{bmatrix} x + \begin{bmatrix} 0 \\ 1 \\ 0 \end{bmatrix} u \\ y = \begin{bmatrix} 1 & 0 & 0 \\ -1 & 1 & 0 \end{bmatrix} x \end{cases} \tag{6.143}$$

Exercise 1053. *Let* M *be a system over* $\mathbf{A} = \mathbf{K}[\mathbf{q}; \boldsymbol{\alpha}]$ *where* $\boldsymbol{\alpha}$ *is an automorphism of* $\mathbf{K}$ *such that* $\boldsymbol{\alpha}^N = 1$ $(N > 1)$. *Let* $R(\mathbf{q}) \in {}^q\mathbf{A}^k$ *be a matrix of definition of* M.

(i) Prove that the corresponding matrix of definition of $\tilde{M}$ *is* $\tilde{R}(\Delta) \in {}^{Nq}\mathbf{S}^{Nk}$ *given by*

$$(e_N(\mathbf{q}) \otimes I_k)\, R(\mathbf{q}) = \tilde{R}(\Delta)\, (e_N(\mathbf{q}) \otimes I_k)\,.$$

(ii) Assume that $R(\mathbf{q}) = f(\mathbf{q}) = \sum_{0 \le i \le 7} a_i \mathbf{q}^i$ $(a_7 \neq 0)$ *and that* $N = 2$. *Check that*

$$\tilde{R}(\Delta) = \begin{bmatrix} f_0(\Delta) & f_1(\Delta) \\ f_1^{\boldsymbol{\alpha}}(\Delta) & f_0^{\boldsymbol{\alpha}}(\Delta) \end{bmatrix}$$

where

$$f_0(\Delta) = a_0 + a_2\Delta + a_4\Delta^2 + a_6\Delta^3,$$
$$f_1(\Delta) = a_1 + a_3\Delta + a_5\Delta^2 + a_7\Delta^3.$$

Exercise 1054. *Let M be a system over $\mathbf{A} = \mathbf{K}[\mathbf{q};\boldsymbol{\alpha}]$ where $\boldsymbol{\alpha}$ is an automorphism of $\mathbf{K}$ such that $\boldsymbol{\alpha}^N = 1$ $(N > 1)$. Let $R(\mathbf{q}) \in {}^q\mathbf{A}^k$ be a matrix of definition of M.*

(i) Prove that the corresponding matrix of definition of $\tilde{M}$ is $\tilde{R}(\Delta) \in {}^{Nq}\mathbf{S}^{Nk}$ given by

$$(e_N(\mathbf{q}) \otimes I_k) R(\mathbf{q}) = \tilde{R}(\Delta)(e_N(\mathbf{q}) \otimes I_k).$$

(ii) Assume that $R(\mathbf{q}) = f(\mathbf{q}) = \sum_{0 \le i \le 7} a_i \mathbf{q}^i$ $(a_7 \neq 0)$ and that $N = 2$. Check that

$$\tilde{R}(\Delta) = \begin{bmatrix} f_0(\Delta) & f_1(\Delta) \\ f_1^{\boldsymbol{\alpha}}(\Delta) & f_0^{\boldsymbol{\alpha}}(\Delta) \end{bmatrix}$$

where

$$f_0(\Delta) = a_0 + a_2\Delta + a_4\Delta^2 + a_6\Delta^3,$$
$$f_1(\Delta) = a_1 + a_3\Delta + a_5\Delta^2 + a_7\Delta^3.$$

6.10 Notes and References

The algebraic approach to define the poles of an LTV system was recently introduced by the authors: [239], [240] and [241]. It is based on the formal local theory of the differential equations presented in [286] and on the theory of non-commutative polynomials introduced in [265]. A key notion is the minimal polynomial and the Wedderburn polynomials studied in [203] and related references. This factorization is similar to the one previously obtained by Gelfand et al. in [135] using quasideterminants techniques. The first attempts to characterize the stability from the factors of the differential operator of an LTV system are due to Kamen in [174] and Zhu and Johnson in [379], [381], [380]. The transfer matrix of an LTV system was introduced in the form used here in [117]. The poles and zeros along with the relations among them have been given in this algebraic approach in the case of LTI systems in [44].

N-piled forms were used by several authors for the study of N-periodic discrete-time systems, notably Grasselli *et al.* [145], [146], El Mrabet and Bourlès [108], for systems given by a Rosenbrock description. In [108], the system coefficients are assumed to belong to a ring with zero-divisors. The present account, where the coefficients are assumed to belong to a field, is clear and general; it is based on modules and the functor $\mathfrak{C}$. The price to pay is that no reasonable injective cogenerator is presently known in the category ${}_{\mathbf{T}}\mathbf{Mod}$ (that in (**5.4.2.3**) is valid for LTV discrete-time systems with coefficients in $\mathbf{K} = \mathbb{C}(t)$, thus for a class of systems which are non-periodic).

There were attempts to extend the notions of pole and zero to nD systems. Wood et al. [363] considered the case of nD systems systems with constant coefficients, and Pommaret ([278], Sect. V.2)–see, also, Pommaret and Quadrat [279]–the case of nD systems systems with varying coefficients. An nD system with varying coefficients can be viewed as a filtered module M over a filtered ring $\mathbf{A}$. A graded module M^g over a graded ring $\mathbf{A}^g$ is canonically associated with this filtered module. To clarify ideas, assume that $n = 1$ and that the ring of operators is $B_1 = \mathbf{K}[\partial, \boldsymbol{\delta}]$ ($\boldsymbol{\delta} = d/dt$, $\mathbf{K} = \mathbb{C}(t)$); B_1 is a filtered ring and the associated graded ring B_1^g is deduced from B_1 by making the indeterminate ∂ commute with the coefficients, i.e., the commutation rule $\partial a = a\partial + a^{\boldsymbol{\delta}}$ in B_1 is replaced in B_1^g by $Xa = aX$ (denoting by X the new indeterminate), so that $B_1^g \cong \mathbf{K}[X]$. The first step of the approach in [278] consists in replacing the filtered module M by the associated graded module M^g, i.e., roughly speaking, in replacing the initial time-varying system by the corresponding "frozen system". Then, the approach in [363] and that in [278] become similar. Consider an LTI control system (M, u, y) over $\mathbf{A} = \mathbb{C}[\partial]$. The module of system poles is M/M_{in} where $M_{in} = [u]_{\mathbf{A}}$. Consider its annihilator $\mathfrak{a}$ and the radical $\sqrt{\mathfrak{a}}$ of $\mathfrak{a}$. Then, the system poles are defined in [363], [278] as being the points of the algebraic set $\mathcal{Z}_{\mathfrak{a}}$ (2.10.1.**2**). Consider now an LTV control system (M, u, y) over B_1, with equation $(\partial^3 + 3a\partial^2 + 3a^2\partial + a^3)y = u$, and let $M_{in} = [u]_{B_1}$. The graded module associated with M is the module M^g over $\mathbf{K}[X]$ given by the equation $(X + a)^3 y^g = u^g$ where the indeterminate X and the coefficient t commute. The graded module associated with M_{in} is the free $\mathbf{K}[X]$-module $M_{in}^g = [u^g]_{\mathbf{K}[X]}$. Consider the torsion module $(M/M_{in})^g \cong M^g/M_{in}^g$ ([278], Sect. IV.3). Its annihilator is the ideal $\left((X + a)^3\right)$ in $\mathbf{K}[X]$, the radical of which is $(X + a)$, thus $\mathcal{Z}_{\mathfrak{a}} = \{a\}$. This algebraic set, which is the "characteristic set" of M/M_{in}, is called the *set of system poles* in [278]. In fact, it is only a subset of the set of "frozen system poles": that obtained when the multiplicities are not taken into account. Moreover, it is not connected to the solutions of the equations which are related the full sets of zeros of $P(\partial)$ or the fundamental sets of roots of $P(\partial)$ used in this section. Therefore, unfortunately, this set $\mathcal{Z}_{\mathfrak{a}}$ cannot be considered as the set of system poles in a consistent way.

7 Structure at Infinity and Impulsive Behaviors

7.1 Introduction

The topic of this chapter is the structure at infinity of discrete and continuous linear time-varying systems in a unified approach. When interconnecting two subsystems, "impulsive motions" may arise. In the continuous-time case, those impulsive motions are linear combinations of the Dirac distribution δ and its derivatives; they were first studied by Verghese [346]. In the discrete-time case, those impulsive behaviors are backward solutions with finite support; they were revealed by Lewis [216]. The space spanned by all impulsive motions of a system is called its "impulsive behavior" and is denoted as $\mathfrak{B}_{\infty}$. The structure of this impulsive behavior must be studied, and, for the integrity of the system resulting from the interconnection, all impulsive motions must be avoided.

Before entering into details, let us consider the following continuous-time system with constant coefficients, in "descriptor form":

$$(E\partial - A)\,\mathbf{x} = B\mathbf{u}, \quad t \in \mathbb{T}_0 \tag{7.1}$$

where the function $\mathbf{u}$ is the "system input", assumed to be known, $\mathbf{x}$ is the "descriptor vector" and $\partial = d/dt$, where the time-derivative is understood in the sense of distributions; E, A and B are matrices belonging to $\mathrm{Mat}_q(\mathbf{k}), \mathrm{Mat}_q(\mathbf{k})$ and ${}^q\mathbf{k}^m$, respectively ($\mathbf{k} = \mathbb{R}$ or $\mathbb{C}$). Suppose that $\mathbb{T}_0 = [0, +\infty)$, which means that the system is *formed* at time $t = 0$ (as a result, for example, of switching or of component failure in some other system; such events are frequent in electrical circuits, mechanics, hydraulics, etc. [133]). Therefore, let us call (7.1) a "temporal system" (to point out the difference with the classic situation where $\mathbb{T}_0 = \mathbb{R}$, and where system (7.1) is thus perpetually existing). Assume that the matrix pencil $Es - A$ is regular (i.e., that the polynomial $|Es - A|$ is nonzero [132]) to ensure that (7.1) has solutions [179]. If E is singular, the restrictions to $\mathbb{T}_0$ of the components $\mathbf{x}_i$ of $\mathbf{x}$ contain impulsive motions with coefficients only depending on the

H. Bourlès and B. Marinescu: Linear Time-Varying Systems, LNCIS 410, pp. 453–496.
springerlink.com

"initial values" $\mathbf{x}_i(0^-)$, when the latter are incompatible with the equation $(E\partial - A)\,\mathbf{x} = B\mathbf{u}$. These impulsive motions, which are said to be "uncontrollable" due to their complete dependence on initial conditions, span the "autonomous part" $\mathfrak{B}_{\infty,a}$ of $\mathfrak{B}_{\infty}$. To know what event arose at time $t = 0$, the *values* of the above-mentioned coefficients are not significant, as opposed to the *structure* of $\mathfrak{B}_{\infty,a}$. Setting $\mathbb{T} = \mathbb{R}$, the temporal system (7.1) can be written in the more general form

$$\begin{cases} R(\partial)\,\mathbf{w}(t) = \mathbf{e}(t)\,,\ t \in \mathbb{T} \\ \quad \mathbf{e}(t) = 0,\ t \in \mathbb{T}_0 \end{cases} \tag{7.2}$$

where $R(\partial)$ is a $q \times k$ matrix $(k = q + m)$ with entries in $\mathbf{k}[\partial]$ and $\mathbf{w}$ is the column, the entries of which are the system variables (here the components $\mathbf{x}_i$ and $\mathbf{u}_i$ of $\mathbf{x}$ and $\mathbf{u}$, respectively); the function $\mathbf{e}$ has any restriction to the complement $\mathbb{T} \setminus \mathbb{T}_0$ of $\mathbb{T}_0$ in $\mathbb{T}$. It is known that the structure of $\mathfrak{B}_{\infty,a}$ is completely determined by the structure of the "zeros at infinity" of the matrix $R(\partial)$ [343]–a notion which is explained below. Therefore, the characterization of the structure of $\mathfrak{B}_{\infty,a}$ is *not* an analytic problem (involving derivations, integrations, etc., in the framework of the theory of distributions), but an *algebraic* one, which makes the calculations much simpler and easier to computerize.

A similar problem is posed by discrete-time systems [216], [217]. The variables are now sequences (denoted as functions defined on the set of integers $\mathbb{Z}$). Let $\mathbf{q}$ be the usual shift forward operator $\mathbf{w}(t) \mapsto \mathbf{w}(t+1)$, define the discrete-time derivative $\partial = \mathbf{q} - 1$, and with this notation consider the discrete-time system with constant coefficients (7.1). Assume that the sequence $\mathbf{u}$ (again called the input) is known and that the matrix pencil $Ez - A$ is regular. Suppose that the matrix E is singular (which means that the system is noncausal) and that $\mathbb{T}_0 = \{..., -2, -1, 0\}$, i.e., that the system exists only up to the "final time" $t = 0$ (a phenomenon which arises in various fields: for example the "Leontief model", in economy, describes the time pattern of production in several interrelated production sectors; it is of the form (7.1), possibly noncausal, and valid up to a finite final time [220]). For the same reason as above, let us call (7.1) (or (7.2) which is the most general form) a "temporal system". Due to the fact that (7.1) is noncausal, the restrictions to $\mathbb{T}_0$ of the variables $\mathbf{x}_i$ contain backward solutions with finite support (i.e., impulsive motions), with coefficients only depending on the "final values" $\mathbf{x}_i(1)$. As in the continuous-time case, these impulsive motions, said to be uncontrollable due to their complete dependence on final conditions, span the "autonomous part" $\mathfrak{B}_{\infty,a}$ of $\mathfrak{B}_{\infty}$. Considering the temporal system (7.2), where $\mathbb{T} = \mathbb{Z}$ and where the sequence $\mathbf{e}$ has any restriction to $\mathbb{T} \setminus \mathbb{T}_0$, the structure of $\mathfrak{B}_{\infty,a}$ is again a key problem. As will be shown below, this structure is determined by the "structure at infinity" of the matrix $R(\partial)$ by the structure of the *zeros at infinity* of $R(\partial)$, exactly as in the continuous-time case. The complete structure at infinity of $R(\partial)$ consists of its poles at infinity and of its zeros at infinity.

The notion of "temporal interconnection" is useful for the sequel. Any system may be considered as resulting from the interconnection of subsystems [369], [303]. In the continuous-time case, a switching, a component failure, etc., as mentioned above, are interconnections *starting at a given initial time* (assumed to be zero without loss of generality, since the origin of time can be freely chosen), i.e., only effective on $\mathbb{T}_0 = [0, +\infty) \subsetneq \mathbb{T}$; such an interconnection is said to be "temporal" in what follows. In the discrete-time case, a temporal interconnection is an interconnection *valid up to a given final time* (also assumed to be zero), i.e., only effective on $\mathbb{T}_0 = \{..., -2, -1, 0\} \subsetneq \mathbb{T}$. A temporal system results from the temporal interconnection of subsystems. This is clear when considering (7.2) which is obtained by interconnecting the system $R(\partial)\,\mathbf{w} = \mathbf{e}$ with the trivial system $\bar{\mathbf{e}} = 0$ through the temporal interconnection $\mathbf{e}(t) = \bar{\mathbf{e}}(t)\,, t \in \mathbb{T}_0$.

7.2 Structure at Infinity

7.2.1 Matrices over S

Let $\mathbf{K}$ be an Ore domain, let $\mathbf{A} = \mathbf{K}[\partial; \boldsymbol{\alpha}, \boldsymbol{\delta}]$ (Definition 355), where $\boldsymbol{\alpha}$ an automorphism and $\boldsymbol{\delta}$ an $\boldsymbol{\alpha}$-derivation of $\mathbf{K}$, and let $\boldsymbol{\beta} = \boldsymbol{\alpha}^{-1}$. Consider the ring $\mathbf{S} = \mathbf{K}[[\sigma; \boldsymbol{\alpha}, \boldsymbol{\delta}]]$, where $\sigma = \partial^{-1}$ (Subsect. 2.9.1). We know that $\mathbf{S}$ is a domain and that if $\mathbf{K}$ is Noetherian, $\mathbf{S}$ is again a Noetherian (Theorem 381). If $\mathbf{K}$ is local with maximal ideal $\mathfrak{m}$, then $\mathbf{S}$ is local with maximal ideal $(\mathfrak{m}, \mathfrak{a})$ where $\mathfrak{a} = (\sigma)$ (Theorem 383).

Example 1055. *(i) Let* $\mathbf{K} = \mathbf{k}\{t\}$ **(2.9.1.1)** *and* $\mathbf{A} = \mathbf{K}[\partial; \boldsymbol{\delta}]$, $\boldsymbol{\delta} = d/dt$. *Then* $\mathbf{A} = A_{1c}(\mathbf{k})$ *(Subsect. 2.7.5) is a local ring with maximal ideal* (t, σ) *and, as shown by Corollary 365, this ring is a simple Noetherian domain.*
(ii) Consider the first Weyl algebra $A_1(\mathbf{k}) = \mathbf{k}[t][\partial; \boldsymbol{\delta}]$*; this ring is a simple Noetherian domain but is not local, for* $\mathbf{k}[t]$ *is not. Nevertheless,* $\mathbf{k}[t]$ *can be embedded in* $\mathbf{k}\{t\}$*, and therefore* $A_1(\mathbf{k})$ *can be embedded in the local ring* $A_{1c}(\mathbf{k})$.

The operator σ can be viewed as a kind of "integration operator" (see below). Let us study the general linear group $\mathrm{GL}_n(\mathbf{S})$.

Proposition 1056. *(i) Let*

$$U = \sum_{i \geq 0} \Gamma_i\, \sigma^i \in \mathrm{Mat}_n(\mathbf{S}),$$

where $\Gamma_i \in \mathrm{Mat}_n(\mathbf{K})\,, i \geq 0$. *The matrix* U *belongs to* $\mathrm{GL}_n(\mathbf{S})$ *if, and only if* $\Gamma_0 \in \mathrm{GL}_n(\mathbf{K})$
(ii) Let $U \in \mathrm{GL}_n(\mathbf{S})$ *and* $k \in \mathbb{Z}$. *There exist two matrices* $U_k, U'_k \in \mathrm{GL}_n(\mathbf{S})$ *such that* $\sigma^k\, U = U_k\, \sigma^k$ *and* $U\, \sigma^k = \sigma^k\, U'_k$.

Proof. (i) If Γ_0 is invertible, U can be written in the form $\Gamma_0 (I_n - X), X \in \sigma \operatorname{Mat}_n(\mathbf{S})$. The matrix $I_n - X$ is invertible with inverse $\sum_{i \geq 0} X^i$. Conversely, if U is invertible, there exists $L = \sum_{i \geq 0} \Lambda_i \sigma^i \in \operatorname{Mat}_n(\mathbf{S})$ such that $U L = I_n$. This implies $\Gamma_0 \Lambda_0 = I_n$, thus Γ_0 is invertible.

(ii) Let $U = \sum_{i \geq 0} \Gamma_i \sigma^i \in \operatorname{GL}_n(\mathbf{S})$. By the commutation rule (2.44), with $Z = \sigma$, $\sigma U = \left(\sum_{i \geq 0} \Theta_i \sigma^i \right) \sigma$ with $\Theta_0 = \Gamma_0^\beta$. The matrix Γ_0^β (the entries of which are the images of the entries of Γ_0 by the automorphism β) is invertible, therefore $\sum_{i \geq 0} \Theta_i \sigma^i \in \operatorname{GL}_n(\mathbf{S})$ by (i). By induction, for any $k \in \mathbb{N}$, there exist $U_k, U_k' \in \operatorname{GL}_n(\mathbf{S})$ such that $\sigma^k U = U_k \sigma^k$ and $U \sigma^k = \sigma^k U_k'$. By the first equality, $U \sigma^{-k} = \sigma^{-k} U_k$, by the second equality, $\sigma^{-k} U = U_k' \sigma^{-k}$, thus (ii) is proved. ■

Proposition 1057. *Assume that* $\mathbf{K}$ *is a division ring.*

(i) The ring $\mathbf{S}$ *is a principal ideal domain and is local with maximal left ideal* $\mathbf{S} \sigma = \sigma \mathbf{S} = (\sigma)$.
(ii) The units of $\mathbf{S}$ *are the power series of order zero; any nonzero element* $a \in \mathbf{S}$ *can be written in the form* $a = \upsilon \sigma^{\omega(a)} = \sigma^{\omega(a)} \upsilon'$, *where* υ *and* υ' *are units of* $\mathbf{S}$.
(iii) Let a *and* b *be nonzero elements of* $\mathbf{S}$*; then* $b \parallel a$ *(i.e.,* b *is a* total divisor *of* a*) if, and only if* $\omega(b) \leq \omega(a)$. *Therefore, every nonzero element of* $\mathbf{S}$ *is invariant.*
(iv) The residue class division ring of $\mathbf{S}$ *(Lemma and Definition 343) is* $\mathbf{K}$.

Proof. If $\mathbf{K}$ is a division ring, its maximal ideal is (0), therefore $\mathbf{S}$ is a local Noetherian domain with maximal ideal $\mathfrak{a} = (\sigma)$. The only nonzero elements of $\mathbf{S}$ are the powers $\sigma^i, i \geq 0$, and their associates (Corollary 384), and this proves (i), (ii) and (iii). The residue class division ring of $\mathbf{S}$ is $\mathbf{S}/(\sigma) \cong \mathbf{K}$. ■

When $\mathbf{K}$ is a division ring, every matrix over $\mathbf{S}$ has a Smith normal form Σ (Lemma and Definition 655) by Proposition 1057 and Theorem 658. Therefore, in that case, for any matrix $R \in {}^q\mathbf{S}^k$, there exist matrices $U \in \operatorname{GL}_q(\mathbf{S})$ and $V \in \operatorname{GL}_k(\mathbf{S})$ such that $U R V^{-1} = \Sigma$ where

$$\boxed{\Sigma = \begin{bmatrix} \operatorname{diag}(\sigma^{\mu_i})_{1 \leq i \leq r} & 0 \\ 0 & 0 \end{bmatrix}, \quad 0 \leq \mu_1 \leq \ldots \leq \mu_r}. \tag{7.3}$$

Let $\mu_i, 1 \leq i \leq s$ be the zero elements in the list $\{\mu_i, 1 \leq i \leq r\}$ (if any). The following is an obvious consequence of Corollary and Definition 662:

Proposition 1058. *The noninvertible invariant factors* σ^{μ_i} $(1 \leq i \leq s)$ *of* R *coincide with its elementary divisors.*

Remark 1059. *Assuming that* $\mathbf{K}$ *is a division ring, the Smith normal form of a matrix* $R \in {}^{q}\mathbf{S}^{k}$ *can be obtained using only elementary operations (Subsect. 2.10.3), i.e., secondary operations (Definition 649) are not necessary. Indeed, the Smith normal form is obtained using the following procedure (referring to the kinds of elementary operations specified in Subsect. 2.10.3):*

(a) Using row and column operations of the third kind, put an element of least order (Definition 380) in position (1,1).
(b) Using row and column operations of the second kind, reduce that element to a power of σ*, say* σ^{μ_1}*.*
(c) Using row and column operations of the first kind, annihilate all elements of the first row and the first column of R*, except the element in position (1,1). Then,* R *is put into the form*

$$\begin{bmatrix} \sigma^{\mu_1} & 0 \\ 0 & R_1^{+} \end{bmatrix}$$

and σ^{μ_1} *is a total divisor of all entries of* R_1*. By induction, using the same operations, the desired Smith normal form is obtained.*

7.2.2 Smith-MacMillan form at Infinity

Let $\mathbf{K}$ be an Ore domain, let $\mathbf{A} = \mathbf{K}[\partial; \boldsymbol{\alpha}, \boldsymbol{\delta}]$, let $\mathbf{B} = \mathbf{K}\left[\sigma^{-1}, \sigma; \boldsymbol{\alpha}, \boldsymbol{\delta}\right]$, and let $\mathbf{L} = \mathbf{K}((\sigma; \boldsymbol{\alpha}, \boldsymbol{\delta}))$; $\mathbf{B}$ (resp., $\mathbf{L}$) is the ring of Laurent polynomials (resp., of formal Laurent series) in σ, equipped with the commutation rule (2.44) with $Z = \sigma$. Let $\hat{\mathbf{K}}$ be the division ring of fractions of $\mathbf{K}$, let $\mathbf{Q} = \hat{\mathbf{K}}(\partial; \boldsymbol{\alpha}, \boldsymbol{\delta})$, $\hat{\mathbf{S}} = \hat{\mathbf{K}}\left[\left[X^{-1}; \boldsymbol{\alpha}, \boldsymbol{\delta}\right]\right]$, and $\hat{\mathbf{L}} = \hat{\mathbf{K}}((\sigma; \boldsymbol{\alpha}, \boldsymbol{\delta}))$. The relations between these rings, already established in Theorem 385, are summarized in the table below, where $\longrightarrow$ is used for $\subseteq$.

$$\begin{array}{ccccc} & & \mathbf{B} & & \\ & \nearrow & & \searrow & \\ \mathbf{A} & \longrightarrow & \mathbf{S} & \rightarrow & \mathbf{L} \\ \updownarrow & & \downarrow & & \downarrow \\ \mathbf{A} & & \hat{\mathbf{S}} & \longrightarrow & \hat{\mathbf{L}} \\ & \searrow & & \nearrow & \\ & & \mathbf{Q} & & \end{array}$$

If $\mathbf{K}$ is a division ring, then $\mathbf{S} = \hat{\mathbf{S}}$, $\mathbf{L} = \hat{\mathbf{L}}$, and $\mathbf{B} \subseteq \mathbf{Q}$.

Theorem and Definition 1060. *(1) Let* $R \in \hat{\mathbf{L}}^{p \times m}$ *be a matrix of rank* r*. There exist matrices* $U \in \mathrm{GL}_p\left(\hat{\mathbf{S}}\right)$ *and* $V \in \mathrm{GL}_m\left(\hat{\mathbf{S}}\right)$ *such that*

$$U\,R\,V^{-1} = \begin{bmatrix} \operatorname{diag}\left(\sigma^{\nu_i}\right)_{1 \leq i \leq r} & 0 \\ 0 & 0 \end{bmatrix}, \quad \nu_1 \leq \ldots \leq \nu_r, \tag{7.4}$$

and the integers $\nu_i \in \mathbb{Z}$ $(1 \leq i \leq r)$ *are uniquely determined from* R. *The integer*

$$c(R) = -\sum\nolimits_{1 \leq i \leq r} \nu_i$$

is called the content *of* R.
(2) The matrix in the right-hand side of the equality in (7.4) is called the Smith-MacMillan (normal) form *of* R *over* $\hat{\mathbf{L}}$. *If* $R \in {}^p\mathbf{Q}^m$ *and* $\mathbf{Q}$ *is embedded in* $\hat{\mathbf{L}}$, *(7.4) is called the* Smith-MacMillan form at infinity *of* R.
(3) Let $\mathbf{R}$ *be* $\mathbf{A}$ *or* $\mathbf{B}$, *or let* $\mathbf{R} = \mathbf{Q}$ *if* $\mathbf{K}$ *is a division ring, and let* $R(\partial) \in {}^q\mathbf{R}^k$ *be a matrix of rank* r. *If there exist matrices* $U \in \mathrm{GL}_q(\mathbf{S})$ *and* $V \in \mathrm{GL}_k(\mathbf{S})$ *such that*

$$U(\sigma)\, R(\partial)\, V^{-1}(\sigma) = \begin{bmatrix} \mathrm{diag}\,(\sigma^{\nu_i})_{1 \leq i \leq r} & 0 \\ 0 & 0 \end{bmatrix} \in {}^q\mathbf{L}^k, \tag{7.5}$$
$$\nu_1 \leq \ldots \leq \nu_r, \quad \nu_i \in \mathbb{Z}\ (1 \leq i \leq r)$$

then the matrix on the right of the equality in (7.5) is called the Smith-MacMillan form at infinity *of* $R(\partial)$.
(4) If $R(\partial) \in {}^q\mathbf{R}^k$ *(*$\mathbf{R} = \mathbf{A}$ *or* $\mathbf{B}$, *or* $\mathbf{R} = \mathbf{Q}$ *if* $\mathbf{K}$ *is a division ring) has a Smith-MacMillan form at infinity, this form is unique, and* $R(\partial)$ *is called* regular at infinity. *Consider the finite sequences* $(\bar{\varsigma}_i)_{1 \leq i \leq r}$ *and* $(\bar{\pi}_i)_{1 \leq i \leq r}$ *defined as follows:* $\bar{\varsigma}_i = \max(0, \nu_i)$ *and* $\bar{\pi}_i = \max(0, -\nu_i)$. *Among the nonnegative integers* $\bar{\varsigma}_i$ *(resp.,* $\bar{\pi}_i$*), those which are nonzero (if any) are called the* structural indices *of the* zero (resp., of the pole) at infinity *of the matrix* $R(\partial)$*; they are put in increasing (resp., decreasing) order and denoted by* ς_i $(1 \leq i \leq \rho)$ *(resp.,* π_i $(1 \leq i \leq s)$*). If* $\varsigma_1 > 0$, $R(\partial)$ *is said to have a* blocking zero at infinity of order ς_1. *The natural integer* $\#(TP_\infty) = \sum_{1 \leq i \leq s} \pi_i$ *(resp.,* $\#(TZ_\infty) = \sum_{1 \leq i \leq \rho} \varsigma_i$*) is called the* degree *of the pole (resp., the zero) at infinity of* $R(\partial)$. *If* $\bar{\varsigma}_i = 0$ *(resp.,* $\bar{\pi}_i = 0$*), then* $R(\partial)$ *is said to have no zero (resp., no pole) at infinity. The integer*

$$c_\infty(R) = -\sum\nolimits_{1 \leq i \leq r} \nu_i$$

is called the content at infinity *of* R.

Proof. (1) Let σ^k be a least common denominator of all entries of G and $A^+ = \sigma^k G \in \hat{\mathbf{S}}^{p \times m}$. According to Proposition 1058, there exist matrices $\bar{U} \in \mathrm{GL}_p\left(\hat{\mathbf{S}}\right)$ and $V \in \mathrm{GL}_m\left(\hat{\mathbf{S}}\right)$ such that $\bar{U}\, A^+\, V^{-1} = \Sigma$, where Σ is given by (7.3). Therefore, $\bar{U}\, \sigma^k\, G\, V^{-1} = \Sigma$, and by Proposition 1056, $\bar{U}\, \sigma^k = \sigma^k\, U$ where $U = \bar{U}'_k \in \mathrm{GL}_p\left(\hat{\mathbf{S}}\right)$. Thus, the equality in (7.4) holds with $\nu_i = \mu_i - k$ $(1 \leq i \leq r)$.

(3) Assume that $R(\partial) \in {}^q\mathbf{R}^k$ has a Smith-MacMillan form at infinity. Embedding $\mathbf{R}$ in $\hat{\mathbf{L}}$ and $\mathbf{S}$ in $\hat{\mathbf{S}}$, we still have (7.5) with $R \in \hat{\mathbf{L}}^{q \times k}$,

$U \in \mathrm{GL}_q\left(\hat{\mathbf{S}}\right)$ and $V \in \mathrm{GL}_k\left(\hat{\mathbf{S}}\right)$, thus the right-hand member of (7.5) is the Smith-MacMillan form of R over $\hat{\mathbf{L}}$. This form is unique by (1). ∎

Lemma 1061. *Assume that* $\mathbf{K}$ *is a division ring. Let* $\mathbf{D}\left(\mathbf{L}^{\times}\right)$ *be the derived group of the multiplicative group* $\mathbf{L}^{\times} \triangleq \mathbf{L}\backslash\{0\}$ *(Subsect. 1.6.7), let* $\mathbf{L}^{\times ab}$ *be the abelianization of* $\mathbf{L}^{\times}$ *(Definition 170), and let* $\mathbf{U}(\mathbf{S})$ *be the group of units of* $\mathbf{S}$.

(i) $\mathbf{D}\left(\mathbf{L}^{\times}\right) \subseteq \mathbf{U}(\mathbf{S})$.
(ii) For any $a \in \mathbf{L}^{\times}$, *let* $\overline{a}$ *be the canonical image of* a *in* $\mathbf{L}^{\times ab}$. *Let* G *be the group consisting of all* $\overline{\sigma^i}, i \in \mathbb{Z}$. *The map* ψ :

$$\mathbb{Z} \to G : i \mapsto \overline{\sigma^i}$$

is a group-isomorphism.
(iii) Let $\varphi : \mathbf{L}^{\times} \to \mathbf{L}^{\times ab} = \mathbf{L}^{\times}/\mathbf{D}\left(\mathbf{L}^{\times}\right)$ *be the canonical epimorphism, and let* $\overline{\mathbf{U}} = \varphi(\mathbf{U}(\mathbf{S}))$. *Then* $G \cap \overline{\mathbf{U}} = \left\{\overline{1}\right\}$.
(iv) Let $U \in \mathrm{GL}_n(\mathbf{S})$. *The Dieudonné determinant* $\det(U)$ *of* U *(Lemma and Definition 402) exists since* $\mathbf{S}$ *can be embedded in* $\mathbf{L}$, *and there exists a unit* $\upsilon \in \mathbf{S}$ *such that* $\det(U) = \overline{\upsilon} \triangleq \varphi(\upsilon)$.
(v) Let $R \in \mathrm{GL}_n(\mathbf{L})$. *Then the Dieudonné determinant of* R *can be uniquely written in the form*

$$\det(R) = \overline{\upsilon}\,\overline{\sigma^{-c}} = \varphi(\upsilon)\,\varphi\left(\sigma^{-c}\right) \tag{7.6}$$

where υ *is a unit of* $\mathbf{S}$ *and* $c = c(R)$ *is the content of* R.
(vi) Let $R_1, R_2 \in \mathrm{GL}_n(\mathbf{L})$, *and let* $R = R_1 R_2$. *The contents of these matrices are related by* $c(R) = c(R_1) + c(R_2)$.

Proof. (i): Any element $x \in \mathbf{S}^{\times}$ can be uniquely written in the form $\upsilon\sigma^i = \sigma^i\upsilon'$ $(i \in \mathbb{N})$, thus any element $x \in \mathbf{L}^{\times}$ can be uniquely written in the form $\upsilon\sigma^i = \sigma^i\upsilon'$ $(i \in \mathbb{Z})$. Let $x_1 = \upsilon_1\sigma^{i_1}$ and $x_2 = \upsilon_1\sigma^{i_2}$ be elements of $\mathbf{L}^{\times}$. Their commutator is, according to Definition 163,

$$(x_1, x_2) = (x_2\, x_1)^{-1}\,(x_1\, x_2) = \upsilon\,\sigma^{-(i_1+i_2)}\sigma^{(i_1+i_2)}\,\upsilon'$$

where υ, υ' are units of $\mathbf{S}$; thus $(x_1, x_2) \in \mathbf{U}(\mathbf{S})$. Therefore, $\mathbf{D}\left(\mathbf{L}^{\times}\right) \subseteq \mathbf{U}(\mathbf{S})$.

(ii): The map ψ is obviously a group-epimorphism. If $\psi(i) = 1$, then $\sigma^i \in \mathbf{U}(\mathbf{S})$ by (i), therefore $i = 0$, and this proves that ψ is injective.

(iii): If $\overline{\sigma^i} \in \overline{\mathbf{U}}$ for some integer i, then there exists a unit υ of $\mathbf{S}$ and an element $d \in \mathbf{D}\left(\mathbf{L}^{\times}\right)$ such that $\sigma^i = \upsilon d$. Since $\upsilon d \in \mathbf{U}(\mathbf{S})$ by (i), $\sigma^i \in \mathbf{U}(\mathbf{S})$, which implies $\sigma^i = 1$.

(iv): According to Remark 1059, U is the product of elementary matrices, the Dieudonné determinant of which is $\overline{1}$, of permutation matrices, the Dieudonné determinant of which is $\overline{-1}$, and of diagonal matrices with units of $\mathbf{S}$ on the diagonal, the determinant of which is $\overline{\upsilon}$ for some unit $\upsilon \in \mathbf{S}$.

(v): Let $R \in \mathrm{GL}_n(\mathbf{L})$. By Theorem and Definition 1060(1), there exist matrices $U, V \in \mathrm{GL}_n(\mathbf{S})$ such that $R = U^{-1} \operatorname{diag}(\sigma^{\nu_i})_{1 \leq i \leq n} V$. Therefore, by (iv) $\det R = \overline{\upsilon}\, \overline{\sigma^{-c}}$ where $\overline{\upsilon} = \det(V) / \det(U)$. Assume that $\det R = \overline{\upsilon'}\, \overline{\sigma^{-c'}}$. Then $\overline{\sigma^{c-c'}} = \overline{\upsilon\, \upsilon'^{-1}} \in \overline{\mathbf{U}}$, thus $\overline{\sigma^{c-c'}} = \overline{1}$ by (iii), and $c = c'$ by (ii). As a consequence, $\overline{\upsilon} = \overline{\upsilon'}$, and the uniqueness of (7.6) is proved.

(vi): We have $\det(R) = \det(R_1) \det(R_2)$, and by (v) $\det(R_i) = \overline{\upsilon_i}\, \overline{\sigma^{-c_i}}$ $(i = 1, 2)$ where each υ_i is a unit of $\mathbf{S}$. Therefore,

$$\det(R) = \overline{\upsilon_1\, \upsilon_2}\, \overline{\sigma^{-(c_1+c_2)}}$$

and again by (v), $c(R) = c(R_1) + c(R_2)$. ■

We now consider the general situation in Theorem and Definition 1060.

Lemma 1062. *Let* $D = \operatorname{diag}(\sigma^{\mu_i})_{1 \leq i \leq n}$ $(\mu_i \in \mathbb{Z})$ *and let* $U \in \mathrm{GL}_n(\mathbf{S})$. *There exists* $U' \in \mathrm{GL}_n(\mathbf{S})$ *such that* $U\, D = D\, U'$.

Proof. Let $I_i \in \mathrm{Mat}_n(\mathbb{Z})$ be the matrix, all entries of which are zero except the (i, i) entry which is equal to 1. Then

$$D = \sum_{1 \leq i \leq n} \sigma^{\mu_i} I_i,$$

therefore by Proposition 1056, there exist $U_i \in \mathrm{GL}_n(\mathbf{S})$ such that

$$U\, D = U \sum_{1 \leq i \leq n} \sigma^{\mu_i} I_i = \sum_{1 \leq i \leq n} \sigma^{\mu_i} U_i\, I_i = D\, U'$$

where U' is the matrix, the ith column of which is the ith column of U_i. Thus, we have obtained $U\, D = D\, U'$ with $U' \in \mathrm{Mat}_n(\mathbf{S})$. Embedding $\mathbf{S}$ in $\hat{\mathbf{L}}$, there exists by Lemma 1061 a unit υ of $\hat{\mathbf{S}}$ such that $\det(U') = \overline{\upsilon}$, therefore $U' \in \mathrm{GL}_n\left(\hat{\mathbf{S}}\right) \cap \mathrm{Mat}_n(\mathbf{S})$. Since, $U' = D^{-1} U\, D$, we have $U'^{-1} = D^{-1} U^{-1} D$ where $U^{-1} \in \mathrm{GL}_n(\mathbf{S})$, thus the above rationale proves that $U'^{-1} \in \mathrm{GL}_n\left(\hat{\mathbf{S}}\right) \cap \mathrm{Mat}_n(\mathbf{S})$. Therefore, $U' \in \mathrm{GL}_n(\mathbf{S})$. ■

Theorem 1063. *Let* $R_1 \in {}^p\mathbf{R}^r$ *and* $R_2 \in {}^r\mathbf{R}^m$ *be two matrices of rank* r *and regular at infinity. Then* $R_1 R_2$ *is regular at infinity and* $c_\infty(R_1 R_2) = c_\infty(R_1) + c_\infty(R_2)$.

Proof. There exist matrices $U_1 \in \mathrm{GL}_p(\mathbf{S})$, $V_1, U_2 \in \mathrm{GL}_r(\mathbf{S})$, and $V_2 \in \mathrm{GL}_p(\mathbf{S})$ such that

$$U_1^{-1} R_1\, V_1 = \begin{bmatrix} D_1 \\ 0 \end{bmatrix}, \quad D_1 = \operatorname{diag}(\sigma^{\mu_i})_{1 \leq i \leq r},$$
$$U_2^{-1} R_2\, V_2 = \begin{bmatrix} D_2 & 0 \end{bmatrix}, \quad D_2 = \operatorname{diag}(\sigma^{\nu_i})_{1 \leq i \leq r}.$$

Setting $R = R_1 R_2$, we obtain

$$R = U_1 \begin{bmatrix} D_1 \\ 0 \end{bmatrix} U \begin{bmatrix} D_2 & 0 \end{bmatrix} V_2^{-1}$$

where $U = V_1^{-1} U_2 \in \mathrm{GL}_n(\mathbf{S})$. Let $W_2 = V_2^{-1}$ and

$$U_1 = \begin{bmatrix} U_{11} & U_{12} \end{bmatrix}, \quad W_2 = \begin{bmatrix} W_{21} \\ W_{22} \end{bmatrix}.$$

Then,

$$R = \begin{bmatrix} U_{11} D_1 U D_2 W_{21} & 0 \\ 0 & 0 \end{bmatrix};$$

by Lemma 1062, there exists $U' \in \mathrm{GL}_n(\mathbf{S})$ such that $U D_2 = D_2 U'$, therefore

$$R = \begin{bmatrix} U_{11} D U' W_{21} & 0 \\ 0 & 0 \end{bmatrix}$$

where $D = D_1 D_2$. This can be written

$$R = U_1 \begin{bmatrix} D & 0 \\ 0 & 0 \end{bmatrix} \begin{bmatrix} U' & 0 \\ 0 & I \end{bmatrix} V_2^{-1}$$

where $D = \mathrm{diag}\left(\sigma^{\mu_i + \nu_i}\right)_{1 \leq i \leq r}$ and $\begin{bmatrix} U' & 0 \\ 0 & I \end{bmatrix} \in \mathrm{GL}_p(\mathbf{S})$. Therefore, R is regular at infinity and $c_\infty(R) = -\sum_{1 \leq i \leq n} (\mu_i + \nu_i) = c_\infty(R_1) + c_\infty(R_2)$. ■

Definition 1064. *Assuming that* $\mathbf{K}$ *is a division ring, let* $G(\partial)$ *be the transfer matrix of a control system* (M, u, y) *over* $\mathbf{A}$ *(Corollary and Definition 853(i)). The order of* (M, u, y) *is* $\dim_{\mathbf{K}}(M / [u]_{\mathbf{A}})$ *in that case (Theorem and Definition 861(4)). Assume that* (M, u, y) *is both controllable and observable.*

(i) The pole (resp., the zero) at infinity of $G(\partial)$ *are called the* transmission pole *(resp., the* transmission zero*)* at infinity *of* (M, u, y).
(ii) The natural integer $\dim_{\mathbf{K}}(M / [u]_{\mathbf{A}}) + \#(TP_\infty)$ *is called the* MacMillan degree *of* $G(\partial)$*, and is denoted by* $\boldsymbol{\delta}_{\mathbf{M}}(G)$.

Assuming that $\mathbf{K}$ is a division ring, the transfer matrix $G(\partial)$ of the control system (M, u, y) over $\mathbf{A}$ can be expanded as:

$$G(\partial) = \sum_{i \geq \nu_1} \Theta_i \sigma^i, \quad \Theta_{\nu_1} \neq 0.$$

Definition 1065. *The transfer matrix $G(\partial)$ (or the control system (M,u,y)) is said to be* proper *(resp.,* strictly proper*) if $\nu_1 \geq 0$ (resp., $\nu_1 \geq 1$). It is said to be* biproper *if $G(\partial)$ is invertible, proper, and with a proper inverse.*

7.2.3 Modules over S

7.2.3.1 A Canonical Structure

Let $\mathbf{A} = \mathbf{K}[\partial; \boldsymbol{\alpha}, \boldsymbol{\delta}]$ where $\mathbf{K}$ is an Ore domain, $\boldsymbol{\alpha}$ is an automorphism and $\boldsymbol{\delta}$ an $\boldsymbol{\alpha}$-derivation of $\mathbf{K}$. Let $R(\partial) \in {}^q\mathbf{A}^k$ be a matrix of rank r.

Theorem 1066. *Assume that $R(\partial)$ is regular at infinity.*

(1) There exists a strongly left-coprime factorization $(D(\sigma), R^+(\sigma))$ of $R(\partial)$ over $\mathbf{S}$ *(Definition 518(i) and Exercise 954), i.e., a pair of matrices $(D(\sigma), R^+(\sigma))$ such that $D(\sigma) \in \mathrm{Mat}_q(\mathbf{S})$ is invertible over* $\mathbf{L}$*, $R^+(\sigma) \in {}^q\mathbf{S}^k$, $R(\partial) = D^{-1}(\sigma) R^+(\sigma)$, and $(D(\sigma), R^+(\sigma))$ is strongly left-coprime.*
(2) For any strongly left-coprime factorization $(D(\sigma), R^+(\sigma))$ of $R(\partial)$ over $\mathbf{S}$ *the following properties hold:*
(a) $D(\sigma)$ is equivalent (Definition 280(i)) to

$$\begin{bmatrix} \mathrm{diag}\left(\sigma^{\pi_i}\right)_{1 \leq i \leq s} & 0 \\ 0 & I_{q-s} \end{bmatrix}$$

where the π_i $(1 \leq i \leq s)$ are the structural indices of the pole at infinity of $R(\partial)$;
(b) $R^+(\sigma)$ is equivalent to

$$\begin{bmatrix} I_s & 0 & 0 \\ 0 & \mathrm{diag}\left(\sigma^{\bar{\varsigma}_i}\right)_{s+1 \leq i \leq r} & 0 \\ 0 & 0 & 0 \end{bmatrix}$$

where $\bar{\varsigma}_i = 0$ for $s+1 \leq i \leq r-\rho$ and the $\bar{\varsigma}_{r-\rho+i}$ $(1 \leq i \leq \rho)$ are the structural indices $\varsigma_i \geq 1$ of the zero at infinity of $R(\partial)$.
(3) Let $M^+ = \mathrm{coker}_{\mathbf{S}}(\bullet R^+(\sigma))$. This left $\mathbf{S}$*-module is uniquely determined by $R(\partial)$, and*

$$M^+ = \mathcal{T}\left(M^+\right) \oplus \Phi^+ \tag{7.7}$$

where $\Phi^+ \cong M^+ / \mathcal{T}(M^+) \cong \mathbf{S}^{k-r}$ and

$$\mathcal{T}\left(M^+\right) \cong \bigoplus\nolimits_{1 \leq i \leq \rho} \tilde{C}_{\varsigma_i}, \quad \tilde{C}_{\varsigma_i} \triangleq \frac{\mathbf{S}}{(\sigma^{\varsigma_i})}. \tag{7.8}$$

The torsion submodule $\mathcal{T}(M^+)$ of M^+ is determined up to $\mathbf{S}$*-isomorphism by the zero at infinity of $R(\partial)$.*

Proof. (1) Since $R(\partial)$ is regular at infinity, this matrix has a Smith-MacMillan form at infinity, and there exist matrices $U(\sigma) \in \mathrm{GL}_q(\mathbf{S}), V(\sigma) \in \mathrm{GL}_k(\mathbf{S})$ such that

$$R(\partial) = U^{-1}(\sigma) D_1^{-1}(\sigma) D_2(\sigma) V(\sigma),$$
$$D_1(\sigma) = \begin{bmatrix} \mathrm{diag}(\sigma^{\bar{\pi}_i})_{1\leq i\leq s} & 0 & 0 \\ 0 & I_{r-s} & 0 \\ 0 & 0 & I_{q-r} \end{bmatrix}$$
$$D_2(\sigma) = \begin{bmatrix} I_s & 0 & 0 \\ 0 & \mathrm{diag}(\sigma^{\bar{\varsigma}_i})_{s+1\leq i\leq r} & 0 \\ 0 & 0 & 0 \end{bmatrix},$$

thus $R(\partial) = D^{-1}(\sigma) R^+(\sigma)$ with $D = D_1 U$ and $R^+ = D_2 V$.

Using elementary column operations, the matrix $\begin{bmatrix} D_1(\sigma) & D_2(\sigma) \end{bmatrix}$ is easily seen to be right-equivalent to $\begin{bmatrix} I_q & 0 \end{bmatrix}$. Therefore, there exists a matrix

$$Q(\sigma) = \begin{bmatrix} Q_1(\sigma) \\ Q_2(\sigma) \end{bmatrix} \in \mathrm{GL}_k(\mathbf{S})$$

such that $\begin{bmatrix} D_1 & D_2 \end{bmatrix} Q(\sigma) = I_q$, i.e., the following Bézout identity holds:

$$D_1 Q_1 + D_2 Q_2 = I_q.$$

This is equivalent to

$$D U^{-1} Q_1 + R^+ V^{-1} Q_2 = I_q,$$

therefore the pair $(D(\sigma), R^+(\sigma))$ is strongly left-coprime.

(2) Let $(D'(\sigma), R'^+(\sigma))$ be any strongly left-coprime factorization of $R(\partial)$ over $\mathbf{S}$. Then $R(\partial) = D'^{-1}(\sigma) R'^+(\sigma)$ and there exist matrices $X'(\sigma), Y'(\sigma)$ over $\mathbf{S}$ such that $D'X' + R'^+Y' = I_q$. Left-multiplying this expression by D'^{-1} one obtains $X' + D'^{-1}R'^+Y' = D'^{-1}$, i.e., $X' + D^{-1}R^+Y' = D'^{-1}$. Left-multiplying this equality by D yields

$$D X' + R^+ Y' = D D'^{-1} \in \mathrm{Mat}_q(\mathbf{S}).$$

By symmetry, $D'D^{-1} \in \mathrm{Mat}_q(\mathbf{S})$, thus $W \triangleq D'D^{-1} \in \mathrm{GL}_q(\mathbf{S})$. We now have $D' = W D$ and $R'^+ = W R^+$. Therefore, $D' \equiv D_1$ and $R'^+ \equiv D_2$.

(3) The torsion submodule $\mathcal{T}(M^+)$ is determined up to $\mathbf{S}$-isomorphism by the structural indices of the zero at infinity of $R(\partial)$. ■

Extending the terminology introduced in Subsect. 3.6.2 we are led to the following.

Definition 1067. *(i) The elements* σ^{ς_i} $(1 \leq i \leq \rho)$*–or the ideals generated by them–are the* nonzero invariant factors *of the left* $\mathbf{S}$*-module* M^+*, and they coincide with its* nonzero elementary divisors*; the number of times a same*

element σ^{ς_i} *is encountered in the list* $\{\sigma^{\varsigma_i}, 1 \leq i \leq \rho\}$ *is the* multiplicity of that elementary divisor*;* $\operatorname{rk} \Phi^+ = k - r$ *is the* multiplicity of the elementary divisor 0.
(ii) The non-negative integer $\# (M^+) = \sum_{1 \leq i \leq \rho} \varsigma_i$ *is called the* degree *of* M^+.

Remark 1068. *Not all f.g. modules over* $\mathbf{S}$ *have the structure in Theorem 1066(3), except when* $\mathbf{K}$ *is a division ring. The full subcategory of* $_{\mathbf{S}}\mathbf{Mod}$*, all objects of which have the structure of the torsion module* $\mathcal{T}(M^+)$ *in Theorem 1066(3), is denoted by* $_{\mathbf{S}}\mathbf{Mod}^{struc}$ *in the sequel.*

7.2.3.2 A Canonical Cogenerator

For any non-negative integer μ, let $\tilde{C}_\mu = \frac{\mathbf{S}}{(\sigma^\mu)}$, and let $\tilde{\delta}^{(\mu-1)}$ be the canonical image of $1 \in \mathbf{S}$ in $\tilde{C}_\mu$; the $\mathbf{S}$-module $\tilde{C}_\mu$ is isomorphic to a submodule of $\tilde{C}_{\mu+1}$ under right multiplication by σ and $\tilde{\delta}^{(\mu)}\sigma = \sigma + (\sigma^{\mu+1}) = \sigma\tilde{\delta}^{(\mu)}$; identifying $\tilde{\delta}^{(\mu-1)}$ with $\sigma\tilde{\delta}^{(\mu)} = \tilde{\delta}^{(\mu)}\sigma$, $\tilde{C}_\mu$ is embedded in $\tilde{C}_{\mu+1}$, and for any $\mu \geq 1$

$$\tilde{C}_\mu = \oplus_{i=1}^{\mu} \mathbf{K}\tilde{\delta}^{(i-1)}. \tag{7.9}$$

Let $\tilde{\Delta}$ be the directed union of all modules $\tilde{C}_\mu, \mu \geq 1$ (Example 19), i.e.,

$$\tilde{\Delta} \triangleq \varinjlim_{\mu} \tilde{C}_\mu = \oplus_{\mu \geq 0} \mathbf{K}\tilde{\delta}^{(\mu)}. \tag{7.10}$$

The left $\mathbf{S}$-module $\tilde{\Delta}$ becomes a left $\mathbf{L}$-module, setting $\sigma^{-1}\tilde{\delta}^{(\mu)} = \tilde{\delta}^{(\mu+1)}$, thus $\tilde{\Delta}$ becomes also a left $\mathbf{A}$-module by restriction of the ring of scalars. Considering σ and ∂ as operators on $\tilde{\Delta}$, σ is a left inverse of ∂, but σ has no left inverse since $\sigma\tilde{\delta} = 0$. Note that

$$\tilde{C}_\mu = \ker_{\tilde{\Delta}} (\sigma^\mu \bullet). \tag{7.11}$$

Theorem 1069. *(1) Assume that* $\mathbf{K}$ *is a division ring. Then* $\tilde{\Delta}$ *is the canonical cogenerator of the category* $_{\mathbf{S}}\mathbf{Mod}$ *(Corollary and Definition 643), and it is a large injective cogenerator (Definition 814).*
(2) Let $\mathbf{K}$ *be any Ore domain; then* $\tilde{\Delta}$ *is a cogenerator for the subcategory* $_{\mathbf{S}}\mathbf{Mod}^{struc}$ *of* $_{\mathbf{S}}\mathbf{Mod}$ *introduced in Remark 1068.*

Proof. (1) (a) Let us prove that for any $\mu \geq 1$, $\tilde{\Delta} = E\left(\tilde{C}_\mu\right)$ (the injective hull of $\tilde{C}_\mu$). This proof is similar to that of Proposition 603, with some simplifications, although $\mathbf{S}$ is not assumed to be commutative.

(i) $\tilde{\Delta}$ is divisible. Indeed, let $0 \neq \tilde{x} \in \tilde{\Delta}$ and $0 \neq r \in \mathbf{S}$. There exist an integer $\mu \geq 0$ and a unit υ such that $r = \sigma^\mu \upsilon$. Let ν be the least positive integer such that $\tilde{x} \in \tilde{C}_\nu$. There exists $\tilde{y} \in \tilde{C}_{\mu+\nu}$ such that $\tilde{x} = \sigma^\mu \tilde{y}$, therefore $\tilde{x} = \sigma^\mu \upsilon \upsilon^{-1} \tilde{y} = r\, \tilde{z}$ where $\tilde{z} = \upsilon^{-1} \tilde{y}$.

(ii) Since $\mathbf{S}$ is a principal ideal domain, $\tilde{\Delta}$ is injective.

(iii) A divisible $\mathbf{S}$-module containing $\tilde{\delta}^{(\mu)} = \sigma \tilde{\delta}^{(\mu+1)}$ must contain an element $\tilde{x}$ such that $\sigma \tilde{\delta}^{(\mu+1)} = \sigma \tilde{x}$, thus that module must contain $\tilde{\delta}^{(\mu+1)}$. Therefore, by induction, $E\left(\tilde{C}_\mu\right) \supseteq \tilde{\Delta}$, and by (ii), $E\left(\tilde{C}_\mu\right) = \tilde{\Delta}$.

(b) $\tilde{C}_1$ is the only simple $\mathbf{S}$-module (Definition 313(i)), therefore, by Corollary and Definition 643, $\tilde{\Delta} = E\left(\tilde{C}_1\right)$ is the canonical cogenerator of the category ${}_{\mathbf{S}}\mathbf{Mod}$, and is injective.

(c) All proper ideals in $\mathbf{S}$ are two-sided since they are of the form (σ^μ), $\mu \geq 1$. They are easily seen to be meet-irreducible (**3.5.2.4**). For every $\mu \geq 1$, $\tilde{C}_\mu \subseteq \tilde{\Delta}$, thus the injective cogenerator $\tilde{\Delta}$ is large by Theorem 815(1).

(2) Let $T = \bigoplus_{1 \leq i \leq \rho} \tilde{C}_{\varsigma_i} \in {}_{\mathbf{S}}\mathbf{Mod}^{struc}$. Then $T = \mathbf{S}^\rho / N$ where $N = \operatorname{im}_{\mathbf{S}} \left(\bullet R^+ (\sigma)\right)$, $R^+ (\sigma) = \operatorname{diag}\left(\sigma^{\varsigma_i}\right)_{1 \leq i \leq \rho}$. Setting

$$N^\perp = \operatorname{Hom}_{\mathbf{S}} \left(T, \tilde{\Delta}\right) = \ker_{\tilde{\Delta}} \left(R^+ (\sigma) \bullet\right),$$

we obtain

$$\begin{aligned} N^\perp &= \left\{ \tilde{\mathbf{w}} \in {}^\rho \tilde{\Delta} : R^+ (\sigma)\, \mathbf{w} = 0 \right\} \\ &= \left\{ \tilde{\mathbf{w}} = \left[\, \tilde{\mathbf{w}}_1 \; \cdots \; \tilde{\mathbf{w}}_\rho \,\right]^T : \tilde{\mathbf{w}}_i \in \tilde{C}_{\varsigma_i} \;\; (1 \leq i \leq \rho) \right\}; \end{aligned}$$

therefore, $N^{\perp\perp} = \operatorname{Ann}_l^{\mathbf{S}} \left(N^\perp\right) = N$. By Theorem 825(2), $\tilde{\Delta}$ is a cogenerator for ${}_{\mathbf{S}}\mathbf{Mod}^{struc}$. ∎

Proposition 1070. *Let* $R(\partial) \in {}^q\mathbf{A}^k$.

(1) The matrix $R(\partial)$ *is regular at infinity if, and only if Conditions (i)-(iii) below hold:*
(i) there exists a strongly left-coprime factorization $\left(D(\sigma), R^+(\sigma)\right)$ *of* $R(\partial)$ *over* $\mathbf{S}$*;*
(ii)

$$D(\sigma) \equiv \begin{bmatrix} \operatorname{diag}\left(\sigma^{\pi_i}\right)_{1 \leq i \leq r} & 0 \\ 0 & I_{q-r} \end{bmatrix}$$

over $\mathbf{S}$*, where* $\pi_i \geq \pi_{i+1} \geq 0 \; (1 \leq i \leq r-1)$*;*
(iii) $R^+(\sigma) \equiv \begin{bmatrix} \operatorname{diag}\left(\sigma^{\varsigma_i}\right)_{1 \leq i \leq r} & 0 \\ 0 & 0 \end{bmatrix}$ *over* $\mathbf{S}$*, where* $0 \leq \varsigma_i \leq \varsigma_{i+1}$ $(1 \leq i \leq r-1)$.
(i) and (ii) can be replaced by (i') and (ii') below:
(i') there exists a strongly right-coprime factorization $\left(R^+(\sigma), D'(\sigma)\right)$ *of* $R(\partial)$ *over* $\mathbf{S}$*;*

(ii')

$$D'(\sigma) \equiv \begin{bmatrix} \operatorname{diag}(\sigma^{\pi_i})_{1 \le i \le r} & 0 \\ 0 & I_{k-r} \end{bmatrix}$$

with the same condition as in (ii) on the elements π_i.
(2) Then, $R(\partial)$ is regular at infinity, and its Smith-MacMillan form at infinity is

$$\begin{bmatrix} \operatorname{diag}(\sigma^{\varsigma_i - \pi_i})_{1 \le i \le r} & 0 \\ 0 & 0 \end{bmatrix}.$$

Proof. (1) The condition is necessary by Theorem 1066. Let us prove that it is sufficient, considering a strongly left-coprime factorization (for a strongly right-coprime factorization, the proof is similar, with obvious changes).

(a) We have $R(\partial) = D^{-1}(\sigma) R^{+}(\sigma)$.

(b) There exist matrices $U_1(\sigma), V_1(\sigma), U_2(\sigma) \in \mathrm{GL}_q(\mathbf{S})$ and a matrix $V_2(\sigma) \in \mathrm{GL}_k(\mathbf{S})$ such that

$$U_1(\sigma) D(\sigma) V_1^{-1}(\sigma) = \begin{bmatrix} \operatorname{diag}(\sigma^{\pi_i})_{1 \le i \le r} & 0 \\ 0 & I_{q-r} \end{bmatrix} = D_1(\sigma),$$

$$U_2(\sigma) R^{+}(\sigma) V_2^{-1}(\sigma) = \begin{bmatrix} \operatorname{diag}(\sigma^{\varsigma_i})_{1 \le i \le r} & 0 \\ 0 & 0 \end{bmatrix} = D_2(\sigma).$$

Therefore, $R = V_1^{-1} D_1^{-1} U\, D_2 V_2$ where $U = U_1 U_2^{-1}$. By Lemma 1062, there exists a matrix $U' \in \mathrm{GL}_q(\mathbf{S})$ such that $D_1^{-1} U = U' D_1^{-1}$. This yields $R = V_1^{-1} U' D_1^{-1} D_2 V_2$. Thus, $R(\partial)$ is regular at infinity and its Smith-MacMillan form at infinity is $D_1^{-1}(\sigma) D_2(\sigma)$. ∎

The ring $\mathbf{S}$ is Hausdorff and complete in the (σ)-adic topology (Theorem 381(i)). When $\mathbf{K} = \mathbf{k}$, $\mathbf{S}$ is commutative, thus, according to Matlis' theory ([199], §3I; [33], n°X.8.3), $\mathbf{S}$ is the endomorphism ring $\tilde{\mathbf{E}}$ of $\tilde{\Delta}$. In the general case, we have the following result:

Lemma 1071. *Let $\mathbf{K}$ be any Noetherian domain, and let $\tilde{\mathbf{E}}$ be the endomorphism ring of the left $\mathbf{S}$-module $\tilde{\Delta}$ (Example 230(ii)). The ring $\mathbf{S}$ is a subring of $\tilde{\mathbf{E}}$, and therefore, $\tilde{\Delta}$ is an $(\mathbf{S}, \mathbf{S})$-bimodule.*

Proof. Let $\mu \in \mathbf{S}$; the action $\bullet\mu : \tilde{\Delta} \to \tilde{\Delta} : \tilde{\mathbf{x}} \mapsto \tilde{\mathbf{x}}\mu$ is well-defined and is obviously $\mathbb{Z}$-linear. In addition, for any $\nu \in \mathbf{S}$ and for any $\tilde{\mathbf{x}} \in \tilde{\Delta}$, $(\nu\tilde{\mathbf{x}})\mu = \nu(\tilde{\mathbf{x}}\mu)$, therefore the action $\bullet\mu$ is $\mathbf{S}$-linear. This proves that $\bullet\mu \in \tilde{\mathbf{E}}$, therefore $\mathbf{S} \subseteq \tilde{\mathbf{E}}$, and $\tilde{\Delta}$ is an $(\mathbf{S}, \mathbf{S})$-bimodule. ∎

The corollary below is a consequence of Theorem 236(i).

Corollary 1072. *Let $R^{+}(\sigma) \in {}^{q}\mathbf{S}^{k}$, and let $M^{+} = \operatorname{coker}_{\mathbf{S}}(\bullet R^{+}(\sigma))$. Then $\ker_{\mathbf{S}}(R^{+}(\sigma)\bullet) = \operatorname{Hom}_{\mathbf{S}}\left(M^{+}, \tilde{\Delta}\right)$ and $\operatorname{Hom}_{\mathbf{S}}\left(\mathcal{T}(M^{+}), \tilde{\Delta}\right)$ are right $\mathbf{S}$-modules.*

7.3 Impulsive Systems and Behaviors

7.3.1 Temporal Systems

7.3.1.1 The Framework

In this subsection, we give a provisional and heuristic definition of a "temporal system". Continuous-time systems and the discrete-time systems are considered. These systems are defined over the ring $\mathbf{A} = \mathbf{K}[\partial; \boldsymbol{\alpha}, \boldsymbol{\delta}]$. We assume that $\mathbf{K}$ is a subring of the commutative domain $\mathcal{O}(\mathbb{R})$ consisting of all $\mathbf{k}$-valued analytic functions in $\mathbb{R}$ and that $\boldsymbol{\alpha}$ is an automorphism of $\mathcal{O}(\mathbb{R})$.

In the continuous-time case, $\boldsymbol{\alpha} = 1$, $\boldsymbol{\delta} = d/dt$ is usual time-derivative, and ∂ is the time-derivative in the sense of distributions. In the discrete-time case, $\boldsymbol{\alpha}$ is the shift-forward operator, $\boldsymbol{\delta} = \boldsymbol{\alpha} - 1$, $\mathbf{q}$ extends $\boldsymbol{\alpha}$ and $\partial = \mathbf{q} - 1$ (Subsect. 2.7.3). In both cases, $\mathbf{k}$ is a field of constants (Subsect. 2.3.3).

7.3.1.2 Generators and Relations

Let $M = \operatorname{coker}_{\mathbf{A}}(\bullet R(\partial))$ be a system, where $R(\partial) \in {}^{q}\mathbf{A}^{k}$. The system equations can be written

$$\begin{cases} R(\partial)\, w = e, \\ \quad e = 0. \end{cases} \tag{7.12}$$

This is the definition of a finitely presented module M by generators and relations (Definition 512). The equations (7.12) correspond to the exact sequence

$$\mathbf{A}^{q} \overset{\bullet R(\partial)}{\longrightarrow} \mathbf{A}^{k} \overset{\varphi}{\longrightarrow} M \to 0. \tag{7.13}$$

The module of generators is $\mathbf{A}^{k} = [\varepsilon]_{\mathbf{A}}$ where $\varepsilon = (\varepsilon_i)_{1 \le i \le k}$ is the canonical basis of $\mathbf{A}^{k}$; the module of relations is $\operatorname{im}_{\mathbf{A}}(\bullet R(\partial)) = [r]_{\mathbf{A}}$ where $r = (r_j)_{1 \le j \le q}$ and r_j is the jth row of $R(\partial)$. Let w_i and e_j be the canonical image of ε_i and r_j, respectively, in the quotient $M = \mathbf{A}^{k} / [r]_{\mathbf{A}}$ $(1 \le i \le k, 1 \le j \le q)$. The equations (7.12) are satisfied, and the second one (i.e., $e = 0$) expresses the fact that the relations existing between the system variables are active.

7.3.1.3 Continuous-Time Temporal System

In the continuous-time case, let $\mathbb{T} = \mathbb{R}$ and $\mathbb{T}_0 = [0, +\infty)$. In place of (7.12), consider the equations (7.2), i.e.,

$$R(\partial)\, \mathbf{w}(t) = \mathbf{e}(t), \quad t \in \mathbb{T}, \tag{7.14}$$

$$\mathbf{e}(t) = 0, \quad t \in \mathbb{T}_0. \tag{7.15}$$

where the entries of the column $\mathbf{w}$ belong to the appropriate signal space W (to be specified in the sequel). By (7.15), the relations between the system variables are now active only during the time period $\mathbb{T}_0$, i.e., the system is *formed at initial time zero* (due, e.g., to a failure or a switch). On the complement $\mathbb{T} \setminus \mathbb{T}_0$ of $\mathbb{T}_0$ in $\mathbb{T}$, we assume that $\mathbf{e}$ can be any C^∞ function.

Definition 1073. *The system of differential equations* (7.14), (7.15) *is the* temporal system with matrix of definition $R(\partial)$.

7.3.1.4 Discrete-Time Temporal System

In the discrete-time case, let $\mathbb{T} = \mathbb{Z}$ and $\mathbb{T}_0 = \{..., -2, -1, 0\}$. Definition 1073 still makes sense, and (7.15) means that the relations between the system variables are active only *up to final time zero.* On $\mathbb{T} \setminus \mathbb{T}_0$, the sequence $(\mathbf{e}(t))_{t \in \mathbb{T}}$ can have any values.

7.3.2 *Signal Spaces and Their Relations*

7.3.2.1 Continuous-Time Case

From the analytic point of view, the temporal system $\mathbf{\Sigma}$ defined by (7.14), (7.15) is formed as follows: take for $\mathbf{e}$ in (7.14) any C^∞ function $\mathbb{R} \to {}^q\mathbf{k}$; then multiply $\mathbf{e}$ by $1 - \Upsilon$, where Υ is the Heaviside function (i.e., $\Upsilon(t) = 1$ for $t > 0$ and 0 otherwise).

For any positive integer n, let $I_n = \left(-\frac{1}{n}, +\infty\right)$ and $W_n = \mathcal{E}(I_n)$. Then $W \triangleq \underrightarrow{\lim} W_n$ is the space of germs of all $\mathbf{k}$-valued C^∞ functions defined in an open neighborhood of $[0, +\infty)$ (Example 20). Let

$$\Delta = \oplus_{\mu \geq 0} \mathbf{K} \delta^{(\mu)} \tag{7.16}$$

where δ is the Dirac distribution. The $\mathbf{A}$-module generated by $S_0 \triangleq (1 - \Upsilon) W$ is (as $\mathbf{K}$-module)

$$S = S_0 \oplus \Delta. \tag{7.17}$$

For any distribution $T \in \Delta$ (with support $\subseteq \{0\}$) and for any neighborhood Ω of 0 in $\mathbb{R}$, the integral $\int_\Omega T$ is well-defined ([322], p. 88). Abusing the language, the punctual distribution T and the above integral are denoted by $t \mapsto T(t)$ and by

$$\int_\Omega T(\varsigma)\, d\varsigma,$$

respectively. In particular, for any $t < 0$,

$$\int_{+\infty}^{t} \delta^{(\mu)}(\varsigma)\, d\varsigma = \begin{cases} \delta^{(\mu-1)}(t) & \text{if } \mu \geq 1, \\ \Upsilon(t) - 1 & \text{if } \mu = 0. \end{cases}$$

whereas for any $t > 0$ and any $\mu \geq 0$,

$$\int_{+\infty}^{t} \delta^{(\mu)}(\varsigma)\, d\varsigma = 0.$$

For any distribution $\mathbf{w} \in S$, there exists $n \geq 1$ such that $\mathbf{w} \in (1 - \Upsilon) W_n \oplus \Delta$, thus

$$(\sigma \mathbf{w})(t) \triangleq \int_{+\infty}^{t} \mathbf{w}(\varsigma)\, d\varsigma$$

is well-defined for any $t \in I_n \setminus \{0\}$. Setting

$$(\sigma \mathbf{w})(0) = (\sigma \mathbf{w})\left(0^{-}\right) \triangleq \lim_{t \to 0, t<0} (\sigma \mathbf{w})(t),$$

σ becomes a well-defined $\mathbf{k}$-linear operator $S \to S$. Since the inverse of σ is ∂, both ∂ and σ are automorphisms of the $\mathbf{k}$-vector space S. The latter is a left $\mathbf{L}$-module (thus a left $\mathbf{S}$-module which is a left $\mathbf{A}$-module, by restriction of the ring of scalars), and S_0 is an $\mathbf{S}$-submodule of S. The left $\mathbf{A}$-module Δ is not an $\mathbf{S}$-module, but there exists by (7.17) an isomorphism of $\mathbf{K}$-module

$$\tau : \Delta \tilde{\longrightarrow} S/S_0 \triangleq \bar{\Delta} \tag{7.18}$$

and $\bar{\Delta} = S/S_0$ has a canonical structure left $\mathbf{S}$-module. The projection $\pi_{\Delta} : S \twoheadrightarrow \Delta$ parallel to S_0 is $\mathbf{K}$-linear.

7.3.2.2 Discrete-Time Case

Let Υ be the sequence defined by $\Upsilon(t) = 1$ for $t > 0$ and 0 otherwise. From the analytic point of view, the temporal system $\mathbf{\Sigma}$ defined by (7.14), (7.15) is formed as follows: take for $\mathbf{e}$ in (7.14) any sequence $\mathbb{Z} \to {}^{q}\mathbf{k}$; then multiply $\mathbf{e}$ by Υ. Set $W = \mathbf{k}^{\mathbb{Z}}$ and $S_0 = \Upsilon W$. Let Δ be defined as in (8.7), but where $\delta \triangleq \partial \Upsilon$ is the "Kronecker sequence", such that $\delta(t) = 1$ for $t = 0$ and 0 otherwise (thus, Δ is the $\mathbf{A}$-module consisting of all sequences with left and finite support). The $\mathbf{A}$-module generated by S_0 is (as $\mathbf{K}$-module) $S = S_0 \oplus \Delta$. The operator ∂ is an automorphism of the $\mathbf{k}$-vector space S, and $\sigma = \partial^{-1}$ is the $\mathbf{k}$-linear operator defined on S by

$$(\sigma \mathbf{w})(t) = \sum_{j=-\infty}^{t-1} \mathbf{w}(j);$$

the set S is an $\mathbf{L}$-module. The $\mathbf{K}$-isomorphism (7.18) still holds, and $\bar{\Delta} = S/S_0$ has again a canonical structure of left $\mathbf{S}$-module. The projection $\pi_{\Delta} : S \twoheadrightarrow \Delta$ parallel to S_0 is defined as above; it is again $\mathbf{K}$-linear.

7.3.2.3 Structure of $\bar{\Delta}$

The continuous-time case and the discrete-time case are now jointly considered. Let us further detail the structure of $\bar{\Delta}$. For any $\mu \geq 1$,

$$\partial^{\mu} S_0 = S_0 + \left(\bigoplus_{1 \leq i \leq \mu} \mathbf{K} \delta^{(i-1)} \right),$$

as easily shown by induction. For any $\mu \geq 0$, let $\bar{\delta}^{(\mu)}$ (resp., $\overline{\delta^{(\mu)}}$) be the canonical image of $\delta^{(\mu)} \in S$ in $S/\partial^{\mu} S_0$ (resp., S/S_0), i.e.,

$$\bar{\delta}^{(\mu)} = \delta^{(\mu)} + \partial^{\mu} S_0, \quad \overline{\delta^{(\mu)}} = \delta^{(\mu)} + S_0.$$

For $\mu = 0$, $\overline{\delta^{(\mu)}} = \bar{\delta}^{(\mu)}$, and for any $\mu \geq 1$, $\overline{\delta^{(\mu)}} \subsetneq \bar{\delta}^{(\mu)}$ (see Theorem 1074(iv) below for more details).

The map

$$\sigma \bullet : S/\partial^{\mu} S_0 \rightarrow S/\partial^{\mu-1} S_0 : a + \partial^{\mu} S_0 \mapsto \sigma a + \partial^{\mu-1} S_0$$

is **k**-linear monomorphism, under which $\sigma \bar{\delta}^{(\mu)}$ and $\bar{\delta}^{(\mu-1)}$ are identified. We have **k**-linear embeddings

$$S/\partial^{\mu} S_0 \xrightarrow{\sigma \bullet} S/\partial^{\mu-1} S_0 \xrightarrow{\sigma \bullet} \ldots \xrightarrow{\sigma \bullet} S/S_0 = \bar{\Delta}.$$

By restriction of the ring of scalars, the left **S**-module $\bar{\Delta}$ is a left **K**-module given by

$$\bar{\Delta} = \oplus_{\mu \geq 0} \mathbf{K} \bar{\delta}^{(\mu)} \tag{7.19}$$

where

$$\begin{aligned} \sigma \bar{\delta}^{(\mu)} &= \bar{\delta}^{(\mu-1)}, \quad \mu \geq 1, \\ \sigma^{\mu} \bar{\delta}^{(\mu-1)} &= 0, \quad \mu \geq 1. \end{aligned}$$

Theorem 1074. *(i) Every element* $\tilde{\mathbf{w}}$ *of* $\tilde{\Delta}$ *can be* uniquely *written in the form* $\lambda(\partial) \tilde{\delta}$ *for some differential operator* $\lambda(\partial) \in \mathbf{A}$.
(ii) For any $a \in \mathbf{K}$ *and any integer* $\mu \geq 0$,

$$a \delta^{(\mu)} = \sum_{i=0}^{\mu} (-1)^i \binom{\mu}{i} a^{\delta^i \beta^{\mu}}(0) \, \delta^{(\mu-i)}. \tag{7.20}$$

(iii) For every element $\bar{\mathbf{w}}$ *of* $\bar{\Delta}$*, there exists a differential operator* $\lambda(\partial) \in \mathbf{A}$ *such that* $\bar{\mathbf{w}} = \lambda(\partial) \bar{\delta}$*. In addition, for any* $a \in \mathbf{K}$ *and any* $\mu \geq 1$,

$$a \bar{\delta}^{(\mu)} = a^{\beta^{\mu}}(0) \, \bar{\delta}^{(\mu)}, \tag{7.21}$$

therefore the following equalities hold:

$$\mathbf{K}\bar{\delta}^{(\mu-1)} = \mathbf{k}\bar{\delta}^{(\mu-1)}, \quad \bar{\Delta} = \oplus_{\mu\geq 0}\mathbf{k}\bar{\delta}^{(\mu)}, \tag{7.22}$$

$$C_\mu \triangleq \oplus_{1\leq i\leq \mu}\mathbf{K}\delta^{(i-1)} = \oplus_{1\leq i\leq \mu}\mathbf{k}\delta^{(i-1)}, \quad \Delta = \oplus_{i\geq 0}\mathbf{k}\delta^{(i)}. \tag{7.23}$$

$$\bar{C}_\mu \triangleq \oplus_{1\leq i\leq \mu}\mathbf{K}\bar{\delta}^{(i-1)} = \oplus_{1\leq i\leq \mu}\mathbf{k}\bar{\delta}^{(i-1)} = \ker_{\bar{\Delta}}\left(\sigma^\mu \bullet\right), \tag{7.24}$$

For convenience, we set $C_0 = \bar{C}_0 = 0$.
(iv) For any $\mu \geq 0$,

$$\partial^\mu S_0 = S_0 \oplus C_\mu \quad \text{and} \quad \bar{\delta}^{(\mu)} = \overline{\delta^{(\mu)}} \oplus C_\mu. \tag{7.25}$$

(v) For every $\mathbf{w} \in \Delta$ *(resp.,* $\bar{\mathbf{w}} \in \bar{\Delta}$*), there exists a* unique *differential operator* $\bar{\lambda}(\partial) \in \mathbf{k}[\partial]$ *such that* $\mathbf{w} = \bar{\lambda}(\partial)\delta$ *(resp.,* $\bar{\mathbf{w}} = \bar{\lambda}(\partial)\bar{\delta}$*). Both* $\mathbf{k}$*-spaces* Δ *and* $\bar{\Delta}$ *have a canonical structure of* $\mathbf{k}[\partial]$*-module, and the well-defined map* $\varkappa : \bar{\Delta} \to \Delta : \bar{\lambda}(\partial)\bar{\delta} \mapsto \bar{\lambda}(\partial)\delta \ \left(\bar{\lambda}(\partial) \in \mathbf{k}[\partial]\right)$ *is a* $\mathbf{k}[\partial]$*-linear isomorphism.*
(vi) There exists a well-defined $\mathbf{S}$*-linear epimorphism* $\psi : \tilde{\Delta} \twoheadrightarrow \bar{\Delta}$ *given by* $\lambda(\partial)\tilde{\delta} \mapsto \lambda(\partial)\bar{\delta}$ $(\lambda(\partial) \in \mathbf{A})$*, and* ψ *is an isomorphism when* $\mathbf{K} = \mathbf{k}$.

Proof. (i): Let us show by induction that the map $\mathbf{A} \to \tilde{\Delta} : \lambda(\partial) \mapsto \lambda(\partial)\tilde{\delta}$ is a $\mathbf{k}$-linear monomorphism. The $\mathbf{k}$-linearity is obvious.

(a) Let $\lambda_0 \in \mathbf{K}$ and assume that $\lambda_0\tilde{\delta} = 0$. Then $\lambda_0(1 + (\sigma)) \subseteq (\sigma)$, which implies $\lambda_0 = 0$.

(b) Assume that for any polynomial $\lambda(\partial) = \sum_{0\leq i\leq \mu}\partial^i\lambda_i \in \mathbf{A}$, of degree $\leq \mu$, the equality $\lambda(\partial)\tilde{\delta} = 0$ implies $\lambda(\partial) = 0$, and let $\lambda'(\partial) = \sum_{0\leq i\leq \mu+1}\partial^i\lambda_i \in \mathbf{A}$ be a polynomial of degree $\leq \mu+1$, such that $\lambda'(\partial)\tilde{\delta} = 0$. Left-multiplying this equality by σ^μ, this yields $\lambda_{\mu+1}\tilde{\delta} = 0$, therefore, by (a), $\lambda_{\mu+1} = 0$.

(ii): For $\mu = 0$, $a\delta = a(0)\delta$ and (7.20) holds true. Assume that (7.20) holds for $\mu = n$. By the commutation rule (2.37), for any $a \in \mathbf{K}$,

$$a\partial = \partial a^{\boldsymbol{\beta}} - a^{\boldsymbol{\beta\delta}}.$$

Therefore,

$$a\delta^{(n+1)} = a\partial\delta^{(n)} = \left(\partial a^{\boldsymbol{\beta}} - a^{\boldsymbol{\beta\delta}}\right)\delta^{(n)} = \partial a^{\boldsymbol{\beta}} - a^{\boldsymbol{\beta\delta}}\delta^{(n)}$$

We have by assumption

$$\begin{aligned}\partial a^{\boldsymbol{\beta}} &= \partial\left(\sum_{i=0}^{n}(-1)^{i}\binom{n}{i}a^{\boldsymbol{\delta}^{i}\boldsymbol{\beta}^{n+1}}(0)\,\delta^{(n-i)}\right)\\ &= a^{\boldsymbol{\beta}^{n+1}}(0)\,\delta^{(n+1)}+\sum_{j=0}^{n-1}(-1)^{j+1}\binom{n}{i+1}a^{\boldsymbol{\delta}^{j+1}\boldsymbol{\beta}^{n+1}}(0)\,\delta^{(n-j)}\end{aligned}$$

and by the binomial identity

$$\binom{n}{j+1}=\binom{n+1}{j+1}-\binom{n}{j}.$$

Finally, one obtains

$$a\delta^{(n+1)}=\sum_{i=0}^{n+1}(-1)^{i}\binom{n+1}{i}a^{\boldsymbol{\delta}^{i}\boldsymbol{\beta}^{n+1}}(0)\,\delta^{(n+1-i)}$$

and (7.20) is proved by induction.

(iii): (a) By (7.19), for every $\bar{\mathbf{w}}\in\bar{\Delta}$, there exists $\lambda(\partial)\in\mathbf{A}$ such that $\bar{\mathbf{w}}=\lambda(\partial)\,\bar{\delta}$.

(b) Let $a\in\mathbf{K}$ and let $\mu\geq 0$. Then, $a\bar{\delta}^{(\mu)}=a\left(\delta^{(\mu)}+\partial^{\mu}S_0\right)$, and by (7.20), there exists $c_{\mu-1}\in\bigoplus_{0\leq i\leq\mu-1}\mathbf{k}\delta^{(i)}$ such that $a\delta^{(\mu)}=a^{\boldsymbol{\beta}^{\mu}}(0)\,\delta^{(\mu)}+c_{\mu-1}$. Therefore, $a\delta^{(\mu)}-a^{\boldsymbol{\beta}^{\mu}}(0)\,\delta^{(\mu)}\in\partial^{\mu}S_0$, so that

$$a\delta^{(\mu)}+\partial^{\mu}S_0=a^{\boldsymbol{\beta}^{\mu}}(0)\,\delta^{(\mu)}+\partial^{\mu}S_0,$$

which implies (7.21). Last, (7.21) implies (7.22). The equalities in (7.24), as well as those in (7.23), are now clear (they have connections with (7.11)).

(iv): The two equalities of (7.25) are easily established by induction.

(v) is clear.

(vi): We know by (i) that for every $\tilde{\mathbf{w}}\in\tilde{\Delta}$ there exists a unique $\lambda(\partial)\in\mathbf{A}$ such that $\tilde{\mathbf{w}}=\lambda(\partial)\,\tilde{\delta}$. By (iii), or every $\bar{\mathbf{w}}\in\bar{\Delta}$ there exists some $\lambda(\partial)\in\mathbf{A}$ such that $\bar{\mathbf{w}}=\lambda(\partial)\,\bar{\delta}$. Therefore, $\psi:\tilde{\Delta}\to\bar{\Delta}$ is a well-defined surjection which is obviously $\mathbf{S}$-linear. If $\mathbf{K}=\mathbf{k}$, then ψ is an isomophism by (v). ■

7.3.3 Impulsive Behavior

Consider a temporal system $\boldsymbol{\Sigma}$ with matrix of definition $R(\partial)\in{}^{q}\mathbf{A}^{k}$ (Definition 1073) and assume that $R(\partial)$ is regular at infinity (Theorem and Definition 1060(4)).

Proposition 1075. *The following properties are equivalent: (i) For any* $\mathbf{e}\in{}^{q}S_0$*, there exists* $\mathbf{w}\in{}^{k}S$ *such that (7.14) is satisfied. (ii) The matrix* $R(\partial)$ *is full row rank (i.e., of rank* $r=q$*)*

Proof. (i) $\Rightarrow$ (ii): If the matrix $R(\partial)$ is not full row rank, $\bullet R(\partial)$ is not injective, i.e., there exists a nonzero element $\eta(\partial) \in \mathbf{A}^q$ such that $\eta(\partial) R(\partial) = 0$. Therefore, for $\mathbf{w} \in {}^k S$ and $\mathbf{e} \in {}^q S_0$ to satisfy (7.14), $\mathbf{e}$ must satisfy the "compatibility condition" $\eta(\partial) \mathbf{e} = 0$.

(ii) $\Rightarrow$ (i): By Theorem 1066, assuming that $q = r$,

$$U(\sigma) R(\partial) V^{-1}(\sigma) = \begin{bmatrix} \operatorname{diag}\{\sigma^{\nu_i}\}_{1 \le i \le r} & 0 \end{bmatrix}$$

and (7.14) is equivalent to

$$\begin{bmatrix} \operatorname{diag}\{\sigma^{\nu_i}\}_{1 \le i \le r} & 0 \end{bmatrix} \mathbf{v} = \mathbf{h}, \tag{7.26}$$

where

$$\mathbf{v} = V(\sigma) \mathbf{w}, \quad \mathbf{h} = U(\sigma) \mathbf{e} \tag{7.27}$$

and (7.26) is equivalent to $\sigma^{\nu_i} \mathbf{v}_i = \mathbf{h}_i$, $1 \le i \le r$. For any $\nu_i \in \mathbb{Z}$ and any $\mathbf{h}_i \in S_0$, $\mathbf{v}_i = \partial^{\nu_i} \mathbf{h}_i$ belongs to S. Therefore, (i) holds because $\mathbf{h}$ spans ${}^q S_0$ as $\mathbf{e}$ spans the same space (since S_0 is an $\mathbf{S}$-module and $U(\sigma) \in \mathrm{GL}_q(\mathbf{S})$). ■

In the sequel, the matrix $R(\partial)$ is assumed to be full row rank. Considering the Smith-MacMillan form at infinity (7.5) of $R(\sigma)$, let

$$\mathbf{r} = \{1, ..., r\}, \mathbf{r}^+ = \{j \in \mathbf{r} : \nu_j \ge 0\}, \mathbf{r}^- = \{j \in \mathbf{r} : \nu_j < 0\}.$$

Let $\mathfrak{B} \subseteq {}^k S$ be the $\mathbf{k}$-space consisting of all $\mathbf{w} \in {}^k S$ for which there exists $\mathbf{e} \in {}^r S_0$ such that (7.14) holds. In the equalities (7.26), (7.27), let

$$\mathbf{v} = \begin{bmatrix} \mathbf{v}_a \\ \mathbf{v}_c \end{bmatrix}, \quad \mathbf{w}_a = V^{-1}(\sigma) \begin{bmatrix} \mathbf{v}_a \\ 0 \end{bmatrix},$$

where $\mathbf{v}_a$ has r rows, and let

$$Z(\sigma) = V^{-1}(\sigma) \begin{bmatrix} I_r & 0 \\ 0 & 0 \end{bmatrix} V(\sigma) \in \mathrm{Mat}_k(\mathbf{S}).$$

Then, $\mathbf{w}_a = Z(\sigma) \mathbf{w}$. Let

$$\mathfrak{B}_a = Z(\sigma) \mathfrak{B} = \{\mathbf{w}_a \in {}^k S : \mathbf{w} \in \mathfrak{B}\}$$

Definition 1076. *(i) The* $\mathbf{k}$*-space* $\mathfrak{B}_\infty = \pi_\Delta \mathfrak{B} \subseteq {}^k \Delta$ *is the* impulsive behavior *of the temporal system* $\mathbf{\Sigma}$*, where the projection* π_Δ *is extended to* ${}^k S$*, acting on this power of* S *component-wise.*
(ii) The $\mathbf{k}$*-space* $\mathfrak{B}_{\infty,a} = \pi_\Delta \mathfrak{B}_a \subseteq {}^k \Delta$ *is the* autonomous impulsive behavior *of the temporal system* $\mathbf{\Sigma}$.
(iii) $M^+ \triangleq \operatorname{coker}_{\mathbf{S}}(\bullet R^+(\sigma))$ *is the* impulsive system *of the temporal system* $\mathbf{\Sigma}$.

Theorem 1077. *(i)* $V(\sigma) \mathfrak{B} = (\oplus_{i \in \mathbf{r}} \partial^{\nu_i} S_0) \oplus S^{k-r}$;
(ii) $\pi_\Delta V(\sigma) \mathfrak{B} = (\oplus_{i \in \mathbf{r}^+} C_{\nu_i}) \oplus \Delta^{k-r}$;

(iii) $V(\sigma)\mathfrak{B}_a = \begin{bmatrix} I_r & 0 \\ 0 & 0 \end{bmatrix} V(\sigma)\mathfrak{B} = \oplus_{i\in\mathbf{r}}\partial^{\nu_i}S_0$;

(iv) $\pi_\Delta \begin{bmatrix} I_r & 0 \\ 0 & 0 \end{bmatrix} V(\sigma)\mathfrak{B} = \oplus_{i\in\mathbf{r}^+}C_{\nu_i}$.

(v) $M^+ \cong \left(\oplus_{i\in\mathbf{r}^+}\tilde{C}_{\nu_i}\right) \oplus \mathbf{S}^{k-r}$;

(vi) $\mathcal{T}(M^+) \cong \oplus_{i\in\mathbf{r}^+}\tilde{C}_{\nu_i}$;

(vii) $V(\sigma)\ker_{\tilde{\Delta}}(R^+(\sigma)\bullet) = \left(\oplus_{i\in\mathbf{r}^+}\tilde{C}_{\nu_i}\right) \oplus \tilde{\Delta}^{k-r}$;

(viii) $\mathrm{Hom}_\mathbf{S}\left(\mathcal{T}(M^+), \tilde{\Delta}\right) \cong \oplus_{i\in\mathbf{r}^+}\tilde{C}_{\nu_i}$.

Proof. (i): We know that

$$\mathfrak{B} = \left\{\mathbf{w} \in {}^kS : (7.14) \text{ holds for some } \mathbf{e} \in {}^rS_0\right\}$$

and (7.14) is equivalent to

$$U(\sigma)R(\partial)V^{-1}(\partial)V(\sigma)\mathbf{w} = \mathbf{e},$$

i.e., to (7.26) where $\mathbf{v}$ and $\mathbf{h}$ are defined according to (7.27). The space of all $\mathbf{v} \in {}^kS$ for which there exists $\mathbf{h} \in {}^rS$ such that (7.26) holds is isomorphic to $(\oplus_{i\in\mathbf{r}}\partial^{\nu_i}S_0) \oplus S^{k-r}$ according to the proof of Proposition 1075. By the first equality of (7.27), this space is $V(\sigma)\mathfrak{B}$.

(ii) is a consequence of (i) and of the first equality of Theorem 1074(iv).

By (i),

$$\begin{bmatrix} I_r & 0 \\ 0 & 0 \end{bmatrix} V(\sigma)\mathfrak{B} = \oplus_{i\in\mathbf{r}}\partial^{\nu_i}S_0,$$

thus (iii) results from the definition of $\mathfrak{B}_a$.

(iv) is proved similarly.

(v) and (vi) are clear.

(vii) and (viii) are consequences of (7.11). ■

Let us show through an example how the above can be used to determine $\mathfrak{B}_{\infty,a}$ in an algebraic way.

Example 1078. *Consider the temporal system with matrix of definition*

$$R(\partial) = \begin{bmatrix} -1 & \partial^2 + t & 0 & 0 \\ 0 & 0 & \partial^2 & -1 \\ 0 & 1 & -1 & 0 \end{bmatrix} \tag{7.28}$$

over the first Weyl algebra $A_1(\mathbf{k})$ *or over* $\mathbf{k}[t][\partial;\boldsymbol{\alpha},\boldsymbol{\alpha}-1]$ *(Definition 355).*

(1) Setting $\sigma = \partial^{-1}$, *one obtains*

$$R(\partial) = D^{-1}(\sigma)R^+(\sigma)$$

where

$$D(\sigma) = \mathrm{diag}\left(\sigma^2, \sigma^2, 1\right), R^+(\sigma) = \begin{bmatrix} -\sigma^2 & 1+\sigma^2 t & 0 & 0 \\ 0 & 0 & 1 & -\sigma^2 \\ 0 & 1 & -1 & 0 \end{bmatrix}. \tag{7.29}$$

Form the matrix $E(\sigma) = \left[D(\sigma)\ R^+(\sigma) \right]$, *i.e.,*

$$E(\sigma) = \begin{bmatrix} \sigma^2 & 0 & 0 & -\sigma^2 & 1+\sigma^2 t & 0 & 0 \\ 0 & \sigma^2 & 0 & 0 & 0 & 1 & -\sigma^2 \\ 0 & 0 & 1 & 0 & 1 & -1 & 0 \end{bmatrix},$$

and carry out the following elementary column operations (Subsect. 2.10.3):

- subtract from the 5th column both the 3rd one and the 1st one right-multiplied by t*;*
- add to the 1st (resp., 4th) column the (new) 5th one multiplied by $-\sigma^2$ *(resp.,* σ^2*);*
- add to the 6th column the 3rd one;
- add to the 2nd (resp., 7th) column the 6th one multiplied by $-\sigma^2$ *(resp.,* σ^2*);*
- interchange the 1st column and the 5th one, and interchange the 2nd column and the 6th one.

Doing so, $E(\sigma)$ *is found to be right-equivalent (Definition 280) to* $\left[I_3\ 0 \right]$*. Therefore,* $(D(\sigma), R^+(\sigma))$ *is a strongly left-coprime factorization of* $R(\sigma)$ *over* $\mathbf{S}$ *(Theorem 1066).*
(2) After some algebra, one obtains

$$U(\sigma) R^+(\sigma) V^{-1}(\sigma) = \left[\mathrm{diag}\left(1, 1, \sigma^2\right) \quad 0 \right]$$

with

$$U(\sigma) = \begin{bmatrix} 1 & 0 & -1 \\ 0 & 0 & 1 \\ -1 & 1 & 1 \end{bmatrix},$$

$$V^{-1}(\sigma) = \begin{bmatrix} t & t & 1+t\sigma^2 & 1+t\sigma^2 \\ 1 & 1 & \sigma^2 & \sigma^2 \\ 1 & 0 & \sigma^2 & \sigma^2 \\ 0 & 0 & 0 & 1 \end{bmatrix}.$$

Therefore, by Proposition 1070, $R(\partial)$ *is regular at infinity, and its Smith-MacMillan form at infinity is*

$$\left[\mathrm{diag}\left(\sigma^{-2}, \sigma^{-2}, \sigma^2\right)\ 0 \right].$$

(3) Let

$$R^+(\sigma)\, w^+ = 0$$

be the equation of the impulsive system M^+, *where* $w^+ = \left[\left(w_i^+\right)_{1\leq i\leq 4}\right]^T$. *This equation is equivalent to*

$$\left[\operatorname{diag}\left(1,1,\sigma^2\right) \quad 0\right] v^+ = 0$$

where

$$v^+ = V(\sigma)\, w^+ = \begin{bmatrix} -\sigma^2 & t\sigma^2 & 1 & 0 \\ 0 & 1 & -1 & 0 \\ 1 & -t & 0 & -1 \\ 0 & 0 & 0 & 1 \end{bmatrix} w^+. \tag{7.30}$$

The torsion submodule of M^+ *is* $\left[v_3^+\right]_{\mathbf{S}} \cong \mathbf{S}/\left(\sigma^2\right)$, *where*

$$v_3^+ = w_1^+ - t\, w_2^+ - w_4^+.$$

(4) $\mathfrak{B}_{\infty,a}$ *is the* $\mathbf{k}$*-space spanned by the variable*

$$t \mapsto \mathbf{v}_3(t) \quad (t \in \mathbb{T}_0)$$

where $\mathbf{v}_3 = \mathbf{w}_1 - t\,\mathbf{w}_2 - \mathbf{w}_4$, *and* $\mathfrak{B}_{\infty,a} \cong C_2 = \mathbf{k}\delta \oplus \mathbf{k}(\partial\delta)$.

7.3.4 Causal Laplace Transform and Anticausal Z-Transform

The impulsive behavior has been determined in the previous subsection in an algebraic way. The comparison with an analytic approach is instructive. For this, an appropriate causal Laplace transform and an appropriate anticausal Z-transform must be developed.

7.3.4.1 Causal Laplace Transform

We consider the continuous-time case. Every germ $f \in W$ is defined on some interval $I_n = \left(-\frac{1}{n}, +\infty\right)$. Let $f_\alpha : t \mapsto e^{\alpha t} f(t)$ for any $t \in I_n$. Let W_e be the $\mathbf{k}$-space consisting of all germs $f \in W$ for which there exists $\alpha \geq 0$ such that $\Upsilon f_\alpha \in L_1(\mathbb{R}^+)$, with the identification in (**1.8.1.4**); set $Q_e = \Upsilon W_e \oplus \Delta$ and $S_e = (1-\Upsilon)\, W_e \oplus Q_e$.

Lemma and Definition 1079. *(i) For any* $\mathbf{w} \in S_e$, *there exists a unique decomposition of the form*

$$\mathbf{w} = f + \mathbf{w}_+, \quad f \in (1-\Upsilon)\, W_e, \quad \mathbf{w}_+ \in Q_e. \tag{7.31}$$

(ii) The elements of Q_e *are Laplace-transformable.*
(iii) The causal Laplace transform $\mathcal{L}_+$ *on* S_e *is defined by*

$$\mathcal{L}_+(\mathbf{w}) = \mathcal{L}(\mathbf{w}_+)$$

where $\mathcal{L}$ *is the two-sided Laplace transform.*

(iv) There exist canonical **k***-linear isomorphisms*

$$\mathcal{L}_+(S_e) \cong \frac{S_e}{(1-\Upsilon)\,W_e} \cong Q_e. \tag{7.32}$$

Proof. (i) and (ii) are clear.

(iv): Since $\ker \mathcal{L}_+ = (1-\Upsilon)\,W_e$, the first isomorphism in (7.32) is obtained by the first Noether isomorphism theorem (Theorem 52(1)). The second isomorphism in (7.32) is a consequence of Theorem 147(ii). ∎

Let $\mathbf{w} \in S_e$ and consider the decomposition (7.31) of $\mathbf{w}$. There exists an integer $m > 0$ such that f is defined in $\left(-\frac{1}{m}, +\infty\right)$; in addition, there exists a function $g \in W_e$, defined in $\left(-\frac{1}{m}, +\infty\right)$, such that $f = (1-\Upsilon)\,g$, thus

$$\partial f = (1-\Upsilon)\,\partial g - g(0)\,\delta.$$

On the other hand,

$$\mathcal{L}(\partial(\mathbf{w}_+))(s) = s\mathcal{L}(\mathbf{w}_+)(s) = s\,\hat{\mathbf{w}}(s)$$

where $\hat{\mathbf{w}}(s) \triangleq \mathcal{L}_+(\mathbf{w})(s)$. The function g and the distribution $\mathbf{w}$ have the same restriction to $\left(-\frac{1}{m}, 0\right)$ (with the identification in (**1.8.1.4**)), therefore

$$g(0) = \lim_{t\to 0, t<0} \mathbf{w}(t) \triangleq \mathbf{w}\left(0^-\right),$$

and this makes sense although $\mathbf{w}$ is a distribution (see [43] for a more general account). Finally, $\partial\mathbf{w} = \partial(\mathbf{w}_+) + \partial f$, hence

$$(\partial\mathbf{w})_+ = \partial(\mathbf{w}_+) + (\partial f)_+ = \partial(\mathbf{w}_+) - g(0)\,\delta = \partial(\mathbf{w}_+) - \mathbf{w}\left(0^-\right)\delta,$$

so that

$$\mathcal{L}_+(\partial\mathbf{w})(s) = s\,\hat{\mathbf{w}}(s) - \mathbf{w}\left(0^-\right).$$

Corollary and Definition 1080. *(i) The operator* $\partial_0^n : \mathbf{w} \mapsto \partial^n\mathbf{w}(0^-)$ *is well-defined on* S_e *for any integer* $n \geq 0$*.*
(ii) For any $n \geq 0$*,*

$$\mathcal{L}_+(\partial^n\mathbf{w})(s) = s^n\,\hat{\mathbf{w}}(s) - \sum_{i=0}^{n-1} s^{n-1-i}\,\partial_0^i\mathbf{w}.$$

The above expression is equivalent to

$$\boxed{\mathcal{L}_+(\partial^n\mathbf{w})(s) = s^n\,\hat{\mathbf{w}}(s) - \tfrac{s^n-\partial_0^n}{s-\partial_0}\mathbf{w}.} \tag{7.33}$$

(iii) Abusing the language, for any $n \geq 0$*,*

$$\boxed{\mathcal{L}_+(t^n\mathbf{w}(t))(s) = (-1)^n\,\tfrac{d^n}{ds^n}\hat{\mathbf{w}}(s).} \tag{7.34}$$

Proof. (i) and (ii) are easily obtained by induction. (iii) is classical ([369], Sect. A.3, Remark 22). ∎

7.3.4.2 Anticausal Z-Transform

We consider the discrete-time case. Let S_e be the subspace of $\mathbf{k}^{\mathbb{Z}}$ consisting of all sequences with left bounded support. By (7.23), $\Delta = (1 - \Upsilon) S_e$, and

$$S_e = \Delta \oplus \Upsilon S_e.$$

Lemma and Definition 1081. *(i) Every element* $\mathbf{w} \in S_e$ *can be decomposed in a unique way as*

$$\mathbf{w} = \mathbf{w}_- + \mathbf{w}_+,$$

$\mathbf{w}_- \in \Delta$, $\mathbf{w}_+ \in \Upsilon S_e$.
(ii) The anticausal Z*-transform* $Z_-(\mathbf{w})$ *of* $\mathbf{w}$ *is defined as*

$$Z_-(\mathbf{w}) = Z(\mathbf{w}_-)$$

where $Z(\mathbf{w}_-)$ *is the two-sided* Z*-transform of* $\mathbf{w}_-$.
(iii) The image of Z_- *is* $\mathbf{k}\left[z^{-1}\right]$*, its kernel is* ΥS_e*, therefore there exists canonical* $\mathbf{k}$*-linear isomorphisms*

$$\mathbf{k}\left[z^{-1}\right] \cong \frac{S_e}{\Upsilon S_e} \cong \Delta.$$

Let ∂_0^n $(n \geq 0)$ be the operator defined on S_e as follows: for any $\mathbf{w} \in S_e$,

$$\partial_0^0 \mathbf{w} = -\mathbf{w}(1), \quad \partial_0^n \mathbf{w} = \partial_0^0 (\partial^n \mathbf{w})$$

(where $\partial \mathbf{w}(n) = \mathbf{w}(n+1) - \mathbf{w}(n)$); in addition, let $s = z - 1$.

Proposition 1082. *(i) For any* $\mathbf{w} \in S_e$ *and any integer* $n \geq 0$,

$$Z_-(\partial^n \mathbf{w}) = s^n Z_-(\mathbf{w}) - \sum_{i=0}^{n-1} s^{n-1-i}\, \partial_0^i \mathbf{w},$$

i.e.,

$$\boxed{Z_-(\partial^n \mathbf{w}) = s^n Z_-(\mathbf{w}) - \frac{s^n - \partial_0^n}{s - \partial_0} \mathbf{w}.} \tag{7.35}$$

(ii) For any $\mathbf{w} \in S_e$ *and any integer* $n \geq 0$*, abusing the language,*

$$\boxed{Z_-\left(\frac{(t+n-1)!}{(t-1)!} \mathbf{w}(t)\right) = (-1)^n z^n \frac{d^n}{dz^n} Z_-(\mathbf{w}(t)).} \tag{7.36}$$

Proof. (i): We proceed by induction. For $n = 1$,

$$\begin{aligned} Z_{-}(\partial \mathbf{w}) &= Z_{-}\left((\partial \mathbf{w})_{-}\right) = \textstyle\sum_{t \leq 0} (\mathbf{w}(t+1) - \mathbf{w}(t))\, z^{-t} \\ &= \mathbf{w}(1) + (z-1)\, Z_{-}(\mathbf{w}). \end{aligned}$$

Assuming that (7.35) holds for $n = k$,

$$\begin{aligned} Z_{-}\left(\partial^{k+1} \mathbf{w}\right) &= Z_{-}\left(\partial\left(\partial^{k} \mathbf{w}\right)\right) \\ &= s^{k+1} Z_{-}(\mathbf{w}) - \sum_{i=0}^{k} s^{k-i}\, \partial_0^i \mathbf{w}. \end{aligned}$$

(ii) For any integer $n \geq 0$,

$$Z_{-}\left(\frac{(t+n-1)!}{(t-1)!}\, \mathbf{w}(t)\right) = \textstyle\sum_{t \leq 0} \mathbf{w}(t)\, \dfrac{(t+n-1)!}{(t-1)!}\, z^{-t};$$

the equality

$$\frac{d^n}{dz^n} z^{-t} = (-1)^n\, \frac{(t+n-1)!}{(t-1)!}\, z^{-(t+n)}$$

yields

$$\frac{(t+n-1)!}{(t-1)!} z^{-t} = (-1)^n\, z^n \frac{d^n}{dz^n} z^{-t}.$$

■

Note that (7.33) and (7.35) are completely similar.

Example 1083. *(Example 1078 cont'd.)*

(1) Consider the continuous-time case. Then, this temporal system is defined over the first Weyl algebra $A_1[\mathbf{k}]$, $\mathbb{T} = \mathbb{R}$ and $\mathbb{T}_0 = [0, +\infty[$. For $t \geq 0$ we obtain

$$\begin{aligned} -\hat{\mathbf{w}}_1(s) + s^2 \hat{\mathbf{w}}_2(s) - \left(s\partial_0^0 + \partial_0^1\right) \mathbf{w}_2 - \frac{d\hat{\mathbf{w}}_2}{ds}(s) &= 0, \\ s^2 \hat{\mathbf{w}}_3(s) - \left(s\partial_0^0 + \partial_0^1\right) \mathbf{w}_3 - \hat{\mathbf{w}}_4 &= 0, \\ \hat{\mathbf{w}}_2(s) - \hat{\mathbf{w}}_3(s) &= 0. \end{aligned}$$

Note that the last equation does not imply

$$\left(s\partial_0^0 + \partial_0^1\right) \mathbf{w}_2 = \left(s\partial_0^0 + \partial_0^1\right) \mathbf{w}_3$$

since in general, for $i = 1, 2$

$$\partial_0^i \mathbf{w}_2 = \lim_{t \to 0, t<0} \partial^i \mathbf{w}_2(t) \neq \lim_{t \to 0, t<0} \partial^i \mathbf{w}_3(t) = \partial_0^i \mathbf{w}_3.$$

Therefore,

$$-\hat{\mathbf{w}}_1(s) - \frac{d\hat{\mathbf{w}}_2}{ds}(s) + \hat{\mathbf{w}}_4(s) = \left(\partial_0^1 + s\partial_0^0\right)(\mathbf{w}_2 - \mathbf{w}_3)$$

which yields

$$-\mathbf{w}_1(t) + t\,\mathbf{w}_2(t) + \mathbf{w}_4(t) = \alpha_0(\partial\delta) + \alpha_1\delta, \quad t \geq 0,$$

$$\alpha_i = \partial_0^i(\mathbf{w}_2 - \mathbf{w}_3), \quad i = 1, 2.$$

The $\mathbf{k}$*-space* $\mathfrak{B}_{\infty,a}$ *is generated by the variable* $t \mapsto -\mathbf{w}_1(t) + t\,\mathbf{w}_2(t) + \mathbf{w}_4(t)$*, and* $\mathfrak{B}_{\infty,a} \cong C_2 = \mathbf{k}\delta \oplus \mathbf{k}(\partial\delta)$*. This is consistent with the result obtained in Example 1078.*
(2) In the discrete-time case, the temporal system is defined over $\mathbf{k}[t][\partial; \boldsymbol{\alpha}, \boldsymbol{\alpha} - 1]$*,* $\mathbb{T} = \mathbb{Z}$ *and* $\mathbb{T}_0 = \{..., -2, -1, 0\}$*. The calculations are similar to those in case (1), but* $\mathfrak{B}_{\infty,a} \cong C_2 = \mathbf{k}\delta \oplus \mathbf{k}(\partial\delta)$ *is a* $\mathbf{k}$*-space spanned by backward solutions with finite support.*

7.3.5 Temporal Interconnections

7.3.5.1 Interconnection of Linear Systems

The interconnection of linear systems is defined in [118]. In the case of several systems, one may first interconnect two of them, then interconnect a third one with the system resulting from the interconnection of the two first ones, etc. Therefore, it is sufficient to consider the case of two linear systems M_1 and M_2. Their interconnection is a fibered sum [206]. Let us briefly explain this point.

Let G be a finite free $\mathbf{A}$-module and assume that there exist two morphisms $\bullet h_i : G \to M_i, i = 1, 2$. Let H be the submodule of $M = M_1 \oplus M_2$ generated by the elements of the form $gh = (gh_1, -gh_2), g \in G$, i.e., $H = \operatorname{im}_G(\bullet h)$, where $h = (h_1, -h_2)$. The quotient module $\breve{M} = M/H$, written $M_1 \coprod_G M_2$, is the fibered sum of M_1 and M_2 over G (with respect to the morphisms h_1, h_2); from the point of view of systems theory, it is the *interconnected system* (with respect to h_1, h_2).

Let $\bullet\xi : M \to \breve{M}$ be the canonical epimorphism and set $\breve{h} = h\xi$, so that

$$g\breve{h} = 0, g \in G. \tag{7.37}$$

The system $\breve{M}$ is defined by an equation consisting of the equations of the subsystems M_i, plus the *interconnection equation* (7.37). More specifically, let us assume that M_i is defined by the equation $R_i(\partial)\,w^i = 0$ $(i = 1, 2)$, where $w^i = \left(w_1^i, ..., w_{k_i}^i\right)$. The interconnection equation can be written $J_1(\partial)\,\breve{w}^1 = J_2(\partial)\,\breve{w}^2$, where $J_1(\partial)$ and $J_2(\partial)$ are matrices over $\mathbf{A}$, with the same number of rows and with, respectively, k_1 and k_2 columns; $J_1(\partial)$ and $J_2(\partial)$ are the *interconnection matrices*. Then, $\breve{M} = \operatorname{coker}_{\mathbf{A}}(R(\partial)\,\bullet)$ where

$$R(\partial) = \begin{bmatrix} R_1(\partial) & 0 \\ 0 & R_2(\partial) \\ J_1(\partial) & -J_2(\partial) \end{bmatrix} \tag{7.38}$$

7.3.5.2 Temporal Interconnections

Consider two temporal systems

$$\begin{cases} R_i(\partial)\,\mathbf{w}^i(t) = \mathbf{e}^i(t),\ t \in \mathbb{T} \\ \mathbf{e}^i(t) = 0,\ t \in \mathbb{T}_0 \end{cases}$$

$(i = 1, 2)$. Let these temporal systems be interconnected through interconnection matrices J_1 and J_2 with coefficients in $\mathbf{k}$. We are led to the following definition.

Definition 1084. *The interconnected temporal system is defined by*

$$\begin{cases} R(\partial)\,\mathbf{w}(t) = \mathbf{e}(t),\ t \in \mathbb{T}, \\ \mathbf{e}(t) = 0,\ t \in \mathbb{T}_0, \end{cases}$$

where

$$R(\partial) = \begin{bmatrix} R_1(\partial) & 0 \\ 0 & R_2(\partial) \\ J_1 & -J_2 \end{bmatrix}.$$

One has the following result.

Theorem 1085. *Assume that $R_1(\partial)$ and $R_2(\partial)$ are regular at infinity. Then $R(\partial)$ is regular at infinity and the impulsive system M^+ of the interconnected temporal system is given by $M^+ = \operatorname{coker}_{\mathbf{S}}(\bullet R^+(\sigma))$ where*

$$R^+(\sigma) = \begin{bmatrix} R_1^+(\sigma) & 0 \\ 0 & R_2^+(\sigma) \\ J_1 & -J_2 \end{bmatrix} \tag{7.39}$$

and $(D_i(\sigma), R_i^+(\sigma))$ is any strongly left coprime factorization over $\mathbf{S}$ of $R_i(\partial)$ $(i = 1, 2)$.

Proof. Let $R^+(\sigma)$ be given by (7.39) and

$$R^+(\sigma) = \begin{bmatrix} R_1^+(\sigma) & 0 \\ 0 & R_2^+(\sigma) \\ J_1 & -J_2 \end{bmatrix}, \quad D(\sigma) = \begin{bmatrix} D_1(\sigma) & 0 & 0 \\ 0 & D_2(\sigma) & 0 \\ 0 & 0 & I_p \end{bmatrix}$$

where p is the number of rows of the matrices J_1 and J_2. Then, using elementary operations, $(D(\sigma), R^+(\sigma))$ is easily shown to be a strongly left coprime factorization over $\mathbf{S}$ of $R(\partial)$. The theorem is now a consequence of Proposition 1070. ∎

By Theorem 1085 and (7.38), M^+ can be written as a fibered sum of the impulsive systems $M_i^+ = \operatorname{coker}_{\mathbf{S}}(\bullet R_i^+)$ with interconnection matrices J_1 and J_2. In other words, *the impulsive system of the interconnected temporal system is obtained by interconnecting the impulsive systems of the temporal subsystems.*

Example 1086. *(Example 1078 cont'd.) Set* $w_1 = u_1$*,* $w_2 = y_1$*,* $w_3 = u_2$*,* $w_4 = y_2$*. The temporal system with matrix of definition (7.28) can be viewed as resulting from the temporal series interconnection of Temporal System 1, with input* u_1*, output* y_1 *and equation* $\left(\partial^2 + t\right) y_1 - u_1 = 0$*, with Temporal System 2, with input* u_2*, output* y_2 *and equation* $\partial^2 u_2 - y_2 = 0$*; the interconnection equation is* $\mathbf{u}_2(t) = \mathbf{y}_1(t)$*,* $t \in \mathbb{T}_0$*. The equation of the impulsive system* M_1^+*, associated with Temporal System 1, is*

$$\sigma^2 \left(u_1^+ - t y_1^+\right) - y_1^+ = 0,$$

that of the impulsive system M_2^+*, associated with Temporal System 2, is*

$$u_2^+ - \sigma^2 y_2^+ = 0;$$

the temporal interconnection equation yields

$$u_2^+ = y_1^+.$$

Therefore, from the two last equations, $y_1^+ = \sigma^2 y_2^+$*; substituting this value of* y_1^+ *in the first equation, one obtains*

$$\sigma^2 \left(u_1^+ - t y_1^+ - y_2^+\right) = 0,$$

in accordance with the result in Example 1078.

7.3.6 The Axiomatic Characterization of Temporal System

7.3.6.1 Introduction

As already said, Definition 1073 is heuristic and provisional. Our aim is now to characterize a temporal system in an axiomatic way. We assume that either

(I) the framework is that specified in (**7.3.1.1**)

or

(II) $\mathbf{K}$ is a division ring, $\mathbf{A} = \mathbf{K}[\partial; \boldsymbol{\alpha}, \boldsymbol{\delta}]$ where $\boldsymbol{\alpha}$ is an automorphism and $\boldsymbol{\delta}$ an $\boldsymbol{\alpha}$-derivation of $\mathbf{K}$.

Without loss of generality, a control system over the ring $\mathbf{A}$ can be given by an admissible PMD (Lemma & Definition 856, and Remark 858(3)), and two PMDs characterize the same control system if, and only if they are strictly equivalent (Definition 859 and Theorem 860).

7.3.6.2 Full Equivalence

The equation (5.28) of an admissible PMD can be put into the usual form

$$R(\partial)\, w = 0,$$
$$w = \begin{bmatrix} \xi \\ u \\ y \end{bmatrix},$$
$$R(\partial) = \begin{bmatrix} D(\partial) & -N(\partial) & 0 \\ Q(\partial) & W(\partial) & -I_p \end{bmatrix}. \tag{7.40}$$

Assume that $R(\partial)$ is regular at infinity, let $(A(\sigma), R^+(\sigma))$ be a strongly left-coprime factorization of $R(\partial)$ over $\mathbf{S}$, and write

$$R^+(\sigma) = \begin{bmatrix} D^+(\sigma) & -N^+(\sigma) & Z^+(\sigma) \\ Q^+(\sigma) & W^+(\sigma) & Y^+(\sigma) \end{bmatrix} \tag{7.41}$$

according to the sizes in (7.40). The impulsive system associated with $R(\partial)$ is $M^+ = \operatorname{coker}_{\mathbf{S}}(\bullet R^+(\sigma))$, and $M^+ = [\xi^+, u^+, y^+]_{\mathbf{S}}$ where

$$R^+(\sigma) \begin{bmatrix} \xi^+ \\ u^+ \\ y^+ \end{bmatrix} = 0. \tag{7.42}$$

Definition 1087. *Consider two admissible PMD's $\{D_i, N_i, Q_i, W_i\}$ with matrices over $\mathbf{A}$ and denote by M_i^+ the associated impulsive systems $(i = 1, 2)$. These PMD's are* fully equivalent *if (i) they are strictly equivalent (Definition 859) and (ii) there exists an $\mathbf{S}$-isomorphism $M_1^+ \cong M_2^+$.*

Consider case (II) in (**7.3.6.1**). Let $R_i(\partial)$ $(i = 1, 2)$ be the matrix of definition associated with the PMD $\{D_i, N_i, Q_i, W_i\}$ following (7.40)). We know that $R_1(\partial), R_2(\partial)$ satisfy a comaximal relation

$$R_1 T_2 = T_1 R_2,$$

i.e.,

$$\begin{bmatrix} R_1(\partial) & T_1(\partial) \end{bmatrix} \begin{bmatrix} -T_2(\partial) \\ R_2(\partial) \end{bmatrix} = 0$$

(see the proof of Theorem 860). Assume that $\begin{bmatrix} D_i & N_i \end{bmatrix}$, or equivalently R_i, is left-regular $(i = 1, 2)$.

Definition 1088. *Let $A(\partial) \in {}^r\mathbf{A}^{k_1}$ and $B(\partial) \in {}^r\mathbf{A}$ be such that $\operatorname{rk}_{\mathbf{A}} \begin{bmatrix} A(\partial) & B(\partial) \end{bmatrix} = r$. The matrices $A(\partial), B(\partial)$ are said to be* left-coprime at infinity *if $\begin{bmatrix} A(\partial) & B(\partial) \end{bmatrix}$ has no zero at infinity (Theorem and Definition 1060(4)).* Right-coprimeness at infinity *is likewise defined.*

Theorem 1089. *Let* $\left(A_1(\sigma), \left[R_1^+(\sigma)\; T_1(\sigma)\right]\right)$ *be a left-coprime factorization of* $\left[R_1(\partial)\; T_1(\partial)\right]$ *and let* $\left(\begin{bmatrix} -T_2^+(\sigma) \\ R_2^+(\sigma) \end{bmatrix}, A_2(\sigma)\right)$ *be a right-coprime factorization of* $\begin{bmatrix} -T_2(\partial) \\ R_2(\partial) \end{bmatrix}$ *over* $\mathbf{S}$. *The two PMDs* $\{D_1, N_1, Q_1, W_1\}$ *are* fully equivalent *if, and only if:*

(i) they are strictly equivalent;
(ii) (R_1, T_1) *and* (T_2, R_2) *are, respectively, left-coprime at infinity and right-coprime at infinity;*
(iii) $\boldsymbol{\delta}_{\mathbf{M}}(R_1) = \boldsymbol{\delta}_{\mathbf{M}}\left(R_1\; T_1\right)$ *and* $\boldsymbol{\delta}_{\mathbf{M}}(R_2) = \boldsymbol{\delta}_{\mathbf{M}}\begin{pmatrix} -T_2 \\ R_2 \end{pmatrix}$.

Proof. (1) Condition (i) is obviously necessary. Assume that (i) holds. Then by Theorem 860, there exists matrices M_1, X_1, M_2, X_2 over $\mathbf{A}$, of appropriate size, such that

$$\begin{bmatrix} M_2 & 0 \\ -X_2 & I_p \end{bmatrix}\begin{bmatrix} D_1 & -N_1 \\ Q_1 & W_1 \end{bmatrix} = \begin{bmatrix} D_2 & -N_2 \\ Q_2 & W_2 \end{bmatrix}\begin{bmatrix} M_1 & X_1 \\ 0 & I_m \end{bmatrix}, \tag{7.43}$$

where (D_2, M_2) is left-coprime, and (D_1, M_1) is right-coprime. Since $\mathbf{K}$ is a division ring, $\mathbf{A}$ is a principal ideal domain (Theorem 467), thus strong left (resp., right) coprimeness of a pair of matrices over $\mathbf{A}$ is equivalent to its left (resp., right) coprimeness (Theorem 653(iii)). The relation (7.43) is equivalent to the comaximal relation

$$\left[R_2\; T_2\right]\begin{bmatrix} -T_1 \\ R_1 \end{bmatrix} = 0$$

where

$$R_i = \begin{bmatrix} D_i & -N_i & 0 \\ Q_i & W_i & -I_p \end{bmatrix}, \quad i = 1, 2,$$

$$T_1 = \begin{bmatrix} M_1 & X_1 & 0 \\ 0 & I_m & 0 \\ 0 & 0 & -I_p \end{bmatrix}, \quad T_2 = \begin{bmatrix} M_2 & 0 \\ -X_2 & I_p \end{bmatrix}.$$

(2) Let

$$\left(A_1(\sigma), \left[R_1^+(\sigma)\; T_1^+(\sigma)\right]\right) \text{ and } \left(\begin{bmatrix} -T_2^+(\sigma) \\ R_2^+(\sigma) \end{bmatrix}, A_2(\sigma)\right)$$

be a left-coprime factorization of $\left[R_1(\partial)\; T_1(\partial)\right]$ over $\mathbf{S}$ and a right-coprime factorization of $\begin{bmatrix} -T_2(\partial) \\ R_2(\partial) \end{bmatrix}$ over $\mathbf{S}$, respectively. We have

$$A_1^{-1}\left[R_1^+\; T_1^+\right]\begin{bmatrix} -T_2^+ \\ R_2^+ \end{bmatrix} A_2^{-1} = 0,$$

therefore

$$\left[\, R_1^+ \; T_1^+ \,\right] \begin{bmatrix} -T_2^+ \\ R_2^+ \end{bmatrix} = 0. \tag{7.44}$$

(2) We will show that $R_i^+(\sigma)$ is a matrix of definition of the impulsive system M_i^+ $(i = 1, 2)$ if, and only if (iii) holds.

We have $R_1(\partial) = A_1^{-1}(\sigma) R_1^+(\sigma)$. By Proposition 1070, $R_1^+(\sigma)$ is a matrix of definition of the impulsive system M_1^+ if, and only if $\left(A_1(\sigma), R_1^+(\sigma)\right)$ is a left-coprime factorization of $R_1(\partial)$ over $\mathbf{S}$. This holds true if, and only if

$$c_\infty\left(\left[\, A_1 \; R_1^+ \,\right]\right) = c_\infty\left(A_1 \left[\, I_q \; R_1 \,\right]\right) = 0.$$

By Theorem 1063,

$$c_\infty\left(A_1 \left[\, I_q \; R_1 \,\right]\right) = c_\infty(A_1) + c_\infty\left(\left[\, I_q \; R_1 \,\right]\right),$$

and

$$c_\infty\left(\left[\, I_q \; R_1 \,\right]\right) = \boldsymbol{\delta}_{\mathbf{M}}\left(\left[\, I_q \; R_1 \,\right]\right) = \boldsymbol{\delta}_{\mathbf{M}}(R_1)$$

(Exercise 1110(iii)). Therefore, the above condition holds if, and only if $\boldsymbol{\delta}_{\mathbf{M}}(R_1) = -c_\infty(A_1)$.

Since $\left(A_1, \left[\, R_1^+ \; T_1^+ \,\right]\right)$ is a left-coprime factorization of $\left[\, R_1 \; T_1 \,\right]$,

$$c_\infty\left(\left[\, A_1 \; R_1^+ \; T_1^+ \,\right]\right) = c_\infty(A_1) + c_\infty\left(\left[\, R_1 \; T_1 \,\right]\right) = 0,$$

therefore $\boldsymbol{\delta}_{\mathbf{M}}\left(\left[\, R_1 \; T_1 \,\right]\right) = -c_\infty(A_1)$.

Therefore, $R_1^+(\sigma)$ is a matrix of definition of M_1^+ if, and only if

$$\boldsymbol{\delta}_{\mathbf{M}}\left(\left[\, R_1 \; T_1 \,\right]\right) = \boldsymbol{\delta}_{\mathbf{M}}(R_1).$$

By a similar rationale, the same holds with the index 1 changed to 2.

(3) Last, the following conditions are equivalent: (a) the relation (7.44) is comaximal, (b) $\operatorname{coker}_{\mathbf{S}}\left(\bullet R_1^+\right) \cong \operatorname{coker}_{\mathbf{S}}\left(\bullet R_2^+\right)$, (c) Condition (ii) holds. Indeed, (a)⇔(b) by Theorem 520, and (a)⇔(c) by Exercise 1109(ii). ∎

7.3.6.3 An Algebraic Definition of a Temporal System

Definition 1090. *A* temporal control system $\mathbf{\Sigma}$ *is an* equivalence class of fully equivalent admissible PMDs. *The* impulsive system M^+ *of* $\mathbf{\Sigma}$ *is defined (up to* $\mathbf{S}$*-isomorphism) in accordance with Definition 1076, and its* control system (M, u, y) *is defined in accordance with Lemma and Definition 856 (thus,* M *is defined up to* $\mathbf{A}$*-isomorphism).*

7.4 Poles and Zeros at Infinity

In this section, $\mathbf{K}$ is a division ring (except when otherwise stated) and $\mathbf{A} = \mathbf{K}[\partial; \boldsymbol{\alpha}, \boldsymbol{\delta}]$ where $\boldsymbol{\alpha}$ is an automorphism and $\boldsymbol{\delta}$ an $\boldsymbol{\alpha}$-derivation of $\mathbf{K}$.

We consider a temporal control system $\mathbf{\Sigma}$ (Definition 1090) with matrix of definition $R(\partial)$; there exists a PMD $\{D, N, Q, W\}$ belonging to $\mathbf{\Sigma}$, the system matrix $P(\partial)$ of which (Definition 855(1)) is connected to $R(\partial)$ by (7.40). The transfer matrix of $\mathbf{\Sigma}$ is $G(\partial) = Q(\partial) D^{-1}(\partial) N(\partial) + W(\partial)$. The transmission poles and zeros at infinity of have already been defined (Definition 1064). The other kinds of poles and zeros at infinity are defined in what follows from the impulsive system M^+ (Definition 1076(iii)) in the same way than the "finite" poles and zeros have been defined in Chapter 6 from the system M (see, also, [43]). Nevertheless, the case of poles and zeros at infinity is simpler, for the principal ideal domain $\mathbf{S}$ is rigid (**2.13.2.3**), whereas the principal ideal domain $\mathbf{A}$ is not.

7.4.1 Uncontrollable Poles at Infinity

Definition 1091. *The module of* uncontrollable poles at infinity *(also called the module of input–decoupling zeros at infinity, and denoted by IDZ_∞) is $\mathcal{T}(M^+)$. Its elementary divisors are the nonzero elementary divisors of the matrix $R^+(\sigma)$.*

Example 1092. *(Example 1078 cont'd.). Embedding $\mathbf{k}[t]$ into its field of fractions $\mathbf{k}(t)$, the transfer function of System 1 is $G_1(\partial) = \left(\partial^2 + t\right)^{-1}$, that of System 2 is $G_2(\partial) = \partial^2$, and the transfer function of the interconnected temporal system is $G(\partial) = \partial^2 \left(\partial^2 + t\right)^{-1}$. Therefore, System 1 has a transmission zero at infinity with structural index 2, and System 2 has a transmission pole at infinity with structural index 2. The interconnected system has no transmission pole and no transmission zero at infinity. Abusing the language, a pole/zero cancellation at infinity arises due to the temporal interconnection. The temporal system has a "hidden mode at infinity" which turns out to be an uncontrollable pole at infinity with set of elementary divisors $\{\sigma^2\}$ (for $\mathcal{T}(M^+) \cong \mathbf{S}/(\sigma^2)$).*

7.4.2 System Poles at Infinity

Let $\breve{M}^+ = \mathbf{L} \bigotimes_{\mathbf{S}} M^+$ and let $\breve{\varphi} : M^+ \to \breve{M}$ be the canonical map defined by $\breve{\varphi}(w^+) = \breve{w}^+ = 1_{\mathbf{L}} \bigotimes_{\mathbf{S}} w^+$, where $1_{\mathbf{L}}$ is the unit-element of $\mathbf{L}$.

Lemma 1093. *(i) The $\mathbf{L}$-vector space $\breve{M}^+$ is of dimension m and $\breve{M}^+ = [\breve{u}^+]_{\mathbf{L}}$.*
(ii) Let u^+, y^+ be the columns defined by (7.42). The module $[u^+]_{\mathbf{S}}$ is free of rank m and $M^+/[u^+]_{\mathbf{S}}$ is torsion.
(iii) There exists a free submodule Φ^+ of M^+ such that (7.7) holds and

$$[u^+, y^+]_{\mathbf{S}} = \mathcal{T}\left([u^+, y^+]_{\mathbf{S}}\right) \oplus \left(\Phi^+ \cap [u^+, y^+]_{\mathbf{S}}\right).$$

In particular, $[u^+]_{\mathbf{S}} \triangleleft \Phi^+$.

Proof. The module M^+ is defined by $R^+(\sigma)\, w^+ = 0$, therefore $A^{-1}(\sigma)\, B^+(\sigma)\, \breve{w}^+ = 0$, i.e., $R(\partial)\, \breve{w}^+ = 0$. Thus (using the notation in Subsect. 7.2.2 with $\mathbf{L} = \hat{\mathbf{L}}$ and $\mathbf{S} = \hat{\mathbf{S}}$) $\breve{M}^+ = \mathbf{L} \bigotimes_{\mathbf{Q}} \hat{M} = \mathbf{L} \bigotimes_{\mathbf{Q}} [\hat{u}]_{\mathbf{Q}} = [\breve{u}^+]_{\mathbf{L}}$, and (i) is proved.

By (i), $[u^+]_{\mathbf{S}}$ is of rank m, and $[u^+]_{\mathbf{S}}$ is free since $[u^+]_{\mathbf{S}}$ is generated by m elements (Corollary and Definition 341(1)); in addition, $\mathbf{L} \bigotimes_{\mathbf{S}} (M^+ / [u^+]_{\mathbf{S}}) = 0$, thus $M^+ / [u^+]_{\mathbf{S}}$ is torsion.

(iii) is a consequence (ii) and of Corollary 661. ∎

In what follows, Φ^+ denotes the free $\mathbf{S}$-module in Lemma 1093(iii).

Definition 1094. *The module of* system poles at infinity, *denoted by* SP_∞, *is* $M^+ / [u^+]_{\mathbf{S}}$.

The elementary divisors of $M^+ / [u^+]_{\mathbf{S}}$ are those of the submatrix $\begin{bmatrix} D^+(\sigma) & Z^+(\sigma) \\ Q^+(\sigma) & Y^+(\sigma) \end{bmatrix}$ of $R^+(\sigma)$.

Example 1095. *(Example 1078 cont'd.) Let us transform the matrix* $R^+(\sigma)$ *in (7.29), using permutations of rows and columns, in such a way that the impulsive control system* M^+ *be described by an equation of the form (7.42) where* $u^+ = u_1^+$ *and* $y^+ = y_2^+$. *The new matrix* $R^+(\sigma)$ *is given by*

$$R^+(\sigma) = \begin{bmatrix} 1+\sigma^2 t & 0 & -\sigma^2 & 0 \\ 1 & -1 & 0 & 0 \\ 0 & 1 & 0 & -\sigma^2 \end{bmatrix}.$$

Therefore,

$$\begin{bmatrix} D^+(\sigma) & Z^+(\sigma) \\ Q^+(\sigma) & Y^+(\sigma) \end{bmatrix} = \begin{bmatrix} 1+\sigma^2 t & 0 & 0 \\ 1 & -1 & 0 \\ 0 & 1 & -\sigma^2 \end{bmatrix} \equiv \begin{bmatrix} 1 & 0 & 0 \\ 0 & 1 & 0 \\ 0 & 0 & \sigma^2 \end{bmatrix}.$$

with the notation in Definition 280(i) (the equivalence is obtained using elementary operations, and using the fact that $1+\sigma^2 t$ *is a unit of* $\mathbf{S}$ *by Theorem 381(i)). Therefore,* $SP_\infty \cong \tilde{C}_2$, *and the set of elementary divisors of* SP_∞ *is* $\{\sigma^2\}$.

Remark 1096. *The calculations in the above example are still valid if* $\mathbf{k}[t]$ *is not embedded into* $\mathbf{k}(t)$.

7.4.3 Hidden Modes at Infinity

The module of uncontrollable poles at infinity is also called the module of *input-decoupling zeros at infinity* and is denoted by IDZ_∞.

Definition 1097. *(i) The module of* output-decoupling zeros at infinity *(denoted by ODZ_∞) is $M^+ / [y^+, u^+]_{\mathbf{S}}$.*
(ii) The module of input-output decoupling zeros at infinity (denoted by $IODZ_\infty$) is $\mathcal{T}(M^+) / (\mathcal{T}(M^+) \cap [y^+, u^+]_{\mathbf{S}})$.
(iii) The module of hidden modes at infinity (denoted by HM_∞) is $M^+ / (\Phi^+ \cap [y^+, u^+]_{\mathbf{S}})$.

The elementary divisors of ODZ_∞ are those of the submatrix $\begin{bmatrix} D^+(\sigma) \\ Q^+(\sigma) \end{bmatrix}$.

Example 1098. *(Example 1078 cont'd.)*
(1) We have

$$\begin{bmatrix} D^+(\sigma) \\ Q^+(\sigma) \end{bmatrix} = \begin{bmatrix} 1+\sigma^2 t & 0 \\ 1 & -1 \\ 0 & 1 \end{bmatrix} \equiv \begin{bmatrix} 1 & 0 \\ 0 & 1 \\ 0 & 0 \end{bmatrix},$$

therefore $ODZ_\infty = 0$.
(2) In addition, $\mathcal{T}(M^+) = [v^+]_{\mathbf{S}}$ where $v^+ = u^+ - t\xi_1^+ - y^+$, and $\xi_1^+ = \xi_2^+ = \sigma^2 y^+$, therefore $v^+ = u^+ - (t\sigma^2 + 1) y^+ \in [y^+, u^+]_{\mathbf{S}}$, $\mathcal{T}(M^+) \triangleleft [y^+, u^+]_{\mathbf{S}}$, therefore $\mathcal{T}(M^+) \cap [y^+, u^+]_{\mathbf{S}} = \mathcal{T}(M^+)$ and $IODZ_\infty = 0$. Note that Remark 1096 is still valid in this context.

Lemma 1099. *(i) Let T_1^+ and T_2^+ be submodules of a f.g. $\mathbf{S}$-module, such that T_1^+ and T_2^+ are torsion and $T_1^+ \cap T_2^+ = 0$. Then*

$$\varepsilon(T_1^+ \oplus T_2^+) = \varepsilon(T_1^+) \overset{\bullet}{\cup} \varepsilon(T_2^+)$$

*where $\overset{\bullet}{\cup}$ is the disjoint union (**1.2.1.3**) and where $\varepsilon(T^+)$ is the set of elementary divisors of the f.g. torsion $\mathbf{S}$-module T^+.*
(ii) Let M_1^+, M_2^+, M_3^+ be f.g. $\mathbf{S}$-modules such that $M_1^+ \triangleleft M_2^+ \triangleleft M_3^+$ and M_{i+1}^+/M_i^+ is torsion $(i = 1, 2)$. Then

$$\#(M_3^+/M_1^+) = \#(M_3^+/M_2^+) + \#(M_2^+/M_1^+).$$

Proof. This is a consequence of Corollary 663. ∎

Theorem 1100. *The following equality holds:*

$$\varepsilon(HM_\infty) = \varepsilon(IDZ_\infty) \overset{\bullet}{\cup} \varepsilon(ODZ_\infty) \setminus \varepsilon(IODZ_\infty).$$

Proof. We have

$$\frac{M^+}{\Phi^+ \cap [y^+, u^+]_{\mathbf{S}}} = \frac{\mathcal{T}(M^+) \oplus \Phi^+}{\Phi^+ \cap [y^+, u^+]_{\mathbf{S}}} \cong \mathcal{T}(M^+) \oplus \frac{\Phi^+}{\Phi^+ \cap [y^+, u^+]_{\mathbf{S}}}, \tag{7.45}$$

this isomorphism holding because $(\Phi^{+} \cap [y^{+}, u^{+}]_{\mathbf{S}}) \cap \mathcal{T}(M^{+}) = 0$. In addition,

$$\frac{M^{+}}{[y^{+}, u^{+}]_{\mathbf{S}}} = \frac{\mathcal{T}(M^{+}) \oplus \Phi^{+}}{(\mathcal{T}(M^{+}) \cap [y^{+}, u^{+}]_{\mathbf{S}}) \oplus (\Phi^{+} \cap [y^{+}, u^{+}]_{\mathbf{S}})}$$
$$\cong \frac{\mathcal{T}(M^{+})}{\mathcal{T}(M^{+}) \cap [y^{+}, u^{+}]_{\mathbf{S}}} \oplus \frac{\Phi^{+}}{\Phi^{+} \cap [y^{+}, u^{+}]_{\mathbf{S}}} \tag{7.46}$$

by (2.9). The theorem is a consequence of (7.45), (7.46) and of Lemma 1099. ∎

Example 1101. *(Example 1078, cont'd.) We obtain by Theorem 1100*

$$\varepsilon(HM_{\infty}) = \{\sigma^2\} \overset{\bullet}{\cup} \varnothing \setminus \varnothing = \{\sigma^2\}.$$

Remark 1096 is still valid.

7.4.4 Invariant Zeros at Infinity

Definition 1102. *The module of* invariant zeros at infinity *(denoted by IZ_{∞}) is $\mathcal{T}(M^{+}/[y^{+}]_{\mathbf{S}})$.*

The elementary divisors of IZ_{∞} are those of the submatrix $\begin{bmatrix} D^{+}(\sigma) & -N^{+}(\sigma) \\ Q^{+}(\sigma) & W^{+}(\sigma) \end{bmatrix}$.

Example 1103. *(Example 1078 cont'd.) We have*

$$\begin{bmatrix} D^{+}(\sigma) & -N^{+}(\sigma) \\ Q^{+}(\sigma) & W^{+}(\sigma) \end{bmatrix} = \begin{bmatrix} 1+\sigma^2 t & 0 & -\sigma^2 \\ 1 & -1 & 0 \\ 0 & 1 & 0 \end{bmatrix} \equiv \begin{bmatrix} 1 & 0 & 0 \\ 0 & 1 & 0 \\ 0 & 0 & \sigma^2 \end{bmatrix},$$

therefore $IZ_{\infty} \cong \tilde{C}_2$, and the set of elementary divisors of IZ_{∞} is $\{\sigma^2\}$. Remark 1096 is still valid.

7.4.5 System Zeros at Infinity

The module of system zeros at infinity is not defined. However, the *degree of the system zeros at infinity* (denoted by $\#(SZ_{\infty})$ is defined as follows:

Definition 1104. $\#(SZ_{\infty}) = \#(TZ_{\infty}) + \#(HM_{\infty})$.

Example 1105. *(Example 1078 cont'd.) We have*

$$\#(SZ_{\infty}) = 0 + 2 = 2.$$

7.4.6 Transmission Poles and Transmission Zeros at Infinity

Lemma and Definition 1106. *(1) Consider the two torsion modules* **S***-modules TP_∞ and TZ_∞ below:*

$$TP_\infty = \frac{\Phi^+ \cap [y^+, u^+]_{\mathbf{S}}}{[u^+]_{\mathbf{S}}}, \quad TZ_\infty = \mathcal{T}\left(\frac{\Phi^+ \cap [y^+, u^+]_{\mathbf{S}}}{\Phi^+ \cap [y^+]_{\mathbf{S}}}\right).$$

These modules TP_∞ and TZ_∞ are called the module of transmission poles at infinity *and the* module of transmission zeros at infinity, *respectively.*
(2) The set $\varepsilon(TP_\infty)$ (resp., $\varepsilon(TZ_\infty)$) of all elementary divisors of TP_∞ (resp., TZ_∞) is given by $\{\sigma^{\pi_i}, 1 \le i \le s\}$ (resp., $\{\sigma^{\varsigma_i}, 1 \le i \le \rho\}$) where $\{\pi_i : 1 \le i \le s\}$ (resp., $\{\varsigma_i : 1 \le i \le \rho\}$) is the set of structural indices of the transmission pole (resp., the transmission zero) at infinity (Definition 1064).

Proof. (2) (a) The matrix $R^+(\sigma)$ can be put into the form $L(\sigma) R_0^+(\sigma)$ where $R_0^+(\sigma)$ has no elementary divisor. Then, Φ^+ is described by the equation

$$R_0^+(\sigma) \begin{bmatrix} \xi^+ \\ u^+ \\ y^+ \end{bmatrix} = 0.$$

To simplify the notation, assume that $R^+(\sigma)$ has no elementary divisor, i.e., that $M^+ = \Phi^+$. Then,

$$TP_\infty = \frac{[y^+, u^+]_{\mathbf{S}}}{[u^+]_{\mathbf{S}}}, \quad TZ_\infty = \mathcal{T}\left(\frac{[y^+, u^+]_{\mathbf{S}}}{[y^+]_{\mathbf{S}}}\right)$$

and $R_0^+(\sigma) = R^+(\sigma)$.

(b) The equation (7.42) of $M^+ = \Phi^+$ can be written

$$\begin{bmatrix} D^+(\sigma) \\ Q^+(\sigma) \end{bmatrix} \xi^+ + \begin{bmatrix} -N^+(\sigma) & Z^+(\sigma) \\ W^+(\sigma) & Y^+(\sigma) \end{bmatrix} \begin{bmatrix} u^+(\sigma) \\ y^+(\sigma) \end{bmatrix} = 0, \tag{7.47}$$

thus $M^+ / [y^+, u^+]_{\mathbf{S}}$ is given by the equation

$$\begin{bmatrix} D^+(\sigma) \\ Q^+(\sigma) \end{bmatrix} \bar{\xi}^+ = 0.$$

This module is zero (i.e., $M^+ = [y^+, u^+]_{\mathbf{S}}$) if, and only if $\begin{bmatrix} D^+(\sigma) \\ Q^+(\sigma) \end{bmatrix}$ is left-invertible, i.e., $D^+(\sigma)$ and $Q^+(\sigma)$ are right-coprime. If this does not hold, let $R(\sigma)$ be a gcrd of $D^+(\sigma)$ and $Q^+(\sigma)$; we can write $D^+(\sigma) = D_0^+(\sigma) R(\sigma)$ and $Q^+(\sigma) = Q_0^+(\sigma) R(\sigma)$ where $D_0^+(\sigma)$ and $Q_0^+(\sigma)$ are right-coprime, i.e., $\begin{bmatrix} D_0^+(\sigma) \\ Q_0^+(\sigma) \end{bmatrix}$ is left-invertible. Then (7.47) can be written

$$\begin{bmatrix} D_0^+(\sigma) \\ Q_0^+(\sigma) \end{bmatrix} \xi'^+ + \begin{bmatrix} -N^+(\sigma) & Z^+(\sigma) \\ W^+(\sigma) & Y^+(\sigma) \end{bmatrix} \begin{bmatrix} u^+(\sigma) \\ y^+(\sigma) \end{bmatrix} = 0$$

where $\xi'^+ = R(\sigma)\xi^+$. To simplify the notation, assume that $D^+(\sigma)$ and $Q^+(\sigma)$ are right-coprime. Then, $\begin{bmatrix} D^+(\sigma) \\ Q^+(\sigma) \end{bmatrix} = \begin{bmatrix} D_0^+(\sigma) \\ Q_0^+(\sigma) \end{bmatrix}$ is left-invertible and $\xi^+ = \xi'^+$ is such that $[\xi^+]_{\mathbf{S}} \lhd [y^+, u^+]_{\mathbf{S}}$.

(c) The module $[y^+, u^+]_{\mathbf{S}}$ has an equation of the form

$$A^+(\sigma)\, y^+ = B^+(\sigma)\, u^+.$$

This module is free, therefore $A^+(\sigma)$ and $B^+(\sigma)$ are left-coprime. In addition, with the notation in Subsect. 7.4.2, $[\breve{y}^+]_{\mathbf{L}} \lhd [\breve{u}^+]_{\mathbf{L}}$, and since $\breve{u}^+$ is a basis of $[\breve{u}^+]_{\mathbf{L}}$, there exists a unique matrix $G^+(\sigma)$ such that $\breve{y}^+ = G^+(\sigma)\,\breve{u}^+$. Therefore, $A^+(\sigma)$ is invertible over $\mathbf{L}$ and $G^+(\sigma) = A^+(\sigma)^{-1} B^+(\sigma)$. In addition, $(A(\sigma), R^+(\sigma))$ is a left-coprime factorization of $R(\partial)$ over $\mathbf{S}$ (**7.3.6.2**), therefore $\breve{y}^+ = G(\partial)\,\breve{u}^+$, which proves that $G^+(\sigma) = G(\partial)$.

(d) The module TP_∞ is described by the equation $A^+(\sigma)\,\overline{y^+} = 0$, where the entries of $\overline{y^+}$ are the canonical images of the entries of y^+ in $[y^+, u^+]_{\mathbf{S}} / [u^+]_{\mathbf{S}}$, therefore the elementary divisors of TP_∞ are those of $A^+(\sigma)$. Likewise, the elementary divisors of TZ_∞ are those of $B^+(\sigma)$, and (2) is proved since the left fraction $G(\partial) = A^+(\sigma)^{-1} B^+(\sigma)$ is irreducible over $\mathbf{S}$. ∎

Example 1107. *(Example 1078 cont'd.) The module M^+ is described by $\begin{bmatrix} \operatorname{diag}(1,1,\sigma^2) & 0 \end{bmatrix} v^+ = 0$ where v^+ and w^+ are related by (7.30) and $w^+ = \begin{bmatrix} u^+ & \xi_1^+ & \xi_2^+ & y^+ \end{bmatrix}$. The equations can be written into the form*

$$\begin{cases} \sigma^2 v_3^+ = \sigma^2\left(u^+ - t\xi_1^+ - y^+\right) = 0, \\ \sigma^2 y^+ = \xi_1^+, \\ \xi_1^+ = \xi_2^+. \end{cases}$$

The second equation implies $t\sigma^2 y^+ = t\xi_1^+$. We have $\Phi \cong M^+/\mathcal{T}(M^+)$ where $\mathcal{T}(M^+) = [v_3^+]_{\mathbf{S}}$. Let $\bar{\xi}_1^+, \bar{\xi}_2^+, \bar{u}^+$ and $\bar{y}^+$ be the canonical images of ξ_1^+, ξ_2^+, u^+ and y^+, respectively, in $M^+/\mathcal{T}(M^+)$. We obtain

$$\begin{cases} \bar{u}^+ - \left(1 + t\sigma^2\right)\bar{y}^+ = 0, \\ \sigma^2 \bar{y}^+ = \bar{\xi}_1^+, \\ \bar{\xi}_1^+ = \bar{\xi}_2^+, \end{cases}$$

therefore $\bar{\xi}_1^+ = \bar{\xi}_2^+$ belongs to $[\bar{y}^+, \bar{u}^+]_{\mathbf{S}}$, and this module is defined by the equation $\bar{u}^+ - \left(1 + t\sigma^2\right)\bar{y}^+ = 0$. Therefore, $TP_\infty \cong [\bar{y}^+, \bar{u}^+]_{\mathbf{S}} / [\bar{u}^+]$, the equation of which is $\left(1 + t\sigma^2\right)\breve{y}^+$ where $\breve{y}^+$ is the canonical image of $\bar{y}^+$ in $[\bar{y}^+, \bar{u}^+]_{\mathbf{S}} / [\bar{u}^+]$. Since $1 + t\sigma^2$ is a unit of $\mathbf{S}$, $TP_\infty = 0$. Likewise, $ZP_\infty \cong [\bar{y}^+, \bar{u}^+]_{\mathbf{S}} / [\bar{y}^+] = 0$.

Remark 1096 is still valid with this definition of TP_∞ and ZP_∞ (whereas it is not with Definition 1064(i)).

7.4.7 Relations between the Various Poles and Zeros at Infinity

We have already seen how $\varepsilon(HM_\infty)$, $\varepsilon(IDZ_\infty)$, $\varepsilon(ODZ_\infty)$ and $\varepsilon(IODZ_\infty)$ are related (Theorem 1100). In addition, we have:

Theorem 1108. *(1) In the general case, the following equalities hold:*
(i)

$$\#(SP_\infty) = \#(HM_\infty) + \#(TP_\infty).$$

(ii)

$$\#(TZ_\infty) + \#(IODZ_\infty) \leq \#(IZ_\infty) \leq \#(SZ_\infty).$$

(2) Assume that $\mathbf{K}$ *is a division ring, let* $G(\partial) \in {}^p\mathbf{Q}^m$ *be the transfer matrix of the temporal system, and let* $r = \mathrm{rk}_{\mathbf{Q}}(G(\partial))$.
(iii) If $r = p$ *(i.e., if* $G(\partial)$ *is right-invertible), then*

$$\varepsilon(TZ_\infty)\,\dot{\cup}\,\varepsilon(IDZ_\infty) \subseteq \varepsilon(IZ_\infty)$$

and in particular

$$\#(TZ_\infty) + \#(IDZ_\infty) \leq \#(IZ_\infty)$$

(iv) If $r = m$ *(i.e., if* $G(\partial)$ *is left-invertible), then*

$$\#(TZ_\infty) + \#(ODZ_\infty) \leq \#(IZ_\infty).$$

(v) If $r = p = m$ *(i.e., if* $G(\partial)$ *is square and invertible), then*

$$\#(TZ_\infty) + \#(HM_\infty) = \#(IZ_\infty) = \#(SZ_\infty).$$

Proof. (1), (i) We have $[u^+]_{\mathbf{S}} \lhd \Phi^+ \cap [u^+, y^+]_{\mathbf{S}} \lhd M^+$, therefore by Lemma 1099(ii),

$$\#(M^+/[u^+]_{\mathbf{S}}) = \#(M^+/(\Phi^+ \cap [u^+, y^+]_{\mathbf{S}})) + \#((\Phi^+ \cap [u^+, y^+]_{\mathbf{S}})/[u^+]_{\mathbf{S}}).$$

(ii), (a): Let $T^+ = \mathcal{T}(M^+)$. We have by Lemma 1093(ii) and by (2.9)

$$\begin{aligned}\frac{M^+}{[y^+]_{\mathbf{S}}} &= \frac{T^+ \oplus \Phi^+}{(T^+ \cap [y^+]_{\mathbf{S}}) \oplus (\Phi^+ \cap [y^+]_{\mathbf{S}})} \\ &\cong \frac{T^+}{(T^+ \cap [y^+]_{\mathbf{S}})} \oplus \frac{\Phi^+}{(\Phi^+ \cap [y^+]_{\mathbf{S}})}.\end{aligned}$$

In addition,

$$\frac{\Phi^+ \cap [y^+, u^+]_{\mathbf{S}}}{(\Phi^+ \cap [y^+]_{\mathbf{S}})} \lhd \frac{\Phi^+}{(\Phi^+ \cap [y^+]_{\mathbf{S}})},$$

therefore

$$\#\left(\mathcal{T}\left(\frac{\Phi^+}{(\Phi^+ \cap [y^+]_{\mathbf{S}})}\right)\right) \geq \#\left(TZ_\infty\right).$$

Since $T^+ \cap [y^+]_{\mathbf{S}} \lhd T^+ \cap [y^+, u^+]_{\mathbf{S}} \lhd T^+$, we have by Lemma 1099(ii)

$$\begin{aligned}\#\left(\frac{T^+}{T^+ \cap [y^+]_{\mathbf{S}}}\right) &= \#\left(\frac{T^+}{T^+ \cap [y^+, u^+]_{\mathbf{S}}}\right) + \#\left(\frac{T^+ \cap [y^+, u^+]_{\mathbf{S}}}{T^+ \cap [y^+]_{\mathbf{S}}}\right) \\ &\geq \#\left(\frac{T^+}{T^+ \cap [y^+, u^+]_{\mathbf{S}}}\right) = \#\left(IODZ_\infty\right),\end{aligned}$$

and the left equality of (ii) is proved.

(b): We have

$$\frac{M^+}{[y^+, u^+]_{\mathbf{S}}} = \frac{\Phi^+}{\Phi^+ \cap [y^+, u^+]_{\mathbf{S}}} \oplus \frac{T^+}{T^+ \cap [y^+, u^+]_{\mathbf{S}}}$$

therefore, putting $T_1^+ = \Phi^+ / (\Phi^+ \cap [y^+, u^+]_{\mathbf{S}})$,

$$\#\left(T_1^+\right) = \#\left(ODZ_\infty\right) - \#\left(IODZ_\infty\right).$$

In addition,

$$\begin{aligned}\mathcal{T}\left\{\frac{M^+}{[y^+]_{\mathbf{S}}}\right\} &= \mathcal{T}\left\{\frac{\Phi^+}{\Phi^+ \cap [y^+]_{\mathbf{S}}}\right\} \oplus \frac{T^+}{T^+ \cap [y^+]_{\mathbf{S}}}, \\ \#\left(\mathcal{T}\left\{\frac{\Phi^+}{\Phi^+ \cap [y^+]_{\mathbf{S}}}\right\}\right) &= \#\left(\frac{\Phi^+}{\Phi^+ \cap [y^+, u^+]_{\mathbf{S}}}\right) + \#\left(\frac{\Phi^+ \cap [y^+, u^+]_{\mathbf{S}}}{\Phi^+ \cap [y^+]_{\mathbf{S}}}\right),\end{aligned}$$

therefore

$$\begin{aligned}\#\left(IZ_\infty\right) &= \#\left(T_1^+\right) + \#\left(TZ_\infty\right) + \#\left(\frac{T^+}{T^+ \cap [y^+]_{\mathbf{S}}}\right) \\ &= \#\left(ODZ_\infty\right) - \#\left(IODZ_\infty\right) + \#\left(TZ_\infty\right) + \#\left(\frac{T^+}{T^+ \cap [y^+]_{\mathbf{S}}}\right).\end{aligned}$$

Since $\#\left(T^+/T^+ \cap [y^+]_{\mathbf{S}}\right) \leq \#\left(IDZ_\infty\right)$, the right inequality of (ii) is proved.

(2) Let us embed $\mathbf{Q}$ in $\mathbf{L}$. With the notation in the proof of Lemma and Definition 1106, we have $\breve{y}^+ = G(\partial)\,\breve{u}^+$.

(iii) If $r = p$, then $\dim [\breve{y}^+]_{\mathbf{L}} = p$, therefore $[y]_{\mathbf{S}}$ is free of rank p by Corollary and Definition 341, so that

$$\mathcal{T}\left\{\frac{M^+}{[y^+]_{\mathbf{S}}}\right\} \cong \mathcal{T}\left\{\frac{\Phi^+}{[y^+]_{\mathbf{S}}}\right\} \oplus T^+ \supseteq \mathcal{T}\left\{\frac{\Phi^+ \cap [y^+, u^+]_{\mathbf{S}}}{[y^+]_{\mathbf{S}}}\right\} \oplus T^+.$$

(iv) If $r = m$, then $M^+ / [y^+]_{\mathbf{S}}$ is torsion. By Lemma 1099(ii),

$$\begin{aligned} \#\left(\frac{M^+}{[y^+]_{\mathbf{S}}}\right) &= \#\left(\frac{M^+}{[y^+, u^+]_{\mathbf{S}}}\right) + \#\left(\frac{[y^+, u^+]_{\mathbf{S}}}{[y^+]_{\mathbf{S}}}\right) \\ &\geq \#\left(\frac{M^+}{[y^+, u^+]_{\mathbf{S}}}\right) + \#\left(TZ_\infty\right). \end{aligned} \tag{7.48}$$

(v) If $r = m = p$, then $M^+ / [y^+]_{\mathbf{S}}$ is torsion and $[y]_{\mathbf{S}}$ is free. Therefore,

$$\frac{[y^+, u^+]_{\mathbf{S}}}{[y^+]_{\mathbf{S}}} \cong \left(T^+ \cap [y^+, u^+]_{\mathbf{S}}\right) \oplus \frac{\Phi^+ \cap [y^+, u^+]_{\mathbf{S}}}{[y^+]_{\mathbf{S}}}.$$

In addition, $0 \lhd \left(T^+ \cap [y^+, u^+]_{\mathbf{S}}\right) \lhd T^+$, therefore by Lemma 1099(ii),

$$\begin{aligned} \#\left(T^+ \cap [y^+, u^+]_{\mathbf{S}}\right) &= \#\left(T^+\right) - \#\left(\frac{T^+}{T^+ \cap [y^+, u^+]_{\mathbf{S}}}\right) \\ &= \#\left(IDZ_\infty\right) - \#\left(IODZ_\infty\right), \end{aligned}$$

so that

$$\#\left(\frac{[y^+, u^+]_{\mathbf{S}}}{[y^+]_{\mathbf{S}}}\right) = \#\left(TZ_\infty\right) + \#\left(IDZ_\infty\right) - \#\left(IODZ_\infty\right).$$

(v) is a consequence the above equality and of (7.48). ■

7.5 Exercises

Exercise 1109. *Let $A(\partial) \in {}^r\mathbf{A}^{k_1}$, $B(\partial) \in {}^r\mathbf{A}^{k_2}$ where $\mathbf{A} = \mathbf{K}[\partial; \boldsymbol{\alpha}, \boldsymbol{\delta}]$ and $\mathbf{K}$ is a division ring. Let $R(\partial) = \begin{bmatrix} A(\partial) & B(\partial) \end{bmatrix}$. Assuming that $\operatorname{rk} R(\partial) = r$, let $(D(\sigma), R^+(\sigma))$ be a left-coprime factorization of $R(\partial)$ over $\mathbf{S}$, and set $R^+(\sigma) = \begin{bmatrix} A^+(\sigma) & B^+(\sigma) \end{bmatrix}$, $A^+(\sigma) \in {}^r\mathbf{S}^{k_1}$, $B^+(\sigma) \in {}^r\mathbf{S}^{k_1}$.*

(i) Prove that the Smith-MacMillan form of $R^+(\sigma)$ over $\mathbf{S}$ is $\begin{bmatrix} I_r & 0 \end{bmatrix}$ if, and only if $R(\partial)$ has no zero at infinity.
(ii) Deduce that $A^+(\sigma), B^+(\sigma)$ are left-coprime if, and only if $\begin{bmatrix} A(\partial) & B(\partial) \end{bmatrix}$ has no zero at infinity.
(iii) Generalize the above to the case when $\mathbf{K}$ is a Noetherian domain and $R(\partial)$ is regular at infinity.

Exercise 1110. *Let $\mathbf{K}$ be a division ring, and let $\mathbf{A} = \mathbf{K}(\partial; \boldsymbol{\alpha}, \boldsymbol{\delta})$. Let $A(\partial) \in {}^q\mathbf{Q}^m$ and $B(\partial) \in {}^q\mathbf{Q}^p$ be two matrices of rank r, and set $F(\partial) = \begin{bmatrix} A(\partial) \\ B(\partial) \end{bmatrix}$.*

(i) Assuming that the Smith-MacMillan form at infinity of $A(\partial)$ and of $B(\partial)$ are $\operatorname{diag}(\sigma^{\nu_1}, ..., \sigma^{\nu_r}, 0, ..., 0)$ and $\operatorname{diag}(\sigma^{\lambda_1}, ..., \sigma^{\lambda_r}, 0, ..., 0)$, respectively, show that the Smith-MacMillan form at infinity of $F(\partial)$ is $\operatorname{diag}(\sigma^{\varepsilon_1}, ..., \sigma^{\varepsilon_r}, 0, ..., 0)$ with $\varepsilon_i = \min\{\nu_i, \lambda_i\}, 1 \leq i \leq r$.

(ii) Deduce from (i) that if $F(\partial)$ has no pole at infinity, then $A(\partial)$ and $B(\partial)$ have the same property, and that $c_\infty(F) \geq \max\{c_\infty(A), c_\infty(B)\}$.
(iii) Show that $\boldsymbol{\delta}_{\mathbf{M}}\left(\begin{bmatrix} A(\partial) \\ I_m \end{bmatrix}\right) = \boldsymbol{\delta}_{\mathbf{M}}(A(\partial))$.
(Hint: for (i), $B(\partial)$ and $B(\partial)V$ have the same Smith-MacMillan form over $\mathbf{S}$ if $V \in \mathbf{GL}_m(\mathbf{S})$; use row elementary operations once $A(\partial)$ and $B(\partial)V$ have been reduced to their Smith-MacMillan form over $\mathbf{S}$, where V is a suitable element of $\mathbf{GL}_m(\mathbf{S})$.)

Exercise 1111. *Assume that $\mathbf{K} = \mathbf{k}$. Let $\mathbf{\Sigma}_1$ be the system $(\partial + 1)y = u$, $\mathbf{\Sigma}_2$ be the system $\bar{y} = \partial^2 \bar{u}$, and consider the interconnection $\bar{y} = u$ (i.e., $\mathbf{\Sigma}_2 \to \mathbf{\Sigma}_1$), active only for $t \in \mathbb{T}_0$. The resulting temporal system is denoted by $\mathbf{\Sigma}$.*

(1) Uncontrollable poles at infinity: *(i) What kind of "pole-zero cancellation at infinity" does arise when $\mathbf{\Sigma}$ is formed? (ii) Check that $\mathcal{T}(M^+) \cong \tilde{C}_1 = \mathbf{k}\tilde{\delta}$. (iii) "Where", in $\mathbf{\Sigma}$, does the uncontrollable impulsive behavior $\mathbf{k}\delta$ arise?*
(2) System poles at infinity: *Show $\mathbf{\Sigma}$ has one transmission pole at infinity of order 1 and that $SP_\infty \cong \tilde{C}_2$.*
(3) Unobservable poles at infinity: *Consider the same systems $\mathbf{\Sigma}_1$ and $\mathbf{\Sigma}_2$ as above, and the temporal interconnection $y = \bar{u}$ (i.e., $\mathbf{\Sigma}_1 \to \mathbf{\Sigma}_2$), active only for $t \in \mathbb{T}_0$. The resulting temporal system is denoted by $\mathbf{\Sigma}'$. (i') What kind of "pole-zero cancellation at infinity" does arise when $\mathbf{\Sigma}'$ is formed? (ii') Check that $M^+/[y^+, u^+]_{\mathbf{S}} \cong \tilde{C}_1 = \mathbf{k}\tilde{\delta}$. (iii') "Where", in $\mathbf{\Sigma}'$, does the "unobservable impulsive behavior" $\mathbf{k}\delta$ arise?*

Exercise 1112. *Assume that $\mathbf{K} = \mathbf{k}$. Let $\mathbf{\Sigma}_1$ be the system $y = \partial^2 u$, let $\mathbf{\Sigma}_2$ be the system $\partial^3 \bar{y} = \partial \bar{u}$, and consider the interconnection $y = \bar{u}$ (i.e., $\Sigma_1 \to \Sigma_2$), active only for $t \in \mathbb{T}_0$. (i) What kind of "pole-zero cancellation at infinity" does arise when the temporal system is formed? (ii) Calculate IDZ_∞, ODZ_∞, $IODZ_\infty$, HM_∞, SP_∞ and IZ_∞. (Answers: $IDZ_\infty = IODZ_\infty = 0$, $ODZ_\infty \cong HM_\infty \cong SP_\infty \cong IZ_\infty \cong \tilde{C}_2$.)*

Exercise 1113. *Assume that $\mathbf{K} = \mathbf{k}[t]$. Consider the temporal system, the matrix of definition $R(\partial)$ of which is given by (7.40) with*

$$D(\partial) = \begin{bmatrix} 1 & 0 \\ t\partial^3 & \partial^2 \end{bmatrix}, \quad N(\partial) = \begin{bmatrix} 0 \\ (t-1)\partial \end{bmatrix},$$
$$Q(\partial) = \begin{bmatrix} t\partial & t^2\partial \end{bmatrix}, \quad W(\partial) = t^2\partial.$$

(i) Is $R(\partial)$ is regular at infinity?
(ii) Calculate SP_∞, IDZ_∞, ODZ_∞, $IODZ_\infty$, HM_∞, IZ_∞, TP_∞, TZ_∞ and $\#(SZ_\infty)$. Interpretation?
(Answers: (i): Yes. (ii): $SP_\infty \cong \tilde{C}_1 \oplus \tilde{C}_1$, $IDZ_\infty \cong ODZ_\infty \cong IODZ_\infty \cong HM_\infty \cong IZ_\infty \cong TP_\infty \cong \tilde{C}_1$, $TZ_\infty = 0$ and $\#(SZ_\infty) = 1$.

7.6 Notes

The "structure at infinity" of matrix pencils goes back to Kronecker and Weierstrass (see [132], Chap. XII) who introduced the notion of infinite elementary divisor. The connection between this notion and the impulsive motions which may arise in a system was pointed out by Verghese in his Ph.D. thesis [346] and in several papers, some of them written in collaboration with coauthors–especially his advisor T. Kailath [348], [350], [347], [349]. In these papers, the notion of MacMillan degree (which goes back to MacMillan [248]) was also clarified The following result was proved in [349]: let $\boldsymbol{\delta}_{\mathrm{M}}$ (resp., $\boldsymbol{\delta}_{\mathbf{z}}$) be the number of poles (resp., of zeros), finite and infinite, of a transfer matrix $G(s)$ ($\boldsymbol{\delta}_{\mathrm{M}}$ is the MacMillan degree); then, $\boldsymbol{\delta}_{\mathrm{M}} - \boldsymbol{\delta}_{\mathbf{z}} \geq 0$ in general, with equality when $G(s)$ is square and nonsingular; $\boldsymbol{\delta}_{\mathbf{P}} - \boldsymbol{\delta}_{\mathbf{z}}$ is called the *defect* of $G(s)$ ([170], Subsect. 6.5.4). Usual elementary row and column operations (over the ring $\mathbf{A}$) do not preserve the structure at infinity, as emphasized by Pugh and Ratcliffe [284]. Based on this observation, the study of the various kinds of poles and zeros at infinity (including hidden modes at infinity) was completed by Vardulakis [341], by van der Weiden and Bosgra [358], and by Ferreira [112]. The Smith-MacMillan form at infinity of a general polynomial matrix (with constant coefficients) was studied by Vardulakis *et al.* [344]. These results are presented in Vardulakis' monograph [345].

Full equivalence of two PMDs was defined by Hayton *et al.* [159], in the case $\mathbf{K} = \mathbf{k}$, and for continuous-time systems, by the necessary and sufficient condition in Theorem 1089. This is a refinement of *restricted equivalence* and of *strong equivalence*, introduced by Verghese [346] and by Bosgra and van der Weiden [25], respectively. An interpretation of full equivalence in terms of the existence of a bijective map between the finite and infinite solution sets of the differential equations describing two systems was given later by Pugh *et al.* [283]. The characteristics of a system left invariant by full equivalence were studied by Karampetakis and Vardulakis [178]. See also, e.g., Karampetakis *et al.* [179], [180], [177].

The extension of poles, zeros and hidden modes at infinity to time-varying systems over $\mathbf{A} = \mathbf{K}[\partial, \delta]$, where $\mathbf{K}$ is a differential field, was carried out by Bourlès and Marinescu [46] using module theory. They (implicitly) introduced the notion of impulsive system. The notions of temporal system, of temporal interconnection and of regularity at infinity are due to Bourlès [40], [42], who showed that the continuous-time case and the discrete-time case can be jointly treated. Theorem 1108 was announced in [46] by analogy with the result on the relations between finite poles and zeros obtained by Bourlès and Fliess [44] (this result was slightly improved in [43]). Theorem 1085 was obtained in [40]. The causal Laplace transform and the anticausal Z-transform in Subsect. 7.3.4 were developed in [39]. Definition 1087 (i.e., the general definition of full equivalence) and Theorem 1089 were obtained in [42].

Part III
Applications

8 Analysis of LTV Systems

8.1 Introduction

Many engineering problems need a time-varying modeling and some examples are shown in this section. The most common class of LTV systems is the one of periodic systems. They are encountered in signal processing and communications as, for example, filters which incorporate modulators in the signal path (see, *e.g.*, [252] in which spectral characterizations of linear periodically time-varying systems are used to the study of gated phased-locked loops), as well as in control.

However, the physical time-dependance of some parameters is not the only reason for studying LTV systems. Indeed, LTV representations are obtained when simplifying complicated models as it is the case of the linearization of the nonlinear dynamics around a given trajectory. In [255] LTV models are used for the linear analysis of electrical circuits while [130] shows that a simplified observer design can be carried out using an LTV model.

Linear Parameter-Varying (LPV) systems form a special class of LTV systems in which the coefficients of the system depend on a time-varying parameter. LPV systems arise in the gain scheduling ([310], [212]). Again, despite the supplementary difficulty introduced by the time-dependance of the parameters, this kind of modeling can lead to important simplifications of the problem. In [287], the LPV modeling applied to the voltage control of the power systems leads to a decentralized control scheme. Other industrial LPV applications are reported in [15] and [365].

We continue here by presenting three industrial applications for which the time-varying modeling is mandatory. It is also shown how the algebraic approach presented in Part II is used to analyse the structural properties of some industrial systems.

H. Bourlès and B. Marinescu: Linear Time-Varying Systems, LNCIS 410, pp. 499–514.
springerlink.com

8.2 Need of Time-Varying Models

8.2.1 *Current-Mode Control of a Converter*

As power electronics has been increasingly used in the control of electrical drives in the last decades, modeling of current-mode controlled converters has been a topic of interest for the control and power electronics communities.

A boost converter as considered in [338] is shown in Fig. 8.1 where D defines the impulse modulation.

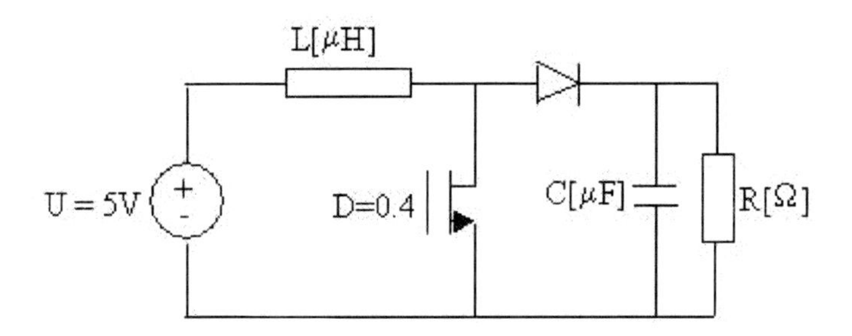

Fig. 8.1 Example boost converter

Under normal operating conditions, the switch is turned on every T seconds, and is turned off when the inductor current $i_L(t)$ reaches a peak value of the control signal $i_p(t)$ minus a compensation ramp. For the control design, the relation between a perturbation $\widehat{i}_p(t)$ in the control signal and the resulting perturbation $\widehat{i}_L(t)$ is investigated. The modeling usually adopted (see, *e.g.*, [293], [338]) is based on the assumption that the input and output voltages do not vary significantly and, as a consequence, the relation between the perturbation in control and the resulting current perturbation can be approximated by a sample-and-hold system as given in Fig. 8.2.

This leads to a linear time-invariant discrete-time model

$$H(z) = \frac{\Delta i_L(z)}{\Delta i_p(z)} = \frac{(M_1 + M_2)z}{(M_c + M_1)z - (M_c - M_2)} \tag{8.1}$$

where M_1, M_2 and M_c are the slope magnitudes of the rising inductor current, falling inductor current, and slope-compensation ramp.

The main effects not modeled by the sample-and-hold approximation are the variation in sampling time and the finite slope of the current perturbation transition.

This modeling is correct at low-frequencies. Its flaws at high frequencies were pointed out in [272] where it is shown that the system in Fig. 8.2 is time-varying for control perturbations approaching half the switching frequency.

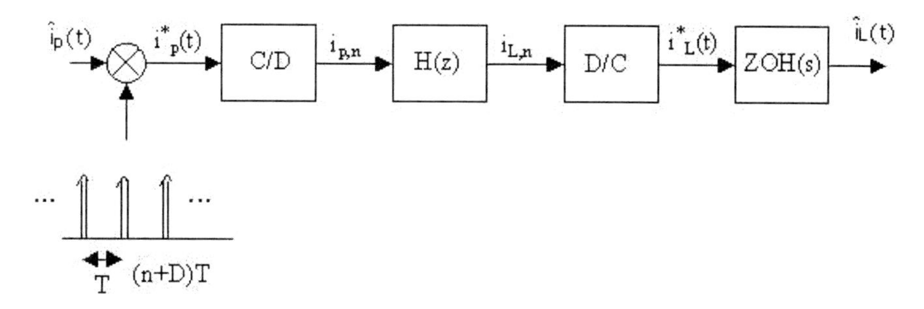

Fig. 8.2 System for modeling the relation between perturbations in control and perturbations in current

Indeed, the sampling process, which is modeled by impulse modulation, generates replicas of the input frequency spectrum centered at multiples of the sampling (switching) frequency, $f_{sw} = 1/T$:

$$\widehat{i_p^*}(s) = \frac{1}{T} \sum_{n=-\infty}^{\infty} \widehat{i}_p(s + \frac{2\pi nj}{T}). \tag{8.2}$$

The effects of the replicas are ignored in (8.1) where the approximation $\widehat{i_p^*}(s) \approx \frac{1}{T}\widehat{i}_p(s)$ is used instead of (8.2). If for low-frequency this approximation is justified since the low-pass zero-order hold in Fig. 8.2 filters the high order frequency replicas, it is more critical for higher frequency perturbations since the zero-order holder filtering is less efficient and thus contributions due to the replicas are present at the output. The response of the system in Fig. 8.2 at these frequencies depends on the position of the control signal with respect to the sampling points. It follows that a time-varying modeling is necessary for the analysis in this case.

8.2.2 Modeling Highway Vehicles with Time-Varying Velocity

In an automated highway system the steering control of a vehicle is based on a model which mainly captures the lateral dynamics of the vehicle and, in particular, the interaction between the tires and the road surface. Traditionally, a linear time-invariant model is obtained by linearizing the dynamics assuming that the vehicle is traveling at a constant velocity. This model does not cover the situation when a vehicle changes highway lanes while accelerating. In [261] such a different approach is proposed on the basis of a linear time-varying model developed by linearizing the nonlinear dynamics for a specified velocity trajectory. This way of modeling and the main differences with a traditional time-invariant approach are summarized here as an example.

Consider a time-varying velocity defined by

$$\dot{v}_x(t) = \begin{cases} t, & 0 \le t < 2 \\ 2, & 2 \le t < 7.5 \\ 2 - (t - 7.5), & 7.5 \le t < 9.5 \\ 0, & 9.5 \le t < 12 \end{cases} \tag{8.3}$$

or, in other words, a given acceleration profile. The LTV model consists in the linearization of the full nonlinear model around the trajectory (8.3).

The nonlinear model consists in the equations of the mass center of the vehicle and the tire/road interface and is schematically given in Fig. 8.3.

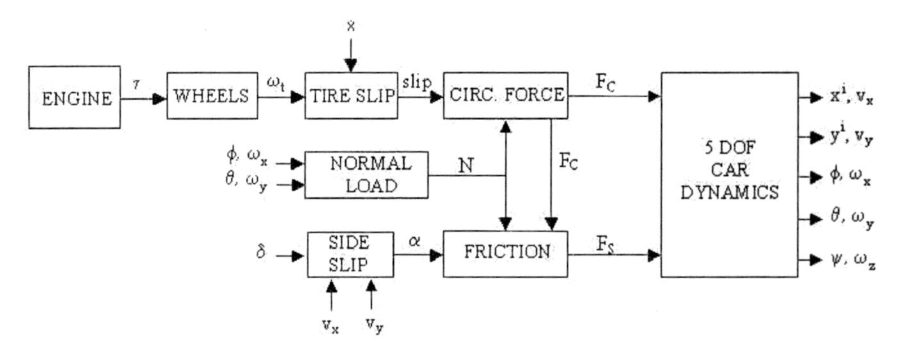

Fig. 8.3 Block diagram of the nonlinear model of the highway vehicle

The inputs are τ the engine/braking torque and δ the front wheel steering angle. The tire/road interface is described by ω_t the wheel angular velocity, SLIP the longitudinal (tire) slip, α the side slip angle, F_C the circumferential (tire) force, and F_S the side (tire) force. The motion of the center of mass is described in the inertial frame by x^i the longitudinal position, y^i the lateral position, and ψ the yaw angle. In the body frame, the motion of the center of mass is descibed by the following five degrees of freedom: ϕ the roll angle, ω_x the roll rate, θ the pitch angle, ω_y the pitch rate and ω_z the yaw rate. The equations of the center of the mass are

$$\begin{bmatrix} \dot{x}^i \\ m\dot{v}_x \\ \dot{y}^i \\ m\dot{v}_y \\ \dot{\phi} \\ J_x\dot{\omega}_x \\ \dot{\theta} \\ J_y\dot{\omega}_y \\ \dot{\psi} \\ J_z\dot{\omega}_z \end{bmatrix} = \begin{bmatrix} v_x cos\phi - v_y sin\phi \\ m v_y \omega_y \\ v_x sin\phi + v_y cos\phi \\ -m v_x \omega_z \\ \omega_x \\ (J_y - J_z)\omega_y\omega_z \\ \omega_y \\ (J_z - J_x)\omega_x\omega_z \\ \omega_z \\ (J_x - J_y)\omega_x\omega_y \end{bmatrix} + \begin{bmatrix} 0 \\ \Sigma F_x^b \\ 0 \\ \Sigma F_y^b \\ 0 \\ \Sigma M_x^b \\ 0 \\ \Sigma M_y^b \\ 0 \\ \Sigma M_z^b \end{bmatrix} \tag{8.4}$$

where J is the moment of inertia about each axis, ΣF^b is the sum of forces along each axis in the body frame, and ΣM^b is the sum of the moments about each axis in the body frame.

The interaction between the tire and the road generates the forces and the moments in (8.4). Their expressions depend on the assumptions made on the time-dependancy of the velocity. For example, the normal load N of each tire is the fraction of the vehicle weight supported by each tire. For a constant velocity, the normal load distribution is static and given by

$$N_f^0 = \frac{mgl_2}{2(l_1 + l_2)}, \quad N_r^0 = \frac{mgl_1}{2(l_1 + l_2)} \tag{8.5}$$

where N_f^0 (respectively N_r^0) is the normal load on the front (respectively rear) tires and the length l_1 (respectively l_2) represents the distance from the front (respectively rear) axle to the center of mass. Indeed, at constant velocity, there is no roll motion and, as a result, the normal load is equally distributed between the right and left side of the vehicle. This is no longer valid if one considers time-varying velocity. In this case, the normal load depends on the vehicle's suspension. The suspension is modeled as a mass-spring-damper system with a spring constant K and a damper constant β. The normal force is

$$\begin{aligned} N_{FL} &= N_f^0 - (Kz_{FL} + \beta \dot{z}_{FL}) \\ N_{FR} &= N_f^0 - (Kz_{FR} + \beta \dot{z}_{FR}) \\ N_{RL} &= N_r^0 - (Kz_{RL} + \beta \dot{z}_{RL}) \\ N_{RR} &= N_r^0 - (Kz_{RR} + \beta \dot{z}_{RR}) \end{aligned} \tag{8.6}$$

where z is the suspension deflection at each tire. This deflection depends on the pitch and roll angles.

The angular velocity ω_t is given by the torque produced by the engine or braking T_e, the circumferential force F_C and the friction force F_f:

$$J_t \dot{\omega}_t = T_e - h(F_C + F_f) \tag{8.7}$$

where J_t is the moment of inertia of the tire and $F_f = \mu N$.

The slips α for each tire are computed using algebraic relations from the velocities in the ground plane v_x^g, v_y^g and the front wheel steering angle δ. The computations are skipped here and the reader can directly refer to [261] for these details.

The equations (8.4), (8.7) and (8.6) define a nonlinear dynamics which can be linearized for the acceleration trajectory defined by (8.3) with initial values for the velocity and the front wheel steering angle δ. In the linearization process in [261] five states were retained from the nonlinear process:

$$x = \left[y \; v_y \; \psi \; \dot{\psi} \; \delta \right]^T \tag{8.8}$$

The first four states describe the motion of the center of mass: relative lateral position y, the yaw angle ψ and their derivatives. The last state represents the steering actuator modeled by the first order transfer function $\frac{1}{1+0.2s}$. With the choice of the state (8.8), the linear model has the state representation

$$A = \begin{bmatrix} 0 & 1 & v_x & 0 & 0 \\ 0 & a_{11} & 0 & a_{12} & b_1 \\ 0 & 0 & 0 & 1 & 0 \\ 0 & a_{21} & 0 & a_{22} & b_2 \\ 0 & 0 & 0 & 0 & -b \end{bmatrix}, \quad B = \begin{bmatrix} 0 \\ 0 \\ 0 \\ 0 \\ b \end{bmatrix} \tag{8.9}$$

where

$$\begin{aligned} a_{11} &= -\frac{\widetilde{C}_{S_f}+\widetilde{C}_{S_r}}{mv_x} & a_{12} &= \frac{\widetilde{C}_{S_r}l_2-\widetilde{C}_{S_f}l_1}{mv_x} - v_x \\ a_{21} &= \frac{\widetilde{C}_{S_r}l_2-\widetilde{C}_{S_f}l_1}{I_z v_x} & a_{22} &= -\frac{\widetilde{C}_{S_r}l_2^2+\widetilde{C}_{S_f}l_1^2}{I_z v_x} \\ b_1 &= \frac{\widetilde{C}_{S_f}+F_{C_f}-aN_f}{m} & b_2 &= \frac{(\widetilde{C}_{S_f}+F_{C_f}-aN_f)l_1}{m} \end{aligned} \tag{8.10}$$

where $a,\ b \in \mathbb{R}$ and $\widetilde{C}_{S_f}$ and $\widetilde{C}_{S_r}$ are the effective cornering stiffness values of the front and rear wheels:

$$\begin{aligned} \widetilde{C}_{S_f} &= \frac{N_f}{N_f^0} C_{S_f} \sqrt{1-(\frac{F_{C_f}}{0.8N_f})^2} \\ \widetilde{C}_{S_r} &= \frac{N_r}{N_r^0} C_{S_r} \sqrt{1-(\frac{F_{C_r}}{0.8N_r})^2} \end{aligned} \tag{8.11}$$

and the circumferential forces and the normal loads are:

$$\begin{aligned} &for\ 0 \le t < 2, & F_{C_f} &= -9t - 150 & F_{C_r} &= 1853t + 1134 \\ & & N_f &= -690t + 10000 & N_r &= 920t + 7420 \\ &for\ 2 \le t < 7.5, & F_{C_f} &= -168 & F_{C_r} &= 204t + 4440 \\ & & N_f &= 8620 & N_r &= 9260 \\ &for\ 7.5 \le t < 9.5, & F_{C_f} &= 9t - 235.5 & F_{C_r} &= -1665t + 18458 \\ & & N_f &= 690t + 3445 & N_r &= -920t + 16160 \\ &for\ 9.5 \le t < 12, & F_{C_f} &= -150 & F_{C_r} &= 2640 \\ & & N_f &= 10000 & N_r &= 7420. \end{aligned} \tag{8.12}$$

The state-space representation (8.9) is time-varying since the entries of the matrix A depend on the time-variant speed v_x (and the coefficients (8.10)).

8.2.3 *Polynomial Time-Varying Models*

The servo systems are a particular class of control systems. They usually control the position of second or third order systems as quick as possible and in the presence of some external disturbances. Such systems are implemented in several domains like aeronautics, rocket techniques, antenna positioning, computer hard disk drive control, etc.

A bang-bang control, known to move in minimal time an object from one place to another, cannot be used in practical applications concerned with overshoots/undershoots, residual vibrations and disturbance rejection problems. A model-following control method is usually used as a tradeoff between fast movement and accurate settling as shown in [340] for the control of a hard disk drive. The control is of polynomial type

$$u_n = a_1 n + a_2 n^2 + a_3 n^3 + u_0 \tag{8.13}$$

where n is the discrete time. The desired trajectories are chosen as the output of the model when the control input is generated by a polynomial function of time of type (8.13).

Moreover, to capture the evolution of the operating point it is useful to model the system with time-varying coefficients. In [253], in order to solve the problem of an antenna positioning in presence of varying loads according to weather conditions, the following second order linear dynamic polynomial type model

$$A\ddot{y} + B\dot{y} + Cy = u \tag{8.14}$$

with the time-varying coefficients

$$\begin{aligned} A &= at^2 + bt + c \\ B &= dt + e \\ C &= f \end{aligned} \tag{8.15}$$

where $a,\ b,\ c,\ d,\ e,\ f\ \in \mathbb{R}$ was considered along with a polynomial control (of type (8.13)).

8.3 Analysis of Structural Properties

The algebraic framework presented in Part II is used here to analyse two industrial systems.

8.3.1 Controllability and Observability Analysis

In this section it is shown how the analysis of the controllability and observability of a power system is used for the reduction of the order of the simulation models of such systems.

8.3.1.1 Synchrony in Power Systems

A power system is the structure needed to supply consumers with electrical energy as schematically represented in Fig. 8.4. It is mainly composed by generators (G) which are rotating machines which produce the energy, loads (L),

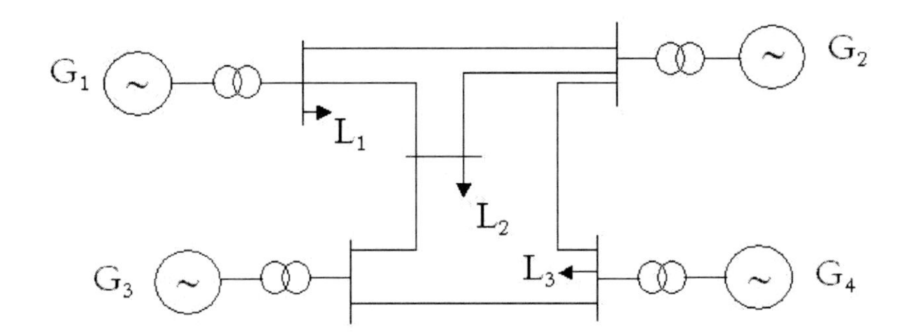

Fig. 8.4 A power system

which are the consumers (of various kinds) and a grid used to transport electricity from producers to the loads. The grid consists in several components, mainly transmission lines and transformers.

The dynamic model of such a system consists in the set of equations which represent the physical laws of the behavior of each component. They are of differential type as, for example, for the rotating machines and their regulations, or of algebraic type, as in the case of transmission lines. This results in a huge size system which is not easy to handle even for simulation purposes. For that reason, the reduction of the order of a dynamic model of a power system is a topic of interest in power systems analysis.

With the usual techniques of reduction based on a balanced realization one is faced with some difficulties in this case. Indeed, a reduced model with physical meaning should be provided, i.e., a model whose variables should preserve their initial meaning like voltages, machines rotating speeds, etc. A mathematical object like, e.g., a transfer function, is not acceptable. This pushed the researchers in the electrical engineering field to imagine specific reduction methods which exploit *redundancy* in a given interconnected power system. Classes of similarity of the rotating machines have been defined. Based on this notion, only one representative of each class is retained in the reduced model while the other machines are replaced by static compensations.

Synchrony, is up to date the most powerful similarity concept [291], [292]. Let $\Lambda = \{\lambda_1, ..., \lambda_p\}$ be a class of interesting modes of a given power system to be reduced. This means that the reduced model is expected to reproduce at least the phenomena linked to these dynamics. The set Λ is called *chord* in the sequel.

Definition 1114. *Two generators* i *and* j *are said* exactly (approximatively) Λ-synchronous *if their angular variations are exactly (approximatively) in constant proportion for any transient in which only the modes in the chord* Λ *are excited:*

$$\delta_i = K\delta_j \tag{8.16}$$

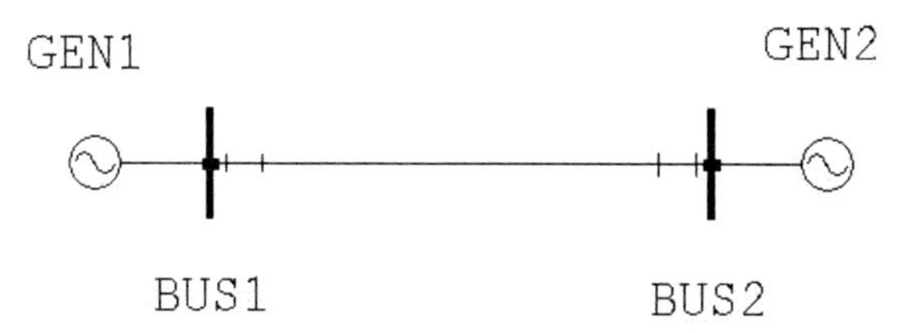

Fig. 8.5 Two-machines power system

8.3.1.2 Representation of the Power System

Synchrony in power systems is usually investigated with the so-called *classical model* for generators [191] which consists only in the swing equation. For the simple example of a power system consisting only of two machines in Fig. 8.5 this leads to

$$M_i\ddot{\delta}_i = P_{m_i} - P_{e_i}, \tag{8.17}$$

$i \in \{1,2\}$ where $M_i = \frac{2H_i}{\omega_0}$ and H_i, P_{m_i}, P_{e_i} and δ_i denote the inertias, the mechanical powers, the electrical powers and respectively the angles of the machines (ω_0 is the nominal speed, generally equal to 1).

If a linear approximation is used around a given operating point, the grid equations are

$$\begin{aligned} P_{e_1} &= P_s(\delta_1 - \delta_2) \\ P_{e_2} &= P_s(\delta_2 - \delta_1) \end{aligned} \tag{8.18}$$

where $P_s = \frac{E_{10}E_{20}}{X}\cos(\delta_{10} - \delta_{20})$ and X is the sum of the reactance of the line which connects the two machines and the transient reactances of the two machines, δ_{10}, δ_{20} are the equilibrium values of the angles of the machines and E_{10}, E_{20} are the terminal voltages which are supposed constant. Thus, P_s is constant if X is constant. However, X may vary in time. Indeed, it should be noted that, in the context of synchrony and model reduction, the model above is an aggregated representation of a real power system, which is usually an interconnection of several sub-systems. The lines between the machines are in fact a model of the real grid of the system. Roughly speaking, it provides the impedance "seen" from the stator of each machine. This impedance varies with the operating conditions of the system. This includes the topology modifications which can be amplified by the control of the FACTS present on the system (see, e.g., the case of a Thyristor Controlled Switched Capacitor (TCSC) presented in Section 9.4). In this case X is no longer constant, $X = X(t)$ and the system is thus LTV.

If the power system is balanced, the mechanical powers of the two machines are not independent variables. If, moreover, the loads are supposed constant, the relation among the powers is

$$P_{m_1} + P_{m_2} = 0. \tag{8.19}$$

Equations (8.17), (8.18) and (8.19) constitute an *analytical model* of the system, i.e., a set of physical variables δ_1, δ_2, P_{m_1}, P_{m_2}, P_{e_1} and P_{e_2} which satisfy a set of equations, closely related to the so-called Differential and Algebraic Equations (DAE) form of the power system (see, e.g., [244]). This form is close to physics and also widely used in simulation and modal analysis because of its sparsity properties.

Notice that this representation can be compacted if variables P_{e_1}, P_{e_2}, P_{m_1} are eliminated from equations (8.19) and (8.18) into equations (8.17):

$$\begin{aligned} M_1\ddot{\delta}_1 &= P_{m_1} - P_s(\delta_1 - \delta_2) \\ M_2\ddot{\delta}_2 &= -P_{m_1} - P_s(\delta_2 - \delta_1). \end{aligned} \tag{8.20}$$

8.3.1.3 Analogy with a Mechanic System

Consider the system of two masses m_1 and m_2 connected by a spring of elasticity constant k in Fig. 8.6 where x_1 and x_2 denote the horizontal positions of the two masses respectively. If, $f_1 = -f_2$, the system is described by the equations

$$\begin{aligned} m_1\ddot{x}_1 &= f_1 - k(x_1 - x_2) \\ m_2\ddot{x}_2 &= -f_1 - k(x_2 - x_1). \end{aligned} \tag{8.21}$$

Notice that the model (8.21) is the same as the one provided by (8.20) with the following correspondences among the variables:

- $M_i \leftrightarrow m_i$: the inertia of the rotating machines correspond to the masses of the two objects.
- $\delta_i \leftrightarrow x_i$: the angular positions of the two machines correspond to the linear positions of the two objects.
- $P_{m_i} = f_i$: the mechanical powers correspond to the external forces.
- $P_s \leftrightarrow k$: the line reactance corresponds to the stiffness of the spring (more precisely, k corresponds to P_s, but in the expression of P_s, the most structural parameter is X).

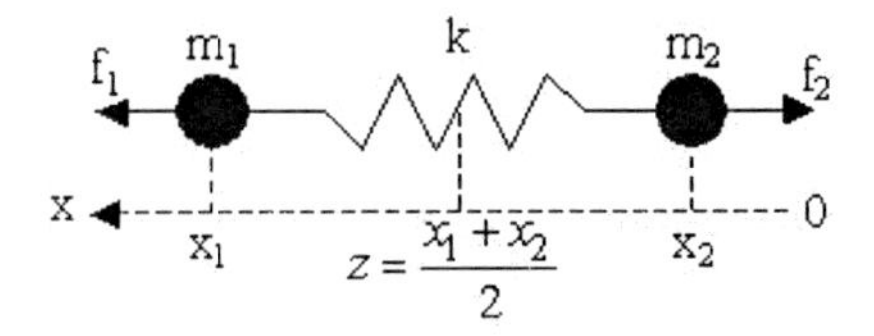

Fig. 8.6 Two masses with a spring

8.3.1.4 Lack of Controllability

The DAE form (8.20) for the two-machines system can be written

$$S(\partial)w = 0, \tag{8.22}$$

where

$$S(\partial) = \begin{bmatrix} M_1\partial^2 + P_s & -P_s & -1 \\ -P_s & M_2\partial^2 + P_s & 1 \end{bmatrix}, \quad w = \begin{bmatrix} \delta_1 \\ \delta_2 \\ P_{m_1} \end{bmatrix}. \tag{8.23}$$

Equation (8.22) defines the module associated with the linear representation of the two-machines system. The Smith normal form of its presentation matrix $S(\partial)$ is $diag\{1, \partial^2\}$, from which follows that the system is not controllable with a double input-decoupling zero at $\partial = 0$. As a matter of fact, if one adds the two rows of (8.22), one obtains

$$\partial^2(M_1\delta_1 + M_2\delta_2) = 0 \tag{8.24}$$

which implies that $M_1\delta_1 + M_2\delta_2$ is a nonzero *torsion element* Δ, such that $\partial^2\Delta = 0$. The variable Δ, which is the *mean speed* of the system, evolves independently of the rest of the system with an *uncontrollable* dynamics. In the mechanical analogy, if the system is driven in a symmetric way by taking $f_1 = -f_2$, the position of the *mass center* $z = m_1x_1 + m_2x_2$ cannot be controlled.

If one looks now for solutions in $W = C^\infty(\mathbb{R})$ (see Theorem 831), i.e., trajectories of the system, (8.24) leads to $\delta_1(t) = -\frac{M_2}{M_1}\delta_2(t) + c_1t + c_2$ where $c_1,\ c_2 \in \mathbb{R}$. As physically the angles of the machines remain bounded as $t \to +\infty$, $c_1 = 0$. Also, as the origin of the axis over which the angles evolve can be arbitrarily chosen, c_2 can be assumed to be zero, which leads to

$$\delta_1 = -\frac{M_2}{M_1}\delta_2 \tag{8.25}$$

which is of type (8.16) with $K = -\frac{M_2}{M_1}$. Thus, lack of controllability leads to synchrony. Fig. 8.7 shows the response $\delta_1 - \delta_2$ of the autonomous system ($P_{m_1} = 0$) with initial conditions $\delta_{1_0} = \dot{\delta}_{1_0} = 1$, $\delta_{2_0} = \dot{\delta}_{2_0} = 0$ which excite mode $\partial = 0$ only (solid line). With $\delta_{1_0} = 1$, $\dot{\delta}_{1_0} = \delta_{2_0} = \dot{\delta}_{2_0} = 0$ only the other (controllable) mode $s = \pm j\sqrt{\frac{P_s(M_1+M_2)}{M_1M_2}}$ of the system is excited which leads to the response in the dotted line which is not of synchrony type.

Remark 1115. *From equation (8.25) follows that $\Delta = 0$ which is apparently incompatible with the torsion equation (8.24). However, (8.25) and (8.24) are obtained with two different modelizations and, from the algebraic point of view, with two different systems. Indeed, for the DAE form (8.17), (8.18) and (8.19), no angle reference is chosen for the angles of the*

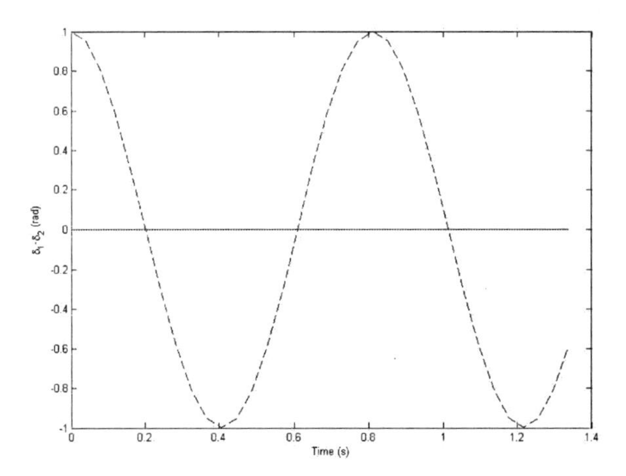

Fig. 8.7 Responses of $\delta_1 - \delta_2$ to different initial conditions

machines. This is the usual manner to proceed in power systems since the resulting representation is sparse *(i.e., use sparse matrices) and facilitates thus numerical analysis and simulation. To derive (8.25) an angle reference have been chosen.*

Notice that the above conclusion about the lack of the controllability is independent of the choice of the control variables. Indeed, no a priori distinction has been made among the physical variables δ_1, δ_2 and P_{m_1} which means that this conclusion is intrinsic, i.e., does not depend on the way in which the system is controlled.

8.3.1.5 Lack of Observability

Choice of inputs and outputs: Observability depends on the input u and the output y which are chosen. Therefore, starting from the system M given by (8.22) one should construct a control system according to Definition 851, i.e., to chose both a vector of inputs u and a vector of outputs y. A well-defined input for the two-machines example is $u = P_{m_1}$. Indeed, $M/[P_{m_1}]_{\mathbf{R}}$ is torsion. Let now $y = \delta_2 - \delta_1$. The equations of the control system (M, u, y) defined in this way are

$$\begin{bmatrix} M_1\partial^2 + P_s & -P_s & -1 & 0 \\ -P_s & M_2\partial^2 + P_s & 1 & 0 \\ -1 & 1 & 0 & -1 \end{bmatrix} \begin{bmatrix} \delta_1 \\ \delta_2 \\ P_{m_1} \\ y \end{bmatrix} = 0. \tag{8.26}$$

Unobservable modes: According to Definition 1019, the nonobservable modes of (M, u, y) are the Smith zeros of a matrix of definition of the module $M/[u, y]_{\mathbf{R}}$. From the computational point of view, this means to eliminate from (8.26) u, y and the columns which correspond to u and y: a presentation of $M/[u, y]_{\mathbf{R}}$ is thus

$$\underline{S}(\partial)\begin{bmatrix}\underline{\delta}_1\\ \underline{\delta}_2\end{bmatrix} = 0,\quad \underline{S}(\partial) = \begin{bmatrix} M_1\partial^2 + P_s & -P_s \\ -P_s & M_2\partial^2 + P_s \\ -1 & 1 \end{bmatrix}. \tag{8.27}$$

Notice that $\underline{S}(\partial) = S(\partial)^T$. Thus, the Smith normal form of $\underline{S}(\partial)$ is still $diag\{1, \partial^2\}$. The only nonobservable mode is $\partial^2 = 0$. To find the nonobservable dynamics let $\overline{S}(\partial) = U_2(\partial)U_1(\partial)S(\partial)V_1(\partial)$ with the unimodular matrices

$$U_1(\partial) = \begin{bmatrix} 1 & 0 & P_s \\ 0 & 1 & -(M_2\partial^2 + P_s) \\ 0 & 0 & 1 \end{bmatrix},\ U_2(\partial) = \begin{bmatrix} 1 & 0 & 0 \\ -\frac{M_2}{M_1} & 1 & 0 \\ 0 & 0 & 1 \end{bmatrix}, \tag{8.28}$$
$$V_1(\partial) = \begin{bmatrix} 1 & 0 \\ 1 & 1 \end{bmatrix}.$$

This transforms (8.27) into

$$\overline{S}(\partial)\begin{bmatrix}\overline{\delta}_1\\ \overline{\delta}_2\end{bmatrix} = 0 \tag{8.29}$$

where $\overline{S}(\partial) = \begin{bmatrix} M_1\partial^2 & 0 \\ 0 & 0 \\ 0 & 1 \end{bmatrix}$ and $\begin{bmatrix}\overline{\delta}_1\\ \overline{\delta}_2\end{bmatrix} = V_1^{-1}$ with V_1 given by (8.28). The first row in (8.29) leads to $\partial^2\overline{\delta}_1 = 0$ which gives the *nonobservable dynamics.* As $\overline{\delta}_1 = \underline{\delta}_1$, the trajectories of the latter nonobservable dynamics, over the solutions space $W = C^\infty(\mathbb{R})$, are $\underline{\delta}_1 = c_1 t + c_2$. For the same reasons mentioned in the preceding paragraphe, $c_1 = 0$, thus $\underline{\delta}_1 = c_2$. Together with the third relation in (8.27) this leads to

$$\underline{\delta}_1 = \underline{\delta}_2 = const\ (not\ necessarily\ 0) \tag{8.30}$$

and obviously to $y \equiv 0$: when the nonobservable mode $\partial = 0$ is excited, no difference between the angles of the machines is observed. Relation (8.30) is again of type (8.16) which means that the lack of observability leads also to synchrony. Notice that, as for the lack of controllability, the apparent gap between the nonobservable dynamics and the synchrony relation (8.30) is explained by Remark 1115.

8.3.1.6 The Zero Mode

The investigations above revealed a somewhat dual situation. The same double mode $\partial = 0$ and the same dynamics have been found both for the uncontrollable and nonobservable parts of the system. This mode is now further investigated. Let skip the noncontrollable subsystem $\mathcal{T}(M)$ of the system M, i.e., the factor ∂^2 in (8.24). A presentation of the remaining controllable quotient $M_c = M/\mathcal{T}(M)$ of the system M is thus

$$\overline{S}(\partial)\begin{bmatrix}\overline{\delta}_1\\ \overline{\delta}_2\\ \overline{P}_{m_1}\end{bmatrix} = 0, \quad \overline{S}(\partial) = \begin{bmatrix} M_1 & M_2 & 0\\ M_1\partial^2 + P_s & -P_s & -1\end{bmatrix}. \tag{8.31}$$

If the same inputs/outputs as before are now chosen for the controllable quotient (8.31), the module $M_c/[\overline{u}, \overline{y}]_{\mathbf{R}}$ of the output-decoupling zeros of M_c is given by $\widetilde{S}(\partial)\begin{bmatrix}\widetilde{\delta}_1\\ \widetilde{\delta}_2\end{bmatrix} = 0, \quad \widetilde{S}(\partial) = \begin{bmatrix} M_1 & M_2\\ M_1\partial^2 + P_s & -P_s\\ -1 & 1\end{bmatrix}$. All the invariant factors of $\widetilde{S}(\partial)$ are equal to 1, thus M_c has no output-decoupling zeros. Following Definition 1023, $\partial = 0$ is a double *input-output decoupling zero* of M.

The double input-output decoupling zero $\partial = 0$ has a clear significance in power systems. One of the zero mode is connected to the absence of angle reference in the model (8.17), (8.18) (see, e.g., [65]). Indeed, δ_1 and δ_2 are not independent which explains the singularity of all considered presentation matrices. The second zero mode is the so-called *frequency mode* of the system (see, e.g., [191]) and expresses a structural similarity in the response of the machines of any power system. When a detailed model of the system is used, the frequency mode gives the dynamics of the response of *all* machines of the system at the first moments after a power imbalance of the system: the speeds of all machines of a power system decrease with the same slope which is the well-known *gradient* of the system [191]. The case of the interconnected European power system can be seen in Fig. 8.8 which contains the responses of some machines to a generation outage in France.

8.3.1.7 Exact Synchrony/Approximate Synchrony/Model Reduction

The analysis above clarified some points:

- exact synchrony is related to the simultaneous lack of controllability and observability, i.e., to the input-output decoupling-zeros.
- the lack of controllability is a "built-in" property of the system (thus independent of the choice of the input variables of the system).

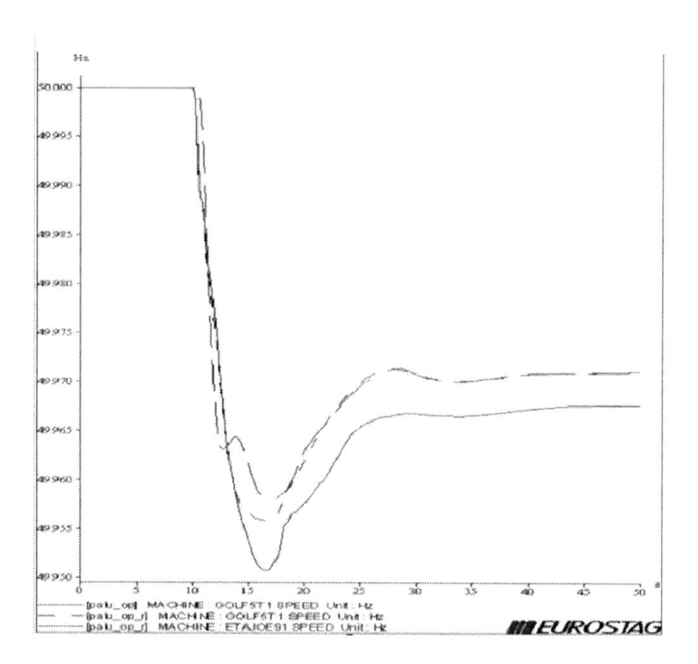

Fig. 8.8 European frequency gradient

- the only exact similar behavior of the machines of a power system is the one given by the frequency gradient, i.e., by the frequency mode which is $\partial = 0$ when computed using the classical model for the machines.

This provides a link between systems theory and the specific analysis widely used by the experts in power systems. Also, these results are a basis for model reduction techniques. It has been proved that the only exact synchrony is related to the zero mode and this is not constructive. However, the link with the lack of controllability and observability is fundamental since it opens the way to reduction methods based on balanced-realizations. This has been exploited in [242], [233] [243] to provide structure-preserving reduction methodologies for power systems.

8.3.2 Stability Analysis

A bandwidth adaptation of linear filters is sometimes needed to fulfil the specifications of several classes of industrial applications. Such an example is treated in [382] where a time-varying bandwidth filter of the form

$$P(\partial)u_{out} = u_{in}, \; P(\partial) = \partial^2 + [2\xi\omega_n(t) - \frac{d\omega_n}{dt}(\omega_n(t))^{-1}]\partial + (\omega_n(t))^2 \quad (8.32)$$

is used in the design of a missile autopilot. In order to cope with the requirements on the tracking performance and on the actuator rate limit, the control is filtered by a block of type (8.32) which is supposed to reduce the acceleration and the rate of the control in case of tracking of abrupt trajectories and, on the contrary, to have have little influence on smooth trajectories which can be tracked without limiting the actuator. These two requirements cannot be achieved with a fixed-parameter filter and a well-chosen time dependance $t \mapsto \omega_n(t)$ in (8.32) of the bandwidth should be considered. Consider the autonomous part

$$P(\partial)y = 0 \tag{8.33}$$

of (8.32). For $0 < \xi < 1$, the solutions of (8.33) in $\mathcal{E}_d$ (see Theorem 837) can be easily found to be $y(t) = \alpha_1 y_1(t) + \alpha_2 y_2(t)$, $y_{1,2}(t) = e^{(-\xi \pm i\sqrt{1-\xi^2}) \int \omega_n(t)dt}$ where $i^2 = -1$ and α_1, $\alpha_2 \in \mathbb{C}$. According to the elementary equations (6.6), to the fundamental set of solutions $\{y_1(t), y_2(t)\}$ above corresponds the fundamental set of roots $\{\gamma_1(t), \gamma_2(t)\}$ of P: $\gamma_{1,2} = -\xi\omega_n(t) \pm i\omega_n(t)\sqrt{1-\xi^2}$. As the bandwidth variation is necessarily such that $\omega_n(t) > 0$, $\forall t$, the condition of Theorem 998 is satisfied and the filter (8.32) is thus exponentially stable.

9 Modeling: Choice of the Input Variables of an LTV System

9.1 Introduction

A linear system has been defined as a module, or roughly speaking, as a set of variables which satisfy a set of differential and algebraic equations. This definition has a strong connection with the construction of the model of a dynamic systems in practice. As a matter of fact, one of the most common way of building models for dynamic systems is the so-called *analytical modeling* which consists of writing the equations of the basic principles of physics in accordance to the nature of the system (electrical energy conservation laws, Lagrange formalism for mechanical systems, thermodynamics heat exchange laws, etc.). Examples for this way of doing are the models presented in Chapter 8.

The analytical models are thus obtained in a sort of generalized form in which neither the control variables (the *inputs*) nor the measured variables (the *outputs*) are distinguished among the variables of the system at this stage of the model construction. The choice of the inputs and of the outputs is an important step of the modeling process which can be non trivial. Of course, the class of variables which are good candidates to be inputs is limited by practical considerations like the type and availability of the actuators. However, theoretical considerations which must be made in a systematic way on the model at several levels are discussed in this section. For this, the framework in Chapter 7 is used. Let M be an LTV system, i.e., a $\mathbf{R}$-module M where $\mathbf{R} = \mathbf{K}[\partial; \alpha, \delta]$ and $\mathbf{K}$ is the Noetherian domain to which the coefficients of the system belong. It is assumed that M is given by a general representation (5.1) in which the matrix of definition $R(\partial)$ is regular at infinity (according to Theorem and Definition 1060(4)). Consider a *temporal system* M_t which has the same matrix of definition as M along with its *impulsive system* M^+. M^+ is an $\mathbf{S}$-module where $\mathbf{S} = \mathbf{K}[[\sigma; \alpha, \delta]]$, $\sigma = \partial^{-1}$.

The criteria needed for the choice of the inputs are discussed in Section 9.2 while a recursive form for the general algorithm is given in Section 9.3.

H. Bourlès and B. Marinescu: Linear Time-Varying Systems, LNCIS 410, pp. 515–522.
springerlink.com

This algorithm is applied in Section 9.4 to the case of an electrical circuit which is a model used in power engineering for power transmission lines.

9.2 Criteria for the Choice of the Inputs

The following points should be analyzed when a control system (M, u, y) (according to Definition 851) is designed starting from a given system M:

(i) first, the system should be stabilizable.

(ii) the minimal number of the input variables must be determined. For the given linear system M, the set of inputs $u = \{u_1, \ldots, u_m\}$ should be such that the quotient module $M/[u]_{\mathbf{R}}$ is a torsion module (see Chapter 5). The inputs u are independent if, and only if the module $[u]_{\mathbf{R}}$ is free of rank m. It follows that the number of independent inputs is the rank of the module which defines the system:

$$m = rank(M) \tag{9.1}$$

Roughly speaking, u should be chosen of minimal size and in order to annihilate the "degrees of freedom" of the system M.

(iii) one may also choose a vector y of outputs such that the transfer $u \longrightarrow y$ is *proper*. As shown in Section 7.2.2, properness can be directly characterized in terms of the pole of the system at infinity. This characterization is more convenient also from a computational point of view and, as a consequence, is used in the following chapter to state an algorithm for the systematic choice of the inputs.

(iv) the dynamic of the system should be also free from impulsive motions due to the initial conditions. Indeed, this may happen with temporal systems as in the following example:

Example 1116. *Let the condenser in Fig. 9.1 be short-circuited at time $t = 0$ by closing the switch in the same figure. If the capacitance C is supposed to be the unit for simplification, the equations which describe the behavior of this circuit for $t \geq 0$ are*

$$\begin{cases} x_1 = 0 \\ \dot{x}_1 = x_2 + x_3 \end{cases} \tag{9.2}$$

where x_2, x_3 denote the short-circuit current and the current injected by the ideal current source S respectively, and x_1 is the voltage on the condenser. The autonomous impulsive behavior $\mathfrak{B}_{\infty,a}$ (Definition 1076) is given by

$$\begin{cases} x_1(t) = 0 \\ x_2(t) + x_3(t) = -x_1(0^-)\delta \end{cases} \tag{9.3}$$

where $x_1(0_-)$ is the initial condition, i.e., *a measure of the initial load of the condenser and δ is the Dirac distribution. If $x_1(0^-) \neq 0$, an impulsive*

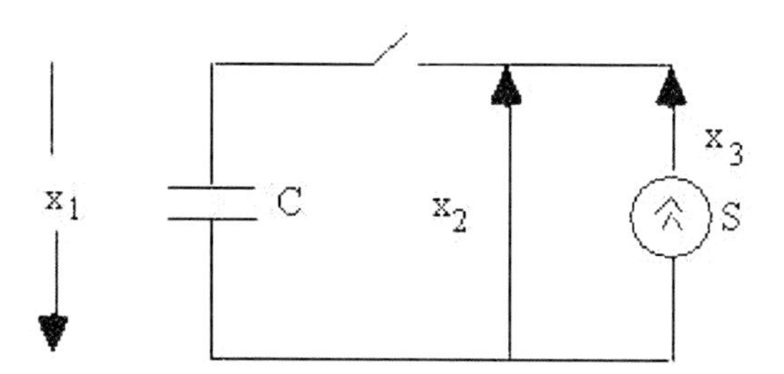

Fig. 9.1 Switched electrical circuit

motion x_2+x_3 is generated. This is the physical interpretation of an input-decoupling zero at infinity as defined in Section 7.4.3. It can be evaluated starting from the matrix of definition $R(\partial)=\begin{pmatrix}1&0&0\\ \partial&-1&-1\end{pmatrix}$ of the temporal system M_t defined by (9.2). A strongly left-coprime factorization of $R(1/\sigma)$ is

$$R(1/\sigma)=D^{-1}(\sigma)R^{+}(\sigma),$$
$$D(\sigma)=\begin{bmatrix}1&0\\0&\sigma\end{bmatrix},\quad R^{+}(\sigma)=\begin{bmatrix}1&0&0\\1&-\sigma&-\sigma\end{bmatrix}. \tag{9.4}$$

The Smith normal form of $R^{+}(\sigma)$ exists and is $\begin{bmatrix}1&0&0\\0&\sigma&0\end{bmatrix}$ ($R(\partial)$ is thus regular at infinity). It follows that $R^{+}(\sigma)$ is a matrix of definition of the module of the input-decoupling zero at infinity M_t or, in other words, of the impulsive system M^{+}. The input-decoupling zeros at infinity can be investigated on the above Smith normal form from which follows that the temporal system M_{+} given by 9.2 has an input-decoupling zero at infinity of degree one, i.e., $\sharp(IDZ_\infty)=1$ (according to Definition 1091 and Definition 1060).

Initial conditions can also excite the output-decoupling zero at infinity, if any, in which case the resulting impulsive motions are not observable from the chosen input u and output y.

The conditions given before Example 1116 are of a different nature: conditions (i) and (ii) concern the finite dynamics of the system while the last two conditions concern the dynamics at infinity. Also, from another point of view, conditions (i) and the part of (iv) concerning the input-decoupling zero at infinity are independent of the choice of the input u. Condition (ii) obviously concerns the inputs u and conditions (iii) and the part of (iv) on the output-decoupling zero at infinity depends both on u and y.

Conditions (iii) and (iv) can be merged into a single condition concerning the system pole at infinity

(iii') $\sharp(SP_\infty)=0$

i.e., the system is required to have no pole at infinity. Notice that condition (iii') is independent of the choice of y and this presents a practical advantage: at the modeling stage the designer is concerned only with the choice of the input u. Afterwards, whatever the measured output y will be, it will lead to proper transfer functions. This is an important advantage when analyzing systems with many variables for which it is difficult to check both u and y because of the increasing combinatorics. This is the case of the example presented in Section 9.4.

9.3 General Algorithm

Consider a temporal system M_t and the LTV system M which has the same matrix of definition as M_t. If M is stabilizable and M_t has no input-decoupling zero at infinity, one should look for u which satisfies conditions (ii) and (iii') formulated in the preceding section. To obtain an efficient algorithm for the choice of u two transformations of the above conditions are needed. First, condition (ii) can be also formulated using the impulsive system M^+. Next, both conditions can be written in recursive forms.

Using Lemma 1093 (ii), condition (ii) can be written in an equivalent manner

(ii') $[u^+]_\mathbf{S}$ is free of rank $m = Card(u^+)$
(ii") $M^+/[u^+]_\mathbf{S}$ is torsion, *i.e.*, $m = rank(M^+)$.

To obtain an iterative form for (ii'), (ii") and (iii'), the following notations are considered: $u^+ = \{u_1^+, ..., u_m^+\}$, the set $u_1^+, ..., u_k^+$ $(k \leq m)$ is denoted by u^{+^k} and u^{+^0} is the empty set: $u^{+^0} = \{\varnothing\}$.

Conditions (ii') and (ii") can be easily put in an iterative form. Indeed, these conditions are satisfied if, and only if

$$rank(M^+/[u^{+^k}]_\mathbf{S}) = m - k, \quad 0 \leq k \leq m. \tag{9.5}$$

Condition (iii") can be also checked iteratively from the following results:

Lemma 1117. *Let M_1^+, M_2^+ and M^+ be* **S***-modules such that $M_1^+ \subseteq M_2^+ \subseteq M^+$, M_1^+ and M_2^+ are free and M_1^+ is a direct summand of M_2^+. Then $\sharp(M^+/M_1^+) \leq \sharp(M^+/M_2^+)$.*

Proof. $\mathcal{T}(\frac{M^+}{M_2^+}) \cong \mathcal{T}(\frac{M^+/M_1^+}{M_2^+/M_1^+}) \supseteq \frac{\mathcal{T}(M^+/M_1^+)}{\mathcal{T}(M_2^+/M_1^+)}$, where $\mathcal{T}(.)$ denotes the torsion submodule. As M_1^+ is a direct summand of M_2^+, $\mathcal{T}(M_2^+/M_1^+) = 0$ which leads to the result. ■

Denote by $\sharp(M^+/[u^+]_\mathbf{S})$ the degree of the system pole at infinity (i.e., $\sharp(M^+/[u^+]_\mathbf{S}) = \sharp(SP_\infty))$ and, by analogy, $\sharp(M^+/[u^{+^k}]_\mathbf{S})$ the similar degree obtained when the input u^{+^k} is used instead u^+.

Proposition 1118. $\sharp(M^+/[u^+]_{\mathbf{S}}) = 0$ *if, and only if*

$$\sharp(M^+/[u^{+^k}]_{\mathbf{S}}) = 0, \ k = 1, \ ..., \ m. \tag{9.6}$$

Proof. From Lemma 1117 with $M_1^+ = [u^{+^{k-1}}]_{\mathbf{S}}$ and $M_2^+ = [u^{+^k}]_{\mathbf{S}}$, it follows that $\sharp(M^+/[u^{+^{k-1}}]_{\mathbf{S}}) \leq \sharp(M^+/[u^{+^k}]_{\mathbf{S}})$, $k = 1, \ ..., \ m$ which proves the result. ∎

The following iterative algorithm can now be stated:

Algorithm 1119

Step 0:

- *check if M is stabilizable (compute the finite input-decoupling zero). If not, stop (the problem has no solution).*
- *compute the degree of the input-decoupling zero at infinity. If it is grater than zero, stop (no solution)*
- *compute the number of independent inputs: $m = rank(M) = rank(M^+)$.*

***Step k** ($k \geq 1$):*

It is supposed that $u^{+^{k-1}}$ has been chosen at step $k-1$ such that

- $rank(M^+/[u^{+^{k-1}}]) = m - (k-1)$
- $\sharp(M^+/[u^{+^{k-1}}]_{\mathbf{S}}) = 0.$

Then, u^{+^k} is chosen such that conditions (9.5) and (9.6) are satisfied.

Notice that using the result of Lemma 1117 the combinatorics of the algorithm above is much reduced. Indeed, the input variable which does not respect condition (9.6) at step k has never to be tested again since it will always provide a pole of the system at infinity.

9.4 Control Model of an Electrical Circuit

The algorithm presented above is applied to the electrical circuit in Fig. 9.2. It can be considered as a simply RLC circuit but it represents also, under certain conditions, the so-called π-model used for high-voltage transmission lines in voltage control and simulation of power systems (see, *e.g.*, [191]). Moreover, if the value of the reactance X is supposed to be continuously adaptable by the law

$$X(t) = e^{\alpha(t-t_0)}, \ \alpha > 0, \ t_0 \leq t \leq t_0 + \Delta \tag{9.7}$$

the same circuit provides a model for a Thyristor Controlled Switched Capacitor which is a FACTS often used in the power systems industry to control

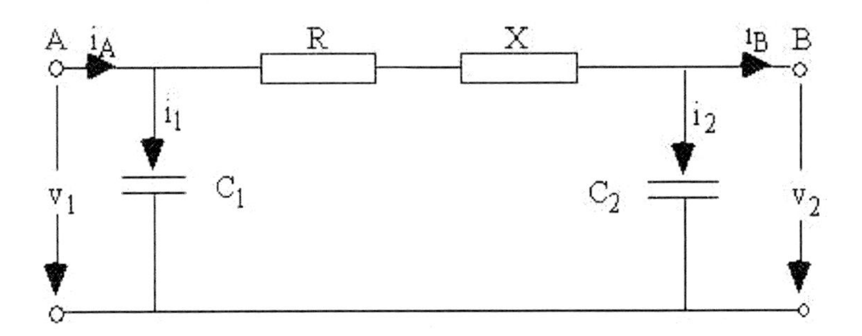

Fig. 9.2 RLC electrical circuit

the power flow on a transmission line of an electrical grid [191]. Roughly speaking, it consists in a thyristor controlled reactance placed in series with the electrical line on which the power flow must be controlled.

The equations describing the circuit in Fig. 9.2 are:

$$\begin{aligned} &C\dot{v}_1 = i_1 \\ &C\dot{v}_2 = i_2 \\ &v_1 = R(i_A - i_1) + X(t)\tfrac{d}{dt}(i_A - i_1) + v_2 \\ &i_A - i_1 = i_B + i_2 \end{aligned} \tag{9.8}$$

where $X(t)$ is given by (9.7).

A matrix of definition of the module M defined by (9.8) is

$$R(\partial) = \begin{bmatrix} \partial C_1 & 0 & -1 & 0 & 0 & 0 \\ 0 & \partial C_2 & 0 & -1 & 0 & 0 \\ 1 & -1 & R + \partial X(t) & 0 & -(R + \partial X(t)) & 0 \\ 0 & 0 & -1 & -1 & 1 & -1 \end{bmatrix} \tag{9.9}$$

A strongly left-coprime factorization of $R(1/\sigma)$ over $\mathbf{S}$ is $R(1/\sigma) = D^{-1}(\sigma)R^{+}(\sigma)$ where

$$\begin{aligned} D(\sigma) &= \begin{bmatrix} \sigma & 0 & 0 & 0 \\ 0 & \sigma & 0 & 0 \\ 0 & 0 & \sigma & 0 \\ 0 & 0 & 0 & 1 \end{bmatrix}, \\ R^{+}(\sigma) &= \begin{bmatrix} C_1 & 0 & -\sigma & 0 & 0 & 0 \\ 0 & C_2 & 0 & -\sigma & 0 & 0 \\ \sigma & -\sigma & P(\sigma) & 0 & -P(\sigma) & 0 \\ 0 & 0 & -1 & -1 & 1 & -1 \end{bmatrix} \end{aligned} \tag{9.10}$$

and $P(\sigma) = R\sigma + X(t)$.

Step 0: The matrix of definition of M^+ can be transformed using elementary row and column operations:

$$R^+(\sigma) \sim \begin{bmatrix} 1 & 0 & 0 & 0 & 0 & 0 \\ 0 & C_1 & 0 & -\sigma & 0 & 0 \\ 0 & 0 & C_2 & 0 & -\sigma & 0 \\ 0 & \sigma & -\sigma & P(\sigma) & 0 & -P(\sigma) \end{bmatrix} \sim$$

$$\sim \begin{bmatrix} 1 & 0 & 0 & 0 & 0 & 0 \\ 0 & 1 & 0 & 0 & 0 & 0 \\ 0 & 0 & C_2 & 0 & -\sigma & 0 \\ 0 & 0 & -\sigma & P(\sigma)+\frac{1}{C_1}\sigma & 0 & -P(\sigma) \end{bmatrix} \sim$$

$$\sim \begin{bmatrix} 1 & 0 & 0 & 0 & 0 & 0 \\ 0 & 1 & 0 & 0 & 0 & 0 \\ 0 & 0 & 1 & 0 & 0 & 0 \\ 0 & 0 & 0 & P(\sigma)+\frac{1}{C_1}\sigma & \frac{1}{C_2}\sigma & -P(\sigma) \end{bmatrix} \sim$$

$$\sim \begin{bmatrix} 1 & 0 & 0 & 0 & 0 & 0 \\ 0 & 1 & 0 & 0 & 0 & 0 \\ 0 & 0 & 1 & 0 & 0 & 0 \\ 0 & 0 & 0 & 1 & 0 & 0 \end{bmatrix}$$

because $P(\sigma)$ is a unit of $\mathbf{S}$ since $X(t) \neq 0$ in (9.7). It follows that $rank(R^+(\sigma)) = 4$ and $m = rank(M) = rank(M^+) = rank(R^+(\sigma)) - dim\{w\} = 4 - 2 = 2$, where $w = \{v_1,\ v_2,\ i_1,\ i_2,\ i_A,\ i_B\}$. So, $m = 2$ and two input variables have to be found in the following two steps.

Step 1 (choice of the first input variable): Consider v_1^+ as a candidate for the first control variable. A matrix of definition of the quotient module $M^+/[v_1^+]_{\mathbf{S}}$ is obtained from $R^+(\sigma)$ by eliminating the column corresponding to v_1^+: $R^+_{v_1^+}(\sigma) = \begin{bmatrix} 0 & -\sigma & 0 & 0 & 0 \\ C_2 & 0 & -\sigma & 0 & 0 \\ -\sigma & P(\sigma) & 0 & -P(\sigma) & 0 \\ 0 & -1 & -1 & 1 & -1 \end{bmatrix}$. As for $R^+(\sigma)$, it can be shown that $R^+_{v_1^+}(\sigma) \sim \begin{bmatrix} 1 & 0 & 0 & 0 & 0 \\ 0 & 1 & 0 & 0 & 0 \\ 0 & 0 & 1 & 0 & 0 \\ 0 & 0 & 0 & \sigma & 0 \end{bmatrix}$ from which follows that $\sharp(M^+/[v_1^+]_{\mathbf{S}}) = 1$. As a result, condition (9.6) is violated and v_1^+ cannot be a control variable. $R^+_{v_2^+}(\sigma)$ has the same property, hence v_2^+ is also inadequate. However, $\sharp(M^+/[i_1^+]_{\mathbf{S}}) = 0$ and $rank(M^+/[i_1^+]_{\mathbf{S}}) = 1$, thus i_1^+ can be a control variable.

Step 2 (choice of the second input variable): From Proposition 1118 neither v_1^+ nor v_2^+ can be the second control variable. A candidate is i_2^+

and a matrix of definition of $M^+/[i_1^+ \; i_2^+]_\mathbf{S}$ is $R^+_{i_1^+,\, i_2^+}(\sigma) \sim \begin{bmatrix} 1\,0\,0 & 0 \\ 0\,1\,0 & 0 \\ 0\,0\,1 & 0 \\ 0\,0\,0 & -P(\sigma) \end{bmatrix}$.

As $P(\sigma)$ is a unit of $\mathbf{S}$ it follows that $R^+_{i_1^+,\, i_2^+}(\sigma) \sim \begin{bmatrix} 1\,0\,0\,0 \\ 0\,1\,0\,0 \\ 0\,0\,1\,0 \\ 0\,0\,0\,1 \end{bmatrix}$ is nonsingular at $\sigma = 0$. In addition, $M^+/[i_1^+,\; i_2^+]_\mathbf{S}$ is torsion, so that the final solution can be $u^+ = \{i_1^+,\; i_2^+\}$.

Notice that the solution is not unique; one can verify that $\{i_A^+,\; i_B^+\}$ is also a solution to the problem. As a result, the electrical circuit, considered as an independent system or in connection with other systems, must be controlled by currents; the voltage control is not possible.

9.5 Notes and References

The modeling problems treated in this Chapter arise naturally in the intrinsic algebraic framework presented in Chapter 5. Indeed, in this formalism, no distinction is made a priori among the variables which define the system M. The definition of the input variables u and the condition that $M/[u]_\mathbf{R}$ must be a torsion submodule of M were first given in [115]. The extension to the topics concerning the structure at infinity was done in [45]. The recursive algorithm for the choice of the input variables of a control system has been used in the analysis of electrical circuits in [234].

10 Open-Loop Control by Model-Matching

10.1 Introduction

The model-matching problem consists in assigning the transfer matrix of a system. Many control problems reduce to model-matching and several approaches have been developed. The model-matching can be *exact* when a given transfer matrix is assigned or *approximate* when the design is such that the transfer matrix of the controlled system gets a specified structure (as is the case for decoupling control) or gets as close as possible to a reference transfer matrix. In the latter case, the distance between the two transfers is expressed in terms of a chosen norm of the difference between the two transfer matrices like, *e.g.*, the $\|.\|_\infty$ norm for the H_∞ control. The *exact* model-matching (EMM) consists in precisely assigning all entries of the transfer matrix to given transfer functions. It can be done by *feedforward* (*pre* or *postcompensation*) or by *feedback* (*static* or *dynamic*).

Both continuous- and discrete-time cases are addressed in a unified way using directly the transfer matrix formalism introduced in Section 5.5.2. In Section 10.2 a parametrization of the class of proper solutions is given using the Smith-MacMillan form at infinity of a transfer matrix. In Section 10.3 it is shown that, from an algebraic point of view, most of the EMM-type problems mentioned above can be reduced to an exact feedforward one. Related problems like the output-disturbance decoupling and system decoupling ones are treated in the same framework in Sections 10.4 and 10.5 respectively. In the last two sections applications to two physical systems are discussed.

10.2 Feedforward Exact Model-Matching

10.2.1 Problem Formulation

Let $\mathbf{K}$ be a differential field and $\mathbf{R} = \mathbf{K}[\partial; \delta]$ denote the ring of generalized differential operators as introduced in Section 2.7.2. Let $\mathbf{Q} = \mathbf{K}(\partial; \alpha, \delta)$ be

H. Bourlès and B. Marinescu: Linear Time-Varying Systems, LNCIS 410, pp. 523–543.
springerlink.com

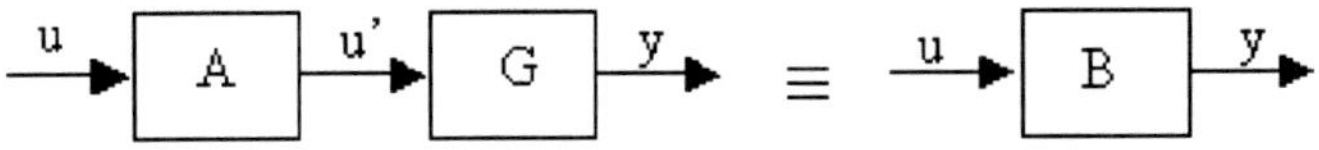

Fig. 10.1 Exact left feedforward model-matching

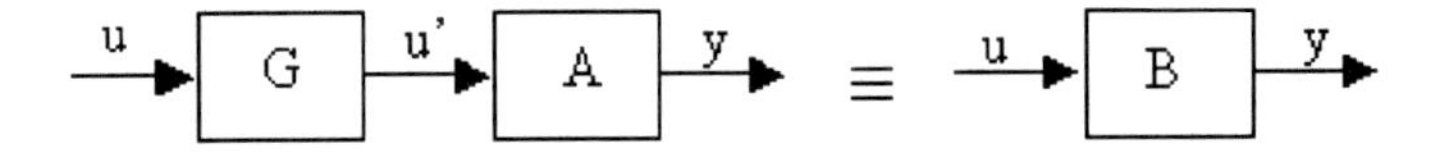

Fig. 10.2 Exact right feedforward model-matching

the quotient field of $\mathbf{R}$. Set $\sigma = \partial^{-1}$ and $\beta = \alpha^{-1}$ and let $\mathbf{S} = \mathbf{K}[[\sigma; \beta, \delta]]$ be the ring of formal power series in σ (see Section 2.9.1). Let $\mathbf{L} = \mathbf{K}((\sigma; \beta, \delta))$ be the quotient field of $\mathbf{S}$, i.e., the field of Laurent series in σ. It is equipped with the commutation rule (2.44) (with $Z = \sigma$, $\alpha = 1$ and $(.)^{\delta} = \frac{d}{dt}(.)$). Every element of $\mathbf{Q}$ can be considered as an element of $\mathbf{L}$, which is of the form $\sum_{i \geq \nu} a_i \sigma^i$, $\nu \in \mathbb{Z}$, $a_{\nu} \neq 0$. A transfer matrix is a matrix with entries in $\mathbf{Q}$ or, alternatively, in $\mathbf{L}$ (Section 5.5.2).

The feedforward exact model matching for a given plant with (proper) transfer matrix $A(\partial)$ consists in finding a compensator with transfer matrix $G(\partial)$ such that the series connection of the initial plant with the compensator has a given desired transfer matrix $B(\partial)$. If a postcompensator is used as in Fig. 10.1, the so-called *left* feedforward exact model-matching problem is to be solved:

$$G(\partial) A(\partial) = B(\partial). \tag{10.1}$$

In case of precompensation as in Fig. 10.2, the *right* feedforward exact model-matching formulation is obtained:

$$A(\partial) G(\partial) = B(\partial). \tag{10.2}$$

The feedforward EMM problems lead to a linear system of equations over $\mathbf{Q}$ or over $\mathbf{L}$ if $\mathbf{Q}$ is embedded in $\mathbf{L}$. The usual conditions for the existence of solutions can thus be used:

Proposition 1120. *Equation (10.1) (respectively equation (10.2)) has a solution if, and only if*

$$\begin{gathered} rank(A(\partial)) = rank\left(\begin{bmatrix} A(\partial) \\ B(\partial) \end{bmatrix}\right) \\ (respectively\ rank(A(\partial)) = rank\left(\begin{bmatrix} A(\partial) & B(\partial) \end{bmatrix}\right)). \end{gathered} \tag{10.3}$$

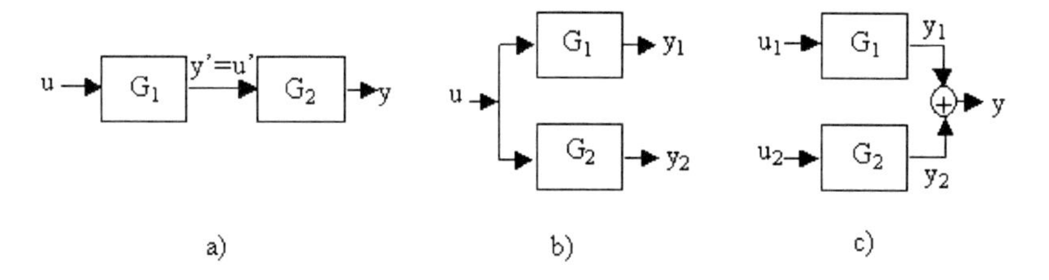

Fig. 10.3 Types of interconnections

10.2.2 Transmission Poles and Zeros at Infinity of Composite Transfer Matrices

The results presented in this section are technical and will be used in the proofs of the main results in the sequel. We are mainly interested here in evaluating the degrees of the poles and zeros at infinity of transfer matrices obtained by multiplying, appending or inverting initially given transfer matrices.

In the more general framework of the interconnection of LTV systems the first situation mentioned above corresponds to cancelations which may occur between poles and zeros, *i.e.*, to hidden modes. The system poles and the hidden modes at infinity introduced in Section 7 as **S**-modules are used here along with their degrees to conclude about the poles and zeros at infinity of the product of two transfer matrices. For this, systems are constructed from the input-output mappings defined by each of the transfer matrices as follows: let $\sum_{G_1}$ and $\sum_{G_2}$ be the systems defined by the input-output mappings given by the transfer matrices G_1 and G_2 respectively. The system $\sum_{G_2G_1}$ of the interconnection in 10.3.a is described by the set of equations of $\sum_{G_1}$ and $\sum_{G_2}$, plus the interconnection equation $y' = u'$. The transfer function of $\sum_{G_2G_1}$ is obviously $S = G_2G_1$. Then, one has the following

Proposition 1121. Series interconnection*: For the series interconnection in Figure 10.3.a,* $\sharp(TP_\infty(S)) = \sharp(TP_\infty(G_1)) + \sharp(TP_\infty(G_2)) - \sharp(HM_\infty(\sum_{G_2G_1}))$.

Proof. From Theorem 1108 1.(i), $\sharp(SP_\infty(\sum_{G_2G_1})) = \sharp(TP_\infty(S) + \sharp(HM_\infty(\sum_{G_2G_1})$. It follows that

$$\sharp(TP_\infty(S)) = \sharp(SP_\infty(\Sigma_{G_2G_1})) - \sharp(HM_\infty(\Sigma_{G_2G_1})) \tag{10.4}$$

Next, $\sharp(SP_\infty(\sum_{G_2G_1}))$ can be evaluated from the degrees of the system poles at infinity of each subsystem $\sum_{G_1}$, $\sum_{G_2}$: let $G_i(\partial) = D_i^{-1}(\partial)N_i(\partial)$, $B_i(\partial) = \left[-D_i(\partial)\ N_i(\partial)\right]$ and

$$B_i(\sigma) = B_i(\partial)\mid_{\partial=\sigma^{-1}} = A_i^{-1}(\sigma)\left[-B_{i_D}^{+}(\sigma)\ B_{i_N}^{+}(\sigma)\right], \quad i \in \{1,2\}$$

be left-coprime factorizations of B_1 and B_2 over $\mathbf{S}$. $B^+_{1_D}$ and $B^+_{2_D}$ are matrices of definition of the modules of system poles at infinity of $\sum_{G_1}$ and $\sum_{G_2}$ respectively. A matrix B^+_{12} of definition of the module of the system poles at infinity of $\sum_{G_2G_1}$ can be deduced from $B^+_{1_D}$ and $B^+_{2_D}$ taking into account the equation of interconnection of the two systems:

$$B^+_{12} = \begin{bmatrix} -B^+_{1_D}(\sigma) & 0 & 0 \\ 0 & -B^+_{2_D}(\sigma) & B^+_{2_N}(\sigma) \\ I & 0 & -I \end{bmatrix}.$$

Using elementary row and column transformations, one obtains $B^+_{12} \sim \begin{bmatrix} 0 & \overline{B}^+_{12} \\ I & 0 \end{bmatrix}$ with $\overline{B}^+_{12} = \begin{bmatrix} B^+_{1_D}(\sigma) & 0 \\ B^+_{2_N}(\sigma) & B^+_{2_D}(\sigma) \end{bmatrix}$. By taking the Dieudonné determinant of the last expression of B^+_{12}, $|B^+_{12}| = |\overline{B}^+_{12}| = |B^+_{1_D}||B^+_{2_D}|$ or, equivalently, $|\sigma^{\Sigma_i \nu_i^{B^+_{12}}}| = |\sigma^{\Sigma_j \nu_j^{B^+_{1_D}}}||\sigma^{\Sigma_k \nu_k^{B^+_{2_D}}}|$ where $\nu_i^{B^+_{12}}$, $\nu_j^{B^+_{1_D}}$ and $\nu_k^{B^+_{2_D}}$ denote the structural indices at infinity of matrices B^+_{12}, $B^+_{1_D}$ and $B^+_{2_D}$ respectively. As $B^+_{1_D}$ and $B^+_{2_D}$ define the modules of system poles at infinity of $\sum_{G_1}$ and $\sum_{G_2}$, it follows that $\sharp(SP_\infty(\sum_{G_2G_1})) = \sharp(SP_\infty(\sum_{G_1})) + \sharp(SP_\infty(\sum_{G_2}))$. Also, as $\sum_{G_1}$ and $\sum_{G_2}$ are built from the input/output mappings given by G_1 and G_2, their only system poles at infinity are the transmission ones and coincide with the transmission poles at infinity of G_1 and G_2 respectively. Thus $\sharp(SP_\infty(\sum_{G_1G_2})) = \sharp(TP_\infty(G_1)) + \sharp(TP_\infty(G_2))$ which exploited in (10.4) leads to the conclusion. ■

Proposition 1122. Parallel interconnection*: Let P_b be the transfer matrix of the interconnection in Figure 10.3.b and P_c the one of the interconnection in Figure 10.3.c (in each case the dimensions of the transfer matrices G_1 and G_2 are supposed to be compliant). Then $\sharp(TP_\infty(P_i)) \leq \sharp(TP_\infty(G_1)) + \sharp(TP_\infty(G_2))$ where $i \in \{b, c\}$.*

Proof. $P_b = \begin{bmatrix} G_1 \\ G_2 \end{bmatrix}$ and $P_c = \begin{bmatrix} G_1 \vdots G_2 \end{bmatrix}$. The result follows from the computation of the Smith-MacMillan form at infinity of P_b and P_c respectively and it is an extension of the well-known in the LTI case characterization of the degree of the transmission pole at infinity of a transfer matrix in terms of the lowest negative power of σ which occurs in any minor of the transfer matrix written over $\mathbf{L}$ (see, *e.g.*, [170]). ■

Consider first a full column rank $m \times r$ matrix G. A left inverse transfer matrix G^{-L} can be constructed and investigated using the Smith-MacMillan form at infinity of G given in Section 7.2.2. Let $P^{-1}GU = \begin{bmatrix} diag\{\sigma^{\nu_1}, ..., \sigma^{\nu_r}\} \\ 0 \end{bmatrix}$ be a Smith-MacMillan factorization at infinity of G. $G^{-L}G = I_r$ leads to $G^{-L}P \begin{bmatrix} diag\{\sigma^{\nu_1}, ..., \sigma^{\nu_r}\} \\ 0 \end{bmatrix} = U$, or, if one partitions $G^{-L}P = \begin{bmatrix} \widetilde{G}_1 \vdots \widetilde{G}_2 \end{bmatrix}$

according to the zero bloc structure of the Smith-MacMillan form at infinity of G, to $\widetilde{G}_1 = Udiag\{\sigma^{-\nu_1}, ..., \sigma^{-\nu_r}\}$. It follows that the set of the left inverses of G is

$$G^{-L} = \left[Udiag\{\sigma^{-\nu_1}, ..., \sigma^{-\nu_r}\} \vdots \Omega \right] P^{-1} \tag{10.5}$$

where Ω is any $r \times (m - r)$ matrix, ν_i are the structural indices at infinity of G and $U \in GL_r(\mathbf{S})$ and $P \in GL_m(\mathbf{S})$ are transformations which provide the Smith-MacMillan form at infinity of G. Parametrization (10.5) can also be written over $\mathbf{Q}$ with $\sigma = \partial^{-1}$.

Similarly, in the case where G is full row rank, the set of its right inverses is:

$$G^{-R} = U \begin{bmatrix} diag\{\sigma^{-\nu_1}, ..., \sigma^{-\nu_m}\}P^{-1} \\ \Omega \end{bmatrix} \tag{10.6}$$

Remark 1123. *This approach avoids direct matrix inversions. Even the inversion of P can be avoided: indeed, as indicated in Section 7.2.2, the Smith-MacMillan form at infinity of G is given by the Smith normal form over $\mathbf{S}$ of $\overline{G}$ (where $G = \sigma^{-k}\overline{G}$): $P^{'-1}\overline{G}U = diag\{\sigma^{\mu_1}, ..., \sigma^{\mu_r}, 0, ..., 0\}$ and thus $P^{'-1}$ is directly obtained from the elementary transformations which lead to the Smith form of $\overline{G}$ and not by a matrix inversion. Next, from Proposition 1056 it follows that there exists an invertible matrix P such that $\sigma^k P^{-1} = P^{'-1}\sigma^k$ from which P^{-1} can be obtained from P'^{-1} without any matrix inversion.*

With this parametrization of the sets of the inverse matrices, relations between the degree of the transmission poles at infinity of the transfer matrix and the degree of the transmission zeros at infinity of its inverses can be given as a direct consequence of Proposition 1122:

Corollary 1124. *Let G be a transfer matrix. Then:*

- *if G is full column (respectively row) rank, then $\sharp(TP_\infty(G^{-L})) \geq \sharp(TZ_\infty(G))$ (respectively $\sharp(TP_\infty(G^{-R})) \geq \sharp(TZ_\infty(G))$)*
- *if G is square, full (row or column) rank and has structural indices at infinity ν_i, $i = 1, ..., n$, then the structural indices at infinity of G^{-L} and G^{-R} are $-\nu_i$, $i = 1, ..., n$ and $\sharp(TP_\infty(G^{-L})) = \sharp(TP_\infty(G^{-R})) = \sharp(TZ_\infty(G))$.*

10.2.3 Existence of Proper Solutions

To be implementable, the solution $G(\partial)$ must be *proper*. Properness is characterized in terms of the transmission pole at infinity as defined by Theorem and Definition 1060. The following result is necessary to state conditions for the existence of proper EMM solutions:

Proposition 1125. *Let $A(\partial) \in \mathbf{Q}^{m \times r}$ and $B(\partial) \in \mathbf{Q}^{l \times r}$ be full column rank transfer matrices. If the pole and the zero at infinity of $F(\partial) = \begin{bmatrix} A(\partial) \\ B(\partial) \end{bmatrix}$ have degree zero, then the poles at infinity of A and B have degree zero.*

Proof. Set $\sigma = 1/\partial$ and let $F(1/\sigma) = \begin{bmatrix} A(1/\sigma) \\ B(1/\sigma) \end{bmatrix}$. The Smith-MacMillan form at infinity of A (respectively of B) is of the form $\sigma^{-k'} P'^{-1} \underline{\underline{A}} U'$ (respectively $\sigma^{-k"} P"^{-1} \underline{\underline{B}} U"$) where k' (respectively $k"$) is the least common denominator (up to similarity) of all entries of $\underline{A}$ (respectively of $\underline{B}$) and $\underline{A}(\sigma) = \sigma^{-k'} \underline{\underline{A}}(\sigma)$ where $\underline{\underline{A}}$ is a matrix with entries in $\mathbf{S}$ (respectively $\underline{B}(\sigma) = \sigma^{-k"} \underline{\underline{B}}(\sigma)$ where $\underline{\underline{B}}$ is a matrix with entries in $\mathbf{S}$). P' and U' (respectively $P"$ and $U"$) are unimodular matrices over $\mathbf{S}$ which give the Smith normal form of $\underline{\underline{A}}$ (respectively $\underline{\underline{B}}$): $P'^{-1} \underline{\underline{A}} U' = \begin{bmatrix} \Lambda \\ 0 \end{bmatrix}$ with

$$\Lambda = \begin{bmatrix} \sigma^{\nu_1} & \dots & 0 \\ \vdots & \ddots & \vdots \\ 0 & \dots & \sigma^{\nu_r} \end{bmatrix} \text{ (resp. } P"^{-1} \underline{\underline{B}} U" = \begin{bmatrix} \Gamma \\ 0 \end{bmatrix} \text{ with } \Gamma = \begin{bmatrix} \sigma^{\mu_1} & \dots & 0 \\ \vdots & \ddots & \vdots \\ 0 & \dots & \sigma^{\mu_r} \end{bmatrix})$$

(see Lemma and Definition 655). Let $k = max\{k', k"\}$. Consider the case $k = k' > k"$. Then, the Smith-McMillan form at infinity of F is $\begin{bmatrix} \Upsilon \\ 0 \end{bmatrix}$ with

$$\Upsilon = \begin{bmatrix} \sigma^{\alpha_1} & \dots & 0 \\ \vdots & \ddots & \vdots \\ 0 & \dots & \sigma^{\alpha_r} \end{bmatrix},$$

where $\alpha_i = min\{\nu_i, \mu_i + k - k"\}$, $1 \leq i \leq r$. As the pole and the zero at infinity of F have degree zero, it follows $\alpha_i = k$, $1 \leq i \leq r$. One obtains that $k = min\{\nu_i, \mu_i + k - k"\}$, $1 \leq i \leq r$ from which $k \leq \nu_i$ and $k" \leq \mu_i$, $1 \leq i \leq r$, *i.e.*, the poles at infinity of A and B, have degree zero. The same rationale can be made in the case $k = k" > k'$. ■

An elegant interpretation of the necessary and sufficient condition for the feedforward EMM problem to have proper solution(s) was given in the time-invariant case in [346] and [349] using the notion of content at infinity of a transfer matrix G denoted by $c_\infty(G)$ and generalized to the time-varying case in Theorem and Definition 1060. The following property is the key point for providing conditions for the existence of proper solutions of the feedforward EMM. The formulation in the time-varying case is the same as in the invariant one (see, *e.g.*, [170], [346]), but the proof is slightly different.

Theorem 1126. *Suppose that $rank(A) = rank(F)$, where $F = \begin{bmatrix} A(\partial) \\ B(\partial) \end{bmatrix}$. The left feedforward EMM problem has proper solutions if, and only if $c_\infty(A) = c_\infty(F)$.*

Proof. Let factorize F as $F(\partial) = \overline{F}(\partial) Q(\partial)$, $\overline{F}(\partial) = \begin{bmatrix} \overline{A(\partial)} \\ \overline{B(\partial)} \end{bmatrix}$ where the pole an the zero at infinity of $\overline{F}(\partial)$ has degree zero and $Q(\partial)$, $\overline{F}(\partial)$ are full rank. Then (10.1) can be written as

$$G(\partial) \overline{A}(\partial) = \overline{B}(\partial) \tag{10.7}$$

where $\overline{A}(\partial)$ is left invertible; from (10.7) one obtains

$$G(\partial) = \overline{B}(\partial)\overline{A}(\partial)^{-L} \tag{10.8}$$

where $\overline{A}(\partial)^{-L}$ is a left inverse of $\overline{A}$.

By Proposition 1125, the poles at infinity of $\overline{A}(\partial)$ and $\overline{B}(\partial)$ have degree zero. From (10.8), one deduces that the pole at infinity of G has degree zero if, and only if the zero at infinity of $\overline{A}(\partial)$ has degree zero, or equivalently, if, and only if $c_\infty(\overline{A}) = 0$. Now, $\overline{A}(\partial)Q(\partial) = A(\partial)$, where $\overline{A}(\partial)$ and $Q(\partial)$ are full column rank and full row rank, respectively. Using Theorem 1063, one obtains:

$$c_\infty(A) = c_\infty(\overline{A}) + c_\infty(Q). \tag{10.9}$$

As Q and F have the same structure at infinity, (10.9) yields $c_\infty(A) = c_\infty(\overline{A}) + c_\infty(F)$. As a result, G is proper if, and only if $c_\infty(A) = c_\infty(F)$. ■

Remark 1127. *Usually it is assumed that A has full column rank (see,* e.g.*, [63] and [170]). Here we treat the general case; from a practical point of view, considering that A is not full column rank means that the input variables are not independent. The plant model can then be changed in order to reduce the number of physical input variables,* i.e.*, to obtain a full column rank transfer matrix.*

Obviously, the same kind of result can be established for the right feedforward EMM problem.

10.2.4 Parametrization of the Class of Proper Solutions

For what follows, the transfer matrices are considered over $\mathbf{L}$. Consider as before the left feedforward EMM problem (10.1) with $A(\sigma) \in \mathbf{L}^{m \times r}$, $B(\sigma) \in \mathbf{L}^{l \times r}$ and A full column rank. The situation $m < r$ is trivial since (10.1) has in this case no solution or a unique one. The solution (if exists) is given by a matrix inversion. The interesting case is $m > r$ in which a class of solutions may exist. As for the demonstration of Proposition 1125, let σ^k be the least common denominator of all entries of A and let

$$P^{-1}\overline{A}U = \begin{bmatrix} \Lambda \\ 0 \end{bmatrix} \tag{10.10}$$

be the Smith form of $\overline{A}(\sigma) = \sigma^k A(\sigma)$ over $\mathbf{S}$. One can thus write (10.1) as $G(\sigma)\sigma^{-k}P\begin{bmatrix} \Lambda \\ 0 \end{bmatrix}U^{-1} = B$. From Proposition 1056(i), there exist an invertible matrix $\overline{P}$, such that

$$\sigma^{-k}P = \overline{P}\sigma^{-k}. \tag{10.11}$$

As a consequence, (10.1) can be further written as $G(\sigma)\overline{P}\sigma^{-k}\begin{bmatrix}\Lambda\\0\end{bmatrix}U^{-1}=B$. Let $\overline{G}=G\overline{P}$ and consider the partition $\overline{G}=\begin{bmatrix}\overline{G}_1\ \overline{G}_2\end{bmatrix}$ according to Λ and the zero block of the Smith form of $\overline{A}$. It follows that $\overline{G}_1\sigma^{-k}\Lambda=BU$ or finally $\overline{G}_1=BU\Lambda^{-1}\sigma^k$.

$\overline{G}_1$ is thus fixed by A and B and $\overline{G}_2$ is free to take any value. If A and B satisfy the condition of Theorem 1126, $\overline{G}_1$ is proper. If, moreover, $\overline{G}_2$ is chosen to be proper, the solution of the EMM problem is proper. It follows:

Theorem 1128. *The proper solutions of the left feedforward EMM problem (10.1) with $A(\sigma)\in\mathbf{L}^{m\times r}$, $m>r$, $rank(A)=r$, $B(\sigma)\in\mathbf{L}^{l\times r}$, if exist, can be parameterized as*

$$G(\sigma)=\begin{bmatrix}BU\Lambda^{-1}\sigma^k\ \Omega\end{bmatrix}\overline{P}^{-1} \tag{10.12}$$

where k, Λ, U are such that $\sigma^{-k}P^{-1}AU=\begin{bmatrix}\sigma^{-k}\Lambda\\0\end{bmatrix}$ is the Smith-MacMillan of A over $\mathbf{L}$, $\overline{P}$ is given by (10.11) and Ω is any proper transfer matrix of dimension $l\times(m-r)$.

A similar parametrization can be of course given for the proper solutions of the right feedforward EMM problem.

From a computational point of view, one can notice that, despite the fact that the expression of the solutions (10.12) contains two inverses of matrices, no special effort is needed to provide them. Indeed, the first one is trivial since Λ is diagonal: $\Lambda=\begin{bmatrix}\sigma^{\nu_1}&\cdots&0\\\vdots&\ddots&\vdots\\0&\cdots&\sigma^{\nu_r}\end{bmatrix}\Longrightarrow\Lambda^{-1}=\begin{bmatrix}\sigma^{-\nu_1}&\cdots&0\\\vdots&\ddots&\vdots\\0&\cdots&\sigma^{-\nu_r}\end{bmatrix}$. Also, $\overline{P}$ need not to be inverted since from (10.11) one obtains $P^{-1}\sigma^k=\sigma^k\overline{P}^{-1}$ so that $\overline{P}^{-1}$ can be computed from P^{-1} which is directly obtained when constructing the Smith form of $\overline{A}$.

Example 1129. *Consider the following continuous-time ($\partial=d/dt$ and $\alpha=1$) transfer matrices: $A(\partial)=\begin{bmatrix}\partial^{-1}\\(\partial+1)^{-1}t\end{bmatrix}$, $B(\partial)=(\partial+2)^{-1}t$. One can easily check that*

$$\begin{bmatrix}1&0\\-1&(t-\sigma)^{-1}(1+\sigma)\end{bmatrix}A(\sigma)=\begin{bmatrix}\sigma\\0\end{bmatrix} \tag{10.13}$$

which gives the Smith-MacMillan form at infinity of A. It follows that $c_\infty(A)=1$. Similarly it can be shown that $c_\infty\left(\begin{bmatrix}B\\A\end{bmatrix}\right)=1$. Thus, from Theorem 1126, the left feedforward EMM problem has proper solutions.

From (10.13) one deduces that $k = 0$, $\Lambda = \sigma$, $U = I_2$ *and* $\overline{P}^{-1} = P^{-1} = \begin{bmatrix} 1 & 0 \\ -1 & (t-\sigma)^{-1}(1+\sigma) \end{bmatrix}$. *It follows from (10.12) that all proper solutions to the left feedforward EMM problem are in this case*

$$G(\sigma) = \left[\, (\sigma^{-1}+2)t\sigma^{-1} - \Omega(\sigma) \;\; \Omega(\sigma)(t-\sigma)^{-1}(1+\sigma) \,\right] \tag{10.14}$$

where $\Omega(\sigma)$ *is any proper transfer function. The class of solutions can be further written as* $G(\sigma) = \left[\, (1+2\sigma)^{-1}\sigma t\sigma^{-1} - \Omega(\sigma) \;\; \Omega(\sigma)(t-\sigma)^{-1}(1+\sigma) \,\right]$, *or, using the commutation rule (2.44) (with* $Z = \sigma$, $\alpha = 1$ *and* $(.)^\delta = \frac{d}{dt}(.)$*),*

$$G(\sigma) = [(1+2\sigma)^{-1}(t-\sigma) - \Omega(\sigma) \quad \Omega(\sigma)(t-\sigma)^{-1}(1+\sigma)]. \tag{10.15}$$

To check this result, let pull-back the solution (10.15) in $\boldsymbol{Q}$: $G(\partial) = [(\partial+2)^{-1}(\partial t - 1) - \Omega(\partial) \quad \Omega(\partial)(\partial t - 1)^{-1}(1+\partial)]$. *Using the commutation rule (2.33) (with* $X = \partial$ *and* $(.)^\delta = \frac{d}{dt}(.)$*),*

$$G(\partial) = \left[\, (\partial+2)^{-1}t\partial - \Omega(\partial) \;\; \Omega(\partial)(t\partial)^{-1}(\partial+1) \,\right]$$

and $G(\partial)A(\partial) = (\partial+2)^{-1}t - \Omega(\partial)\partial^{-1} + \Omega(\partial)(t\partial)^{-1}t = (\partial+2)^{-1}t$. *Thus,* $G(\partial)A(\partial) = B(\partial)$, $\forall \Omega(\partial)$.

Obviously, for any $\Omega(\partial)$ *proper,* $G(\partial)$ *is proper. As an example, the particular solution obtained for* $\Omega(\partial) = 0$, $G(\partial) = \left[\, (\partial+2)^{-1}t\partial \; 0 \,\right]$ *is proper.*

10.3 Feedback Exact Model-Matching

The EMM problem occurs in many forms according to the type of regulator used (static/dynamic), the configuration used for the implementation (feedback/feedforward) and the variables used for the control (state/output). The solutions proposed in the literature are often dedicated to one of this situations but some of them are equivalent from an algebraic point of view. In this section it is shown how some EMM problems can be reduced to the exact feedforward one.

10.3.1 Dynamic Output Feedback EMM

Consider the situation in Figure 10.4 where the input of the controller G is switched to y. The output feedback EMM problem is to find a proper transfer matrix G such that the transfer matrix $v \longmapsto y$ in Figure 10.4 matches exactly a given transfer matrix B.

This problem can be written

$$[I - A(\partial)G(\partial)]^{-1}A(\partial) = B(\partial) \tag{10.16}$$

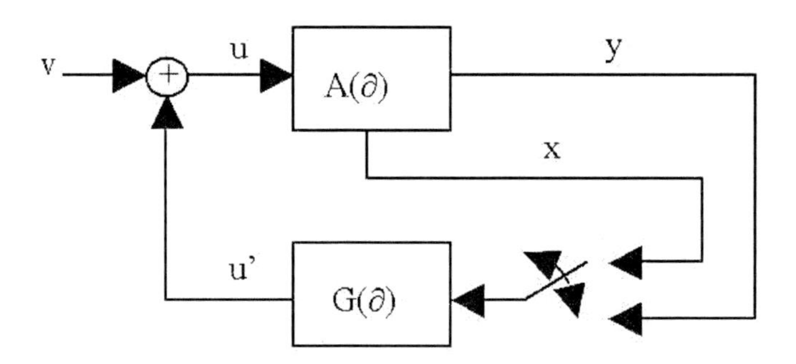

Fig. 10.4 Feedback EMM

and can be solved in two steps:

- first, find $X(\partial)$ (not necessarily proper) such that $X(\partial)B(\partial) = A(\partial)$. This is a left feedforward EMM type problem.
- next, find $G(\partial)$ proper such that $A(\partial)G(\partial) = I - X(\partial)$. This is a right feedforward EMM type problem.

Example 1130. *Consider the discrete-time ($\partial = z - 1$, $\alpha = q$, $\delta = q - 1$ where q is the forward shift operator) transfer matrices: $A(z) = (1+z)^{-2}tz^2$, $B(z) = (t+z)^{-1}z$. As A is a single input/single output system, the solutions of the two EMM problems are trivial. First, $X(z)B(z) = A(z) \Longrightarrow X(z) = (1+z)^{-2}tz(t+z)$. Using the commutation rule (2.33) (with $X = z-1$, $\alpha = q$ and $(.)^\delta = (q-1)(.)$), the solution of the first step is $X(z) = (1+z)^{-2}[-2t + t^2 + (t+t^2)z + tz^2]$. At the second step, from $A(z)G(z) = 1 - X(z)$ one obtains $G(z) = z^{-2}[t^{-1} + 2 - t + (2t^{-1} - 1 - t)z + (t^{-1} - 1)z^2]$ which is the proper solution to the problem.*

This can be easily checked if one uses the equations of the closed-loop in Figure 10.4

$$\begin{cases} y = (1+z)^{-1}tz^2u \\ u' = z^{-2}[t^{-1} + 2 - t + (2t^{-1} - 1 - t)z + (t^{-1} - 1)z^2]y \\ u = u' + v \end{cases}$$

to compute the transfer function from v to y: $y = [(1+z)^2 - 1 - 2t + t^2 - (2 - t - t^2)z + (1-t)z^2]^{-1}tz^2v = [t(-2 + t + z + tz + z^2)]^{-1}tz^2v$ or, using again the commutation rule (2.33), $y = [tz(t+z)]^{-1}tz^2v$ which gives the desired transfer function $B(z)$.

10.3.2 Dynamic State Feedback EMM

Consider now a state x feedback in Figure 10.4 along with a state representation of the strictly proper system A (as defined in Section 5.5.3):

$$\begin{cases} \partial x(t) = Fx(t) + \Gamma u(t) \\ y(t) = Hx(t) \end{cases} \tag{10.17}$$

where $F \in \mathbf{K}^{n\times n}$, $\Gamma \in \mathbf{K}^{n\times r}$, $H \in \mathbf{K}^{m\times n}$. Let $B(\partial)$ be the desired transfer matrix. State feedback solutions are sometimes attractive so that we consider a control law $u = v + K(\partial)x(t)$. If one consider Γ of full column rank (in other words, independent inputs), the dynamic state feedback is equivalent to a dynamic precompensation. More precisely, the closed-loop transfer matrix, *i.e.*, the transfer $v \longmapsto y$ is $A(\partial)[I_r - G(\partial)(\partial I_n - F)^{-1}\Gamma]^{-1}$. The EMM problem is thus to find a proper compensator with transfer matrix $G(\partial)$ such that $A(\partial)[I_r - G(\partial)(\partial I_n - F)^{-1}\Gamma]^{-1} = B(\partial)$. This problem can be further solved in two steps:

- first, find $X(\partial)$ (not necessarily proper) such that $A(\partial)X(\partial) = B(\partial)$. This is a right feedforward EMM type problem.
- next, find $G(\partial)$ proper such that $G(\partial)\overline{A}(\partial) = \overline{B}(\partial)$, where $\overline{A} = (\partial I_n - F)^{-1}\Gamma X(\partial)$ and $\overline{B} = X(\partial) - I_r$. This is a left feedforward EMM type problem.

10.4 Output Disturbance Decoupling Problem

Several control problems can be treated as model-matching problems. In this section we investigate how the output disturbance decoupling one can be fitted into the EMM framework.

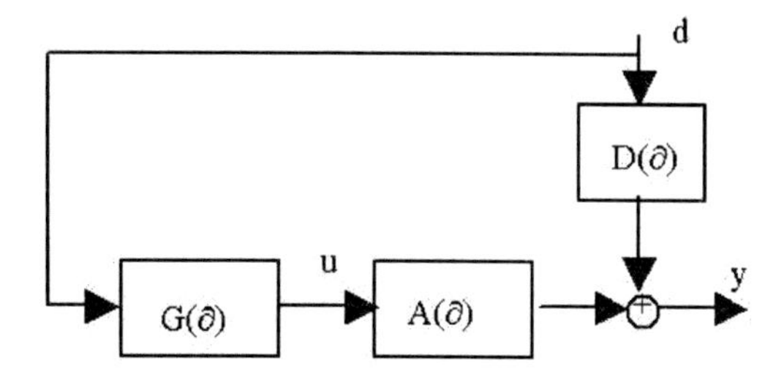

Fig. 10.5 Output disturbance decoupling

Consider the case in Figure 10.5 where the output of system is polluted by a disturbance signal d. It is assumed that d is measurable and its model (transfer matrix D) is known.

The output disturbance problem consist in finding a control law to cancel the effect of d on y. If a precompensator G is chosen for this, the problem reduces to a right feedforward EMM problem: given $A(\partial)$ and $D(\partial)$, find $G(\partial)$ proper such that

$$A(\partial)G(\partial) = -D(\partial). \tag{10.18}$$

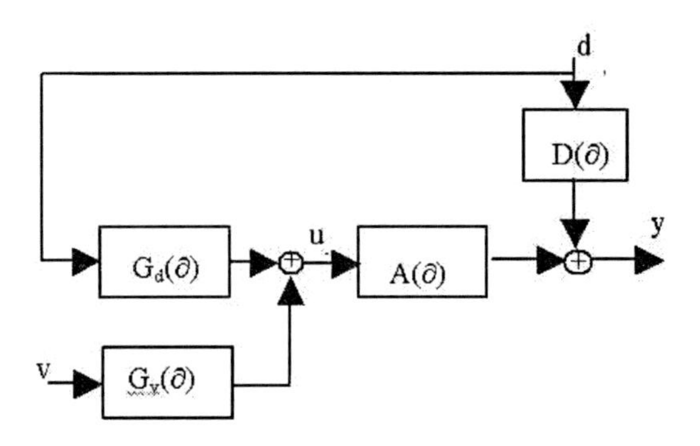

Fig. 10.6 Two-degrees of freedom EMM

With the control law given by (10.18) the output is exactly zero. However, the output disturbance decoupling can be integrated in a general control problem. A two-degrees of freedom regulator can be used like in Figure 10.6 to reject the disturbance d and to achieve another control objective like, *e.g.*, to assign a desired transfer matrix B from v to y.

The output y is given by

$$y = (D(\partial) + A(\partial)G_d(\partial))d + A(\partial)G_v(\partial)v. \tag{10.19}$$

To exactly cancel the effect of disturbance d on the output y one has to choose G_d such that

$$A(\partial)G_d(\partial) = -D(\partial). \tag{10.20}$$

Next, to ensure the desired B transfer from v to y it is sufficient to choose G_v such that

$$A(\partial)G_v(\partial) = B(\partial). \tag{10.21}$$

Obviously, to be implementable G_d and G_v should be proper. Problems (10.20) and (10.21) are right feedforward EMM independent problems.

10.5 System Decoupling Problem

The techniques of decoupling have been extensively investigated for linear time-invariant systems in the purpose of noninteracting control. The problem is to find a proper compensator such that the corrected system has a diagonal structure (purely diagonal or block diagonal). In this way, a subset of inputs affect only a subset of outputs and the regulation problem can be decomposed in parallel independent subproblems.

If a dynamic precompensation is chosen, from an algebraic point of view, the decoupling problem can be reduced to a right feedforward EMM problem if a desired decoupled transfer matrix is specified. This is in general too restrictive and only the structure (dimensions of the blocks) of the decoupled transfer matrix has to be specified.

10.5.1 Necessary and Sufficient Condition for Block-Decoupling

The necessary and sufficient condition for general block decoupling by dynamic precompensation formulated in [158] can be also given in the (continuous- and discrete-time) time-varying case.

Definition 1131. *Let B be a $m \times q$ transfer matrix and $m_1, ..., m_k \in \mathbb{N}$ such that $\sum_{i=1}^{k} m_i = m$, $m_i \neq 0$, $i = 1, ..., k$. B is said $(m_1, ..., m_k)$* decoupled *if there exist positive integers $q_1, ..., q_k$ such that $\sum_{i=1}^{k} q_i = q$ and $B = diag\{B_1, \; ..., \; B_k\}$, where B_i, $i = 1, ..., k$ are $m_i \times q_i$ transfer matrices.*

Definition 1132. *A compensator G is called admissible for a given plant A if the following rank condition is satisfied: $rank(AG) = rank(A)$.*

This assumption avoids trivial solutions to the decoupling problem, in particular $G = 0$.

Theorem 1133. *Given an $m \times r$ transfer matrix A and $m_i \in \mathbb{N}$ such that $m_i \neq 0$, $i = 1, ..., k$ and $m = \sum_{i=1}^{k} m_i$, there exists an admissible compensator $\overline{G}$ such that $A\overline{G}$ is $(m_1, ..., m_k)$ decoupled if, and only if $rank(A) = \sum_{i=1}^{k} q_i$, where $q_i = rank(A_i)$ and A_i are the row blocks induced on A by the partition $(m_1, ..., m_k)$:*

$$A = \begin{bmatrix} A_1 \\ \vdots \\ A_k \end{bmatrix} \begin{matrix} \}m_1 \\ \\ \}m_k. \end{matrix} \tag{10.22}$$

Proof. *The if part:* Consider the Smith-MacMillan form at infinity of each block A_i of the partition (10.22) of A: there exist matrices $P_i \in \mathbf{GL}_{m_i}(\mathbf{S})$ and $U_i \in \mathbf{GL}_r(\mathbf{S})$ such that $P_i^{-1} A_i U_i = diag\{\sigma^{\nu_1}, ..., \sigma^{\nu_{q_i}}, 0, ..., 0\}$, where ν_j, $j = 1, ..., q_i$ are the structural indices at infinity of A_i. It follows that $A_i = P_i \begin{bmatrix} \overline{A}_i \\ 0 \end{bmatrix}$ where $\overline{A}_i = \begin{bmatrix} diag\{\sigma^{\nu_1}, ..., \sigma^{\nu_{q_i}}\} \vdots 0 \end{bmatrix} U_i^{-1}$ is full row rank. Since $rank(A) = \sum_{i=1}^{k} q_i$ it follows that

$$\overline{A} = \begin{bmatrix} \overline{A}_1 \\ \vdots \\ \overline{A}_k \end{bmatrix} \tag{10.23}$$

is right invertible, so there exists $\overline{G} \in \mathbf{L}^{r \times q}$, where $q = \sum_{i=1}^{k} q_i$, such that

$$\overline{A}\,\overline{G} = I_q. \tag{10.24}$$

Let

$$\overline{G} = \left[\overline{G}_1 \vdots \cdots \vdots \overline{G}_k\right] \tag{10.25}$$

be the $(q_1, ..., q_k)$ column partition of $\overline{G}$. From (10.24) results

$$\overline{A}_i\overline{G}_j = \begin{cases} I_{q_i} \; if \; i = j \\ 0 \; if \; i \neq j \end{cases}, \quad i, j = 1, ..., k. \tag{10.26}$$

Next, $A = diag\{P_1, ..., P_k\}\begin{bmatrix} \begin{bmatrix} \overline{A}_1 \\ 0 \end{bmatrix} \\ \vdots \\ \begin{bmatrix} \overline{A}_k \\ 0 \end{bmatrix} \end{bmatrix}$. Using this factorization of A and (10.26), one obtains $\overline{B} = A\overline{G} = diag\{P_1, ..., P_k\}diag\{\begin{bmatrix} I_{q_1} \\ 0 \end{bmatrix}, ..., \begin{bmatrix} I_{q_k} \\ 0 \end{bmatrix}\} = diag\{\overline{B}_i\}$, where $\overline{B}_i = P_i\begin{bmatrix} I_{q_i} \\ 0 \end{bmatrix}$, $i = 1, ..., k$. Obviously, $\overline{B}$ is a $(m_1, \ldots, m_k)$ decoupled transfer matrix. Also, since $P_1, \ldots, P_k$ are invertibles, we have $rank(A\overline{G}) = \sum_{i=1}^{k} q_i = rank(A)$ which means that $\overline{G}$ is admissible.

The only if part: Suppose $\overline{G}$ is an admissible decoupling precompensator for A, *i.e.*, $A\overline{G} = diag\{B_1, ..., B_k\}$, where B_i, $i = 1, ..., k$ are transfer matrices with m_i rows respectively. The rows of the transfer matrix A of the plant and the columns of the transfer matrix G of the compensator are partitioned according to the dimensions of the blocks B_i: $A = \begin{bmatrix} A_1 \\ \vdots \\ A_k \end{bmatrix}$ and $G = \left[\overline{G}_1 \vdots \ldots \vdots \overline{G}_k\right]$. Then $A_i\overline{G}_i = B_i$, $i = 1, ..., k$. It follows that $rank(B_i) \leq rank(A_i) = q_i$, $i = 1, ..., k$ and $rank(B) = \sum_{i=1}^{k} rank(B_i) \leq \sum_{i=1}^{k} q_i$. But also $rank(B) = rank(A\overline{G}) = rank(A) = q$, since $\overline{G}$ is admissible. One deduces that $rank(B_i) = q_i$, $i = 1, ..., k$ and $q = \sum_{i=1}^{k} q_i$. ■

10.5.2 Parametrization of the Solutions

The solution $\overline{G}$ provided in the proof of Theorem 1133 is not unique. We now give the whole class of admissible compensators G which $(m_1, ..., m_k)$ decouple the given $m \times r$ transfer matrix A. For this, from A construct $\overline{A}$ as defined by (10.23) and consider the set of right inverses of $\overline{A}$ given by (10.6)

$$\overline{G} = \overline{A}^{-R} = \overline{U}\begin{bmatrix} diag\{\sigma^{-\mu_1}, ..., \sigma^{-\mu_q}\}\overline{P}^{-1} \\ \Omega \end{bmatrix} \tag{10.27}$$

where μ_i, $i = 1, ..., q$, $q = rank(\overline{A}) = rank(A)$ are the structural indices at infinity of $\overline{A}$ and Ω is any $(r-q) \times q$ transfer matrix. Consider again the $(q_1, ..., q_k)$ partition of $\overline{G}$ (10.25), where $q_i = rank(A_i)$, $i = 1, ..., k$ and A_i are the blocks of A induced by the $(m_1, ..., m_k)$ partition (10.22). Then, any compensator

$$G = \left[G_1 \vdots \ldots \vdots G_k \right], \quad G_i = \overline{G}_i \Theta_i, \quad i = 1, ..., k \tag{10.28}$$

where Θ_i is any transfer matrix with q_i rows ($q_i = rank(A_i)$), provides the same type of decoupling as $\overline{G}$: $AG = B$ where

$$B = diag\{B_i\}, \quad B_i = P_i \begin{bmatrix} \Theta_i \\ 0 \end{bmatrix}, \quad i = 1, ..., k. \tag{10.29}$$

As for the EMM problem, the Smith-MacMillan form at infinity of A provides not only the characterization of the class of all decoupling solutions but also a pragmatic way of computing which avoids direct matrix inversions.

10.5.3 Proper and Minimal Delay Solutions

Θ_i can always be chosen to obtain a proper compensator G. This can be trivially achieved by assigning to $\Theta_i(\sigma)$ sufficiently large powers of σ. However, one can ask for the simplest structure of the decoupled system B which allows a proper decoupling solution G. The complexity of B is measured by the degree of its transmission zeros at infinity or, roughly speaking, in terms of the degrees of its denominators when written over $\mathbf{Q}$ as shown below. This is known as *the minimum delay problem* for the decoupling in the continuous LTI case and it is now addressed in the general LTV case. Suppose that the $m \times r$ transfer matrix A can be $(m_1, ..., m_k)$ block-decoupled ($m = \sum_{i=1}^{k} m_i$, $m_i \in \mathbb{N}$, $m_i \neq 0$, $i = 1, ..., k$). Let $G_i = \overline{G}_i \Theta_i$, $i = 1, ..., k$ be the decoupling solution as found in Section 10.5.2.

First, we are interested in finding necessary and sufficient conditions to obtain proper decoupling solutions. The transfer matrices G_i's given by (10.28) are proper if, and only if $\sharp(TP_\infty(G_i) = 0$, or, using Proposition 1121, if, and only if $\sharp(TP_\infty(\overline{G}_i)) + \sharp(TP_\infty(\Theta_i)) = \sharp(HM_\infty(\sum_{\overline{G}_i \Theta_i}))$. As $B = AG = diag\{B_i\}$, with B_i given by (10.29) in which $P_i \in \mathbf{GL}_{m_i}(\mathbf{S})$, must also be proper, it follows that Θ_i's are proper. Thus, the necessary and sufficient condition for the block-decoupling compensator G to be proper is

$$\sharp(TP_\infty(\overline{G}_i)) = \sharp(HM_\infty(\sum_{\overline{G}_i \Theta_i})) = \sharp(TZ_\infty(\Theta_i)), \quad i = 1, ..., k \tag{10.30}$$

which means that if $\overline{G}_i$ has transmission poles at infinity, these poles must be compensated by hidden modes at infinity introduced by Θ_i in $\sum_{\overline{G}_i \Theta_i}$, or, equivalently, by the transmission zeros at infinity of Θ_i. If a sum of (10.30) is performed following i, it results: $\Sigma_{i=1}^{k}\sharp(TP_{\infty}(\overline{G}_i)) = \Sigma_{i=1}^{k}\sharp(TZ_{\infty}(B_i)) = \sharp(TZ_{\infty}(B))$. From Proposition 1122, $\Sigma_{i=1}^{k}\sharp(TP_{\infty}(\overline{G}_i)) \geq \sharp(TP_{\infty}(\overline{G}))$ and, from Corollary 1124, $\sharp(TP_{\infty}(\overline{G})) \geq \sharp(TZ_{\infty}(A))$ (the latter equality is satisfied, for instance, when Ω in (10.27) is 0). Thus, the proper decoupling compensators G are such that the resulting decoupled transfer matrix B satisfies $\sharp(TZ_{\infty}(B)) \geq \sharp(TZ_{\infty}(A))$. Notice that if the inequality in the latter formula is strict, the transmission zeros at infinity of Θ should compensate more than the transmission zeros at infinity of A. Indeed, this is because supplementary transmission poles at infinity may be brought by Ω in parametrization (10.27).

Next, we investigate the minimal delay solution which is obtained when Ω in (10.27) does not introduce supplementary transmission poles at infinity. This is trivially achieved with $\Omega = 0$ and, in that case, the minimal delay solution is provided by

$$\Theta_i = \overline{U}_i diag\{\sigma^{-\nu_1}, ..., \sigma^{-\nu_p}, 1, ..., 1\},\ i = 1, ..., k \tag{10.31}$$

where $\overline{P}_i^{-1}\overline{G}_i\overline{U}_i = \begin{bmatrix} diag\{\sigma^{\nu_1}, ..., \sigma^{\nu_p}, \sigma^{\nu_{p+1}}, ..., \sigma^{\nu_{q_i}}\} \\ 0 \end{bmatrix}$ is a Smith-MacMillan factorization at infinity of $\overline{G}_i$ with $\nu_l < 0,\ l = 1, ..., p$ and $\nu_j \geq 0,\ j = p+1, ..., q_i$. Obviously, $G_i = \overline{G}_i\Theta_i = \overline{P}_i \begin{bmatrix} diag\{1, ..., 1, \sigma^{\nu_{p+1}}, ..., \sigma^{\nu_{q_i}}\} \\ 0 \end{bmatrix}$ and it is thus proper.

Example 1134. *Consider the following time-varying transfer matrix:* $A(\partial) = \begin{bmatrix} \partial^{-1}t & 0 & 0 & \partial^{-2} \\ 0 & \partial^{-1} & 0 & \partial^{-3} \\ -\partial^{-1} & -\partial^{-1} & -\partial^{-2} & -\partial^{-2} - \partial^{-3} \end{bmatrix}$ *and consider the partition* $m_1 = 2$, $m_2 = 1$. *Let* $A_1(\sigma) = A_1(\partial)\mid_{\partial=\sigma^{-1}} = \begin{bmatrix} (t-\sigma)\sigma & 0 & 0 & \sigma^2 \\ 0 & \sigma & 0 & \sigma^3 \end{bmatrix}$. *The Smith-MacMillan form at infinity of* A_1 *is* $\left[diag\{\sigma, \sigma\} \vdots 0\right]$ *and the one of* A *is* $\left[diag\{\sigma, \sigma, \sigma^2\} \vdots 0\right]$ *from which it follows that* $rank(A_1) = 2$, $rank(A) = 3$ *and the condition of Theorem 1133 is fulfilled:* $rank(A) = rank(A_1) + rank(A_2)$ *thus the (2, 1) decoupling problem has a solution in this case. As both* A_1 *and* A_2 *are full row rank,* $A_1 = \overline{A}_1$ *and* $A_2 = \overline{A}_2$, *so* $A = \overline{A}$ *is right invertible with* $A^{-R} = \overline{G}$ *given by (10.27) with* $\Omega = 0$:

$$\overline{G} = \begin{bmatrix} 0 & -\sigma^{-1} & -\sigma^{-1} \\ -1 & \sigma + \sigma^{-1} - t & \sigma - t \\ -\sigma^{-2} & \sigma^{-1} - \sigma^{-2}t & \sigma^{-1} - \sigma^{-2}t \\ \sigma^{-2} & -\sigma^{-1} + \sigma^{-2}t & -\sigma^{-1} + \sigma^{-2}t \end{bmatrix}. \tag{10.32}$$

Let $\overline{G} = \left[\overline{G}_1 \vdots \overline{G}_2\right]$ *be the partition (10.25) of* $\overline{G}$. *The minimal delay solution (10.31) can be computed from the following Smith-MacMillan factorizations at infinity of* $\overline{G}_1$ *and* $\overline{G}_2$: $\overline{P}_i^{-1}\overline{G}_i\overline{U}_i = \begin{bmatrix} \overline{\Lambda}_i \\ 0 \end{bmatrix}$, $i \in \{1,2\}$ *with* $\overline{\Lambda}_1 = diag\{\sigma^{-2},\ \sigma^{-1}\}$, $\overline{U}_1 = \begin{bmatrix} 1 & \sigma - t \\ 0 & 1 \end{bmatrix}$, $\overline{\Lambda}_2 = \sigma^{-2}$, $\overline{U}_2 = 1$ *which leads to* $\Theta_1(\sigma) = \begin{bmatrix} \sigma^2 & \sigma^2 - t\sigma \\ 0 & \sigma \end{bmatrix}$ *and* $\Theta_2(\sigma) = \sigma^2$. *The solution provided by (10.28) is* $G(\sigma) = \begin{bmatrix} 0 & -1 & -\sigma \\ -\sigma^2 & 1 & \sigma^3 - t\sigma^2 \\ -1 & 0 & -t-\sigma \\ 1 & 0 & t+\sigma \end{bmatrix}$ *which is proper and* $(2,1)-$*block decoupling: using the multiplication rule (2.44) (with* $Z = \sigma$, $\alpha = 1$ *and* $(.)^{\delta} = \frac{d}{dt}(.)$*) one could check that* $A(\sigma)G(\sigma) = \begin{bmatrix} \sigma^2 & -\sigma t & 0 \\ 0 & \sigma & 0 \\ 0 & 0 & \sigma^2 \end{bmatrix} = B(\sigma)$ *which is proper and (2,1) decoupled. Also,* $rank(AG) = 3 = rank(A)$, *thus* G *is admissible. Moreover, the Smith-MacMillan form at infinity of* B *is* $diag\{\sigma, \sigma^2, \sigma^2\}$, *thus* $\sharp(TZ_\infty(B)) = 5$. *Notice that this solution has less "delay" than the one obtained by simply dividing the columns of* $\overline{G}$ *(given by (10.32)), the right inverse of* A, *by large enough powers of* σ *in order to render each entry of* G *proper. Doing so, each column would be multiplied by at least* σ^2 *which leads to a solution* B *with* $\sharp(TZ_\infty(B))$ *equal at least to 6. In fact, 6 is the minimal "delay" which can be achieved for the system in this example by row-by-row decoupling,* i.e., *for the* (1, 1, 1) *decoupling.*

10.6 Case of a FACTS in a Power System

Consider again the electrical circuit in Figure 9.2. As mentioned in Section 9, this model is used to take into account the evolution of the voltages and the currents of the electrical transmission lines for an upper control level of the hierarchical grid control of a power system, *i.e.*, the Electrical High Voltage (EHV) control which is, for instance, the control of 225kV and 400kV lines in France. When the branch reactance X is time-varying one obtains the model of a Thyristor Controlled Switched Capacitor (TCSC) which is a power electronics device controlled to adapt the impedance of the branch.

It has been shown in Section 9 that in order to control this model without impulsive motions, two currents have to be chosen as inputs. Let us take $u = [i_A\ i_B]^T$ and $y = [v_1\ v_2]^T$. Let us normalize C_i, setting $C_1 = C_2 = 1$ and assume that the value of R is negligible ($R = 0$). Using the equations (9.8), one can compute the transfer matrix $A(s)$ from u to y

$$A(s) = \begin{bmatrix} D^{-1}(s)N(s) & D^{-1}(s) \\ D^{-1}(s)N(s) + s^{-1} & D^{-1}(s) - s^{-1} \end{bmatrix} \tag{10.33}$$

where $D(s) = (2+\dot{X}s+Xs^2)s$, $N(s) = 1+\dot{X}+Xs^2$ and $X(t)$ is given by (9.7). We wonder whether there exists a proper transfer matrix $G(s)$ satisfying the left feedforward EMM problem (10.1) with $B(s)$ of the form: $B(s) = \begin{bmatrix} k_1 s^{-1} & 0 \\ 0 & k_1 s^{-1} \end{bmatrix}$. In other words, we try to find a feedforward compensator to decouple the transfer $u \mapsto y$ and to retrieve the usual time invariant relation between voltage and current as in the first two equations of (9.8). Set $\sigma = 1/s$ and compute first the Smith-MacMillan form at infinity of A:

$$A(1/\sigma) = \begin{bmatrix} D^{-1}(1/\sigma)N(1/\sigma) & D^{-1}(1/\sigma) \\ D^{-1}(1/\sigma)N(1/\sigma) + \sigma & D^{-1}(1/\sigma) - \sigma \end{bmatrix} \tag{10.34}$$

where $D(1/\sigma) = 2/\sigma + \dot{X}/\sigma^2 + X/\sigma^3$, $N(1/\sigma) = 1 + \dot{X}/\sigma + X/\sigma^2$. Using the commutation rule (2.44) (with $Z = \sigma$, $\alpha = 1$ and $(.)^\delta = d/dt(.)$) one obtains

$$A(1/\sigma) = \tau^{-1}(\sigma) \begin{bmatrix} \gamma(\sigma)\sigma & -\sigma^3 \\ \varepsilon(\sigma)\sigma & \pi(\sigma)\sigma \end{bmatrix} \tag{10.35}$$

where

$$\begin{aligned} \tau(\sigma) &= 2\alpha - \beta - (\alpha - \beta)t + (\alpha - \beta)\sigma + 2\sigma^2 \\ \gamma(\sigma) &= \alpha - (\alpha - \beta)t + 2(\alpha - \beta)\sigma + 2\sigma^2 \\ \varepsilon(\sigma) &= \alpha - \beta - (\alpha - \beta)\sigma + \sigma^2 \\ \pi(\sigma) &= -2\alpha + \beta + (\alpha - \beta)t - (\alpha - \beta)\sigma - \sigma^2 \end{aligned} \tag{10.36}$$

are units of $\mathbf{S}$ if $\alpha \neq 0$, $\alpha \neq \beta$, $\beta \neq 2\alpha$. Using the three classic elementary column and row operations defined in Section 2.10.3, one can further transform (10.35) to finally obtain the Smith MacMillan form:

$$P^{-1}(\sigma)A(1/\sigma)U(\sigma) = \Lambda \tag{10.37}$$

where

$$\begin{aligned} P^{-1}(\sigma) &= \begin{bmatrix} 1 & \gamma^{-1}(\sigma)(\sigma)^2\xi^{-1}(\sigma) \\ 0 & \xi^{-1}(\sigma) \end{bmatrix} \begin{bmatrix} \tau(\sigma) & 0 \\ -\tau(\sigma) & \tau(\sigma) \end{bmatrix}, \\ U(\sigma) &= \begin{bmatrix} \gamma^{-1}(\sigma) & 0 \\ 0 & \varepsilon^{-1}(\sigma) \end{bmatrix}, \ \Lambda = \begin{bmatrix} \sigma & 0 \\ 0 & \sigma \end{bmatrix} \end{aligned} \tag{10.38}$$

and $\xi(\sigma) = \varepsilon^{-1}(\sigma)\pi(\sigma) - \sigma\gamma^{-1}(\sigma)\sigma$ is another unit of $\mathbf{S}$. Hence $c_\infty(A) = -2$. Similar calculations yield $c_\infty(\begin{bmatrix} B \\ A \end{bmatrix}) = -2$. Therefore, by Theorem 1126, the EMM problem admits proper solutions in this case. The parametrization of the whole class of proper solutions is given by (10.10) with $k = 1$, P, U and Λ given by (10.38) and Ω any proper transfer matrix with 2 rows and 2 columns.

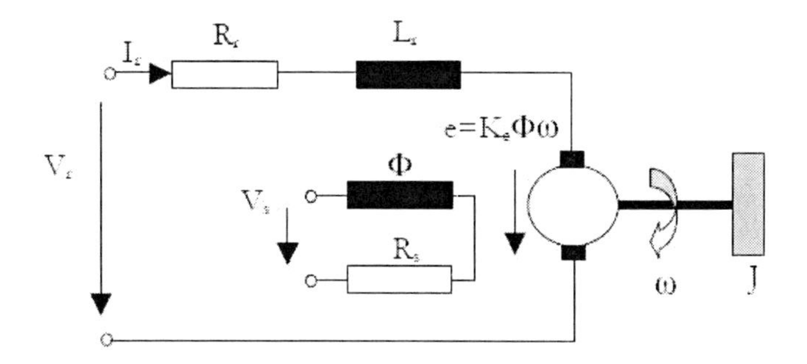

Fig. 10.7 Equivalent circuit of the DC motor

10.7 Case of a DC Motor

Consider a separated excitation variable flux DC motor (see, *e.g.*, [213]) described by the equations:

$$\begin{array}{l} L_s \frac{dI_s}{dt} + R_s I_s = V_s(t), \\ L_s I_s(t) = \Phi(t) \\ L_r \frac{dI_r(t)}{dt} + R_r I_r(t) = V_r(t) - K_e \Phi(t)\omega(t) \\ J\frac{d\omega(t)}{dt} + f\omega(t) = K_m \Phi(t) I_r(t) \end{array} \tag{10.39}$$

where the indices s and r denote respectively the stator and rotor variables as follows: R the resistances, L the inductances, I the currents and V the voltages. J is the rotor and load inertia and Φ is the flux produced by the stator current. The equivalent circuit of the motor is given in Fig. 10.7. By elimination of I_r from the last two equations of (10.39) one can obtain the relation between the rotor velocity and the rotor voltage:

$$\frac{d^2\omega}{dt^2} + \alpha_1(t)\frac{d\omega}{dt} + \alpha_0(t)\omega = \beta(t)V_r \tag{10.40}$$

where

$$\begin{array}{l} \alpha_0(t) = \frac{fR_r}{L_r J} + \frac{K_m K_e f}{L_r J}\Phi^2 \\ \alpha_1(t) = \frac{f}{J} - \frac{\dot{\Phi}(t)}{\Phi(t)} \\ \beta(t) = \frac{K_m \Phi(t)}{J L_r}. \end{array} \tag{10.41}$$

To simplify, it is assumed that the variation in the stator coil voltage induces a periodic time-varying flux given by:

$$\Phi(t) = \Phi_0(1 + \gamma sin(\pi t)). \tag{10.42}$$

The time-varying transfer function from V_r to ω is thus:

$$\mathcal{H}(s) = [s^2 + \alpha_1(t)s + \alpha_0(t)]^{-1}\beta(t). \tag{10.43}$$

The EMM problem for this system has been considered in [189] under the assumption of constant flux in which case coefficients (10.41) are constant. A desired transfer function

$$B(s) = (s^2 + 2w_\nu s + w_\nu^2)^{-1} \tag{10.44}$$

is assigned to the closed-loop. This problem can now be addressed considering the time dependance (10.42) of the flux: $A(\sigma) = [\sigma^{-2} + \alpha_1\sigma^{-1} + \alpha_0]^{-1}\beta = [1 + \sigma\alpha_1 - \sigma^2\dot{\alpha}_1 + \sigma^2\alpha_0]^{-1}\sigma^2\beta$ from which results that the Smith-MacMillan form of A over $\mathbf{L}$ is σ^2 so that $c_\infty(A) = -2$. Similarly it can be shown that the Smith-MacMillan form of $[A \;\; B]$ over $\mathbf{L}$ is $[\sigma^2 \;\; 0]$ so that $c_\infty([A \;\; B]) = -2 = c_\infty(A)$ and, from Theorem 1126, it follows that the right feedforward EMM problem has here proper solutions. As A and B are scalar transfer functions, the solution to the equation $AG = B$ is unique and can be trivially obtained by inverting A: $G(s) = A^{-1}(s)B(s) = \beta^{-1}[s^2 + \alpha_1 s + \alpha_0](s^2 + 2w_r s + w_r^2)^{-1}$, which is obviously proper.

Remark 1135. *The EMM topics were addressed here for the LTV systems with coefficients defined over a differential* field $\boldsymbol{K}$. *However, the techniques used avoid matrix inversions and allow thus the extension of these results to systems defined over* domains *which have matrices of definition* regular at infinity *according to Theorem and Definition 1060 (4). For example, this is the case for the DC motor presented in the section above if* $|\gamma| < 1$. *Indeed, then* $\Phi(t) > 0$ *and the system is defined over the ring* $\boldsymbol{K}[\partial]$ *where* $\boldsymbol{K} = O(\mathbb{R})$ *is the domain of the real analytic functions.*

10.8 Notes and References

The model matching problem, firstly formulated in the late seventies, was next studied for LTI systems in several forms according to the nature of the compensator (static or dynamic compensator) and the type of the compensation (feedback or feedforward) (see, *e.g.*, [4], [228], [342]). Although a lot of work has been carried out on these topics in the time-invariant case, very few results exist in the time-varying one, maybe because of the difficulty to characterize the input-output behavior for this kind of systems. In [271] the Silverman's inversion algorithm is used to solve the problem in the continuous-time case while in [267] and [8] the solution is obtained solving a system of first order differential equations obtained considering identical *weighting pattern matrices* for the model and for the closed-loop system considered also in the continuous-time only. Input-output representations for LTV system have been introduced in the late '80 in [173], [110] and [117]. The series developments proposed in [173] and [110] has been used in [109] and [110] to address the model matching problem for LTV systems. The definition of transfer matrices for LTV systems over the field of fractions of the two-sided Ore domain $\mathbf{K}[s]$, where $\mathbf{K}$ is the differential field of definition to which the system's coefficients are supposed to belong, allowed a more direct and constructive (from the implementation point of view) approach for the model matching topics in [235]. The presentation in this chapter follows this approach.

11
Closed-Loop Control by Output Feedback Pole Placement

11.1 Introduction

The pole placement consists of tuning the natural response of a given system. The input-output representation is convenient for engineers and thus widely used in industrial applications. For this reason, the output feedback problem is of particular interest. It is addressed here using the transfer matrix formalism given in Section 5.5.2 and the stability characterization by the notions of poles introduced in Chapter 6. We consider continuous-time systems (i.e., ∂ stands exclusively for d/dt) which are supposed controllable and observable. These results allow the closed-loop control of LTV systems but also to drive nonlinear systems around a given trajectory.

In Section 11.2 a typical LTV closed-loop system is analyzed. Its poles are defined and used to provide conditions for internal stability. The pole placement procedure is presented in Section 11.3 along with issues for the implementation of the regulators. A Youla-type parameterization of the class of all stabilizing regulators is given in Section 11.4. In Section 11.5 it is treated the case of the well-known two-degrees of freedom control structure while Section 11.6 is devoted to the application of these techniques to the control of nonlinear systems around a given trajectory.

11.2 Closed-Loop Transfer Matrices and Stability

Let $\mathbf{R}$ be an Ore domain and $\mathbf{F} = \mathrm{Q}(\mathbf{R})$ its division ring of fractions. Consider a typical closed-loop system encountered in the output-feedback control of linear systems as given in Figure 11.1 where $(\mathcal{H}, u, y)$ and $(\mathcal{G}, w, v)$ are strongly observable and controllable control systems over $\mathbf{R}$, whose transfer matrices are denoted by $\mathcal{H}(\partial)$ and $\mathcal{G}(\partial)$.

Let $\tilde{u} = \begin{bmatrix} r \\ d \end{bmatrix}$ and $\tilde{y} = \begin{bmatrix} y \\ v \end{bmatrix}$. Assume that $\mathcal{H}$ (resp., $\mathcal{G}$) has both a left-coprime factorization (i.e., of type (i) as introduced in Section 6.6.3)

H. Bourlès and B. Marinescu: Linear Time-Varying Systems, LNCIS 410, pp. 545–580.
springerlink.com

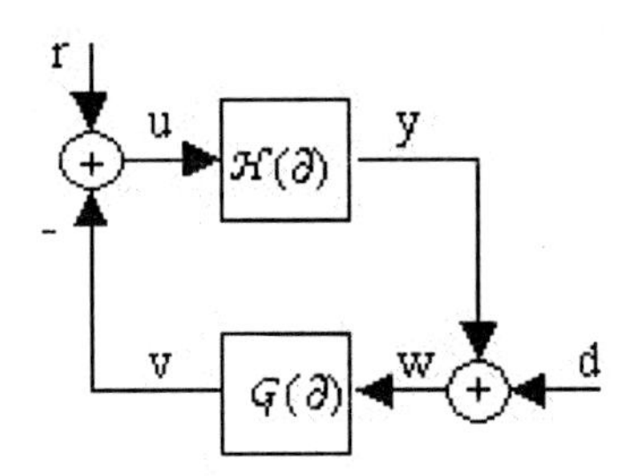

Fig. 11.1 LTV multi input/multi output closed-loop

$\mathbf{F}^{p\times m} \ni \mathcal{H}(\partial) = A_L(\partial)^{-1}B_L(\partial)$ (resp. $\mathbf{F}^{m\times p} \ni \mathcal{G}(\partial) = S_L(\partial)^{-1}R_L(\partial)$) and a right-coprime factorization (i.e., of type (ii)) $\mathcal{H}(\partial) = B_R(\partial)A_R(\partial)^{-1}$ (resp. $\mathcal{G}(\partial) = R_R(\partial)S_R(\partial)^{-1}$).

The latter factorizations are obviously related by (6.64) and by

$$S_L(\partial)R_R(\partial) - R_L(\partial)S_R(\partial) = 0. \tag{11.1}$$

Let

$$A_{cl}(\partial) = A_L(\partial)S_R(\partial) + B_L(\partial)R_R(\partial) \tag{11.2}$$

and

$$\overline{A}_{cl}(\partial) = S_L(\partial)A_R(\partial) + R_L(\partial)B_R(\partial). \tag{11.3}$$

Let M be the system resulting from the interconnection in Figure 11.1.

Theorem and Definition 1136. *(1) The following conditions are equivalent:*
(i) A_{cl} is invertible over $\mathbf{F}$.
(ii) $\overline{A}_{cl}$ is invertible over $\mathbf{F}$.
(iii) $(M, \tilde{u}, \tilde{y})$ is a control system.
(2) The closed-loop in Figure 11.1 is said to be well-posed *if the above equivalent conditions hold.*

Proof. Let $\hat{u}$ and $\hat{y}$ be the columns whose entries are the images of the entries of $\tilde{u}$ and $\tilde{y}$ by the Laplace functor. As easily checked, condition (ii) is necessary and sufficient for a matrix $\mathcal{T}_{\mathcal{H},\mathcal{G}}$ to exist, such that $\hat{y} = \mathcal{T}_{\mathcal{H},\mathcal{G}}\hat{u}$. So is also condition (iii). This matrix $\mathcal{T}_{\mathcal{H},\mathcal{G}}$ is then the transfer matrix of the closed-loop system, and therefore this system is a control system. (See the proof of the theorem below for more details.) ■

Theorem 1137. *Assume that the control system is well-posed.*

(i) Then $\mathcal{T}_{\mathcal{H},\mathcal{G}}$ admits two bicoprime factorizations $(*, A_{cl}, *, *)$ and $(*, \overline{A}_{cl}, *, *)$.
(ii) Therefore, the closed-loop control system is such that $M/[\tilde{u}] \cong$ $\operatorname{coker}_{\mathbf{R}}(\bullet A_{cl}) \cong \operatorname{coker}_{\mathbf{R}}(\bullet \overline{A}_{cl})$.

Proof. As easily checked, we have

$$\mathcal{T}_{\mathcal{H},\mathcal{G}}(\partial) = \begin{bmatrix} \mathcal{H}_{ry}(\partial) & \mathcal{H}_{dy}(\partial) \\ \mathcal{H}_{rv}(\partial) & \mathcal{H}_{dv}(\partial) \end{bmatrix} \tag{11.4}$$

where

$$\begin{aligned} \mathcal{H}_{ry}(\partial) &= [I_p + \mathcal{H}(\partial)\mathcal{G}(\partial)]^{-1}\mathcal{H}(\partial) = \mathcal{H}(\partial)[I_m + \mathcal{G}(\partial)\mathcal{H}(\partial)]^{-1} \\ \mathcal{H}_{dy}(\partial) &= -[I_p + \mathcal{H}(\partial)\mathcal{G}(\partial)]^{-1}\mathcal{H}(\partial)\mathcal{G}(\partial) \\ \mathcal{H}_{dv}(\partial) &= [I_m + \mathcal{G}(\partial)\mathcal{H}(\partial)]^{-1}\mathcal{G}(\partial) \\ \mathcal{H}_{rv}(\partial) &= I_m - [I_m + \mathcal{G}(\partial)\mathcal{H}(\partial)]^{-1}. \end{aligned} \tag{11.5}$$

Next,

$$\begin{aligned} &[I_m + \mathcal{G}(\partial)\mathcal{H}(\partial)]^{-1} = \\ &= [I_m + S_L(\partial)^{-1}R_L(\partial)B_R(\partial)A_R(\partial)^{-1}]^{-1} = \\ &= A_R(\partial)[S_L(\partial)A_R(\partial) + R_L(\partial)B_R(\partial)]^{-1}S_L(\partial) = \\ &= A_R(\partial)\overline{A}_{cl}^{-1}S_L(\partial), \end{aligned}$$

where $\overline{A}_{cl}$ is given by (11.3).

Also, it can be directly checked that

$$[I_m + \mathcal{G}(\partial)\mathcal{H}(\partial)]^{-1} = I_m - \mathcal{G}(\partial)[I_p + \mathcal{H}(\partial)\mathcal{G}(\partial)]^{-1}\mathcal{H}(\partial)$$

from which follows that

$$\begin{aligned} &[I_m + \mathcal{G}(\partial)\mathcal{H}(\partial)]^{-1} = \\ &= I_m - R_R(\partial)S_R(\partial)^{-1}[I_p + A_L(\partial)^{-1}B_L(\partial)R_R(\partial)S_R(\partial)^{-1}]^{-1}A_L(\partial)^{-1}B_L(\partial) = \\ &= I_m - R_R(\partial)A_{cl}(\partial)^{-1}B_L(\partial), \end{aligned}$$

where $A_{cl}(\partial)$ is given by (11.2).

The latter deductions lead to the following expressions of the closed-loop transfer $r \mapsto v$:

$$\mathcal{H}_{rv}(\partial) = R_R(\partial)A_{cl}(\partial)^{-1}B_L(\partial) = I_m - A_R(\partial)\overline{A}_{cl}(\partial)^{-1}S_L(\partial). \tag{11.6}$$

Using similar developments, one can obtain for the other transfers of the closed-loop:

$$\begin{aligned} \mathcal{H}_{dy}(\partial) &= I_p - S_R(\partial)A_{cl}(\partial)^{-1}A_L(\partial) = -B_R(\partial)\overline{A}_{cl}(\partial)^{-1}R_L(\partial) \\ \mathcal{H}_{ry}(\partial) &= S_R(\partial)A_{cl}(\partial)^{-1}S_L(\partial) = B_R(\partial)\overline{A}_{cl}(\partial)^{-1}S_L(\partial) \\ \mathcal{H}_{dv}(\partial) &= R_R(\partial)A_{cl}(\partial)^{-1}A_L(\partial) = A_R(\partial)\overline{A}_{cl}(\partial)^{-1}R_L(\partial). \end{aligned} \tag{11.7}$$

The expressions (11.6) and (11.7) provide the following bicoprime factorizations of (11.4):

$$\begin{aligned} &\mathcal{T}_{\mathcal{H},\mathcal{G}}(\partial) = \\ &= \begin{bmatrix} 0 & -I_p \\ 0 & 0 \end{bmatrix} + \begin{bmatrix} S_R(\partial) \\ R_R(\partial) \end{bmatrix} A_{cl}(\partial)^{-1} \begin{bmatrix} B_L(\partial) & A_L(\partial) \end{bmatrix} = \\ &= \begin{bmatrix} 0 & 0 \\ I_m & 0 \end{bmatrix} + \begin{bmatrix} B_R(\partial) \\ -A_R(\partial) \end{bmatrix} \overline{A}_{cl}(\partial)^{-1} \begin{bmatrix} S_L(\partial) & -R_L(\partial) \end{bmatrix}. \end{aligned} \tag{11.8}$$

Indeed, on one hand, as $R_R(\partial)$ and $S_R(\partial)$ are right-coprime, it follows that $S_R(\partial)$ and $A_{cl}(\partial)$ are right-coprime and, furthermore, that $\begin{bmatrix} S_R(\partial) \\ R_R(\partial) \end{bmatrix}$ and $A_{cl}(\partial)$ are right-coprime. On the other hand, as $A_L(\partial)$ and $B_L(\partial)$ are left-coprime, it follows that $A_{cl}(\partial)$ and $\begin{bmatrix} B_L(\partial) & A_L(\partial) \end{bmatrix}$ are left-coprime. In the same way it can be concluded that $\begin{bmatrix} B_R(\partial) \\ -A_R(\partial) \end{bmatrix}$ and $\overline{A}_{cl}(\partial)$ are right-coprime and $\overline{A}_{cl}(\partial)$ and $\begin{bmatrix} S_L(\partial) & -R_L(\partial) \end{bmatrix}$ are left-coprime.

(ii) is now an obvious consequence of Theorem 893. ■

Definition 1138. *Consider that $A_{cl}(\partial)$ or, equivalently, $\overline{A}_{cl}(\partial)$ admits a direct sum decomposition (6.54) (resp. (6.52)) over an Ore field $\breve{\boldsymbol{K}}$. Then, the* poles (resp. quasi-poles) of the closed-loop system *in Fig. 11.1 are the conjugacy classes of the Smith zeros of $A_{cl}(\partial)$ which coincide with those of $\overline{A}_{cl}(\partial)$.*

The result above has the same formulation as in the LTI case. However, its significance is quite different: in the LTV case, the four transfers of the closed-loop system do not have the same *dynamic*. Indeed, as $A_{cl}(\partial) \neq \overline{A}_{cl}(\partial)$, the inputs r and d do not impose the same *transients* on u and y. If the poles of the closed-loop system are placed over the field K^{∞} defined by (6.46), $A_{cl}(\partial) \sim \overline{A}_{cl}(\partial)$ means that the dynamics of u and y are only *asymptotically* similar, i.e., for $t \to +\infty$. This is however sufficient to conclude about stability.

Definition 1139. *The closed-loop in Fig. 11.1 is said* internally exponentially stable *if all transfers $r \mapsto y$, $r \mapsto v$, $d \mapsto y$ and $d \mapsto v$ are exponentially stable.*

Corollary 1140. *If $\mathcal{G}(\partial)$ is chosen such that a fundamental set of roots of $A_{cl}(\partial)$ (or, equivalently, of $\overline{A}_{cl}(\partial)$) belongs to an Ore field and satisfies the stability condition of Theorem 998, then the closed-loop system in Fig. 11.1 is internally exponentially stable.*

Notice that the internal stability of the closed-loop system ensures disturbance rejection and tracking properties. Indeed, first, the transfer $d \mapsto y$ given by the first line in (11.7) is exponentially stable. Next, one can compute the transfer $r \mapsto e$, where $e = r - y$ is the *tracking error*: $\mathcal{H}_{re}(\partial) = I_m - H_{ry}(\partial)$ which is exponentially stable since $H_{ry}(\partial)$ is.

11.3 Regulator Synthesis and Implementation

11.3.1 Pole Placement

The pole placement problem: For the closed-loop system in Fig. 11.1 one has to find the compensator transfer matrix $\mathcal{G}(\partial)$ such that the closed-loop has a desired fundamental set of poles $\Lambda_{cl} = \{\lambda_1, ..., \lambda_n\}$ which belong to an Ore field and satisfy the stability condition of Theorem 998.

Let $P_{cl}(\partial) = [\partial - \lambda_i;\ i = 1, ..., n]_l$ be the *minimal polynomial* of Λ_{cl} (see Section 4.3.4). Following Definition 767, $P_{cl}(\partial)$ is of degree n since Λ_{cl} is P-idependent. Let $A_{cl}(\partial) \in \mathbf{R}^{m\times m}$ and with the Smith normal form given by Λ_{cl}, i.e., such that $A_{cl}(\partial) \sim diag\{1, ..., 1, P_{cl}(\partial)\}$ (see Lemma and Definition 655). The problem is thus to find $S_R(\partial)$ and $R_R(\partial)$ which satisfy (11.2). The latter is a diophantine equation which has solutions since $A_L(\partial)$ and $B_L(\partial)$ are left-coprime and $\mathbf{R}$ is Euclidean. The whole class of solutions can directly be obtained by solving the system of *differential* equations obtained from (11.2). The solutions of interest for the pole placement problem are the ones for which $S_R(\partial)$ is nonsingular. Notice that, as mentioned in Section 11.2, alternatively (11.3) can be solved (as $A_R(\partial)$ and $B_R(\partial)$ are right-coprime). Compared to (11.2), when directly solved for $S_L(\partial)$ and $R_L(\partial)$, (11.3) leads to a system of *algebraic* equations. Following the discussion in Section 11.2, the above pole placement is different from the one in the LTI case: the same dynamics cannot be assigned for all the closed-loop transfer matrices by choosing $\mathcal{G}(\partial)$[1]. Only a common *asymptotic* behavior is imposed in terms of the chosen poles of the closed-loop system.

Equations (11.2), (11.3), (6.64) and (11.1) lead to

Corollary 1141. *Let $A_L(\partial) \in \boldsymbol{R}^{p\times p}$, $B_L(\partial) \in \boldsymbol{R}^{p\times m}$ be left-coprime and $A_{cl}(\partial) \in \boldsymbol{R}^{p\times p}$. Then there exist 7 matrices $A_R(\partial) \in \boldsymbol{R}^{m\times m}$, $B_R(\partial) \in \boldsymbol{R}^{p\times m}$, $S_R(\partial) \in \boldsymbol{R}^{p\times p}$, $R_R(\partial) \in \boldsymbol{R}^{m\times p}$, $R_L(\partial) \in \boldsymbol{R}^{m\times p}$, $S_L(\partial) \in \boldsymbol{R}^{m\times m}$ and $\overline{A}_{cl}(\partial) \sim A_{cl}(\partial)$ such that the following* generalized diophantine identity *holds:*

$$\begin{bmatrix} A_L(\partial) & B_L(\partial) \\ R_L(\partial) & -S_L(\partial) \end{bmatrix} \begin{bmatrix} S_R(\partial) & B_R(\partial) \\ R_R(\partial) & -A_R(\partial) \end{bmatrix} = \begin{bmatrix} A_{cl}(\partial) & 0 \\ 0 & \overline{A}_{cl}(\partial) \end{bmatrix}. \tag{11.9}$$

11.3.2 Canonical Forms of Transfer Functions

Factorizations (i) and (ii) of Proposition 1010 lead to the right and left canonical forms, well-known for the LTI systems. They are important for the practical implementation of the regulator since they provide a systematic way to ensure causality in the input/output relations. For any entry $\mathcal{G}_{ij}(\partial)$ of $\mathcal{G}(\partial)$,

[1] This is the case even when constant poles are placed by A_{cl}. Indeed, in that case, the poles of the other closed-loop transfers are not necessarily constant.

consider a factorization of type (i): $\mathcal{G}_{ij}(\partial) = a_L(\partial)^{-1} b_L(\partial)$ where $a_L(\partial) = \partial^n + a_{n-1}\partial^{n-1} + ... + a_0$, $b_L(\partial) = b_n \partial^n + b_{n-1}\partial^{n-1} + ... + b_0$ with $a_i, b_j \in \mathbf{K}$. Using the commutation rule (2.33) with $X = \partial$ and $(.)^\delta = d/dt(.)$ one can write $a_L(\partial) = \partial^n + \partial^{n-1}\overline{a}_{n-1} + ... + \overline{a}_0$ and $b_L(\partial) = \partial^n \overline{b}_n + \partial^{n-1}\overline{b}_{n-1} + ... + \overline{b}_0$ from which follows $v = \overline{b}_n w + \partial^{-1}(\overline{b}_{n-1}w - \overline{a}_{n-1}v) + ... + \partial^{-n}(\overline{b}_0 w - \overline{a}_0 v)$ which is a causal description of $\mathcal{G}_{ij}(\partial)$. It can be implemented with the scheme in Fig. 11.2 which does not contain derivators.

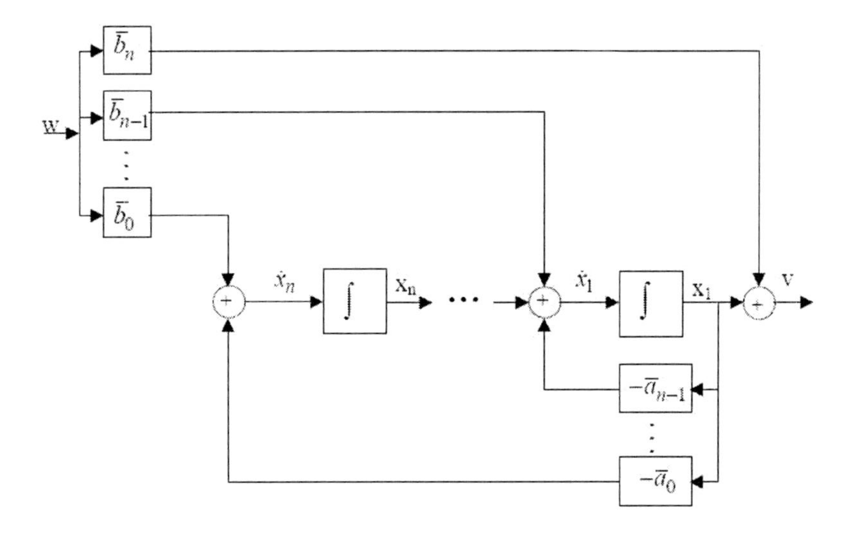

Fig. 11.2 Left canonical form implementation

This corresponds to an observable state-space realization, called the *left canonical form* of $\mathcal{G}_{ij}(\partial)$, which can be associated to (i):

$$\begin{cases} \dot{x} = A_l x + B_l w \\ v = C_l x + D_l w \end{cases}$$
$$A_l = \begin{bmatrix} -\overline{a}_{n-1} & 1 & \cdots & 0 \\ \vdots & \ddots & \ddots & \vdots \\ -\overline{a}_1 & \cdots & \ddots & 1 \\ -\overline{a}_0 & \cdots & \cdots & 0 \end{bmatrix}, \; B_l = \begin{bmatrix} \overline{b}_{n-1} \\ \vdots \\ \overline{b}_0 \end{bmatrix} \tag{11.10}$$
$$C_l = \begin{bmatrix} 1 \; 0 \cdots 0 \end{bmatrix}, \; D_l = \overline{b}_n.$$

A dual form can be obtained for a factorization (ii) of $\mathcal{G}_{ij}(\partial) = b_R(\partial) a_R^{-1}(\partial)$ where $a_R(\partial) = \partial^n + a_{n-1}\partial^{n-1} + ... + a_0$, $b_R(\partial) = b_n\partial^n + b_{n-1}\partial^{n-1} + ... + b_0$. Then a causal description is $\xi^{(n)} = w - a_{n-1}\xi^{(n-1)} - ... - a_0\xi$ which leads to

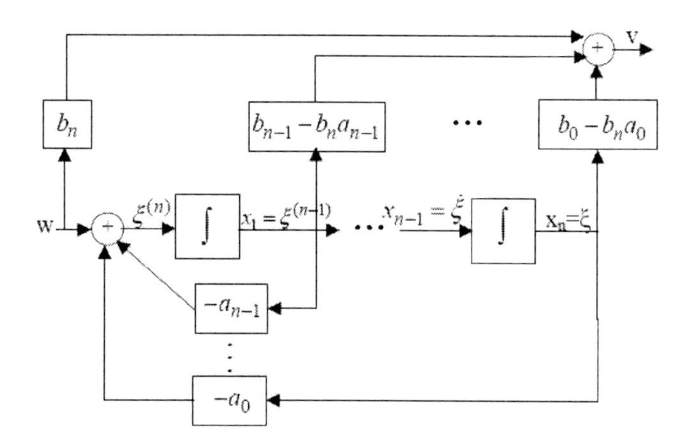

Fig. 11.3 Right canonical form implementation

the implementation in Fig. 11.3 associated to the controllable *right canonical form*

$$\begin{cases} \dot{x} = A_r x + B_r w \\ v = C_r x + D_r w \end{cases}$$

$$A_r = \begin{bmatrix} -a_{n-1} & -a_{n-2} & \cdots & -a_0 \\ 1 & \ddots & \ddots & \vdots \\ \vdots & \ddots & \ddots & 0 \\ 0 & \cdots & 1 & 0 \end{bmatrix}, \; B_r = \begin{bmatrix} 1 \\ 0 \\ \vdots \\ 0 \end{bmatrix} \tag{11.11}$$

$$C_r = \begin{bmatrix} b_{n-1} - b_n a_{n-1} & \cdots & b_0 - b_n a_0 \end{bmatrix}, \; D_r = b_n.$$

11.3.3 Regulator Implementation

To implement $\mathcal{G}(\partial)$, one has to find canonical forms of each entry of $\mathcal{G}(\partial)$ starting from $S_R(\partial)$ and $R_R(\partial)$. The inversion of $S_R(\partial)$, which is a difficult task in the non commutative case, can be replaced by a diagonalization of $S_R(\partial)$: let $U(\partial)$ and $V(\partial)$ be unimodular matrices which lead to the Smith normal form of $S_R(\partial)$:

$$U(\partial)S_R(\partial)V(\partial) = diag\{1, ..., 1, P_S(\partial)\}.$$

Then, $S_R(\partial) = U(\partial)^{-1} diag\{1, ..., 1, P_S(\partial)\} V(\partial)^{-1}$ from which follows

$$U(\partial)^{-1} diag\{1, ..., 1, P_S(\partial)\} V(\partial)^{-1} \mathcal{G}(\partial) = R_R(\partial)$$

and finally $\mathcal{G}(\partial) = V(\partial) diag\{1, ..., 1, P_S(\partial)^{-1}\} U(\partial) R_R(\partial)$. The latter expression provides factorizations (not necessarily coprime) for each entry $\mathcal{G}_{ij}(\partial)$

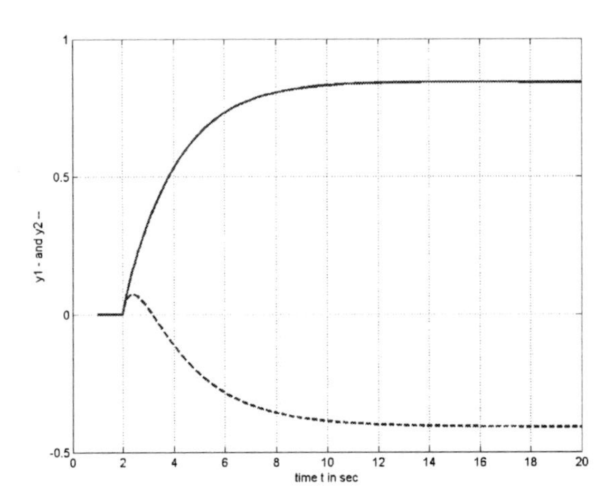

Fig. 11.4 Output responses for Example 1142

without any matrix inversion. The latter are implemented with one of the causal canonical forms presented in Section 11.3.2.

Example 1142. *Let* $\mathcal{H}(\partial) = \begin{bmatrix} (\partial - 1)^{-1} \\ (\partial + 2 + t^{-1})^{-1} \end{bmatrix}$ *which is obviously unstable because of its first entry. The transfer matrix can also be trivially factorized:* $\mathcal{H}(\partial) = A_L(\partial)^{-1} B_L(\partial)$, $A_L(\partial) = \begin{bmatrix} \partial - 1 & 0 \\ 0 & \partial + 2 + t^{-1} \end{bmatrix}$, $B_L(\partial) = \begin{bmatrix} 1 \\ 1 \end{bmatrix}$.

To place the poles $\Lambda_{cl} = \{-2, -2\}$ *one can solve (11.2) to get* $S_R(\partial) = \begin{bmatrix} \partial + \frac{6t+5}{3t+1} & \frac{t}{3t+1} \\ \frac{-9t}{3t+1} & \frac{t}{3t+1} \end{bmatrix}$, $R_R(\partial) = \begin{bmatrix} \frac{9t}{3t+1}\partial + \frac{9(6t^2+5t+2)}{(3t+1)^2} & \frac{-t}{3t+1}\partial + \frac{3t^2+t-1}{(3t+1)^2} \end{bmatrix}$. *Next,* $U(\partial) S_R(\partial) V(\partial) = diag\{1, \partial + 5\}$ *is the Smith normal form of* $S_R(\partial)$ *obtained with*

$$U(\partial) = \begin{bmatrix} 3 + t^{-1} & 0 \\ 1 & -1 \end{bmatrix} \text{ and } V(\partial) = \begin{bmatrix} 0 & 1 \\ 1 & -(3 + t^{-1})\partial - 6 - 5t^{-1} \end{bmatrix}.$$

Thus, $\mathcal{G}(\partial) = V(\partial) diag\{1, (\partial + 5)^{-1}\} U(\partial) R_R(\partial)$. *The closed-loop responses* y *to a unitary step on* r *shown in Fig. 11.4 are stable and compatible with the closed-loop poles* Λ_{cl}. *Notice that, provided that a factorization (ii) is available,* $\mathcal{G}(\partial)$ *can be computed by solving (11.3) instead of (11.2). This has the advantage to lead to a system of algebraic equations. A factorization (ii) of* $\mathcal{H}(\partial)$ *is of the form* $\mathcal{H}(\partial) = B_R(\partial) A_R(\partial)^{-1}$ *where* $A_R(\partial)$ *is a right least common multiple of* $\partial - 1$ *and* $\partial + 2 + t^{-1}$, *i.e.,* $A_R(\partial) = \partial^2 + \frac{(1+2t)(1+3t)-t^2}{t(1+3t)} - \frac{1+2t}{1+3t}$. *It follows that* $B_R(\partial) = \begin{bmatrix} \frac{1+2t}{1+3t} + \frac{t}{1+3t}\partial \\ \frac{t}{1+3t} - \frac{t}{1+3t}\partial \end{bmatrix}$.

11.4 Parametrization of the Class of Stabilizing Controllers

The whole class of stabilizing controllers can be given. For that, $\mathcal{G}(\partial)$ must be written into another form. Let $X(\partial)$ and $Y(\partial)$ be such that

$$A_L(\partial)X(\partial) + B_L(\partial)Y(\partial) = I_p \tag{11.12}$$

and let $\mathcal{K}(\partial) = A_R^{-1}(\partial)[R_R(\partial)A_{cl}^{-1}(\partial) - Y(\partial)]$ which, on the one hand, leads to

$$Y(\partial) + A_R(\partial)\mathcal{K}(\partial) = R_R(\partial)A_{cl}^{-1}(\partial). \tag{11.13}$$

On the other hand,

$$\begin{aligned}
&X(\partial) - B_R(\partial)\mathcal{K}(\partial) = \\
&= X(\partial) - B_R(\partial)A_R^{-1}(\partial)R_R(\partial)A_{cl}^{-1}(\partial) + B_R(\partial)A_R^{-1}(\partial)Y(\partial) = \\
&= X(\partial) - A_L^{-1}(\partial)B_L(\partial)R_R(\partial)A_{cl}^{-1}(\partial) + A_L^{-1}(\partial)B_L(\partial)Y(\partial) = \\
&= A_L^{-1}(\partial)[A_L(\partial)X(\partial) + B_L(\partial)Y(\partial) - B_L(\partial)R_R(\partial)A_{cl}^{-1}(\partial)].
\end{aligned}$$

Using (11.12) and (11.2), it follows that

$$\begin{aligned}
&X(\partial) - B_R(\partial)\mathcal{K}(\partial) = \\
&= A_L^{-1}(\partial)[A_L(\partial)S_R(\partial) + B_L(\partial)R_R(\partial) - B_L(\partial)R_R(\partial)]A_{cl}^{-1}(\partial)
\end{aligned}$$

and thus, finally,

$$X(\partial) - B_R(\partial)\mathcal{K}(\partial) = S_R(\partial)A_{cl}(\partial)^{-1}. \tag{11.14}$$

Moreover, $X(\partial) - B_R(\partial)\mathcal{K}(\partial)$ is invertible since $S_R(\partial)$ is invertible. From (11.13) and (11.14) it can be easily deduced the following *right factorization* of the controller:

$$\mathcal{G}(\partial) = [Y(\partial) + A_R(\partial)\mathcal{K}(\partial)][X(\partial) - B_R(\partial)\mathcal{K}(\partial)]^{-1}. \tag{11.15}$$

Notice that $A_L(\partial)[X(\partial) - B_R(\partial)\mathcal{K}(\partial)] + B_L(\partial)[Y(\partial) + A_R(\partial)\mathcal{K}(\partial)] = A_L(\partial)X(\partial) + B_L(\partial)Y(\partial) + [B_L(\partial)A_R(\partial) - A_L(\partial)B_R(\partial)]\mathcal{K}(\partial)$ and, using (11.12) and (6.64), it follows that $A_L(\partial)[X(\partial) - B_R(\partial)\mathcal{K}(\partial)] + B_L(\partial)[Y(\partial) + A_R(\partial)\mathcal{K}(\partial)] = I_p$ thus (11.15) is right-coprime.

Let now $X'(\partial)$ and $Y'(\partial)$ be such that

$$X'(\partial)A_R(\partial) + Y'(\partial)B_R(\partial) = I_m. \tag{11.16}$$

If one additions (11.14) left multipied by $Y'(\partial)$ to (11.13) left multiplied by $X'(\partial)$ it is obtained a relation which leads to the following factorization of $\mathcal{K}$:

$$\mathcal{K}(\partial) = [X'(\partial)R_R(\partial) - Y'(\partial)S_R(\partial)]A_{cl}(\partial)^{-1} + Y'(\partial)X(\partial) - X'(\partial)Y(\partial). \tag{11.17}$$

To obtain a left factorization of $\mathcal{G}(\partial)$, let $\overline{\mathcal{K}}(\partial)$ be such that

$$Y'(\partial) + \overline{\mathcal{K}}(\partial)A_L(\partial) = \overline{A}_{cl}(\partial)^{-1}R_L(\partial). \tag{11.18}$$

As before, it follows that

$$X'(\partial) - \overline{\mathcal{K}}(\partial)B_L(\partial) = \overline{A}_{cl}(\partial)^{-1}S_L(\partial) \tag{11.19}$$

which, along with (11.18) lead to the dual *left factorization* of the controller:

$$\mathcal{G}(\partial) = [X' - \overline{\mathcal{K}}(\partial)B_L(\partial)]^{-1}[Y' + \overline{\mathcal{K}}(\partial)A_L(\partial)]. \tag{11.20}$$

It is easy to check that factorization (11.20) is left-coprime.

As before, from (11.18) and (11.19) it can be obtained a factorization of $\overline{\mathcal{K}}(\partial)$:

$$\overline{\mathcal{K}}(\partial) = \overline{A}_{cl}(\partial)^{-1}[S_L(\partial)Y(\partial) - R_L(\partial)X(\partial)] + Y'(\partial)X(\partial) - X'(\partial)Y(\partial). \tag{11.21}$$

Moreover, if $X'(\partial)$ and $Y'(\partial)$ are chosen to satisfy

$$X'(\partial)Y(\partial) = Y'(\partial)X(\partial) \tag{11.22}$$

in addition to (11.16)[2], the two factorizations (11.17) and (11.20) lead to $\mathcal{K}(\partial) = \overline{\mathcal{K}}(\partial)$. Relations (6.64), (11.12), (11.16) and (11.22) lead to a *doubly coprime factorization* of the plant and of the regulator which is the following generalized diophantine equation:

$$\begin{bmatrix} A_L(\partial) & B_L(\partial) \\ Y'(\partial) & -X'(\partial) \end{bmatrix} \begin{bmatrix} X(\partial) & B_R(\partial) \\ Y(\partial) & -A_R(\partial) \end{bmatrix} = \begin{bmatrix} I_p & 0 \\ 0 & I_m \end{bmatrix}. \tag{11.23}$$

The *Youla-Kučera parametrization* introduced for LTI systems in [367] holds also for LTV systems:

Theorem 1143. *The class of exponentially stabilizing controllers for the closed-loop system in Fig. 11.1 is given by the right-coprime factorization (11.15) or by the left one (11.20) where $\mathcal{K}(\partial)$, respectively $\overline{\mathcal{K}}(\partial)$, is an exponentially stable transfer matrix (according to Definition 1013).*

Proof. First, (11.17) can be written also

$$\mathcal{K}(\partial) = \{X'(\partial)R_R(\partial) - Y'(\partial)S_R(\partial) + [Y'(\partial)X(\partial) - X'(\partial)Y(\partial)]A_{cl}(\partial)\}A_{cl}(\partial)^{-1}$$

from which follows that $\mathcal{K}(\partial)$ is a stable transfer matrix if $A_{cl}(\partial)$ is chosen to provide stable closed-loop poles. The same argument is valid for the form

[2] From a procedural point of view, this can be achieved by choosing first $Y'(\partial) = A_R^{-1}(\partial)Y(\partial)A_L(\partial)$ and, next, $X'(\partial) = [I_m - Y'(\partial)B_R(\partial)]A_R^{-1}(\partial)$.

(11.21). Next, let $\mathcal{K}(\partial)$ be an exponentially stable transfer matrix and let $\mathcal{G}(\partial)$ be given by (11.15). Then,

$$y = A_L(\partial)^{-1}B_L(\partial)\{r - [Y(\partial) + A_R(\partial)\mathcal{K}(\partial)][X(\partial) - B_R(\partial)\mathcal{K}(\partial)]^{-1}y\}$$

so

$$\{I_p + A_L(\partial)^{-1}B_L(\partial)[Y(\partial) + A_R(\partial)\mathcal{K}(\partial)][X(\partial) - B_R(\partial)\mathcal{K}(\partial)]^{-1}\}y = \\ = A_L(\partial)^{-1}B_L(\partial)r.$$

The latter equality left multiplied on both sides by $A_L(\partial)$ leads to $[A_L(\partial) + B_L(\partial)(Y(\partial) + A_R(\partial)\mathcal{K}(\partial))(X(\partial) - B_R(\partial)\mathcal{K}(\partial))^{-1}] = B_L(\partial)r$ or, moreover to $[A_L(\partial)X(\partial) + B_L(\partial)Y(\partial) + (B_L(\partial)A_R(\partial) - A_L(\partial)B_R(\partial))\mathcal{K}(\partial)](X(\partial) - B_R(\partial)\mathcal{K}(\partial))^{-1} = B_L(\partial)r$ from which, using (11.12) and (6.64), follows that $y = [X(\partial) - B_R(\partial)\mathcal{K}(\partial)]B_L(\partial)r$, thus

$$\mathcal{H}_{ry}(\partial) = [X(\partial) - B_R(\partial)\mathcal{K}(\partial)]B_L(\partial). \tag{11.24}$$

Consider now a factorization (ii) of $\mathcal{K}(\partial)$: $\mathcal{K}(\partial) = U_R(\partial)V_R(\partial)^{-1}$. The poles of $\mathcal{K}(\partial)$ are the roots of the polynomial $V_R(\partial)$ and they respect the stability condition of Theorem 998 because $\mathcal{K}(\partial)$ is exponentially stable. $\mathcal{H}_{ry}(\partial) = [X(\partial)V_R(\partial) - B_R(\partial)U_R(\partial)]V_R(\partial)^{-1}B_L(\partial)$. As $V_R(\partial)$ and $U_R(\partial)$ are right-coprime, $V_R(\partial)$ and $X(\partial)V_R(\partial) - B_R(\partial)U_R(\partial)$ are also right-coprime and thus exist $\overline{V}_R(\partial)$ and $S(\partial)$ left-coprime such that

$$[X(\partial)V_R(\partial) - B_R(\partial)U_R(\partial)]V_R(\partial)^{-1} = \overline{V}_R(\partial)^{-1}S(\partial)$$

and $\overline{V}_R(\partial) \sim V_R(\partial)$. It follows $\mathcal{H}_{ry}(\partial) = \overline{V}_R(\partial)^{-1}S(\partial)B_L(\partial)$, thus exponentially stable. Following Theorem 1137, the closed-loop system is internally exponentially stable. The same conclusion can be obtained with factorization (11.21). ∎

11.5 Two-Input Regulators

A more general type of feedback system is obtained if the regulator in Fig. 11.1 is enriched by a second input as in Fig. 11.5. The new regulator $\mathcal{G}(\partial) = \begin{bmatrix} \mathcal{G}_1(\partial) & \mathcal{G}_2(\partial) \end{bmatrix}$ is indeed now directly driven by the reference signal r and by the feedback y and may ensure more flexibility in the synthesis of the control law.

11.5.1 The Two-Degrees-Of-Freedom (2DOF) Control Structure

A two-input regulator can be implemented with the 2DOF structure in Fig. 11.6. Widely used in industrial applications, it consists of three distinct transfers $\mathcal{R}(\partial)$, $\mathcal{S}(\partial)$ and $\mathcal{T}(\partial)$ located on the feedback path, the direct path and,

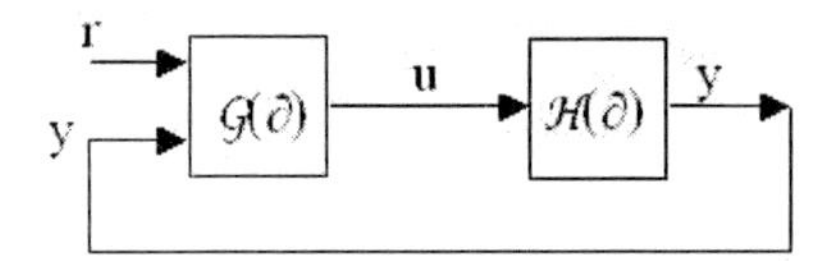

Fig. 11.5 General two-input regulator

respectively, on the feedforward path. The 2DOF control loop is studied here for a single input/single output LTV plant $\mathcal{H}(\partial)$. Let $\mathcal{T}(\partial) = V^{-1}(\partial)T(\partial)$, $\mathcal{S}(\partial) = S_L^{-1}(\partial)V(\partial)$, $\mathcal{R}(\partial) = V^{-1}(\partial)R_L(\partial)$ where $V(\partial)$ is an a priori given polynomial with coefficients in $\mathbb{R}$ and having roots in the left-half complex plane. Its only role is to ensure proper transfers for the three blocs. The regulator synthesis consists in choosing left-coprime polynomials $S_L(\partial)$ and $R_L(\partial)$ to place the poles of the closed-loop system at desired locations $\Lambda_{cl} = \{\lambda_1, ..., \lambda_n\}$ or, equivalently, to impose A_{cl} and to determine $T(\partial)$ to ensure tracking properties of the overall closed-loop system as explained in the preceding sections.

11.5.1.1 Closed-Loop Stability

Following Definition 1139, the closed-loop system in Fig. 11.6 is internally exponentially stable if the six transfers $r \mapsto u$, $r \mapsto y$, $d_1 \mapsto u$, $d_1 \mapsto y$, $d_2 \mapsto u$ and $d_2 \mapsto y$ are exponentially stable. It is now shown that the same kind of pole placement as introduced in Section 11.3.1 ensures the exponential stability of the closed-loop system of the 2DOF structure. The closed-loop relation with $d_1 = d_2 = 0$ is $y = \mathcal{H}(\partial)\mathcal{S}(\partial)(\mathcal{T}(\partial)r - \mathcal{R}(\partial)y)$, or, moreover, $[1 + \mathcal{H}(\partial)\mathcal{S}(\partial)\mathcal{R}(\partial)]y = \mathcal{H}(\partial)\mathcal{S}(\partial)\mathcal{T}(\partial)r$ which, using a right-coprime factorization for the plant $\mathcal{H}(\partial) = B_R(\partial)A_R^{-1}(\partial)$ leads to

$$[1 + B_R(\partial)A_R^{-1}(\partial)S_L^{-1}(\partial)R_L(\partial)]y = B_R(\partial)A_R^{-1}(\partial)S_L^{-1}(\partial)T(\partial)r.$$

Because $B_R(\partial)$ is invertible in this case, the latter relation left multiplied by $A_R(\partial)B_R^{-1}(\partial)$ leads to

$$[A_R(\partial)B_R^{-1}(\partial) + S_L^{-1}(\partial)R_L(\partial)]y = S_L(\partial)^{-1}T(\partial)r$$

or, equivalently, to

$$[S_L(\partial)A_R(\partial) + R_L(\partial)B_R(\partial)]B_R^{-1}(\partial)y = T(\partial)r$$

from which one deduces the closed-loop transfer function

$$y = \mathcal{H}_{ry}(\partial)r, \quad \mathcal{H}_{ry}(\partial) = B_R(\partial)\overline{A}_{cl}^{-1}(\partial)T(\partial) \tag{11.25}$$

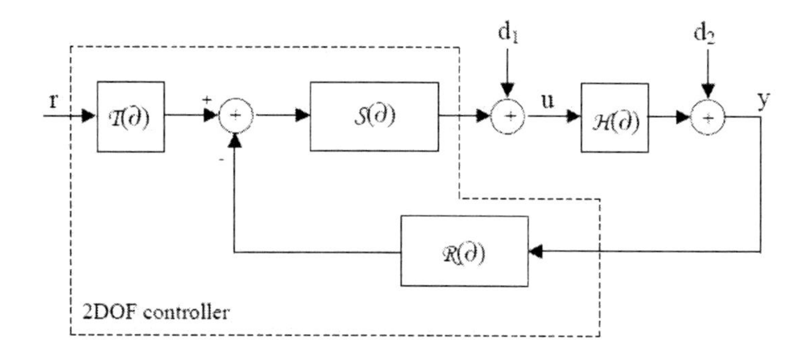

Fig. 11.6 2DOF closed-loop system

where $\overline{A}_{cl}(\partial)$ is given by (11.3). The desired poles Λ_{cl} can thus be placed by solving (11.3) where $\overline{A}_{cl}(\partial)$ is obtained from Λ_{cl} as explained in Section 11.3.

For the same closed-loop, one can now write $u = \mathcal{S}(\partial)(\mathcal{T}(\partial)r - \mathcal{R}(\partial)\mathcal{H}(\partial)u)$ from which follows $[1 + S_L^{-1}(\partial)R_L(\partial)B_R(\partial)A_R^{-1}(\partial)]u = S_L^{-1}(\partial)T(\partial)r$ or, moreover,

$$S_L^{-1}(\partial)(S_L(\partial)A_R(\partial) + R_L(\partial)B_R(\partial))A_R^{-1}(\partial)u = S_L^{-1}(\partial)T(\partial)r$$

thus

$$u = \mathcal{H}_{ru}(\partial)r, \quad \mathcal{H}_{ru}(\partial) = A_R(\partial)\overline{A}_{cl}^{-1}(\partial)T(\partial). \tag{11.26}$$

Let now $r = d_2 = 0$: $u = d_1 - \mathcal{S}(\partial)\mathcal{R}(\partial)\mathcal{H}(\partial)u$ from which, using the same type of calculations, follows that

$$u = \mathcal{H}_{d_1u}(\partial)d_1, \quad \mathcal{H}_{d_1u}(\partial) = A_R(\partial)\overline{A}_{cl}^{-1}(\partial)S_L(\partial). \tag{11.27}$$

Next, $\mathcal{H}_{d_1y}(\partial) = \mathcal{H}(\partial)\mathcal{H}_{d_1u}(\partial)$ from which follows

$$y = \mathcal{H}_{d_1y}(\partial)d_1, \quad \mathcal{H}_{d_1y}(\partial) = B_R(\partial)\overline{A}_{cl}^{-1}(\partial)S_L(\partial). \tag{11.28}$$

If $r = d_1 = 0$, $u = -\mathcal{S}(\partial)\mathcal{R}(\partial)(d_2 + \mathcal{H}(\partial)u)$ which leads to

$$u = \mathcal{H}_{d_2u}(\partial)u, \quad \mathcal{H}_{d_2u}(\partial) = -A_R(\partial)\overline{A}_{cl}^{-1}(\partial)R_L(\partial). \tag{11.29}$$

Also, $\mathcal{H}_{d_2u}(\partial) = -\mathcal{S}(\partial)\mathcal{R}(\partial)\mathcal{H}_{d_2y}(\partial)$ from which

$$\begin{aligned}
&\mathcal{H}_{d_2y}(\partial) = R_L^{-1}(\partial)S_L(\partial)A_R(\partial)\overline{A}_{cl}^{-1}(\partial)R_L(\partial) = \\
&= R_L^{-1}(\partial)[A_R^{-1}(\partial)S_L^{-1}(\partial)]^{-1}[S_L(\partial)A_R(\partial) + R_L(\partial)B_R(\partial)]^{-1}R_L(\partial) = \\
&= R_L^{-1}(\partial)[1 + R_L(\partial)B_R(\partial)A_R^{-1}(\partial)S_L^{-1}(\partial)]R_L(\partial) = \\
&= \{[1 + R_L(\partial)B_R(\partial)A_R^{-1}(\partial)S_L^{-1}(\partial)]R_L(\partial)\}^{-1}R_L(\partial) = \\
&= \{R_L(\partial)[1 + B_R(\partial)A_R^{-1}(\partial)S_L^{-1}(\partial)R_L(\partial)]\}^{-1}R_L(\partial) = \\
&= [1 + \mathcal{H}(\partial)S_L^{-1}(\partial)R_L(\partial)]^{-1}.
\end{aligned}$$

If one choses now a left-coprime factorization for the plant $\mathcal{H}(\partial) = A_L^{-1}(\partial)B_L(\partial)$ and a right-coprime one for the $\mathcal{R}$ part of the regulator (i.e., $R_R(\partial)$ and $S_R(\partial)$ right-coprime which verify (11.1), it follows

$$\begin{aligned}\mathcal{H}_{d_2y}(\partial) &= [1 + A_L(\partial)B_L(\partial)R_R(\partial)S_R^{-1}(\partial)] = \\ &= \{A_L^{-1}(\partial)[A_L(\partial)S_R(\partial) + B_L(\partial)R_R(\partial)]S_R^{-1}(\partial)\}^{-1},\end{aligned}$$

thus

$$y = \mathcal{H}_{d_2y}(\partial)d_2, \quad \mathcal{H}_{d_2y}(\partial) = S_R(\partial)A_{cl}^{-1}(\partial)A_L(\partial) \tag{11.30}$$

where $A_{cl}(\partial)$ is given by (11.2).

Proposition 1144. *The closed-loop system in Fig. 11.6 is internally exponentially stable if $A_{cl}(\partial)$ (or, equivalently, $\overline{A}_{cl}(\partial)$) has a fundamental set of roots which belongs to an Ore field and satisfies the stability condition of Theorem 998.*

Proof. As $A_R(\partial)$ and $B_R(\partial)$ are right-coprime, $\overline{A}_{cl}(\partial)$ and $B_R(\partial)$ and, respectively, $\overline{A}_{cl}(\partial)$ and $A_R(\partial)$, are also right-coprime. It follows that the closed-loop transfers (11.25), (11.28) and, respectively, (11.26), (11.27) and (11.29) are exponentially stable if $\overline{A}_{cl}(\partial)$ has a fundamental set of roots which belongs to an Ore field and satisfies the stability condition of Theorem 998. Similarly, $A_{cl}(\partial)$ and $S_R(\partial)$ are right-coprime because $R_R(\partial)$ and $S_R(\partial)$ are right-coprime and thus (11.30) is exponentially stable if $A_{cl}(\partial)$ has the same property above. ■

11.5.1.2 Reference Tracking

A basic system performance requirement is that the output y follows the reference r.

Definition 1145. *The exponentially stable closed-loop system in Fig. 11.6 is said to* asymptotically track *the reference r if, with $d_1(t) \equiv d_2(t) \equiv 0$, $lim_{t\to+\infty}e(t) = 0$ where $e(t) = y(t) - r(t)$ is the* tracking error.

From (11.25) one can easily deduce the transfer function of the tracking error: $e = \mathcal{H}_{re}(\partial)r$, $\mathcal{H}_{re}(\partial) = B_R(\partial)\overline{A}_{cl}^{-1}(\partial)T(\partial) - 1$. As $\overline{A}_{cl}(\partial)$ and $B_R(\partial)$ are right-coprime (as shown in the preceding paragraph), there exist $\widetilde{A}_{cl}(\partial)$ and $\widetilde{B}_R(\partial)$ left-coprime such that

$$\widetilde{A}_{cl}(\partial)B_R(\partial) = \widetilde{B}_R(\partial)\overline{A}_{cl}(\partial). \tag{11.31}$$

Thus, $\mathcal{H}_{re}(\partial) = \widetilde{A}_{cl}^{-1}(\partial)P(\partial)$ where $P(\partial) = \widetilde{B}_R(\partial)T(\partial) - \widetilde{A}_{cl}(\partial)$.

The most current class of reference signals is the constant ones, i.e., the reference set-points: $r = const$. Obviously, a sufficient condition for asymptotic tracking of a constant reference is $P(0) = 0$ or, equivalently,

$$\widetilde{A}_{cl}(0) = (\widetilde{B}_R T)(0). \tag{11.32}$$

Proposition 1146. *A constant reference r is asymptotically tracked if condition (11.32) is satisfied.*

In practice, once $A_{cl}(\partial)$ (or, equivalently, $\overline{A}_{cl}(\partial)$) is chosen and consequently $S_L(\partial)$ and $R_L(\partial)$ computed, the feedforward term $T(\partial)$ is determined such that (11.32) is satisfied. Thus, the property of asymptotic tracking can be ensured independent from the pole placement of the closed-loop.

When the feedforward term is not used in the 2DOF structure, i.e., when $T(\partial) = 1$, asymptotic tracking of constant references can still be ensured if the poles of the closed-loop are placed such that

$$\widetilde{A}_{cl}(0) = \widetilde{B}_R(0). \tag{11.33}$$

Remark 1147. *The left-coprime factorisation used to deduce (11.31) is not unique. Thus, a different left-coprime pair $\widetilde{\widetilde{A}}_{cl}(\partial)$ and $\widetilde{\widetilde{B}}_R(\partial)$ might satisfy $\widetilde{\widetilde{A}}_{cl}(\partial)B_R(\partial) = \widetilde{\widetilde{B}}_R(\partial)\overline{A}_{cl}(\partial)$. However, $\widetilde{A}_{cl}(0)/\widetilde{B}_R(0) = \widetilde{\widetilde{A}}_{cl}(0)/\widetilde{\widetilde{B}}_R(0)$ thus condition (11.33) as well as (11.32) are consistent.*

11.5.1.3 Disturbance Rejection

One usual additional requirement of system performance is that the output y tracks the reference r even in the presence of disturbances.

Definition 1148. *The exponentially stable closed-loop in Fig. 11.6 with $d_1 \neq 0$ and $d_2(t) \equiv 0$ (resp. $d_2 \neq 0$ and $d_1(t) \equiv 0$) is said to* asymptotically reject *disturbance d_1 (resp. d_2) if $lim_{t \to +\infty} e(t) = 0$, where $e(t) = y(t) - r(t)$ is the* tracking error.

As $\mathcal{H}_{d_1 y}(\partial) = B_R(\partial)\overline{A}_{cl}^{-1}(\partial)S_L(\partial) = \widetilde{A}_{cl}^{-1}(\partial)\widetilde{B}_R(\partial)S_L(\partial)$, where $\widetilde{A}_{cl}^{-1}(\partial)$ and $\widetilde{B}_R(\partial)$ satisfy (11.31), the same arguments as in the preceding paragraph can be used. If, moreover, one uses the evaluation of a product of polynomials formula of Lemma 758, we have

Proposition 1149. *The exponentially stable closed-loop in Fig. 11.6 asymptotically rejects* constant *input disturbances $d_1 = const$ if*

$$\begin{gathered} S_L(0) = 0 \ (i.e., \ s_0 = 0 \ where \ S_L(\partial) = s_n\partial^n + ... + s_1\partial + s_0) \\ or \\ \widetilde{B}_R({}^{S_L(0)}0) = 0 \ (if \ S_L(0) \neq 0). \end{gathered} \tag{11.34}$$

Notice that the situation is different from the LTI case where condition $S_L(0) = 0$, which means that the controller in the direct chain of the closed-loop must have an integral effect, is necessary and sufficient for asymptotic input disturbance rejection. In the LTV case, the same condition is only a sufficient condition.

Consider now the case of the output disturbance d_2. Using dual factorizations for the plant and the regulator, the transfer function from d_2 to y (11.30) can be written $\mathcal{H}_{d_2 y}(\partial) = \widehat{A}_{cl}^{-1}(\partial)\widehat{S}_R(\partial)A_L(\partial)$ where $\widehat{A}_{cl}(\partial)$ and $\widehat{S}_R(\partial)$ are left-coprime such that $\widehat{A}_{cl}(\partial)S_R(\partial) = \widehat{S}_R(\partial)A_{cl}(\partial)$. Thus, we have the following result

Proposition 1150. *The exponentially stable closed-loop in Fig. 11.6 asymptotically rejects* constant *output disturbances* $d_2 = const$ *if*

$$A_L(0) = 0 \ (i.e.,\ a_0 = 0 \ where \ A_L(\partial) = a_n\partial^n + ... + a_1\partial + a_0)$$
$$or \tag{11.35}$$
$$\widehat{S}_R({}^{A_L(0)}0) = 0 \ (if \ A_L(0) \neq 0).$$

As mentioned in Remark 1147 for asymptotic tracking, conditions (11.34) and (11.35) are also coherent. Indeed, if an equivalent factorization $\widetilde{A}_{cl}(\partial)$ and $\widetilde{\widetilde{B}}_R(\partial)$ (respectively $\widehat{\widehat{A}}_{cl}(\partial)$ and $\widetilde{\widetilde{S}}_R(\partial)$) is used, $\widetilde{\widetilde{B}}_R(\partial)$ and $\widetilde{B}_R(\partial)$ (respectively $\widetilde{\widetilde{S}}_R(\partial)$ and $\widetilde{S}_R(\partial)$) have the same roots since they differ by a multiplicative constant (unimodular factor).

11.5.2 Application to the Control of a Variable Flux DC Motor

The separated excitation variable flux DC motor presented in Section 10.7 can now be controlled in closed-loop. For that, the LTV transfer function $\mathcal{H}(\partial)$ from the rotor voltage V_r to the speed ω of the motor given by (10.43) is considered. If the variation in the stator coil voltage induces a periodic time-varying flux (10.42) with $\Phi_0 = 2$, $\gamma = 0.5$, $k = 4$, the open-loop response of ω to a unitary step on V_r is oscillatory as shown in solid-line in Fig. 11.7. This dynamic can be changed if the constant poles $\Lambda_{cl} = \{-1, -1, -1, -1\}$ are placed. For that, a right-coprime factorization is used for $\mathcal{H}(\partial) = B_R(\partial)A_R(\partial)^{-1}$ with $A_R(\partial) = \beta^{-1}\partial^2 + \beta^{-1}\alpha_1\partial + \beta^{-1}\alpha_0$ and $B_R(\partial) = 1$. Consider (11.3) with $\overline{A}_{cl}(\partial) = (\partial+1)^4$. Notice that in this case the asymptotic tracking condition (11.32) is satisfied because $A_{cl}(\partial) = \overline{A}_{cl}(\partial) = \widetilde{A}_{cl}(\partial) = (\partial+1)^4$ and $\widetilde{B}_R(\partial) = B_R(\partial) = 1$ and both have free terms equal to 1. Thus, there is no need for a feedforward term and it is taken $T(\partial) = 1$. An integral type regulator is demanded: $S_L(\partial) = s_2(t)\partial^2 + s_1(t)\partial$, $R_L(\partial) = r_2(t)\partial^2 + r_1(t)\partial + r_0(t)$. Thus condition (11.34) for input disturbance rejection is satisfied. Equation (11.3) lead to a system of 5 algebraic equations. A set of solutions is

$$s_2(t) = \tfrac{K_m\Phi(t)}{JL_r}$$

$$s_1(t) = \tfrac{K_m}{J^2L_r}[-\Phi(t)f + 3J\dot{\Phi} + 4J\Phi(t)]$$

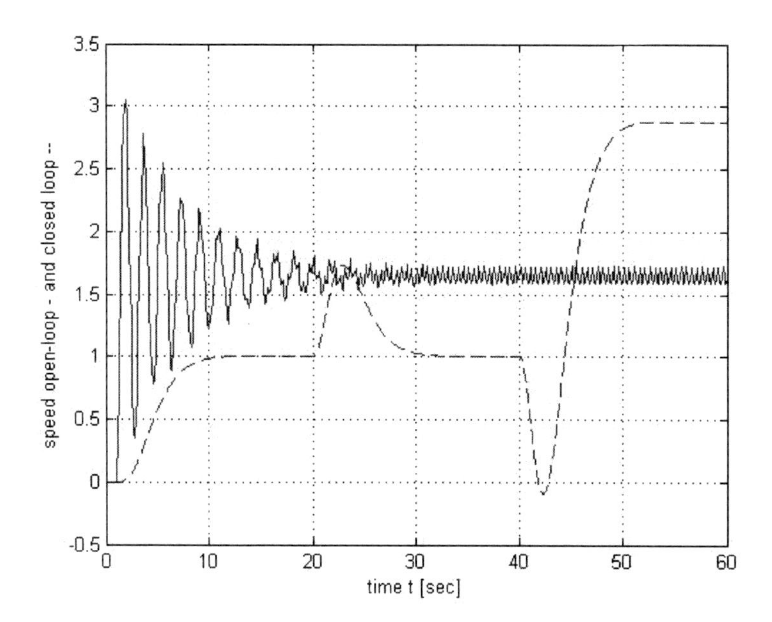

Fig. 11.7 Speed responses for the DC motor

$r_2(t) = \frac{-1}{J^2 L_r \Phi}(-L_r f^2 \Phi + 3LfJ\frac{d\Phi}{dt} + 4LFJ\Phi - 8LJ^2\frac{d\Phi}{dt} + fRJ\Phi + K_e fJ\Phi^3 - 3LJ^2\frac{d^2\Phi}{dt^2} - 6LJ^2\Phi)$

$r_1(t) = \frac{-1}{J^2 L_r \Phi^2}(-\Phi^2 f^2 R + f\frac{d\Phi}{dt}R\Phi J + 4fRJ\Phi^2 - \Phi^4 K_e f^2 + 5fK_e J\frac{d\Phi}{dt}\Phi^3) + 4K_e fJ\Phi^4 + Lf^2\frac{d\Phi}{dt}\Phi - 3fLJ(\frac{d\Phi}{dt})^2 - 4L\Phi fJ + 3\frac{d^2\Phi}{dt^2}\frac{d\Phi}{dt}J^2 L - 4L\Phi\frac{d^2\Phi}{dt^2}J^2 + 8L(\frac{d\Phi}{dt})^2 J^2 - \Phi\frac{d^3\Phi}{dt^3}J^2 L - 4\Phi^2 J^2 L$

$r_0(t) = \frac{1}{J^2 L_r \Phi^2}(-f^2 R\frac{d\Phi}{dt}\Phi + fRJ(\frac{d\Phi}{dt})^2 + 4RfJ\frac{d\Phi}{dt}\Phi + K_e f^2\frac{d\Phi}{dt}\Phi^3 - 3fK_e J(\frac{d\Phi}{dt})^2\Phi^2 - 4fK_e\frac{d\Phi}{dt}\Phi^3 + fRJ\Phi\frac{d^2\Phi}{dt^2} - fJK_e\Phi^3\frac{d^2\Phi}{dt^2} + J^2 L\Phi^2)$.

Transfer matrices $\mathcal{S}(\partial) = S_L(\partial)^{-1}V(\partial)$ and $\mathcal{R}(\partial) = V(\partial)^{-1}R_L(\partial)$ are implemented with one of the canonical forms presented in Section 11.3.2. The dotted line in Fig. 11.7 gives the speed response to a unitary step input on r, the rotor voltage reference, obtained with the loop closed by the LTV regulator above. It can be seen that the response is aperiodic and the dynamic corresponds to the multiple pole $\partial = -1$ placed by output feedback. The steady-state error is zero. At time $t = 20sec$, an input step disturbance has been simulated (d_1 in Fig. 11.6). It can be seen that this disturbance is also fully rejected in steady-state. This is no longer the case with a step output disturbance (d_2 in Fig. 11.6) simulated at time $t = 40sec$. Indeed, condition

$S(0) = 0$ is not sufficient for asymptotic output disturbance rejection as in the LTI case. As a matter of fact, the above regulator does not satisfy condition (11.35).

11.6 Application to the Control of Nonlinear Systems

Nonlinear systems can be approximated by LTV systems. Thus, the LTV approach developed here allows us to control nonlinear systems without going into a purely nonlinear control formalism.

11.6.1 Differential Field Extensions

Let $\mathbf{K}$ be a differential field and $\mathbf{L}/\mathbf{K}$ a differential field extension. As for the non-differential case, for an element $\xi \in \mathbf{L}$ two cases are possible:

- either ξ is $\mathbf{K}$-*differentially algebraic* if it satisfies an algebraic differential equation $P(\xi, \dot{\xi}, ..., \xi^{(n)}) = 0$ where P is a polynomial with coefficients in $\mathbf{K}$ and with indeterminates ξ and its n first derivatives. The extension $\mathbf{L}/\mathbf{K}$ is said to be *differentially algebraic* if any element of $\mathbf{L}$ is $\mathbf{K}$-differentially algebraic.
- either ξ is $\mathbf{K}$-*differentially transcendental* if it is not $\mathbf{K}$-differentially algebraic, i.e., if it does not satisfy *any* algebraic differential equation.

A set $\{\xi_i,\ i \in I\}$ of elements in $\mathbf{L}$ is said to be $\mathbf{K}$-*differentially algebraically dependent* if the set $\{\xi_i^{(\nu_i)},\ i \in I,\ \nu_i = 0, 1, 2, ...\}$ of derivatives of any order is $\mathbf{K}$-algebraically dependent.

A set which is not $\mathbf{K}$-differentially algebraically dependent is called $\mathbf{K}$-*differentially algebraically independent.* A $\mathbf{K}$-differentially algebraically independent set which is maximal with respect to inclusion is called a *differential transcendence basis* of $\mathbf{L}/\mathbf{K}$.

Two such bases have the same cardinality which is the *differential transcendence degree* of $\mathbf{L}/\mathbf{K}$, denoted by $deg.tr.diff_{\mathbf{K}}\mathbf{L}$. Obviously, $\mathbf{L}/\mathbf{K}$ is a differential algebraic extension if, and only if $deg.tr.diff(\mathbf{L}/\mathbf{K}) = 0$.

Let $z = \{z_i \in \mathbf{L}),\ i \in I\}$ be a set of elements of $\mathbf{L}$. $\mathbf{K} < z >$ denotes the differential field generated by $\mathbf{K}$ and the set z, i.e., the smallest differential subfield of $\mathbf{L}$ which contains $\mathbf{K}$ and z. The set z is a *set of generators* of $\mathbf{K} < z >$ and, if I is finite, $\mathbf{K} < z >$ is said *finitely generated.*

Among all possible choices of sets $\xi = \{\xi_1, ..., \xi_m\}$ of m $\mathbf{K}$-differentially algebraically independent elements of $\mathbf{L}$ where $m = deg.tr.diff_{\mathbf{K}}\mathbf{L}$, take one such that $deg.tr.diff_{\mathbf{K}<\xi>}\mathbf{L} < \xi >$ is maximum, say δ. The integer δ is called the *defect* of $\mathbf{L}/\mathbf{K}$.

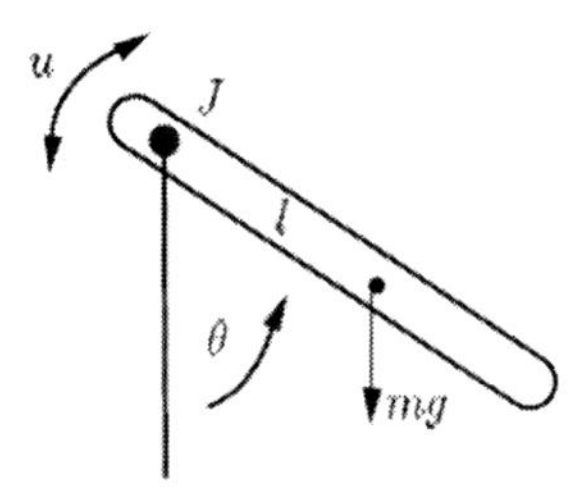

Fig. 11.8 Robot arm

11.6.2 Nonlinear Systems

11.6.2.1 Nonlinear Systems as Field Extensions

In the Fliess's algebraic approach ([114], [119] and related references) a nonlinear system is defined as a finite set of variables related by algebraic differential equations. The following intrinsic characterization was given using field extensions:

Definition 1151. *A nonlinear system Σ over a differential ground field* $\boldsymbol{K}$ *is a finitely generated extension* $\mathbf{L}/\boldsymbol{K}$.

Example 1152. *Consider the robot arm in Fig. 11.8 which pivots in the vertical plan. Its movement is described by the nonlinear equation*

$$J\ddot{\theta}(t) + mglsin\theta(t) = u(t) \tag{11.36}$$

where θ is the angular position, J, m and l are constant parameters of the robot (respectively the inertia, the mass and the length of the arm) which belong to the field $\mathbb{R}$ and u is the torque provided by the motor which drives the arm. With the change of variable $x = tg(\theta/2)$, i.e., $sin\theta = \frac{2x}{1+x^2}$, $\dot{\theta} = \frac{2\dot{x}}{1+x^2}$ and $\ddot{\theta} = \frac{2(1+x^2)-2\dot{x}^2\ddot{x}}{(1+x^2)^2}$, (11.36) is replaced by

$$2J\frac{2(1+x^2)-2\dot{x}^2\ddot{x}}{(1+x^2)^2} + mgl\frac{2x}{1+x^2} = u \tag{11.37}$$

which defines a differential extension of the field $\mathbb{R}$ generated by u and x.

Let $u = \{u_1, ..., u_m\}$ be a finite set of differential quantities, i.e., of variables of the system Σ.

Definition 1153. *A* dynamic nonlinear system *is a finitely generated differential algebraic extension* $\mathbf{L}/\boldsymbol{K} < u >$.

Notice that if u is a differential transcendence basis of $\mathbf{L}/\mathbf{K}$, then $\mathbf{L}/\mathbf{K} < u >$ is a differential algebraic extension. An *input* of the system Σ is thus a vector u such that $deg.tr.diff_{\mathbf{K}<u>}\mathbf{L} = 0$. An *output* of the system Σ is a finite set $y = \{y_1, ..., y_p\}$ of elements of $\mathbf{L}$. A system Σ for which an input u and an output y have been defined form an *nonlinear control system* (Σ, u, y).

11.6.2.2 State-Representations of Nonlinear Systems

Definition 1151 is intrinsic in the sense that it makes no a distinction among the variables of definition of the system (neither state, nor input variables). However, a *state-space representation* can be obtained for a nonlinear control system (Σ, u, y) as follows: let $x = \{x_1, ..., x_n\}$ be a (non differential) transcendence basis of $\mathbf{L}/\mathbf{K} < u >$. Thus, every element w of $\mathbf{L}$ is $\mathbf{K} < u >$-algebraic dependent on the components of x, i.e., satisfies a polynomial equation $P(w, x, u, \dot{u}, ..., u^{(\nu)}) = 0$. In particular, the components of x and y verify

$$\begin{cases} F_i(\dot{x}_i, x, u, \dot{u}, ..., u^{(\alpha_i)}) = 0, \ i = 1, ..., n \\ H_j(\dot{y}_j, x, u, \dot{u}, ..., u^{(\beta_j)}) = 0, \ j = 1, ..., p \end{cases} \tag{11.38}$$

where F_i and H_j are polynomials with coefficients in $\mathbf{K}$. The form (11.38) is called a *generalized state representation* of the nonlinear system Σ. The set x is a (minimal, as n is the degree of transcendence of $\mathbf{L}/\mathbf{K} < u >$) generalized state and x_i, $i = 1, ..., n$ are the generalized state variables of Σ.

Under the hypotheses of the theorem of implicit functions, the following *explicit* local form can be obtained from (11.38):

$$\begin{cases} \dot{x}_i = f_i(x, u, ..., u^{(\alpha_i)}), \ i = 1, ..., n \\ y_j = h_j(x, u, ..., u^{(\beta_j)}), \ j = 1, ..., p. \end{cases} \tag{11.39}$$

A state representation (11.39) with $\alpha_i = 0$ is called a *Kalman state representation* and contains thus derivatives of the inputs only for the output equations

$$\begin{cases} \dot{x}_i = f_i(x, u), \ i = 1, ..., n \\ y_j = h_j(x, u, ..., u^{(\beta_j)}), \ j = 1, ..., p. \end{cases} \tag{11.40}$$

A nonlinear analogue of the notion of properness of transfer matrices of linear systems introduced by Definition 1065 is obtained when $\beta_j = 0$ in (11.40). If, moreover, $\mathbf{K}$ is not a field of constants, (11.40) reduces to the so-called *classical nonlinear state representation*, used by most of the nonlinear analysis and control approaches. The standard notation is

$$\begin{cases} \dot{x} = f(t, x, u), \ \ t \geq t_0 \\ y = g(t, x, u). \end{cases} \tag{11.41}$$

Conversely, if the coefficients of the set of the differential equations which define the nonlinear system Σ are functions *which satisfy algebraic differential equations*, it is possible to transform the equations of Σ into a polynomial form (11.38) as shown for Example 1152. For systems which coefficients do not satisfy condition above, an intrinsic definition of Σ was given in [122] and related references in a geometric differential approach which uses objects of type (11.41).

11.6.3 Linearization

11.6.3.1 Kähler Differential

The Kähler differential, denoted by $d_{\mathbf{L}/\mathbf{K}}$, provides a connection between the nonlinear and linear algebraic structures used to define systems. It associates to a differential extension $\mathbf{L}/\mathbf{K}$ a left $\mathbf{L}[\frac{d}{dt}]$-module $\Omega_{\mathbf{L}/\mathbf{K}}$. More precisely, the Kähler differential is a $\mathbf{K}$-linear application $d_{\mathbf{K}/\mathbf{k}} : \mathbf{L}/\mathbf{K} \to \Omega_{\mathbf{L}/\mathbf{K}}$ which has the following properties [169]:

$$\begin{array}{l} \forall x, y \in \mathbf{L}, d_{\mathbf{L}/\mathbf{K}}(x+y) = d_{\mathbf{L}/\mathbf{K}}x + d_{\mathbf{L}/\mathbf{K}}y \\ \forall x, y \in \mathbf{L}, d_{\mathbf{L}/\mathbf{K}}(xy) = (d_{\mathbf{L}/\mathbf{K}}x)y + x d_{\mathbf{L}/\mathbf{K}}y \\ \forall x \in \mathbf{L}, d_{\mathbf{L}/\mathbf{K}}\dot{x} = \frac{d}{dt}(d_{\mathbf{L}/\mathbf{K}}x) \\ \forall a \in \mathbf{K}, d_{\mathbf{L}/\mathbf{K}}a = 0. \end{array} \tag{11.42}$$

Definition 1154. *Let Σ be the nonlinear system defined by the differential field extension $\boldsymbol{L}/\boldsymbol{K}$. The* generic tangent linear system *Σ_T associated with the nonlinear system Σ is the $\boldsymbol{K}[\partial]$-module $\Omega_{\boldsymbol{L}/\boldsymbol{K}}$ spanned by the Kähler differential of the elements of $\boldsymbol{L}$. To a dynamics $\boldsymbol{L}/\boldsymbol{K} < u >$ it is associated the generic tangent linear dynamics $\Omega_{\boldsymbol{L}/\boldsymbol{K}}$ with input $d_{\boldsymbol{L}/\boldsymbol{K}}u$ and for the nonlinear control system (Σ, u, y), the output of the generic tangent input-output system is $d_{\boldsymbol{L}/\boldsymbol{K}}y$.*

Example 1155. *Consider the nonlinear system given by the variables x and u related by $\dot{x}+x^2 = u$. Using the properties of the Kähler differential (11.42), the generic linear tangent of this system is*

$$\frac{d}{dt}(d_{\boldsymbol{L}/\boldsymbol{K}}x) + 2x(t)d_{\boldsymbol{L}/\boldsymbol{K}}x = d_{\boldsymbol{L}/\boldsymbol{K}}u. \tag{11.43}$$

11.6.3.2 Linearization around a Given Trajectory

The generic linear tangent system Σ_T can be parameterized if one considers the trajectories of Σ, i.e., time dependence expressions for the variables of Σ which satisfy the equations of definition of Σ. For Example 1155, one trajectory T^* is $u^*(t) = 0$, $x^*(t) = t^{-1}$ since, obviously, $x^*(t) = t^{-1}$ verifies $\dot{x}^*(t) + (x(t))^2 = 0$. Thus, $x(t) = x^*(t)$ in (11.43) leads to

$$\dot{d}_x + 2x^*(t)d_x = d_u \Leftrightarrow \dot{d}_x - t^{-1}d_x = d_u \tag{11.44}$$

which is the linear tangent of Σ around trajectory T^*, denoted by Σ_{T^*}.

The tangent systems Σ_{T^*} can also be achieved using the classical (Fréchet) differential. If Σ is given by a state representation (11.41), a trajectory T^* of Σ is $T^* = (x^*(t), u^*(t), y^*(t))$ such that

$$\begin{cases} \dot{x}^*(t) = f(t, x^*(t), u^*(t)), \quad t \geq t_0 \\ y^*(t) = g(t, x^*(t), u^*(t)). \end{cases} \tag{11.45}$$

Let $d_u(t)$ be a fixed input and note by $\widehat{x}(t) = x^*(t) + d_x(t)$ the state trajectory in response to $\widehat{u}(t) = u^*(t) + d_u(t)$. A first order development of f and g lead to

$$\begin{cases} \dot{\widehat{x}}(t) = f(t, x^*(t), u^*(t)) + \frac{\partial f}{\partial x}(t, x^*(t), u^*(t))d_x(t) + \frac{\partial f}{\partial u}(t, x^*(t), u^*(t))d_u(t) \\ \widehat{y}(t) = g(t, x^*(t), u^*(t)) + \frac{\partial g}{\partial x}(t, x^*(t), u^*(t))d_x(t) + \frac{\partial g}{\partial u}(t, x^*(t), u^*(t))d_u(t) \end{cases} \tag{11.46}$$

where $\widehat{y}(t) = y^*(t) + d_y(t)$ is the output of Σ in response to $\widehat{u}(t) = u^*(t) + d_u(t)$. Using (11.45) along with the notations

$$\begin{cases} A(t) = \frac{\partial f}{\partial x}(t, x^*(t), u^*(t)), \quad B(t) = \frac{\partial f}{\partial u}(t, x^*(t), u^*(t)) \\ C(t) = \frac{\partial g}{\partial x}(t, x^*(t), u^*(t)), \quad D(t) = \frac{\partial g}{\partial u}(t, x^*(t), u^*(t)), \end{cases} \tag{11.47}$$

(11.46) leads to the following LTV state representation of Σ_{T^*}:

$$\begin{cases} \dot{d}_x = A(t)d_x + B(t)d_u \\ d_y = C(t)d_x + D(t)d_u. \end{cases} \tag{11.48}$$

Clearly, (11.46) is an approximation of Σ valid in the vicinity of trajectory T^*, i.e., the band around the trajectory in Fig. 11.9. For Example 1155, $A(t) = -2x^*(t)$, $B(t) = 1$ and the resulting state form of the linearization around trajectory $(x^*(t) = t^{-1}, u^*(t) = 0)$ obviously corresponds to (11.44).

If f and g of (11.41) are analytic (i.e., $f, g \in \mathcal{O}(\mathbb{R})$), then the poles of the LTV system (11.48) can be placed as shown in Section 11.3. Also, the entries of $A(t)$, $B(t)$, $C(t)$, $D(t)$ in (11.48) belong also to $\mathcal{O}(\mathbb{R})$ and they can be embedded in the field of meromorphic functions $\mathcal{M}(\mathbb{R})$. A regulator can be synthesized as shown in Section 11.3 to place the poles of the LTV closed-loop system at exponentially stable locations over an Ore field. As a consequence, the resulting control d_u is also analytic and, from Lemma 974 it follows that d_x and d_y of the aforementioned LTV closed-loop are also analytic. In Section 11.6.6 it is shown that this stabilizing LTV controller is suitable for the control of the initial nonlinear system Σ.

There exists an important particular case of the linearization above:

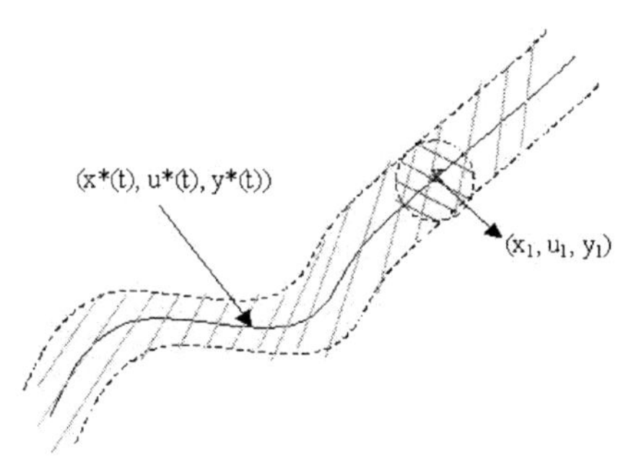

Fig. 11.9 Linearization around a given trajectory/operation point

11.6.3.3 Linearization around an Equilibrium Point

Trajectory (11.45) reduces to an *equilibrium point* when $f(t, x_e, u_e) = 0$, $\forall t \geq t_0$ and thus $x^*(t) = x_e = const.$ and $x^*(t) = u_e = const.$. When Σ is time-invariant, i.e., given by

$$\begin{cases} \dot{x} = f(x, u) \\ y = g(x, u) \end{cases} \tag{11.49}$$

instead of (11.41). Its tangent linear approximation around the equilibrium point (x_e, y_e) is the following LTI system

$$\begin{cases} \dot{d}_x = Ad_x + Bd_u \\ d_y = Cd_x + Dd_u \end{cases} \tag{11.50}$$

where $d_x = \widehat{x} - x_e$, $d_u = \widehat{u} - u_e$, $d_y = \widehat{y} - y_e$ ($y_e = g(x_e, u_e)$) and A, B, C and D are matrices with entries in $\mathbf{K}$ which is now a field of constants.

The linearization around an equilibrium point has obviously the advantage to lead to an LTI system with more control facilities. However, the LTV approximation (11.48) can be controlled using the results of the Section 11.3. Before this, a special class of nonlinear systems, called flat systems, is investigated especially due to the fact that trajectories can be easily generated for this class of systems.

11.6.4 Trajectory Planning

11.6.4.1 Flat Systems

Definition 1156. *A nonlinear system Σ is called* differentially flat *or simply* flat *if there exists a vector of variables $z(t)$ such that:*

- *the components of z are differentially independent, i.e., not related by any differential equation.*
- *the components of z are differential functions of the system definition variables and a finite number of their derivatives.*
- *any system variable is a differential function of z, i.e., is a function of the components of z and a finite number of their derivatives.*

If Σ is given by a state representation of the form (11.49), then it is flat if there exists

$$z(t) = h(x(t), u(t), \dot{u}(t), ..., u^{(\delta)}(t)) \tag{11.51}$$

with components differentially independent and two functions $E(.)$, $F(.)$ such that

$$\begin{cases} x(t) = E(z(t), \dot{z}(t), ..., z^{(\alpha)}(t)) \\ u(t) = F(z(t), \dot{z}(t), ..., z^{(\beta)}(t)). \end{cases} \tag{11.52}$$

z is called the *flat output* of Σ. (11.52) shows that a flat system can be inverted (from the flat output z towards the input u) without integrating any equation.

The system of Example 1152 is flat with the flat output $y = \theta$ (respectively $y = x$). Indeed, $u = J\ddot{y} - mglsin(y)$ (respectively $u = 2J\frac{(1+x^2)\ddot{x}-\dot{x}^2}{(1+x^2)^2} + 2mgl\frac{x}{1+x^2}$).

11.6.4.2 Computation of $x^*(t)$, $u^*(t)$

In motion planning of nonlinear systems it is of interest to obtain a whole trajectory $(x^*(t), u^*(t), y^*(t))$ compliant with a desired output $y^*(t) = y_d(t)$. This provides in fact an open-loop control: given an output tracking objective $y_d(t)$, $t \in [t_0\ t_f]$, under the hypothesis of perfect model and no disturbances, it is sufficient to excite the system with the input $u^*(t)$ on the time interval $[t_0\ t_f]$ to obtain as response of the system the desired output $y_d(t)$. Of course, the above hypothesis of perfect model is not realistic and, for a practical implementation, a closed-loop control should be provided. However, as shown in the next section, to synthesize such a control, it is important to know not only the desired output, but also state and input trajectory $x^*(t), u^*(t)$ of the system, compliant with $y_d(t)$. This problem, called *trajectory planning*, is investigated in three distinct situations:

- *Σ is flat and the trajectory is to be planned on the flat output:* the trajectory planning problem can in this case be trivially solved since $x^*(t)$ and $u^*(t)$ are given by (11.52) with $z(t) = y(t) = y_d(t) = y^*(t)$.
- *Σ is flat but $y(t) \neq z(t)$:* $z^*(t)$ can be constructed from $y^*(t)$ following the steps:

1. compute the transient points from $y^*(t)$ and its derivatives

$$\begin{aligned} y^*(t_i) &= y_i, \ i = 1, ..., N-1 \\ y^{*(j)}(t_i) &= y_{ij}, \ j = 1, ..., M. \end{aligned} \tag{11.53}$$

2. from $y = \phi(z, \dot{z}, ..., z^{(q)})$ compute $y^{(j)} = \phi_j(z, \dot{z}, ..., z^{(q+j)})$, $j = 1, ..., M$.
3. solve the system (11.53) for $z^*(t_i)$ and $z^{*(j)}(t_i)$.
4. (linearly) interpolate $z^*(t)$ between the points t_i.
5. compute the error $e(t) = y^*(t) - \phi(z^*, \dot{z}^*, ..., z^{*(q)})$. If e satisfies a tolerance criterium, then $z^*(t)$ is declared compliant with $y^*(t)$ and it is used to generate $x^*(t)$ and $u^*(t)$ by (11.52) with $z(t) = z^*(t)$. If not, supplementary points t_i are inserted at step 1 and the procedure is run again.

- *Σ is not flat:* the problem may not always have a solution. More precisely, one has to *invert* the general control system (Σ, u, y).

Definition 1157. *Consider the state representation (11.49) of Σ with $u \in \mathbb{R}^m$, $y \in \mathbb{R}^p$, $x \in \mathbb{R}^n$. Let $\mathcal{U}$ denote the set of analytic applications $t \mapsto u(t)$ from $\mathbb{R}$ to $\mathbb{R}^m$ and $\mathcal{Y}$ the set of analytic applications $t \mapsto y(t)$ from $\mathbb{R}$ to $\mathbb{R}^p$. For a given initial condition x_0 of the state equation of (11.49), let $y(t) = H_{x_0}(u(t))$ be the solution of (11.49) with entry $u \in \mathcal{U}$ and initial condition x_0. Then, Σ is said* left *(respectively* right*)* invertible *if application H_{x_0} is injective (respectively surjective).*

Necessary and sufficient conditions for invertibility have been provided in [114] in the algebraic framework adopted here using the *differential output rank of the system.*

Definition 1158. *The* differential output rank, *denoted by ρ, of Σ with respect to the output y is the differential transcendence degree of the differential field extension generated by this output: $\rho = deg.tr.diff(\boldsymbol{K} < y > /\boldsymbol{K})$.*

Proposition 1159. *Σ is left (respectively right) invertible if, and only if $\rho = m$ (respectively $\rho = p$).*

Remark 1160. *In the case of* linear systems, *the differential output rank is the rank of the module generated by y: $\rho = rank([y]_{\mathbf{R}})$. The following are equivalent:*

- *the linear control system Γ is left (respectively right) invertible.*
- *the transfer matrix of Γ is left (respectively right) invertible.*
- *$\rho = m$ (respectively $\rho = p$) where m is the number of inputs (respectively p is the number of outputs) of Γ.*

When Σ is left invertible, the solution is provided by the so-called *structure algorithm* given firstly in [328] for linear systems and extended in

[160] and [218] to the nonlinear ones. The output of this algorithm is the inverse system with input y, output u and state x. To compute $x^*(t)$, $u^*(t)$ compliant with $y^*(t)$ one has to integrate the latter system.

11.6.5 Flatness-Based Trajectory Tracking

11.6.5.1 System Differential Equivalence

Definition 1161. *Two nonlinear systems* $\mathbf{L}/\mathbf{K}$ *and* $\overline{\mathbf{L}}/\mathbf{K}$ *are said* differentially equivalent *or* equivalent by endogenous feedback *if any element of* $\mathbf{L}$ *(respectively* $\overline{\mathbf{L}}$*) is* $\mathbf{K}$*-differentially algebraic over* $\overline{\mathbf{L}}$ *(respectively* $\mathbf{L}$*). Two dynamics* $\mathbf{L}/\mathbf{K} < u >$ *and* $\overline{\mathbf{L}}/\mathbf{K} < \overline{u} >$ *are differentially equivalent if the corresponding systems are so.*

System $\overline{\mathbf{L}}/\mathbf{K}$ can be created from system $\mathbf{L}/\mathbf{K}$ by transforming the original (endogenous) variables of the latter system without creation of new (exogenous) variables. This transformation is a particular type of *dynamic feedback*. If a state representation of type (11.49) is used both for the system $\mathbf{L}/\mathbf{K}$ and for its feedback, the latter can be written

$$\begin{cases} \dot{v} = a(x, v, w) \\ u = k(x, v, w) \end{cases} \tag{11.54}$$

which leads to the closed-loop system

$$\begin{cases} \dot{x} = f(x, k(x, v, w)) \\ \dot{v} = a(x, v, w) \\ y = g(x, v, w) \end{cases} \tag{11.55}$$

where w is the new input. The feedback (11.54) is thus endogenous if (11.55) and the open-loop system $\mathbf{L}/\mathbf{K}$ are differentially equivalent. In this case, the state variable v is endogenous to the initial system $\mathbf{L}/\mathbf{K}$. Obviously, differentially equivalent systems does not have the same state dimension but they have the same number of independent inputs. Also, they have the same generic tangent system: $\Omega_{\mathbf{L}/\mathbf{K}} \sim \Omega_{\overline{\mathbf{L}}/\mathbf{K}}$ (see [121] for a complete proof).

11.6.5.2 Connections with Linear Systems

From Definition 1156 immediately follows

Proposition 1162. *A nonlinear system* $\boldsymbol{L}/\boldsymbol{K}$ *if differentially flat if, and only if its defect is zero.*

Notice also that the generic tangent linear system of a nonlinear flat system Σ is controllable. Indeed, if $\mathbf{K} < y > /\mathbf{K}$ is differentially transcendental, the module $\Omega_{\mathbf{K}} < y > /\mathbf{K}$ spanned by $d_{\mathbf{K}<y>/\mathbf{K}} y_1, ..., d_{\mathbf{K}<y>/\mathbf{K}} y_p$ is necessarily free. Flatness may be thus viewed as a nonlinear extension of the notion of linear controllability. Indeed, consider the case of a linear system Γ and let $\Gamma = \mathcal{T}(\Gamma) \oplus \Phi$ be a decomposition (3.61). A basis of Φ plays the role of the flat output when Γ is free. When $\mathcal{T}(\Gamma) \neq 0$, the differential field extension $\mathcal{T}/\mathbf{K}$ generated by $\mathcal{T}(\Gamma)$ is differentially algebraic and its (non-differential) transcendence degree is equal to the dimension of $\mathcal{T}(\Gamma)$ as $\mathbf{K}$-vector space. It follows that Γ is flat if, and only if it is controllable.

Indeed, consider a controllable linear system Γ. Suppose for the moment that it has only one input and consider its right canonical form (11.11). The loop consisting of Γ fed back by the control

$$u = a_{n-1}x_1 + a_{n-2}x_2 + ... + a_0 x_n + v \tag{11.56}$$

has the following cascade of n integrators state representation $\dot{z}_1 = v$, $\dot{z}_2 = z_1$, ..., $\dot{z}_n = z_{n-1}$ where $z_1 = x_1$, ..., $z_n = x_n$ are the flat outputs. This can be further written,

$$z_n^{(n)} = v. \tag{11.57}$$

Control (11.56) is obviously endogenous, thus Γ is differentially equivalent to (11.57) which is called the *Brunowsky (controllability) canonical form* of a linear system. In the case where Γ has m inputs, its Brunowsky form consists of m independent systems of the form (11.57)

$$z_i^{(\nu_i)} = v_i, \ i = 1, ..., m \tag{11.58}$$

where the ν_i's are the controllability indices of Γ (see [170] and [120] for details).

The same equivalence can be obtained for flat nonlinear systems.

Proposition 1163. *A flat nonlinear system Σ with m (independent) outputs is differentially equivalent to system (11.58) where $z = (z_1, ..., z_m)$ is the flat output.*

Proof. Using the second relation in (11.52), one can propose the control law

$$u(t) = F(z(t), \dot{z}(t), ..., z^{(\beta)}(t), v(t)) \tag{11.59}$$

where $v(t)$ is the new control of the resulting closed-loop. If $\frac{\partial F}{\partial z^{(\beta)}}$ is locally invertible, control (11.59) leads to the linear decoupled system (11.58). ∎

For Example 1152, $u = mglsin(z) + v$ transforms the initial nonlinear system into the differentially equivalent LTI one $J\ddot{\theta} = v$.

Thus both flat nonlinear systems and controllable linear systems can be brought to the linear canonical form (11.58). It follows that flat systems are the class of nonlinear systems which are differentially equivalent to the particular controllable realization (11.58). Flatness can thus be interpreted like a generalization of the well-known static feedback linearization results (see, e.g., [166]) to the case of dynamic feedback. However, several main differences between the two approaches must be pointed out. First, linearization (11.58) is global and not valid only in a (sufficiently small) vicinity of an operating point. Next, as any component of u and x can be retrieved from the flat output z without integrating any differential equation, the flat dynamics has a trivial zero-dynamics. Both mentioned facts are important advantages for the flatness-based control which is studied in the next two sections.

11.6.5.3 Flatness-Based Trajectory Tracking

The problem now is to track a specified desired output trajectory $y^*(t)$.

Definition 1164. *A closed-loop control* asymptotically tracks *the desired output $y^*(t)$ if $lim_{t\to+\infty}[y^*(t) - y(t)] = 0$.*

As shown in Section 11.6.4.2.2, a full trajectory $(x^*(t), u^*(t), y^*(t))$ compliant with the desired output $y^*(t)$ can be computed starting from $y^*(t)$. Control $u^*(t)$ cannot be directly applied to a real system subject to model errors and perturbations. A *closed-loop* control should be provided. For that, two distinct situations can be encountered which are now studied from a control point of view.

Tracking of a flat output:

This first case corresponds to the situation of a flat system with $z(t) \equiv y(t)$. The linearizing control (11.59) can be supplemented by a standard linear feedback with *constant* coefficients

$$v_i(t) = z_i^{(\nu_i)}(t) + \sum_{j=0}^{\nu_i - 1} K_{ij}[z_i^{*(j)}(t) - z_i^{(j)}(t)], \; i = 1, ..., m \tag{11.60}$$

to stabilize the linear system (11.57) (obviously $z_i^{*(\nu_i)}(t) = y_i^{*(\nu_i)}(t)$). More specifically, the *constant* gains K_{ij}, $j = 0, ..., \nu_i - 1$, $i = 1, ..., m$ are chosen such that the roots of polynomials $A_{bf_i}(\partial) = \partial^{\nu_i} + \sum_{j=0}^{\nu_i - 1} K_{ij}\partial^j$ are in the left half of the complex plane. As a consequence, the closed-loop dynamic is $A_{bf}(\partial)_i[z_i^*(t) - z_i(t)] = 0$, $i = 1, ..., m$ and thus $lim_{t\to+\infty}[z^*(t) - z(t)] = 0$, i.e., $lim_{t\to+\infty}y(t) = y^*(t)$.

The overall nonlinear control is thus (11.59) with $v(t)$ given by (11.60).

Example 1165. Control of a synchronous alternator [331]: *A simplified model of a synchronous alternator, the so-called classical model with voltage dynamics, is (see, e.g., [213]):*

$$\begin{cases} \dot{\theta} = \omega \\ \dot{\omega} = \frac{1}{J}P_e - \frac{d}{J}\omega + \frac{P_m}{J} \\ T'_{d_0}\dot{E}' = -\frac{x_d}{x'_d}E' + \frac{x_d - x'_d}{x'_d}Ecos(\theta) + E_{fd} \end{cases} \tag{11.61}$$

where θ is the machine angle, ω the machine speed, E' the voltage behind the reactance x'_d and P_e the electrical power. Under the assumption of constant and nominal (thus unitary) grid-side voltage, $P_e = \frac{E'}{x'_d x_{tfo}} sin(\theta)$, where x_{tfo} is the reactance of the transformer which is a constant. P_m is the mechanical power and it is considered constant (this is a usual hypothesis for the voltage regulation); it becomes thus a parameter. The other parameters of the machine are: J the inertia, d the damping coefficient, x_d the direct axis transient reactance, x'_d the direct axis subtransient reactance. Grouping the constant terms, (11.61) can be written as the following state LTI model

$$\begin{cases} \dot{x}_1 = x_2 \\ \dot{x}_2 = -b_1 x_3 sin(x_1) - D x_2 + P \\ \dot{x}_3 = b_3 cos(x_1) - b_4 x_3 + E - u \end{cases} \tag{11.62}$$

where the state variables are $x_1 = \theta$, $x_2 = \omega$ and $x_3 = E'$. The electric part of the machine is driven by the excitation system, thus the control variable in this case is the excitation voltage E_{fd}: $u = E_{fd}$. The model (11.62) is flat with the flat output $z = x_1$. Indeed, simple transformations of (11.62) lead to

$$\begin{cases} x_1 = z \\ x_2 = \dot{z} \\ x_3 = \frac{P - D\dot{z} - \ddot{z}}{b_1 sin(z)} \\ u = E - \frac{b_4}{b_1}\frac{P - D\dot{z} - \ddot{z}}{sin(z)} + b_3 cos(z) - \frac{(-D\ddot{z} - z^{(3)})sin(z) - (P - D\dot{z} - \ddot{z})\dot{z}cos(z)}{b_1 sin^2(z)}. \end{cases} \tag{11.63}$$

Moreover the machine angle is systematically measured in practice. Thus, if $x_1^*(t)$ is a desired trajectory to be imposed to the machine angle (usually step references). We are in the case $z = y$ and a direct control can be synthesized based on the measurement of the flat output. The control law (11.59) is in the case of the alternator

$$\begin{aligned} u = E - \frac{b_4}{b_1}\frac{P - D\dot{z} - \ddot{z}}{sin(z)} + b_3 cos(z) - \\ -\frac{(-D\ddot{z} - v^{(3)})sin(z) - (P - D\dot{z} - \ddot{z})\dot{z}cos(z)}{b_1 sin^2(z)} \end{aligned} \tag{11.64}$$

and leads to $z^{(3)}(t) = v^{(3)}(t)$. The second control v is chosen

$$v(t) = z^{*(3)}(t) - k_2[z^{(2)}(t) - z^{*(2)}(t)] - k_1[\dot{z}(t) - \dot{z}^*(t)] - k_0[z(t) - z^*(t)] \tag{11.65}$$

with k_0, k_1, k_2 such that the roots of the polynomial $P(\partial) = s^3 + k_2 s^2 + k_1 s + k_0$ are in the left half complex plane. This ensures an exact tracking: $lim_{t \to +\infty} z(t) = z^*(t)$.

Tracking of a non flat output:

Consider now the situation when the system is flat with flat output $z(t)$ but the desired trajectory $y^*(t)$ is specified on an output which is not flat, i.e., $y(t) \neq z(t)$. It has been shown in Section 11.6.4.2.2 that a desired flat output $z^*(t)$ compliant with $y^*(t)$ can be constructed exploiting the flatness properties of the system. However, if $z(t)$ is not measurable, a control of type (11.59) cannot be implemented. Moreover, even when $z(t)$ is measurable, (11.59) ensures *only* $lim_{t \to +\infty} z(t) = z^*(t)$ and not necessarily $lim_{t \to +\infty} y(t) = y^*(t)$. In this case, a better solution is to implement a LTV control law as explained in the following section.

11.6.6 Trajectory Tracking by LTV Control

We are interested now in tracking the output trajectory for a general nonlinear (not necessarily flat) system.

Proposition 1166. *Let Σ be a nonlinear system, T^* one of its trajectories $(x^*(t), u^*(t), y^*(t))$ and Σ_{T*} the linear tangent system around this trajectory. Suppose that Σ_{T*} is exponentially stabilizable. Then, $\mathcal{G}(s)$ can be chosen to exponentially stabilize the linear closed-loop system in Fig. 11.1 where $\mathcal{H}(s)$ is the transfer matrix of Σ_{T*} and, when applied to the nonlinear system Σ as in Fig. 11.10, to provide an asymptotic tracking of the desired output trajectory $y^*(t)$.*

Proof. Let

$$\begin{cases} \dot{\overline{x}} = \overline{f}(t, \overline{x}, \overline{u}) \\ \overline{y} = \overline{g}(t, \overline{x}, \overline{u}) \end{cases} \tag{11.66}$$

where $\overline{x} = x - x^*$, $\overline{u} = u - u^*$, $\overline{y} = y - x^*$, $\overline{f}(t, \overline{x}, \overline{u}) = f(t, x, u) - f(t, x^*, u^*)$, $\overline{g}(\overline{x}, \overline{u}, t) = g(t, x, u) - g(t, x^*, u^*)$ be the *variational* system relative to trajectory $(x^*(t), u^*(t), y^*(t))$ (see also Section 12.4.1). Obviously, $(\overline{x} = 0, \overline{u} = 0, \overline{y} = 0)$ is an equilibrium point of (11.66). Moreover, $\frac{\partial \overline{f}}{\partial \overline{x}}|_{\overline{x}=0,\overline{u}=0} = \frac{\partial f}{\partial x}|_{x=x^*,u=u^*}$, $\frac{\partial \overline{f}}{\partial \overline{u}}|_{\overline{x}=0,\overline{u}=0} = \frac{\partial f}{\partial u}|_{x=x^*,u=u^*}$, $\frac{\partial \overline{g}}{\partial \overline{x}}|_{\overline{x}=0,\overline{u}=0} = \frac{\partial g}{\partial x}|_{x=x^*,u=u^*}$, $\frac{\partial \overline{g}}{\partial \overline{u}}|_{\overline{x}=0,\overline{u}=0} = \frac{\partial g}{\partial u}|_{x=x^*,u=u^*}$ and $d_{\overline{x}} = d_x$, $d_{\overline{u}} = d_u$, $d_{\overline{y}} = d_y$ thus, the linear tangent E_{T^0} of (11.66) around the origin is (11.48) and (11.47), the same as the linear tangent of Σ around $(x^*(t), u^*(t), y^*(t))$. It follows that the regulator $\mathcal{G}(\partial)$ stabilizes also the closed-loop system Fig. 11.1 where $\mathcal{H}(\partial)$ is the transfer matrix of E_{T^0}. If a state representation of this controller is

$$\begin{cases} \dot{x}_r = A_r(t) x_r + B_r(t) d_y \\ -d_u = C_r(t) x_r + D_r(t) d_y \end{cases} \tag{11.67}$$

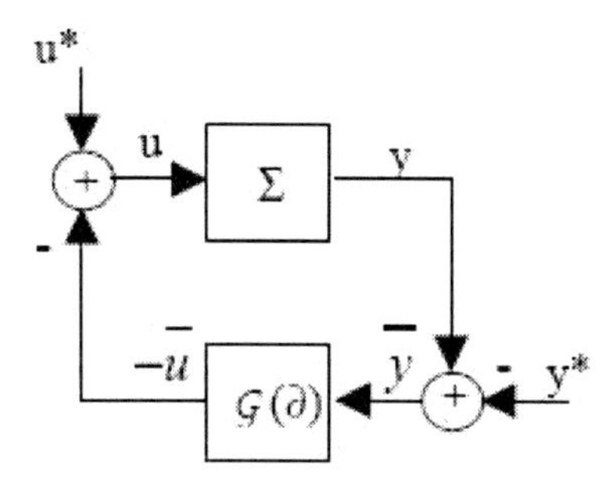

Fig. 11.10 General nonlinear closed-loop system

then, the equations of the closed-loop are thus (11.48) and (11.67) which, after elimination of d_u and d_y lead to the following *state representation* of the LTV closed-loop system:

$$\dot{X}_{LTV} = A_{cl}(t)X_{LTV}, \; X_{LTV} = \begin{bmatrix} d_x \\ x_r \end{bmatrix},$$
$$A_{cl}(t) = \begin{bmatrix} A - B(I + D_r D)^{-1} D_r C & -B(I + D_r D)^{-1} C_r \\ B_r C - B_r D(I + D_r D)^{-1} D_r C & A_r - B_r D(I + D_r D)^{-1} C_r \end{bmatrix}. \tag{11.68}$$

If, moreover, A_r, B_r, C_r, D_r are chosen such that $A_{cl}(t)$ is bounded and Lipschitz in a neighboring of the origin[3], from Proposition 1196 follows that the origin is an asymptotically stable equilibrium point of the nonlinear closed-loop system in Fig. 11.10 and thus $lim_{t\to+\infty}(y(t) \to y^*(t)) = 0$. ■

The following example has been used several times to test and demonstrate gain-scheduling approaches (see, e.g., [310]). It is now treated with the LTV methodology presented above.

Example 1167. *Consider the nonlinear system defined by $\dot{x} = -x + u$, $y = tanh(x)$. It is flat with the flat output y. Indeed, from the latter equation one can easily express x and u in function of y and its derivatives:*

$$x = \tfrac{1}{2} ln \tfrac{1+y}{1-y}, \quad u = \dot{x} + x. \tag{11.69}$$

Equations (11.69) can be exploited to generate a full system trajectory $(x^*(t), u^*(t), y^*(t))$ from a desired output $y^*(t)$. If a first-order type response is desired, let

$$y^*(t) = \alpha(1 - e^{-t/T}), \quad T = 1, \; \alpha = 0.9 \; . \tag{11.70}$$

[3] Notice that these conditions are trivially satisfied if $A_{cl}(t)$ has constant entries, i.e., if constant poles are placed for the closed-loop system.

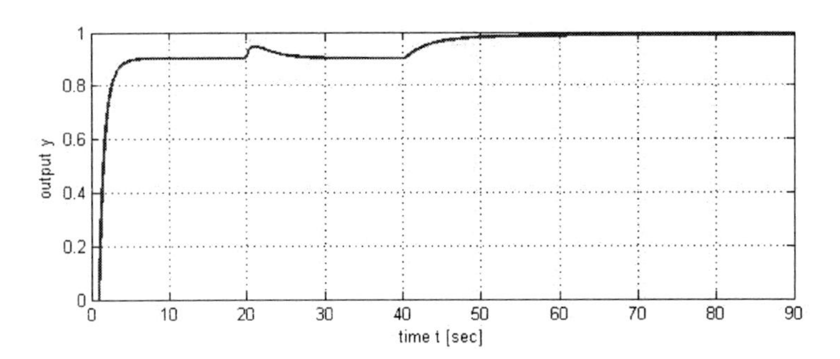

Fig. 11.11 Output response of the nonlinear closed-loop

The associated tangent linear system around the trajectory (11.70) Σ_{T^*} *is* $\dot{d}_x = -d_x + d_u$, $d_y = [1 - y^{*2}(t)]d_x$. *The transfer function of* Σ_{T^*} *is thus* $\mathcal{H}(\partial) = [1 - y^{*2}(t)](\partial + 1)^{-1}$ $(d_y = \mathcal{H}(\partial)d_u)$. *An 2DOF regulator as presented in Section 11.5 which places the closed-loop poles to* $\Lambda_{cl} = \{-1, -1\}$ *is* $S(\partial) = [1 - y^{*2}(t)]\partial$,

$$R(\partial) = (1 - 4\tfrac{y^*\dot{y}^*}{1-y^{*2}})\partial + 1- \\ -2y^*\tfrac{\dot{y}^* - 2\dot{y}^*y^{*2} + \dot{y}^*y^{*4} + \ddot{y}^* - \ddot{y}^*y^{*2} + 2\dot{y}^{*2}y^*}{1-y^{*2}} - 2(1 + y^{*2}).$$

When trajectory (11.70) is used, the response y *of the closed-loop system in Fig. 11.10 with zero initial conditions is given in Fig. 11.11 from* $t = 0$ *to* $t = 40s$. *A unitary step input disturbance applied at time* $t = 20sec$ *is also rejected with good transients and without steady-state error because of the integral effect of the regulator (no free term for* $S(\partial)$*).*

The control (11.67), plotted in Fig. 11.12, stabilizes the closed-loop in Fig. 11.10 only locally around trajectory $(x^*(t), u^*(t), y^*(t))$ of Σ. The domain of validity of such a control (i.e., the width of the zone around the trajectory in Fig. 11.9) depends on the application (i.e., on the nature of Σ) and it is difficult to estimate. In general, the closed-loop system can be driven around the linearization trajectory by including a supplementary reference signal. For Example 1167, as an 2DOF regulator is used, this transforms the closed-loop in Fig. 11.10 into the one in Fig. 11.13. In Fig. 11.11 it is shown from time $t = 40sec$ to $t = 60sec$ the response y to a step of magnitude 0.09 on reference r in Fig. 11.13.

Remark 1168. *The proposed LTV output feedback allows stable operation for the whole range* $r \in (-1, 1)$. *This is not the case with the classical gain-scheduling schemes. Indeed, in [310] it is shown that the closed-loop response diverges for* $r > 0.8$ *if precautions are not taken during the linearization to counteract the action of the so-called* hidden coupling terms. *One existing solution for this is the* velocity-based linearization, *discussed in the next section in comparison with the linearization along a given trajectory.*

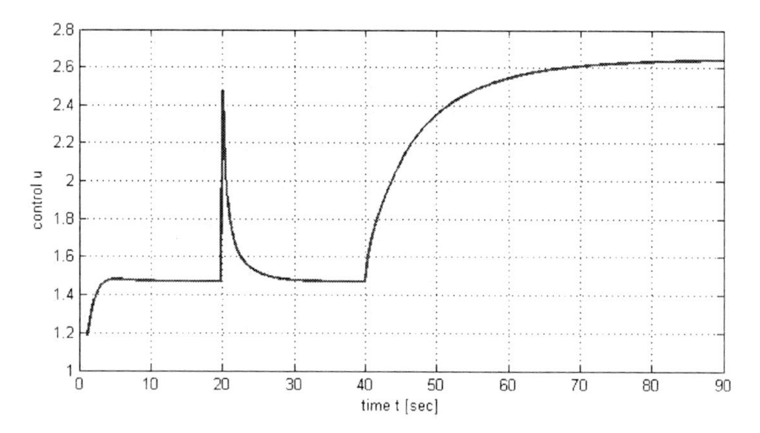

Fig. 11.12 Control of the nonlinear closed-loop

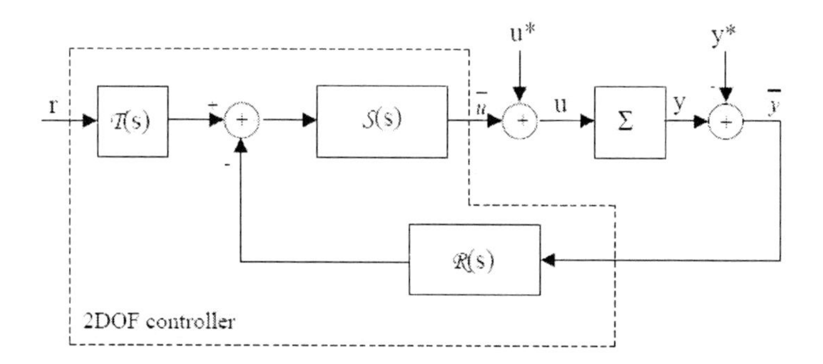

Fig. 11.13 Nonlinear 2DOF closed-loop

Remark 1169. *Notice that the solution presented in Section 11.6.6 is general, i.e., does not require Σ to be flat. First, trajectory planning can be done in the general case as exposed in Section 11.6.4.2 but, of course, flatness would facilitate this task and should be exploited at this stage if ever the system is flat. Next, control law (11.67) is not based on flatness properties and can thus be used in the case where the system is not flat or when it is flat but the output to track is not flat. Moreover, it presents also a more robust alternative to the control presented in Section 11.6.5 for tracking flat outputs of flat systems. Indeed, control law (11.60) is not only a feedback of the flat output $z(t)$ but also of a number of its derivatives. Although solutions have been proposed to compute those derivatives (see, e.g., [331] and [246]), their presence in the closed-loop still diminish robustness and noise filtering (for a detailed analysis the reader should refer to a background textbook on linear robust control like, e.g., [333]).*

11.6.7 Connections and Differences with the Gain-Scheduling Control

The gain-scheduling is a control technique used to overcome the limitations due to the linearization around a single equilibrium point and thus to extend the domain of validity of the controller. The nonlinear system is linearized around several equilibrium points and for each obtained LTI model an LTI controller is synthesized. The final control law consists in an adaptation of the control parameters inside the family previously computed and for that a measured variable, called the *scheduling variable* is used (see [212], [310] for a survey of the topic). The keypoint is to parameterize a set of equilibrium points of the given nonlinear system Σ by the scheduling variable σ. If a state representation (11.49) is used, let $(x_o(\sigma), u_o(\sigma), y_o(\sigma))$ be this family. Then, the family of linear systems around those equilibrium points is also parameterized by σ:

$$\begin{cases} \dot{d}_x = A(\sigma)d_x + B(\sigma)d_u \\ d_y = C(\sigma)d_x + D(\sigma)d_u \end{cases} \tag{11.71}$$

where $d_x(t) = x(t) - x_o(\sigma)$, $d_u(t) = u(t) - u_o(\sigma)$, $d_y(t) = y(t) - y_o(\sigma)$.

11.6.7.1 Scheduling on a Reference Trajectory

The scheduling variable must be measurable and, preferably exogenous. Its choice is, of course, application dependent, but the existing gain-scheduling approaches differ mainly in the way in which (11.71) is written. One encountered situation is the scheduling on a reference trajectory [318] where σ is in fact an a priori given trajectory $(x^*(t), u^*(t), y^*(t))$ of Σ. The LTV approach presented above allows one to better asses the closed-loop stability and to skip thus the usual hypothesis of slow varying input and scheduling variable under which the reference trajectory scheduling was previously done.

11.6.7.2 Velocity-Based Linearization

In Section 11.6.3, the linearization around an equilibrium point has been presented as a particular case of the linearization around a trajectory. The *velocity-based linearization*, another important particular case, is now introduced. It has been shown that the state $\widehat{x}(t)$ in (11.46) is a first order approximation of the state $x(t)$ of the nonlinear system Σ. More precisely, the state equation of (11.48) can be written

$$\dot{\widehat{x}}(t) = \dot{x}^*(t) + A(t)(\widehat{x}(t) - x^*(t)) + B(t)(\widehat{x}(t) - x^*(t)). \tag{11.72}$$

Let now $t_1 > t_0$ and $x_1 = x^*(t_1)$, $u_1 = u^*(t_1)$, $y_1 = y^*(t_1)$ be one point of the trajectory (11.45) of Σ. Notice that (x_1, u_1, y_1) is not necessarily an equilibrium point of Σ. A second approximation in the evaluation of Σ is to consider in (11.46) the trajectory $(x^*(t), u^*(t), y^*(t))$ frozen to its point (x_1, u_1, y_1). If, moreover, Σ is time-invariant, i.e., $\mathbf{K}$ is a field of constants, this leads to the following approximation of the state x of Σ:

$$\dot{\widehat{x}}(t) = f(x_1, u_1) + A(\widehat{x}(t) - x_1) + B(\widehat{u}(t) - u_1) \tag{11.73}$$

where A and B are matrices with constant entries. A similar development leads to

$$\widehat{y}(t) = y_1 + C(\widehat{x}(t) - x_1) + D(\widehat{u}(t) - u_1). \tag{11.74}$$

The constant terms in (11.73) and (11.74) can be skipped if one differentiate both equations which leads to the LTI state representation

$$\begin{cases} \dot{\widehat{x}}(t) = \widehat{w} \\ \dot{\widehat{w}} = A\widehat{w} + B\dot{\widehat{u}}(t) \\ \dot{\widehat{y}} = C\widehat{w} + D\dot{\widehat{u}}(t) \end{cases} \tag{11.75}$$

called the *velocity-based linearization* of Σ around the constant operating point (x_1, u_1, y_1). Compared to (11.50), (11.75) has the advantage to be computable in every operating point, not only in an equilibrium one. Also, approximation (11.75) is stronger than (11.48) and thus valid in a smaller region (the disc centered to (x_1, u_1, y_1) instead of the band around trajectory $(x^*(t), u^*(t), y^*(t))$ in Fig. 11.9).

The velocity-based linearization was proposed in [211] as a way to improve the efficiency of the gain scheduling scheme. More precisely, it has been shown that a velocity-based implementation of the gain-scheduling controller avoids the so-called *hidden coupling terms* which may cause the instability of the overall nonlinear closed-loop system. Notice that the velocity-based linearization may be interpreted as a frozen realization of the leariration around a trajectory at a given point (x_1, u_1, y_1) of that trajectory. Indeed, the state matrices of (11.75) are $A(t_1)$, $B(t_1)$, $C(t_1)$, $D(t_1)$ in (11.47) and (11.48). However, the comparison stops here as (11.75) is of higher order than (11.48) and their variables have also different signification. For Example 1155, the velocity-based linearization is obtained with the state $X = \begin{bmatrix} x \ \dot{x} \end{bmatrix}^T$:

$$\dot{X} = A(t_1)X + B(t_1)U, \quad A(t_1) = \begin{bmatrix} 0 & 1 \\ 0 & -2x(t_1) \end{bmatrix}, \quad B(t_1) = \begin{bmatrix} 0 \\ 1 \end{bmatrix}, \quad U = \dot{u}. \tag{11.76}$$

The LTI transfer function $U \mapsto y = x$ is $\mathcal{H}(\partial) = (s+2x_1)^{-1}$, with $x_1 = x^*(t_1)$ whereas the LTV transfer matrix of system (11.44) is $\mathcal{H}'(\partial) = (s+2x^*(t))^{-1}$.

11.7 Notes and References

The output-feedback pole placement problem has been widely studied in the literature and a survey is given in [196]. Factorizations over Euclidean rings are used as, for example, in [354]. The latter are valid also on noncommutative rings as used for LTV systems. However, the exposal is adapted to treat the closed-loop stability item in the LTV case based on the definitions given for finite poles in the LTV case in Chapter 6; from this point of view, the results in Sections 11.2, 11.3 and 11.4 are original and have been recently published in [236] and [237]. The topic has been addressed in the LTV case in [281] and [110] for systems with coefficients in a *ring*. For the LTV systems defined over *fields*, the transfer matrix formalism [117] and the characterization of the stability using the notion of poles given in Chapter 6 were used here to better assess computations and closed-loop stability for the output feedback pole placement problem. This provides a general solution of the problem tackled in some particular cases of LTV systems in [377], [151].

Nonlinear systems have been studied in the differential algebraic approach introduced by Fliess in [114], [119] and further extended to the differential geometric one in [122]. An overview of the state representation construction and input-output inversion problems is given in [86]. In this intrinsic framework, flatness has been introduced in [121] and further developed in [122]. An overview is given in [305].

Part of the computations have been done with the Maple packages OreModules [67] and Janet [20].

Part IV
Complements

12
Analytic Theory of LTV Systems

12.1 Introduction

We recall here some known facts about the stability of LTV systems given by their state representation

$$\dot{x}(t) = A(t)x(t) + B(t)u(t) \tag{12.1}$$

$$y(t) = C(t)x(t) + D(t)u(t) \tag{12.2}$$

where $A(t)$, $B(t)$, $C(t)$ and $D(t)$ are matrices which entries are real-valued piecewise continuous functions of t and the dimension of the state x is n. The structural properties of such an LTV system are analyzed using the *solutions* of the system of differential equations (12.1) in Section 12.2. The Lyapunov analysis of stability is recalled in Section 12.4. Definitions of the poles and zeros are given in Section 12.5 using Lyapunov transformations of the initial system. This analytic approach is a complement of the intrinsic algebraic one presented in Chapter 6.

12.2 Solutions and Trajectories

Definition 1170. *The control system given by (12.1) and (12.2) is* well-defined *if its equations of definition aforementioned satisfy the conditions of existence and uniqueness of the solutions of the associated initial condition problem.*

A system is thus well-defined if, for a given set of initial conditions and inputs, it has a unique *trajectory*. Since $A(.)$ is assumed to be continuous, for any initial state x_0 and any u, there exists a unique solution to (12.1), thus the state LTV system above is well-defined.

H. Bourlès and B. Marinescu: Linear Time-Varying Systems, LNCIS 410, pp. 583–603.
springerlink.com

Consider first the *autonomous part* of the system (12.1), i.e., the system obtained when $u = 0$:

$$\dot{x} = A(t)x. \tag{12.3}$$

As (12.1) is well-defined, the autonomous system (12.3) is also well-defined. The space of solutions of (12.3) is of dimension n.

Definition 1171. *A basis of the space of solutions of the autonomous system (12.3), i.e., a set $x_1(t), ..., x_n(t)$ of linearly independent solutions, is called a* fundamental set of solutions. *A matrix $\Psi(t) = [x_1(t)...x_n(t)]$ whose columns are the vectors of a basis of the solution space of (12.3) is called a* fundamental matrix *of (12.3).*

A fundamental matrix of (12.3) is solution of the matrix equation

$$\frac{d}{dt}\Psi(t) = A(t)\Psi(t) \tag{12.4}$$

and, conversely, any nonsingular solution of (12.4) is a fundamental matrix of (12.3).

Definition 1172. *Let $\Psi(t)$ be a fundamental matrix of (12.3). Then*

$$\Phi(t, t_0) = \Psi(t)\Psi^{-1}(t_0), \ t \geq t_0 \tag{12.5}$$

is called the state transition matrix *of (12.3).*

Notice that the above definition is consistent in the sense that $\Phi(t, t_0)$ is uniquely defined by $A(t)$ and independent of the particular choice of $\Psi(t)$. Indeed, for two different fundamental matrices $\Psi_1(t)$ and $\Psi_2(t)$, there exists $P(t)$ nonsingular such that $\Psi_2(t) = \Psi_1(t)P(t)$. Thus, following the Definition 1172, $\Phi(t, t_0) = \Psi_1(t)\Psi_1^{-1}(t_0) = \Psi_2(t)P^{-1}(t)P(t)\Psi_2^{-1}(t_0) = \Psi_2(t)\Psi_2^{-1}(t_0)$.

The transition matrix has the following properties:

Lemma 1173. *If $\Phi(t, t_0)$ is the state transition matrix of the LTV system (12.1), then:*

$$\begin{aligned} &(i)\ \Phi(t, t) = I_n \\ &(ii)\ \Phi(t_2, t_0) = \Phi(t_2, t_1)\Phi(t_1, t_0) \\ &(iii)\ \Phi^{-1}(t, t_0) = \Phi(t_0, t). \end{aligned}$$

Proof. Properties (i) and (ii) follow directly from Definition (12.5). (iii) is a special case of (ii) with $t_1 = t_2$ and $t_3 = t_0$. ■

The solution of (12.3) with initial conditions $x(t_0) = x_0$ is

$$x(t) = \Phi(t, t_0)x_0. \tag{12.6}$$

Formula (12.6) can be directly checked using the definition relation (12.5). It shows that the state transition matrix is a linear transformation which maps the initial condition x_0 into the state x at time t.

Notice that, in the particular case of LTI systems, $\Psi(t) = e^{At}$ and $\Phi(t, t_0) = e^{A(t-t_0)}$.

The solution of the system (12.1) with initial condition $x(t_0) = x_0$ can be given using the state transition matrix:

$$\begin{aligned} x(t) &= \Phi(t, t_0)x_0 + \int_{t_0}^{t} \Phi(t, \tau)B(\tau)u(\tau)d\tau = \\ &= \Phi(t, t_0)\left[x_0 + \int_{t_0}^{t} \Phi(t_0, \tau)B(\tau)u(\tau)d\tau\right], t \geq t_0 \ . \end{aligned} \tag{12.7}$$

It can be directly checked that both expressions above verify (12.1). The first terms in (12.7) give the so-called *unforced response* (i.e., the response due exclusively to the initial conditions) while the second ones are the contributions of the input u.

12.3 Periodic Systems

When the entries of $A(t)$ are piecewise continuous functions of period T, (12.3) defines a *periodic* autonomous system. Such kind of differential equations were first investigated by Floquet [123] (see also [164], [3]). Any fundamental matrix $\Psi(t)$ of a periodic system is periodic with the same period: $\Psi(t+T) = \Psi(t)\Psi^{-1}(t_0)\Psi(T)$, $\forall t \in \mathbb{R}$.

For each matrix B (possibly with complex entries) such that $e^{TB} = \Psi^{-1}(0)\Psi(T)$, there is a periodic (of period T) matrix function $t \mapsto P(t)$ such that

$$\Psi(t) = P(t)e^{tB}, \ \forall t \in \mathbb{R}. \tag{12.8}$$

Also, there exists a real matrix R and a real periodic (of period $2T$) matrix function $t \mapsto Q(t)$ such that

$$\Psi(t) = Q(t)e^{tR}, \ \forall t \in \mathbb{R}. \tag{12.9}$$

The mapping (12.8) leads to a change of coordinates under which the original periodic system becomes a linear system with real constant coefficients.

Proposition 1174. *A linear system with periodic coefficients is reducible to a system with constant coefficients.*

Proof. The change of coordinates

$$x(t) = P(t)y(t) \tag{12.10}$$

where $P(t)$ is given by (12.8) transforms (12.3) into $\dot{y} = C(t)y$ where $C(t) = P(t)^{-1}A(t)P(t) - P(t)^{-1}\dot{P}(t)$. Using $P(t) = \Psi(t)e^{-Bt}$ and $\dot{\Psi}(t) = A(t)\Psi$, the latter is written $C(t) = e^{Bt}\Psi^{-1}(t)A(t)\Psi(t)e^{-Bt} - e^{Bt}\Psi^{-1}(t)A(t)\Psi(t)e^{-Bt} + e^{Bt}\Psi^{-1}(t)\Psi(t)e^{-Bt}B = B$. ∎

In Section 12.5.1 it is shown that the change of coordinates (12.10) is a Lyapunov transformation and, as a consequence, the LTI system in the new coordinates has the same stability properties as the original periodic one.

Definition 1175. • *the representation (12.8) is called the* Floquet normal form *for the fundamental matrix* $\Psi(t)$.
- *the eigenvalues of* e^{TB} *are called the* characteristic multipliers *of the periodic system defined by (12.3).*
- *a* Floquet exponent *of the periodic system is a complex* μ *such that* $e^{\mu T}$ *is a characteristic multiplier of the system.*

12.4 Stability

Stability is a seminal problem in the analysis of dynamic systems. The object of a stability study is to draw conclusions about the behavior of the system without computing its solutions or trajectories. Starting with the theory of the differential equations, the stability items evolved in parallel with the different engineering disciplines and lead to formulations and problematics some time quite different according to the nature and specificities of each discipline. A unified approach, extending the notion of energy coming from mechanics, was developed by the Russian mathematician Lyapunov [222] to analyze in the same way the properties of any type of system modeled by differential equations. This formalism, intensively developed and extended since, is today an indispensable tool in the analysis and synthesis of dynamic systems. We will first give the basics of this approach in the general case of nonlinear systems and, next, we will focus of the advances of the analysis of the LTV case.

12.4.1 Stability Definitions

Let Σ be a nonlinear system given by

$$\dot{x} = f(t, x) \tag{12.11}$$

for $t \geq t_0$ and $x(t_0) = x_0$. Let $\|.\|$ denote a (any) norm of $\mathbb{R}^n$. Then Σ is well-defined in the sense of Definition 1170 if $f(x,t)$ is *continuous* in a set $S = \{\|x - x_0\| \leq a,\ t_0 \leq t \leq t_0 + b\}$ and *Lypschitz* (i.e., $\exists K > 0$ such that $\|f(t,x_1) - f(t,x_2)\| \leq K\|x_1 - x_2\|$ for all (t,x_1), $(t,x_2) \in S$)[1]. All the systems considered below are supposed well-defined.

Lyapunov theory of stability is concerned with the analysis of the asymptotic (i.e., as $t \to +\infty$) behavior of the trajectories of Σ. Let a solution of

[1] Notice that the two conditions are sufficient, but continuity of f is not necessary for the existance of a continuous solution and there exist solutions which do not met the Lypschitz condition ([69], [164]).

(12.11) for $t \geq t_0$ which verifies the initial condition $x(t_0) = x_0$ be denoted by $x(t; t_0, x_0)$. In the Lyapunov's original work [222] it is called the *initial (or non perturbed) trajectory*. It is of interest the behavior of other trajectories, called *perturbed trajectories*, initiated *close* to the initial trajectory.

Definition 1176. *Trajectory $x(t; t_0, x_0)$ of system Σ is said* stable *if for every $\epsilon > 0$, there exists a $\delta = \delta(\epsilon; f, t_0) > 0$ such that any perturbed trajectory $x(t; t_0, x_1)$ with $\|x_1 - x_0\| \leq \delta$ satisfies $\|x(t; t_0, x_1) - x(t; t_0, x_0)\| \leq \epsilon$ for all $t_0 \leq t < +\infty$. The same trajectory it is said* unstable *if it is not stable.*

In the definition above the value of δ depend both on ϵ and t_0 but the notion of stability itself is independent of the choice of t_0. This corresponds to the notion of *uniform stability* formalized by the following

Definition 1177. *The solution $x(t; t_0, x_0)$ is said* uniformly stable *if for any $\epsilon > 0$ there exists $\delta = \delta(\epsilon; f) > 0$ such that any perturbed trajectory $x(t; t_0, x_1)$ with $\|x_1 - x_0\| \leq \delta$ satisfies $\|x(t; t_0, x_1) - x(t; t_0, x_0)\| \leq \epsilon$ for all $t_0 \leq t < +\infty$.*

The above definitions of stability can be interpreted as properties of continuity of the solution of (12.11) with respect to the initial conditions. Moreover, the initial trajectory $x(t; t_0, x_0)$ may attract the perturbed trajectories which leads to a stronger notion of stability:

Definition 1178. *Trajectory $x(t; t_0, x_0)$ of system Σ is said (uniformly)* asymptotically stable *if it is (uniformly) stable and there exists a $\delta = \delta(\epsilon; f, x_0) > 0$ such that any perturbed trajectory $x(t; t_0, x_1)$ with $\|x_1 - x_0\| \leq \delta$ satisfies $\|x(t; t_0, x_1) - x(t; t_0, x_0)\| \to 0$ as $t \to +\infty$.*

Notice that an atrractive trajectory is not necessarily stable (see [355], 5.1.32 for an example). The ball $\|x_1 - x_0\| \leq \delta$ is called the *domain of attraction* of trajectory $x(t; t_0, x_0)$. All the concepts of stability above are *local* in the sense that they pertain only to the behavior of perturbed trajectories starting close to the initial trajectory. If $\delta = +\infty$, trajectory $x(t; t_0, x_0)$ is said *globally* asymptotically stable. If the the *rate of attraction* is concerned, we have

Definition 1179. *Trajectory $x(t; x_0, t_0)$ is said* exponentially stable *if there exist constants $m > 0$, $\lambda > 0$ and $\delta = \delta(\epsilon; f, x_0) > 0$ such that any perturbed trajectory $x(t; t_0, x_1)$ with $\|x_1 - x_0\| \leq \delta$ satisfies $\|x(t; x_0, t_0) - x(t; t_0, x_0)\| \leq me^{-\lambda(t-t_0)}, t \geq t_0$.*

A particular type of trajectories are the *equilibrium points* x_e which satisfy $f(x_e, t) = 0$, $\forall t$. Thus, a trajectory of Σ initialized in an equilibrium point x_e is $x(t; t_0, x_e) = x_e$, $\forall t$. The analysis of the stability of a general trajectory $x^*(t) = x(t; t_0, x_0)$ reduces to the analysis of the stability of the origin (i.e., of the equilibrium point $x_e = 0$) of some new system derived from Σ. Indeed, the change of variables

$$\overline{x}(t) = x(t) - x^*(t) \tag{12.12}$$

leads to Σ_V given by

$$\dot{\overline{x}}(t) = F(t, \overline{x}) \tag{12.13}$$

where $F(t, \overline{x}) = f(t, x^*(t) + \overline{x}(t)) - f(t, x^*(t))$. Σ_V is called the *variational system* with respect to trajectory $x^*(t)$. As $F(t, 0) = 0$, the origin is an equilibrium point for Σ_V. The stability of $x^*(t)$ with respect to Σ is reduced to the stability of $\overline{x}(t) = 0$ with respect to Σ_V. Therefore, without loss of generality, we will assume that $f(t, 0) = 0$ and we will study the stability of the origin $x = 0$ for (12.11).

12.4.2 Lyapunov's Direct Method

Lyapunov's theory generalizes ideas which link stability to the energy of a system. For an isolated system like (12.11), i.e. without exogenous variables, the *total energy* is a function which is zero at the origin (which is an equilibrium) and positive elsewhere. If the system is initially at the equilibrium point $x = 0$ and a perturbation brings it into a state close to 0 where the energy is positive, its behavior is characterized depending on the way in which its energy varies. If the system dynamics are such that the energy of the system is nonincreasing with time, the system is *stable*. If the energy decreases with time till zero, the system is *asymptotically stable* and if the energy increases, the system is said *unstable*. This energetic interpretation, pioneered by Lagrange for mechanical systems, was extended by Lyapunov to generalized systems defined by any differential equations. The energy function was replaced by a function with some prespecified properties.

Definition 1180. *Let $x = 0$ be an equilibrium point for the nonlinear system Σ given by (12.11). A* Lyapunov function *for Σ is a continuously differentiable function $V : [t_0, +\infty) \times \mathbb{R}^n \to \mathbb{R}$ for which exists $r > 0$ such that*

$$V(t, 0) = 0, \ \forall t \geq t_0 \text{ and } V(t, x) > 0, \forall t \geq t_0, \ \forall x \text{ such that } \|x\| < r, \tag{12.14}$$

$$\dot{V}(t, x) \leq 0, \ \forall t \geq t_0, \ \forall x \text{ such that } \|x\| < r \tag{12.15}$$

where $\dot{V}$ is evaluated along the trajectories of Σ.

The derivative in the definition above is dependent on the nature of Σ, i.e., on f: $\dot{V}(t, x) = \frac{\partial V}{\partial t} + \sum_{i=1}^{n} \frac{\partial V}{\partial x_i} \dot{x}_i = \frac{\partial V}{\partial t} + \frac{\partial V}{\partial x} f(x)$.

Proposition 1181. *The equilibrium $x = 0$ of system Σ is stable if there exists a Lyapunov function V for Σ.*

Proof. Since V is a Lyapunov function for Σ, there exists a class K function α[2] and $s > 0$ such that

$$\alpha(\|x\|) \leq V(t, x), \ \forall t \geq 0, \forall x \text{ such that } \|x\| \leq s. \tag{12.16}$$

[2] i.e., $\alpha : [0, \ a) \to [0, \ +\infty)$, strictly increasing and $\alpha(0) = 0$.

Let $\epsilon > 0$, $t_0 > 0$,

$$\epsilon_1 = min\{\epsilon, r, s\} \tag{12.17}$$

where r is the one in (12.14). Choose $\delta > 0$ such that

$$\beta(t_0, \delta) < \alpha(\epsilon_1) \tag{12.18}$$

where $\beta(t_0, \delta) = sup_{\|x\| \leq \delta} V(t_0, x)$. This choice satisfies the condition of Definition 1176. Indeed, let x_0 such that $\|x_0\| < \delta$. Then $V(t_0, x_0) \leq \beta(t_0, \delta) < \alpha(\epsilon_1)$.

Moreover, with a such choice of x_0, x stays in the ball $\|x\| < r$: suppose that there exists x such that $\|x\| \geq \epsilon_1$ and let T be such that

$$\|x(t; t_0, x_0)\| \leq \epsilon_1, \ \forall t \in [t_0, T) \tag{12.19}$$

and

$$\|x(T; t_0, x_0)\| = \epsilon_1. \tag{12.20}$$

Since $\epsilon_1 \leq r$, it follows from (12.15) that

$$V(T; x(T; t_0, x_0)) \leq V(t_0, x_0) < \alpha(\epsilon_1). \tag{12.21}$$

But, as $\epsilon_1 \leq s$, from (12.20) and (12.16) we have $V(T; x(T; t_0, x_0)) \geq \alpha(\|x(T; t_0, x_0)\|) = \alpha(\epsilon_1)$ which contradicts (12.21).

Thus, from (12.15) follows that $V(t, x(t; t_0, x_0)) \leq V(t_0, x_0) < \alpha(\epsilon_1)$, $\forall t \geq t_0$. From the latter inequality and (12.16) follows that $\alpha(\|x(t; t_0, x_0)\|) < \alpha(\epsilon_1)$. Finally, as α is strictly increasing, $\|x(t; t_0, x_0)\| < \epsilon_1 \leq \epsilon$, $\forall t \geq t_0$, thus 0 is a stable equilibrium point of Σ. ∎

Example 1182. *Consider a simple pendulum as in Fig. 12.1. Using Newton's second law of motion, we can write the equation of motion in the tangential direction as* $ml\ddot{\theta} = -mgsin\theta - kl\dot{\theta}$. *If the frictional resistance is neglected* ($k = 0$) *and after suitable normalization, this leads to* $\ddot{\theta} + sin\theta = 0$. *The state representation which corresponds to the choice of the state variables* $x_1 = \theta$, $x_2 = \dot{\theta}$ *is*

$$\begin{cases} \dot{x}_1 = x_2 \\ \dot{x}_2 = -sinx_1. \end{cases} \tag{12.22}$$

The origin $x_1 = 0$, $x_2 = 0$ *(the vertical low equilibrium position) is an equilibrium of the pendulum. The total energy of the pendulum is the sum of the potential and kinetic energies, i.e.,* $V(x_1, x_2) = (1 - cosx_1) + \frac{1}{2}x_2^2$. *Function* V *satisfies the conditions of Definition 1180, so it is a Lyapunov function for the pendulum. Thus, the origin is a stable equilibrium.*

Similar results exist for the other types of stability introduced above. They are mentioned here without proof (for which the reader should refer, for example, to [186] or [355]).

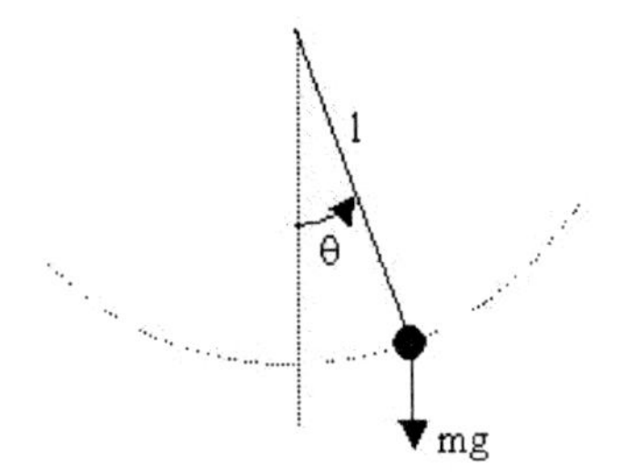

Fig. 12.1 Simple pendulum

Proposition 1183. *The equilibrium $x = 0$ of system Σ is*

- uniformly stable *if there exists a* descrescent *Lyapunov function for Σ.*
- uniformly asymptotically stable *if there exists a descrescent Lyapunov function for Σ such that $-\dot{V}$* is a locally positive definite function.
- exponetially stable *if there exists constants a, b, c, $r > 0$, $p \geq 1$ such that $a\|x\|^p \leq V(x,t) \leq b\|x\|^p$, $\forall t \geq 0$, $\forall x$ such that $\|x\| \leq r$.*

Propositions 1181 and 1183 provide only sufficient conditions for stability but the converse results are also true (see [154], [186]). They basically provide Lyapunov functions for each case of stability. However, they are not constructive since the construction of the Lyapunov functions always uses the solution of the system. Their interest stays theoretical. An example of such converse result is the "only if" part of Proposition 1196 in the case of nonlinear systems and Proposition 1193 for linear systems.

12.4.3 Linear Systems

12.4.3.1 Autonomous Systems

Consider the LTV autonomous (i.e., without input signals) system Γ defined by (12.3) where $A(t)$ is continuous. Thus, Γ is well-defined in the sense of Definition 1170. Clearly, $x = 0$ is an equilibrium point of (12.3). It is an isolated equilibrium point (i.e., there are no other equilibrium points in its vicinity) if $A(t)$ is nonsingular for some $t \geq 0$.

The stability of Γ can be characterized using the state transition matrix $\Phi(t; t_0)$ of (12.3) as introduced by Definition 1172.

Proposition 1184. *The equilibrium point $x = 0$ of Γ is stable if, and only if for each $t_0 \geq 0$*

$$sup_{t \geq t_0} \|\Phi(t, t_0)\| = m(t_0) < \infty \tag{12.23}$$

where $\|\Phi(t, t_0)\|$ is the matrix induced norm[3].

[3] For $A \in \mathbf{R}^n$, $\|A\| = sup_{x \neq 0, x \in \mathbf{R}^n} \frac{\|Ax\|}{\|x\|} = sup_{\|x\|=1} \|Ax\| = sup_{\|x\| \leq 1} \|Ax\|$. When a specific norm p ($p = 1, 2, \infty$) is used on $\mathbf{R}^n$, the resulting induced norm is denoted by $\|.\|_p$.

Proof. *The "if" part:* Suppose (12.23) holds, and let $\epsilon > 0$, $t_0 \geq 0$ be specified. If $\delta = \epsilon/m(t_0)$, then $\|x(t_0)\| \leq \delta \Rightarrow \|x(t)\| = \|\Phi(t,t_0)x(t_0)\| \leq \|\Phi(t,t_0)\|\|x(t_0)\| \leq m(t_0)\delta = \epsilon$.

The "only if" part: Let $\epsilon, \delta > 0$ and δ_1 be such that $0 < \delta_1 < \delta$. If (12.23) is false, then $\|\Phi(t,t_0)\|$ in an unbounded function of t for some $t_0 \geq 0$, i.e., there exists $t \geq t_0$ such that $\|\Phi(t,t_0)\| > \epsilon/\delta_1$. Next, select a vector v of norm one such that $\|\Phi(t,t_0)v\| = \|\Phi(t,t_0)\|$ and $x(t_0) = \delta_1 v$. It follows that $\|x(t_0)\| \leq \delta_1 < \delta$. Moreover, $\|x(t)\| = \|\Phi(t,t_0)x(t_0)\| = \|\delta_1\Phi(t,t_0)v\| = \delta_1\|\Phi(t,t_0)\| > \epsilon$. Thus, following Definition 1176, 0 is un unstable equilibrium. ■

Notice that there is a difference in the meaning of stability between linear and nonlinear autonomous systems. In the linear case, instability is synonymous of unboundedness of some solution as shown in the necessity part of the proof of Proposition 1184. This is not the case for nonlinear systems where the instability can be accompanied by the boundedness of all solutions as in the following example.

Example 1185. *Consider the Van der Pol oscillator given by*

$$\begin{cases} \dot{x}_1 = x_2 \\ \dot{x}_2 = -x_1 + (1 - x_1^2)x_2. \end{cases} \tag{12.24}$$

The origin $x_1 = 0$, $x_2 = 0$ is an equilibrium of (12.24). The response to any nonzero initial conditions is oscillatory and this results in the limit cycle in Fig. 12.2 (obtained for $x_{1_0} = x_{2_0} = 0.1$). Following Definition 1176 the origin is unstable: by choosing $\epsilon > 0$ sufficiently small such that the ball $\|x(t)\| < \epsilon$ is contained in the limit cycle in Fig. 12.2, trajectories starting in this ball will leave it and so no $\delta > 0$ can be found to satisfy the condition of Definition 1176. However, all trajectories are bounded by the limit cycle.

Proposition 1186. *The equilibrium point $x = 0$ of Γ is (globally) uniformly asymptotically stable if, and only if the state transition matrix satisfies the inequality*

$$\|\Phi(t,t_0)\| \leq me^{-\lambda(t-t_0)}, \ \forall t \geq t_0 \geq 0 \tag{12.25}$$

for some positive constants m and λ.

Proof. Due to the linear dependence (12.6) of $x(t)$ on $x_0 = x(t_0)$, if the origin is uniformly asymptotically stable, it is globally so.

The "if" part: $\|x(t)\| \leq \|\Phi(t,t_0)\|x_0 \leq m\|x_0\|e^{-\lambda(t-t_0)}$, thus (12.25) is sufficient for asymptotic stability.

The "only if" part: If $x = 0$ is uniformly asymptotically stable, then there exist finite constants μ and T such that

$$\|\Phi(t,t_0)\| \leq \mu, \ \forall t \geq t_0 > 0 \tag{12.26}$$

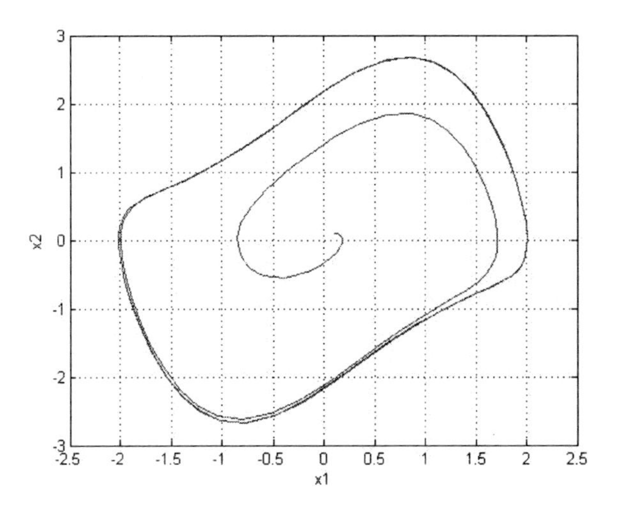

Fig. 12.2 Limit cycle of the Van der Pol oscillator

and

$$\|\Phi(t_0 + t, t_0)\| \leq 1/2, \ \forall t \geq T, \ \forall t_0 \geq 0. \tag{12.27}$$

For a given $t \geq t_0$, let k be an integer such that $t_0 + kT \leq t < t_0 + (k+1)T$. It follows, using Lemma 1173 (ii), that $\Phi(t, t_0) = \Phi(t; t_0 + kT)\Phi(t_0 + kT; t_0 + kT - T)...\Phi(t_0 + T; t_0)$ thus $\|\Phi(t, t_0)\| \leq \|\Phi(t; t_0 + kT)\| \prod_{j=1}^{k} \|\Phi(t_0 + jT; t_0 + jT - T)\|$. Repeated application of (12.26) and (12.27) gives $\|\Phi(t, t_0)\| \leq \mu 2^{-k} \leq (2\mu) 2^{-(t-t_0)/T}$. Hence (12.25) is satisfied with $m = 2\mu$ and $\lambda = \frac{log 2}{T}$. ■

Thus, for linear systems, uniform asymptotic stability is equivalent to exponential stability. Also, notice that in the general case of nonlinear systems, an equilibrium point is *globally* uniformly or exponentially stable if it is the *only* equilibrium of the system. This is the case for nondegenerate LTV systems (i.e., with nonsingular state matrix). This makes a major difference between the two classes of systems and, for linear systems, we will speak shortly about stability *of the autonomous system* (instead of stability of an equilibrium point or trajectory).

Another important particularity of LTV systems is that condition (12.25) is not easy to check in practice because knowledge of the state transition matrix $\Phi(t, t_0)$ requires the integration of equations (12.3). Propositions 1184 and 1186 have just a theoretical interest.

12.4.3.2 Periodic Systems

If the entries of the state matrix $A(t)$ in (12.3) are periodic functions, the stability can be evaluated from the Floquet exponents introduced by Definition 1180.

Proposition 1187. *The periodic system (12.3) is*

- *exponentially stable if the real parts of all its Floquet exponents are negative.*
- *stable if the real parts of all its Floquet exponents are nonpositive*
- *unstable otherwise.*

12.4.3.3 Nonautonomous Systems

When the influence of the inputs is taken into account, the *nonautonomous system* (12.1) is to be studied. Let $e(t) = x(t; t_0, x_0) - x(t; t_0, x_1)$ be the *error* between the initial and perturbed trajectories. From (12.7) it follows that $e(t) = \Phi(t, t_0)(x_0 - x_1)$ and thus $e(t)$ is the solution of the autonomous system with the initial condition $e(t_0) = x_0 - x_1$ no matter what $u(t)$ is. As a consequence, the question of stability of the solution $x(t; x_0, t_0)$ of the nonautonomous system reduces to: for a given $\epsilon > 0$, does exist $\delta > 0$ such that $\|e(t_0)\| < \delta \Rightarrow \|e(t)\| < \epsilon$, $\forall t \geq t_0$. This is the problem of stability of the trivial solution ($x(t) = 0$) of the autonomous system (12.3). A similar rationale can be made for the asymptotic and exponential stability, which leads to

Proposition 1188. *A linear nonautonomous system is stable (asymptotically or exponentially stable) if, and only if the associated linear autonomous system is stable (asymptotically or exponentially stable).*

However, in practice a stronger notion of stability is needed. Indeed, following Proposition 1188, trajectories of a stable nonautonomous system are not necessarily bounded as shown by the following example.

Example 1189. *The linear nonautonomous system given by*

$$\dot{x} = -x + t \tag{12.28}$$

is exponentially stable since its autonomous part $\dot{x} = -x$ is exponentially stable. The solutions of (12.28) from $t_0 = 0$ are $x(t) = -1 + t + x_0 e^{-t}$. Thus $e(t) = [x_0 - x_1]e^{-t} \to 0$ as $t \to +\infty$. However, $x(t) \to +\infty$.

To ensure also bounded trajectories, i.e., *input to state stability*, other supplementary conditions must be added for the nonautonomous part $f(t) = B(t)u(t)$:

$$\|f(t)\| \leq b < \infty, \; t \geq t_0. \tag{12.29}$$

Also, the following result is used:

Lemma 1190. *(Banach-Steinhaus Lemma) If the vector u given by*

$$u(t) = \int_{t_0}^{t} \Phi(t,\tau) f(\tau) d\tau, \ t, \tau \geq t_0 \tag{12.30}$$

is bounded as $t \to +\infty$ for every vector f satisfying (12.29), then the matrix $\Phi(t,\tau)$ satisfies the condition

$$\int_{t_0}^{t} \|\Phi(t,\tau)\| d\tau \leq c < +\infty \text{ for some } c \text{ and all } t \geq \tau \geq t_0 \tag{12.31}$$

See [16], [383] for a proof.

Proposition 1191. *When B and u are such that the non autonomous part $f(t) = B(t)u(t)$ satisfies (12.29), a necessary and sufficient condition for (12.1) to have bounded solutions on $[t_0, +\infty)$ is (12.31), where $\Phi(t, t_0)$ is the state transition matrix of the autonomous part of (12.1).*

Proof. *The "if part"* : suppose (12.31) holds. The solution of the non autonomous system (12.1) is given by (12.7). Using (12.31) and (12.29), it follows that $\|x\| \leq (\|\Phi(t,t_0)x_0\| + b)c$. Again, from (12.31) follows that $\Phi(t,t_0)x_0$, the solution of the autonomous part, is bounded for $t \to +\infty$ from which follows that x is bounded.

The "only if" part: suppose x is bounded when $t \to +\infty$. Then, $\Phi(t,t_0)x_0$, the autonomous part of the solution is bounded as a particular case of x (obtained with $u = 0$). From (12.29), it follows that $\int_{t_0}^{t} \|\Phi(t,\tau)B(\tau)u(\tau)\| d\tau$ is bounded and, using Lemma 1190 with $f(t) = B(t)u(t)$, one can conclude that $\int_{t_0}^{t} \|\Phi(t,\tau)\| d\tau$ is bounded. ■

Condition (12.29) is natural and usually satisfied in practice. On the one hand, physical systems have continuous and bounded coefficients over $[t_0, +\infty)$, the interval over which the model is considered; they do not exhibit any finite-time escape phenomena. On the other hand, to investigate (input to state or input/output) stability it is natural to consider bounded inputs, i.e., to excite the system with physical signals of finite energy.

The last two propositions above give the basis of stability analysis for linear systems: it is necessary and sufficient to study the stability of the autonomous system in order to conclude on the stability of the general (nonautonomous) system. This is the main difference with the nonlinear case, and allows one to speak for the linear systems about the (asymptotic or exponential) stability *of the system* without any danger of confusion.

12.4.3.4 The Case of LTI Systems

If all the coefficients of the system are constant, (12.3) particulizes to

$$\dot{x} = Ax \tag{12.32}$$

where the entries of A are constants. In this particular case of linear systems, the results are much simpler and constructive. First, the state transition matrix reduces to $\Phi(t, t_0) = e^{A(t-t_0)}$ which can be computed in practice without integrating (12.32). Next, stability is directly characterized by the eigenvalues of A:

Proposition 1192. *The equilibrium $x = 0$ of (12.32) is*

- *(globally) exponentially stable if, and only if all eigenvalues of A have negative real parts.*
- *stable if, and only if all eigenvalues of A have nonpositive real parts and, in addition, every eigenvalue of A having a zero real part is a simple zero of the minimal polynomial of A.*

The result above is a well-known fact and it is recalled without proof.

12.4.3.5 Lyapunov Functions

In the case of linear systems, Lyapunov functions can be directly provided using the state transition matrix.

Proposition 1193. *It is assumed that the origin is an uniformly asymptotically (i.e., also exponentially) stable equilibrium of the linear system (12.3) and its state matrix is bounded in the sense that*

$$m_0 = sup_{t \geq 0} \|A(t)\| < \infty. \tag{12.33}$$

Then for each bounded function $Q(t)$ which is a symmetric positive definite matrix for each $t > 0$ and such that there exists a constant $\alpha > 0$ such that

$$\alpha x^T x \leq x^T Q(t) x, \ \forall t \geq 0, \forall x \in \mathbb{R}^n, \tag{12.34}$$

the function

$$V(t, x) = x^T P(t) x \tag{12.35}$$

where

$$P(t) = \int_t^\infty \Phi^T(\tau, t) Q(\tau) \Phi(\tau, t) d\tau \tag{12.36}$$

is a Lyapunov function for (12.3).

Proof. Notice first that $P(t)$ is well-defined by (12.36). Indeed, as (12.3) is exponentially stable, there exist positive constants m, λ such that $\|\Phi(\tau, t)\| \leq me^{-\lambda(\tau - t)}$, $\forall \tau \geq t \geq 0$. Also, $Q(t)$ is bounded.

Next, $P(t)$ is positive defined for each $t \geq 0$: from (12.36) follows that $x^T P(t)x = \int_t^\infty x^T \Phi^T(\tau,t)Q(\tau)\Phi(\tau,t)x d\tau = \int_t^\infty s^T(\tau;t,x)Q(\tau)s(\tau;t,x)d\tau$ where $s(\tau;t,x)$ is the solution of (12.3) evaluated at time τ with the initial condition x at the initial time t. From (12.34) follows that $x^T P(t)x \geq \alpha \int_t^\infty \|s(\tau;t,x)\| d\tau$. As (12.3) is exponentially stable, roughly speaking, the right term of the latter inequality can be minorated by the initial conditions x of $s(\tau;t,x)$ which leads to $V(t,x) = x^T P(t)x \geq \beta x^T x$ for some $\beta > 0$, thus $V(t,x) > 0$. Moreover, $\dot{V}(t,x) = x^T[\dot{P}(t) + A^T(t)P(t) + P(t)A(t)]x = -x^T Q(t)x$, thus $\dot{V}(t,x) < 0$. It follows that V satisfies the conditions of Definition 1180. ■

In the particular case of LTI systems, the Lyapunov functions are of the form $V(x) = x^T Px$ where $P \in \mathbf{C}^{n \times n}$ is the real symmetric matrix solution of the so-called *Lyapunov matrix equation*

$$A^T P + PA = -Q \tag{12.37}$$

where Q is any positive definite matrix. As a matter of fact,

Proposition 1194. *Given a matrix with constant entries, the following statements are equivalent:*

- *all the eigenvalues of A have negative real parts.*
- *there exists some positive definite matrix Q such that (12.37) has a unique solution P, and this solution is positive definite.*
- *for every positive definite matrix Q, (12.37) has a unique solution P, and this solution is positive definite.*

12.4.4 Lyapunov's Indirect Method

The Lyapunov's indirect method, called also the *linearization method*, uses the tangent linear system of the initial nonlinear system Σ to draw stability conclusions on Σ. Let Σ be given by (12.11). As shown in Section 11.6.3, the tangent linear system Σ_{T^0} of Σ around the origin $x = 0$ is the LTV system (12.3) where $A(t) = \frac{\partial f}{\partial x}(t,x) \mid_{x=0}$. It is assumed that Σ_{T^0} is well-defined in the sense of Definition 1170 (i.e., $A(t)$ is continuous). Supplementary properties are needed on the Jacobian matrix of Σ.

Lemma 1195. *Let $f : \mathbb{R}_+ \times \mathbb{R}^n \to \mathbb{R}^n$ be continuous and differentiable with respect to t. Consider that the Jacobian $\frac{\partial f}{\partial x}(t,x)$ is* Lipschitz *in a neighboring $\|x\| \leq r$, $r > 0$ of the origin,* uniformly *in t. Let the function $f_1(t,x)$ be defined by*

$$f_{1_i}(t,x) = [\frac{\partial f_i}{\partial x}(t,z_i) - \frac{\partial f_i}{\partial x}(t,0)]x \tag{12.38}$$

for some z_i. Then, there exists a point z_i on the line segment connecting x to the origin such that

$$\|f_1(x,t)\| \leq L\|x\|^2 \tag{12.39}$$

for some $L > 0$.

Proof. The Lipschitz condition of $\frac{\partial f}{\partial x}(t,x)$ is

$$\|\frac{\partial f_i}{\partial x}(t,x_1) - \frac{\partial f_i}{\partial x}(t,x_2)\| \leq L_i\|x_1 - x_2\| \tag{12.40}$$

for $\|x_1\|$, $\|x_2\| \leq r$, $\forall t \geq 0$, $L_i > 0$ and $i = 1,...,n$. By the mean value theorem[4], $f_i(x,t) = f_i(0,t) + \frac{\partial f_i}{\partial x}(t,z_i)x$ where z_i is a point on the line segment connecting x to the origin and $i = 1,...,n$. As $f(t,0) = 0$ for all $t \geq 0$, it follows, using (12.38), that $f_i(t,x) = \frac{\partial f_i}{\partial x}(t,0)x + f_{1_i}(t,x)$, thus

$$f(t,x) = A(t)x + f_1(t,x). \tag{12.41}$$

Moreover, from (12.40) one has $\|\frac{\partial f_i}{\partial x}(x,t) - \frac{\partial f_i}{\partial x}(0,t)\| \leq L_i\|z\| \leq L_i\|x\|$ from which the conclusion follows with $L = n\ max\{L_1,...,L_n\}$. ∎

Proposition 1196. *Let $x = 0$ be an equilibrium point of Σ and assume that Σ_{T^0} is well-defined and, moreover, the Jacobian $\frac{\partial f}{\partial x}(x,t)$ is* bounded *and* Lipschitz *in a neighboring $\|x\| \leq r$, $r > 0$ of the origin,* uniformly *in t. Then, $x = 0$ is an exponentially stable equilibrium point for the nonlinear system Σ if, and only if it is an exponentially stable equilibrium point for the LTV system (12.3).*

Proof. *The "if" part:* Suppose $x = 0$ is an exponentially stable equilibrium point for the LTV system (12.3). Since $A(t)$ is bounded, it follows, as in the proof of Proposition 1193 that

$$P(t) = \int_t^\infty \Phi^T(\tau,t)\Phi(\tau,t)d\tau \tag{12.42}$$

is well-defined for all $t \geq 0$ and exist positive constants α and β such that

$$\alpha x^T x \leq x^T P(t)x \leq \beta x^T x,\ \forall x \in \mathbb{R}^n,\ \forall t \geq 0. \tag{12.43}$$

As Σ_{T^0} is exponentially stable, it follows that the function $x^T P(t)x$ is descrescent positive definite, thus a suitable Lyapunov candidate: $V(t,x) = x^T P(t)x$. Moreover, $\dot{V}(t,x) = x^T\dot{P}(t)x + f^T(t,x)P(t)x + x^T P(t)f(t,x) = x^T[\dot{P}(t) + A^T(t)P(t) + P(t)A(t)]x + 2x^T P(t)[f(t,x) - A(t)x]$. From (12.42) follows that $\dot{P}(t) + A^T(t)P(t) + P(t)A(t) = -I_n$, thus

[4] For a function $f : \mathbb{R}^n \to \mathbb{R}$ continuously differentiable at each point x of an open set $S \subset \mathbb{R}^n$ let x, $y \in S$ such that $L(x,y) \subset S$ where $L(x,y) = \{z \mid z = \theta x + (1-\theta)y,\ 0 < \theta < 1\}$ is the line segment joining x and y. Then, there exists a point $z \in L(x,y)$ such that $f(y) - f(x) = \frac{\partial f}{\partial x}\mid_{x=z}(y-x)$. (see, e.g., [5]).

$$\dot{V}(t,x) = -x^T x + 2x^T P(t)[f(t,x) - A(t)x]. \tag{12.44}$$

From Lemma 1195, one can chose $r > 0$ such that $L = \frac{s}{\beta}$ in (12.39) with $s < \frac{1}{2r}$. Thus, $\|f(t,x) - A(t)x\| \leq \frac{1}{2\beta r}\|x\|^2$, $\forall t \geq 0$, $\forall\ x$ such that $\|x\| \leq r$. The latter inequality in conjunction with (12.43) leads to $|\ 2x^T P(t)[f(t,x) - A(t)x]\ | \leq 2srx^T x$, $\forall t \geq 0$, $\forall\ x$ such that $\|x\| \leq r$. Thus, (12.44) leads to $\dot{V}(t,x) \leq -(1-2s)x^T x$, $\forall t \geq 0$, $\forall\ x$ such that $\|x\| \leq r$ which, as $s < \frac{1}{2r}$, means that $-\dot{V}$ is a locally positive definite function. Thus, following Proposition 1183, $x = 0$ is an exponentially stable equilibrium for Σ.

The only if part: Suppose $x = 0$ is an exponetially stable equilibrium for Σ. Then, there exist positive constants m, λ and c such that

$$\|x(t)\| \leq m\|x(t_0)\|e^{-\lambda(t-t_0)},\ \forall t \geq t_0 \geq 0,\ \forall \|x_{t_0}\| < c. \tag{12.45}$$

Consider the function $V(t,x) = \int_t^{t+\delta} s^T(\tau;t,x)s(\tau;t,x)d\tau$ where δ is a constant to be chosen and $s(\tau;t,x)$ is the solution of (12.11) that starts at t in x, i.e., $s(t) = x$. It is now shown that δ can be chosen such that V is a Lyapunov function for Σ:

- *there exist positive constants c_1, c_2 such that $c_1\|x\|^2 \leq V(t,x) \leq c_2\|x\|^2$:* First, from (12.45) follows that

$$V(t,x) \leq \int_t^{t+\delta} m^2 e^{-2\lambda(\tau-t)}d\tau\|x\|^2 = \frac{m^2}{2\lambda}(1-e^{-2\lambda\delta})\|x\|^2.$$

  Thus, $c_2 = \frac{m^2}{2\lambda}(1-e^{-2\lambda\delta})$. Next, as the Jacobian is bounded, let $\|\frac{\partial f}{\partial x}(t,x)\| \leq T$, for $\|x\| \leq r$. It follows $\|f(t,x)\| \leq T\|x\|$ from which follows that $s(\tau;t,x)$ satisfies the lower bound $\|s(\tau;t,x)\|^2 \geq \|x\|^2 e^{-2T(\tau-t)}$. Hence $V(t,x) \geq \int_t^{t+\delta} e^{-2T(\tau-t)}d\tau\|x\|^2 = \frac{1}{2T}(1-e^{-2T\delta})\|x\|^2$, thus $c_1 = \frac{1}{2T}(1-e^{-2T\delta})$.
- *there exists c_3 such that $\dot{V}(t,x) \leq -c_3\|x\|^2$:* Let $s(\tau;t,x) = \frac{\partial}{\partial t}s(\tau;t,x)$, $s_x(\tau;t,x) = \frac{\partial}{\partial x}s(\tau;t,x)$. Then, $\dot{V}(t,x) = \frac{\partial V}{\partial t} + \frac{\partial V}{\partial x}f(t,x) = s^T(t+\delta;t,x)s(t+\delta;t,x) - s^T(t;t,x)s(t;t,x) + \int_t^{t+\delta} 2s^T(\tau;t,x)s_t(\tau;t,x)d\tau + \int_t^{t+\delta} 2s^T(\tau;t,x)s_x(\tau;t,x)d\tau f(t,x) = s^T(t+\delta;t,x)s(t+\delta;t,x) - \|x\|_2^2 + \int_t^{t+\delta} 2s^T(\tau;t,x)[s_t(\tau;t,x) + s_x(\tau;t,x)f(t,x)]d\tau$. As

$$s_t(\tau;t,x) + s_x(\tau;t,x)f(t,x) \equiv 0,\ \forall \tau \geq t,$$

  it follows that $\dot{V}(t,x) = s^T(t+\delta;t,x)s(t+\delta;t,x) - \|x\|_2^2 \leq -(1 - k^2 e^{-2\lambda\delta})\|x\|_2^2$. By choosing $\delta = \frac{ln(2k^2)}{2\lambda}$ and using the equivalence of norms it follows that $c_3 = \frac{1}{2}$.
- *there exists c_4 such that $\|\frac{\partial V}{\partial x}\| \leq c_4\|x\|$:* similar calculations lead to $c_4 = \frac{2k}{\lambda-T}[1-e^{(\lambda-T)\delta}]$.

Let now compute the derivative of V but along the trajectories of the linear system (12.3): $\frac{\partial V}{\partial t} + \frac{\partial V}{\partial x}A(t)x = \frac{\partial V}{\partial t} + \frac{\partial V}{\partial x}f(t,x) - \frac{\partial V}{\partial x}f_1(t,x)$, where $f_1(t,x)$ is given by (12.41). Using Lemma 1195 and the second item above, $\frac{\partial V}{\partial t} + \frac{\partial V}{\partial x}A(t)x \le -c_3\|x\|^2 + c_4T\|x\|^3 < -(c_3 - c_4T\rho)\|x\|^2$, $\forall\ \|x\| < \rho$. The choice $\rho < min\{r, \frac{c_3}{c_4T}\}$ ensures that the derivative of V along the trajectories of (12.3) is negative definite in $\|x\| < \rho$, thus the origin is an exponentially stable equilibrium point for the linear system (12.3). ∎

12.5 Poles of the System

12.5.1 Lyapunov Transformations

Consider again an LTV system given by the space representation (12.3). A change of variables

$$x = L(t)y, \tag{12.46}$$

where L is a nonsingular and continuously differentiable for $t \ge t_0$ matrix, brings (12.3) to

$$\dot{y} = B(t)y, \tag{12.47}$$

where $B(t) = L^{-1}(t)A(t)L(t) - L^{-1}(t)\dot{L}(t)$. We are interested in the class of transformations which preserve stability properties of (12.3).

Definition 1197. *The transformation (12.46) is called a* Lyapunov transformation *if $L(t)$ is a nonsingular and continuously differentiable for $t \ge t_0$ matrix and $L(t)$, $\dot{L}(t)$, $L^{-1}(t)$ are bounded for $t \ge t_0$.*

Let $f(t)$ be a complex-valued function defined on the interval $[t_0,\ +\infty)$.

Definition 1198. *The characteristic or Lyapunov exponent of f is*

$$\chi[f] = limsup_{t\to+\infty}\frac{ln \mid f(t) \mid}{t}. \tag{12.48}$$

$\chi[f]$ determines the growth of the absolute value of f with respect to the exponential scale. As a matter of fact, the Lyapunov characteristic of $e^{\alpha t}$ is α. Moreover, $\mid f(t) \mid = te^{\frac{ln|f(t)|}{t}}$.

Vector-functions $x_1, ..., x_n$ defined on $[t_0,\ +\infty)$ and having different finite Lyapunov exponents are linearly independent. It follows that the solutions of a linear system (12.3) cannot have more that n different Lyapunov exponents.

Definition 1199. *The set of all different Lyapunov exponents of solutions of a linear system is called its* Lyapunov spectrum.

The Lyapunov transformations do not change the Lyapunov spectrum:

Proposition 1200. *Systems (12.3) and (12.47) have the same Lyapunov spectra.*

Proof. On the one hand, from (12.46), obviously $\|x\| \leq \|L(t)\|\|y\| \leq K\|y\|$, thus $\chi[x] \leq \chi[y]$. On the other hand, $y = L^{-1}(t)x$ and, $L^{-1}(t)$ is also bounded, it follows that $\|y\| \leq K'\|x\|$, thus $\chi[y] \leq \chi[x]$. Finally, $\chi[x] = \chi[y]$ for any solutions of (12.3) and (12.47). ■

12.5.2 Reduction of a Linear System to a Triangular Form

System (12.3) can always be reduced to a state form having an upper-triangular state matrix. We are interested in the case in which this kind of reduction, called *Perron transformation* [273], can be achieved with a Lyapunov change of coordinates.

The result below is widely known and used in linear algebra. It is the consequence of the orthogonalization of the basis of the state transition matrix of (12.3) (see, e.g., [3]).

Lemma 1201. *Any fundamental matrix $\Psi_A(t)$ of (12.3) can be written*

$$\Psi_A(t) = U(t)R(t) \tag{12.49}$$

where $U(t)$ is unitary (i.e., $U^T(t)U(t) = I_n$) and $R(t)$ is an upper triangular matrix with positive elements on the diagonal.

Proposition 1202. *System (12.3) can be reduced to a system with an* upper triangular *state matrix with real diagonal coefficients by a* unitary *change of variable (12.46). Moreover, if the initial system has* bounded *coefficients, the coefficients of the reduced system are also* bounded *and the unitary transformation is* Lyapunov.

Proof. The change of variables (12.46) with $L(t) = U(t)$ where $U(t)$ is the unitary matrix in (12.49) brings the initial system (12.3) to (12.47) where

$$B(t) = U^{-1}(t)A(t)U(t) - U^{-1}(t)\dot{U}(t). \tag{12.50}$$

A fundamental matrix of the reduced system is $\Psi_B(t) = U^{-1}(t)\Psi(t)$ where $\Psi(t)$ is a fundamental matrix of the initial system. From (12.49) follows that $\Psi(t) = R(t)$, thus $\Psi_B(t)$ is upper triangular. Also, $\dot{\Psi}_B(t) = B(t)\Psi_B(t)$ from which $B(t) = \dot{\Psi}(t)\Psi^{-1}(t)$, thus $B(t)$ is upper triangular also. Moreover,

$$b_{ii} = \dot{r}_{ii}(t)r_{ii}^{-1}(t), \; i = 1, ..., n \tag{12.51}$$

which, following Lemma 1201, are positive. Suppose now the coefficients of the initial system are bounded: $sup_{t \geq t_0}\|A(t)\| \leq M$. Then $\|U^{-1}(t)A(t)U(t)\| \leq M$ for $t \geq t_0$ since $U(t)$ is unitary. Using (12.50), it follows that

$$\|B(t)\| \leq M + \|V(t)\| \tag{12.52}$$

where $V(t) = U^{-1}(t)\dot{U}(t)$. It can be easily checked that $V(t) = -V^T(t)$. As $b_{ii} > 0$, $i = 1, ..., n$, it follows from (12.50) that $V_{ii}(t) \leq (U^{-1}(t)A(t)U(t))_{ii}$, $i = 1, ..., n$. Thus, $\|V(t)\| \leq M$ which, combined with (12.52) leads to $\|B(t)\| \leq 2M$ for $t \geq t_0$. Moreover, $\|U(t)\|$ and $\|U^{-1}(t)\|$ are bounded since U is unitary. The boundedness of $\|\dot{U}(t)\|$ follows from the boundedness of $V(t)$ since $\dot{U}(t) = U(t)V(t)$. ■

12.5.3 Poles and Stability

Poles can be defined for LTV systems with bounded coefficients. Indeed, following Proposition 1202, the transformation towards the upper triangular form is in this case Lyapunov and preserves the stability properties of the initial system.

Definition 1203. *Consider the system defined by (12.3) and suppose $A(t)$ is bounded. A set $\mathcal{P}$ of system poles for (12.3) is the diagonal of the state matrix $B(t)$ given by (12.50) of the upper triangular reduced system: $\mathcal{P} = \{b_{11}, ..., b_{nn}\}$.*

Proposition 1204. *The system defined by (12.3) where $A(t)$ is bounded is exponentially stable if, and only if*

$$limsup_{t \to +\infty} b_{ii}(t) < 0, \ i = 1, ..., n. \tag{12.53}$$

Proof. We will use the same notations as in Proposition 1202.

The "if" part: Suppose (12.53) is satisfied. As shown in the proof of the mentioned proposition, transformation $U(t)$ brings also a fundamental matrix to an upper triangular form: $\Psi_B(t) = R(t)$. A fundamental set of solutions of the reduced system (12.47) is formed by the columns $y_i(t) = [r_{1i}(t)...r_{n-1i}(t) \ r_{ii}(t) \ 0...0]^T$ of $\Psi_B(t)$ or, equivalently, $R(t)$. From the elementary equations (12.51) follows that $r_{ii}(t) = \int b_{ii}(t)dt$, $i = 1, ..., n$. Obviously $r_{ii}(t) \to 0$ exponentially when $t \to +\infty$ because of (12.53). The same property holds for the other components of each y_i of the fundamental set: the $i-1$ row of $\dot{y}_i(t) = B(t)y_i(t)$ is

$$\dot{r}_{i-1i}(t) = b_{i-1i-1}(t)r_{i-1i}(t) + f(t) \tag{12.54}$$

where $f(t) = b_{i-1i}(t)r_{ii}(t)$. As $b_{i-1i}(t)$ is bounded and $r_{ii}(t)$ is exponentially decreasing to zero, f satisfies (12.29), thus, from Proposition 1191, the state of (12.54), i.e., $r_{i-1i}(t)$ exponentially decreases to zero. Following in the same manner, it can be concluded that all the entries of $\Psi(t)$ exponentially decrease to zero, i.e., that the reduced system (12.47) is exponentially stable. It follows that the initial system (12.3) is also exponentially stable since the transformation $L(t) = U(t)$ is Lyapunov.

The "only if" part: Suppose (12.3) is exponentially stable. Then (12.47) is also exponentially stable. As shown in the proof of the "if" part, r_{ii}, $i = 1, ..., n$ are elements of the vectors of a fundamental set of solutions of (12.47). Thus they exponentially decrease to zero. From the elementary equations (12.51) it follows (12.53). ■

It is important to notice that the stability characterization above holds only for systems with bounded coefficients. Indeed, this is a key point in the demonstration above. Without this condition, the transformation $L(t) = U(t)$ is no longer Lyapunov. Moreover, $U(t)$ is built from a fundamental matrix of the initial system which can be determined only after computation of all solutions to the system (12.3). Thus, stability characterization (12.53) is not constructive even in the particular case of systems having bounded coefficients.

When $B(t)$ has the particular form $B(t) = diag\{b_1, ..., b_n\}$, the pole set $\mathcal{P}$ corresponds to the *parallel D-spectrum* introduced in [378] and related references. It furthermore corresponds to a fundamental set of poles defined in Section 6.6 for the particular case of linear systems defined by a W-polynomial.

Example 1205. *Consider the third order Euler differential equation:*

$$t^3 y^{(3)}(t) - t^2 \ddot{y}(t) - 2t\dot{y}(t) - 4y(t) = 0. \tag{12.55}$$

A state representation is (12.3) with $x(t) = [y(t)\ \dot{y}(t)\ \ddot{y}(t)]^T$ *and* $A(t) = \begin{bmatrix} 0 & 1 & 0 \\ 0 & 0 & 1 \\ -4t^{-3} & -2t^{-2} & t^{-1} \end{bmatrix}$. *Notice that* $A(t)$ *is bounded for* $[t_0,\ +\infty)$, $t_0 > 0$. *A set of poles which is also a parallel D-spectrum is* $\gamma_1(t) = 4t^{-1}$, $\gamma_2(t) = it^{-1}$, $\gamma_3(t) = -it^{-1}$ *where* $i^2 = -1$. *Equation (12.55) can be written* $P(\partial)y = 0$ *where* $\partial = d/dt$ *and* $P(\partial) = t^3\partial^3 - t^2\partial^2 - 2t\partial - 4$. *The* γ*'s given above form a fundamental set of roots of the skew polynomial* $P(\partial)$ *and can be computed directly from the factors of* $P(\partial)$.

12.6 Exercises

Exercise 1206. *For the LTV system (12.1) and (12.2) consider the matrix*

$$W_o(t_0, t_1) = \int_{t_0}^{t_1} \Phi^T(\tau, t_0) C^T(\tau) C(\tau) \Phi(\tau, t_0) d\tau,$$

called the observability grammian of (12.1), (12.2), where $\Phi(\tau, t_0)$ *is the state transition matrix of (12.1). Show that (12.1), (12.2) is observable at* t_0 *if, and only if it exists* $t_1 > t_0$ *such that* $W_o(t_0, t_1)$ *is nonsingular.*

Exercise 1207. *Show that the Lyapunov transformations form a group.*

12.7 Notes and References

Overviews of the presentation and analysis of an LTV system from its state representations are given in [63], [170], [304], [151]. Stability is derived from the original approach of Lyapunov [222]. Our presentation follows mainly the ones in [355], [186] and [61]. Lyapunov transformations and reducibility of state representations of LTV systems are treated in [3]. In [262], Lyapunov transformations were used to obtain the Perron's upper triangular form [273] of the state matrix, on which the notion of *pole set of the state matrix* was defined.

References

[1] Abramczuk, W.: A Class of Surjective Convolution Operators. Pacific J. of Math. 110, 1–7 (1984)

[2] Abramov, S.A.: Rational solutions of linear differential and difference equations with polynomial coefficients. U.S.S.R. Comput. Maths. Math. Phys. 29, 7–12 (1989)

[3] Adrianova, L.Y.: Introduction to Linear Systems of Differential Equations. American Mathematical Society, Providence (1995)

[4] Anderson, B.D.O., Scott, R.W.: Parametric Solution of the Stable Exact Model Matching Problem. IEEE Trans. on Automatic Control, 137–138 (February 1977)

[5] Apostol, T.M.: Mathematical Analysis. Addison-Wesley, Reading (1957)

[6] Artin, E.: Geometric Algebra. Interscience (1957)

[7] Artin, E.: Galois Theory. 2nd supplemented edn. University of Notre Dame Press (1942) (6th printing, 1971)

[8] Arvanitis, K.G., Paraskevopoulos, P.N.: Uniform exact model matching for a class of linear time-varying analytic systems. Systems and Control Letters 19, 313–323 (1992)

[9] Baer, R.: Abelian Groups that are Direct Summands of Every Containing Abelian Group. Bull. Amer. Math. Soc. 46, 800–806 (1940)

[10] Barkatou, M.A., Pflügel, E.: Formal solutions of linear differential and difference equations. Programmirovanie 1 (1997)

[11] Barkatou, M.A.: On rational solutions of systems of linear differential equations. J. Symb. Comput. 28(4/5), 547–568 (1999)

[12] Bass, H.: Finistic dimension and homological generalization of semi-primary rings. Trans. Math. Amer. Soc. 95, 466–488 (1960)

[13] Bastida, J.R., Lyndon, R.: Field Extensions and Galois Theory. Addison-Wesley Publishing Company, Menlo Park (1984)

[14] Bedoya, H., Lewin, J.: Ranks of free matrices over Ore domains. Proc. Amer. Math. Soc. 62, 233–236 (1977)

[15] Beck, G., Packard, A.K.: Robust Performance of Linear Parametrically Varying System Using Parametrically Dependent Linear Feedback. Systems Control Letters 23(3), 205–215 (1994)

[16] Bellman, R.: An an application of a Banach-Steinhaus theorem to the study of the boundedness of solutions of nonlinear differential and difference equations. Annals of Mathematics 49(3), 515–522 (1948)

[17] Berenstein, C.A., Dostal, M.A.: The Ritt theorem in several variables. Ark. Mat. 12, 267–280 (1974)
[18] Birkhoff, G.: Lattice Theory, 3rd edn. American Mathematical Society, Providence (1984)
[19] Björk, J.-E.: Rings of Differential Operators, North Holland Mathematics Library (1979)
[20] Blinkov, Y.A., Cid, C.F., Gerdt, V.P., Plesken, V.P., Robertz, D.: The Maple Package Janet: I. Polynomial Systems. II. Linear Partial Differential Equations. In: Proc. 6th Int. Workshop on Computer Algebra in Scientific Computing, Passau, Germany, pp. 31–40, resp. 41-54 (2003), `http://wwwb.math.rwth-aachen.de/Janet`
[21] Blomberg, H., Ylinen, R.: Algebraic Theory for Multivariable Linear Systems. Academic Press, London (1983)
[22] Bode, H.W.: Network Analysis and Feedback Amplifier Design. Van Nostrand, New York (1945)
[23] Borel, A.: Linear Algebraic Groups. W.A. Benjamin, Inc. (1969)
[24] Borel, A., et al.: Algebraic D-modules. Academic Press, London (1987)
[25] Bosgra, O.H., van der Weiden, A.J.J.: Realization in generalized state-space form for polynomial matrices and the definition of poles, zeros and decoupling zeros at infinity. Internat. J. Control 33, 393–411 (1981)
[26] Bourbaki, N.: Théorie des ensembles. Hermann, Parris (1970); English translation: Theory of Sets. Springer, Heidelberg (2004)
[27] Bourbaki, N.: Algèbre, Chapitres 1 à 3. Hermann, Paris (1970); English translation: Algebra I. Springer, Heidelberg (1989)
[28] Bourbaki, N.: Algèbre, Chapitres 4 à 7. Masson (1981); English translation: Algebra II. Springer, Heidelberg (1990)
[29] Bourbaki, N.: Algèbre, Chapitre 8. Hermann, Paris (1973)
[30] Bourbaki, N.: Algèbre, Chapitre 10 – Algèbre homologique. Masson (1980)
[31] Bourbaki, N.: Algèbre commutative, Chapitres 1 à 4 & 5 à 7, Hermann, Paris (1985); English translation: Commutative Algebra, ch. 1-7. Springer, Heidelberg (1989)
[32] Bourbaki, N.: Algèbre commutative, Chapitres 8 à 9. Masson (1983)
[33] Bourbaki, N.: Algèbre commutative, Chapitre 10. Masson (1998)
[34] Bourbaki, N.: Topologie générale, Chapitres 1 à 4. Hermann, Paris (1974)
[35] Bourbaki, N.: Topologie générale, Chapitres 5 à 10, pp. 5–10. Hermann, Paris (1971); English translation: General Topology, ch. 5-10. Springer, Heidelberg (1989)
[36] Bourbaki, N.: Espaces vectoriels topologiques. Masson (1981); English translation: Topological Vector Spaces. Springer, Heidelberg (1987)
[37] Bourbaki, N.: Fonctions d'une variable ré elle. Hermann, Paris (1976); English translation: Functions of a Real Variable. Springer, Heidelberg (2004)
[38] Bourbaki, N.: Groupes et algèbres de Lie, Chapitres 1 et 2-3. Hermann, Paris (1971-1972); English translation: Lie groups and Lie algebras, ch. 1-3. Springer, Heidelberg (1989)
[39] Bourlès, H.: A New Look on Poles and Zeros at Infinity in the Light of System Interconnection. In: Proc. 41st Conf. on Decision and Control, Las Vegas, Nevada, December 10-13, pp. 2125–2130 (2002)
[40] Bourlès, H.: Impulsive systems and behaviors in the theory of linear dynamical systems. Forum. Math. 17, 781–808 (2005)

[41] Bourlès, H.: Structural properties of discrete and continuous linear time-varying systems: a unified approach. In: Lamnabhi-Lagarrigue, F., Loria, A., Panteley, E. (eds.) Advanced topics in control systems theory – Lecture Notes from FAP 2004. LNCIS, vol. 311, ch. 6, pp. 225–280. Springer, Heidelberg (2005)

[42] Bourlès, H.: Structural properties of linear systems – Part II: Structure at Infinity. In: Loria, A., Lamnabhi-Lagarrigue, F., Panteley, E. (eds.) Advanced topics in control systems theory – Lecture Notes from FAP 2005. LNCIS, vol. 328, ch.7, pp. 259–284. Springer, Heidelberg (2006)

[43] Bourlès, H.: Linear Systems. ISTE-Wiley (2010)

[44] Bourlès, H., Fliess, M.: Finite poles and zeros of linear systems: an intrinsic approach. Internat. J. Control 68, 897–922 (1997)

[45] Bourlès, H., Marinescu, B.: Infinite Poles and Zeros: a Module Theoretic Standpoint with Application. In: Proc. of the 35th Conf. on Decision and Control, Kobe, Japan (1996)

[46] Bourlès, H., Marinescu, B.: Poles and Zeros at Infinity of Linear Time-Varying Systems. IEEE Trans. on Automat. Control 44, 1981–1985 (1999)

[47] Bourlès, H., Oberst, U.: Duality for differential-difference systems over Lie groups. SIAM J. Control Optim. 48, 2051–2084 (2009)

[48] Bourlès, H., Oberst, U.: Elimination, fundamental principle and duality for analytic linear systems of partial differential-difference equations with constant coefficients. In: Proc. MTNS 2010, Budapest, pp. 1551–1555 (2010)

[49] Bourlès, H., Oberst, U.: Fundamental principle for analytic linear systems of partial differential-difference equations with constant coefficients. Math. Control Signal Systems (submitted)

[50] Bréthé, D., Loiseau, J.J.: Stabilization of linear time-delay systems. JESA-RAIRO-APII 6, 1025–1047 (1997)

[51] Bruhat, F.: Sur les représentations induites des groupes de Lie. Bull. Soc. Math. de France 84, 97–205 (1956)

[52] Bruhat, F.: Distributions sur un groupe localement compact et applications à l'étude des représentations des groupes p -adiques. Bull. Soc. Math. de France 89, 43–75 (1961)

[53] Bowtell, A.J., Cohn, P.M.: Bounded and invariant elements in 2-firs. Proc. Camb. Phil. Soc. 69, 1–12 (1971)

[54] Bronstein, M., Petkovšek, M.: An introduction to pseudo-linear algebra. Theoretical Computer Science 157, 3–33 (1996)

[55] Buchsbaum, D.A.: Exact Categories and Duality. Trans. Amer. Math. Soc. 80(1), 1–34 (1955)

[56] Byrnes, J., Ostheimer, G. (eds.): Computational Noncommutative Algebra and Applications. Kluwer Academic Publishers, Dordrecht (2004)

[57] Callier, F.M., Desoer, C.A.: Multivariable Feedback Systems. Springer, Heidelberg (1982)

[58] Cartan, H.: Théorie de Galois pour les corps non commutatifs. Annales de l'ENS, 3ème série 64, 59–77 (1947)

[59] Cartan, H., Eilenberg, S.: Homological Algebra. Princeton University Press, Princeton (1956)

[60] Cerezo, A., Rouvière, F.: Résolubilit é locale d'un opérateur différentiel invariant du premier ordre. Ann. Sc. ENS 4, 21–30 (1971)

[61] Cesari, L.: Asymptotic behavior and stability problems in ordinary diffrential equations, 3rd edn. Academic Press, New York (1971)

[62] Chase, S.U.: Direct Product of Modules. Trans. Amer. Math. Soc. 97, 457–473 (1960)
[63] Chen, C.-T.: Linear Systems theory and Design. Holt, Rinehart and Wilson (1984)
[64] Chirikjian, G.S., Kyatkin, A.: An Operational Calculus for the Euclidean Motion Group with Applications in Robotics and Polymer Science. The Journal of Fourier Analysis and Applications 6, 583–606 (2000)
[65] Chow, J.H.: Time-scale Modelling of Dynamic Networks with Applications to Power Systems. LNCIS, vol. 46. Springer, Heidelberg (1982)
[66] Chyzak, F.: Fonctions holonomes en calcul formel. Ph.D Thesis, Ecole polytechnique, France (Mai 27, 1998)
[67] Chyzak, F., Quadrat, A., Robertz, D.: OreModules: A symmetric package for the study of multidimensional linear systems. In: Proc. 16th Int. Symp. on Mathematical Theory on Networks and Systems (MTNS 2004), Leuven, Belgium, July 5-9 (2004), `http://wwwb.math.rwth-aachen.de/OreModules`
[68] Chyzak, E., Quadrat, A., Robertz, D.: Effective algorithms for parametrizing linear control systems over Ore algebras. Applicable Algebra in Engineering, Communications and Computing 16, 319–376 (2005)
[69] Coddington, E.A., Levinson, N.: Theory of Ordinary Differential Equations. McGraw-Hill, New York (1955)
[70] Cohn, P.M.: Noncommutative Unique Factorization Domains. Trans. Amer. Math. Soc. 109, 313–331 (1963); Corr. ibid. 119, 552 (1965)
[71] Cohn, P.M.: Free ideal rings. Journal of Algebra 1, 47–69 (1964)
[72] Cohn, P.M.: Bézout rings and their subrings. Proc. Cambridge Phil. Soc. 1, 251–264 (1968)
[73] Cohn, P.M.: The embedding of firs in skew fields. Proc. London Math. Soc. 23(3), 193–213 (1971)
[74] Cohn, P.M.: Unique Factorization Domains. The American Mathematical Monthly 80, 1–18 (1973)
[75] Cohn, P.M.: The similarity reduction of matrices over a skew field. Math. Zeits. 132, 151–163 (1973)
[76] Cohn, P.M.: Universal Algebra, 2nd edn. D. Reidel, Dordrechtz (1981)
[77] Cohn, P.M.: Free Rings and Their Relations, 2nd edn. Academic Press, London (1985)
[78] Cohn, P.M.: Skew fields –Theory of general division rings. Cambridge University Press, Cambridge (1995)
[79] Cohn, P.M.: Basic Algebra –Groups, Rings and Fields. Springer, Heidelberg (2003)
[80] Cohn, P.M.: Further Algebra and Applications. Springer, Heidelberg (2003)
[81] Cohn, P.M.: Some remarks on projective-free rings. Algebra Univers. 49, 159–164 (2003)
[82] Cohn, R.M.: Difference Algebra. John Wiley, Chichester (1965)
[83] Commault, C., Dion, J.M., Torres, J.: Invariant Spaces of Linear Systems Application to Bloc Decoupling. In: Proc. of the 27th Conference on Decision and Control, Austin, Texas (December 1988); Also in IEEE Trans. on Automatic Control 35(5), 618–623 (1990)
[84] Conte, G., Perdon, A.M.: Infinite zero module and infinite pole module. In: Proc. of the Conference on Analysis and Optimization of Systems, Nice and Analysis and Optimization of Systems, Session 6 Linear Systems I, vol. 62, pp. 302–315. Springer, Heidelberg (1984)

[85] Coutinho, S.C.: A Primer of Algebraic $\mathcal{D}$-modules. London Math. Soc. Texts, vol. 33. Cambridge University Press, Cambridge (1995)
[86] Delaleau, E.: Sur les dérivées de l'entrée en réprésentation et commande des systèmes non linéaires, Ph.D. Dissertation, Univ. Paris-Sud., Orsay (1993)
[87] Delaleau, E., Rudolph, J.: An intrinic characterization of properness for linear time-varying systems. Journal of Mathematical Systems, Estimation and Control 5, 1–18 (1995)
[88] Delenclos, J., Leroy, A.: Noncommutative Symmetric Functions and W-polynomials. J. Algebra Appl. 6(5), 815–837 (2007)
[89] Dicks, W., Sontag, E.D.: Sylvester Domains. Journal of Pure and Applied Algebra 13, 243–275 (1978)
[90] Dicks, W.: Free Algebras Over Bézout Domains are Sylvester Domains. Journal of Pure and Applied Algebra 27, 15–28 (1983)
[91] Dieudonné, J.: Les déterminants sur un corps non commutatif. Bulletin de la Soc. Math. de France 71(2), 27–45 (1943)
[92] Dieudonné, J.: Eléments d'analyse, vol. I-VI. Gauthier-Villars, Paris (1969-1975); English translation: Treatise on analysis, vol. I-VI. Academic Press, London (1969-1978)
[93] Dieudonné, J.: Introduction to the Theory of Formal Groups. M. Dekker, New York (1973)
[94] Dixmier, J.: Algèbres enveloppantes. Gauthier-Villars, Paris (1994); English translation: Enveloping Algebras. AMS Bookstore (1996)
[95] Duflo, M., Wigner, D.: Sur la résolubilité des équations différentielles invariantes sur un groupe de Lie. Séminaire Equations aux dérivées partielles(Ecole Polytechnique) 9, 1–7 (1977-1978)
[96] Dunford, N., Schwartz, J.T.: Linear Operators–Part I: General Theory. Wiley, Chichester (1958)
[97] Eckmann, B., Schopf, O.: Über injecktiv Moduln. Archiv. der Mathematik 4, 75–78 (1953)
[98] Edwards, H.M.: Fermat's Last Theorem –A Genetic Introduction to Algebraic Number Theory, 5th printing. Springer, Heidelberg (2000)
[99] Ehrenpreis, L.: Solutions of some problems of division III. American J. of Mathematics 78, 685–715 (1956)
[100] Ehrenpreis, L.: Solutions of some problems of division IV. American J. of Mathematics 82, 148–170 (1960)
[101] Ehrenpreis, L.: A fundamental principle for systems of linear differential equations with constant coefficients and some of its applications. In: Proc. Int. Symp. on Linear Spaces, Jerusalem (1960)
[102] Ehrenpreis, L.: Fourier analysis in several complex variables. Wiley, Chichester (1970)
[103] Eijndhoven, S.J.L., Habets, L.C.G.J.M.: Equivalence of Convolution Systems in a Behavioral Framework. Math. Contr., Sign., Syst. 16, 175–206 (2003)
[104] Eilenberg, S., Steenrod, N.: Foundations of Algebraic Topology. Princeton University Press, Princeton (1952)
[105] Eisenbud, D.: Commutative Algebra with a View Toward Algebraic Geometry. Springer, Heidelberg (1995)
[106] Eisenbud, D., Robson, J.C.: Modules over Dedekind prime rings. J. Algebra 16, 67–85 (1970)
[107] Eisenbud, D., Robson, J.C.: Hereditary Noetherian prime rings. J. Algebra 16, 86–104 (1970)

[108] El Mrabet, Y., Bourlès, H.: Periodic-polynomial interpretation for structural properties of linear periodic discrete-time systems. Systems & Control Letters 33(4), 241–251 (1998)
[109] Emre, E.: Generalized Model Matching and (F,G)-Invariant Submodules for Linear Systems over Rings. Linear Algebra and Its Applications 50, 133–166 (1983)
[110] Emre, E., Tai, H.M., Seo, J.H.: Transfer Matrices, Realization, and Control of Continuous-Time Linear Time-Varying Systems via Polynomial Fraction Representations. Linear Alg. and its Appl. 141, 79–104 (1990)
[111] Everett, C.: Vector Spaces over Rings. Bull. Amer. Math. Soc. 48, 312–316 (1942)
[112] Ferreira, P.M.G.: Infinite system zeros. Internat. J. Control 32, 731–735 (1980)
[113] Fitting, H.: Über den Zusammenhang zwischen dem Begriff der Gleichartigkeit zweier Ideale und dem Äquivalenzbegriff der Elementarteilertheorie. Math. Ann. 112, 572–582
[114] Fliess, M.: Automatique et corps différentiels. Forum. Math. 1, 227–238 (1989)
[115] Fliess, M.: Some basic structural properties of generalized linear systems. Systems & Control Letters 15, 391–396 (1990)
[116] Fliess, M.: A remark on Willems' trajectory characterization of linear controllability. Systems & Control Letters 19, 43–45 (1992)
[117] Fliess, M.: Une Interprétation Algébrique de la Transformation de Laplace et des Matrices de Transfert. Linear Algebra Appl. 203-204, 429–442 (1994)
[118] Fliess, M., Bourlès, H.: Discussing some examples of linear system interconnections. Systems & Control Letters 27, 1–7 (1996)
[119] Fliess, M., Glad, S.T.: An Algebraic Approach to Linear and Nonlinear Control. In: Trentelman, H.L., Willems, J.C. (eds.) Essays on Control: Perspectives in the Theory and its Applications, pp. 223–267. Birkhäuser, Basel (1993)
[120] Fliess, M.: Some remarks on the Brunovsky canonical form. Kybernetika 29(5), 417–422 (1993)
[121] Fliess, M., Levine, J., Martin, P., Rouchon, P.: Flatness and defect of nonlinear systems: introducing theory and examples. Internat. J. Control 61, 1327–1361 (1995)
[122] Fliess, M., Lévine, J., Martin, P., Rouchon, P.: A Lie-Bäcklund Approach to Equivalence and Flatness of Nonlinear Systems. IEEE Trans. on Autom. Contr. AC-44, 922–937 (1999)
[123] Floquet, G.: Sur les équations différentielles lineaires à coefficients périodiques. Annales Scientifiques de l'ENS, 2ème série, tomme 12, 47–88 (1883)
[124] Fornasini, E., Rocha, P., Zampieri, S.: State-space realization of 2D finite-dimensional behaviours. SIAM J. Control Optim. 31, 1502–1515 (1993)
[125] Forney, G.D.: Minimal Bases of Rational Vector Spaces with Application to Multivariable Linear Systems. SIAM Journal of Control 13, 493–520 (1975)
[126] Franke, C.H.: Picard-Vessiot Theory of Linear Homogeneous Difference Equations. Amer. J. Math. 84, 89–109 (1962)
[127] Freyd, P.: Abelian Categories –An Introduction to the Theory of Functors. Harper & Row, New York (1964)
[128] Fröhler, S.: Linear differential systems with variable coefficients–a duality theorem, Ph.D Thesis, Innsbruck (1997)

[129] Fröhler, S., Oberst, U.: Continuous time-varying linear systems. Systems & Control Letters 35, 97–110 (1998)
[130] Fromion, V., Scorletti, G., Barbot, J.P.: Quadratic Observers for Estimation and Control in Induction Motors. In: Proc. of the American Control Conference, San Diego, California, June 1999, pp. 2143–2147 (1999)
[131] Fuhrmann, P.A.: On strict system equivalence and similarity. Internat. J. Control 25, 5–10 (1977)
[132] Gantmacher, F.R.: The Theory of Matrices, vol. 2. Chelsea Publishing Company (1959)
[133] Ton Geerts, A.H.W., Schumacher, J.M.: Impulsive-Smooth Behavior in Multimode Systems. Part I: State-space and Polynomial Representations. Automatica 32, 747–758 (1996)
[134] Gelfand, S.I., Manin, Y.I.: Methods of Homological Algebra, 2nd edn. Springer, Heidelberg (2003)
[135] Gelfand, I., Retakh, V., Wilson, R.L.: Quadratic linear algebras associated with factorizations of noncommutative polynomials and noncommutative differentials polynomials. Sel. math., New Ser. 7, 493–523 (2001)
[136] Gentile, E.R.: On rings with one-sided field of quotients. Proc. Amer. Math. Soc. 11(13), 380–384 (1960)
[137] Gluesing-Luerssen, H.: A behavioral approach to delay-differential systems. SIAM J. Control Optim. 35, 480–499 (1997)
[138] Gluesing-Luerssen, H.: Linear Delay-Differential Systems with Commensurate Delays: An Algebraic Approach. Springer, Heidelberg (2002)
[139] Gluesing-Luerssen, H., Vettori, P., Zampieri, S.: The algebraic structure of DD systems: a behavioral perspective. Kybernetika 37, 397–426 (2001)
[140] Godement, R.: Topologie algébrique et théorie des faisceaux. Hermann, Paris (1958)
[141] Godement, R.: Cours d'algèbre. Hermann, Paris (1966); English Translation: Algebra. Houghton Mifflin Co. (1968)
[142] Goldie, A.W.: Localization in noncommutative Noetherian rings. J. Algebra 5, 89–105 (1967)
[143] Goldie, A.W.: Non-commutative Dedekind rings. Séminaire Dubreil. Algèbre et théorie des nombres 21(2), exp. 20, 1–4 (1967-1968)
[144] Gondran, M., Minoux, M.: Graphs and Algorithms. Wiley Interscience, Hoboken (1984)
[145] Grasselli, O.M., Tornambè, A., Longhi, S.: A Polynomial Approach to Deriving a State-space Model of Periodic Process Prescribed by Diference Equations. Cir. Signal Process. 13(2-3), 373–384 (1994)
[146] Grasselli, O.M., Longhi, S., Tornambè, A.: System equivalence for periodic models and systems. SIAM J. Control Optim. 33(2), 455–468 (1995)
[147] Grothendieck, A.: Sur les espaces $(\mathcal{F})$ et $(\mathcal{DF})$. Summa Brasiliensis Mathematicae 3(6), 57–123 (1954)
[148] Grothendieck, A.: Sur quelques points d'alg èbre homologique. Tohuku Mathematical Journal 9, 119–221 (1957)
[149] Grothendieck, A.: Topological Vector spaces. Gordon and Breach (1973)
[150] Grothendieck, A., Dieudonné, J.: Eléments de géométrie algébrique I. Springer, Heidelberg (1971)
[151] Guglielmi, M.: Systèmes linéaires variables dans le temps. In: de Larminat, P. (ed.) Commande des systèmes linéaires, ch. 8, Lavoisier, Paris, vol. 8, pp. 271–284 (2002)

[152] Gurevič, D.I.: Counterexamples to a problem of L. Schwartz. Funct. Anal. Appl. 9, 116–120 (1975)
[153] Habets, L.C.G.J.M.: System equivalence for AR-systems over rings –with an application to delay-differential systems. Math. Contr., Sign., Syst. 12, 219–244 (1999)
[154] Hahn, W.: Stability of motion. Springer, Heidelberg (1967)
[155] Hall, M.: The Theory of Groups. The MacMillan Comp. (1959)
[156] Hartshorne, R.: Algebraic Geometry. Springer, Heidelberg (1977)
[157] Hautus, M.L.J.: Controllability and observability conditions for linear autonomous systems. Ned. Akad. Wetenschappen, Proc. Ser. A 72, 443–448 (1969)
[158] Hautus, M.H., Heymann, M.: Linear Feedback Decoupling - Transfer Function Analysis. IEEE Trans. on Automatic Control AC-28(8), 823–832 (1983)
[159] Hayton, G.E., Pugh, A.C., Fretwell, P.: Infinite elementary divisors of a matrix polynomial and implications. Internat. J. Control 47, 53–64 (1988)
[160] Hirschorn, R.M.: Invertibility of multivariable nonlinear control systems. IEEE Trans. on Autom. Contr. AC-24, 855–865 (1979)
[161] Hörmander, L.: Linear Partial Differential Operators, 3rd edn. Springer, Heidelberg (1969)
[162] Hörmander, L.: The Analysis of Partial Differential Operators, I and II. Springer, Heidelberg (1990-1993)
[163] Ilchmann, A., Nürnberger, I., Schmale, W.: Time-varying polynomial matrices. Int. J. Control 40, 329–362 (1984)
[164] Ince, E.L.: Ordinary differential equations. Dover Publications, Inc., New-York (1956)
[165] Ireland, K., Rosen, M.: A Classical Introduction to Modern Number Theory, 2nd edn. Springer, Heidelberg (1990)
[166] Isidori, A.: Nonlinear Control Systems. Springer, Heidelberg (1993)
[167] Jacobson, N.: On Pseudo-Linear Transformations. Proc. Nat. Acad. Sci. 21, 667–670 (1935)
[168] Jacobson, N.: Pseudo-Linear Transformations. Annals of Mathematics 38(2), 484–507 (1937)
[169] Johnson, J.: Kähler Differentials and Differential Algebra. The Annals of Mathematics, 2nd Ser. 89(1), 92–98 (1969)
[170] Kailath, T.: Linear Systems. Prentice-Hall, Englewood Cliffs (1980)
[171] Kalman, R.E.: On the general theory of control systems. In: Proc. 1st IFAC Congress, Moscow (1960)
[172] Kalman, R.E.: Mathematical description of linear dynamical systems. SIAM J. Control 1, 152–192 (1963)
[173] Kamen, E.W., Khargonekar, P.P., Poolla, K.R.: A Transfer-Function Approach to Linear Time-Varying Discrete-Time Systems. SIAM Journal of Control and Optimization 23(4) (July 1985)
[174] Kamen, E.W.: The poles and zeros of a linear time-varying system. Linear Algebra and its Applications 98, 263–289 (1998)
[175] Kaplansky, I.: Elementary Divisors and Modules. Trans. Amer. Math. Soc. 66, 464–491 (1949)
[176] Kaplansky, I.: Projective modules. Ann. Math. 68, 372–377 (1958)
[177] Karampetakis, N.P.: On the solution space of discrete time AR-representations over a finite time horizon. Linear Algebra Appl. 382, 83–116 (2004)

[178] Karampetakis, N.P., Vardulakis, A.I.: Generalized state-space system matrix equivalents of a Rosenbrock system matrix. IMA J. of Math. Control & Information 10, 323–344 (1993)
[179] Karampetakis, N.P., Vardulakis, A.I.: On the Solution Space of Continuous Time AR Representations. In: Proc. ECC 1993, pp. 1784–1789 (1993)
[180] Karampetakis, N.P., Vologiannidis, S., Vardulakis, A.I.G.: A new notion of equivalence for discrete time AR representations. Internat. J. Control 77, 584–597 (2004)
[181] Kashiwara, M.: Analytic study of partial differential equations. Master's Thesis, Tokyo University (December 1970); French translation: Mémoires de la Société Mathé matique de France, Série 2 63, 1–72 (1995)
[182] Kashiwara, M., Kawai, T., Kimura, T.: Foundations of Algebraic Analysis. Princeton University Press, Princeton (1986)
[183] Kashiwara, M., Schapira, P.: Categories and Sheaves. Springer, Heidelberg (2006)
[184] Kato, G., Struppa, D.C.: Fundamentals of Algebraic Microanalysis. Marcel Dekker, New York (1999)
[185] Kelleher, J.-J., Taylor, B.-A.: Closed ideals in locally convex algebras of analytic functions. J. Reine Angew. Math. 255, 190–209 (1972)
[186] Khalil, H.K.: Nonlinear systems, 3rd edn. Pearson Education, London (2002)
[187] Kolchin, E.R.: Differential Algebra and Algebraic Group. Academic Press, London (1973)
[188] Köthe, G.: "Dualität in der Funktionentheorie. J. Reine Angew. Math. 191, 30–49 (1953)
[189] Koumboulis, F.N., Skarpetis, M.G.: Robust Exact Model-Matching via P-D Feedback with Application to DC Servo Motor. In: Proc. of ICECS 1996, pp. 602–605 (1996)
[190] Köthe, G.: Topological Vector Spaces I and II. Springer, Heidelberg (1969-1979)
[191] Kundur, P.: Power System Stability and Control. In: Balu, N.J., Lauby, M.G. (eds.) Electrical Power Research Institute. McGraw-Hill, New York (1993)
[192] Kung, S., Kailath, T.: Some notes on valuation theory in linear systems. In: Proc. IEEE Conference on Decision and Control, San-Diego, CA (1979)
[193] Kreindler, E., Sarachik, P.E.: On the Concepts of Controllability and Observability of Linear Systems. IEEE Trans. on Automat. Control. 9, 129–136 (1964)
[194] Krull, W.: Galoissche Theorie der unendlichen Erweiterungen. Math. Ann. 100, 687–698 (1928)
[195] Krull, W.: Allgemeine Bewertungstheorie. J. Reine Angew. Math. 167, 169–196 (1932)
[196] Kučera, V.: Diophantine Equations in Control - A Survey. Automatica 29(6), 1361–1375 (1993)
[197] Lam, T.Y.: A general theory of Vandermonde matrices. Expositiones Mathematicae 4, 193–215 (1986)
[198] Lam, T.Y.: A First Course in Noncommutative Rings, 2nd edn. Springer, Heidelberg (2001)
[199] Lam, T.Y.: Lectures on Modules and Rings. Springer, Heidelberg (1999)
[200] Lam, T.Y.: Serre's Problem on Projective Modules. Springer, Heidelberg (2006)

[201] Lam, T.Y., Leroy, A.: Vandermonde and Wronskian Matrices over Division Rings. J. of Algebra 119, 308–336 (1988)
[202] Lam, T.Y., Leroy, A.: Principal One-Sided Ideals in Ore Polynomial Rings. Contemporary Mathematics 259, 333–352 (2000)
[203] Lam, T.Y., Leroy, A.: Wedderburn Polynomials over Division Rings. J. of Pure and Applied Algebra 186, 43–76 (2004)
[204] Lam, T.Y., Leroy, A., Ozturk, A.: Wedderburn Polynomials over Division Rings, II. In: Jain, S.K., Parvathi, S. (eds.) Noncommutative Rings, Group Rings, Diagram Algebras and their Applications, pp. 73–98. AMS, Providence (2008)
[205] Lang, S.: Introduction to Algebraic Geometry, 4th printing. Addison Wesley, Reading (1973)
[206] Lang, S.: Algebra, corrected, 3rd edn. Addison-Wesley, Reading (1999)
[207] Lang, S.: Algebraic Number Theory, 2nd edn. Springer, Heidelberg (1994)
[208] Lang, S.: Complex Analysis, 4th edn. Springer, Heidelberg (1999)
[209] Larsen, M.D., Lewis, W.J., Shores, T.S.: Elementary Divisor Rings and Finitely Presented Modules. Trans. Amer. Math. Soc. 187(1), 231–248 (1974)
[210] LaSalle, J.P.: The time optimal control problem. In: Contributions to the Theory of Nonlinear Oscillations V, pp. 1–24. Princeton Univ. Press, Princeton (1960)
[211] Leith, D.J., Leithead, W.E.: Input-output linearization velocity-based gain scheduling. Int. J. of Contr. 72(3), 229–246 (1999)
[212] Leith, D.J., Leithead, W.E.: Survey of Gain-Scheduling Analysis & Design. Internat. J. Control 73(11), 1001–1025 (2000)
[213] Leonhard, W.: Control of Electrical Drives, 3rd edn. Springer, Heidelberg (2001)
[214] Leroy, A.: Pseudo linear transformations and evaluation in Ore extensions. Bull. Belg. Math. Soc. 2, 321–347 (1995)
[215] Levy, L.: Torsion-free and divisible modules over non-integral domains. Canadian J. of Math. 15, 132–151 (1963)
[216] Lewis, F.L.: Descriptor Systems: Decomposition Into Forward and Backward Subsystems. IEEE Trans. on Automat. Control 29, 167–170 (1984)
[217] Lewis, F.L., Mertzios, B.G.: On the Analysis of Discrete Linear Time-Invariant Singular Systems. IEEE Trans. on Automat. Control 35, 506–511 (1990)
[218] Li, C.W., Feng, Y.-K.: Functional Reproductibility of General Multivariable Analytic Nonlinear Systems. Int. J. Control 45, 255–268 (1987)
[219] Lissner, D.: Outer product rings. Trans. Amer. Math. Soc. 116, 526–535 (1965)
[220] Luenberger, D.G.: Dynamic Equations in Descriptor Form. IEEE Trans. on Automat. Control 22, 312–321 (1977)
[221] Loiseau, J.J.: Invariant Factors Assignment for a Class of Time-Delay Systems. Kybernetika 3, 265–275 (2001)
[222] Lyapunov, A.M.: Problème Générale de la Stabilité du Mouvement (1902) (french translation); reproduced in: Annals of Mathematics. Princeton Univ. Press, Princeton (1949)
[223] MacLane, S.: Homology. Springer, Heidelberg (1963)
[224] MacLane, S.: Categories for the Working Mathematician, 2nd edn. Springer, Heidelberg (1998)
[225] MacLane, S., Birkhoff, G.: Algebra, 2nd edn. Collier-Macmillan Pub. (1979)

[226] Magid, A.R.: Lectures on Differential Galois Theory. American Mathematical Society, Providence (1997); edition reprinted with corrections
[227] Maisonobe, P., Sabbah, C. (eds.): $\mathcal{D}$ -modules cohérents et holonomes. Hermann, Paris (1993)
[228] Malabre, M., Kučera, V.: Infinite Structure and Exact Model Matching Problem: A Geometric Approach. IEEE Trans. on Automatic Control AC-29(3), 266–268 (1984)
[229] Malcev, A.I.: On the immersion of an algebraic ring in a skew field. Math. Ann. 113, 686–691 (1937)
[230] Malgrange, B.: Existence et approximation des solutions des équations aux dérivées partielles et des é quations de convolution. Ann. Inst. Fourier 6, 271–355 (1956)
[231] Malgrange, B.: Sur les systèmes différentiels à coefficients constants. Séminaire Jean Leray (Collège de France, Paris), 7, 1–13 (1961-1962)
[232] Malgrange, B.: Systèmes linéaires à coefficients constants. Séminaire Bourbaki 246, 1–11 (1962-1963)
[233] Mallem, B., Marinescu, B., Rouco, L.: Structure-Preserving Dynamic Equivalents for Large-Scale Power Systems using Border Synchrony. In: Proc. CIGRE Symposium, Zagreb, Croatia (2007)
[234] Marinescu, B.: Analyse et synthèse des systèmes linéaires généralisés avec application au réglage secondaire de la tension des réseaux électriques, Ph. D. dissertation, Université de Paris-Sud (1997)
[235] Marinescu, B.: Model-matching and decoupling for continuous- and discrete-time linear time-varying systems. International Journal of Control 82(6), 1018–1028 (2009)
[236] Marinescu, B.: Output Feedback for Linear Time-Varying Systems. In: Proc. IASTED, Intelligent Systems and Control, Cambridge, MA-US (2009)
[237] Marinescu, B.: Output feedback pole placement for linear time-varying systems with application to the control of nonlinear systems. Automatica 46, 1524–1530 (2010)
[238] Marinescu, B., Bourlès, H.: The Exact Model-Matching Problem for Linear Time-Varying Systems: An Algebraic Approach. IEEE Trans. on Automatic Control 48(1), 166–169 (2003)
[239] Marinescu, B., Bourlès, H.: Poles of Linear Time-Varying Systems Revisited. In: Proc. of the IFAC Workshop SSSC 2007, Iguasu Falls, Brazil, October 17-19 (2007)
[240] Marinescu, B., Bourlès, H.: Poles and stability of linear time-varying systems. In: Proc. of the MTNS, Blacksburg, Virginia, USA (2008)
[241] Marinescu, B., Bourlès, H.: An intrinsic algebraic setting for poles and zeros of linear time-varying systems. Systems and Control Letters 58, 248–253 (2009)
[242] Marinescu, B., Mallem, B., Rouco, L.: Model Reduction of Interconnected Power Systems via Balanced State-Space Representation. In: Proc. European Control Conference, Kos, Greece (2007)
[243] Marinescu, B., Mallem, B., Rouco, L.: Large-Scale Power System Dynamic Equivalents Based on Standard and Border Synchrony. IEEE Trans. on Power Systems 25(4) (2010)
[244] Marinescu, B., Rouco, L.: A Unified Framework for Nonlinear Dynamic Simulation and Modal Analysis for Control of Large-Scale Power Systems. In: Proc. PSCC, Liège, Belgium (2005)

[245] Martineau, A.: Les hyperfonctions de M. Sato. Séminaire Bourbaki, Exposé n° 314 (1960-1961)
[246] Martins, J.R.R.A., Sturdza, P., Alonso, J.J.: The Complex-Step Derivation Approximation. ACM Tran. on Mathematical Software 29(3), 245–262 (2003)
[247] McConnell, J.C., Robson, J.C.: Noncommutative Noetherian Rings. American Mathematical Society, Providence (2001)
[248] McMillan, B.: Introduction to formal realizability theory. Bell Syst. Tech. J. 31, 217–279, 541–600 (1952)
[249] Mitchell, B.: Theory of Categories. Academic Press, New York (1965)
[250] Morimoto, M.: An Introduction to Sato's Hyperfunctions. American Mathematical Society, Providence (1993)
[251] Morse, A.S.: Ring models for delay-differential systems. Automatica 12, 529–531 (1976)
[252] Mosquera, C., Scalise, S., Taricco, G.: Spectral Characterization of Feedback Linear Periodically Time-Varying Systems. In: Proc. of the International Conference on Acoustics, Speech and Signal Processing, pp. 1209–1212 (2002)
[253] Mukaetov, T., Jing, Y., Dimirovski, G.: Variable Load at Antenna Servosystem: Analytical Approach with Time Varying Models. In: Proc. of the International Conference on Telecommunications, Modern Satellites, Cables and Broadcasting Services (1999)
[254] Nakayama, T.: A Note on the Elementary Divisor Theory in Non-Commutative Domains. Bull. Amer. Math. Soc. 44(10), 719–723 (1938)
[255] Neerhoff, F.L., van der Kloet, P., van der Staveren, A., Verhoeven, C.J.M.: Time-Varying Small-Signal Circuits for Nonlinear Electronics. In: Proc. NDES 2000, Catania, Italy, May 18-20, pp. 85–89 (2000)
[256] Northcott, D.: Finite Free Resolutions. Cambridge University Press, Cambridge (1976)
[257] Oberst, U.: Duality Theory for Grothendieck Categories and Linearly Compact Rings. J. of Algebra 15, 473–542 (1970)
[258] Oberst, U.: Multidimensional Constant Linear Systems. Acta Applicandae Mathematicae 20, 1–175 (1990)
[259] Oberst, U.: On the Minimal Number of Trajectories Determining a Multidimensional System. Math. Control Signal Systems 6, 264–288 (1993)
[260] Oberst, U.: Variations on the Fundamental Principle for Linear Systems of Partial Differential and Difference Equations with Constant Coefficients. Applicable Algebra in Engineering, Communication and Computing 6, 211–243 (1995)
[261] O'Brien, R.T.: Modeling Highway Vehicles with Time-Varying Velocity. In: Arlington, V.A. (ed.) Proc. of the American Control Conference, June 25-27 (2001)
[262] O'Brien, R.T., Iglesias, P.A.: On the poles and zeros of linear, time-varying systems. IEEE Trans. on Circuit and Systems-I: Fund. Theory and Applications 48(5), 565–577 (2001)
[263] Ojanguren, M., Sridharan, R.: Cancellation of Azumaya Algebras. J. Algebra 18, 501–505 (1971)
[264] Ore, O.: Linear equations in non-commutative fields. Annals of Math. 32, 463–477 (1931)
[265] Ore, O.: Theory of non-commutative polynomials. Annals of Math. 34, 480–508 (1933)

[266] Palamodov, V.P.: Linear Differential Operators with Constant Coefficients. Springer, Heidelberg (1970)

[267] Paraskevopoulos, P.N.: Exact Model Matching of linear time-varying systems. Proc. IEE 125(1) (January 1978)

[268] Parreau, F., Weit, Y.: Schwartz's Theorem on Mean Periodic Vector-Valued Functions. Bull. Soc. Math. France 117, 319–325 (1989)

[269] Passman, D.S.: A course in ring theory. American Mathematical Society, Providence (2004)

[270] Perbeno, L.: Notes on strict system equivalence. Internat. J. Control 25, 21–38 (1977)

[271] Perdon, A.M., Conte, G., Moog, C.H.: Structural Properties in the Control of Linear Time-Varying System. In: Proc. of the ECC 1991 European Control Conference, Grenoble, France, pp. 2–5 (1991)

[272] Perreault, D.J., Verghese, G.C.: Time-Varying Effects in Models for Current-Mode Control. In: Proc. of PESC 1995, pp. 621–628 (1995)

[273] Perron, O.: Die Ordungszahlen Linear Differentialgleichungssystems. Math. Zeits. 31, 748–766 (1930)

[274] Petkovšek, M., Wilf, H.S., Zeilberger, D.: A=B (1997), `http://www.cis.upenn.edu/~wilf/AeqB.html`

[275] Pillai, H., Shankar, S.: A behavioral approach to control of distributed systems. SIAM J. Control Optim. 37, 388–408 (1998)

[276] Polderman, J.W., Willems, J.C.: Introduction to Mathematical Systems Theory–A Behavioral Approach. Springer, Heidelberg (1998)

[277] Pommaret, J.-F.: Dualité différentielle et applications. C.R. Acad. Sci. Paris, t. 320, Série I, pp. 1225–1230 (1995)

[278] Pommaret, J.-F.: Partial differential control theory, vol. I-II. Kluwer Academic Publishers, Dordrecht (2001)

[279] Pommaret, J.-F., Quadrat, A.: Algebraic analysis of linear multidimensional control systems. IMA J. of Math. Control & Information 16, 275–297 (1999)

[280] Pommaret, J.-F., Quadrat, A.: Localization and parametrization of linear multidimensional control systems. Systems & Control Letters 37, 247–260 (1999)

[281] Poola, K., Khargonekar, P.: Stabilizability and stable-proper factorizations for linear time-varying systems. SIAM J. of Control and Optimization 25(3), 723–736 (1987)

[282] Popescu, N.: Abelian Categories with Applications to Rings and Modules. Academic Press, London (1973)

[283] Pugh, A.C., Karampetakis, N.P., Vardulakis, A.I.G., Hayton, G.E.: A Fundamental Notion of Equivalence for Linear Multivariable Systems. Internat. J. Control 39, 1141–1145 (1994)

[284] Pugh, A.C., Ratcliffe, P.A.: On the zeros and poles of a rational matrix. Internat. J. Control 30, 213–226 (1979)

[285] van der Put, M., Singer, M.F.: Galois Theory of Difference Equations. Springer, Heidelberg (1997)

[286] van der Put, M., Singer, M.F.: Galois Theory of Linear Differential Equations. Springer, Heidelberg (2003)

[287] Qiu, W., Vittal, V., Khammash, M.: Decentralized Power System Stabilizer Design Using Linear Parameter Varying Approach. IEEE Transactions on Power Systems 19(4), 1951–1960 (2004)

[288] Quadrat, A.: The Fractional Representation Approach to Synthesis Problems: An Algebraic Analysis Viewpoint. Part I (Weakly) Doubly Coprime Factorizations. SIAM J. Control Optim. 42, 266–299 (2003)
[289] Quadrat, A., Robertz, D.: Computation of bases of free modules over the Weyl algebras. J. Symbolic Computation 42, 1113–1141 (2007)
[290] Quillen, D.: Projective modules over polynomial rings. Invent. Math. 36, 167–171 (1976)
[291] Ramaswamy, G.N., Verghese, G.C., Rouco, L., Vialas, C., Demarco, C.L.: Synchrony, Aggregation and Multi-Area Eigenanalysis. IEEE Trans. on Power Systems 10(4), 1986–1993 (1995)
[292] Ramaswamy, G.N., Rouco, L., Filatre, O., Verghese, G.C., Panciatici, P., Lesieutre, B.C., Peltier, D.: Synchronic Modal Equivalencing (SME) for Structure-Preserving Dynamic Equivalents. IEEE Trans. on Power Systems 11(1), 19–29 (1996)
[293] Ridley, R.: A New, Continuous-Time Model for Current-Mode Control. IEEE Transactions on Power Electronics 6(2), 271–280 (1991)
[294] Ritt, J.F.: Differential Algebra. American Mathematical Society, Providence (1950)
[295] Robba, P.: Lemme de Hensel pour les opérateurs différentiels (I). Groupe de travail d'analyse ultranumérique, tome 2, exp. no. 16, 1–11 (1975)
[296] Robba, P.: Lemme de Hensel pour les opérateurs différentiels (II). Groupe de travail d'analyse ultranumérique, tome 6, exp. no. 23 23, 1–8 (1979)
[297] Robson, J.C.: Non-commutative Dedekind rings. J. of Algebra 9, 249–265 (1968)
[298] Rocha, P., Willems, J.C.: Behavioral controllability of delay-differential systems. SIAM J. Control Optim. 35, 254–264 (1997)
[299] Roos, J.-E.: Locally noetherian categories and linearly compact rings. Applications. Springer Lecture Notes, vol. 92, pp. 197–277 (1969)
[300] Rosenberg, A., Zelinsky, D.: On the finiteness of the injective hull. Math. Zeit. 70, 372–380 (1959)
[301] Rosenbrock, H.H.: State-space and Multivariable Theory, Nelson (1970)
[302] Rosenbrock, H.H.: A comment on three papers. Internat. J. Control 25, 1–3 (1977)
[303] Rosenbrock, H.H., Pugh, A.C.: Contributions to a hierarchical theory of systems. Internat. J. Control 19, 845–867 (1974)
[304] Rotella, F.: Systèmes linéaires non stationnaires. Techniques de l'ingénieur, traité Informatique industrielle R7185, 1–16 (2003)
[305] Rotella, F., Zambettakis, I.: Commande des systèmes par platitude. Téchniques de l'ingénieur, traité Informatique industrielle S7450, 1–18 (2008)
[306] Rotman, J.J.: An Introduction to Homological Algebra. Academic Press, NewYork (1979)
[307] Rubin, H., Rubin, J.E.: Equivalents of the Axiom of Choice, II. North-Holland, Amsterdam (1985)
[308] Rudin, W.: Real and Complex Analysis, 3rd edn. McGraw-Hill, New York (1987)
[309] Rudolph, J.: Duality in Time-Varying Linear Systems: A Module Theoretic Approach. Linear Algebra Appl. 245, 83–106 (1996)
[310] Rugh, W.J., Shamma, J.S.: Research on gain scheduling. Automatica 36, 1401–1425 (2000)

[311] Rump, W.: Almost abelian categories. Cahiers de topologie et géométrie différentielle catégoriques 42(3), 163–225 (2001)
[312] Ruth, W.J., Shamma, J.S.: Research on gain scheduling. Automatica 36, 1401–1425 (2000)
[313] Samuel, P.: Théorie algébrique des nombres, 2nd edn. Hermann, Parris (1971); English translation: Algebraic Number Theory. Dover Publications, New York (2008)
[314] Sato, K.: Theory of hyperfunctions. J. Fac. Sci. Univ. Tokyo 1(8), 139-193, 387–437 (1959-1960)
[315] Sattinger, D.H., Weaver, O.L.: Lie Groups and Algebras with Applications to Physics, Geometry, and Mechanics. Springer, Heidelberg (1986)
[316] Schaefer, H.H.: Topological Vector Spaces, 2nd edn. Springer, Heidelberg (1999)
[317] van der Schaft, A.J.: Duality for linear systems: External and state space characterization of the adjoint system. In: Bonnard, B., Bride, B., Gauthier, J.P., Kupka, I. (eds.) Analysis of Controlled Dynamical Systems, pp. 393–403. Birkhäuser, Basel (1991)
[318] Shamma, J.S., Athans, M.: Analysis of Gain Scheduled Control for Nonlinear Plants. IEEE Trans. on Autom. Contr. AC-35(8), 898–907 (1990)
[319] Schapira, P.: Microdifferential Systems in the Complex Domain. Springer, Heidelberg (1985)
[320] Schneiders, J.-P.: An Introduction to $\mathcal{D}$-modules. Bull. Soc. Royale des Sciences de Liège 63(3-4), 223–295 (1994)
[321] Schwartz, L.: Théorie géné rale des fonctions moyenne-périodique. Ann. of Math. 48, 857–928 (1947)
[322] Schwartz, L.: Théorie des distributions, 3rd edn. Hermann, Paris (1966)
[323] Schwartz, L.: Analyse IV. Hermann, Paris (1993)
[324] Serre, J.P.: Faisceaux algébriques cohérents. Ann. of Math. 61, 197–278 (1955)
[325] Serre, J.P.: Géométrie algébrique et g éométrie analytique. Ann. Inst. Fourier 6, 1–142 (1955-1956)
[326] Serre, J.P.: Modules projectifs et espace fibré à fibre vectorielle. Sém. Dubreil-Pisot 23 (1957-1958)
[327] Serre, J.P.: Lie Algebras and Lie Groups. Benjamin (1965)
[328] Silverman, L.M.: Inversion of multivariable linear systems. IEEE Trans. on Autom. Contr. AC-14, 270–276 (1969)
[329] Silverman, L.M., Meadows, H.E.: Controllability and observability in time-variable linear systems. SIAM J. Control 5, 64–73 (1967)
[330] Singer, M.F.: Liouvillian Solutions of Linear Differential Equations with Liovillian Coefficients. J. Symbol. Comput. 11, 251–273 (1991)
[331] Sira-Ramírez, H., Fliess, M.: On the output feedback control of a synchronous generator. In: Proc. of the 43rd IEEE Conf. on Decision and Control, Atlantis, Bahamas, USA, pp. 4459–4464(2004)
[332] Sizer, W.S.: Similarity of sets of matrices over skew fields. Ph. D. Thesis, London University (1975)
[333] Skogestaad, S., Postlethwaite, I.: Multivariable feedback control - Analysis and design. John Wiley, Chichester (1996)
[334] Smith, M.C.: On Stabilization and the Existence of Coprime Factorizations. IEEE Trans. on Automat. Control 34(9), 1005–1007 (1989)
[335] Sontag, E.D.: Mathematical Control Theory. Springer, Heidelberg (1990)

[336] Stafford, J.T.: Module structure of Weyl algebras. J. London Math. Soc. 18, 429–442 (1978)
[337] Suslin, A.A.: Projective modules over a polynomial ring are free. Soviet. Math. Dokl. 17, 1160–1164 (1976)
[338] Tan, F., Middlebrook, R.: Unified Modeling and Measurement of Current-Programmed Converters. In: IEEE Power Electronic Specialists Conference, pp. 380–387 (1993)
[339] Trèves, F.: Topological Vector Spaces. Academic Press, New York (1967)
[340] Uchida, H., Semba, T.: Reference Model Generation Using Time-Varying Feedback for Model-Following Control and Application to a Hard Disk Drive. In: Proc. of the American Control Conference, Anchorage, AK, May 8-10 (2002)
[341] Vardulakis, A.I.G.: On infinite zeros. Internat. J. Control 32, 849–866 (1980)
[342] Vardulakis, A.I.G.: Proper Rational Matrix Diophantine Equations and the Exact Model Matching Problem. IEEE Trans. on Automatic Control AC-29(5), 475–477 (1984)
[343] Vardulakis, A.I.G., Fragulis, G.: Infinite Elementary Divisors of Polynomial Matrices and Impulsive Motions of Linear Homogeneous Matrix Differential Equations. Circuits Systems Signal Process 8, 57–373 (1989)
[344] Vardulakis, A.I.G., Limebeer, D.J.N., Karkanias, N.: Structure and Smith-MacMillan form of a rational matrix at infinity. Internat. J. Control 35, 701–725 (1982)
[345] Vardulakis, A.I.G.: Linear Multivariable Control. Wiley, Chichester (1991)
[346] Verghese, G.C.: Infinite-Frequency Behaviour in Generalized Dynamical Systems, Ph. D. dissertation, Electrical Engineering Department, Stanford University (1979)
[347] Verghese, G.C.: Comments on 'Properties of the system matrix of a generalized state-space system. Internat. J. Control 31, 1007–1009 (1980)
[348] Verghese, G.C., van Dooren, P., Kailath, T.: Properties of the system matrix of a generalized state-space system. Internat. J. Control 30, 235–243 (1979)
[349] Verghese, G.C., Kailath, T.: Rational Matrix Structure. IEEE Trans. on Automat. Control 26, 434–439 (1981)
[350] Verghese, G.C., Lévy, B.C., Kailath, T.: A Generalized State-Space for Singular Systems. IEEE Trans. on Automat. Control 26, 811–830 (1981)
[351] Verlet, J.L.: Introduction à la théorie des hyperfonctions. Séminaire Lelong. Analyse, tome 7, exposé 5, 1–7 (1966-1967)
[352] Vettori, P., Zampieri, S.: Some Results on Systems Described by Convolution Equations. IEEE Trans. on Automat. Control 46, 793–797 (2001)
[353] Vettori, P., Zampieri, S.: Module theoretic approach to controllability of convolutional systems. Linear Algebra Appl. 351-352, 739–759 (2002)
[354] Vidyasagar, M.: Control System Synthesis –A Factorization Approach. MIT Press, Cambridge (1985)
[355] Vidyasagar, M.: Nonlinear Systems Analysis, 2nd edn. SIAM, Philadelphia (2002)
[356] Wang, S.H., Davison, E.J.: A Minimisation Algorithm for the design of Linear Multivariable Systems. IEEE Transactions on Automatic Control 18, 220–225 (1973)
[357] Wedderburn, J.H.M.: On division algebras. Trans. A.M.S. 22, 129–135 (1921)
[358] van der Weiden, A.J.J., Bosgra, O.H.: The determination of structural properties of a linear multivariable system by operations of system similarity–2. Non-proper systems in generalized state-space form. Internat. J. Control 32, 489–537 (1980)

[359] Weiss, L.: The Concepts of Differential Controllability and Differential Observability. J. Math. Analysis Appl. 10, 442–449 (1965)
[360] Willems, J.C.: From time series to linear systems – Part I: Finite-dimensional linear time invariant systems. Automatica 22, 675–694 (1986)
[361] Willems, J.C.: Models for dynamics. In: Dynamics Reported, vol. 2, pp. 171–269. Wiley, Chichester (1989)
[362] Willems, J.C.: Paradigms and Puzzles in the Theory of Dynamical Systems. IEEE Trans. on Automat. Control 36, 259–294 (1991)
[363] Wood, J., Oberst, U., Rogers, E., Owens, D.: A Behavioural Approach to the Pole Structure of $1D$ and nD Linear Systems. SIAM J. Control Optim. 38, 627–661 (2000)
[364] Wood, J., Rogers, E., Owens, D.: Controllable and Autonomous nD Linear Systems. Multidimensional Signal Processing 10, 33–69 (1999)
[365] Wu, F., Yong, X.H., Packard, A.K., Becker, G.: Induced L_2 norm control for LPV systems with bounded parameter variation rates. International Journal of Robust Nonlinear Control 6(9/10), 205–215 (1994)
[366] Yazici, B.: Stochastic Deconvolution over Groups. IEEE Trans. on Information Theory 50, 494–510 (2004)
[367] Youla, D.C., Jabr, H.A., Bongiorno, J.: Modern Wiener-Hopf Design of Optimal Controllers - Part II: The Multivariable Case. IEEE Trans. on Autom. Contr. AC-21(3) (1976)
[368] Youla, D.C., Gnavi, G.: Notes on n-Dimensional System Theory. IEEE Trans. Circuits Syst. 26, 105–111 (1979)
[369] Zadeh, L.A., Desoer, C.A.: Linear System Theory: The State Space Approach. McGraw-Hill, New York (1963)
[370] Zariski, O., Samuel, P.: Commutative Algebra, vol. I & II. D. Van Nostrand Company, Inc., New York (1958-1960)
[371] Zerz, E.: Primeness of multivariate polynomial matrices. Systems & Control Letters 29, 139–145 (1996)
[372] Zerz, E.: On strict equivalence for multidimensional systems. Internat. J. Control 73, 495–504 (2000)
[373] Zerz, E.: Topics in Multidimensional Linear Systems Theory. Springer, Heidelberg (2000)
[374] Zerz, E.: Algebraic Systems Theory, Lehrstuhl D für Mathematik. RWTH, Aachen (2006)
[375] Zerz, E.: An Algebraic Analysis Approach to Linear Time-varying Systems. IMA Journal of Mathematical Control and Information 23, 113–126 (2006)
[376] Zerz, E.: Discrete multidimensional systems over $\mathbb{Z}_n$, Systems & Control Letters, pp. 702–708 (2007)
[377] Zhu, J., Johnson, C.D., Thompson, P., Wendt, R.: Stabilization Of Time-Varying Linear Systems Using A New Time-Varying Eigenvalue Assignement Technique. In: Proc. of the IEEE Southeastcon, pp. 212–218 (1990)
[378] Zhu, J.J., Johnson, C.D.: Unified canonical forms for matrices over a differential ring. Linear Algebra and its Applications 147, 201–248 (1991)
[379] Zhu, J.J.: A unified spectral theory for linear time-varying systems-progress and chalanges. In: Proc. of the 34th Conference on Decision and Control, New-Orleans (LA-US) (1995)

[380] Zhu, J.J.: Series and parallel d-spectra for multi-input-multi-output linear time-varying systems. In: Proc. of the South-Eastern Symposium on Systems Theory, Baton Rouge (LA-US) pp. 125–129 (1995)
[381] Zhu, J.J.: A necessary and sufficient stability criterion for linear time-varying systems. In: Proc. of the South-Eastern Symposium on Systems Theory, Baton Rouge (LA-US) pp. 115–119 (1996)
[382] Zhu, J.J., Mickle, M.C.: Missile Autopilot design Using a New Linear Time-Varying Control Technique. Journal of Guidance, Control and Dynamics 20(1), 150–157 (1997)
[383] Zygmund, A.: Trigonometrical series. Chelsea Publishing Company, New York (1952)

Index

Lecture Notes in Control and Information Sciences

Edited by M. Thoma, F. Allgöwer, M. Morari

Further volumes of this series can be found on our homepage:
springer.com

Vol. 410: Bourlès, H., Marinescu, B.;
Linear Time-Varying Systems
637 p. 2011 [978-3-642-19726-0]

Vol. 409: Xia, Y., Fu, M., Liu, G.-P.;
Analysis and Synthesis of
Networked Control Systems
198 p. 2011 [978-3-642-17924-2]

Vol. 408: Richter, J.H.;
Reconfigurable Control of
Nonlinear Dynamical Systems
291 p. 2011 [978-3-642-17627-2]

Vol. 407: Lévine, J., Müllhaupt, P.:
Advances in the Theory of Control,
Signals and Systems with
Physical Modeling
380 p. 2010 [978-3-642-16134-6]

Vol. 406: Bemporad, A., Heemels, M.,
Johansson, M.:
Networked Control Systems
appro. 371 p. 2010 [978-0-85729-032-8]

Vol. 405: Stefanovic, M., Safonov, M.G.:
Safe Adaptive Control
appro. 153 p. 2010 [978-1-84996-452-4]

Vol. 404: Giri, F.; Bai, E.-W. (Eds.):
Block-oriented Nonlinear System Identification
425 p. 2010 [978-1-84996-512-5]

Vol. 403: Tóth, R.;
Modeling and Identification of
Linear Parameter-Varying Systems
319 p. 2010 [978-3-642-13811-9]

Vol. 402: del Re, L.; Allgöwer, F.;
Glielmo, L.; Guardiola, C.;
Kolmanovsky, I. (Eds.):
Automotive Model Predictive Control
284 p. 2010 [978-1-84996-070-0]

Vol. 401: Chesi, G.; Hashimoto, K. (Eds.):
Visual Servoing via Advanced
Numerical Methods
393 p. 2010 [978-1-84996-088-5]

Vol. 400: Tomás-Rodríguez, M.;
Banks, S.P.:
Linear, Time-varying Approximations
to Nonlinear Dynamical Systems
298 p. 2010 [978-1-84996-100-4]

Vol. 399: Edwards, C.; Lombaerts, T.;
Smaili, H. (Eds.):
Fault Tolerant Flight Control
appro. 350 p. 2010 [978-3-642-11689-6]

Vol. 398: Hara, S.; Ohta, Y.;
Willems, J.C.; Hisaya, F. (Eds.):
Perspectives in Mathematical System
Theory, Control, and Signal Processing
appro. 370 p. 2010 [978-3-540-93917-7]

Vol. 397: Yang, H.; Jiang, B.;
Cocquempot, V.:
Fault Tolerant Control Design for
Hybrid Systems
191 p. 2010 [978-3-642-10680-4]

Vol. 396: Kozlowski, K. (Ed.):
Robot Motion and Control 2009
475 p. 2009 [978-1-84882-984-8]

Vol. 395: Talebi, H.A.; Abdollahi, F.;
Patel, R.V.; Khorasani, K.:
Neural Network-Based State
Estimation of Nonlinear Systems
appro. 175 p. 2010 [978-1-4419-1437-8]

Vol. 394: Pipeleers, G.; Demeulenaere, B.;
Swevers, J.:
Optimal Linear Controller Design for
Periodic Inputs
177 p. 2009 [978-1-84882-974-9]

Vol. 393: Ghosh, B.K.; Martin, C.F.;
Zhou, Y.:
Emergent Problems in Nonlinear
Systems and Control
285 p. 2009 [978-3-642-03626-2]

Vol. 392: Bandyopadhyay, B.;
Deepak, F.; Kim, K.-S.:
Sliding Mode Control Using Novel Sliding
Surfaces
137 p. 2009 [978-3-642-03447-3]

Vol. 391: Khaki-Sedigh, A.; Moaveni, B.:
Control Configuration Selection for
Multivariable Plants
232 p. 2009 [978-3-642-03192-2]

Vol. 390: Chesi, G.; Garulli, A.;
Tesi, A.; Vicino, A.:
Homogeneous Polynomial Forms for
Robustness Analysis of Uncertain
Systems
197 p. 2009 [978-1-84882-780-6]

Vol. 389: Bru, R.; Romero-Vivó,
S. (Eds.):
Positive Systems
398 p. 2009 [978-3-642-02893-9]

Vol. 388: Jacques Loiseau, J.; Michiels, W.;
Niculescu, S-I.; Sipahi, R. (Eds.):
Topics in Time Delay Systems
418 p. 2009 [978-3-642-02896-0]

Vol. 387: Xia, Y.;
Fu, M.; Shi, P.:
Analysis and Synthesis of
Dynamical Systems with Time-Delays
283 p. 2009 [978-3-642-02695-9]

Vol. 386: Huang, D.;
Nguang, S.K.:
Robust Control for Uncertain
Networked Control Systems with
Random Delays
159 p. 2009 [978-1-84882-677-9]

Vol. 385: Jungers, R.:
The Joint Spectral Radius
144 p. 2009 [978-3-540-95979-3]

Vol. 384: Magni, L.; Raimondo, D.M.;
Allgöwer, F. (Eds.):
Nonlinear Model Predictive Control
572 p. 2009 [978-3-642-01093-4]

Vol. 383: Sobhani-Tehrani E.;
Khorasani K.;
Fault Diagnosis of Nonlinear Systems
Using a Hybrid Approach
360 p. 2009 [978-0-387-92906-4]

Vol. 382: Bartoszewicz A.;
Nowacka-Leverton A.;
Time-Varying Sliding Modes for Second
and Third Order Systems
192 p. 2009 [978-3-540-92216-2]

Vol. 381: Hirsch M.J.; Commander C.W.;
Pardalos P.M.; Murphey R. (Eds.)
Optimization and Cooperative Control Strategies:
Proceedings of the 8th International Conference
on Cooperative Control and Optimization
459 p. 2009 [978-3-540-88062-2]

Vol. 380: Basin M.
New Trends in Optimal Filtering and Control for
Polynomial and Time-Delay Systems
206 p. 2008 [978-3-540-70802-5]

Vol. 379: Mellodge P.; Kachroo P.;
Model Abstraction in Dynamical Systems:
Application to Mobile Robot Control
116 p. 2008 [978-3-540-70792-9]

Vol. 378: Femat R.; Solis-Perales G.;
Robust Synchronization of Chaotic Systems
Via Feedback
199 p. 2008 [978-3-540-69306-2]

Vol. 377: Patan K.
Artificial Neural Networks for
the Modelling and Fault
Diagnosis of Technical Processes
206 p. 2008 [978-3-540-79871-2]

Vol. 376: Hasegawa Y.
Approximate and Noisy Realization of
Discrete-Time Dynamical Systems
245 p. 2008 [978-3-540-79433-2]

Vol. 375: Bartolini G.;
Fridman L.; Pisano A.; Usai E. (Eds.)
Modern Sliding Mode Control Theory
465 p. 2008 [978-3-540-79015-0]

Vol. 374: Huang B.; Kadali R.
Dynamic Modeling, Predictive Control
and Performance Monitoring
240 p. 2008 [978-1-84800-232-6]

Vol. 373: Wang Q.-G.; Ye Z.; Cai W.-J.;
Hang C.-C.
PID Control for Multivariable Processes
264 p. 2008 [978-3-540-78481-4]

Vol. 372: Zhou J.; Wen C.
Adaptive Backstepping Control of
Uncertain Systems
241 p. 2008 [978-3-540-77806-6]

Vol. 371: Blondel V.D.; Boyd S.P.;
Kimura H. (Eds.)
Recent Advances in Learning and Control
279 p. 2008 [978-1-84800-154-1]

Vol. 370: Lee S.; Suh I.H.;
Kim M.S. (Eds.)
Recent Progress in Robotics:
Viable Robotic Service to Human
410 p. 2008 [978-3-540-76728-2]

Vol. 369: Hirsch M.J.; Pardalos P.M.;
Murphey R.; Grundel D.
Advances in Cooperative Control and
Optimization
423 p. 2007 [978-3-540-74354-5]

Vol. 368: Chee F.; Fernando T.
Closed-Loop Control of Blood Glucose
157 p. 2007 [978-3-540-74030-8]

CPSIA information can be obtained at www.ICGtesting.com
227919LV00002B/1/P

9 783642 197260